Lecture Notes in Networks and Systems

Volume 203

The series "Lecture Notes in Networks and Systems" publishes the latest developments in Networks and Systems—quickly, informally and with high quality. Original research reported in proceedings and post-proceedings represents the core of LNNS.

Volumes published in LNNS embrace all aspects and subfields of, as well as new challenges in, Networks and Systems.

The series contains proceedings and edited volumes in systems and networks, spanning the areas of Cyber-Physical Systems, Autonomous Systems, Sensor Networks, Control Systems, Energy Systems, Automotive Systems, Biological Systems, Vehicular Networking and Connected Vehicles, Aerospace Systems, Automation, Manufacturing, Smart Grids, Nonlinear Systems, Power Systems, Robotics, Social Systems, Economic Systems and other. Of particular value to both the contributors and the readership are the short publication timeframe and the world-wide distribution and exposure which enable both a wide and rapid dissemination of research output.

The series covers the theory, applications, and perspectives on the state of the art and future developments relevant to systems and networks, decision making, control, complex processes and related areas, as embedded in the fields of interdisciplinary and applied sciences, engineering, computer science, physics, economics, social, and life sciences, as well as the paradigms and methodologies behind them.

Indexed by SCOPUS, INSPEC, WTI Frankfurt eG, zbMATH, SCImago.

All books published in the series are submitted for consideration in Web of Science.

More information about this series at http://www.springer.com/series/15179

Pradeep Kumar Singh · Sławomir T. Wierzchoń ·
Sudeep Tanwar · Maria Ganzha ·
Joel J. P. C. Rodrigues
Editors

Proceedings of Second International Conference on Computing, Communications, and Cyber-Security

IC4S 2020

Springer

Editors
Pradeep Kumar Singh
Department of Computer Science
and Engineering
ABES Engineering College
Ghaziabad, India

Sudeep Tanwar
Department of Computer Science
and Engineering
Institute of Technology
Nirma University
Ahmedabad, Gujarat, India

Joel J. P. C. Rodrigues
Instituto de Telecomunicações
Federal University of Piauí (UFPI)
Teresina, Piauí, Brazil

Sławomir T. Wierzchoń
Institute of Computer Science
Polish Academy of Sciences
Warsaw, Poland

Maria Ganzha
Faculty of Mathematics and Information
Science
Warsaw University of Technology
Warsaw, Poland

ISSN 2367-3370 ISSN 2367-3389 (electronic)
Lecture Notes in Networks and Systems
ISBN 978-981-16-0732-5 ISBN 978-981-16-0733-2 (eBook)
https://doi.org/10.1007/978-981-16-0733-2

This Springer imprint is published by the registered company Springer Nature Singapore Pte Ltd.
The registered company address is: 152 Beach Road, #21-01/04 Gateway East, Singapore 189721, Singapore

Preface

This book features selected research papers presented at the Second International Conference on Computing, Communications, and Cyber-Security (IC4S 2020), organized in Krishna Engineering College (KEC), Ghaziabad, India, along with Academic Associates: Southern Federal University, Russia; IAC Educational, India; and ITS Mohan Nagar, Ghaziabad, India, during October 3–4, 2020. It includes papers from the area of communication and network technologies, advanced computing technologies, data analytics and intelligent learning, the latest electrical and electronics trends, and security and privacy issues.

The number of presented works and the enthusiasm of its participants allow us to hope that the conference will turn into an intellectual event that will bring together scientists from various fields of communication and network technologies on a common platform.

We are grateful to all authors for their contributions and the members of the Technical Program Committee for their tremendous support and motivation to make IC4S 2020 a great success. We are also grateful to the session leaders and keynote speakers: Dr. Richard Evans, Brunel University, London, UK; Prof. Kusum Deep, IIT Roorkee, India; Dr. Pelin Angin, Ankara Middle East Technical University, Turkey; Dr. Anand Nayyar, Duy Tan University, Da Nang, Vietnam; Prof. (Dr.) Madhuri Bhavsar from Nirma University, Ahmedabad, for sharing their technical lectures and enlightening the conference delegates. In addition, we express our gratitude to Aninda Bose, Senior Editor of Springer, for his continued help and guidance to further improve the quality of the proceedings.

Ghaziabad, India — Dr. Pradeep Kumar Singh
Warsaw, Poland — Dr. Sławomir T. Wierzchoń
Ahmedabad, India — Dr. Sudeep Tanwar
Warsaw, Poland — Dr. Maria Ganzha
Teresina, Brazil — Dr. Joel J. P. C. Rodrigues
October 2020

Contents

Communication and Network Technologies

Smart Aging Wellness Sensor Networks: A Near Real-Time Daily Activity Health Monitoring, Anomaly Detection and Alert System 3
Sharnil Pandya, Mayur Mistry, Ketan Kotecha, Anirban Sur, Asif Ghanchi, Vedant Patadiya, Kuldeep Limbachiya, and Anand Shivam

Ant Colony Optimization for Traveling Salesman Problem with Modified Pheromone Update Formula 23
Rahil Parmar, Naitik Panchal, Dhruval Patel, and Uttam Chauhan

Face Mask Detection Using Deep Learning During COVID-19 39
Soham Taneja, Anand Nayyar, Vividha, and Preeti Nagrath

Compact Millimeter-Wave Low-Cost Ka-Band Antenna for Portable 5G Communication Gadgets 53
Raqeebur Rehman, Javaid A. Sheikh, Khurshed A. Shah, Zahid A. Bhat, Shabir A. Parah, and Shahid A. Malik

Lightweight De-authentication DoS Attack Detection Methodology for 802.11 Networks Using Sniffer 67
Zakir Ahmad Sheikh and Yashwant Singh

Power Distribution Control for SIMO Wireless Power Transfer Systems 81
Sanjog Ganotra

Performance Analysis of WRAN in Light of Full Duplex Capability 97
Khusali Obhalia, Mayur M. Vegad, and Prashant B. Swadas

Design of Planer Wide Band Micro-Strip Patch Antenna for 5G Wireless Communication Applications: Review 109
Praveen Tiwari and Praveen Kumar Malik

Planar UWB Antenna for MIMO/Diversity Applications 121
Pramod Singh and Rekha Agarwal

Design and Analysis of Wearable Textile UWB Antenna for WBAN Communication Systems 141
Bhawna Tiwari, Sindhu Hak Gupta, and Vipin Balyan

Advanced Computing Technologies

Stock Prices Prediction from Financial News Articles Using LSTM and XAI 153
Shilpa Gite, Hrituja Khatavkar, Shilpi Srivastava, Priyam Maheshwari, and Neerav Pandey

Precision Agriculture: Methodologies, Practices and Applications 163
Sharnil Pandya, Mayur Mistry, Pramit Parikh, Kashish Shah, Gauravsingh Gaharwar, Ketan Kotecha, and Anirban Sur

Predicting Customer Spent on Black Friday 183
Ashish Arora, Bhupesh Bhatt, Divyanshu Bist, Rachna Jain, and Preeti Nagrath

Analyzing the Need of Edge Computing for Internet of Things (IoT) 203
Ajay Pratap, Ashwani Kumar, and Mohit Kumar

A Mobile-Based Farm Machinery Hiring System 213
Oluwasefunmi Arogundade, Rauf Qudus, Adebayo Abayomi-Alli, Sanjay Misra, JohnBosco Agbaegbu, Adio Akinwale, and Ravin Ahuja

Cloud Computing Offered Capabilities: Threats to Software Vendors 227
Oluwasefunmi Arogundade, Funmilayo Abayomi, Adebayo Abayomi-Alli, Sanjay Misra, Christianah Alonge, Taiwo Olaleye, and Ravin Ahuja

The Sentimental Analysis of Social Media Data: A Survey 241
Vartika Bhadana and Hitendra Garg

A Review Study on IoT-Based Smart Agriculture System 253
Mir Saqlain Sajad and Farheen Siddiqui

Secure Group Data Sharing with an Efficient Key Management without Re-Encryption Scheme in Cloud Computing 265
Lalit Mohan Gupta, Hitendra Garg, and Abdus Samad

Energy-Efficient Bonding Based Technique for Message Forwarding in Social Opportunistic Networks 279
Satbir Jain, Ritu Nigam, Deepak Kumar Sharma, and Shilpa Ghosh

Threat Modelling and Risk Assessment in Internet of Things: A Review 293
Mahapara Mahak and Yashwant Singh

Deep Learning-Based Attack Detection in the Internet of Things 307
Parushi Malhotra and Yashwant Singh

Data Analytics and Intelligent Learning

An Improved Dictionary Based Genre Classification Based on Title and Abstract of E-book Using Machine Learning Algorithms 323
Vrunda Thakur and Ankit C. Patel

A Novel Multicast Secure MQTT Messaging Protocol Framework for IoT-Related Issues 339
Sharnil Pandya, Mayur Mistry, Ketan Kotecha, Anirban Sur, Pramit Parikh, Kashish Shah, and Rutvij Dave

Experiential Learning Through Web-Based Application for Peer Review of Project: A Case Study Based on Interdisciplinary Teams 361
Pallavi Asthana, Sudeep Tanwar, Anil Kumar, Ankit Yadav, and Sumita Mishra

Machine Learning Applications for Computer-Aided Medical Diagnostics 377
Parita Oza, Paawan Sharma, and Samir Patel

Music Genre Classification ChatBot 393
Rishit Jain, Ritik Sharma, Preeti Nagrath, and Rachna Jain

Detection of COVID-19 by X-rays Using Machine Learning and Deep Learning Models 409
Yash Varshney, Piyush Anand, Achyut Krishna, Preeti Nagrath, and Rachna Jain

Implication of Machine Learning Models Toward Education Loan Repayment Rate Analysis 423
Anushree Bansal and Shikha Singh

Predicting the Result of English Premier League Matches 435
Ashutosh Ranjan, Vishesh Kumar, Devansh Malhotra, Rachna Jain, and Preeti Nagrath

Comment Filtering Based Explainable Fake News Detection 447
Dilip Kumar Sharma and Sunidhi Sharma

Comparative Analysis of Intelligent Solutions Searching Algorithms of Particle Swarm Optimization and Ant Colony Optimization for Artificial Neural Networks Target Dataset 459
Abraham Ayegba Alfa, Sanjay Misra, Adebayo Abayomi-Alli, Oluwasefunmi Arogundade, Oluranti Jonathan, and Ravin Ahuja

An Online Planning Agent to Optimize the Policy of Resources Management 471
Aditya Shrivastava, Aksha Thakkar, and Vipul Chudasama

CNN-Based Approach to Control Computer Applications by Differently Abled Peoples Using Hand Gesture 485
Hitesh Kumar Sharma, Prashant Ahlawat, Manoj Kumar Sharma, Md Ezaz Ahmed, J. C. Patni, and Sahil Taneja

A Frequency-Based Approach to Extract Aspect for Aspect-Based Sentiment Analysis 499
Rahul Pradhan and Dilip Kumar Sharma

Sentiment Analysis Techniques on Food Reviews Using Machine Learning 511
Shilpa Gite, Abhishek Udanshiv, Rajas Date, Kartik Jaisinghani, Abhishek Singh, and Prafful Chetwani

Parts of Speech (POS) Tagging for Dogri Language 529
Shivangi Dutta and Bhavna Arora

Deep Learning-Based 2D and 3D Human Pose Estimation: A Survey 541
Pooja Parekh and Atul Patel

Deepfake: An Overview 557
Anupama Chadha, Vaibhav Kumar, Sonu Kashyap, and Mayank Gupta

Instinctive and Effective Authorization for Internet of Things 567
Nidhi Sinha, Meenatchi Sundaram, and Abhijit Sinha

Methodical Analysis and Prediction of COVID-19 Cases of China and SAARC Countries 581
Sarika Agarwal and Himani Bansal

Coronary Artery Disease Prediction Techniques: A Survey 593
Aashna Joshi and Maitrik Shah

IoT Supported Healthcare (Or: Computer Aided Healthcare)

COVID-19 a "BIG RESET"—Role of GHRM in Achieving Organisational Sustainability in Context to Asian Market 607
Meenu Chaudhary and Loveleen Gaur

Anemia Multi-label Classification Based on Problem Transformation Methods 627
Bhavinkumar A. Patel and Ajay Parikh

Automated Disease Detection and Classification of Plants Using Image Processing Approaches: A Review 641
Shashi and Jaspreet Singh

Heart Disease Prediction Using Machine Learning 653
Jaydutt Patel, Azhar Ali Khaked, Jitali Patel, and Jigna Patel

Future of Augmented Reality in Healthcare Department 667
Gouri Jha, Lavanya shm. Sharma, and Shailja Gupta

E-health in Internet of Things (IoT) in Real-Time Scenario 679
Gourav Jha, Lavanya Sharma, and Shailja Gupta

Diagnosis of Heart Disease Using Internet of Things and Machine Learning Algorithms 691
Amit Kishor and Wilson Jeberson

Diabetes Prediction Using Machine Learning 703
Harsh Jigneshkumar Patel, Parita Oza, and Smita Agrawal

Myocardial Infarction Detection Using Deep Learning and Ensemble Technique from ECG Signals 717
Hari Mohan Rai, Kalyan Chatterjee, Alok Dubey, and Praween Srivastava

BL_SMOTE Ensemble Method for Prediction of Thyroid Disease on Imbalanced Classification Problem 731
Rajshree Srivastava and Pardeep Kumar

Computer-Aided-Diagnosis System for Symptom Detection of Breast and Cervical Cancer 743
Piyushi Jain, Drashti Patel, Jai Prakash Verma, and Sudeep Tanwar

Blockchain Adoption for Trusted Medical Records in Healthcare 4.0 Applications: A Survey 759
Umesh Bodkhe, Sudeep Tanwar, Pronaya Bhattacharya, and Ashwin Verma

The Amalgamation of Blockchain and IoT: A Survey 775
Jignasha Dalal

C2B-SCHMS: Cloud Computing and Bots Security for COVID-19 Data and Healthcare Management Systems 787
Vivek Kumar Prasad, Sudeep Tanwar, and Madhuri Bhavsar

FemtoCloud for Securing Smart Homes—An Edge Computing Solution for Internet of Thing Applications 799
Abhinav Rawat, Avani Jindal, Akshat Singhal, and Abhirup Khanna

Security and Privacy Issues

Global Intrusion Detection Environments and Platform for Anomaly-Based Intrusion Detection Systems 817
Jyoti Snehi, Abhinav Bhandari, Manish Snehi, Urvashi Tandon, and Vidhu Baggan

Image Steganography Using Bit Differencing Technique 833
Mudasir Rashid and Bhavna Arora

Detection and Prevention of DoS and DDoS in IoT 845
Meetu Sharma and Bhavna Arora

Approach for Ensuring Fragmentation and Integrity of Data in SEDuLOUS 857
Anand Prakash Singh and Arjun Choudhary

Efficient Classification of True Positive and False Positive XSS and CSRF Vulnerabilities Reported by the Testing Tool 871
Monika Shah and Himani Lad

A Survey on Hardware Trojan Detection: Alternatives to Destructive Reverse Engineering 885
Archit Saini, Gahan Kundra, and Shruti Kalra

Comparative Study of Various Intrusion Detection Techniques for Android Malwares 899
Leesha Aneja and Jaspreet Singh

Error Detection Using Syntactic Analysis for Air Traffic Speech 909
Narayanan Srinivasan and S. R. Balasundaram

Road Segmentation from Satellite Images Using Custom DNN 927
Harshal Trivedi, Dhrumil Sheth, Ritu Barot, and Rainam Shah

Performance Analysis of SoC and Hardware Design Flow in Medical Image Processing Using Xilinx Zed Board FPGA 945
Neol Solanki, Chintan Patel, Neel Tailor, and Nadimkhan Pathan

SDN Firewall Using POX Controller for Wireless Networks 967
Sulbha Manoj Shinde and Girish Ashok Kulkarni

Latest Electrical and Electronics Trends

Output Load Capacitance Scaling-Based Energy-Efficient Design of ROM on 28 nm FPGA 987
Pankaj Singh, Bishwajeet Pandey, Neema Bhandari, Shilpi Bisht, and Neeraj Bisht

Image Correction and Identification of Ishihara Test Images for Color Blind Individual 997
Himani Bansal, Lalit Bhagat, Satyam Mittal, and Ayush Tiwari

Gradient Feature-Based Classification of Patterned Images 1007
Divya Srivastava, B. Rajitha, and Suneeta Agarwal

Corrosion Estimation of Underwater Structures Employing Bag of Features (BoF) 1017
Anant Sinha, Sachin Kumar, Pooja Khanna, and Pragya

On Performance Enhancement of Molecular Dynamics Simulation Using HPC Systems 1031
Tejal Rathod, Monika Shah, Niraj Shah, Gaurang Raval, Madhuri Bhavsar, and Rajaraman Ganesh

Author Index 1045

Editors and Contributors

About the Editors

Dr. Pradeep Kumar Singh is currently working as a Professor in the Department of CSE at ABES Engineering College, Ghaziabad, India. He has completed his Ph.D. in Computer Science and Engineering from Gautam Buddha University (State Government University), Greater Noida, UP, India. He received his M.Tech. (CSE) with Distinction from GGSIPU, New Delhi, India. Dr. Singh is having life membership of Computer Society of India (CSI), a life member of IEI and promoted to Senior Member Grade from CSI and ACM. He is an associate editor of IJAEC, IJISMD journals IGI Global USA, Security and Privacy, WILEY, and section editor of Discover IoT journal from Springer. He has published almost 100 research papers in various international journals and conferences of repute. He has received three sponsored research projects grant from Government of India and Government of HP worth Rs. 25 Lakhs. He has edited total 14 books from Springer and Elsevier and also edited several special issues for SCI and SCIE Journals from Elsevier and IGI Global. He has total 829 google scholars, h-index 17, and i-10 Index 31 in his account.

Prof. Sławomir T. Wierzchoń received M.Sc. and Ph.D. degrees in Computer Science from Technical University of Warsaw, Poland. He holds Habilitation (D.Sc.) in Uncertainty Management from Polish Academy of Sciences. In 2003, he received the title of Professor from the President of Poland. Currently, he is Full Professor at the Institute of Computer Science of Polish Academy of Sciences. His research interests include computational intelligence, uncertainty management, information retrieval, machine learning, and data mining. He is an author/co-author of over 100 peer-reviewed papers in international journals and international conferences. He published, as an author/co-author, 11 monographs from the field of machine learning. In the period 2000–2013, he co-organized 13 international conferences on intelligent information systems. Co-authored proceedings from these conferences were published by Springer. He coedited two volumes of proceedings of the International Conference on Computer Information Systems and Industrial Management, and he has served as a guest co-editor of three special issues of Information and

Control journal. Currently, he is a member of the editorial board for some international journals, as well as a member of many program committees for international conferences. He cooperated with medical centers in the area of statistical data analysis and knowledge discovery in databases.

Dr. Sudeep Tanwar is working as a Professor in the Computer Science and Engineering Department at Institute of Technology, Nirma University, Ahmedabad, Gujarat, India. He is visiting Professor in Jan Wyzykowski University in Polkowice, Poland and University of Pitesti in Pitesti, Romania. He received B.Tech. in 2002 from Kurukshetra University, India, M.Tech. (Honor's) in 2009 from Guru Gobind Singh Indraprastha University, Delhi, India and Ph.D. in 2016 with specialization in Wireless Sensor Network. He has authored or coauthored more than 200 technical research papers published in leading journals and conferences from the IEEE, Elsevier, Springer, Wiley, etc. Some of his research findings are published in top cited journals such as IEEE TNSE, IEEE TVT, IEEE TII, Transactions on Emerging, Telecommunications Technologies, IEEE WCM, IEEE Networks, IEEE Systems Journal, IEEE Access, IET Software, IET Networks, JISA, Computer Communication, Applied Soft Computing, JPDC, JNCA, PMC, SUSCOM, CEE, IJCS, Software: Practice and Experience, MTAP, Telecommunication System. He has also edited/authored 14 books with International/National Publishers like IET, Springer. One of the edited textbook entitled, *Multimedia Big Data Computing for IoT Applications: Concepts, Paradigms, and Solutions* published in Springer in 2019 is having 3.7 million downloads till 20 May 2021. It attracts the attention of the researchers across the globe. He has guided many students leading to M.E./M.Tech. and guiding students leading to Ph.D. He is Associate Editor of IJCS, Wiley, PHYCOM, Elsevier, COMCOM, Elsevier and Security and Privacy Journal, Wiley. His current interest includes Wireless Sensor Networks, Fog Computing, Smart Grid, IoT, and Blockchain Technology. He was invited as Guest Editors/Editorial Board Members of many International Journals, invited for keynote Speaker in many International Conferences held in Asia and invited as Program Chair, Publications Chair, Publicity Chair, and Session Chair in many International Conferences held in North America, Europe, Asia and Africa. He has been awarded best research paper awards from IEEE GLOBECOM 2018, IEEE ICC 2019, and Springer ICRIC-2019.

Dr. Maria Ganzha is Associate Professor in the Faculty of Mathematics and Information Science. She has an M.S. and a Ph.D. degrees in Mathematics from the Moscow State University, Russia, and a Doctor of Science degree (in Computer Science) from the Polish Academy of Sciences. Maria has published more than 200 research papers, is on editorial boards of 6 journals and a book series, and was invited to program committees of more than 150 conferences. She is also Technical Coordinator of the European project ASSIST-IoT (Architecture for Scalable, Self-*, human-centric, Intelligent, Secure, and Tactile next generation IoT) project. She has 1797 google citations, h-index 20, and i-10 index 62 in her account. Her area of interest includes computational intelligence, distributed systems, agent-based computing, and semantic data processing.

Joel J. P. C. Rodrigues [S'01, M'06, SM'06, F'20] is a professor at the Federal University of Piauí, Brazil; senior researcher at the *Instituto de Telecomunicações*, Portugal; and collaborator of the Post-Graduation Program on Teleinformatics Engineering at the Federal University of Ceará (UFC), Brazil. Prof. Rodrigues is the leader of the Next Generation Networks and Applications (NetGNA) research group (CNPq), an IEEE Distinguished Lecturer, Member Representative of the IEEE Communications Society on the IEEE Biometrics Council, and the President of the scientific council at ParkUrbis—Covilhã Science and Technology Park. He was Director for Conference Development—IEEE ComSoc Board of Governors, Technical Activities Committee Chair of the IEEE ComSoc Latin America Region Board, a Past-Chair of the IEEE ComSoc Technical Committee on eHealth, a Pastchair of the IEEE ComSoc Technical Committee on Communications Software, a Steering Committee member of the IEEE Life Sciences Technical Community and Publications Co-chair. He is the editor-in-chief of the International Journal of E-Health and Medical Communications and editorial board member of several high-reputed journals. He has been general chair and TPC Chair of many international conferences, including IEEE ICC, IEEE GLOBECOM, IEEE HEALTHCOM, and IEEE LatinCom. He has authored or coauthored over 950 papers in refereed international journals and conferences, 3 books, 2 patents, and 1 ITU-T Recommendation. He had been awarded several Outstanding Leadership and Outstanding Service Awards by IEEE Communications Society and several best papers awards. Prof. Rodrigues is a member of the Internet Society, a senior member ACM, and Fellow of IEEE.

Contributors

Adebayo Abayomi-Alli Federal University of Abeokuta, Abeokuta, Ogun State, Nigeria

Funmilayo Abayomi Federal University of Abeokuta, Abeokuta, Ogun, Nigeria

Rekha Agarwal Department of Electronics and Communication, Amity School of Engineering and Technology, Noida, India

Sarika Agarwal Department of CSE/IT, Jaypee Institute of Information Technology, Noida, India

Smita Agrawal CE Department, Institute of Technology, Nirma University, Ahmedabad, India

Suneeta Agarwal Professor, Motilal Nehru National Institute of Technology, Allahabad, Prayagraj, India

JohnBosco Agbaegbu Federal University of Abeokuta, Abeokuta, Ogun, Nigeria

Prashant Ahlawat Manipal University, Jaipur, India

Md Ezaz Ahmed CS Department, Saudi Electronic University, Al Madina, Kingdom of Saudi Arabia

Ravin Ahuja Shri Viskarma Skill University, Gurgaon, India

Adio Akinwale Federal University of Abeokuta, Abeokuta, Ogun, Nigeria

Abraham Ayegba Alfa Kogi State College of Education, Ankpa, Nigeria

Christianah Alonge Federal University of Abeokuta, Abeokuta, Ogun, Nigeria

Piyush Anand Department of ECE, Bharati Vidyapeeth's College of Engineering (Aff. To IPU), New Delhi, India

Leesha Aneja GD Goenka University, Gurugram, India

Oluwasefunmi Arogundade Federal University of Abeokuta, Abeokuta, Ogun State, Nigeria

Ashish Arora Department of ECE, Bharati Vidyapeeth's College of Engineering, New Delhi, India

Bhavna Arora Department of Computer Science and Information Technology, Central University of Jammu, Jammu, Jammu and Kashmir, India

Pallavi Asthana Amity University, Noida, Uttar Pradesh, India

Vidhu Baggan Institute of Engineering and Technology, Chitkara University, Chitkara University, Punjab, India

S. R. Balasundaram National Institute of Technology, Tamil Nadu, Tiruchirappalli, India

Vipin Balyan Department of Electrical, Electronics and Computer Engineering, Cape Peninsula University of Technology, Cape Town, South Africa

Anushree Bansal Department of Computer Science and Engineering, Amity School of Engineering and Technology, Amity University, Lucknow, India

Himani Bansal Department of CSE & IT, Jaypee Institute of Information Technology, Noida, India

Ritu Barot LD Engineering College, Ahmedabad, India

Vartika Bhadana GLA University, Mathura, UP, India

Lalit Bhagat Department of CSE & IT, Jaypee Institute of Information Technology, Noida, India

Abhinav Bhandari Department of Computer Science and Engineering, Panjabi University, Patiala, India

Neema Bhandari G.B. Pant University of Agriculture and Technology, Pantnagar, Uttarakhand, India

Zahid A. Bhat Department of Electronics and Instrumentation Technology, University of Kashmir, Srinagar, India

Bhupesh Bhatt Department of ECE, Bharati Vidyapeeth's College of Engineering, New Delhi, India

Pronaya Bhattacharya Department of Computer Science and Engineering, Institute of Technology, Nirma University, Ahmadabad, Gujarat, India

Madhuri Bhavsar Computer Science Department, Institute of Technology, Nirma University, Ahmedabad, India

Neeraj Bisht Birla Institute of Applied Sciences, Bhimtal, Uttarakhand, India

Shilpi Bisht Birla Institute of Applied Sciences, Bhimtal, Uttarakhand, India

Divyanshu Bist Department of ECE, Bharati Vidyapeeth's College of Engineering, New Delhi, India

Umesh Bodkhe Department of Computer Science and Engineering, Institute of Technology, Nirma University, Ahmadabad, Gujarat, India

Anupama Chadha MRIIRS, Faridabad, Haryana, India

Kalyan Chatterjee Department of Electrical Engineering, Indian Institute of Technology (ISM), Dhanbad, India

Meenu Chaudhary Amity International Business School, Amity University, Noida, India

Uttam Chauhan Vishwakarma Government Engineering College, Ahmedabad, Gujarat, India

Prafful Chetwani Symbiosis Institute of Technology, Pune, Symbiosis International (Deemed University), Pune, Maharashtra, India

Arjun Choudhary Sardar Patel University of Police Security and Criminal Justice, Jodhpur, India

Vipul Chudasama Department of Computer Science and Engineering, Institute of Technology, Nirma University, Ahmedabad, India

Jignasha Dalal K. J. Somaiya Institute of Engineering and Information Technology, Mumbai University, Mumbai, India

Rajas Date Symbiosis Institute of Technology, Pune, Symbiosis International (Deemed University), Pune, Maharashtra, India

Rutvij Dave Department of Computer Science Engineering, Ganpat University, Ahmedabad, Gujarat, India

Alok Dubey Department of ECE, Krishna Engineering College, Ghaziabad, India

Shivangi Dutta Central University of Jammu, Jammu and Kashmir, India

Gauravsingh Gaharwar Department of Computer Science Engineering, Navrachana University, Vadodara, Gujarat, India

Rajaraman Ganesh Institute of Plasma Research, Gandhinagar, India

Sanjog Ganotra Department of Electronics and Communication, Knowledge Park III, Uttar Pradesh, India

Hitendra Garg GLA University, Mathura, UP, India

Loveleen Gaur Amity International Business School, Amity University, Noida, India

Asif Ghanchi Symbiosis Center for Applied Artificial Intelligence and Symbiosis Institute of Technology (SIT), Symbiosis International (Deemed) University, Pune, India

Shilpa Ghosh Division of Information Technology, University of Delhi (Netaji Subhas Institute of Technology), New Delhi, India

Shilpa Gite Symbiosis Institute of Technology, Symbiosis International (Deemed University), Pune, India

Lalit Mohan Gupta Dr. A.P.J. Abdul Kalam Technical University, Lucknow, India

Mayank Gupta MRIIRS, Faridabad, Haryana, India

Shailja Gupta Department of Computer Science and Technology, Manav Rachna University, Faridabad, India

Sindhu Hak Gupta Department of Electronics and Communications, Amity University, Noida, India

Piyushi Jain Institute of Technology, Nirma University, Ahmedabad, Gujarat, India

Rachna Jain Department of CSE, Bharati Vidyapeeth's College of Engineering (Aff. To IPU), New Delhi, India

Rishit Jain Department of ECE, Bharati Vidyapeeth's College of Engineering, New Delhi, India

Satbir Jain Department of Computer Engineering, Netaji Subhas University of Technology (Formerly Netaji Subhas Institute of Technology), New Delhi, India

Kartik Jaisinghani Symbiosis Institute of Technology, Pune, Symbiosis International (Deemed University), Pune, Maharashtra, India

Wilson Jeberson Department of Computer Science and Information Technology, Sam Higginbottom University of Agriculture, Technology and Sciences, Allahabad, U.P., India

Gourav Jha Amity Insititute of Information Technology, Amity University, Noida, Uttar Pradesh, India

Gouri Jha Amity Insititute of Information Technology, Amity University, Noida, Uttar Pradesh, India

Avani Jindal University of Petroleum and Energy Studies (UPES),, Dehradun, Uttarakhand, India

Oluranti Jonathan Covenant University, Otta, Nigeria

Aashna Joshi L.D. College of Engineering, Ahmedabad, India

Shruti Kalra Jaypee Institute of Information Technology, Noida, India

Sonu Kashyap MRIIRS, Faridabad, Haryana, India

Azhar Ali Khaked Institute of Technology Nirma University, Ahmedabad, India

Abhirup Khanna University of Petroleum and Energy Studies (UPES),, Dehradun, Uttarakhand, India

Pooja Khanna Amity University, Lucknow Campus, Lucknow, India

Hrituja Khatavkar Symbiosis Institute of Technology, Symbiosis International (Deemed University), Pune, India

Amit Kishor Department of Computer Science and Information Technology, Sam Higginbottom University of Agriculture, Technology and Sciences, Allahabad, U.P., India

Ketan Kotecha Symbiosis Center for Applied Artificial Intelligence and Symbiosis Institute of Technology (SIT), Symbiosis International (Deemed) University, Pune, India

Achyut Krishna Department of ECE, Bharati Vidyapeeth's College of Engineering (Aff. To IPU), New Delhi, India

Girish Ashok Kulkarni Department of Electronics & Telecommunication Engineering, HSM'S Shri Sant Gadge Baba College of Engineering & Technology, Bhusawal, India

Deepak Kumar Sharma Department of Information Technology, Netaji Subhas University of Technology (Formerly Netaji Subhas Institute of Technology), New Delhi, India

Anil Kumar Amity University, Noida, Uttar Pradesh, India

Ashwani Kumar Sreyas Institute of Engineering and Technology, Hyderabad, Telangana, India

Mohit Kumar Cambridge Institute of Technology, Ranchi, Jharkhand, India

Pardeep Kumar JUIT, Himachal Pradesh, Waknaghat, Solan, India

Sachin Kumar Amity University, Lucknow Campus, Lucknow, India

Vaibhav Kumar MRIIRS, Faridabad, Haryana, India

Vishesh Kumar Department of Electronics and Communication, Engineering, Bharati Vidyapeeth College of Engineering, New Delhi, India

Gahan Kundra Jaypee Institute of Information Technology, Noida, India

Himani Lad Nirma University, Ahmedabad, Gujarat, India

Kuldeep Limbachiya Symbiosis Center for Applied Artificial Intelligence and Symbiosis Institute of Technology (SIT), Symbiosis International (Deemed) University, Pune, India

Mahapara Mahak Department of Computer Science and Information Technology, Central University of Jammu, Samba, Jammu and Kashmir, India

Priyam Maheshwari Symbiosis Institute of Technology, Symbiosis International (Deemed University), Pune, India

Devansh Malhotra Department of Electronics and Communication, Engineering, Bharati Vidyapeeth College of Engineering, New Delhi, India

Parushi Malhotra Department of Computer Science and Information Technology, Central University of Jammu, Samba, Jammu and Kashmir, India

Praveen Kumar Malik Lovely Professional University, Jalandhar, India

Shahid A. Malik Department of Electronics and Instrumentation Technology, University of Kashmir, Srinagar, India

Sumita Mishra Amity University, Noida, Uttar Pradesh, India

Sanjay Misra Covenant University, Otta, Nigeria

Mayur Mistry Department of Computer Science Engineering, Ganpat University, Ahmedabad, Gujarat, India

Satyam Mittal Department of CSE & IT, Jaypee Institute of Information Technology, Noida, India

Preeti Nagrath Department of Computer Science and Engineering, Bharati Vidyapeeth's College of Engineering (Aff. To IPU), New Delhi, Delhi, India

Anand Nayyar Graduate School, Faculty of Information Technology, Duy Tan University, Da Nang, Vietnam

Ritu Nigam Division of Computer Engineering, University of Delhi (Netaji Subhas Institute of Technology), New Delhi, India

Khusali Obhalia Computer Engineering Department, BVM Engineering College, Vallabh Vidyanagar, Gujarat, India

Taiwo Olaleye Federal University of Abeokuta, Abeokuta, Ogun, Nigeria

Parita Oza CE Department, Institute of Technology, Nirma University, Ahmedabad, India;
Pandit Deendayal Petroleum University, Gandhinagar, India

Naitik Panchal Vishwakarma Government Engineering College, Ahmedabad, Gujarat, India

Bishwajeet Pandey Birla Institute of Applied Sciences, Bhimtal, Uttarakhand, India

Neerav Pandey Symbiosis Institute of Technology, Symbiosis International (Deemed University), Pune, India

Sharnil Pandya Symbiosis Center for Applied Artificial Intelligence and Symbiosis Institute of Technology (SIT), Symbiosis International (Deemed) University, Pune, India

Shabir A. Parah Department of Electronics and Instrumentation Technology, University of Kashmir, Srinagar, India

Pooja Parekh Charotar University of Science and Technology, Changa, Gujarat, India

Ajay Parikh Department of Computer Scince, Gujarat Vidyapith, Ahmedabad, Gujarat, India

Pramit Parikh Department of Computer Science Engineering, Navrachana University, Vadodara, Gujarat, India

Rahil Parmar Vishwakarma Government Engineering College, Ahmedabad, Gujarat, India

Vedant Patadiya Symbiosis Center for Applied Artificial Intelligence and Symbiosis Institute of Technology (SIT), Symbiosis International (Deemed) University, Pune, India

Ankit C. Patel L.D. College of Engineering, Ahmedabad, India

Atul Patel Charotar University of Science and Technology, Changa, Gujarat, India

Bhavinkumar A. Patel Department of Computer Scince, Gujarat Vidyapith, Ahmedabad, Gujarat, India

Chintan Patel Birla Vishvakarma Mahavidyalaya, Vallabh Vidyanagar, Gujarat, India

Dhruval Patel Vishwakarma Government Engineering College, Ahmedabad, Gujarat, India

Drashti Patel Institute of Technology, Nirma University, Ahmedabad, Gujarat, India

Harsh Jigneshkumar Patel CE Department, Institute of Technology, Nirma University, Ahmedabad, India

Jaydutt Patel Institute of Technology Nirma University, Ahmedabad, India

Jigna Patel Institute of Technology Nirma University, Ahmedabad, India

Jitali Patel Institute of Technology Nirma University, Ahmedabad, India

Samir Patel Pandit Deendayal Petroleum University, Gandhinagar, India

Nadimkhan Pathan Birla Vishvakarma Mahavidyalaya, Vallabh Vidyanagar, Gujarat, India

J. C. Patni Department of Cybernetics, School of Computer Science, University of Petroleum and Energy Studies, EnergyAcres, Dehradun, Uttarakhand, India

Rahul Pradhan GLA University, Mathura, India

Pragya MVPG College, Lucknow University, Lucknow, India

Vivek Kumar Prasad Computer Science Department, Institute of Technology, Nirma University, Ahmedabad, India

Ajay Pratap AIIT, Amity University Uttar Pradesh, Lucknow, India

Rauf Qudus Federal University of Abeokuta, Abeokuta, Ogun, Nigeria

Hari Mohan Rai Department of ECE, Krishna Engineering College, Ghaziabad, India

B. Rajitha Assistant Professor, Motilal Nehru National Institute of Technology, Allahabad, Prayagraj, India

Ashutosh Ranjan Department of Electronics and Communication, Engineering, Bharati Vidyapeeth College of Engineering, New Delhi, India

Mudasir Rashid Central University of Jammu, Jammu, Jammu and Kashmir, India

Tejal Rathod Institute of Technology, Nirma University, Ahmedabad, India

Gaurang Raval Institute of Technology, Nirma University, Ahmedabad, India

Abhinav Rawat University of Petroleum and Energy Studies (UPES),, Dehradun, Uttarakhand, India

Raqeebur Rehman Department of Electronics and Instrumentation Technology, University of Kashmir, Srinagar, India

Archit Saini Jaypee Institute of Information Technology, Noida, India

Mir Saqlain Sajad Department of Computer Science, School of Engineering Sciences and Technology, Jamia Hamdard University, New Delhi, India

Abdus Samad Women's Polytechnic, Aligarh Muslim University, Aligarh, India

Kashish Shah Department of Computer Science Engineering, Navrachana University, Vadodara, Gujarat, India

Khurshed A. Shah Department of Physics, S.P College, Cluster University, Srinagar, India

Maitrik Shah L.D. College of Engineering, Ahmedabad, India

Monika Shah Institute of Technology, Nirma University, Ahmedabad, Gujarat, India

Niraj Shah Institute of Technology, Nirma University, Ahmedabad, India

Rainam Shah Gandhinagar Institute of Technology, Ahmedabad, India

Dilip Kumar Sharma Department of Computer Engineering and Applications, GLA University, Mathura, Uttar Pradesh, India

Hitesh Kumar Sharma Department of Cybernetics, School of Computer Science, University of Petroleum and Energy Studies, EnergyAcres, Dehradun, Uttarakhand, India

Lavanya Sharma Amity Insititute of Information Technology, Amity University, Noida, Uttar Pradesh, India

Lavanya shm. Sharma Amity Insititute of Information Technology, Amity University, Noida, Uttar Pradesh, India

Manoj Kumar Sharma Manipal University, Jaipur, India

Meetu Sharma Department of Computer Science and Information Technology, Central University of Jammu, Jammu, Jammu and Kashmir, India

Paawan Sharma Pandit Deendayal Petroleum University, Gandhinagar, India

Ritik Sharma Department of ECE, Bharati Vidyapeeth's College of Engineering, New Delhi, India

Sunidhi Sharma Department of Computer Engineering and Applications, GLA University, Mathura, Uttar Pradesh, India

Shashi G D Goenka University, Gurugram, India

Javaid A. Sheikh Department of Electronics and Instrumentation Technology, University of Kashmir, Srinagar, India

Zakir Ahmad Sheikh Department of Computer Science and Information Technology, Central University of Jammu, Jammu, Jammu and Kashmir, India

Dhrumil Sheth Softvan Pvt. Ltd, Ahmedabad, India

Sulbha Manoj Shinde Department of Electronics & Telecommunication Engineering, HSM'S Shri Sant Gadge Baba College of Engineering & Technology, Bhusawal, India

Anand Shivam Symbiosis Center for Applied Artificial Intelligence and Symbiosis Institute of Technology (SIT), Symbiosis International (Deemed) University, Pune, India

Aditya Shrivastava Department of Computer Science and Engineering, Institute of Technology, Nirma University, Ahmedabad, India

Farheen Siddiqui Department of Computer Science, School of Engineering Sciences and Technology, Jamia Hamdard University, New Delhi, India

Akshat Singhal University of Petroleum and Energy Studies (UPES),, Dehradun, Uttarakhand, India

Abhishek Singh Symbiosis Institute of Technology, Pune, Symbiosis International (Deemed University), Pune, Maharashtra, India

Anand Prakash Singh Sardar Patel University of Police Security and Criminal Justice, Jodhpur, India

Jaspreet Singh GD Goenka University, Gurugram, India

Pankaj Singh Birla Institute of Applied Sciences, Bhimtal, Uttarakhand, India

Pramod Singh USIT, Guru Govind Singh Indraprastha University, New Delhi, India

Shikha Singh Department of Computer Science and Engineering, Amity School of Engineering and Technology, Amity University, Lucknow, India

Yashwant Singh Department of Computer Science and Information Technology, Central University of Jammu, Jammu, Jammu and Kashmir, India

Abhijit Sinha Garden City University, Bangalore, India

Anant Sinha Amity University, Lucknow Campus, Lucknow, India

Nidhi Sinha Garden City University, Bangalore, India

Jyoti Snehi Institute of Engineering and Technology, Chitkara University, Chitkara University, Punjab, India

Manish Snehi Engineering Services, Infosys Limited, Chandigarh, India

Neol Solanki Birla Vishvakarma Mahavidyalaya, Vallabh Vidyanagar, Gujarat, India

Narayanan Srinivasan National Institute of Technology, Tamil Nadu, Tiruchirappalli, India

Divya Srivastava Assistant Professor, Bennett University, Greater Noida, UP, India

Praween Srivastava Department of ECE, Krishna Engineering College, Ghaziabad, India

Rajshree Srivastava Department of Computer Science and Engineering, JUIT, Himachal Pradesh, Waknaghat, Solan, India

Shilpi Srivastava Symbiosis Institute of Technology, Symbiosis International (Deemed University), Pune, India

Meenatchi Sundaram Garden City University, Bangalore, India

Anirban Sur Symbiosis Institute of Technology (SIT), Symbiosis International (Deemed) University, Pune, India

Prashant B. Swadas Birla Vishvakarma Mahavidyalaya (BVM) Engineering College, Vallabh Vidyanagar, Gujarat, India

Neel Tailor Birla Vishvakarma Mahavidyalaya, Vallabh Vidyanagar, Gujarat, India

Urvashi Tandon Chitkara Business School, Chitkara University, Punjab, India

Sahil Taneja Department of Cybernetics, School of Computer Science, University of Petroleum and Energy Studies, EnergyAcres, Dehradun, Uttarakhand, India

Soham Taneja Bharati Vidyapeeth's College of Engineering, New Delhi, Delhi, India

Sudeep Tanwar Department of Computer Science and Engineering, Institute of Technology, Nirma University, Ahmadabad, Gujarat, India

Aksha Thakkar Department of Computer Science and Engineering, Institute of Technology, Nirma University, Ahmedabad, India

Vrunda Thakur L.D. College of Engineering, Ahmedabad, India

Ayush Tiwari Department of CSE & IT, Jaypee Institute of Information Technology, Noida, India

Bhawna Tiwari Amity University, Noida, Uttar Pradesh, India

Praveen Tiwari Lovely Professional University, Jalandhar, India

Harshal Trivedi Softvan Pvt. Ltd, Ahmedabad, India

Abhishek Udanshiv Symbiosis Institute of Technology, Pune, Symbiosis International (Deemed University), Pune, Maharashtra, India

Yash Varshney Department of ECE, Bharati Vidyapeeth's College of Engineering (Aff. To IPU), New Delhi, India

Mayur M. Vegad Birla Vishvakarma Mahavidyalaya (BVM) Engineering College, Vallabh Vidyanagar, Gujarat, India

Ashwin Verma Department of Computer Science and Engineering, Institute of Technology, Nirma University, Ahmadabad, Gujarat, India

Jai Prakash Verma Institute of Technology, Nirma University, Ahmedabad, Gujarat, India

Vividha Bharati Vidyapeeth's College of Engineering, New Delhi, Delhi, India

Ankit Yadav Amity University, Noida, Uttar Pradesh, India

Communication and Network Technologies

Smart Aging Wellness Sensor Networks: A Near Real-Time Daily Activity Health Monitoring, Anomaly Detection and Alert System

Sharnil Pandya, Mayur Mistry, Ketan Kotecha, Anirban Sur, Asif Ghanchi, Vedant Patadiya, Kuldeep Limbachiya, and Anand Shivam

Abstract In the growing automation of existing world, activity modeling is being used in the field of technology to serve various purposes. One such field, which will be majorly benefited from daily activity modeling and life- living activities analysis, is monitoring of seasonal behavior pattern of elderly people, which can be further utilized in their remote health analysis and monitoring. Today's demand is to develop a system with minimum human interaction and automatic anomaly detection and alert system. The proposed research work emphasizes to diagnose elderly persons daily behavioral patterns by observing their day-to-day routine activities with respect to time, location and context. To grow the accurateness of the structure, numerous sensing as well as actuator units have been deployed in elderly homes. Popular this research paper, we have recommended a unique sensing fusion technique to monitor seasonal, social, weather related and wellness observations of routine tasks. A novel daily activity learning model has been proposed which can record contextual data observations of various locations of a smart home and alert caretakers in the case of anomaly detection. We have analyzed monthly data of two old-aged smart homes with more than 5000 test samples. Results acquired from the investigation validate the accuracy and the efficiency of the proposed system which are recorded for 20 activities.

Keywords Activity modeling · Anomaly detection · Activities of daily living · Cognitive computing

S. Pandya (✉) · K. Kotecha · A. Ghanchi · V. Patadiya · K. Limbachiya · A. Shivam
Symbiosis Center for Applied Artificial Intelligence and Symbiosis Institute of Technology (SIT), Symbiosis International (Deemed) University, Pune, India
e-mail: sharnil.pandya@scaai.siu.edu.in

M. Mistry
Department of Computer Science Engineering, Ganpat University, Ahmedabad, Gujarat, India
e-mail: mayur.mtechbda1703@ict.gnu.ac.in

A. Sur
Symbiosis Institute of Technology (SIT), Symbiosis International (Deemed) University, Pune, India
e-mail: sur@sitpune.edu.in

P. K. Singh et al. (eds.), *Proceedings of Second International Conference on Computing, Communications, and Cyber-Security*, Lecture Notes in Networks and Systems 203,
https://doi.org/10.1007/978-981-16-0733-2_1

1 Introduction

In the today's growing world of smart houses across the globe and automation of daily living activities in people's living, the need of remote health monitoring has arisen rapidly in what is called Web 2.0, where two-way communication has become commonplace, allowing for the participation of users, remote health monitoring is aimed to target a vast number of audiences in elderly homes, which are currently deprived of quality health services [1]. In conventional remote health monitoring activity, majorly two approaches are used: firstly, utilization of on-body sensors to measure body parameters, and second is using sensors system in smart house for activity modeling, which sends data and activity report to medical experts for their views and decisions [24–29]. Whereas the main problem with first approach is that it also contributes toward inconvenience of the residents majorly elderly people and the second approach always relies on an external health specialist, which also increases the routine health expenses [30–32]. Hence, today's demand is to develop a system with minimum human interaction and automatic anomaly detection and alert system [2, 33–36]. The proposed system mainly aims toward diagnosis of behavioral pattern of elderly people by observation of their day-to-day activities monitored daily, weekly, monthly and yearly manner thorough a pattern recognition approach-based machine learning models and reinforcement learning methodology. In the proposed research work, a novel approach has been introduced to not only carry our routine health monitoring of elderly persons but to identify, detect and recognize anomaly conditions by providing immediate notification to the caregivers or health experts [3, 37–43]. The proposed research paper is further divided into four sections: (i) Design and Experimental Setup, (ii) Activity Modeling Methodology, (iii) Results and Discussions and (iv) Future Enhancements.

1.1 Proposed Model: Layered Model

See Fig. 1.

2 System Architecture

As depicted in the below diagram, the whole health monitoring system architecture is divided into five different layers, each having its own functional importance [44–48]. Architecture starts from the physical layer which comprises of different hardware setups, and going through edge layer, cloud layer and processing layer, we reach the end and topmost layer, the application layer which finally interacts with the caretakers and medical specialists [49–55].

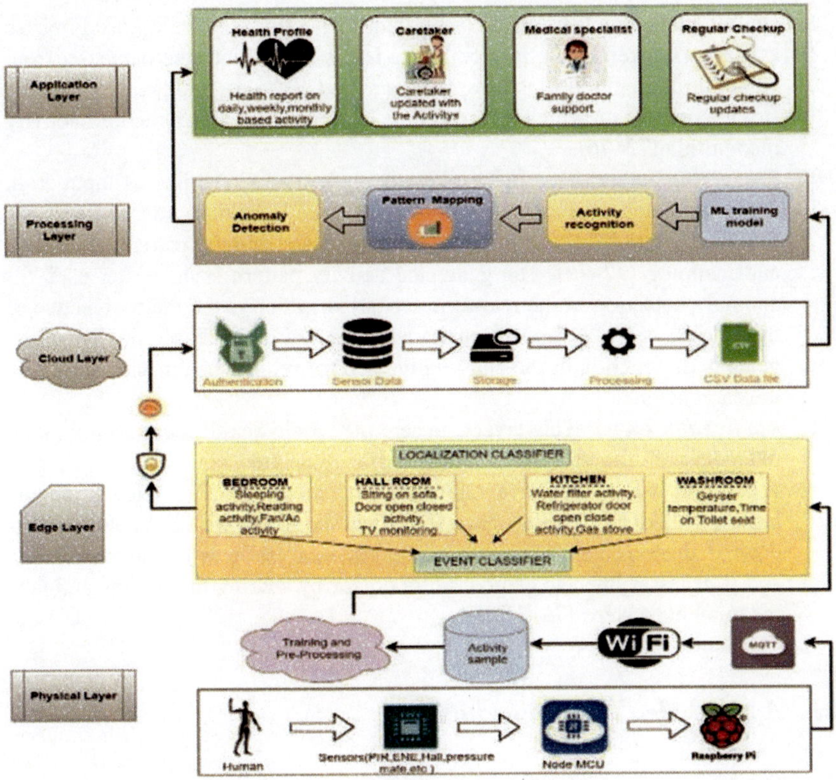

Fig. 1 System architecture [3]

(A) ***Physical Layer***: This is the bottommost layer, which contains the individual hardware components like sensors, micro-controllers like ESP8266, conditioning circuits and micro-computers, e.g., Raspberry Pi 3B+ . This layer is responsible for data generation after sensing the environment and passing it on to the edge layer [44–48, 56–60].

(B) ***Edge Layer***: This layer receives data from the sensors over Wi-Fi and performs the data pre-processing on the micro-computer, before sending it to the cloud hence utilizing the concept of edge computing [61–65]. This layer comprises various sensing units inside each room and activity samples provided form the basis of activity identification by the event classifier. Data from event classifier is then transferred over Internet to the cloud layer [66–69]. In here we use MQTT over SSL (Secure Socket Layer) to encrypt and secure the information between MQTT clients and MQTT broker, which in our case is Raspberry Pi 3B+.

(C) ***Cloud Layer***: This layer is comprised of different modules for the function of data authentication, real-time database, file storage and ML Kit for applying machine learning on generated CSV [70–73]. Here we use Firebase as BaaS (Backend as a Service). Data is first stored into the database, from where through cloud functions, we generate CSV files, and are stored in the storage module on which later machine learning model is applied for activity recognition [73–76].

(D) ***Processing Layer***: This layer takes the generated CSV files as input from cloud layer and applies the developed machine learning model on the CSV files, leading to results of activity recognition and activity pattern generation and mapping [77–79]. The generated activity pattern is then later used for anomaly detection in the real-time scenarios, which is the main objective of the whole system. A report is prepared at the end of day based on the pattern generated, which is in the end finally used for remote health monitoring by the caretakers.

(E) ***Application Layer***: This layer is where the whole health monitoring system interacts with the caretakers, based on the generated report. Regular report generation helps respective caretakers to remotely monitor elderly people health, without regular assistance of medical specialists with also results in cut-down in their spending. System also alerts caretakers on detection of anomaly, and if necessary he/she can consult a doctor by sharing the reports and take recommended steps [79–81].

2.1 *Activity Health Monitoring*

The objectives defined under this research, range from the provision of low-level data to sensors then to high-quality information addition and information transmission via together data-driven and information-driven methods. [4]. New publications have to do with the recognition of the work itself, as an income of mining high-quality data. Here, a range of actions, nonetheless general rule for altogether is that they would be seen through non-professional (e.g., "making foods," "bathing" or "watching TV while sitting on the couch") [5, 79–81]. When our activities are properly identified and automatically installed, a wide range of services are available, for instances, emergency medical emergencies, early diagnosis, and expert advice on general lifestyles, health monitoring, and physician assistance. Some practical samples of these applications can be found in the mobile emergency response system, fall detection system and related to daily health monitoring activities and recommend active lifestyle resources [6, 79–82]. The new data gathered, from the sensors, caters to a larger cause, helping professionals with valuable information to detect anomalies and facilitate patients.

2.2 Anomaly Detection

In addition to behavioral modeling, finding behavior change "anomaly detection" is equally important and challenging task. Changes in normal living behavior are tracked through anomaly detection, and the complexity of abnormal data is not considered to be the same as abnormal isolation behavior. [7, 83], and this has more to do with the problem of phase 1 classification than the traditional phase 2 (binary) problem. Detection changes can include changes in various aspects of a situation, for example, design, temporary, period imperative, work order, fitness position so on [6, 84]. Essentially, behavioral discovery is the challenge of measuring human behavior from sensor numbers and anomaly detection to the test of in what way to detect behavioral modifications that contain the trial (creative) of a model that may best detention the normal lifestyle (anomaly). These are two techniques to sense behavior change, comment and divider [8]. The comment puts a strain on the general behavior and considers any new entries that are inconsistent with this classical modal by way of an anomaly. Discrimination reads inconsistency information on or after past figures as well as examining aimed at the same design as new contribution result figures to reflect differences. The comment plan is additional accessible by way of the irregularity facts are less visible trendy actual lifetime, providing sample learning examples.

Anomaly detection, still being in a rookie among other methods, works in a clever household and is highly established cutting-edge additional areas, for instance, intervention recognition, fake recognition, agent recognition, industrialized feature recognition, image processing, text data, etc. [9, 85]. A range of machine learning methods have been used for anomaly detection, for instances, segmentation (derived law, neural network, Bayesian, SVM), Nearest-Neighbor, clustering, computation, information theory, viewer, etc. [10, 86]. Now an insolent home-based environment, change detection, has been used extensively in the security awareness of elder people and assistive technology for brain injury as dementia patients. Data can be determined in three ways, (1) an anomaly, (2) a contextual anomaly and (3) an integrated anomaly [11, 87].

2.3 Activities of Daily Living

The idea of ADL alludes to the things that we do in our everyday life and day-to-day self-care activities, for example, feeding, bathing, dressing, grooming, work, homemaking and recreation. These exercises define the capacity near living popular private households autonomously.

ADL are sorted interested in twice primary gatherings, fundamental activities of daily living also devices ADL. ADL is the essential residential and important human exercises for people's day-by-day schedules including portability, eating, drinking, dozing, dressing, washing and heading off to the restroom, and so on.

IADLs are different assignments that are not significant forever [12, 88]. Notwithstanding, IADLs comfort the old and impeded people, for example, housework, getting ready food, drug, working out, shopping, pressing, clearing and calling. The acknowledgment of human exercises in indoor situations is a troublesome undertaking to accomplish. ADL centers on tending to the difficulties and discovering answers for understanding human movement in keen conditions [13, 88]. Pervasive conditions like Smart homes have encouraged the everyday exercises discovery, helped members with housekeeping projects just as helped the disabled people and old to live serenely and freely [14, 88].

2.3.1 Activity Sensors for Daily Living

There are three types of sensors together of devices for the person action acknowledgment with wearable-based, physical condition-based and several sensors as following:

Physical Environment Based Sensors

Sorts of sensors, for example, RFID, proximity, pressure, ZigBee, and Wi-Fi can be utilized toward recognizing the association among the individual also nature nearby them. Ecological variable-based sensors utilize the crude information after detected items to expect action attempted via people [15, 89]. The conveyed devices identify action through substances as well as their connection with objects. The information stands gathered through universal sensors then referred to a neighborhood server for additional handling.

Wearable Sensors

Wearable sensors, for example, inertial sensors, accelerometers, electromechanical switches, goniometers and pedometers gyroscopes, are body-appended sensors as well as remain viewed as widely recognized devices aimed by people action acknowledgment. They are utilized toward perceive movements of human figure besides in the direction of help the people acknowledgment developments just, for example, drops recognition utilizing remote observing frameworks [16, 90]. The gadgets are intended to constantly quantify physiological information (fundamental indications of the human body) and biomechanical information. Investigating such information will assist with distinguishing human exercises in day-by-day living and make an interpretation of them into an important structure utilizing design acknowledgment.

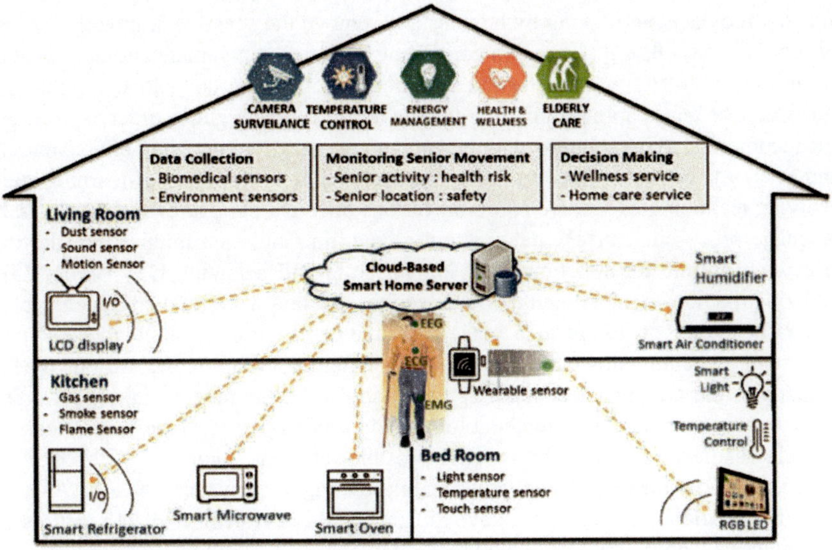

Fig. 2 Architecture of cognitive computing [17, 18]

Remaining Activity Recognition Sensors

Camera-grounded devices remain broadly rummage sale for social action acknowledgment inside a limited detecting inclusion; these sensors depend proceeding the camera footage as well as audiovisual arrangements toward perceived humanoid movement utilizing PC image calculations [17, 91]. Audiovisual instruments for instances, RGB video, RGB-D video as well as penetration pictures measuring device remain basic kinds of chromatic devices also largely recognized in mortal movement acknowledgment by decent acknowledgment proportions. Be that as it may, they are expensive, have high vitality utilization and additionally require visit upkeep and stand exposed near protection-linked alarms [18, 92].

Cognitive Computing

Cognitive computing term is generally used to elaborate the technological principles and platforms that are basically built upon the emerging fields of artificial intelligence along with signal processing. These platforms comprise of all latest machine learning and deep learning areas ranging from natural language processing, digital signal processing for audio classification to speech recognition, virtual reality, and robotics intelligence till the field of computer vision which incorporates vast application in this era [17, 84, 87, 91] (Fig. 2).

Cognitive computing in general sense refers to all the latest hardware and software setups that are aimed at stimulating the overall human brain functioning, in order to assist the decision-making by humans and remove the sense of ambiguity in the system [17, 58, 78, 84]. Cognitive computing systems amalgamate and harmonize the information from various technical sources, keeping the contextual weights in consideration and keeping conflicting sources and data in thought, in order to propose and recommend a best solution in the end of processing. To achieve such effectiveness and accuracy, cognitive systems incorporate all different types of self-learning and evolving technologies that are based on various different categories such as natural language processing (NLP), data pattern recognition and data mining in order to effectively mimic the functioning of human brain and behavior. [18, 49, 61, 75] Utilizing the non-living computer systems to resolve and workout the problems that generally humans are assigned to arises the unavoidable requirement of large amount of structured & non-structured data & information which are fed to various types of machine learning and deep learning algorithms. Over the time, cognitive systems have evolved and refined the methodology of identifying patterns in order to become capable of foresee and model respective possible solutions accordingly [19, 30, 40, 52, 65, 83]. To achieve such evolving capabilities, cognitive systems' five key/major attributes are defined which must have for any cognitive systems and are mentioned by Cognitive Computing Consortium [21–23]:

Contextual: As having a deep knowledge of context of the problem is integral part of solving the problem and critical in thought process, they should understand, mine and identify the contextual data, recognize, identify and fetch/extract various elements such as processes, user profile, the respective domain, rules and regulations, assigned goals and tasks. [20, 26, 34, 48] They may fetch and utilize various types of data such as structured, unstructured, sensory, audio and visual and imagery data [57, 68, 70, 79].

Interactive: One of the key components of cognitive systems is human–computer interactions (HCI). Hence, one of the integral requirements is that the users must be able to communicate with the systems easily and effectively, in both as an input and feedback manner, based on the evolving conditions of environment under consideration [19, 46–54, 61–70, 80–84]. Different technologies incorporated inside the system must be able to interact with other attached devices, micro-processors and remote cloud platforms too [20, 49, 59, 74].

ADAPTIVE: Cognitive systems are aimed at assisting the decision-making systems. Hence, they must be built in a fashion that allows them to iteratively collect information and have that much flexibility integrated into them that allows them to learn as the informational data changes and goals of the overall systems evolve. Hence, systems must work mainly on the real-time data and make adjustments to the pre-defined systems accordingly [20, 90–92].

Stateful and Iterative: Cognitive computing systems and technologies can also be trained and modified to recognize and approach problems by collecting surplus data by asking questions, and sensory real-time inputs in case available state information are insufficient or vague in nature to reach any final decision. In such situations, systems can approach the problem by using any similar event that has previously occurred in history [20, 83, 90–92].

3 Experimental Setup

In the experimental setup, each sensor is placed in different positions in the house, with a single ESP8266 attached to it, where the basic data processing takes place. ESP8266 comes with a Wi-Fi module attached to it, using which the processed data is then transferred and logged into Raspberry Pi B+ server. To avoid data duplication and erroneous data, pre-processing of sensor data is done at the ESP8266 level, thus ensuring that Raspberry Pi is only used as local server where clean data is stored in the whole setup. Logged data is periodically sent to the cloud storage to ensure backup of data is kept in case of any system failure, along with the logs of whole systems working, which can also be later on utilized for system's debugging (Fig. 3).

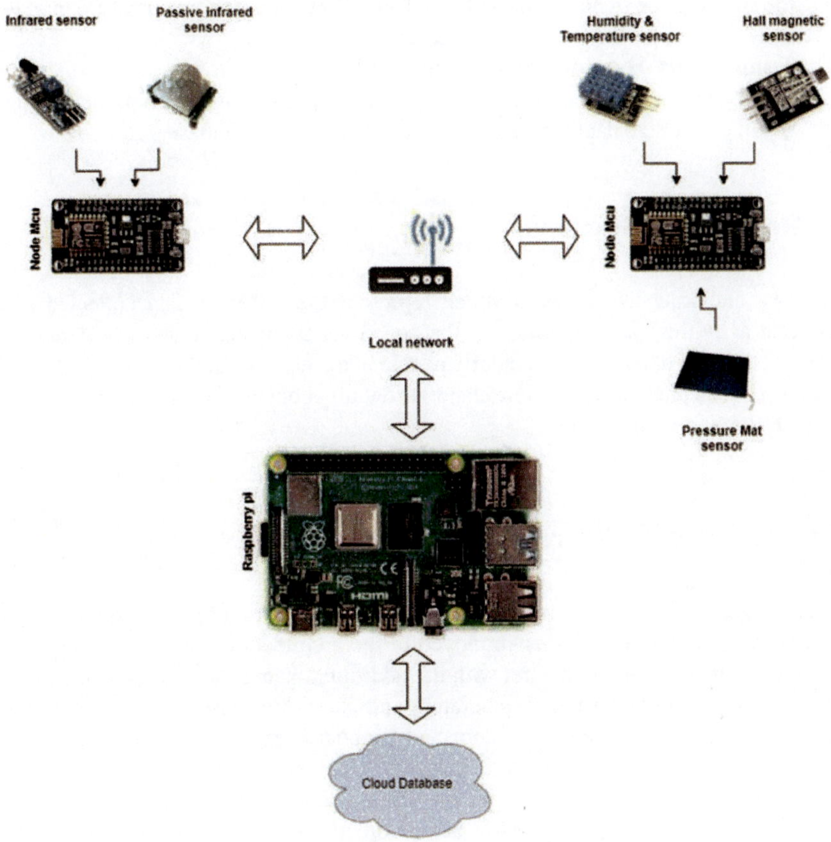

Fig. 3 Experimental setup

1. **PIR Sensor**:
PIR sensor is used to detecting the activity of the human in a room; it basically detects the motion of the human; it sends the data when it detects some motion. With the help of local network, it sends the data to cloud database.
2. **IR-Sensor (IR)**:
The IR sensor is similar to human's visionary detects, it is utilized to recognize the deterrents, and it sends the information to cloud database we a few snags is identify like when the entryway is close it distinguish the obstruction.
3. **Hall Magnetic Sensor**:
Hall magnetic sensor detects the magnetic field around the sensors. When it detects the magnetic field, it gets active and sends data to cloud database.
4. **Pressure Mat Sensor**:
Pressure mat is use to detect the pressure on any surface basically is used on the sitting place in house like sofa, bed, etc. When it gets some pressure, it gets active and sends the data to cloud database.
5. **Temperature-Humidity Sensor (dht-11)**:
The dht11 sensor detects the temperature and humidity in the room; it takes the 1 min of average data and sends the data to the cloud database.

4 Results and Discussion

Here we elaborate two datasets which represent the different data collected for different activities being recorded in the house via the installed sensory units for ADL (activity of daily living) of elderly people in the homes, for the cross-validation of the proposed methodology. More elaborate details about the datasets are mentioned in the following sections too.

4.1 Activities of Different Home Sensors Dataset

Data is being gathered on a single elderly resident house for a fixed period of about 1 week of time period. Sensors employed in the architecture are ambient sensors, being non-intrusive and in nature, with the additional benefit of having availability in low cost in the market too. Temperature and Humidity sensor DHT11, pressure mats on sofas and resting places, door and cupboard sensors are deployed on the scene for the respective activity data collection.

Table 1 reveals the activity mappings of a single elderly resident house, where activities as mentioned in the table are recorded; all these activities have their individual sensors, which for a period of 1 week, map the individual's data.

This Fig. 4 depicts the measure of humidity in atmosphere outside the house, and due to regular raining conditions, in ending days the humidity reaches the extreme level of 95% (Fig. 5).

Table 1 Sample of collected different home sensors data

Activity	Start time	End time
Humidity in atmosphere outside the house	26/07/2019	01/08/2019
Climate temperature	13/07/2019	19/07/2019
Bedroom	13/07/2019	19/07/2019
Internal temperature of house	26/07/2019	01/08/2019
Hall Temperature	13/07/2019	19/07/2019

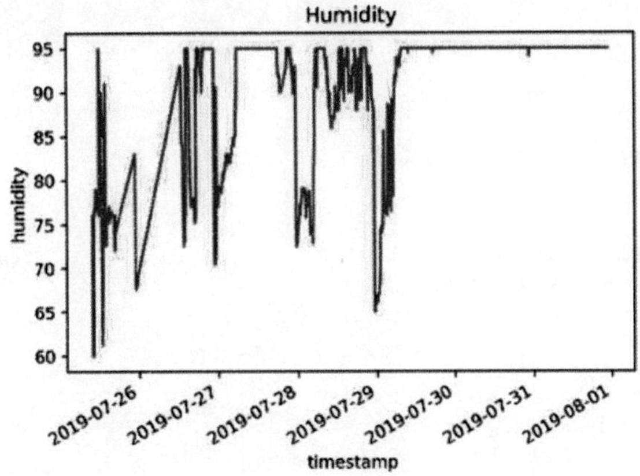

Fig. 4 Humidity temperature

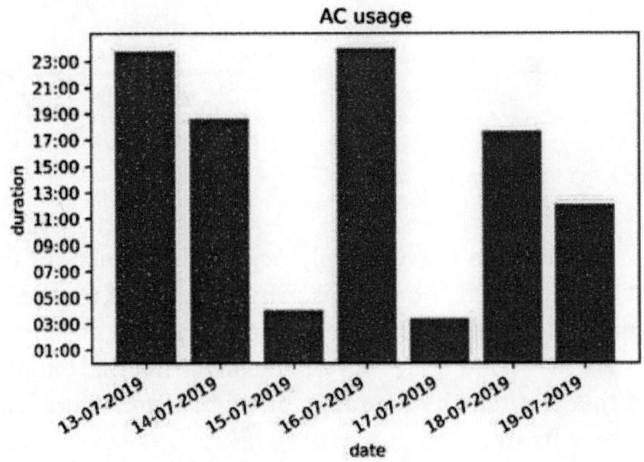

Fig. 5 Climate temperature

It depicts maximum usage on 13th and 16th of August, and minimum usage on 15th and 17th, depending on various factors like climate temperature, humidity, and time of presence in house and health condition on the respective weekdays.

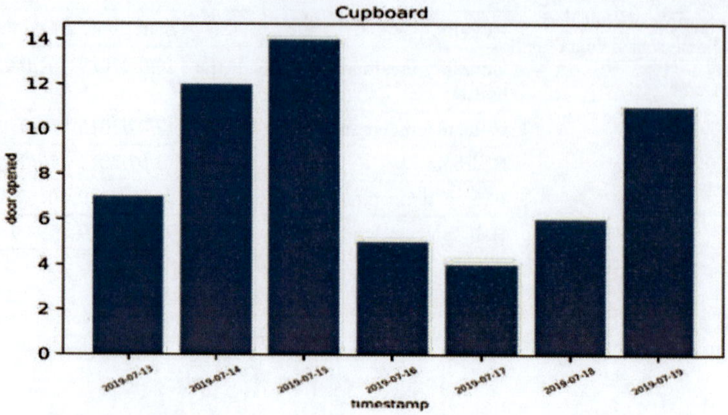

Fig. 6 Bedroom temperature

Fig. 7 Variation in the internal temperature of house

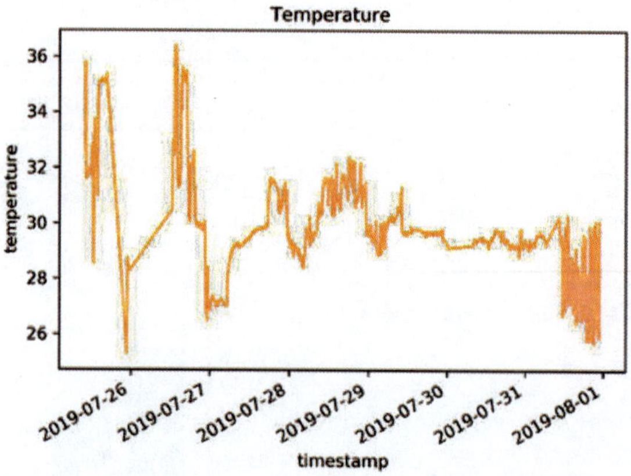

As shown in Fig. 6, simply depicts the number of times, the cupboard is opened in that day. In real time, this activity helps us to determine whether any activity is going on in the bedroom or person is reading or sleeping, etc.

As shown in Fig. 7, the variation in the internal temperature of house. And analysis of this plot tells us the range of room temperature suitable for the elderly people normal living.

4.2 *Activities of Daily Living Dataset*

Data generated by these sensors are both digital and analog in nature. For example, data generated by Hall magnetic sensor is digital and binary in nature and only 1

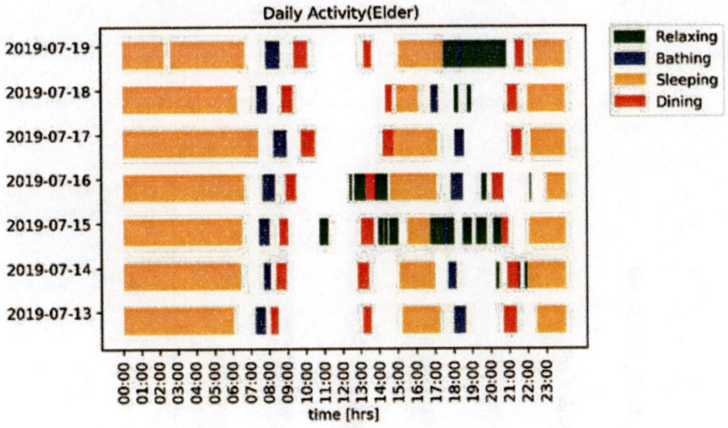

Fig. 8 Daily activity of elder people

Table 2 Sample of collected ADL data

Activity	Start time	End time
Relaxing	13/07/2019	19/07/2019
Bathing	13/07/2019	19/07/2019
Sleeping	13/07/2019	19/07/2019
Dining	13/07/2019	19/07/2019

or 0, indicating active and non-active states respectively, similarly data generated by pressure mat is also digital and binary in nature. Activities performed by the elderly people in the house are inferred from the sensor readings only with some applied suitable logic varying for each activity measurement. Figure 8 depicts the set of activities which are aimed for recognition by the network of sensors installed in the site location (elderly person house). Activities being covered/ recognized by the sensory network setup includes sleeping, bathing, reading, cooking, watching television, opening and closing of cupboards and doors of house, relaxing and sitting on the sofas and chairs, lavatory usage taking the total number of covered activities up to 14 in count. Each recorded activity has start time, location, end time, duration, attached to it as depicted in Table 2.

4.3 Comparisons of Elder and Young People Dataset

As compared to others, when elder people health deteriorates, their lavatory visits show a significant deviation from normal observed pattern. As you can see during 1 week, when elder person in the house was ill, there is observable change in their amount of lavatory visits (Table 3).

Table 3 Sample of collected ADL data

Activity	Start time	End time
Comparison of elder and younger people	13/07/2019	19/07/2019

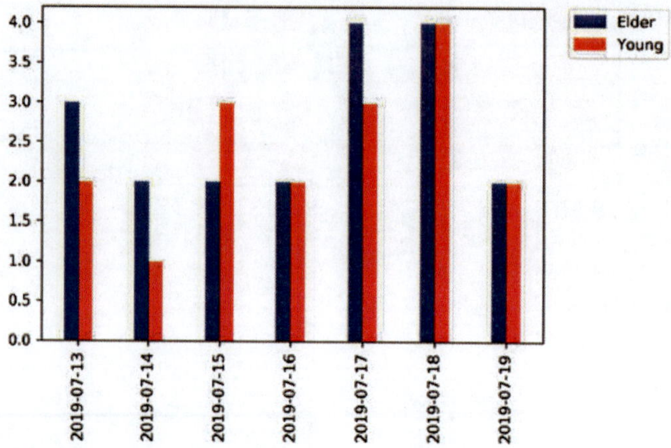

Fig. 9 Comparison of elder and younger data

As compared to others, when elder people health deteriorates, their lavatory visits show a significant deviation from normal observed pattern. As you can see during 17th and 18th August, when elder person in the house was ill, there is observable change in their amount of lavatory visits. In Fig. 9, there is a comparison made between the data of a elder person and the data of a youngster over a week of time.

5 Conclusion

This article presented report on the existing state of Internet of things (IoT) research by examining the currently relatable, and available chunks of literature, underlining current trends in the ongoing research field. We also aim toward elaborating the current challenges faced in developing the remote health monitoring systems, for predictive health monitoring and how concerns of privacy invasion has led down the systems based on visual imagery systems based on real-time computer vision applications and has pushed the researchers to move towards the health monitoring using non-invasive sensor networks. We propose a remote health monitoring system based on a wireless sensor network deployed on a house of an elderly person, with the aim of remote health monitoring of the elderly people inside the house and predict any abnormality based on the predictive analysis of behavioral pattern of the concerned person or patient, which can be done utilizing various machine learning models such as predictive time-series analysis and advanced hidden Markov models. Such systems can be effectively deployed in various locations such as homes consisting of elderly persons with their children staying in remote location, old age homes which are having very less caretakers in comparison to the strength of elderly people residing there, in hospitals for patient health monitoring, etc. In each of these locations, remote and predictive health monitoring can be a boon and that too based on the concept of edge computing and non-body sensors; hence, no need for person under consideration to regularly wear sensor on them. Using such systems, medical personal can also continuously see the behavioral reports generated by the system on his smartphones or laptops and give his prescriptions based on history of the patient or the person under surveillance. Some of the shortcomings of the current system are the erroneous data generated by the wrongly calibrated sensors, weak predictive power of the currently existing models and requirement of significant amount of data before the model starts to give its predictive results. But all of these shortcomings can be effectively overcome using some advanced machine learning models, making models capable of evolving themselves based on the real-time scenarios, and utilizing data collected in one location to be utilized for analysis for same group person until significant amount of data is available for current scenario. However, we aim toward further exploring this field of remote health monitoring and continuously making improvements in the current system, and further evolve the remote health monitoring systems by incorporating existing technologies in novel manner to further enhance the capability that IoT has to reshape or world.

References

1. Cad avid H, Garson W (2018) Towards a smart farming platform: from IoT-based crop sensing to data analytics. In: Radford T (ed) The Guardian [Internet] 2005. Springer. https://doi.org/10.1007/978-3-319-98998-3_19. Available from: https://www.theguardian.com/science/2005/mar/30/environment.research.
2. Nandurkar S, Thool V, Thool R (2014) Design and development of precision agriculture system using wireless sensor network. In: International conference on automation, control, energy and systems (ACES). Hooghly
3. Andrew R, Malekian R, Bogatinoska D (2018) IoT solutions for precision agriculture. In: MIPRO. Opatija
4. Benyezza H, Bouhedda M (2018) Smart irrigation system based thing speak and Arduino. International conference on applied smart systems. In: ICASS. Médéa
5. Mat I, Kassim M, Harun A (2015) Precision agriculture applications using wireless moisture sensor network. In: IEEE 12th Malaysia international conference on communications. Kuching
6. Fountas S, Aggelopoulou K, Gemtos T (2016) Precision agriculture: crop management for improved productivity and reduced environmental impact or improved sustainability "supply chain management for sustainable food networks. In: Supply chain management for sustainable food networks
7. Miles C (2019) The combine will tell the truth: on precision agriculture and algorithmic rationality. In: Big Data & Society, 1–12
8. Zhang L, Dabipi I, Brown W (2018) Internet of things applications for agriculture. Wiley
9. Ghayvat H, Mukhopadhyay S, Gui X, Suryadevara N (2015) WSN-and IoT-based smart homes and their extension to smart buildings. Sensors 15:10350–10379

10. Patil K, Kale N (2016) A model for smart agriculture using IoT. In: International conference on global trends in signal processing, information computing and communication. IEEE
11. Ashwini BV (2018) A study on smart irrigation system using IoT for surveillance of crop-field. Int J Eng Technol 7:370–373
12. Ananthi N, Divya J, Divya M, Janani V (2017) IoT based smart soil monitoring system for agricultural production. IEEE
13. González-Teruel J, Torres-Sánchez R, Blaya-Ros P, Toledo-Moreo A, Jiménez-Buendía M, Soto-Valles F (2019) Design and calibration of a low-cost SDI-12 soil moisture sensor. Sensors. 491. https://doi.org/10.3390/s19030491:19
14. Cambra C, Sendra S, Lloret J, Lacuesta R (2018) Smart system for bicarbonate control in irrigation for hydroponic precision farming. Sensors. 18
15. Kumar R, Dharwadkar N (2018) IoT based low-cost weather station and monitoring system for precision agriculture in India. IEEE;
16. Bhakta I, Phadikar S, Majumder K (2019) State of the art technologies in precision agriculture: a systematic review. J Sci Food Agricul
17. Cloudscene. Cloudscene. [Internet]. 2018 Available from: https://cloudscene.com/news/2018/05/internet-of-things-iot/
18. Pflaum A, Gölzer P (2018) The IoT and digital transformation: toward the data-driven enterprise. IEEE Comput Soc 18(1536–1268):5
19. Gupta B, Quamara M (2018) An overview of internet of things (IoT): architectural aspects, challenges, and protocols. Wiley
20. Balafoutis A, Beck B, Fountas S, Vangeyte J, Van der Wal T, Soto I, Gómez-Barbero M, Barnes A, Eory V Precision agriculture technologies positively contributing to GHG emissions mitigation, farm productivity and economics. Sustainability
21. Phupattanasilp P, Tong S (2019) Augmented reality in the integrative internet of things (AR-IoT): application for precision farming. Sustainability. 2658. https://doi.org/10.3390/su11092658:11
22. Ahmed N, De D, Hussain I (2018) Internet of things (IoT) for smart precision agriculture and farming in rural areas. IEEE Internet Things J 5 https://doi.org/10.1109/JIOT.2018.2879579
23. Naha R, Garg S, Georgakopoulos D, Jayaraman P, Gao L, Xiang Y, Ranjan R (2016) Fog computing: survey of trends, architectures, requirements, and research directions. IEEE Access 4:2169–3536
24. Sarker V, Queralta J, Gia T, Tenhunen H, Westerlund T (2019) A Survey on LoRa for IoT: integrating edge computing. In: Fourth international conference on fog and mobile edge computing
25. Raza U, Kulkarni P, Sooriyabandara M (2016) Low power wide area networks: an overview. IEEE
26. Ismail D, Rahman M, Saifullah A (2019) Low-power wide-area networks: opportunities, challenges, and directions. IEEE
27. Wixted A, Kinnaird P, Larijani H, Tait A, Ahmadinia A, Strachan N (2016) Evaluation of LoRa and LoRaWAN for wiereless sensor Network. IEEE, 16 (978-1-4799-8287-5).
28. Shilpa A, Muneeswaran V, Rathinam D (2019) A precise and autonomous irrigation system for agriculture: IoT based self propelled center pivot irrigation system. In: 5th international conference on advanced computing & communication systems
29. Chen W, Lin Y, Lin Y, Chen R, Liao J (2018) AgriTalk: IoT for precision soil farming of turmeric cultivation. IEEE
30. Elijah O, Rahman A, Orikumhi I, Leow C (2018) An overview of internet of things (IoT) and data analytics in agriculture: benefits and challenges. IEEE Internet Things J 5:2327–4662
31. Premkumar A, Monishaa P, Thenmozhi K, Amirtharajan R, Praveenkumar P (2018) IoT assisted automatic irrigation system using IoT assisted automatic irrigation system using wireless sensor nodes. In: International conference on computer communication and informatics. IEEE
32. Olatinwo S, Joubert T (2019) Enabling communication networks for water quality monitoring applications: a survey. IEEE 7:100332

33. Dholu M, Ghodinde K (2018) Internet of Things (IoT) for precision agriculture application. In: International conference on trends in electronics and informatics. IEEE
34. Naik N, Shete V, Danve S (2016) Precision agriculture robot for seeding function. IEEE
35. Chang C, Srirama S, Buyya R (2019) Internet of things (iot) and new computing paradigms. Wiley
36. Shin S, Chuang C, Huang H (2016) A Security framework for MQTT. In: IEEE conference on communications and network security
37. García S, Larios D, Barbancho J, Personal E, Mora-Merchán J, León C (2019) Heterogeneous LoRa-based wireless multimedia sensor network multiprocessor platform for environmental monitoring. Sensors 19 https://doi.org/10.3390/s19163446:3446
38. Linh AN P, Kim T (2018) A Study of the Z-wave protocol: implementing your own smart home gateway. IEEE, 18 (978-1-5386-6350-9)
39. Leikanger T, Schuss C, Häkkinen J (2017) Near field communication as sensor to cloud service interface. IEEE, 17 (978-1-5090-1012-7)
40. Liu Y, Qian K (2016) A novel tree-based routing protocol in ZigBee wireless networks. IEEE, 16 (978-1-5090-1781-2)
41. Martínez R, Pastor J, Álvarez B, Iborra A (2016) A testbed to evaluate the FIWARE-based IoT platform in the domain of precision agriculture. Sensors. 197.9 https://doi.org/:10.3390/s16111979:16
42. Carnevale L, Galletta A, Fazio M, Celesti A, Villari M (2018). Designing a FIWARE cloud solution for making your travel smoother: the FLIWARE experience. In: IEEE 4th international conference on collaboration and internet computing
43. Yu S, Park K, Park Y (2019) A secure lightweight three-factor authentication scheme for IoT in cloud computing environment. Sensors 19. https://doi.org/10.3390/s19163598:3598
44. Shirazi S, Gouglidis A, Farshad A, Hutchison D (2017) The extended cloud: review and analysis of mobile edge computing and fog from a security and resilience perspective. IEEE 35(0733–8716):11
45. Sarangi S, Naik V, Choudhury S, Jain P, Kosgi V, Sharma R, Bhatt P, Srinivasu P (2019) An affordable IoT edge platform for digital farming in developing regions. IEEE
46. Satyanarayanan M (2017) The emergence of edge computing. IEEE Comput Soc. 17:0018–9162
47. Math A, Ali L, Pruthviraj U (2018) Development of smart drip irrigation system using IoT. IEEE, 18 (978-1-5386-5323-4)
48. Pandithurai O, Aishwarya S, Aparna B, Kavitha K. Agro-tech: a digital model for monitoring soil and crops using internet of things (IoT). IEEE, 17. (978-1-5090-4855-7)
49. Aagaard A, Presser M, Andersen T (2019) Applying Iot as a leverage for business model innovation and digital transformation. IEEE, 19. (978-1-7281-2171-0)
50. Chandra N, Khatri S, Som S (2019) Business models leveraging IoT and cognitive computing. IEEE, 19. (978-1-5386-9346-9)
51. Whitmore A, Agarwal A, Da Xu L (2014) The internet of things—a survey of topics and trends. Springer
52. Pandya S., Sur. A., Kotecha K (2020) Smart epidemic Tunnel-IoT based sensor-fusion assistive technology for COVID19 disinfection. Emerald
53. Patel NR, Kumar, S (2017) Enhanced clear channel assessment for slotted CSMA/CA in IEEE 802.15.4. Wireless Pers Commun 95:4063–4081
54. Patel NR, Kumar S (2018) Wireless sensor networks' challenges and future prospects. In: 2018 international conference on system modeling & advancement in research trends (SMART). Moradabad, India, pp 60–65
55. Ghayvat H, Awais M, Pandya S, Ren H, Akbarzadeh S, Chandra Mukhopadhyay S, Chen C, Gope P, Chouhan A, Chen W (2019) Smart aging system: uncovering the hidden wellness parameter for well-being monitoring and anomaly detection, smart aging system: uncovering the hidden wellness parameter for well-being monitoring and anomaly detection. Sensors 19:766
56. Saket, S, Sharnil P (2016) An overview of partitioning algorithms in clustering techniques

57. Jaimeel MS, Ketan K, Sharnil P, Choksi DB, Joshi N (2017) Load balancing in cloud computing: methodological survey on different types of algorithm. IN: 2017 international conference on trends in electronics and informatics (ICEI). https://doi.org/10.1109/ICOEI.2017.8300865
58. Ghayvat H, Pandya S, Shah S, Mukhopadhyay SC, Yap MH, Wandra KH (2016) Advanced AODV approach for efficient detection and mitigation of wormhole attack in MANET. In: 2016 10th international conference on sensing technology (ICST).
59. Pandya S, Shah J, Joshi N, Ghayvat H, Mukhopadhyay SC, Yap MH (2016) A novel hybrid based recommendation system based on clustering and association mining. IN: 2016 10th international conference on sensing technology (ICST)
60. Patel S, Singh N, Pandya S (2017) IoT based smart hospital for secure healthcare system, 2017/5. Int J Recent and Innov Trends Comput Commun
61. Pandya SP, Prajapati MR, Thakar KP Assessment of training needs of farm women. Guj J Ext Edu 25(2):169–171
62. Pandya S, Ghayvat H, Sur A, Awais M, Kotecha K, Saxena S, Jassal N, Pingale G (2020) Pollution weather prediction system: smart outdoor pollution monitoring and prediction for healthy breathing and living. Sensors 20:5448
63. Pandya S, Ghayvat H, Kotecha K, Awais M, Akbarzadeh S, Gope P Smart home anti-theft system: A novel approach for near real-time monitoring and smart home security for wellness protocol. Appl Syst Innov 1(4):42
64. Patel RR, Pandya SP, Patel PK Characterization of farming system in north west agro climatic Zone of Gujarat State. Guj J Ext. Edu 27(2):206–208
65. Pandya S, Ghayvat H, Kotecha K, Yep MH, Gope P (2018) Smart home anti-theft system: a novel approach for near real-time monitoring. In: Smart home security and large video data handling for wellness protocol
66. Joshi N, Kotecha K, Choksi DB, Pandya S (2018) Implementation of novel load balancing technique in cloud computing environment … on computer communication and informatics (ICCCI)
67. Patel W, Pandya S, Mistry V (2016) i-MsRTRM: Developing an IoT based intelligent medicare system for real-time remote health monitoring-2016. In: 8th international conference on computational
68. Wandra KH, Pandya S (2012) A survey on various issues in wireless sensor networks. Int J Sci Eng
69. Swarndeep Saket J, Pandya S Implementation of extended K-Medoids algorithms to increase efficiency and scalability using large dataset. Int J Comput Appl
70. BholaYO, Socha BN, Pandya SB, Dubey RP, Patel MK (2019) Molecular structure, DFT studies, Hirshfeld surface analysis, energy frameworks, and molecular docking studies of novel (E)-1-(4-chlorophenyl)-5-methyl-N′-((3-methyl-5-phenoxy-1-phenyl-1H-pyrazol-4-yl) methylene)-1H-1, 2, 3-triazole-4-carbohydrazide. Molecul Crystals Liquid Crystals
71. Patel WD, Pandya S, Koyuncu B, Ramani B, Bhaskar S (2019) NXTGeUH: LoRaWAN based NEXT generation ubiquitous healthcare system for vital signs monitoring & falls detection. In: 2018 IEEE Punecon
72. Dandvate HS, Pandya S (2016) New approach for frequent item set generation based on Mirabit hashing algorithm. In: 2016 international conference on inventive
73. Swarndeep SJ, Pandya S (2016) Implementation of extended k-medoids algorithm to increase efficiency and scalability using large datasets. Int J Comput Appl
74. Wandra K, Pandya S (2014) Centralized timestamp based approach for wireless sensor networks. Int J Comput Appl
75. Garg D, Goel P, Pandya S, Ganatra A, Kotecha K (2002) A deep learning approach for face detection using YOLO. In: 2018 IEEE Punecon
76. Sur A, Pandya S, Sah RP, Kotecha K, Narkhede S (2020) Influence of bed temperature on performance of silica gel/methanol adsorption refrigeration system at adsorption equilibrium. Particul Sci Technol
77. Sur S, Sah RP, Pandya S (2020) Milk storage system for remote areas using solar thermal energy and adsorption cooling. Mater Today Proc

78. Cohen JM, Pandya S, Tangirala K, Krasenbaum LJ (2020) Treatment patterns and characteristics of patients prescribed AJOVY Emgality, or Aimovig,—Headache
79. Cohen JM, Pandya S, Krasenbaum LJ, Thompson SF (2020) A real-world perspective of patients with episodic migraine or chronic migraine prescribed AJOVY in the United States. Headache
80. Barot V, Kapadia V, Pandya S (2020) QoS enabled IoT based low cost air quality monitoring system with power consumption optimization. Cybernet Inform Technol
81. Ghayvat H, Pandya S, Patel S (2019) Proposal and preliminary fall-related activities recognition in indoor environment. In: 2019 IEEE 19th international conference on
82. Akbarzadeh S, Ren H, Pandya S, Chouhan A, Awais M (2019) Smart aging system
83. Ghayvat H, Pandya S (2018) Wellness sensor network for modeling activity of daily livings–proposal and off-line preliminary analysis. In: 2018 4th international conference on computing
84. Awais M, Kotecha K, Akbarzadeh S,Pandya S (2018) Smart home anti-theft system
85. Patel M, Pandya S, Patel S (2017) Hand gesture based home control device using IoT. Int J Adv Res
86. Pandya S, Yadav AK, Dalsaniya N, Mandir V Conceptual study of agile software development
87. Samani MD, Karamta M, Bhatia J, Potdar MB (2016) Intrusion detection system for DoS attack in cloud. Int J Appl Informat Syst
88. Review on various security threats & solutions and network coding based security approach for VANET
89. Bhatia J, Shah B (2013) Int J Adv Eng
90. Review on variants of reliable and security aware peer to peer content distribution using network coding
91. Patel P, Bhatia J (2012) Nirma University International Conference on
92. Bhatia J, Kakadia P, Bhavsar M, Tanwar S (2019) SDN-enabled network coding based secure data dissemination in VANET environment IEEE Internet Things J

Ant Colony Optimization for Traveling Salesman Problem with Modified Pheromone Update Formula

Rahil Parmar, Naitik Panchal, Dhruval Patel, and Uttam Chauhan

Abstract Traveling Salesman Problem is a combinatorial problem from which various other problems have been derived in the real-world application. It is a well-known NP-complete problem. Its instances are used in various fields around the globe. There have been various optimization techniques that are used to solve this problem. The Ant Colony Optimization (ACO) is an optimization method that is very useful in solving various artificial intelligence problems and obtaining the optimized solution. There have been methods proposed after its introduction in 1991. When using the traditional ACO pheromone update formula on the large dataset of Traveling Salesman Problem, one might get an optimal solution at the cost of a great amount of time. In this paper, we have proposed a modification in the basic Ant Colony Optimization pheromone update formula for discovering the optimized solution for the Traveling Salesman Problem using the probability from the pheromone value from succeeding nodes. This updated formula also helps in reducing the time to obtain the optimal solution as compared to the traditional formula.

Keywords Artificial intelligence · Ant colony optimization · Traveling salesman problem

1 Introduction

Artificial Intelligence (AI) can be described as intelligence in machines that can resemble human intelligence and can perform some tasks, which requires logical thinking to solve the problem. Some of the tasks can be categorized from everyday tasks like speech, translation, generation of language to expert level tasks like manufacturing planning, scientific analysis, etc. Now for solving these tasks, the AI system needs some dataset, which it can process and find the solution. These processes require searching and optimizing. For example, the AI system which is designed to solve the single-player tic tac toe game, two-player chess or to get the shortest path

R. Parmar (✉) · N. Panchal · D. Patel · U. Chauhan
Vishwakarma Government Engineering College, 382424 Ahmedabad, Gujarat, India

P. K. Singh et al. (eds.), *Proceedings of Second International Conference on Computing, Communications, and Cyber-Security*, Lecture Notes in Networks and Systems 203,
https://doi.org/10.1007/978-981-16-0733-2_2

in the graph, the system needs to search the input dataset to achieve the required outcome. For searching in the dataset, the AI system uses various search algorithms like Depth First Search (DFS), Breadth First Search (BFS), Iterative Deepening DFS, bi-directional search or A* search, etc. [8].

The efficiency of these algorithms may depend on the size of the search state space generated from the given input. As the size of the obtained search space state becomes larger, these algorithms can take a very long time to generate the solution [2]. Besides, the solutions may become less accurate and less efficient. Heuristic methods are used for increasing the efficiency of these algorithms by optimizing the generated path to a solution. It can be done by finding such solutions that decrease the size of the search state space and the time required to achieve these results. Heuristic techniques often generate good-enough solutions, but it does not guarantee the optimal solution [4].

Optimization algorithms are important to the various fields in real life. Examples of the practical implementation of the optimization algorithm include train scheduling, telecommunication, shape formation, routing technique, drone pathfinding, and many more.

1.1 Meta-Heuristic Methods in AI

The term meta-heuristic is composed of two Greek words, the suffix meta means "more organized" and the heuristics mean "to find". A meta-heuristic procedure is used to find the best solutions from the set of feasible solutions. Meta-heuristic can be used in combinatorial optimization to generate a better solution with more efficiency than simpler heuristic methods. Combinatorial optimization is to find the near-optimum solution from the set of feasible solutions. The Traveling Salesman Problem is a good example of NP-hard problems in combinatorial optimization [4].

The Particle Swarm Optimization (PSO), Genetic Algorithm (GA), and Simulated Annealing are the fields where the meta-heuristic methods are used widely. Genetic Algorithms are the search-based optimization techniques that are based on the concept of natural selection and genetics. The Genetic Algorithms can be briefly summarized as follows [10]:

1. First, a node is selected at random from the given nodes.
2. Then the pool of possible solutions is evaluated from the current node.
3. These solutions then undergo various recombinations and mutations which produce the new children.
4. This process is repeated for various generations. The fitness value is assigned to each child solution obtained.
5. The node with a better solution is more likely to produce an optimized path.

This way the Genetic Algorithm is implemented on various problem domains.

The Simulated Annealing has been developed by taking inspiration from the metal annealing. The annealing method involves the heating and cooling of the metal to change its physical property by changing its inner structure. As the metal cools, its new structure becomes fixed and retains its new property. Suppose S is a set of obtainable solutions. The initial start point is selected at random from the set of obtainable solutions. Now from the currently selected solution, the cost function to other solutions is calculated using the gradient descent which is, in this case is performed on the temperature. If the calculated cost function is reduced, then the current solution is replaced by the generated solution. Otherwise, the generated solution is rejected. This process is carried out until the optimal solution is obtained. This technique is used in various other fields like Machine Learning and Deep Learning to train the model to predict the result and learn to optimize.

Particle Swarm Intelligence is an algorithm that is derived from the social behavior of the flock of birds, schools of fish, and the ant colony [17]. This algorithm emulates the interaction between the members to share the information among them. The individual solution in these methods is considered as a particle. Each individual in PSO flies with the velocity that is dynamically adjusted according to its own experience and the experience of its companion [22]. Given the solution set S with the position of each solution in sample space, we can obtain the optimal solution using PSO. From the current position of the swarm, the fitness of the particle can be calculated by using the objective function. The solution with a better fitness value is selected, and the optimal solution is created. While creating the optimal solution, the previous best solutions are remembered for the backtracking.

This paper is further organized as follows: Sect. 2 explains the famous Traveling Salesman Problem, it also involves how to solve Traveling Salesman Problem using Ant Colony Optimization. Section 3 describes the variants in the Ant Colony Optimization. Following this, we have listed the current state-of-the-art techniques in Sect. 4. Next, we have presented modified formula along with the necessary pseudocode in Sect. 5. In Sect. 6, we have showcased our experimental results in the form of a graph, where we have compared the proposed approach with Ant Colony System. Finally, we present some concluding remarks in Sect. 7.

2 Traveling Salesman Problem (TSP)-NP-Hard

The TSP is a very well-known NP-hard problem in computer science. The TSP is an NP-Hard problem so it means that to solve this problem, there is no direct efficient way. If the solution of TSP is found, then it is possible to find the solution of the famous P versus NP problem. Besides, there are other applications of TSP, so we decided to apply Ant Colony Optimization (ACO) on TSP.

The problem can be described as there is a list of cities which are to be visited by the traveling salesman, the distance between the cities are given. The traveling salesman has to visit every city on the list and return to the original city, but he can only visit each city only once and after visiting a city he cannot re-visit the city. The main aim of the TSP is to find the shortest route to visit each city once. TSP is a problem of combinatorial optimization [19]. The TSP can be defined by the Hamiltonian cycle problem.

In a directed or undirected graph, there may exist a path that visits each node in the graph exactly once this type of path is called the Hamiltonian path, when the Hamiltonian path forms a cycle it is called the Hamiltonian cycle. In TSP, we need to find the Hamiltonian cycle with the smallest cost. Many problems are similar or related to the original TSP such as Generalized TSP, Bottleneck TSP, Steiner TSP, etc. The generalized TSP is known as the SetTSP [1]. In the generalized TSP, there are sets of nodes of the graph and from each set, we have to visit at least one node and we need to visit each set to find the smallest cost Hamiltonian cycle. The TSP can be called the specialized case of the Generalized TSP in which each set contains exactly one city [12]. In the figure given below, each node represents the city in a graph. The ACO is implemented on the symmetric TSP here in this case. In symmetric TSP, the distance from city A to city B and from city B to city A is the same. So, it can be represented as $d_{AB} = d_{BA}$, where d is the distance between these two cities, which forms an undirected graph. In asymmetric TSP, the distance from city A to city B and from city B to city A can be different because the graph for asymmetric TSP is directed graph so the path between two cities may be in one direction only (Fig. 1).

The TSP is a problem of combinatorial optimization. It is observed that TSP having 20 or fewer numbers of cities can be solved using the specific techniques like dynamic programming or branch and bound method. It can provide the optimal solutions efficiently [20], but when the number of cities increases, the combinations to search the feasible solution also increases. So for a larger number of cities, AI search optimization techniques like Swarm Intelligence, Artificial Bee Colony (ABC), ACO, or genetic algorithm can be applied to solve the TSP efficiently.

TSP is directly applied in logistics and transportations. Additionally, TSP can be applied in many areas such as the production of ICs and PCBs, where it is used in drilling machines for decreasing the time of these processes. TSP is used in computing the DNA sequencing. In finding the shortest path between airports, the TSP can be

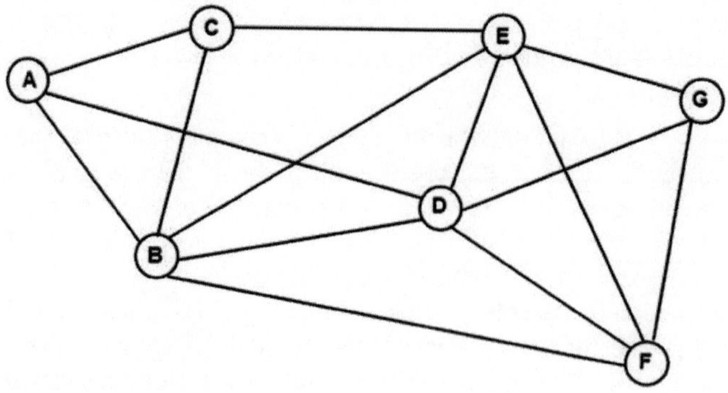

Fig. 1 Traveling salesman problem

applied. Moreover, The TSP can be applied to deliver the power to home using the fiber optics network designing.

There are many state-of-the-art methods to solve the TSP. These methods include some complete algorithms for solving the TSP, most of the algorithms are based on branch and cut methods. The Concorde TSP solver has solved almost all of the TSPLIB instances. The current largest solved TSPLIB instance includes 85,900 cities. Other state-of-the-art methods are the type of Stochastic Local Search (SLS) algorithms, these are Construction heuristic, Hybrid methods, Population-based methods. Population-based algorithms are mostly used as a basis for developing more efficient TSP algorithms. There are a few limitations of SLS algorithms. SLS algorithms can sometimes get into the Stagnation situation, it is difficult to know if the algorithm is in stagnation or not. It is also not guaranteed to find a solution.

2.1 TSP using ACO

ACO was developed from the foraging behavior of the real ants [5]. The first application of ACO that successfully had an advantage over Simulated Annealing and Genetic Algorithm approaches that were aimed to solve this type of dynamic problems. The efficient ACO algorithms adjust between search intensification and diversification.

But in ACO, there exist problems of stagnation situation and premature convergence. Furthermore, the convergence speed of ACO is slow, and as the size of the problem increases, these problems become more obvious. Hence, the ACO algorithm needs more improvement [18]. Initially, the ants move in a random direction in search of the food near their nest. The ants leave behind the special type of chemical known as pheromone due to ant's weak perception about the environment, to communicate to the other ants. The ant travels from its nest to the food source by finding the optimal path and leaving the traces of pheromone for the other ants to follow. As more amount of pheromone is accumulated on the path, the ants will start following the same path [9]. In the ACO algorithm, ants are the agents that search for the solution in the available solutions and then try to optimize the path to reach the solution. The biologists have described the foraging behavior of ants composed of three mechanisms. (1) Selection Mechanism: If more information is on the path, then the probability of choosing that path is higher. (2) Update Mechanism: The amount of pheromone increases with the number of ants and decreases with time. (3) Co-ordination Mechanism: The communication between the ants is carried out in coordination [21] (Fig. 2).

While traveling through the solutions, the (artificial) ants deposit the trail of (artificial) pheromone on the edges. The solution of the problem depends on the quality and distance of the solution of the previous solution [19]. After certain iteration, the ants travel through the same path and show the little deviation in the path, providing the optimized solution to the problem. The Ant System algorithm was originally a set of three algorithms based on their implementation. Those algorithms were Ant-Cycle,

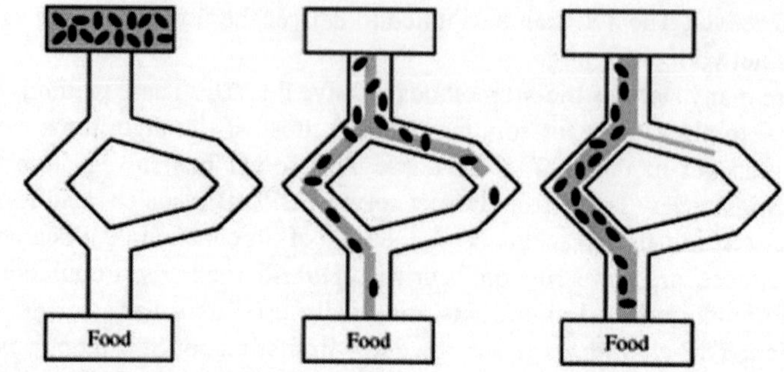

Fig. 2 Ant colony system

Ant-Density, and Ant-Quantity. The difference between these three was the updating of the pheromone at the nodes. In the Ant-Cycle algorithm, the pheromone update was done at the end of each ant's tour while in the latter two, the pheromone was updated at each subsequent step. In Ant-Cycle, the pheromone is updated depending on the length of the tour completed by the individual ant, while in the other two algorithms the pheromone is updated according to distance between two cities [3].

Given the weighted graph $G = (V, E)$ of n cities, the TSP can be stated as the smallest tour length of the agent based on the Hamiltonian distance. The Hamiltonian distance between the two cities city i and city j can be given by d_{ij}. The ant k at the city i chooses the city j through probability p_{ij}^{k}, which is calculated as the function of the city distance and the pheromone amount at that city [15]. The probability for choosing the next city j can be given as:

$$P_{ij}^{(t)} = \begin{cases} \frac{\left(\tau_{ij}^{\alpha}\right)\left(\eta_{ij}^{\beta}\right)}{\sum \tau_{ij}^{\alpha}\big)\left(\eta_{ij}^{\beta}\right)}, & \text{if } j \notin \text{ tabu list} \\ 0, & \text{otherwise} \end{cases} \tag{1}$$

Here, τ is the pheromone value at city j and η is the visibility of the city j from city i (i.e., the inverse of the distance between the cities). The pheromone value τ is updated at every edge according to Eq. (2).

$$\tau_{ij}(t+1) = \rho\left(\tau_{ij}(t)\right) + \tau_{ij}^{k}(t) \tag{2}$$

where ρ is the rate of evaporation and its range is in $0 < \rho \leq 1$. Here the evaporation rate ρ is used for avoiding the unlimited deposition of the pheromone. It allows the ants to improvise the path rather than being stuck up in the local best option. Here $\Delta\tau_{ij}$ is the pheromone the previously visited ants accumulated on the node and it is given as follow:

$$\Delta\tau_{ij}^{k}(t) = \begin{cases} \frac{Q}{L^{k}}, & \text{if } j \text{ is visited} \\ 0, & \text{otherwise} \end{cases} \tag{3}$$

where L^k is the distance traveled by the ant so far. Q is the positive constant. From Eq. (3), it can be seen that better the tour of the ant, the more amount of pheromone is deposited by the ant at the node. After some iterations, the path chosen by many ants will contain more amount of pheromone and therefore it has more probability of being chosen by the ants in upcoming iterations.

3 Variants in ACO

Originally, the ACO algorithm was known as Ant System (AS) which was presented by Dorigo et al. in the early 90 s. Later there were some improvements introduced over the AS. These improvements include the Elitist Ant System (EAS), Rank-Based Ant System (RBAS), MAX–MIN Ant System(MMAS), and Ant Colony System (ACS). In these improvements, the methods for solution generation and pheromone evaporation were similar but the pheromone update and pheromone trail management were improved. In all the systems, the evaporation factor is used which evaporates the pheromone at the nodes at the predetermined rate, which in turn helps the algorithm to explore new paths and enables it to escape the premature convergence to the optimal solution. Over the years, many improvements are added to the original algorithm.

Elitist AS: It was the first improvement over the AS. In EAS, the best optimal tour traveled obtained since the start is dispensed with additional reinforcements. This is done by pheromone sublimated by the additional ant called best-so-far ant and the tour is called tbs (best-so-far-tour) [6].

Rank Based AS: The rank-based Ant System improvement was done by Bullnheimer et al. In RBAS each individual ant deposits a certain amount of pheromone which decreases with its range and the ant with the best-so-far tour will deposit the colossal amount of pheromone in each iteration. As a result, RBAS provides a moderately better result than EAS and significantly better results than AS [6].

MMAS: The MIN–MAX Ant System was introduced by Stützle and Hoos. The MMAS has four modifications over the AS. The first modification is that either only the ant which has produced the best tour in the current iteration, which is called iteration-best ant or the best-so-far ant is allowed to change or add the pheromone to the nodes. This may direct result in a stagnation situation. In this situation, all the ants follow only one path or tour because it has an excessive amount of pheromone deposited. This tour may be good or suboptimal. The stagnation situation may halt the exploration of another path that can be optimal. Hence to counteract the stagnation situation, the second modification was implemented in MMAS [6].

As per the second modification, it limits the pheromone trails' possible range in the interval $[\tau_{\min}, \tau_{\max}]$. The third modification was to initialize the pheromone value to the upper limit. Along with a compact pheromone evaporation rate, the exploration rate of other tours from the beginning of the search is increased. The fourth modification was that the pheromone values are to be reinitialized whenever the system is stuck in a stagnation situation or when no enhanced tour is obtained for some consecutive iterations.

Based on experimental results, it is seen that for TSP having a small number of cities, the iteration-best ant should modify the pheromone values and for TSP having a larger number of cities pheromone on the best-so-far tour is updated only.

ACS: Ant Colony System differs from AS in three respects. Primarily, ACS uses more belligerent action choice rule than AS. Subsequently, global pheromone update and local pheromone update. In tour construction, ACS ants use the pseudorandom-proportional action choice rule, it is given by:

$$j = \begin{cases} -\text{argmax}_{L \in N_i^k} \tau_{il} \left(\eta_{il}^{\beta} \right), & \text{if } q \leq q_0 \\ 0, & \text{otherwise} \end{cases} \tag{4}$$

where q is randomly distributed variable in the range [0, 1]. q_0 $(0 \leq q_0 \leq 1)$ is a parameter, J is calculated by $(\alpha = 1)$. If the probability is q_0, then the ant selects the best possible solution learned from previous pheromone trails information. Here the ant makes use of the previously acquired knowledge. The tour is constructed based on best-so-far tours or to reconnoiter other tours. If the probability is $(1\text{–}q_0)$ the biased exploration is done. In global pheromone update, only the ant with a best-so-far tour is allowed to update the pheromone over the best-so-far tour. In the local pheromone update, it is applied immediately after each arc is crossed. After an arc is visited, the pheromone value over that arc is reduced, so the exploration rate of other arcs is increased and it does not result in the stagnation situation.

4 State-of-the-art

Since its development, various research papers and the literature have been published for the improvements in the ACO. The ACO has been applied to various domains in the different forms and with the hybridization with various other techniques, the improvements have been made according to the need. In the latest ACO research, authors mainly focus on the areas of integrating various algorithms with the ACO and the application of ACO in various other disciplines.

4.1 Hybrid Method Based on ACO and 3-Opt Algorithm

One of the most common problems faced during the implementation of the ACO is premature stagnation. The stagnation can be defined as the condition in which all the ants traverse through the same track and construct the same repercussion, again and again considering there is no other optimal path. To overcome this problem, the hybrid algorithm which contains multiple colonies of ants shares the global best tours occasionally to guide to the better solution. If the colony is stuck in the stagnation condition, the other colonies extricate it from the condition. Thus, this algorithm provides better and more robust solutions. This algorithm requires the colonies to be executed independently at the same time to find the optimal tour [11]. This algorithm considers the tour completed by ant and finds the optimal solution. But it fails to look at the tour which is yet to be performed. For a very large dataset, we might need large number of iteration for the maximum number of tours to be exploited.

4.2 *Annealing Elitist Ant System with Mutation Operator*

In this algorithm proposed by Abdulqader M. Mohsen, the ant has the two options for selecting the next city to be visited. Given *n* ants which are to be traveled through m cities, the ants will be distributed randomly across the cities. Initially, the pheromone level at all the cities is initialized to a small positive integer. At every iteration, an ant will have to choose either mutation operation or simulated annealing, based on the multifariousness of elitist ant system, to improve the performance of the algorithm. Mutation Operator is an algorithm in which each ant is provided with the chance to amend according to the predetermined probability. This operator helps the algorithm to survey different scope in the search space. If the diversity is greater than some value, the algorithm needs intensification which can be achieved through annealing on the ratio of solution pool. If the diversity is less than some value, it means the algorithm is losing its diversity and there is a probability to be stuck in local minima. Therefore, the algorithm needs to increase diversity, which can be attained by applying the mutation operator [16]. The variousness in the fitness of the ants in the algorithm can be obtained by the Euclidean Distance (ED) which is given as:

$$\mathrm{ED} = \frac{\overline{d} - d_{\min}}{d_{\max} - d_{\min}} \tag{5}$$

Here,

Here $\overline{d}$ is the mean of the fitness value of the best ant and fitness value of other prevailing ants in the solution. $d_{\min}$ is the distances of the worst ant fitness and $d_{\max}$ is the second-best ant fitness from the best ant respectively. In this algorithm, we need to tune diversity parameter in order to decide whether the algorithm requires diversification or intensification based on the dataset provided. This makes the algorithm very tedious to execute on different datasets as we need to reset the value for different datasets.

4.3 Greedy–Levy Flight ACO

There is always a perplexity in the reinforcement learning of further exploitation and exploration. To address this dilemma, Greedy–Levy Flight ACO was developed. The Greedy–Levy ACO comprises two algorithms, Epsilon Greedy Algorithm and Levy Flight Algorithm. The Epsilon-Greedy policy is technique of exploration which is used in ACS (Ant Colony System) algorithm [14]. The exploitation in Epsilon-Greedy policy is done with the probability of epsilon while selecting the best node available and the exploration is done with the probability of 1-epsilon. In case of 1-epsilon probability, the Levy Flight technique is used to improve the results. In this algorithm, initially a random number P is generated such that $0 < P < 1$. If the $P \leq \varepsilon$, the candidate solution with maximum probability is selected. If $P > \varepsilon$, then a candidate is selected randomly by using the Levy Flight algorithm.

The Greedy–Levy ACO tries to achieve the balance between local search and global search for generating more optimal results which is very crucial in increasing efficiency of the algorithm and it is done by implementing the epsilon-greedy method and Levy Flight technique [13]. The probability of choosing the next node in the Greedy–Levy algorithm can be given as follows:

$$P_{ij}^{(t)} = \begin{cases} \operatorname{argmax}\left\{\left(\tau_{ij}^{\alpha}\right)\left(\eta_{ij}^{\beta}\right)\right\}, & \text{if } P \leq \epsilon \\ 1 - A \times \frac{1-P_{\text{levy}}}{1-P_{\text{threshold}}} \times \left(1 - \frac{\left(\tau_{ij}^{\alpha}\right)\left(\eta_{ij}^{\beta}\right)}{\sum \tau_{ij}^{\alpha}\right)\left(\eta_{ij}^{\beta}\right)}\right) & \text{if } P > \epsilon, \text{ if } P_{\text{levy}} \geq P_{\text{threshold}} \\ \frac{\left(\tau_{ij}^{\alpha}\right)\left(\eta_{ij}^{\beta}\right)}{\sum \tau_{ij}^{\alpha}\right)\left(\eta_{ij}^{\beta}\right)}, & \text{else} \end{cases} \tag{6}$$

Here, A = altering ration of Levy Flight.

P_{levy} = Probability of turning on/off levy Flight altering, $0 < P_{\text{levy}} < 1$.

$P_{\text{threshold}}$ = Parameter for Levy Flight threshold, $0 < P_{\text{threshold}} < 1$.

The Greedy–Levy algorithm improves the program by tuning the exploration and exploitation of the ants, and hence, it requires more iteration to get to the optimal solution.

5 Proposed Solution

Ant Colony Optimization is one of the most popular search methods among artificial intelligence based on the real behavior of ants. There have been various modifications to the algorithm since its formulation by Dorigo. In this paper, we propose a modification in the basic ACO's pheromone update formula which improves the convergence speed of ACO in comparison to the basic ACO model and improves the time required to get the optimal solution. When an ant travels from city i to city j, the pheromone value of the city j is updated according to the distance traveled by that ant so far. We propose to add a γ amount of pheromone of the next city k from the city j, where city k has the maximum amount of pheromone value from the city j. Hence, the pheromone value at the city j will be updated by the Eq. (5) given below:

$$\tau_{ij}(t+1) = \rho\left(\tau_{ij}(t)\right) + \Delta\tau_{ij}^{k}(t) + \gamma\left(\tau_{jk}(t)\right) \tag{7}$$

Here τ_{jk} is the pheromone value at the city k from the city j and *gamma* is the factor which decides the amount of pheromone value to be added. The city k is selected based on the following equation:

$$k = \operatorname{argmax}\left\{\tau_{jl}\right\} \tag{8}$$

The city k is determined based on the next city from city j which has the maximum pheromone value. By adding this modification to the original algorithm, the convergence speed increases, and the amount of time required to reach the optimal solution decreases greatly. The pseudocode for the ACO is given below:

Algorithm 1: Pseudocode for ACO

```
Distribute ants at random cities;
for i in number of ants do
    while all the cities are not visited do
        Add city to tabu list;
        Calculate the probability for next city;
        Calculate pheromone value;
        increment k by 1;
    end
end
Increment i by 1;
Calculate optimized path;
```

The reason for this convergence speed and the more optimal solution lies in the future path of the solution. In the proposed modification, we are adding a certain portion of the pheromone from the next nearest solution which adds the probability of the next succeeding path to the current solution. As the number of iteration increases and the number of ants visiting the nodes increases, the pheromone value from the succeeding optimal path is accumulated at the node, which increases the probability of choosing the node with the global best solution.

While in the original Ant Colony System, the ant chooses the next node based on the local optimal solution. Moreover, as the number of ants increases beyond some point, the pheromone value from various nodes is repeatedly accumulated at some nodes, which in turn leads the ants to choose the path with maximum pheromone amount which is the result of the various repeated pheromone from the succeeding nodes.

6 Experiments and Results

In this section, we provide the experiments and their results obtained on the execution the algorithm. The algorithm has been developed in python and has been implemented on the Windows 10 64-bit Operating System, Intel Core i7 2.5 GHz processor and 8 GB RAM. We adjusted the parameters for the comparison of the original algorithm with the proposed algorithm. Here we have varied the number of cities and the number of ants to present the difference in the optimal solution and time provided by the original and the proposed algorithm. In the implementation of the algorithm, we have used the real data of the TSPlib benchmark libraries which is available on the web: https://comopt.ifi.uniheidelberg.de/software/TSPLIB95/XML-TSPLIB/instances/.

According to the survey paper published by Dorigo and Stutzle, in 2019, minmax ACO and Ant Colony System are still the state-of-the-art techniques [7].

Hence we've compared the results obtained with Ant Colony System and used it as a benchmark. To verify the performance of the modified formula, it has been applied to various TSP instances. The algorithm provides the promising results when used with the modified pheromone update formula. We have implemented and compared the results obtained by the ACS algorithm with the traditional pheromone update formula and the modified pheromone update formula. For the experimentation, we have initialized the variables as follows: $\alpha = 1$, $\beta = 2$, $\rho = 0.5$ and $\gamma = 0.4$ (Table 1; Figs. 3 and 4).

It can be perceived from the results that the proposed modification generates an optimized path with a fewer number of ants and less time than the original ACO. It is due to the consideration of the pheromone value of the next node from the traveled node which adds the pheromone value of the succeeding node with an optimal solution from the graph. With the increasing number of iteration, the pheromone value from the succeeding nodes gets accumulated at the node which provides the optimal solution according to the global optimal solution. While in traditional ACO, the next node is selected based on the local best solution. Because of this, the next node

Table 1 Statistical comparison with benchmark algorithm

TSP instances	No. of cities	ACS with original formula (s)	ACS with proposed formula (s)	Percentage change (%)
berlin52	52	8.13	6.29	22.63
eil76	76	17.01	10.3	39.45
kroA100	100	37.27	22.0	40.97
pr124	124	89.41	41.15	53.98
rat195	195	265.34	156.62	40.97
a280	280	813.57	472.56	41.92
rd400	400	2669.75	1464.34	45.15
ali535	535	6385.28	3184.64	50.12

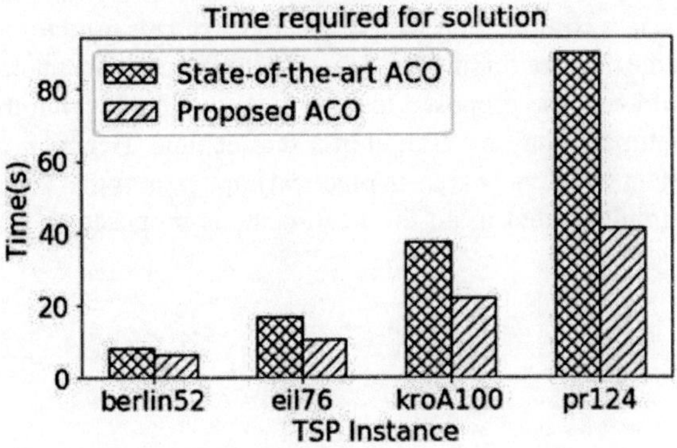

Fig. 3 Comparison of algorithm for less than 150 cities

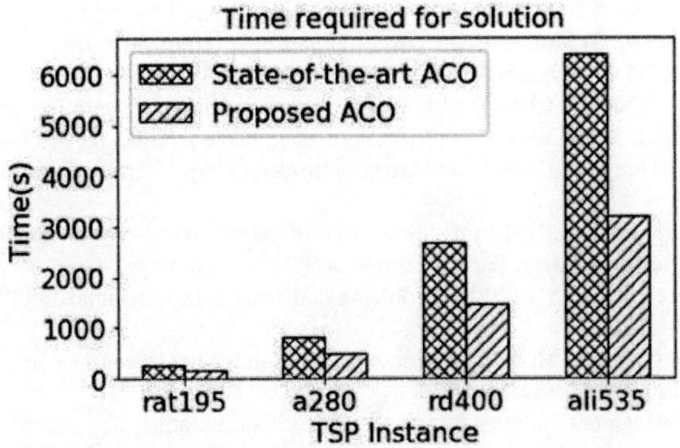

Fig. 4 Comparison of algorithm for more than 150 cities

selected by the ants contains the pheromone value from the optimal node succeeding that node and ant select the node according to the global best solution. As a result, the optimal solution obtained in the proposed modification is better than the traditional ACO algorithm. Also, the experimental results vary when we change the parameters used in the algorithm. Here, the rate of evaporation also plays an important role in providing the optimal solution.

7 Conclusion

In this paper, we are proposing the modification in the pheromone update formula of the ACO. The new pheromone update formula updates the pheromone value of current node with a certain amount of the pheromone value from the next node with maximum pheromone from the current node. It can be deduced that the proposed method performs less computation with less number of ants used for a large number of cities. Hence, the time for finding the optimal solution decreases notably by 41.9%. The advantage of using the proposed method is that with a low number of ants and iterations, the optimal solution is found in a shorter time. Hence, it can be used in solving the problem which may arise in practical applications that are combinatorial. Also, it can be implemented in all the new methods proposed to find the optimal solution.

References

1. Generalized tsp.https://en.wikipedia.org/wiki/Set_TSP_problem. Accessed 14 May 2020
2. State space search. https://en.wikipedia.org/wiki/State_space_search. Accessed 15 Apr 2020
3. Asmar D, Elshamli A, Areibi S (2005) A comparative assessment of aco algorithms within a tsp environment. Dyn Contin Discr Impul Syst Series B Appl Algorithms 1:462–467
4. Blum C, Roli A (2003) Metaheuristics in combinatorial optimization: overview and conceptual comparison. ACM Computing Surveys (CSUR) 35(3):268–308
5. Chen H, Tan G, Qian G, Chen R (2018) Ant colony optimization with tabu table to solve tsp problem. In: 2018 37th Chinese Control Conference (CCC). IEEE, pp 2523–2527
6. Dorigo M, and Thomas S (2004) Ant colony optimization algorithms for the traveling salesman problem
7. Dorigo M, Thomas S (2019) Ant colony optimization: overview and recent advances. In: Handbook of metaheuristics. Springer, pp 311–351
8. Knight K, Rich E, Nair SB (2009) Problems and search. In: Artificial Intelligence, 3rd edn. Tata McGraw-Hill
9. Gao W (2020) New ant colony optimization algorithm for the traveling salesman problem. Int J Computat Intell Syst 13(1):44–55
10. David EG (1989) Genetic algorithms in search. In: Optimization, and Machine Learning
11. Şaban G, Mostafa M, Ömer KB, Halife K (2018) A parallel cooperative hybrid method based on ant colony optimization and 3-opt algorithm for solving traveling salesman problem. Soft Comput 22(5):1669–1685
12. Wu J, Ouyang A (2012) A hybrid algorithm of aco and delete-cross method for tsp. In: 2012 International Conference on Industrial Control and Electronics Engineering. IEEE, pp 1694–1696
13. Liu Y, Cao B (2020) A novel ant colony optimization algorithm with levy flight. IEEE Access 8:67205–67213
14. Liu Y, Cao B, Li H (2020) Improving ant colony optimization algorithm with epsilon greedy and levy flight. JSP 24(25):54
15. Mavrovouniotis M, Yang S (2013) Ant colony optimization with immigrants schemes for the dynamic travelling salesman problem with traffic factors. App Soft Comput 13(10):4023–4037
16. Mohsen AM (2016) Annealing ant colony optimization with mutation operator for solving tsp. In: Computational intelligence and neuroscience
17. Ester ME (2017) Ant colony optimization for predicting gene interactions from expression data
18. Raghavendra BV (2015) Solving traveling salesmen problem using ant colony optimization algorithm. J Appl Comput Math JACM 4(6):260
19. Stützle T, Dorigo M et al (1999) Aco algorithms for the traveling salesman problem. Evolut Algorithms Eng Comput Sci 4:163–183
20. Supaporn S, Deacha P (2012) Solving traveling salesman problems via artificial intelligent search techniques. In: Proceedings of the 11th WSEAS international conference on artificial intelligence, knowledge engineering and data bases. World Scientific and Engineering Academy and Society (WSEAS), pp 137–141

21. Wang J, Yang X (2016) Application of improved ant colony algorithm on travelling salesman problem. In: 28th Chinese control and decision conference (CCDC)
22. Liu Y, Hou Z, Jiang C (2005) Unit commitment by binary particle swarm optimization. In *Proceedings of the 7th WSEAS International Conference on Mathematical Methods and Computational Techniques In Electrical Engineering*, pages 372–377. Citeseer, 2005.

Face Mask Detection Using Deep Learning During COVID-19

Soham Taneja, Anand Nayyar, Vividha, and Preeti Nagrath

Abstract With the onset of the COVID-19 pandemic, the entire world is in chaos and is talking about novel ways to prevent virus spread. People around the world are wearing masks as a precautionary measure to prevent catching this infection. While some are following and taking this measure, some are not still following despite official advice from the government and public health agencies. In this paper, a face mask detection model that can accurately detect whether a person is wearing a mask or not is proposed and implemented. The model architecture uses MobileNetV2, which is a lightweight convolutional neural network, therefore requires less computational power and can be easily embedded in computer vision systems and mobile. As a result, it can create a low-cost mask detector system that can help to identify whether a person is wearing a mask or not and act as a surveillance system as it works for both real-time images and videos. The face detector model achieved high accuracy of 99.98% on training data, 99.56% on validation data, and 99.75% on testing data.

Keywords Computer vision · Deep learning · Object detection · Face detection · Convolutional neural network · Face mask detection

1 Introduction

The COVID-19 pandemic is an ongoing pandemic caused by severe acute respiratory syndrome coronavirus 2 [1]. The virus is transmitted primarily among people in close contact, most often by small droplets formed by sneezing, coughing, and talking.

S. Taneja · Vividha · P. Nagrath
Bharati Vidyapeeth's College of Engineering, New Delhi, Delhi 110063, India
e-mail: sohamtaneja.cse1@bvp.edu.in

P. Nagrath
e-mail: preeti.nagrath@bharatividyapeeth.edu

A. Nayyar (✉)
Graduate School, Faculty of Information Technology, Duy Tan University, Da Nang 550000, Vietnam
e-mail: anandnayyar@duytan.edu.vn

P. K. Singh et al. (eds.), *Proceedings of Second International Conference on Computing, Communications, and Cyber-Security*, Lecture Notes in Networks and Systems 203,
https://doi.org/10.1007/978-981-16-0733-2_3

Instead of moving large distances through air, droplets typically fall on the ground or surfaces [2]. 7.03 million COVID-19 cases have been registered to date in more than 188 countries and territories resulting in more than 403,000 deaths [3]. On May 21, there were 100,000 new infections worldwide, the most from the outbreak of the pandemic, while total 5 million cases have surpassed [4]. India currently has the highest number of confirmed cases in Asia with a total of reported cases crossing the mark of 100,000 on May 19 and 200,000 on June 3, 2020 [5–7]. The reproductive number by which COVID-19 spread is extremely high due to which it is difficult to control this virus [8]. Since there is no effective vaccine to prevent COVID-19 till date, it is crucial to take important precautionary measures. In this ongoing pandemic, one of the recommendations received from different public health agencies including WHO and governments is wearing face masks [9]. In June 2020, the WHO modified its policy on wearing face masks, saying they should be worn in public places to help prevent the spread of COVID-19 [10]. It is mandatory to wear a mask in more than 75 countries, and 88% of the world's total population lives in these countries [11]. Researchers have also shown that wearing masks helps to avoid contracting and further spread of virus [12]. In all these circumstances, it becomes necessary to build a surveillance system that can detect whether a person is wearing a mask or not to prevent the further spread of this contagious virus. Surveillance systems like these can be installed at the entrances of offices and shops to check whether the employees and the customers are wearing a mask or not and let in only those who are detected with a mask. Using this as the motivation, this paper tells how computer vision can be used to achieve this.

In recent years, deep learning has made significant breakthroughs in many fields of computer vision, including general object detection and facial recognition. Unlike the previous face recognition algorithms, deep learning does not have to manually build features, as the convolutional neural networks would automatically learn useful features from the training images [13]. The face mask detector primarily comes under object detection and face detection. In this project, the face mask detector model is trained using transfer learning with MobileNetV2 architecture due to its light-weighted design and state-of-the-art performance [14]. It is primarily used as a classifier that classifies the image into two classes—face with mask and without mask after training. To apply this detector on to real-time images and videos, Haar feature-based cascade classifier is employed which detects the faces in the real-time images and passes the coordinates of the detected face to the model which then classifies the detected face as with mask or without mask. The primary reason to opt for a light-weighted model was to help in building low cost and maintenance face detector system that can easily work in mobiles and embedded computer vision systems.

The objectives of this paper are:

- Training the face detector model using transfer learning with MobileNetV2
- Evaluating the model on validation and training set
- Extracting the face (or faces) from real-time images and videos using Haar feature-based cascade classifier

- Applying the face mask detector to the extracted face (or faces) to get the results
- The whole paper is divided into VI sections. Section 1 is the introduction. In Sect. 2, related works in the area of object detection and face detection are mentioned along with works on how computer vision is contributing in controlling COVID-19. In Sect. 3, methodology of the face mask detector is discussed. In Sect. 4, the experimental setup and results are mentioned which describes the dataset and the experimental results. Lastly, Sect. 5 concludes the paper and discusses future work in the given field.

2 Related Works

Due to huge progress in network architecture, such as inception and ResNet, object detectors based on the convolutional neural networks (CNNs) have become more and more established and have achieved great success in recent years [15–17]. Object detection can be divided into two groups: one-stage detector and two-stage detectors. In two-stage detectors, first the region of interest (RoI) is detected by select search or regional proposal framework. Then these regions are passed onto a classifier. R-CNN and FPN use two-stage detectors [18, 19]. In one-stage detectors, the model works directly over a dense sample of potential sites instead first extracting the regions of interest. YOLO and SSD models work on one-stage detector principle [20–22]. As two-stage detectors work in two steps, they are comparatively slower to one-stage detectors but more accurate than them. YOLO is an intelligent, convolutional neural network (CNN) for real-time detection of objects. The algorithm basically applies a single neural network to the image, then divides the image into regions and predicts bounding boxes and probabilities for each region. Such bounding boxes are weighted by the probabilities predicted. Although one-stage detectors have simple architecture and high speed, they cannot compete with state-of-the-art performance of two-stage detectors. However, RetinaNet is a one-stage detector that achieves similar performance to two-stage detectors [23]. In this paper, a two-stage detector framework is implemented.

The CNN-based methods have lately dominated face detection. It can be said as a building block in computer vision-related tasks. Some of the CNN-based detection methods such as cascade CNN enhance detection performance by training a significant interleaved CNN models [24]. Hongwei Qin et al. propose the joint training of cascaded CNNs to perform end-to-end optimization [25]. Zhu et al. propose an estimated max overlapping score for determining anchor matching efficiency [26]. MTCNN proposes joint resolution of face detection and alignment using multiple multitask CNNs [27].

According to Ulhaq et al., computer vision has enjoyed recent success in solving various complex healthcare problems and has the potential to contribute to the COVID-19 control fight [28]. For the diagnosis of COVID-19, many deep learning-based computer vision models are proposed. The best model till date is COVID-Net proposed by Darwin [29]. Yunlu Wang used depth camera and deep learning as a classifier for abnormal respiratory patterns that can contribute accurately and unobtrusively to wide screening of people infected with the virus [30]. Somboonkaew et al. have proposed a mobile platform for an automatic fever screening system based on the temperature of the infrared forehead [31]. Jun Chen et al. proposed a CT image dataset of 46,096 images of both healthy and infected patients, classified by expert radiologists. Of 106 admitted patients with 51 confirmed COVID-19 pneumonia and 55 control patients, it was obtained. The study used deep-learning segmentation models only to distinguish the infected area between healthy and infected patients in CT images [32]. In the same manner, this paper hopes to contribute to healthcare and governance center among this pandemic by proposing a light-weighted face detector model which can act as a surveillance system.

3 Methodology

3.1 Face Detection

For the detection of faces with or without masks, the primary approach is to detect the number of faces in the given image. A lot of study work is proposed in this field which made detecting of faces in an image easy using different machine learning and image processing techniques. Various classification methods are proposed for face detection. These methods can be broadly categorized into four groups [33]. First is knowledge-based in which face detection depends on a predefined set of rules, for instance, the distance between different features of the face like the distance between eyes and nose. The major flaw with this method was the definition of the rules governing around the method. If the set of rules is detailed, it would tend to cause overfitting and would provide a lot of false negatives; whereas, if the rules are not detailed, the data would be insufficient to most of the faces. To overcome this problem, the second type known as feature-based methods came into play. The feature-based method detects the faces by feature extraction. Firstly, instead of defining rules, it is trained as a classifier, for instance, whether features such as eyes or nose are present or not. If they are present, then it further differentiates the facial and non-facial regions. The third type was methods of template matching which used predefined face templates to detect face. It employed the correlation between the image given and the predefined template. However, these methods did not take changes in shape, scale, and pose into account.

The most revolutionary research in face detection was the Viola–Jones face detection algorithm [34]. It helped in performing real-time face detection by analyzing the pixels of full-frontal faces. It is based on machine learning, where using a cascade function, it is trained on many positive and negative images. The algorithms consist of four steps. First is the feature selection. Some similar characteristics are shared by all human faces like the region being darker than other parts and the nose being shinier than the rest. Moreover, the location of these features is also the same. These features are matched by the pixel intensities and the respective distance between them. The image frame is split into a grid of rectangles, and the feature selection uses these rectangles to identify features using the detection window in the frame. From the sum of the values in the rectangular subset of the grid, the integral image is created. The next step is the implementation of classification function using the learning algorithm which constructs a 'strong' classifier as a linear combination of weighted simple 'weak' classifiers. The last step is the training classifier, i.e., the cascade classifier which is used to detect faces in the image using parameters such as the maximum acceptable false-positive rate per layer and the minimum acceptable detection rate per layer where the layer describes the stage of classifiers applied one by one on different features. In this paper, Haar feature-based cascade classifier is to detect face (or faces) in real-time images.

3.2 *MobileNetV2 Architecture*

In this project, MobileNetV2 is used which is a lightweight model for image classification. This model is suitable for any device with low computational power as it reduces the need for main memory access. It can be implemented efficiently in any modern system using standard operations, without impacting its performance. MobileNetV2's basic building block is a depth-separable bottleneck convolution with residuals [14]. The model uses depth-wise separable convolutions to replace a fully convolutional layers operator with a factorized version that splits the convolution into two separate layers and is an important feature for many robust neural network architectures [35–37]. The depth-separable convolution is so given the name because it deals with the depth dimension, i.e., the number of channels as well as with the spatial dimensions. Inverted residuals and linear bottlenecks are the two primary ideas of MobileNetV2 architecture. The need for inverted residuals and linear bottlenecks is that firstly, feature maps can be encoded in low-dimensional subspaces, and secondly, nonlinear activations result in a loss of information given their ability to increase representational complexity. The model includes the initial fully convolution layer with 32 filters, followed by 19 layers of residual bottlenecks. There are two types of blocks where one block has a 1 stride and the second block has a stride of 2. Each block is composed of three layers. The first layer is a layer of 1×1 convolution with the activation function of Relu6. It is implemented with low precision computation owing to its robustness. The second layer is a layer with 3×3 depth-wise convolution. By applying a single convolutional kernel per input channel, it performs lightweight filtering. The third layer is again a 1×1 convolution layer, the difference being that instead of nonlinear, it has a linear function. MobileNetV2 is a major improvement over MobileNetV1 and drives state-of-the-art visual recognition including classification, detection, and segmentation of object classes [38, Fig. 1].

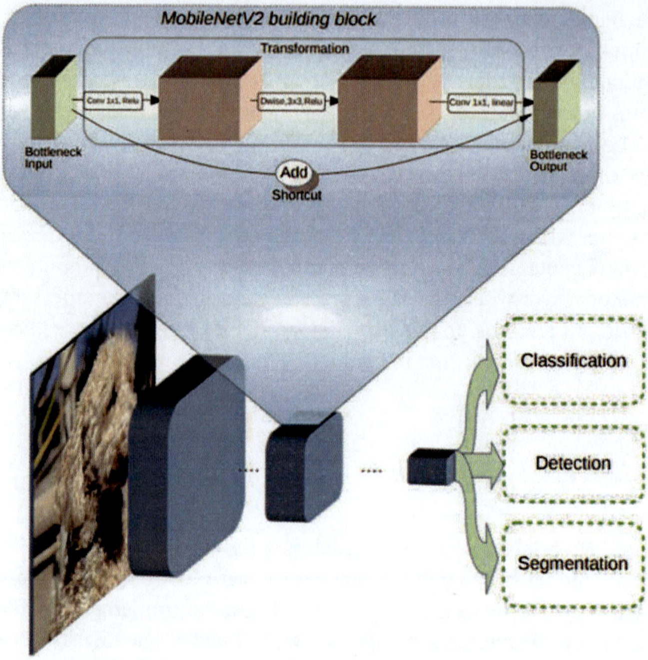

Fig. 1 MobileNetV2 architecture where the blue blocks represent composite convolutional building blocks

3.3 *Transfer Learning*

Transfer learning (TL) is a research problem in machine learning (ML) which focuses on storing the gained knowledge while solving one problem and applying it to a related but different problem [39]. It is an effective approach in deep learning where pre-trained models are used as the reference point for computer vision and natural language processing tasks due to the enormous computational and time resources required to build neural network models on these issues and the huge skill leaps they provide on related issues.

In this paper, transfer learning using MobileNetV2 is used due to a lack of large datasets on the given topic and vast computational resources. Furthermore, creating a model with low computational power is one of the primary objectives of this study.

3.4 *Model Architecture*

The architecture of the face mask detector is based on two-stage detector framework. Figure 2 shows phase 1 of the model workflow. The model is trained using transfer

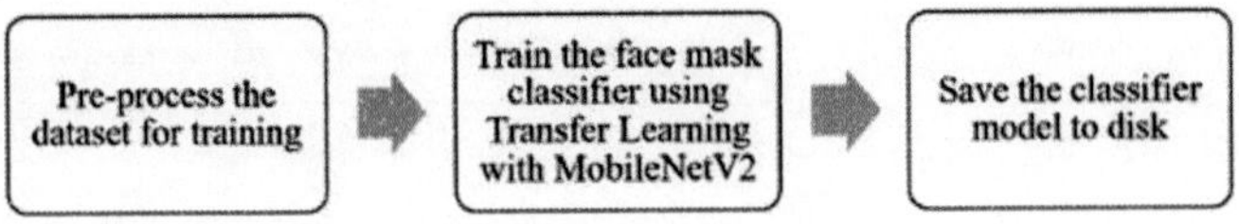

Fig. 2 Training the face mask detector

Fig. 3 Apply the face mask classifier on real-time images

learning with MobileNetV2 as a face mask classifier that classifies the image into two classes—face with a mask and without a mask. The model is then saved as a classifier to the disk. Figure 3 shows phase 2 of the model workflow. The face mask classifier is loaded from the disk to detect real-time images fed to the model. For the face detection in real-time images, Haar feature-based cascade classifiers are used. As it works on only single channeled images, the images are first converted into grayscale images. Then, the cascade classifier is applied which returns the coordinates of the detected face (or faces). Those coordinates stored as an array are passed to the face mask classifier to determine whether the face detected in the image is wearing a mask or not.

4 Experimental Setup and Results

This section contains the experimental setup and results obtained by training and testing the face mask detector on the dataset used. The dataset contains 11,792 images with 10,000 images in the training set, 800 images in the validation set, and 992 images in the training set [40]. Each set has two classes of images—faces with mask and without mask. The performance on each set is given in Table 1. The training accuracy obtained is 99.98%. The validation accuracy came out to be 99.56% suggesting that the model is not overfitted. For further evaluation, the performance on testing set was also evaluated which came out to be 99.75%. Figure 4 displays the graphs between the number of epochs versus accuracy and loss, respectively.

Table 1 Performance on training, validation, and testing set

Data	Accuracy (%)
Training	99.98
Validation	99.56
Testing	99.75

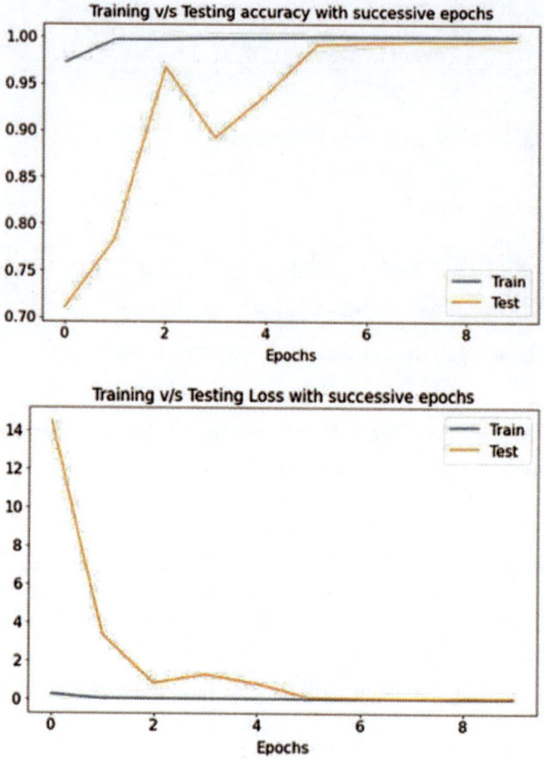

Fig. 4 Training versus testing accuracy and loss graph, respectively

It is visible that after the last epoch, the testing accuracy was almost similar to the training accuracy. The graph verifies that the model is not overfitted as it is achieving high accuracy even on the testing set implying that the model can work on real-life images as well thus proving its efficiency and feasibility. To effectively evaluate the model, confusion matrix was obtained for each of the sets. Confusion matrix is a table with four different combinations of actual values and predicted values. These are true positive, false positive, true negative, and false negative. True positive (TP) is when the actual value is positive, and it is consequently predicted as positive by the model, whereas false positive (FP) is when the actual value is negative but it is predicted positive by the model. Similarly, true negative (TN) is when the actual value is negative as well as it is predicted as negative, and false negative (FN) is when the actual value is positive but is predicted as negative. Clearly, therefore, for an accurate model, the true positives should be greater than false positives. Figure 5 clearly shows less number of false positive and greater number of true positive suggesting that the model has accurately classified majority of the images. Figures 6 and 7 are some examples of face detector applied on real-world images.

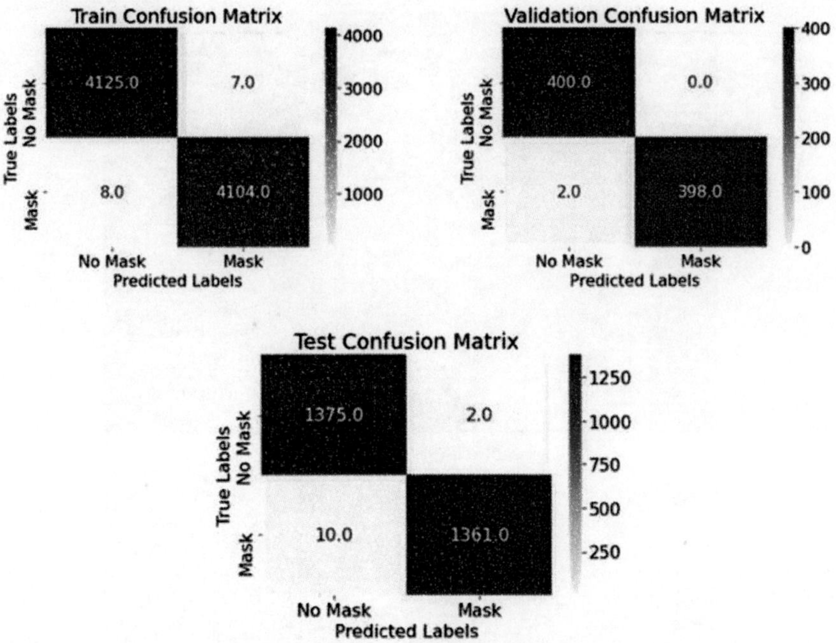

Fig. 5 Confusion matrix of training, validation, and testing set, respectively

5 Conclusion

In this paper, a face mask detector is proposed and implemented that can detect whether the person is wearing a mask or not among this COVID-19 pandemic. This model used a two-stage detector framework. For the training of the model, it used transfer learning with MobileNetV2 framework on dataset of 11,800 images. In order to extract the RoI, it used Haar feature-based cascade classifier which extracted the regions and passed to the trained classifier to get the desired results. Experimental results showed training accuracy of 99.9% and testing accuracy of 99.75%. The model displayed remarkable results when it was used to detect a person wearing a mask or not in real-time images and video stream. The face mask detector proposed and implemented in this paper can be used as a surveillance system on shops and offices to ensure that people are wearing face masks to prevent the spreading of the virus. Due to its low computational power yet state-of-the-art performance, it can be easily embedded on computer vision systems thus cutting the cost of making the detector.

For future research work, one-stage detector framework like YOLO and SSD can be used for face mask detection which would reduce the computational time significantly hence making the detection faster. Furthermore, different frameworks and datasets can be used to train and evaluate the face detector model to enhance the

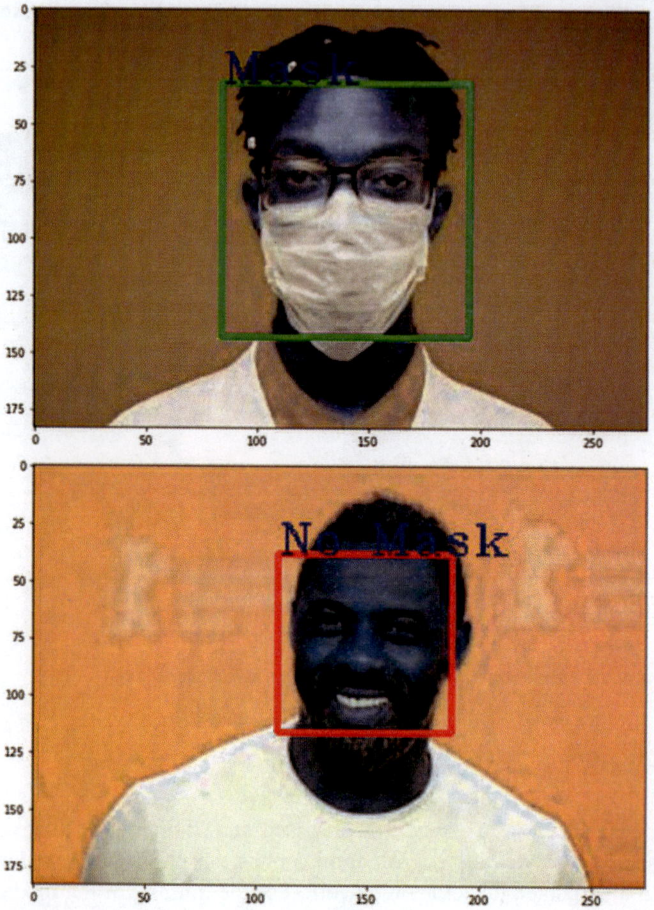

Fig. 6 Face mask detector applied on real-world images

performance of the model over real-time images and videos. For instance, people often use scarf and cloth as masks. Due to the same features as that of a face mask, the model detects it as a face mask. Therefore, training the detector with datasets that contain such images and with more accurate feature selection, the model would be more feasible for real-world applications.

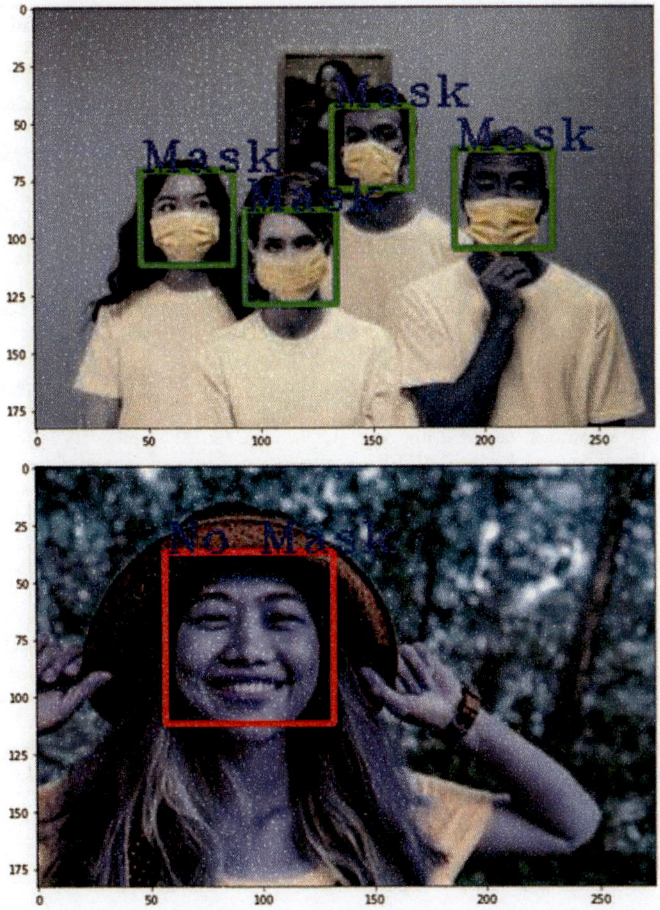

Fig. 7 Face mask detector applied on real-world images

References

1. WHO Naming the coronavirus disease (COVID-19) and the virus that causes it, 201, [Online]. Available: https://www.who.int/emergencies/diseases/novel-coronavirus-2019/technical-guidance/naming-the-coronavirus-disease-(covid-2019)-and-the-virus-that-causes-it
2. CDC (2020) How COVID-19 Spreads, 2020, [Online]. Available: https://www.cdc.gov/coronavirus/2019-ncov/prevent-getting-sick/prevention.html
3. COVID-19 Dashboard by the Center for Systems Science and Engineering (CSSE) at Johns Hopkins University (JHU). ArcGIS. Johns Hopkins University. Retrieved 8 June 2020.
4. Live updates: Global cases top 5 million as WHO reports worst day yet for new infections. The Washington Post. Retrieved 21 May 2020.
5. India most infected by Covid-19 among Asian countries, leaves Turkey behind. Hindustan Times. 29 May 2020. Retrieved 30 May 2020.

6. India's case count crosses 100,000, Delhi eases restrictions: Covid-19 news today. Hindustan Times. 19 May 2020. Retrieved 20 May 2020.
7. Daily COVID-19 bulletin. PIB India (@PIB_India) on Twitter. Twitter. Retrieved 3 June 2020.
8. Liu Y, Gayle AA, Wilder-Smith A, Rocklöv J (2020) The reproductive number of Covid-19 is higher compared to SARS coronavirus. J Travel Med
9. Feng S, Shen C, Xia N, Song W, Fan M, Cowling BJ (2020) Rational use of face masks in the Covid-19 pandemic. In: The Lancet Respiratory Medicine
10. Coronavirus: WHO advises to wear masks in public areas, reversing policy 5 June 2020, bbc.com
11. Masks4All, What countries require masks in public or recommend masks? 2020, [Online]. Available: https://masks4all.co/what-countries-have-mask-laws/
12. Leung NHL, Chu DKW, Shiu EYC, Kwok-Hung C, McDevitt JJ, Hau BJP, Hui-Ling Y, Li Y, Dennis KMI, Peiris JSM et al (2020) Respiratory virus shedding in exhaled breath and efficacy of face masks. Nature Medicine, pp 1–5
13. Zou Z, Shi Z, Guo Y, Ye J (2019) Object detection in 20 years: A survey. arXiv preprint arXiv: 1905.05055
14. Sandler M, Howard A, Zhu M, Zhmoginov A, Chen L-C (2018) MobileNetV2: inverted residuals and linear bottlenecks. In: The IEEE conference on computer vision and pattern recognition (CVPR)
15. Szegedy C, Liu W, Jia Y, Sermanet P Reed SE, Anguelov D, Vanhoucke V, Rabinovich A (2015) Going deeper with convolutions. In: CVPR Dumitru Erhan
16. Szegedy C, Vanhoucke V, Ioffe S, Shlens J, Wojna Z (2016) Rethinking the inception architecture for computer vision. In: CVPR
17. He K, Zhang X, Ren S, Sun J (2016) Deep residual learning for image recognition. In: CVPR
18. Ren S, He K, Girshick RB, Sun J (2017) Faster R-CNN: towards real-time object detection with region proposal networks. TPAMI
19. Lin T, Dollár P, Girshick RB, He K, Hariharan B, Belongie SJ (2017) Feature pyramid networks for object detection. In: CVPR
20. Redmon J, Divvala SK, Girshick RB, Farhadi A (2016) You only look once: Unified, real-time object detection. In: CVPR
21. Redmon J, Farhadi A (2016) YOLO9000: better, faster, stronger. CoRR
22. Liu W, Anguelov D, Erhan D, Szegedy C, Reed SE, Fu C, Berg AC (2016) SSD: single shot multibox detector. In: ECCV
23. Lin T, Goyal P, Girshick RB, He K, Dollár P (2017) Focal loss for dense object detection. In: ICCV
24. Li H, Lin Z, Shen X, Brandt J, Hua G (2015) A convolutional neural network cascade for face detection. In: CVPR
25. Qin H, Yan J, Li X, Hu X (2016) Joint training of cascaded CNN for face detection. In: CVPR
26. Zhu C, Tao R, Luu K, Savvides M (2018) Seeing small faces from robust anchors perspective. In: CVPR
27. Zhang K, Zhang Z, Li Z, Qiao Y (2016) Joint face detection and alignment using multitask cascaded convolutional networks. SPL
28. Ulhaq A, Khan A, Gomes D, Paul M (2020) Computer vision for COVID-19 control: a survey. In arXiv:2004.09420
29. Gáal G, Maga B, Lukács A (2020) Attention u-net based adversarial architectures for chest x-ray lung segmentation. arXiv preprint arXiv:2003.10304
30. Wang Y, Hu M, Li Q, Zhang X, Zhai G, Yao N (2020) Abnormal respiratory patterns classifier may contribute to large-scale screening of people infected with Covid-19 in an accurate and unobtrusive manner
31. Somboonkaew A, Prempree P, Vuttivong S, Wetcharungsri J, Porntheeraphat S, Chanhorm S, Pongsoon P, Amarit R, Intaravanne Y, Chaitavon K, Sumriddetchkajorn S (2017) Mobile-platform for automatic fever screening system based on infrared forehead temperature. In: 2017 Opto-electronics and communications conference (OECC) and photonics global conference (PGC), pp 1–4

32. Chen J, Wu L, Zhang J, Zhang L, Gong D, Zhao Y, Hu S, Wang Y, Hu X, Zheng B et al (2020) Deep learning-based model for detecting 2019 novel coronavirus pneumonia on high-resolution computed tomography: a prospective study. medRxiv
33. Yang M-H, Kriegman DJ, Ahuja N (2002) Detecting faces in images: a survey. IEEE Trans Pattern Anal Mach Intell 24(1):34–58, January 2002, [Online]. Available: https://doi.org/10.1109/34.982883
34. Viola P, Jones MJ (2004) Robust real-time face detection. Int J Comput Vision 57(2):137–154, May 2004, [Online]. Available: https://doi.org/10.1023/B:VISI.0000013087.49260.fb
35. Zhang X, Zhou X, Lin M, Sun J (2018) ShuffleNet: an extremely efficient convolutional neural network for mobile devices. In: The ieee conference on computer vision and pattern recognition (CVPR)
36. Chollet F (2017) Xception: deep learning with depthwise separable convolutions. In: The IEEE conference on computer vision and pattern recognition (CVPR)
37. Howard AG, Zhu M, Chen B, Kalenichenko D, Wang W, Weyand T, Andreetto M, Adam H (2017) MobileNets: efficient convolutional neural networks for mobile vision applications. CoRR, vol. abs/1704.04861, 2017.
38. Overview of MobileNetV2 Architecture, [Online]. Available: (https://ai.googleblog.com/2018/04/mobilenetv2-next-generation-of-on.html)
39. Jeremy W, Ventura D, Warnick S (2007)Spring research presentation: a theoretical foundation for inductive transfer. Brigham Young University, College of Physical and Mathematical Sciences. Archived from the original on 2007–08–01. Retrieved 2007–08–05.
40. Dataset." [Online]. Available: https://www.kaggle.com/ashishjangra27/face-mask-12k-images-datas

Compact Millimeter-Wave Low-Cost Ka-Band Antenna for Portable 5G Communication Gadgets

Raqeebur Rehman, Javaid A. Sheikh, Khurshed A. Shah, Zahid A. Bhat, Shabir A. Parah, and Shahid A. Malik

Abstract This paper presents design and simulation of a low-cost 5G Millimeter-wave planar antenna with defected ground structure operating in the Ka-band portions of millimeter-wave. The antenna resonates at multiple frequencies of Ka-band especially at 27.80 GHz, 30.86 GHz and 33.74 GHz with a return loss of −21.41 dB, −24.03 dB and −22.27 dB, respectively, and has an impedance bandwidth of 53.5%. The presented antenna has been designed on a low cost FR4 substrate with a dielectric *K* value of 4.4 and a dissipation value of 0.004. The overall profile of the designed structure is $30 \times 40 \times 0.8\ \text{mm}^3$. The antenna proposed is a compact structure with a peak gain achievement of 3.79 dBi and suits best for the employment with 5G mobile devices and gadgets. Other parameters such as radiation pattern, VSWR, polar plot and surface current density have also been discussed. The well performance of the presented antenna with reference to the return loss (S_{11}), peak gain and associated radiation pattern makes it a sterling and compact design antenna for use in the 5G Millimeter-Wave mobile devices.

Keywords Millimeter-wave · Ka-band · Impedance bandwidth · Peak gain · Radiation pattern

1 Introduction

The wide spectrum available in Millimeter-wave frequency proves out to be a promising hope for fifth-generation mobile communication [1, 2]. High propagation

R. Rehman (✉) · J. A. Sheikh · Z. A. Bhat · S. A. Parah · S. A. Malik
Department of Electronics and Instrumentation Technology, University of Kashmir, Srinagar, India
e-mail: raqeeb.scholar@kashmiruniversity.net

J. A. Sheikh
e-mail: sheikhjavaid@uok.edu.in

K. A. Shah
Department of Physics, S.P College, Cluster University, Srinagar, India

P. K. Singh et al. (eds.), *Proceedings of Second International Conference on Computing, Communications, and Cyber-Security*, Lecture Notes in Networks and Systems 203,
https://doi.org/10.1007/978-981-16-0733-2_4

loss of millimeter-waves due to atmospheric absorption should also be considered which proves to be a challenge for design of Millimeter-wave 5G Networks. So, the solution will be to incorporate antennas with higher directivity to make the high capacity and high-speed Millimeter-wave 5G communication possible [3, 4]. Also, the size of antennas at high Millimeter-wave frequencies shows a sharp reduction up to a few millimeters which proves out to be much challenging for the utilization of traditional antennas like horns, Yagi, sector antennas and dipole arrays in spite of their higher directivities and gains at some specified geometries [5–10]. Further, the implementation cost of such antennas increases to a much higher extent. So, the requirement is to incorporate high gain and efficiently compact antennas for 5G Millimeter-wave links. The compactness in the size of antennas at higher frequencies enables us to design planar antennas of different shapes to felicitate higher spectral efficiency, better coverage of signal, less interference and much higher gains. Along with this, the concept of specific feeding techniques for antenna arrays and half-wavelength inter-element spacing can be incorporated for the achievement of much higher gains and better directivity. But in order to maintain a high front-to-back ratio of antenna field patterns, the geometries of the Millimeter-wave antennas need to have a keen care in the design process to circumvent the formation of high side lobes. Also, for the millimeter-wave antennas, the emerging need will be for wideband, low cost, compact size, easy implementation, low-complexity and efficient integration ability with other microwave or millimeter-wave circuitry. Yagi and Horn antennas can furnish much higher directivity and bandwidth at millimeter-wave frequency bands, but their huge and bulky shapes restrict them to easily integrate with planar microstrip circuitry. Some novel antennas with unique techniques of miniaturization have been reported recently, but most of fail to achieve the high directivity requirement of millimeter-wave communication. Besides it, a very good contribution to millimeter-wave antennas has been reported in the literature. Fan et al. [11] have proposed a conical-beam omni-directional antenna with wideband horizontal polarization for Millimeter-wave applications. There is 22.9% impedance bandwidth achievement in the design within 39–49.3 GHz band with a better radiation pattern. The gain in the prescribed band extends from 4.6 to 6 dBi. Due to the incorporation of substrate integrated waveguide (SIW) radial power divider and a conical reflector, it proves as good prototype for Millimeter-wave communication with minimum losses. But in comparison with the conventional millimeter-wave antennas with planar structures, the design proves to be a little bit complicated. Other antennas are also presented with wide angle scan characteristics. These antennas are usually incorporated with a meta-material surface on which the specific geometries of radiating microstrip structures are printed to form the unit cells. These meta-material unit cells direct the E-field of antennas to propagate in specified directions. In [12], a multi-beam characteristic antenna array with tapered slots has been reported for Massive MIMO mmWave communication. There is a low complexity regarding the antenna geometry computation. With the incorporation of SIW feeding geometry, the integration ability of the planar circuits with the given antenna has been eased up to a greater extent. The antenna section elements have been spaced efficiently according to the half-wavelength criteria. The gain ranges from 8.2 to 9.6 dBi in the operating frequency of 24–32 GHz for every antenna element. There is a very good contribution to Massive MIMO mmWave systems in the presented design. Kumar et al. [13] have proposed a compact square CPW-fed strip and loaded slots antenna with circular polarization for satellite communications. In the concerned antenna, a grounded pair of spiral-shaped slots, a grounded L-strip and a slot in rectangular shape in lower left CPW ground plane are responsible for the gain of dual CP. A specific perturbation in ground plane and CPW structure also adds to the improvement in dual CP. A detailed parametric study of the coplanar antenna has been carried out, and the effect of specific geometries regarding the performance of antenna is recorded. The effect of slot cut outs and spiral M position variation on the axial ratio (AR) bandwidth and impedance bandwidth has been analyzed. The antenna comes up a gain with a peak of 6.36 dBic, a dual-band response and 3-dB AR bandwidths in both dual bands when provided a frequency range from 3 to 14 GHz. This antenna suits best for the downlink Ku-band frequency and wideband wireless but can't be used for higher frequencies due to the increase in the loss for coplanar structure proposed in this antenna leading to much lesser gains and higher losses. Though all the reported antennas [15–20] have some specific and special characteristics, most of them show a limit to attain the demanded gain and radiation characteristics like the polarization diversity and better directivity. Moreover, these antenna structures fail to attain the required compactness, multi-band operation characteristic, easy mounting, etc. Though their implementation turns to be a little bit complicated task, the authors have attained a good milestone and have tried their best to provide a trade-off and a better solution for of the limitations of 5G millimeter-wave communication.

Keeping the above constraints in view, a compact low-cost 5G millimeter-wave Ka-band planar antenna is presented which consists of a pistol-shaped radiating patch and a ground structure with a cut out (DGS) to make the overall structure highly miniaturized to fit into mobile or portable 5G devices. The microstrip patch and ground of the presented antenna are defected or slotted in order to disturb the surface current distribution to have inductive and capacitive effects which makes the presented antenna to resonate at multiple frequencies of Ka-band. In addition to this, it helps in the size reduction of the presented antenna to a much greater extent. The antenna comes up with an impedance bandwidth of 53.5% with a gain of peak value 3.79 dBi. Due to the low-cost substrate used in the presented antenna, it proves to be very economical millimeter-wave antenna for application point of view. Hence, the antenna can serve as a sterling candidate for portable 5G gadgets utilizing the millimeter-wave frequency spectrum.

2 Antenna Conformation and Analysis

2.1 Antenna Design

The full conformation of antenna is brought up in Fig. 1. The antenna proposed has got a ground plane (defected) which consists of a horizontal rectangular slot at the top. The radiating structure is a pistol-shaped patch comprising of a vertical slot cut toward the left, a triangular slot in the middle and five circular cut outs; three toward the top and two toward the right of patch, respectively, to modify the surface distribution current of ground plane and patch which makes the antenna to have a high impedance bandwidth and multi-band operation. The antenna proposed has been designed on 0.8 mm FR4 square-shaped dielectric of value 4.4 and tan $\delta = 0.004$. The full structural shape profile of antenna is $30 \times 40 \times 0.8\ mm^3$. The proposed antenna construction makes it to have a nominal gain of 3.79 dBi for both E and H planes. The design variables of the suggested antenna structure are explained below and brought forward in Table 1.

The EM simulator HFSS V.15 has been utilized to carry on the full-fledged parametric analysis of the presented antenna for the optimized values regarding the

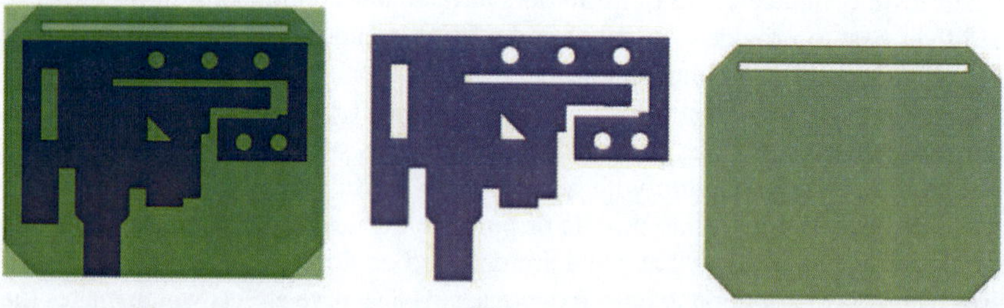

Fig. 1 Structure of the low-cost millimeter-wave antenna proposed, **a** complete view; **b** top view; **c** bottom view (structure designed in ANSYS EM simulator HFSS V.15)

Table 1 Design variables of the presented antenna

Parameter	Value (mm)	Parameter	Value (mm)
L_{gnd}	36	L_{vsp}	8
W_{gnd}	30	W_{vsp}	2
L_{sub}	40	$rad_{circle\text{-}upper}$	1
W_{sub}	30	T_{hyp}	4.24
L_{patch}	32	$T_{bas} = T_{per}$	3
W_{patch}	20	$rad_{circle\text{-}right}$	1
L_{slhor}	28	L_{feed}	6
W_{slhor}	1	D_{hsp}	1

dimensions of substrate. A detailed look over the different values for the dimensions of substrate and its impact over the various characteristics of antenna has been studied. The substrate used here is 0.8 mm with a permittivity value of 4.4 and dissipation factor value of 0.004. The return loss values in dB and VSWR values of the concerned antenna for the different substrate dimensions are given in Fig. 2a and b, respectively. The full parametric setup has been organized, and the finalized values of $30 \times 40\ \text{mm}^2$ have been taken into consideration. Thus, it can be concluded from the respective figures that the given antenna attains the best substrate width dimension at 40 mm due to which the antenna attains efficient multi-band operation and a high impedance bandwidth of 53.5% and a gain with a peak of 3.79 dBi. Moreover, the compact structure obtained using the full parametric analysis results in the application of antenna particularly for the 5G millimeter-wave mobile devices as the

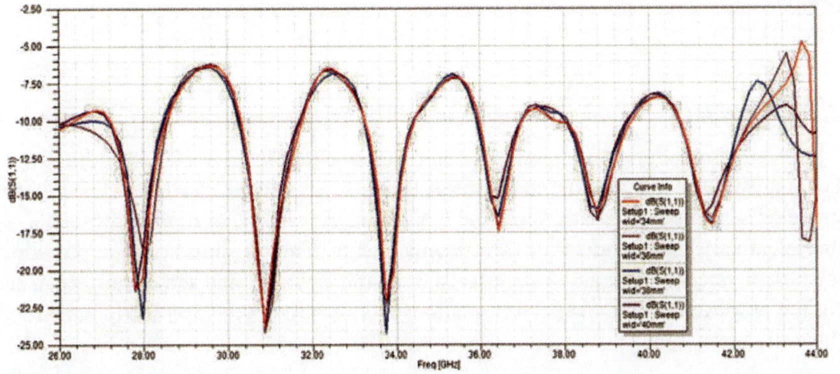

(a). Parametric variations of Return Loss (dB) for varied substrate widths. (Results obtained with the simulations performed in HFSS V.15)

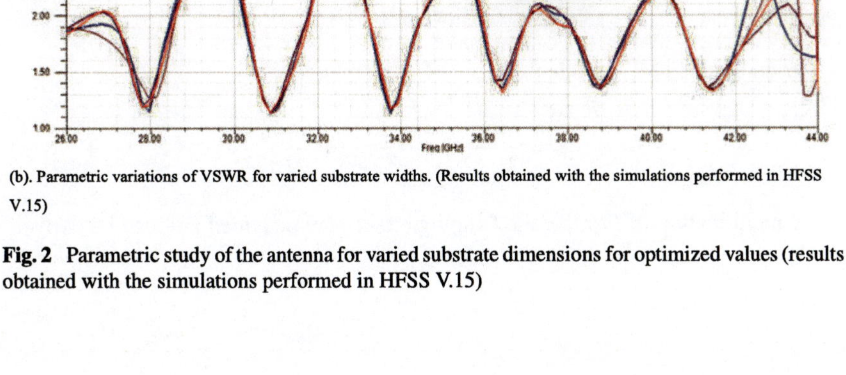

(b). Parametric variations of VSWR for varied substrate widths. (Results obtained with the simulations performed in HFSS V.15)

Fig. 2 Parametric study of the antenna for varied substrate dimensions for optimized values (results obtained with the simulations performed in HFSS V.15) concerned antenna comes up with a planar and low-cost structure with miniaturized dimensions.

Here, L_{gnd} and W_{gnd} depict the length and associated width of the ground plane. L_{sub} and W_{sub} show the length and associated width of the dielectric material. L_{patch} and W_{patch} denote the length and associated width of radiating patch of the presented antenna. L_{slhor} and W_{slhor} denote the length and associated width of the upper horizontal slot of ground plane. The vertical slot length and associated width to left of radiating patch are denoted by L_{vsp} and W_{vsp} respectively. The radii of upper three circular cut outs on the patch is shown by $rad_{circle\text{-}upper}$ and the $rad_{circle\text{-}right}$ shows the radii of the two circular cut outs to right of radiating patch. The hypotenuse and the two sides of the inner triangular cut out on the pistol-shaped patch are denoted by T_{hyp} and T_{bas}, T_{per} respectively. L_{feed} depicts the length of feed line and the D_{hsp} denotes the distance of the horizontal slot from the top of the concerned radiating patch of the antenna proposed.

2.2 *Analysis*

For analysis of planar microstrip antennas, there are a number of methods. The FDTD model, cavity model, transmission line model and the method of moments model are the selectively used models for the antenna design. The transmission line provides the best physical insight and is the easiest of all the methods, and same has been used for the proposed design also. The equations from (1) to (5) are applied to construct the design of antenna proposed.

$$\varepsilon_{r_{eff}} = \frac{\varepsilon_r + 1}{2} + \frac{\varepsilon_r - 1}{2}\left[\frac{1}{\sqrt{1 + 12\frac{h}{w}}}\right] \quad (1)$$

$$L_{eff} = L + 2\Delta L$$

where

$$\Delta L = 0.412h\frac{(\varepsilon_{reff} + 0.3)\left(\frac{w}{h} + 0.264\right)}{(\varepsilon_{reff} - 0.258)\left(\frac{w}{h} + 0.8\right)} \quad (2)$$

For dominant TM_{010} mode f_r is

$$(f_r)_{010} = \frac{1}{2L\sqrt{\varepsilon_r}} * \frac{1}{\sqrt{\mu_0 \varepsilon_0}} = \frac{\vartheta_0}{2L\sqrt{\varepsilon_r}} \quad (3)$$

By modification of Eq. (3), the fringing effect may be added and can be derived as

$$(f_{rc})_{010} = \frac{1}{2L_{\text{eff}}\sqrt{\varepsilon_{\text{reff}}}} * \frac{1}{\sqrt{\mu_0\varepsilon_0}} = \frac{1}{2(L+\Delta L)\sqrt{\varepsilon_{\text{reff}}}} * \frac{1}{\sqrt{\mu_0\varepsilon_0}} = q\frac{\vartheta_0}{2L\sqrt{\varepsilon_r}} \quad (4)$$

$$q = \frac{(f_{rc})_{010}}{(f_r)_{010}}$$

Equation (4) denotes the factor regarding length reduction and is known as fringe factor

$$W = \frac{1}{2f_r\sqrt{\mu_0\varepsilon_0}}\sqrt{\frac{2}{\varepsilon_r+1}} = \frac{\vartheta_0}{2f_r}\sqrt{\frac{2}{\varepsilon_r+1}} \quad (5)$$

Equation (5) corresponds to the width of radiating patch and obviously depends upon the resonant frequency and the dielectric permittivity of the material (FR4) employed to construct the profile of the antenna presented.

3 Antenna Simulation Results and Discussion

The Electromagnetic Simulator ANSYS HFSS v.15 has been utilized to accomplish the design process of the proposed Millimeter-wave antenna. A deep parametric analysis has been carried on during the design steps for the optimized selection of particular values regarding the dimensions of antenna and especially for the for the length and associated width of the horizontal and vertical slots as well as radii of the circular cut outs and substrate dimensions for the achievement of multi-band operation, better radiation performance and impedance bandwidth. When the antenna is given an excitation at the port with a frequency sweep of 26–44 GHz with 28 GHz as setup frequency, a response of antenna is depicted for return loss and associated VSWR as demonstrated in Fig. 3a and b.

3.1 Return Loss, VSWR and Impedance Bandwidth

Figure 3 shows the simulated S_{11} and VSWR values of the presented antenna. It can be inferred that the return loss value is below −10 dB for the multiple frequencies of 27.80, 30.86 and 33.74 GHz and has achieved the best level of −24.03 dB with a good impedance matching corresponding to the VSWR of 1.13. Also, it can be interpreted from the RL versus frequency curve that the antenna proposed has a much better impedance bandwidth of 53.5%.

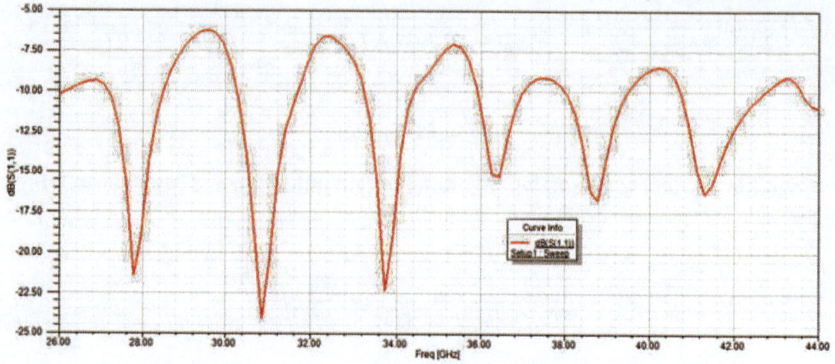

(a). S_{11} of the Antenna Presented (Results obtained with the simulations done in HFSS V.15)

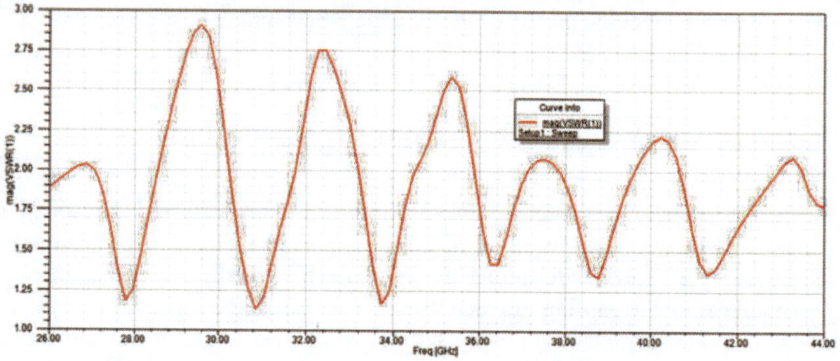

(b). VSWR of the Antenna Presented (Results obtained with the simulations done in HFSS V.15)

Fig. 3 Return loss and associated VSWR of the antenna presented (results obtained with the simulations done in HFSS V.15)

3.2 Radiation Performance

Gain. The antenna proposed is further examined for the peak azimuthal and elevation gains which is illustrated from Fig. 4a and b. From the respective figures, it can be noted that the presented antenna has almost equal E and H plane peak gain corresponding to the varied Phi values (305°, 326° and 338°) and Theta values (20°, 23° and 44°). This feature of the presented antenna makes it be horizontal and vertical polarization independent.

Radiation Pattern. The radiation pattern (normalized) characteristic related to E plane and H plane of the antenna presented is illustrated in Fig. 5. (Results obtained with the simulations done in HFSS V.15.)

From the radiation patterns relevant to the E and H planes of the antenna, it can be inferred that the antenna proposed has a good directivity suitable for the 5G millimeter-wave communication. In addition to that, it can also be noticed from the

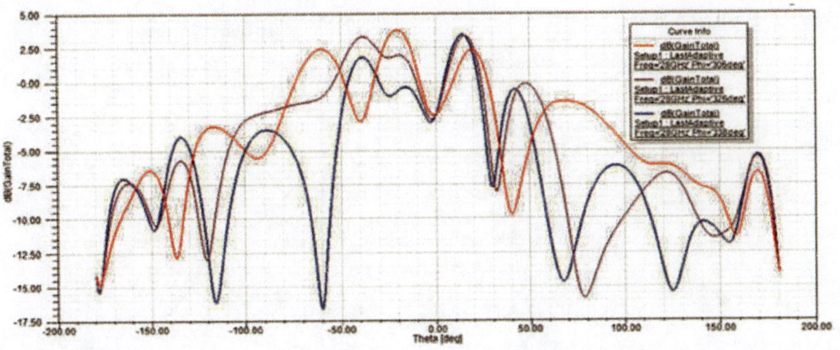

(A). Azimuth gain of antenna presented for Phi= 305^0, 326^0 and 338^0 (Results obtained with the simulations done in HFSS V.15)

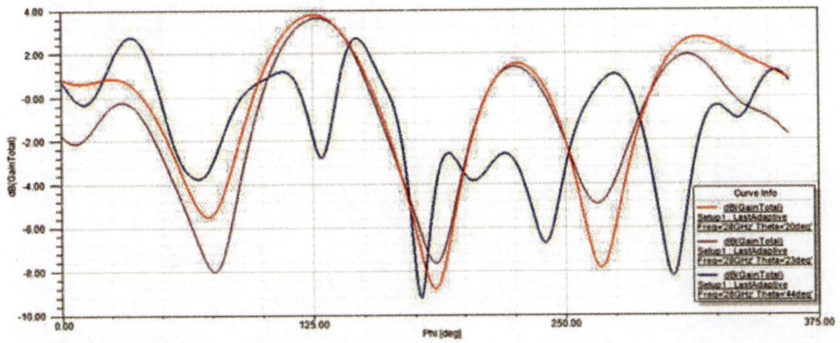

(B). Elevation gain of antenna presented for Theta= 20^0, 23^0 and 44^0. (Results obtained with the simulations done in HFSS V.15)

Fig. 4 Azimuthal and elevation gain of the presented antenna (results obtained with the simulations done in HFSS V.15)

radiation patterns that the antenna proposed has also a better field scanning ability which is an essential need for portable 5G millimeter-wave gadgets.

3D Polar-Radiation Pattern and Surface current dissemination. Figure 6a and b shows the far-field polar 3-D radiation characteristic pattern of the antenna presented for the variation in both the theta and phi angular sweeps and the logarithmic surface current dissemination of both the ground plane and the radiating patch.

From the 3D radiation pattern, it can be inferred that the design proposed has a good field coverage which is an essential need in the millimeter-wave 5G devices. Also, it shows that the density of the field is almost same for both the E plane and H plane. Moreover, the logarithmic surface current dissemination of both the ground plane and the pistol-shaped radiating patch signifies that the antenna presented has a good sensing capability for the incoming EM fields.

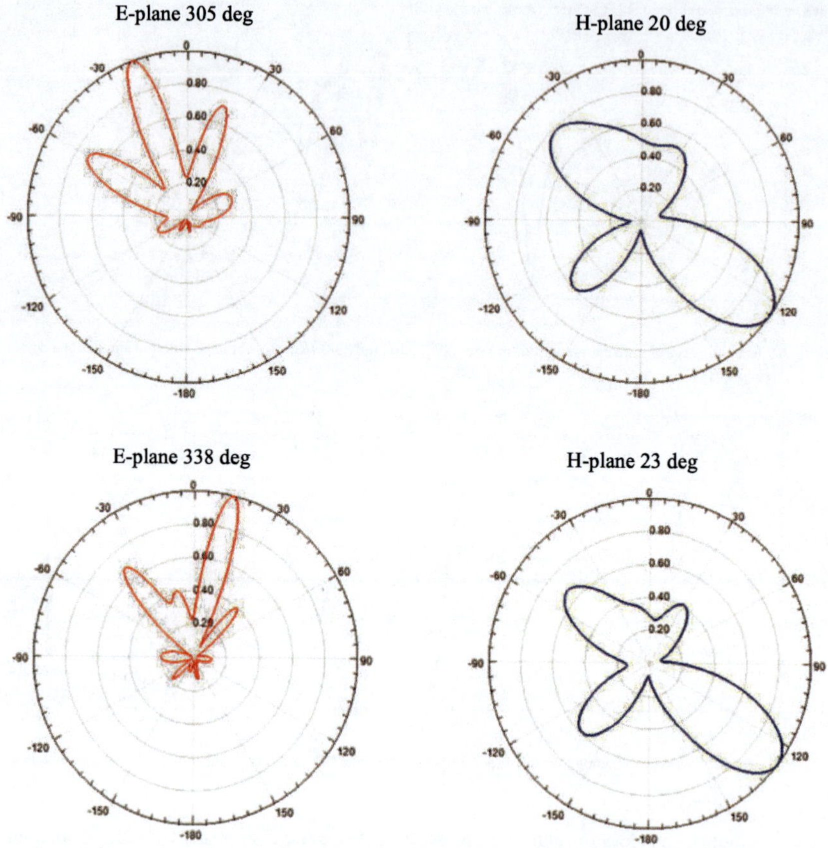

Fig. 5 Radiation patterns of the antenna presented (results obtained with the simulations done in HFSS V.15)

4 State-of-the-Art Comparison

For the support and validity of this work, a comparison is made with previous reported antennas and is depicted in Table 2. From the table, it can be noticed that the antenna presented has much better characteristics as compared with the other antennas in terms of the operating frequency band, impedance bandwidth, simple planar geometry, better peak gain and most importantly its utilization in the portable 5G millimeter-wave gadgets due to its miniaturized structure and easy integration with the millimeter-wave frequency circuits.

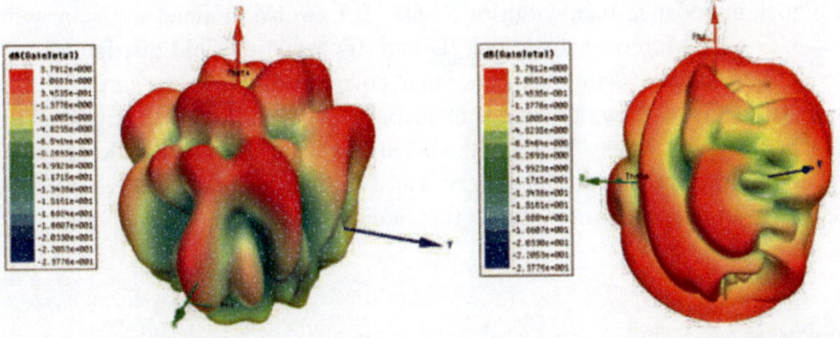

(a). 3D Radiation Pattern of the antenna presented for separate sweeps of phi and theta angles. (Results obtained with the simulations done in HFSS V.15)

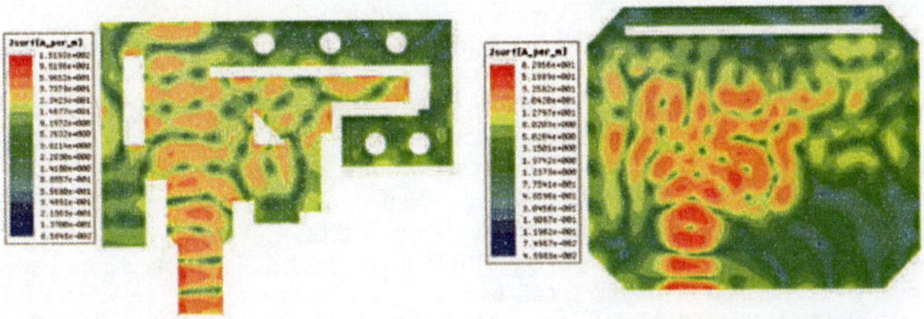

(b). Surface current dissemination of pistol patch and defected ground plane. (Results obtained with the simulations done in HFSS V.15)

Fig. 6 Normalized 3D radiation pattern and the logarithmic surface current distribution (results obtained with the simulations done in HFSS V.15)

Table 2 Proposed antenna compared with the previous reported antennas

Reference/year	Freq. band (GHz)	Impedance bandwidth (%)	Peak gain (dBi)	Type
[11]/2018	39–49.3	22.90	5.5	SIW/reflector omni
[12]/2017	24–32	47	9	SIW/tapered slot
[14]/2016	57–67	–	2.2	Loop microstrip
Proposed work	26–40	53.50	3.79	Planar/low-cost slotted

5 Conclusion

A low-cost planar millimeter-wave antenna operating in the Ka-band was designed and simulated successfully. The antenna proposed has a simple and well miniaturized structure suitable for its incorporation in the 5G millimeter-wave portable devices. The antenna successes in achieving the multi-band characteristics and especially much higher impedance bandwidth of 53.5%. It has also attained the gain with a peak of 3.79 dBi. Moreover, from the 2D and 3D radiation field distribution, it is inferred that the antenna achieves a better field coverage and from the surface current dissemination, it is quite evident that antenna proposed has a good sensing capability for the incoming EM waves. Keeping the discussed features into consideration, it is concluded that the antenna proposed serves as a demanding and excellent candidate for the 5G millimeter-wave communication devices.

References

1. Ayanoglu E, Swindlehurst AL, Heydari P, Capolino F (2014) Millimeter-wave massive MIMO: the next wireless revolution. IEEE Commun Mag 52(9):56–62
2. Rappaport TS, Mayzus RH, Zhao S (2013) Millimeter-wave mobile communications for 5G: it will work! IEEE Access 1(1):335–349
3. Roh W, Park J, Park JH, Seol JY (2014) Millimeter-wave beam-forming as an enabling technology for 5G cellular communications: theoretical feasibility & prototype results. IEEE Commun Mag 52(2):106–113
4. Kim Y, Lee H (2016) Feasibility of mobile cellular communications at millimeter wave frequency. IEEE J Sel Topics Signal Process 10(3):589–599
5. Wang H, Fang DG, Zhang B, Che WQ (2010) Dielectric loaded SIW H-plane horn antennas. IEEE Trans Antennas Propag 58(3):640–647
6. Li M, Luk KM (2015) Wideband magneto-electric dipole antenna for 60-GHz millimeter-wave communications. IEEE Trans Antennas Propag 63(7):3276–3279
7. Zhang Y, Qing X, Chen ZN, Hong W (2011) Wideband mmWave SIW slotted narrow-wall fed cavity antennas. IEEE Trans Antennas Propag 59(5):1488–1496
8. Yang TY, Hong W, Zhang Y (2014) Wideband mmWave SIW cavity-backed rectangular patch antenna. IEEE Antennas Wirel Propag Lett 13(13):205–208
9. Wu K, Djerafi T (2012) Corrugated SIW antipodal antenna array linearly tapered slot fed by quasi-triangular power divider. Propag Electron Res 26(5):139–151
10. Ghiotto A, Parment F, Wu K, Vuong TP (2016) Millimeter-wave air-filled substrate integrated waveguide anti-podal linearly tapered slot antenna. IEEE Antennas Wirel Propag Lett 24(5):1–4
11. Fan K (2018) Wideband horizontally polarized Omni-directional antenna with a conical beam for millimeter-wave applications. IEEE Trans Antennas Propag 66(9):4437–4448
12. Yang B (2017) Compact tapered slot millimeter-wave antenna array for massive MIMO 5G systems. IEEE Trans Antennas Propag 65(12):6721–6727
13. Kumar A, Mahendra MS, Rajendra PY (2019) Dual coplanar waveguide fed circular polarized compact slotted square antenna for wireless and satellite uses. AEU---Int J Electron Commun 108:181–188
14. Ghazizadeh MH, Fakharzadeh M (2016) 60 GHz Omni-directional segmented loop antenna. In: IEEE international symposium on antennas and propagation, pp 1653–1654
15. Tiwari RN, Singh P, Kanaujia BK, Barman PB (2019) Wideband monopole planar antenna with stepped ground plane for WLAN/WiMAX applications. In: Singh P, Paprzycki M, Bhargava B, Chhabra J, Kaushal N, Kumar Y (eds) FTNCT 2018, communications in computer and information science, vol 958. Springer, Singapore, pp 253–264
16. Zhou Z, Wei Z, Tang Z, Yin Y (2019) Design and analysis of a high isolation wideband multiple-microstrip antenna dipole. IEEE Antennas Wirel Propag Lett 18(4):722–726
17. Tang MC, Li D, Chen X, Wang Y, Ziolkowski RW, Hu K (2019) Compact tri-polarization diversity wideband, reconfigurable and wideband filtenna. IEEE Trans Antennas Propag 67(8):5689–5694

18. Wang J, Lu WB, Liu ZG, Zhang AQ, Chen H (2019) Graphene-based microwave antennas with reconfigurable pattern. IEEE Trans Antennas Propag 68(4):2504–2510
19. Hussain S, Qu SW, Zhou WL, Zhang P, Yang S (2020) Design and fabrication of wideband dual-polarized dipole array for 5G wireless systems. IEEE Access 8:65155–65163
20. Wu GB, Zeng YS, Chan KF, Chen BJ, Qu SW, Chan CH (2020) High-gain filtering reflect-array antennas for millimeter-wave applications. IEEE Trans Antennas Propag 68(2):805–812

Lightweight De-authentication DoS Attack Detection Methodology for 802.11 Networks Using Sniffer

Zakir Ahmad Sheikh and Yashwant Singh

Abstract Wireless networks are prone to many types of attacks. The open access of wireless technology puts the whole wireless environment into a threat. The new wireless technologies like wireless sensor networks (WSN) and the Internet of Things (IoT) face the challenges of security. The Internet of Things is yet to be standardized because of the unavailability of security solutions to certain wireless threats. Internet of Things uses different technologies for communication such as Bluetooth, Wi-Fi, Li-Fi, and 6LoPAN. Of all, Wi-Fi is mainly preferred. But Wi-Fi itself possesses certain security issues. One of the issues of Wi-Fi is the nature of Management Frames in its 802.11 standard. The Management Frames in 802.11 standard are unencrypted, hence easily interpretable. This very nature of Wi-Fi leads to many attacks like De-authentication attacks and Dis-association attacks. This paper provides an approach to detect De-authentication DoS attacks in Wi-Fi-based IoT networks, using the sniffing concept. For network sniffing, a Python-based tool known as Scapy is used. Attacks and Detection are done in a real-time environment. We used Kali Linux tools and Scapy for real-time attacks. The De-authentication DoS attacks are performed on a Wi-Fi-based IoT network, and the detections are triggered on a Scapy-based Sniffer program. Our algorithm also detects spoofed De-authentication DoS attacks.

Keywords De-authentication attack · DoS attacks · Wi-Fi attacks · Packet sniffer · Scapy · Internet of things · Wireless sensor networks · 802.11 standard

Z. A. Sheikh (✉) · Y. Singh
Department of Computer Science and Information Technology, Central University of Jammu, Jammu, Jammu and Kashmir, India

Y. Singh
e-mail: yashwant.csit@cujammu.ac.in

P. K. Singh et al. (eds.), *Proceedings of Second International Conference on Computing, Communications, and Cyber-Security*, Lecture Notes in Networks and Systems 203,
https://doi.org/10.1007/978-981-16-0733-2_5

1 Introduction

De-authentication DoS attacks induce some sort of hindrance to the network connectivity by either de-authenticating a Wi-Fi-based device or dis-associating it. For the critical networks like wireless sensor networks (WSN) and the Internet of Things (IoT), any sort of connection interruption, and disconnection can be dangerous for the overall network performance. Connection interruptions and disconnection in critical situations may result in the loss of important data. Thus, there is a need to develop solutions that protect the network connectivity.

Our paper provides a methodology in the form of the de-authentication DoS Attack Detection Algorithm (DDADA), for the detection of De-authentication DoS attacks. We use Scapy for creating a Sniffer program that monitors the network. Based on the network traffic, our sniffer program marks the frames as Normal or Abnormal (as a part of De-authentication DoS Attack). We used three thresholds for the detection of De-authentication DoS attacks that are Time_Difference, *Frame_Count*, and *Gap*. *Time_Difference* indicates a difference in times from disconnection to reconnection. The *Frame_Count* indicates number of De-authentication frames from a particular device that is the level of tolerance for the number of De-authentication frames to be treated as Normal. We also used an equation for *Frame_Count* which changes the tolerance level dynamically based on the number of De-authentication DoS attacking devices. Our algorithm ensures that the Spoofed De-authentication DoS attacks are detected by using the *Gap* threshold. We ran our Algorithm on a Kali Linux-based environment. The rest of the paper is organized as: Sect. 2 contains Background, Sect. 3 discusses the Proposed Scheme that is followed by Methodology in Sect. 4. Results and Discussion are discussed in Sect. 5. Finally, the Conclusion and future work are discussed in Sect. 6.

2 Background

Wireless LANs are very famous for their ability to provide mobility to connected nodes. Wi-Fi has better mobility than Bluetooth because of the smaller range in Bluetooth. 802.11 is the standard for Wi-Fi technology. As per the 802.11 standard, there are three types of frames in Wi-Fi, namely Data Frames, Control Frames, and Management Frames. The standard also mentions that only Data Frames are encrypted. The unencrypted nature of Control Frames and Management Frames lets humans interpret the frames. De-authentication Frames are Management Frames and are hence unencrypted. As per the 802.11 standard, De-authentication is a Notification and Not Request, and any node receiving De-authentication must acknowledge it. The unencrypted nature and the Notification nature of the De-authentication frame lead to a De-authentication attack loophole in Wi-Fi networks. There exist three possible states of a Wi-Fi device. The Wi-Fi device is initially in State 1, and is hence *Not Authenticated* and *Not Associated*. For connectivity, the node needs to complete authentication. Any node *Authenticated* but *Not Associated* is in State 2. Afterward, for data communication, the device needs to be *Authenticated* and *Associated*. Any device *Authenticated* and *Associated* is in State 3 [1].

The De-authentication attack leads to connection termination between two connected devices/nodes. When connectivity is terminated by the De-authentication attack, the data could not be sent or received unless Re-authentication is done [1, 2]. The impact of the De-authentication attack on communication is shown in Fig. 1. The figure depicts that the data is blocked after the De-authentication attack. The data is continuously blocked unless Re-authentication is done. The impact of the De-authentication attack is severe than the Dis-association attack because the former needs both Re-association and Re-authentication but the later needs only Re-association [1].

The De-authentication is mainly done by Spoofing MAC address of Access Point (AP) and/or MAC address of station (STA). The following are the different scenarios for De-authentication attack using Spoofed MAC [3]:

- Craft a Frame and set MAC address of victim STA as Source Address and MAC address of AP as Destination Address. When AP receives this frame, the AP assumes the frame as genuine and acknowledges it, which leads the STA to be De-authenticated with AP.

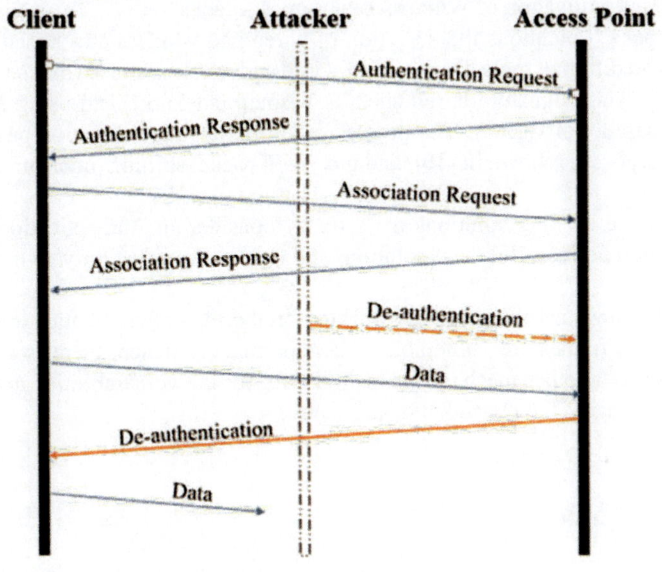

Fig. 1 De-authentication attack and data communication

- Craft a Frame and set MAC address of victim AP as Source Address and MAC address of STA as Destination Address. When STA receives this frame, the STA assumes the frame as genuine and acknowledges it, which leads the AP to be De-authenticated with STA.
- Craft a Frame and set MAC address of victim AP as source address and broadcast MAC address, i.e., (FF:FF:FF:FF:FF:FF) as destination address. This broadcasts the frame to all the connected STAs. When STAs receive this frame, the STAs assume the frame as genuine and acknowledge it, which leads all the STAs to be De-authenticated. This type of attack is severe in nature.

Scapy is a powerful and interactive Python-based tool used for Packet Sniffing, Packet Manipulation, Packet Cloning, and many other tasks. A brief introduction about the usage of the tool, its different commands, and types of operations is given in [4]. The paper [5] analyzed different Packet Sniffing tools. Wireshark, NMap, ZenMap, Tcpdump, Kismet, Ntop, Dsniff, Caspa, Cain and Abel, Etherape, and Ethereal are the tools analyzed. In paper [6], the author(s) demonstrated the usage of Packet Sniffing in switched and not switched networks.

A De-authentication Denial-of-Service attack Detection and Prevention methodology is given in [3]. The paper proposes an Intrusion Detection System (IDS) and Intrusion Prevention System (IPS) for Detection and Prevention of De-authentication Denial-of-Service (DoS) attacks in 802.11 networks.

An automated approach for Detection and Prevention of De-authentication and Dis-association DoS attacks in 802.11 networks is mentioned in [7]. This paper also analyzed these attacks using Wireshark. Paper [8] also provides a solution for the Detection and Mitigation of Wireless Link Layer Attacks.

The paper [9] mentions the detection of wired and wireless attacks using three concepts and different tools. Detection of wired and wireless attacks manually using Wireshark, using Signature-based such as Kismet and Snort, and using Machine Learning-based tool WEKA. The procedure to filter De-authentication packets for analysis purposes is shown in [10], and the Wi-Fi packet sniffing procedure is given in [11].

Most of the existing solutions try a fix by considering Authentication of De-authentication Frames. But these solutions still lack complete security, as mentioned in Table 1.

Table 1 shows that the solutions working on the idea of Authentication of De-authentication frames are vulnerable to certain attacks. Hence, there is a need to incorporate some other mechanisms or find a fix for the vulnerabilities in existing solutions, to handle the De-authentication attacks.

Table 1 Comparative analysis of existing solutions for De-authentication attack

Existing solution	Problem
R-ARP for De-authentication [12]	The clients IP address can be spoofed
Frame sequence number based De-authentication [13–15]	Prone to sniffing (i.e., prediction) attack
1-bit authentication of management frames [16, 17]	50% chances of a correct guess
Random bit scheme [18]	Uses 20-bit keys that can be brute-forced
UUID mechanism for authentication of De-authentication packets [19]	UUID can be brute-forced
Encrypted authentication keys [16, 19]	Not all encryption algorithms are suitable for IoT environments

3 Proposed Scheme

Our proposed scheme is based on Python-based Sniffer. The frames are monitored, sniffed, and analyzed for the Detection of De-authentication DoS attacks. The architectural diagram of our proposed solution is shown in Fig. 2.

The frames flow continuously in an active Wi-Fi network. Our Sniffer performs the following operations in an active Wi-Fi network:

- **Frame Filtering**: Filters frames which contains all of the below-mentioned attributes:

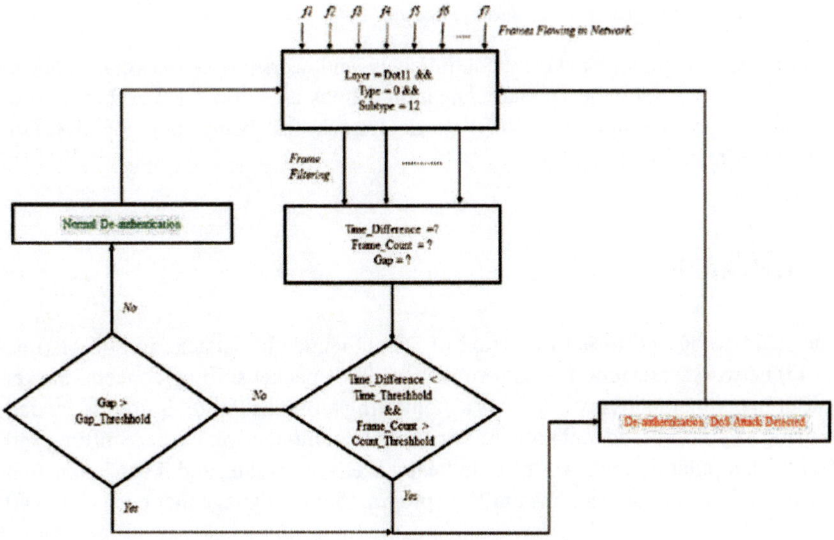

Fig. 2 De-authentication sniffer for 802.11 networks using Scapy

  - Dot11 Layer
  - Frame Control *type* of value 0
  - Frame Control *subtype* of value 12

- **Time Difference and Frame Count**: For every filtered frame, the *Time_Difference* and *Frame_Count* are calculated. *Time_Difference* is the difference of times between the previous filtered frame and the current frame, of the user possessing a particular Source address. *Frame_Count* is the number of filtered frames (De-authentication frames) received in *Time_Difference* time.
- **Sequence Number Rate Analysis (SNRA)** [20]: It is a lightweight algorithm used for Spoofing Detection [21]. It uses the following logic for the detection of spoofing:

$$\text{Gap} = \frac{S(i) - S(i-1)}{T(i) - T(i-1)} \tag{1}$$

  where S(i) and S(i-1) are the Sequence Numbers (SN) of ith and $(i-1)$th frame, respectively. $T(i)$ and $T(i-1)$ are the time of reception of ith and $(i-1)$th frame, respectively.
- **Analysis**: The Time_Difference and *Frame_Count* of individual Source Address are compared with the *Time_Threshold* and *Count_Threshold* respectively. Also, Sequence Number (SN) Gap is calculated as given in Eq. 1. The De-authentication DoS Attack event is triggered if any one of the following condition satisfies:

  i. Time_Difference is less than *Time_Threshold* AND *Frame_Count* is greater than *Count_Threshold.*
  ii. The *Gap* is greater than *Gap_Threshold.*

The process in Fig. 2 works continuously and works as a network monitor. It continuously marks the De-authentication frames as Normal De-authentication frames or De-authentication DoS attack frames (as being part of the De-authentication DoS attack).

4 Methodology

Our methodology performs Detection of De-authentication attacks in 802.11 (i.e., Wi-Fi) networks. Detection is performed by using a packet sniffing concept. Packet Sniffer in our methodology is used as a monitoring tool. The Packet Sniffer is created by using a Python-based tool known as Scapy. It is an Interactive, Packet Sniffing, and Packet Manipulating tool. A Sniffer in Scapy is easy to create, and it can be created in a single line of code which is unlike C programming language that needs 30 to 40 lines of code for the creation of simple packet Sniffer. The network needs to be in Monitor mode for our Sniffer Program, De-authentication Program, and aircrack-ng utility to work. The aircrack-ng can be used for performing De-authentication DoS attacks. We also created a program for De-authentication DoS Attacks using Scapy. For changing the network mode from Managed to Monitor, we use the following commands:

- sudo airmon-ng check
- airomon-ng check kill
- sudo airmon-ng start wlan0

The first command is used for checking the currently running processes, and the second command accordingly stops the running processes. Finally, the third command changes the network mode of wlan0 to Monitor mode. It also renames the wlan0 interface name to wlan0mon. Our Sniffer filter frames based on the following credentials. The frames filtered contain three following properties:

i. The frame contains Dot11 layer, AND
ii. The frame has Frame Control type value 0, AND
iii. The frame has a Frame Control subtype value 12.

Frame Control type of 0 means that the Frame is Management Frame, and its subtype value of 12 means that the Frame is De-authentication Frame. Based on the above-mentioned properties, the De-authentication packets are filtered and rest are ignored. This procedure is done using Scapy Program as shown in Algorithm 1.

Algorithm 1 De-authentication DoS Attack Detection Algorithm (DDADA)

Algorithm 1: De-authentication DoS Attack Detection Algorithm (DDADA)

Input: De-authentication Requests from Devices

Output: Mark De-authentication Frame as Normal or De-authentication DoS Attack Frame.

Initialization: BSSIDs=[ ], Time_Difference=0, Time_Threshold=X, Frame_Count=0, Count_Threshold=Y, Gap=0, Gap_Threshold=Z, stime=0, etime=0, sold=0,snew=0,num=0,mycount=defaultdict(int), openMonitorMode()

```
procedure detectDeauth(frm)
 if (frm.haslayer(Dot11) and frm.type==0 and frm.subtype==12) then
   if (frm.getlayer(Dot11).addr2 not in BSSIDs) then
     BSSIDs.append(frm.getlayer(Dot11).addr2)
     addr2 ← frm.getlayer(Dot11).addr2
     mycount[addr2] ← 1
     num ← num + 1
     Count_Threshold ← int(ceil((num/num^1.5)*10))
     snew ← SequenceNumber(frm.SC)
   else
     etime ← time.clock()
     Time_Difference ← etime-stime
     sold ← snew
     snew ← SequenceNumber(frm.SC)
     Gap ← (snew-sold)/etime-stime
     addr2 ← frm.getlayer(Dot11).addr2
     Frame_Count ← mycount[addr2] ← mycount[addr2]+1
     if (Time_Difference < Time_Threshold and Frame_Count > Count_Threshold) then
       print ("De-authentication DoS Attack Detected")
     else
       if (Gap > Gap_Threshold) then
         print ("De-authentication DoS Attack Detected")
       else
         print ("Normal De-authentication")
       end if
     end if
   end if
   stime ← time.clock()
 else
   print("Packet Ignored")
 end if
end procedure
while true do
sniff(iface="Wi-Fi", prn=detectDeauth())
end while
```

We used three thresholds in our Algorithm. The *Time_Threshold* means for how much time and *Count_Threshold* means how many De-authentication frames to be accepted as Normal in *Time_Threshold* time. If the number of De-authentication frames exceeds the value of *Count_Threshold* in time less than *Time_Threshold*, then the De-authentication DoS attack is Detected. Also, if Gap is greater than *Gap_Threshold*, the De-authentication DoS Attack message is triggered. Choosing a higher value of *Time_Threshold* reduces the Attack Detection Rate and its lower value

Table 2 Impact of altered values of thresholds

Attribute	Value	Impact
Time_Threshold	Lower	False positives
Time_Threshold	Higher	Low detection rate
Count_Threshold	Lower	False positives
Count_Threshold	Higher	Low detection rate
Gap_Threshold	Lower	False positives
Gap_Threshold	Higher	Low detection rate

triggers a higher rate of False Positives. Further, the lower value of *Count_Threshold* increases False Positives, and its higher value reduces Attack Detection Rate as mentioned in Table 2.

The *Gap_Threshold* detects Spoofing and is based on Sequence Number and Time Difference. *Gap_Threshold* is the number of frames sent in a Normal Communication by an 802.11 device in one second. For effectiveness, this value can be calibrated, and accordingly, the accuracy is detected. For different networks, different values of *Gap_Threshold* could be effective, because of different values of inter variable.

5 Results and Discussion

Our algorithm uses three threshold values. There is no fixed value for the thresholds to effectively work on all the networks but needs to be adjusted in different environments. The *Gap_Threshold* depends on the number of frames flowing per second in an 802.11-based network. The impact on performance by different values of *Time_Threshold*, *Count_Threshold*, and *Gap_Threshold* is given in Table 2.

A single De-authentication frame is not always enough for successfully De-authenticating a device. Sometimes more than one de-authentication frames are needed if as in the case when the devices are in a distance and possessing a low signal strength. As it has been seen that on an average of 2 De-authentication frames, 96% of the devices gets successfully De-authenticated. With a single De-authentication frame, about 88% of devices get disconnected. More than 4 De-authentication frames even touch 99% De-authentication success [3]. Keeping given this fact that a single De-authentication frame is not enough, we have introduced a threshold, i.e., *Count_Threshold* with a dynamic value calculated by Eq. 2:

$$Count\_Threshold = \mathrm{int}\left(\mathrm{ceil}\left(\frac{\mathrm{num}}{\mathrm{num}^{1.5}} * 10\right)\right) \quad (2)$$

where num is the number of devices notifying for De-authentication (NDD). Equation 2 changes the threshold values based on the number of devices. The more the number of devices the smaller its value is, and the smaller the number of devices the more its value is. Our equation is one example of how the *Count_Threshold* can be

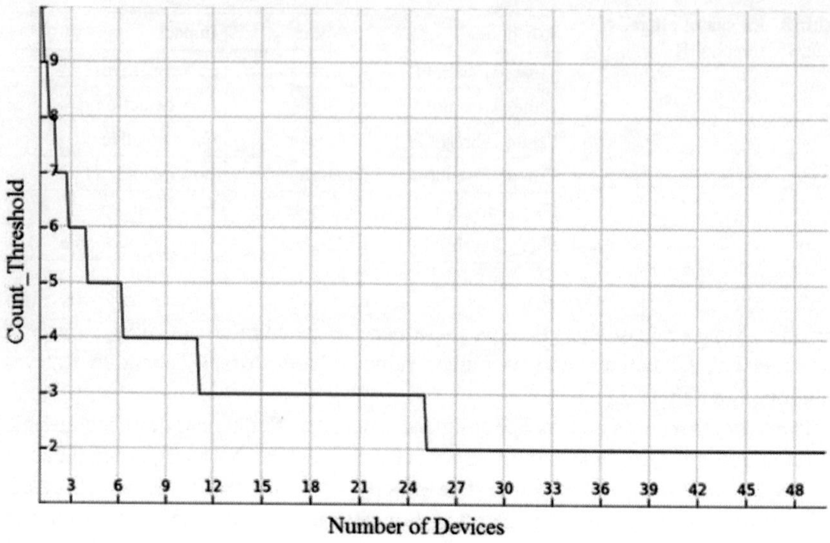

Fig. 3 Count_Threshold value of networks with different number of devices

used to have dynamic values depending upon the network situations. For networks containing devices ranging from 1 and to 50, this equation gives values of thresholds from 10 to 2 (known to be a tolerance level) as shown in Fig. 3.

As Fig. 3 clearly shows that severe the attack in the form of the number of attacking devices, the smaller the value of *Count_Threshold* and vice versa. Our DDADA detects a De-authentication DoS attack when either *Count_Threshold* is crossed by *Frame_Count* and *Time_Difference* is below its threshold *Time_Threshold* or *Gap_Threshold* is crossed by Gap variable. This is shown by two simple examples in Figs. 4 and 5.

The different scenarios of Fig. 4 depict whether the *Frame_Count* has crossed its threshold or not in situations where Time_Difference is below its *Time_Threshold*. If *Time_Difference* is below its *Time_Threshold* and *Frame_Count* crosses its *Count_Threshold*, De-authentication DoS Attack message is triggered, as in Scenario 2 and 4. Figure 5 depicts different scenarios of the relation between *Gap* and *Gap_Threshold*. In situations where the *Gap* variable crosses *Gap_Threshold*, De-authentication DoS Attack message is triggered, as in Scenario 3 and 5. The creation of the Sniffer and De-authentication Attacker program has been done using Scapy which is a Python-based tool. We executed our algorithm (DDADA) on a Linux-based environment. We took a fixed threshold *Time_Threshold* with value 1 s and *Gap_Threshold* of value 20 frames. We generated the value of the third threshold, i.e., *Count_Threshold* dynamically using Eq. 2. We ran our Sniffer and performed De-authentication attacks various times. The Detection Rates are shown in the Performance graph in Fig. 6 and the Detection Rate is calculated by the formula shown in Eq. 3:

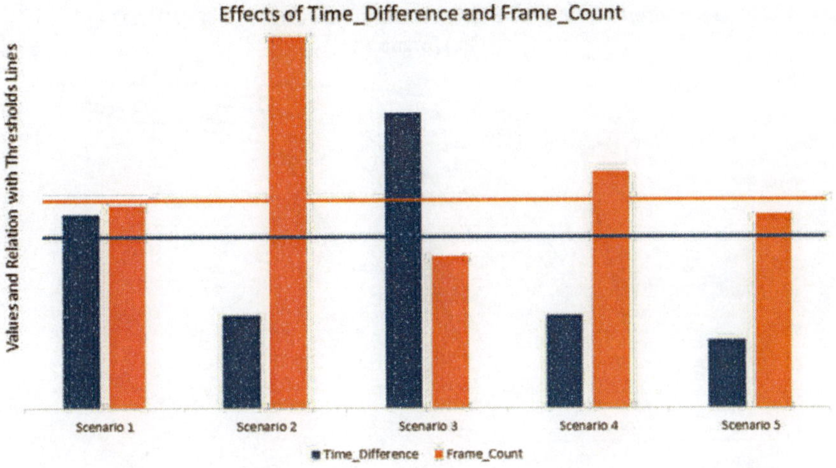

Fig. 4 Time_Threshold and Frame_Count values and scenarios of De-authentication DoS attack detection

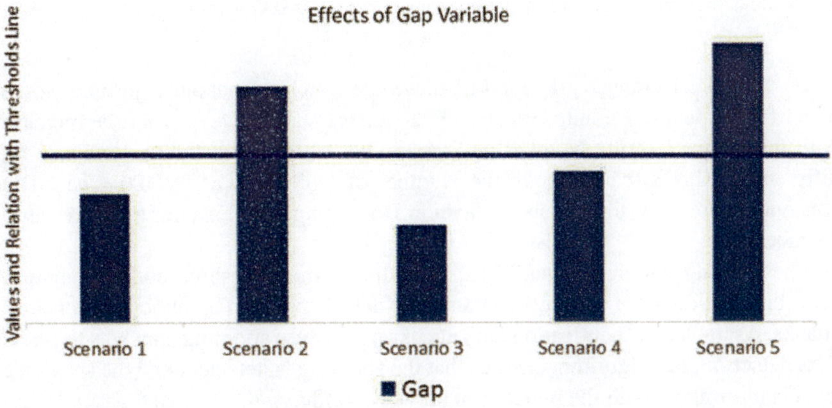

Fig. 5 Gap threshold values and scenarios of De-authentication DoS attack detection

$$\text{Detection Rate} = \left(\frac{\text{TP}}{\text{TP} + \text{FN}}\right) \tag{3}$$

where TP is True Positive and FN is False Negative. A TP occurs when a real De-authentication attack is treated as attack and FN occurs when a De-authentication attack is treated as Normal.

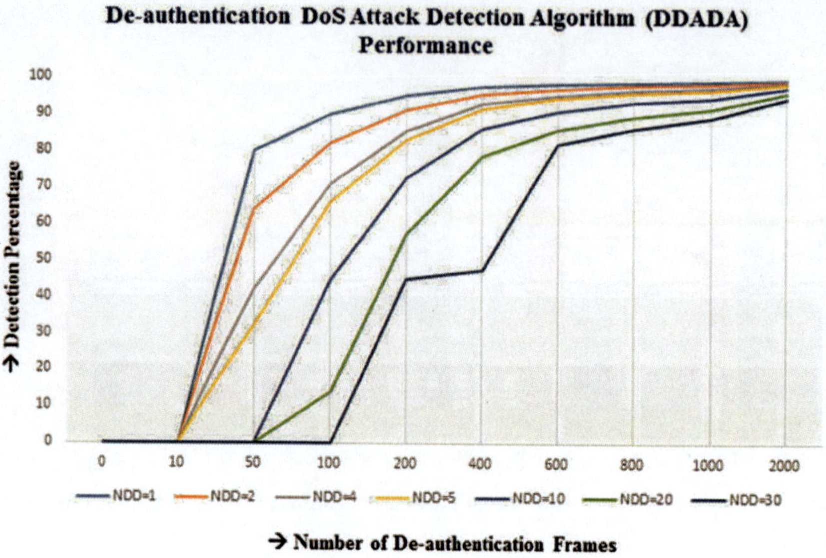

Fig. 6 De-authentication DoS attack detection algorithm (DDADA) performance

From the Performance graph in Fig. 6, we can conclude that our algorithm works better for the severe De-authentication DoS attacks and the Detection Rate increase with the increase in the Number of Frames. We have shown the performance on different scenarios that differ in the Number of DoS Devices (NDD). The NDD concept ensures that the De-authentication DoS using MAC spoofing is taken into consideration.

The existing solution in paper [3] uses the Frame Threshold and Throughput Threshold but does not use any mechanism to detect spoofed De-authentication DoS attacks. As the De-authentication DoS attacks mainly use spoofing concepts to evade DoS detection, our algorithm ensures that the spoofing is detected, and the tolerance level is minimized with the increase in network traffic.

6 Conclusion and Future Work

De-authentication DoS attack is severe in nature. For wireless sensor networks (WSNs) and the Internet of Things (IoT), real-time communication is very important. So, these networks cannot compromise with security issues. De-authentication DoS attack interrupts the data communication. WSN and IoT networks mainly use 802.11 standard for communication and are also limited in terms of computational power and storage. We have come up with a lightweight solution for Detection of De-authentication DoS attacks in 802.11 networks which possesses limited attributes. Our algorithm is lightweight in terms of the linearity of the algorithm itself and the thresholds used. We have particularly selected the Sequence Number Rate Analysis (SNRA) algorithm for spoofing detection, which is an effective lightweight algorithm in comparison to the others available. For networks, our DDADA works effectively but needs the proper calibration of three thresholds. In the future, we tend to have a lightweight prevention methodology for De-authentication DoS attacks, and also the inclusion of some more attributes for detection will be focused. Also, the machine learning approaches could be used for the dynamic estimation of three thresholds. Moreover, Machine Learning and Deep Learning based IDS could enhance the performance and reduce the overhead of manual attribute selection.

References

1. Cheema R (2011)Deauthentication/disassociation attack: implementation and security in wireless mesh networks. Int J Comput Appl 23(7):7–15
2. Noman HA, Abdullah SM, Kama N, Noman SA (2016) A lightweight scheme to mitigate deauthentication and disassociation DoS attacks in wireless 802.11 networks
3. Agarwal M, Biswas S, Nandi S (2013) Detection of de-authentication denial of service attack in 802.11 networks
4. Rohith Raj S (2018) SCAPY—a powerful interactive packet manipulation program. In: 2018 international conference on networking, embedded and wireless systems (ICNEWS), pp 1–5
5. Kaur I, Kaur H, Singh EG (2014) Analysing various packet sniffing tools. Int J Electr Electron Comput Sci Eng 1(5):65–69
6. Verma A, Singh A (2013) An approach to detect packets using packet sniffing. Int J Comput Sci Eng Surv 4(3):21–33
7. Noman HA, Abdullah SM, Mohammed HI (2015) An automated approach to detect deauthentication and disassociation dos attacks on wireless 802.11 networks
8. Aung MAC, Thant KP (2017) Detection and mitigation of wireless link layer attacks. In: 2017 IEEE 15th international conference on software engineering research, management and applications (SERA), pp 173–178
9. Kaur J (2020) Wired LAN and wireless LAN attack detection using signature based and machine learning tools
10. Analyzing deauthentication packets with wireshark (2020), pp 1–11
11. Linux H, Paython W, Point A (2020) Wireless sniffing: how to build a simple WiFi sniffer in Python
12. Sweigart C (2003) Interested in learning more? Institute, Author retain full rights. Security 401:1–39
13. Jo C, Barros P, Tavares M Vulnerabilities in IoT Devices for smart home environment
14. Visoottiviseth V, Akarasiriwong P, Chaiyasart S, Chotivatunyu S (2017) PENTOS: penetration testing tool for Internet of Thing devices, pp 2279–2284
15. Williams R, Mcmahon E, Samtani S, Patton M, Chen H (2017) Identifying vulnerabilities of consumer Internet of Things (IoT) devices : a scalable approach, pp 179–181
16. Arora A Preventing wireless deauthentication attacks over
17. Divya, Kumar S (2010) Analysis of denial of service attacks in IEEE 802.11s wireless mesh networks. Lecture notes of the institute for computer sciences, social informatics and telecommunications engineering (LNICST), vol 41, pp 1–10
18. Achalakul TBig data analytics. How data fuel the world. www.BigDataExperience.org
19. Kamble A, Bhutad S (2018) Survey on Internet of Things (IOT). In: 2018 2nd International Conference on Inventive Systems and Control (ICISC), pp 307–312
20. Madory D (2020) New methods of spoof detection in 802.11b wireless networking

21. Vaidya A, Jaiswal S, Motghare M (2016) A review paper on spoofing detection methods in wireless LAN. In: Proceedings of the 10th international conference on intelligent systems and control (ISCO 2016)

Power Distribution Control for SIMO Wireless Power Transfer Systems

Sanjog Ganotra

Abstract Transmission of power wirelessly has emerged as a prominent technology in recent decades; however, the selective flow of power in systems with multiple receivers has always been a conflicting issue. This paper proposes a favorable method to control power distribution among multiple receivers using different resonating frequencies for transmitter and receiver coils. A multiple load system with different parameters has been analyzed theoretically, using Advanced Design System (ADS), and experimentally by practical implementation. Upon detailed analysis, it is found that power transfer efficiency reaches its maximum value at the resonant frequency of the receiver under consideration, irrespective of the resonant frequency of the common transmitter. It has also deduced that the driving efficiency of a coil has a great emphasis on the probable amount of power received, lesser the difference between driving frequency and the resonant frequency of the coil, less isolated is the coil.

Keywords Power transfer · Efficiency · Resonant frequency · Multiple-load

1 Introduction

Transferring power wirelessly has transpired to be a revolutionizing approach in the recent years [1–3]. Along the course of time, wireless nature of power transfer has been widely implemented in laptops, mobiles and other PDA devices [4], in bionic implants such as pacemakers [5]. Solar panels, fuel cells and combustion engines also use WPT systems to source power [6]. However, WPT systems are susceptible to weak transmission efficiency, with increase in distance between the transmitter and receiver, the power transmitted diminishes. Chaidee et al. [7] clearly show that the transmission efficiency achieved at 20 cm is far greater than that achieved at 30 cm.

Many prior works have been proposed which show promising results in optimizing the power transfer efficiency over a longer range. Use of optimum load to

S. Ganotra (✉)
Department of Electronics and Communication, Delhi Technical Campus, 28/1, Knowledge Park III, Uttar Pradesh, India

P. K. Singh et al. (eds.), *Proceedings of Second International Conference on Computing, Communications, and Cyber-Security*, Lecture Notes in Networks and Systems 203,
https://doi.org/10.1007/978-981-16-0733-2_6

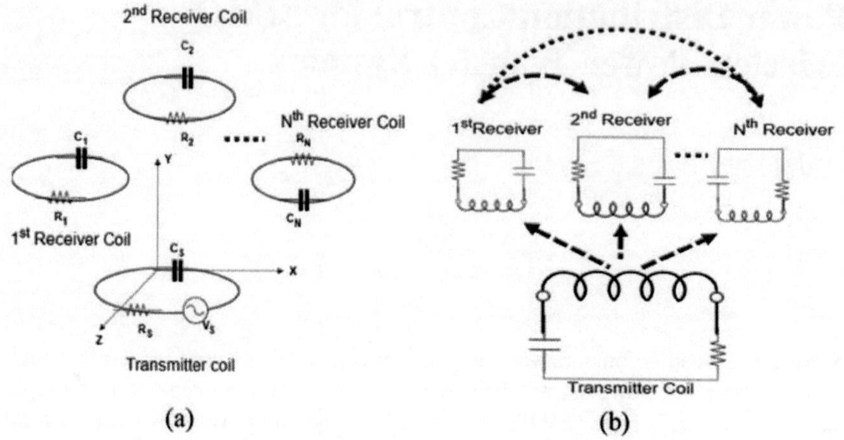

Fig. 1 Configuration of a SIMO system: **a** geometry and **(b)** equivalent circuit

achieve peak efficiency was highlighted from the beginning. This approach is pursued by deducing optimum values for source and load impedances from the conjugated matching conditions as in [8]. Many unique system designs build solely for optimization purposes at higher frequency have also been encountered. Hafeez et al. [9] propose a system design that employs offset reflectors fed by conical horns at 6 GHz frequency for long-range wireless power transfer. A special design implementation using parameters for a four-coil geometry has been proposed in [10], which shows a maximum efficiency of 92.9% at 6.5 cm coupling distance.

Other than optimizing power transfer efficiency, an approach to replicate receivers in WPT systems has been a profound research topic. Single-input multiple-output (SIMO) configuration refers to a single source of power transmission with multiple destinations as receivers. SIMO or single-input multiple-output configuration is a curtailment to MIMO or multiple-input multiple-output. For a general MIMO system, power transfer efficiency is calculated using Z matrix formulation based on set of circuit design equations [9]. Although studies like [11] which show a detailed scalability analysis for SIMO resonant wireless power systems have been conducted, applications for SIMO WPT scenarios are limited. Figure 1 shows general configuration of a SIMO WPT system with N receivers.

2 Formulating Problem Statement

In SIMO WPT systems, selective power transmission to a specific receiver is still a sore subject. The transmitter tends to transmit power to all the receivers in vicinity, which provides involuntary power to the receivers who are at standby or do not require the same amount of power as the other receivers. This flaw could lead electrical instruments to malfunction. The ability to control power distribution is a salient feature for a power distributing circuit. The primary function of a power distribution circuit is to transmit a predetermined amount of power to any end of receiver. A fine grade distribution technique is one which is cost effective requires minimum number of power supplies.

In this paper, a method using resonant frequency of the receivers has been demonstrated, which can be a solution to the above stated flaw. Resonance is the phenomenon for a system to oscillate with larger amplitudes at some frequencies as compared to others, these are called its resonant frequencies. Every system has its own natural resonant frequency. In this paper we validate that, when the driving frequency of a system overlaps the resonant frequency, peak efficiency is attained. To vindicate our statement, a detailed analysis comparing the experimental and theoretical values has been conducted. The transmitter and receiver coils have different parameters all together.

3 Single-Input Single-Output Power Transfer

In this section, we will discuss the power transfer at different resonant frequencies when only one load is present. Figure 2a and b shows the abstract model and the schematic diagram for single load power transfer. The Tx and Rx coils both have different parameters (*R*, *L*, *C*) denoted by subscript 1 and 2, respectively. Driving angular frequency is denoted by ω.

Since the transmitter and receiver coils have different parameters, their resonant frequencies vary, which are denoted by F_1 and F_2.

We know that

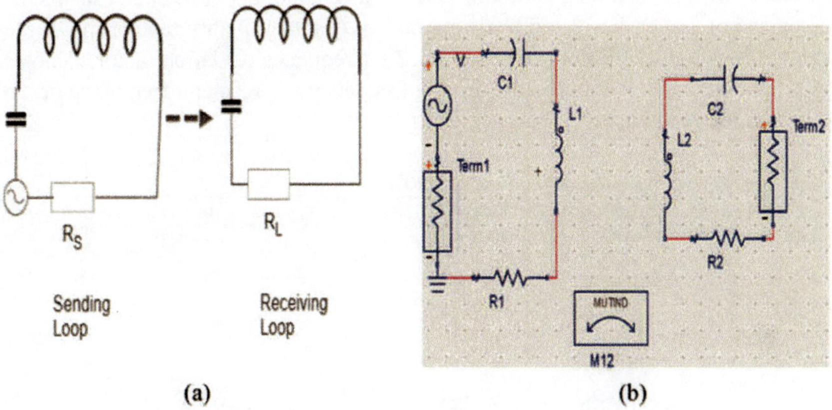

Fig. 2 **a** Single-load abstract model and **b** single-load schematic diagram

$$F_1 = \frac{1}{2\pi\sqrt{L_1 C_1}}, \quad F_2 = \frac{1}{2\pi\sqrt{L_2 C_2}}. \tag{1}$$

Due to the receiving loop, the sending loop shall experience a reflected impedance which is expressed as

$$\mathrm{Re}(Z_{\mathrm{ref}}) = \frac{\omega^4 C_1^2 M^2 (R_1 + R_2)}{\left(\omega^2 L_{\mathrm{term1}} - 1\right) + \omega^2 C_1^2 (R_1 + R_2)^2} \tag{2}$$

$$\mathrm{Im}(Z_{\mathrm{ref}}) = \frac{-\omega^3 C_1 M^2 \left(\omega^2 C_1 L_1 - 1\right)}{\left(\omega^2 C_1 L_S - 1\right) + \omega^2 C_1^2 (R_1 + R_2)^2} \tag{3}$$

At resonance Z_{ref} becomes

$$Z_{\mathrm{ref}} = \frac{\left((\omega M_{12})^2\right)\left(R_2 + R_{\mathrm{term2}} - j\omega L_2 - \frac{1}{j\omega C_2}\right)}{(R_2 + R_{\mathrm{term2}})^2 + (\omega L_2 - 1/\omega C_2)^2} \tag{4}$$

Power transfer efficiency (PTE), η, is represented by the ratio of power consumed by the load to the power available with source. The mathematical expression for PTE

$$\eta = \frac{\frac{(\omega M_{12})^2 (R_2 + R_{\mathrm{term2}})}{(R_2 + R_{\mathrm{term2}})^2 + \left(\omega L_2 - \frac{1}{\omega C_2^2}\right)}}{R_{\mathrm{term1}} + R_2 + \frac{(\omega M_{12})^2 (R_2 + R_{\mathrm{term2}})}{\left((R_2 + R_{\mathrm{term2}})^2 + \left(\omega L_2 - \frac{1}{\omega C_2}\right)^2\right)}} \cdot \frac{R_{\mathrm{term2}}}{R_1 + R_{\mathrm{term2}}} \tag{5}$$

For better assessment of power transfer, with different resonant frequencies we employ the following parameters as depicted in [12]. L_M (load matching factor) represents the load matching conditions. S_M (source matching impedance) shows the percentage of power dissipated from source. T_Q (transfer quality factor) shows the extent of how tightly the coils are coupled. F_D (frequency deviation factor) denotes the extent of deviation between the driving frequency and resonant frequency of the coil. They are mathematically expressed as,

$$L_M = \frac{R_{\mathrm{term2}}}{R_2}, \quad S_M = \frac{R_{\mathrm{term1}}}{R_1}, \quad T_Q = \frac{\omega M_{12}}{\sqrt{(R_1 R_2)}},$$
$$F_{D_1} = \frac{\omega L_1 - \frac{1}{\omega C_1}}{R_1}, \quad F_{D_2} = \frac{\omega L_1 - \frac{1}{\omega C_2}}{R_2} \tag{6}$$

Taking (6) into consideration,

$$\eta = \frac{T_Q^2}{(1 + S_M)(1 + L_M) + \frac{1+S_M}{1+L_M} F_{D2}^2 + T_Q^2} \cdot \frac{L_M}{1 + L_M} \tag{7}$$

From this we infer that the transfer quality factor is directly proportional to the PTE. Therefore, power is transferred more efficiently if the coils are more tightly coupled. A lower S_M leads to higher efficiency, which is obvious because greater loss of power at the source end would lead to a smaller amount of power available for transmission.

4 Single-Input Multiple-Output Power Transfer

In the previous section, we discussed that the PTE reaches its peak value, when the operating frequency of receiver is equal to the resonant frequency, irrespective of the resonant frequency of the transmitter This phenomenon can be used to control the power distribution in a WPT system with multiple receivers, as a higher amount of power is transferred to the load showing lesser deviation from its resonant frequency. Figure 3a and b shows schematic diagram for a two-load WPT system. For ease of calculation, mutual inductance between the receiver coils is ignored. The resonant frequencies for transmitter and the two receivers are denoted by F_1, F_2, F_3 respectively. They are mathematically denoted as,

$$F_1 = \frac{1}{2\Pi\sqrt{(L_1C_1)}},\quad F_2 = \frac{1}{2\Pi\sqrt{(L_2C_2)}},\quad F_3 = \frac{1}{2\Pi\sqrt{(L_3C_3)}} \tag{8}$$

The transmitting loop experiences reflective impedance due to receiving loop 1 which is mathematically expressed as

$$Z_{\text{ref1}} = \frac{(\omega M_{12})^2}{R_2 + R_{\text{term2}} + j\omega L_2 + \frac{1}{j\omega C_2}}$$

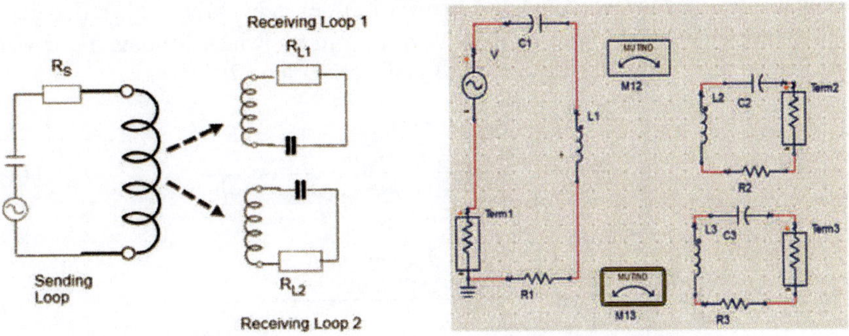

(a) (b)

Fig. 3 **a** Two-load abstract model and **b** two-load schematic diagram

$$= \frac{(\omega M_{12})^2\left(R_2 + R_{\text{term2}} - J\omega L_2 - \frac{1}{J\omega C_2}\right)}{(R_2 + R_{\text{term2}})^2 + \left(\omega L_2 - \frac{1}{(\omega c_2)^2}\right)} \tag{9}$$

Similarly, the transmitter experiences a reflective impedance from receiving loop 2 which is mathematically expressed as,

$$Z_{\text{ref2}} = \frac{(\omega M_{13})^2}{R_3 + R_{\text{term3}} + j\omega L_3 + \frac{1}{j\omega C_3}}$$
$$= \frac{(\omega M_{12})^2\left(R_2 + R_{\text{term2}} - J\omega L_2 - \frac{1}{J\omega C_2}\right)}{(R_3 + R_{\text{term3}})^2 + \left(\omega L_3 - \frac{1}{(\omega c_3)^2}\right)} \tag{10}$$

We shall assess the two-load WPT system with the same parameters as described in Sect. 2. The mathematical representations for load matching factor (L_m), transfer quality factor (T_Q) and frequency deviation factor (F_D) for receiver loops 1 and 2 are shown in (11), (12) and (13), respectively.

$$L_{M1} = \frac{R_{\text{term2}}}{R_2}, \quad L_{M2} = \frac{R_{\text{term3}}}{R_3} \tag{11}$$

$$T_{Q12} = \frac{\omega M_{12}}{\sqrt{(R_1 R_2)}}, \quad T_{Q13} = \frac{\omega M_{13}}{\sqrt{(R_1 R_2)}} \tag{12}$$

$$F_{D2} = \frac{\omega L_2 - \frac{1}{\omega C_2}}{R_2}, \quad F_{D3} = \frac{\omega L_3 - \frac{1}{\omega C_3}}{R_3} \tag{13}$$

S_M (source matching factor) for single-load and two-load WPT system will be same, as the transmitter loop remains unchanged. Taking the above parameters into account, efficiency for receiver 1 and receiver 2 is shown in (14) and (15). Equation (16) shows a simplified ratio of η_1 with respect to η_2, which denotes the power distribution ratio between receiver loop 1 and receiver loop 2

$$\eta_1 = \frac{\frac{T_{Q12}^2 L_{M1}}{F_{D2}^2 + (1+L_{M1})^2}}{1 + S_M + \frac{T_{Q12}^2(1+L_{M1})}{(1+L_{M1})^2 + F_{D2}^2} + \frac{T_{Q13}^2(1+L_{M2})}{(1+L_{M2})^2 + F_{D3}^2}} \tag{14}$$

$$\eta_2 = \frac{\frac{T_{Q12}^2 L_{M2}}{F_{D3}^2 + (1+L_{M2})^2}}{1 + S_M + \frac{T_{Q12}^2(1+L_{M1})}{(1+L_{M1})^2 + F_{D2}^2} + \frac{T_{Q12}^2(1+L_{M1})}{(1+L_{M1})^2 + F_{D2}^2}} \tag{15}$$

$$\frac{\eta_1}{\eta_2} = \frac{F_{D3}^2 + T_{Q12}^2 L_{M1}(1 + L_{M2})^2}{F_{D2}^2 + T_{Q13}^2 L_{M2}(1 + L_{M1})^2} \tag{16}$$

For identical receivers,

$$\frac{\eta_1}{\eta_2} = \frac{F_{D3}^2 + (1 + L_{M2})^2}{F_{D2}^2 + (1 + L_{M1})^2} \tag{17}$$

F_{D2} will be 0 at F_2 and F_{D3} will be 0 at F_3. If the difference between the F_2 and F_3 is large, then F_{D3}^2 will be at peak at F_2 and F_{D2}^2 will be at peak at F_3. Therefore, maximum PTE is achieved only at resonant frequency of each receiver.

5 Performance Evaluation

A 21 gauge wire was used to make helical coils for both receivers and transmitter, with 12 turns and a 0.01 m pitch. The inductance and resistance were maintained at 20.4 μH and 0.138 Ω, respectively. To achieve different resonating frequencies, 2.8 nF capacitors were connected in parallel to attain driving frequency of 181, 191, 203 kHz. The source and load impedance were 0.5 Ω each. The entire system is driven by a power meter supplying a constant supply of 50 W.

A. **For single-input single-output power transfer**

The analysis for power transfer efficiency was done at 0.2 and 0.3 m distance between the transmitter and the receiver. When the distance between the transmitter and the receiver loop is 0.2 m, the mutual inductance was 1.05 μH; for 0.3 m separation, the mutual inductance was 0.34 μH. Figures 4 and 5 show comparison between experimental and calculated values for 0.2 m and 0.3 m, respectively. Here, the resonant frequencies of the receiver loops are 181 kHz Figs. 6 and 7 show the same analysis as Figs. 5 and 6, but here the resonant frequency was changed to 203 kHz. The transmitter was sustained at 191 kHz.

From Figs. 4, 5, 6, and 7, we infer that only when the resonant frequency was set to be 181 kHz, maximum efficiency, i.e., 0.794 and 0.35 were achieved when the driving frequency was close or equal to their resonant frequencies. Similarly, when the resonant frequency was changed to 203 kHz, peak efficiencies 0.43 and 0.78, were also attained at their resonant frequency.

B. **For single-input multiple-output power transfer**

For the ease of practical implementation, in case of multiple receivers only two receiver loops were considered. Figures 8 and 9 show comparison between calculated and experimental results for two load system, when the resonant frequency for receiver 1 is 181 kHz and the resonant frequency for receiver 2 is 203 kHz, respectively. The mutual inductance is 0.574 μH.

Figures 10 and 11 show calculated versus experimental results for two-load system when F_2 was changed to 191 kHz and F_3 was changed to 203 kHz, respectively. The distance between the transmitter and load coils was maintained at 0.25 m. From

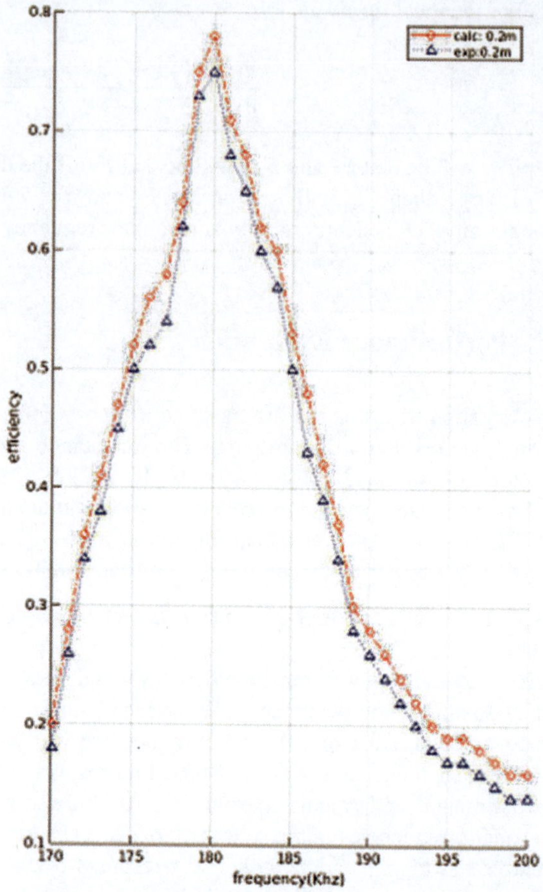

Fig. 4 Calculated versus experimental results for single-load transfer at 0.2 m apart, when $F_2 = 181$ kHz

Figs. 7 and 8, we infer that every receiver receives maximum power, when their loads are operated at their own resonant frequency irrespective of the transmitter.

6 Conclusion

In this paper, a detailed analysis for a technique to control power distribution in SIMO WPT system was conducted. It can be inferred that each load shows maximum power transfer efficiency when it is operated at a frequency equal or close to its resonant frequency, irrespective of the resonant frequency of the transmitter. When the difference between resonant frequencies of the loads is large enough as in case of Fig. 7, it is preferred to use individual power supplies. When the difference between resonant frequencies is not large enough, simultaneous power feeding is preferred.

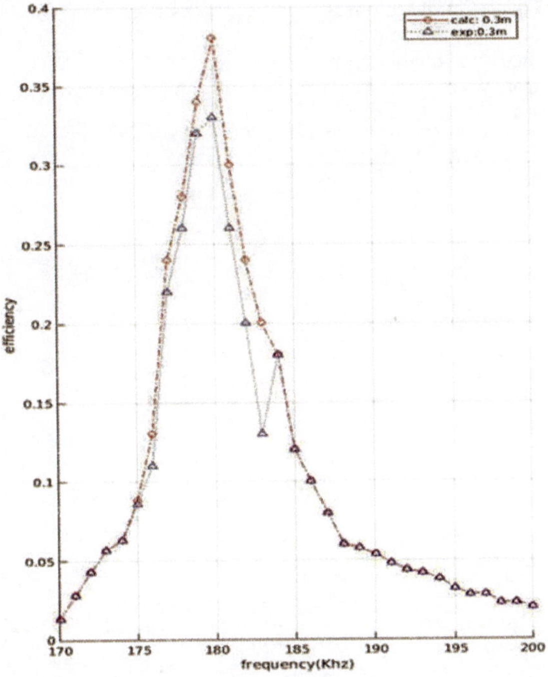

Fig. 5 Calculated versus experimental results for single-load transfer at 0.3 m apart, when $F_2 = 181$ kHz

By closing in the difference between operating and resonant frequencies, multiple receivers can be also be used together, with variable amount of power being fed from a single source. Therefore, a cost-effective method to control selective power transfer in SIMO WPT systems was proposed that showed promising results.

7 Other Related Works

Although the question for controlling power distribution has been long prevailing, some prior works have also been published. The traditional method for the same is to vary the size for receiver coils, according to the need of the instrument. A receiver which requires more power will have a coil with a bigger diameter as compared to those receivers with low power requirements.

Other works such as [13–15] use different innovative methods to differentiate designated coils according to power and timing requirements. In [13], Shotaru Kuga et al. have used heterodyne detection method for use in PSIM simulation, which is a costly method. In [14], Wen Ding et al. use in-wheel outer rotor switched reluctance motor (ORSRM) drives for selective power transfer. ORSRM drives use resonant frequencies to match the designated receiver to the desired rotor's signal. However, it is again an expensive practice because it also requires a DC/DC converter to

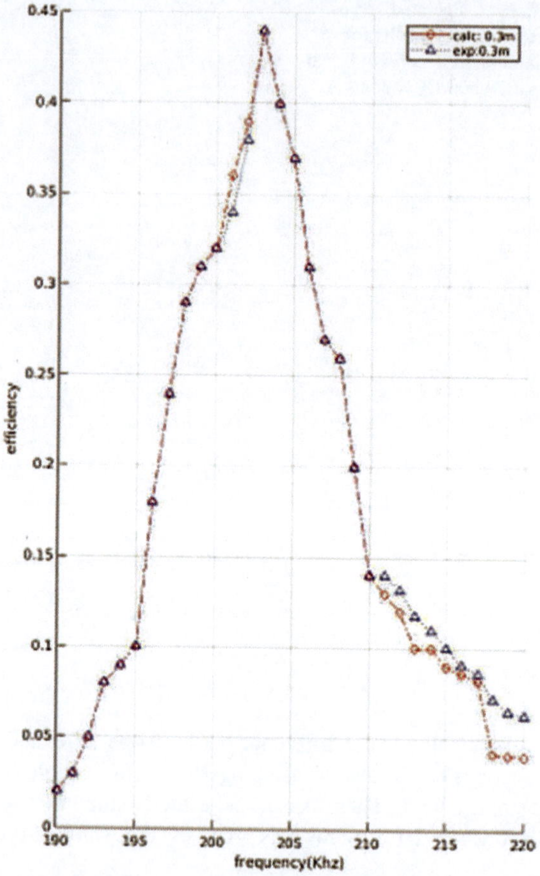

Fig. 6 Calculated versus experimental results for single-load transfer 0.3 m apart, when $F_2 = 203$ kHz

regulate the input voltage. In [15], Takashiro Nozaki et al. use magnetic resonant coupling method to couple the designated receiver with transmitter by keeping other receivers shut. The method proposed in [15] is cost-effective; however, it does not allow simultaneous activation of power coils.

In a nutshell, the technique proposed in this paper is a better solution to the selective power distribution issue in SIMO WPT systems because it is cost-effective and allows power regulation with simultaneously activating all receiver coils.

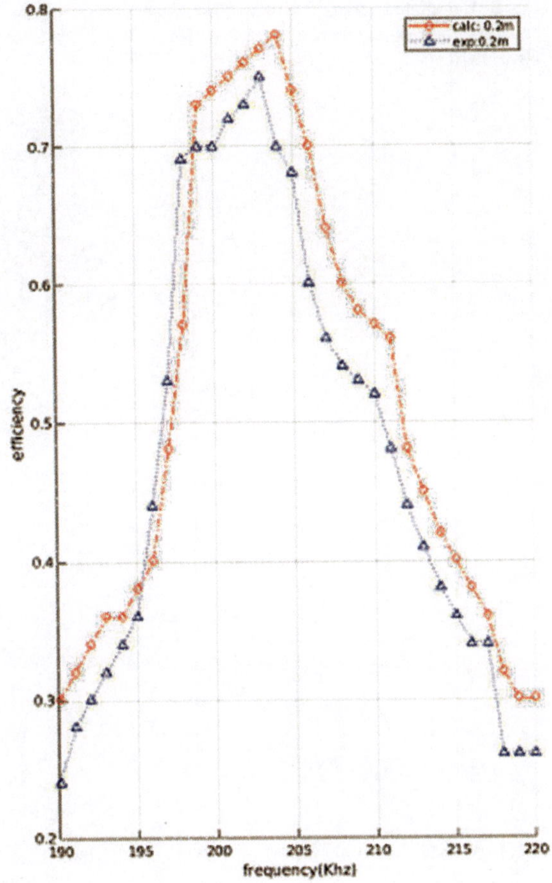

Fig. 7 Calculated versus experimental results for single-load transfer at 0.2 m apart, when $F_2 = 203$ kHz

Fig. 8 Calculated versus experimental results when, $F_2 = 181$ kHz

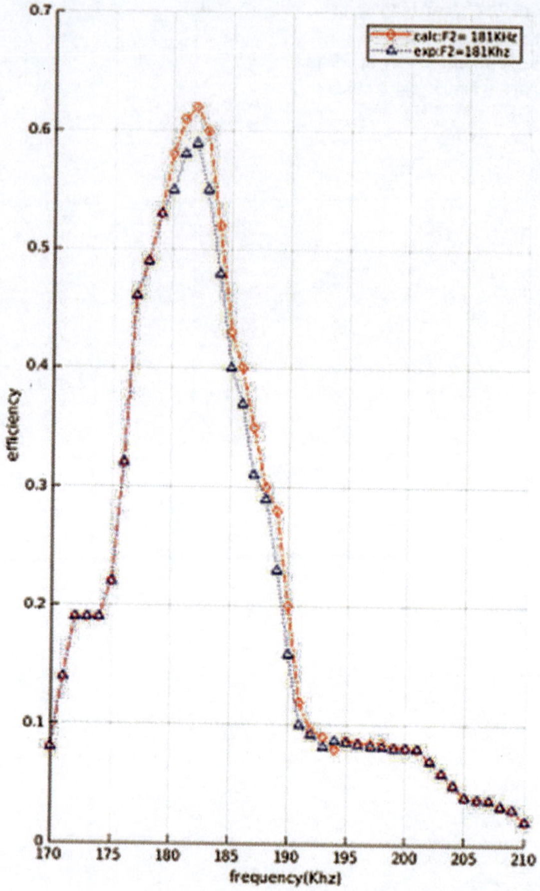

Fig. 9 Calculated versus experimental results when, $F_2 = 191$ kHz

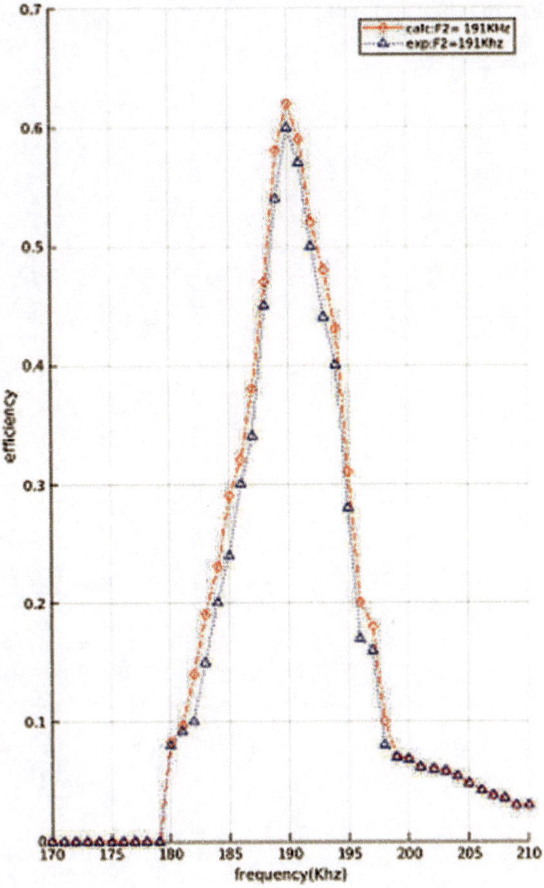

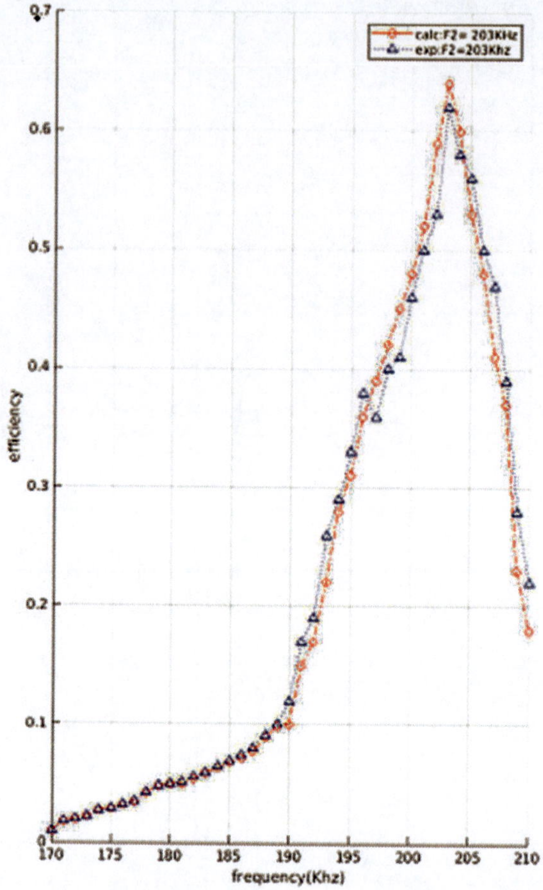

Fig. 10 Calculated versus experimental results when $F_2 = 203$ kHz

Fig. 11 Calculated versus experimental results when, $F_3 = 203$ kHz

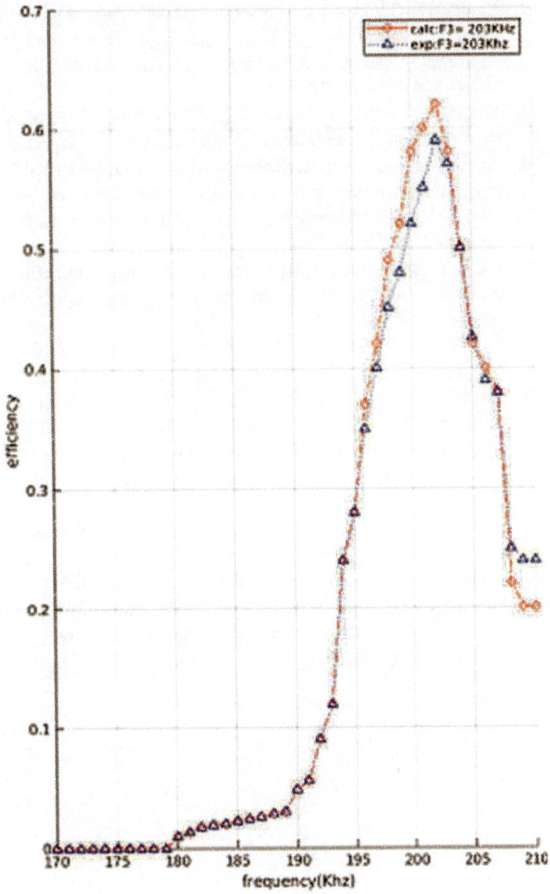

References

1. Hui SYR, Zhong W, Lee CK (2014) A critical review of recent progress in mid-range wireless power transfer. IEEE Trans Power Electron 29(9):4500–4511
2. Nguyen T, Li S, Li W, Mi CC (2014) Feasibility study on bipolar pads for efficient wireless power chargers. In: Annual IEEE applied power electronics conference and exposition (APEC), Fort Worth, TX (542), pp 1676–1682
3. Zhang Y, Zhao Z, Chen K (2014) Frequency decrease analysis of resonant wireless power transfer. IEEE Trans Power Electron 29(3):1058–1063
4. Mao H, Yang B, Li Z, Song S, Zhao X (2017) Flexible and efficient 6.87 MHz wireless charging for metal cased mobile devices using controlled resonance power architecture. In: IEEE wireless power transfer conference (WPTC), Taipei, Taiwan, pp 1–4
5. Lee B, Kiani M, Ghovanloo M (2016) A triple-loop inductive power transmission system for biomedical applications. IEEE Trans Biomed Circ Syst 10(1):138–148
6. Sah, A (2013) Design of wireless power transfer system via magnetic resonant coupling at 13.56 MHz. In: Proceedings of IOE graduate conference, Kathmandu, Nepal, vol 1

7. Chaidee E, Sangswang A, Naetiladdanon S, Mujjalinvimut E (2017) Influence of distance and frequency variations on wireless power transfer. In: 14th international conference on electrical engineering/electronics, computer, telecommunications and information technology (ECTI-CON), Phuket, pp 572–575
8. Yuan Q, Chen Q, Li J, Sawaya K (2012) Optimum load of WPT system analysed by S-parameter. In: 6th European conference on antennas and propagation (EUCAP), Prague, pp 3604–3608
9. Hafeez MA, Yousef K, Raheem MA, Khaled M (2019) Design of 6 GHz high efficiency long range wireless power transfer system using offset reflectors fed by conical horns. In: International conference on innovative trends in computer engineering (ITCE), Aswan, Egypt, pp 365–370
10. Liu B, Chen Z, Hsu H (2018) Implementation of high efficiency coupling coil in wireless power transfer system. In: IEEE wireless power transfer conference (WPTC), Montreal, QC, Canada, pp 1–4
11. Kim S, Lee B (2016) Analysis of efficiencies for multiple-input multiple-output wireless power transfer systems. J Electromagn Eng Sci 16(4):126–133
12. Zhang Y, Zhao Z (2014) Frequency splitting analysis of two-coil resonant wireless power transfer. IEEE Antennas Wirel Propag Lett 13:400–402
13. Suetsugu T, Kuga S (2015) Selective wireless transfer system using heterodyne detection. In: IEEE international telecommunications energy conference (INTELEC), Nankai, Osaka, Japan, pp 1–4
14. Li Y, Ding W, Song K, Bian H (2019) A new type of in-wheel outer rotor switched reluctance motor drive based on selective wireless power transfer technology. In: 22nd international conference on electrical machines and systems (ICEMS), Harbin, China, pp 1–5
15. Nakagawa T, Furusato K, Nozaki T, Murakami T, Imura T (2017) Selective wireless power transfer for multiple receivers by using magnetic resonance coupling. In: International symposium on antennae and propagation (ISAP), Phuket, pp 1–3

Performance Analysis of WRAN in Light of Full Duplex Capability

Khusali Obhalia, Mayur M. Vegad, and Prashant B. Swadas

Abstract IEEE 802.22 is a wireless regional area network (WRAN) standard that works on cognitive radios, allowing sharing of unused spectrum allocated to television services on a non-interfering basis. This standard is helpful in providing data services to rural areas with less population. IEEE 802.22 has three operations: transmission, reception and sensing of data. This standard operates on half duplex mode where not only transmission and reception by a participating station occurs at different times, but sensing of a channel is also not allowed while transmission is going on at any node. This reduces the spectrum usage. Recent developments have shown that full duplex communication, i.e., concurrent transmission and reception by a node, is possible even in wireless networks. The full duplex operation helps in increasing the throughput and reduces collisions. In this paper, our attempt is to explore the possibility of exploiting full duplex capabilities of wireless nodes in IEEE 802.22 WRAN by allowing a node to perform the process of channel sensing while it is engaged in transmission of some data. The simulation results show noticeably good performance enhancement of WRAN in terms of throughput by 9% which leads to efficient usage of spectrum without inviting increased interference.

Keywords Cognitive radio networks · Full duplex communication · IEEE 802.22 · Wireless regional area network (WRAN)

K. Obhalia (✉)
Computer Engineering Department, BVM Engineering College,
Vallabh Vidyanagar, Gujarat, India

Mayur M. Vegad · Prashant B. Swadas
Birla Vishvakarma Mahavidyalaya (BVM) Engineering College,
Vallabh Vidyanagar, Gujarat, India
e-mail: mayurmvegad@bvmengineering.ac.in

Prashant B. Swadas
e-mail: pbswadas@bvmengineering.ac.in

P. K. Singh et al. (eds.), *Proceedings of Second International Conference on Computing, Communications, and Cyber-Security*, Lecture Notes in Networks and Systems 203,
https://doi.org/10.1007/978-981-16-0733-2_7

1 Introduction

IEEE 802.22 is the first cognitive radio wireless regional area network (WRAN) standard. This standard uses geolocation and spectrum sensing to detect available and occupied channels of a spectrum. It uses frequencies that are not used by television (TV) stations (primary users PUs) [2]. An incumbent or secondary user (called as Customer Premises Equipment or CPE) is a WRAN device that can access the spectrum when primary users are inactive. WRAN standard achieves higher spectrum utilization by using the otherwise unutilized TV channels. Such channels are also called spectrum holes or white spaces. Particularly, in remote rural areas, TV channels are largely unoccupied. The CPEs of WRAN are relatively low-cost devices as licensing is not required. Further, due to characteristics of the portion of the spectrum allocated for TV bands, IEEE 802.22-based network can provide affordable service in large area with data as well as voice services with Quality-of-Service (QoS) support [2].

Wireless networks demand more spectral resources to support increasing number of users. The increase in number of wireless devices and data traffic has boost the misconception that there is scarcity in wireless commodity. But spectrum usage analyses have revealed that large portions of the spectrum are not efficiently used. Cognitive radio (CR) [7, 9] promises for enhancing the efficiency of spectrum usage. It uses dynamic spectrum access strategies instead of traditional static spectrum. Dynamic spectrum access strategies allow for opportunistic exploitation of unused spectrum.

IEEE 802.22 conventionally uses half duplex (HD) communication method, i.e., at a given time either sensing or transmission occurs, exclusively. WRAN allocates a dedicated period for sensing the channel, thereby ceasing all the transmissions from secondary users during that period. During our analysis, we came to know that the sensing of channel appears for fixed intervals thus missing the available empty channels otherwise. This increases the chances of missing the empty channels during "no-sensing" periods. This leads to following drawbacks for HD cognitive radio networks (CRNs):

- In HD mode, spectrum sensing and normal communication (secondary transmission and reception) take place separately. The secondary users are obliged to sacrifice substantial fraction of their available share of communication time for sensing. Also, there is a possibility of not detecting a PU transmission during SU transmission. The secondary users are obliged to sacrifice substantial fraction of their available share of communication time for sensing. Also, there is a possibility of not detecting a PU transmission during SU transmission. This situation may lead to loss of data and harmful interference to primary users.
- Further, HD cognitive radio devices use two separate channels for transmission and reception of data. This increases communication latency as both of these channels are required to be sensed for white spaces.

Recent advancements in technology have made full duplex communication possible in wireless networks too. This allows a wireless node to have concurrent transmission and reception of data in a single time frequency channel, thus improving the spectral efficiency and over all throughput of the system [1]. The cognitive radio technology used in IEEE 802.22 WRAN allows its secondary users to shift their communication to new channel within stipulated time, thereby avoiding interference to the primary users. The use of full duplex in cognitive radio-based WRAN would result in following advantages:

- Increase in throughput
- Reduced probability of false alarm (i.e., wrong results of sensing)
- Regular scheduling of quiet period can be avoided because of the possibility of continuous sensing, along with the transmission in full duplex mode
- Minimize data loss
- Increase overall network capacity
- Improve spectrum utilization.

Considering the effectiveness of cognitive radio in spectrum usage and dynamic spectrum access strategy, we have studied the possible impact of using Full Duplex in IEEE 802.22 standard. In this work, we have implemented the transmission of data and sensing of a channel simultaneously. This leads to improved spectrum usage. The simulation results show that full duplex cognitive radio networks would have a huge impact on existing and future applications of CRNs.

Remaining paper is organized as follows. Section 2 discusses related works in the area of exploiting full duplex for wireless communications. Section 3 describes the implementation of full duplex in WRAN. Simulations and the discussion on the results obtained are covered in Sect. 4. Finally, Sect. 5 concludes the paper while dictating some possible future works.

2 Related Works

Many works have explored the possibility of using full duplex communication mode in different wireless standards. The authors in [8] suggest an in-band full duplex method that overcomes the problem of self-interference. Febrianto et al. [5] demonstrate improvements in terms of the average throughput of both primary users and secondary users for some specific transmission schemes. They have proposed FD-MAC protocol and analyzed it for full duplex cooperative spectrum sensing. In [4], a wireless full duplex spectrum sensing (FD-SS) scheme has been proposed for secondary users in multichannel non-time-slotted cognitive radio networks. Tan et al. [10] have proposed a full duplex cognitive MAC (FDC-MAC) protocol that does not require any synchronization among secondary users. They have shown that their approach achieves higher throughput than half duplex MAC protocol and mitigates self-interference.

In the approach proposed by Afifi et al. in [3], the secondary users are equipped with Self-Interference Suppression (SIS) capable radios. To detect the primary users, these radios can operate in a simultaneous transmission and sensing (TS) mode or simultaneous transmission and reception (TR) mode. This approach achieves about 50% reduction in the collision probability and 100% increase in the throughput as compared to the half duplex case. Exploiting full duplex mode in CRNs, in [11], the authors use frame fragmentation during data transmission phase for timely detection of active PUs. Each data packet is divided into multiple fragments and active SU makes sensing detection at the end of each data fragment. Self-interference management and sensing overhead control is achieved by an algorithm to configure MAC protocol. Liao et al. [6] propose Listen and talk (LAT) protocol to simultaneously sense and access the vacant spectrum. Spectrum utilization efficiency and secondary throughput under the LAT protocol has been provided in closed-form, based on which a unique tradeoff between the secondary transmit power and the secondary throughput has been reported.

3 Full Duplex in WRAN

Cognitive sensing is the key feature of a WRAN Base Station (BS). In this, CPEs sense the spectrum and sends a period report to BS informing about what they sense. From this report, BS decides whether any change is necessary to the channel being used by the CPEs. In WRAN, PHY layer helps in flexible jump to different channel without errors or losing CPEs connections and MAC layer helps in performing sensing. BS manages all the activities within its cell and the CPEs connected to it. MAC layer has two structures: Frame and Superframe [11]. Frames are responsible for sensing of spectrum by scheduling quiet period and transmission of data between CPE and BS. CPE transmits data to BS using Upstream Burst. BS sends data to CPE using Downstream Burst.

Wireless regional area network works on half duplex communication. In WRAN, superframe allocates quiet period within frames that performs sensing. When quiet period initiates, all the transmissions in progress by CPEs are ceased, i.e., during quiet period no transmission will take place. This time is only allocated for sensing the channels. Figure 1 shows intra-frame quiet period scheduled in single superframe. In this superframe, frames 0, 5 and 11 contains quiet period coloured grey. The rest of the frames are colored white that indicates transmission process. This means that in frames 0, 5 and 11 from Fig. 1 transmission does not occur. If there is an incumbent present in the frame 0, 5 and 11, then the transmission in the upcoming frames will not occur. If there is no incumbent detected (in sensing frame 5), then the transmission occurs in the next frame (frame 6).

Now consider Fig. 2 where an incumbent occurs in frames 2, 6, 8 and 12 (colored black). But quiet period is only scheduled in frames 0, 4 and 7. That means there will be transmission occurring in rest of the frames. This will cause interference to incumbent.

There might be a possibility that sensing gives incorrect result. It may say that incumbent is present even if there is no active incumbent and vice versa, also called false alarm. This results in interference or no transmission of data.

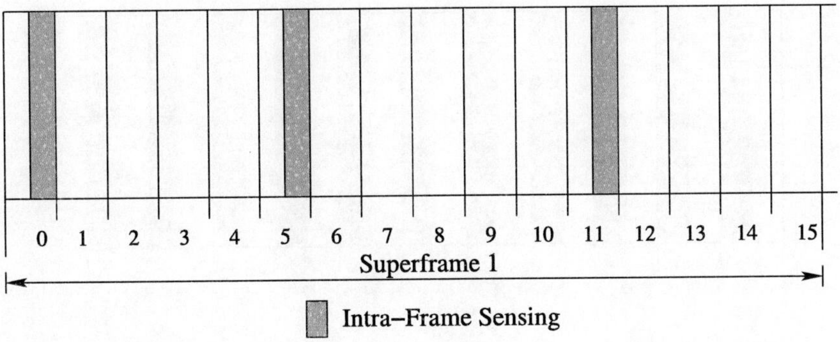

Fig. 1 Superframe sensing

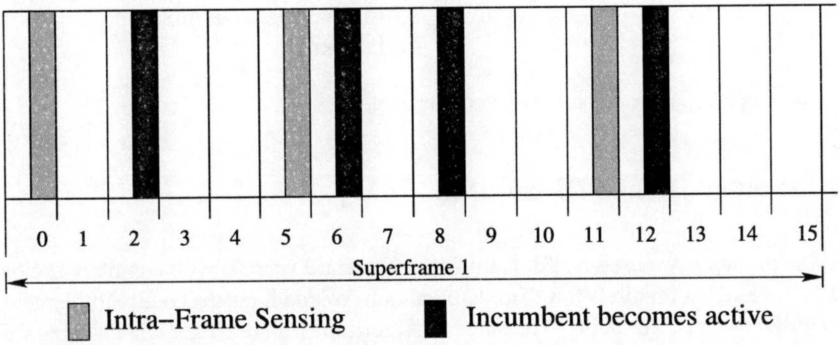

Fig. 2 Superframe sensing with active primary user

3.1 Advantages of Allowing Transmission and Sensing Simultaneously

If this is considered, then whenever an upstream subframe starts, there will be transmission and sensing done in parallel. Figure 3 depicts simultaneous transmission and sensing in upstream. If there is no incumbent detected during this period, the transmission continues anyways. If an incumbent is encountered during this phase, then the transmission needs to stop and incumbent is allowed to use the channel because primary users have the first access to spectrum. If we implement full duplex in IEEE 802.22 standard, then

- No requirement to schedule quiet period, i.e., no transmission will be ceased or interrupted for sensing
- Protection to incumbent also less interference
- Reduced probability of false alarm
- Increased throughput.

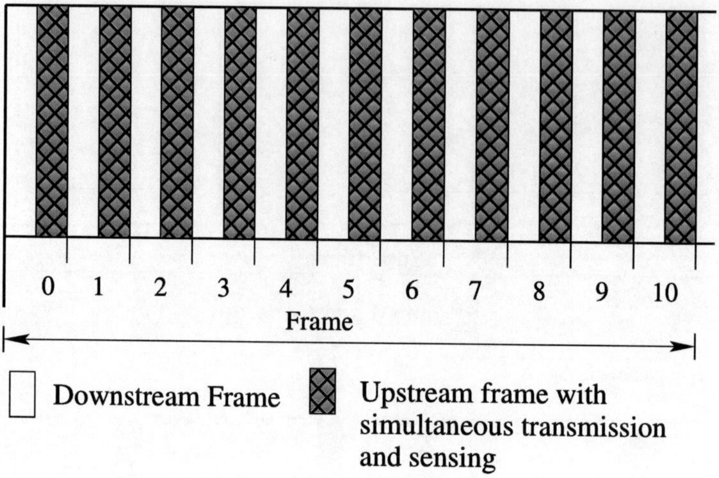

Fig. 3 Simultaneous sensing and transmission in a single frame

4 Simulations and Discussions

For simulations, we used NetSim simulator (Standard ver 9.1) with cognitive radios [12]. NetSim has inbuilt WRAN implementation. We modified the code to implement the full duplex functionality. The steps in Algorithm 1 dictate the main changes we incorporated for implementing the full duplex functionality:

Algorithm 1: Steps to Incorporate Full Duplex Mechanism into WRAN Implementation of NetSim

---

Reset Intra-Frame-Quiet-Period to *Zero* in the headers of Frame-Control and Superframe-Control
Start Spectrum-Sensing-Function
Calculate the Keep-out-Distance
Let $numIncumb \leftarrow$ Number of Incumbents Present
if $numIncumb > 0$ **then**
    $incumbPresent \leftarrow True$
    $signalType \leftarrow 0$ {No transmissions allowed}
else
    Create Form-Upstream-Burst-Event {Allow Data Transmission}
    Create Transmit-Upstream-Burst-Event
end if

Table 1 lists the different parameter settings. *intraFrameQuietPeriodDuration* is the amount of time in terms of frames when no transmission of data from secondary user takes place. $keepOutDistance$ is the maximum distance at which a CPE can detect an incumbent. In our study to implement full duplex in WRAN, we set the

Table 1 Simulation parameters

Parameter	Value
Intra-frame-quiet-period-cycle-length	0
Intra-frame-quiet-period-duration	0 s
Keep-out-distance	100 m
Operational-interval	10 s
Operational-time	10 s
Operating-frequency-start	54 MHz
Operating-frequency-end	132 MHz
Simulation-time	100 s

$intraFrameQuietPeriodDuration$ to 0. This means there will not be any quiet period involved. We let a CPE count the number of incumbents within the keep out distance of the BS using Spectrum Sensing Function. If any incumbent is found active the flag for no-transmission is set. Otherwise, Upstream Bursts are transmitted to BS from the CPE that transmits data. In this way, sensing and transmission are both allowed to perform concurrently.

Figure 4 shows a scenario of a WRAN network that includes two BSs, seven CPE_{CR}'s (Cognitive Radio Customer Premises Equipment) and six incumbents. CPE C, D, J, K and L are affiliated to the BS A and CPE F and G are affiliated to the BS E. CPE_{CR}s communicate with each other via BS. The BS is connected to the CPEs via 802.22 wireless links.

There are five applications set in the shown scenario as follows: App1 (Application 1): CPE C sends data to CPE D; App2: CPE F sends data to CPE G; App3: CPE G sends data to CPE D; App4: CPE K sends data to CPE G; App5: CPE L sends data to CPE F.

For our study, we created five different scenarios. In all the scenarios, the number of BSs and CPEs are kept constant. However, the number of incumbent is increased by one in each increasing-numbered scenario. We set different operating frequency range for incumbents and BS in which they will operate. Table 2 gives further details of these five scenarios. Note that, Fig. 4 shows the scenario no. (5) from Table 2. Followings are the common parameters for all five scenarios. The keepOutDistance is fixed to 100 m. Any incumbent is assumed to be active for 10 s. This period is referred to operationalTime. The duration between two successive incumbent operations, called $operationalInterval$ is kept 10 s. All the scenarios are simulated for 100 s.

In the results extracted, we have noticed that there is a little increase in the number of errored packets with full duplex case, though the amount of increase is very negligible.

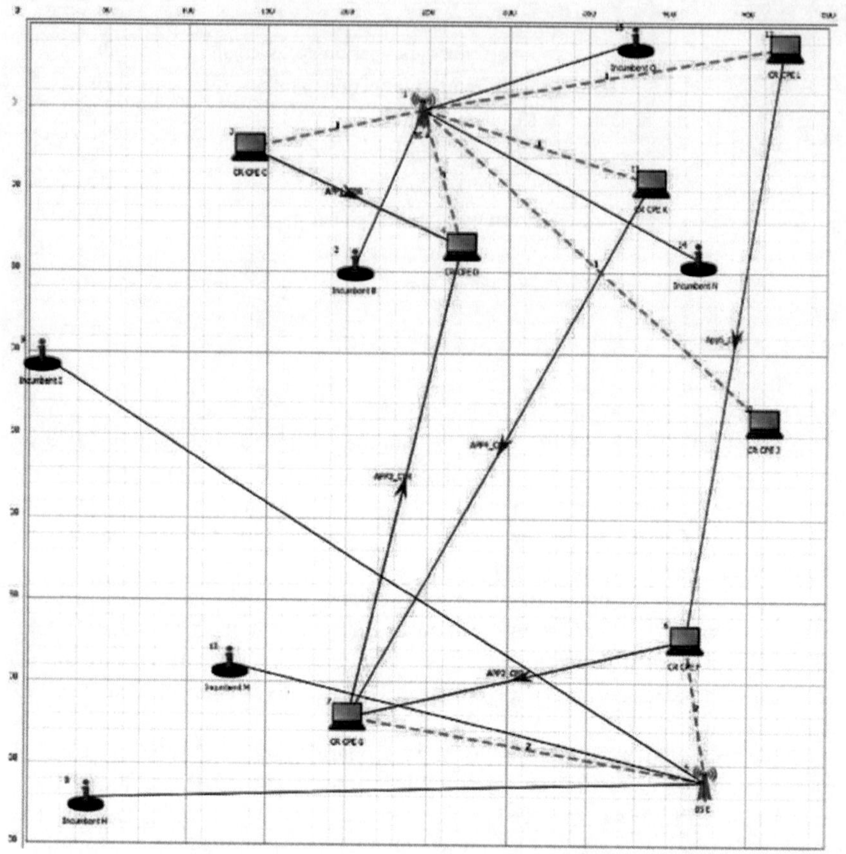

Fig. 4 Screenshot of one of the network scenarios simulated

Figure 5 shows comparison of throughput for full duplex and half duplex cases for different network scenarios listed in Table 2. As it can be noticed, with full duplex implementation, the throughput obtained is higher for all five scenarios, with about 9% improvement over the half duplex case. The improvement is due to the allowance of performing sensing and transmission simultaneously. It is worth noticed here that this improvement has been obtained just by allowing sensing while transmitting. The performance can be significantly improved, by much higher percentage gain if reception is also allowed along with the transmission. However, this would require significant changes in the standard. We have noticed that the interference in both the cases are almost same. That means, with the same amount of interference, more packets were received with full duplex implementation. Further, this argument remains valid for all scenarios, i.e., for all the different operating frequencies of TV band.

Table 2 Operating frequencies for BSs and incumbents

Scenario No.	BS with operating frequency	Incumbent	Incumbent operating frequency
1	A:54–84	B	54–60
	E:54–84	H	66–72
2	A:54–96	B	54–60
	E:54–96	H	66–72
		I	78–84
3	A:54–108	B	54–60
	E:54–108	H	66–72
		I	78–84
		M	90–96
4	A:54–120	B	54–60
		N	102–108
	E:54–120	H	66–72
		I	78–84
		M	90–96
5	A:54–132	B	54–60
		N	102–108
		O	114–120
	E:54–132	H	66–72
		I	78–84
		M	90–96

5 Conclusion

In this work, we have investigated the impact of exploiting full duplex capabilities of modern wireless radios in IEEE 802.22 WRAN. We studied how upstream subframe can be used for both transmission of data and sensing of channel. This phenomenon has resulted in more effective usage of the spectrum while not increasing the amount of interference to the incumbents. The simulation results show that with full duplex implementation, for a given amount of interference, there is about 9% improvement in terms of throughput as compared to the half duplex case.

There is still a scope of further exploiting full duplex phenomenon leading to significant improvement in WRAN throughput performance. In our work, the full duplex implementation allowed a CPE, only sensing while it is transmitting. It seems plausible to allow reception as well while transmission is going on at a CPE. While this change seems so much promising in offering great throughput benefits, one needs to ensure that it does not lead to unprecedented increase in the interference. Further, this change would also require significant modifications in the standard.

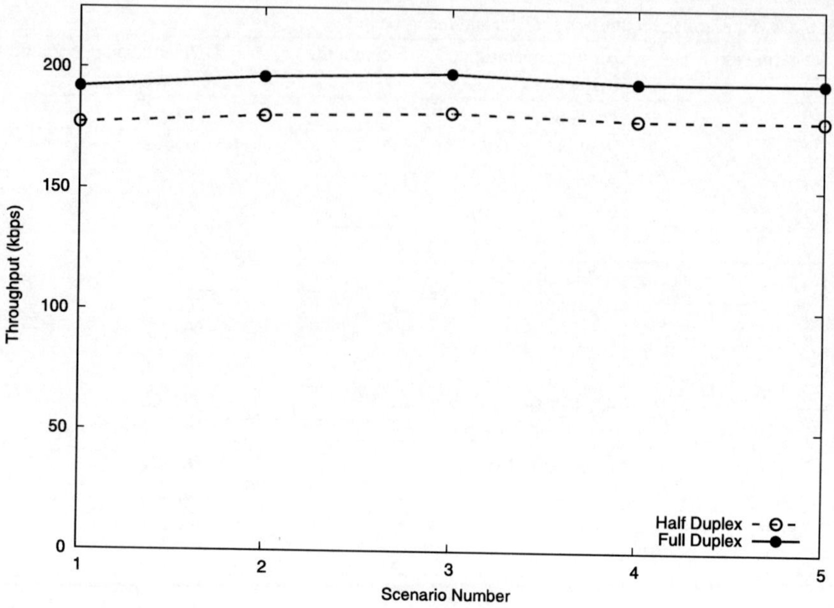

Fig. 5 Simulation results: performance improvement in terms throughput

References

1. IEEE ComSoc full duplex wireless communications emerging technologies initiatives. https://fdc.committees.comsoc.org
2. IEEE standard for information technology—local and metropolitan area networks—specific requirements—Part 22: Cognitive wireless ran medium access control (MAC) and physical layer (PHY) specifications: Policies and procedures for operation in the TV bands (2011). IEEE Std 802.22-2011, pp 1–680
3. Afifi W, Krunz M (2014) Adaptive transmission-reception-sensing strategy for cognitive radios with full-duplex capabilities. In: 2014 IEEE international symposium on dynamic spectrum access networks (DYSPAN), pp 149–160
4. Cheng W, Zhang X, Zhang H (2015) Full-duplex spectrum-sensing and MAC-protocol for multichannel nontime-slotted cognitive radio networks. IEEE J Sel Areas Commun 33(5):820–831
5. Febrianto T, Hou J, Shikh-Bahaei M (2017) Cooperative full-duplex physical and MAC layer design in asynchronous cognitive networks. Wirel Commun Mob Comput 2017:1–14. https://doi.org/10.1155/2017/8491920
6. Liao Y, Wang T, Song L, Han Z (2017) Listen-and-talk: protocol design and analysis for full-duplex cognitive radio networks. IEEE Trans Vehic Technol 66(1):656–667
7. Mitola J, Maguire GQ (1999) Cognitive radio: making software radios more personal. IEEE Pers Commun 6(4):13–18
8. Sabharwal A, Schniter P, Guo D, Bliss DW, Rangarajan S, Wichman R (2014) In-band full-duplex wireless: challenges and opportunities. IEEE J Sel Areas Commun 32(9):1637–1652. https://doi.org/10.1109/JSAC.2014.2330193
9. Setoodeh P, Haykin S (2017) Fundamentals of cognitive radio. Wiley, New York

10. Tan LT, Le LB (2015) Design and optimal configuration of full-duplex MAC protocol for cognitive radio networks considering self-interference. IEEE Access 3:2715–2729
11. Tan LT, Le LB (2015) Distributed MAC protocol design for full-duplex cognitive radio networks. In: 2015 IEEE global communications conference (GLOBECOM), pp 1–6
12. Tetcos, Bangalore, India: NetSim Network Simulator, Standard Version 9.1

Design of Planer Wide Band Micro-Strip Patch Antenna for 5G Wireless Communication Applications: Review

Praveen Tiwari and Praveen Kumar Malik

Abstract In this paper, wide band antenna with reduced size and irregular coplanar micro-strip strip took care of fast correspondence applications. Two planned antennas with the radiator patches have been picked as elliptical and semi- elliptical structure. The designs of antenna radiation designs are almost omnidirectional radiation over the ultra-wide band scope with appropriate gain. For the 4th and 5th generation application an antenna wideband structure is examined. Antenna is created on ease effectively accessible FR-4 substrate. The burrowing impact of ENZ thin network with coordinating two microstrip lines along with various impedances attributes have been examined. An exceptionally minimized gaping loop band pass planer channel through unequal frequency result and covers the range of 2.5–2.6 GHz for 4th generation and 3.6–3.7 GHz range for 5th generation purposes have been discussed. To accomplish increased sharp cut-off frequencies, one limitless plus three-limited transmission zeros are effectively produced on the upper and lower ends of the 4th, 5th generation passbands. The 5th generation communication involves a position of lightweight, increased gain. In this article, for 28 GHz, the basic structure of patch antenna have been used to guarantee unwavering quality, versatility, high proficiency, a good shape, low profile, high gain and high efficiency. For the 5G application in Wi Fi, Wi Max array of broadband printed dipole antenna, 5G compact MIMO antenna, Impedance Matching Using ENZ Metamaterials, Planer UWB Antenna for High Speed Communications and UWB slot antenna may be used.

Keywords Ultra wide band (UWB) antennas · Slot antennas · 5G communications · ACS-fed · Monopole antenna · Bandwidth · Radiation

1 Introduction

A polarization reconfigurable aperture coupled magneto-electric (ME) dipole antenna is introduced. A completely useful model is created and tried, exhibiting

P. Tiwari (✉) · P. K. Malik
Lovely Professional University, Jalandhar, India

P. K. Singh et al. (eds.), *Proceedings of Second International Conference on Computing, Communications, and Cyber-Security*, Lecture Notes in Networks and Systems 203,
https://doi.org/10.1007/978-981-16-0733-2_8

the antenna having an overlap bandwidth of 16% ranging from 5.07 to 5.95 GHz for axial ratio ≤3 dB and return loss (S11) ≤ −10 dB. A deliberate radiation efficiency of around 85% over the overlap band and gain of roughly 8.2 dBi are gotten for all polarization states. In addition, because of favorable circumstances of the ME dipole, the proposed structure can realize stable radiation designs together with both back radiation levels and cross polarization lower than 13 dB [1]. To understand a wide recurrence scope of activity, the projected antenna is supplied by a balun, it comprises of a rectangular slot and folded microstrip line. To achieve smallness, the engraved dipole is taken at 45 degree. The solo component antenna results a gain of 4.5–5.8 dBi, bandwidth of 36.2% (26.5–38.2 GHz) and return loss (S11) <−10-dB. In this design of two printed-dipole antenna, a stub was inserted and it yielded for a 4.8-mm center to center distancing (0.42–0.61λ at 26–38 GHz), a low mutual coupling of <−20 dB. Due to nearness with the stubs, a higher gain, a lower side-lobe level, wider scanning angle in the low-frequency region resulted for the array [2]. Dependent on vector synthesized system, a dual polarized antenna base is exhibited in the article. It consists of a reflector, a feed structure and four radiators. All the existing radiators take part in the analysis, and practically there is not any cross-polarization current at the radiators. Consequently, proposed antenna component might cover 3.3–3.6 GHz with high-level port separation and at a low level cross polarization. Meanwhile model of the antenna component have additionally been considered. At that point, the antenna component is applied towards an array. By allocating diverse weighting element besides stage for every component's port, the array of antenna be able to produce distinctive pattern to meet up the multi-mode necessities for communication system. By using such type of techniques, low cross polarization, low side lobes, high front to back ratio and high isolation can be achieved. Therefore, by giving such merits, the Multi Input Multi Output antenna apparatus will locate its versatile application for 5G support station application [3]. The 2 × 2 planer MIMO antenna system of framework is explicitly intended to point lower fifth generation working for operating groups running between 2 and 12 GHz. The proposed band likewise contains the IEEE 802.11 a/b/g/n/ac standards. The geometry of antenna array has been simulated by utilizing CST software. The model is incredibly scaled down by complete structure size of 13 × 25 × 0.254 mm^3. Return loss is under −10 dB across the working band and lowest values of −32.5 dB for 9.2 GHz and −35 dB at 5.2 GHz have seen, while the return loss is −25.2 dB at center frequency. The measured maximum gain is as 4.8 dB. Likely aftereffects of Envelope Correlation Coefficient and addition decent variety with plan is accomplished. The fractional bandwidth estimated as 143.2% thus it fulfills its UWB requirement [4].

2 Need for Literature Review

Wireless Communication having tremendous interest of antenna framework and their small size design, antenna configuration turns out to be increasingly challenging and required. As of now microstrip patch antennas have been broadly utilized in satellite, military, aviation communication, radars, biomedical and for mobile communication on account of its natural parameters. The literature survey has been done for the study of various microstrip antenna parameter and its characteristics for future wireless communication application. The recent demand of ultra wide band for future application attracted researchers for miniaturization of antenna. These antennas are useful for various types of wireless applications. It is very essential for an antenna to have minimum size and weight along with its performance capabilities to different functions and multiband operations. The microstrip antenna have more advantages over conventional microstrip antenna. The design of a microstrip antenna is such that is consists of a ground plane on one side and a radiating patch on the other side. The patch is made up of conducting material having shapes of rectangle, circle, square, triangle and hexagonal etc. The property of substrate is such that its dielectric constant should be minimum and radiation is maximum.

3 Survey Methodology

Table 1 shows the work carried out in various research articles by using different kind of antenna design.

4 Review of the Existing Literature

Ibrahim [5] presented an antenna as Compact Planer UWB Antenna for High Speed Communication. In this paper a 2D design with intended UWB monopole antenna with asymmetric coplanar strip took care of as outlined in Fig. 1a. A past FR4 substrate is utilized for the antenna plan along with 1.6 mm of thickness and permittivity of 4.4. Additionally, it is apparent from the structure as in Fig.1a that the radiator of the planned antenna is quasi- elliptical associated through 50 Ω asymmetric coplanar strip taking care of line. The components of the asymmetric coplanar strip taking care of line approaches, 8 mm in stature, hole rises to 0.3, 3 mm of width to guarantee coordinating with 50 Ω SMA connector. The ground plane has twisted side to accomplish an ideal transfer speed reasonable for ultra-wide band correspondences. Therefore, the setup has the favourable minimal size than main design. The size is diminished with the greater part as a result of the ACS—took care of arrangement. The reception apparatus size equivalents $28 \times 11.5 \times 1.6$ mm^3. The created photograph of the projected ultra-wide band antenna with ACS is appeared in Fig. 1b. The projected antenna have been estimated via utilizing R&S vector arrange analyser ZVA40. Figure 2 displays the simulated and estimated consequences of reflection coefficient [5]. It can be seen from the reproduced outcomes that antenna is worked in recurrence band between 3.2 and 11.7 GHz with return loss (S_{11}) lower than 10 dB. Additionally, unmistakably the deliberate outcomes concur well with the reproduced ones. An anechoic chamber utilizing the NSI 800F-30 framework was

Table 1 Comparison of various antenna designs

S. No.	Methodology	Remark
1	Magneto-electric (ME) dipole antenna	Overlap bandwidth is 16% for 5.07 and 5.95 GHz, Return Loss is ≤ -10 decibels, Axial Ratio <3 decibels, Radiation Efficiency is 80% and Gain is 8.2 dBi
2	Printed-dipole with array of 8 elements	Return Loss <–10 decibels and BW of 36.2% for the range 26.5–38.2 GHz, Gain is 4.5–5.8 dB wrt isotropic antenna, mutual coupling of <–20 dB, lower side lobe
3	Antenna with dual-polarization using vector synthetic technique	For 3.3–3.6 GHz with increased port isolation and lower cross polarization. Its size is very minimum
4	Compact 2 × 2 planar MIMO antenna	For 2–12 GHz, structure size of $13 \times 25 \times 0.254$ mm^3 Return loss is less than −10 dB, mutual coupling is less that −20 dB and fractional bandwidth is 143.2%
5	Micro-strip patch antenna which is filled by air with recessed ground	For frequency 60 GHz, The substrate is alumina ceramic with height 0.127 mm having relative permittivity = 9.8. An increase in impedance bandwidth (9.48%) for 58.2 to 65 GHz and enhancement of radiation efficiency by 24.97%, gain increment is 2.64 decibels
6	The defected ground plane is used with rectangular inset feed for micro-strip antenna	Return loss less than −12 dB and VSWR is less than 2, return loss is −10 dB for 11.2 GHz
7	Asymmetric coplanar strip (ACS) compact planer antenna	The Bandwidth is analysed for 3.2 GHz and greater than 11.7 GHz. The size of the antenna is 28×11.5 mm^2 and radiation pattern is omnidirectional
8	Wideband antenna design	For band number ranging from 33 to 43 for 5G New Radio n78 band the techniques of Time division duplex long-term evolution bands is used. Operating in the frequency ranging from 1.85 to 3.8 GHz, Its results are bandwidth percentage (69.02%), scattering loss (−42 dB) and the input impedance of the antenna is 50.22 Ω
9	Compact planer openloop bandpass filter (BPF)	Covering the 2.5–2.6 GHz and 3.6–3.7 GHz, microstrip BPF employs four open-loop ring resonators with 50 Ω, implemented on a Rogers RO3010 substrate with a relative dielectric constant of 10.2 and a very compact size of $11 \times 9 \times 1.27$ mm^3
10	Single element patch antenna which is having shape of umbrella	For 28 GHz, The substrate used is Rt-5880 with size of 25 × 19 × 0.5 mm, gain of 7.88 dBi, Impedance bandwidth ranging from 27.673 to 28.118 GHz. With return loss ≤ −10 dB

(continued)

Table 1 (continued)

S. No.	Methodology	Remark
11	Slot antennas	For the range of frequency from 24 to 53 GHz, dielectric substrate—the Roger's 5880 is used, |S11| <−10 dB, gain ranges from 3.7 to 6.7 dB

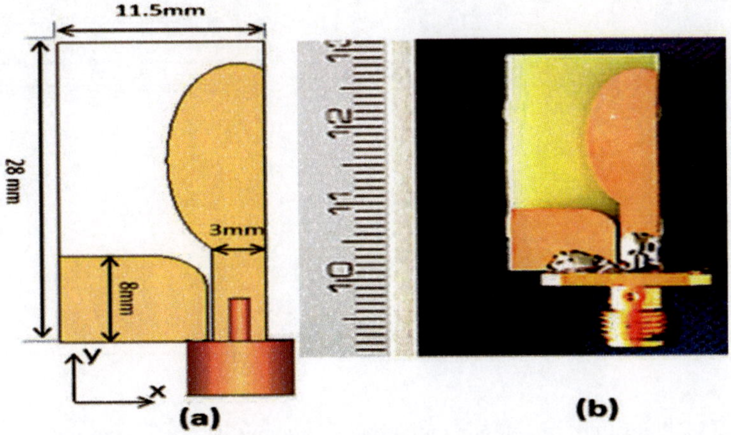

Fig. 1 Antenna fed geometry with ACS **a** 2 D layout, **b** photo of the fabricated antenna [5]

utilized to quantify the projected antenna radiation designs. The projected antenna with ACS-took care of radiation designs in the (x-y plane) ($\theta = 90°$), (y-z plane) ($\varphi = 90°$) and (x-z plane) ($\varphi = 0°$), are estimated. The normal proficiency of the projected antenna is about 90% inside the ultra-wide band recurrence band.

Singh et al. [6] presented a wideband antenna for 4th and 5th generation applications in this article. The designed antenna in this article have been explored with time division duplex long term evolution bands development groups for wideband tasks from band number 33–43 for 5th generation communication. The antenna configuration is reproduced utilizing CST Microwave Studio and is manufactured on effectively accessible FR-4 substrate. The outcomes were reproduced and estimated for 4th and 5th generation groups for gain, minimum reflected power, radiation characteristics and bandwidth [6]. The better tunable outcomes have been shown for different size dipole element upon a similar dipole, as given in Fig. 2.

The design of antenna configuration comprises of a radiated folded length to scale down the antenna elements. The radiating components are equipped for controlling parameters with the utilization of controlling boundaries. The antenna designed was fit for working in the sphere of the scope of 1.85–3.8 GHz, which covers eleven TDD LTE groups 5G NR n78 band and from Band No. 33–43. A wide transmission capacity of 1.95 GHz is acquired along with a minimum scattering loss of −42 dB and data transfer capacity level of 69.02 %. The antenna is having input impedance

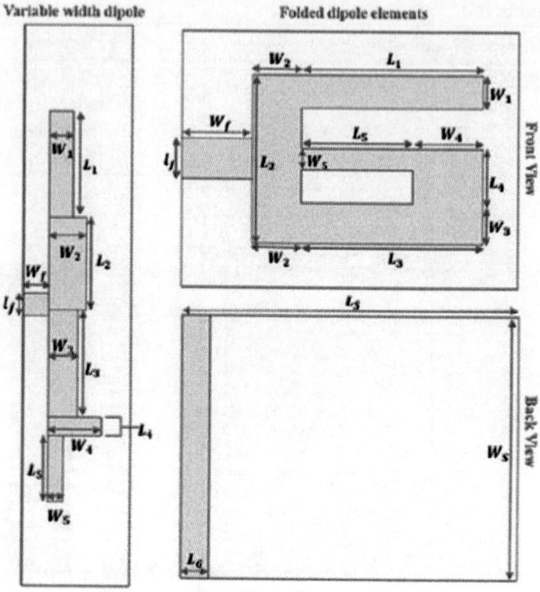

Fig. 2 Different size dipole element [6]

of 50.22 Ω, the pattern of antenna radiation is doughnut pattern; therefore, it is most appropriate with versatile wireless mobile applications. Additionally, the structure was compact and planer to fit any mobile device.

Yasir et al. [7] presented the article which proposes an exceptionally reduced planer open loop bandpass channel by unequal frequency response and covers the 3.6–3.7 GHz and 2.5–2.6 GHz range for 4th and 5th generation applications, separately. The microstrip band pass filter utilizes design of a four open-loop ring resonators with 50 Ω tapped lines for the ports of input and output. The cross-coupling coefficients parameter among the resonators are advanced to resonate at the required frequency by appropriate data transfer capacity [7, 10]. The announced BPF is structured and upgraded by utilizing CST tool, and is fabricated on a Rogers RO3010 substrate. A structure for an exceptionally minimized 4-pole microstrip BPF was introduced as well as executed during the article with asymmetric frequency response. Three limited transmission zeroes were effectively produced on the upper boundary of the passbands to expand selectivity of the proposed BPF.

Ahmad et al. [8] studied that 5th generation communication system requires position of safety, lightweight, high gain, and basic structure of antenna towards guarantee dependability, versatility and high efficiency, thus an innovative shape having property of low profile, high gain and high efficiency patch antenna for 5th generation wireless communication at the frequency 28 GHz. The proposed antenna has been fabricated on a compact Rogers Substrate Rt-5880. The proposed plan gives an increase of 7.88 dBi at 28 GHz and efficiency is 92%. Impedance BW of the projected antenna varies from 27.673 to 28.118 GHz with return loss (S11) ≤ -10 dB [8]. Thus proposed antenna is fabricated and utilized the CST microwave studio

2018. Therefore, this antenna may be used for 5G MIMIO Applications because of the basic plan arrangement, ease in creation with minimal effort the suggested structure. The Geometry of proposed antenna is given in Fig. 3a Front view 3b Back view

The suggested antenna is having following optimized dimensions as given in Table 2.

The antenna design which is having the properties of low profile with high gain and high efficiency operating in 28 GHz with an Umbrella shape is analysed without

Fig. 3 **a** Front view [8], **b** Back view [8]

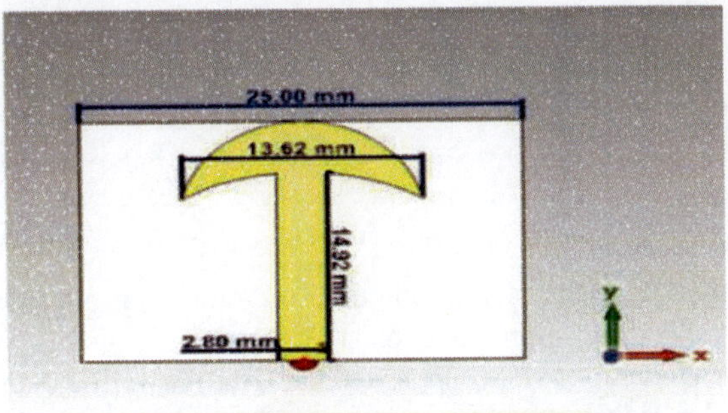

(a)

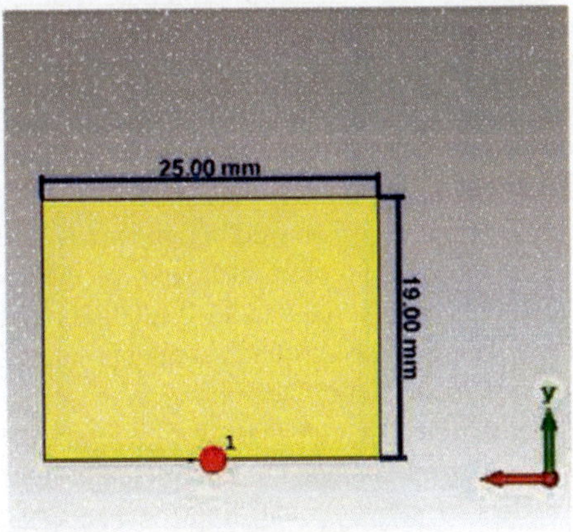

(b)

Table 2 Dimensions of the suggested antenna [8]

Parameters	sx	sy	st	ct	fw	fl
Values (mm)	25	19	0.5	0.035	2.8	15
Parameters	pur	pvr	purc	pvrc	qu	qv
Values (mm)	7	8	11	10	21	18

Table 3 Geometric values of the proposed antenna [9]

Parameter values (mm)	Antenna prototypes				
	PA	PC	PD1	PD2	PD3
r_{dx}	1.053	1.053	1.053	1.15	1.15
r_{dy}	0.963	0.963	0.963	1.07	1.07
r_{ux}	–	–	0.316	0.25	0.25
r_{uy}	–	–	0.289	0.3	0.3
g	–	–	0.01	0.03	0.03
r_{sx}	–	0.632	0.632	0.7	0.7
r_{sy}	–	0.578	0.578	0.55	0.55
X_s	–	1.030	1.030	1	1.2
Y_s	–	1.256	1.256	1.47	1.47
Lin	–	–	–	0.3	0.3

using any complex techniques. Therefore, it is appropriate for milli-meter wave 5th generation communication.

Ioannis et al. [9] presented, an Ultra-Wide-Band antenna of slot type, which was planned proposed for the future 5th generation wireless systems. The antennas are having property of planar and compact, thus has extremely little size and correct arrangement of their structure prompted models accompanied by powerful activity for a ultra-wide frequency range of 24–53 GHz. Meanwhile proposed antenna is having geometric values as given in Table 3.

The antennas fed with CPW feeding of slot type with small size. The antenna shows ultra-wide band results having impedance BW, to be specific $|S11| < -10$dB, for the range of 20–55 GHz and thus covers most of the bands in 5th generation groups. The antenna displayed high gain which fluctuate from 3.7 to 6.7 dB [9, 10] .

Jaiswal et al. [11] presented a paper on the impacts of limited air filled depressed land all things considered, radiation qualities of the microstrip antenna at 60 GHz are explored. The recessed ground antenna is examined utilizing 2-Dimensional capacitance prototype to comprehend impact measurements of the recessed ground plane at compelling dielectric constant. Definite parametric investigation, measurements of the recessed ground plane has completed to improve presenting antenna at 60 GHz. The models of conventional, improved recessed ground antenna at 60 GHz were created with alumina artistic substrate of height 0.127 mm having relative permittivity of 9.8. Estimated findings show that with recessed ground plane, an upgrade in −10 dB, impedance BW capacity by 9.48% for the frequency range of 58.2–65 GHz and in radiation efficiency by 24.97% over ordinary antenna can be accomplished at 60.1 GHz. The gain increase of 2.64 dB was accomplished with recessed ground plane in the estimations. The decent understanding among hypothetical and estimated results affirms the advantages of utilizing recessed ground plane [11]. The image of traditional antenna patch is displayed in Fig. 4.

Fig. 4 Antenna patch image [11]

Thus impacts of various boundaries of recessed ground all things being equal and radiation qualities of antennas are examined by full wave simulation along with semi static capacitance pattern.

Mousavi Khaleghi et al. [12] investigated that the possibility of article was to broaden the burrowing impact of ENZ limited channel for coordinating two microstrip lines with various impedance characteristics. The primary bit of leeway of this technique to plan a channel with sub-wavelength electrical scope to acquire comparative coordinating situation when compared through a regular λ/4-transformer. The structure's bandwidth is straight forwardly identified with the bandwidth of the ENZ-metamaterial. The proposed matching circuit is contained a metallic wall and an ENZ limited channel. To understand the ENZ channel, a rectangular waveguide that works for TE_{10} mode was planned and it was executed by utilizing a substrate integrated waveguide technology. Many methods are additionally required instead of imitating the metallic wall to lessen the ENZ channel cross section. The intended structure for various impedances estimations with 50, 100, and 150 Ω was planned, simulated, manufactured, and tested. Also, for utilization of the matching of network, a patch antenna was matched over the required frequency range. Reproduction of result dependent on CST software had great concurrence to the estimations. It was demonstrated that the bandwidth of the circuit is ranging from 8 to 15% [12]. The prototype matching circuit by using ENZ material is shown in Fig. 5.

Chauhan et al. [13] discussed in the document about a rectangular inset feed microstrip antenna with abandoned ground plane for ultra- wide band application. Thus, author discovered about the creating even Shape openings right underneath transmitting patch between the ground plane outcomes ultra- wide band parameters. Freeloading impact was limited to shape with streamlining width and length of the slot, with the goal of uniform return loss have been achieved for whole UWB extend.

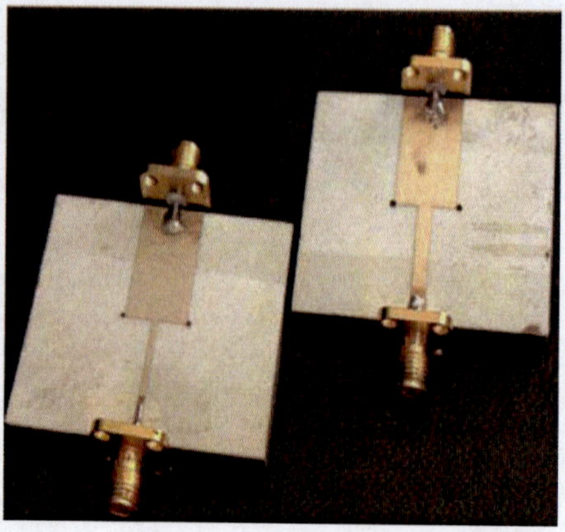

Fig. 5 Matching circuit using ENZ material [12]

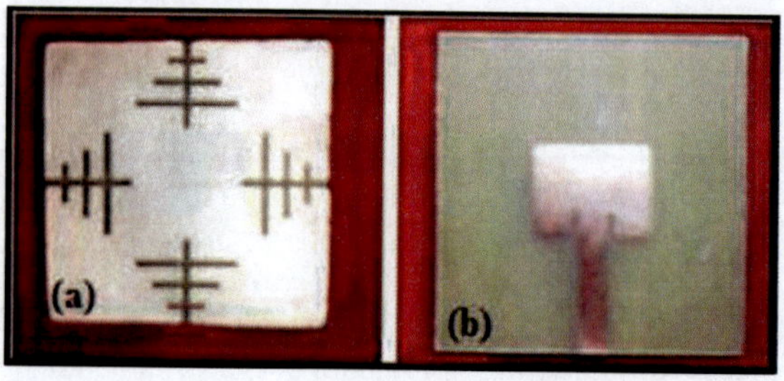

Fig. 6 Design of UWB antenna **a** ground plane, **b** patch antenna [13]

The designed antenna configuration shows return loss beneath −12 dB and VSWR under 2 in the whole ultra-wide band extend. An impedance bandwidth capacity of 11.2 GHz at −10dB return loss was achieved. Further, band notching property is used to reduce the signal interruption of Wi-Max is achieved by variation of the ground. The impact of first, T-cut space in ground structure was seen to result band notching from 3.28 to 4.58 GHz. In this way, the projected antenna conquers the signal obstruction issue with existing Wi-Max, IEEE 802.16 protocol framework and can be appropriately utilized for ultra-wide band applications [13]. The design of the fabricated antenna is shown in Fig. 6.

5 Conclusion

In the recent years, there is rapid development in the wireless technologies like Bluetooth, WLAN, GPS, WiFi and Wi Max. Some of these have fabricated in the mobile handset to fulfil the requirement of radio frequency (RF) by using wideband characteristics of antenna technology for different frequency band. Meanwhile the futuristic wireless communication applications such as Bluetooth, WLAN, WiFi, WiMax etc. uses circularly polarised radiation. The antenna configuration is the simplest for micro strip antenna. The wideband microstrip antenna provides overall improvement in the antenna parameters. This paper examines about the different kinds of antennas regarding their parameters like design, gain, simulation methods and their results. The significance of wide band and UWB antenna is examined for executing in 5G wireless communication applications. The study included investigation of the antennas where return loss is found to be $|S11| \leq -10$. The UWB antenna should be grown with the end goal that it surpasses the enhancement given by the current antennas along with expanding various communication applications. The UWB ends up being promising for giving inclusion to the fifth-generation applications such as Wi-Fi and Wi Max.

Acknowledgements The author is thankful to Dr. Praveen Kumar Malik for his appreciation, specialized suggestions and good technological help. Sincerely, I would like to thank Lovely Professional University to support for composing the review article. The writer is very much obliged to friends for motivating regarding research article publication.

References

1. Ge L, Yang X, Zhang D, Li M, Wong H (2017) Polarization-reconfigurable magnetoelectric dipole antenna for 5G Wi-Fi. IEEE Antennas Wirel Propag Lett 16:1504–1507. https://doi.org/10.1109/LAWP.2016.2647228
2. Ta SX, Choo H, Park I (2017) Broadband printed-dipole antenna and its arrays for 5G applications. IEEE Antennas Wirel Propag Lett 16:2183–2186. https://doi.org/10.1109/LAWP.2017.2703850
3. Huang H, Li X, Liu Y (2018) 5G MIMO antenna based on vector synthetic mechanism. IEEE Antennas Wirel Propag Lett 17(6):1052–1055. https://doi.org/10.1109/LAWP.2018.2830807
4. AL-Saif HT, Usman H, Chughtai M, Muhammad Tajammal Nasir J (2018) Compact ultra-wide band MIMO antenna system for lower 5G bands. Wirel Comm Mobile Comput Hindawi 2018:6. https://doi.org/10.1155/2018/2396873
5. Ibrahim A (2019) Compact planer UWB antenna for high speed communications. In: 2019 international conference on innovative trends in computer engineering (ITCE), Aswan, Egypt, pp 266–269. https://doi.org/10.1109/ITCE.2019.8646646
6. Singh H (2020) Designing and analysis of wideband antenna for 4G and 5G Applications. J Sci Ind Res 79:297–301
7. Al-Yasir YIA, OjaroudiParchin N, Abdulkhaleq A, Hameed K, Al-Sadoon M, Abd-Alhameed R (2019) Design, simulation and implementation of very compact dual-band microstrip bandpass filter for 4G and 5G Applications. In: 2019 16th international conference on synthesis, modeling, analysis and simulation methods and applications to circuit design (SMACD), Lausanne, Switzerland, pp 41–44. https://doi.org/10.1109/SMACD.2019.8795226
8. Ahmad I, Houjun S, Ali Q, Samad A (2020) Design of umbrella shape single element patch antenna with high gain and high efficiency for 5G wireless communication in 28 GHz. In: 2020 17th international Bhurban conference on applied sciences and technology (IBCAST), Islamabad, Pakistan, pp 710–713. https://doi.org/10.1109/IBCAST47879.2020.9044577
9. Ioannis G, Katherine S (2018) Design of ultra wide band slot antennas for future 5G mobile communication applications. In: 2018 7th international conference on modern circuits and systems technologies (MOCAST), Thessaloniki, pp 1–4. https://doi.org/10.1109/MOCAST.2018.8376611

10. Tiwari P, Malik PK (2020) Design of UWB Antenna for the 5G mobile communication applications: a review. In: 2020 international conference on computation, automation and knowledge management (ICCAKM), Dubai, United Arab Emirates, pp 24–30. https://doi.org/10.1109/ICCAKM46823.2020.9051556
11. Malik PK, Singh M (July, 2019) Multiple bandwidth design of micro strip antenna for future wireless communication. Int J Recent Technol Eng 8(2):5135–5138. ISSN: 2277-3878. https://doi.org/10.35940/ijrte.B2871.078219
12. Kaur A, Malik PK (2020) Tri state, T shaped circular cut ground antenna for higher 'X' band frequencies. In: 2020 International Conference on Computation, Automation and Knowledge Management (ICCAKM), Dubai, United Arab Emirates, pp 90–94
13. Chauhan A, Sharma S, Kalyan R, Tripathi CC (2013) Design and fabrication of ultra wideband antenna with band notching property using T-shaped defected ground structure. In: Fifth international conference on advances in recent technologies in communication and computing (ARTCom 2013), Bangalore, pp 229–235. https://doi.org/10.1049/cp.2013.2216

Planar UWB Antenna for MIMO/Diversity Applications

Pramod Singh and Rekha Agarwal

Abstract Ultra-wideband (UWB) communication technology is found to be suitable for short-range high-speed data transfer. But due to the limitation of maximum transmit power, the range and channel capacity are the matter of concern. If we use MIMO technology in UWB-based system, the channel capacity and rate of data transmission can be made better. After 2002 when Federal Communication Commission (FCC) declared a dedicated frequency range to UWB system, a lot of research has been carried out on MIMO antenna for UWB system. Although there are several challenges like the issue of size and compactness appeared, when a number of antenna elements were increased. There was another problem of mutual coupling between closely spaced radiating elements. Use of printed antenna technology was found to be helpful in the reduction in size and area of antenna significantly. Similarly, for the reduction of mutual coupling, several isolation techniques had been incorporated for significant reduction of mutual coupling as well as correlation coefficient. An extensive survey is presented here to appreciate and acknowledge the research carried out in design of UWB antenna for MIMO applications.

Keywords EMI · Planar · UWB · Mutual coupling

1 Introduction

Researchers are working to develop wideband communication devices with several constraints like: compact size, less power requirement, reduced cost, high speed of data transmission, and wide bandwidth. In these devices for interacting with outside world, antenna is very essential component and it must also have small size, low cost, conformal shape, and stable gain as well as radiation pattern. Since the evolution of

P. Singh (✉)
USIT, Guru Govind Singh Indraprastha University, New Delhi, India

R. Agarwal
Department of Electronics and Communication, Amity School of Engineering and Technology, Noida, India

P. K. Singh et al. (eds.), *Proceedings of Second International Conference on Computing, Communications, and Cyber-Security*, Lecture Notes in Networks and Systems 203,
https://doi.org/10.1007/978-981-16-0733-2_9

communication system, we can see that the process of development is never ending starting from 1G, 2G to 5G, and so on. In each subsequent generation, bandwidth, data rate, and features are increasing while size, area, and power consumption is reducing.

UWB systems are known for data transfer at very high rate at very low power level. These systems are also known as time domain, carrier free, impulse, and baseband systems used in the field of wireless communication. Due to UWB technology, it is possible to provide high-speed data transfer between portable wireless devices like digicam, laptop, mobile, and printer within small range. The performance of devices based on UWB technology can be made better by choosing an appropriate antenna. It is observed that planar antenna, made using printed circuit technology, is of low cost, compact size, and adjustable according to the shape of the structure on which it has to be mounted. Due to these features, planar antenna is the most frequently used antenna in current wireless systems. Microstrip patch antenna is a type of planar antenna which is most widely used. Although it has an inherent drawback of low impedance bandwidth, which can be compensated by using several improving its bandwidth using various methods.

Due to small power requirement for signal transmission in UWB system, the probability of interference with other signals is very low. However, the low transmitted power; the UWB system can be used only in short-range communication [1]. This limitation can be removed by the use of MIMO technology.

The block diagram representation of MIMO system is shown in Fig. 1. The MIMO technology enables us to increase the rate of data transmission as well as capacity of channel is also improved along with optimum usage of frequency spectrum [2]. But the major challenge in the implementation of this technique is to maintain isolation between nearby radiators due to compact nature of devices.

In MIMO system, isolation coefficient is very significant parameter in evaluating the performance of the system. It is defined as signal or energy received by a nearby

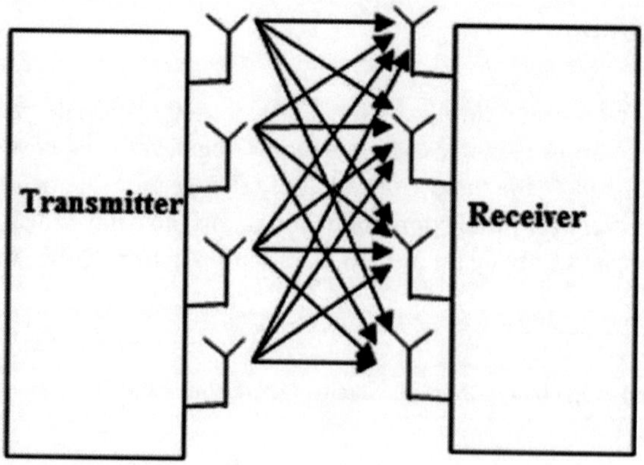

Fig. 1 General block diagram of MIMO system

antenna when another antenna is active. This coupling will affect input impedance, return loss, and radiation pattern of the elements present in the MIMO system. An empirical model of mutual coupling was presented in [3], as shown in Eq. (1),

$$C_{mn} = \exp\left(\frac{-2d_{mn}}{\lambda}(\alpha + j\pi)\right), \quad m \neq n$$
$$C_{nm} = 1 - \frac{1}{N}\sum_{m}\sum_{n \neq m} C_{mn} \tag{1}$$

where

C_{mn} Mutual coupling between *m*th and *n*th radiator.
d_{mn} Distance between *m*th and *n*th radiator.
N Number of elements in MIMO system.
α Parameter controlling coupling.

There are several challenges in design of UWB antenna for MIMO applications. Enormous work has been carried out by researchers in this field starting from a wideband antenna for MIMO to ultra-wideband antenna for MIMO applications. Here, a study is presented showing work of various researchers.

2 Literature Review of Wideband as Well as Ultra-Wideband Antenna for MIMO Applications

2.1 Wideband Antenna for MIMO Applications

In 2005, Y. Ge, K.P. Esselle, and T.S. Bird developed E-shaped diversity antenna with corrugated wings for handheld wireless communication devices. This design was able to operate over wireless communication bands and standards, WLAN 802.11a and HiperLAN2 between 5 and 6 GHz with good diversity performance (Fig. 2).

Sajad Mohammad Ali Nezhad et al. suggested that it is possible to deliver good performance in MIMO system with small elemental spacing. The proposed structure consists of monopole antenna of E shape with two slots a delivered multiband operation and good MIMO performance. Five different configurations by placing two monopoles of MIMO system in different ways were tested. The results showed that when the antenna elements were orthogonal mutual coupling was lower than when elements are parallel. A particular orientation of this system was found to be useful for MIMO applications for three resonance frequencies at 2.4, 5.4, and 5.8 GHz (Fig. 3).

Marko Sonki and Erkki Salon presented a simple structure with two monopole antenna elements with very low mutual coupling at 2.45 GHz center frequency. Here,

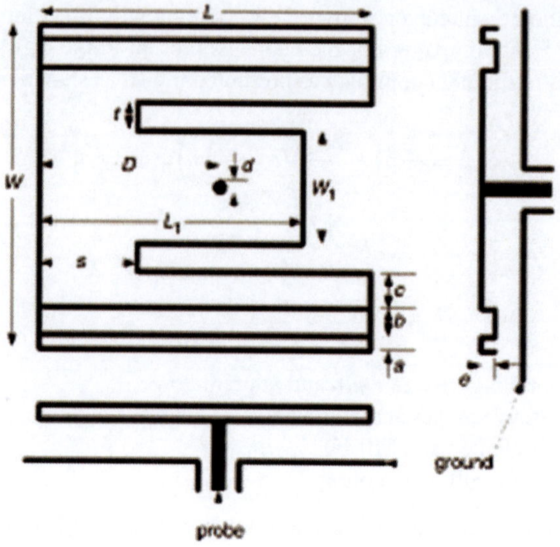

Fig. 2 E-shaped diversity antenna [4]

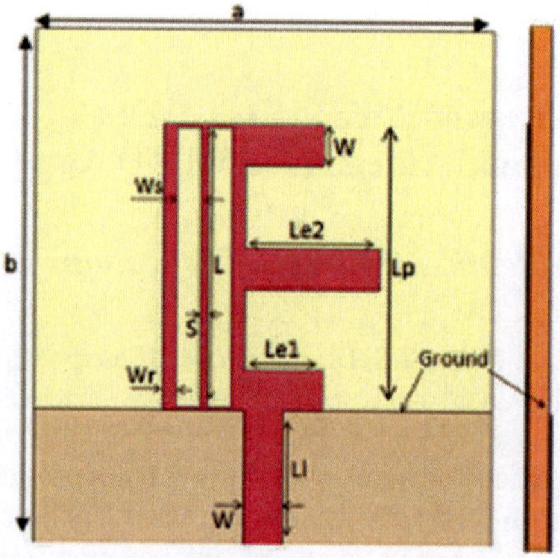

Fig. 3 E-shaped monopole with two slots [5]

for isolation between radiating elements, two λ/2 slots are placed into the ground plane. Very promising results were found with reduced mutual coupling and good radiation properties (Fig. 4).

MinseokHan et al. developed a MIMO system with polarization diversity for 4G handset applications. In this structure, a layer of radiators was made one above the other on a single substrate with spacing of only 0.8 mm. This structure provided

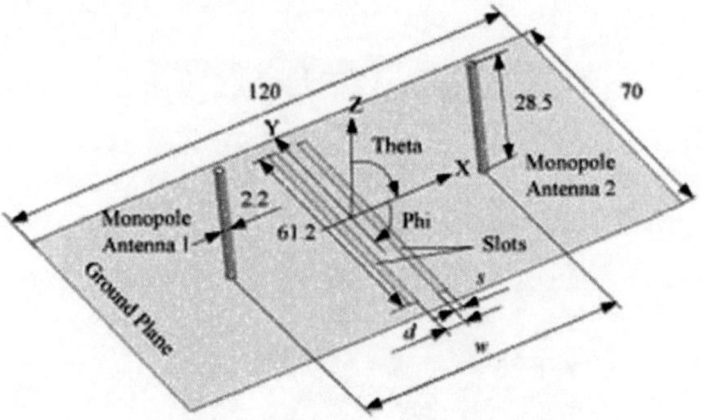

Fig. 4 Dual monopole MIMO antenna [6]

dual band operation 0.7/2.5 GHz with polarization diversity. The surface current distribution at 0.75 and 2.5 GHz can be used to explain the analysis of the proposed structure (Fig. 5).

A dual band monopole antenna for UWB application was proposed by H. U. Iddi, M.R. Kamarudin, T.A. Rahman, and R. Dewan, UTM Skudai. Here, B-shaped monopole elements operating for ISM band and WLAN with and impedance bandwidth of 29.9% and 33.8%, respectively. Both the radiating elements were placed $\lambda/12$ apart with a mutual coupling of −22 and −30 dB (Fig. 6).

Manoj K. Meshram et al. presented a MIMO antenna having two antennas for 4G-LTE and Wi-Fi applications in electronic gadgets. A DGS plane is used for reduction of mutual coupling. In this design, there were two meandered line planer inverted F-shaped antennas with a folded patch with an inter-digital capacitive strip. In this research, the effect of user proximity on MIMO antenna performance was also studied. Two commonly arising conditions were studied, one during messaging, hand is close to device and second during calling, head is closed to device. The FEKO simulation models of articulated hand and head were used to study the effect on performance of MIMO system. From the results, it was found that the effect of hand and head is almost negligible on both, return loss and mutual coupling between two antennas (Fig. 7).

Sajad Mohammad et al. designed and analyzed an UWB antenna with additional WLAN at 2.4 GHz found its applications in MIMO systems. The proposed structure is combination of a patch, monopole, and defected ground structure. Four different configurations of two-element MIMO systems had been tested using current distribution and it had good current distribution as well as satisfactory isolation between two elements (Fig. 8).

In MIMO system, the major concern is mutual coupling between closely spaced multiple radiating elements. There were many techniques applied by different researchers for improved isolation or reduced coupling. In 2013, Mohammad S. Sharawi1 et al. used reactive impedance on top side of structure for higher band (2.3–2.98 GHz) isolation while complementary capacitive loaded loops on ground

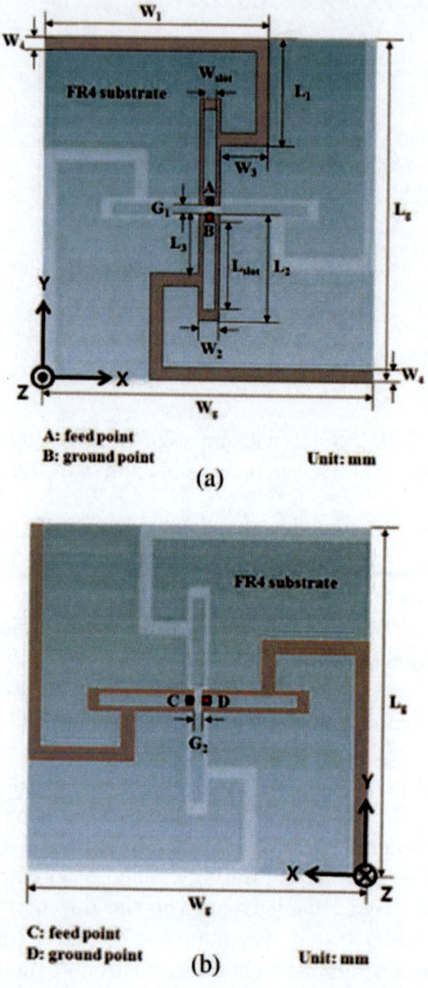

Fig. 5 Multiband MIMO antenna for using polarization diversity, **a** top view, **b** bottom view [7]

for medium and lower band (827–853 MHz) isolation. However, this isolation was obtained at the cost of 5% reduction in efficiency (Fig. 9).

Hari S. Singh et al. proposed dual band MIMO radiator with good isolation. The proposed system was operating over WLAN bands (2.4–2.485 and 5.15–5.85 GHz). The mutual couplings are −28 dB for WLAN 11.b/g band and lower than −26 dB for WLAN 11.a. These structures were found to have good MIMO performance and useful in mobile handset applications (Fig. 10).

Saber Soltani and Ross D. Murch proposed a MIMO antenna system with arbitrary even number of ports. In this approach, a canonical 2-port antenna was used which could be replicated together to form a MIMO antenna system up to 22 antenna

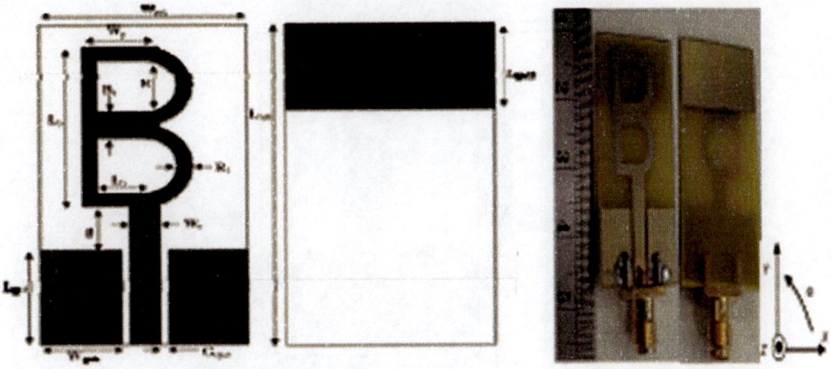

Fig. 6 B-shaped monopole antenna [8]

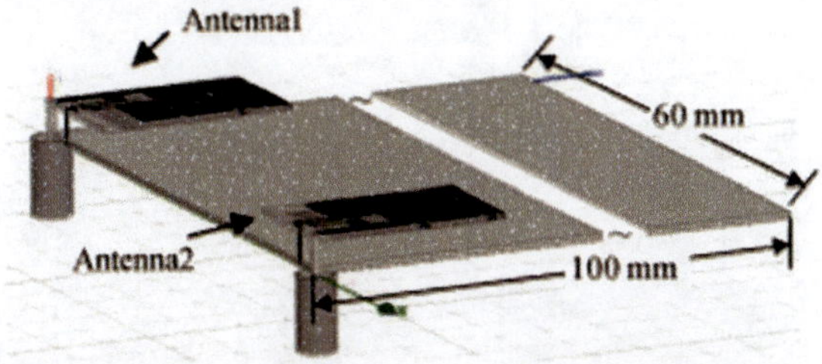

Fig. 7 Configuration of the proposed antenna for mobile handset application [9]

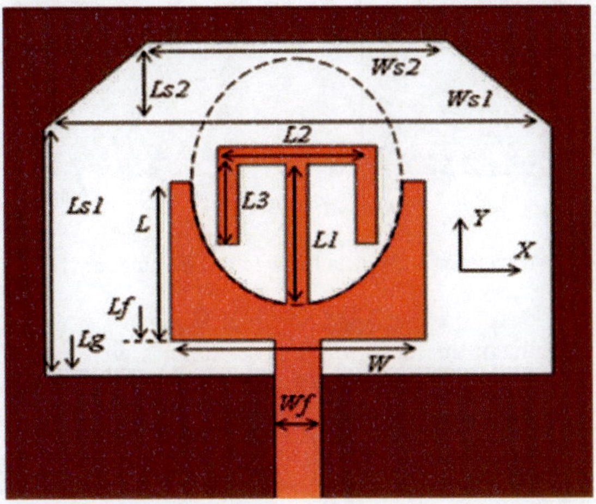

Fig. 8 Printed slot antenna with UWB/WLAN bands [10]

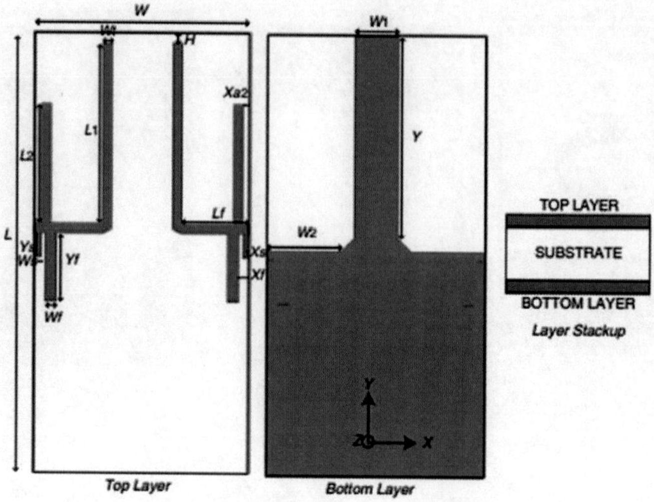

Fig. 9 2 × 1 four-shaped MIMO antenna [11]

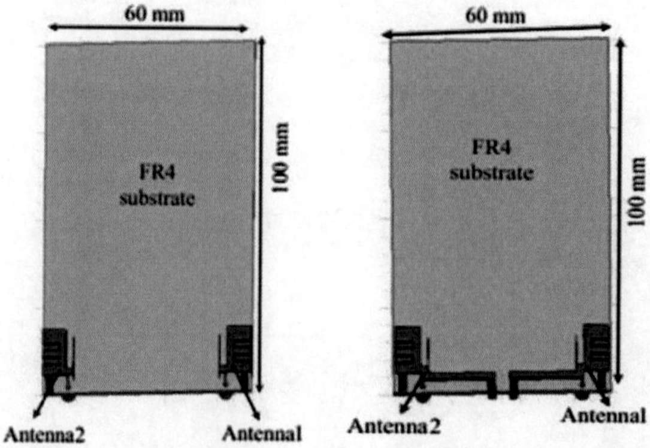

Fig. 10 Dual band MIMO antenna without and with folded shorting strip [12]

per square wavelength. In this design, 20 port antennas were used operating over frequency of 2.6 GHz and bandwidth of 100 MHz. Here, the isolation better than 10 dB was achieved for the proposed design (Fig. 11).

2.2 UWB Antenna for MIMO Applications

Seokjin Hong et al. presented a two-element UWB antenna for MIMO applications. Here, two Y-shaped radiators were used with three stubs on the ground plane to

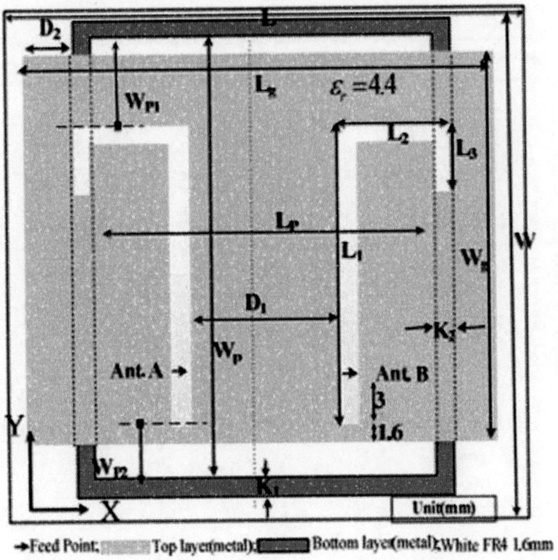

Fig. 11 Two-port MIMO antenna [13]

reduce coupling between radiating elements. By comparing surface results in two structures, the use of stubs was justified. It was concluded from the results that isolation characteristics were improved and it also reduced the lowest resonant frequency (Fig. 12).

In 2009, Terence S. P. See and Zhi Ning Che devised an ultra-wideband MIMO antenna for handheld portable devices. In this design, two triangular notched radiating elements were there with modified ground plane to ensure lower mutual coupling. From the results, it was clear that the structure was able to have satisfactory gain, efficiency, isolation, and return loss in the range of 3.1–5 GHz. In this research, to compute energy emerging from antenna, transfer function approach is used. In order to facilitate researchers in the field of antenna designing, a parametric analysis was carried out to study the effect of spacing between the radiators, length of the notch on the radiator, position of feed point, etc., on the return loss and mutual coupling (Fig. 13).

A. Najam, Y. Duroc et al. proposed a pair of circular monopole antenna for an UWB MIMO system. The isolation below −15 dB between the antennas elements was achieved by inverted Y-shaped and Y-shaped stub. Equivalent circuit analysis of this structure was presented by LC band stop filter. In order to test the ability of this system for UWB applications, time domain response of the system is analyzed. Here, the radiating elements were fed by fifth derivative of Gaussian pulse and response was good in time domain which was further justified by group delay response also. In this research, the effectiveness of stub was justified by comparing its performance with

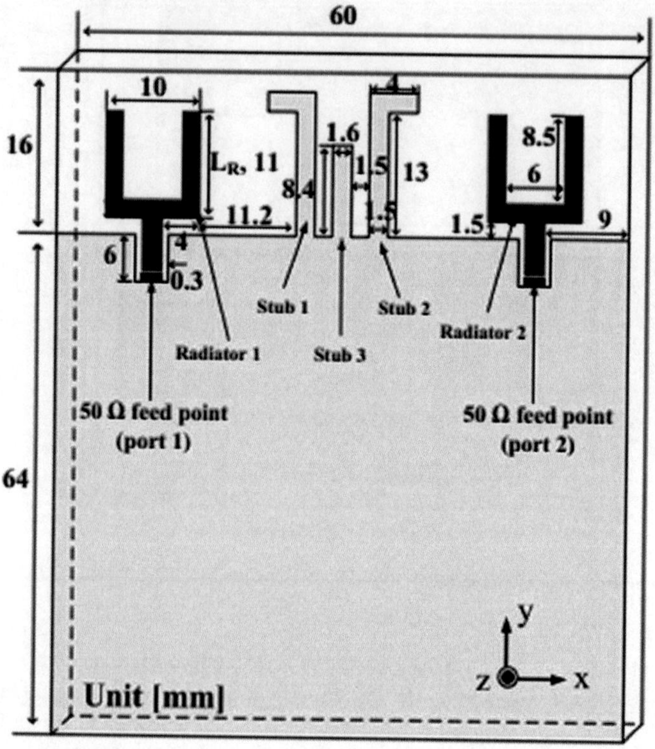

Fig. 12 Geometry of diversity antenna for UWB [14]

and without stub. The plotted s parameter curves of UWB MIMO antenna clearly showed that isolation increases significantly by the insertion of stub especially in the higher frequency (Fig. 14).

In 2012, M. R. Kamarudin et al. discussed spatial design effect on radiation pattern in ultra-wideband MIMO system. The proposed design consisted of two radiating elements where each element was a combination of seven small circles surrounding a single middle circle. This system was developed on Taconic substrate and covers the entire UWB with three resonant frequencies at 3, 6, and 10 GHz. The results showed that the antenna has relative stable radiation pattern at above-mentioned frequencies (Fig. 15).

Bybi P. et al. proposed a compact UWB, dual band antenna for MIMO systems. This design is composed of two orthogonal U-shaped radiating elements with cross-shaped strip for isolation between elements. The operating range of this structure was 2.6–11 GHz with mutual coupling better than 15 dB except lower end of frequency range (Fig. 16).

Another MIMO antenna for UWB applications of very compact size was proposed by Jian-Feng Li, Qing-Xin Chu, Zhi-Hui Li, and Xing-Xing Xia. There were two

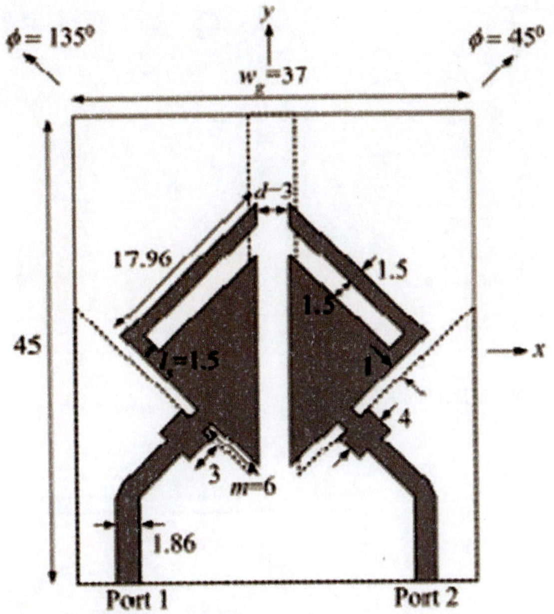

Fig. 13 Compact UWB diversity antenna [15]

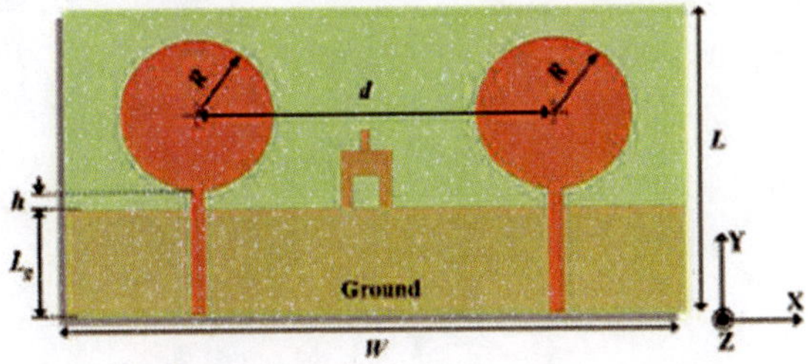

Fig. 14 UWB MIMO antenna with Y-shaped stub [16]

antenna elements of dimension 5.5 × 11 mm^2 connected to two protruded ground parts, respectively. These protruded ground parts were connected by a metal strip to reduce mutual coupling between 3 and 4 GHz. This structure was operated over UWB range with two stop bands from 3.3 to 3.7 GHz and 5.15–5.85 GHz (Fig. 17).

Nayab Gogosh et al. proposed a novel H-shaped isolating structure for shared and separate ground planes of radiating patches and it suppressed mutual coupling significantly. In this design, two ellipse-shaped antennas are placed at certain angle and a circular slot at center. The mutual coupling was found to be −20dB for shared and −25dB for isolated ground planes. This proposed design found its application in wireless personal area network and high-speed WLAN modems (Fig. 18).

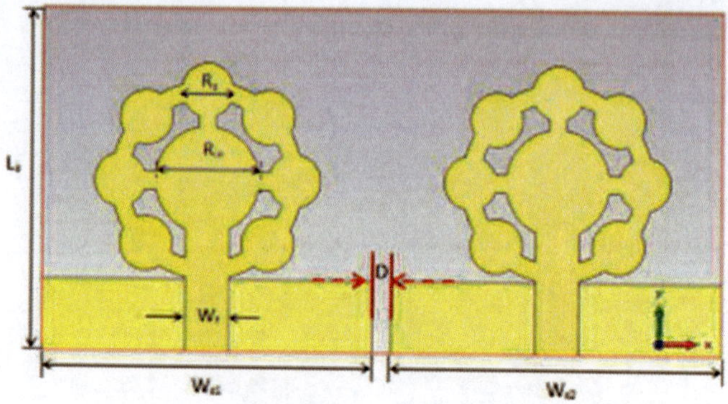

Fig. 15 Proposed UWB MIMO antenna [17]

Fig. 16 Dual band MIMO antenna [18]

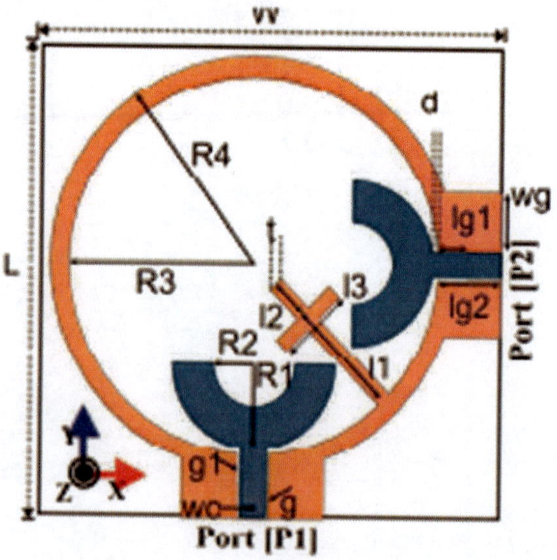

Jian Ren et al. proposed a compact MIMO antenna for UWB systems. In this design, there were two L-shaped antennas placed perpendicular to each other to ensure minimum mutual coupling and a narrow slot at the ground plane. The results showed that this structure is operating over entire UWB with mutual coupling below −15 dB and envelope correlation coefficient better than 0.02 (Fig. 19).

Most of the MIMO system used linear structure for isolation between radiating elements. In 2014, Muhammad Bilal et al. proposed a compact MIMO antenna doublet which used sin curve-based T-shaped decoupling structure which ensured isolation of at least −20 dB. This structure consisted of a pair of semielliptical antenna elements with impedance matching and isolation arrangements. For the enhanced

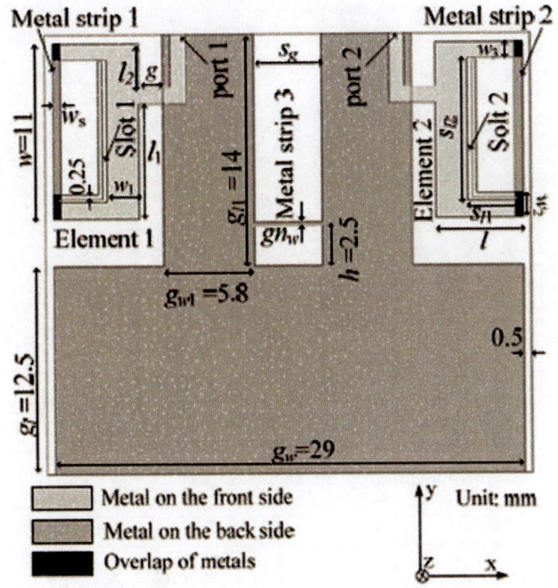

Fig. 17 Dual band notched UWB MIMO antenna [19]

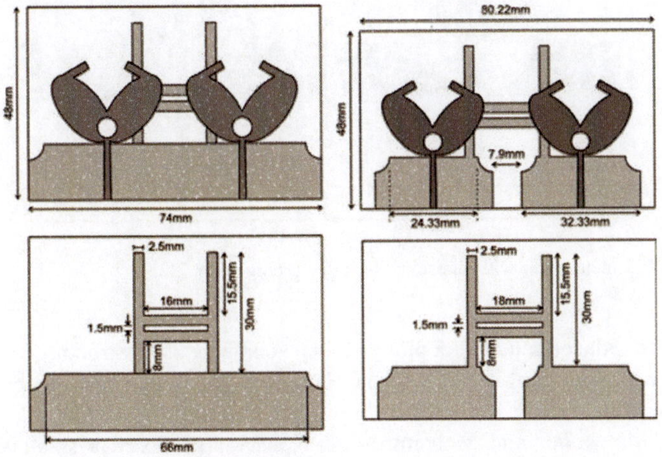

Fig. 18 MIMO antenna with shared ground plane and separate ground plane [20]

impedance matching to ensure UWB coverage, DGS planes, tapered feed lines, and meander line parasitic structures were used (Fig. 20).

Lihong Wang et al. proposed a compact flexible polarization antenna for wideband system. This MIMO system has two radiating elements placed perpendicular to each

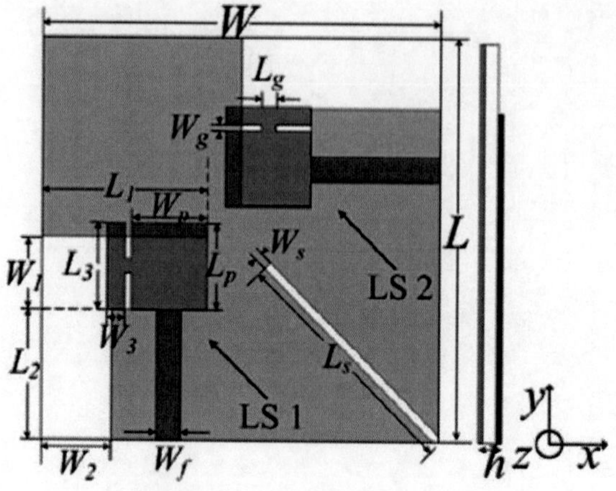

Fig. 19 UWB/WLAN MIMO antenna top and side view [21]

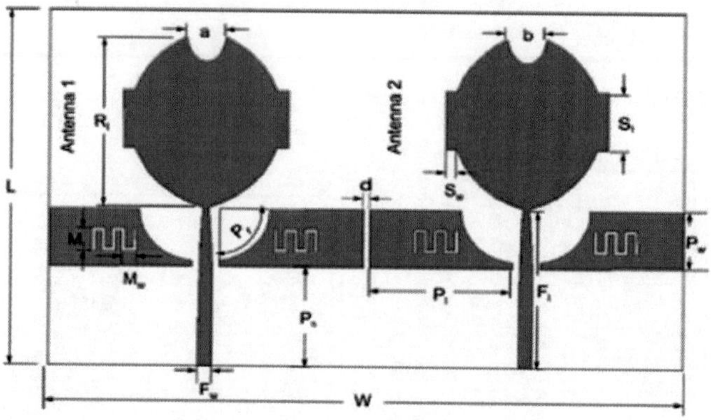

Fig. 20 UWB antenna pair with sinusoidal isolating arrangement [22]

other ensure reduced mutual coupling. It was a compact sized structure of 26 × 38 mm^2 dimension which covers 3.1–10.6 GHz range with mutual coupling better than 30 dB (Fig. 21).

Narges Malekpour and Mohammad A. Honarvar proposed a small size 2 × 2 MIMO antenna system for ultra-wideband communication applications. Here, monopole radiating elements were used and comb-line structure on the ground plane to ensure good impedance matching and reduced mutual coupling. Mutual coupling between two radiating monopoles is below −25 dB and mutual coupling across the entire ultra-wideband frequency range (Fig. 22).

Shrivishal Tripathi et al. proposed a compact UWB antenna with WLAN notch for MIMO applications. Here, two circular monopoles orthogonally place with microstrip feed were used. A fractal slot was introduced on the ground for enhanced

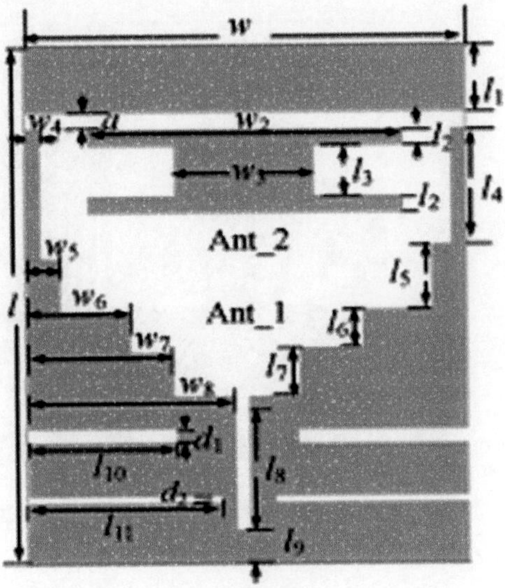

(a)

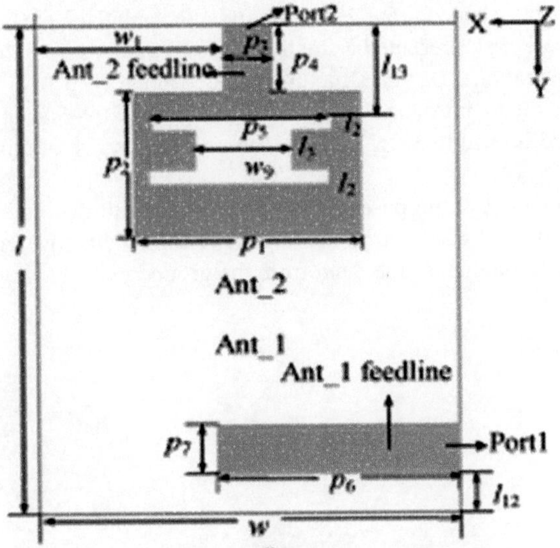

(b)

Fig. 21 Compact UWB diversity antenna **a** ground plane and **b** patch [23]

Fig. 22 UWB MIMO antenna with comb-line structure on ground plane [24]

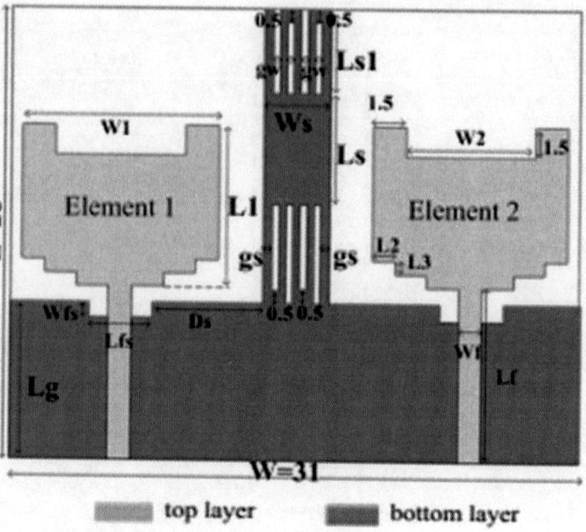

bandwidth. To introduce stopband at WLAN, elliptical split ring resonator (ESRR) was used at radiator. Isolation better than 21 dB between radiating elements was ensured L-shaped, I-shaped grounded stubs and a rectangular slot (Fig. 23).

Richa Chandel et al. presented a small size UWB antenna with dual band notch for MIMO applications. In this structure, tapered microstrip line feed was used with L-shaped slit to introduce stopband at WLAN (5.09–5.8 GHz) and IEEE INSAT/extended C band (6.3–7.27 GHz). The operating range of this structure was 2.9–20 GHz (Fig. 24).

Kamel S. Sultan et al. proposed an UWB antenna with two notch. Here, quasi self-symmetry method was used to achieve wide bandwidth coverage from 2.4 to 12 GHz. In this technique, a slot is cut from the ground plane having shape similar

Fig. 23 UWB MIMO antenna [25]

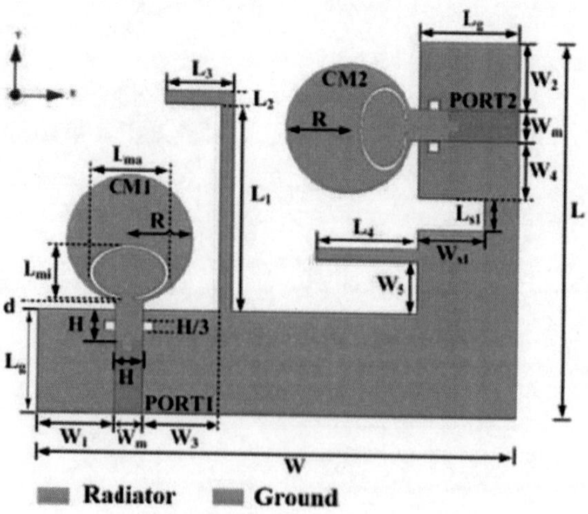

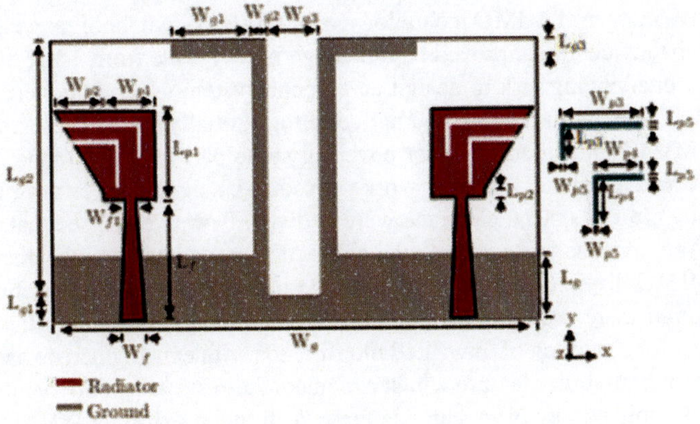

Fig. 24 MIMO antenna with L-shaped slit [26]

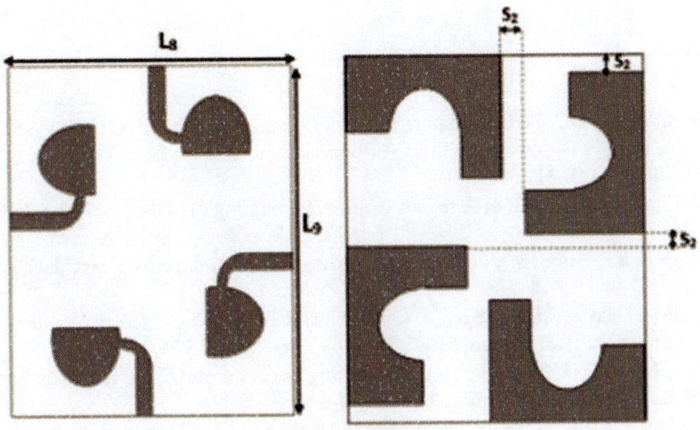

Fig. 25 MIMO antenna with four radiating elements [27]

the patch on the top of the substrate. Dual band notch was due to square ring on the ground plane and C-shaped slot on the patch. To evaluate the system performance, four different parameters: envelope correlation coefficient, directive gain, total active reflection coefficient, and channel capacity loss were used (Fig. 25).

3 Conclusion

UWB technology is need for modern handheld wireless electronics devices because of their high speed of data transmission, low power transmission level (which is not causing interference to other existing narrow band applications), and secure communication. The inherent drawback of these system is lower range and low channel capacity. By merger of MIMO technology with UWB system, both range as well as channel capacity can be improved. UWB range is very wide from 3.1 to 10.6 GHz. It is a very challenging task to design an antenna system which completely covers entire UWB range with reduced mutual coupling. Initially, several researchers had designed MIMO antenna system for covering some part of UWB range but those were not UWB antennas. But their work proved to be significant as different ideas for isolating closely spaced elements were evolved. Now, we can see that there are several UWB antenna systems are available with very compact size which can be used in MIMO/diversity applications because of their good isolation characteristics. Although there are several things that can be incorporated like: insertion of notch bands for reduction of unwanted interference with existing narrowband applications, reconfigurability feature which can make these systems more flexible to use in different applications. If we can use these features in exiting UWB antenna for MIMO, then our system becomes more versatile, flexible, and effective.

References

1. FCC, Washington, DC, FCC 1st report and order on ultra-wideband technology, Feb 2002
2. Kaoser T, Zheng F, Dimitrov E (2009) An overview of ultra wideband systems with MIMO. Proc IEEE 97(2):285–312
3. Chen X, Zhang S, Li Q (2018) A review of mutual coupling in MIMO systems. IEEE Access, Apr 2018
4. Ge Y, Esselle KP, Bird TS (2005) Compact diversity antenna for wireless devices. Electron Lett 41(2)
5. Nezhad SMA, Hassani HR (2010) A novel triband E-shaped printed monopole antenna for MIMO application. IEEE Antennas Wirel Propag Lett 9:576–579
6. Sonkki M, Salon E (2010) Low mutual coupling between monopole antennas by using two slots. IEEE Antennas Wirel Propag Lett 9:138–141
7. Han M, Choi J, Dong H, Gu S, Seou (2011) Multiband MIMO antenna using orthogonally polarized dipole elements for mobile communications. Microw Opt Technol Lett 53(9)
8. Iddi HU, Kamarudin MR, Rahman TA, Dewan R, Skudai UTM (2012) Design of dual-band B-shaped monopole antenna for MIMO application. In: Proceedings of the IEEE international symposium on antennas and propagation
9. Meshram MK, Animeh RK, Pimpale AT, Nikolova NK (2012) A novel quad-band diversity antenna for LTE and Wi-Fi applications with high isolation. IEEE Trans Antennas Propag 60(9)
10. Sajad Mohammad A, Hassani HR, Foudazi A (2013) A dual-band WLAN/UWB printed wide slot antenna for MIMO/diversity applications. Microwave Opt Technol Lett 55(3)
11. Sharawi MS, Numan AB, Aloi DN (2013) Isolation improvement in a dual band dual element MIMO antenna system using capacitive loaded loops. Progr Electromagn Res 134:247–266
12. Singh HS, Meruva B, Pandey GK, Bharti PK, Meshram MK (2013) Low mutual coupling between MIMO antennas by using two folded shorting strips. Progr Electromagn Res B 53:205–221
13. Soltani S, Murch RD (2015) A compact planar printed MIMO antenna design. IEEE Trans Antennas Propag 63(3)
14. Hong S, Chung K, Lee J, Jung S, Lee S-S, Choi J (2008) Design of a diversity antenna with stubs for UWB applications. Microwave Opt Technol Lett 50(5)
15. See TSP, Chen ZN (2009) An ultra-wideband diversity antenna. IEEE Trans Antennas Propag 57(6)

16. Najam A, Duroc Y, Tedjni S (2011) UWB-MIMO antenna with novel stub structure. Progr Electromagn Res C 19:245–257
17. Jusoh PM, Jamlos MF, Kamarudin MR, Ahmad ZA, Romli MA, Ronald SH (2012) A UWB MIMO spatial design effect on radiation. In: PIERS proceedings, Kuala Lumpur, MALAYSIA, 27–30 Mar 2012
18. Chacko BP, Augustin G, Denidni TA (2012) Uniplanar UWB antenna for diversity applications. In: Proceedings of the 2012 IEEE international symposium on antennas and propagation, 12 Nov 2012
19. Li J-F, Chu Q-X, Li Z-H, Xia X-X (2013) Compact dual band-notched UWB MIMO antenna with high isolation. IEEE Trans Antennas Propag 61(9)
20. Gogosh N, Farhan Shafique M, Saleem R, Usman I, Faiz AM (2013) An UWB diversity antenna array with a NOVEL H-type decoupling structure. Microwave Opt Technol Lett 55(11)
21. Ren J, Hu W, Yin Y, Fan R Compact printed MIMO antenna for UWB applications. IEEE Antennas Wirel Propag Lett 13:1517–1520
22. Bilal M, Saleem R, Shafique MF, Khan HA (2014) MIMO application UWB antenna doublet incorporating a sinusoidal decoupling structure. Microwave Opt Technol Lett 56(7)
23. Wang L, Xu L, Chen X, Yang R, Han L, Zhang W A compact ultra-wideband diversity antenna with high isolation. IEEE Antennas Wirel Propag Lett 13:35–38
24. Malekpour N, Honarvar MA (2016) Design of high-isolation compact MIMO antenna for UWB application. Progr Electromagn Res C 62:119–129
25. Tripathi S, Mohan A, Yadav S (2017) A compact MIMO/diversity antenna with WLAN band-notch characteristics for portable UWB applications. Progr Electromagn Res C 77:29–38
26. Chandel R, Gautam AK, Rambabu K (2018) Tapered fed compact UWB MIMO-diversity antenna with dual band-notched characteristics. IEEE Trans Antennas Propag 66(4)
27. Sultan KS, Abdullah HH (2019) Planar UWB MIMO-diversity antenna with dual notch characteristics. Progr Electromagn Res C 93:119–129

Design and Analysis of Wearable Textile UWB Antenna for WBAN Communication Systems

Bhawna Tiwari, Sindhu Hak Gupta, and Vipin Balyan

Abstract In present scenario, with application-centric approach of all the modern communication devices, integration of electronic gadgets with wearable accessories is in high demand. This research paper presents design and simulation of wearable antenna using flexible textile substrate and analysis of various performance antenna parameters. The design and simulation of proposed jeans substrate-based textile antenna have been performed using CST 2018 Microwave Studio. Main emphasis of the current research work is to demonstrate a small-sized UWB antenna design having overall antenna size 20 mm × 22 mm × 1.07 mm. The resonant frequencies of the designed dual-band antenna are observed to be 4.45 and 8.75 GHz and have a wide fractional bandwidth of 103.5% in the ultra-wide band range. The presented microstrip patch textile antenna is compact, robust and flexible that makes it perfect choice to be utilized as body worn antenna for WBAN communication systems for wireless health monitoring systems.

Keywords WBAN · UWB · Textile antenna · Gain · VSWR · Efficiency

1 Introduction

Wireless body area networks (WBAN) is research trend nowadays that makes wireless health monitoring systems feasible. The design and implementation of wide variety of flexible textile fabric-based wearable antennas are preferred in recent research trend that are light weight, malleable and conveniently portable in

B. Tiwari (✉)
Amity University, Noida, Uttar Pradesh, India

S. H. Gupta
Department of Electronics and Communications, Amity University, Noida, India
e-mail: shak@amity.edu

V. Balyan
Department of Electrical, Electronics and Computer Engineering, Cape Peninsula University of Technology, Cape Town, South Africa
e-mail: balyanv@cput.ac.za

P. K. Singh et al. (eds.), *Proceedings of Second International Conference on Computing, Communications, and Cyber-Security*, Lecture Notes in Networks and Systems 203,
https://doi.org/10.1007/978-981-16-0733-2_10

comparison to rigid substrate-based traditional antennas for wireless communication systems. The antenna's impedance bandwidth and therefore antenna performance can be significantly improved by using textile fabric materials having low values of dielectric constant which helps in reduction of antenna losses associated with the antenna's proximity electromagnetic waves. On-body antenna radio frequency wave propagation and radiation is explained that finds application in different fields such as WBANs, medical sensor networks, personal and industrial communication, etc. The frequency range from 3.1 to 10.6 GHz has been authorized for implementing ultra-wide band (UWB) systems for commercial utilities. Wearable textile UWB antennas incorporating different designs patterns using fabric materials were implemented and discussed. Fabric material finds perfectly suitable to be used for flexible textile antenna for on body worn applications as these can be easily embedded into person's clothing [1, 2].

Motivation behind the current work is to incorporate microstrip patch antenna utilizing the advantage of dielectric properties of jeans fabric substrate for WBAN communication systems as body worn, flexible UWB antenna. The major aim of this presented work is based on development and simulation of a compact dual-band UWB textile-based antenna using jeans fabric which provides superior antenna performance and analysis of various antenna performance parameters when worn on human body.

The remaining paper is categorized as described. Related work in context to antenna designs as mentioned in references is discussed in Sect. 2. Proposed textile antenna design structure with dimensions and the CST simulation model is demonstrated in Sect. 3. Section 4 characterizes designed antenna simulation results and discussion. Section 5 concludes the current research work.

2 Related Work

A novel technique of monitoring, recognition and classification of different human activities that involves hand movements is described in [3]. Transmission and reflection coefficients parameter of on body antennas are observed and extract different characteristics, and machine learning approach is applied for various hand movement classifications with high classification accuracy. The application of on body antenna parameters S_{11} and S_{21} to classification of various human activities like sitting, boxing, etc. by using deep convolution neural network approach (DCNN) is also presented.

Martin Frank et al. in [4] demonstrate highly reliable in body UWB antenna communication and its effect on human tissues by utilizing different layer of human tissues embedded in biomedical human model. A small-sized UWB antenna design in terms of electrical dimensions and its electrical equivalent model is described in [5]. Analysis of jeans-based textile antenna is performed by considering both the antenna design theory and circuit theory analysis. Fabricated antenna results are compared with simulation result in context to antenna performance as well as specific absorption rate (SAR) characteristics. A low-profile low-weight UWB antenna optimized for WBAN purpose is fabricated on printed circuit board (PCB) and proposed in [6]. The tremendous improvement in antenna performance in terms of very high impedance bandwidth (21.5 GHz) shown in antenna simulation results. A compact circular patch antenna having dimensions in micrometers with terahertz band operation optimized for WBAN applications is designed and presented. A highly efficient and flexible wearable body worn antenna is designed using CST 2016 software. Specific absorption ratio (SAR) analysis is also performed for the proposed electro textile-based antenna [7–10]. In a review paper [11], mathematical modeling analysis of wide-band and ultra-wideband antennas is shown by considering WB patch, WB monopole, slotted antennas, tapered slotted and many other design geometries. Enabling technologies of WBAN applications are discussed in by Cao et al. [12] A novel technique to demonstrate the properties of textile materials with low values of electrical permittivity for flexible antennas has been explained. Mishra B. et al. in the research work [13] proposed a rectangular-shaped slotted circular patch antenna geometry designed for Wi-Fi/WLAN wireless utilities.

A novel design of the high-performance textile antenna with three resonant frequency bands and large fractional bandwidth has been designed and analyzed in [14] which is optimized for various wireless communication operations. Textile antennas can be used in wide variety of application which includes military, medical fields, etc. The proposed jeans- and copper-based antenna is preferred as wearable textile antenna as this fabric substrate material possesses many properties like its washability, wearability, flexibility, cost effectiveness, and very less maintenance requirements. The implemented natural rubber substrate-based compact and flexible UWB antenna is presented in [15] and preferred as suitable option as flexible antenna for WBAN applications. Its flexibility and less weight can incorporate any bending or twisting of human body without deforming antenna characteristics while wearing it on any part of the human body. Wang et al. in [16] demonstrate the various antenna geometries of multi-band UWB body worn antennas with jeans textile material as antenna substrate. Different design technical issues and associated challenges of wearable antennas for WBAN utility are also highlighted in their research work. The research work in [17] considers fabrication and simulation results of the novel design of compact textile antenna with different optimization techniques. The presented antenna for UWB wireless applications is operable on multiple channels that can incorporate all bending, twisting, curves, etc. conditions of human body for body communication.

3 Proposed Antenna Design

Wearable antennas can be classified into two categories. One is non-textile antennas that provide in-built antennas embedded in accessories like glasses, etc., and other is textile antennas that can be perfectly integrated into clothing. Current works focus on design and simulation of wearable textile antenna in which jeans fabric is taken as substrate material and copper is utilized as conducting part. Textile material exhibits relatively lower value of loss tangent and lower value of electric permittivity that leads to reduction of surface wave losses and bandwidth enlargement. Because of its flexibility and light weight, textile antennas are preferred as wearable antenna integrated into clothing without compromising wearability and user comfort. Table 1 represents applicable notations used in the paper and its description.

Table 2 describes the characteristics of the substrate material used in proposed antenna design. Radiating patch, substrate and partial ground with defective ground structure (DGS) with optimized dimensions are presented in Fig. 1, and antenna parameters values are depicted in Table 3.

The designed geometry of the proposed wearable textile antenna consists of a radiating patch with an asymmetric microstrip feed line and a partial ground plane with DGS structure as shown in Fig. 2. Defects or slots are inserted on the ground plane of antenna to improve various antenna parameters like enhancement of gain, impedance bandwidth, etc. The optimized dimensions are attained through optimization performed in CST Microwave Studio simulator. Three slots in radiating patch are introduced with dimensions $W_3 \times L_3$, $W_4 \times L_4$ and $W_5 \times L_5$, defective ground structure is utilized with slots dimensions $W_1 \times L_1$, $W_2 \times L_2$ in partial ground and adjustment of feed dimensions $W_f \times L_f$ leads to improved impedance matching thereby enhancing antenna performance.

Table 1 Table of notations

Notation	Description	Notation	Description
L_{sub}	Length of substrate	L_p	Length of patch
W_{sub}	Width of substrate	W_p	Width of patch
L_f	Length of feed line	h_s	Height of substrate material
W_f	Width of feed line	S_{11}	Return loss or reflection coefficient
L_g	Length of ground	Tanδ	Loss tangent of substrate material
W_g	Width of ground	εr	Relative permittivity of substrate material

Table 2 Substrate material properties

Textile substrate material	(εr)	(Tanδ)	(h_s)
Jeans	1.7	0.025	1

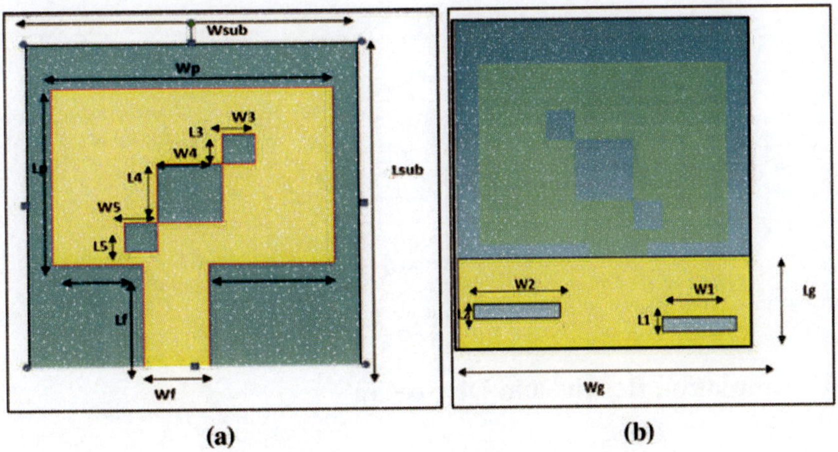

Fig. 1 Proposed antenna design geometry with dimensions. **a** Radiating patch and substrate. **b** Partial ground with DGS structure

Table 3 Dimensions of proposed antenna geometry

S. No.	Antenna parameter	Dimension (mm)	S. No.	Antenna parameter	Dimension (mm)
1	L_{sub}	22	8	W_p	17
2	W_{sub}	20	9	h_s	1
3	L_f	7	10	$L1 \times W1$	1×6
4	W_f	4	11	$L2 \times W2$	1×5
5	L_g	6	12	$L3 \times W3$	2×2
6	W_g	20	13	$L4 \times W4$	4×4
7	L_p	16	14	$L5 \times W5$	2×2

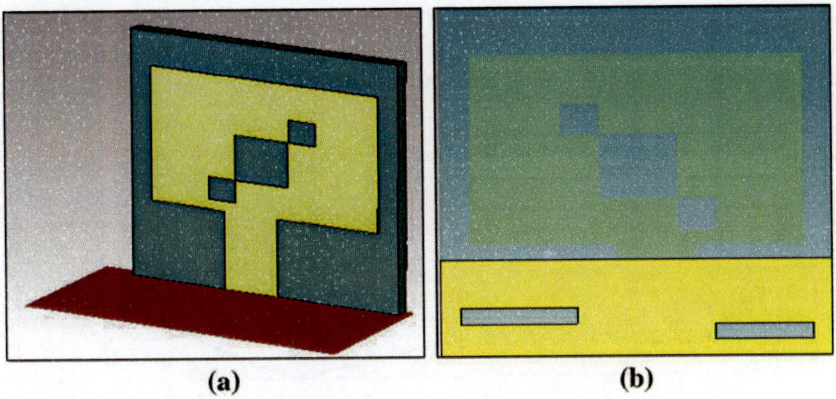

Fig. 2 CST model of proposed antenna. **a** Radiating patch design. **b** Partial DGS structure ground design

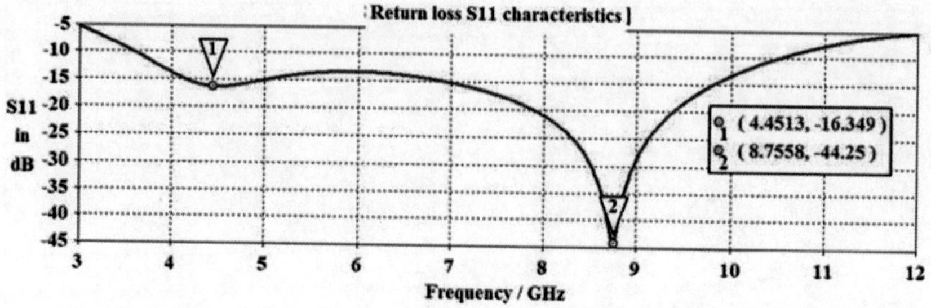

Fig. 3 Return loss characteristics (S_{11} vs. frequency plot)

4 Simulation Results and Discussion

4.1 Return Loss Characteristics

Return loss characteristics signify reflection coefficient or S_{11} value which indicates the quantity of power which gets reflected by the antenna. It has been observed from simulation result that values of S_{11} at resonant frequency 4.45 GHz and 8.75 GHz is -16.349 dB and -44.25 dB, respectively. Figure 3 demonstrates return loss characteristics for developed antenna having jeans as substrate material which confirms its suitability for UWB operation as S_{11} lies below -10 dB for the whole UWB frequency band as desired.

4.2 VSWR Characteristics

Voltage wave standing Ratio (VSWR) is an indication of how properly an antenna is matched with its attached feed line. It is ascertained from Fig. 4 that the observed values of VSWR lies in its desired value range that is 2 or lesser than 2. The simulation

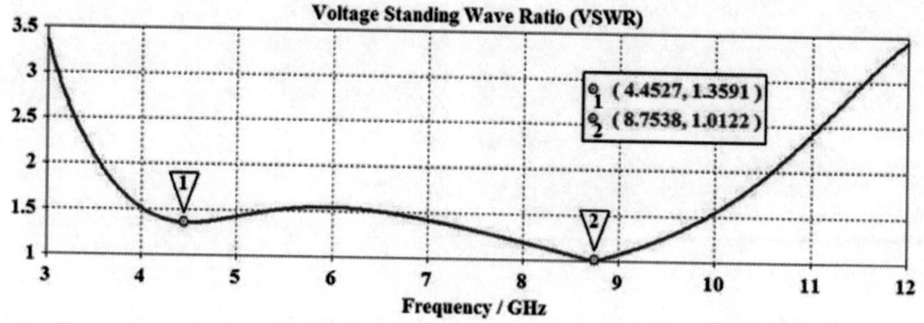

Fig. 4 VSWR characteristics (VSWR vs. frequency plot)

result shows VSWR values of 1.35 and 1.0122 at resonant frequency 4.45 GHz and 8.75 GHz, respectively.

4.3 Radiation Pattern Characteristics

Radiation patterns describing antenna radiation characteristics in terms of gain and directivity 3D plot for designed textile antenna at resonant frequencies are shown in Fig. 5. The simulation results show that the observed gain and directivity at resonant frequency 4.45 GHz are 1.89 dB and 3.4 dBi, respectively. At resonant frequency

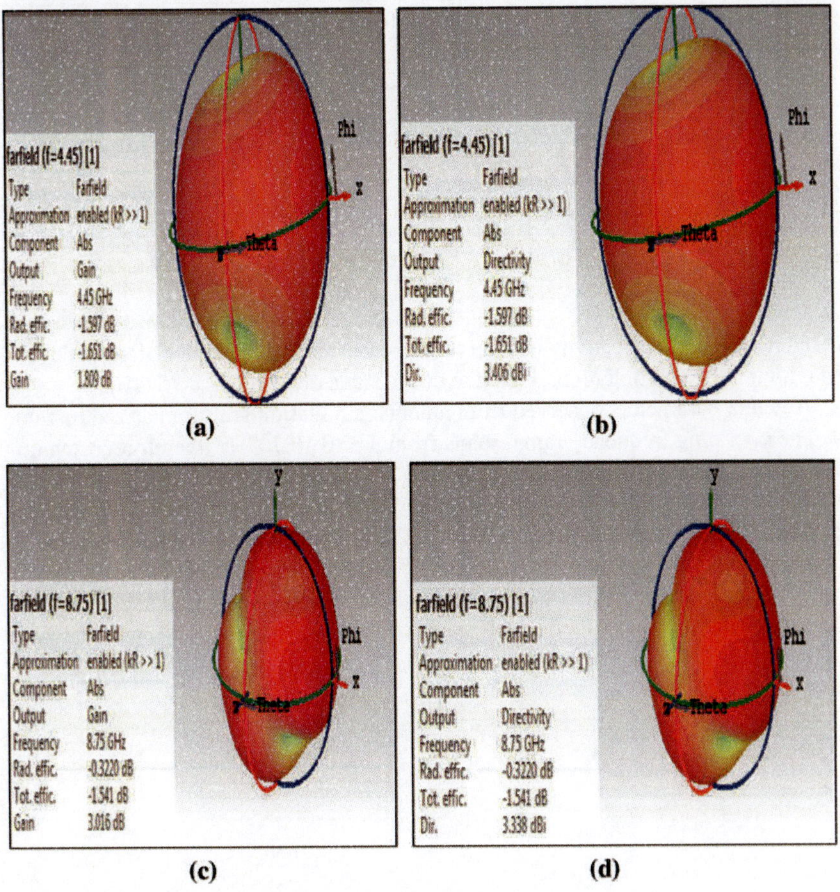

Fig. 5 Radiation pattern characteristics. **a** Gain plot at resonant frequency 4.45 GHz. **b** Directivity plot at 4.45 GHz. **c** Gain plot at resonant frequency 8.75 GHz. **d** Directivity plot at 8.75 GHz

Table 4 Antenna performance parameters of proposed antenna design

Resonant frequency (GHz)	Return loss S_{11} (dB)	Gain (dB)	VSWR	Directivity (dBi)	Efficiency (%)	Impedance bandwidth (GHz)	Fractional bandwidth (%)
$F1 = 4.45$ $F2 = 8.75$	−16.349 −44.25	1.89 3.0166	1.35 1.01	3.4 3.338	55.8 90.3	3.4–10.7 (7.2 GHz)	103.5

Table 5 Comparison of antenna geometry dimensions of proposed antenna design with comparable antennas mentioned in reference

Ref. No.	Substrate used	Overall antenna size (in mm)	Substrate dimension (in mm)	Ground dimension (in mm)	Patch dimension (in mm)
[14]	Jeans	86 × 90 × 1.06	86 × 90	86 × 30	Circular with radius 14 mm
[15]	Rubber and FR4	–	42 × 34	42 × 10	17 × 21.4
[16]	Jeans	45 × 35	45 × 35	38 × 28	20 × 10 × 1
[17]	Felt	39 × 42 × 3.34	39 × 42 ×	39 × 42	32 × 39
Proposed	Jeans	20 × 22 × 1.07	20 × 22 × 1	20 × 6 × 0.035	17 × 16 × 0.035

8.75 GHz gain and directivity is observed to be 3.016 dB and 3.338 dBi, respectively. Maximum antenna radiation efficiency of 90.3% is observed at 8.75 GHz.

Antenna parameters observed from antenna simulation are summarized in Table 4. The operating frequency range spans from 3.4 to 10.7 GHz, therefore, providing 7.2 GHz impedance bandwidth and 103.5% fractional bandwidth in UWB range.

The comparative study of the antenna dimensions and antenna performance parameters of presented antenna with existing comparable antenna designs is presented in Tables 5 and 6, respectively. Results demonstrate improvement of proposed antenna in context to compactness, impedance bandwidth and return loss characteristics and thus superiority in comparison to other antenna designs presented in literature.

Table 6 Comparison of antenna performance characteristics of proposed antenna design with comparable antennas mentioned in reference

Ref. No.	Resonant frequency (GHz)	S_{11} (dB) at resonant frequency	Gain (dB)	Directivity (dBi)	Impedance bandwidth
[14]	2.1366, 4.756, 11.49	22.23, −38.10, −20.79	–	3.3, 4.23, 5.19	4–6 GHz 11–12 GHz
[15]	5.5, 7.5	−18, −24	–	–	–
[16]	3, 7, 9	−12, −18, −35	2.7, 4.17, 4.074	–	6.4 GHz (86.48%)
[17]	3.5, 8.3	−17, −15	–	–	3.6–4.3 GHz 6.3–10.1 GHz
Proposed	4.45, 8.75	−16.349, −44.25	1.89, 3.0166	3.4, 3.338	3.4–10.7 GHz (103.5%)

5 Conclusion

The major objective of this work is to design and simulate a novel small-sized wearable antenna based on jeans textile substrate in UWB frequency range. The slotted radiating patch and partial ground with DGS structure is designed using adhesive copper, and jeans fabric is utilized as substrate material for antenna design. The dual-band textile UWB antenna using jeans fabric has been designed and simulated that is compact and possess large impedance bandwidth and high efficiency. The research paper also presented the comparative analysis of the antenna performance characteristics like reflection coefficient, VSWR, directivity, impedance fractional bandwidth, gain and radiation efficiency. The simulation results of presented jeans textile substrate-based antenna show return loss S_{11} values of −16.349 dB and −44.25 dB at resonating frequency 4.45 GHz and 8.75 GHz, respectively. The highest gain of 3.016 dB and peak efficiency of 90.3% is observed at the resonant frequencies 8.875 GHz, respectively. The proposed compact antenna has overall dimensions of 20 mm × 22 mm × 1.07 mm. The return loss characteristic shows that the designed antenna operating frequency lies from 3.4 to 10.7 GHz, therefore providing 103.5% fractional bandwidth in UWB range. Integration of flexibility, light weight, compact size, mechanically robust, efficient with the desired antenna performance characteristics are the key features that confirm its suitability for on body wearable application for wireless body area network communication systems.

References

1. Balanis CA (2004) Antenna theory: analysis and design. Wiley, New York
2. Osman MAR, Rahim MKA, Azfar M, Samsuri NA, Zubir F, Kamardin K (2011) Design, implementation and performance of ultra wide band textile antenna. Progr Electromagn Res 27:307–325
3. Gupta SH, Sharma A, Mohta M, Rajawat A (2020) Hand movement classification from measured scattering parameters using deep convolutional neural network. Measurement 151
4. Frank M, Lurz F, Kempf M, Rober J, Weigel R, Koelpin A (2020) Miniaturized ultra-wideband antenna design for human implants. IEEE Radio Wirel Symp
5. Yadav A, Kumar Singh V, Kumar Bhoi A, Marques G, Garcia-Zapirain B, de la Torre Díez I (2020) Wireless body area networks: UWB wearable textile antenna for telemedicine and mobile health systems. Micromachines 11:558
6. Yang D, Jianzhong Hu, Liu S (2018) A low profile UWB antenna for WBAN applications. IEEE Access 6:25214–25219
7. Guraliuc AR, Barsocchi P, Potorti F, Nepa P (2011) Limb movements classification using wearable wireless transceivers. IEEE Trans Inf Technol Biomed 15(3):474–480
8. Mukhopadhyay SC (2015) Wearable sensors for human activity monitoring: a review. IEEE Sens J 15(3):1321–1330
9. Rubani Q, Gupta SH, Kumar A (2019) Design and analysis of circular patch antenna for WBAN at terahertz frequency. Optik-Int J Light Electron Opt:529–536
10. Ahmed MI, Ahmed MF, Shaalan A-E (2018) Novel electro textile patch antenna on jeans substrate for wearable applications. Progr Electomagn Res 83:255–265
11. Saeidi T, Ismail I, Wen WP, Alhawari ARH, Mohammadi A (2019) Ultra-wideband antennas for wireless communication applications. Int J Antennas Propag:1–25
12. Cao H, Leung V, Chow C, Chan H (2009) Enabling technologies for wireless body area networks: a survey and outlook. IEEE Commun Mag 47(12):84–93
13. Mishra B, Singh V, Diwedi A, Pandey AK, Sarwar A, Singh R (2017) Slots loaded multilayered circular patch antenna for WiFi/WLAN applications, computing and network sustainability. Springer, pp 49–59
14. Khan S, Singh VK, Naresh B (2015) Textile antenna using jeans substrate for wireless communication application. Int J Eng Technol Sci Res 2(11)
15. Lakshmanan R, Sukumaran SK (2016) Flexible ultra wide band antenna for WBAN application. Proc Technol 24:880–887
16. Wang JC, Lim EG, Leach M, Wang Z, Man KL, Huang Y (2016) Conformal wearable antennas for WBAN applications, vol II. In: Proceedings of the international multi conference of engineers and computer scientists. Hong Kong (Mar 2016)
17. Samal PB, Soh PJ, Zakaria Z (2019) Compact microstrip-based textile antenna for 802.15.6 WBAN-UWB with full ground plane. Int J Antennas Propag:1–12

Advanced Computing Technologies

Stock Prices Prediction from Financial News Articles Using LSTM and XAI

Shilpa Gite, Hrituja Khatavkar, Shilpi Srivastava, Priyam Maheshwari, and Neerav Pandey

Abstract The stock market is very complex and volatile. It is impacted by positive and negative sentiments which are based on media releases. The scope of the stock price analysis relies upon the ability to recognize the stock movements. It is based on technical fundamentals and understanding the hidden trends which the market follows. Stock price prediction (Vachhani et al in Mach Learn-Based Stock Market Anal Short Surv (2020) [1]) has consistently been an extremely dynamic field of exploration and research work. However, arriving at the ideal degree of precision is still an enticing challenge. In this paper, we are proposing a combined effort of using efficient machine learning techniques coupled with a deep learning technique—long short-term memory (LSTM) to use them to predict the stock prices with a high level of accuracy. Sentiments derived by users from news headlines have a tremendous effect on the buying and selling patterns of the traders as they easily get influenced by what they read. Hence, fusing one more dimension of sentiments along with technical analysis should improve the prediction accuracy. LSTM networks have proved to be a very useful tool to learn and predict temporal data having long-term dependencies. In our work, the LSTM model uses historical stock data along with sentiments from news items to create a better predictive model

Keywords Recurrent neural network (RNN) · Long short-term memory (LSTM) · Explainable AI (XAI) · Stock price

S. Gite (✉) · H. Khatavkar · S. Srivastava · P. Maheshwari · N. Pandey
Symbiosis Institute of Technology, Symbiosis International (Deemed University), Pune, India
e-mail: shilpa.gite@sitpune.edu.in

H. Khatavkar
e-mail: hrituja.khatavkar@sitpune.edu.in

S. Srivastava
e-mail: shilpi.srivastava@sitpune.edu.in

P. Maheshwari
e-mail: priyam.maheshwari@sitpune.edu.in

N. Pandey
e-mail: neerav.pandey@sitpune.edu.in

P. K. Singh et al. (eds.), *Proceedings of Second International Conference on Computing, Communications, and Cyber-Security*, Lecture Notes in Networks and Systems 203,
https://doi.org/10.1007/978-981-16-0733-2_11

1 Introduction

Stock market investment can be a very tricky and tedious job, stock market prediction has been an object of studies for the past many decades and is a very difficult task because of its complexity, disordered information and dynamism. There are several technical indicators and sources of information which affect the stock prices, but due to the substantial amount of data present, it becomes difficult to predict the prices. However, with the advancement in technology, the opportunity to procure profit from the stock market has increased, which helps experts to make a better prediction.

There is a famous hypothesis in finance called the Efficient Market Hypothesis [2], which states that asset prices cannot fully depend on obsolete information and market prices react to new information, for example, financial news articles, social media blogs, etc. Given the major rules of financial institutes and their information gathering convention we believe it is important to focus on financial news at news events. With the advancement in artificial intelligence information coming from both financial time series and textual data can be employed to forecast stock prices [3].

In this paper, we are suggesting a combined effort involving multiple machine learning techniques and deep learning techniques for providing visual representations through tables for enabling us to get a clear outline which helps the regular public with amateur knowledge in this domain, to anticipate their future moves. We are using the data of the National Stock Exchange (NSE and the news headlines aggregated from Pulse. Pulse has aggregated 210,000+ Indian finance news headlines from various news websites like Business Standard, The Hindu Business, Reuter and many more eminent journal websites).

Recurrent neural network (RNN) has proved to be a powerful model for processing context information from textual data. We propose using LSTM for news sentiment classification employing simulating the interactions of words during the compositional process [4]. LSTM incorporates a memory cell which is a unit of computation that supersedes the traditional deep learning neurons in the hidden layer of the network. To trust the behaviour of the model proposed, we also intend to make our model more explainable. XAI aims to create a collection of machine learning techniques that produce more explainable models. Using XAI techniques, we wish to provide knowledge about the prediction made by the model so that the user can make an educated decision while trading.

2 State-of-the-Art (Past Work in the Field)

Kalyani et al. [5] in their research are using supervised machine learning as classification and other text mining techniques to check news polarity. The news articles with its polarity score and text converted to TF-IDF vector space are fed to the classifier. Three different classification algorithms (SVM, Random Forest and Naïve Bayes) are implemented to check and improve classification accuracy. Results of all the three algorithms are compared based on accuracy, precision, recall and other model evaluation methods. When comparing the results of all classifiers, SVM classifier performs well for unknown data. The Random Forest algorithm also worked well compared to the Naive Bayes algorithm. Finally, the relationship between news articles and stock prices data is plotted.

Nayak et al. [6] used the historical data from 2003 obtained from Yahoo Finance and used two models to predict the stock trend. One model was built for the prediction of daily stock by considering all the data available on a daily basis. The second model that was built was for monthly prediction of stocks and it considered data available on a monthly basis. Also, two different datasets were used for each model. A historical price dataset was used for the Daily Prediction model and historical data from 2003 obtained from Yahoo finance is used for a monthly prediction model. The dataset was modelled using various models like boosted decision tree, logistic regression and support vector machine. Up to 70% accuracy was observed using the Support Vector Machine.

Hiransha et al. [7] used four types of deep learning architectures, i.e. convolutional neural network (CNN), recurrent neural networks (RNNs), multilayer perceptron (MLP) and long short-term memory (LSTM) for predicting the stock price of a company based on the historical prices available. Here closing price of two different stock markets, National Stock Exchange (NSE) of India and New York Stock Exchange (NYSE) by daily basis, is considered. The model was trained with the stock price of a single company from NSE. The prediction was done for five different companies from both NSE and NYSE. Although the network was trained with NSE data, it was able to predict for NYSE also. This was because both NSE and NYSE markets share some common inner dynamics. The results were then compared with the results of the ARIMA model, and it was inferred that the neural networks outperform the existing linear model (ARIMA).

Vargas et al. [9] in their paper propose an RCNN model to forecast intraday directional movements of the S&P 500 index. The model inputs financial new titles published the day before the prediction day and makes use of seven technical indicators which are extracted from the target series. Each news goes through a two-step process—first, a word2vec model to generate a vector representation of each word and then an average of all the word vectors of that same title is performed. The RCNN model takes advantage of deep learning models: CNN and RNN. The CNN model is used to extract semantic information from texts, while the RNN LSTM model is used to catch the context information and to interpret the stock data features for prediction purposes.

3 LSTM

LSTM stands for Long Short-Term Memory. Hochreiter and Schmidhuber [11] introduced LSTMs that used memory cells and gates to store information for long

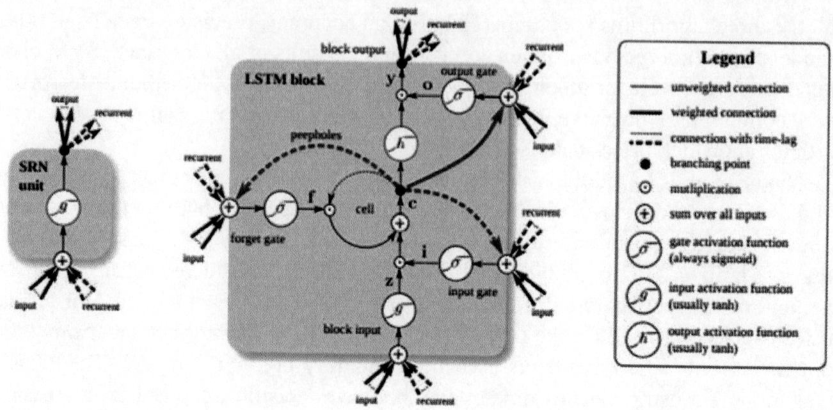

Fig. 1 Detailed schematic of the long short-term memory block (right) as used in the hidden layers of a neural network

periods or to forget unnecessary information [10]. They are capable of learning long-term dependencies, and LSTMs are explicitly designed to handle tasks involving historical context. The ability to learn comes from the memory cells. LSTMs have a chain-like structure making it easier to pass on information. The information is passed on as a state of the cell from one memory cell to another. The output of the network is modulated by the state of these cells.

The architecture of LSTM allows for constant error flow with the help of constant, self-connected units [11]. This flow of error and states is propagated with the help of the three gates: input gate, output gate and forget gate that each LSTM memory cell block is composed of input gates which modulate how much new information is received by the cell. The forget gates determine what amount of information from the previous cell is passed on to the current cell, and they determine what information is relevant and what information needs to be forgotten (Fig. 1).

3.1 Forward Pass

Let *xt* be the input vector at time *t*, *N* be the quantity of LSTM blocks and *M* be the quantity of inputs. Then we get the subsequent weights for an LSTM layer:

- Input weights: $Wz, Wi, Wf, Wo \in R\ N \times M$
- Recurrent weights: $Rz, Ri, Rf, Ro \in R\ N \times N$
- Peephole weights: $pi, pf, po \in R\ N$
- Bias weights: $bz, bi, bf, bo \in R\ N$

Then the vector formulas for a vanilla LSTM layer forward pass can be written as [11]:

$$
\begin{aligned}
&zt = Wzxt + Rzyt - 1 + bz \\
&zt = g(zt) && \text{block input} \\
&it = Wixt + Riyt - 1 + pict - 1 + bi \\
&it = \sigma(it) && \text{input gate} \\
&ft = Wfxt + Rfyt - 1 + pfct - 1 + bf \\
&ft = \sigma(ft) && \text{forget gate} \\
&ct = ztit + ct - 1ft && \text{cell} \\
&ot = Woxt + Royt - 1 + poct + bo \\
&ot = \sigma(ot) && \text{output gate} \\
&yt = h(ct)\,ot && \text{block output}
\end{aligned}
$$

where σ, g and h are pointwise nonlinear activation functions. The logistic sigmoid ($\sigma(x) = 1/1 + e - x$) is employed as a gate activation function and also the hyperbolic tangent ($g(x) = h(x) = \tanh(x)$) is typically used because the block input and output activation function. Pointwise multiplication of two vectors is denoted by.

Backpropagation

The deltas inside the LSTM block are then calculated as [11]:

$$
\begin{aligned}
&\delta yt = \Delta t + RzT\delta zt + 1 + RiT\delta it + 1 + RTf\delta ft + 1 + RTo\delta ot + 1 \\
&\delta ot = \delta yth(ct)\sigma'(ot) \\
&\delta ct = \delta ytoth'(ct) + po\delta ot + pi\delta it + 1 + pf\delta ft + 1 + \delta ct + 1ft + 1
\end{aligned}
$$

Here, Δt is the vector of deltas passed down from the layer above. If E is the loss function it formally corresponds to $\partial E\ \partial yt$, but not including the recurrent dependencies. Then [12]:

$$
\begin{aligned}
&\delta ft = \delta ctct - 1\sigma'(ft) \\
&\delta \bar{i}t = \delta ctzt\sigma'(\bar{i}t) \\
&\delta zt = \delta ctitg'(zt)
\end{aligned}
$$

3.2 Why LSTM?

LSTM networks have internal contextual state cells. These cells act as short- and long-term memory cells, and therefore, the output of the LSTM model network relies on the state of these cells. These memory cells help the model remember historical context as predictions made by the LSTM network are conditioned by past experiences of inputs to the network. This helps us make better predictions. LSTM networks manage to keep the context of information fed by inputs by integrating

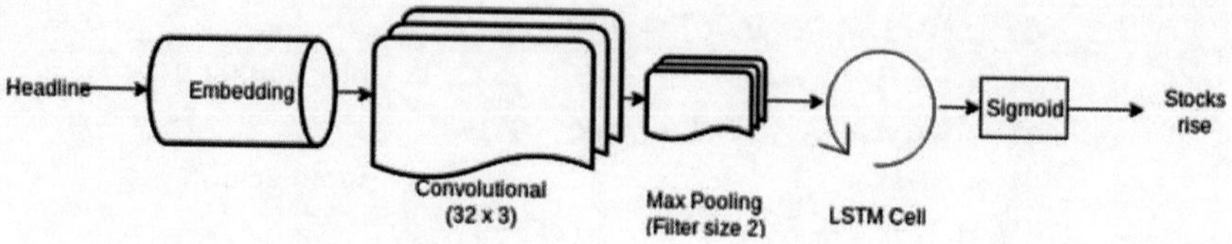

Fig. 2 Internal architecture of the LSTM is

a loop that allows information to flow, in one direction, i.e. from one step to the following.

Similar to humans that read one word after another and can make sense from a sentence, these networks are conditioned by the past experiences of the network's input (Fig. 2).

3.3 *LSTM Versus RNN*

All RNNs have feedback loops within the recurrent layer. This lets them maintain information in 'memory' over time. But due to the gradient of the loss function decays exponentially with time (called the vanishing gradient problem), it becomes difficult to train standard RNNs to help solve problems involving long-term temporal dependencies. In contrast to RNNs, LSTM networks include a 'memory cell' that helps it remember information for longer periods. LSTMs use an additive method, compared to the RNNs multiplicative method, to update the cell states and help us preserve these long-term dependencies. Additionally, the three gates present in an LSTM memory cell help us control the flow and mixing of previous states and inputs. Although the complexity to train the model is a little higher and it has a higher operating cost, it still gives us better controllability. This controllability of tuning our gates as per our requirements helps us in getting more desired results.

4 XAI

We want our model to output not only the prediction part but also the explanation as to why the prediction turned out that way. If our machine makes a prediction why should the user trust the prediction made by the machine? Today, machine learning and artificial intelligence (AI) are exploited to make decisions in many fields like medical, finance and sports. There are cases where machines aren't 100% accurate. And while dealing with a sensitive computation of stock market prediction, utmost care must be taken. So, the user should blindly trust the choice of the machine? How can the user trust AI systems that derive inferences on probable unfair grounds? To solve the problem of trust between the user and artificial intelligence, XAI can be employed. It gives us the reasoning for a prediction made by the model. Mainly, XAI is employed to resolve the black box problem. This 'black box' phase is interpreted by XAI and explained to the users. The user cannot completely depend on the model without a clear understanding of the model and the way the output is achieved. XAI provides a clear understanding of how the model achieved a certain prediction or result. XAI gives a human-understandable explanation for the prediction. Current models enable interpretation but leave it to the user to apply their own knowledge, bias and understanding. After studying various XAI [13] tools, we realized the one which will determine the transparency of our model in the best possible way was LIME.

LIME, short for Local Interpretable Model-Agnostic Explanations [14], is a tool which is used to explain what machine learning model is performing. It is applied to the models which perform behaviour but we do not know how it works. Machine learning models can be explained using plots, for example, linear regression, decision trees; these are some of the classifiers that can be explained using plots (It can only give an explanation for the global, but it cannot give an explanation of any local observation). By using LIME, we can understand what model does, which features it picks on to create or develop prediction. LIME is model-agnostic.

4.1 *Use of XAI*

XAI is used vastly in many fields. But there are only temporal models for XAI [15]. Some examples of XAI is Google's Inception neural network on some arbitrary images. In this case, we keep as explanations the parts of the image that are most positive towards a certain class (Fig. 3).

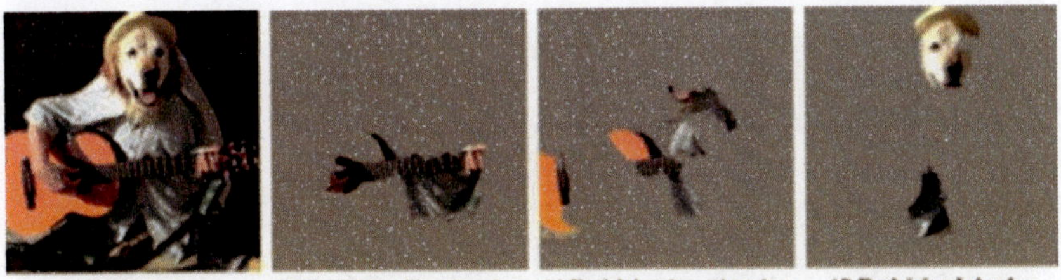

(a) Original Image (b) Explaining *Electric guitar* (c) Explaining *Acoustic guitar* (d) Explaining *Labrador*

Fig. 3 Explaining an image classification made by Google's Inception network, highlighting positive pixels

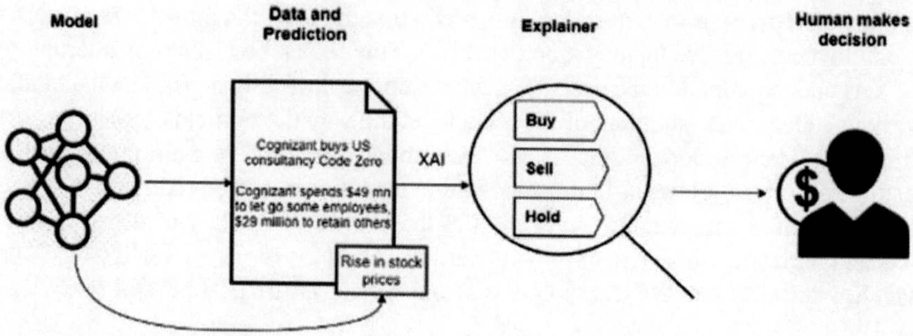

Fig. 4 XAI explaining predictions for every data sample

4.2 *Use in Our Software*

Currently, all models provide a prediction of stock prices and enable interpretation but leave it to the user to apply their knowledge, bias and understanding. This could prove dangerous and lead to a financial loss as supported by the above factors, different people might come up with different explanations for the identical decision. In our paper, we introduce an approach where the model after providing the prediction, also explains the prediction by looking for buzzwords like 'buys', 'sells', 'profit', 'loss', etc., in the news headlines, and also finding out how repeatedly these words have occurred. With this information about the rationale behind the model, the user is now empowered to trust the model and can make a knowledgeable decision about buying, selling or holding the stock. As with LIME, we get a cumulative explanation for features [16]. Our model is not just a black box now and the customers know the insides of the system through just one graph. Thus, we achieve the explanations along with the insights from the data (Fig. 4).

5 Conclusion and Discussion

This paper describes a technique for stock predictions using financial headlines that are modelled over LSTM networks. Our goal was to produce a critical review of the key ideas in the fields of LSTM and XAI that align with stock market prediction. We hope that the references cited cover the major theoretical issues and provide access to the major branches of the literature that deal with similar methodologies and will guide the researcher in interesting research directions. Also, comparisons with RNN give us better inferences for LSTM being a better choice of the network over RNN. Finally, our paper reviews a solution to generate an explanatory prediction along with the predicted values to justify as to why the model has forecasted the given output with the help of XAI tools.

References

1. Vachhani H, Obiadat M, Thakkar A, Shah V, Sojitra R, Bhatia J, Tanwar S (2020) Machine learning-based stock market analysis: a short survey. https://doi.org/10.1007/978-3-030-38040-3_2
2. Titan A (2015) The efficient market hypothesis: review of specialized literature and empirical research. Proc Econ Fin 32:442–449
3. Roondiwala M, Patel H, Varma S (2017) Predicting stock prices using LSTM 6(4):2015–2017
4. Mathews SM (2017) Dictionary and deep learning algorithms with applications to remote health monitoring systems. Univ. Delaware, Newark, DE, USA, Technical Report
5. Kalyani J, Bharathi HN, Rao J (2016) Stock trend prediction using news sentiment analysis. Int J Comput Sci Inf Technol (IJCSIT) 8(3):67–76
6. Nayak A, Pai MM, Pai RM (2016) Prediction models for Indian stock market. Proc Comput Sci 89:441–449
7. Hiransha M, Gopalakrishnan EA, Menon VK, Soman KP (2018) NSE stock market prediction using deep-learning models. Proc Comput Sci 132:1351–1362.https://doi.org/10.1016/j.procs.2018.05.050
8. Qun Z, Xu L, Zhang G (2017) LSTM neural network with emotional analysis for prediction of stock price. Eng Lett 25(2):167–175
9. Vargas MR, de Lima BSLP, Evsukoff AG (2017) Deep learning for stock market prediction from financial news articles. In: 2017 IEEE international conference on computational intelligence and virtual environments for measurement systems and applications (CIVEMSA). IEEE, June 2017
10. Elman JL (1990) Finding structure in time. Cogn Sci 14:179–211
11. Hochreiter S, Schmidhuber J (1997) Long-short term memory. Neural Comput
12. Greff K, Srivastava RK, Koutník J, Steunebrink BR (2017, Oct 4) LSTM: a search space odyssey
13. Radwan S What does explainable AI really mean?
14. Shrikumar A, Greenside P, Shcherbina A, Kundaje A (2016) Not just a black box: learning important features through propagating activation differences
15. Doran D, Schulz S, Besold TR (2017) What does explainable ai really mean? A new conceptualization of perspectives. arXiv:1710.00794
16. Gite S, Khatavkar H, Kotecha K, Srivastava S, Maheshwari P, Pandey N (2021) Explainable stock prices prediction from financial news articles using sentiment analysis. PeerJ Comput Sci 7:e340. https://doi.org/10.7717/peerj-cs.340

Precision Agriculture: Methodologies, Practices and Applications

Sharnil Pandya, Mayur Mistry, Pramit Parikh, Kashish Shah, Gauravsingh Gaharwar, Ketan Kotecha, and Anirban Sur

Abstract IoT-enabled modern agricultural methodologies can change the current agriculture practices by automating the entire process of agriculture from crop management, water irrigation to making better decisions based on real-time monitoring of environmental conditions, soil conditions and landscape conditions. In the recent times, technology-enabled precision agriculture solutions have enabled a paradigm shift from static and manual agriculture methodologies to automated precision-oriented agricultural methodologies using the latest technologies such as Internet of agricultural things, AI-based agricultural analytics, cloud computing and WSN-enabled crop monitoring and control. In the proposed review work, a rigorous and detailed assessment has been conducted to identify the research gaps and analyze the latest technology-enabled PA methodologies and applications. Furthermore, in the proposed review work, we have presented IoAT-based PA model, which is comprised of five layers. The first layer depicts the physical layer devices, second layer describes security protocols, third layer highlights efficient data management practices, fourth layer provides effective irrigation models, and the final layer discusses technology-enabled water management services. In the end, along with

S. Pandya · K. Kotecha
Symbiosis Center for Applied Artificial Intelligence and Symbiosis Institute of Technology (SIT), Symbiosis International (Deemed) University, Pune, India
e-mail: sharnil.pandya@scaai.siu.edu.in

K. Kotecha
e-mail: head@sccai.siu.edu.in

M. Mistry (✉)
Department of Computer Science Engineering, Ganpat University, Ahmedabad, Gujarat, India
e-mail: mayur.mtechbda1703@ict.gnu.ac.in

P. Parikh · K. Shah · G. Gaharwar
Department of Computer Science Engineering, Navrachana University, Vadodara, Gujarat, India

A. Sur
Symbiosis Institute of Technology (SIT), Symbiosis International (Deemed) University, Pune, India
e-mail: sur@sitpune.edu.in

P. K. Singh et al. (eds.), *Proceedings of Second International Conference on Computing, Communications, and Cyber-Security*, Lecture Notes in Networks and Systems 203,
https://doi.org/10.1007/978-981-16-0733-2_12

future directions, we have represented categorical analysis of the conducted review work in the form of graphical results along with gained experiences and learnt lessons.

Keywords Crops irrigation · Precision agriculture (PA) · Internet of agricultural things (IoAT)

1 Introduction and Related Work

By the year 2050, the population of planet Earth is estimated to rise up to a large sum of 9.1 billion inhabitants. The current population of 7.3 billion has looted up almost 2/3rd of Earth's resources [1]. In the last 60 years, due to high demand for commodities, such as food, fuel, fiber, timber and freshwater, more land has been assigned to agriculture than eighteenth and nineteenth century era. The facts tell us that there is currently a clear deficit in natural resources and if the ongoing depletion of resources is not dealt with in the right way, the future generations will have nothing to foster [2]. Water irrigation is a primary concern for the development of agriculture in developing countries. To resolve the issue of water irrigation, fellow researchers have made some efforts by proposing innovative methodologies such as drip irrigation, automated sprinkler technology and smart farming. Smart farming method (also known as precision agriculture (PA)) makes the use of latest technologies such as Internet of agricultural things, big data analytics, cyber–physical systems, intelligent sensing and systems and machine learning methodologies to address the issue of resource management by proving efficient management of available resources [3, 4].

Furthermore, latest technologies have also contributed immensely in solving the problem of water shortage and in the development of innovative irrigation methodologies such as development of a digital system for agriculture, which notifies the farmer of certain uncertainties in advance and helps them tackle these undesirable situations [5, 6]. A fundamental component of the PA system is field mapping. The concept of precision agriculture is not very old, and it was started before 25 years when innovative technologies such as wireless sensor networks and GPS came into existence [7].

PA is an innovative method of collecting, analyzing and reporting data, which plays a crucial role in making crop management decisions [6]. Modern agriculture is currently facing a new era with increasing production needs and reducing availability. The exponential increase in population brings with it the need of a greater amount of production; the ability to control resources accurately allows greater control of expenses/profits and helps to curb the depletion of natural resources such as rivers and soil nutrients [8]. PA allows monitoring of environmental variables in order to have a precise response and detailed visualization of plantation behavior to solve specific problems in each case [9]. As shown in Fig. 1, four main aspects of precision agriculture cycle which comprise all the different techniques used are sense, analyze, predict and act. In the given figure, which is the adaptation of a cloud platform service, is integrated with sensors, which in turn predicts the result with the help of different

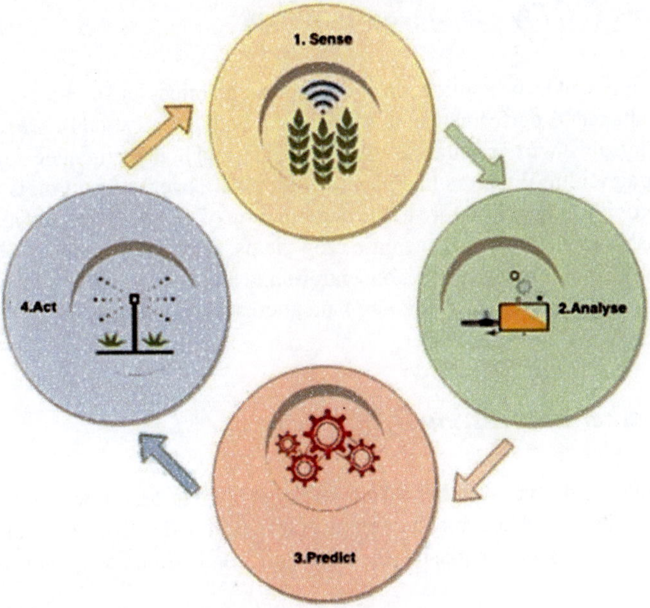

Fig. 1 Architecture of precision agriculture life cycle

models and eventually notifies the farmer which helps them to take necessary actions and further their technology enriched farm. As described in Table 1, four aspects of the precision agriculture cycle define how the processes are carried out and below we see what each individual step clearly achieves, and Table 2 represents a list of used terminologies.

2 1. Precision Agriculture Methodologies and Practices

2.1 Irrigation System

As far as agricultural production is concerned, innovative water irrigation systems are very important in the entire crop management process, which includes preparation of seedbeds, utilization of nutrients, pesticides and insecticides, germination, yield and growth and re-growth of plantation. Highly precise and variable irrigation systems play a critical role in the entire process of PA [10].

2.2 *Land Quality Management System*

Land management is the application of operations, practices and treatments to protect soil and enhance its performance. It is important, both directly and indirectly, to crop productivity and environmental sustainability [11], and management helps us control important agricultural factors like plant health, plant cover and soil moisture which helps to provide a bigger picture to the combination of technologies and agriculture-based sensors to cut costs and increase crop yields, such as pH sensor [12]. The pH sensor is capable to measure the concentration levels of oxygen, ion and hydrogen which will assist farmers in predicting land altercation conditions [13].

2.3 *Weather Forecasting System*

The accurate field level weather information helps in assessing the workability of the farmer's fields and becomes more efficient in pest and disease control, avoiding destruction of crops and increasing the workability and efficiency in their operations [14, 15].

3 Challenges

The increasing need for IoT to interconnect all the components of the visible and virtual world has resulted in businesses and consumers to heavily rely on IoT-based technologies to make advancements in their respective fields.

3.1 *Data Insight*

Increase in efficiency and productivity is the main aim of IoT-induced technologies. Collecting ample amount of data to use it as a utensil to fuel consumer experience is the Holy Grail, which all businesses rely on. The collection of sensor analytics to increase the data expansion promises many benefits in farming, mainly reduction in delays due to external communication, which eventually helps farms situated in rural areas with poor connectivity [16, 17].

3.2 Sustainable Computing

With the growing need for sustainability of resources, the goal of large-scale organizations is to make the whole world a better place. After exploring many applications of IoT technologies, the World Economic Forum (WEF) found that 84% were addressing or could potentially address the UN's Sustainable Development Goals. Sustainable computing helps to conquer some of the world's growing hurdles and tackles those problems by giving solutions, such as smart grid. A smart grid consists of all the latest developing/developed IoT sustainability technologies. It detects and reacts to local changes in usage through an electric supply network that uses digital communications technology.

3.3 Information Security

Information security has a central role in these developments. The spread of knowledge and skills that it engenders worldwide helps foster the essential communication between all those concerned with food production and the environment from research through to the farmers and field workers in the general sense, everywhere [18]. Performing some of the data analytics directly within the farm also has number of benefits from the information security viewpoint. Only a summary of the farm-monitoring data will be shared by third party. This will complicate obtaining specific sensitive information from the data shared.

3.4 Data Analytics

Data analytics assists in making impactful, proactive and profitable decisions that can increase opportunities and efficiencies in the entire process of smart farming. However, data analytics require good data to be successful, and data that is incomplete or is incorrect will provide insights that are not fully analyzed. Data from in-field sensors, collection of input data at each level and economic functions of decisions will continue to be critical for success of data analytics [19].

3.5 Cyber–Physical Systems

Cyber–physical systems will form an integrated environment of cyber and physical objects which assists farmers in real-time monitoring of farms, crop management, soil management by proving variety of information such as requirement of water, required concentration levels of oxygen, hydrogen and ion, harsh environmental conditions and many more.

4 Legacy Systems

4.1 *IoAT-Based Crop Monitoring System*

The 3D coordinates node, connected to the crop identification module, represents various IoAT systems. This system deals with the application characteristic of IoT and exposes the potential it has to improve the precision agriculture techniques. The setback the proposed model possesses is the inclusion of soil, climate and water sensor in the same category as crop identification, nutrient control and energy control.

4.2 *Camera Chessboard Pattern Arrangement*

The main issue with this kind of arrangement is it helps to deal with the superimposing of IoT information with each and every object on the views of the camera. This chessboard pattern arrangement goes through a certain procedure to find out whether any projection errors exist in the placement of cameras. The software, which is used to calibrate these cameras, is a pre-existing MATLAB toolbox code, which helps in finding the camera projection errors. After processing their codes and creating virtual block diagrams of their display and interactions, a graphic of the plant through camera visualization is created, which helps the farmers to virtually glance at their plants every day.

4.3 *Rural Architectural Network*

Rural architecture network is a WSN-enabled network designed to improve crop management practices to take just-in-time crop management decisions [20, 21]. The proposed architecture provides scalability and covers large geographical regions. It uses 6LoWPAN-based WSN network for sensing and actuating related operations.

Advancing from our previous review of a multi-camera arrangement on a compact farm, here we see the scale increase and observe multiple sensors nodes scattered across a large farmland area. These nodes perform the sensing and actuating operations simultaneously, increasing the scalability and the coverage range. The use of IPv6 for this setup has been essential. Not only does it solve the address space and auto-configuration complication, but is also combined with the adaption layer to introduce the 6LoWPAN protocol stack. Small packet size, low bandwidth (between

250/40/20 kbps) and other mesh topology drawbacks are only the start of the disadvantages of the 6LoWPAN network. Although it has its pros, such as energy saving through usage of low-powered listening nodes, the main disadvantage of 6LoWPAN is the lack of application that utilizes 6LoWPAN because of the extensive training that it requires to get accustomed to the technology and the knowledge of IPv6 protocol. Knowledge of stack and the workability of IPv6 is a must for the end users handling this architecture. Routing solutions, performance analysis, the architectural complexity of the proposed CLAD protocol, effects of weather on the 6LoWPAN-based network and the assurance of quality of service (QoS) are all thoroughly provided in the conducted survey to legitimately implement fog and cloud computing technologies in their proposed framework. In the recent times, the usage of IoAT-based applications has increased exponentially due to its low-energy requirements and coverage of large geographical regions. In the last decade, LoRaPAN and LoRaWAN technologies have replaced conventional communications protocols such as Wi-Fi and classic Bluetooth [22–25].

Generally speaking, in the fog computing architectures, IoAT-based devices are connected to heterogeneous devices in close proximity to IoAT users. The major challenges in running IoAT applications in fog computing environment are management of available resources and scheduling of routine tasks [21, 26, 27].

4.4 Aerial View and Central Pivot Irrigation System

Even though this framework is the closest thing, we have to maximize IoT implementation, and many times the final applications of these technologies are not user-friendly. The understanding of the vast possibilities of these technologies is a very important factor in discovering and applying them. Mitigation of the knowledge gap is very important in order to use the experimented technology efficiently and appropriately. With this, privacy issues, data loss and manipulation over longer transmissions are also outstanding issues, which have to be dealt with. This kind of methodology is unique in the uniform distribution of the quantity of water in the entire farmland area [28, 29]. According to the conducted research work, each proposed system has its advantages and disadvantages, and a few major advantages and disadvantages that come with the implementation of this system are for irrigating larger geographical fields. The proposed architecture demands only 60% of water as compared to the conventional irrigation systems with low cost and higher efficiency.

It is imperative that regular servicing and maintenance should be provided to such systems to avoid the possibility of glitches, which requires huge initial capital investment [28, 30]. Again, if the outstanding problems of this proposed model are mitigated and necessary steps such as inclusion of subsidies to farmers are taken to motivate them to make revolutionary changes to their farm and that is when the agriculture sector will start to cultivate in the right direction. Further, we look at a smart water management system, which combines various other technologies, with the goal to further the outreach of IoT technologies and deduce any ongoing problems with the existing architectures. Smart water management system can provide real-time notification of the irrigation and crop conditions to farmers, which may assist farmers in taking just-in-time decisions.

4.5 Proposed Approach

This section discussed an IoAT system, which has been proposed by us to provide an overview of an IoT-induced precision agriculture system, which levels through a hierarchy of steps, customizable services and services which can be replicated for common IoT systems such as this. From minute sensor installation on the farmland to final gross profit of end users, this architecture provides an overview or a blueprint, which can be followed for similar IoT systems.

Layer 1: Physical Layer

IoT physical layer devices such as sensors, actuators and controllers interact and automate the process of fertilization, water irrigation and the utilization of pesticides and insecticides for better crop growth management [31]. The IoAT system combines several conventional technologies such as RFID, NFC, cloud computing, WSNs to automate the crop management processes [32]. WSN and WMSN architectures share information through wireless channels consisting of thousands of interconnected nodes which are used to sense, process and communicate with IoT-integrated technologies, which play a key role in any IoT-induced experiments [33].

These architectures are very precise as they follow a specific blueprint of all the required components, which have to be included for successful transmission of signals [34]. WSN/WMSN has their advantages and disadvantages, and although these modules are the pioneers of node processing, increase in efficiency, ease of use and consistency is always a concern [35]. Moreover, to understand the management of proposed IoT Information models, it is essential to dig deeper into protocol independent object management, which occurs at a conceptual level. In WSN-based systems, WQM is a major concern. Such systems demand reliability, sustainability and timely quality of service to communicate WQM data over longer distances [36–38].

In general, IoAT-enabled WSN systems consist of numerous communication systems which are required for routing, sensors and actuators for transmitting crop-related information, power source backup and user-friendly GUI-based design [39]. The processing block consists of a microcontroller, which helps in logical and arithmetic operations, whose results are stored in the memory. Further, the outcome of the processing block is used by the communication block via a third-party application, so the farmer can make better decisions depending upon their area of interest. The sensor block contains different types of sensors such as humidity sensor, temperature sensor, pH sensor, which ensures that the data taken is reliable and can be analyzed systematically using mathematical algorithms [40]. Generally speaking, the common IoAT systems include three major technologies: (i) sensors and actuators (ii) fog computing and edge computing.

(iii) cloud computing storage. These technologies are capable to provide relevant information to GUI-based devices, and it also interconnects the heterogeneous IoAT architecture [41].

Layer 2: Network Layer

This layer deals with data collection, management, data privacy and security and IoAT device management [42, 43]. Data acquisition, security and management: Protocol and software components for data acquisition are the key characteristics of this layer in addition to security and device management functions. This layer utilizes message brokers such as mosquito. Mosquito is capable to implement IoT-based protocols such as MQTT. MQTT is a lightweight public IoT protocol, which assists IoAT systems in interconnecting the controllers, middle wares, routers, gateways, servers and sensing devices. MQTT is a widely used protocol for resource and energy constraint IoAT applications [44, 45]. In the recent times, technologies such as LoRaWAN have enabled long-range wireless communications up to the range of 25–30 km. It requires less power and lower bandwidth for complex wireless communications. However, it is a difficult task for LoRaWAN to initiate real-time multimedia streaming over lower bandwidth for applications such as weather monitoring [46, 47]. LoRaWAN makes the use of IPV6 protocol to communicate with IoAT-based systems. Furthermore, in the last decade, with the advent of numerous communication protocols Z-Wave, ZigBee, NFC, RFID, Thread, SigFox and IFTTT, interconnection of cyber and physical objects has become easier. Z-Wave and Thread are smart communication protocol which is being used in smart homes which requires to interconnect home appliances such as freeze, fans, lights, ACs, dish washer and many more [48, 49]. NFC is a tag-based technology which is used for identifying various devices and short-range communications. NFC is capable to interconnect and control sensing devices directly via smart phones [50]. ZigBee is a two-way communication protocol which is widely used for applications such as tree routing, vector routing and low-rate data communications [51].

Layer 3: Data Management Layer

This layer deals with data distribution, processing and storage. It utilizes a software management firmware platform such as FIWARE. FIWARE is a platform for the development and global deployment of Internet applications of the future. It provides a totally open, public and free architecture as well as a set of specifications that allow developers, service providers, companies and other organizations to develop products [52, 53]. FIWARE COSMOS is the reference implementation of the FIWARE's Big Data Analysis Generic Enabler. The Big Data Analysis Generic Enabler is intended to deploy means for analyzing both batch and stream data [54, 55].

FIWARE CYGNUS and FIWARE ORION are apache flume-based middle wares which are responsible for managing persistent data storage, which is widely used for managing historical data. It mainly deals with data collection, data processing and context aware information exchange [52]. FIWARE ORION is a c + + implementation of FIWARE platform, which is capable to manage the entire life cycle of data management. It includes functions such as managing queries, routine updates and subscriptions. FIWARE QUANTUM LEAP is a mechanism which can store data in the form of time series such as ngsi-tsdb [52].

Layer 4: Water Irrigation and Distribution Methodologies

Conventional agriculture designs need to make the use of drones and soil sensors to acquire contextual information related to soil conditions. The conventional models also utilize computation intelligence techniques for water quality management and cloud computing analytics for safety or performance sensitive applications [22]. However, in the proposed approach, we have proposed model designs, which can be easily implemented by agricultural practitioners. These approaches utilize numerous analytical methodologies and tools to provide real-time contextual information to agricultural practitioners [56, 57].

(a) **Edge computing:** Edge computing is an efficient data acquiring approach, which is widely used in automation-based IoAT applications to provide real-time analytical computations and notifications related to soil conditions, environmental conditions and landscape-related information [47, 49, 58]. It is a mobile sensing mechanism which provides real-time monitoring of remote locations which will assist agricultural practitioners' in making farmland-related decisions [59]. In the last decade, fellow researchers have proposed diverse methodologies to implement fog and edge computing-based IoAT systems [60].

b) **Irrigation Models**: An irrigation model consists of different irrigation methods, which are used by agriculture practitioners in the recent times. There are five methods to the irrigation procedure: (i) surface irrigation (ii) sprinkler irrigation (iii) drip irrigation (iv) subsurface irrigation and (v) drone-assisted irrigation. In the case of surface irrigation, the water is uniformly distributed over the soil surface due to the effect of gravitational flow. Sprinkle irrigation resembles a raindrop type action in which water is sprayed across the farm field. Drip irrigation is an approach in which water is distributed in the form of drops or small streams. Smart drip irrigation systems are feasible to implement for farmers and incur very less initial cost. In such systems, due to automation, no human intervention is required which also results in mitigation of water wastage [61–63]. Farmers can manage and control such system using a smart phone. In the case of subsurface irrigation approach, the water is supplied below the soil surface. The water is usually given within the plant root zone. Drone-assisted irrigation is an approach which uses IoAT-assisted technologies such as WSNs [64–66]. Agriculture-specific drones help keep a bird's eye view on crop production and help monitor crop growth. Improving the overall farm efficiency, the drones also provide a setup where farmers can monitor their crops periodically, according to their liking. The main issue drones help tackle is that of pesticides [67, 68]. Pests are detected, and multiple areas of land are secured within minutes. The only concern drone-assisted irrigation creates is that of privacy. With no access authority needed for drones, flying over someone's property with a mic and camera attached to it is as easy as said and could a potential privacy violation.

Layer 5: Water Application Services

User-friendly GUI-based designs are provided to farmers and distributors for better understanding and faster access to water application services. After the hierarchy of IoT in agriculture is followed, the final layer brings the distribution of these cultivated services, as well as how well the proposed system meets user requirement and lives up to developer and user expectations [69]. The developers use the final data gathered from this experiment to further improve their IoT ecosystem components, and the users (farmers) which are physically working with this technology on a day-to-day basis benefit from the application of these services and help provide important user experience data towards the implemented experiment. Due to energy constraints of water quality management sensing devices, a reliable energy harvesting methodology is required to manage the power backup requirements. However, the efficient implementation of energy harvesting methodology is still an open issue for fellow researchers [36]. An efficient routing protocol is required to manage smooth routing of information between sensing devices, controllers, routers, gateways and cloud computing-based data storage servers [70]. In the recent times, it has been observed that the land is shrinking gradually which increases the pressure on the natural resources. In such scenarios, latest technologies can assist farmers in getting contextual information of variety of crops for just-in-time crop management decisions [70].

Furthermore, the statistical representation of all reviewed data is shown in the further sections, which gives insight into the detailed conducted survey of latest IoAT-based articles [71, 72]. After reviewing all the processing layers and deriving how an IoT system should function for specific tasks, analysis of the intrinsic connection between IoAT-based systems and data is represented [73]. Analysis is the most valuable step in the decision-making process. All the data generated from IoT devices is only of value if it is subjected to analysis. Actionable insights and meaningful conclusions are what IoAT systems aim to extract whilst implementing a data analytics (DA) process. As shown in Fig. 2, we have done the distribution and classification of the IoAT articles with respect to different categories. These articles are classified into following categories:

(i) technology (ii) applications (iii) challenges (iv) business models (v) future directions and (vi) overview of the conducted survey. Furthermore, the surveyed articles have been divided into three subsections: (i) major category, (ii) sub-category and (iii) sub-sub-category. These classifications will assist in observing intrinsic applications, identifying latest trends and provide a broader sense of architecture distribution. It also assists in identifying the challenges which threaten IoT dispersion and open research issues which steer technology in the right direction for the future [74, 75].

(a) **Technology**—IoT, which stands for Internet of things, is a network, which is embedded with technology and has the ability to collect and exchange data [75]. Like and ecosystem, it links multiple systems with the common denominator of IoT embedded technologies and helps us extend IoT technologies further than our personal computers and mobile devices.

(b) **Applications**—The application layer of this architecture has a lot of potential for expansion. IoT devices are likely to be useful in connection between people themselves and the devices around them [76]. From tracking day-to-day activities of an individual, to monitoring their sleep for more efficient and healthy living, the applications of IoT are limitless and always have a scope of further growth [77, 78].

(c) **Challenges**—The challenges in the IoT field are numerous. In order to successfully implement IoT practices, these challenges have to be dealt with and minimal or completely removed [79, 80]. From security, privacy and legal challenges to challenges, which in general affect all three of these aspects, are to be sought out from this architecture. Encryption issues, identity management, communication compatibility, global accountability issues and ownership of data are the relevant challenges that have to be tackled [81].

(d) **Business Models**—With changes in IoT technologies to increase productivity and efficiency in day-to-day lives, business models also have to be altered to meet IoT expectations [82]. For sustainable growth and future developments in the field of precision agriculture, agricultural-based industries are required to identify sustainable technology-based agriculture designs which can provide real-time monitoring of crops and provide latest soil conditions. The conducted survey identifies the gaps between the conventional agriculture-based systems and innovations adopted by agriculture-based industries [83].

(e) **Future Directions**—Even though forecasting the future of agriculture-based architecture is debatable, the vision of its advancement is the driving force for the cultivation of the Web of things [84, 85]. For integration of smart objects with the real world, the ease of the process of development, increase in integration between devices and overcoming challenges to increase the feasibility of ideas are the prime factors, which have to be met with.

(f) **Overview/Survey**—In this section, we have discussed detailed categorical analysis of IoAT technologies as represented in Fig. 2.

Figure 3 represents a graphical view of the number of papers used under two different publications, respectively, journal papers and conference papers. Figure 4 shows the distribution of all the different sources from which the analysis and review work has been carried out.

This display of data gives us a lot of insight into the existing research which progresses, whose data we have used for analysis and all the concurrent reference models from which our analysis has been derived. Figure 5 represents a world map representation of the conduced analysis of the latest research works.

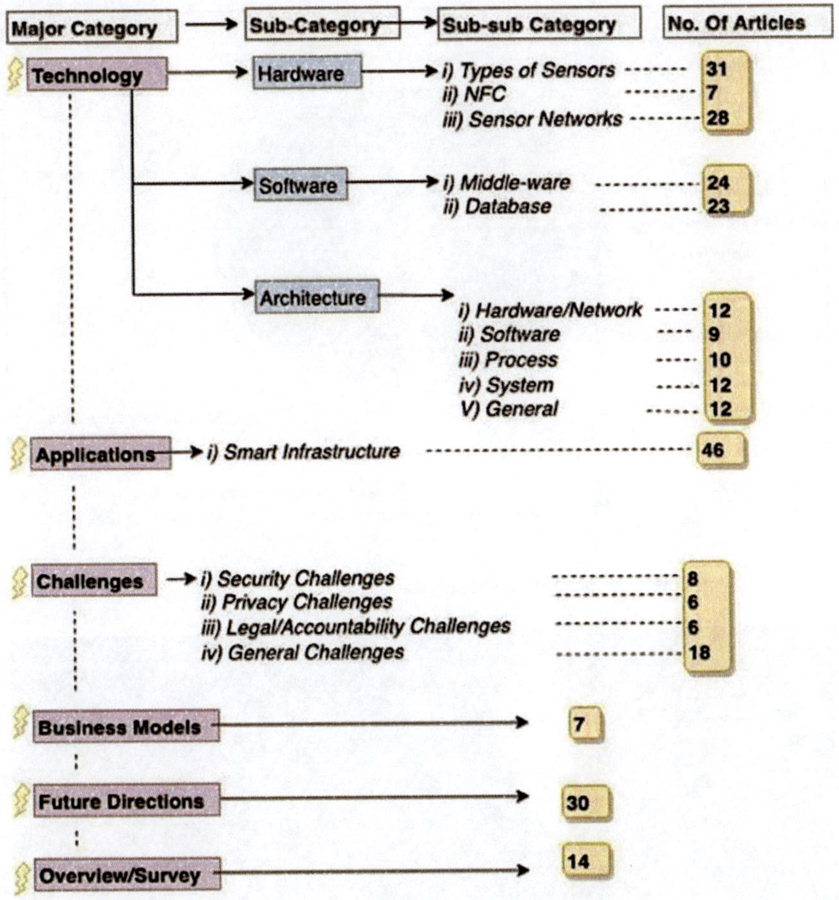

Fig. 2 Statistical display of articles according to their research fields

Fig. 3 Reviewed literature differentiated by publication type

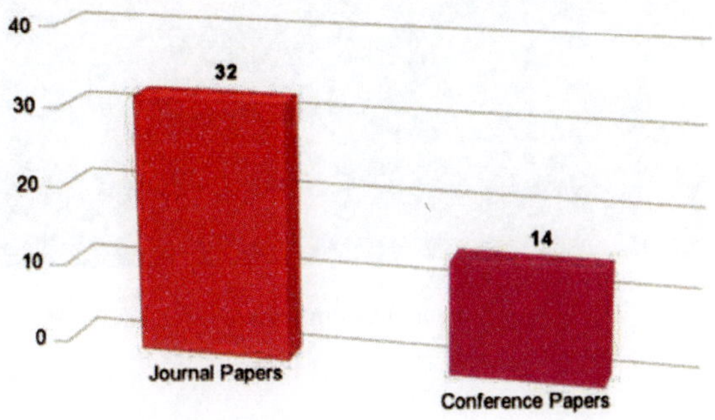

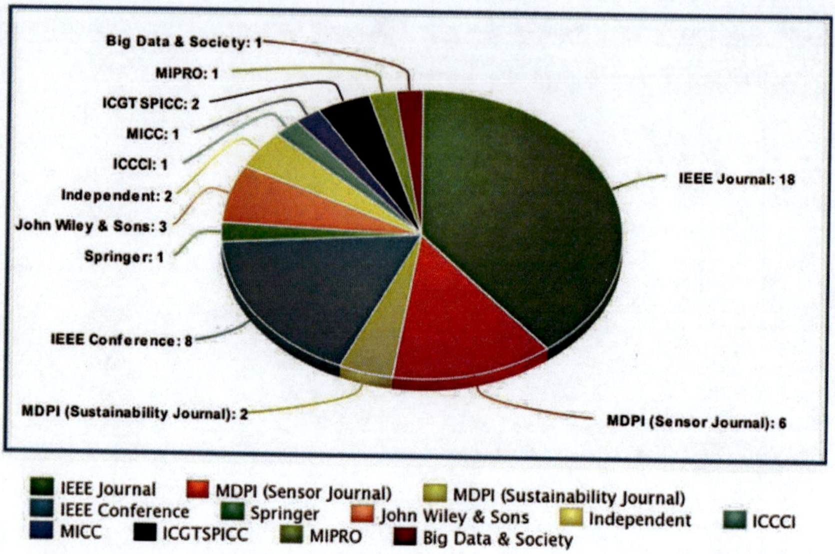

Fig. 4 Pie chart representation of different publications

Fig. 5 Distribution of paper locations based on review work

5 Conclusions

Precision agriculture practices have a significant influence on the modern agricultural practices, techniques and applications. The latest research findings also suggest that appropriate usage of precision-based practices can create a great impact on the productivity incline and sustainability factors [86–88]. This research paper aims to discuss the numerous technology-based irrigation systems, recent industrial practices and innovations, various precision agriculture-based platforms along with the discussion of a newly introduced approach. In this research work, numerous case studies were also discussed to assess the effectiveness of the legacy systems.

Most of the research done in this domain aims to focus on the remote monitoring and control of various agricultural crops, its resource requirements, such as desired water level, quantity of insecticides and pesticides, usage of fertilizers to identify the vegetation indices. However, we have made a genuine attempt to not only analyze the existing methodologies but have proposed an effective model to achieve better results.

Acknowledgements **Conflicts of Interest** The authors declare no conflicts of interest.

Authors Contribution S. Pandya, K. Shah, P. Parikh designed and conceptualized the manuscript. G. Gaharwar did the proof-reading, editing and English grammar check. The manuscript was written jointly by all the authors.

References

1. Cadavid H, Garson W (2018) Towards a smart farming platform: from IoT-based crop sensing to data analytics. Springer, Berlin. https://doi.org/10.1007/978-3-319-98998-3_19. Radford T. The Guardian. [Internet]. 2005 Available from: https://www.theguardian.com/science/2005/mar/30/environment.research
2. Nandurkar S, Thool V, Thool R (2014) Design and development of precision agriculture system using wireless sensor network. In: International conference on automation, control, energy and systems (ACES), Hooghly
3. Andrew R, Malekian R, Bogatinoska D (2018) IoT solutions for precision agriculture. In: MIPRO, Opatija
4. Benyezza H, Bouhedda M (2018) Smart irrigation system based ThingSpeak and Arduino. In: International conference on applied smart systems, ICASS, Médéa
5. Mat I, Kassim M, Harun A (2015) Precision agriculture applications using wireless moisture sensor network. In: IEEE 12th Malaysia international conference on communications, Kuching
6. Fountas S, Aggelopoulou K, Gemtos T (2016) Precision agriculture: crop management for improved productivity and reduced environmental impact or improved sustainability. In: Supply chain management for sustainable food networks
7. Miles C (2019) The combine will tell the truth: on precision agriculture and algorithmic rationality. Big Data Soc, pp 1–12
8. Zhang L, Dabipi I, Brown W (2018) Internet of things applications for agriculture. Wiley
9. Ghayvat H, Mukhopadhyay S, Gui X, Suryadevara N (2015) WSN-and IoT-based smart homes and their extension to smart buildings. Sensors 15:10350–10379

10. Patil K, Kale N (2016) A model for smart agriculture using IoT. In: International conference on global trends in signal processing, information computing and communication. IEEE, New York
11. Ananthi N, Divya J, Divya M, Janani V (2017) IoT based smart soil monitoring system for agricultural production. IEEE, New York
12. González-Teruel J, Torres-Sánchez R, Blaya-Ros P, Toledo-Moreo A, Jiménez-Buendía M, Soto-Valles F (2019) Design and calibration of a low-cost SDI-12 soil moisture sensor. Sensors 19:491. https://doi.org/10.3390/s19030491
13. Cambra C, Sendra S, Lloret J, Lacuesta R (2018) Smart system for bicarbonate control in irrigation for hydroponic precision farming. Sensors 18
14. Kumar R, Dharwadkar N (2018) IoT based low-cost weather station and monitoring system for precision agriculture in India. IEEE, New York
15. Bhakta I, Phadikar S, Majumder K (2019) State of the art technologies in precision agriculture: a systematic review. J Sci Food Agric
16. Cloudscene. Cloudscene. [Internet]. 2018 Available from: https://cloudscene.com/news/2018/05/internet-of-things-iot/
17. Pflaum A, Gölzer P (2018) The IoT and digital transformation: toward the data-driven enterprise. IEEE Comput Soc 18(1536–1268):5
18. Gupta B, Quamara M (2018) An overview of Internet of Things (IoT): architectural aspects, challenges, and protocols. Wiley
19. Balafoutis A, Beck B, Fountas S, Vangeyte J, Van der Wal T, Soto I, Gómez-Barbero M, Barnes A, Eory V (2017) Precision agriculture technologies positively contributing to GHG emissions mitigation, farm productivity and economics. Sustainability
20. Ahmed N, De D, Hussain I (2018) Internet of things (IoT) for smart precision agriculture and farming in rural areas. IEEE Internet Things J 5. https://doi.org/10.1109/JIOT.2018.2879579
21. Naha R, Garg S, Georgakopoulos D, Jayaraman P, Gao L, Xiang Y, Ranjan R (2016) Fog computing: survey of trends, architectures, requirements, and research directions. IEEE Access 4:2169–3536
22. Sarker V, Queralta J, Gia T, Tenhunen H, Westerlund T (2019) A survey on LoRa for IoT: integrating edge computing. In: Fourth international conference on fog and mobile edge computing
23. Raza U, Kulkarni P, Sooriyabandara M (2016) Low power wide area networks: an overview. IEEE, New York
24. Ismail D, Rahman M, Saifullah A (2019) Low-power wide-area networks: opportunities, challenges, and directions. IEEE, New York
25. Wixted A, Kinnaird P, Larijani H, Tait A, Ahmadinia A, Strachan N (2016) Evaluation of LoRa and LoRaWAN for wireless sensor Network. IEEE, New York. 16. 978-1-4799-8287-5
26. Patel P, Bhatia J (2012) Nirma University International Conference on
27. Bhatia J, Kakadia P, Bhavsar M, Tanwar S (2019) SDN-enabled network coding based secure data dissemination in VANET environment. IEEE Int Things J
28. Shilpa A, Muneeswaran V, Rathinam D (2019) A precise and autonomous irrigation system for agriculture: IoT based self propelled center pivot irrigation system. In: 5th international conference on advanced computing & communication systems
29. Bhatia J, Shah B (2013) Int J Adv Eng
30. Review on variants of reliable and security aware peer to peer content distribution using network coding
31. Chen W, Lin Y, Lin Y, Chen R, Liao J (2018) AgriTalk: IoT for precision soil farming of turmeric cultivation. IEEE, New York
32. Elijah O, Rahman A, Orikumhi I, Leow C (2018) An overview of internet of things (IoT) and data analytics in agriculture: benefits and challenges. IEEE Internet Things J 5:2327–4662
33. Premkumar A, Monishaa P, Thenmozhi K, Amirtharajan R, Praveenkumar P (2018) IoT assisted automatic irrigation system using IoT assisted automatic irrigation system using wireless sensor nodes. In: International conference on computer communication and informatics. IEEE, New York

34. Patel S, Singh N, Pandya S (2017) IoT based smart hospital for secure healthcare system. Int J Recent Innov Trends Comput Commun
35. Pandya SP, Prajapati MR, Thakar KP, Assessment of training needs of farm women. Guj J Ext Edu 25(2):169–171
36. Olatinwo S, Joubert T (2019) Enabling communication networks for water quality monitoring applications: a survey. 7:100332 (IEEE, New York)
37. Pandya S, Sur A, Kotecha K (2020) Smart epidemic Tunnel-IoT based sensor-fusion assistive technology for COVID-19 Disinfection. Emerald
38. Pandya S, Ghayvat H, Sur A, Awais M, Kotecha K, Saxena S, Jassal N, Pingale G (2020) Pollution weather prediction system: smart outdoor pollution monitoring and prediction for healthy breathing and living. Sensors 20:5448
39. Dholu M, Ghodinde K (2018) Internet of things (IoT) for precision agriculture application. In: International conference on trends in electronics and informatics. IEEE, New York
40. Naik N, Shete V, Danve S (2016) Precision agriculture robot for seeding function. IEEE, New York
41. Chang C, Srirama S, Buyya R (2019) Internet of things (IoT) and new computing paradigms. Wiley
42. Cohen JM, Pandya S, Krasenbaum LJ, Thompson SF (2020) A real-world perspective of patients with episodic migraine or chronic migraine prescribed AJOVY in the United States. In: Headache
43. Barot V, Kapadia V, Pandya S (2020) QoS enabled IoT based low cost air quality monitoring system with power consumption optimization. Cybernet Inform Technol
44. Shin S, Chuang C, Huang H (2016) A security framework for MQTT. In: IEEE conference on communications and network security
45. Ghayvat H, Pandya S, Patel A (2019) Proposal and preliminary fall-related activities recognition in indoor environment. In: 2019 IEEE 19th international conference on communication technology (ICCT)
46. García S, Larios D, Barbancho J, Personal E, Mora-Merchán J, León C (2019) Heterogeneous LoRa-based wireless multimedia sensor network multiprocessor platform for environmental monitoring. Sensors 19:3446. https://doi.org/10.3390/s19163446
47. Samani MD, Karamta M, Bhatia J, Potdar MB (2016) Intrusion detection system for DoS attack in cloud. Int J Appl Inform Syst
48. Linh An P, Kim T (2018) A study of the Z-wave protocol: implementing your own smart home gateway. IEEE, New York, vol 18. 978-1-5386-6350-9
49. Review on various security threats & solutions and network coding based security approach for VANET
50. Leikanger T, Schuss C, Häkkinen J (2017) Near field communication as sensor to cloud service interface, vol 17. IEEE, New York, 978-1-5090-1012-7
51. Liu Y, Qian K (2016) A novel tree-based routing protocol in ZigBee wireless networks, vol 16. IEEE, New York. 978-1-5090-1781-2
52. Martínez R, Pastor J, Álvarez B, Iborra A (2016) A Testbed to evaluate the FIWARE-based IoT platform in the domain of precision agriculture. Sensors 16:1979. https://doi.org/10.3390/s16111979
53. Akbarzadeh S, Ren H, Pandya S, Chouhan A, Awais M (2019) Smart aging system
54. Carnevale L, Galletta A, Fazio M, Celesti A, Villari M (2018) Designing a FIWARE cloud solution for making your travel smoother: the FLIWARE experience. In: IEEE 4th international conference on collaboration and internet computing
55. Ghayvat H, Pandya S (2018) Wellness sensor network for modeling activity of daily livings–proposal and off-line preliminary analysis. In: 2018 4th international conference on computing
56. Yu S, Park K, Park Y (2019) A secure lightweight three-factor authentication scheme for IoT in cloud computing environment. Sensors 19:3598. https://doi.org/10.3390/s19163598
57. Awais M, Kotecha K, Akbarzadeh S, Pandya S (2018) Smart home anti-theft system
58. Shirazi S, Gouglidis A, Farshad A, Hutchison D (2017) The extended cloud: review and analysis of mobile edge computing and fog from a security and resilience perspective, vol 35, p 11. IEEE, New York. 0733-8716

59. Sarangi S, Naik V, Choudhury S, Jain P, Kosgi V, Sharma R, Bhatt P, Srinivasu P (2019) An affordable IoT edge platform for digital farming in developing regions. IEEE, New York
60. Satyanarayanan M (2017) The emergence of edge computing. IEEE Comput Soc 17. 0018-9162
61. Math A, Ali L, Pruthviraj U (2018) Development of smart drip irrigation system using IoT, vol 18. IEEE, New York. 978-1-5386-5323-4
62. Patel M, Pandya S, Patel S (2017) Hand gesture based home control device using IoT. Int J Adv Res
63. Pandya S, Yadav AK, Dalsaniya N, Mandir V, Conceptual study of agile software development
64. Wandra K, Pandya S (2014) Centralized timestamp based approach for wireless sensor networks. Int J Comput Appl
65. Sur A, Pandya S, Sah RP, Kotecha K, Narkhede S (2020) Influence of bed temperature on performance of silica gel/methanol adsorption refrigeration system at adsorption equilibrium. Particulate Sci Technol
66. Cohen JM, Pandya S, Tangirala K, Krasenbaum LJ (2020) Treatment patterns and characteristics of patients prescribed AJOVY, Emgality, or Aimovig. In: Headache
67. Garg D, Goel P, Pandya S, Ganatra A, Kotecha K (2002) A deep learning approach for face detection using YOLO. In: 2018 IEEE Punecon
68. Sur A, Sah RP, Pandya S (2020) Milk storage system for remote areas using solar thermal energy and adsorption cooling. Mater Today: Proc
69. Swarndeep SJ, Pandya S (2016) Implementation of extended K-Medoids algorithm to increase efficiency and scalability using large datasets. Int J Comput Appl
70. Pandithurai O, Aishwarya S, Aparna B, Kavitha K (2017) Agro-tech: a digital model for monitoring soil and crops using internet of things (IOT), vol 17. IEEE, New York. 978-1-5090-4855-7
71. Bhola YO, Socha BN, Pandya SB, Dubey RP, Patel MK (2019) Molecular structure, DFT studies, Hirshfeld surface analysis, energy frameworks, and molecular docking studies of novel (E)-1-(4-chlorophenyl)-5-methyl-N′-((3-methyl-5-phenoxy-1-phenyl-1H-pyrazol-4-yl) methylene)-1H-1, 2, 3-triazole-4-carbohydrazide. Mol Crystals Liquid Crystals
72. Patel WD, Pandya S, Koyuncu B, Ramani B, Bhaskar S (2019) NXTGeUH: LoRaWAN based NEXT generation ubiquitous healthcare system for vital signs monitoring & falls detection. In: 2018 IEEE Punecon
73. Dandvate HS, Pandya S (2016) New approach for frequent item set generation based on Mirabit hashing algorithm. In: 2016 International conference on inventive
74. Ghayvat H, Pandya S, Shah S, Mukhopadhyay SC, Yap MH, Wandra KH (2016) Advanced AODV approach for efficient detection and mitigation of wormhole attack in MANET. In: 2016 10th international conference on sensing technology (ICST)
75. Pandya S, Shah J, Joshi N, Ghayvat H, Mukhopadhyay SC, Yap MH (2016) A novel hybrid based recommendation system based on clustering and association mining. In: 2016 10th international conference on sensing technology (ICST)
76. Shah JM, Kotecha K, Pandya S, Choksi DB, Joshi N (2017) Load balancing in cloud computing: methodological survey on different types of algorithm. In: 2017 international conference on trends in electronics and informatics (ICEI).https://doi.org/10.1109/ICOEI.2017.8300865
77. Wandra KH, Pandya S (2012) A survey on various issues in wireless sensor networks. Int J Sci Eng
78. Swarndeep Saket J, Pandya S (2016) Implementation of extended K-Medoids algorithms to increase efficiency and scalability using large dataset. Int J Comput Appl
79. Joshi N, Kotecha K, Choksi DB, Pandya S (2018) Implementation of novel load balancing technique in cloud computing environment on computer communication and informatics (ICCCI)
80. Patel W, Pandya S, Mistry V (2016) i-MsRTRM: developing an IoT based intelligent medicare system for real-time remote health monitoring. In: 2016 8th International Conference on Computational Intelligence and Communication Networks (CICN)
81. Aagaard A, Presser M, Andersen T (2019) Applying Iot as a leverage for business model innovation and digital transformation, vol 19. IEEE, New York. 978-1-7281-2171-0

82. Saket S, Pandya S (2016) An overview of partitioning algorithms in clustering techniques
83. Chandra N, Khatri S, Som S (2019) Business models leveraging IoT and cognitive computing, vol 19. IEEE, New York. 978-1-5386-9346-9
84. Pandya S, Ghayvat H, Kotecha K, Awais M, Akbarzadeh S, Gope P (2018) Smart home anti-theft system: a novel approach for near real-time monitoring and smart home security for wellness protocol. Appl Syst Innov 1(4):42
85. RR Patel, SP Pandya, PK Patel (2016) Characterization of farming system in North West agro climatic Zone of Gujarat State. Guj J Ext Edu 27(2):206–208
86. Patel NR, Kumar S (2017) Enhanced clear channel assessment for slotted CSMA/CA in IEEE 802.15.4. Wireless Pers Commun 95:4063–4081
87. Patel NR, Kumar S (2018) Wireless sensor networks' challenges and future prospects. In: 2018 International conference on system modeling & advancement in research trends (SMART), Moradabad, India, pp 60–65
88. Ghayvat H, Awais M, Pandya S, Ren H, Akbarzadeh S, Chandra Mukhopadhyay S, Chen C, Gope P, Chouhan A, Chen W (2019) Smart aging system: uncovering the hidden wellness parameter for well-being monitoring and anomaly detection. Sensors 19:766
89. Ashwini BV (2018) A study on smart irrigation system using IoT for surveillance of crop-field. Int J Eng Technology7:370–373
90. Phupattanasilp P, Tong S (2019) Augmented reality in the integrative internet of things (AR-IoT): application for precision farming. Sustainability 11:2658. https://doi.org/10.3390/su11092658
91. Whitmore A, Agarwal A, Da Xu L (2014) The internet of things—a survey of topics and trends. Springer, Berlin
92. Pandya S, Ghayvat H, Kotecha K, Yep MH, Gope P (2018) Smart home anti-theft system: a novel approach for near real-time monitoring. In: Smart home security and large video data handling for wellness protocol

Predicting Customer Spent on Black Friday

Ashish Arora, Bhupesh Bhatt, Divyanshu Bist, Rachna Jain, and Preeti Nagrath

Abstract The paper provides insights into the expenditures of consumers on a Black Friday. On the day of Black Friday, most of the retail shops are hugely crowded, therefore it becomes very difficult to control the crowd and to have proper stocks of a variety of products. This study analyzes the shopping pattern based on various categories like age, occupation, marital status, city, etc. This study centers around the field of forecast models to build up an exact and efficient calculation to dissect the client spending before and yield the future going through of the clients with the same highlights. In this study, a forest regressor is used to predict the expenditure of consumers. Further, this study talks about the information prehandling and perception procedures utilized to achieve the ideal outcomes. With the help of this study, any retail store participating in Black Friday can improve its efficiency and prepare himself to handle consumers.

Keywords Root mean square error · Machine learning · Regression · USA

A. Arora (✉) · B. Bhatt · D. Bist
Department of ECE, Bharati Vidyapeeth's College of Engineering, New Delhi 110063, India
e-mail: ashisharora111122@gmail.com

B. Bhatt
e-mail: bhupeshbhatt8222@gmail.com

D. Bist
e-mail: dbist411@gmail.com

R. Jain · P. Nagrath
Department of CSE, Bharati Vidyapeeth's College of Engineering, New Delhi 110063, India
e-mail: rachna.jain@bharatividyapeeth.edu

P. Nagrath
e-mail: preeti.nagrath@bharatividyapeeth.edu

P. K. Singh et al. (eds.), *Proceedings of Second International Conference on Computing, Communications, and Cyber-Security*, Lecture Notes in Networks and Systems 203,
https://doi.org/10.1007/978-981-16-0733-2_13

1 Introduction

The day after Thanksgiving, otherwise called Black Friday, started from the USA and now it is celebrated in different parts of the world [1]. This deal comes just once consistently and marks the Christmas shopping season that pursues from the Friday Thanksgiving Day and proceeds until December 24, the day preceding Christmas [2]. The day is among the busiest day of shopping. The retailer gives an enormous measure of markdown and offers on a wide range of things [3]. This term was initially used to describe a financial crisis that occurs in the USA gold market on September 24, 1869. The most usually perennial story behind the Black Friday tradition links it to retailers. Here 'black' signifies 'profit' and 'red' means 'loss'. During the whole year, the retailers operate in losses but on this day they make huge profits [4]. Deals are so high for Black Friday that it has become an indispensable day for stores and the economy by and generally 30% of all the yearly retail deals happen as of now [5]. This prediction is primarily based on different factors like age group, marital status, occupation, etc. [6]. ShopperTrak reports that over the last six weeks (Oct. 3 - Nov. 13) shopper traffic has increased 3.2 percent compared to the same period last year as earlier sales and promotions prompted consumers to visit various retail locations and spend. Additionally, retailers received a rather unexpected lift from Veteran's Day – which accounted for a 10 percent rise in enclosed malls compared to 2009 – helping boost overall traffic performance early in the holiday season [7]. "Despite various economic pressures, consumers have remained resilient and found a way to spend throughout 2010, although not at the level retailers experienced prior to the depths of the recession," said Bill Martin, co-founder of ShopperTrak.

This study is done in this manner: Section 1 is the introduction of the paper, In Sect. 2 there is data preprocessing technique. Section 3 is the visualization of data and finding out hidden patterns present in the data. Section 4 demonstrates the model applied to the dataset. Segment 5 shows the distinctive machine learning techniques utilized alongside its accuracy and execution (performance). Section 6 is the literature review of the paper.

2 Related Work

There is a lot of related work on the dataset. Wu [8] has applied various models and compared the performance, where complex models like neural networks as well as simple models like regression both are used. The XG boost gave the better result on this dataset than others. Majumder [5] has also used a random forest regressor to predict the price with good visualization of data for finding interesting data trends. Trung [9] has performed Black Friday sale prediction through extreme gradient boosted trees and our model has better accuracy than their random forest model. The author has concluded that the utilizations of bagging and boosting techniques can accomplish incredible performance and be additionally improved by a legitimate combination of models' hyperparameters tuning. Thomas et al. [10] did an exploratory analysis of Black Friday consumption rituals and data was collected by 38 interviews. The discoveries of this examination show that Black Friday customers plan for the ritual by advertisements and deliberately outlining their arrangements for the day [3]. Sumit Kalra had used four models and compared their performance nicely which are Xgboost T fidfT transformer, Xgboost + T fidf Transformer, Extra TreesRegressor, and various test sizes of the test case are taken. Maharjan [11] has given full investigation of on Black Friday sale featuring using Apriori algorithm. The strategy found interesting patterns with regard to client's purchasing behaviors from the datasets executing the affiliation rules for buyer's Web-based purchasing conduct which will help the retailers to configure suitable advertising techniques for selling their items on the Web. Milavec [4] analyzed consumer's misbehavior on Black Friday, and various factors of consumer misbehavior were included in the study [12]. Mixed methods research here to refer to all procedures collecting and analyzing both quantitative and qualitative data in the context of a single study (sensu lato Tashakkori and Teddlie 2003) and this method used to better analysis of data [13]. In recent years, the Linked Open Data approach initiated by the World Wide Web Consortium has received increasing attention for Data analysis and Data visualization [14]. Researcher used Different models and technique for better data analysis and visualization [15]. Ability to recognize and track patterns in data help businesses shift through the layers of seemingly unrelated data for meaningful relationships. Through this analysis it becomes easy for the online retailers to determine the dimensions that infl uence the uptake of online shopping and plan effective marketing strategies. This paper builds a roadmap for analyzing consumer's online buying behaviour with the help of Apriori algorithm.

3 Proposed Approach

3.1 Preprocessing

It is perceived as the beginning of the busiest shopping seasons in a year. There are 12 columns and 537,578 rows in the dataset. The dataset gives the information about the money the customer spent on the sales depending on various features like age, occupation, stay in the current city, marital status, etc. The age group in the data is given in some range. The marital status is either 0 or 1. 0 represents unmarried and 1 represents married. There are three different types of product category and the purchases column gives the detail of the money spent. There are some empty data in the dataset [5].

During this analysis, it is found that the age group between 25 and 35 purchased more as compared to any other group [8].

Data

Variable	Definition
	User ID
Product_ID	Product ID
Gender	Sex of User
Age	Age in bins
Occupation	Occupation (Masked)
City_Category	Category of the City (A,B,C)
Stay_In_Current_City_Years	Number of years stay in current city
Marital_Status	Marital Status
Product_Category_1	Product Category (Masked)
Product_Category_2	Product may belongs to other category also (Masked)
Product_Category_3	Product may belongs to other category also (Masked)
Purchase	Purchase Amount (Target Variable)

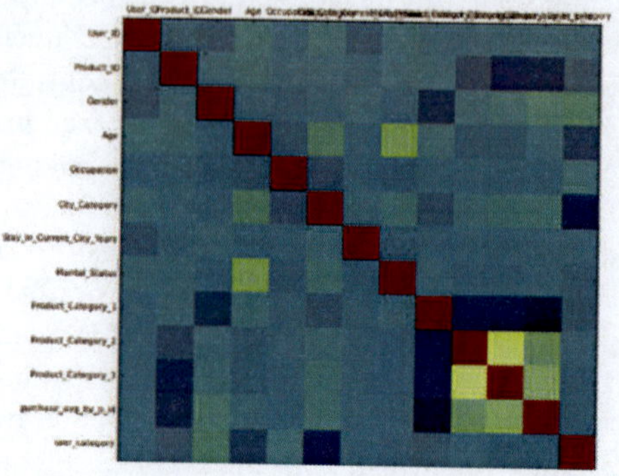

As there was a lot of empty data in the dataset, they were filled with the average of their column's data. The column where data which was in a range (age) was converted to classes by label encoding. The user ID was removed as it was not helpful in the prediction of the prices. The data table after preprocessing looks like this. With the help of pandas and sklearn libraries, we modified our dataset [3].

City_category	Stay_In_Current_City_Years	Marital_Status	Product_Category_1 Product
0	2	0	2
0	2	0	0
0	2	0	11
0	2	0	11
2	4	0	7

The data present in the dataset usually contains multiple labels. All labels need not be numerical. Some labels also contain non-numerical data which can be proved to be very difficult to work with. This includes words, range, etc. They are generally used to make the data understandable for humans. So, it is important to change these non-numerical data to numerical mapping so that processing can be done easily, that is, where label encoding comes handy.

3.1.1 Example

An attribute having output classes 0–17, 17–35, 20–30. On applying Label Encoding to this column, 0–17 is replaced with 0, 17–35 is replaced with 1, and 20–30 is replaced with 2. With this, it very well may be concluded that 20–30 class has higher priority than 17–35 or 0–17. But in reality, there is no connection between these classes here.

Now our data is ready on which we can apply the different supervised algorithm to predict the result.

Data mapping to numbers [2]		
Gender	M	1
Gender	F	2
City	A	1
City	B	2
City	C	3
Married	M	1
Unmarried	U	2
Age	0–17	17
Age	18–25	25
Age	26–35	35
Age	36–45	45
Age	46–50	50
Age	51–55	55
Age	55+	57

3.2 *Data Visualization*

We have used jupyter notebook for training, testing, and analysis of the dataset. Pandas library is utilized for reading and altering CSV records. With the help of matplotlib library, visualization of data becomes very easy. Matplotlib is the library of Python that is used to plot various types of charts like scatter plots, bar charts, heatmap Histograms, power spectra, errorcharts [3], etc. it is easy to use and has a variety of functionality. Seaborn is also used for the analysis of data and used for statistics and graphical plotting in Python. These libraries are important for data visualization and to know hidden patterns from data.

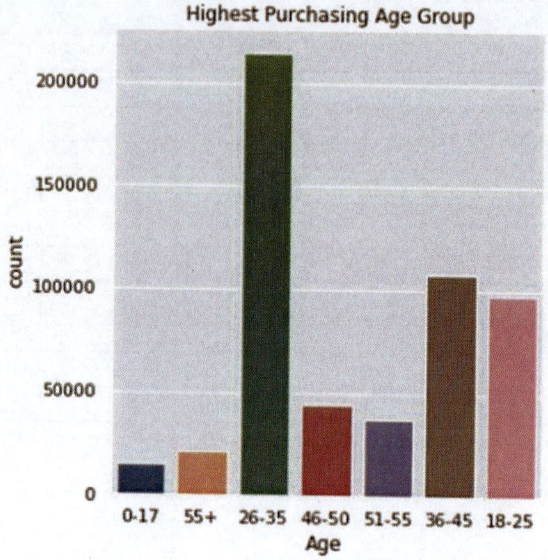

From the analysis of the data, we get that the highest purchasing age group is 26–35 and there was. Also, the number of purchases done by my male is greater than women.

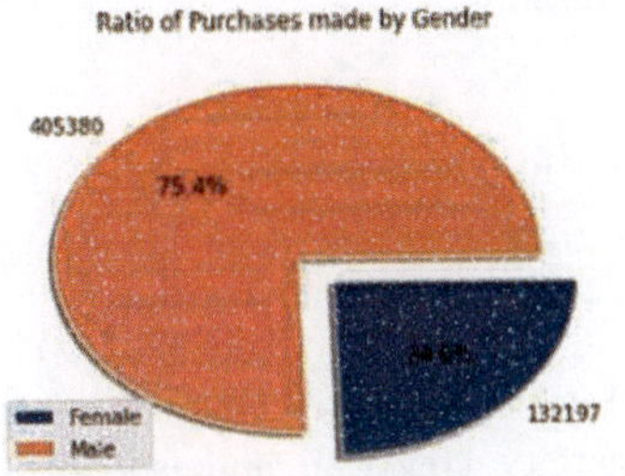

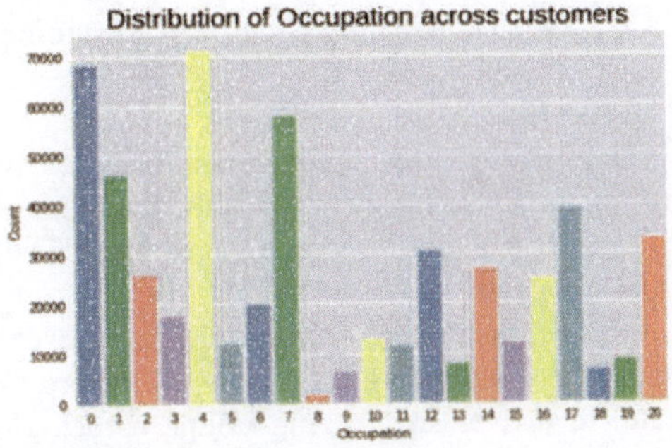

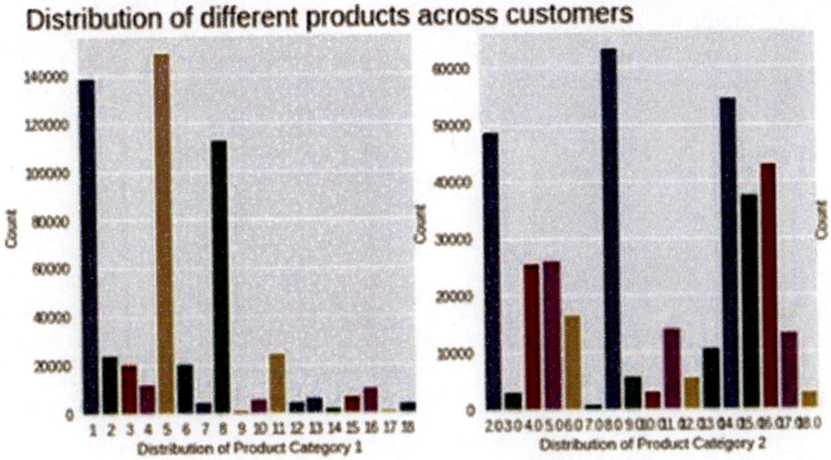

3.3 *Random Forest Regressor Model*

The model which we applied to this dataset is a random forest regressor. Although other models were applied (like linear regression, Decision Tree, Gradient Boosting Algorithm and SVM), the best possible model with best accuracy from them is Random Forest Regressor. A random forest is an outfit method equipped for performing both regression and grouping errands with the utilization of various choice trees with the use of **bagging**. Bagging, in the random forest, is a technique, includes preparing every choice tree on an alternate information test where examining is finished with substitution [5].

3.3.1 Two Mainstream Groups of Ensemble Methods

BAGGING
A few estimators have assembled autonomously on subsets of the information and their expectations have arrived at the midpoint of [9].

Bagging can diminish variance with practically zero impact on inclination.

BOOSTING

Base estimators are fabricated consecutively. Each resulting estimator centers around the shortcomings of the past estimators. Generally, a few frail models 'collaborate' to deliver an amazing troupe model. Boosting can decrease bias without bringing about a higher difference.

E.g., Gradient boosted trees, AdaBoost.

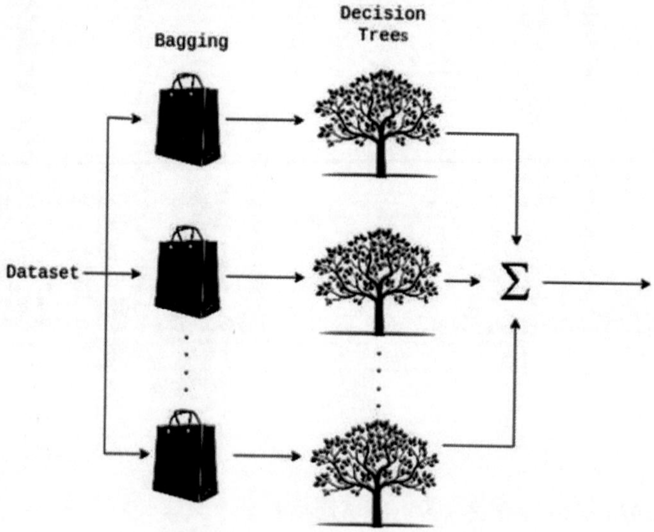

Basic Idea of Random Forest algorithm

The main motive behind this is to add multiple decision trees rather than getting output from a single tree.

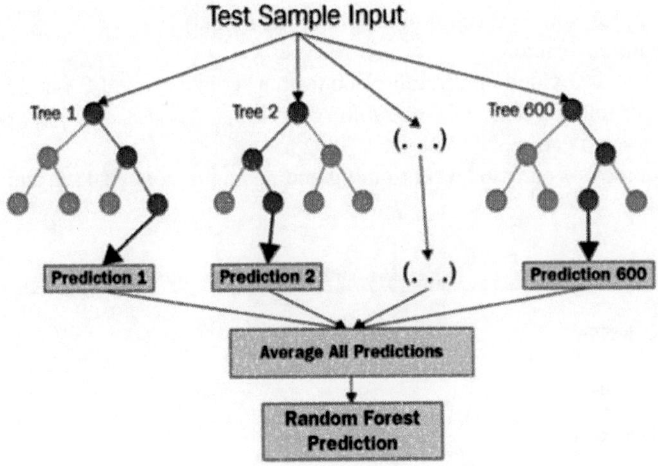

Random Forest Regressor

3.3.2 Main Features of the Algorithm

Decrease in overfitting: by averaging a few trees, the danger of overfitting is the exact moment.

Less variance: When multiple trees are used, it reduces the chance of slipping across a single tree that may not do well due to the relationship between the test and train data.

3.3.3 Implementation:

1. Loading the data using pandas and importing necessary libraries.

2. Preprocessing the data.

(2a). Filling the null values.

(2b). Removing the unnecessary columns.

(2c). Encoding the categorical data.

3. Splitting the dataset between the training and test data.

4. Applying the random forest regressor using ensemble class in sklearn, & creating the Random Forest Regressor class.

5. Fitting the training data.

6. Testing the data on test data.

After tuning the parameter, we get the best result at this value of the parameters.

n_estimator = 10 (no of trees in forest model).

max_depth = 500 (maximum depth of each tree of the model).

min_samples_leaf = 3 (minimum leaf nodes of each tree of the model).

min_samples_spilt = 3 (minimum spilt of the model).
max_features = auto.
The dataset is divided into training and testing set in a ratio of 0.8 and 0.2.
RMSE for training data = 2032.9908970869153.
RMSE for test data = 2817.386128678023.
The accuracy of our model on training and testing data is 78.42% and 69.41%, respectively.

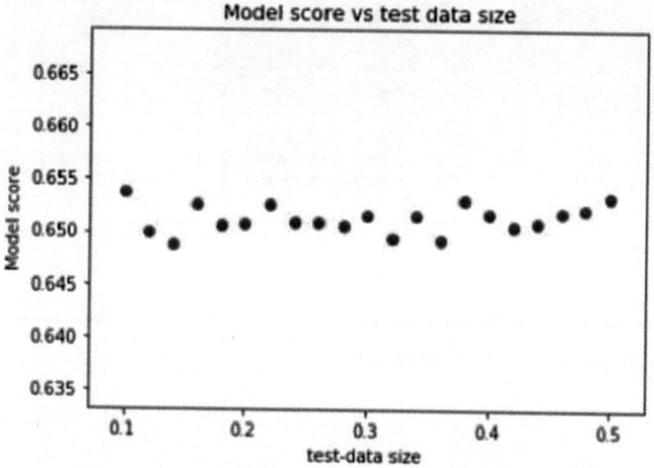

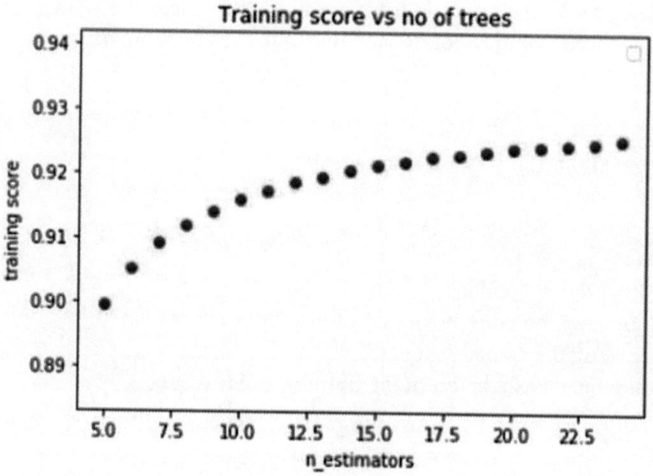

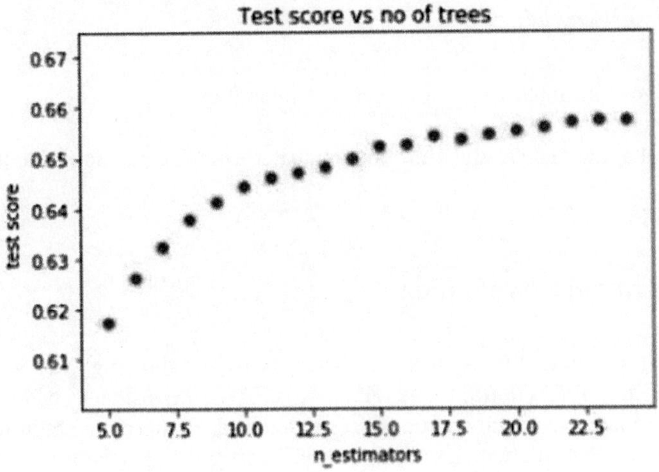

So the optimal choice of **n_estimators** is 10 keeping the factor time complexity in mind (as on increasing the no of the tree the complexity would have increased).

The accuracy of the training size goes to 92.53%. It is also observed that the result gets improved slightly on increasing the number of trees, and on the test size it is around 65.7%.

3.4 *Why This Model is Best Suited for This*

- **No feature scaling needed**: No component scaling (normalization and standardization) required in the event of random forest as it utilizes rule-based methodology rather than separation count.
- Random forest can consequently deal with **missing qualities**.
- Random forest is nearly **less affected by noise**.
- Linear relapse overimproves true issues by **expecting a direct relationship** among the factors.
- The mistake if there should arise an occurrence of random forest regressor is less when contrasted with other models.
- It is the best-fit model because there is no overfitting and underfitting in the model.

In Sumit Kalra's paper [3], they have made four models and their respective performance is found out based on the size of training and test data. The RMSE calculated there, on test data is around 3200 in every algorithm, but we have got a better result.

Apart from the random forest regressor, there were other models applied to the data- Linear regression, SVR, and decision tree, but the best results were produced from forest regressor.

3.5 Dermits of Model

- This algorithm makes a ton of trees and joins their output, so complexity is a factor here.
- It cannot extrapolate at all to information that are outside the range that they have seen.

4 Performance Evaluation

This section describes the experimental setup used for validating the performance of Black Friday Prediction Dataset. The motive of the experiment conducted is to provide the context in which a particular regressor produces the comparable result, as, in practice, the classifier with low computational cost and high accuracy is preferred.

4.1 Discussion on Result

We have used the various model to compare the performance of all the models so that we can get to know the best-suited model for our dataset. The model which we have used are:

4.1.1 Linear Regression

The linear regression is a method that finds the relationship between a dependent variable/feature, denoted as y, and 1 or more independent variables, denoted as x [8].

The linear model has fit pretty well in the data and the results were quite good.

RMSE on training data = 4653.425309837938.

RMSE on test data = 4649.916318697418.

Linear regression is a regression technique in which the independent variable has a linear relationship with the dependent variable. However, this fact is an assumption here. The time complexity of linear regression overtraining is $\mathbf{O(p^2n + p^3)}$ and on prediction is $\mathbf{O(p)}$, where n is the number of the training sample is the number of features. However, the complexity of models, time, and space complexity is mostly proportional to the size of the input data.

There are several disadvantages of linear regression model:

The linear regression model is very delicate to outliers, e.g., if most of your data lives in the range (5,10) on the x-axis, but you have one or two points out at $x = 100$, this could significantly change our regression results. Here, in our case, there is less likely to give out this error as we have large no. of training data and our predictors will be near to the training data.

Linear regression is an incredible device to break down the connections among the factors yet it is not suggested for most reasonable applications since it overimproves true issues by expecting a straight relationship among the factors.

4.1.2 SVR

Support vector machine can also be used as a regression method, keeping all the main features that characterize the algorithm (maximum margin). The fundamental idea is: to limit mistake, individualizing the hyperplane which boosts the edge, remembering that piece of the error is tolerated.

To use SVR, we used ***SVM's svm*** class. The kernel used was default one ***rbf***. The RBF kernel on two samples x and x', represented as feature vectors in some input space, is defined [16] as.

However, the results were not showing after applying the SVR. Even after waiting for one hour, there was no output. Even after the feature scaling (StandardScaling from skleanrn. preprocessing).

This model does not do well when the dataset is large as it needs more training time. It also does not perform very well, when our dataset contains noise, i.e., target classes are mixing.

4.1.3 Decision Tree

The prime objective of this algorithm is to find a training model that can be used to predict or value the target variable by **learning simple decision guidelines** deduced from prior data.

The tree is looked from the root to predict a class label in the decision tree. We look at the estimations of the root attribute with the record's attribute. Based on the examination, we follow the branch comparing to that worth and hop to the following node.

RMSE training data = 3298.560612.

RMSE test data = 3786.9045400893683.

There is overfitting of data which gave good accuracy in training data set but low accuracy in the testing dataset.

4.1.4 Gradient Boosting Algorithm

Gradient boosting is one of the most remarkable strategies for making predictive models. Boosting came out of whether a powerless learner can be changed to turn out to be better [9]. The main acknowledgment of boosting that saw extraordinary achievement in the application was Adaptive Boosting or AdaBoost for short.

The parameter max_depth was taken 2, n_estimators3, and learning_rate were 1.

However, there was a pretty good result but not better than the forest regressor model.

RMSE training data = 3261.0803035857753.

RMSE test data = 3262.5107685140533.

In above models there were some demerits moving to the forest regressor is a good choice. If we have only used the decision tree then the overfitting problem would have been raised. The complexity on training is $\mathbf{O(n^2pn_{trees})}$ and over the test is $\mathbf{O(n^2pn_{trees})}$.

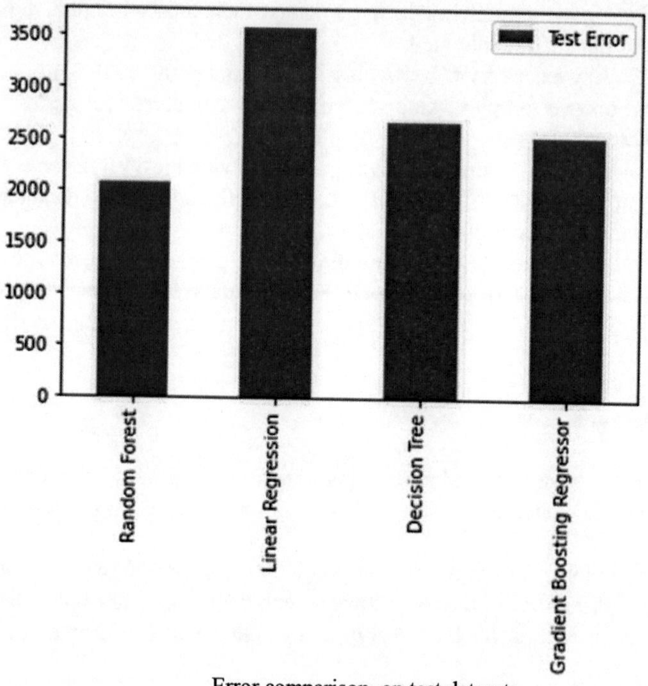

Error comparison on test dataset

The test error is less compared to the other model which makes it more suitable for our dataset.

4.2 *Discussion on Random Forest Regressor Model Result*

After selecting the model, now we have tuned the parameter to get best accuracy of random forest model.

1. Changing the n_estimator: we have seen that we get low training score at 10 value of n_estimator

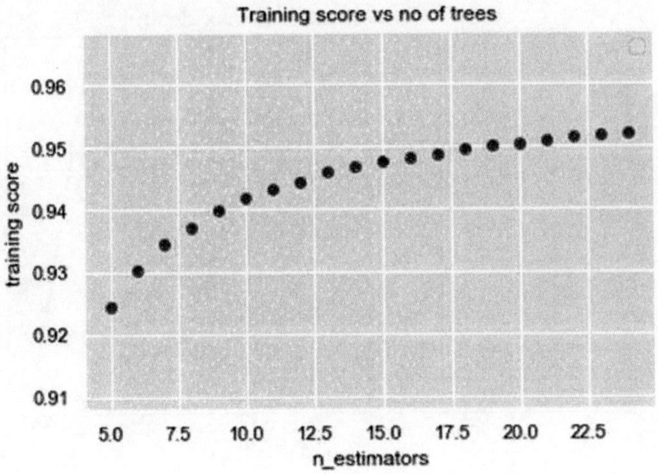

2. Changing the min leaf nodes : we have seen that we get low score at min_leaf_nodes value of 3.

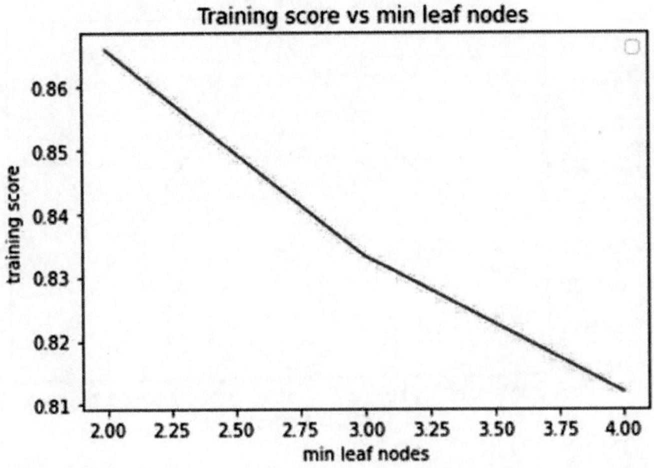

3. Changing the max_depth: we have seen that we get a low score at max_depth of the value of 700.

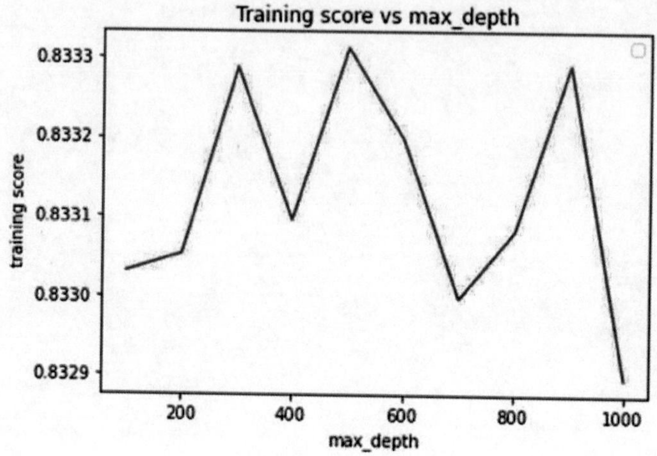

4. Changing the min_sampling: we have seen that we get a low score at min_sampling of the value of 3.

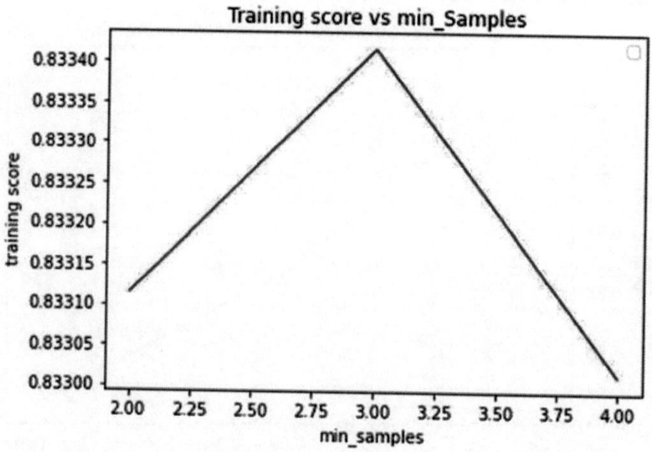

5. Changing the max_features: we have seen that we get a low score at max_features of the value of 'auto'.

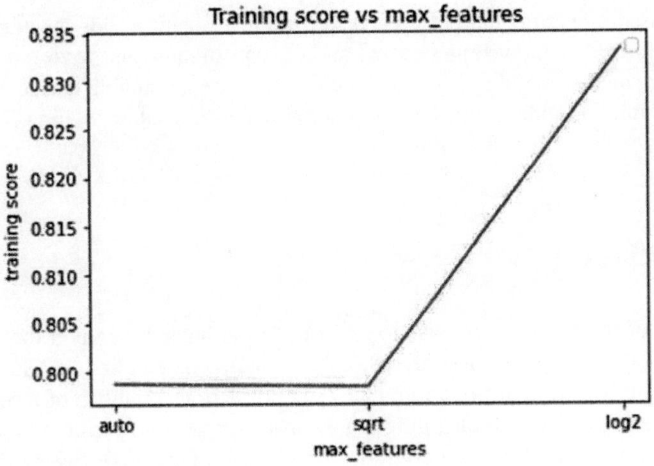

After tuning all parameter we get best accuracy at:
n_estimator = 10 (no of trees in forest model).
max_depth = 700 (maximum depth of each tree of the model).
min_samples_leaf = 3 (minimum leaf nodes of each tree of the model).
min_samples_spilt = 3 (minimum spilt of the model).
max_features = auto.
Root mean squared error:
RMSE for training data = 2032.9908970869153.
RMSE for test data = 2817.386128678023.

5 Literature Review

There is a great deal of related work on the dataset. The dataset consists of around 5.5 lakhs records, where each record comprises of 12 highlights they have applied different models and looked at the performance. Some have given full investigation of on Black Friday sale featuring using the Apriori algorithm. But our model is better than this because we have used random forest regressor model and tuned the parameter to get better efficiency of that model, we have seen some other researcher also work on this dataset by random forest regressor model but our model has good accuracy than others researcher. We also reduce overfitting to get better result. We have calculated the RMSE score then compared our model with other models. This dataset is from Analytics Vidhya. The data was analyzed using various techniques (KalraSumit2020) like Xgboost and Tfidf Transformer. The data was analyzed using various techniques (KalraSumit2020) like Linear Regression, SVM, Gradient Boosting Algorithm and XgBoost Regression. Different cases of these models were taken in which the training and testing data is varied. There were five cases in the splitting of data as follows: Case1 (90%, 10%), Case2 (70%, 30%), Case3 (50%,

50%), Case4, preparing dataset was taken from investigation Vidhya which is additionally utilized as the testing dataset, and Case5, training, and testing data is also taken from analytics Vidhya. It is concluded that Case4, planning dataset was taken from examination Vidhya which is also used as the testing dataset (Ching-Seh (Mike) Wu 2018).

6 Conclusion

The goal of the examination was to investigate the buys done upon the arrival of Black Friday. An input dataset of 788,639 passages, where at first 550,069 sections comprise of preparing information, and 233,600 sections establish of testing information. Various supervised learning algorithms were used to predict the price, i.e., linear regression, SVR, and random forest regressor. In our study, we firstly did the data preprocessing and data visualization is also performed by plotting sufficient bar graphs, pie charts, etc. The main thing is that the random forest regressor model outperforms many other methods and which was further improved by tuning the parameters. By using this algorithm, there is reduction in overfitting and variance is also less. We have got a better result by putting the max_depth parameter to 700, however complexity is the issue. We have put the value of the various parameter where the accuracy is maximum.

References

1. [Online]. Available: https://www.thebalance.com/what-is-black-friday-3305710
2. Smith O, Raymen T (2017) Shopping with violence: black friday sales in the British context. J Consum Culture 17(3): 677–694
3. Kalra S et al (2020) Analysing and predicting the purchases done on the day of black friday. In: 2020 international conference on emerging trends in information technology and engineering (ic-ETITE). IEEE, New York
4. Milavec B (2012) An analysis of consumer misbehavior on Black Friday. Diss. University of Delaware
5. Majumder G (2019) Analysis and prediction of consumer behaviour on black friday sales. J Gujarat Res Soc 21(10s):235–242
6. Shopper Trak Reports Positive Response to Early Holiday Promotions Boosts Projections or 2010 Holiday Season. Shopper Trak, Press Release (2010).
7. Holiday Watch (2010) Media Guide 2006 Holiday Facts and Figure International Council of Shopping Centers
8. Wu C-SM, Patil P, Gunaseelan S (2018) Comparison of different machine learning algorithms for multiple regression on black friday sales data. In: 2018 IEEE 9th international conference on software engineering and service science (ICSESS). IEEE, New York
9. Trung ND et al, Black friday sale prediction via extreme gradient boosted trees
10. Thomas JB, Peters C (2011) An exploratory investigation of black friday consumption rituals. Int J Retail Distrib Manage
11. Maharjan M (2019) Analysis of consumer data on black friday sales using Apriori algorithm. SCITECH Nepal 14(1):17–21

12. Tashakkori A, Teddlie C, Teddlie CB (1998) Mixed methodology: combining qualitative and quantitative approaches, vol 46. Sage
13. Ma X (2017) Linked geoscience data in practice: where W3C standards meet domain knowledge, data visualization and OGC standards. Earth Sci Inform 10(4):429
14. Badie B, Berg-Schlosser D, Morlino L (eds) International encyclopedia of political science, vol 1. Sage
15. AI-Malaise AS (2013) Implementation of Apriori algorithm to analze organization data: building decision support system. Int J Comput Appl 66(9):27
16. Ganapathi A (2009) Predicting and optimizing system utilization and performance via statistical machine learning. Diss. UC Berkeley

Analyzing the Need of Edge Computing for Internet of Things (IoT)

Ajay Pratap, Ashwani Kumar, and Mohit Kumar

Abstract Whenever we need to monitor something or sense the data such as traffic, temperature, or may be pollution then Internet of things (IoT) comes in picture. IoT devices can only collect and transfer data for analysis. Today's increased computer literacy enables them to perform complex real-time computations. The concept of edge computing means running fewer processes in cloud and transferring rest of the processes to the user's computer, IoT device, or server side. In this paper, recent advances in computer technology and their impact on IoT have been discussed. I have also created computer market classification by analyzing and classifying current prose, thus revealing the secrecy and supportive features of various IoT computational models. I have also highlighted the important requirements to increase the use of the edge of computing in IoT. Few implementation case studies examine the overhead caused by edge calculations and present a possible implementation of the computing edge paradigm.

Keywords Edge computing · Internet of things (IoT)

1 Introduction

IoT devices sense the data, take it out from the sensors, and then send it to the global monitoring place. Numbers of IoT devices used are growing rapidly and resulting data mean that new technologies are needed to be explored to meet customer requirements and ensure efficient management. Existing IoT model for sending all data to the cloud for processing requires alternatives due to high volume of data to be

A. Pratap (✉)
AIIT, Amity University Uttar Pradesh, Lucknow Campus, Lucknow, India
e-mail: apratap@lko.amity.edu

A. Kumar
Sreyas Institute of Engineering and Technology, Hyderabad, Telangana, India

M. Kumar
Cambridge Institute of Technology, Ranchi, Jharkhand 834001, India

P. K. Singh et al. (eds.), *Proceedings of Second International Conference on Computing, Communications, and Cyber-Security*, Lecture Notes in Networks and Systems 203,
https://doi.org/10.1007/978-981-16-0733-2_14

processed and the cost of centering and processing each piece of data. This is mainly important for IoT services because they can produce large amounts of new data for analysis. Cloud computing performs weaker as compared to edge computing because of its centralized nature on processing, storage, and security [1]. Many cloud applications are powered by the user, resulting in comprehensive data analytics opportunity. Nevertheless, using the cloud as central server only increments the frequency of communication between user devices, what we refer to as edge devices and geographically distant cloud data centers. Edge computing speaks to computing topology that is placing contents, computing and processing closer to the edge of the networking [2]. Edge computing is a deployment model intended to push critical data processing and storage characteristics where the device is located. Edge computing is the practice of processing data near the edge of your network, where the edge computing is a distributed and open information technology architecture [3]. This means that the data can be processed more efficiently. The main motive of this approach is to discover the possibilities of computing performance on edge through which network load is directed.

This paper also explores the benefits of edge computing and some diverse use cases where it can be implemented and identifies some potential next steps to be taken by the industry [4].

2 Literature Survey

Paper [5] discusses the problems existing IoT model related to data access delays, traffic, and bandwidth of Internet access channels and the proposed edge computing model with a solution case study based on education. Edge computing is gaining more and more popularity in the IoT field. In 2018, it was one of the major technological developments that formed the basis of the next generation of digital commerce. At the same time, given the huge amount of data and the need to maximize computer resources, we are also seeing a growing tendency to send data into the cloud and handling the challenges of synchronization and connectivity related to edge computing [6]. Bob O'Donnell founder and chief analyst of Technalysis Research said, "Finally, the world is heading towards a more distributed model, in which cloud-based computing plays a critical, but not solitary, role in managing the capabilities of both connected consumer devices and critical devices for enterprise and industry". This, in turn, will lead to increased interest for devices with computing and storage capabilities on the edge of the network [7].

Gartner Group suggests: "About 10% of the data generated by the enterprise is created and processed outside a traditional data center or a cloud. By 2022, Gartner predicts that figure will reach 75%" [8]. Sun et al. [9] discussed that how big IoT data is generated from distributed IoT devices and they are transmitted to cloud which is called brain of big data processing. A model has been proposed in [10] in which migration time can be estimated when average transmission data rate is given.

Sun et al. [11] proposed a hierarchical architecture for fog computing where each fog node provides flexible IoT services and maintains user privacy also. Discussion on three typical edge computing techniques which are mobile edge computing, cloudlets, and fog computing is being provided in [12]. Current edge computing architectures and their challenges are examined in [13].

Pratap et al. [14] have explored the work done in the field of BDasS implementation and its challenges related to Indian banking system. In [15], comprehensive review of various methods is applied for the big data dimension reduction processes and redundancy elimination. Kumar [16] has proposed scheme for secure transmission of digital media between a service provider and consumer with cloud storage.

3 Problem Identification

Cloud computing technologies implement a variety of IoT services for the storage of data, processing of data, and remote access such as Amazon Web Services IoT Platform or Google Cloud IoT Platform. This approach simplifies the development of applications for IoT bot that have some disadvantages also which are as followings:

- Restrictions on the intensity of the data flow for storing on Cloud stores
- Significant delay in access to data stored on Cloud stores
- Permanent access of Internet in case of growing number of IoT sites
- Need of increased bandwidth.

In cloud computing, users perform their encoding and install it in the cloud. The cloud company then decides where the computer issues should take place in the cloud. Users have zilch or part of the awareness of the program. This is one of the advantages of cloud computing that communication is clear to the user. Typically, the application is written in a single programming code and compiled for a specific platform where the application runs only in the cloud. Conversely the processing is removed from the cloud and sent to the edge of the nodes, and since the junctions of these nodes are different from each other, the programmer therefore faces great difficulty in writing applications that can be applied to the edge of a computer science model.

4 Edge Computing as a Solution

Concept of edge computing is capable to solve the problems related to connectivity and latency challenges, bandwidth constraints and restriction on intensity of data flow. Edge computing describes a computing topology in which information processing and content collection and delivery are placed closer to the sources of this information.

For solving the problem with the computer programming edge of computer science, we set forth the theory of computer stream, which is defined as a series of operations on the data along dissemination of data. The calculation can be performed anywhere on the URL till the time the application defines where the processing should take place. In this case, the computer stream can help user to determine where the computer issues should be done and how the data is increased after calculations happened on the beach. By distributing the computer stream, we expect the data to be calculated as closely as possible.

Data permissions and data transfer costs can be reduced. In a computer feed you can switch the function and it should also be data and status with functionality reallocation. In addition, cooperation issues should be addressed by numerous laws in the calculation model of the edge.

5 Architecture of Edge Computing

Edge computing enables technology to perform data processing on the periphery of the network. For example, a smart phone is the edge in between user and cloud. Figure 1 is the illustration of bilateral computing streams in computing. In the edge of computing things not only behave like data consumers, but also as data producers. On the perimeter, things can not only request specific services and content from the cloud, but also work on cloud computing. Edge can handle computer storage of data, offloading and computing, as well as distributed request and delivery services from cloud to user. With these online jobs, the edge should be well-planned to meet the

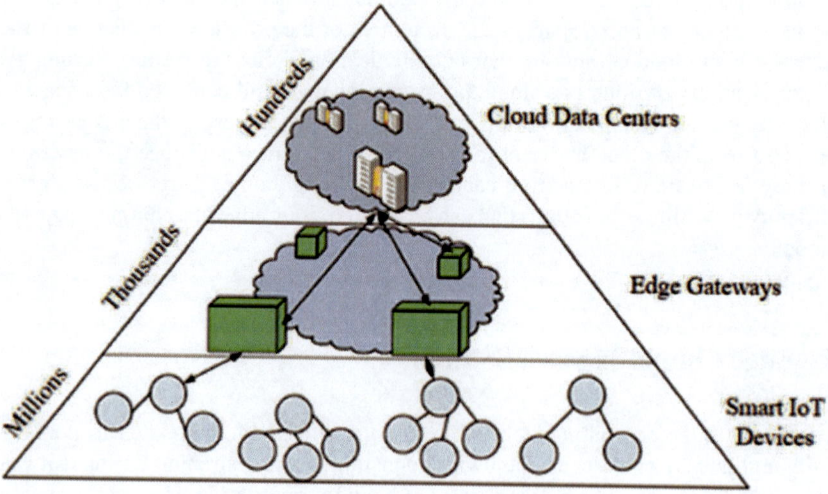

Fig. 1 Layered model for cloud edge-based IoT service delivery

requirements effectively in services such as consistency, safety, and confidentiality [17].

6 Benefits of Edge Computing

The data is mostly generated on the network edge, so it will be more resourceful to process data on the side. Cloud computing, nowadays, is not considered a good option for processing data due to latency problems and security reasons when data is happening on the edge. In this part of the paper, we have discussed mainly why edge computing is considered better than cloud computing.

Low latency: Compared to the cloud, the edge is closer to the IoT device. This means that communication has to travel a short distance to obtain local processing power, which speeds up data transfer and processing.

Longer battery life for IoT devices: Due to improved latency, the communication channels are being able to open for a shorter period of time, thereby increasing the battery life of a powered IoT device.

More efficient data management: Processing data on the edge server makes simplifying quality management such as filtering and prioritization more efficient. Completing this data management aside means that you can set a cleaner cloud dataset for further analysis (Fig. 2).

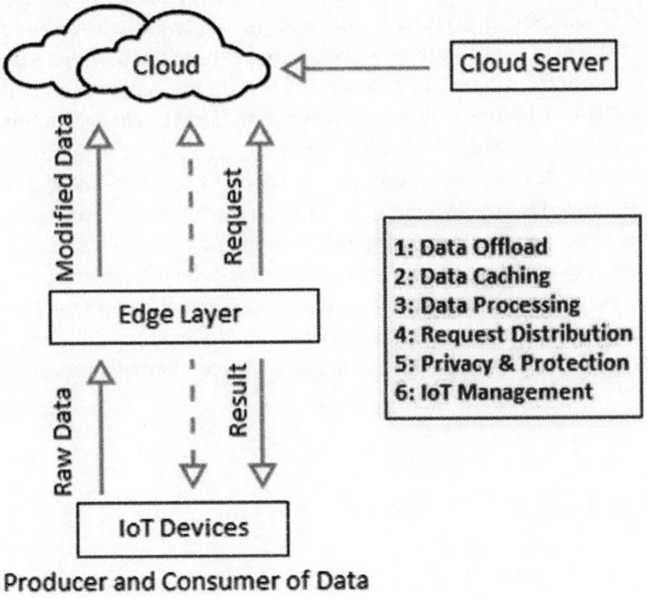

Fig. 2 Computing request processing

Access to data analytics and AI: Edge computing comes in the way that analytics and AI require very fast response time or involve the processing of large "real-time" data that cannot be transmitted to central systems.

Resilience: More possible communication paths are provided by the edge than a centralized model. If there is a failure at the edge, other resources are also present to provide continuous operation.

Scalability: Since the processing is decentralized with the edge model, the low load should eventually be placed on the network. This means that scaling IoT devices should have less impact of resource on the network, especially if the application and control planes are located sideways with data.

7 Applications of IoT Devices

In this particular part of the paper, we will come across different case studies to showcase our idea of edge computing.

7.1 Cloud Offloading

In the case of cloud computing, almost all the processing is performed on the cloud, resulting in time delays, which weakens the user experience. In edge computing, edge contains computing resources, which shifts the workload from cloud. In the case of IoT, the data is not only produced but also consumed at the edge. Possible area that can profit from the edge of computing is shopping online service. A user may often make changes to the shopping cart. By default, these changes to his shopping cart are made in the cloud, and further, the updated view of shopping cart comes to the customer's device. Depending on the speed of the network and the load on servers, this procedure could take long time and even longer for mobile phones. Nowadays, online shops are becoming increasingly popular, so it becomes a priority to improve the customer experience by solving time-related problems. In such cases, the computer edge of a computer comes to the fore, by uploading the shopping cart upgrade from servers to online edges, it is possible to temporarily reduce the delay. With the help of a desktop computer, we can enhance interactive service quality in the early stages by reducing the delay [18].

7.2 Video Analytics

Extensive use of mobile phones and network cameras makes video analysis an assembly tool. Cloud computing is no longer suitable for apps that require video analysis due to transmission latency and confidentiality issues. Currently, various kinds of cameras are extensively installed in the city areas and in almost all the vehicles. Suppose a child gets missing, it is possible that the camera installed in that particular area might capture the missing child. But due to traffic cost and privacy issues the data from the camera is generally not uploaded to the cloud and even if the data is available on the cloud, it will take considerably longer time to search and upload such huge quantity of data which also puts the life of the missing child at stake. Conversely, with the edge computing model, the cloud can itself generate a request of searching the missing child and transfer it to the equipment in a particular target area [18].

7.3 Smart Homes

Nowadays, even domestic environments benefit from IoT. Some products such as intelligent lighting, smart TVs, and vacuum cleaner robots have already been developed and are available on the market. However, a smart home: smart home environments are primarily a combination of connected devices, with inexpensive wireless sensors and controllers positioned for the floor, pipes, and even rooms and walls. Each day huge amount of data is being produced by these devices which should be processed locally in order to provide privacy protection which makes it clear that cloud computing is not suitable for smart homes. However, the edge computation is considered ideal for building a smart home: with a portal that runs a special edge OS operating system at home, you can connect things and manage your home easily. Data can be processed locally and continue to load Internet bandwidth and service [18].

7.4 Augmented Reality Devices

The word virtual reality is definitely more common among most people, whereas augmented reality (AR) is a recent development and has more practical presentations. Instead of creating a virtual world, AR lights up with digital elements in real-world environments. Sustainable AR devices such as glasses and headphones are sometimes used to create this effect, but AR is mostly experienced by the users through their smartphone screens. Anyone with games like Pokemon GO or uses a filter on Snapchat or Instagram after using AR. The technology behind AR requires processing of visual data devices and visual elements in real time. Without edge computer architecture, it will be necessary to return this visual data to the central cloud servers where digital items can be added before they are returned to the device leading to a larger latency. Retail chains use AR technology to add an extra set of data to your shopping experience. AR can easily view product information and sales alerts giving customers a reason to shop in person instead of using online retailers.

A computer terminal is critical to providing minimal intelligence to these services [18].

7.5 *Automobile Industry*

The automotive industry has already invested billions of dollars in technology development. In order to operate safely, these vehicles need to collect and analyze large amounts of data relating to their environment, instructions, and weather conditions, not to mention the communication with other vehicles on the road. They will also need to transfer data back to manufacturers to monitor usage and maintenance notifications as well as connections to local networks.

Unfortunately, this stream of transmitted data will go through the same flow of traffic produced by mobile phones, personal computers, and various other connected devices. With so many additional vehicles that collect and transmit data, bandwidth is inevitable if manufacturers do not adopt new computer solutions. It is one thing for an office computer to experience an uncomfortable delay when accessing a network; it is completely different for a self-propelled car to tilt when driving at 65 mph on an open highway. Edge computing model enables independent vehicles to collect, process, and share data between vehicles and wider networks in real time. Compared to the network edge of geographically located data centers to gather and distribute important raw information to local authorities, crisis response services and car manufacturers, peripheral vehicles will give supreme consistency without compromising the network's infrastructure [18].

8 Findings

In today's era, data is power. People are hungry for business data. And this hunger is gaining maximum outlook over the data that has recorded IoT at a massive rate. As more data is collected, merged, processed, and analyzed, they help to make better business decisions, streamline processes, increase customer engagement, and gain a competitive edge. We are already aware of Gartner's forecast of IoT which clearly states that by 2020, the Internet of things will reach over 20.4 billion devices. Therefore, we can imagine the amount of data that will be collected and the processing power it needs to analyze. Many IoT experts have reported that the implementation of IoT is imminent, but they need a solution that addresses growing concerns due to the processing of moderate amounts of data. Typically, data is collected from devices connected to the IoT ecosystem and sent to data or clouds, where they are further processed and analyzed [19]. This method has proven to be reliable, but it takes a considerable amount of time, especially when talking about real-time updates. A better and faster approach is the edge of the computer. Every day, tools become more effective at reducing data center stress and developing the ability to improve—and in some cases even move—cloud capabilities, which allow IoT companies to take the initiative instead of affecting costs and delays. By sending data to a data center or on-premises cloud, companies are now prioritizing information processing and decision-making closer to data, each an IoT device. Audio Statistics is a great way to power IoT ecosystems without a doubt, but the edge of computing can prove to be a game changer for the Internet of things, providing the ultimate in computing and analytics power in an increasingly connected world [20].

9 Conclusion

Better understanding of the network hierarchies, different levels of interactions among network elements and sensor devices can lead to the better competency to IoT developers. Almost all the services are being sent from the cloud to the edge of the network as they process data in them and ensure lesser reaction times and consistency also. If high volume data is processed on the edge instead of being loaded into the cloud, then bandwidth can also be saved. Some situations are listed where sophisticated computing can thrive from cloud downloads in smart environments. The tools of rapid prototyping with model oriented or process-oriented approach can be next research domain which allow us to bring IoT projects to the level of ready-made solutions.

References

1. Liu ZY, Ding W, Atiquzzaman M (2019) A survey on secure data analytics in edge computing. IEEE Internet Things J
2. Gartner top 10 strategic technology trends for 2018. https://www.gartner.com/smarterwithgartner/gartner-top-10-strategic-technology-trends-for-2018
3. Chiang M, Ha S, Chih-Lin I, Risso F, Zhang T (2017) Clarifying fog computing and networking: 10 questions and answers. IEEE Commun Mag 55(4)
4. Sahni Y, Cao J, Zhang S, Yang L (2017) Edge mesh: a new paradigm to enable distributed intelligence in internet of things. IEEE Access 5:16,441–16,458
5. Plakhteyev A, Perepelitsyn A, Frolov V (2018) Edge computing for IoT: an educational case study. In: 9th IEEE international conference on dependable systems, services and technologies, DESSERT'2018, 24–27 May, 2018, Kyiv, Ukraine
6. Mass J, Chang C, Srirama SN (2019) Edge process management: a case study on adaptive task scheduling in mobile IoT. Internet Things 6:100051
7. ESP8266 Arduino Core Documentation. Release 2.4.0. Ivan Grokhotkov. May 14, 2017. https://media.readthedocs.org/pdf/arduino-esp8266/docs_to_readthedocs/arduino-esp8266.pdf
8. Marković D, Koprivica R (2015) Application of IoT in monitoring and controlling agricultural production. ActaAgriculturae Serbica XX(40):145–153
9. Sun X, Ansari N (2016) Optimizing resource utilization of a data center. IEEE Commun Surv Tutorials. https://doi.org/10.1109/COMST.2016.25582032016
10. Khan MA et al (2016) Moitree: a middleware for cloud-assisted mobile distributed apps. In: 4th IEEE international conference on mobile cloud computing, services, and engineering, Oxford, UK, Mar. 29–Apr. 1, 2016, pp 21–30

11. Sun X, Ansari N (2016) Edge IoT: mobile edge computing for the internet of things. IEEE Commun Mag. https://doi.org/10.1109/MCOM.2016.1600492CM
12. Ai Y, Peng M, Zhang K (2017) Edge computing technologies for internet of things: a primer. Digital Commun Network. https://doi.org/10.1016/d.dcan.2017.07.001
13. Mostafavi SA, Dawlatnazar MA, Paydar F (2019) Edge computing for IoT: challenges and solutions. J Commun Technol Electron Comput Sci 26
14. Pratap A, Sharma SK, Dev H (2019) Challenges and issues related to big data as a service platform-survey on Indian banking system. In: National conference on recent trends in electronics and electrical engineering (NCRTEEE-2019), India, 10–11 June, 2019, pp 65–70
15. Pratap A, Dwivedi A, Dev H (2019) Review of dimensionality reduction techniques in data mining from big data. Int J Res Appl Sci Eng Technol (IJRASET) 7(V)
16. Kumar A (2019) Design of secure image fusion technique using cloud for privacy-preserving and copyright protection. Int J Cloud Appl Comput (IJCAC) 9(3):22–36
17. Mohan N, Kangasharju J, Edge-fog cloud: a distributed cloud for internet of things computations. https://www.cs.helsinki.fi/u/nmohan/documents/2016/EF_Nitinder_Jussi_UH_Final.pdf
18. Botta A, de Donato W, Persico V, Pescapè A (2015) Integration of cloud computing and internet of things: a survey. J Future Gener Comput Syst, pp 1–54
19. Yazed MSM, F Mahmud (2016) The development of IoT based BBT charting and monitoring using Thingspeak. In: International conference on engineering, science and nanotechnology
20. Yi S, Hao Z, Qin Z, Li Q (2015) Fog computing: platform and applications. In: Third IEEE workshop on hot topics in web systems and technologies, pp 73–78

A Mobile-Based Farm Machinery Hiring System

Oluwasefunmi Arogundade, Rauf Qudus, Adebayo Abayomi-Alli, Sanjay Misra, JohnBosco Agbaegbu, Adio Akinwale, and Ravin Ahuja

Abstract Agriculture as the oldest profession dates back to the Stone Age. Advances in technologies like diesel engine tractors and other tools with hydrostatic capability and control brought about agricultural mechanization which increased food productivity and industrialization. The aim of this research work is to design a mobile application for distributing or leasing agricultural machineries to farmers using locations-based services. The design also took into consideration the configuration of the various topologies and other factors that could enhance the flexibility of a mobile application of this nature. The user platform is categorized into three sections: the presentation layer, the business layer, and the data layer. The presentation layer is focused on the design logic and the navigational tools used in locating the right hiring type. It is also responsible for choosing the right data format using data validation technique to protect the app from invalid data entry. The business layer is responsible for logging in, authentication, exception handling, and security matters. While the data layer is focused on facilitating secure data transactions, the beauty of it is that it can be rescaled over time to meet the challenges of the time. The application was developed using JavaScript and MySQL with Phonegap/Cordova, XAMMP, and PHP for the backend. It was validated using formative evaluation which was conducted using interviews and open-ended questionnaires. The results of usability test obtained are promising.

O. Arogundade · R. Qudus · A. Abayomi-Alli · J. Agbaegbu · A. Akinwale
Federal University of Abeokuta, Abeokuta, Ogun, Nigeria
e-mail: arogundadeot@funaab.edu.ng

A. Abayomi-Alli
e-mail: abayomiallia@funaab.edu.ng

A. Akinwale
e-mail: akinwaleat@funaab.edu.ng

S. Misra (✉)
Covenant University, Ota, Nigeria
e-mail: Sanjay.misra@covenantuniversity.edu.ng

R. Ahuja
Shri Viskarma Skill University, Gurgaon, India

P. K. Singh et al. (eds.), *Proceedings of Second International Conference on Computing, Communications, and Cyber-Security*, Lecture Notes in Networks and Systems 203,
https://doi.org/10.1007/978-981-16-0733-2_15

Keywords Mobile application · Market · Information system · Agriculture · Mobile phones

1 Introduction

Three decades from now around 2050, the world will be in dire need to increase agricultural production by 60–110% in order to overcome food challenges and meet the increasing demand for food [1]. As expected, this will pile pressure on resources such as water, land, and energy. To increase agricultural production will be one of the greatest problems to face humanity in the near future [2, 3]. Agriculture is divided into two categories which are: industrialized agriculture and subsistence agriculture. Research has shown that small-, medium-, and large-scale farmers have issues regarding accessibility to modern farm machinery. Small-scale farmers face difficulties in having access to those tools expected to boost their farm activities and eventually affect productivity. However, according to a research conducted by [4] using Abuja as a case study revealed that the major factor limiting the farmers residing in this part of the country (Abuja) is the lack of access to a major farm machinery, tractor in this case, to work on their farms. The motivation behind this study is how so many farmers have gone out of business and the remaining few are suffering from increasingly reduced farm productivity due to lack of access to farm machines that could make farm operations less tedious with improved productivity. The proposed idea is to provide a farm machinery hiring system for these farmers at a subsidized rate. The contributions of this study include: guaranteeing reliable up-to-date information on machines acquisition and hire, offering real-time information on mobile hiring system with respect to availability based on database record and provision of an interface that indicates the nearest hiring point to the customer through GPS.

The paper is organized as follows: Section 2 dealt with related work in the area of machine hiring; Sect. 3 explains the methodology of the proposed system and analysis of the system; Sect. 4 discusses the implementation and evaluation of the mobile hiring system. The paper concludes in Sect. 5.

2 Related Works

The share of agriculture in Africa is declining, partly due to low productivity and limited value addition [5]. Africa is a major importer of agricultural products, with imports ranging from rice, maize, and wheat including livestock products contributing to food security. The contributing factor to this is the mismatch between increase in consumption and increase in production [6]. Agriculture mechanization has drastically increased labor productivity in crop production, by playing a significant role in industrialization, freeing up labor for industry and services [7]. It originally depended on human effort, with the advent of mechanical advances such as steam engine and diesel tractors and other mechanical tools with hydrostatic power that needed control.

The way forward for most unresolved challenges in agriculture lies on more advances that will compel the replacement with human intelligence to meet the needs of superior autonomy in an infinite and unstructured domain [8]. Technology in the twenty-first century plays the role of an enabler to agriculture [9, 10]. Components of mechatronics, like actuators and sensors, play important parts in our farms for seeding, cropping, cleaning, fertilizing, and monitoring of our vegetation. Various approaches have also been applied to aid agricultural processes, for instance using robot arms that nurture the roots of plants and revolving machines to seed, they can also collect, and clean produce [11]. Information and communication technology (ICT)-enabled services sector potentially links markets and farmers and enables a two-way communication with farmers. It has enhanced the digitalization of land records and management, using software to improve efficiency and participation, offering expert advice through mobile phones form some of the potential applications [12]. The authors in [13] developed a mobile-based marketspace app for marketing organic farm products by leveraging automated geo-location services. In [14], the authors used an estimated procedure to show how the adoption of animal traction tractors can affect economies of scope (EOS) for rice and non-rice grains and legumes or seeds as these are the most common crop group widely grown with these methods in Nigeria. In [15], the authors presented an overview of the agricultural ecosystem with respect to DR Congo, their emphasis was major stakeholders within the system. They also showed how mobile technologies were being used in the agricultural sector of the economy.

In [16], the study gave a roadmap for building a collaborative concept connected to a mobility model for sharing agricultural machines. The authors in [17] analyzed the implication of agricultural mechanization on wages by using a unique dataset based on monthly wages with rice price tag for a period spanning from 1995 to 2015. They employed a dynamic panel model calculated by generalized methods of moments. They discovered that increase in rural agricultural wage is a function of increase in agricultural mechanization, for the short and long term.

In [18], the goal of the study was to end hunger, achieve food security, improve nutrition, and promote sustainable agriculture which must be urgently pursued. It was discovered that agricultural mechanization plays a pivotal role in this process. In [19], the index of mechanization and other productivity functions were used as indicators in assessing the impact of mechanization on agricultural production in Umuahia North LGA of Abia State, Nigeria. In [20], the goal of the article is farming mechanized to boost the overall produce and production with the lowest cost of production. The studies have indicated that there was a significant increase in cropping intensity due to the use of farm machinery. Some previous studies [21–24] present some basic concepts of mobile-based applications which the present study may leverage upon though they are not particularly related to farm or agricultural sector.

3 Methodology

The basic motivation for this study is to develop farm machinery hiring system built on realistic viewpoint to offer concrete solutions and results, the study adopts the methods strategy according to [25]. It begins by leaning toward inductive methodology to understand the challenges within the context of applicability and provide the requisite control and guide while collecting requirements [26]. Before continuing, we summarize mobile-based farm machinery hiring system and process methodology. In tune with our realistic method, we developed the framework using computing research methods [25]. This study took only the features that have relationship to our work. The study considered the following:

3.1 *Framework for the Study*

A. **What is our goal?**

- (Discover events)—The authors desired to have a better knowledge of mobile-based farm machinery hiring system in Nigeria.
- (Design a working tool)—The authors will design a mobile tool, for hiring farm machinery to aid and enhance farmers' efficiency.

B. **The data will originate from where?**

- (Explore)—The authors will leverage on existing literatures on mobile apps for farming and general agriculture.
- (Detect, Request)—The authors will interact with stakeholders like extension workers and farmers
- (Prototypical)—The authors will use the UML to prototype real-world design tool encapsulating client operation and admin operation.

C What became of the data?

- (Find subjects, ascertain leanings)—The collected data will be used to better appreciate farm machinery hiring operations and the use of ICT.

D Did they attained their objective?

- (Appraise effects, take decisions)—The authors, armed with understanding and designed a working tool, are now in a better position to perfect the design and develop the software.
- (ascertain constraint)—From the design, authors will be able to ascertain their limitations and see how existing studies can be improved upon.

Arising from the framework for computing research method, this study passed through basis of views from mechanized farming experts, analysis of mechanized farmers in Nigeria and beyond through the use of the Internet, the use of questionnaire administered to farmers and some articles from the Web. The graphs used are

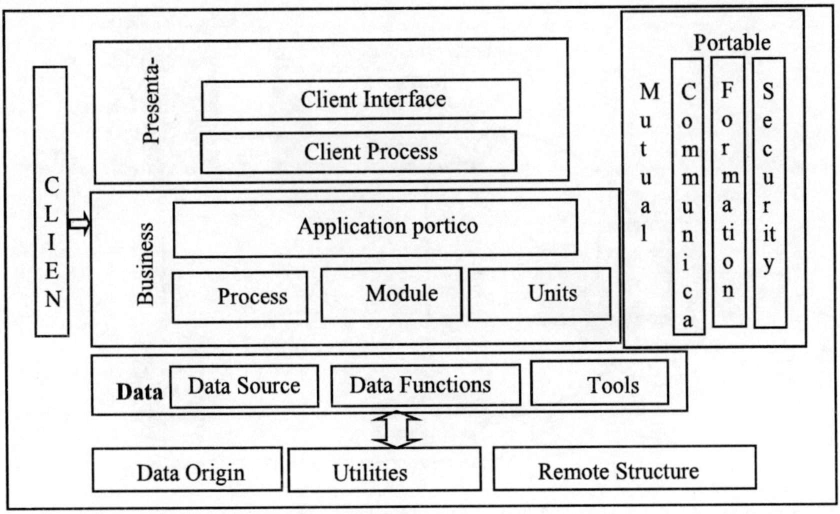

Fig. 1 Architectural design of the hiring system for farmers

for precise reason based on the client's perspective of the system. The different interpretations are referred to as architectural views. These views enhance the structuring of information, while the graphs provide for the propagation of knowledge. Figure 1 shows the architectural design of the hiring system for the study.

3.2 Architectural Design of the Farm Machinery Hiring System Based on Mobile Technology

The design of the architecture was extensible and scalable so that it can contain all parts of the system. Farm machinery hiring system is an application for smart phones that supports an Android operating system and uses GPS function to locate the nearest service point for farmers to hire machines at ease. The architectural design of the hiring system for farmers consists of multiple layers, as shown in Fig. 1.

The mobile app user model contains the presentation, business data, and data storage layers. This framework takes into consideration the case of security and communication angles since these factors determine the flexibility and reliability of the system. For this study, the presentation layer houses the user interface and the rules to traverse the interface. It translates information into human readable format while the business layer takes care of information transfer among the user interface and database section of the system.

The mobile hiring system is connected to the mobile middleware and third-party services (API) before it is sent to the enterprise system for approval and store into database as shown in Fig. 2.

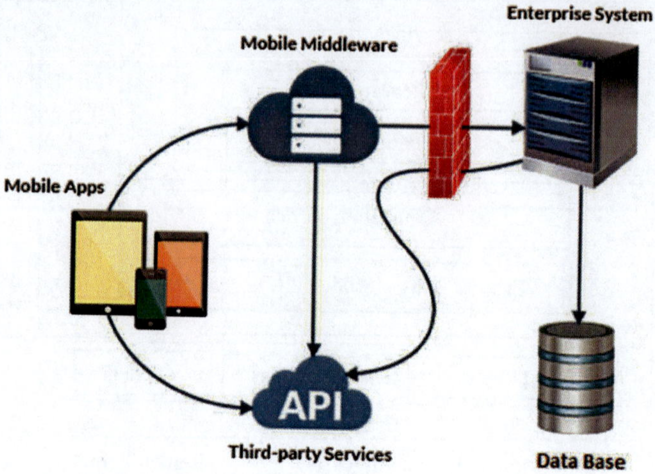

Fig. 2 Connection of the mobile hiring system to the database

Mobile middleware: It is a software that connects disparate mobile applications programs. It essentially hides the complexities of working in mobile environment, allowing for smother device-to-device interaction on the hiring system.

Third-party (API): It is a set of function and procedures allowing the creation of application that access the features or data on the hiring system.

Enterprise system: It is where all request or complains are directed to for approval or attended to.

Database: Farm machinery hiring system is developed with MySQL database because it supports every platform for designing such as Web, mobile, and desktop applications. It also supports remote connection through the Internet and is the most widely used database in the world because of its efficiency.

3.3 *Use Case and Activity Design Considerations*

The design process for the mobile application is illustrated in a graphical notation that shows the interactions between the users and the application. Use case diagram in Fig. 3 is a representation of a user's interaction with the system that shows the relationship between the user and the different actions in which the user is involved. The activity diagrams in Figs. 4 and 5 show how an admin and user (Farmer) flow of activities on the hiring system relates.

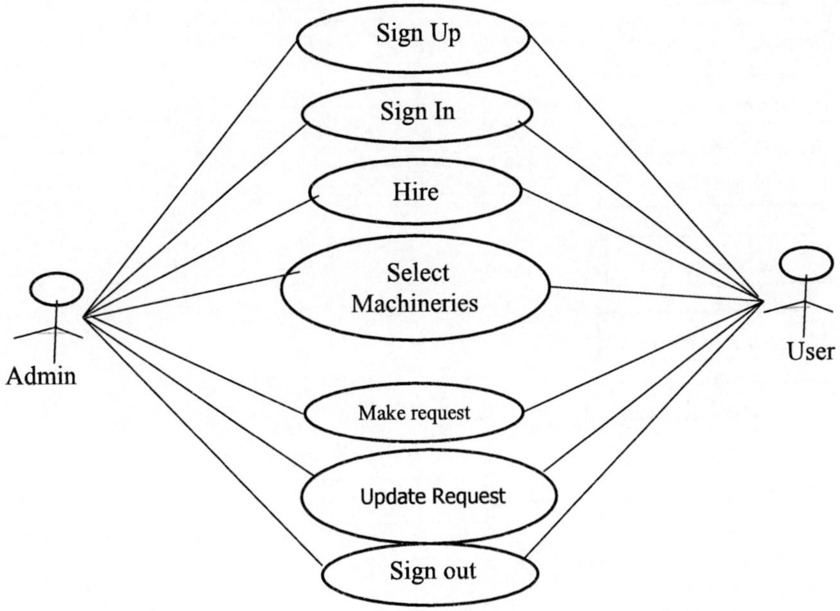

Fig. 3 Use case diagram

3.4 *Business Model Canvas of the Farm Machinery Hiring System*

A business model can be explained by its basic building blocks that presents the way for earning finance. There are four main pillars and these include: customers, offer, infrastructure, and financial viability. System for business model is shown in Table 1.

3.5 *Hiring Strategy*

Several strategies are put in place to ensure smooth relationship between our service and the customers. They include:

i. Any farmer can order for the delivery of any machinery/equipment
ii. If equipment is not currently available in store, the date it would return to store will be made known to the customer. Provide the customer chooses to proceed with the order; the newly available date will then be assigned to the new customer.
iii. Farmer can also request for a technician along with the machinery/implement if they do not have technical skill to use them.

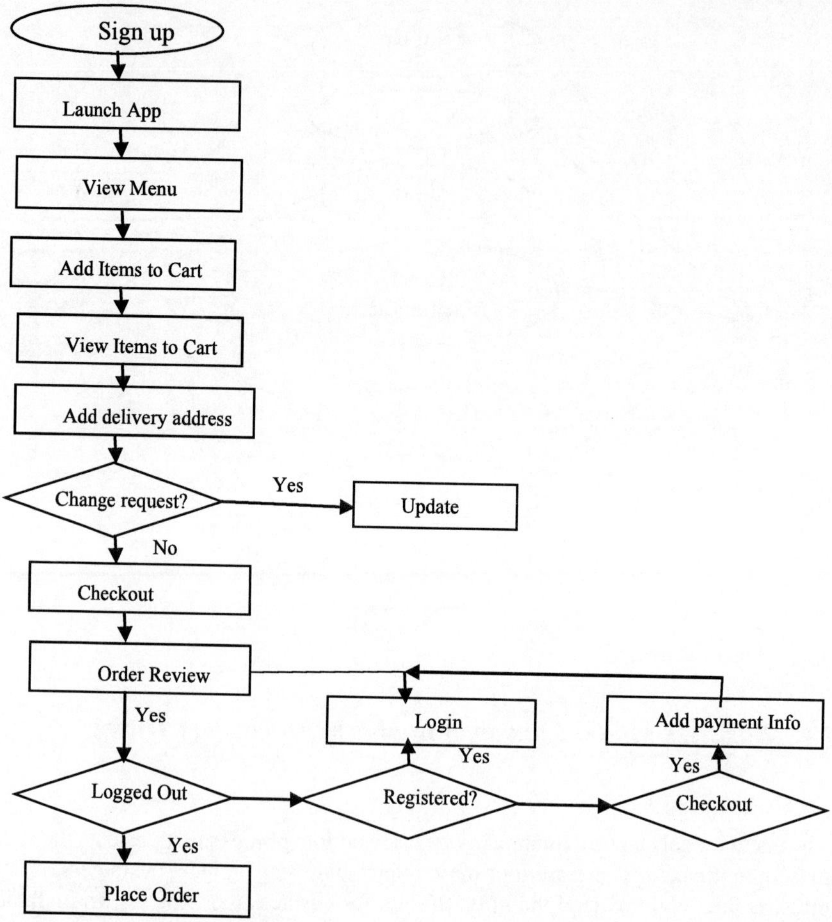

Fig. 4 User activity diagram

4 Result and Discussion

4.1 Implementation

The application is designed from a user point of view. This application supports graphical user interface which enables the users to interact and accomplish their task with ease. The design was simple and understandable. The database used is MySQL where all the information related to markets and users is stored. PHP was used as a server-side scripting language in connecting the application with the database with the files located on the XAMPP Server. GPS and Google Maps were integrated into the application to enable locations and machinery search as shown in Fig. 6.

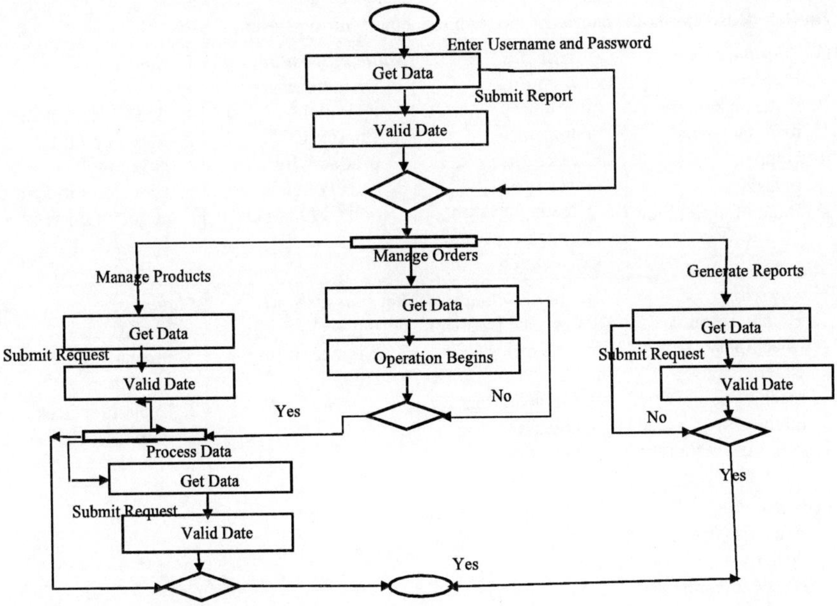

Fig. 5 Admin activity diagram

The program requirements for the implementation of the proposed application are MySQL, JavaScript, PHP, Phonegap/Cordova, and XAMMP. These are suitable and flexible for end customers to use.

4.2 *Evaluation and Discussion*

Formative evaluation was used in this research. It is primarily qualitative in nature; the formative evaluation was conducted through interviews and open-ended questionnaires. This is to ensure quality of service and find a better way of delivering them while having our farmers in mind. The analysis of the questionnaires was done using technology acceptance models (TAMs).

Perception evaluation for farm machinery hiring system

The user perception evaluation for farm machinery hiring system was carried out using 16 farmers—12 male famers and 4 female farmers. Questionnaires were administered and filled by these farmers. The questions raised in the user perception questionnaire were based on the technology acceptance model (TAM). Four factors from TAM were used, which are:

1. Perceived Usefulness (PU): To access farmer's perception of the usefulness of farm machinery hiring system

Table 1 Business model canvas of the farm machinery hiring system

Key partnership	*Key activities*	*Value proposition*	*Customer's SEGMENT*
1. Farmers 2. Farm machinery manufacturing company 3. Investors 4. Bank of agriculture	1. Reach for machinery companies 2. Search for investors 3. Creating awareness through farmer's association	1. Reduced cost of operations 2. Improved productivity 3. Timely machinery delivery on request	1. Based on farm size 2. Based on farm proximity 3. Based on farmer's common interest
Key resources 1. Human resources (machinery experts, auto mechanic engineers) 2. Financial resources capital *Cost structure* 1. After service visitation 2. Physical reach out to machinery companies for inquiry 3. Meeting investors 4. Purchase of equipment (mower, tractor, sprayer, combine harvester)	*Customer relation*1. After service feedback (farmer's hiring) *ship* 2. Farm visitation (farmer's acquisition)	*Revenue stream*1. Hiring 2. Engineer hiring 3. Training	*Channels* 1. Social medias(digital marketing) 2. Radio broadcast 3. Farm association

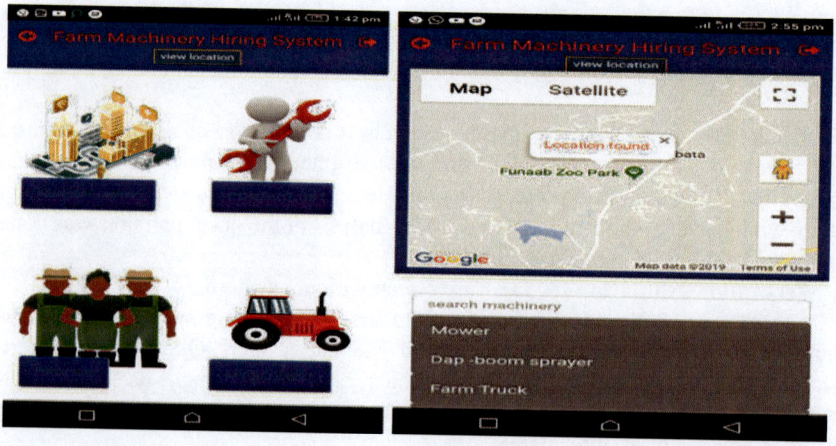

Fig. 6 Interface of farmers' profile and the present location of the farmer to hire machines through GPS

2. Perceived Ease of Use (PEOU): To determine what degree of ease is associated with the usage of the farm machinery hiring system.
3. Attitude toward Using (AH): Used to determine if the use of farm machinery hiring system is a wise idea?
4. Behavioral Intention (INT): Used to determine the degree to which a farmer has formulated plans to use a farm machinery hiring system.

A total of 14 questions was formulated and measured on male farmers and female farmers. This questionnaire was given to 16 respondents. Table 2 summarizes the overall usability feedback of farm machinery hiring system. The feedback received from the participants showed that farmer perception is evidenced by the 4.02 mean score for perceived usefulness, 4.25 mean score for perceived ease of use, 4.56 mean score for attitude toward using, and 4.94 mean score for behavioral intention. Perceived usefulness (PU) results are given in Table 3.

Four items were used to measure the perceived ease of use (PEOU) TAM construct, as shown in Table 4. Further investigation of participants' feedback for PEOU suggests that the farm hiring machinery system is convenient to use than the manual approach to machinery hiring. This finding is supported by the high mean score for PEOU1 and PEOU2 with 5.00 and 3.63, respectively. Considering all accumulated feedback, the PEOU construct also received a median score of 4.40, which evidences 94% agreement on the ease of use of farm machinery hiring system as perceived by farmers.

Attitude toward using (ATU) was measured through three questions, AH1, AH2, and AH3 as shown in Table 5. Our findings indicate agreement on participants' idea to the farm machinery hiring system, with a mean score of 4.88 for AH1 and embraces the software as a good innovation with a high mean score of 3.81 for AH2.

Table 2 Usability evaluation result

Factors	Overall mean	Overall median
Perceived usefulness (PU)	4.02	4.25
Perceived ease of use (PEOU)	4.25	4.40
Attitude toward using (AH)	4.56	4.67
Behavioral intention (INT)	4.94	5.00
Overall average	4.44	4.58

Table 3 Perceived usefulness (PU) results

Factors	Overall mean	Overall median
PU1_Modern approach	5.00	5.00
PU2_Prefer farm machinery hiring system	3.81	4.00
PU3_Accurate Information	3.63	4.00
PU4_ResponseToQuestions	3.63	4.00
PU_Overall	4.02	4.25

Table 4 Perceived ease of use (PEOU) results

Factors	Overall mean	Overall median
PEOU1_Comfort	5.00	5.00
PEOU2_Convenience	3.63	4.00
PEOU3_LessTime	4.94	5.00
PEOU4_UserFriendly	3.94	4.00
PEOU5_HighSpeed	3.75	4.00
PEOU_Overall	4.25	4.40

Table 5 Attitude toward using (AH) results

Factors	Overall mean	Overall median
AH1_LikeTheIdea	4.88	5.00
AH2_GoodInnovation	3.81	4.00
AH3_GoodExperience	5.00	5.00
AH_Overall	4.56	4.67

In general, an overall median score of 4.67 indicates that most farmers like the idea of using farm hiring system.

Finally, behavioral intention (INT) was measured through two questions, INT1 and INT2 shown in Table 6.

Our findings indicate agreement on participants' intention to use a farm machinery hiring system, with mean score 4.88 for INT1. The participants also agreed or strongly agreed that they expect to use a farm machinery hiring system with a mean score of 5.00 for INT2. In all, a median score of 5.00 was also recorded from the behavioral intention assessment, evidencing a high level of agreement among participants with respect to the BI construct. Finally, by assessing the overall result as shown in Table 2, we observe that attitude toward using has the highest average mean score of 4.56 and an average median score of 4.67 when compared to other constructs. This implies that a farm machinery hiring system is a wise idea and the farmers have a positive attitude toward the use of the hiring system.

Table 6 Behavioral intention to use (INT) results

Factors	Overall mean	Overall median
INT1_IntentionToUse	4.88	5.00
INT2_ExpectToUse	5.00	5.00
INT_Overall	4.94	5.00

5 Conclusion

The fundamental goal of the research work is to design a hiring system based on mobile technology for farmers to hire machines with flexibility and ease at their convenience and at a lower price rate. The new system made use of well secured, optimized, and structured database management system for proper storing and retrieval of information. The capabilities of the system are user friendly; the database is capable of storing and retrieving farmers' information quickly. In conclusion, the implementation of this work by any farming association or farmers nationwide as a means of hiring machines would enhance the quality and productivity of their farming and harvest.

In the future, the authors intend to have a planned longitudinal study with economic concerns for a period not more than five years in order to develop an ontology based on the architectural model to assist researchers and farmers.

References

1. Ray DK, Mueller ND, West PC, Foley JA (2013) Yield trends are insufficient to double global crop production by 2050. PLoS ONE 8(6):e66428
2. Godfray HCJ, Beddington JR, Crute IR, Haddad L, Lawrence D, Muir JF, Toulmin C (2010) Food security: the challenge of feeding 9 billion people. Science 327(5967):812–818
3. Licker R, Johnston M, Foley JA, Barford C, Kucharik CJ, Monfreda C, Ramankutty N (2010) Mind the gap: how do climate and agricultural management explain the "yield gap" of croplands around the world? Glob Ecol Biogeogr 19(6):769–782
4. Ajah J (2014) Factors limiting small-scale farmers' access and use of tractors for agricultural mechanization in Abuja, North Central Zone Nigeria. Euro J Sustain Dev 3(1):115–124
5. WEF-World Bank-OECD (2015) Africa competitiveness report Geneva
6. OECD-FAO (2014) OECD-FAO agricultural outlook. OECD, Paris
7. Aguilera E, Guzmán GI, González de Molina M, Soto D, Infante-Amate J (2019) From animals to machines. The impact of mechanization on the carbon footprint of traction in Spanish agriculture: 1900–2014. J Clean Prod 221:295–305
8. Azeta J, Bolu CA, Alele F, Daranijo EO, Onyeubani P, Abioye AA (2019) Application of mechatronics in agriculture: a review. J Phys: Conf 1378(3)
9. Tyagi A, Gupta N, Navani JP, Tiwari MR, Gupta MA (2017) Smart irrigation system. Int J Innov Res Sci Technol 3(10)
10. Ramli NL, Mohd Yamin N, Ab Ghani S, Saad NM, Md Sharif S (2015) Implementation of passive infrared sensor in street lighting automation system. ARPN J Eng Appl Sci 10(22):17120–17126
11. Hessel L, Bar-On D (2003) U.S. Patent No. 6,508,033. Washington, DC: U.S. Patent and Trademark Office
12. Srinivas KR (2018) Cooperation in agriculture in AAGC: innovations and agro-processing. RIS/AAGC-DP, vol 221
13. Arogundade OT, Abayomi-Alli A, Adesemowo K, Bamigbade T, Odusami M, Olowe V (2020) An intelligent marketspace mobile application for marketing organic products. In Responsible design, implementation and use of information and communication technology. I3E 2020. Lecture notes in computer science, vol 12066. Springer, Cham, pp 276–287
14. Takeshima H, Hatzenbuehlerb PL, Edeh HO (2020) Effects of agricultural mechanization on economies of scope in crop production in Nigeria. Agricu Syst 177 (102691)

15. Matiyabu I, Ndayizigamiye P (2019) Enhancing agricultural practices through mobile technology interventions: a case of the democratic republic of Congo. In: IEEE global humanitarian technology conference (GHTC)
16. Pongsuwan W, Pongsuwan H (2019) Social media for smart farmer-shared farming equipment model. Inform Manage Bus Rev 11(2):1–9
17. Hassan F, Kornher L (2019) Let's get mechanized—labor market implications of structural transformation in Bangladesh, pp 1–26
18. Sims BG, Hilmi M, Kienzle J (2016) Agricultural mechanization: a key input for sub-Saharan Africa smallholders. FAO, Rome (Italy). Plant Production and Protection Div. 23
19. Bello RS, Onyeanula P, Saidu MJ, Bello MB (2015) Mechanization and agricultural productivity functions in Umuahia North LGA, Abia, Nigeria. Sci J Bus Manage 3(5-1):21–25
20. Verma SR (2006) Impact of agricultural mechanization on production, productivity, cropping intensity income generation and employment of labour. Status Farm Mech India, pp 133–153
21. Jonathan O, Ogbunude C, Misra S (2018) Design and implementation of a mobile-based personal digital assistant (MPDA). Commun Comput Inform Sci 1031:15–28
22. Jonathan O, Misra S, Ibanga E, Maskeliunas R, Damasevicius R, Ahuja R (2019) Design and Implementation of a mobile webcast application with google analytics and cloud messaging functionality. J Phys Conf Ser 1235(1):012023
23. Abayomi-Zannu TP, Odun-Ayo I, Tatama BF, Misra S (2020) Implementing a mobile voting system utilizing blockchain technology and two-factor authentication in Nigeria. Lect Notes Network Syst 121:857–872
24. Oluwagbemi O, Adewumi A, Misra S, Leon M (2020) MAFODKM: mobile application framework for the management of Omics data and knowledge mining. J Pyhs Conf Ser 1566(1):012132
25. Holz HJ, Applin A, Haberman B, Joyce D, Purchase H, Reed C (2006) Research methods in computing. In: Working Group Reports on ITiCSE on Innovation and Technology in Computer Science Education—ITiCSE-WGR '06
26. Saunders MNK, Lewis P, Thornhill A (2019) Understanding research philosophy and approaches to theory development. In: Research methods for business students. Pearson, Harlow, pp 128–170

Cloud Computing Offered Capabilities: Threats to Software Vendors

Oluwasefunmi Arogundade, Funmilayo Abayomi, Adebayo Abayomi-Alli, Sanjay Misra, Christianah Alonge, Taiwo Olaleye, and Ravin Ahuja

Abstract Cloud computing has permeated and penetrated every aspect of our lives both personal and professional. This technology has gained a phenomenal acceptance as it provides opportunities for organizations, by offering large collections of easily accessible data. This has transformed the way businesses are done. The ever-changing demands of users, fierce competitive activities, and rapidly evolving technology pose a great challenge to software vendors to be innovative and constantly seek to be on top of their game. Cloud computing allows business to scale efficiently as they constantly adjust their operations to the new realities that help them to optimize cost, quality, and time. The main objective of this project is to ascertain if cloud computing offered capability is a threat to software vendors, and the extent of the threat, thus proposing solutions. This study used multi-stage research technique and the research instrument being questionnaire. The completed questionnaire forms were collated, coded, and analyzed using both descriptive and parametric statistics. Quantitative data was coded and entered into statistical packages for social scientists and analyzed using descriptive statistics for expressing the demographic characteristics of the respondents. Qualitative data was analyzed based on the content of the responses. Responses with common themes or patterns were grouped together into coherent categories. Chi-square analysis and logistic analysis were used to show dependencies and relationships between the variables and also to determine the probability that software vendors are aware of cloud computing threats and the severity of those

O. Arogundade · F. Abayomi · A. Abayomi-Alli · C. Alonge · T. Olaleye
Federal University of Abeokuta, Abeokuta, Ogun, Nigeria
e-mail: arogundadeot@funaab.edu.ng

F. Abayomi
e-mail: abayomiallia@funaab.edu.ng

S. Misra (✉)
Covenant University, Ota, Nigeria
e-mail: sanja.misra@covenantuniversity.edu.ng

R. Ahuja
Shri VikarmShilla University, Gurgaon, India

P. K. Singh et al. (eds.), *Proceedings of Second International Conference on Computing, Communications, and Cyber-Security*, Lecture Notes in Networks and Systems 203,
https://doi.org/10.1007/978-981-16-0733-2_16

threats. Quantitative data was rendered in tables and explanation was presented in logical prose.

Keywords Cloud computing · Software vendors · Threats · Information technology · Measurement · Maintainability

1 Introduction

Cloud computing has been a mainstay of the IT world for several years now. Many authorities and practitioners have described it as the backbone of IT which has enjoyed a lot of acceptance from all and sundry in the technology industry across the globe and the impact on business will only continue to grow in the future.

Cloud computing extends computing from the remote of the desktop to the World Wide Web, whereas maintenance and management of resources are out of the bother of the end user. It is likewise the delivery of applications, platforms, data management, operating systems, and other computing functionalities over the Internet instead of on-the-premises infrastructure systems [1]. User of smart phones could easily be a large datacenter with the advent of cloud computing as an extended form of distributed computing, equivalent computing, and network computing [2]. With cloud computing, companies need no huge investment in hardware purchases nor high level personnel recruitment for on-site management; subscription is simply made to access similar services and needed computing needs and only pay for required services [2]. In the work of [3] it was noted that cloud computing provides computing resources, applications as facilities to end users through the Internet. Similarly, it was pointed out that it is a new organizational approach to infrastructure management to form a huge reserve pool and provide flexible application of hardware and software resources upon demand. Cloud computing can also provide easy access to data, projects, and vital business software from any location in the world that provides a fast Internet connection as it also removes the worry about replacing and configuring lost or stolen application software since the software is on the cloud. With these advancements in computing however comes with opportunities, innovations, advancements, productivity, promptness, high return on investments and possibly challenges and threats.

The motivation for this research is established on the way that cloud computing has offered various assistance and innovation to create information and figure requesting equal applications with substantially more conservative costs contrasted with customary equal registering approach. However, it has thus made independent software vendors to change their activities and modify their business structures to accomplish more prominent effectiveness, compelling and adjust to the new wonder of lower normal deals, costs and lower margins, which may be viewed as representing a huge danger to them.

The main thrust of this project is to check if cloud computing offered capabilities are threats to software vendors/merchants and then propose solutions.

Session 2 of this work studies literatures associated therewith the concept, scope, and area of application of cloud computing with associated challenges and threats. Session 3 presents the methodology of this study in carrying out the objectives while session 4 presents the result and discussion. Session 5 is the concluding part with recommendations.

2 Related Works

The subject of cloud computing has been severally discussed in literature with divergent views on germane issues related therewith in addendum to its areas of applications and adoption of the information technology by experts, most especially software development experts. In the work of [1], issues involved in integrating the huge benefits of cloud computing with Internet of things (IoT) were discussed which was referred to as cloud of things. The paper unraveled the components of the proposed cloud of things which addresses several issues including security, protocol support, energy efficiency, identity management, resource allocation, and location of data storage.

In [2] the authors mentioned the role of cloud computing in the educational sector and the issues related to it. The paper underscores the huge burden of responsibilities in terms of quality service delivery in the education subsector while understudying the benefits cloud computing could offer in changing the course of direction in the application of computing in education management. It unravels the stability and the convenient swiftness of query libraries that cloud computing could offer.

While understudying the impact of cloud computing on enterprise software vendors' business models, Boillat and Legner [4] unraveled the challenges this advancement poses to enterprise vendors who might need to adopt new logical dimensions in their quest to adjust to the new realities cloud computing offers. From many case studies covering traditional cloud providers, the work discovers that moving from on-premise software to cloud services has an adverse effect on all business model components including the end user value proposition, resource base, value structure, and monetary flows. It then reinforces cloud computing's unsettling nature in the enterprise software business domain.

The work done in [5] opinionated that the evolution of the SaaS business model was considered impressive and catches the fancy of both scholars and IT practitioners as SaaS vendors deliver on-demand information enterprise services to users, and thus offer a bouquet of computing utility rather than the stand-alone software by vendors. The author showed that a software application that is modulated, open, and standardized will always command the needed market attention. Furthermore, under certain market situations, offers that give users an easy exit option through the software bond will help to upsurge the SaaS vendors' competitiveness.

While understudying a comprehensive survey on cloud computing, Patidar et al. [6] opine different categorization of clouds to depend on where the owner of the cloud data center is located. The authors gave different descriptions to private and public cloud.

A note of caution was the essence of the expository work of [7] while trying to uncover the gray areas in cloud computing that could endanger end user enterprises who are trying to migrate to the cloud computing facilities in the life of their business enterprise. Privacy concerns were the main thrust of the work of [8] which unravels the various levels of concerns in the adoption of cloud computing by enterprises.

Advantages of scalability, resilience, flexibility, efficiency, and outsourcing of cloud computing notwithstanding, security and vulnerability associated therewith were the concern of [9] in his work. The paper noted security as one of the germane issues that could make or mar the acceptance or otherwise of cloud computing noting that the idea of handing over sensitive data of organizations to third parties calls for concern such that the end users need to be watchful in understanding the risks of data breaches in this new platform.

In trying to find a lasting solution to the security concerns mentioned in other literatures, Sun et al. [10] indeed surveyed several security threats militating against adoption of cloud computing toward encouraging more migration of end users by categorizing noted threats into tangible and intangible versions. In [11] the research work understudied various factors that has continued to endeared small enterprises toward the adoption of cloud computing in their daily business processes. It is noted that the provision of the services of cloud computing on pay-as-you-go way makes the enterprise unique compared with payment modalities adopted by traditional computing services. The paper argues that cloud computing tends to prove commercially viable for many small and medium outfits (SMEs) due to its flexibility and convenient payment structure, particularly in the prevalent global cash crunch.

Authors of [12] note the degree of risk consciousness among users of cloud computing with respect to the uniqueness of their desired cloud computing functionality. The outcome of the research is a pointer to the creature of a risk management system prior to the adoption of cloud computing by Switzerland business corporations.

Some school of thoughts like [13] opines that there has never been a more disruptive innovation in the IT landscape such as cloud computing brought ever since the advent of the concept of Internet. The author believes the advent of cloud computing has altered corporate cultures and modus operandi in their bid to advance the course of their service delivery models. The paper further did a survey of the empowering and democratizing benefits of cloud computing which has continued to challenge norms and status quo across public and private organizations.

A forensic enquiry into the cloud computing environment is the basis of the work of [14] which is an expository attempt on the germane issues associated therewith the subject of cloud computing.

The authors in [15] propose a FT Cloud, for building fault-tolerant applications developed through component ranking model; the model is in twofold; the first deployed component entreaty structures and invocation rates for making noteworthy component grading. The authors in [16, 17], and [18] discussed cloud computing issues that concern security and privacy as well as quality models for evaluating platform as a service among other. In [18], a secured private cloud computing system was proposed.

Considering the volume of prevalent global economic slump, Aljabre [19] noted cloud computing fills the void for an affordable and reliable technology to meet daily data processing and storage needs with flexible price regime but observed that cloud computing is not necessarily needed by all businesses. The research work by [20] emphasizes the dire need for understanding the business-related issues around cloud computing including the industry (SWOT) analysis—strengths, weaknesses, opportunities, and threats and then classifies the various issues that affect different practitioners of cloud computing industry.

While there are so many literatures research related to cloud computing and its many interesting applications in business and other human endeavor, the inexhaustible research domain has been an eye opener to other germane undercurrent issues which has further tightened competitiveness of local service providers in IT retailing and cloud computing service providers. In tandem with this observation is the observation that customer service and support has expectedly become a 24×7 commitment response time that customers expect from IT vendors due to the same largesse provided by the cloud computing service providers (PWC Report. Global 100 Software Leaders Report, 2016).

3 Methodology

This section describes the tools and processes involved in the determination of threats that cloud computing has on the activities and businesses of independent software vendors through three main stages of instrument pretest and validation, questionnaire administration, and data analysis. The sequence diagram of the model is as presented in Fig. 1 which encapsulates all stages aforementioned.

3.1 Instrument and Data Collection

The target population for this study comprised of 200 male and female independent software vendors from various departments in Federal University of Agriculture, Abeokuta. The sample size of this study was the 200 independent software vendors. The same number of questionnaires was administered to the independent software vendors who were also the participants because they are deemed to have the relevant information regarding the threats that cloud computing poses to them. The 200 ISVs were randomly sourced from various departments to answer some questions regarding the possible threats that cloud computing poses. The research instrument used for the collection of data was questionnaire. The questionnaire had two basic

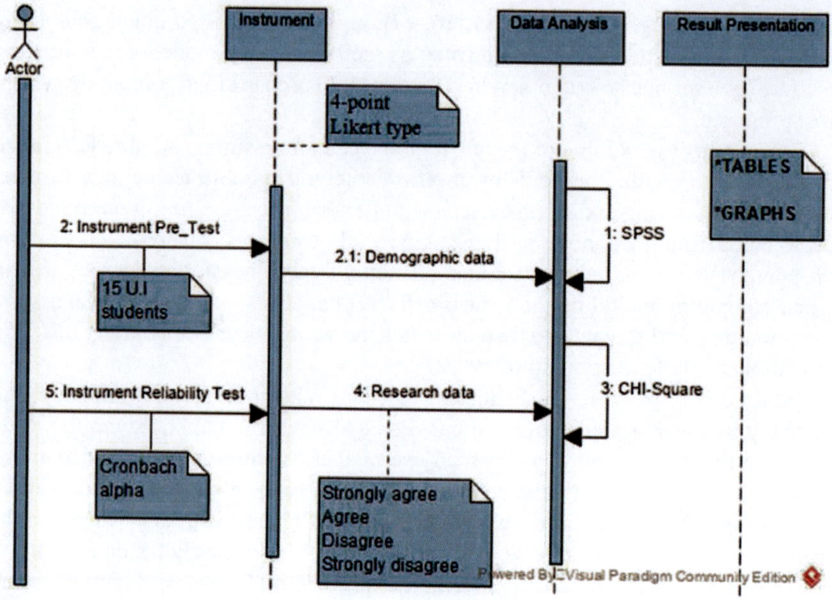

Fig. 1 Sequence diagram of the study

sections: Section A receives the demographic data of the respondents while Section B has the gathered information regarding the study. A 4-point Likert-type format using: Strongly Agree, Agree, Disagree, and Strongly Disagree was used for the rating parameter.

The first stage questionnaire examines the fact that cloud computing offered capability is a threat to software vendors. The second stage questionnaire was administered to know the extent to which the software vendors are being threatened.

3.2 *Instrument Reliability and Validation*

Pretesting of an instrument enables the researcher to discover ambiguities contained in the questions which might have constituted other limitations not envisaged during the study. The pretest of the instrument was carried out in the University of Ibadan with 15 students from who were not part of the study so as to ensure that the validity of the instrument. It was subjected to principal component analysis (PCA) and also it was given to my supervisor, colleagues, and other experts in research to test on the content and face validity. In order to confirm the reliability of the instrument, the questionnaires were subjected to overall reliability analysis of internal consistency. This was measured using Cronbach alpha as a coefficient of internal consistency. Castillio (2009) made available the following rules of the thumb: > 0.9—Excellent,

> 0.8—Good, > 0.7—Acceptable, > 0.6—Questionable, >0.5—Poor, and <0.5—Unacceptable. The acceptable value of 0.7 was used as a cut–off for unwavering quality of this examination.

3.3 *Data Analysis*

The completed questionnaire forms were collated, coded, and analyzed using both descriptive and parametric statistics. Quantitative data was coded and entered into statistical packages for social scientists and analyzed using descriptive statistics for explaining the demographic characteristics of the respondent. Chi-square analysis and logistic analysis were used to show dependencies and relationships between the variables and also to determine the probability that software vendors are aware of cloud computing threats and the severity of those threats. Quantitative data was presented in tables and the explanation was presented in prose.

4 Results and Discussion

The respondents' court across ISVs from the college of physical sciences and those from the non-physical sciences of the university and their distribution is as presented on Table 1. From the table presented, respondents representing physical sciences have the highest frequency of 138 (69%) participation while respondents representing college of non-physical sciences have a frequency participation of 62 (31%). The respondents' year of expertise is likewise presented on Table 2 where 108 (54%) respondents have less than 3 years of expertise, 65 (32.5%) respondents have 3 years of expertise, while 27 (13.5%) have more than 3 years of expertise. On Table 3, the distribution of respondents' area of expertise is presented with 53

Table 1 Distribution of respondents by colleges

College	Frequency	Percentage (%)
Physical sciences	138	69
Non-physical sciences	62	31
Total	200	100

Table 2 Distribution showing respondents' years of expertise

Years of expertise	Frequency	Percentage (%)
Less than 3 years	108	54
3 years	65	32.5
Greater than 3 years	27	13.5
Total	200	100

Table 3 Distribution of respondents' field of expertise

Field of expertise	Frequency	Percentage (%)
Android/Web/graphics design	53	26.5
IT support/DBMS/networking/data science/cybersecurity	32	16
Software development	115	57.5
Total	200	100

Table 4 Frequency 1

Threat awareness						
Variables	Sa	A	I	D	Sd	Total
	N (%)	N (%)	N (%)	N (%)	N (%)	
You do have a prior knowledge of cloud computing	22 (11.0)	158 (79.0)	8 (4.0)	0 (0.0)	12 (6.0)	200
You are aware of some of the threats that cloud computing poses	40 (20.0)	145 (72.5)	3 (1.5)	0 (0.0)	12 (6.0)	200
Cloud computing might really be a threat to software vendors	40 (20.0)	130 (65.0)	4 (2.0)	22 (11.0)	4 (2.0)	200
Software as a service is a main threat to independent software vendors	36 (18.0)	112 (56.0)	15 (7.5)	35 (17.5)	2 (1.0)	200
Do you see yourself as a software vendor	40 (20.0)	160 (80.0)	0 (0.0)	0 (0.0)	0 (0.0)	200

(26.5%) respondents involved in Android/Web/graphics design, 32 (16%) involved in IT support/DBMS/networking/data science/cybersecurity, while 115 (57.5%) are involved in software development.

Tables 4, 5, and 6 show the relationship between the socio-demographic characteristics and awareness of the threat cloud computing poses to the software vendor. Years of expertise, gender, and colleges are found to be statistically dependent with *p-values* 0.002, 0.001, and 0.012, respectively. Gender however was also found to be statistically dependent $p = 0.001$ but with a cell count less than 5 which violates the fourth assumption if chi-square analysis of large expected frequencies, hence, Fisher's exact test *p-value* was reported.

All other variables are found to be independent and we fail to reject the null hypothesis and as such it was concluded that there is no relationship between the variables and threat awareness.

Tables 7, 8, 9 show that those in the field of IT support, database management, networking, data science, and cybersecurity are 1.3 times more likely to feel the severity of the threat than those in the Android development, Web development, UI/UX design, and graphics design fields, which is not statistically significant with *p-value* = 0.619. Software developers are 3.6 times more likely to feel the severity of the

Table 5 Frequency 2

Threat severity						
Variables	SA	A	I	D	SD	Total
	n (%)	n (%)	n (%)	n (%)	n (%)	
Cloud computing has allowed independent software vendors to reduce sales price	29 (14.5)	127 (63.5)	20 (10.0)	20 (10.0)	4 (2.0)	200
Independent software vendors have had to deal with high expectation and pressure from customers due to the agility that cloud computing offers	58 (29.0)	125 (62.5)	15 (7.5)	2 (1.0)	0 (0.0)	200
Independent software vendors have had to become data centers operators or at least contract with a data center for efficiency and for ensuring satisfaction from customers	57 (28.5)	128 (64.0)	12 (6.0)	3 (1.5)	0 (0.0)	200
Independent software vendors are forced to release continuous innovations as customers using cloud, demands continuous innovations	46 (23.0)	140 (70.0)	4 (2.0)	8 (4.0)	2 (1.0)	200
Independent software vendors have had to frequently upgrade their software	69 (34.5)	119 (59.5)	12 (6.0)	0 (0.0)	0 (0.0)	200
Independent software vendors have had to quickly respond to business demands in order not to lose customers	46 (23.0)	131 (65.5)	12 (6.0)	7 (3.5)	4 (2.0)	200
Independent software vendors are challenged by evolving technologies, including cloud computing	39 (19.5)	135 (67.5)	12 (6.0)	12 (6.0)	2 (1.0)	200

threat posed by cloud computing capabilities to software vendors than their Android development, Web development, UI/UX design, and graphics design counterparts, and this is statistically significant with *p-value* = 0.005.Threat awareness was also evaluated and those with good awareness or those who are well aware of the threat posed by cloud computing capabilities are 5.81 times more likely to feel the threat severity than those who are not well aware of the threat posed.

Table 7 describes the relationship between the socio-demographic characteristics and severity of the threat cloud computing poses to the software vendor. Both fields of expertise and threat awareness are found to be statistically dependent with *p-value* 0.001,*and* $p < 0.01$, respectively.

Table 6 Crosstabs (Chi-square analysis) 1

Threat awareness					
Variables	Good	Poor	Total	X^2	*p-value*
	n (%)	n (%)			
Age group					
18–24 years	123 (67.6)	59 (32.4)	182	1.06	0.302
>24 years	10 (55.6)	8 (44.4)	18		
Gender					
Male	106 (62.0)	65 (38.0)	171	–	0.001[a]
Female	27 (93.1)	2 (6.9)	29		
College					
Non-physical sciences	49 (79.0)	13 (21.0)	62	6.34	0.012*
Physical sciences	84 (60.9)	54 (39.1)	138		
Field of expertise					
Android/Web/graphics design	31 (58.5)	22 (41.5)	53	5.25	0.073
IT/DBMS/networking/data science/cybersecurity	18 (56.3)	14 (43.8)	32		
Software development	84 (73.0)	31 (27.0)	115		
Years of expertise					
<3 years	71 (65.7)	37 (34.3)	108	12.25	0.002*
3 years	51 (78.5)	14 (21.5)	65		
>3 years	11 (40.7)	16 (59.3)	27		

[a]Fisher's Exact
**p-value* < 0.05 [b]Likelihood ratio

All other variables are found to be independent and we fail to reject the null hypothesis and as such we conclude that there is no relationship between the variables and threat awareness.

Further investigation on the level of dependence from the chi-square was carried out using logistic regression analysis and the table above reports the odds ratio (OR) value of each variable. The table above shows that the respondents of the feminist gender are 8.23 times more likely to be aware of the threat that cloud computing poses to software and it is statistically significant with *p-value* = 0.007 than their male counterparts. Although those in non-physical science colleges were reported to be 1.53 times more likely to be aware of the cloud computing capability threat to software vendors than those in physical science college, this however, is not significant statistically with *p-value* = 0.275. Respondents who have 3 years and less than 3 years of expertise in their respective fields are 4.66, 3.12 times more likely to be aware of the threat posed by cloud computing capabilities than those who have spent less more 3 years in their various fields, respectively, this is significant statistically with *p-values* = 0.004, 0.016, respectively.

Table 7 Crosstabs (Chi-square analysis) 2

Threat severity					
Variables	High	Low	Total	X^2	*p-value*
	n (%)	n (%)			
Age group					
18–24 years	150 (82.4)	32 (17.6)	182	2.64	0.104
>24 years	12 (66.7)	6 (33.3)	18		
Gender					
Male	139 (81.3)	32 (18.7)	171	0.06	0.802
Female	23 (79.3)	6 (20.7)	29		
College					
Non-physical sciences	54 (87.1)	8 (12.9)	62	2.17	0.141
Physical sciences	108 (78.3)	30 (21.7)	138		
Field of expertise					
Android/Web/graphics design	36 (67.9)	17 (32.1)	53	13.10	0.001*
IT/DBMS/networking/data science/cybersecurity	23 (71.9)	9 (28.1)	32		
Software development	103 (89.6)	12 (10.4)	115		
Years of expertise					
<3 years	85 (78.7)	23 (21.3)	108	3.10	0.212
3 years	57 (87.7)	8 (12.3)	65		
>3 years	20 (74.1)	7 (25.9)	27		
Threat awareness					
Good awareness	121 (91.0)	12 (9.0)	133	25.68	0.000*
Poor awareness	41 (61.2)	26 (38.8)	67		

**p-value* < 0.05

Table 8 Logistic regression 1

Threat awareness			
Variables	Or	95% C. I	*P-value*
Gender			
Male	1	–	–
Female	8.23	1.8–37.66	0.007*
College			
Physical sciences	1	–	–
Non-physical sciences	1.53	0.71–3.30	0.275
Years of expertise			
>3 years	1	–	–
3 years	4.66	1.62–13.46	0.004*
<3 years	3.12	1.24–7.87	0.016*

**P-Value* < 0.05

Table 9 Logistic regression 2

Variables	OR	95% C. I	*p-value*
Field of expertise			
Android/Web/graphics design	1	–	–
IT/DBMS/networking/data science/cybersecurity	1.30	0.46–3.69	0.619
Software development	3.60	1.49–8.70	0.005*
Threat awareness			
Poor awareness	1	–	–
Good awareness	5.81	2.63–12.81	0.000*

**p-value* < 0.05

5 Conclusion

Respondents' responses on variables of study using chi-square revealed that there is no significant dependency on variables and threat awareness, although using chi-square analysis and logistic regression to measure the threat awareness that cloud computing poses to software vendors shows a level of dependency between gender, college, and field of expertise, but it shows no concrete relationship among other variables. We can therefore conclude that the rate of awareness is high, especially among those with 3 years and above years of expertise. Software vendors can investigate the work of cloud computing by utilizing hybrid approach with a straightforward start. Leveraging and modernizing existing foundation acknowledges adequacy through the discussion of static servers, stockpiling and systems administration into a virtualized pool of resources. It can begin with a division with scarcely any applications with a couple of clients. Obviously, there will be a move to the cloud. Training of staff and furthermore helping them to understand the ability required for the organization to be proactive and explore emerging opportunities 1 on the grounds that it pushes ahead in terms of innovation and competitiveness. Cloud computing undoubtedly changes our lives in a big way, from users to content to applications and policies. Software vendors should be customer-centric by always thinking of the users before they think about data center, network, infrastructure, software, and hardware. It is immensely important to always be prepared to profile their end users in terms of their needs, contents, applications and industry and/or business specific stated or implied need and weave this into their operating policies.

Acknowledgements The authors acknowledge the sponsorship provided by Covenant University through the Centre for Research, Innovation and Discovery (CUCRID).

References

1. Aazam M, Khan I, Alsaffar AA, Huh EN (2014, January) Cloud of things: integrating internet of things and cloud computing and the issues involved. In: Proceedings of 2014 11th international Bhurban conference on applied sciences & technology (IBCAST) Islamabad, Pakistan, 14th–18th January, 2014, pp 414–419. https://doi.org/10.1109/IBCAST.2014.6778179
2. Chandra DG, Malaya DB (2012, March) Role of cloud computing in education. In: 2012 international conference on computing, electronics and electrical technologies (ICCEET), pp 832–836
3. Ma W, Zhang J (2012, July) The survey and research on application of cloud computing. In: 2012 7th International conference on computer science & education (ICCSE), pp 203–206. https://doi.org/10.1109/ICCSE.2012.6295057
4. Boillat T, Legner C (2013) From on-premise software to cloud services: the impact of cloud computing on enterprise software vendors' business models. J Theor Appl Electron Commerce Res 8(3):39–58. ISSN 0718-1876
5. Ma D (2007, July) The business model of "software-as-a-service". In: IEEE international conference on services computing (scc 2007), pp 701–702. https://doi.org/10.1109/SCC.2007.118
6. Patidar S, Rane D, Jain P (2012, January) A survey paper on cloud computing. In: 2012 second international conference on advanced computing & communication technologies, pp 394–398
7. Onwudebelu U, Chukuka B (2012, October) Will adoption of cloud computing put the enterprise at risk? In: 2012 IEEE 4th international conference on adaptive science & technology (ICAST), pp 82–85
8. Arnold S (2009) Cloud computing and the issue of privacy. KM World, 14–22
9. Kuyoro SO, Ibikunle F, Awodele O (2011) Cloud computing security issues and challenges. Int J Comput Networks (IJCN) 3(5):247–255
10. Sun D, Chang G, Sun L, Wang X (2011) Surveying and analyzing security, privacy and trust issues in cloud computing environments. Proc Eng 15:2852–2856. https://doi.org/10.1016/j.proeng.2011.08.537
11. Sultan NA (2011) Reaching for the "cloud": how SMEs can manage. Int J Inf Manage 31(3):272–278
12. Brender N, Markov I (2013) Risk perception and risk management in cloud computing: results from a case study of Swiss companies. Int J Inf Manage 33(5):726–733. https://doi.org/10.1016/j.ijinfomgt.2013.05.004
13. Sultan N (2013) Cloud computing: a democratizing force? Int J Inf Manage 33(5):810–815. https://doi.org/10.1016/j.ijinfomgt.2013.05.010
14. Taylor M, Haggerty J, Gresty D, Lamb D (2011) Forensic investigation of cloud computing systems. Netw Secur 2011(3):4–10
15. Zheng Z, Zhou TC, Lyu MR, King I (2011) Component ranking for fault-tolerant cloud applications. IEEE Trans Serv Comput 5(4):540–550. https://doi.org/10.1109/TSC.2011.42
16. Odun-Ayo I, Ajayi O, Misra S (2018) Cloud computing security: issues and developments. Lecture notes in engineering and computer science, 2235.
17. Olokunde T, Misra S, Adewumi A (2017) Quality model for evaluating platform as a service in cloud computing. Commun Comput Inform Sci 756:280–291
18. Olowu M, Yinka-Banjo C, Misra S (2018) A secured private-cloud computing, system, communications in computer and information. Science 1051:373–384
19. Aljabre A (2012) Cloud computing for increased business value. Int J Bus Soc Sci 3(1)
20. Marston S, Li Z, Bandyopadhyay S, Zhang J, Ghalsasi A (2011) Cloud computing—the business perspective. Decis Support Syst 51(1):176–189. https://doi.org/10.1016/j.dss.2010.12.006

The Sentimental Analysis of Social Media Data: A Survey

Vartika Bhadana and Hitendra Garg

Abstract Nowadays, machine learning plays a very important role in every field. For recommendation systems, user feedback is relevant because they contain different forms of emotional details that may affect the reliability or consistency of the recommendation. Online reviews, comments are very helpful in selecting the said items and services as it gives real feedback about the quality of these items and services. The categorizations of these items based on feedback provided by actual users are known as sentimental analysis. In this study, we described various machine learning techniques and parameters used for the sentimental analysis of reviews, comments, and feedback available on health care, Facebook, Twitter, and other social media networking sites. The study reveals that the most commonly used approaches are machine learning and deep learning.

Keywords Machine learning techniques · Lexicon-based approaches · Corpus-based approaches · Deep learning · Neural networks

1 Introduction

As sentiment analysis is a major problem in today's world, people are facing difficulty in predicting things. As people, we still seem to attract like-minded people. Also, surveys show that they are not confident socializing with individuals with common values, what people should believe, and who will help us accomplish those goals. Etymologically, individuals prefer to connect to similar-minded groups. Many clusters make up a group. Modularity is one of the key factors considered during the quantity calculation of populations. If the features of the clusters are studied in-depth, it may be helpful to define the unique character profile of particular clusters or groups of individuals with like-minds. As sentiment analysis can be in many forms,

V. Bhadana · H. Garg (✉)
GLA University, Mathura, UP, India

V. Bhadana
e-mail: vartika.bhadana_mtcs19@gla.ac.in

P. K. Singh et al. (eds.), *Proceedings of Second International Conference on Computing, Communications, and Cyber-Security*, Lecture Notes in Networks and Systems 203,
https://doi.org/10.1007/978-981-16-0733-2_17

it can be in health care, social media, e-commerce Website, and language prediction. Machine learning techniques are very helpful in predicting sentiments. As social media has many reviews, individuals cannot predict that the reviews or comment is of sentiment good, bad, excellent, or average. Some reviews are in different languages which cannot be understood by an individual. If citizens have to book the hotel for accommodation, there also sentiments are major issues that time, also it is difficult to predict which hotel is best, and many reviews can be fake also which are best for that particular hotel. So, to overcome these problems, some techniques are been used which will predict these issues from the different datasets of different types.

The proposed survey paper aims to give knowledge about the sentiment in different fields. Sentiment analysis uses different techniques to find the best accuracy among the sentiments.

2 Related Work

These sentiment analyses have been discussed under the following headings.

2.1 Sentiments Polarity Improvement

As the sentiment detection polarity describes that there are so many long reviews or comments that can be computed in the short meaningful sentences, firstly, they follow five policies like.

2.1.1 Most Occurring First (MOF)

This approach defines the most occurring goal as the overall review's main aim and then measures the review's polarity dependent on the review's main objective. This technique is analogous to the voting process, which is effective when the analysis has a dominant purpose. Nevertheless, in the study, two or three targets may be set for the same frequencies. The following methods should be used instead, in these situations.

2.1.2 Most General First (MGF)

Under this method, the most general objective is known as the main objective of the overall analysis, and therefore, the analysis polarity is determined based on the main objective. An ontology-based approach is used to identify the most general goal.

2.1.3 Most Specific First (MSF)

This approach defines the most important target as the overall review's main target and then measures the review's polarity based on the review's main objective.

2.1.4 First Occurring First (FOF)

It takes the first target listed in the analysis as the overall review's key target and then determines the review's polarity based on that goal.

2.1.5 Last Occurring First (LOF)

This approach considers the last goal found in the analysis as the key objective of the whole study and then measures the polarity of the review based on this objective.

The next step is the part of speech (POS) tags in which reviews are been compressed and the lexicon approach is been applied to it. Basiri et al. focused on the Persian language and have used the dataset of hotels and movies. As the Persian language has some drawback also as the lexicon are not precise as they are for the English language. There is no large dataset for the Persian language to train and test for ML techniques that will affect the polarity. The Persian language can have grammatical mistakes as compared to English. As they choose for the lexicon approach because it is simpler than the ML techniques as lexicon does not require training data, it is domain-independent.

Ontology is a systematic classification of definitions and the interaction between them. That may be as basic as definition taxonomy, which may include axioms and constraints to describe multiple facets of the natural world. The former is commonly considered a lightweight ontology, while the latter is regarded as an ontology of high weight. Chi et al. developed a lightweight ontology from the ground up, as there was no ontology existing for Persian principles relating to their datasets [1].

The system proposed typically has four major steps. The first section, pre-processing, involves phases of tokenization, standardization, and POS marking that are implemented sequentially on analysis sentences. The next step is called "extraction of possible words" and is responsible for separating future words from the statements. Typically, such words include adjectives, adverbs, and negations. After identifying the goals of each paragraph, the third stage called "target recognition" is to determine the key target for the whole study. The final stage is the classification process, which is responsible for determining the review's final polarity [2].

In the Persian language, certain pre-processing measures are identical to the Arabic language. For example, in both languages, the tokenization stage is more difficult than in English since they do not use capital letters and do not have specific punctuation rules. Another example is the stemming method in the Persian language, like the Arabic language, where in addition to adding prefix and postfix to the stems to produce a new grammatical form, it is also possible to attach infixes to the stem which makes the stemming method more complex than English in the Arabic and Persian languages [2]. Rathi et al., info on the mining of thoughts, how the importance of polarity deals with positive and negative and how to cope with Roman language reviews and journals.

Two pre-processing steps are expected to be performed on the dataset prior to the main process: tokenization and normalization.

Boundaries of words are defined in the tokenization process which is necessary for the next steps. Use functional units like phrase, sentence, and separators like space and line, and there suggested method tokenizes reviews. Normalization is a typical stage in text processing in which the different representations of a single word are combined and translated into the same type.

Basiri et al. have taken the preview dataset which is a labelled dataset, so analysis has two labels in it: one to indicate its polarity and the other to define its principal goal. The target of reviews of unknown destinations is defined as "implied." For example, the target is implied in the analysis seen in Example 1.

Example 1 Hello, it looks great, but now you can purchase two Android phones with the same or better specifications without restriction when you update the device and more. Please dismiss mark prejudice. Something costlier cannot automatically be easier.

Best accuracy was measured by the lexicon approach and $F1$ measure, and among the five identification techniques, MGF scores the best rank [1].

A hybrid ML-based approach is been implemented to integrate SVM and Naïve Bayes in the online Persian film review data collection to identify consumer comments as either favourable or negative. As a result, an increase compared with previous studies has been identified. Hajmohammadi et al. studied the close association between the extraction process of the characteristic and the findings obtained [3].

2.2 *Sentiments in the User Rating on Social Media Sites*

It is another issue that describes the deep leaning technique. The proposed work is a profound learning model for processing user comments and creating a potential user rating for user suggestions. First, the program uses sentiment analysis as the input points to create a vector of the function. Next, the device implements dataset noise reduction to improve user rating classification. Finally, for the suggestions, a deep belief network and sentiment analysis (DBNSA) should achieve data learning. The experimental results suggested greater precision of this device than conventional methods. Analysis of the feelings is based on a lexicon of thought. An opinion lexicon is a word dictionary that communicates the polarity of words by positive or negative emotions such as happy, good, bad or disgusting. In sentiment analysis, these opinion terms are used as the primary predictor for measuring the user's opinions. Many public lexical sets have become accessible in recent years, such as SentiWordNet. The first important task in sentiment analysis is to define the opinion objectives (aspects, persons, and issues of recognition of topics) on which opinions are expressed. Next, the lexicon of opinion has to be built.

The paper introduces three steps of noise reduction. The program extracts user comments in the first step that contain just ten words or fewer, and for which the user comments do not contain a word from the opinion lexicon. The second step is the identification of negative terms used in the lexicon of opinion. For example, "not bad" is a negative term that is included with "evil" in the positive opinion lexicon. Use the Glove algorithm to measure the sense of similarity for the negative term opinion lexicon [4]. In the third step, they delete all remarks which are categorized as good impressions, making the lexicon target the negative impression, and vice versa. Using a DBN and sentiment analysis, they propose a novel approach to predict user ratings from user comments. The sentiment analysis generates the feature vector based on the good and bad opinions of products or services as recorded by users in their comments. They also enforce noise reduction procedures that remove short comments, comments with no speech in them, and false rating comments. The DBNSA model outperforms other baseline models in the experimental section and outperforms baseline models in training failure performance, accuracy, and recall on Yelp and Amazon datasets [5]. A strategy is proposed by Samuel et al. which clusters and then indexes the tweets based on the emotions and emoticons present in the tweet [6].

Reviews also help us recognize market dynamics and tactics, as that may be achieved by nostalgic research as it encourages one to recognize popular items as it enables companies, enterprises, to use and grow accordingly. It can also be used by people themselves in general to search for which movie to watch, which laptop to purchase, but when we find spam reviews, a person does not know whether they are false or not in fact; however, they change our perspective. The author goes over this in a step-by-step style with various articles and outlines how a person can recognize the right feelings for other readers and distinguish between the true and the false reviews [7].

2.3 *Sentiment Analysis Challenges and Techniques*

Anees et al. mention the reviews of the product, the user got confused whether the reviews are right or wrong. If they are not sure about the reviews, they will take reviews from the people who have bought that product or they will assume whether the product is good or bad by seeing the rating of product [4]. The main objective is to find the suitable techniques to get the correct reviews. The information can be divided into subjective and objective forms. Objective means facts, and subjective means emotions which are used in sentiment analysis. Suggested machine leaning approaches, lexicon-based approaches, and hybrid approaches are been used [8]. The lexicon approaches divide the document into lexemes which are used to examine the sentences it is further divided into corpus-based and dictionary-based. Corpus finds out the positive, negative, and neutral polarity of the sentences. Machine learning is divided into supervised and unsupervised learning. E-commerce dataset is used by Anees. Supervised learning requires the desired output with the actual output. In machine learning approach, they have used the Naïve Bayes. The related work is done in which the dataset is taken as 70% training data and 30% testing data.

$$P\left(\frac{c}{x}\right) = \frac{P(x/c)P(c)}{P(x)}$$

$$P(c|x) = P(x1|c)^{*}P(x2|c) \ldots^{*} P(xn|c)^{*}P(c).$$

The above equation is Naïve Bayes which is been commonly used to predict the accuracy of sentiments.

2.4 Sentiments Also Play a Major Role in Health Care

Abualigah Laith, et al. tell us about the sentiment in health care. There is some technique which can improve healthcare quality. The healthcare sentiments are related to the hospital performance, which hospital is best, and recovery process which can be taken from other patients. The technique tells that there is no need to ask other patients whether the hospital is good or bad. Past there were surveys of the different hospital, but they consume a lot of money and time. Abualigah Laith, et al. have taken a dataset in which the hospital list is there, and it has some option that reviews which is the best hospital for medication. Natural language processing involves a technique called emotion analysis mining; it recognizes the emotional meaning behind a paper picture. It is a common way for organizations to define and categorize an aggregate of feelings about a commodity, service or concept. Thus, the polarity of ambiguous words (context-dependent words) needs to be decided efficiently and effectively, and then, the aspect-based description produced. The author used k-nearest neighbour classifier in this paper to evaluate the polarity of the context-dependent terms [9].

Some resources are needed for this function such as polarized lexicon. Sentiment data mining in health care is not well studied, partially because patients are given some trust and an interpretation of their emotions, and often patients are using social media. These are inspired by the mining of product ratings in sentiment analysis. Next, they describe the root of the lexicon, using terms from the general realm of sentiment analysis and their polarity, and then they create a lexicon of medical emotion analysis based on a sample of drug feedback. Most views include terms of thought and have the same polarity in all circumstances. But there are several words of opinion called context-dependent terminology that in various situations have distinct polarities [10].

In this, the Arabic languages are been used the dataset of different categories in which different techniques are been implemented. The transmission of the natural language has many problems that may alter the nature of the expression of emotions in many aspects. Some of the problems are linked to data form, while others are apparent to some sort of text analysis. In this many techniques are been applied to the Naïve Bayes, for example, is powerful and fast computing, without being affected by trivial features. It does presume individual characteristics, however [11].

2.5 Sentiment Analysis of Social Media, Politics, etc., Can Be Predicted by Deep Learning Techniques

The World Wide Web, such as social networks, groups, web pages, and blogs, creates vast volumes of data in the form of thoughts, feelings, viewpoints, and claims regarding different world activities, goods, brands, and policies. User emotions shared on the Internet have a great impact on readers, sellers of goods and policymakers. The unstructured type of social media data is expected to be processed and well-structured, and much focus has been paid to sentiment analysis for this reason [12]. Examination of emotion is referred to as an as text entity which is used to identify our emotions conveyed in various ways such as negative, good, favourable, and unfavourable. In the field of natural language processing (NLP), the problems for trend analysis are the lack of adequate labelled data. And the sentiment analysis and deep learning approaches have been combined to solve these problems, and deep learning models are successful because of their automatic learning ability.

As deep learning models can be a deep neural network, recursive neural network and many more other neural networks are been used. Via the use of deep learning models, this study presented adequate studies relevant to sentiment analysis. After reviewing all of these experiments, it is known that the interpretation of emotions can be done more effectively and reliably by using deep learning methods. As the study of emotions is used to forecast consumer attitudes, and deep learning models are more about modelling or mimicking the human mind, and they have more precision than shallow models. Deep learning networks are good than SVMs and normal neural networks because they have more hidden layers than normal neural networks with one or two hidden layers. The author introduces a system that focuses on clustering and indexing tweets, based on their geographical and temporal characteristics. The X-means clustering was used, which does not enable the user to enter the cluster number, but rather takes enter from the index of the tweets-created specified functions [13].

Many different forms of neural network are there, i.e. convolutional neural network, recursive neural network, deep neural network, recurrent neural network, deep belief network, hybrid neural network, and another neural network [14]. One is to capture people's opinions worldwide, which is called opinion mining or nostalgic research. It encourages them to consider the consumer need and help to produce the stuff people want and to isolate the items that are undesired or to solve a particular problem. This makes development at a much faster rate [15].

Deep learning networks perform automated extraction of features which does not require human interaction because it will save time and there is no need for

Table 1 Various approaches used for sentiment analysis

Authors	Approaches used	Dataset
Basiri et al. [1]	Lexicon approach	Hotel and movies dataset
Chen et al. [5]	Deep brief network and sentiment analysis	Yelp and Amazon dataset
Anees et al. [16]	Naïve Bayes	E-commerce dataset
Abualigah et al. [13]	NLP	Heath dataset
Zhang et al. [14]	CNN	Twitter dataset

software engineering. Sentiment analysis consists of various kinds of comments about problems. The ability to resolve differences in the process by making few modifications in the program itself requires a deep learning basic power feather. This approach often has some drawbacks relative to previous versions such as SVM. It needs massive datasets and is incredibly costly to train. These sophisticated models will train for weeks using computers fitted with costly GPUs [11] (Table 1).

3 Various Parameters

Lexicon approach: This approach uses a sentiment lexicon to explain the polarity of a textual material (positive, negative, and neutral). This methodology is more intuitive and can be applied quickly, as opposed to algorithms based on deep learning. Nonetheless, the downside is that it requires human intervention in the process of text analysis. The more popular the amount of material, the more notable the task would be for sifting through the noise, recognizing the meaning and separating valuable details from various sources of knowledge (Table 2).

Naïve Bayes: Easy term description based on 'theorem of Bayes.' This is a 'Bag of Words' technique for contextual interpretation of a substance (text described as set of its words, rejecting grammar, and word order while retaining multiplicity).

Table 2 Various parameters for sentimental analysis

Year	Approach use	References
2019	Lexicon	[1]
2020	Lexicon	[4]
2020	Naïve Bayes	[2]
2017	Deep learning	[16]
2019	DBNSA	[5]

Deep learning: Deep learning tailors a multilayer solution to the neural network's hidden layers. In conventional approaches to machine learning, features are described and extracted either manually or by using methods for selecting features. Nonetheless, features are automatically taught and extracted in deep learning models, gaining greater precision and efficiency.

Deep belief network and sentiment analysis (DBNSA): The DBN and sentiment analysis (DBNSA) approach is a discriminative classifier, which predicts the likelihood of rating from a WordNet sentiment analysis created word vector and then uses deep learning to train a model that predicts ranking.

4 Dataset

Three datasets are been used in which, preview dataset is manually labelled, and two other datasets of hotels and movies are been introduced [1].

The first collection of data is comprised of user comments on movies that are obtained from the Website of Naghdefarsi.com [17], and the second set of data includes comments on hotels reported on different Websites concerned. Aside from these latest datasets, they have named the current per view dataset [18] recently published in the Persian language for document-level SA. This data collection contains reviews of wireless appliances and will be compiled at Digikala.com in 2017 [16].

Next, this also contains three datasets, and these three show different results. Amazon dataset is used which is been collected by crawling the Amazon Website and camera category and its comment. 70% of the data for the deep learning model is for training and 30% for testing. Chen et al. use the Academic Challenge 5th round of Yelp data collection, which consists of over 1.5 million reviews, 36 600 users, and 61,000 companies [5].

There are 212,983 customer ratings for the hotel category in the Trip-Advisor data collection. The data collection is composed of 12,773 hotel travellers in tourism areas. The Trip-Advisor dataset is a Xml structure. The data collection of Trip-Advisor is obtained by browsing the Trip-Advisor Website and collecting only details from hotels and their reviews. The data collection for the Trip-Advisor is identical to that for the Amazon dataset [5].

The data is been collected from the e-commerce Website. Web scraper has been used to scrape comments from Amazon product URL and store them in spreadsheet form. The scraped comments are pre-processed to save time and energy for the computation [4].

The Arabic language suffers the lack of immense open databases for applications of AI and emotion analysis. This study launched a massive dataset, called BRAD, which is Arabic Dataset's biggest book reviews [19]. This dataset contains 490,587 inn surveys obtained from the Booking.com site record that includes the message of the survey in the Arabic language, the assessment by the commentator on a scale of 1–10 stars, and various attributes of the hosting/analyst. They make the complete unequal dataset available just like a good subset. Six prevalent classifiers are used using Modern Standard Arabic (MSA) for evaluating the datasets [11].

ATwitter dataset containing 1269 images is chosen for experimental work, and back propagation is introduced. The photos are marked with Amazon Mechanical Turk (MTurk) and common crowd intelligence. Five employees were involved in creating sentiment mark for each graphic. On this dataset, the proposed model was evaluated and got better performance than existing systems. Results show that the device proposed achieves high efficiency without fine tuning on Flickr dataset [11].

5 Conclusion

As the analysis suggests about many different techniques and by which they get many different accuracies. An article, come across many different results, first is the decomposes of a long review into its constituent sentences and then detects the main target of each sentence. Finally, using the POS tags, the proposed method filters out all words except the potential terms, considering a comprehensive sentiment lexicon, and computes the polarity of the sentence. Moreover, five target identification strategies, including MOF, MGF, MSF, FOF, and LOF, are proposed to come up with the main target of the review. The author suggests a novel approach for estimating user ratings from user feedback by using a DBN and sentiment analysis. The sentiment analysis generates the function vector based on the good and bad opinions of products or services as stated by users in their comments. They also enforce noise reduction procedures that remove short posts, comments with no speech in them, and false rating comments. In this paper, author used a lexicon-based approach to measure feedback sentiment. Lexicon-based approach provides more specificity because they use a word. Analysis of emotions is a good way to help people get a decision and learn information. This approach attempts to examine the social Web, anywhere an identified issue will only reach the appropriate authority if they quickly find it. The best advice cannot be accessed via social media and different consumer materials. Sentiment analysis automates this process. Sentiment analysis is geared at gathering more knowledge to help consumers make the correct judgement about the analysed. To change this challenge, the techniques of sentiment analysis applied to data mining and machine learning. Author after reviewing all of these experiments, like convolutional neural network, deep belief neural network and many more, it is known that interpretation of emotions can be done more effectively and reliably way by using deep learning methods. As the study of emotion is used to forecast consumer attitudes, deep learning models are more about anticipating or imitating the human spirit, and deep learning models have greater precision than shallow models. Deep learning networks are different than SVMs and normal neural networks because they have more layers hidden than normal neural networks with one or two layers revealed. Deep learning networks can deliver training in both supervised and no supervised ways.

6 Limitations and Challenges

Chen et al. have not discussed social relationships and corresponding timeline comments, as user comments are influenced by the social relationships of the users. As consumers are influenced by their past interaction with related goods or services, timeframe is also a significant consideration [5]. Anees et al. [4] show some challenges that there can be multiple language input, fake input, emoticons, and sarcastic reviews. The transmission of natural language has many problems that in several ways will affect the presentation of the emotion analysis. Some of the problems are connected to data type, while others are obvious to any kind of text analysis. Compared to previous versions such as SVM, this system still has certain drawbacks. It needs massive datasets and is incredibly costly to train. These complex models will practice for weeks using fitted machines with costly GPUs [11].

References

1. Basiri ME et al (2019) Improving sentiment polarity detection through target identification. IEEE Trans Comput Soc Syst
2. Duwairi R, El-Orfali M (2014) A study of the effects of preprocessing strategies on sentiment analysis for Arabic text. J Inform Sci 40(4):501–513
3. Hajmohammadi MS, Ibrahim R (2013) A SVM-based method for sentiment analysis in Persian language. In: International conference on graphic and image processing (ICGIP 2012), vol 8768. International Society for Optics and Photonics
4. Anees AF et al (2020) Survey paper on sentiment analysis: techniques and challenges, No. 2389. EasyChair
5. Chen R-C (2019) User rating classification via deep belief network learning and sentiment analysis. IEEE Trans Comput Soc Syst 6(3):535–546
6. Samuel A, Sharma DK (2018) A novel framework for sentiment and emoticon-based clustering and indexing of tweets. J Inform Knowl Manage 17(02):1850013
7. Agarwal Y, Sharma DK, Katarya R (2019)Sentiment/opinion review analysis: detecting spams from the good ones! In: 2019 4th international conference on information systems and computer networks (ISCON), Mathura, India, 2019, pp 557–563. https://doi.org/10.1109/ISCON47742.2019.9036249
8. Digikala Internet Shop (2019) Accessed 23 Feb 2019. [Online]. Available: https://www.digikala.com/
9. Garg S, Sharma DK (2015)Feature based clustering considering context dependent words. In: 2015 1st international conference on next generation computing technologies (NGCT), Dehradun, 2015, pp 713–718. https://doi.org/10.1109/NGCT.2015.7375214
10. Garg S, Sharma DK (2016) Sentiment classification of context dependent words. In: Satapathy S, Joshi A, Modi N, Pathak N (eds) Proceedings of international conference on ICT for sustainable development. Advances in intelligent systems and computing, vol 408. Springer, Singapore. https://doi.org/10.1007/978-981-10-0129-1_73
11. Abualigah L et al (2020) Sentiment analysis in healthcare: a brief review. In: Recent advances in NLP: the case of Arabic language. Springer, Cham, pp 129–141
12. Maynard D, Funk A (2011) Automatic detection of political opinions in tweets. In: Extended semantic web conference. Springer, Berlin, Heidelberg
13. Samuel A, Sharma DK (2017) A spatial, temporal and sentiment based framework for indexing and clustering in twitter blogosphere, p361

14. Zhang Y et al (2016) Sentiment classification using comprehensive attention recurrent models. In: 2016 international joint conference on neural networks (IJCNN). IEEE, New York
15. Agarwal Y, Katarya R, Sharma DK (2019)Deep learning for opinion mining: a systematic survey. In: 2019 4th international conference on information systems and computer networks (ISCON), Mathura, India, 2019, pp 782–788. https://doi.org/10.1109/ISCON47742.2019.9036187
16. Nasr FM, Mohamed SE, Shaaban M, Hafez TAM (2017) Building sentiment analysis model using Graphlab. Int J Sci Eng Res 8:11551160
17. Naghdefarsi (2019) Accessed 23 Feb 2019. [Online]. Available: https://naghdefarsi.com/
18. Basiri ME, Kabiri A (2018) Words are important: improving sentiment analysis in the Persian language by lexicon refining. In: ACM transactions on Asian and low-resource language information processing (TALLIP), vol 17.4, p 118
19. Alayba AM et al (2018) Improving sentiment analysis in Arabic using word representation. In: 2018 IEEE 2nd international workshop on Arabic and derived script analysis and recognition (ASAR). IEEE, New York

A Review Study on IoT-Based Smart Agriculture System

Mir Saqlain Sajad and Farheen Siddiqui

Abstract Agriculture has been going on in every country for ages, and it is also the economic system's backbone, which is declining due to overpopulation and urbanization, hence smart agriculture is necessary for all. Smart agriculture includes the incorporation of sensors, automated monitoring, networking, and data processing capability. The arrival of the accelerated growth of IoT and site-specific farming activities is difficult to increase crop yield and the productive utilization of assets associated with the agriculture procedure. It gives a real-time evaluation of the different crops and environmental conditions by determining the agricultural land and the volume of fertilizers, water, and other inputs needed. This approach empowers our farmers to attain productivity within the right time frame and thus ensures the production of secure and stable non-toxic crops. We studied different wireless network technologies that can be used in agricultural farms for smart purposes. These technologies are used to monitor the parameters of agriculture, i.e., moisture in the soil, humidity, and temperature, etc. Agriculture issues have frequently hampered the development of the region. The main solution for this issue is smart farming by changing the current conventional strategies for farming. The objective of the task is accordingly to make farming smart utilizing, automation, and IoT advancements. These two activities ought to be done by a remote smart unit or Internet-associated computer, and the process is conducted by interfacing sensors, ZigBee, Wi-Fi devices, cameras, etc. This paper represents numerous possibilities and difficulties in the field of IoT-based smart farming.

Keywords Agriculture · Internet of things (IoT) · Smart agriculture · Arduino Mega-2560 · DS18B20-temperature sensor · Smart farming

M. S. Sajad (✉) · F. Siddiqui
Department of Computer Science, School of Engineering Sciences and Technology, Jamia Hamdard University, New Delhi 110019, India

F. Siddiqui
e-mail: fsiddiqui@jamiahamdard.ac.in

P. K. Singh et al. (eds.), *Proceedings of Second International Conference on Computing, Communications, and Cyber-Security*, Lecture Notes in Networks and Systems 203,
https://doi.org/10.1007/978-981-16-0733-2_18

1 Introduction

Farming is the fundamental level of the people in India. The existence of humans depends on agriculture. Production of the crops is important as that of the world population explosion. The reasons responsible for degrading crop production in agriculture can be affected by water wastage, poor soil fertility, inadequate supply of fertilizer, changes in climate, etc. Automation and IoT technologies are needed to mitigate these effects and maximize farm productivity. IoT technology is capable of making successful agricultural production. The optimization of agricultural resources is needed through the integration of the IoT wireless sensor network [1]. As the world is moving toward emerging technology and modern applications, there is also a need for an agricultural upward trend. IoT has a major role to play in smart farming. IoT systems empower in collecting data on agricultural lands. This IoT-based farm surveillance system uses wireless sensor networks that gather information from various sensors mounted at different nodes and relay it via a wireless protocol. This smart agriculture using the IoT system is operated by Arduino, consisting of a temperature sensor, a moisture sensor, a water level sensor, a DC motor, and a GPRS board.

1.1 The Need for IoT in Agriculture

As per the United Nations Food and Agriculture Organization report, global production of food is projected to rise by 70% in the year 2050 for a changing inhabitant. Farming is the foundation of the human race because it is the major food supply and is necessary to increase the world's economy. It also offers people enough opportunities for jobs. Farmers still use traditional farming methods which result in lower yields of fruit and crops.

Other causes have the main impact on efficiency. Factors include attacks on mosquitoes and rodents which can be achieved by insecticide and pesticides, as well as attacks on birds and wild animals as the crop grows. Crop yields are decreasing due to erratic monsoon rains, water shortages, and inefficient use of water [2]. The data collected provides insights into the numerous environmental factors. Tracking environmental conditions is not a complete remedy for increasing crop yields. There are a variety of other factors that reduce efficiency to a larger degree. Therefore, to provide a result of all these issues, it is important to build an interconnected framework that will look after all variables that impact efficiency at each point, for reference we can see Fig. 1 showing the uses of IoT which will bring better infrastructure in agriculture. However, due to different problems, the full automation of agriculture is not achieved. While applied at the research level, farmers are not provided as a commodity to benefit from the capital. It then deals with the development of smart farming using IoT and supplied it to farmers [3, 4].

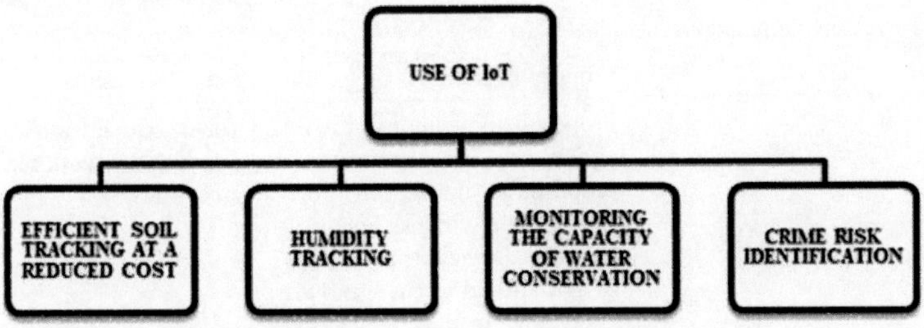

Fig. 1 Use of IoT in smart farming

The goal of smart farming is to increase the quality and quantity of agricultural production by using sensing technologies to make farmers more intelligent and more connected. Farmers are now doing a lot of hard work to grow crops and to make a living alongside their hard work, neither the quantity nor the quality of crops is getting better, this leads to their losses, these losses are happening due to natural calamity as well lack of availability of technology, and because of this farmers are dying day by day. We can quickly solve the major economic and social problems of farmers. Smart farming is an agricultural production method that uses advanced technologies to improve the amount and efficiency of farming products [5, 6].

2 Literature Review

Many researchers have worked for a smart farming technique to get the maximum yield by limiting the wastage of resources to have the optimization of resources in terms of quantity and time. A detailed review of previously done research is presented in Table 1.

3 Proposed Work

Earlier agriculturists have been evaluating the freshness of the soil and to control decisions on the type of crop to be grown. Agriculturists did not even care for the moisture, the water level, and, in particular, the temperature, which was terrible to the farmer. Pesticides are used in the light of a few concerns that have contributed to a real effect on yield if the conclusion is not correct. Productivity is dependent on the last period of harvest in which the farmer relies on.

In the proposed work, a smart agriculture model will describe a real-time monitoring system for properties of soil such as humidity, temperature, water level, and crop disease detection. It will also be possible to monitor the specific process of the field remotely from anywhere, at any time, via mobile and web applications. The

Table 1 Recent study analysis

Author	Proposed solution	Expected results
Zhao et al. [7]	Proposed applications of IoT to agriculture. The system was proposed for monitoring Internet and wireless sensor networks to include the results of agricultural science, and the knowledge management system has been developed. The authors developed field management methods such as field data collection, templates for data processing, and modules for device configuration	The developed software provides the work for greenhouses
Agrawal and Das [8]	They proposed the potential uses and obstacles faced by IoT technologies. Many of the major issues posed in IoT implementations were quality, coordination, protection, efficiency, identity authentication, ownership, confidence, integration, and regulation	Using the mobile communication technologies and radio frequency identification, wireless sensor network would decrease the distance between the practical and theoretical implementations and deployment of IoT applications
Li et al. [9]	Proposed IoT's Wi-Fi-based and smart network with integrated sensors. The author proposed IoT-based WSN, Wi-Fi, and smart grid technologies. Smart grid offers an integrated information collection system, improves data collection reliability, and delivers quality information. IoT is an automated weather forecasting application; air and water information is collected by sensors and transmitted to the cloud for further review	The principle of agricultural accuracy has been implemented. The authors have argued that ZigBee is not better than the new WSN
Fan [10]	Proposed intelligent farming focused on cloud computing. The author proposed the smart farming program depends on the idea of cloud computing and the Internet of Things	Agriculture knowledge cloud has been combined with the IoT to create a global information exchange and a balancing of loads

(continued)

Table 1 (continued)

Author	Proposed solution	Expected results
Chen et al. [11]	Discussed digital agriculture focused on IoT. The automated agriculture research is divided into two steps: The stage first gathers data on wind, temperature, quality, soil, etc. from different sensors. In stage second, ZigBee translates data	The EPC code was used to classify agricultural products. The commodity standard is denoted by EPC code
Prasad et al. [12]	Discussed an expert system for the treatment of mango birds, illnesses, and disorders. The architecture for text animation was introduced in expert system shell (ESTA). The first phase in the discussed approach is awareness development; the second stage is the data-based treatment of illness	Based on the visual signs, a short description of the variety of mango disease and suggestions for illness control is made
Sarma et al. [13]	Discussed an expert system for the detection of sickness in the plants of rice. The network of experts aims to help the farmers solve problems. The first step is to build the base of information in the context of the situational rules	The new software is simple to use and will help individuals who are reluctant to obtain any agriculture expert's assistance
Kaura et al. [14]	Put forward the system that can detect and diagnose diseases in the leaves of cereals. Without the help of experts, it is very hard for farmers to identify diseases in leaves. JAVA proposed comparative imaging methods to identify pathogens	That is why methods such as affine transformation and edge identification were used. It is a web-based community of experts, meaning that it can be accessed from any web-enabled network

(continued)

Table 1 (continued)

Author	Proposed solution	Expected results
Gondchawar et al. [15]	Proposed a smart approach to optimize farming that can be considered Smart agriculture. By indicating an automatic device threat, unintended risks to crops may be avoided by human interference. Real-time control of the climate is an important element in smart farming. GUI-based software will be provided to monitor the hardware device, and the device will be fully independent, fitted with sensors like temperature sensor, humidity sensor, and photo emitter	This program will incorporate a clever approach for agriculture that will solve the challenges of farmers successfully. The climate will not be an obstacle to the production, and development of any plant and will be able to resolve the lack of agricultural resources
Lakshmisudha [16]	Proposed smart precision-based farming that uses wireless sensor networks to track the farming environment. Raspberry pi and ZigBee-based farming tracking systems act as a safe and effective tool for controlling farming parameters. Wireless field tracking not only allows users to minimize human resources, but also allows users to make real improvements. This focuses on the development of apps and software to control, view, and alert users leveraging the benefits of a wireless network sensor system	A smart network focused on precision farming will set the stage for a newer development in farming
Gayatri et al. [17]	Proposed low cost and effective wireless network sensor technologies for the acquisition of moisture of soil and temperature from different farm locations and the decision to make irrigation ON or OFF as needed by the crop controller	Make farming more effective

(continued)

Table 1 (continued)

Author	Proposed solution	Expected results
Gutiérrez et al. [18]	Proposed climate change and flooding have been unpredictable in the past few decades. As a result of this in the recent period, many Indian farmers have implemented smart environment practices called smart agriculture. Smart farming is an advanced and focused information system introduced by IoT (Internet of Things)	IoT is growing quickly and commonly implemented in all cellular settings. The key goal is to gather real-time data to minimize the amount of water wasted in the irrigation cycle to decrease the time spent on the farm
Prathiba et al. [19], Galgalikar [20], Patil et al. [21]	Proposed intelligent irrigation systems that use the Internet of Things. Some wireless sensors are expected to measure soil moisture and water levels. Such sensed information was sent to a smart gateway via a network utilizing the standard IoT border router wireless Br 1000 gateway. Through the router, the information is then transmitted to a web server via a network	A study on smart farming irrigation systems was performed to get a deeper understanding of IoT-based innovations in cloud infrastructure agriculture

various wireless systems used for this purpose include SIGFOX, ZigBee, Wi-Fi, and LPWAN. The cost factor for the deployment of the sensors can be minimized with the use of motors to use a small number of sensors across a wide area of agriculture with rapidly moving sensors around the whole field with the aid of motors.

The use of a wireless network has been used in many fields as of:

a. Flood detection.
b. Smart grid.
c. Home automation.
d. Vehicle flow monitoring.

The use of wireless network technology has been used to produce healthy and productive ranches by predicting the beginning of the disease and by determining improved nutrition, automating irrigation, and detecting foreign matter like weeds in crops. This study focuses on IoT and its devices with wireless network technologies in the agricultural field since agriculture is the primary source of population survival listing some of them below:

a. ZigBee technology: The network technology of the ZigBee wireless has a range of 50–100 m operating at a frequency of 868 MHz. The power consumption and complexity are low. It operates on the AES encryption technique of 128 bits.

b. Bluetooth/BLE technology: This technology has a range of 10–100 m operating at a frequency of 2.4. The power consumption is medium, and the complexity is high.
c. Wi-Fi technology: This technology has a range of 50–100 m operating at a frequency of 2.4 and 5 GHZ. The power consumption and complexity are high.
d. Lora technology: This technology has a range of 15–30 miles/up to 25 km operating in license-free sub-GHz frequency 433 MHz. The Lora technology consumes less power can elevate the battery lifetime of 20 years, supports high capacity, i.e., can add millions of messages per base station, and enables interoperability. The use of Lora technology can lead to cost-effective installation.

Wireless technologies are used as these are cost effective used for long transmission utilizes less power to sense the parameters like temperature, humidity, and moisture to increase the yield by providing accurate information to the farmers. The general steps involved in the proposed methodology are:

a. Soil, humidity, and temperature data are collected by temperature and environmental sensors embedded.
b. The data collected from the sensor nodes is sent to the wireless sensor gateway which may be static or mobile.
c. The data is sent to farmers with alert messages by the application to maximize the yield by limiting and optimizing the resources.
d. The wireless gateway sends data to the cloud using an Internet facility where processing and analysis of data are observed. It is a bidirectional link between gateway and cloud.
e. The software Arduino-ide is used for writing, compiling, and uploading the code in the Arduino device.

4 Benefits of IoT in Agriculture

IoT is viewed as a core component of smart agriculture to provide smart applications and reliable sensors, with agriculturists projected to expand food yield by 70% by 2050, as specialists describe. It permits the processing, and control of loads of information gathered from sensors and the application of cloud infrastructure resources such as map fields, distributed storage, etc., information can be viewed live from everywhere, allowing live tracking and throughout communication for all gathering involved. Some benefits of this technology are as follows:

a. With IoT manufacturing expenses can be lowered to a phenomenal amount, which consequently would improve productivity.
b. Through IoT, the level of productivity will be improved in the utilization of water, soil, pesticides, fertilizers, etc.

c. Enhanced livestock farming–Sensors and devices may be used to identify reproduction and health activities in livestock in the past. Geofencing location tracking will also enhance the control and handling of livestock.
d. Accurate plant and land assessment–Accurate analysis of land output levels over time allows for accurate forecasts of potential crop yields and plant performance.
e. Improved production performance–Analyzing the level of production and the effects of treatment interaction will help farmers to change procedures to increase the quality of the crop.
f. Lower running costs–Automated procedures for planting, manufacturing, and collecting can decrease resource usage, human inaccuracy, and total costs.
g. Real-time analysis and market intelligence–Farmers can imagine growth patterns, soil moisture, sunshine strength, and more progressively and remotely to improve decision making. Water conservation–Climate forecasts and soil moisture sensors enable water to be used only when and where it is needed.
h. Improved productivity–Optimized management of crops such as effective planting, irrigation, fertilizer application, and collecting directly impacts productivity.
i. Equipment tracking–Cultivating gear can be followed and overseen as indicated by creation level, work profitability, and disappointment prediction. Smart farming has an incredible potential to deliver increasingly serious and practical agrarian creation dependent on a progressively solid and asset effective methodology.

The current study would evaluate the existing features of the previous work on an intelligent agricultural device and incorporate the agriculture wireless sensor network to improve the battery life of the nodes and ensure continuous transmission. The cost of deployment of the sensor is minimized due to the use of motors to shift the small number of nodes continuously over large fields.

5 Future Scope

This concept has an immense scope that can be applied in many other areas due to its cost-effective nature. The ability of our farming remains unexplored, but there are still miles to go in this field of research as there are various types of soil texture in different regions of our state. Farmers would benefit from the successful implementation of this proposed project. The main problems that have been encountered and that are yet to be solved in practice are the Internet functioning of the nodes in the field of agriculture and the creation of a user-friendly program that is readily understood by farmers. In the future, we can have several applications too as the use of IoT technology in agriculture is beneficial. Also, collecting environmental criteria for crop growth at a set position to assist farmers to detect issues in time, the innovative advancement of smartphone applications had led to the growth of the promotion of agricultural technologies and a specialist web FAQ.

6 Conclusion

Wireless network technologies (WSN) are used for smart monitoring due to low cost and simple end devices in the field. The implementation of IoT-based devices for broad acres of land may be expanded for potential projects. In turn, the device can be optimized to track soil quality and crop development in growing soil. Microcontrollers and sensors are viably interfacing, and communication between nodes is achieved. All the study and preliminary experiments indicate how this initiative provides a comprehensive solution to the problems of field operations and irrigation. The introduction of this technology in the field will undoubtedly help to increase crop yields and overall performance. The IoT-based design additionally provides ongoing information and interpretation of information that can be utilized far and wide in corresponding with the parameter being tracked in certain areas of the world to identify the irregular activity of a particular crop type. Smart farming would also revolutionize the world of agriculture and maximize productivity, boost efficiency, and save farmers' lives. There is an overwhelming need for a program that makes the farming cycle simpler and burden-free for farmers. India is a completely agriculture-driven system. The capacity to preserve natural assets and give a magnificent boost to crop development is one of the key objectives of integrating such innovation into the world's agricultural domain. Power and time are the most critical factor to save the farmer's effort.

References

1. Anusha A, Guptha A, Rao GS, Tenali RK (2019) A model for smart agriculture using IoT. Int J Innov Technol Explore Eng 8(6):1656–1659
2. Baker N (2005) ZigBee, and bluetooth—strengths and weaknesses for industrial applications. Comput Control Eng 16:20–25
3. Garg SK, Buyya R (2020) Green cloud computing and environmental sustainability. Harnessing Green IT: principles and practices, pp 315–340
4. Tanaka K, Suda T, Hirai K, Sako K, Fuakgawa R, Shimamura M, Togari A (2009) Monitoring of soil moisture and groundwater levels using ultrasonic waves to predict slope failures. Sensors, pp 617–620
5. Ballena K, Satyanvesh D, Sampath NVSSP, Varma KTN, Baruah PK (2014) Agpest: an efficient rule-based expert system to prevent pest diseases of rice and wheat crops. In: Intelligent systems and control ISCO, 2014 IEEE 8th international conference on. IEEE, New York, pp 262–268
6. Negid NK (2014) Expert system for wheat yields protection in Egypt ESWYP. Int J Innov Technol Explor Eng IJITEE. ISSN: 2278-3075
7. Zhao J-C, Zhang J-F, Feng Y, Guo J-X (2010) The study and application of IOT technology in agriculture. In: 2010 3rd IEEE international conference on computer science and information technology ICCSIT, vol 2. IEEE, New York, pp 462–465
8. Agrawal S, Das ML (2011) Internet of things—a paradigm shift of future Internet applications. In: 2011 Nirma University international conference on engineering (NUiCONE). IEEE, New York, pp 1–7
9. Li L, Xiaoguang H, Ke C, Ketai H (2011) The applications of WiFi-based wireless sensor network in the internet of things and smart grid. In: 2011 6th IEEE conference on industrial electronics and applications ICIEA. IEEE, New York, pp 789–793

10. Fan T (2013) Smart agriculture based on cloud computing and IoT. J Convergence Inf Technol 8(2)
11. Chen X-Y, Jin Z-G (2011) Research on key technology and applications for the internet of things. Phys Proc 33:561–566
12. Prasad R, Ranjan KR, Sinha AK (2006) AMRAPALIKA: an expert system for the diagnosis of pests, diseases, and disorders in Indian mango. Knowl-Based Syst 19(1):9–21
13. Sarma SK, Singh KR, Singh A (2010) An expert system for diagnosis of diseases in rice plant. Int J Artif Intell 1(1):26–31
14. Kaura R, Dina S, Pannub PPS (2013) Expert system to detect and diagnose the leaf diseases of cereals. Int J Current Eng Technol 3(4)
15. Gondchawar N, Kawitkar RS (2016) IoT based smart agriculture. Int J Adv Res Comput Commun Eng (IJARCCE) 5(6)
16. Lakshmisudha K, Hegde S, Kale N, IyerS (2011) Smart precision based agriculture using sensors. Int J Comput Appl 146(11) ISSN: 0975-8887
17. Gayatri K, Jayasakthi J, Anandhamala GS, Providing smart agriculture solutions to farmers for better yielding using IoT. In: IEEE international conference on technological innovations in ICT for agriculture and rural development (TIAR2015)
18. Gutiérrez J, Villa-Medina JF, Nieto-Garibay A, Porta-Gándara MÁ (2013) Automated irrigation system using a wireless sensor network and GPRS module. IEEE Trans Instrum Meas ISSN: 0018-9456
19. Prathibha SR, Hongal A, Jyothi MP, IoT based monitoring system in smart agriculture. In: 2017 International conference on recent advances in electronics and communication technology
20. Galgalikar MM (2010) Real-time atomization of agricultural environment for social modernization of Indian agricultural system. IEEE, New York
21. Patil KA, Kale NR, A model for smart agriculture using IoT. In: 2016 International conference on global trends in signal processing, information computing, and communication

Secure Group Data Sharing with an Efficient Key Management without Re-Encryption Scheme in Cloud Computing

Lalit Mohan Gupta, Hitendra Garg, and Abdus Samad

Abstract Cloud computing becomes an essential tool for internet users which provide various computing resources over the cloud. Providing on-demand storage is being one of the major services which is effective to meet out the expectations of an organization when utilizes the computational resources while accessing large amount of data. Storing the data on cloud is being a big challenge for the researchers in terms of maintaining the protection and security of data. Sharing of the data among a group can lead to insecurity of data from external and internal threats. This paper introduces a protected data sharing framework in the cloud storage that maintains the privacy and confidentiality of data. The proposed method uses the MD5 hash function that is relatively faster than the SHA256 hash function for checking the data integrity. The framework also provides a check on accessing and sharing of data. Exhaustive re-encryption computations are avoided and a single encryption key is used to encrypt the entire file. There are two distinct key shares for each of the users and user is allowed to share one at a time get access data. In this way having access to a single portion of a key permit the framework to safeguard the data against internal threats. Cryptographic server is used to store the other key share and performs all the expensive computations which are considered as a trusted third party. Proposed work also measure the efficiency of the model based on the time taken to execute the various operations.

Keywords Access control · Cloud computing · Cryptography · Security · Confidentiality

L. M. Gupta
Dr. A.P.J. Abdul Kalam Technical University, Lucknow, India

H. Garg (✉)
GLA University, Mathura, India

A. Samad
Women's Polytechnic, Aligarh Muslim University, Aligarh, India

P. K. Singh et al. (eds.), *Proceedings of Second International Conference on Computing, Communications, and Cyber-Security*, Lecture Notes in Networks and Systems 203,
https://doi.org/10.1007/978-981-16-0733-2_19

1 Introduction

Cloud computing is becoming a notable research topic for the researchers, academicians, organizations, governments, and industries due to its various services such as elasticity, on-demand storage service, flexibility, and computing resource services provided to its users [1]. Organizations take benefits of cloud due to its on-demand storage services with low cost maintenance in developing the infrastructure [2, 3]. In the cloud based systems, users outsource their data on the cloud and loss their physical control over the data. Due to loss of control, data may be at a risk and outsource data can be leaked to malicious users. Storing and maintaining outsources data to the cloud and the computation on the data is performed by a third-party auditor (TPA), also known as a cloud service provider. TPA is considered to be a partially authentic entity in a cloud environment. Therefore, the untrustworthy nature of the cloud service provider (CSP) raises several security concerns for the industries, organizations, and academicians.

If TPA works on at compromising fashion i.e. same storage server is being used by several users to store and access the data then one's data may be accessible by the malicious users who are not the authorized to access the specific data. Therefore, storage on the cloud leads to major concerns for the organization to maintain the control access over the data [4]. So, the framework must be designed in such a way that it permits only legal users to access the individual or group data. Moreover, the confidentiality and privacy of the data are also suggested to be interested in by the users [5]. To enhance security over the cloud, one of the simplest approaches for any organization to adopt the cryptography tools to keep the data confidential and private. Cryptography tools mainly perform two major operations known as the encryption process and the decryption process. The recent trends in technology uses the DNA based cryptography techniques in which De-oxyribo Neucleic Acid (DNA) can be represented either at biological DNA or with genetic databases (digital DNA). Several DNA cryptographic techniques have been proposed in literature [6–9]. In the encryption phase, it changes the actual meaning of the message using a specific pattern that patterns only known by the intended receiver and decrypts the encoded message into original form known as the decryption process. In a traditional framework, the access restriction on the data, key formation, encryption, and decryption of files is done by the data owner to ensure data privacy and security [10]. In group data, the number of users may be varied time to time i.e. users may be sometimes increased or decreased as per their authorization conditions. Therefore, there is a need to design a flexible cryptographic framework that can handle the varying nature in the number of users (newly joining or leaving user) during the data is being shared among a group and also capable to handle the key management efficiently [11]. There may be a possibility that the existing, leaving, and newly admitting group users can involve doing the mischievous activities to breach the data confidentiality and privacy [11] of group data. Such an internal attack can be more devastating than the external attack that violates data security. In most of the cases, lots of researchers trust the internal entities and only focus to prevent the data from outside attackers. Nevertheless, due to the involvement of multiple users in a group there are various security issues that need to be addressed. We pointed out some of the following important issues which are caused by the involvement of different users in a group during data share.

- Use of single key share among all the users in a group will permit the newly admitted user to access all the past data also known as backward access control
- Similarly, leaving user can have access to all the future communicated data also known as forward access control is also a major concern.

In such situations the chances of penetration of system control or violation of limited allowable access to data may break the security cover [12]. In this paper, we pointed out the above said security concerns of shared group data over the cloud.

The proposed model consists of three components named as the user; a crypto system (CS), and the cloud. The owner of actual data upload data along with the list of authorized users, and the required parameters that are used by the CS to create an access control list (ACL). In our framework, the CS is a fully trusted entity and responsible for all computations like key-creation, encryption, decryption, and access control. The CS create the separate key for each user and the corresponding symmetric key is used to encrypts or decrypt the data. To enhance the security of the system, CS splits each key into two parts for each user of the group such that one part alone cannot form the whole key. Subsequently, one portion of the key is assigned to the corresponding user, whereas the second part of the key is retained by the CS within the access list. Continuously, the source key is deleted from the system using protected overlapping [13].The architecture of the proposed model is similar to standard cloud architecture shown in Fig. 1. The working of the model is described in details in Sect. 3.

This paper contributes in the following ways.

- Retaining only part of the key by every group user keep the data secures against the nasty user.

Fig. 1 Standard cloud server architecture

- This framework ensures the security of data against the backward and forward access control that comes from internal threats.
- Data sharing over the cloud is done securely without the use of the elliptic curve or bilinear Diffie Hellman problem (BDH) cryptographic re-encryption.
- The proposed method ensures the privacy and integrity of data on the cloud using symmetric key encryption.

2 Related Work

Various architectures and security mechanisms have been developed in the last few decades [15–17] that can be utilized on various applications but these methods are not appropriate and more effective for portable devices due to rise in computational cost for key creation. Xu et al. [18] presented a framework named certificate-less proxy re-encryption (CL-PRE) schemes that use symmetric key encryption techniques. These techniques are effectively used to share the data confidentially within a particular group in the public cloud. The key used for encryption is also used in encryption carried out using the public key of the data owner. Subsequently, the encrypted data as well as encrypted key both are to be uploaded to the cloud. With the use of Proxy re-encryption approach the encrypted key is again re-encrypted by the cloud. In this model public–private key pair is formed based on the user's identity and no involvement of certificate is taken into consideration. The computational cost of the standard operation in finite fields is cheaper as compared to bilinear pairing but the proxy re-encryption uses the BDH and bilinear pairing that makes the model computationally intensive.

Seo et al. [19] presented a technique for reducing the computation overhead of the bilinear pairing method. The researcher avoids the use of bilinear pairing in his mediated certificate-less encryption model to share the data in the public cloud. The given model reduces the burden of the data owner by delegating all the task to the cloud such as generate public key and private key pairs for each users and forward the public keys to all of the participating users. This model makes revocation of the user easy to handle due to partial decryption and key management is done by the cloud. This model assumes the public cloud to semi-trusted entities. So, it is not considered safe to transfer the key creation process to the shared multitenant cloud system. Besides, two times the decryption process makes this system less efficient.

Khan et al. [11] introduced a system that uses the concepts of incremental cryptography in which data is partitioned into the number of blocks and subsequently each block is encrypted. The proposed model uses a private cloud to perform the expensive computation of key generation. This model uses El-Gamal Cryptosystem and bilinear pairing to share secret data to the cloud and not able to reduce the computation cost of bilinear pairing.

Chen and Tzeng [12] introduced another secured framework to share the data in a group based on the shared key derivation methods that use the binary tree for key formation. Due to re-keying methodology, the computation cost of the scheme is too high and easy to breakable. In this paper, we designed a framework that can be used to secure data sharing among a group without using the BDH, bilinear pairing, and El-Gamal cryptosystem. The framework eliminates the need of re-encryption and uses symmetric key cryptography tp encrypt the data. In this way computational intensive operations could be minimized that further improve the performance of overall system. Another important feature of the proposed frame work is that it allows user to access only a portion of the key which is helpful to protect data from forward and backward access control attacks.

Rewadkar et al. [20] developed an approach to hire a third-party auditor (TPA) which takes care about the confidentiality of cloud data. In this method, homomorphic encryption techniques are used to encrypt the data before it is shared with the TPA.

Gupta et al. [21] introduced an efficient security framework named TBHM using fully homomorphic properties with MD5 integrity algorithm on health records. Researcher has improved the key generation time, encryption as well as decryption time in his model but the fully homomorphic approach is more expensive as compare to others encryption techniques.

Shen et al. [22] designed a framework to share the data among multiple participants environment using symmetric balanced incomplete block design (SBIBD). The proposed framework suitable to extend the number of users flexibly over the cloud however, security of share data is always being a major concerned.

Garg et al. [23] proposed an auditing protocol for data integrity in cloud computing which reduced the computational cost during system setup phase. Authors protocol incorporate the use of bilinear pairings to verify and perform dynamic operations on data. Security of the system is achieved by the use of and Diffie Hellman Techniques.as well as to retain the privacy of data. Fan et al. [24] introduced another data integrity protocol based on secure identity aggregate signatures (SIBAS). This protocol also provide secure key management in trusted execution environment through (t, n) threshold scheme.

3 The Proposed Model

This section will describe the proposed techniques that able to share the data in a group in a secure manner without re-encryption in the cloud environment. Our model mainly consists of three entities that are described below in details.

3.1 Entities

The Cloud: When owner of actual data outsources his data, it is needed to store anywhere in the cloud storage. For this, the data owner must take storage services from the cloud service provider. To make sure the confidentiality and security of the data, data must be uploaded in encrypted form over the cloud because of the cloud required to be protected against privacy breaches. Our method uses two basic cloud operations: File Uploading and File Downloading.

The CS: In the proposed model, the CS is considered as a trusted entity and is fully responsible to perform all the security operations like access control, key generation handling, encryption and decryption processes, and managing the list of ACL to provide data privacy and confidentiality. Initially, the new users are required to register themselves with the CS to get the security services.

The group Users: The group users are responsible to upload the data file on the cloud storage. There is only one owner for each of the data files and remaining will act as a consumer of the data file. The owner of the file will decide the access rights such as either can be granted for the file or revoked to the file or both to the other group members. The records of constraints related to the access control rights are retained by the CS in the form of an ACL file. It is to be noted that a separate ACL is maintained for each of the data files.

Formation of encryption or decryption Keys: The proposed approach uses a single key encryption technique for each of the data files. This single key is not fully accessed by any of the group members and CS i.e., after performing encryption/decryption process, the encrypted key is divided into two parts, one part is forwarded to the respective user in the group while the other part of the key is retained by the CS. The encrypted data can only be decrypted using both parts of the key, while a single part of the key is not sufficient to decrypt the data.

Formation of Key (Key): The CS generates the random secret keyof 128 bits length for every data files. Key of the model is computed using two-stage process. In the first stage, length of 128-bits random number R is generated such that $R = \{0, 1\}^{128}$ and passes through message digest 5 (MD5) hash function with a 128-bit output. The second stage of the key completely randomizes the initial user-derived random number *R*. The result of the hash function works as a *Key* for data file. The data file is encrypted using standard symmetric key encryption technique such as AES algorithm.

Formation of CS Key Share (*Key_k*): The CS is responsible to generate distinct Key_k for each of the users in the group, such that $Key_k = \{0, 1\}^{128}$. Now, Key_k acts as the *CS* portion of the key and is used to figure out *Key* whenever an encryption/decryption request is received by the CS. Moreover, it is ensured by the comparison that the distinct Key_k is generated for every file user.

Formation of User Key Share (*Key_k'*): Key_k' is evaluated by performing Ex-OR operation between the Key with the corresponding CS key share for each user '*k*' in the group as follows:

$$Key'_k = Key\ Ex - OR\ Key_k$$

Algorithm 1: Key Formation and Encryption of data file
Input: The algorithm take input data file F, list of authorized users, the *Key* and MD5 hash function M_f
Compute: Compute a random number R with length of 128-bits
$R = \{0, 1\}^{128}$
By the use of MD5 hash function compute $Key = M_f(R)$
Cipher File (C) = SKA (*F, Key*)
for each user *k*in the access list (ACL), do
$Key_k = \{0, 1\}^{128}$
$Key_k' = Key \oplus Key_k$
AddKey_k'foruser*k*intheACL
SendKey_k'foruser*k*
end for
remove (Key_k')
remove (*K*)
return the value of *C* to the owner or
upload the value of C to the cloud

4 Design of Proposed Framework

This section describes the design of the proposed framework which uses various cryptographic key operations to achieve the security of shared data among the group. The model performs the following basic operations:

File Upload: To upload a file in a group, the data owner required to encrypt the data for security concerns and send encryption request along with the data file (F) and a list (L) of authorized users with the access constraint parameters to the CS.

For access control Read-only access or Read–write access or a combination of both are used. The CS uses list to create an access control list (ACL) to access the data for each user of a particular group. The owner of data sends list to the CS only if the data has to be shared in a new group. If the group already exists, there is no need to send list with encryption request, rather, sends the group id of the already reachable group. When an encryption request is received the CS generates the ACL from the list and forms a separate group of the users. The ACL is generated independently for each of the file. The ACL comprises the records about the file such as owner ID, file ID, file size, the list of the access user IDs, and other metadata. In case of an already existing group, for each file only the ACL is being generated. Further, the CS creates a key using the above said approach and encrypts the file. The resultant file is named as cipher file (*C*). Consequently, the CS creates Key_k and Key_k' for each user and deletes the key by secure overwriting. In the next step the entry of Key_k is incorporated into ACL for each user. For securing the file, the CS applies hash-based message authentication algorithm (MD5) on every encrypted file. By using a secure channel the encrypted data file, group ID, and Key_k' of the owner are sent to the requesting data owner while group ID and Key_k' are sent to the rest of the users in the group. To forward the user part of the key the public keys of the group users can be used. Algorithm 1 describes the use for the key creation and encryption process at the CS.

File Download: In the download phase, the group member sends a downloading request for the data file to the CS. The CS checks the authenticity of the requested user and permits authenticated user to download the file from the cloud. After downloading is complete the user again sends the decryption request to the CS. The CS at this stage checks the authentication of the member corresponding to the existing ACL. The decryption request incorporates the user part of the key, i.e. Key_k' and handles other authentication credentials. All the details regarding the computation of key are described in the decryption algorithm.

File Update: Updating a file on the cloud has a similar approach as used in uploading the file with a smaller difference. Activities related to the creation of the ACL and key generation are not required during the process of updating a file. Simply, the user who wants to make any changes in the file transmits an update request along with the security parameter to the CS. Then, the requested user is verified by the CS and ACL and accordingly permission is granted.

Algorithm 2 Decryption Process
Input: Cipher file (C), the ACL, the Symmetric Key Algorithm
Compute:
Obtain1[st]portion of the Key (Key_k')from the wishing user k
ObtainCipher file (*C*) from the wishing user or download from the cloud
Retrieve 2nd portion of the Key (Key_k) from the ACL
If $Key_k$is not in ACL,
Then report a access denial message to the user
else
    calculate $Key = Key_k \oplus Key_k'$
    $F = \text{SKA}(C, K)$
Send *F* to the user
end if
remove (*K*)
remove (Key_k').

New Group User Inclusion: In the future, the System may be required to add new users to the existing group. For joining the existing group, each user should make a request to the data owner along with their user ID, group ID, and the access control parameters which are to be included in the ACL. After successful joining, users can perform uploading and downloading files in a similar way discussed above.

Leaving/Revoked Group User: The group owner will inform about the leaving/revoking member to the CS. CS in response to the request removes all records related to the specific leaving user from the ACL of the corresponding files.

5 Motivation Towards Proposed Work

There are several challenges while designing and adopting new methods of maintaining security on clouds. A number of methods have been proposed and reported in this direction; however, no methodology claims completely elimination of the security lapses on shared data among the group. In the proposed work effort is made to minimize security threats faced from both inside/outside. This methodology completely eliminates the possibility of any kind of outsider or insider threats on shared data. This is achieved by introducing partial key sharing as partial key cannot decrypt the shared data alone. All the expensive computation such as key generation, encryption, and decryption are carried out by CS. By employing MD5 hash function data integrity and verification can be incorporated in the proposed method. MD5 is used in the proposed framework because it is claimed in [25] that the execution time of MD5 is better than SHA256 with the same computation complexity Θ(N).

6 Experimental Analysis

Our framework comprises majorly three entities: The users, the CS, and the cloud. The experimental analysis has been carried out for the proposed framework on Windows-10 (64-bits) Operating System, Intel i3 Processor with 2.4 GHz and 8 GB RAM as system configuration. The method was implemented in visual studio 2015 C# using the.Net framework. Amazon S3 was used as a cloud server in our implementation that is communicated by the Amazon Web Services software development kit and.Net application programming interfaces. The CS was used as a trusted third party. Client application which incorporates the functionality of the user was used to communicate with the CS to get the services. .Net libraries are used to establish connections between entities. The class MD5 Crypto Service Provider was used to access all of the features related to MD5 hash function.

Key Generation Time: The proposed method generates a separate and unique key for each file. However, file uploading time and the key sharing are distinctly evaluated for all the participants in the group. The time for key formation for the proposed work is calculated at the interval of 10 units such as 10–100 cloud users (CUs) as shown in Fig. 2 and compared with the results of SeDaSC [14].

It is observed from the Fig. 2 that the number of users increases, then the key generation time automatically increases and the proposed method take less time in key formation as compared to SeDaSC.

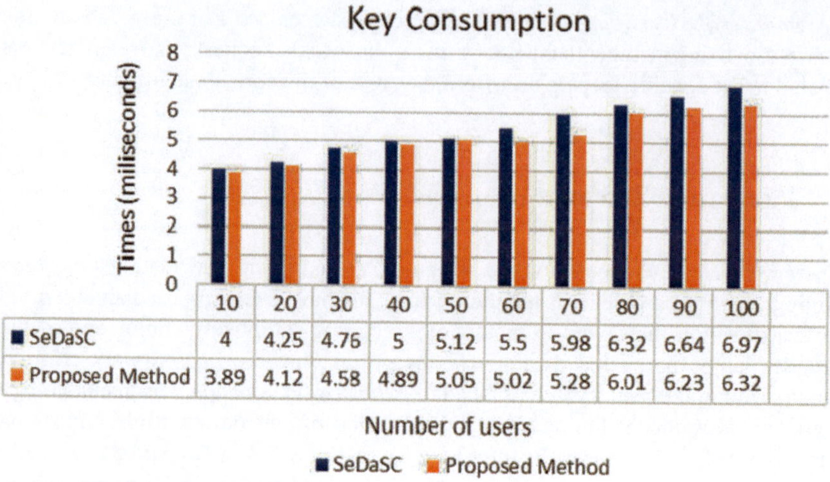

Fig. 2 Key generation time for SeDaSC and proposed method

Encryption and Decryption: We run our encryption/decryption algorithms to evaluate the encryption and decryption times on various file sizes. The results are obtained on various file sizes of 0.1, 0.5, 1, 10, 100, and 500 MBs. In the proposed framework, first the CS computes *key* then encrypts and decrypts the file with the obtained key. Therefore, generation time for a *key* is compared with the execution time of encryption and decryption. This is done to find out the overhead time to generate a key over the total encryption and decryption times. The execution time required in key generation, encryption and decryption evaluated separately and outputs are analyzed which are shown in Figs. 3 and 4, respectively.

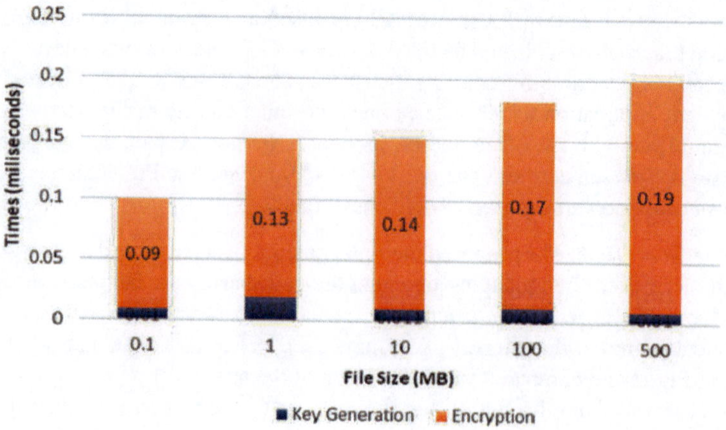

Fig. 3 Execution times for various file sizes

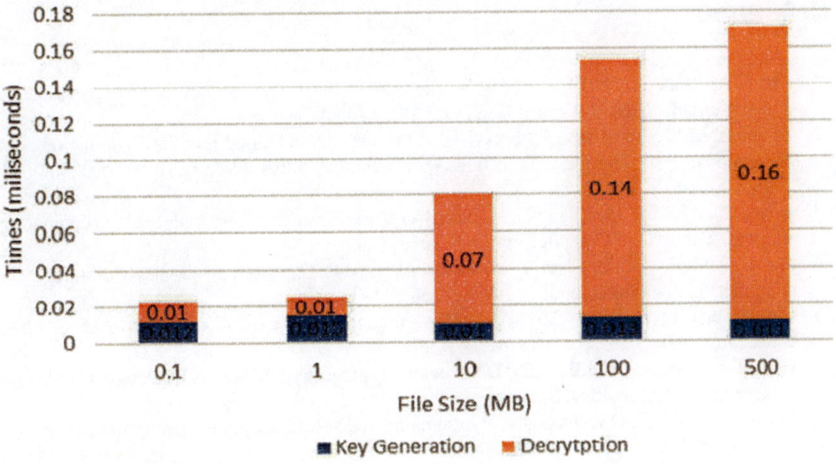

Fig. 4 Decryption times for various file sizes

Figure 3 demonstrates the outcome for encryption time on different file sizes. The result shows that the execution time for encryption increases with the increase in the file sizes. However, the generation time for key on various file sizes almost remains constant with a minor change. This is because the computation time of the key does not depend on the file size.

Figure 4 demonstrates the outcomes for the decryption time on various file sizes. The results for decryption show similar behavior as compared to encryption. The generation time for the key is larger than the decryption time for smaller file sizes. Therefore, for a large file key computation it takes negligible time for decryption.

7 Conclusion

This paper introduces a storage security framework over the cloud that ensures the security and confidentiality of the shared data among the multiple participants in a group without re-encryption techniques. Additionally, the method also provides access control, safeguard against forward and backward access control, and guaranteed deletion by removing all the records used for the decryption of a file. It reduces the burden of the data owner because of all the computation to be carried out by the CS. To evaluate the performance of the proposed model, various parameters are used. The effectiveness of the proposed method is checked by evaluating by key generation, file uploading and downloading time. Conclusively, the methodology could be improved by the trust level in the CS. System security is further enhanced by handling the inside threats in an effective way. Moreover, the results of the proposed method with different key sizes can further be examined. Based on the results obtained it could be argued that the proposed method could be employed in cloud based applications.

References

1. Abbas A, Khan SU (2014) A review on the state-of-the-art privacy preserving approaches in e-health clouds. IEEE J Biomed Health Inf 18(1):1431–1441
2. Alhamazani K et al (2014) An overview of the commercial cloud monitoring tools: research dimensions, design issues, state-of-the-art. Computing. https://doi.org/10.1007/s00607-014-0398-5
3. Khan AN, Kiah MLM, Khan SU, Madani SA (2013) Towards secure mobile cloud computing: a survey. Future Gen Comput Syst 29(5):1278–1299
4. Wei L, Zhu H, Cao Z, Chen Y, Vasilakos AV (2014) Security and privacy for storage and computation in cloud computing. Inf Sci 258:371–386
5. Cloud Security Alliance (2011) Security guidelines for critical areas of focus in cloud computing v3.0
6. Gehani A, LaBean T Reif J (2000) DNA-based cryptography. DIMACS DNA based computers V. American Mathematical Society
7. Gehani A et al (2004) DNA-based cryptography. Lect Notes Comput Sci 2950:167–188
8. Anam B, Sakib K, Hossain MA, Dahal K (2010) Review on the advancements of DNA cryptography. arXiv:1010.0186. 1 Oct 2010
9. Gupta LM, Garg H, Samad A (2019) An improved DNA based security model using reduced Cipher text technique. Int J Comput Netw Inf Secur (IJCNIS) 11(7):13–20. https://doi.org/10.5815/ijcnis.2019.07.03
10. Chen D et al (2014) Fast and scalable multi-way analysis of massive neural data. IEEE Trans Comput. https://doi.org/10.1109/TC.2013.2295806 (to be published)
11. Khan AN, Kiah MM, Madani SA, Ali M, Shamshir-band S (2014) Incremental proxy re-encryption scheme for mobile cloud computing environment. J Super comput 68(2):624–651
12. Chen Y, Tzeng W (2012) Efficient and provably-secure group key management scheme using key derivation. In: Proceedings of IEEE 11th international conference on trust, security, and privacy in computing and communications, pp 295–302
13. Gutmann P (1996) Secure deletion of data from magnetic and solid-state memory. In: Proceedings of 6th USENIX Security Symposium. Focusing Application Cryptography, p 8
14. Ali M et al (2017) SeDaSC: secure data sharing in clouds. IEEE Syst J 11(2):395–404. https://doi.org/10.1109/JSYST.2014.2379646
15. Liu DL, Chen YP, Zhang HP (2010) Secure applications of RSA system in the electronic commerce. In: International conference on future information technology and management engineering, vol 1, pp 86–89
16. Gola KK, Gupta B, Iqbal Z (2014) Modified RSA digital signature scheme for data confidentiality. Int J Comput Appl 106(13)
17. Arora R, Parashar A (2013) Secure user data in cloud computing using encryption algorithms. Int J Eng Res Appl 3(4):1922–1926
18. Xu L, Wu X, Zhang X (2012) CL-PRE: A certificateless proxy re- encryption scheme for secure data sharing with public cloud. In: Proceedings of the 7th ACM symposium on information, computer and communication security, pp 87–88
19. Seo S, Nabeel M, Ding X, Bertino E (2013) An efficient certificate- less encryption for secure data sharing in public clouds. IEEE Trans Knowl Data Eng 26(9):2107–2119
20. Rewadkar DN, Ghatage SY (2014) Cloud storage system enabling secure privacy preserving third party audit. In: International conference on control, instrumentation, communication and computational technologies (ICCICCT). IEEE, New York
21. Gupta LM, Samad A, Garg H (2020) TBHM: a secure threshold-based encryption combined with homomorphic properties for communicating health records. Int J Inf Technol Web Eng (IJITWE) 15(3):1–17
22. Shen J, Zhou T, He D, Zhang Y, Sun X, Xiang Y (2019) Block design-based key agreement for group data sharing in cloud computing. In: IEEE transactions on dependable and secure computing, vol 16(6), pp 996–1010. https://doi.org/10.1109/TDSC.2017.2725953

23. Garg N, Bawa S, Kumar N (2020) An efficient data integrity auditing protocol for cloud computing. Future Gener Comput Syst 109. https://doi.org/10.1016/j.future.2020.03.032
24. Fan Y, Lin X, Tan G, Zhang Y, Dong W, Lei J (2019) One secure data integrity verification scheme for cloud storage. Future Gener Comput Syst 96:376–385
25. Rachmawati D, Tarigan JT, Ginting ABC (2018) A comparative study of message digest 5 (MD5) and SHA256 algorithm. J Phys: Conf Ser 978(1):012116 (IOP Publishing)

Energy-Efficient Bonding Based Technique for Message Forwarding in Social Opportunistic Networks

Satbir Jain, Ritu Nigam, Deepak Kumar Sharma, and Shilpa Ghosh

Abstract In Social opportunistic networks where no end to end path exists between a sender and a destination node, the nodes usually perform message routing using the store-carry and forward pattern. Nodes consume a significant amount of energy during the node discovery phase and the message transmission phase. Therefore, it is challenging to design an energy-efficient message forwarding protocol. This paper improves an existing scheme named bonding based technique for message forwarding in social opportunistic network (BBFT) concerning energy consumption. The newly introduced energy-efficient BBFT approach proposes an energy estimation model that estimates the amount of energy consumed for transmitting, receiving, and scanning in the message forwarding process by the sender node. This energy consumption estimation model reduces the message flooding in the network, which results in conservation in the nodes' residual energy, hence increasing the network's lifetime. Simulation has been performed using ONE simulator to assess the EBBFT algorithm's performance on energy parameters such as average residual energy, dead nodes, and other parameters overhead and dropped messages and compared against the BBFT and the energy-efficient SEIR protocols. On average, EBBFT is 90.65% better than BBFT and 27.18% better than SEIR concerning average residual energy while varying the message generation interval and had no dead nodes.

S. Jain
Department of Computer Engineering, Netaji Subhas University of Technology
(Formerly Netaji Subhas Institute of Technology), New Delhi, India

R. Nigam
Division of Computer Engineering, University of Delhi (Netaji Subhas Institute of Technology), New Delhi, India

D. Kumar Sharma (✉)
Department of Information Technology, Netaji Subhas University of Technology
(Formerly Netaji Subhas Institute of Technology), New Delhi, India

S. Ghosh
Division of Information Technology, University of Delhi (Netaji Subhas Institute of Technology), New Delhi, India

P. K. Singh et al. (eds.), *Proceedings of Second International Conference on Computing, Communications, and Cyber-Security*, Lecture Notes in Networks and Systems 203,
https://doi.org/10.1007/978-981-16-0733-2_20

Keywords Energy estimation model · Transmission energy · Scan energy · Receiving energy · ONE simulator · Direct and indirect bonding

1 Introduction

Social Opportunistic networks [1, 2], i.e., Social Oppnets, is a type of Delay Tolerant Networks (DTNs) [3] where the social behavior of the nodes is used to enhance the message delivery through peer to peer approach in a short-range wireless interfaces. The extensive use of handheld mobile gadgets by people, like tablets, smartphones with internet competencies among users, and their environments, results in gaining significance in these networks. Social Oppnets [4, 5] takes on the store-carry and forward pattern to conduct routing when a sender node has a message to communicate to its target node. So, when the relays are not available to transmit the message currently, the nodes reserve the message in their buffer up to any useful relay nodes are obtained. The selection of the useful relay nodes follows the examination of nodes' forwarding capabilities against a few performance metrics. But there is a possibility that the selected relay node may not have sufficient energy [6, 7] to forward the message to the next relay or the destination, which can cause transmission failure and decreases network reliability. So the energy constraint is a critical issue to handle. Therefore this paper investigates the finite energy of nodes while designing the forwarding protocol for Social OppNets [8]. As the previously designed BBFT [9] protocol works well regarding delivery rate, dropped data packets, and overhead, but takes a large amount of energy. Therefore this paper takes energy into consideration to propose a new energy estimation model to build BBFT energy efficient, which also maximizes the lifetime of nodes in the social OppNets.

The remaining part of the paper is set up as follows. Section 2 discusses the literature review accomplished in the field of energy-based routing in the Social OppNets. Next, Sect. 3 explains the system model and proposed EBBFT scheme; later, Sect. 4 presents the simulations and results. Finally, the conclusion part is discussed in Sect. 5.

2 Literature Review

This section covers a quick overview of the previously introduced BBFT protocol and some current energy-aware routing protocols for OppNets.

BBFT [9] protocol uses the average detachment period and variance between the nodes of the network to calculate the bonding matrics between source and neighbor nodes. The scheme considers both direct Bonding and indirect Bonding with the available neighbor nodes to transmit the data packet from source to destination node. The weakest direct bonded link is replaced with the strongest indirect bonded links to deliver the message, which increases the delivery probability of the BBFT protocol.

Dhurandher et al. introduced EHBPR [10], an energy-aware protocol that works on History-Based Prediction for message forwarding in OppNets. EHBPR proposes a utility function that incorporates the four energy-related factors named Perpendicular, transmission, sparse constant and, residual energy to determine the most efficient relay node to transmit the message. The GAER [11] scheme proposed by Deepak et al. exploits the nature-inspired genetic algorithm for message transmission. This work uses two factors, i.e., mean and place, as a fitness function to check the relay nodes' fitness. The node with the highest fitness is selected as the relay node to transmit the message to subsequent next-hop. Anshuman et el. presents SEIR [11] protocol, which brings energy reduction and incentivization ideas in the message routing strategy. SEIR uses the Stackelberg game-theoretic model, which works on eliminating the node's selfishness by assigning the optimal reward to these relay nodes to increase the probability of successful message transmission. Annalisa et al. proposed a routing protocol [12] (ECF) based on the Energy-aware Centrality for mobile social OppNets. The above-discussed protocols are based on the nodes' residual energy, but our proposed algorithm estimates the energy required for message transmission to increase the network's lifetime.

3 System Model and Proposed Approach

3.1 Energy Estimation Model

The energy estimation model represents the evaluation of a node's capability concerning energy available for message transmission. Whenever nodes transmit or receive messages, they consume a significant amount of energy for this process. If this process consumes more energy than the node's current energy than that node dies or its battery drained. For a successful message transmission routing, nodes must have enough power for performing scanning, message sending, and receiving. Signal Processing, message forwarding, and hardware operation require some amount of energy to complete the task in the Social opportunistic networks scenario. This proposed message forwarding scheme EBBFT specifically focuses:

Scan Energy: The scan energy shows the minimum energy requirements in each scan with other nodes.

$$\mathrm{ER}_{\mathrm{sc}} = e_{\mathrm{sc}} \times \frac{t}{T} \tag{1}$$

where e_{sc} represents consumed energy per scan and T is the scan period.

Transmission Energy: The energy required to forward a message from one node to another node.

$$\mathrm{ER}_{\mathrm{tr}} = e_{\mathrm{tr}} \times p_{\mathrm{tr}} \tag{2}$$

where e_{tr} represents the energy consumed per message transmission and p_{tr} depicts the amount of transmitted messages.

Receiving Energy: The energy required to receive a message by the node.

$$\mathrm{ER}_{rc} = e_{rc} \times p_{rc} \tag{3}$$

where e_{rc} indicates the energy consumed per message reception, and p_{rc} depicts the amount of received messages.

From Eqs. (1)–(3), we estimate the total energy required E_{rq} by a node for successful message delivery. The proposed formula is given below:

$$\mathrm{E}_{rq} = e_{rc} \times \sum_{i=1}^{r} p_{rc} + n \times \left(e_{sc} \times \frac{t}{T} + e_{tr} \sum_{j=1}^{tr} p_{tr} \right) \tag{4}$$

3.2 *Proposed Work*

The proposed scheme assumes that the Social opportunistic networks consist of N social nodes cooperate with each other in performing the message transmission. Nodes participating in data transmission have a sufficient amount of buffer space to hold the messages, and there is no malicious behavior present by the nodes. In the proposed EBBFT protocol, relay nodes selection is based on the node's required energy. EBBFT scheme is an extended version of BBFT protocol, which estimates the node's energy requirement to perform a message transmission. This estimation considers scan energy, transmission energy, and receiving energy of the node and employs a threshold E_{th} condition for next-hop neighbor nodes. Algorithm 1 gives the details of EBBFT approach.

$$E_{th} = \frac{\text{Sum of neighbors available energy}}{\text{number of neighbors}} \tag{5}$$

The essential steps of EBBFT are described below, which mimics the message routing between nodes a and b.

1. First, the source node performs the energy estimation model. Suppose the source node's current energy (E_{avail}) is more than the total energy required (E_{rq}), and neighbor nodes contain more energy than threshold energy (E_{th}). In this case, neighbor nodes are considered for relay nodes selection, and the sender node can perform forwarding.
2. Second, measure the direct bonding $DB_{(a,b)}$ of the source node a with its neighbor nodes by calculating the average detachment period and variance. After that, measure the indirect bonding $IB_{(a,b)}$ by using implicit two-hop paths information between nodes a and b.

Algorithm 1 **EBBFT algorithm for message forwarding**

Begin
sender node a generates message x for destination b
if $E_{avail}(a\ or\ intermediatenode) >= E_{rq}(a\ or\ intermediate\ node)\ AND\ E_{avail}(N_i) >= E_{th}$ **then**
  Calculate direct $DB_{(a,b)}$ and indirect $IB_{(a,b)}$ bonded nodes.
else drop the message x.
  for Each Neighbor Node N_i of a **do**
    Measure $DB_{(N_i,b)}$
    $Q_n[] \leftarrow DB_{(N_i,b)}$ /* assign direct connections */
  end for
  for Each Neighbor Node N_i of sender node a **do**
    Measure $IB_{(N_i,b)}$
    $RB_n[] \leftarrow IB_{(N_i,b)}$ /* assign indirect connections */
  end for
  for Each directly linked Node $Q_n[]$ **do**
    if $DB_{(N_i,b)} > DB_{(a,b)}$ **then**
      $L_n[i] \leftarrow DB_{(N_i,b)}$
    end if
  end for
  for Each $L_n[i]$ **do**
    find $\rho_a \leftarrow \min[DB_{(k,b)}]$ where k $\in N_i$
  end for
  for Each $RB_n[i]$ **do**
    if $I_{(RB_n[i],b)} \geq \rho_a$ **then**
      $M_n[i] \leftarrow RB_n[i]$
    end if
  end for
  Take the nodes of L_n and M_n sets and forward the message copies to them.
end if
End

3. In the third step, select the candidate node with better direct bonding weight compared to the source node concerning destination b. Also, select the indirect neighbors as the candidate node with better indirect bonding weight than the weakest direct bonding weight.
4. In the forth step, source node forwards the message copy to selected direct and indirect bonding nodes.

4 Simulation and Results

4.1 Simulation Setup

EBBFT's performance is evaluated on ONE simulator, and it is compared in contrast with the BBFT scheme and energy-efficient SEIR protocol. The simulation is performed by altering the number of nodes in each group as 40, 80, 120, 180 in each

Table 1 Default simulation parameters

Parameters	Value
Total time of simulation	4300 s
Simulation area	3400 m * 3400 m
Range of speed	0.5–1.5 m
Message TTL	100 min
Total number of nodes	126
Buffer space	10 MB
Message size	1 kB–5 MB
Transmission speed	250 kBps
Transmission range	10 m
Message generation interval	25–35 s
Initial energy	4800 units
Transmit energy	0.5 units
Scan energy	0.1 units
Scan response energy	0.01 units
Threshold energy	300 units
Charging coefficient	20 units

simulation run, and altering the message generation interval values by incrementing it to 10 s as 5–15, 15–25, 25–35, and 35–45 seconds in each simulation run. The default parameters are specified in Table 1, and the mobility model used is the shortest path and map route. The simulation setup used in EBBFT is the same as BBFT; only the changed default settings are mentioned in this paper. The EBBFT algorithm is compared with BBFT and SEIR protocols by using the following matrices.

1. Average residual energy (Joule): This metric represents the average unused energy remained when the simulation ends.
2. Dead nodes: This metric represents the number of drained nodes whose residual energy is zero.
3. Overhead ratio: This metric represents the average number of replicas forwarded per data packet to deliver it to the destination.
4. Dropped messages: This metric represents the number of declined messages from the nodes buffers.

4.2 Results

In this section, results obtained through regressive simulations by using the ONE simulator are discussed.

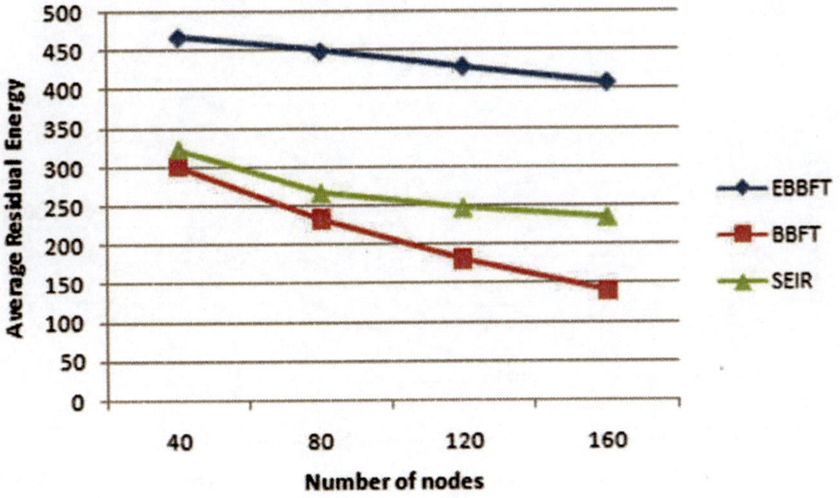

Fig. 1 Number of nodes versus average residual energy

Altering the number of nodes: Figures 1, 2, 3 and 4 illustrates the response of altering the group's number of nodes in the increasing order to assess the EBBFT's performance compared with BBFT and SEIR protocols. It is examined from Fig. 1 that the rise in the number of group nodes corresponds to a decrement in the nodes' average residual energy as the interaction among the nodes becomes more. These high interactions result in additional device scans and responses, reducing the nodes' average residual energy. It is indicated that EBBFT outperforms all the other protocols. On average, EBBFT is 105.04% better than BBFT and 63.84% better than SEIR in regard to average residual energy. Figure 2 depicts the EBBFT produces no dead nodes as compared to the BBFT and SEIR protocols. EBBFT uses an energy estimation method that discards the low energy nodes to get selected as next-hop. Therefore, the probability of node dying in the network is extremely low in EBBFT. Overall, BBFT has 67.75% dead nodes, and SEIR has 39% of dead nodes. Figure 3 indicates that the EBBFT produces the lowest overhead ratio with a significant margin compared to BBFT and SEIR. It causes fewer message copies because of its efficient energy model limiting the number of message replicas forwarded by a node. On average, EBBFT is 99.69% lower than BBFT and 99.82% lower than SEIR in terms of overhead ratio. Figure 4 inspects that the number of dropped messages in EBBFT is remarkably low as compare to BBFT and SEIR protocols. On average, EBBFT is 96.00% lower than BBFT and 96.39% lower than SEIR in terms of dropped message. The hike in dropped messages is due to the more message passing among nodes but limited buffer size.

Altering the message Generation Interval (MGI) Figures 5, 6, 7 and 8 displays the impact of changing the message generation interval in the rising order to assess the performance of EBBFT in comparison with BBFT and SEIR protocols. It is shown

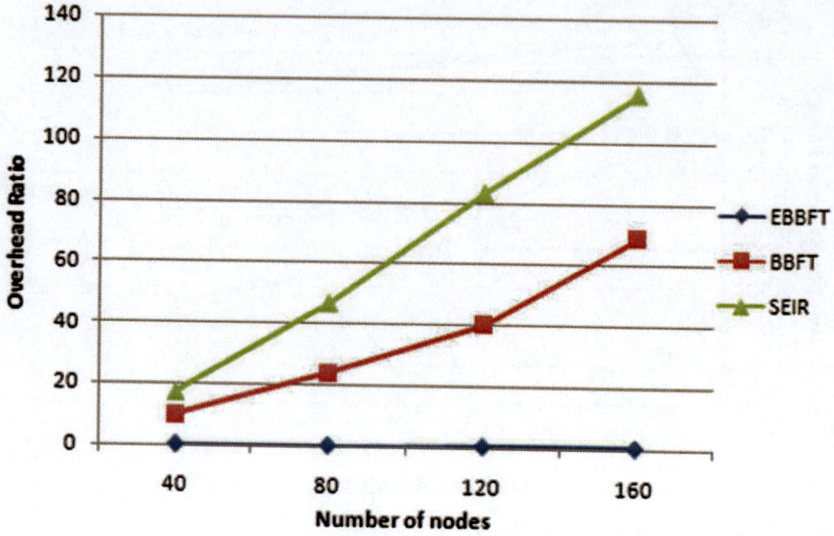

Fig. 2 Number of nodes versus overhead ratio

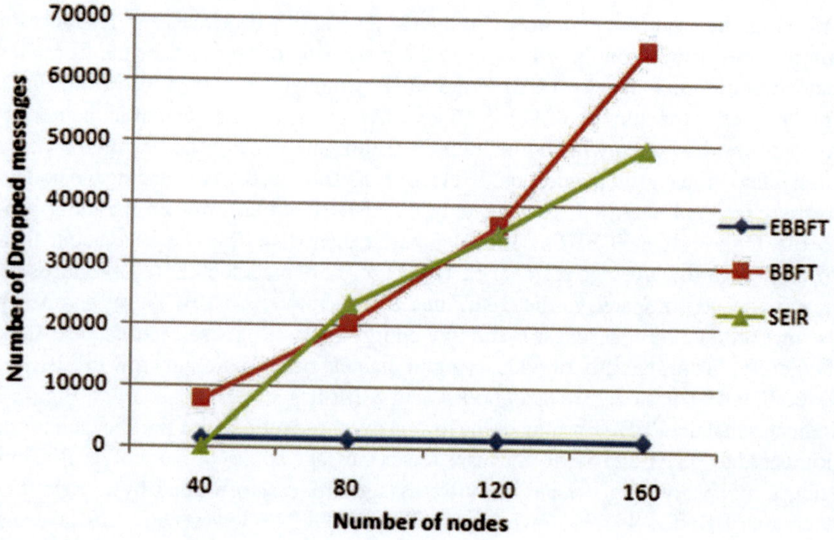

Fig. 3 Number of nodes versus dropped messages

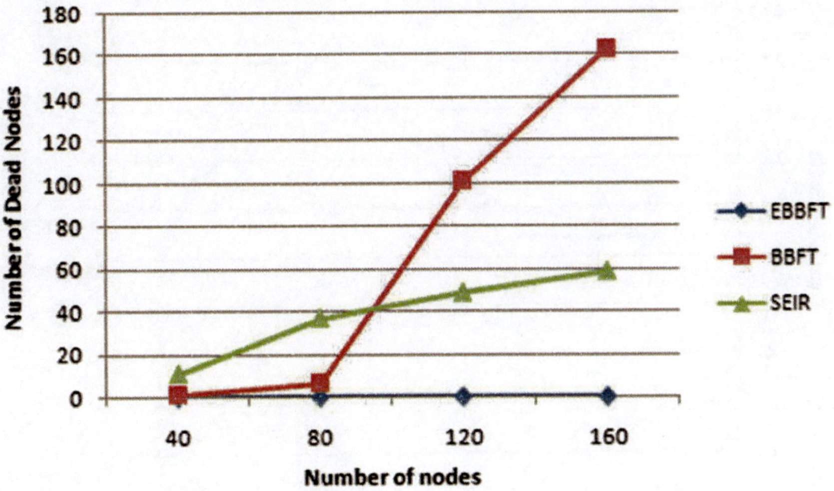

Fig. 4 Number of nodes versus number of dead nodes

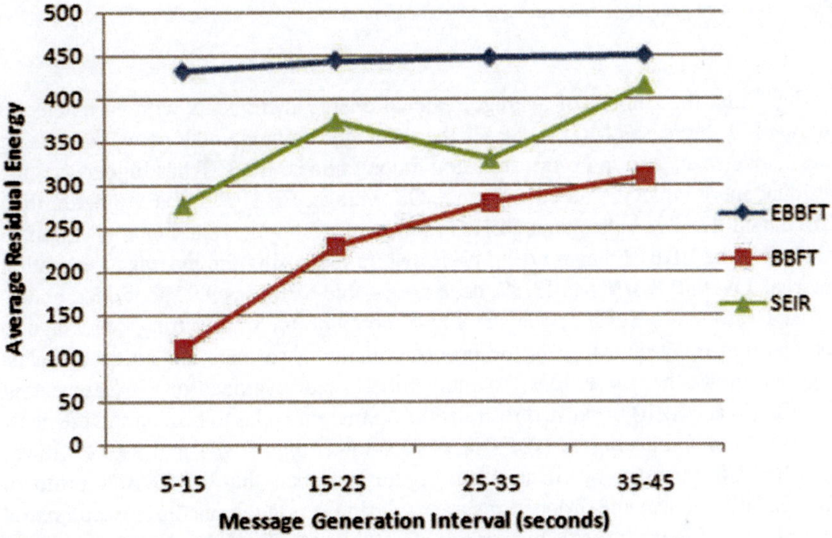

Fig. 5 Message generation interval versus average residual energy

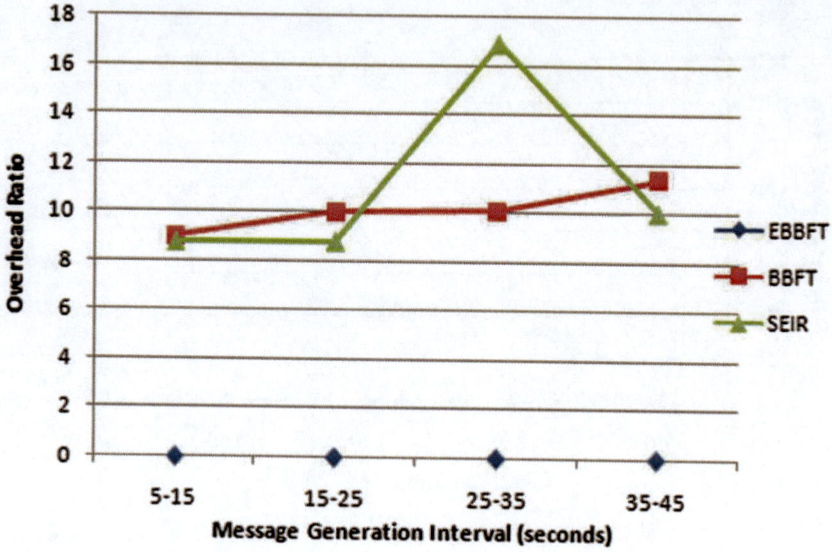

Fig. 6 Message generation interval versus overhead ratio

in Fig. 5 that the amount of average residual energy increments with an extension in the MGI. Increased MGI increases the messages flowing in the network, causing more node scans and more transmission among nodes. EBBFT has higher average residual energy than the other protocols. On average, EBBFT is 90.65% better than BBFT and 27.18% better than SEIR in terms of average residual energy. Figure 6 illustrates the EBBFT has no dead nodes while increasing the message generation interval. Overall, BBFT has 13.5% dead nodes, and SEIR has 9.97% of dead nodes. EBBFT estimates a sender node's residual energy before transmitting a message to its next-hop relay nodes using the proposed method, so there are no dead nodes. Figure 7 shows that the EBBFT maintains the lowest overhead to a greater extent than BBFT and SEIR because it restricts the inefficient nodes to take part in the next-hop selection. On average, EBBFT is 99.91% lower than BBFT and 99.92% lower than SEIR in terms of overhead ratio. Figure 8 inspects that the EBBFT protocol significantly reduces the dropped messages during message transmission compared to BBFT and SEIR protocols by performing efficient buffer utilization. On average, EBBFT is 82.25% lower than BBFT and 76.62% lower than SEIR based on the dropped message criteria.

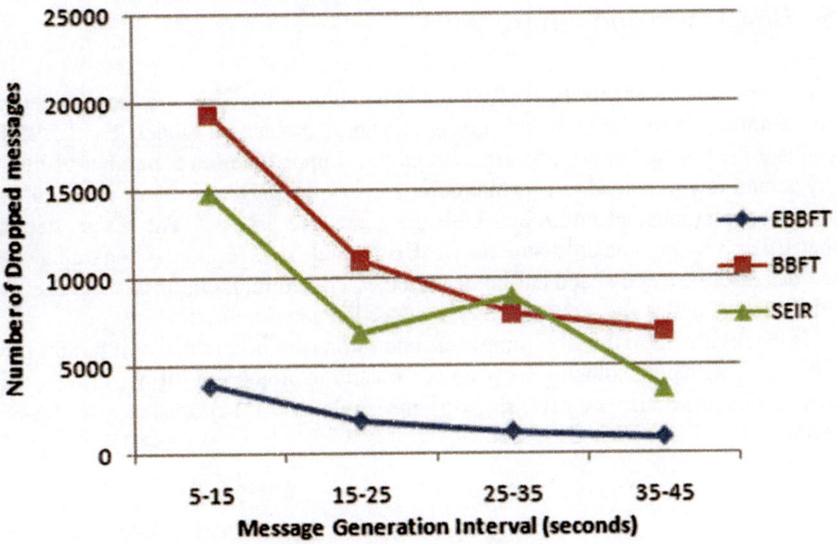

Fig. 7 Message generation interval versus dropped messages

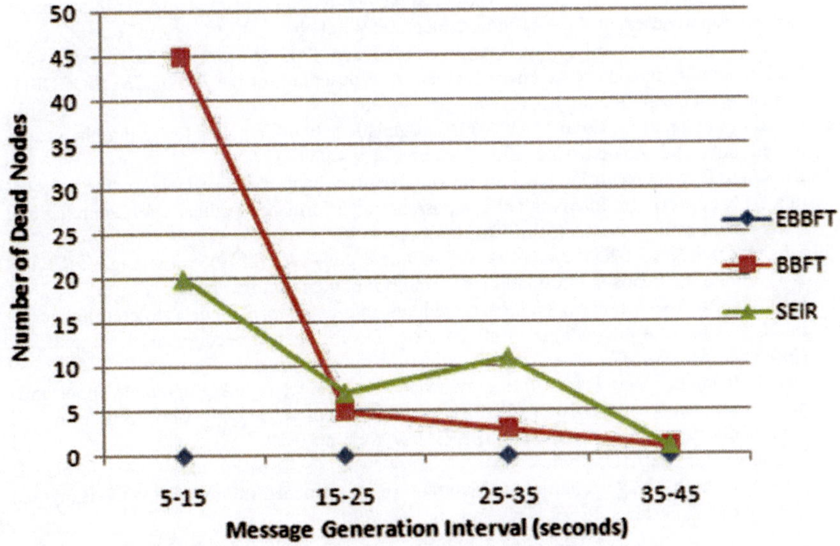

Fig. 8 Message generation interval versus number of dead nodes

5 Conclusion and Future Work

This work showcases that the EBBFT scheme improves the BBFT protocol's energy consumption performance by proposing an energy estimation model. The EBBFT yields more residual energy, fewer dead nodes, dropped the fewer number of messages, and low overhead ratio while comparing with BBFT and SEIR protocol by altering the number of nodes and message generation interval. The results depict that EBBFT is significantly superior to BBFT and SEIR protocols regarding the metrics discussed above and has no dead nodes. Therefore, it diminishes the energy utilization of nodes and, consequently, extends the network lifetime.

In the future, EBBFT will explore more energy constraints to enhance it further and compare it with many other energy-based forwarding protocols [10]. We also would like to work on security aspects using trust mechanisms [13, 14], and incentive-based techniques [15].

References

1. Wang Y, Vasilakos AV, Jin Q, Ma J (2014) Survey on mobile social networking in proximity (MSNP): approaches, challenges and architecture. Wirel Netw 20(6):1295–1311
2. Hui P, Chaintreau A, Scott J, Gass R, Crowcroft J, Diot C (2005) Pocket switched networks and human mobility in conference environments. In: Proceedings of the 2005 ACM SIGCOMM workshop on Delay-tolerant networking, pp 244–251
3. Pelusi L, Passarella A, Conti M (2006) Opportunistic networking: data forwarding in disconnected mobile ad hoc networks. IEEE Commun Mag 44(11):134–141
4. De Rango F, Monteverdi F (2012) Social and dynamic graph-based scalable routing protocol in a DTN network. In: International symposium on performance evaluation of computer and telecommunication systems (SPECTS). IEEE, pp 1–8
5. Singh PK, Panigrahi BK, Suryadevara K, Sharma SK, Singh AP (2019) Proceedings of ICETIT 2019: Emerging trends in information technology, vol 605. Springer Nature, Berlin
6. Lu X, Hui P (2010) An energy-efficient n-epidemic routing protocol for delay tolerant networks. In: 2010 IEEE fifth international conference on networking, architecture, and storage. IEEE, New York, pp 341–347
7. Gao S, Zhang L, Zhang H (2010) Energy-aware spray and wait routing in mobile opportunistic sensor networks. In: 2010 3rd IEEE international conference on broadband network and multimedia technology (IC-BNMT). IEEE, New York, pp 1058–1063
8. Chilipirea C, Petre A-C, Dobre C (2013) Energy-aware social-based routing in opportunistic networks. In: 2013 27th international conference on advanced information networking and applications workshops. IEEE, New York, pp 791–796
9. Nigam R, Sharma DK, Jain S, Gupta S, Ghosh S (2019) Bonding based technique for message forwarding in social opportunistic network. Scalable Comput Pract Experience 20(1):1–15
10. Dhurandher SK, Sharma DK, Woungang I, Saini A (2017) An energy-efficient history-based routing scheme for opportunistic networks. Int J Commun Syst 30(7):
11. Dhurandher SK, Sharma DK, Woungang I, Gupta R, Garg S (2014) GAER: genetic algorithm-based energy-efficient routing protocol for infrastructure-less opportunistic networks. J Supercomput 69(3):1183–1214
12. Socievole A, De Rango F (2015) Energy-aware centrality for information forwarding in mobile social opportunistic networks. In: International wireless communications and mobile computing conference (IWCMC). IEEE, New York, pp 622–627

13. Li N, Das SK (2013) A trust-based framework for data forwarding in opportunistic networks. Ad Hoc Netw 11(4):1497–1509
14. Lilien L, Kamal ZH, Bhuse V, Gupta A (2007) The concept of opportunistic networks and their research challenges in privacy and security. In: Mobile and wireless network security and privacy. Springer, Berlin, pp 85–117
15. Mantas N, Louta M, Karapistoli E, Karetsos GT, Kraounakis S, Obaidat MS (2017) Towards an incentive-compatible, reputation-based framework for stimulating cooperation in opportunistic networks: a survey. IET Netw 6(6):169–178

Threat Modelling and Risk Assessment in Internet of Things: A Review

Mahapara Mahak and Yashwant Singh

Abstract The Internet of things (IoT) plays an important role in our daily lives. They have been used in our homes, hospitals and industries for so long. IoT has also been used to monitor and inform the changes in the environment. Furthermore, the data exchanged within and among IoT devices are increasing aggressively, and the permeative of such systems take them to come in the tenancy of very delicate information, as a result, there are huge risks of security and privacy. Many research works have been conducted to secure the Internet of things. In this paper, we provide an introduction to the Internet of things and reassess some threat models and risk assessment methodologies of IoT.

Keywords IoT · STRIDE · LINDDUN · Risk assessment · DREAD

1 Introduction

The basic definition of "Internet of things" is the assemblage of various objects including sensors which proclaim directly with each other without human involvement [1]. Sethi et al. [25] define IoT as a set of physical devices, vehicles, home appliances and other items embedded with software, sensors, actuators and network connectivity. Each object of the system has a unique identity because of embedded computing system, but they are compatible with already existing Internet infrastructure. Vermesan et al. [30] define the Internet of things as an interrelationship between physical and digital worlds. The digital world communicates with the physical world through different sensors and actuator.

IoT has numerous advantages. It improves customer engagement by optimizing the technology. It reduces waste by providing effective resource management and

M. Mahak (✉) · Y. Singh
Department of Computer Science and Information Technology, Central University of Jammu, Samba, Jammu and Kashmir 181143, India

Y. Singh
e-mail: yashwant.csit@cujammu.ac.in

P. K. Singh et al. (eds.), *Proceedings of Second International Conference on Computing, Communications, and Cyber-Security*, Lecture Notes in Networks and Systems 203,
https://doi.org/10.1007/978-981-16-0733-2_21

improves the data collection as it analyses real data. But IoT is facing some challenges too, and security is the biggest challenge.

In nutshell, IoT is producing a paradigm transformation in infrastructure and services. Whilst this paradigm transformation is occurring, security and privacy are important requirements to tackle a variety of threats, attacks and destructive impact of these on community. So, there is a need to organize the threat and reassess the risk of threats to overcome these challenges. To do this, we create a threat model and reassess the risk through various risk assessment techniques. Here is a brief description of what threat modelling and risk assessment means in IoT [24].

Threat modelling is an approach that identifies, quantify, and address the security risks associated with an IoT system. To understand the risk associated with IoT systems, threat modelling is regarded as the great starting point, this can also help to know how those risks can be mitigated and this computing diagram can provide a starting point for further risks modelling associated with an IoT system. Risk assessment, in general, is understood as the series of actions for identifying, estimating and then prioritizing risks to the assets and operations of an organization. This is a critical activity within risk management and is very important in the Internet of things as it provides the foundation for the risks to be treated that has been identified in an IoT [7, 13, 21, 29].

The remainder of this paper is structured as follows: Sect. 2 presents the related work, Sect. 3 presents various threat models of IoT, Sect. 4 presents the different risk assessment methodologies in IoT, Sect. 5 presents the research challenges and Sect. 6 concludes the paper.

2 Literature Survey

Due to the growing interest in the IoT, there have been diverse publications on threat models and risk assessment methods in various applications of IoT.

In [27], a risk assessment model is proposed for systematic attack propagation depending on the bi-partite graph. Hodo et al. [14] proposed a model for the reduction of threats based on artificial neural networks (ANN). The ANN was verified against the simulated IoT network and was observed the accuracy of 99%. Cagnazzo et al. [8] provide an overview of various threats to mobile health, identified in accordance with STRIDE model. and risk level is determined by means of the DREAD model. Casola et al. [10] proposed a model for threat modelling and risk assessment, using CIA and STRIDE model different threats to assets were determined, and then risk assessment was done by OWASP risk rating methodology. Omotosho et al. [22] proposed a model for IoT-based health monitoring system. Threats to different assets of system and access points were identified using the STRIDE model, and then by DREAD model risk level is determined. This research study is expected to encourage the increase in the security of IoT device infrastructure. Kavallieratos et al. [16] conducted the threat analysis of the smart home, by using the STRIDE threat model. This work was done by taking six different scenarios or instances of the smart home of varying complexity, and it was observed with the increasing complexity of topology, security threats increases. Shakdhe et al.[26] conducted a penetration testing on various IoT applications to find out the vulnerabilities, and then proper countermeasures were proposed to secure these applications. Liu et al. [17] proposed DRAMIA method for dynamic risk assessment of IoT security inspired by an artificial immune system. Costa et al. [11] proposed a method to identify the vulnerabilities in smart home IoT by using open-source tools, as well as provide an example of actual vulnerabilities found in two commercially available devices. Macher et al. [19] proposed an approach called security aware hazard analysis and risk assessment (SAHARA) that is based on the automotive hazard analysis and risk assessment (HARA). This approach is for risk assessment for road vehicles.

From Table 1, we can conclude that IoT systems are vulnerable to risks like DOS attacks. There are lapses in communication protocols. The bridging of multiple architecture-based environment leads to issues like security and interoperability. The threats of cyber-attacks lead to failure of network and data stealing in the multi-dimensional network environment, and security is a critical issue and must be taken into consideration.

To overcome the challenge of security and privacy, there is a need for threat modelling and risk assessment in IoT. We will provide a brief description of various threat models and risk assessment methodologies.

3 Threat Models

Threat modelling is an approach that identifies, quantify and address the security risks associated with an IoT system. The security related to IoT systems can be unusually complex as a huge number of components, imaginably high attack surface and the communication between different parts of the system. To understand the risk associated with IoT systems, threat modelling is regarded as the great starting point, this can also help to know how those risks can be alleviated, and this computing diagram can provide a starting point for additional risk modelling related with an IoT system.

3.1 Stride

One of the most popular threats modelling framework is the STRIDE developed by Microsoft. The base artifact of STRIDE is a data-flow diagram that is presented as the part of software development lifecycle (SDL). These diagrams are used to map a threat with a particular asset. STRIDE is a cipher. (In STRIDE; S stands for Spoofing, T stands for Tampering, R stands for Repudiation, I stands for Information Disclosure, D stands for Denial of Service and E stands for Elevation of privileges). STRIDE consists of various steps;

Table 1 Threat model and risk assessment methodologies

Author	Year	Threat model	Risk assessment methodology	Challenges/issues
Shivraj et al. [27]	2018	STRIDE	Graph theory	Non-self-healing architecture a big requirement to count the DOS, and other attacks poses a large threat
Rak et al.	2019	STRIDE	OWASP risk rating methodology	Lapses in communication protocols like ZigBee, Wi-Fi, etc., leading to severe security issue
Cagnazzo et al. [8]	2018	STRIDE	DREAD	Major challenge is in terms of security and inter-portability needs great insight in the counter measures to be designed
Casola et al. [10]	2019	STRIDE/CIA	OWASP risk rating methodology	Complexity in security requirement solution design
Omotosho et al. [22]	2019	STRIDE	DREAD	There is great challenge to propose counter measures to handle threats in remote devices when integrated together in health services
Kavallieratos et al. [16]	2019	STRIDE	N/A	Detection of threats in a dynamic environment poses a great challenge to devise mechanisms to counter these
Arjun et al. (2019)	2019	Penetration testing	N/A	Misconfiguration of environment as an human error
Liu et al. [17]	2012	N/A	DRAMIA	Need of dynamic real-time risk assessment as most solutions are based on static data
Macher et al. [19]	2015	STRIDE	HARA	Lack of safety that is critical issue in multi-dimensional network environment

(continued)

Table 1 (continued)

Author	Year	Threat model	Risk assessment methodology	Challenges/issues
Juan et al.	2018	N/A	OWASP risk rating methodology	There is need of more privacy in various applications

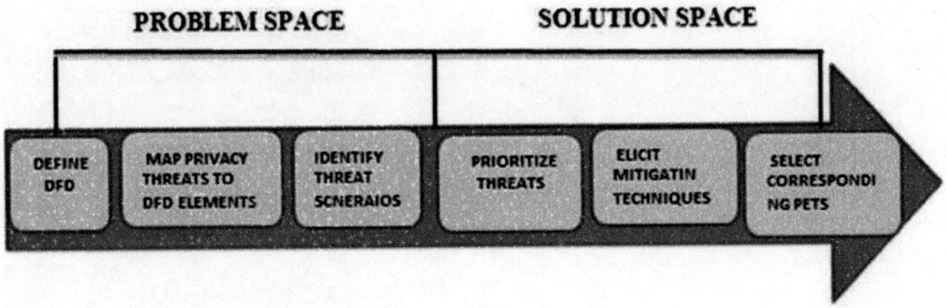

Fig. 1 LINDDUN methodology steps

1. Definition of use scenarios
2. Assemblage of external yokes
3. Define security hypothesis
4. Make records of external security
5. Catalyze DFD of the application being examined
6. Arbitrate type of threat
7. Perceive the vulnerability of system to threats
8. Arbitrate risk
9. Mitigation of plan.

3.2 LINDDUN

It is described as a model to identify privacy threats in software-based systems. The base artifact of this model is also DFD like that of STRIDE. This is the acronym of likability, identifiability, non-repudiation, detectability, disclosure of information, unawareness and non-compliance [4, 20] (Fig. 1).

3.3 CORAS

It operates iteratively among examiners and developers during software developing. The core artefact for CORAS is a UML diagram that involves numerous kinds of

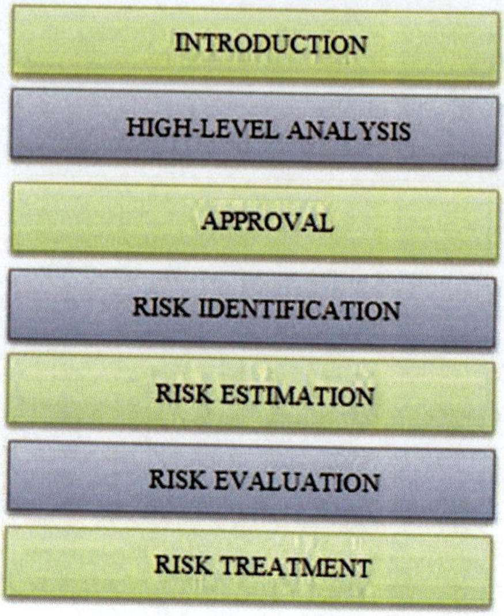

Fig. 2 Steps of CORAS

diagrams viz UML class diagram, UML collaboration diagram and the UML activity diagram. CORAS includes seven steps as [12, 16, 18] (Fig. 2).

3.4 *Attack Trees*

It is another model for modelling the threats. Attack trees are defined as pictorial representations of various scenarios that will occur in a security failure. A tree diagram is used to portray abeyant attacks on a particular system (Fig. 3).

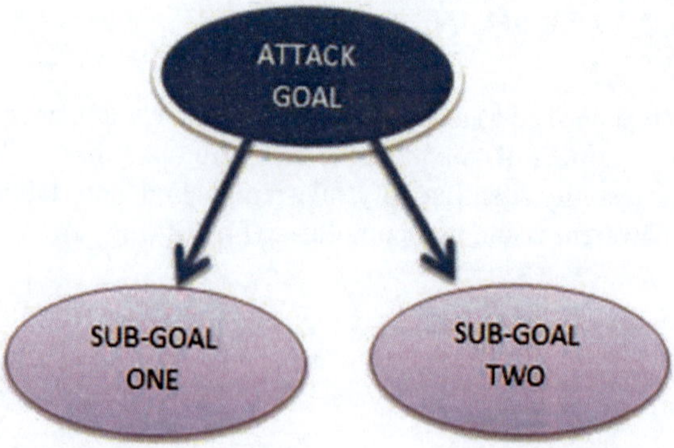

Fig. 3 Attack tree [5]

Table 2 Different threat models

Model	Base artefact used	Tool support requirement
STRIDE	Data-flow diagram	No tool support
LINDDUN	Data-flow diagram	No tool support
CORAS	UML diagrams	Diagram editor
ATTACK TREES	Attack trees	Secular tree

The ultimate goal of attack is represented by root of the tree and leaves represent how we can achieve the goal. For every goal, there is a separate tree. Thus, a set of attack trees is produced by the analysis of system threat [2, 9, 28].

Table 2 shows the threat modelling frameworks as recapitulated in the already existing survey of different threat models [15]. It describes the basic diagram of the model on which threat modelling is done and also provides the information whether any tool support requirement is there are not.

4 Risk Assessment

Risk assessment, in general, is understood as the series of actions for identifying, estimating and then prioritizing risks to the assets and operations of an organization. This is the critical activity within risk management and is very important in the Internet of things as it provides the foundation for the risks to be treated that has been identified in an IoT. Some approaches for risk assessment are as [21].

4.1 OWASP Risk Rating Methodology

This is the approach for estimating the severity of risk. The methodology identifies two factors for calculation of risk, i.e. likelihood and impact. Likelihood and impact depend on various factors like vulnerability factors, threat agent factors, business factors and technical factors. Each factor has a set of options having likelihood rating in the range of 0–9 related with it that is used to calculate the overall likelihood; similarly with the impact factors parameters, there is set of options, and every option is having impact rating from 0–9 which is used for estimating the overall impact. By considering all these parameters for a particular IoT system, we can estimate the severity of risk by taking the mean of the numbers associated with each parameter [23].

After the estimation of likelihood and impact, the values are put together to get the overall severity of a risk. This is done by taking the mean of likelihood estimated and impact factor estimated values and are then classified in classes of low, medium and high according to the value belonging to the specific range, i.e. if between 0 and 3, the severity of risk is low, if between 3 and 6, then medium and if between 6 and 9, it is of high severity. Then after this, the more severe risks are fixed, and countermeasures are taken accordingly (Table 3).

4.2 DREAD

It is another method for risk assessment. In this method, we rate, compare and prioritize the severity of risk. The risk in this method is calculated as shown in Table 4.

In this methodology, every parameter is assigned a number in a range of 1 to 10, and then mean of values is taken after finding the parameters on which threat depends and corresponding values a Higher the number means more serious [3].

5 Research Challenges

The important challenges in ensuring the security and privacy are as follows:

There are lapses in communication protocols leading to severe security issues, and due to the heterogeneity of devices, there are interoperability issues, lack of non-self-healing architecture leads to pose a large threat to the IoT system, lack of detection of threats in a dynamic environment, bridging of multiple architecture-based environment leading to issues like security and inter-portability, and lack of security and safety as there is a threat of cyber-attacks leading to failure of the network. Based on the survey, we conclude that more research must be performed on highlighting possible security threats that harm people and then suggest possible solutions to them.

6 Conclusion

The survey conducted in this paper highlights the technology with brief introspection toward the security aspects concerning it. Starting from the basic introduction of IoT, protocols, applications of IoT and need for security and privacy, the prime attention is paid toward various threat modelling and risk assessment models, namely STRIDE, LINDDUN, CORAS, DREAD and OWASP risk rating methodology. A further listing of research challenges is also provided to narrow the research areas in the field of security and privacy in IoT. This study expected to promote the improvement of the design of the IoT device security infrastructure.

Table 3 Identification of risks in IoT

Identification of risk = likelihood*impact							
Likelihood				Impact			
Threat agent factors		Vulnerability factors		Technical factors		Business factors	
Skill level		Ease of discovery		Loss of confidentiality		Financial damage	
No technical skill	1	Practically impossible	1	Minimal non-sensitive data disclosed	2	Less than the cost to fix the vulnerability	1
Some technical skill	3	Difficult	3	Minimal critical data disclosed	6	Minor effect on annual profit	3
Advanced computer user	5	Easy	7	Extensive non-sensitive data disclosed	6	Significant effect on annual profit	7
Networking and programming skills	6	Automated tools available	9	Extensive critical data disclosed	7	Bankruptcy	9
Security penetration skills	9	Ease of exploit		All data disclosed	9	Reputation damage	
Motive		Theoretical	1	Loss of integrity		Minimal damage	1
Low or no reward	1	Difficult	3	Minimal slightly corrupt data	1	Loss of major accounts	4
Possible reward	4	Easy	5	Minimal seriously corrupt data	3	Loss of goodwill	5
High reward	9	Automated tools available	9	Extensive slightly corrupt data	5	Brand damage	9
Opportunity		Awareness		Extensive serious corrupt data	7	Non-compliance	
Full access or expensive resource required	0	Unknown	1	All data totally corrupt	9	Minor violation	2
Special access Or Resource Required	4	Hidden	4	Loss of availabilty		Clear violation	5
Some access or resource required	7	Obvious	6	Minimal secondary services interrupted	1	High profile violation	7
No access or resource required	9	Public knowledge	9	Minimal primary service interrupted	5	Privacy violation	
Size		Intrusion detection		Extensive secondary services interrupted	5	One individual	3
Developers	2	Active detection In Application	1	All services Completely Lost	9	Hundreds of people	5
System administrators	2	Logged and reviewed	3	Extensive primary services interrupted	7	Thousands of people	7

(continued)

Table 3 (continued)

Identification of risk = likelihood*impact							
Likelihood				Impact			
Threat agent factors		Vulnerability factors		Technical factors		Business factors	
Skill level		Ease of discovery		Loss of confidentiality		Financial damage	
No technical skill	1	Practically impossible	1	Minimal non-sensitive data disclosed	2	Less than the cost to fix the vulnerability	1
Intranet users	4	Logged without review	8	Loss of accountability		Millions people	9
Partners	5	Not logged	9	Fully traceable	1		
Authenticated users	6			Possibly traceable	7		
Anonymous Internet users	9			Completely anonymous	9		

Table 4 Identification of risk in IoT

DREAD RISK = (DAMAGE + REPRODUCIBILITY + EXPLOITABILITY + AFFECTED USERS + DISCOVERABILITY/5)	
DAMAGE POTENTIAL	How much damage a particular threat causes
Nothing	0
Information disclosure	5
Individual non-sensitive user data is compromised	8
Administration non-sensitive data is compromised	9
Data destruction	10
Application unavailability	10
Reproducibility	How easily threat is exploited
Very hard or impossible	0
Authorized user need complex steps	5
Easy steps for authorized user	7.5
Web browser and address bar is enough to exploit without authentication	10
Exploitability	What is needed to exploit this threat
Advanced programming needed and networking knowledge with advanced attack tools	2.5

(continued)

Table 4 (continued)

DREAD RISK = (DAMAGE + REPRODUCIBILITY + EXPLOITABILITY + AFFECTED USERS + DISCOVERABILITY/5)	
DAMAGE POTENTIAL	How much damage a particular threat causes
Already tools are available in public	5
Web application proxy tool in is available	9
Web browser is needed	10
Affected users	How many users will be affected
None	0
Already compromised employer	2.5
Some users of individual or employer privileges, but not all	6
Administration users	8
All users	10
Discoverability	How easy is to discover the threat
Very hard	0
Can discover by monitoring or manipulating HTTP requests	5
If easily discovered by search engine	8
Visibility of information in address of web browser	10

References

1. Alaba FA, Othman M, Abaker I, Hashem T, Alotaibi F (2017) Internet of things security: a survey. J Netw Comput Appl 88:10–28. https://doi.org/10.1016/j.jnca.2017.04.002
2. Albanese M, Probst CW, Workshop I, Goos G (2019) Graphical models for security. https://doi.org/10.1007/978-3-030-36537-0
3. Application Threat Modeling using DREAD and STRIDE (n.d.) Retrieved 26 Nov 2019 from https://haiderm.com/application-threat-modeling-using-dread-and-stride/
4. Atamli AW, Martin A (2014) Threat-based security analysis for the internet of things. In: Proceedings—2014 international workshop on secure internet of things, SIoT, pp 35–43. https://doi.org/10.1109/SIoT.2014.10
5. Attack trees expressed through extended influence diagrams. | Download Scientific Diagram (n.d.) Retrieved 14 Mar 2020 from https://www.researchgate.net/figure/Attack-trees-expressed-through-extended-influence-diagrams_fig3_224373064
6. Aufner P (2020) The IoT security gap: a look down into the valley between threat models and their implementation. Int J Inf Secur 19(1):3–14. https://doi.org/10.1007/s10207-019-00445-y
7. Basics of IoT Security Threat Modelling—Cyber Security Blog (n.d.) Retrieved 24 Mar 2020 from https://www.valencynetworks.com/blogs/basics-of-iot-security-threat-modelling/
8. Cagnazzo M, Hertlein M, Holz T, Pohlmann N (2018) Threat modeling for mobile health systems. In: 2018 IEEE wireless communications and networking conference workshops, WCNCW 2018, pp 314–319. https://doi.org/10.1109/WCNCW.2018.8369033
9. Caldera C (2017) Towards an automated attack tree generator for the IoT
10. Casola V, De Benedictis A, Rak M, Villano U (2019) Toward the automation of threat modeling and risk assessment in IoT systems. Internet of Things 7:100056. https://doi.org/10.1016/j.iot.2019.100056

11. Costa L, Barros JP, Tavares M (2019) Vulnerabilities in IoT devices for smart home environment. In: ICISSP 2019—Proceedings of the 5th International conference on information systems security and privacy, pp 615–622. https://doi.org/10.5220/0007583306150622
12. den Braber F, Brændeland G, Dahl HEI, Engan I, Hogganvik I, Lund MS, Vraalsen F (2006) The CORAS model-based method for security risk analysis. SINTEF, Oslo, (September). Retrieved from https://scholar.google.com/scholar?hl=en&btnG=Search&q=intitle:The+CORAS+Model-based+Method+for+Security+Risk+Analysis#0
13. Getting Started with IoT Security with Threat Modeling | Denim Group (n.d.) Retrieved 24 Mar 2020 from https://www.denimgroup.com/resources/blog/2017/11/getting-started-with-iot-security-with-threat-modeling/
14. Hodo E, Bellekens X, Hamilton A, Dubouilh P-L, Iorkyase E, Tachtatzis C, Atkinson R (2016) Threat analysis of IoT networks using artificial neural network intrusion detection system keywords—internet of things, artificial neural network, denial of service, intrusion detection system and multi-level perceptron, pp 4–9
15. Hussain S, Kamal A, Ahmad S, Rasool G, Iqbal S (2014) Threat modelling methodologies: a survey. Sci. Int. (Lahore) 26(4):1607–1609
16. Kavallieratos G, Gkioulos V, Katsikas SK (2019) Threat analysis in dynamic environments: The case of the smart home. In: 2019 15th international conference on distributed computing in sensor systems (DCOSS), pp 234–240. https://doi.org/10.1109/DCOSS.2019.00060
17. Liu C, Zhang Y, Zeng J, Peng L, Chen R (2012) Research on dynamical security risk assessment for the Internet of Things inspired by immunology. In: Proceedings—International conference on natural computation, (ICNC), pp 874–878. https://doi.org/10.1109/ICNC.2012.6234533
18. Lund MS, Solhaug B, Stølen K (2010) Model-driven risk analysis: the CORAS approach. Springer
19. Macher G, Sporer H, Berlach R, Armengaud E, Kreiner C (2015). SAHARA: a security-aware hazard and risk analysis method. In: Proceedings—Design, automation and test in Europe, DATE, pp 621–624. https://doi.org/10.7873/date.2015.0622
20. Mathematics A (2017) ijpam.eu. 115(8):121–126.
21. Nurse JRC, Creese S, De Roure D (2017) Security risk assessment in internet of things systems. IT Professional 19(5):20–26. https://doi.org/10.1109/MITP.2017.3680959
22. Omotosho A, Ayemlo Haruna B, Mikail Olaniyi O (2019) Threat modeling of internet of things health devices. J Appl Secur Res 1–16. https://doi.org/10.1080/19361610.2019.1545278
23. OWASP Risk Rating Methodology—OWASP. (n.d.). Retrieved 23 Nov 2019 from https://www.owasp.org/index.php/OWASP_Risk_Rating_Methodology
24. Rizvi S, Kurtz A, Pfeffer J, Rizvi M (2018) Securing the internet of things (IoT): a security taxonomy for IoT. In: Proceedings—17th IEEE international conference on trust, security and privacy in computing and communications and 12th IEEE international conference on big data science and engineering, Trustcom/BigDataSE 2018, (December), pp 163–168. https://doi.org/10.1109/TrustCom/BigDataSE.2018.00034
25. Sethi P, Sarangi SR (2017) Internet of things: architectures, protocols, and applications. J Electr Comput Eng 2017. https://doi.org/10.1155/2017/9324035
26. Shakdhe A, Agrawal S, Yang B (2019) Security vulnerabilities in consumer IoT applications. In: Proceedings—5th IEEE international conference on big data security on cloud, BigDataSecurity 2019, 5th IEEE international conference on high performance and smart computing, HPSC 2019 and 4th IEEE international conference on intelligent data and security, pp 1–6. https://doi.org/10.1109/BigDataSecurity-HPSC-IDS.2019.00012
27. Shivraj VL, Rajan MA, Balamuralidhar P (2018) A graph theory based generic risk assessment framework for internet of things (IoT). In: 11th IEEE international conference on advanced networks and telecommunications systems, ANTS 2017, pp 1–6. https://doi.org/10.1109/ANTS.2017.8384121
28. Threat Modeling: 12 Available Methods (n.d.) Retrieved 14 Mar 2020 from https://insights.sei.cmu.edu/sei_blog/2018/12/threat-modeling-12-available-methods.html

29. Threat Modeling | IoT ONE (n.d.) Retrieved 24 Mar 2020 from https://www.iotone.com/term/threat-modeling/t686
30. Tiwary A, Mahato M, Chandrol MK (2018) Internet of Things (IoT): research, architectures and applications. Int J Future Revol Comput Sci Commun Eng. ISSN: 2454-4248. Retrieved from https://www.ijfrcsce.org

Deep Learning-Based Attack Detection in the Internet of Things

Parushi Malhotra and Yashwant Singh

Abstract The exponential growth of the Internet of Things in various domains has escalated the rise of concern in this digital era. The concern is primarily due to the evolution of cyber-attacks leading to the emergence of numerous threats and anomalies. The bottlenecks in traditional security techniques have unclouded the vision of learning techniques for intrusion detection. The use of classical ML techniques for the identification and classification has been around for a long time, but it suffers from the issue of scalability and feature engineering, which limits its usage. In this paper, we have analyzed the use of deep learning for intrusion and anomaly detection. The deep learning algorithms used here are deep neural networks (DNN) and long short-term memory recurrent neural networks (LSTM). Denial of service, malicious control, wrong setup, data type probing, scan, spying, malicious operation are the attacks against which the algorithms are tested. An accuracy of 99.30% is achieved for deep neural networks and 97.50% for the LSTM model.

Keywords Internet of Things · Machine learning · Deep learning · Distributed smart space orchestration system (DS2OS)

1 Introduction

The proliferation in the technological aspects has escalated the presence of IoT in different sectors. These sectors are primarily inclined toward robust and intelligent systems to provide better storage and computing facilities [3]. The core conviction of it revolves around the dynamic interconnection of billions of diverse entities in a wired or wireless fashion. With such growth and advancement, security has become a sizeable concern. The past few years have already recorded some damaging effects of

P. Malhotra (✉) · Y. Singh
Department of Computer Science and Information Technology, Central University of Jammu, Samba, Jammu and Kashmir 181143, India

Y. Singh
e-mail: yashwant.csit@cujammu.ac.in

P. K. Singh et al. (eds.), *Proceedings of Second International Conference on Computing, Communications, and Cyber-Security*, Lecture Notes in Networks and Systems 203,
https://doi.org/10.1007/978-981-16-0733-2_22

it in the form of Mirai botnet and Bashlite attack. Also, besides, numerous scanning, probing, DoS, and other attacks are being launched by the adversaries by exploiting the weaknesses of the existing software. All these together make the entire system vulnerable and raising the concern further. Traditional signature-based approaches require the knowledge of attack for its detection. Hence, it is not suitable for the detection of various zero-day attacks that are being launched against the system. Moreover, such systems need to be updated frequently, which further increases the overhead. Such shortcomings can be overcome by leveraging the potential of learning approaches.

Machine learning is one of the popular learning approaches which provides computationally robust algorithms that have found application in various sectors like pattern recognition, fraud detection, computer vision, and intrusion detection [1]. ML requires a reasonable amount of data to make unambiguous decisions [15]. However, due to issues like scalability and manual feature extraction limits its penetration into the security market mainly because the extracted features may not represent accurate underlying patterns. Moreover, with the increasing growth of IoT technology, more and more devices are being connected to the Internet, which further complicates the usage of classical machine learning approaches. To overcome the problems of traditional methods as well as the classical machine learning algorithms, deep learning approach of data analytics is being applied. Deep learning is an advanced ML approach having the potential of achieving better accuracy in terms of prediction and classification because of the multilayered composition. Deep learning, when combined with intrusion detection, can achieve performance at a superhuman level for the detection of novel attacks and anomalies [13]. The principle benefit of the technology is the omission of manual feature selection and capability to model nonlinear relationships, thereby achieving an edge over ML.

This paper presents a solution for the detection of attack and anomalies via deep learning. Deep neural networks and LSTM algorithms are tested against various attacks that are part of the DS2OS dataset. LSTM networks can remember long-term dependencies, thereby taking into account the previous and recent occurrences to recognize the new ones. Section 2 presents the survey conducted with reference to intrusion detection in IoT, followed by the methodology in Sect. 3. The conclusion and future scope are presented in Sect. 4.

2 Literature Review

Divyatmika and Sreekesh [6] have proposed a two-tier network intrusion detection system (NIDS) using machine learning techniques. The approach is based on TCP/IP data packet features obtained from NSL-KDD DATASET, which commenced by preprocessing the data in wekas and building of an autonomous model using hierarchical agglomerative clustering which clusters the data in two (usual and new patterns) followed by classification of novel traffic using K-nearest neighbor (KNN) which is again followed by using multi-layer perceptron (MLP) for signature-based attacks and reinforcement to classify anomaly detection and subsequently reduce the false alarm rates. A similar approach is presented by [12]. In this, the malignant activities were detected and classified with the application of two-layer dimension reduction and two-tier classification (TDTC) model on the NSL-KDD dataset. Dimensionality reduction was performed to reduce the complexity followed by the application of naïve Bayes, CF-KNN, and KD trees for efficient classification.

Moustafa et al. [10] have proposed an Adaboost ensemble for the detection of attacks in the network by using features of DNS, HTTP protocols in TCP/IP models. It is a three-step structure involving feature extraction via Tcpdump, Bro-ids, and another extractor module. UNSW-NB15 and NIFS datasets were used for the generation of simulated IoT traffic. In another paper by Canedo and Skjellum [4], the authors have conducted suitable experimentation to generate own synthetic data to inspect and carefully scrutinize the usage of ANN in IoT gateway device present in the transport layer to work at the security aspects of the technique. Hasan et al. [8] have compared the anomaly detection mechanism of various ML techniques (LR, SVM, DT, RF ANN) in a virtual environment producing synthetic data. The synthetic data generated is publicly available at Kaggle under the name of DS2OS. The dataset has 357,952 samples and 13 features. Data preprocessing is performed by performing tasks like cleaning missing data, converting categorical data into numeric using encoders succeeded by application of the techniques on the dataset, and comparing their performances in which random forest outperformed with 99.4% accuracy.

Roopak et al. [14] have presented a deep learning approach for DDoS attack detection in the CICIDS2017 dataset. The deep learning model is a combination of 1D-CNN, RNN, LSTM, and a hybrid model of CNN and LSTM. McDermott, Majdani and Petrovski [9] present a novel bidirectional LSTM for sensing of botnet activities in the consumer IoT devices. Rahul et al. [13] have proposed an in-depth deep neural network-based approach to predict attacks on a NIDS. KDD cup 99 was used to train the network. With continuous evaluation and by varying the hidden layer counts, a DNN with three layers, 0.1 learning rate running for 1000 epochs, generated the maximum accuracy. The system was also tested against many shallow ML algorithms.

3 Methodology

3.1 Data Collection and Preprocessing

The overall framework includes steps that involve data collection, preprocessing, implementation of the algorithm followed by evaluation and result analysis. In this paper, the data is collected from the IoT traffic traces available open-source at Kaggle named DS2OS [11]. This dataset contains IOT traffic traces from the application layer. It comprises of four IOT sites with a thermometer, movement sensors,

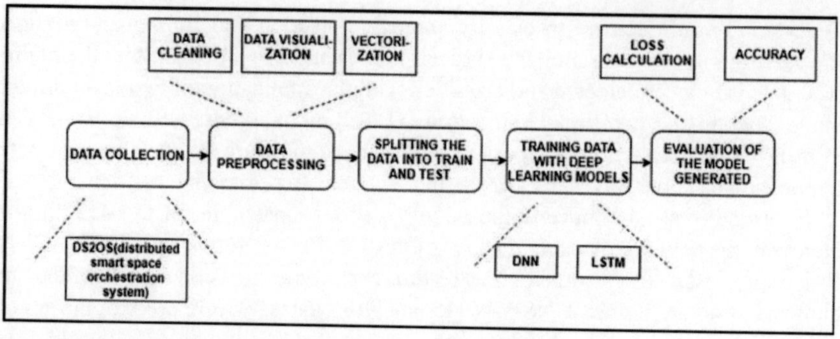

Fig. 1 Process of data collection, cleaning, and classification

washing machines, batteries, thermostats smartphone, light controller, and smart door in different organizations. These traces were collected by Pahl [2]. DS2OS follows a service-oriented architecture where communication between various services takes place via message queuing telemetry transport (MQTT) protocol [11].

The next step in the framework is to preprocess the data, which is extremely necessary to avoid inaccurate results. This step primarily includes data cleaning, which involves dealing with noise and incomplete information followed by data visualization for proper analysis of the data. Figure 1 depicts the data collection, preprocessing, and the analysis structure of the approach followed. The entire dataset has 357,952 data values with 347,935 normal data and 10,017 anomalous data. The attacks mentioned in the dataset are described below.

1. Denial of Service: The attack is performed to flood the server with multiple packets for denying or making the services inaccessible to legitimate users [5].
2. Datatype probing: It refers to the alterations made in the original data type by the malicious node of the network.
3. Scan: It refers to the process of exploiting the vulnerabilities of the system and subsequently harming the system and the data.
4. Malicious control: In this, the adversaries can attack the entire system by capturing the valid network traffic and then causing the damage.
5. Malicious operation: It refers to the task of performing any kind of activity that can infect the system. The intended purpose is to harm the entire setting.
6. Wrong setup: This refers to the occurrence of errors because of the incorrect system configuration.
7. Spying: It refers to a mechanism that can lead to a series of attacks by the adversaries and subsequently cause loss of sensitive data and infrastructure.

Thirteen features describing the dataset include Source ID, Source Address, Source Type, Source Location, Destination Service Address, Destination Service Type, Destination Location, Accessed Node Address, Accessed Node Type, Operations, Value, Timestamp (discrete form), and Normality (seven abnormal and one normal).

3.2 Theoretical Aspect of the Technique with Algorithm

DNN: These networks can process data via the assistance of multiple layers between the input and the output. It can make conclusions with the help of previous experience, thus making the data more abstract. The principle benefit of the technology includes the omission of manual feature extraction and modeling nonlinear relationships. To achieve the nonlinearity activation function plays an important role. The most used activation function is the ReLu activation function. Figure 2 depicts the structure of DNN.

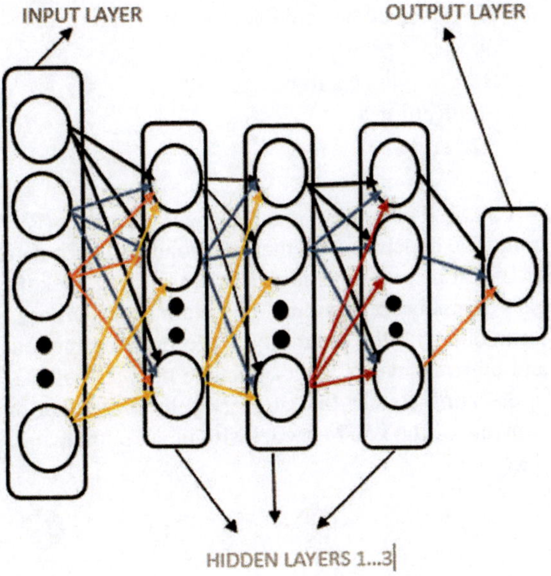

Fig. 2 Deep neural networks

ALGORITHM

Input: Data (D)={ X,Y },{ X,Y }= (X_train, Y_train) , batch_size, epochs, c = class of all features ,X_i = resultant matrix c_1= new class, a=attacks of the dataset

Output: Anomalous or normal

1. **for** each a in D do
2. **for** each class (c) do
3. If anomalous then
4. Use c_1 to build a X_i
5. Compile loss function
6. **end for**
7. **else**
8. Add normal data to X_i
9. **end if**
10. **end for**
11. Return X_i
12. **end**

LSTM: It is a type of recurrent neural network (RNN) with an ability to remember long-time dependencies, thus overcoming the limitations of RNN. The composition of LSTM includes memory cells for keeping back the information along with three gates, namely forget, input, and output for memory orchestration. It is a powerful technique with the capability of representing the relationships between the current and previous events and can handle time-series data. Such skills can be beneficial in detecting attacks that are served by more than one point. Figure 3 depicts the structure of the LSTM model with input, output, and memory blocks.

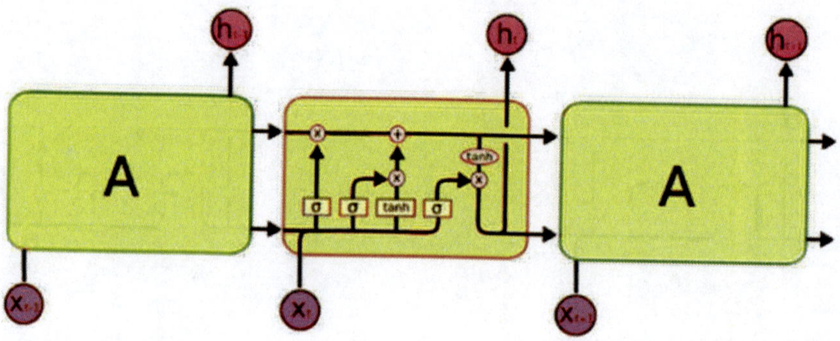

Fig. 3 Lstm architecture *Source* Gulli and Pal [7]

ALGORITHM
Input: Data(D)={ X,Y },){ X,Y }= (X_train, Y_train) Maximun training epoch=20, $\Theta_1, \Theta_2, \Theta_3 \ldots \Theta_{12}$ are the parameters, a=attacks in the dataset, L= LSTM network
Output: Anomalous or normal

```
For each a in D do
            Construct 3D matrix
end
Initialise parameters ϴ1, ϴ2, ϴ3..... ϴ12 in L as 3D input
J← 0
    While J < epoch do
            For each a in D do
            Detect a and report it as anomalous.
    end while,
Compute the time frame.
Optimise the parameter, ϴ1, ϴ2, ϴ3..... ϴ12 in L based on loss funtion.
end
J←J+1
end
```

3.3 Experimental Analysis

The experiment was performed on the ninth generation core i5 processor with GPU, including 2 GB graphic card and 8 GB ram. Along with this, packages like pandas, Numpy, Matplotlib, Scikit-learn, tensor flow, and Keras are used for data cleaning, visualization, and analysis. Before beginning, data was visualized using pandas, Numpy, and Matplotlib packages. This was followed by preprocessing of data in which data was cleaned by filling the missing values and then encoding the data using z-scores for numeric columns and dummy variables for the text values. Two deep learning models, i.e., deep neural networks and LSTM neural networks, were implemented on this cleaned data. DNN was implemented by dropping the timestamp from the features while LSTM was performed with the timestamp. Figure 4 depicts the conversion of timestamp from discrete to time–date form. This conversion was essential for the implementation of the LSTM model. Data was trained by

Fig. 4 Conversion of time from discrete to time-date form

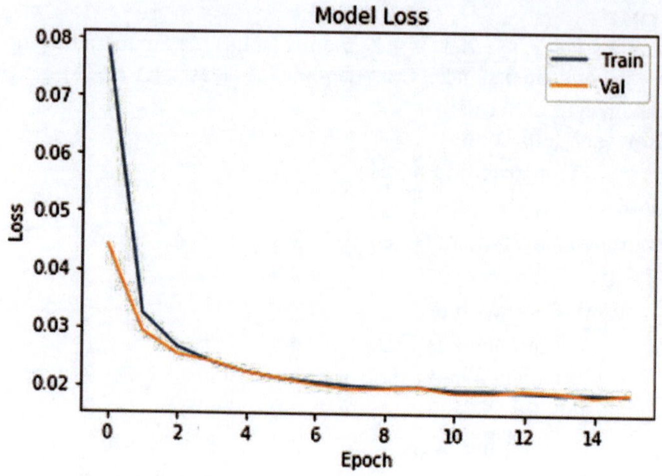

Fig. 5 Line graph depicting the model loss

importing suitable packages followed by splitting the data for training and testing. In the experiment, 75% of data was utilized for training and 25% for testing.

The model comprises three dense layers with 10, 50, 10 nodes in each layer, respectively, with the ReLu activation function. In the output layer, Softmax was used for activation. The training was done with 1000 epochs followed by calculation of loss and accuracy. Figure 5 explains the model loss with respect to the number of epochs. Categorical cross-entropy was used to measure the loss.

The graph depicts the decrease in the loss with the increase of the epoch size.

Figure 6 depicts the accuracy of the model. The model has achieved an accuracy of 99.30%. Thus, conclusions available from the loss and the accuracy graph depict the suitability of the model for this set of data.

After the successful implementation of DNN, the LSTM model was used for training purpose on this data. LSTM models best suited for time series and sequential problems. A large amount of data further improves its suitability. In our model generation, the LSTM network with 100 neurons in the dense layer with a dropout rate of 0.2 was created. Before model generation, the data was split into train and test, and the input was reshaped to be 3D as per the requirement of the LSTM model (samples, timestamp, feature). The model loss was calculated using the mean squared error and Adam optimizer was used for the purpose of optimization. An accuracy of 97.50% was achieved with a minimum loss, which is depicted in Fig. 7.

Figure 8 depicts the accuracy graph of the LSTM model on the simulated traces. The graph presents the accuracy of the model with increasing epoch size.

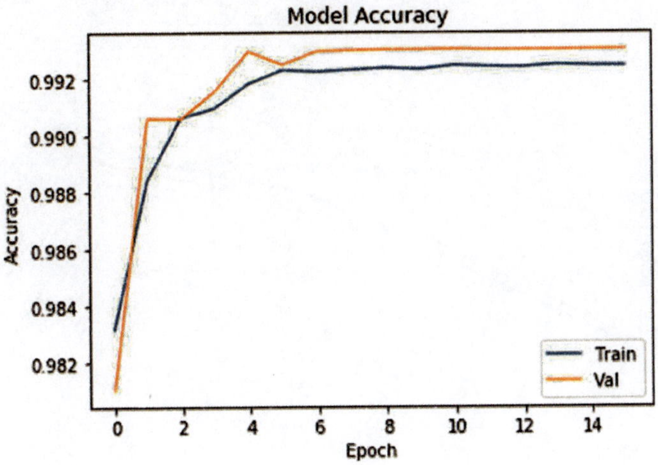

Fig. 6 Line graph depicting accuracy

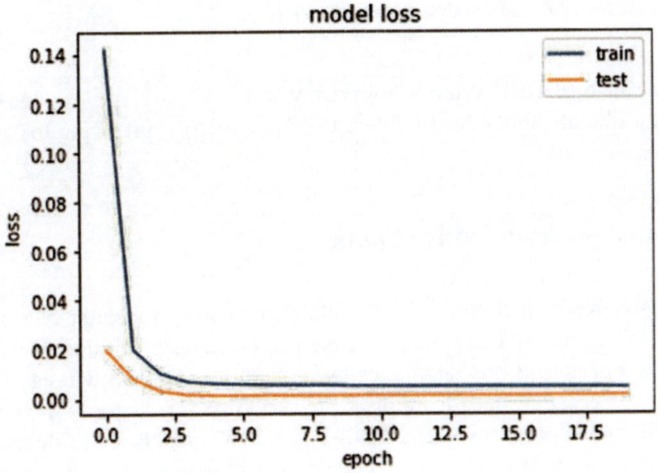

Fig. 7 Model loss with increasing epochs

3.4 Comparative Analysis

A fair amount of analogous work is done by the researchers to come up with a better IDS mechanism to deal with various attacks. Table 1 summarizes such related work to depict the recent trends and approaches.

Both the algorithms were able to achieve a remarkable level of accuracy with minimum loss. However, in the future, LSTM models will be more suitable for real-time IoT traffic due to better speed and better adaptability for collective anomalies.

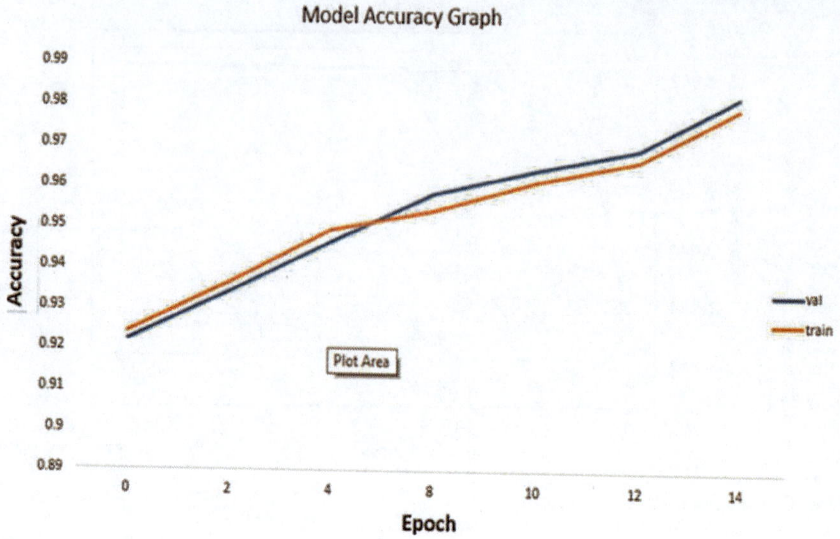

Fig. 8 Accuracy of the model with increasing epoch size

The time taken by the LSTM model for prediction was remarkably less than the DNN, making it a suitable option for future security perspectives involving IoT traffic.

4 Conclusion and Future Scope

The extensive research conducted necessitates the need for a better security mechanism for IoT traffic. In this paper, we have presented deep learning-based IDS for the detection of attacks and anomalies by utilizing the DS2OS dataset. The result implicates the usage of deep learning in IoT security as a better tool compared to other traditional approaches. Here, DNN and LSTM models are evaluated against the simulated IoT traffic. The vast amounts of data generated by IoT devices will require such deep analytics models for better detection of attacks and anomalies. The future scope includes testing of such models for real-time IoT traffic to analyze their applicability in this sector. Hence, a practical application is required to test the algorithms.

Table 1 Detailed analysis

Author	Dataset used	Algorithm	Threats	Challenges	Evaluation
Divyatmika and Sreekesh [6]	NSL-KDD	Clustering + KNN (data classification) + MLP (misuse detection) + reinforcement (anomaly detetion)	DoS, probe, remote-to-local (R2L), user-to-root (U2R)	–	Scan attack: precision-97.7, recall-97.7, f-measure-97.7 SYN: precision-80.8, recall-68.8, f-measure-65.8 UDP: Precision-81, recall-68.8, f-measure-65.8
Moustafa et al. [10]	UNSW-NB15, NIMS	Ensemble model (decision tree + naïve Bayes + ANN)	Analysis, backdoor, DoS, exploit, fuzzers, generic, reconnaissance, worms	Considering other IoT protocols, concentrating on zero-day attacks	Accuracy with DNS data source: 99.54%, accuracy with Http data source: 98.97%
Roopak et al. [14]	CICIDS2017	MLP, 1-d convolutional neural network (CNN), LSTM, CNN + LSTM	DDoS (distributed denial of service)	Lack of deep learning models that can work with highly unbalanced datasets	Accuracy: 1dCNN-95.14%, MLP-86.34%, LSTM-96.24%, CNN + LSTM-97.16%
Rahul et al. [13]	KDD CUP-99	DNN with three layers	DoS, probe, user-to-root (U2R), remote-to-local (R2)	Lack of real-time IoT dataset, evaluation of better deeper networks required	Accuracy: 93%
Hasan et al. [8]	DS2OS	logistic regression (LR), support vector machines (SVM), artificial neural network (ANN), random forest (RF), decision tree (DT)	DoS, data type probing, malicious control, malicious operation, scan, spying, wrong setup control	More robust algorithm required, more inspection required for framework creation, more attention needed for real-time detection	Accuracy: LR-98.3% SVM-98.2% DT-99.4% RF-99.4% ANN-99.4%

(continued)

Table 1 (continued)

Author	Dataset used	Algorithm	Threats	Challenges	Evaluation
Our Study	DS2OS	DNN, LSTM	DoS, data type probing, malicious control, malicious Operation, scan, spying, wrong setup control	Lack of real-time IoT traffic, no multiclass classification	Accuracy: 99.30% for DNN and 97.50% for LSTM

References

1. Alsheikh MA, Lin S, Niyato D, Tan HP (2014) Machine learning in wireless sensor networks: algorithms, strategies, and applications. IEEE Commun Surv Tutor 16(4):1996–2018. https://doi.org/10.1109/COMST.2014.2320099
2. Aubet F (2019) Machine learning-based adaptive anomaly detection in smart spaces machine learning-based adaptive anomaly detection in smart spaces frano (January). https://doi.org/10.13140/RG.2.2.35293.26088
3. Bodkhe U, Mehta D, Tanwar S, Bhattacharya P, Singh PK, Hong WC (2020) A survey on decentralized consensus mechanisms for cyber physical systems. IEEE Access 8:54371–54401. https://doi.org/10.1109/ACCESS.2020.2981415
4. Canedo J, Skjellum A (2016) Using machine learning to secure IoT systems. In: 2016 14th Annual conference on privacy, security and trust, PST 2016, pp 219–222. https://doi.org/10.1109/PST.2016.7906930
5. Deogirikar J, Vidhate A (2017) Security attacks in IoT: a survey. In: Proceedings of the international conference on IoT in social, mobile, analytics and cloud, I-SMAC 2017, pp 32–37. https://doi.org/10.1109/I-SMAC.2017.8058363
6. Divyatmika, Sreekesh M (2016) A two-tier network based intrusion detection system architecture using machine learning approach. In: International conference on electrical, electronics, and optimization techniques, ICEEOT 2016, pp 42–47. https://doi.org/10.1109/ICEEOT.2016.7755404
7. Gulli A, Pal S (2017) Long short term memory-LSTM. https://doi.org/10.1144/GSL.MEM.1999.018.01.02
8. Hasan M, Islam M, Zarif II, Hashem MMA (2019) Internet of things attack and anomaly detection in IoT sensors in IoT sites using machine learning approaches. Internet Things 7:100059. https://doi.org/10.1016/j.iot.2019.100059
9. McDermott CD, Majdani F, Petrovski AV (2018) Botnet detection in the internet of things using deep learning approaches. In: Proceedings of the international joint conference on neural networks, pp 1–8, July 2018. https://doi.org/10.1109/IJCNN.2018.8489489
10. Moustafa N, Turnbull B, Choo KKR (2019) An ensemble intrusion detection technique based on proposed statistical flow features for protecting network traffic of internet of things. IEEE Internet Things J 6(3):4815–4830. https://doi.org/10.1109/JIOT.2018.2871719
11. Pahl MO, Carle G, Klinker G (2016) Distributed smart space orchestration. In: Proceedings of the NOMS 2016—2016 IEEE/IFIP network operations and management symposium, pp 979–984. https://doi.org/10.1109/NOMS.2016.7502936
12. Pajouh HH, Javidan R, Khayami R, Dehghantanha A, Choo KKR (2019) A two-layer dimension reduction and two-tier classification model for anomaly-based intrusion detection in IoT backbone networks. IEEE Trans Emerg Top Comput 7(2):314–323. https://doi.org/10.1109/TETC.2016.2633228

13. Rahul VK, Vinayakumar R, Soman K, Poornachandran P (2018) Evaluating shallow and deep neural networks for network intrusion detection systems in cyber security. In: 2018 9th International conference on computing, communication and networking technologies, ICCCNT 2018, pp 1–6. https://doi.org/10.1109/ICCCNT.2018.8494096
14. Roopak M, Yun Tian G, Chambers J (2019) Deep learning models for cyber security in IoT networks. In: 2019 IEEE 9th Annual computing and communication workshop and conference, CCWC 2019, pp 452–457. https://doi.org/10.1109/CCWC.2019.8666588
15. Tanwar S, Bhatia Q, Patel P, Kumari A, Singh PK, Hong WC (2020) Machine learning adoption in blockchain-based smart applications: the challenges, and a way forward. IEEE Access 8:474–488. https://doi.org/10.1109/ACCESS.2019.2961372

Data Analytics and Intelligent Learning

An Improved Dictionary Based Genre Classification Based on Title and Abstract of E-book Using Machine Learning Algorithms

Vrunda Thakur and Ankit C. Patel

Abstract The amount of digital books or e-books is increasing day by day. Book Assortment is the job of assigning a category or set of appropriate genres to a book. The goal of this research paper is to classify books with related genres. Many existing approaches, like Support Vector Machine (SVM), Neural Text Categorizer (NTC), etc. are available for text mining. We applied existing machine learning algorithms with different datasets and implemented existing feature selection methods to select features. In our proposed dictionary-based approach, we classified books by its attributes like title, description, genre, and author using text mining. In the learning part, we created a dictionary of keywords from the book's description and title and then assigned genres to the keywords. In the classification part, we attributed genres to a book. For classifying the books, we extracted a dataset from web pages using web scraping. Our proposed approach outperforms traditional approaches to reduce the time of training when massive data is considered.

Keywords Genre · Classification · E-book · Dictionary of words · Machine learning · Feature selection · Term frequency · Book categorization

1 Introduction

The virtual form of data is expanding gradually due to the use of cyberspace on a large scale. Most of the information on the Internet is in an unshaped form. It is very tough to relate murky data to each other. This massive data needs to be stored in a particular way. The reason to design algorithms for book categorization is to transform standardized data into business valuable and usable information, which does not work when it comes to unformed data. Book Categorization is the task of assigning predefined categories or sets of appropriate genres to a book [1]. Automatic

V. Thakur (✉) · A. C. Patel
L.D. College of Engineering, Ahmedabad 380015, India

A. C. Patel
e-mail: acpatel@ldce.ac.in

P. K. Singh et al. (eds.), *Proceedings of Second International Conference on Computing, Communications, and Cyber-Security*, Lecture Notes in Networks and Systems 203,
https://doi.org/10.1007/978-981-16-0733-2_23

book categorization is an essential aspect of the research area. We reviewed book categorization methods to derive the techniques that take less time to find a book of interest from a big set of data [2]. Cloud computing models can also play an essential role in enormous data processing for many applications like book categorization [3–5]. The existing classifier needs a numeric form instead of a text.

Today's world faces vast diversification in types of literature, which in turn requires genre classification. For eBooks, there are various areas of interest in the categories of fiction and non-fiction. Fiction is literature that contains a made-up story based on imagination, while non-fiction literature is based on actual events, facts, information, etc. [6]. Both have various subcategories, which in turn necessary to separate the books according to their identification of the topic and genre across multiple subject domains. Literature dives in every time with some new concepts and ideologies, in the work they write. With the initiative of publication of the book, it is essential to have information about the genre of a book. Hence, the correct identification of a book is required. Due to the lack of the technique to decide, up to what extent a book is a subset of a particular genre, many of them end up being poorly classified. This leads to defining a standard approach to assort books and their degree of similarity to a given genre. There is a lack of clearly defined genre taxonomy, predefined rules, or theorems to decide the book genre's correctness. Assigning a book to a particular category is a prime requirement for various applications. The summary of various notations used throughout this paper is listed in Table 1.

2 Background and Related Work

There is a wide range of elaborations and explanations for the term "genre." Bieber says genres can be defined by an external set of rules defined by the targeted audience [7]. According to Swales, texts within a genre may expand over a wide range of linguistic variations. This degree of change depends on how well constrained a genre is and what amount it allows freedom of personal expression [8]. Kessler et al. [9] suggested that genres should be defined narrowly enough so that texts within a genre class possess common structural or linguistic properties. The main concern would be choosing the proper combination of features and classification algorithms. Defining genre properly can be used for recommendation engines for online e-book stores and various other fields. There are multiple genres based on documents, e.g., fantasy, thriller, sci-fi, history, comic, adventure-based autobiographies, novels, crime-based, news, etc.

Existing studies are based on a combination of feature selection methods and machine learning algorithms for genre identification purposes. The best combinations of algorithms lead to an improvement in accuracy and better results. Feature Selection is the phenomenon of selecting relevant features for constructing a model for classification. These methods are used to avoid two major issues, namely overfitting and high dimensionality, and helps in optimizing the models' performance.

Table 1 Summary of notations

Abbreviations	Description
KNN	K nearest neighbor
SVM	Support vector machine
NB	Naïve Baye's classifier
LSTM	Long short term memory
IG	Information gain
RD-TFD	Relative document term frequency difference
TFIG	Term frequency based on information gain
TTFS	t-test based feature selection
IMTFIDF	Improved TFIDF
GININTF	Normal term frequency-based Gini index
EMCFS	Efficient multi-cluster feature selection
MCFS	Multi-cluster feature selection
TS	Term strength
DF	Document frequency
MI	Mutual information
Bi-LSTM	Bidirectional LSTM
RNN	Recurrent neural networks
CNN	Convolutional neural networks
GRU	Gated recurrent unit
HAN	Hierarchical attention network
ABSA	Aspect based sentiment analysis

Liu et al. [10] proposed a new method Relative document-term frequency difference (RD-TFD) based on independent feature space search, which divides features into two separate sets for a dataset. Gupta et al. [11] introduce a new feature selection method, Efficient Multi-Cluster Feature Selection (EMCFS), to obtain relevant feature subset from the original feature space for efficient cluster clustering data. Zheng et al. [12] presented thirteen superior feature selection methods and focused on evaluating its effectiveness of these methods on datasets of different languages. Subsequently, IG and MD feature selection methods gave better performance for Japanese and English Datasets, respectively. Mahnabolis distance emerged out to be the best for multi-class classification. All these methods were tested using different classifiers like SVM, AdaBoost, and KNN, where SVM gave better accuracy than other classifiers. Zhang et al. [13] presented a theory based on the study of Chinese documents. Four Feature Selection Methods (MI, IG, CHI, and DF) and classifiers (KNN, Winnow Classifier, NB, SVM, and Centroid Classifier) have experimented on Chinese Sentiment Corpus datasets. Pederson et al. [14] show the comparative study and evaluation of different feature selection methods in book categorization. Zhao et al. [15] presented an in-depth analysis of feature selection methods (DF, IG, CHI, MI, and TF) on classifiers, namely: NB, ME, Winnow, SVM for sentiment analysis. Sarkar et al. [16] proposed a two-step feature selection method based on univariate selection and clustering. It tends to falsify the myth that Naive Bayes cannot perform better for text classification. The proposed method reduces search space using the univariate feature selection method and then applies clustering to select feature sets. The experiment shows that Naive Bayes turns out to be superior to other classifiers, and IG gives better accuracy. Dey et al. [17] explores the applicability of feature selection methods (Document Frequency, Information Gain, Gain Ratio, Chi-Squared, and Relief-F) and observes their performance of recall, precision, and accuracy.

Classification is the process of predicting the labels of given data points, which is used to categorize data. It is the task of predicting labels for unknown data—for example, Email Spam Detection. A classifier learns from the input data, how data is interrelated, and builds up the relation between data, and thereby predicts the genre. In terms of books, various algorithms are implemented to classify its genre. Table 3 shows the performance of multiple classifiers along with their drawbacks and accuracy measure for genre classification. Various genre classification algorithms along with feature selection methods are listed in Table 2.

Ozsarfati et al. [18] tested five different machine-learning algorithms (RNN, GRU, LSTM, Bi-LSTM, CNN, Naïve Bayes) for classifying genre the 20, 75,575 books dataset with 32 genres. Buczkowski et al. [19] extracted information about books with the change in the domain of predictions. Joseph et al. [20] presented a comparative analysis of various machine learning algorithm's performance on the Gutenberg dataset. Tamara et al. [21] introduced new methods for feature categorization in different kinds of reviews, in the domain of books. Aspect Based Sentiment Analysis was implemented on the Open Editor and Amazon Datasets.

3 Proposed Approach

There are different methodologies and algorithms for genre classification and feature selection. To implement this process, different automatic machine learning algorithms exist. To implement machine-learning algorithms and classify books, we should have a dataset of books. To get the dataset with relevant attributes, we scrapped web pages that had a list of books or detail of books. In this work, we propose a new approach for categorizing the books with precise accuracy than existing algorithms. We implemented based on the process model [22] with existing feature selection methods to select features. Our dataset was retrieved from goodreads and iblist websites, tokenized with an algorithm, and converted to vector representations for every word. Both the datasets consist of information like title, author, and description of the eBooks. Our datasets include ten genres which are listed in Table 3.

Table 2 Analysis of various machine-learning classifiers and feature selection methods

Author	Feature selection methods	Classifier	Datasets	Best performing method
Liu [10]	RD-TFD, CHI, IG, TFIG, TTFS, IMTFIDF, GININTF	NB, SVM	PU123A CSDMC2010, Enron-spam 3	RD-TFD
Gupta [11]	EMCFS, MCFS	KNN	TDT2 Corpus, Reuters 21578	EMCFS
Zheng [12]	IG, MD	SVM, AdaBoost, KNN	Japanese and English language corpus novels	MD (English novels) IG (Japanese novels)
Zhang [13]	DF, MI, IG	KNN, Naive Bayes, winnow classifier	Chinese sentiment document	IG
Pedersen [14]	DF, IG, MI, CSS, TS	KNN linear least square fit	Reuter 22173 collection	IG, CSS
Zhao [15]	DF, IG, CSS, MI	Naive Bayes, max entropy, SVM	Chinese news comments	CSS
Sarkar [16]	IG, MI, CSS, symmetric uncertainty	Naive Bayes, SVM decision trees, KNN	CNAE-9 SMS spam collection	IG
Dey [17]	DF, IG, gain ratio, relief F algorithm	Naive Bayes, SVM, decision trees, AdaBoost classifier	Cornell movie review dataset	Gain ratio

Table 3 Genre distribution in dataset

Goodreads dataset		Iblist dataset	
Genre	Percentile	Genre	Percentile
Adventure	10.85	Cookbooks	9.98
Crime	9.78	Historical	9.23
Children and young adult	8.69	Horror	7.56
Business	6.89	Fantasy	9.87
Comic	9.93	Sci-fi	4.78

3.1 Proposed System

In our new approach for book categorization, we implemented a classifier that gives us more accurate results than other algorithms. We created a dictionary of words and related genres to classify the new coming books and selected a book database with three attributes, viz., Book Title, Book Description, and Book Genre. Our approach is divided into three parts: the data extraction, the training part, and the classification part. In the first part, we scrape the data from the web pages of both websites. After that, various data-cleaning techniques are applied on the dataset to treat missing values and outliers, create a cleaned dataset, and further tokenize the dataset. In the second part, we employ various feature selection methods, they are Document Frequency, Information gain, Chi-square analysis, and keyword frequency respectively on the tokenized dataset, and create vectors of unique keywords as features. We differentiate the title, author, and description accurately and try to gather unique keywords associated with each genre. In the training part, we created a model of keywords and related genres in classification. We considered a one by one book from the dataset, finding words from the book using its title and description. Remove common words called stop words from the word array. In the final step, we merge all book's words in one array, followed by the removal of duplicate words and applying feature selection on word vector. Then, finding out some crucial keywords from a word vector. The whole process results in "Dictionary of Keywords." Now we fetched genres for each keyword. For fetching genres, we took one word, followed by finding that word in the books and storing the books that contain that word. After gathering all books containing that word, we took genres of those books, create an array of unique genres, and assign each genre to that keyword. This process is repeated for every word of word array that results in a table of the word and related genres. The graphical representation of our proposed approach is shown in Fig. 1.

In the classification part, we used a model of word genre created in the training part. We took a book, which needs to be classified. We applied the process of finding keywords from the book. We took a word from unique words and found that word in the table. Considered genres of that word and store it. This process is followed for each term of the new book. Merged the genre of all words, created a vector of it

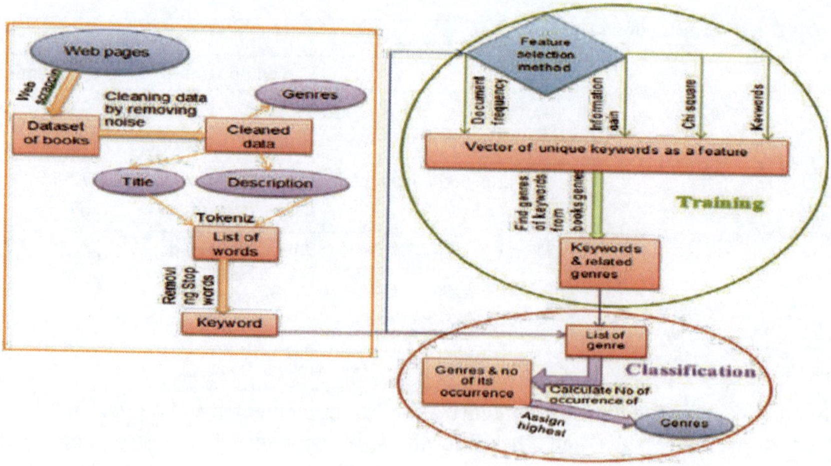

Fig. 1 Proposed system

and computed each genre's frequency. The genre, which has the highest frequency value, was assigned to a new book. Thereby, we get the test book genre, whose genre is determined from the trained dataset.

3.2 Dataset Creation and Finding Keywords

As discussed above, we scrap the dataset from iblist and good reads website, respectively. The information that users can see on web pages is extracted from web pages using a script is known as web scraping. We create our scraper using the Nokogiri gem of the ruby language that hits on the URL and scraps the information. We scrapped 11,128 books from the goodreads dataset containing features like title, genre, and description and 11,000 books from the iblist website. The scrapped dataset contains noisy data, unrecognized symbols, and even the same meaning has different names of genres. The data cleaning process will fill the missing values, treat outliers and mislabels, and remove the dirty data. There are a large number of words in the English dictionary, and stopping words are those that occur commonly in sentences.

Therefore, the removal of those words is necessary. We consider the SQL's stop words list for stop words. To classify a book, we have to select some text as a feature. Selecting the whole text as a feature is not a good idea. Therefore, we split a text into words and select words as features. Stemming is the process of text mining. Book categorization is depending on the number of occurrences of the same words.

For this reason, the classifiers need to remove different forms of words. The simple form of a word is considered as a keyword in book categorization. In the proposed algorithm, we consider the n-character approach for stemming the n-character approach is considering first n characters of a word for matching its plural or past tenses forms. In the n-character method, we create three rules for matching the words. In those rules, we match first n characters of words. If the rules are satisfied, then consider the first n-character of words as a keyword.

1. For each word, do apply the following rules:
   a. If the length of a word is, less than six characters then match the first three characters.
   b. If the length of a word is greater than six and less than or equal to nine characters then match the first four characters.
   c. If the length of a word is greater than 9 characters than match first 5 characters
2. Consider unique keywords obtained after stemming for classification.

3.3 Feature Selection Methods

In book categorization, feature selection is a tedious task due to the enormous bag. Feature defines as an attribute of an object used for classification. We have classified books on the basis of text in books. Therefore, we have to select some text as a feature. To select the whole text as a feature is not a good idea, so split a text into words and select words as features.

3.3.1 Document Frequency

The name document frequency defines the calculation of the rate of documents. In this method, we will be considering words one by one and the number of books that consider that keyword. From the size of the dataset, the threshold value is to be decided. The threshold value is an absolute value that divides words into two parts useful and not useful. The words, which have more document frequency than the threshold, are considered features and other non-features [14]. The term frequency method is the same as a document frequency. The difference between these two methods is in term frequency; we compute the rate of terms from the dataset or genres and in document frequency of document calculated for the term. A systematic process is as shown below:

1. Input the list of books
2. Count the number of books, in which keywords occur a set threshold value depending on dataset size.
3. Find out keywords from all books using defined functions
4. Split the keywords obtained and convert it into an array
5. Take keywords one by one from an array and count the number of books in which the keyword appears.
6. Document Frequency can be computed as the number of boos containing the word divided by the total number of books
7. Output: A training set of books along with the keywords.

3.3.2 Information Gain

The information gain of a feature defines how much information concerning the classification feature gives. It is based on the decrease in entropy after a dataset is split on an attribute. A systematic process is as shown below:

1. Input the list of books
2. Calculate entropy for each feature and add it proportionally to obtain the entire entropy of the split. Entropy is a measurement of impurity due to the feature in an event.
3. Compute entropy for the feature from the class after splitting the dataset.
4. Compute information gain of a feature.

$$\text{Information gain} = \text{entropy} - \text{child entropy} \tag{1}$$

5. After calculating information about each term, we decide the threshold from data of information gain. The term, which has higher information gain than the threshold value, is considered as a feature.

6. Output: Number of features to be considered

3.3.3 Chi-Square Analysis

The Chi-Square test shows a relationship between two variables. Here, it can be termed as an analysis of the association between genre and a word. It is a procedure to find out the statistical significance between the differences between sub-nodes and parent nodes. It can be defined as:

$$X(t, c) = \frac{N * (AD - CB)}{(A + C) * (B + D) * (A + B) * (C + D)} \tag{2}$$

In this method, we decide the threshold value by computing the average chi-square value of each term. Select words, which have a higher average chi-square value than a threshold. A systematic process is as shown below:

Input: List of training books
Keep a track of keywords and their count for each genre and find the value of each keyword using chi-square equation given below:

$$X^2(t, c) = (N * (wz - xy)^2)/((w + y) * (x + z) * (w + x) * (y + z))$$

After chi-square value is obtained we find the average for each keyword
We set a threshold and consider the word as keywords that have high average value than the threshold.
Output: Number of features to be considered.

4 Proposed Algorithm

The following section represents the various functions involved in the proposed approach.

1. Input: List of training books, testing books, and a list of keywords to predict a genre.
2. For each input test-book, find keywords from each book and return a vector of keywords.
3. Perform Algorithms and find keywords using various techniques and feature selection methods.
4. Create numeric vectors of keywords of the test book. Take keywords one by one from all book vector keywords and match if they are present in the current test book.
5. Consider the most occurred keyword from a test book. Calculate the distance between it and the same keywords from the features.

6. Consider minimum distance, and break the distance, if minimum distance == 5. Match the books for each index.
7. Give a list of genres and create a new hash and sort all the genres in descending order by its repetition.
8. Assign the highest occurred genre to a book.

5 Experimental Evaluation

In this section, we evaluate the performance of our proposed algorithm on the scrapped datasets. Most of the algorithms are implemented in ruby language using visual studio code. Python is used for cleaning datasets as well as importing and exporting datasets. We have taken 22,128 books [11,128: goodreads, 11,000: iblist dataset] as dataset and 10 (5 of each dataset) [adventure, crime, children, young-adult, business, comic, cookbooks, horror, historical] as genre.

5.1 *Result Analysis*

We divide the books into various sets to obtain the results. We split the dataset into the slot of 500 books for each set. We take 250 books for training, and the rest of 250 are test dataset, divided into five slots. Each slot contains 50 books for a dataset. We are collecting 500 books for each dataset and collecting 250 books for testing the dataset. We divide 250 testing books into five slots of 50 books. The results for different classifiers with two different datasets shown in this section. After analyzing the effects mentioned above of the proposed approach, we can see that training books separate our records. We consider only 100 training books, then consider 200 training books than 300 and consider up to 500 training books and 50 books for testing. In which we got the correct classification for 39 books and the wrong classification for 11 books. The next 50 books got correct classification with 36 books and wrong classification for 14-books. The meaning of correct classification is the original genre of a book, and the algorithm's occurred genre is similar.

5.1.1 Results of Proposed Approach

The meaning of the wrong classification is the original genre of a book, and the occurred genre by the algorithm is different. The average of the five tests with 50 books each and 100 training data is 76.8% while with 200 training data it is 81.86%, and the average of the proposed approach for this goodreads.com dataset is 82.72%. Detailed results are as shown in Table 4.

From the results, it can be concluded that as the training dataset increases, the algorithm's accuracy is increased. In this dataset when the size of the training books

Table 4 Results of proposed approach on both datasets

Results on datasets of goodreads.com by the proposed approach			Results on datasets of iblist.com by the proposed approach		
No.	Size of training dataset	Accuracy (%)	No.	Size of training dataset	Accuracy (%)
1	100	76.8	1	100	38.0
2	200	81.86	2	200	40.4
3	300	84.8	3	300	50
4	400	85.6	4	400	52.8
5	500	85.8	5	500	54.0
Average		82.72%	Average		47.04%

is 100, we achieved accuracy 76.8%, and as the dataset size increasing, we reached up to 85.6%. By a similar approach, we can calculate the results of the proposed method with datasets of **iblist.com** that are observed to be **47.04%** accordingly.

5.1.2 Results of Existing Algorithms (KNN-SM Model)

As per the results obtained, by applying the proposed algorithm and existing algorithm on our datasets, we can see the increase in accuracy of genre prediction, which increases from 74.96% to 82.72% for goodreads dataset, and 39.84% to 47.04% respectively for iblist dataset (Table 5).

Table 5 Results of KNN-SM model on both datasets

Results on datasets of goodreads.com by the proposed approach			Results on datasets of iblist.com by the proposed approach		
No.	Size of training dataset	Accuracy (%)	No.	Size of training dataset	Accuracy (%)
1	100	72	1	100	39.2
2	200	72.8	2	200	32.4
3	300	74.4	3	300	42.4
4	400	78	4	400	42.4
5	500	77.6	5	500	42.8
Average		74.96%	Average		39.84%

6 Discussion

The feature selection method Chi-square and information gain, term frequency give good results with different classifiers. By decrementing threshold value or incrementing, the number of features is increasing in result accuracy. Expanding the dataset provides better results; classification accuracy can work more accurately on it. Compared with all existing techniques, the proposed approach gives excellent results with 82.7% accuracy for the goodreads.com dataset. By performance analysis, we also observed that when we consider threshold value as 0.5, it provides better efficiency than other benefits. For more accurate results, we need to obtain more keywords, associate with corresponding genres, and perform sentiment analysis on the dataset. As shown in Fig. 2, when the size of the training dataset increases, the accuracy is also increased. We consider the size of the dataset is 100–500, and get good accuracy at the point of 500 around 86%. From the below graph, we observe that efficiency increases gradually as we add up the number of books in the training data sets. From the table, we can see that the chi-square test feature selection method gives better results than all other purposes. It provides an accuracy of around 82.7%. We had implemented all feature selection methods with different values of threshold. In Table 6, we show the results of the proposed approach with the chi-square feature selection method using different threshold values. From the table, we can see that as we decrease the threshold value, we get more accuracy. Therefore, we can say that the algorithms depend on the value of the threshold also. Here we got the highest

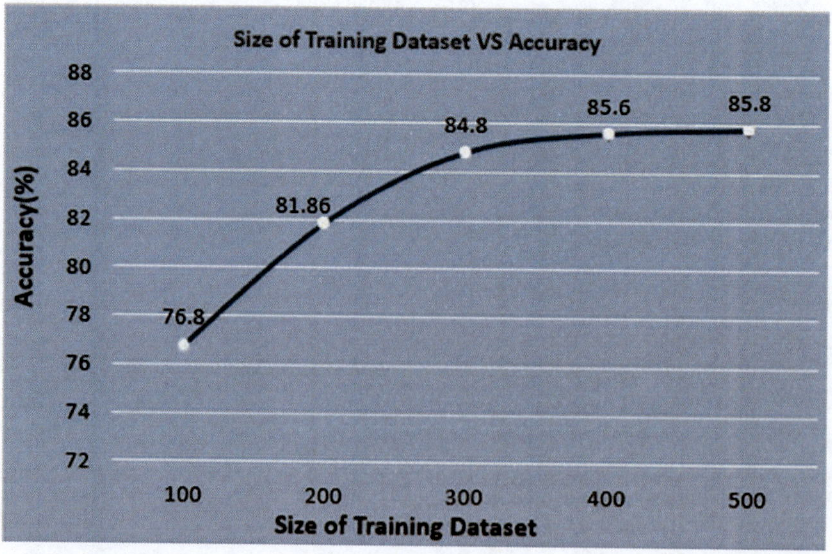

Fig. 2 Comparative analysis of change in accuracy of the proposed approach with different size of the training dataset

Table 6 Accuracy with different threshold value in chi square

Threshold	Accuracy
1.5	75.7
1	78.7
0.5	82.72

Table 7 Performance of proposed approach on feature selection methods

Algorithms	Avg. accuracy on goodreads dataset	Avg. accuracy on iblist dataset
KNN euclidean distance	21.28	20.16
KNN SM model	74.96	39.84
Proposed approach	82.72	47.4

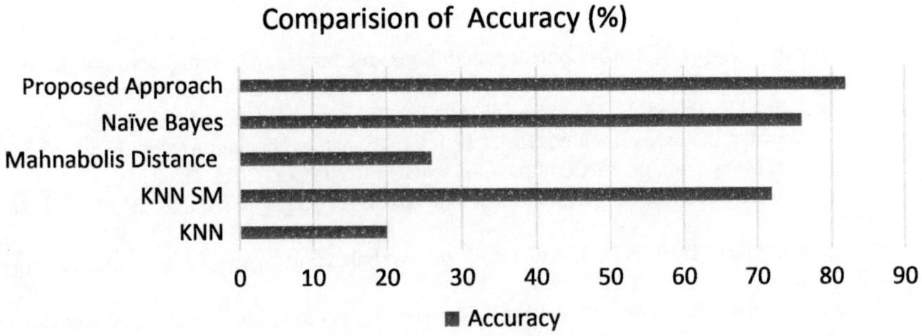

Fig. 3 Comparative analysis of the accuracy of various algorithms

efficiency with threshold value 0.5. Comparative performance of feature selection method is shown in Table 7.

As shown in Fig. 3, the blue bar shows the accuracy obtained by various algorithms. The Mahnabolis Distance Approach algorithm gives worse accuracy in comparison to other algorithms.

While our proposed approach takes less time to train the dataset and gives better accuracy, this training time considers only 50 training books. Therefore, when we reach up to 100,000, or more than that time for training the dataset will be less. Our proposed approach considers that point and gives better accuracy.

7 Conclusion

In this paper, we introduce a new dictionary-based approach for the genre classification of an e-book. The experimental result shows that classification using traditional

KNN algorithms (KNN SM Model and KNN Euclidean Distance) produces not such a high evaluation value, with a satisfaction rate of 74.96 and 21.28% accuracy for good reads dataset. In contrast, our proposed approach gives 82.72% on the same dataset. For the iblist dataset, traditional algorithms provide 20.16 and 39.84% classification accuracy that is lower than the proposed approach. The proposed approach takes less computation time and tends to reasonable accuracy in comparison to traditional KNN algorithms for genre classification. The Chi-square method for feature selection emerged out to be the best among all of the methods considered. Our approach may lead to less accuracy if the appropriate threshold value is not selected. The proposed Approach needs to be more accurate when the size of the dataset is too large. For future work, various combinations of feature selection methods and algorithms can be implemented.

References

1. Mooney RJ, Roy L (2000) Content-based book recommending using learning for text categorization. In: Proceedings of the fifth ACM conference on digital libraries
2. Bhatia J, Patel T, Trivedi H, Majmudar V (2012) HTV dynamic load balancing algorithm for virtual machine instances in cloud. In: 2012 international symposium on cloud and services computing, Mangalore, pp 15–20.https://doi.org/10.1109/ISCOS.2012.25
3. Karimkhan M, Bhatia JB (2014) Sentiment analysis and big data processing. IJCSC 5(1):136–142
4. Bhatia J, Kumhar M (2015) Perspective study on load balancing paradigms in cloud computing. IJCSC 6(1):112–120
5. Bhatia JB (2015) A dynamic model for load balancing in cloud infrastructure. Nirma Univ J Eng Technol (NUJET) 4(1):15
6. MerriamWebster.com. Genre (2014) https://www.merriamwebster.com/dictionary/genre
7. Bieber A (2018) Voices from the interior: reimagining childhood under Janusz Korczak's care. Lion Unicorn 42(3):321–337
8. Swales JM (2019) The futures of EAP genre studies: a personal viewpoint. J English Acad Purposes 38:75–82
9. Kessler B, Nunberg G, Schütze H (1997) Automatic detection of text genre. arXiv preprint cmp-lg/9707002
10. Liu Y et al (2020) A new feature selection method for text classification based on independent feature space search. Math Probl Eng
11. Gupta A, Begum SA (2019) Efficient multi-cluster feature selection on text data. J Inf Optimiz Sci 40(8):1583–1598
12. Zheng W, Jin Z (2020) Comparing multiple categories of feature selection methods for text classification. Dig Scholarship Human 35(1):208–224
13. Liu P et al. (2019) Sentiment analysis of chinese tourism review based on boosting and LSTM. In: 2019 international conference on communications, information system, and computer engineering (CISCE). IEEE
14. Yang Y, Pedersen JO (2017) A comparative study on feature selection in text categorization. ICML 97:412–420
15. Zhao Y, Dong S, Li L (2014) Sentiment analysis on news comments based on a supervised learning method
16. Sarkar SD, Goswami S (2013) Empirical study on filter-based feature selection methods for text classification. Int J Comput Appl 81(6)

17. Sharma A, Dey S (2012) Performance investigation of feature selection methods and sentiment lexicons for sentiment analysis. IJCA Special Issue Adv Comput Commun Technol HPC Appl 3:15–20
18. Ozsarfati E et al (2019) Book genre classification based on titles with comparative machine learning algorithms. In: 2019 IEEE 4th international conference on computer and communication systems (ICCCS). IEEE
19. Buczkowski P, Sobkowicz A, Kozlowski M (2018) Deep learning approaches towards book covers classification. ICPRAM:309–316
20. Worsham J, Kalita J (2018) Genre identification and the compositional effect of the genre in literature. In: Proceedings of the 27th international conference on computational linguistics
21. Álvarez-López T et al (2018) A proposal for book-oriented aspect-based sentiment analysis: comparison over domains. In: International conference on applications of natural language to information systems. Springer, Cham
22. Vachhani H et al (2019) Machine learning-based stock market analysis: a short survey. In: International conference on innovative data communication technologies and application. Springer, Cham

A Novel Multicast Secure MQTT Messaging Protocol Framework for IoT-Related Issues

Sharnil Pandya, Mayur Mistry, Ketan Kotecha, Anirban Sur, Pramit Parikh, Kashish Shah, and Rutvij Dave

Abstract Edge computing and fog computing have emerged as effective and efficient technologies for IoT-related issues. In the recent times, fellow researchers have proposed numerous researches under the area of edge and fog computing. Still, edge and fog computing have remained open research problems. In the proposed research work, we have proposed a MQTT-based broker security mechanism to protect the IoT-based system from a selection of security penetrations such as man-in-the-middle attacks, DDoS, DoS and many more. In general, MQTT broker architecture acts as an intermediator to establish connection between a publisher and subscriber. For secure communication between both the ends, it is essential to establish a novel security protocol which secure communication channel between subscribers and publishers. In the presented research work, we have proposed a novel authentication mechanism which makes the use of MQTT broker in achieving data privacy, authentication and data integrity. Furthermore, a detailed and rigorous analysis of a variety of security attacks has been analyzed and discussed in the end. At last but not the least, we have

S. Pandya (✉) · K. Kotecha
Symbiosis Center for Applied Artificial Intelligence and Symbiosis Institute of Technology (SIT), Symbiosis International (Deemed) University, Pune, India
e-mail: sharnil.pandya@situne.edu.in; sharnil.pandya@scaai.siu.edu.in

K. Kotecha
e-mail: head@sccai.siu.edu.in

M. Mistry (✉) · R. Dave
Department of Computer Science Engineering, Ganpat University, Ahmedabad, Gujarat, India
e-mail: mayur.mtechbda1703@ict.gnu.ac.in

R. Dave
e-mail: rutvij.dave1801@ict.gnu.ac.in

A. Sur
Symbiosis Institute of Technology (SIT), Symbiosis International (Deemed) University, Pune, India

P. Parikh · K. Shah
Department of Computer Science Engineering, Navrachana University, Vadodara, Gujarat, India

P. K. Singh et al. (eds.), *Proceedings of Second International Conference on Computing, Communications, and Cyber-Security*, Lecture Notes in Networks and Systems 203,
https://doi.org/10.1007/978-981-16-0733-2_24

also presented multicast-MQTT secure messaging protocol and compared it with state-of-the-art methodologies such as RSA and advanced encryption standard.

Keywords IoT · M-MQTT · RSA · AES · Attack · DOS · Sensors

1 Introduction

The way that IoT gadgets do not have adequate security highlights has brought about major issue, requesting the requirement for making sure about the correspondence between these gadgets. The expanded dependence of individuals on these systems for playing out their everyday exercises and the reception of existing security plans present open difficulties and genuine worries over access of individual data relating to a gadget and individual protection [1–7]. As indicated by the reports introduced by Helvert Packard, 70% of IoT gadgets are powerless against attacks, similar to secret word security, encryption and general absence of granulator client get to authorizations [2, 7–14]. An as of late detailed, DOS, DDOS, spoofing tempering data attack, expresses that a self-proliferated and tainted botnet infection that contaminates ineffectively ensured IoT gadgets [3, 15–21]. The paper targets understanding DOS attacks on IoT organize and recognizing an appropriate plan dependent on cryptographic strategies to ruin the equivalent [4, 22–29]. In spite of the fact that the experimentation is finished with assortment of cryptographic plans on an IoT arrange-based MQTT correspondence convention the assessment system uncovered that AES and RSA calculation-based cryptography-based plans are most appropriate for upgrading the security of an imparting gadget in Internet of things [5, 14, 30–33]. MQTT depends on a distributer and buy in design. Customers that utilization these conventions will distribute to the specialist to transfer data, and buy in to the dealer to download orders. MQTT will consequently push got messages to every bought in customer. In the event that various customers are bought in to a theme, the intermediary will distribute any got messages to all the customers [34–38]. The upsides of this are the framework that could not care less what number of as well as if any customers are associated, and it will in any case push any got messages to all the bought in customers with no arrangement. Another favorable position is that when another sensor is associated, it basically needs to interface with a system and buy in to the point to get any orders. For distributing data, a sensor can without much of a stretch simply distribute to the server. The other bought in frameworks will at that point get this new data with no requirement for setup.

2 Security Challenges of IoT Layers

Security challenges are connected with verification, classification, and honesty of the information moved through IoT [1, 21, 39–42]. While technological difficulties are more related to the vitality and versatility came about because of the dynamic idea of IoT gadgets [2, 4, 43–48].

2.1 Existing Security Threats in IoT

There are different difficulties, for example, jamming and spoofing attacks and other unauthorized access, which have undermined the trustworthiness of the client's information [49–53]. There are potential arrangements that can assist the person with implementing different safety efforts that can assist with making sure about their IoT gadgets.

2.1.1 Security Threats of IoT Layers

IoT contains a network of profoundly various advanced articles interfaced with one another and with people as well. It gives a sensor network correspondence framework, store and deal with the data, gives get to and likewise handles the privacy protection and data security problems [1, 2, 4, 14, 54–60] (Table 1).

2.1.2 Security Threats

Each IoT layer is vulnerable to security dangers and assaults. These can be dynamic, or inactive, and can begin from outside sources or internal framework owing to an attack by the insider [1–3]. A working ambush direct stops the administration while the latent kind screens IoT orchestrate information without forestalling its administration (Table 2).

2.1.3 Network Layer

The correspondence in the IoT is not exactly equivalent to that of the Web since it is not restricted to machine to human. Regardless, the part of machine-to-machine correspondence that the IoT presents has a security issue of compatibility [1, 2] (Table 3).

Table 1 Existing security threats [1, 2, 4]

Threats	Descriptions
Compatibility	Huge security game plans should not keep the helpfulness of interconnected heterogeneous contraptions in IoT compose structure [6, 61–63]
System constraints	In IoT plan, most of focuses absence of limit breaking point, force and CPU. They generally use low-information move limit correspondence channels. Hereafter, it is ill suited to apply some security techniques, for instance, repeat skipping correspondence and open key encryption estimation. Plan of well-being system is incredibly inconvenient beneath these conditions [61–63]
Data volumes	Sensor-based, coordination's and critical scale structure that have likely outcomes to recollect massive volume of data for central framework or workers [61–63]
Confidentiality	In the interim, a fantastic number of RFID structures are short of reasonable affirmation instrument; anyone can way make and find the personality of the articles doling out them. Interlopers can scrutinize the data, just as change or even eradicate data too [10, 61–63]
Adaptability	The IoT sort out contains a broad number of center points. The prescribed security part on IoT should be versatile [4]
Autonomous control	Traditionalist PCs require customers toward plan and change them to different application situates just as particular correspondence conditions [61–63]. In any case, dissents in IoT framework should develop affiliations quickly, and form/mastermind themselves for acclimating to the stage they are working in. Such a control also incorporates a couple of methods and instruments, for example, self-organizing, self-smoothing out, self-organization, self-repairing and self-making sure about

Table 2 Security challenges

Challenges	Description
Originality	Simply legal customers should be allowed to get to the structure or sensitive information [1, 64–68]
Jurisdiction	The advantages of device portions and applications should be limited as so they can get to simply the resources, they need to do their kept an eye on tasks [2, 64–68]
Privacy	Information transmission between the center points should be protected from intruders [3]
Probity	Related information should not be modified [4, 64–68]
Accessibility and continuation	With a definitive goal to keep up a key decent way from any conceivable operational dissatisfactions and interferences, receptiveness and development in the course of action of security associations ought to be guaranteed [64–68]

Table 3 Threats of network layer

Threats	Description
Selective forwarding	In such attacks, poisonous centers do not propel a couple of messages and explicitly drop them, ensuring that they cannot multiply later on [69–71]
Sybil attack	In such attacks, noxious center points do not advance a couple of messages and explicitly drop them, ensuring that they cannot induce later on. It is clarified as a noxious device wrongly going toward various characters. Sybil attack, an attacker can "be more than one put at once" as a singular Malevolent center. It acquaints various characters with various centers in the framework diminishing the feasibility of accuse lenient plans [72]
Sinkhole attack	It is portrayed in by genuine resource debate between next centers of poisonous center point for the compelled bandwidth and organization get to. It outcomes in stop up and compartment enliven the imperativeness usage of center points included. With sink openings molding in a sensor mastermind, it is exposed against a couple of various types of refusal of organization attacks [13, 72]
Wormhole	This sort of DoS attack starts development of whiles of data since its special area in framework. These developments of data pack are helped prepared over tunneling of snapshots of data done an association of little Idleness [13]
Man-in-the-middle attack	The unapproved amassing can actually, even phony the character of the individual being alluded to and give regularly to acquire data [7]
Hello-flood attack	From an overall perspective, a solitary vindictive focus coordinates a purposeless correspondence which are answering the attacker to make a bigger turn of events [7]
Acknowledgment flooding	Coordinating estimations in sensor-based structures require insistences occasionally [70, 71]. This sort of DoS attack, a noxious center alludes wrong information to appointed next center points by help of these assertions

2.1.4 Application Layer

Since the IoT in spite of everything does not have overall techniques and principles that regulate the collaboration and the headway of employments, there are various issues related to application security [1] [[2]. Different applications have unmistakable approval instruments, which make coordination of all of them difficult to ensure data security and character confirmation [3, 4, 6] (Table 4).

2.1.5 Perception Layer

There are three security issues in IoT discernment layer. First is the nature of far off signs. Generally, the signs are imparted between sensor centers of IoT using far off developments whose capability can be subverted by upsetting waves [1–3] (Table 5).

Table 4 Threats of application layer

Threats	Description
Sniffer/loggers	Aggressors can show sniffer records keen on the structure that accepts fundamental proof as of system improvement. The chief expect to sniffer is to take passwords, reports and E-mail content. Different shows are skewed to sniffing [73, 74]
Injection	Aggressors may show up encryption clearly into demand is performed on worker. It is an outstandingly ordinary attack, easy to abuse, and can cause some horrendous results, for instance, data mishap, data degradation and nonappearance of duty [20]
Session hijacking	This attack reveals singular characters by mishandling security defects in affirmation and meeting organization. Such an attack is amazingly ordinary and effects of attack are incredibly essential. With the character of someone else, aggressor is doing whatever certifiable customer is doing [20, 73, 74]
DDoS	It is the identical standard DOS attack. In any case, it has executed by various aggressors meanwhile
Social engineering	A certifiable peril for demand level some place aggressors could get information after customers by methods for visits, knowing one another

Table 5 Threats of perceptional layer

Threats	Description
Spoofing	It begins with a message passed on to the sensor by the aggressors. [11]
Signal jamming	It has such a DoS snare; it is correspondence network among focus focuses and delays from conversing with each extra
Node capturing	The aggressor gets the sensor community point really replaces the middle with their hurtful focus point. Such an assault usually achieves the assailant extending complete power over the got focus and damages the system
Path-based DoS	This DoS attack, attacker overpowers sensor center points broadened partition missing by flooding a multihop start to finish correspondence route with likewise rehashed packages or injected tricky bundles [69, 73, 74]. Diminished structure openness and exhaustion in arrangement of center points are properties of substantial attack
Node outage	This attack has associated truly framework and they close the handiness of framework portions. Center point organizations, for instance, scrutinizing, assembling and beginning assignments are stopped because of this attack [12]
Eavesdropping	Remote characteristics of RFID structure make it possible that attacker tracks down the characterized information [73, 74], for instance, mystery word or some other information pouring out of tag-to-per client or per client-to-mark making the system weak

3 Application Layer Protocols in IoT

This layer conventions are measures conventions for message going in Internet of things application layer proposed by various normalizations [2]. All-most Internet of things use are Internet Protocol based and they use TCP and UDP for transport. In any case, there are a few message appropriations works that are basic among numerous IoT applications; it is alluring that these capacities be actualized in an interoperable standard manner by various applications.

3.1 MQTT

MQTT is circulated purchase in-based illuminating show. The process cannot be rushed for the TCP/IP meeting to end. Intended for associations with distant areas where less code discovery is required or confined to information move limit [2, 3]. It is disseminating in illuminating plan requires a message authority. MQTT is disseminated in show that licenses edge-of-sort out devices to convey to a mediator. Clients partner with this vendor, which by then intervenes correspondence between the two contraptions. Each contraption can purchase in, or enroll, to explicit topics [2, 4, 5].

3.1.1 MQTT Architecture

See Fig. 1.

3.2 CoAP

Constrained Application Protocol is an electronic route convention that can be utilized by mandatory center points and is mandatorily arranged on the IoT [2, 3, 5]. The convention is expected for M2M applications, for instance, sharp essentialness and building mechanization" [2, 3, 6]. The convention is particularly focused for obliged equipment, for example, 8-bits microcontrollers, low force sensor and comparative gadgets that cannot run on HTTP or TLS [2, 3, 5]. CoAP is a rearranged of the HTTP convention running on UDP that helps spare transmission capacity [2].

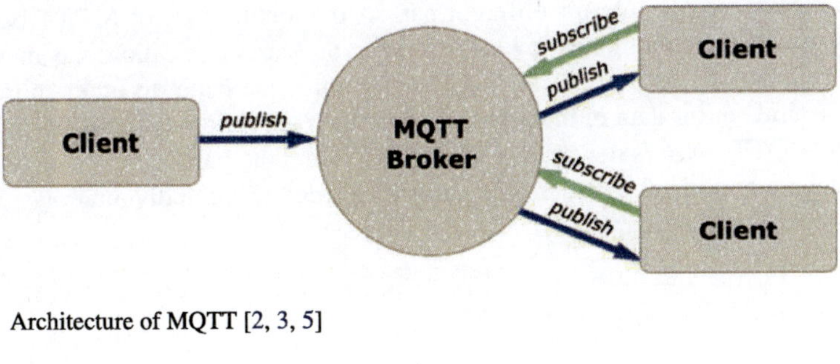

Fig. 1 Architecture of MQTT [2, 3, 5]

3.3 AMQP

Advanced Message Queuing Protocol (AMQP) is a legitimate message-based program for open-source middleware programming [2, 5]. The portraying features of AMQP are message heading, lining, directing (tallying feature point and disperse and purchase in), reliability and security [2, 6].

3.4 XMPP

XMPP is a protocol dependent on Extensible Markup Language (XML) and proposed for texting (IM) and online presence discovery. It works between or among servers and encourages close constant activity [2, 4]. XMPP protocol utilized in IoT. It covers XMPP Core, XMPP addressing, XMPP server and XMPP customer correspondence. XMPP is the short type of Extensible Messaging and Presence Protocol [2]. Messaging: It utilizes short messages as strategy for correspondence between customer (e.g., client) and server.

3.5 HTTP

Hyper Text Transfer Protocol is a TCP/IP-based correspondence show that is used to pass on data (HTML archives, picture records, question results, etc.) on the World Wide Web [2, 6]. The default port is TCP 80, yet various ports can be used too. It gives a standardized way to deal with PCs to talk with each other [2, 4]. HTTP specific decides how clients' requesting data will be constructed and shipped off the worker, and how the workers respond to these sales [2].

4 Attacks on MQTT

Here, attacks are possible for different part of the architecture of MQTT because this is a way for communicating using one-to-one communication and many-to-many communication are transferring the data for 1 publisher to other subscriber using several sensor data in the processing for the encryption and transferring data into the MQTT server/gateway, after that it has found the particular IP addresses in the database and match IP address, then it forwarded to the individual or multiple subscriber to destination part.

4.1 Attack Mitigation Methods in MQTT IoT Environment

See (Table 6).

5 Proposed Framework

IoT communication middleware, utilized for IoT information transmission, is additionally required to be changed, as per the edge registering worldview. The Message Queuing Telemetry Transport (MQTT) Protocol is a broker-based IoT middleware that is generally utilized in IoT frameworks. The area of the broker is significant on the grounds that the MQTT protocol trades information through themes, and all gadgets with a similar subject must be associated with the broker. At the point when a broker is introduced on all edge hubs to utilize the MQTT protocol in edge figuring, the multifaceted nature of the association increments, and the issue of dealing with an enormous number of brokers emerges. IoT gadgets send a determined message to keep up availability with the broker, and this procedure can cause arrange clog because of various brokers introduced on the edge hub. In this manner, a system is expected to oversee brokers introduced on all edge hubs, and to trade information between numerous edge systems without extra association with brokers (Fig. 2).

Table 6 Various security threats on MQTT

Threats	Mitigation
DoS	Utilization of firewall strategies to square DoS assaults particularly TCP-based assaults, rate constraining customers, utilization of circulated MQTT merchants to deal with load, utilization of use layer firewall to screen and square abnormalities
Spoofing	Utilization of authentications to recognize customers and workers, utilization of TLS and utilization of VPN among customer and worker
Fact revelation	Scrambling the information put away in the merchant and IoT gadget, utilization of confided in customer programming on PC and cell phones, start to finish encryption, VPN among customers and dealers
Elevation of privileges	ACL to control access to themes, incapacitating # memberships, utilizing oddity discovery to identify anomalous message substance [15, 16, 72]
Tampering data	Message trustworthiness checks utilizing hashing calculations, utilization of scrambled correspondence advancements, for example, TLS/SSL in MQTT intermediary, X509 endorsements from confided in power [7, 15, 17, 69, 72]

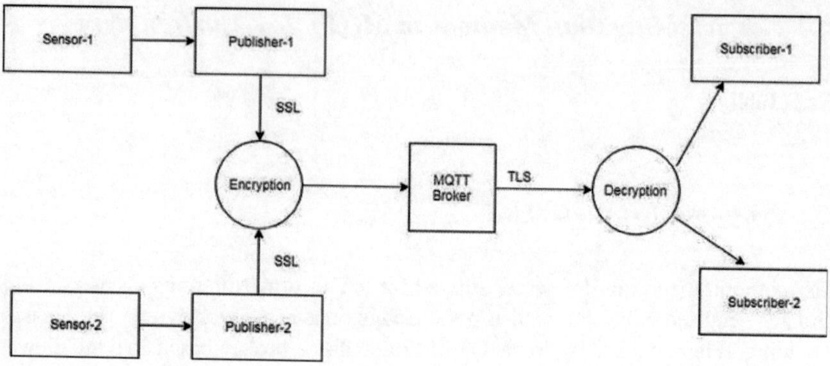

Fig. 2 Many-to-many communication (M-MQTT)

6 Proposed Flowchart

See (Fig. 3).

7 Methodology

7.1 RSA Algorithm

In cryptography, RSA is a calculation implied for open key cryptography which was given by every one of the three researchers Rivest, Shamir and Adleman [16, 17, 75–82]. The RSA calculation is perceived on the numerical part which looks simple yet extreme for useful usage. It essentially implies anybody can undoubtedly locate the prime numbers and even can without much of a stretch various two adequate huge prime numbers together [19, 83–89], however, the real trouble emerges during considering their item. It is colossally hard to factor the result of those two enormous prime numbers. The RSA calculation contains some particular strides for taking care of an issue [16, 17, 19, 22, 90–97].

7.2 AES Algorithm

AES is created to supplant a more seasoned more prohibitive encryption strategy called data encryption standard (DES). This strategy utilizes a similar guideline as AES, however, has a limited key and data square size and is generally speaking more execution substantial [18, 98–105]. AES is known as a square code, otherwise, called a pseudorandom permutation (PRP), where it utilizes a mystery key to scramble and

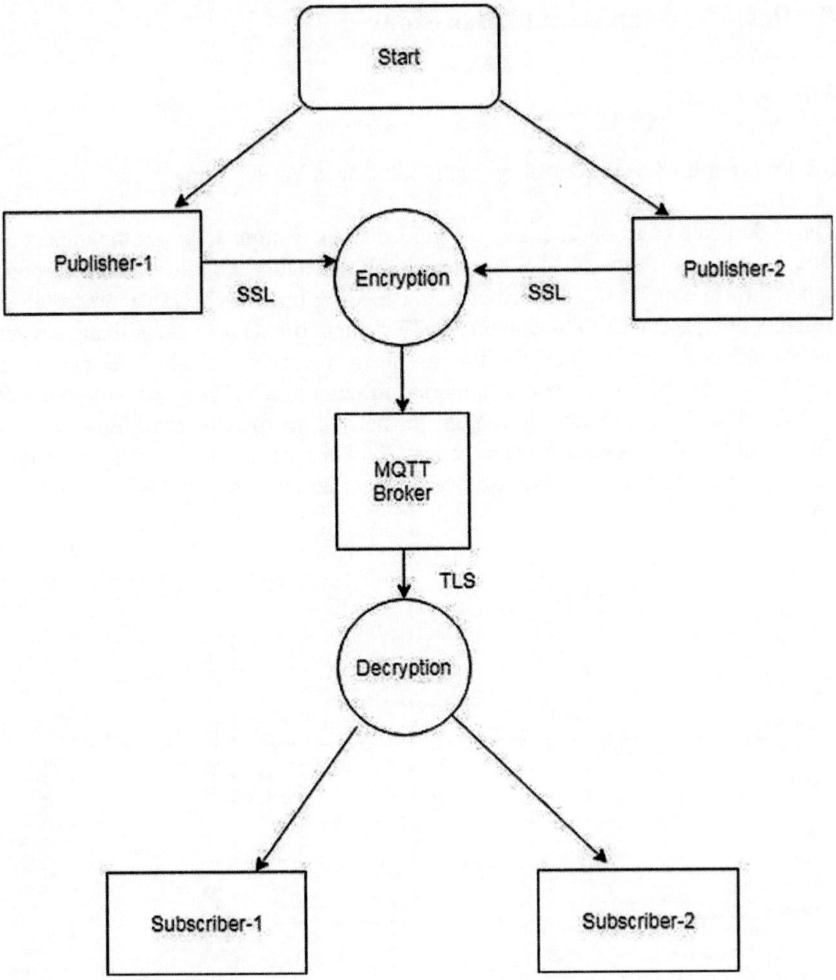

Fig. 3 Flowchart of many-to-many communication (M-MQTT)

unscramble squares of data. Any blend of data sources would get an apparently random permutation of the contribution as yield. The key controls the permutation request and any various keys should bring about apparently random permutations. [20, 106–117] AES can encode 128-piece plain content to 128-piece figure text with control of a 128-, 192-or 256-piece mystery key [18, 20, 113–118]. AES takes a shot at the guideline of substitution-permutation network design with steps considered rounds that are rehashed 9, 11 or multiple times to make the code text [119–123].

8 Results: Graphs and Discussions

8.1 Analysis

8.1.1 Graph of DOS Attack on no. of Messages Versus Time in ns

The underneath chart outlines that the guided investigations to gauge the impact of DoS attacks against the MQTT. Made a MQTT specialist and afterward correspondence information between distributers to supporter utilizing MQTT representative utilizing hub js. The DOS assault on MQTT dealer correspondence utilizing cluster record utilizing hub js. Plays out the assault on intermediary when the exchange messages between distributers to supporter utilizing hub js. With and without DOS assault what amount of time to require for normal postponement of the message. This would made work simpler and we payload encryption utilizing cryptography calculation to make the correspondence safer utilizing RSA and AES calculations (Fig. 4).

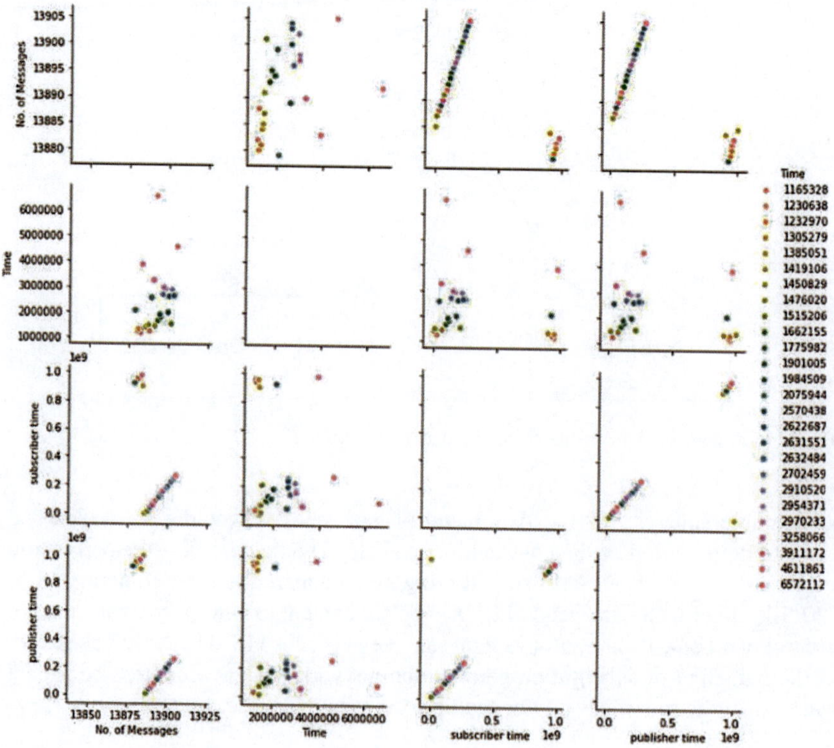

Fig. 4 Graph of DOS attack on no. of messages versus time in ns

8.1.2 The Graph of Difference Between of no. of Messages Versus Publisher Time Versus Subscriber Time in ns of RSA and AES Algorithms

Here, the chart characterizes made a MQTT agent and afterward correspondence information between distributers to supporter utilizing MQTT merchant utilizing hub js. The DOS assault on MQTT specialist correspondence utilizing cluster record utilizing hub js. Plays out the assault on intermediary when the exchange messages between distributers to supporter utilizing hub js. With and without DOS assault what amount of time to require for normal postponement of the message (Fig. 5).

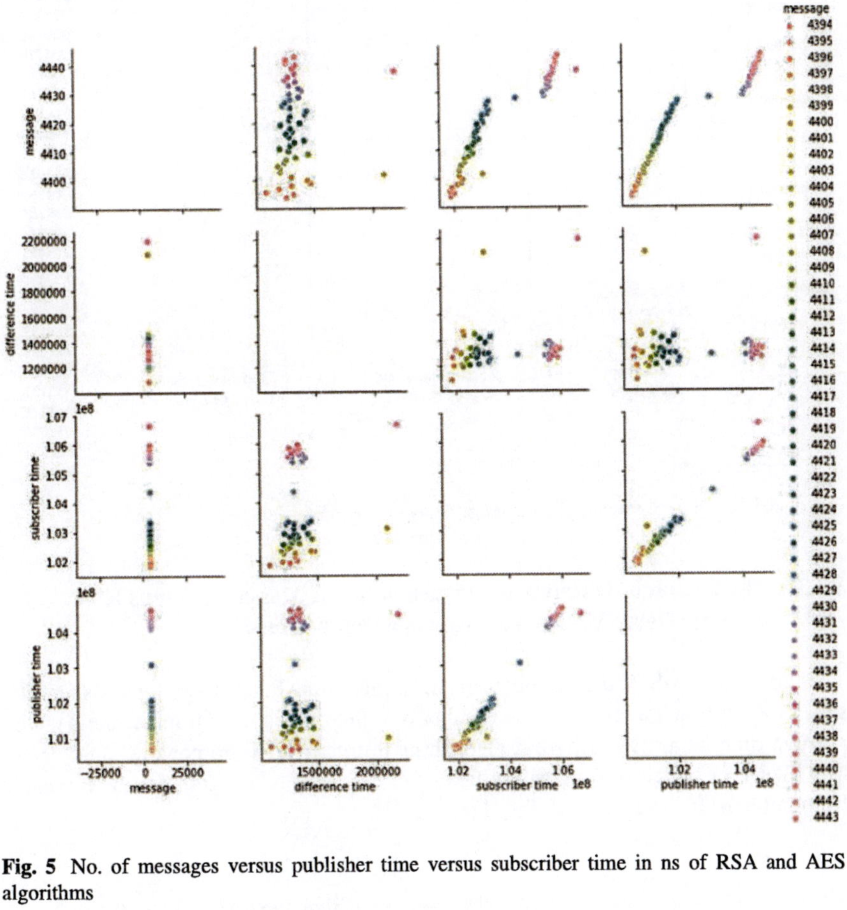

Fig. 5 No. of messages versus publisher time versus subscriber time in ns of RSA and AES algorithms

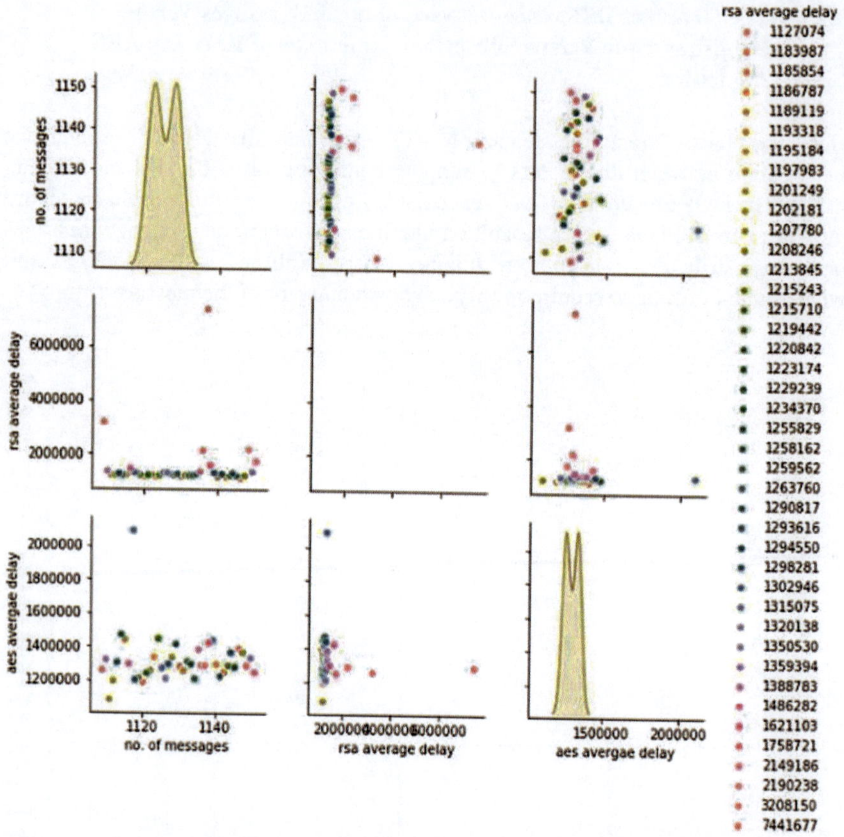

Fig. 6 No. of messages versus RSA average delay versus AES average delay time in ns

8.1.3 The Graph of Difference Between of no. of Messages versus RSA Average Delay Versus AES Average Delay Time in ns

The graph describes the number of messages transfer between publishers to subscriber in how much time. The number of input data is from transfer message in average time using two different algorithms in this MQTT framework. And using MQTT server which one algorithm takes more time to take a transfer the message from publishers' o subscriber side (Fig. 6).

8.1.4 Comparison of Average Delay Between RSA and AES Algorithms

Here, the above graph defines the communicating with the different sensors/devices using MQTT protocol algorithm are comparison between RSA and AES algorithms which is more secure way to communicating to each end user using encryption and

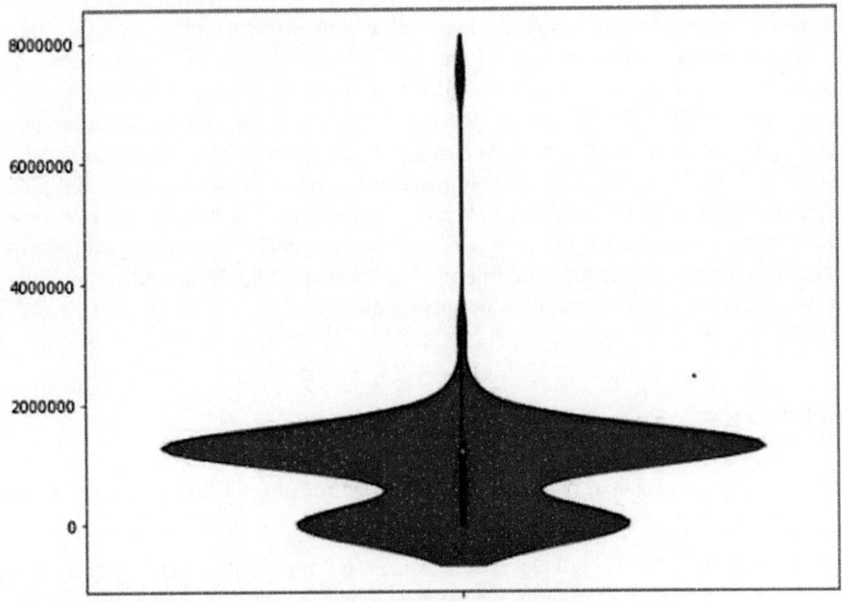

Fig. 7 Comparison of average delay between RSA and AES algorithms

decryption algorithm on application layer. And finds the average delay between using that two algorithms along with which one is faster compare to others. Moreover, we find the AES algorithm faster than RSA algorithm (Fig. 7).

9 Conclusion

This paper has presented edge computing and fog computing that has emerged as effective and efficient technologies for IoT-related issues, and a novel programming-based plan is to keep the assault from the unapproved individual and gives more improved security to information correspondence and information move between the distributor and supporter. We have proposed a novel authentication mechanism which makes the use of M-MQTT broker in achieving data privacy, authentication and data integrity. Cryptographic arrangement is noteworthy instrument of defending delicate information. In this work, this paper presents consolidated methodology of two notable and made sure about calculations RSA and AES. We guided examinations to gauge the impact of DoS attacks against the MQTT. The DOS assault on M-MQTT dealer correspondence utilizing cluster document utilizing node.js. It plays out the assault on intermediary with and without DOS assault what amount of time to require for normal postponement of the message. This would made work simpler and we payload encryption utilizing cryptography calculation to make the correspondence safer utilizing RSA and AES calculations. And it creates the chart of normal postponement among RSA and AES calculations. AES calculation has required significant investment is 0.00133757 s and RSA has required some serious energy is 0.00150854 s. That is the reason AES quicker than RSA calculation and safer multicast correspondence between many-to-numerous distributer and supporter for MQTT. In future, we will execute multicast correspondence at that point plan cryptography calculation at that point tests and investigate result. We will overview the impact of various toxic attacks on IoT contraptions and MQTT message representatives. Additionally, the introduction of a load changed MQTT specialist condition during different attacks is ought to be surveyed.

References

1. Huang X, Craig P, Lin H, Yan Z (2015) SecIoT: a security framework for the internet of things. Secur Commun Netw 9:3083–3094
2. Wind River Systems (2015) Security in the internet of things
3. Suo H, Wan J, Zou C, Liu J (2012) Security in the internet of things: a review. IEEE International Conference on Computer Science and Electronics Engineering, 23–25 March 2012, pp 648–651
4. Nguyen KT, Laurent M, Oualha N (2015) Survey on secure communication protocols for the internet of things. Ad Hoc networks European commission. IoT Privacy, Data Protection, Information Security. https://ec.europa.eu/information_society/newsroom/cf/dae/document.cfm?doc_id=1753ITU-T.Y.2060:OverviewoftheInternetofThings
5. Arseni SC, Halunga S, Fratu O, Vulpe A, Suciu G (2015) Analysis of the security solutions implemented in current internet of things platforms. In: IEEE grid, cloud and high performance computing in science. Romania, 28–30 October 2015, pp 1–4
6. Mattern F, Floerkemeier C (2010) From the internet of computers to the internet of things. In: Sachs K, Petrov I, Guerrero P (eds) From active data management to event-based systems and more. Springer, Berlin Heidelberg, pp 242–259
7. Deng J, Han R, Mishra S (2005) Defending against path-based DoS attacks in wireless sensor networks. In: ACM workshop/security of ad hoc and sensor networks. Alexandria, 7 November 2005, pp 89–96
8. Farooq MU, Waseem M, Khairi A, Mazhar S (2015) A critical analysis on the security concerns of internet of things (IoT). Int J Comput Appl 111:1–6
9. Mitrokotsa A, Rieback MR, Tanenbaum AS (2009) Classifying RFID attacks and defenses. Inf Syst Front 12:491–505
10. Borgohain T, Kumar U, Sanyal S (2015) Survey of security and privacy issues of internet of things. Int J Adv Netw Appl 6:2372–2378
11. Anwar RW, Bakhtiari M, Zainal A, Hanan AA, Qureshi KN (2014) Security issues and attacks in wireless sensor network. World Appl Sci J 30:1224–1227
12. Khoo B (2011) RFID as an enabler of the internet of things: issues of security and privacy. In: IEEE international conferences on internet of things, and cyber, physical and social computing. Dalian, pp 709-712, 19–22 October 2011
13. Ahmed N, Kanhere SS, Jha S (2005) The holes problem in wireless sensor network: a survey. ACM SIGMOBILE Mobile Comput Commun Rev 9:4–18
14. Bhatia J, Govani R, Bhavsar M (2018) Software defined networking: from theory to practice. In: 2018 Fifth international conference on parallel, distributed and grid computing (PDGC). Solan Himachal Pradesh, India, pp 789–794. https://doi.org/10.1109/PDGC.2018.8745762

15. Kalita HK, Kar A (2009) Wireless sensor network security analysis. Int J Next-Gener Netw 1:1–10
16. Karlof C, Wagner D (2003) Secure routing in wireless sensor networks: attacks and countermeasures. Ad hoc networks. Kulshrestha A, Dubey SK (2014) A literature review on sniffing attacks in computer network. Int J Adv Eng Rese Sci, 1(2):22. Hermes Engineering. Security in Web Applications
17. Gupta J, Nayyar A, Gupta P (2015) Security and privacy issues in internet of things (IoT). Int J Res Comput Sci 2:18–22
18. Niruntasukrat CI, Pongpaibool P, Meesublak K, Aiumsupucgul P, Panya A (2016) Authorization mechanism for MQTT-based internet of things. In: 2016 IEEE international conference on communications workshops (ICC), pp 290–295
19. Zeidanloo HR, Manaf AA (2009) Botnet command and control mechanisms. In: 2009 Second international conference on computer and electrical engineering. Dubai, pp 564–568
20. Mektoubi, Hassani HL, Belhadaoui H, Rifi M, Zakari A (1995) New approach for securing communication over MQTT protocol a comparison between RSA and elliptic curve. In: 2016 Third international conference on systems of collaboration (SysCo). Casablanca, pp. 1–6. Rekhter Y, Li T, (1995) A border gateway protocol 4 (BGP-4). In: Internet Engineering Task Force (IETF), RFC 1771
21. Cloudscene (2018) Cloudscene [Internet]. Available from: https://cloudscene.com/news/2018/05/internet-of-things-iot/
22. Lethonen R, Soini J, Majalainen J, Vatiainen H (2003) Multicast control protocol (MCOP). Internet Engineering Task Force (IETF), Internet Draft, draft-lehtonen-magma-mcop-*.txt, (work in progress)
23. Rajvaidya P, Ramachandran KN, Almeroth KC (2003) Detection and deflection of DoS attacks against the multicast source discovery protocol. In: Technical Reports. University of California, Santa Barbara, submitted for publication
24. Deering S (1989) Host extensions for IP multicasting. Internet Engineering Task Force (IETF), RFC 1112
25. Shaheen SH, Yousaf M (2014) Security analysis of DTLS structure and its application to secure multicast communication. 12th International conference on frontiers of information technology
26. Tiloca M (2014) Efficient protection of response messages in DTLS-based secure multicast communication. In: SIN '14 Proceedings of the 7th international conference on security of information and networks
27. Tanwar S, Patel P, Patel K, Tyagi S, Kumar N, Obaidat MS (2017) An advanced internet of thing based security alert system for smart home. In: 2017 International conference on computer, information and telecommunication systems (CITS). Dalian, pp 25–29. https://doi.org/10.1109/CITS.2017.8035326
28. Bhatia J, Kakadia P, Bhavsar M, Tanwar S (2020) SDN-enabled network coding-based secure data dissemination in VANET environment. IEEE Internet Things J 7(7):6078–6087. https://doi.org/10.1109/JIOT.2019.2956964
29. Bhatia J, Shah B (2013) Review on various security threats and solutions and network coding based security approach for vanet
30. Cadavid H, Garson W (2018) Towards a smart farming platform: from IoT-based crop sensing to data analytics. Springer. 2018 https://doi.org/10.1007/978-3-319-98998-3_19. Radford T (2005) The Guardian. [Internet]. Available from https://www.theguardian.com/science/2005/mar/30/environment.research
31. Nandurkar S, Thool V, Thool R (2014) Design and development of precision agriculture system using wireless sensor network. In: International conference on automation, control, energy and systems (ACES). Hooghly
32. Andrew R, Malekian R, Bogatinoska D (2018) IoT solutions for precision agriculture. In: MIPRO. Opatija
33. Benyezza H, Bouhedda M (2018) Smart irrigation system based Thing Speak and Arduino. In: International conference on applied smart systems. In: ICASS. Médéa

34. Zhang L, Dabipi I, Brown W (2018) Internet of things applications for agriculture. Wiley
35. Ghayvat H, Mukhopadhyay S, Gui X, Suryadevara N (2015) WSN-and IoT-based smart homes and their extension to smart buildings. Sensors 15:10350–10379
36. Patil K, Kale N (2016) A model for smart agriculture using IoT. In: International conference on global trends in signal processing, information computing and communication. IEEE
37 Ashwini BV (2018) A study on smart irrigation system using IoT for surveillance of crop-field. Int J Eng Technol 7:370–373
38. Ananthi N, Divya J, Divya M, Janani V (2017) IoT based smart soil monitoring system for agricultural production. IEEE
39. González-Teruel J, Torres-Sánchez R, Blaya-Ros P, Toledo-Moreo A, Jiménez-Buendía M, Soto-Valles F (2019) Design and calibration of a low-cost SDI-12 soil moisture sensor. Sensors 491. https://doi.org/10.3390/s19030491:19
40. Cambra C, Sendra S, Lloret J, Lacuesta R (2018). Smart system for bicarbonate control in irrigation for hydroponic precision farming. Sensors 18.
41. Kumar R, Dharwadkar N (2018) IoT based low-cost weather station and monitoring system for precision agriculture in India. IEEE
42. Bhakta I, Phadikar S, Majumder K (2019) State of the art technologies in precision agriculture: a systematic review. J Sci Food Agric
43. Pflaum A, Gölzer P (2018) The IoT and digital transformation: toward the data-driven enterprise. IEEE Comput Soc 18(1536–1268):5
44. Gupta B, Quamara M (2018) An overview of internet of things (IoT): architectural aspects, challenges, and protocols. Wiley
45. Balafoutis A, Beck B, Fountas S, Vangeyte J, Van der Wal T, Soto I, Gómez-Barbero M, Barnes A, Eory V Precision agriculture technologies positively contributing to GHG emissions mitigation, farm productivity and economics. Sustainability
46. Phupattanasilp P, Tong S (2019) Augmented reality in the integrative internet of things (AR-IoT): application for precision farming. Sustainability 2658 https://doi.org/10.3390/su11092658:11
47. Ahmed N, De D, Hussain I (2018). Internet of things (IoT) for smart precision agriculture and farming in rural areas. IEEE Internet Things J 5. https://doi.org/10.1109/JIOT.2018.2879579
48. Naha R, Garg S, Georgakopoulos D, Jayaraman P, Gao L, Xiang Y, Ranjan R (2016) Fog computing: survey of trends, architectures, requirements, and research directions. IEEE Access 4(2169–3536)
49. Sarker V, Queralta J, Gia T, Tenhunen H, Westerlund T (2019) A survey on LoRa for IoT: integrating edge computing. In: Fourth international conference on fog and mobile edge computing
50. Raza U, Kulkarni P, Sooriyabandara M (2016) Low power wide area networks: an overview. IEEE
51. Ismail D, Rahman M, Saifullah A (2019) Low-power wide-area networks: opportunities, challenges, and directions. IEEE
52. Wixted A, Kinnaird P, Larijani H, Tait A, Ahmadinia A, Strachan N (2016) Evaluation of LoRa and LoRaWAN for wiereless sensor network. IEEE 16. ISBN: 978-1-4799-8287-5
53. Shilpa A, Muneeswaran V, Rathinam D (2019) A precise and autonomous irrigation system for agriculture: IoT based self-propelled center pivot irrigation system. In: 5th international conference on advanced computing and communication systems
54. Elijah O, Rahman A, Orikumhi I, Leow C (2018) An overview of internet of things (IoT) and data analytics in agriculture: benefits and challenges. IEEE Internet Things J 5. ISSN: 2327-4662
55. Premkumar A, Monishaa P, Thenmozhi K, Amirtharajan R, Praveenkumar P (2018) IoT assisted automatic irrigation system using IoT assisted automatic irrigation system using wireless sensor nodes. In: International conference on computer communication and informatics. IEEE

56. Olatinwo S, Joubert T (2019) Enabling communication networks for water quality monitoring applications: a survey. IEEE 7(100332)
57. Dholu M, Ghodinde K (2018) Internet of things (IoT) for precision agriculture application. In: International conference on trends in electronics and informatics. IEEE
58. Naik N, Shete V, Danve S (2016) Precision agriculture robot for seeding function. IEEE
59. Chang C, Srirama S, Buyya R (2019) Internet of things (IoT) and new computing paradigms. Wiley
60. Shin S, Chuang C, Huang H (2016) A security framework for MQTT. In: IEEE conference on communications and network security
61. García S, Larios D, Barbancho J, Personal E, Mora-Merchán J, León C (2019) Heterogeneous LoRa-based wireless multimedia sensor network multiprocessor platform for environmental monitoring. Sensors 19. https://doi.org/10.3390/s19163446:3446
62. Lin H, An P, Kim T (2018) A study of the Z-wave protocol: implementing your own smart home gateway. IEEE 18. (978-1-5386-6350-9)
63. Leikanger T, Schuss C, Häkkinen J (2017) Near field communication as sensor to cloud service interface. IEEE 17. (978-1-5090-1012-7)
64. Liu Y, Qian K (2016) A novel tree-based routing protocol in ZigBee wireless networks. IEEE. 2016;16(978–1- 5090–1781–2).
65. Martínez R, Pastor J, Álvarez B, Iborra A (2016) A testbed to evaluate the FIWARE-based IoT platform in the domain of precision agriculture. Sensors 1979. https://doi.org/10.3390/s16111979:16.
66. Carnevale L, Galletta A, Fazio M, Celesti A, Villari M (2018) Designing a FIWARE cloud solution for making your travel smoother: the FLIWARE experience. In: IEEE 4th international conference on collaboration and internet computing
67. Yu S, Park K, Park Y (2019) A secure lightweight three-factor authentication scheme for IoT in cloud computing environment. Sensors 19. https://doi.org/10.3390/s19163598:3598
68. Shirazi S, Gouglidis A, Farshad A, Hutchison D (2017) The extended cloud: review and analysis of mobile edge computing and fog from a security and resilience perspective. IEEE. 35(0733–8716):11
69. Sharma P, Saluja M, Saluja KK (2012) A review of selective forwarding attacks in wireless sensor networks. Int J Adv Smart Sens Netw Syst 2:37
70. Sarangi S, Naik V, Choudhury S, Jain P, Kosgi V, Sharma R, Bhatt P, Srinivasu P (2019) An affordable IoT edge platform for digital farming in developing regions. IEEE
71. Satyanarayanan M (2017) The emergence of edge computing. IEEE Comput Soc 17. (0018-9162)
72. Douceur JR (2002) The sybil attack. In: Springer international workshop on peer-to-peer systems, Cambridge, 7–8 March 2002, 251–260
73. Math A, Ali L, Pruthviraj U (2018). Development of smart drip irrigation system using IoT. IEEE 18. (978-1-5386-5323-4)
74. Pandithurai O, Aishwarya S, Aparna B, Kavitha K (2017) Agro-tech: a digital model for monitoring soil and crops using internet of things (IOT). IEEE 17. (978-1-5090-4855-7)
75. Aagaard A, Presser M, Andersen T (2019) Applying Iot as a leverage for business model innovation and digital transformation. IEEE 19. (978-1-7281-2171-0)
76. Chandra N, Khatri S, Som S (2019) Business models leveraging iot and cognitive computing. IEEE 19. (978-1-5386-9346-9)
77. Whitmore A, Agarwal A, Da Xu L (2014) The internet of things—a survey of topics and trends. Springer
78. Pandya S, Sur A, Kotecha K (2020) Smart epidemic Tunnel- IoT based sensor-fusion assistive technology for COVID19 disinfection. Emerald
79. Patel NR, Kumar S (2017) Enhanced clear channel assessment for slotted CSMA/CA in IEEE 802.15.4. Wireless Pers Commun 95:4063–4081
80. Patel NR, Kumar S (2018) Wireless sensor networks' challenges and future prospects. In: 2018 International conference on system modeling and advancement in research trends (SMART). Moradabad, India, pp 60–65

81. Ghayvat H, Awais M, Pandya S, Ren H, Akbarzadeh S, Mukhopadhyay SC, Chen C, Gope P, Chouhan A, Chen W (2019) Smart aging system: uncovering the hidden wellness parameter for well-being monitoring and anomaly detection. Sensors 19:766
82. Saket S, Pandya S (2016) An overview of partitioning algorithms in clustering techniques
83. Shah JM, Kotecha K, Pandya S, Choksi DB, Joshi N (2017) Load balancing in cloud computing: methodological survey on different types of algorithm. In: 2017 International conference on trends in electronics and informatics (ICEI). https://doi.org/10.1109/ICOEI.2017.8300865
84. Ghayvat H, Pandya S, Shah S, Mukhopadhyay SC, Yap MH, Wandra KH Advanced AODV approach for efficient detection and mitigation of wormhole attack in MANET. In: 2016 10th international conference on sensing technology (ICST)
85. Pandya S, Shah J, Joshi N, Ghayvat H, Mukhopadhyay SC, Yap MH (2016) A novel hybrid based recommendation system based on clustering and association mining. In: 2016 10th international conference on sensing technology (ICST)
86. Patel S, Singh N, Pandya S IoT based smart hospital for secure healthcare system 2017/5. Int J Recent Innov Trends Comput Commun
87. Pandya SP, Prajapati MR, Thakar KP Assessment of training needs of farm women, Guj J Ext Edu 25(2):169–171
88. Pandya SB, Patel UH, Chaudhary KP, Socha BN, Patel NJ, Bhatt BS DNA interaction, cytotoxicity and molecular structure of cobalt complex of 4-amino-N-(6-chloropyridazin-3-yl) benzene sulfonamide in the presence of secondary ligand pyridine, Appl Organomet Chem 33(12):e5235
89. Pandya S, Ghayvat H, Kotecha K, Awais M, Akbarzadeh S, Gope P Smart home anti-theft system: a novel approach for near real-time monitoring and smart home security for wellness protocol. Appl Syst Innov 1(4):42
90. Patel RR, Pandya SP, Patel PK Characterization of farming system in northwest agro climatic zone of Gujarat state. Guj J Ext Edu 27(2):206–208
91. Pandya S, Ghayvat H, Kotecha K, Yep MH, Gope P (2018) Smart home anti-theft system: a novel approach for near real-time monitoring, smart home security and large video data handling for wellness protocol
92. Joshi N, Kotecha K, Choksi DB, Pandya S (2018) Implementation of novel load balancing technique in cloud computing environment … on computer communication and informatics (ICCCI)
93. Patel W, Pandya S, Mistry V (2016) i-MsRTRM: developing an IoT based intelligent medicare system for real-time remote health monitoring 2016 8th international conference on computational
94. Wandra KH, Pandya S (2012) A survey on various issues in wireless sensor networks. Int J Sci Eng
95. Saket JS, Pandya S Implementation of extended k-medoids algorithms to increase efficiency and scalability using large dataset. Int J Comput Appl
96. Bhola YO, Socha BN, Pandya SB, Dubey RP, Patel MK (2019) Molecular structure, DFT studies, Hirshfeld surface analysis, energy frameworks, and molecular docking studies of novel (E)-1-(4-chlorophenyl)-5-methyl-N′-((3-methyl-5-phenoxy-1-phenyl-1H-pyrazol-4-yl) methylene)-1H-1, 2, 3-triazole-4-carbohydrazide. Mol Cryst Liquid Cryst
97. Patel WD, Pandya S, Koyuncu B, Ramani B, Bhaskar S (2019) NXTGeUH: LoRaWAN based NEXT generation ubiquitous healthcare system for vital signs monitoring & falls detection. 2018 IEEE Punecon
98. Dandvate HS, Pandya S (2016) New approach for frequent item set generation based on Mirabit hashing algorithm. In: 2016 international conference on inventive
99. Swarndeep SJ, Pandya S (2016) Implementation of extended k-medoids algorithm to increase efficiency and scalability using large datasets. Int J Comput Appl
100. Wandra K, Pandya S (2014) Centralized timestamp based approach for wireless sensor networks. Int J Comput Appl

101. Garg D, Goel P, Pandya S, Ganatra A, Kotecha K (2002) A deep learning approach for face detection using YOLO. In: 2018 IEEE Punecon
102. Sur A, Pandya S, Sah RP, Kotecha K, Narkhede S (2020) Influence of bed temperature on performance of silica gel/methanol adsorption refrigeration system at adsorption equilibrium. Particulate Sci Technol
103. Sur A, Sah RP, Pandya S (2020) Milk storage system for remote areas using solar thermal energy and adsorption cooling. Mater Today Proc
104. Cohen JM, Pandya S, Tangirala K, Krasenbaum LJ (2020) Treatment patterns and characteristics of patients prescribed AJOVY, emgality, or Aimovig, HEADACHE
105. Cohen JM, Pandya S, Krasenbaum LJ, Thompson SF (2020) A real-world perspective of patients with episodic migraine or chronic migraine prescribed AJOVY in the United States, HEADACHE
106. Barot V, Kapadia V, Pandya S (2020) QoS enabled IoT based low cost air quality monitoring system with power consumption optimization. Cybernet Inf Technol
107. Ghayvat H, Pandya S, Patel A (2019) Proposal and preliminary fall-related activities recognition in indoor environment. In: 2019 IEEE 19th international conference on
108. Akbarzadeh S, Ren H, Pandya S, Chouhan A, Awais M (2019) Smart aging system
109. Ghayvat H, Pandya S (2018) Wellness sensor network for modeling activity of daily livings–proposal and off-line preliminary analysis. In: 2018 4th International conference on computing
110. Awais M, Kotecha K, Akbarzadeh S, Pandya S (2018) Smart home anti-theft system
111. Patel M, Pandya S, Patel S (2017) Hand gesture based home control device using IoT. Int J Adv Res Eng
112. Pandya S, Yadav AK, Dalsaniya N, Mandir V Conceptual study of agile software development
113. Samani MD, Karamta M, Bhatia J, Potdar MB (2016) Intrusion detection system for DoS attack in cloud. Int J Appl Inf Syst
114. Review on various security threats & solutions and network coding based security approach for VANET
115. Bhatia J, Shah B (2013) Int J Adv Eng
116. Review on variants of reliable and security aware peer to peer content distribution using network coding
117. Patel P, Bhatia J (2012) 2012 Nirma university international conference on
118. Bhatia J, Kakadia P, Bhavsar M, Tanwar S (2019) SDN-enabled network coding based secure data dissemination in VANET environment. IEEE Internet Things J
119. Pooja M, Singh Y (2013) Security issues and Sybil attack in wireless sensor networks. Int J P2P Netw Trends Technol 3:7–13
120. Mat I, Kassim M, Harun A (2015) Precision agriculture applications using wireless moisture sensor network. In: IEEE 12th Malaysia international conference on communications. Kuching
121. Fountas S, Aggelopoulou K, Gemtos T (2016) Precision agriculture: crop management for improved productivity and reduced environmental impact or improved sustainability "supply chain management for sustainable food networks. Supply Chain Manage Sustain Food Netw
122. Miles C (2019) The combine will tell the truth: on precision agriculture and algorithmic rationality. Big Data Soc 1–12
123. Chen W, Lin Y, Lin Y, Chen R, Liao J (2018) AgriTalk: IoT for precision soil farming of turmeric cultivation. IEEE

Experiential Learning Through Web-Based Application for Peer Review of Project: A Case Study Based on Interdisciplinary Teams

Pallavi Asthana, Sudeep Tanwar, Anil Kumar, Ankit Yadav, and Sumita Mishra

Abstract Engineering is collaborative by nature, and this study was conducted to further embed these traits through the experiential learning-based inter-disciplinary project activity spanned over three weeks. Students of three different branches of engineering voluntarily worked on common projects to develop understanding on common topics. They also reviewed the projects of their peer through a Web-based peer review process (AmiPREP) specifically developed to bring technical uniformity in selected projects. After the completion of the projects, participants were interviewed and their responses implicated that the activity proved to be meaningful, interesting and effective to work in inter-disciplinary teams and understand the significance of mathematical modeling through MATLAB/Simulink.

Keywords Inter-disciplinary communication · Experiential learning · Peer review system · Simulink · Firebase app

1 Introduction

Engineering has evolved as a collaborative entity notwithstanding the barriers of specialization. Majority of engineering industry employers solicit graduate engineers based on their critical thinking ability and problem-solving skills. Teamwork

P. Asthana (✉) · A. Kumar · A. Yadav · S. Mishra
Amity University, Noida, Uttar Pradesh, India
e-mail: pasthana@lko.amity.edu

A. Kumar
e-mail: akumar3@lko.amity.edu

S. Mishra
e-mail: smishra3@lko.amity.edu

S. Tanwar (✉)
Nirma University, Ahmedabad, Gujarat, India
e-mail: sudeep.tanwar@nirmauni.ac.in

P. K. Singh et al. (eds.), *Proceedings of Second International Conference on Computing, Communications, and Cyber-Security*, Lecture Notes in Networks and Systems 203,
https://doi.org/10.1007/978-981-16-0733-2_25

and flexibility in technical approach for problem solving and project implementation are the two key words that define the engineering in present era of growth. More than ever, engineers now need to work and learn together [1]. They must have the shared goals of designing and developing as this is the way most of the industries are working now [2]. It is expected that even after choosing a specific engineering discipline, engineering graduate is not 'pigeonholed' into a specific job role. In recent decades, for inclusive education, most engineering course curriculum have been designed consisting of computer simulation and programming languages along with the core subjects to experience the integration of IT in all fields of engineering [3]. But exhaustive academic gaps still exist in different core branches like mechanical engineering, civil engineering, aerospace engineering, electronics engineering, etc., as students are unaware of the importance of interdependence between their courses. All programs have independent program structures with few common subjects like computer programming, basic simulation, engineering mathematics, etc. To initiate the communication related to working on common platform, they must be brought together to work in same team to design and develop applications in cohesive ways [4–6]. This also requires the faculties of different departments of engineering to work in collaborative ways to create better pedagogical strategy for promoting interdisciplinarity [7].

In present times, industries need engineering workforce that can communicate effectively with the people of different engineering background to utilize wider application of theories and principles in innovative ways [8, 9]. Students realize the importance of interdisciplinary education for the first time, when they join jobs or appear for master's degree programs, that require specialized technologies, designed for research and development-related activities. It is imperative to adapt the engineering education with the existing trends of technologies and perspectives and develop the sense of interdisciplinary studies in engineering students.

2 Related Work

Interdisciplinarily in engineering has garnered a lot of attention [10, 11]. Few of the studies have proven the effectiveness by its implementation, within the curriculum of engineering education [12]. It was found that much of the work examines the effect of interdisciplinary study in students. But, more important is to create methods for developing interdisciplinary competencies in students [13, 14]. A thorough need of enhancing these abilities in students can be fulfilled through simple modules where students can explore this dimension of engineering within the realms of the existing course curriculum and assessment process [15]. This work aims to see the effect of an interdisciplinary teamwork on students and check whether such exercise can be useful in developing inherent interdisciplinary nature in them [1, 2, 16]. Because it is still a task to change the curriculum with the pace required by the real-time engineering.

This study was conducted at Amity University, India, with the undergraduate students of mechanical engineering, civil engineering and aerospace engineering. Students of these branches were given the choice of projects based on the topics that they study in common. They also studied mathematical modeling of selected project on MATLAB/Simulink. This was conducted as an experiential learning-based activity for participation of students in interdisciplinary study of projects.

Motivation behind the work:

It is observed that program structure does not provide the scope of interdisciplinary correspondence between the students of different branches in the earlier engineering semesters. This was the motivation behind the development of this Web-based application.

Objective of the Work

This work was properly structured to accommodate two objectives:

1. Experiential learning for students through interdisciplinary study of project for understanding the relationship between different subjects and their implementation through MATLAB/Simulink.
2. To bring uniformity in the selection of projects taken up by students through an automated Web-based peer review system 'AmiPREP' specifically developed for the improvement of projects through peer review process.

3 Experiential Learning Through AmiPREP

In this section, role of peer assessment in the experiential learning has been discussed and also discovered that working in the teams that consist of engineering students of different background add into experiential learning.

(a) **Finalization of Project Title**:

Experiential learning means learning by doing or hands-on learning. Based on four stages of Kolbe's learning cycle [17]:

Concrete experience: Learner must be willing and be actively involved in the experience.

Reflective observation: Learner must be able to reflect on the experience.

Abstract conceptualization: Learner must possess and use analytical skills to conceptualize the experience.

Active experimentation: Learner must possess decision-making and problem-solving skills to apply the new ideas gained from the experience.

Experiential learning is based on reflecting the experiences that has been conceptualized through concrete experience and abstract experimentation.

Teams conceptualized the title of their project study representing the role of each individual member of team. They had to make a detailed study of the existing simulations that they could find from Internet sources, as it is a tough task for the students of the second year to run their own simulations [18–20]. Initially, students finalized the topic based on their current knowledge on courses. In order to refine the topic, they reviewed the projects of other teams through an automated Web-based peer review system [21]. This system was developed by Department of Electronics and Communication and Computer Science and Engineering. It has been observed that most of the students selected the projects based on their limited knowledge, and they did not discuss their project title with other teams. When they reviewed the projects of other teams and received reviews on their own projects, they were encouraged to change the titles of the activity [22]. This enhanced the learning activity and students' reflection as it encouraged the students to select the project at par with other teams [23]. To explain it in better way, two teams selected the advanced topics like finite element analysis on bending of a bracket and analysis of rankin air-cycle, whereas two groups selected the basic topics like mathematical modeling of Bernoulli's equation through MATLAB and modeling of basic equations for mechanical load analysis. They were not interrupted by the concerned faculty members during the selection of the projects; instead, they were asked to review the projects of the peer and upload their own projects on the AMIPREP, a Web-based app to be reviewed by other students. Overview of the process conducted during the study can be explained through a flowchart (Fig. 1).

Tenure of this study was restrained to three weeks only; hence, title of the study was not reviewed after the change, but, if this system is used for major project, then, it could be reviewed again.

(b) **Peer Review of the Project as an Experiential Learning**:

In many experiential learning-based activities, peer review system is implemented to enhance the performance of individuals in teams [24–26]. In this work, students have reviewed the topics of peer teams. Teamwork provides a huge potential for experiential learning. A Web-based peer review system is developed that could present the opportunity for 'Reflective observation.'

This process involved the following steps:

(i) All students uploaded the synopsis of their project.
(ii) They were asked to review the synopsis of projects uploaded by other teams.
(iii) Students could view the feedback received on their project only after they would review at least two projects.

Peer review of projects provided them two opportunities:

(a) By reviewing other projects, they could compare their topics with peers and could infer the opportunity of knowledge creation from their selected topics. Through the feedback, they received on their topics, and they had the idea to change the topics or to continue with the selected topic.
(b) This Web-based application proved to be highly useful as it provided a unique platform to enhance the understanding of students in selecting the project title.

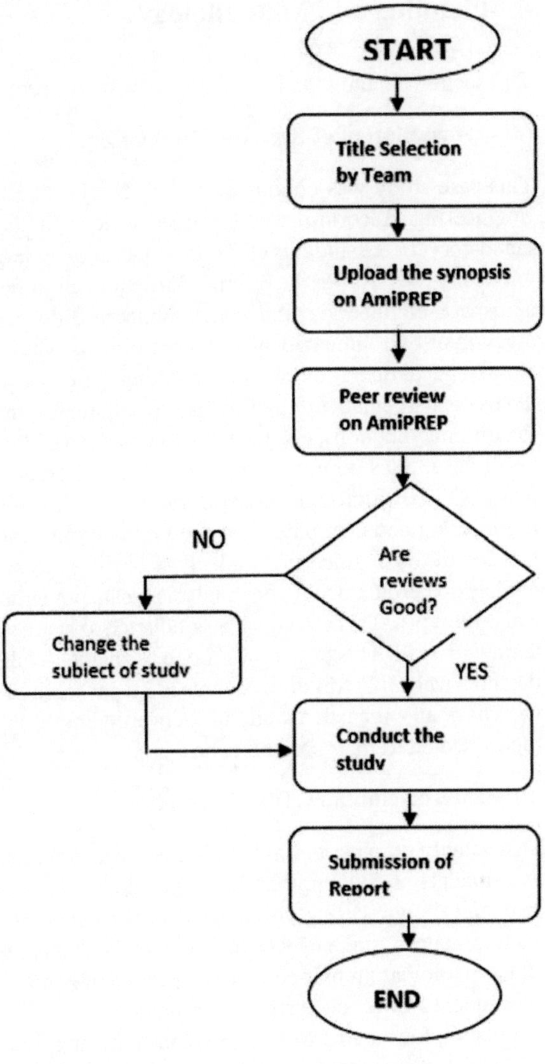

Fig. 1 Flowchart of the complete process adopted for study through AmiPREP

This is an automated process and, hence, did not require the intervention of teaching faculty members [27]. Peer review actuated the students to search for the topics that would be more helpful in knowledge creation and its impact was realized when other two groups who selected the basic topics changed the titles and selected topics named as 'Analysis of load of Bridge' and 'Calculation of Mechanical load for wind turbine.'

4 Planning and Methodology

This section contains the formulation of research problem and formation of the teams.

(a) **Formulation of Research Problem**:

This case study was conducted by the faculty of Electronics and Communication Engineering. According to the course curriculum, the basic simulation laboratory is mandatory for the students of fourth semester of undergraduate engineering courses at university [28]. As per the program structure, all students of mechanical engineering, aerospace engineering and civil engineering do the basic programs that include creation of two-dimensional and three-dimensional arrays and applying logical and mathematical operation on them, creating various types of 2D and 3D plots, plotting the basic mathematical function like trigonometric functions, exponential functions, logarithmic functions, etc. [29]. It was observed by the course instructor that students of all disciplines do not show much interest in this subject, due to mistaken belief that MATLAB/Simulink is useful, mainly for the problem related to the Electronics Engineering and Computer Science Engineering. Knowledge of Simulink is important for all the engineering students as it allows to create appropriate mathematical models of a problem and after analysis create potential solutions without using physical prototypes. This activity was conducted to bring awareness in students and instill teamwork skill. This project was done as an open-ended practical work spanned over the three weeks. Credit of the course is 2 h per week [30]. It was decided that students will make an exhaustive study of their problem through MATLAB/Simulink on the topics related to their core branches.

(b) **Interdisciplinary Team formation**:

This activity of experiential learning is not a compulsory part of assessment; hence, five students of civil engineering, eight students of aerospace and seven students of mechanical engineering volunteered to participate in the project work. During the team formation, students were asked to make the teams that must have students from all the participating branches [31]. Each team had a mix of students having at least one student from each participating branch.

Overall, four teams were formed, each having five students:

Team 1: two students of aerospace engineering, two students of civil engineering and one student of mechanical engineering.

Team 2: three students of aerospace engineering, one student of civil engineering and one student of mechanical engineering.

Team 3: two students of aerospace engineering, one student of civil engineering and two students of mechanical engineering.

Team 4: one student of aerospace engineering, one student of civil engineering and three students of Mechanical Engineering.

5 Development of Web-Based App AMIPREP

This section discusses the design and development of the Web-based peer review system.

AMIPREP stands for Amity Peer Review for Enhancement of Project. Specification of the software used in the designing of AmiPREP.

Software versions: Node Runtime: Version 10, Angular CLI Version: 7.3.3 Angular Version: 7.2.6, Typescript: 3.2.4.

(A) *Angular—Front End of Application*

Front end of the application is built using Angular, a front end mobile and Web app development framework by Google LLC. Google LLC has made Angular an open-source framework which means it is also maintained by developer community around the world, and its code is available for free online.

The app is divided into several reusable components and services for code reusability and better performance. Different components in Angular are rendered in browser's data object model (DOM) using a virtual DOM container. The app loads only once when the main index page sets up itself in the main DOM of browser, and then, this DOM acts as a virtual DOM for rendering all other app components; hence, applications developed using Angular front end can develop as multiple-page application (MPA) or single-page application (SPA). In this project, we have worked on SPAs [32].

Services in Angular are singleton classes which means only one instance of a service class is created. The instance is created when the application loads for the very first time in browser. Services act as a data provider for the application. Each component that utilizes a service needs to inject that service into its constructor through dependency injection.

The authentication system integrated with client application internally uses Firebase Authentication. The client has enabled only email password and Google OAuth2.0 login methods. The user can also register by providing a valid email address and password [33].

(B) *Firebase—Backend for Application*

Backend of this application is developed in Firebase which is a cloud-based service by Google LLC. Firebase is basically a part of Google Cloud Platform [34]. It involves the following steps:

(i) **Authentication**: Firebase Authentication provides a lot of ways to authenticate users, for example, mobile number and OTP login and sign up, email and password, OAuth2.0 login for several providers like Facebook, Apple, etc. AmiPREP uses only email–password login and login with Google account authentication options. In future, other login methods would be implemented as per the user feedbacks and requirements.

(ii) **Database**: The database used is Firebase Firestore Database which is a high-performance, lightweight and lightning fast NoSQL database by Google [32]. Data in Firestore is structure in collections, and each collection in Firestore stores multiple numbers of documents, and each document stores data in key–value pairs like JavaScript Object Notation (JOSN). Firestore provides real-time change detection, and listeners can be implemented on client side to listen for changes in database and update Web app state in real time.

(iii) **Storage**: Images uploaded by users while posting a project or idea on platform are stored in Firebase Storage. Storage allows to store up to 5 GB of data in free tier. The download URL of image linked to a project or post is fetched while writing the project data in Firestore and is added as metadata to it [35–37].

(iv) **Hosting**: AmiPREP is hosted in Firebase using Firebase hosting service. The Web application is served over a secured HTTPS connection and thus provides additional layer of security for its users.

(C) *Application Use Cases*

Figure 2 is the snapshot of the home page of the application having the buttons like My uploads, Upload and quick link to access the new upload for review. A user can login into the application by Google sign in or email password They can also register as a new user on the platform with an active valid email ID. They can view recent active projects.

When student clicks on My Uploads, the page as shown in Fig. 3 will open, and student would enter the basic information about project like title, description, algorithm, etc., and in the output, they can upload a synopsis of the project. They can upload new project ideas and images related to project for the detailed explanation of idea (Fig. 4).

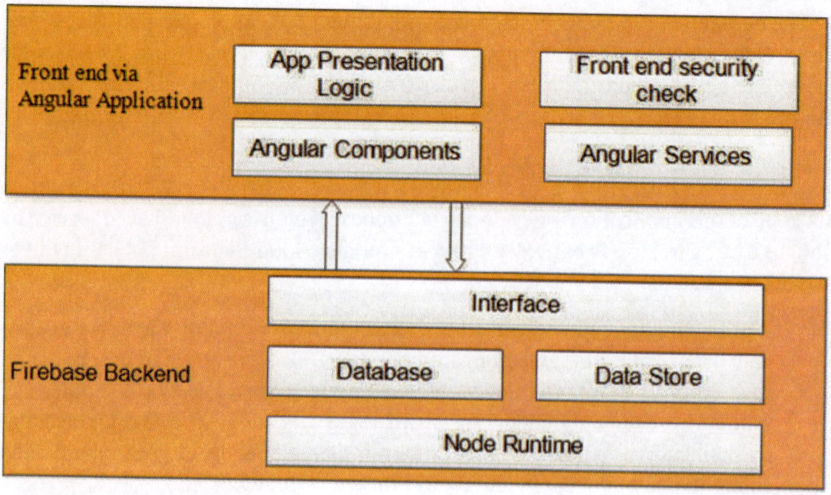

Fig. 2 System architecture diagram

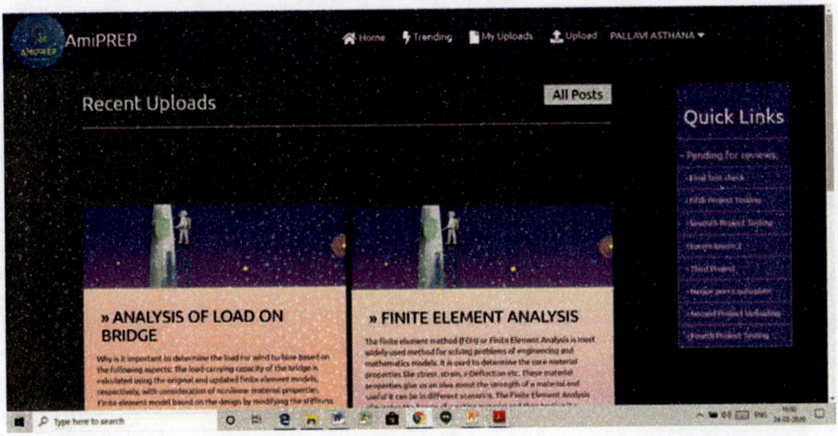

Fig. 3 Appearance of the home page of the application

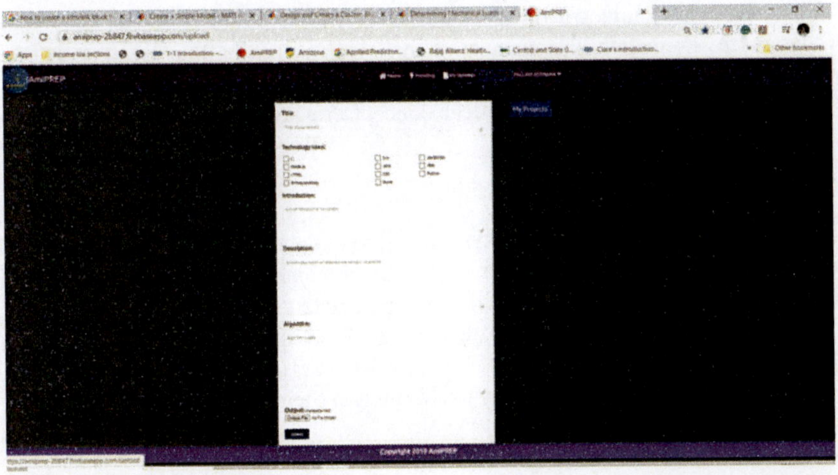

Fig. 4 Description about project and uploading of synopsis

To review the projects of other teams, students will click on the quick links, and in the comment section, a 'Feedback' button will appear. On clicking this button, feedback section, a page will open. On this page, students have to review the project based on simple questions like whether they appreciate the idea of project, is idea innovative, have they earlier heard of this title and how they rate the project idea on the scale of 1–5. They could also provide the suggestion in the comment box. This is anonymous feedback on a project using the feedback form provided in project details section (Fig. 5).

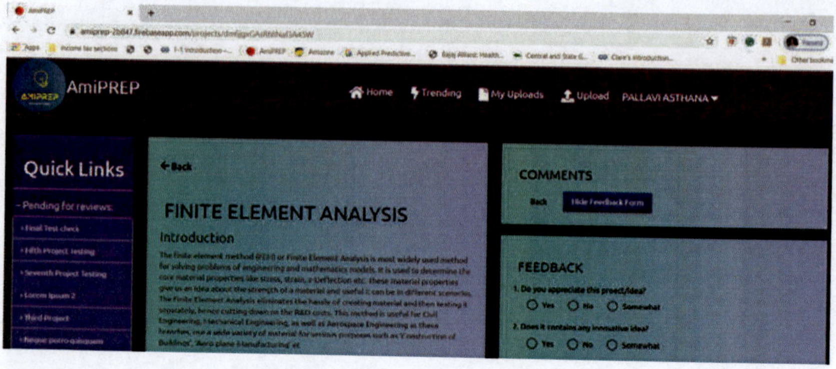

Fig. 5 Feedback for the review of project

6 Project Management

This study was conducted on the students of fourth semester of undergraduate program to strengthen the foundation of MATLAB/Simulink in students and to collaborate the students of varied branches over common topics. Each team was required to study the theoretical concepts related to their topics and their implementation in a Simulink.

(a) **Creating a Model through Simulink**:

Simulink can be used to model a system and to simulate its dynamic behavior. To create a model, few things are required as: block name, block purpose and model purpose. Each block requires its library and parameters.

Each student had to conduct a detailed study of the existing models that are available on the official Web site of MathWorks which is an open access Web site. After studying the simulations, they could run it in their laboratory.

(b) **Implementation of the Study**:

This task was extended over a span of three weeks. Details of the project requirement and students' performance are elaborated below:

Week 1*: Project requirement*: To make the synopsis of work during this week and upload it on AmiPREP for the peer review process.

Students' Performance: All teams uploaded their synopsis on AmiPREP and completed the peer review process.

Week 2: *Project requirement*: Students could analyze the significance of their project in terms of knowledge creation through the feedback process. All team members were expected to explain the relevance of the project with their branch of engineering.

Students' performance: Few students were inclined toward timely finishing of their tasks, while some students submitted the tasks after the reminder of course instructor.

Week 3: *Project requirement*: Students had to make a detailed study of the simulation models as available on the official Web site of MathWorks named as mathswork.com which is an open access Web site. They had to study the modeling of their respective topics or topics related to their work, namely load analysis of bridge, finite element analysis of bending of bracket, analysis of air cycle and load analysis of wind turbine

Students' Performance: All the students of Team 3 submitted the detailed study of their project explaining the mathematical modeling of their project on MATLAB/Simulink. Few students of team 2 and Team 4 submitted the study, but some students found it difficult to understand. They could not explain it well, henceforth, did not submit the document.

Performance of Team-1 was very poor as only one student from the team submitted the final report (Table 1).

(c) Outcome of the work:

This work brought some interesting outcomes.

1. It initiated the interdisciplinary communication among students of three different branches based on the Simulink-based mathematical modeling of their topics.
2. All the student participants volunteered to participate in this activity. At the end of week 2, it was observed that few students were enthusiastic for completing the

Table 1 Branch-wise segregation of the project topics as provided by students

Teams	Aerospace Engg	Mechanical Engg	Civil Engg
Determining mechanical load for wind turbine	Modeling the aerodynamic forces on the wind turbine, including the effects of induced velocity	To accurately predict maximum deflections, and oscillation	Calculation of variable load
Finite element analysis on bending of bracket	Solving a linear elasticity problem is to create a structural analysis model	Damping parameters, body loads, boundary loads, boundary constraints	Stress analysis and its effect of different metals
Analysis of load on bridge	Wind load. Buoyancy effect	Dead load, live load, impact load	Deformation and horizontal effects, erection stresses. Seismic loads
Analysis of rankin air-cycle	To design and understand space propulsion systems, these would have components such as heat exchanges in turbines	As on one of the important aspect of Thermodynamics	For life cycle energy analysis of buildings

task. Few completed it under peer pressure and pressure of the faculty members. Some students do not complete the task.

3. It established the significance of the peer review system for project selection. This Web-based tool can be further used to bring technical uniformity in the major projects in the last semester of engineering program.
4. One important outcome of this work was observed that even in this era of advance and integrated technologies, working in an interdisciplinary atmosphere is still considered an understatement in engineering students, and an awareness in this zone is explicitly required for growth as a competent engineer.

7 Result and Discussion

(A) ***Successful implementation of AmiPREP: Web-based peer review Application***

Topics of the project study were selected by the teams. Through Web-based application, for peer review among the teams, students got the opportunity to review the synopsis of the other teams and received the reviews on their topics as well. This motivated them to change the topic and switch to some advanced topics. Indeed, two out of four teams changed their topics after the review process. This application helped students to select the topics as par with other teams.

Web-based autonomous system created anonymous feedback, and this could be trusted as a powerful tool for the students in all types of project selection in the experiential learning-based models in engineering.

(B) ***Report based on Students' feedback***

This exercise was a voluntary exercise with the consent of students. Therefore, evaluation of the students was not based on the performance in the activity. But, it is also a fact that after this activity, participants generated a sensitivity toward the course of basic simulation laboratory and were able to perform the practical assigned in syllabus with ease. They were also able to help other students having less knowledge about the MATLAB/Simulink.

At the end of week three, an interview was conducted with each team for the feedback of complete process. All the team members were interviewed and were inquired about these basic questions:

Q1—Was this activity meaningful?

Q2—Was this activity interesting?

Q3—Was project review system was effective?

Q4—Were you comfortable to work in a team consisting students from different branches?

Q5—Were you able to understand the significance of mathematical modeling through Simulink?

Q6—Were you able to understand the blocks and libraries of MATLAB/Simulink? (Fig. 6).

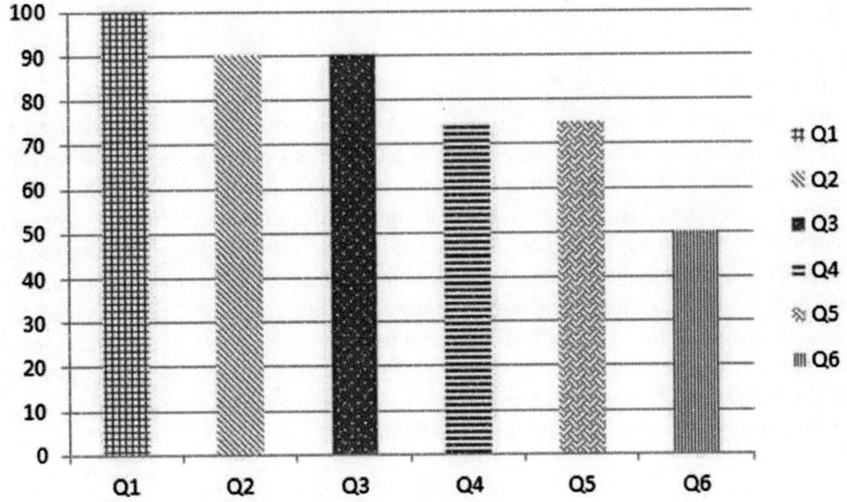

Fig. 6 Response of students based on the feedback interview

Analysis of the responses shows that 100% of the students agreed that this activity was meaningful. It was obvious as all the students volunteered to participate in the activity. Initially, they were briefed about the activity, and later, they were provided with the task of project management. Few students found it cumbersome, and still 90% of the students found it interesting. These 90% agreed to the effectiveness of AmiPREP peer review system. Out of all, 75% students were comfortable in working in an interdisciplinary team.

Most of the students up to 75% understood the significance of mathematical modeling through Simulink and around 50% understood the blocks and libraries of MATLAB Simulink.

(C) ***Discussions based on Students' Feedback***

Overall, these results were found satisfactory when matched with the students understanding of the MATLAB/Simulink. Their performance was also examined and was found at par with their interview responses.

Meanwhile, during the span of this activity, students other than the participants of the course learned the basic programs of MATLAB and to make their functions and utilize the libraries of MATLAB.

This activity proved to be a motivating factor for the knowledge creation by adding one more dimension of interdisciplinary communication through an experiential learning technique.

8 Conclusion

Fundamental prerequisite in today's engineering education is to impart the knowledge, skills and attitudes required to effectively participate in interdisciplinary teams. Teamwork during graduate study provides an excellent opportunity to achieve this objective and enhance the performance of individual team members. This study proved to be useful to inculcate such trait in students through an experiential learning-based model. This paper also established the role of Web-based peer review system to provide better motivation and feedback to the participants. This tool can be further used to enhance the quality of projects through an automated and anonymous peer review system. After the completion of the projects, participants were interviewed and their responses implicated that the activity proved to be meaningful, interesting and effective to work in interdisciplinary teams and understand the significance of mathematical modeling through MATLAB/Simulink.

References

1. Lattuca LR, Knight DB, Ro HK, Novoselich BJ (2017) Supporting the development of engineers' interdisciplinary competence. J Eng Educ 106(1):71–97
2. Borrego M, Newswander LK (2010) Definitions of interdisciplinary research: toward graduate-level interdisciplinary learning outcomes. Rev High Educ 34(1):61–84
3. Rullkoetter P, Whitman R, DeLyser R (2000) Engineering the future: an integrated engineering design experience. In: 30th ASEE/IEEE frontiers in education conference, F1C-12–17, 30th annual volume, Kansas City, 18–21 Oct 2000
4. Desmond A, Jaeger M (2014) Managing the interdisciplinary approach to engineering design. Int J Mech Eng Educ 42(2):174–185
5. Coso AE, Bailey RR, Minzenmayer E (2010) How to approach an interdisciplinary engineering problem: characterizing undergraduate engineering students' perceptions. In: Proceedings—40th frontiers in education conference, Washington, DC, 27–30 Oct 2010
6. Drezek KM, Olson D, Borrego M (2008) Crossing disciplinary borders: a new approach to preparing students for interdisciplinary research. In: 38th proceedings—frontiers in education conference, Saratoga Springs, New York, 22–25 Oct 2008
7. McNair LD, Newswander C, Boden D, Borrego M (2010) Student and faculty interdisciplinary identities in self-managed teams. J Eng Educ 100(2):374–396
8. Coşkun S, Gençay E, Kayıkcı Y (2016) Adapting engineering education to industries: 4.0 vision. In: Proceedings of the 16th production research symposium, pp 258–263, Istanbul, Turkey, 12–14 Oct 2016
9. Bongomin O, Ocen GG, Nganyi EO, Musinguzi A, Omara T (2020) Exponential disruptive technologies and the required skills of industry 4.0. J Eng Educ 2020:1–17
10. National Academy of Engineering (2004) The engineer of 2020: visions of engineering in the new century. National Academies Press, Washington, DC. National Academy of Sciences, Facilitating interdisciplinary research. National Academies Press , Washington, DC
11. The Royal Academy of Engineering (2007) Educating engineers for the 21st century. The Royal Academy of Engineering
12. O'Donnell AM, Derry SJ (1997) Cognitive processes in interdisciplinary groups: problems and possibilities. In: Interdisciplinary collaboration: an emerging cognitive science, Research Monograph No. 5, National Institute for Science Education University of Wisconsin-Madison

13. Taajamaa V, Westerlund T, Liljeberg P, Salakoski T (2013) Interdisciplinary capstone project. In: 41th SEFI annual conference, Leuven, Belgium, 16–20 Sep 2013
14. Renate G (2018) Klaassen,: Interdisciplinary education: a case study. Eur J Eng Educ 43(6):842–859
15. Klein JT, Newell WH (1997) Advancing interdisciplinary studies. In Gaff JG, Ratcliff JL, Associates (eds) Handbook of the undergraduate curriculum: a comprehensive guide to purposes, structures, practices, and change, pp 393–415. Jossey-Bass, San Francisco
16. Newell WH (2008) The intertwined history of interdisciplinary undergraduate education and the association for integrative studies: an insider's view. Issues Integr Stud 26:1–59
17. Kolb AY, Kolb DA (2005) Learning styles and learning spaces: enhancing experiential learning in higher education. Acad Manage Learn Educ 4(2):193–212
18. https://in.mathworks.com/videos/determining-mechanical-loads-for-wind-turbines-81627.html
19. https://in.mathworks.com/discovery/finite-element-analysis.html
20. https://in.mathworks.com/matlabcentral/fileexchange/51815-calculation-of-the-modal-parameters-of-a-suspension-bridge
21. Liu EZ, Lin SS, Chiu CH, Yuan SM (2001) Web-based peer review: the learner as both adapter and reviewer. IEEE Trans Educ 44(3):246–251
22. Topping KE, Ehly SE (2001) Peer-assisted learning. J Educ Psychol Consul 12(2):113–132
23. Purchase HC (2000) Learning about interface design through peer assessment. Assess Eval High Educ 25(4):341–352
24. Macías-Guarasa J et al (2006) A project-based learning approach to design electronic systems curricula. IEEE Trans Educ 49(3):389–397
25. Usher M, Barak M (2018) Peer assessment in a project-based engineering course: comparing between on-campus and online learning environments. Assess Eval High Educ 43(5):745–759
26. Gadola M, Chindamo D (2017) Experiential learning in engineering education: the role of student design competitions and a case study. Int J Mech Eng Educ 47(1):3–22
27. Cornelius S, Gordon C, Harris M (2011) Role engagement and anonymity in synchronous online role play. Int Rev Res Open Dist Learn 12(5):57–73
28. David H (2005) Introduction to matlab for engineering students. Northwestern University
29. https://amizone.net/AdminAmizone/WebForms/naac/DirectProgrammeStructure
30. Schaffer SP, Chen X, Zhu X, Oakes WC (2012) Self-efficacy for cross-disciplinary learning in project-based teams. J Eng Educ 101(1):82–94
31. Kaluža M, Troskot K, Vukelić B (2018) Comparison of front-end frameworks for web applications. J Polytechnic Rijeka 6(1):261–282
32. Dao V(2016) Thesis on 'Development of a front-end application using angular JS: 1UP media company case'. Leppävaara Laurea University of Applied Sciences Leppävaara
33. Chunnu K, Pritam S (2018) Application of firebase in android app development—a study. Int J Comput Appl 179(46):49–53
34. Domes S (2017) Progressive web based apps with react: create lighting fast wed apps with native power using recat and firebase app. Packt Publishing
35. Kim J et al (2018) m'Hybrid mobile-app. on multi-MEC platforms in NFV environment. Int J Eng Technol 7(4):383–386
36. Singh PK, Kar AK, Singh Y, Kolekar MH, Tanwar S (2019) Recent innovations in computing, vol 597. Springer Nature, Switzerland AG. ISBN: 978-3-030-29406-9
37. Singh PK, Panigrahi BK, Suryadevara NK, Sharma SK, Singh AK (eds) (2019) Proceedings of ICETIT 2019, Emerging trends in information technology, Lecture Notes in Electrical Engineering (LNEE), Springer Nature Switzerland AG. https://doi.org/10.1007/978-3-030-30577-2

Machine Learning Applications for Computer-Aided Medical Diagnostics

Parita Oza, Paawan Sharma, and Samir Patel

Abstract Machine learning has made potential developments in biotechnology. Years of medical training are required for correct diagnosis of diseases. Diagnostics is often a very time-consuming process, and it requires strenuous effort. Data generated through varieties of imaging modalities for the diagnoses purpose is very bulky. In the corporate and government hospitals, a high number of patients are visiting per day for the disease diagnosis and treatment. This may cause diagnosis burden on the clinicians and radiologist. For interpretation, overload of image data may produce oversight and observational errors. Machine learning algorithms have recently made huge advancements in automated disease detection and classification. These algorithms can learn to view the patterns in an image similarly the way doctors do by training those using lots of annotated examples. Various machine learning algorithms used for automated diagnosis in medical imaging filed are discussed in the paper. Comparative analysis of these algorithms based on different parameters is also presented. This paper also focused at various applications of machine learning in diagnostic imaging, which can be part of routine clinical work for detection and classification of the process.

Keywords Medical imaging · Machine learning · Computer-aided diagnosis

P. Oza (✉) · P. Sharma · S. Patel
Pandit Deendayal Petroleum University, Gandhinagar, India
e-mail: parita.prajapati@nirmauni.ac.in; parita.ophd19@sot.pdpu.ac.in

P. Sharma
e-mail: Paawan.Sharma@sot.pdpu.ac.in

S. Patel
e-mail: Samir.Patel@sot.pdpu.ac.in

P. Oza
Nirma University, Ahmedabad, India

P. K. Singh et al. (eds.), *Proceedings of Second International Conference on Computing, Communications, and Cyber-Security*, Lecture Notes in Networks and Systems 203,
https://doi.org/10.1007/978-981-16-0733-2_26

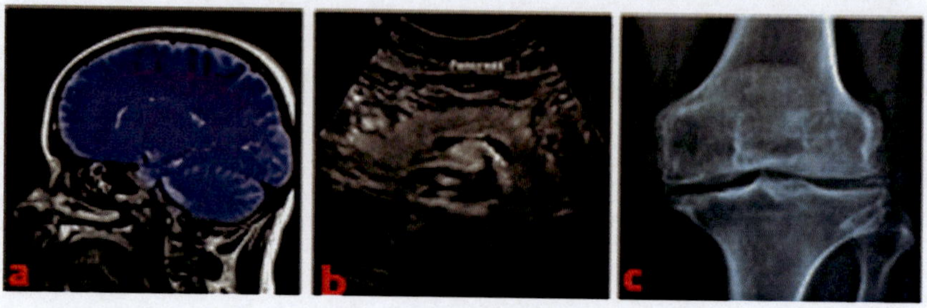

Fig. 1 Medical images by different imaging modalities: **a** MRI, **b** Sonogram, **c** X-ray [2, 3]

1 Introduction to Medical Imaging

Medical imaging is the field of medical science, used to visualize body parts, organs or tissues for clinical diagnosis and continuous disease monitoring. Therefore, it plays an important role to improve public health for all the groups of population. This field includes radiology, optical imaging and nuclear medicine [1]. Images taken through different imaging modalities are presented in Fig. 1.

Radiology: This field is used to find physiological and anatomical details of the human body at very high temporal and spatial resolution. This technique makes use of X-rays and other such agents to produce and process images. Contrast agents are used to enhance medical images. This discipline covers imaging modalities like X-rays, CT scan, ultrasound and MRI [4].

Optical Imaging: This technique makes use of light to interrogate structural and functional information in the living body in real time. This technique is still in the early stages of development [5].

Nuclear Medicine: This imaging uses radioactive substances that are typically injected into the patient's body. It may be inhaled or swallowed, depending on medical condition of the patient. It is used to visualize details of metabolism or molecular function through techniques like PET and SPECT [6].

2 Machine Learning Algorithms

Various machine learning techniques are available and can be applied in medical imaging to classify data, for image analysis, malignancy detection and classification, content-based image retrieval, brain mapping, etc. This section talks about various ML techniques used in medical domain [7, 46].

Supervised learning: This ML technique learns from training dataset with proper label, and based on training, classifier responds to the test data. There are two types: classification and regression. Classification techniques are used for disease prediction and classification in medical imaging. Regression is used to answer the questions like "How much" and "How many".

Unsupervised Learning: In this technique, input provided to the classifier is not labeled. This learning directly infers from the data itself, creates cluster from those data and generates a data-driven decision.

Semi-supervised Learning [7]: This technique is the mixture of both supervised as well as unsupervised techniques. Here, the model is learned from a combination of both annotated and unannotated data. In medical imaging, it is very difficult to obtain properly annotated data. Unlabeled data is usually much easier to obtain in practice.

Deep Learning [8]: Machine learning tends to do feature extraction for the classification process. Images contain many details in it. Hence, extracting number of features for so and so applications may be challenging. The medical field is one of the most noticeable fields where deep learning algorithms can play an important role, especially when it comes to medical imaging. Deep neural networks are designed to learn features from the medical images on their own, and these features can further be used for the process of classification. There is no need of manual feature extraction like in machine learning algorithms. Figure 2 shows types of machine learning and various algorithms used under each category in medical imaging for varieties of applications.

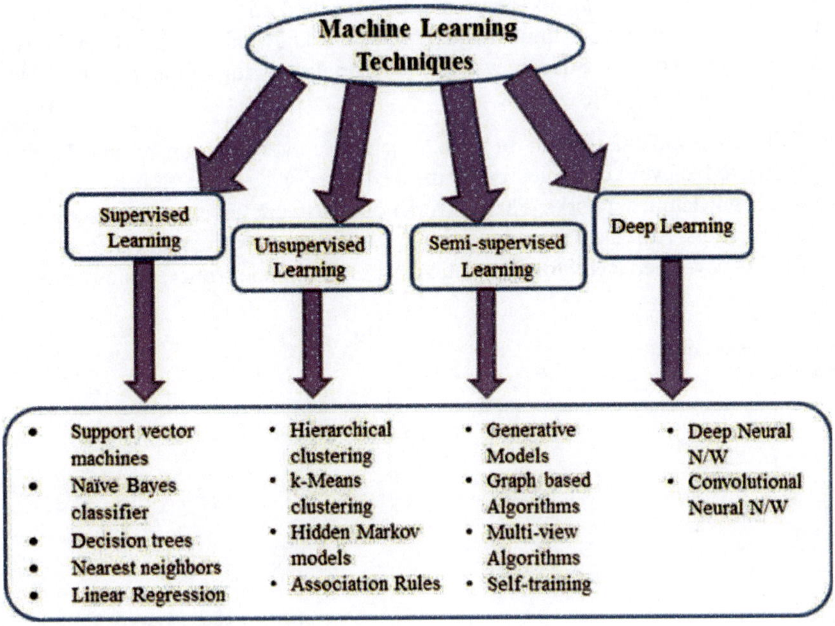

Fig. 2 ML algorithms—taxonomy

2.1 Supervised Approaches

Machine learning algorithms are designed with well-defined learning methods which are best to achieve the objective function of an algorithm. This section presents the review of most commonly used techniques of ML in the field of medical imaging.

Support Vector Machine (SVM): Support vector machine (SVM) is supervised learning algorithm which is best suited for classification as well as regression [9]. The objective of the algorithm is to find out a boundary also known as hyperplane which separates the patterns of the dataset as shown in Fig. 3.

The algorithm chooses the boundary such that the geometric margin can be maximized on the training data while minimizing the training error. Kernel function in the algorithm does mapping of the original data into new space of nonlinearly separable examples. Result of which is a classification of two class problem. SVM supports varieties of kernel function like linear, nonlinear, polynomial, sigmoid and radial basis. SVM classifiers with carefully crafted and selected features have proven to be robust [11].

Neural Networks: Artificial neural networks are biological-inspired networks which are capable of handling varieties of data. The network has large number of "neuron" which is processing elements, and there are weighted connections between them. The process of learning is implemented to acquire knowledge [12]. There are two broad categories of NN. Both are widely used in radiology for segmentation [13] and tissue classification [14].

- Fully connected networks: In such a network, every neuron of one layer is connected to every other neuron on the next layer.
- Recurrent neural networks: This network is used to work with a series of connected information. It is used in radiology for text report classification [15]. In [16], authors have used RNN for automatic disease annotation from radiology reports.

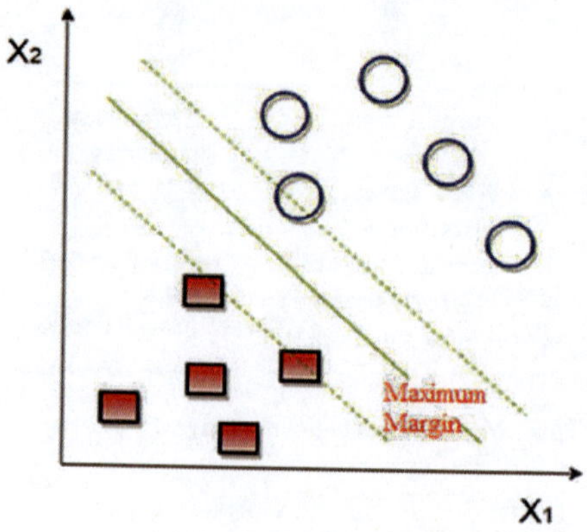

Fig. 3 Support vector machine [10]

Decision Tree: Decision tree is also one of the widely used machine learning approaches for the classification process in medical imaging. Decision tree consists of root node, leaf nodes and internal nodes. The internal nodes use features, and every leaf node represents a class [9]. This approach is used with random forest in medical imaging for classification and prediction of disease. In [17], authors have developed a classification model to distinguish lipomas and lipoma variants using CART analysis. Authors of [18] have used binary decision tree for preoperative cyst screening method using cone beam computed tomography (CBCT).

Random Forest: Random forest is an ensemble learning method which combines many decision trees for prediction or classification of disease. Authors of [19] have developed a new approach called "REPLICA" which is supervised random forest approach for MRI image synthesis. Random forest classifier has proven to be most powerful predictor of machine learning in segmentation of brain images [20].

Table 1 shows comparison of machine learning algorithms discussed above with respect to various parameters like computation cost, processing task, strength and weakness.

2.2 *Evaluation Criteria [9]*

In clinical practice, evaluation criteria rely on label like positive and negative. Healthy cases are termed as negative, and cases with disease are termed as positive.

The computer-aided diagnosis system should be designed such that it results into minimum false positives and false negatives. Classification models can be assessed using many popular performance metrics. Sensitivity and specificity are widely used as evaluation criteria. Sensitivity is true positive rate, and specificity is true negative rate.

$$\text{Sensitivity} = \frac{\text{True Positive}}{\text{True Positive} + \text{False Negative}} \tag{1}$$

$$\text{Specificity} = \frac{\text{True Negative}}{\text{True Negative} + \text{False Positive}} \tag{2}$$

Another assessment method to evaluate model is receiver operating characteristics (ROC) which shows the relationship between sensitivity and specificity shown in Fig. 4. Area under the curve (AUC) measures the entire area under the ROC curve.

3 Need for ML in Medical Imaging

For human, data processing becomes very much tedious if it is in large quantity and not organized properly. The amount of image data acquired during different scans

Table 1 Comparison of machine learning algorithms

Algorithm	Computational cost	Data processing	Strength	Weakness
Support vector machine	High on large data	Classification/regression	• Robust against noisy data • Promising prediction results	• Binary classifier • Slow for large data • Memory intensive
Neural networks	Depends on training Function	Classification	• Fault tolerance • Parallel processing ability	• Hardware dependence
Decision tree	High	Classification/regression	• Simple rules • Handle both numerical and categorical variables • Provide definite clue of important feature for classification	• Biased toward features with many levels • Prone to errors in classification with many class and less data
Random forest	High	Classification/regression	• Efficient for large data • Robust against missing values • Generated forest can be used for future reference	• Complex model as it combines many trees • Computationally much expensive

of patient body is becoming overwhelming for vision of medical practitioner. This image overload for the interpretation process may result into observational error [22]. So, there is a need to build a system which is clinically proven and works as radiology assistance. The objective to build such system is to reduce the false negative rate due to observational oversights. The system is called computer-aided diagnosis (CAD), which is one of the emerging technologies to help radiologists for the interpretation of medical images. The purpose of CAD can be used for both, detection of the anomaly and classification which is likelihood that the anomaly represents a malignancy or not.

Figure 5 shows common steps required to build the CAD system for detection and classification of various irregularities like breast cancer, lung cancer, brain tumor, colon cancer and other similar disease. Once images are processed, they are given

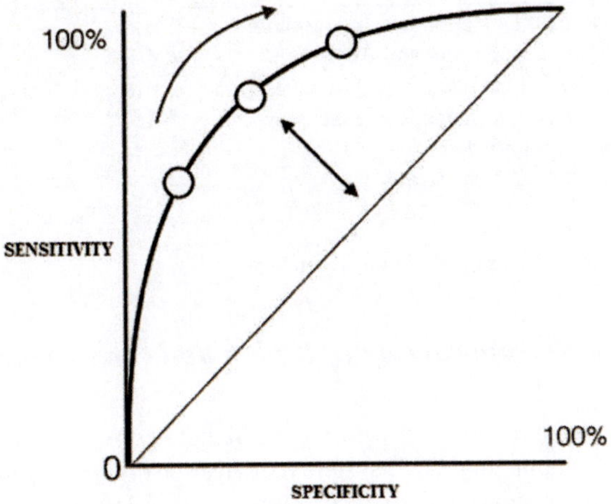

Fig. 4 ROC curve [21]

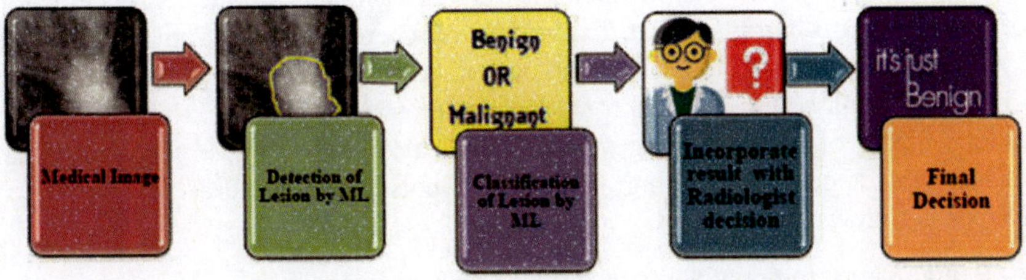

Fig. 5 Computer-aided diagnosis

as an input to the machine learning algorithm which is already well trained to detect and classify anomaly during the training phase. The output generated by this system is then associated with the radiologist's decision for the confirmation of the final result.

4 Machine Learning Applications in Medical Imaging

This section discusses about various application domains where machine learning approaches can be applied in the field of medical imaging. Lots of research have been done in this direction. Few such research applications are presented here. Figure 6 presents categorical view of such applications.

- Automated Detection and Interpretation
- Radiology Reporting and Analysis
- Automatic Localization and Identification of Vertebra
- Content Based Image Retrieval
- Cross Modality Transfer
- Low Dose CT Denoising

Fig. 6 Different applications of ML in medical imaging

4.1 Automated Abnormality Detection and Interpretation

Automated detection of anomaly within given radiology image has more impact in the field of medical imaging. CAD system for detection of anomaly like lung cancer, brain tumor, skin cancer and colon cancer is always helpful to take the second opinion to avoid observational oversight.

Breast cancer is most common type of cancer in women worldwide, which may cause death of the women. So, to increase the survival rate of women, early detection of breast cancer is highly needed. This is the area where machine learning techniques are expected to be used to assist radiology practitioner [23]. A mammogram is an imaging modality which is used for the detection of breast cancer when there is no other evidence of anything being wrong. Micro-calcifications are very small calcium deposits that appear in mammogram as bright spots. There are three types of breast cancer as follows:

- Benign breast tumors: Non-cancerous and cannot spread
- In situ cancers: Cells are within the lobule and have not gone through the basal membrane
- Invasive cancers: Cells have spread into the surrounding tissues by breaking the basal membrane.

Micro-calcifications are the presentation of in situ cancer. When these discrete tiny calcium spots come closer, they may form a lesion which can be malignant. Hence, early detection of micro-calcification from mammogram images can save a women life. Figure 7 shows micro-calcification (left) and breast lesion (right) in mammogram.

Bone age analysis and automated age determination are another application where ML can be applied. This has considerable utility for pediatric radiology. In some medical conditions, child's growth is excessive or stunted in such case where bone age of patient can be assessed with the help of X-rays. Bone has some ossification center that appears and fuses at a specific time interval. Before it fuses if the correct age prediction is done, then by injecting growth hormones, at least enough height can be achieved for a child. Table 2 shows age wise growth of bones, this can be fed to ML algorithm along with annotated palm X-ray images for proper prediction of age, and accordingly, the decision can be taken whether a child needs extra growth hormones or not. Applying deep learning techniques to medical imaging data like

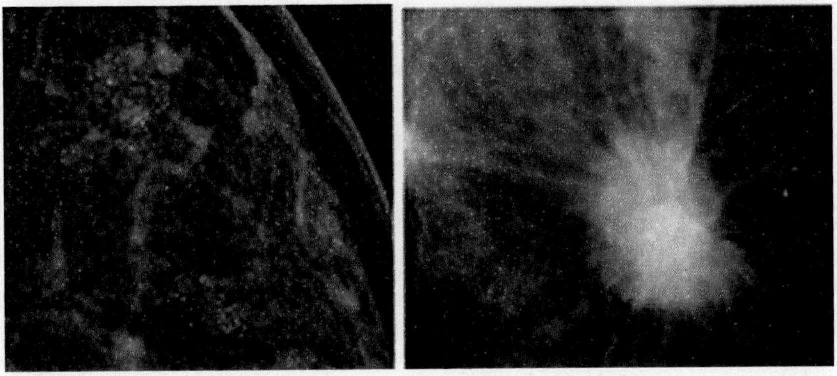

Fig. 7 Mammogram images [24]

Table 2 Bone development stages [25]

Bone type	Age
Distal phalanges	9th fetal week
Metacarpals	10th fetal week
Proximal phalanges	11th fetal week
Middle phalanges	12th fetal week
Middle phalanx	14th fetal week
Capitate	4 months
Hamate	4 months
Triquetral	3 years
Lunate	4–5 years
Trapezium	6 years
Trapezoid	6 years
Scaphoid	6 years
Pisiform	11 years

X-rays, CT and MRI scans has proven itself a powerful method for a broad range of image diagnosis, recognition and classification tasks. One of these tasks is the establishment and use of appropriate and fully automated methods for determining the bone age since it is a crucial parameter for monitoring and assessing the health status of children.

The manual process of bone age prediction is a time-consuming process, whereas its automated prediction using deep learning models has proven better accuracy. An algorithm can be developed to predict the bone age of children given X-ray images [26].

Bone scan is another application where machine learning or deep learning algorithm can be applied. Bone scan is performed by injecting radioactive substance to the patient's body who are taking cancer treatment. It is very difficult and time

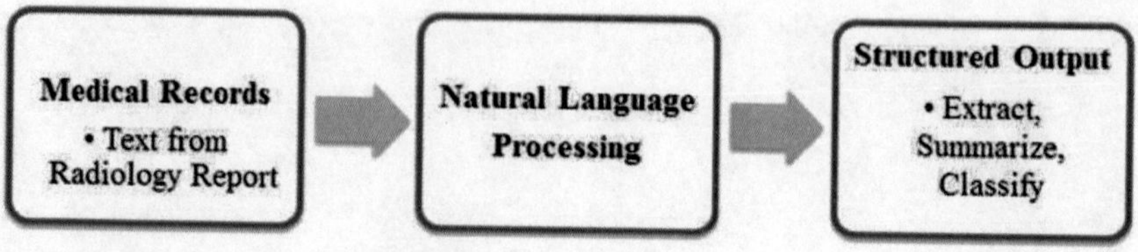

Fig. 8 Information extraction from radiology report

consuming to detect changes between successive bone scan images of a patient. So, the CAD system was built [27] which can take successive bone scans as an input and perform their temporal subtraction to see the effect of radiotherapy. One cold lesion (white circle) and two hot lesions (dark circles) are correctly marked by the CAD system [27, 28]. Cold lesion shows the reduced effect of bone metastasis, wherein hot lesion shows that cancer has spread in the detected region.

4.2 Radiology Reporting and Analytics

In many applications, there is a need to extract important information from the radiology report and to categorize and store extracted information in the form of structured data. Natural language processing can be used to extract such information [29]. Machine learning techniques could also be used to extract terminology used in the report for the purpose of quality improvement and information analytics [30]. In the article radiology informatics [31], the example of information extraction from narrative radiology report is shown. Using supervised algorithm, they have used annotations available in the report to summarize the same. Figure 8 shows steps for information extraction from medical records like radiology reports of CT scan or X-ray.

4.3 Automatic Localization and Identification of Vertebrae in CT Scans [32]

Vertebra is an important anatomical landmark in our body and provides former structure of spinal cord. Automatic and accurate localization and identification of vertebrae from CT and MRI images is substantial for the clinical task like surgical plan and post-operative assessment.

Spinal cord has five sections: C1–C8 cervical, T1–T12 thoracic, L1–L5 lumber, S1–S5 sacrum and coccyx as shown in Fig. 9. Automatic identification of all these vertebrae can reduce manual analysis by the radiologist. This automation can help in linking of radiology reports with corresponding image regions of the cord. For the patient of trauma with multiple injuries in the spinal cord, after his or her scan,

Fig. 9 Spine annotation [33]

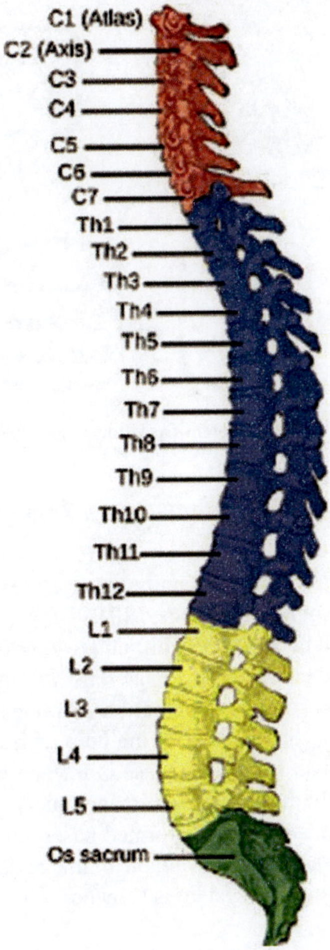

radiologist has to find out exact location of any type of injury that may be fracture or dislocation. This process is manual hence wastes lots of time of the clinicians, and also, it should be done so precisely to avoid any kind of oversight errors. Localizing and identifying vertebrae automatically can be a great help to the medical imaging fraternity. In [32], authors have proposed an algorithm which is based on regression forests and graphical model of probability. The regression forest is used for the detection part, and for the precise localization and identification of specific vertebrae, hidden Markov model is used. The experimentation was performed over 200 labeled CT scans.

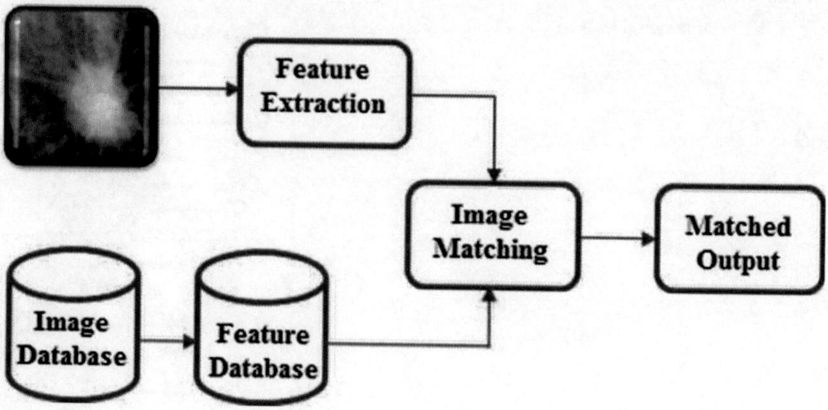

Fig. 10 Content-based image retrieval

4.4 Content-Based Image Retrieval

In medical imaging, images are taken for diagnosis. With the modalities like ultrasounds, CT scans, MRI, X-rays, etc., more and more digitized images are produced. It becomes very difficult to handle this large image database. Hence, there is need for the system for image retrieval for supporting clinical decision making. This system is known as content-based image retrieval (CBIR).

In CBIR, with the help of feature vectors, visual image contents are extracted. The feature database is formed using these multi-dimensional feature vectors. For the retrieval of an image, query image is presented to the system, and then, image matching is performed based on similarity between feature vector of query image and those of available in image database. The block diagram of content-based image retrieval system is shown in Fig. 10 [34].

4.5 Cross Modality Transfer

MR to CT: CT scan exposes to high radiation during an image acquisition which causes side effect to the patients and very much risky for a child and pregnant women. Compare to CT, MR is safe as it causes no radiation free modality. Methods have been developed to estimate CT images for MR for diagnosis purpose. Also, there are several medical conditions where a patient has to undergo various imaging modality for confirmed analysis. So, to reduce the financial burden on the patient, cross modality transfer is needed [35].

CT to MR: MRI has vital role in organ segmentation and for analysis of tissues in a body. However, use of MR is restricted due to high cost and due to increased use of metal implants in the patient's body. There are methods available which can estimate MR images given CT images as an input. Authors in paper [36] suggested an approach to transform two-dimensional brain CT image slices into two-dimensional brain MR image slices using generative adversarial network (GAN) model.

4.6 Low Dose CT Denoising [37]

CT scan is one of the most useful imaging modalities for the analysis of bone structure. But, it exposes to the radiation, and hence, there is a potential risk to the patient especially to child and pregnant women since rays produced by this modality cause genetic damage and increase chances of cancer. One possible solution to this is to reduce radiation dose. Reducing radiation dose may increase the noise in the image, and information needed for proper diagnosis can compromise. So, methods have been developed to design better image reconstruction for low dose CT. Various methods like sonogram filtration before reconstruction [38–40], iterative reconstruction [41, 42] and image post-processing after reconstruction [43–45] are available for image reconstruction to reduce the effect of noise due to low dose CT.

5 Shortcomings and Open Issues

Although machine learning and deep learning technology have made acceptable achievements in the field of medical imaging, there are still some challenges. First is large ground truth labeled data. For a classifier to be well trained, there is a need of large number of proper annotated or labeled data. Image scarcity is the biggest problem in medical field. Second one is imbalance of categories in existing dataset. There should be proper proportion of both types of data to train the model, positive as well as negative cases. Model trained with such imbalanced data may result into classification error. Third is linking medical data with the images. There are certain medical conditions where only images cannot help in diagnosis, and there is a need to link image data with other pathological reports and patient history. Integration of these data is very challenging. Also, it is a matter of human lives, and acceptance of classifier for the diagnosis and treatment monitoring by health professionals is also an important constrained element. There is also a need of extensive collaboration between researchers and health industry.

6 Conclusion

Machine learning in medical analytics has a very wide scope in terms of providing decision support for diagnostics and a line of treatment. It offers methods for making an automated intelligent decision using training data through which complex patterns can be learned well. In the medical field, there are lots of data to be interpreted in the form of images. Hence, making analytical solution is not possible. This image overload may result into observational errors by the radiologist. Computer-aided diagnosis is an important subject of research in radiology diagnostic. The paper focused at various domains applicable for computer-aided diagnostics (CAD) which can be a part of routine clinical work for detection and classification of the process. Various machine learning algorithms used in literature are also presented with their comparative analysis. Some shortcomings and issues which are needed to be addressed for doing research in the field of life using field of machines are also discussed.

References

1. http://www.iambiomed.com/specialization/medical_imagingphp
2. Image is licensed under CC0 1.0 Universal, url: https://pxhere.com/en/photo/992853
3. Image is licensed under CC0 1.0 Universal, url: https://pxhere.com/en/photo/619689
4. https://medlineplus.gov/ency/article/007451.htm
5. https://www.nibib.nih.gov/science-education/science-topics/optical-imaging
6. https://www.radiologyinfo.org/en/info.cfm?pg=gennuclear
7. Pillai R, Oza P, Sharma P (2020) Review of machine learning techniques in health care. In: Singh P, Kar A, Singh Y, Kolekar M, Tanwar S (eds) Proceedings of ICRIC 2019. Lecture Notes in Electrical Engineering, vol 597. Springer, Cham
8. McBee MP et al (2018) Deep learning in radiology. Acad Radiol 25(11):1472–1480
9. Zhang Zhenwei, Sejdić Ervin (2019) Radiological images and machine learning: trends, perspectives, and prospects. Comput Biol Med 108:354–370
10. https://en.wikipedia.org/wiki/File:Support_vector_machine.jpg
11. Torheim T, Malinen E, Kvaal K, Lyng H, Indahl UG, Andersen EKF, Futsaether CM (2014) Classification of dynamic contrast enhanced MR images of cervical cancers using texture analysis and support vector machines. IEEE Trans Med Imag 33(8):1648–1656
12. https://www.nextbigfuture.com/2019/12/what-are-the-limits-of-deep-learning-going-beyond-deep-learning.html
13. Prasoon A, Petersen K, Igel C, Lauze F, Dam E, Nielsen M (2013) Deep feature learning for knee cartilage segmentation using a triplanar convolutional neural network. In: International conference on medical image computing and computer-assisted intervention, pp 246–253
14. Cruz-Roa AA, Arevalo Ovalle JE, Madabhushi A, González Osorio FA (2013) A deep learning architecture for image representation, visual interpretability and automated basal-cell carcinoma cancer detection. In: International conference on medical image computing and computer-assisted intervention, p 403
15. Banerjee I et al (2019) Comparative effectiveness of convolutional neural network (CNN) and recurrent neural network (RNN) architectures for radiology text report classification. Artif Intell Med 97:79–88
16. Lee C et al (2019) Automatic disease annotation from radiology reports using artificial intelligence implemented by a recurrent neural network. Am J Roentgenol 212(4):734–740
17. Shim EJ et al (2020) An MRI-based decision tree to distinguish lipomas and lipoma variants from well-differentiated liposarcoma of the extremity and superficial trunk: classification and regression tree (CART) analysis. Eur J Radiol, p 109012
18. Pitcher B et al (2017) Binary decision trees for preoperative periapical cyst screening using cone-beam computed tomography. J Endod 43(3):383–388
19. Jog A et al (2017) Random forest regression for magnetic resonance image synthesis. Medical image analysis 35:475–488

20. Huynh T et al (2015) Multi-source information gain for random forest: an application to CT image prediction from MRI data. In: International workshop on machine learning in medical imaging, pp 321–329. Springer, Cham
21. https://commons.wikimedia.org/wiki/File:ROC_curve.svg
22. Wernick M, Yang Y, Brankov J, Yourganov G, Strother S (2010) Machine learning in medical imaging signal processing magazine. IEEE 27(4):25–38
23. Polan DF, Brady SL, Kaufman RA (2016) Tissue segmentation of computed tomography images using a random forest algorithm: a feasibility study. PhysMed Biol 61(17):6553–6569
24. https://radiologyassistant.nl/breast/bi-rads-for-mammographyand-ultrasound-2013
25. Sutton D, Textbook of radiology and imaging, 3rd edn.
26. De Sanctis V, Di Maio S, Soliman AT, Raiola G, Elalaily R, Millimaggi G (2014) Hand X-ray in pediatric endocrinology: Skeletal age assessment and beyond. Indian J Endocrinol Metabol 18(7):63
27. Doi K (2007) Computer-aided diagnosis in medical imaging: historical review, current status and future potential. Comput Med Imag Graph 31:198–211
28. Shiraishi J, Li Q, Appelbaum D, Pu Y, Doi K (2006) Development of a computer-aided diagnostic scheme for detection of interval changes in successive whole-body scans. Med Phys (in press [PubMed])
29. Sevenster M, Buurman J, Liu P, Peters JF, Chang PJ (2015) Natural language processing techniques for extracting and categorizing finding measurements in narrative radiology reports. Appl Clin Inform 6(3):600–610
30. Hassanpour S, Langlotz CP, Amrhein TJ, Befera NT, Lungren MP (2017) Performance of a machine learning classifier of knee MRI reports in two large academic radiology practices: a tool to estimate diagnostic yield. AJR Am J Roentgenol 208(4):750–753
31. http://langlotzlab.stanford.edu/machine-learning
32. Glocker B, Feulner J, Criminisi A, Haynor DR, Konukoglu E (2012) Automatic localization and identification of vertebrae in arbitrary field-of-view CT scans. In: International conference on medical image computing and computer assisted intervention
33. Image by CNX OpenStax, licensed under the Creative Commons Attribution 4.0. https://commons.wikimedia.org/wiki/File:Figure_38_01_07.jpg
34. Kumar Ashnil, Kim Jinman, Cai Weidong, Fulham Michael, Feng Dagan (2013) Content-based medical image retrieval: a survey of applications to multidimensional and multimodality data. J Digit Imaging 26(6):1025–1039
35. Wolterink JM, Dinkla AM, Savenije MH, Seevinck PR, van den Berg CA, Išgum I (2017) Deep mr to ct synthesis using unpaired data. In: International workshop on simulation and synthesis in medical imaging. Springer, pp 14–23
36. Jin CB, Kim H, Liu M, Jung W, Joo S, Park E, Ahn YS, Han IH, Lee JI, Cui X (2019) Deep CT to MR synthesis using paired and unpaired data. Sensors 19(10):2361–2379
37. Yang Q, Yan P, Zhang Y, Yu H, Shi Y, Mou X, Kalra MK, Zhang Y, Sun L, Wang G (2018) Low-dose CT image denoising using a generative adversarial network with Wasserstein distance and perceptual loss. IEEE Trans Med Imag 37(6):1348–1357
38. Wang J, Lu H, Li T, Liang Z (2005) Sinogram noise reduction for low-dose CT by statistics-based nonlinear filters. In: Medical imaging 2005: image processing, vol 5747. International Society for Optics and Photonics, pp 2058–2067
39. Wang J, Li T, Lu H, Liang Z (2006) Penalized weighted least-squares approach to sinogram noise reduction and image reconstruction for low-dose x-ray computed tomography. IEEE Trans Med Imag 25(10):1272–1283
40. Manduca A, Yu L, Trzasko JD, Khaylova N, Kofler JM, McCollough CM, Fletcher JG (2009) Projection space denoising with bilateral filtering and CT noise modeling for dose reduction in CT. Med Phys 36(11):4911–4919
41. Beister M, Kolditz D, Kalender WA (2012) Iterative reconstruction methods in x-ray CT. Phys Med Eur J Med Phys 28(2):94–108
42. Hara AK, Paden RG, Silva AC, Kujak JL, Lawder HJ, Pavlicek W (2009) Iterative reconstruction technique for reducing body radiation dose at CT: feasibility study. Am J Roentgenol 193(3):764–771

43. Ma J, Huang J, Feng Q, Zhang H, Lu H, Liang Z, Chen W (2011) Low-dose computed tomography image restoration using previous normal-dose scan. Med Phys 38(10):5713–5731
44. Chen Y, Yin X, Shi L, Shu H, Luo L, Coatrieux J-L, Toumoulin C (2013) Improving abdomen tumor low-dose CT images using a fast dictionary learning based processing. Phys Med Biol 58(16):5803
45. Feruglio PF, Vinegoni C, Gros J, Sbarbati A, Weissleder R (2010) Block matching 3D random noise filtering for absorption optical projection tomography. Phys Med Biol 55(18):5401
46. Singh PK, Kar AK, Singh Y, Kolekar MH, Tanwar S (eds) (2019) Proceedings of ICRIC 2019: recent innovations in computing, vol 597. Springer Nature

Music Genre Classification ChatBot

Rishit Jain, Ritik Sharma, Preeti Nagrath, and Rachna Jain

Abstract Classification of music on the basis of genre is a sub-domain of the multidisciplinary field of music information retrieval (MIR) that is gaining traction among researchers and data scientists. Even though this problem has been extensively researched and tested, the problem still lies in the foundations, as the true definition of genre still lies to the mercy of human subjectivity. In this paper, we have proposed a classification model which employs a convolutional neural network (CNN) to differentiate between audio files by assessing the visual representations of their timbral features [1]. The music genre classification model is outlined by a ChatBot model built using NLTK, which can simulate an intelligent conversation with a user, and it employs a feature that enables it to recognize and process the audio file based on the input from the user. The GTZAN dataset [2] was used for training the music genre classification model, and the so trained model for this purpose yielded an accuracy of nearly 68.9%. The accuracy so obtained is relatively better than several other classification models that we had researched. Through extensive research and constant trials, we can state, with some certainty, that such a system can be extensively used alongside several music streaming services, as it would facilitate the process of automation of the classification of songs.

Keywords Music genre classification · ChatBot · Convolutional neural networks · Machine learning · Natural language processing

R. Jain (✉) · R. Sharma
Department of ECE, Bharati Vidyapeeth's College of Engineering, New Delhi 110063, India
e-mail: rishitjainn@gmail.com

R. Sharma
e-mail: ritiksharma373@gmail.com

P. Nagrath · R. Jain
Department of CSE, Bharati Vidyapeeth's College of Engineering, New Delhi 110063, India
e-mail: preeti.nagrath@bharatividyapeeth.edu

R. Jain
e-mail: rachna.jain@bharatividyapeeth.edu

P. K. Singh et al. (eds.), *Proceedings of Second International Conference on Computing, Communications, and Cyber-Security*, Lecture Notes in Networks and Systems 203,
https://doi.org/10.1007/978-981-16-0733-2_27

1 Introduction

Genres in music are conventional categories that classify different bits of music to different sets of traditions or conventions. Music is a source of entertainment for people around the globe. The different genres of music have an appeal to different kinds of people. The lyrics and the language used kept aside, and characteristics such as instrumentation and the harmonic rhythm can be used to categorize and organize the songs [3].

In this day and age, there are millions of musicians, who release new and unique songs almost every day. Despite the advanced algorithms and sorting techniques that are being used currently, classification of music on the basis of its different aspects still poses a problem. Genre being a paramount characteristic of any audio clip, it is important to have an autonomous classification system to identify the genre of a particular audio file on the basis of certain timbral features. To facilitate this problem, we worked on a project whose primary objective was to develop a technology that would provide a user with an interactive ChatBot that simulates an intelligent conversation with the user and is capable of sorting an audio file according to its genre, through proper assessment of the key features of the audio and the prediction of the genres based on the features extracted. Such a model could be very useful, if implemented alongside an audio streaming service, where it can help a user keep track of his/her favourite music and even get recommendations based on his/her preferred genres.

Prior research suggests that there are several different techniques and methodologies that can be adopted to tackle the problem of classification of music. However, all the models so developed for this purpose have never been implemented further to be used for our convenience. In contrast, this research is an attempt to fill in some gaps in the previous implementations and even add some new features to facilitate the entire process. Through our research,

1. We have created a classification model using a convolutional neural network (CNN), comprising four layers, which ensure an efficient processing of the extracted features and give out a predicted genre for a particular file. This model yielded a training accuracy of about 97.8%.
2. The music genre classification model was convoluted with a ChatBot model, which was built using a deep neural network (DNN) and the natural language tool kit (NLTK), and it was trained using a custom dataset. The ChatBot is capable of simulating an intelligent conversation with a user, and it delivered a testing accuracy of 100%.
3. The ChatBot facilitates the process of classification and prediction of the genre of a particular audio file, by taking a natural language input from the user along with the audio file.

For this project, certain timbral features [4] are extracted from the audio files given as input, and using those features, the audio file will be categorized into either one of the ten genres (blues, classical, country, disco, hip-hop, jazz, metal, jazz, pop, reggae, rock) as per the GTZAN dataset [2]. This classification model is convoluted with a simple ChatBot [5], which converses with the user and takes the input for the audio file in ".wav" format and gives an appropriate response with the predicted genre of the given audio file.

This paper is ordered in a manner, such that, following the introduction, the Sect. 2 comprises the literature review which outlines a detailed comparison of the work done by several authors for music genre classification and the different strategies that were adopted to tackle this problem. In Sect. 3, we introduced the datasets that were employed for training the music genre classification model. Section 4 of this paper comprises a concise explanation of the two deep learning models created for this project in the form of properly laid out algorithms. Section 5 comprises the methodology and materials section which contains the entire account of the project and all the relevant details of the types of libraries, networks and methods that were used for making the music genre classification system and the ChatBot. Following the methodology, Sect. 6 of the paper contains a brief outline of all the observations and results inferred from this project. Section 7 mentions the conclusions drawn from our project and the research conducted for it, along the way. Finally, the last section of the paper contains all the references and citations to previous researches.

2 Literature Review

Over the last two decades, there has been a surge in the application of several different machine learning and deep learning techniques to tackle the problem of classification of music. Numerous studies and researches have been conducted, which employed carrying techniques such as K-nearest neighbours, support vector machines (SVMs), decision trees, convolutional neural networks (CNNs) and several others. These different methodologies can be used to extract certain features from an audio file, and based on which, it can be classified into different categories. Some of those features may include content-based acoustic features, loudness, quality, beat or pitch. Working on these features, typical timbral features [4] such as energy of the signal, spectral centroid, spectral flux, spectral roll-off, zero crossings, linear prediction coefficients and mel-frequency cepstral coefficients (MFCCs) [1, 6].

Several studies were conducted to develop an efficient music genre classification model, as it can be seen in Table 1. For the process of classification of music on the basis of genre, Tzanetakis and Cook [2] proposed a comprehensive set of features which were explored through a model created using K-nearest neighbours and Gaussian mixture models. Lambrou et al. [7] used statistical features to classify music into three primary genres (rock, piano and jazz) in the temporal domain as well as three different wavelet transform domains. Deshpande et al. [8] defined a classification using Gaussian mixtures, support vector machines (SVMs) and K-nearest neighbours (K-NN) based on the typical timbral features to classify the audio files into rock, piano and jazz. Soltau et al. [9] suggested that representation of the temporal structures of an input signal yields a new set of abstract features that can be assessed

Table 1 Tabular comparison of different approaches for classification of music

Author/Year	Topic	Dataset used	Paradigm/Methods	Findings
Soltau et al. (1998) [9]	"Recognition of music types"	Custom dataset (rock, pop, techno and classic)	Explicit time modelling with neural networks	For a small dataset, ETM-NN gives an accuracy of 81.9%
Lambrou et al. (1998) [7]	"Classification of audio signals using statistical features on time and wavelet transform domains"	Custom audio files (rock, pop and jazz)	Statistical analysis of features from time and wavelet transform domains	Statistical analysis of 12 musical signals gave a classification accuracy of more than 90%, but not suitable for all files
Deshpande et al. (2001) [8]	"Classification of music signals in the visual domain"	Custom dataset (MP3 format—rock, jazz, and classical)	Gaussian mixtures, SVMs, K-nearest neighbour	Accuracy of the SVM classifier was 90%, while that of the K-NN model was 75
Tzanetakis et al. (2002) [2]	"Musical genre classification of audio signals"	GTZAN dataset	K-nearest neighbours, Gaussian mixture	An Accuracy of 61% was achieved using K-NN on GTZAN dataset
Vyas et al. (2014) [10]	"Automatic mood detection of indian music using mfccs and k-means algorithm"	Custom dataset (WAV format—happy/sad song)	K-means algorithm	An accuracy of about 90% was achieved to classify music into happy/sad
Asim Ali et al. (2017) [11]	"Automatic music genres classification using machine learning"	GTZAN dataset	K-nearest neighbours, SVMs	SVMs gave an overall accuracy of 77%, and it proved to be a better classifier than K-NN
Ramírez et al. (2019) [12]	"Machine learning for music genre: multifaceted review and experimentation with audioset"	GTZAN dataset, ISMIR 2004, latin music, ballroom, Ismis 2011, million song dataset, etc.	Decision trees, NB classifiers, linear SVMs, DNNs, RNNs	Out of all the methods used, RNNs gave the highest AUC of 0.929
Liu et al. (2019) [13]	"Bottom-up broadcast neural network for music genre classification"	GTZAN, ballroom and extended ballroom datasets	Bottom-up broadcast neural network (BBNN)	On experimentation with different datasets, BBNN gave an average accuracy of 97.2%

(continued)

Table 1 (continued)

Author/Year	Topic	Dataset used	Paradigm/Methods	Findings
Dokania et al. (2019) [14]	"Graph representation learning for audio and music genre classification"	GTZAN and AudioSet datasets	Siamese neural network and graph neural network	Training accuracy for GTZAN dataset was 99.5% and that for AudioSet was 91.3%

and learnt by ANNs, thereby making them capable of music genre classification. Vyas et al. [10] used MFCC for the mood detection of the Indian Music, using K-means algorithm. Asim et al. [11] used a model build using K-nearest neighbours and support vector machines (SVM) to classify the ten genres of GTZAN dataset. Drawing out a multifaceted review, Ramírez et al. [12] used several methodologies such as decision trees, Naïve Bayes classifiers, linear SVMs, deep neural networks (DNNs) and recurrent neural networks (RNNs) to experiment the classification of music with multiple datasets. Liu et al. [13] built a model using a bottom-up broadcast neural network to implement a solution of classification of music on GTZAN, ballroom and extended ballroom dataset. Dokania et al. [14] implemented a Siamese neural network and a graph neural network to explore the structural associations in the features of an audio file.

A ChatBot may be defined as an artificial intelligence (A.I.) software which is capable of simulating a conversation with the user using natural language processing (NLP) [15]. In layman's terms, a ChatBot is a program uses a machine learning algorithm to study and predict a response based on the user's input [5]. Natural language tool kit (NLTK) [16] is a free package/library for Python, which is widely used for the purpose of text manipulation and processing. The NLTK library comprises certain modules and functions that enable the developer to assess the data of the taken dataset. TensorFlow and TFLearn libraries were used to prepare and train a deep neural network for the ChatBot.

There exists another open source library—ChatterBot—which provides an easy to implement framework for ChatBots. The library contains a few trainer classes, which can be used to train a bot for a particular dataset of messages.

3 Dataset

To train the model built for the ChatBot, a custom dataset is comprised of a file with six primary tags ("greeting", "goodbye", "age", "name", "shop" and "hours"). For each tag, there were appropriate patterns and responses that define a conversation and a few possible questions and answers a conversation may have. This dataset is used such that when the user input is somewhat similar to a pattern string, then the ChatBot would predict a relevant response from the dataset.

Table 2 Number of Songs per Genre in the GTZAN dataset

GTZAN genre collection dataset		
S. No.	Genre	Total number of songs
1.	Blues	100
2.	Classical	100
3.	Country	100
4.	Disco	100
5.	Hip-Hop	100
6.	Jazz	100
7.	Metal	100
8.	Pop	100
9.	Reggae	100
10.	Rock	100

Total 1000 audio tracks in the GTZAN dataset

For the purpose of classification on the basis of genre, we made use of the GTZAN genre collection dataset [2], created and used for the well-known paper published by G. Tzanetakis and P. Cook in IEEE Transactions on Musical Genre Classification of Audio Signals. This dataset has been extensively used for research and evaluation for music genre recognition (MGR). As it can be seen in Table 2 drawn below, the GTZAN dataset comprises a total of 1000 songs and ten genres. Each audio file within the dataset is 30 s long, and the tracks are 22,050 Hz mono 16-bit audio files in.wav format. The GTZAN dataset is quite small, containing only a 1000 audio files, thus, it would generally not be suitable for deeper neural networks, but for the model in consideration, the dataset was just enough to produce considerable results.

4 Proposed Model

This section comprises a brief outline and explanation of our proposed model and how we effectively used the small datasets, to produce better results. Algorithm 1 is an outline of the music genre classification model, which was trained using the GTZAN [2] dataset. Algorithm 2 is an outline of the primary ChatBot model which is convoluted with the classification model.

Algorithm 1 Proposed Music Genre Classification Model Algorithm

Check If *Trained Classification Model is not available,* **then**
  Import Libraries
  Load GTZAN Dataset (Total 1000 Audio Files)
  Plot the Spectrogram of an example file for test
  For *audio files in dataset,* **do**
    Calculate Timbral Features for each audio file
    Append the values for all features in GenreData.csv
    Open the .csv file as a list and Label Encode the genres
    Scale the feature columns
    Run train test split (Training Set = 60%, Validation Set = 20%, Test Set = 20%)
  Create the 4 layers (Input, Output & 2 Hidden) of the CNN model using Keras
  Compile the model (Optimizer = 'adam', Loss = 'sparse_categorical_crossentropy')
  Train the model (epochs = 25, batch size = 16)
  Plot the Training Loss (0.0782) & Accuracy (97.8%)
  Run the model for validation set
  Plot Validation Parameters – Test Accuracy (68.9%) & Test Loss (1.48)
  Save the trained model
Else
  Load trained model
  Test model with Test Dataset

Algorithm 2 Proposed ChatBot Model Algorithm

Input : Natural Language Command from the User
Output: Predicted Genre Label

Check If *Trained ChatBot Model is not available,* **then**
  Load Custom Dataset
  Convert the tags and labels in the dataset to a list
  Pass the lists (words and labels) through the Lancaster Stemmer
  New list of labels is One Hot Encoded
  Append the new lists (word and label) to the training and output lists
  Create layers of the Deep Neural Network (Activation = SoftMax)
  Train the DNN model (epochs = 1000, batch size = 8)
  Save the trained model
Else
  Load trained model
  Take natural language input from the user
  Tokenize and Stem the String from the User's Input
  Append the words from the stemmer into a sorted list (labels)
  Compare the list (labels) to the tags in the dataset.
  If *labels = 'shop' or 'music',* **then**
    Load music genre classification model
    Test classification model with test dataset
    Return the Accuracy of the Classifier
    Return the predicted genre of the test set files
  Else
    Check For *the relevant tag in list labels*
      Return random response and ask user for input until 'Exit'

5 Methodology and Materials

This section encapsulates all the specific details about the methods, functions and libraries used for the genre classification model and the NLTK model for the ChatBot. It also contains a brief outline of the two models created and the theory behind it (Fig. 1).

5.1 *Music Genre Classification Model*

A classification model built using convolutional neural networks [17, 18] through the Keras library, was used to classify the audio files according to the ten genres as per the GTZAN dataset [2]. The model we created yielded a test accuracy of approximately 68.9%, which, in comparison with the other models built for the same purpose, is quite efficient.

Spectrogram Generation

A spectrogram is a graphical illustration of the spectrum of a particular signal as it visualizes the signal time and the frequency on the x-axis and the y-axis, respectively. A colormap is an array of colours which is used to quantify the magnitude of a given frequency within a specified timeframe [19]. All the files in the taken dataset were

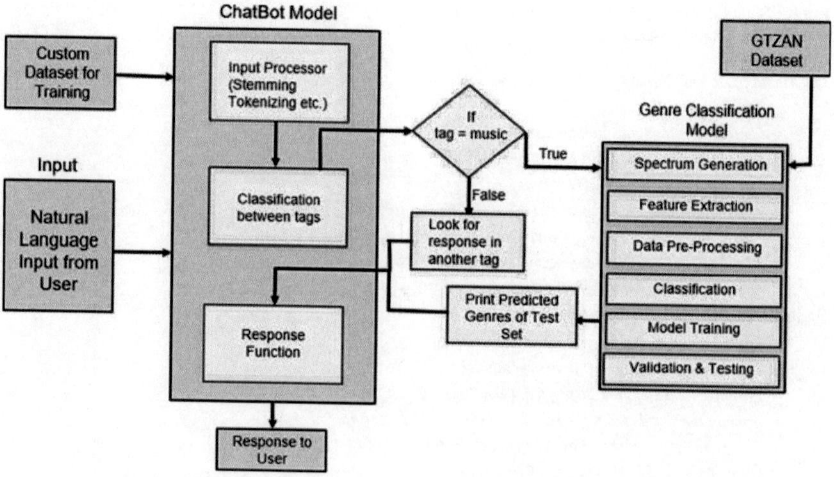

Fig. 1 Data flow representation of the ChatBot model and the working of the genre classifier

converted into .csv files, which in turn, make the process of features extraction, relatively easier.

Features Extraction

Certain timbral features [1, 4] can be drawn out from the audio files, based on which the classification can be performed. There are primarily five features that are used for audio signal processing [20].

Mel-Frequency Cepstral Coefficients (MFCC)

The mel-frequency cepstral coefficients (MFCCs) [6, 3] of a signal are a certain set of features (usually about 10–20 features) that best describe the appearance of the spectral envelope. It illustrates the features of the human voice. Figure 2 illustrates the plot of MFCC for a sample audio file of .wav format.

Spectral Centroid

The spectral centroid can be used to depict the frequency upon which the energy of the spectrum is centred. It may be estimated by taking the weighted mean of the different frequencies present in the audio. Figure 3 represents the representation for the same on the waveform of the audio signal.

Zero-Crossing Rate

The zero-crossing rate is the sum of the signal changes along with the signal, e.g. the rate at which the signal changes from positive to negative [21]. The zero cross-sectional structure can be seen in Fig. 4. This feature is widely used in speech recognition and music information retrieval (MIR). Usually, they have a high amount of very visual effects such as those of metal and rock.

Chroma Frequencies

Chroma frequencies, as illustrated in Fig. 5, are an efficient and clear representation for an audio file, in which the entire spectrum is estimated onto 12 bins depicting a total of 12 different semitones (or chroma) of the music spectrum.

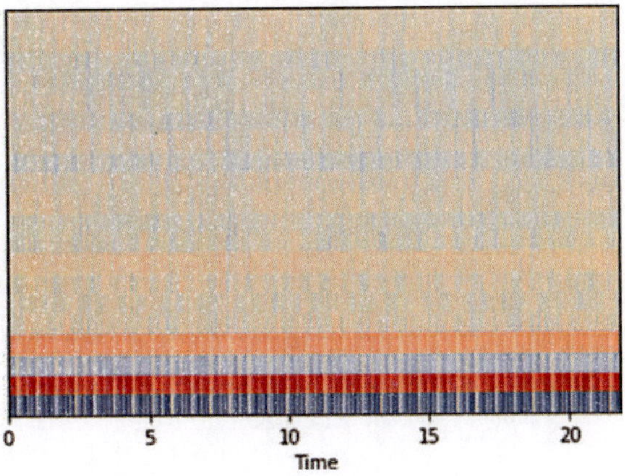

Fig. 2 MFCC of a classical audio file

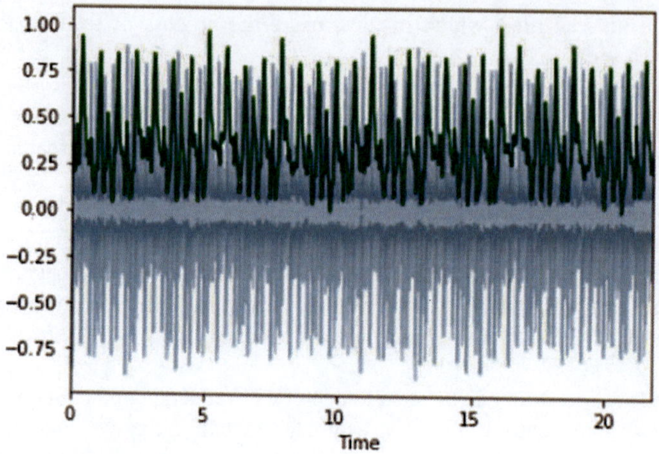

Fig. 3 Spectral centroid of a classical audio file

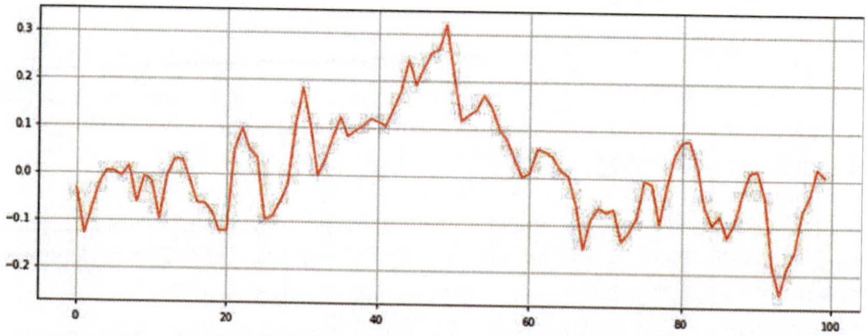

Fig. 4 Zero-crossing rate of a classical audio file

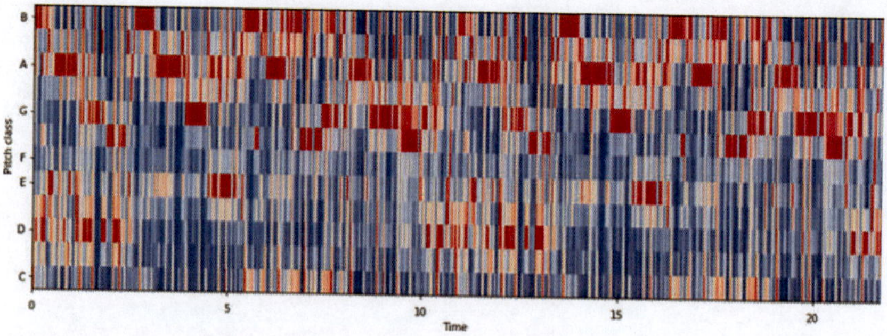

Fig. 5 Chromic frequency of a classical audio file

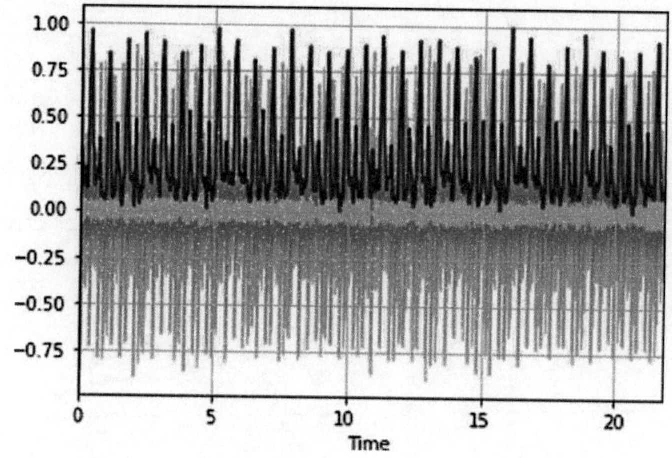

Fig. 6 Spectral roll-off of a classical audio file

Spectral Roll-off

The spectral roll-off can be defined as the frequency under which a part or some percentage of the total spectrum energy is contained. Figure 6 is a representation of the spectral roll-off.

Data Preprocessing

Before training the model, the audio files are assessed appropriately and are converted into meaningful representations. We extract the above-mentioned features from every audio file. All these features are then appended to a new .csv file to make the analysis of data easier, and then, it can be passed on for classification.

Classification

Classification is done using Keras [22] sequential model, which is essentially a convolutional neural network (CNN) model [23]. The model comprises four layers—the first dense layer is an input layer with the "ReLU" [24] activation function, the second and third layers are dense layers with "ReLU" activation function, but fewer number of neurons, and the fourth layer employs the "SoftMax" [24] activation function, through which the output is taken. The value inside dense represents the dimension of output space. The input will have 256 neurons, and the output will have only ten neurons which represent the ten genres of the dataset.

Model Training and Learning

The optimizer used for the learning process of the model is Adam optimizer [25] which is an extension to stochastic gradient descent, and it produces fast result in deep learning algorithm. The loss is calculated using "sparse_categorical_crossentropy" function. We have trained our model using fit function and have trained it for 20 epochs.

Validation

We set apart 200 samples from our training data to be used as validation set. After applying the model on the validation set, we yielded an accuracy of 68.9%. Figure 7

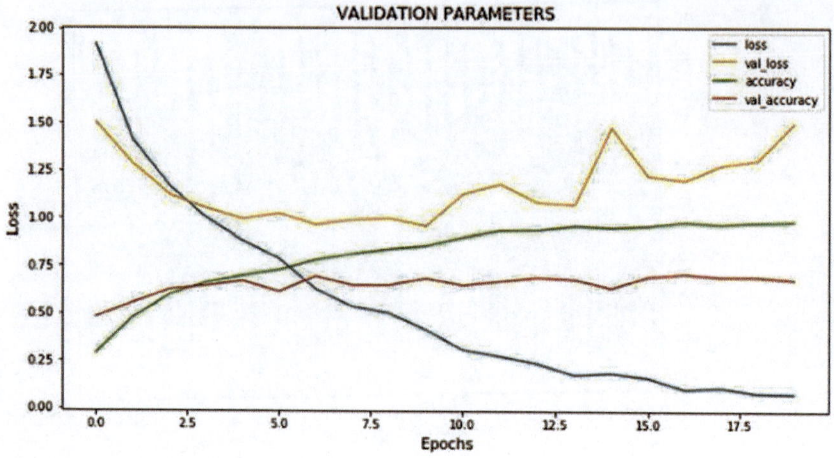

Fig. 7 Graph of different validation parameters

represents a plot for the validation parameters—loss, validation set loss, accuracy and validation set accuracy.

5.2 *ChatBot Model*

The ChatBot being the parent framework for the project is built using a deep neural network [17] through the TensorFlow and TFLearn libraries. A custom dataset of .json format was used to train the model using natural language tool kit (NLTK). The ChatBot simulates a conversation with the user and for a particular message, and it runs and shows the results of the music genre classifier.

Natural Language Tool Kit

NLTK or natural language tool kit [16] is a free plugin/package for Python that provides a framework and several functions which can be used to manipulate the text taken from a dataset or extracted from speech recognition or speech to text conversions. This toolkit is a convenient for the purpose of organization of the text into sentences and then into relevant keywords that would help the ChatBot recognize and process the input efficiently.

ChatBot Logic

The ChatBot is essentially a computer program which mimics an intelligent conversation [26]. In order for the ChatBot to produce a suitable response to the user's natural language text input, the ChatBot model [5] must be trained so that it takes the input from the user, split the strings of input into words and then tag the words with labels which can be further compared with the previously known conversational elements. Figure 7 is a pictorial representation of the data flow of the complete project, which clearly represents the functioning of the ChatBot and the classification model. The model prepared for this project yielded a training accuracy of 100%. The functioning of the ChatBot model can be divided into three sections.

Input Recognition and Model Training

In this part, the preprocessing of the input data takes place. A Lancaster Stemmer was employed, which converts the natural language input to a string of root words. The string, thus obtained, is then tokenized and separated according to the "tag" it may lie in, according to the dataset. The tags and the string obtained are sorted according to labels which are then passed through the one hot encoder, where the data is essentially referred to as by zeroes and ones in a matrix. The so-called bag of words thus obtained is then stored in a "training" array, and their respective labels are stored in an "output" array.

Finally, the model is trained with four layers (input layer, two hidden layers and an output layer). The fourth layer uses the "SoftMax" [24] activation function. The trained model thus created is a deep neural network which is saved to be used further.

Classification

A function is defined which can conveniently stem and tokenize the input given to it and pass them into an array "bag[ ]". This function will be used to convert the input into a form that can be easily understood by the trained model. The input taken from the user is finally run through the classifier which compares the relevant strings with the dataset and produces a response.

Response

Primary "Chat()" function is defined which calls the pre-trained model and takes in an input from the user. The user input is converted processed and is then compared with the elements in the dataset. Based on the input, the ChatBot generates an appropriate response. If the ChatBot is asked to run the genre classifier, then the test accuracy of the genre classification model, it gives an output of the predicted genres for the testing files of the GTZAN dataset [2].

As represented in Fig. 1, if the given input corresponds to a specific tag—"music" or "shop"—in the dataset, then the ChatBot will call the genre classifier, which in turn will predict the genres of a given test set, and the predicted genres will be given as a response by the ChatBot along with the test accuracy of the classification model.

6 Results

A custom dataset of tags and responses was used to train the ChatBot, and the NLTK library functions were used to select the relevant features from the given data. The model for the ChatBot was created using "TensorFlow" and "TFLearn" libraries [27]. Appropriate functions were defined that called the ChatBot and simulated the conversation. The primary feature of the ChatBot is the ability to classify audio files on the basis of genre and return the predicted genre of the test set and the testaccuracy of the model, to the user.

The music genre classification model was created using "TensorFlow" [27] and "Keras" [22] libraries. A convolutional neural network (CNN) was defined for the classification process which was trained using the GTZAN dataset.

Plots for the training accuracy and the training loss have been represented in Figs. 8 and 9. From these representations, it is evident that with increasing number of

Fig. 8 Plot for training accuracy

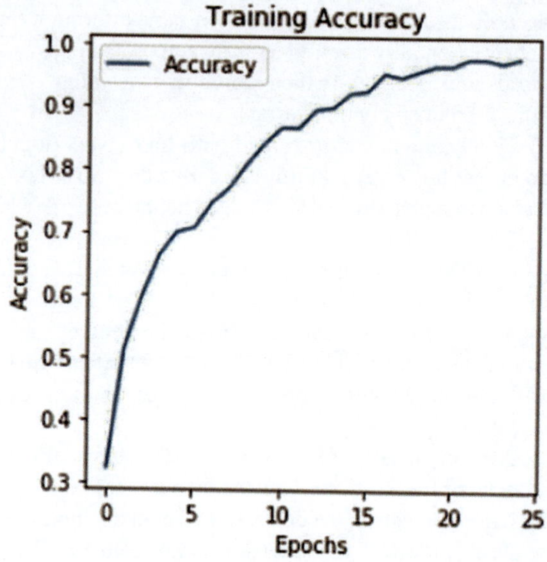

Fig. 9 Plot for training loss

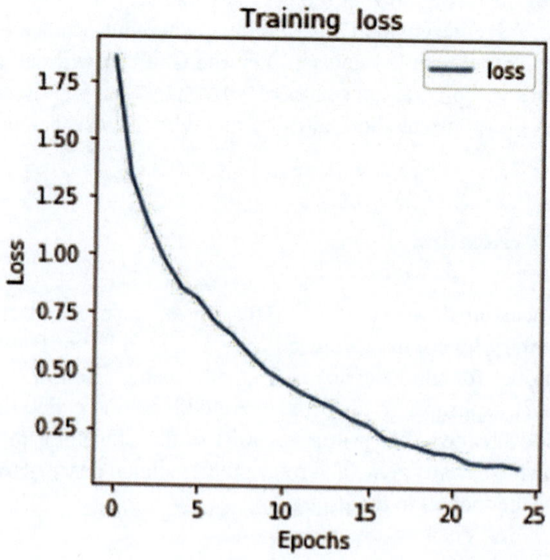

epochs (or iterations), the training loss decreases, and the training accuracy increases. On running the said models, a training accuracy of 97.8% and a test accuracy of approximately 68.9% were obtained from the classification model, with a training loss of 0.0782 and a test loss of 1.48.

Meanwhile on training the ChatBot DNN [17] model, we obtained a training accuracy of 100% and a total loss of 0.01242. This clearly signifies that the ChatBot was successfully trained, and when called, it produces a suitable response for the natural language input from the user.

7 Conclusion

In this paper, we discussed a ChatBot project whose primary function is to classify audio files on the basis of genre, through a convolutional neural network model built using Keras [22]. For the training of the classification model, we can use more elaborate and extensive datasets such as the million song dataset [28], which comprises a million audio files and accounts for almost 250 GB of storage data. Use of the million song dataset for training the model could have produced a more thorough training process, thereby yielding a better accuracy. The idea behind this project was to create a convenient and user-friendly ChatBot that would simulate a conversation with the user and as per the request from the user, run the genre classifier to categorize the audio file according to its genre. Such a technology can be used in audio streaming services which require fast and efficient management of data. Classification of audio tracks on the basis of genre can be useful as the user can access the songs to his/her liking. This particular project can be improved, and it can be implemented into any audio streaming service app, where the ChatBot can take note of the user's preferences and assist the user with recommendations based on his or her preferred genres.

References

1. Rabiner L, Juang B (1993) Fundamentals of speech recognition. Prentice-Hall, NJ
2. Tzanetakis G, Cook P (2002) Musical genre classification of audio signals. IEEE Trans Speech Audio Process 10(5):293–302. https://doi.org/10.1109/TSA.2002.800560
3. MUSIC type classification by spectral contrast feature. Department of Computer Science and Technology, Tsinghua University, China {llu, hjzhang}@ microsoft.com. Database, pp 0–3
4. Caclin A, McAdams S, Smith BK, Winsberg S (2005) Acoustic correlates of timbre space dimensions: a confirmatory study using synthetic tones. J Acoust Soc Am 118:471
5. Følstad, Brandtzaeg PB (2017) Chatbots and the new world of HCI. Interactions 24(4):38–42. https://doi.org/10.1145/3085558
6. Li D, Sethi IK, Dimitrova N, McGee T (2001) Classification of general audio data for content-based retrieval. Pattern Recogn Lett 22(5):533–544
7. Lambrou T, Kudumakis P, Speller R, Sandler M, Linney A (1998) Classification of audio signals using statistical features on time and wavelet transform domains. In: Proceedings of the

1998 IEEE international conference on acoustics, speech and signal processing, ICASSP'98 (Cat. No. 98CH36181), vol 6, pp 3621–3624
8. Deshpande H, Singh R, Nam U (2001) Classification of music signals in the visual domain. In Proceedings of the COST-G6 conference on digital audio effects
9. Soltau H, Schultz T, Westphal M, Waibel A (1998) Recognition of music types. In: Proceedings of the 1998 IEEE international conference on acoustics, speech and signal processing *ICASSP*, vol 2, pp 1137–1140. https://doi.org/10.1109/icassp.1998.675470
10. Vyas G, Dutta MK (2014) Automatic mood detection of indian music using mfccs and k-means algorithm. In: 2014 7th International conference on contemporary computing IC3 2014, pp 117–122. https://doi.org/10.1109/ic3.2014.6897159
11. Asim M, Ahmed Z (2017) Automatic music genres classification using machine learning. Int J Adv Comput Sci Appl 8(8):337–344. https://doi.org/10.14569/ijacsa.2017.080844
12. Ramírez J, Flores MJ (2019) Machine learning for music genre: multifaceted review and experimentation with audioset. J Intell Inf Syst. https://doi.org/10.1007/s10844-019-00582-9
13. Liu C, Feng L, Liu G, Wang H, Liu S (2019) Bottom-up broadcast neural network for music genre classification, pp 1–7. [Online]. Available: http://arxiv.org/abs/1901.08928
14. Dokania S, Singh V (2019) Graph representation learning for audio & music genre classification, no. 2017. [Online]. Available: http://arxiv.org/abs/1910.11117
15. Elhadad M (2010) Natural language processing with python Steven Bird, Ewan Klein, and Edward Loper. University of Melbourne, University of Edinburgh, and BBN Technologies) O'Reilly Media, Sebastopol, CA, xx + 482 pp; paperbound, ISBN 978-0-596-51649-9, $44.99; on-line free of charge at nltk.org/book. Comput Linguist 36:767–771. https://doi.org/10.1162/coli_r_00022
16. Loper E, Bird S (2002) NLTK: the natural language toolkit. [Online]. Available: http://arxiv.org/abs/cs/0205028
17. Salamon J, Bello JP (2017) Deep convolutional neural networks and data augmentation for environmental sound classification. IEEE Signal Process Lett 24(3):279–283. https://doi.org/10.1109/LSP.2017.2657381
18. Liang Y, Zhou Y, Wan T, Shu X, (2019) Deep neural networks with depthwise separable convolution for music genre classification. In: IEEE 2nd international conference on information communication and signal processing, ICICSP 2019, pp 267–270. 10.1109/ICICSP48821.2019.8958603
19. Costa YMG, Oliveria LS, Koerich AL et al (2011) Music genre recognition using spectrograms. In: 2011 18th International conference on systems, signals and image processing, pp 1–4
20. Ng AY (2004) Feature selection, L 1 vs. L 2 regularization, and rotational invariance. In: Proceedings of the twenty-first international conference on machine learning, p 78
21. Bahuleyan H (2018) Music genre classification using machine learning techniques
22. Choi K, Joo D, Kim J (2017) Kapre: on-GPU audio preprocessing layers for a quick implementation of deep neural network models with Keras
23. Chillara S, Kavitha AS, Neginhal SA, Haldia S, Vidyullatha KS (2019) Music genre classification using machine learning algorithms: a comparison. Int Res J Eng Technol 6(5):851–858
24. Avinash SV (2017) Understanding activation functions in neural networks. Medium 4(12):1–10
25. Zhang Z (2018) Improved adam optimizer for deep neural networks. In: 2018 IEEE/ACM 26th international symposium on quality of service (IWQoS), Banff, AB, Canada, pp 1–2. 10.1109/IWQoS.2018.8624183
26. Abdul-Kader S, John D (2015) Survey on Chatbot design techniques in speech conversation systems. Int J Adv Comput Sci Appl 6(7):72–80. https://doi.org/10.14569/ijacsa.2015.060712
27. Tang Y (2016) TF.Learn: TensorFlow's high-level module for distributed machine learning, pp 1–7
28. Bertin-Mahieux T, Ellis DPW, Whitman B, Lamere P (2011) The million song dataset. In: Proceedings of the **12th** international society for music information retrieval conference ISMIR 2011, pp 591–596

Detection of COVID-19 by X-rays Using Machine Learning and Deep Learning Models

Yash Varshney, Piyush Anand, Achyut Krishna, Preeti Nagrath, and Rachna Jain

Abstract In India, test for COVID-19 is very expensive and not everybody can afford it. This document provides knowledge and awareness to the reader on COVID-19 screening of a person using radio chest x-ray images. Here Machine Learning and Deep Learning algorithms like CNN and max-pooling are used. These algorithms identifies different features in the images and help us to distinguish between a COVID-19 and non COVID-19 chest X-ray. This paper also describes the data set of COVID19 open image X-rays.It was created by collecting medical images from websites and publications. Our model accuracy is following a trend of greater than 95% on every run time. Machine learning models can't have 100% accuracy and hence, this is the best one can get.

Keywords COVID-19 · Corona · Image · Kernel · X-Ray recognition · Pneumonia · Network · Matrix · Convolution

1 Introduction

The objective of this research work is to get fast and accurate results cheaply using Machine Learning and Deep Learning models by using X-rays as input training data of large number of patients. IEEE and concerned authorities have managed to develop data sets related to this disease. One of them is chest Xray images and these data sets with some little cleaning process can easily go through deep learning networks to get trained and predict whether a person is corona positive or not with very high accuracy i.e., above 95%. In order to achieve our predictions, Convolution Neural Networks

Y. Varshney (✉) · P. Anand · A. Krishna
Department of ECE, Bharati Vidyapeeth's College of Engineering (Aff. To IPU), New Delhi, India

P. Nagrath · R. Jain
Department of CSE, Bharati Vidyapeeth's College of Engineering (Aff. To IPU), New Delhi, India
e-mail: preeti.nagrath@bharatividyapeeth.edu

R. Jain
e-mail: rachna.jain@bharatividyapeeth.edu

P. K. Singh et al. (eds.), *Proceedings of Second International Conference on Computing, Communications, and Cyber-Security*, Lecture Notes in Networks and Systems 203,
https://doi.org/10.1007/978-981-16-0733-2_28

for processing X-ray images, max pooling for enhancing the quality and reducing dimensions of the image has been used. In order to understand these two terms one should know about Artificial Neural Networks. Data cleaning operations have been performed as the data sets from IEEE consists data of many other diseases other than COVID-19 and we have compared X-rays in this data with X-rays of pneumonia patients as corona also develop similar symptoms. Corona is more dangerous than pneumonia because in addition to pneumonia symptoms it can also affect affected persons kidneys (damaging them completely) leading to his/her death. This effort is just one step ahead towards finishing this pandemic by which whole world has been affected.

The work illustrated in this paper will surely help in the field that focuses on testing the sick for corona as this study provides a cheap way of analyzing human lungs and verify them if they are in a healthy state or not and whether they are corona positive or not. This technique will help doctors to attain test results fast and also on the other poor can also afford this technique.

This paper is organized in the following order. Firstly, the topic is introduced. Secondly, all the related work is discussed .In the third section various algorithms have been explained. Later after that, the architecture, results and finding of this study are discussed. Finally, this study has been concluded for the readers.

2 Related Work

Deep learning is playing an important role in improving today's generation technology creating next generation technology. In the paper [1] it has been explained thoroughly that deep learning has already made a huge improvement in various areas, such as medical field, automatic vehicles, weather prediction and voice recognition. It also covers various types of deep learning architectures such as DCN, DRN, RNN, reinforcement learning etc. In [2], it elaborates the various types of DNN architectures, training algorithms. There are various shortcomings of the training algorithms discussed. It also describes the necessity of optimization of training algorithms which will increase speed and efficiency of the network. There are few architectures, its corresponding algorithms and its implementations has been explained. In the concluding part the paper also emphasis the importance of deep learning and describes that it is still in its early stage. Deep learning is a growing field and various approaches have been used to bring it closer to work like biological neural network and to some extent mankind is successful in doing so. One such astonishing example is image processing through Convolution neural networks [3] and other tools like max pooling. Image processing in turn is acting as a powerful tool for humans to fight COVID-19. Proper screening of patients is major step towards fighting this pandemic.

As in [4] chest radiology gives evidences of Pneumonia and the X-ray of that patient can be distinguished from the uninfected person. So,unlike [4] instead of using COVID-net we have just used convolution and max pooling and still got very high

Fig. 1 Related work

Method	Month-Year	Accuracy (%)
DeTraC	March-2020	95.12
ResNetV2	April-2020	97
COVID-Net	April-2020	96.78
VGG-19	March-2020	86.7
Adam(our model)	2020	97.52

accuracy. As described in [5] CNN can train large amount of data, with millions of parameters. Hence CNN is very useful for detecting features that are not visible with naked eyes. The evolution of image processing or texture recognition is thoroughly explained in [6] and now CNN is accepted widely.

As explained in [7], max pooling is also essential for this classification as it boosts convergence rate by selecting higher-grade invariant features which enhances performance [7]. It also suggests use of soft-max function for representing probability of each category and selecting the class with highest probability as output.

We got motivated from the need of faster interpretation of radiography images so we planned to train a Deep learning Model [8] based on a CNN network and results have shown to be quite promising in terms of accuracy in detecting patients infected with COVID-19 via radiography imaging, we have collected data from two different data sets that are widely available from github [9] and kaggle for pneumonia where other were using only from one of them [4].Our model is trained on 224 images based on Covid and Normal images which were greater than other [10] due to which the model gets better training and we achieve a greater accuracy 97.52% in validation accuracy and 98.8% in training accuracy with adam optimizer where other optimizer like DeTraC [11] got lesser accuracy 97.2% (Fig. 1).

3 Preliminaries

Before computing predictions of the test, one should know how those results were achieved. In this section various algorithms are explained that has been implemented for this purpose.

3.1 *Artificial Neural Networks (ANN)*

The elementary unit of Artificial Neural networks is called Artificial neuron, which is similar to the neurons that are in a biological brain. These artificial neurons pass information from one neuron to another neuron. At every node or neuron, signal from the input or the previous neuron is processed and passed to the next node or neuron. In general, there are three types of layers: input layer, hidden layer and output layer. Input nodes are connected to each node of hidden layer in every possible way to one another. Similarly each node of hidden layer is connected to each node of output/next hidden layer. The hidden layers' node perform some specific function on the input given to it which is the main task of this process. There can be multiple hidden layers in these networks. Also each node to node connection have some weight. To decide efficient value of weights we use a technique called Backward Propagation.

The *Back-propagation* algorithm uses gradient descent to decide weights that minimize the error function of the given neural network. The result of the decided weights gives the minimal value of the error function of the model. In simple words we perform the following steps:

- First, random value of weights are set and the model is propagated forward.
- If, some error is observed so to reduce such error, backward propagation is done and the value of 'W' increased.
- Afterward, if the error increases, the value of weights can't be increased—So, again propagated to backwards and the value of weights are decreased.
- Now, the error has been reduced. Refer to [1–3].

3.2 *Convolution Operation and Convolution Neural Networks*

For identification of some particular object in images with many things and background we have to adopt to a new technique known as Convolution Neural Network or simply CNN. Here we decide a filter through which our image is passed which is a small matrix whose dimensions are decided by us. For example, if we have a 3×3 filter then this filter will slide over each 33×3 matrix of pixels in the image. This sliding is refer to as convolution. When the filter first lands on the 33×3 pixels of the input. The filter dot products itself with the input and the computed result is stored. This is done until we are done with each 33×3 pixels of image. After this process, we'll be left with new representation of our input. This matrix is the output and final result of our convolution operation. We can make this filter as pattern detector and hence can detect objects, shapes etc. after passing our image to many of such filters. These filter may be present in one hidden layer of our neural network or in many. Hence, such type of neural network is called CNN. Refer Figs. 2 and 3.

Convolution Neural Network comes under Deep Learning that deals with the neural networks for image processing. CNN represents huge break-through in the image recognition in machine and deep learning. It analyzes the visual features in

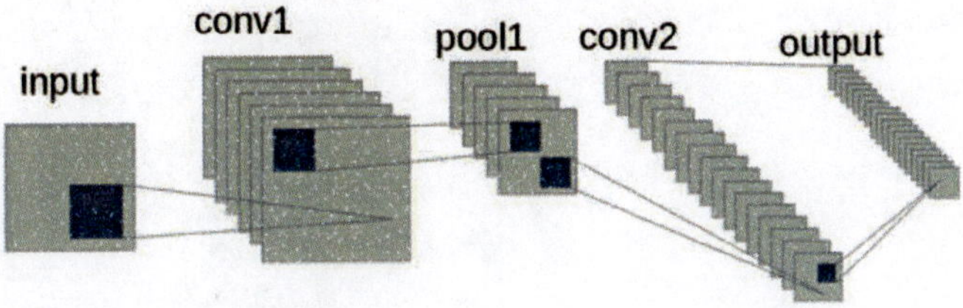

Fig. 2 Convolution neural network

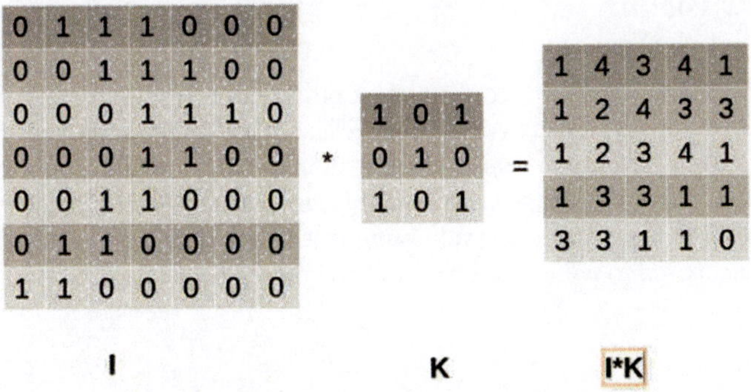

Fig. 3 Convolution operation on an image

an image and classify images on the basis of these features. In this classification process, we take an Input of images and output a categorical result (like "Covid" or "Normal") or the probability that input is from particular class ("there's a 90% probability that this Person is a Normal"). For radiology it is easier but for fast results we can train our machine with a Convolution neural network!

Following is an example of a CNN architecture:

INPUT – Convolution – ReLU – Convolution – ReLU – Pooling – Convolution – ReLU – Pooling – Convolution – ReLU – Pooling – OUTPUT.

This clears that CNN is very good network and it seems to be ideal for processing any type of 2D images as also shown in [5, 6, 12].CNN provides the programmer the advantage of using any kind of filter to detect any kind of features. CNN is used in many applications like face recognition, NLP, recommender systems, video processing and many more. CNN can also be modified with many other algorithms according to the need of the programmer. It also uses very little pre-processing.

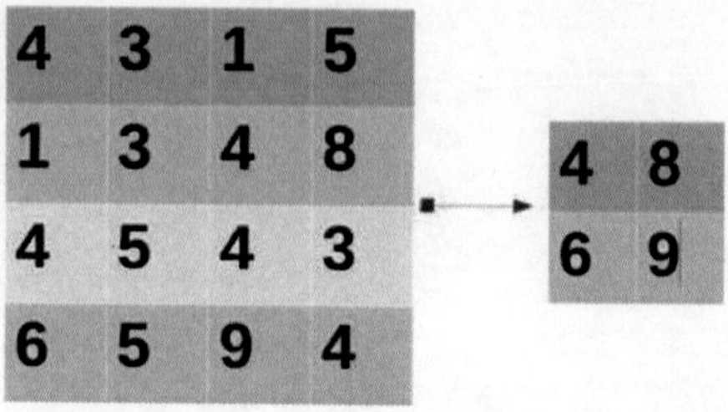

Fig. 4 Max pooling

3.3 *Max Pooling*

Max pooling is a sample-based discretization process. In this process a filter of desired dimensions is passed over the image. The stride i.e. the amount of blocks the filter will move per iteration is also set according to the purpose of implementation. At every iteration on the image at different places selecting different blocks the filter outputs the block with maximum value amongst all blocks that lie within the size of filter. It is also used to reduce variance and computations. Refer to Fig. 4.

4 Datasets

4.1 *Chest X-Ray Images (Pneumonia)*

Pneumonia is a very dangerous disease especially for infants below 2 years and for those who are above 65 years of age. This is because there immune system isn't much active. It causes infection in lungs causing the air sacs to be filled with puss or fluid. This results in abnormal and difficult breathing. Sometimes the patients don't even get to know that he/she is carrying the disease. This condition is called walking pneumonia by doctors. This situation can affection one or both lungs but the patient suffer similarly in both conditions. We are discussing it because one of the symptoms of COVID-19 is pneumonia refer to [9]. This data set is divided into 3 folders train, test, validation inside each of these there are 3 folders named categorically into Pneumonia and Normal (X-Ray Images).The total no. of these images are 5863 (Fig. 5).

4.2 *Ieee8023/Covid-Chestxray-Dataset*

This data set is collected from hospitals, doctors and physicians who treat corona virus patients. It consists of X-rays of patients having diseases like SARS, MERS

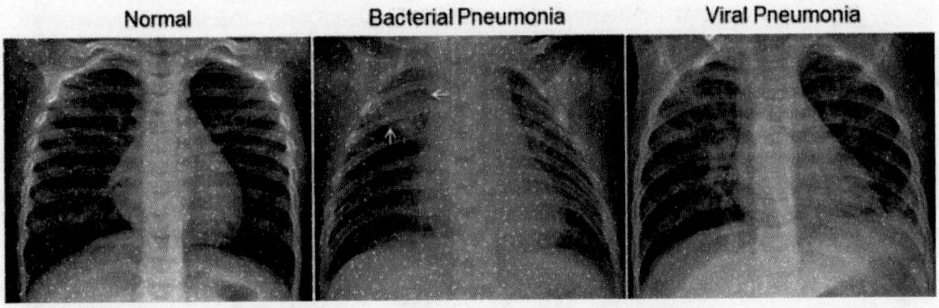

Fig. 5 Convolution neural network

and ARDS or suspected of having them. This data is released on a basis of regular intervals in the Github repository (Fig. 6).

5 Architecture

In this architecture, at first this model takes an X-RAY RGB (coloured) image as input. Padding is same here. Padding means the number of pixels integrated with the image when the kernel of the Convolution neural network processes the image. A 3 $\times$ 3 kernal having 32 filters which will extract the features from this image has been used. Then the no. of filters has been increased from 32 to 64 for better extraction of features of the image. Padding is valid here. After that to discretized, max pooling is done. This pooling will down sample the representation of the input image. There is a dropout (of value $p = 0.25$) layer (Fig. 7).

In the dropout layer, each neuron generates 1 feature map. Since dropout works with every neuron, dropping a neuron implies that the corresponding feature map has been dropped. The dropout layer is generally added after the pooling layer. At the next 2 steps, the first 2 steps of the model will be repeated again, with a change in no. of filters to 64 128 from 32 64 and the pooling remains the same. Again, there is a dropout (of value $p = 0.25$) layer. Now add a flatten layer that will make a vector of the all the connected layers of the processed image, for instance, function used here for flattening is Sequential.add(Flatten()). Now, add 1 fully connected layer with Sequential.add(Dense)) function in the Keras. There is an addition of dropout (of value $p = 0.25$) layer again. The model is trained now. After all these processes, at last the model produces the output. The output is generated with the help of prediction. The output produced is binary in nature i.e. infected or not infected (Fig. 8).

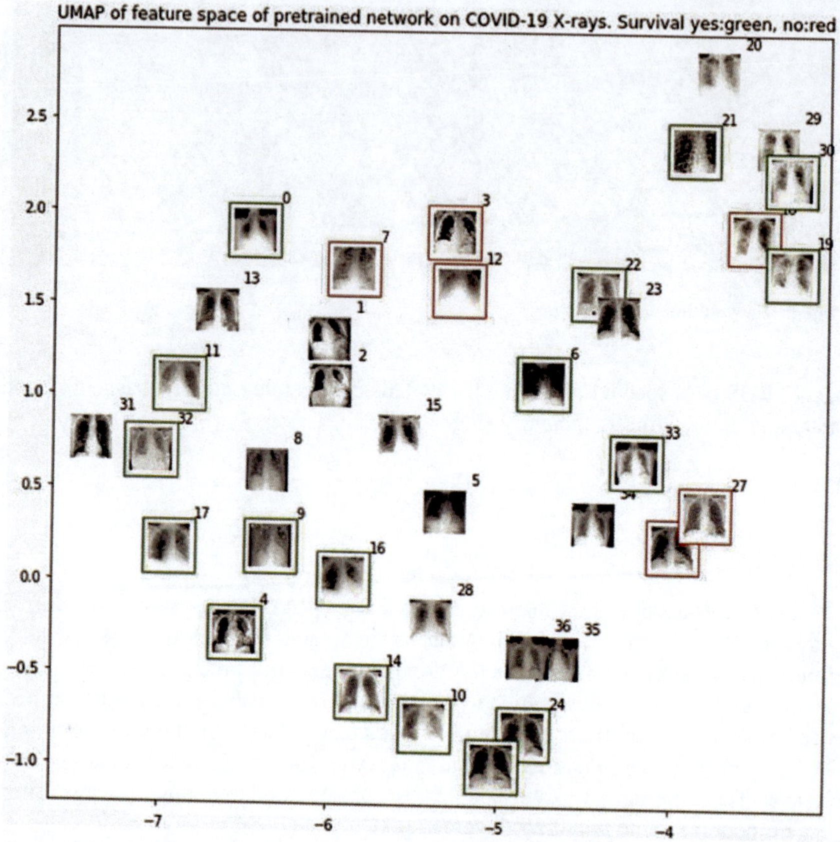

Fig. 6 UMAP of feature space of pretained network on Covid-19 X-rays. Survival yes: green, no: red

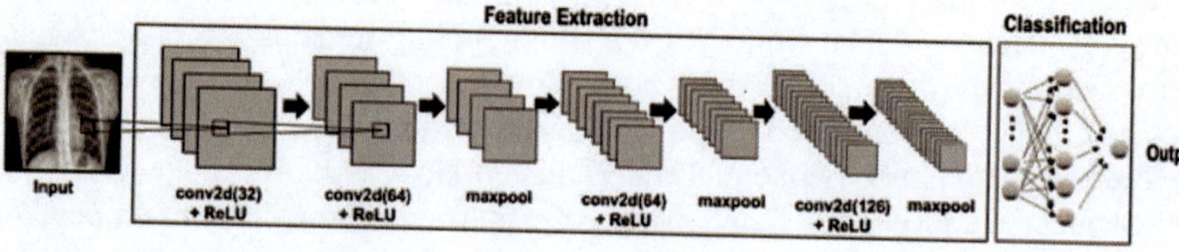

Fig. 7 Architecture of the implemented model

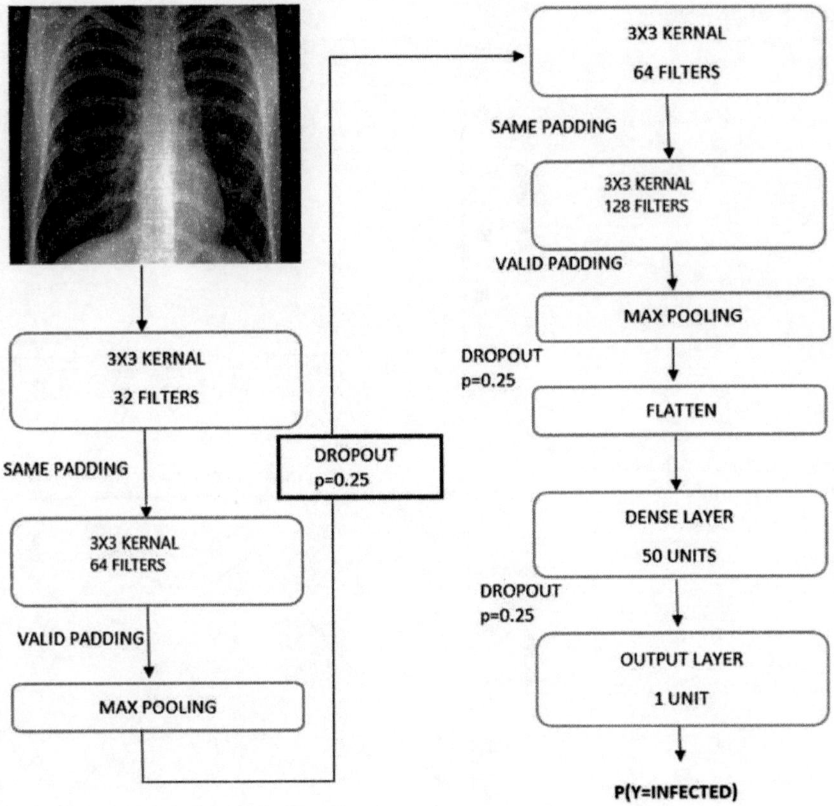

Fig. 8 Flowchart of the implemented model

6 Results and Analysis

6.1 Accuracy Curve

Referencing to [13], Accuracy curves is one of the method to study the progress of deep neural networks. For anyone who has some experience in Deep Learning, using accuracy and loss curves is obvious. A more important curve is the one with both training and validation accuracy. The gap between training and validation accuracy is a clear indication of over fitting. The larger the gap, the higher the over fitting. Hence, it's clear that IEEE data set will show over fitting just after 2 epoch while kaggle data set after epoches greater than 8 converges which is a good thing for our model (Figs. 9 and 10).

Fig. 9 IEEE model (accuracy—97.52%)

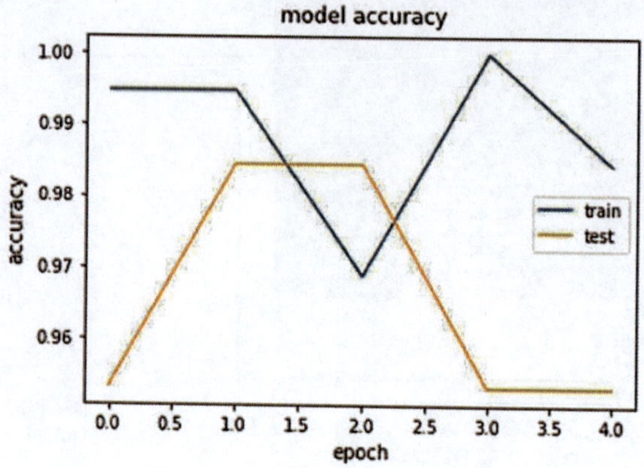

Fig. 10 Kaggle Model (accuracy—96.66%)

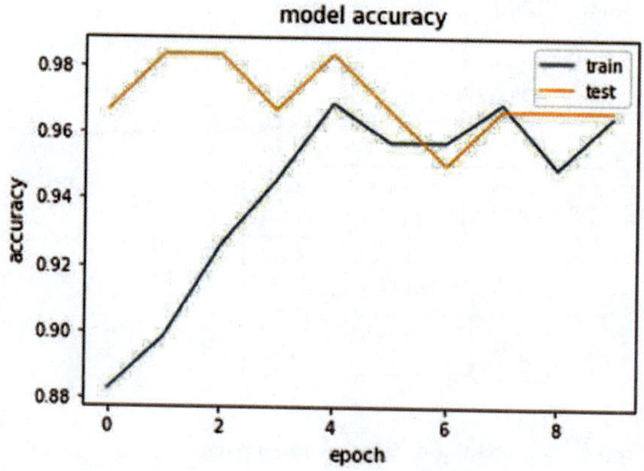

6.2 *Confusion Matrix*

Here in Fig. 7 and in Fig. 8 two confusion matrix are shown one representing dataset of IEEE and other representing that of Kaggle respectively. Here 0 represents Corona positive while 1 represents normal patient. The top left box (0, 0).Here on x-axis represents the actual data while y-axis represents predicted data. The (0,0) box represents people that actually have corona and our model also detected them while (1,1) box represents people who in real don't have corona and also our model predicted about them correctly. Hence, the no. written in these two box represents the no. of people and therefore this no. should be high. Similarly the blue box represents the reverse of this situation and hence no. labelled on it should be low. Our confusion matrix follow these principles and hence our model is working good! Refer to [7] (Figs. 11 and 12).

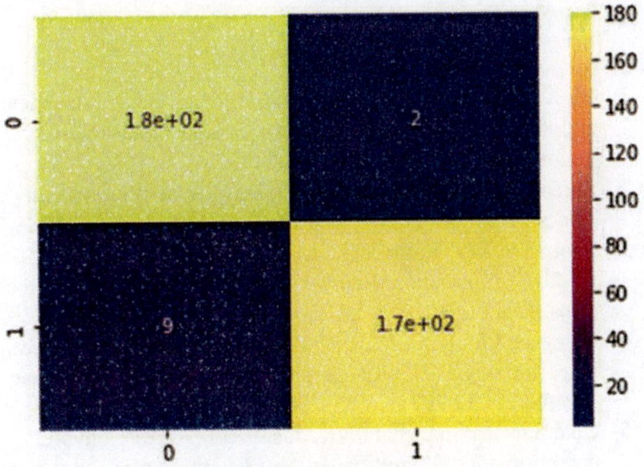

Fig. 11 IEEE Confusion matrix

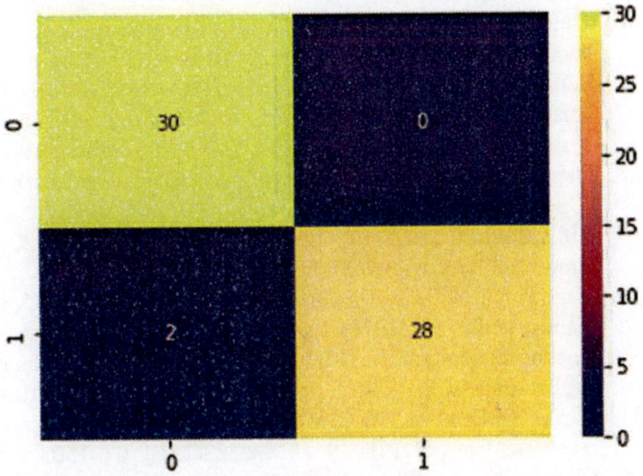

Fig. 12 Kaggle Confusion matrix

7 Conclusion

It can be concluded that Deep learning technology is acting as a boon for the humanity in this pandemic situation. As a programmer who don't know about biology can also predict by just processing X-rays through these type of neural networks. This project must encourage others to implement there knowledge somehow, to improve this situation. Since normal blood tests are expensive, this type of tests are very reliable. But there is a drawback also. As 100% accurate model can't be made so there is chance that positive patients go undetected and hence they can spread the virus. For eg in X-rays of 100 people and with accuracy say, 97% implies a chance of leaving 3 people undetected and hence they can become the virus spreading agents if not quarantined.

References

1. Du X, Cai Y, Wang S, Zhang L (2016) Overview of deep learning. In: 2016 31st Youth academic annual conference of chinese association of automation (YAC), Wuhan, pp 159–164. https://doi.org/10.1109/YAC.2016.7804882
2. Shrestha A, Mahmood A (2019) Review of deep learning algorithms and architectures. IEEE Access 7:53040–53065. https://doi.org/10.1109/ACCESS.2019.2912200
3. Albawi S, Mohammed TA, Al-Zawi S, (2017) Understanding of a convolution neural network. In: 2017 International conference on engineering and technology (ICET), Antalya, pp 1–6. https://doi.org/10.1109/ICEngTechnol.2017.8308186
4. Wang L, Wong A (2020) COVID-net: a tailored deep convolution neural network design for detection of COVID-19 cases from chest radiographyimages. arXiv:2003.09871. https://arxiv.org/abs/2003.09871
5. Chauhan R, Ghanshala KK, Joshi RC (2018) Convolution neural network (CNN) for image detection and recognition. In: 2018 First international conference on secure cyber computing and communication (ICSCCC), Jalandhar, India, pp 278–282. https://doi.org/10.1109/ICSCCC.2018.8703316.
6. Zhu G, Li B, Hong S, Mao B (2018) Texture recognition and classification based on deep learning. In: 2018 Sixth international conference on advanced cloud and big data (CBD), Lanzhou, pp 344–348. https://doi.org/10.1109/CBD.2018.00068
7. Huang J, Ling CX (2005) Using AUC and accuracy in evaluating learning algorithms. IEEE Trans Knowl Data Eng 17(3):299–310. https://doi.org/10.1109/TKDE.2005.50
8. Fang Y et al (2020) Sensitivity of chest CT for covid-19: comparison to RTPCR. Radiology 200432
9. IEEE Covid Chest X-Ray Dataset. Accessed 7 Mar 2020. [Online].Available: https://github.com/ieee8023/covid-chestxray-dataset
10. Nagi J, Ducatelle F, Di Caro GA, Cireşan D, Meier U, Giusti A, Nagi F, Schmidhuber J, Gambardella LM (2011) Max-pooling convolution neural networks for vision-based hand gesture recognition. In: 2011 IEEE international conference on signal and image processing applications, ICSIPA2011, pp 342–347. https://doi.org/10.1109/ICSIPA.2011.6144164
11. Lan L, Xu D, Ye G, Xia C, Wang S, Li Y, Xu H (2020) Positive RT-PCR test results in patients recovered from COVID-19. JAMA 323(15):1502–1503
12. Chen Z. Zhou Y (2019) Research on automatic essay scoring of composition based on CNN and OR. In: 2019 2nd International conference on artificial intelligence and big data (ICAIBD), Chengdu, China, pp 13–18. https://doi.org/10.1109/ICAIBD.2019.8837007
13. Bonettini N, Paracchini M, Bestagini P, Marcon M, Tubaro S (2019) Hyperspectral X-ray denoising: model-based and data-driven solutions. In: 2019 27th European signal processing conference (EUSIPCO), A Coruna, Spain, pp 1–5. https://doi.org/10.23919/EUSIPCO.2019.8903151
14. Wang W, Xu Y, Gao R et al. (2020) Detection of SARS-CoV-2 in different types of clinical specimens. JAMA 323(18):1843–1844. https://doi.org/10.1001/jama.2020.3786
15. Jacobs S, Bean CP (1963) Fine particles, thin films and exchange anisotropy. In: Magnetism GT, Rado, Suhl H (eds), vol 3. Academic, New York, pp 271–350
16. Imran A, Posokhova I, Qureshi HN, Masood U, Riaz S, Ali K, John CN, Nabeel M, Hussain I (2020) AI4COVID-19: AI enabled preliminary diagnosis for COVID-19 from cough samples via an app. arXiv:2004.01275. https://arxiv.org/abs/2004.01275
17. Zhong L, Mu L, Li J, Wang J, Yin Z, Liu D (2020) Early prediction of the 2019 novel coronavirus outbreak in the mainland china based on simple mathematical model. IEEE Access: Pract Innov Open Solutions 8:51761–51769. https://doi.org/10.1109/ACCESS.2020.2979599
18. Beers A, Brown J, Chang K, Campbell JP, Ostmo S, Chiang MF, Kalpathy-Cramer J (2018) High-resolution medical image synthesis using progressively grown generative adversarial networks. arXiv:1805.03144. [Online]. Available: https://arxiv.org/abs/1805.03144

19. Dai W, Doyle J, Liang X, Zhang H, Dong N, Li Y, Xing EP (2017) SCAN: structure correcting adversarial network for chest X-raysorgan segmentation, arXiv:1703.08770. [Online]. Available: https://arxiv.org/abs/1703.08770
20. Mondal S, Agarwal K, Rashid M (2019) Deep learning approach for automatic classification of X-ray images using convolution neural network. In: 2019 Fifth international conference on image information processing (ICIIP), Shimla, India, pp 326–331. https://doi.org/10.1109/ICIIP47207.2019.8985687
21. Ayan E, Unver HM (2019) Diagnosis of pneumonia from chest X-ray image using deep learning. In: 2019 Scientific meeting on electrical-electronics biomedical engineering and computer science (EBBT), Istanbul, Turkey, pp 1–5. https://doi.org/10.1109/EBBT.2019.8741582
22. Bouchahma M, Ben Hammouda S, Kouki S, Alshemaili M, Samara K (2019) An automatic dental decay treatment prediction using a deep convolution neural network on X-ray images. In: 2019 IEEE/ACS 16th international conference on computer systems and applications (AICCSA), Abu Dhabi, United Arab Emirates,pp 1–4. https://doi.org/10.1109/AICCSA47632.2019.9035278
23. Wibisono A, Adibah J, Priatmadji FS, Viderisa NZ, Husna A, Mursanto P (2019) Segmentation-based knowledge extraction from chest X-ray images. In: 2019 4th Asia-Pacific conference on intelligent robot systems (ACIRS), Nagoya, Japan, pp 225–230. https://doi.org/10.1109/ACIRS.2019.8935951.
24. Ren X et al (2019) Regression convolution neural network for automated pediatric bone age assessment from hand radiograph. IEEE J Biomed Health Inform 23(5):2030–2038. https://doi.org/10.1109/JBHI.2018.2876916
25. Marom ND, Rokach L, Shmilovici A (2010) Using the confusion matrix for improving ensemble classifiers. In: 2010 IEEE 26-th convention of electrical and electronics engineers in Israel, Eliat, pp 000555–000559. https://doi.org/10.1109/EEEI.2010.5662159
26. Abbas A, Abdelsamea M, Gaber M (2020).Classification of COVID-19 in chest X-ray images using DeTraC deepConvolution neural network. https://doi.org/10.1101/2020.03.30.20047456

Implication of Machine Learning Models Toward Education Loan Repayment Rate Analysis

Anushree Bansal and Shikha Singh

Abstract Education loan is a significant factor contributing to one's decision to pursue studies. Students now do not hesitate about taking the risk of being in debt of thousands of dollars. They believe in getting the highest quality of academic qualification first without realizing how risky it can get to be in such an enormous amount of debt. Banks nowadays are also wary of approving loans because they face difficulty in analyzing how credible the borrower is. It makes the whole process very tedious and time taking and often proves to be inefficacious. Prediction of education loan repayment rate can make the job easier for both banks and the applicants. This research paper aims at analyzing the education loan repayment rate by the use of machine learning algorithms. Machine learning is extensively used now and finds application in almost every domain. Predictions and analysis carried out using ML helps in making an informed decision and gives an idea of how future trends predict to look. In this research, various features are analyzed and researched thoroughly by the use of Python language. Its extensive set of libraries enables easy manipulation and visualization of the data. The paper contains a description of the analysis and rich visuals to produce a clearer image of the dataset. Various models are implemented, and their accuracy is measured using the R2 score.

Keywords Machine learning · Loan repayment · Prediction · Regression

1 Introduction

Education loans are the loans which are generally taken by the student or their guardians to fund the student's studies. Students tend to invest a colossal amount of money to get a quality education from esteemed universities. In the present era, where self-learning and online training have become so prevalent among youngsters,

A. Bansal (✉) · S. Singh
Department of Computer Science and Engineering, Amity School of Engineering and Technology, Amity University, Lucknow, India

P. K. Singh et al. (eds.), *Proceedings of Second International Conference on Computing, Communications, and Cyber-Security*, Lecture Notes in Networks and Systems 203,
https://doi.org/10.1007/978-981-16-0733-2_29

a large population continues to prefer offline, face-to-face classes and is inclined to achieve a proper degree and recognition from an established institute over online certifications and courses.

Education is a crucial part of every person's life. Students take a loan of a massive amount of money worth thousands of dollars to study at their desired universities without analyzing the chances of whether this loan is repayable or not. Banks also require a proper assessment of various features related to the student to sanction these loans. Some of these features may include—degree opted for, the period of their education, job prospects after completing the degree, etc. These features play a vital role in determining the loan repayment ability of a student.

The USA is known as a hub for opportunities. It is famous for the most esteemed universities located in the country and equally for the investment required to study in them. As of the data available until March 2019, about 43 million Americans hold student loans that were delivered by the chief sector of the education loan market that is the federal government programs [1].

Around 20% of borrowers are in default—which means they have gone without payment for at least 270 days—millions more are late, and more than a million loans go into default every year, as testified by the U.S. Department of Education [2]. Every year millions of students enroll in programs at colleges in the USA alone with a sum of loans in trillions of dollars. This paper deals with studying the education loan repayment rate of students studying in colleges of the USA.

Machine learning and artificial intelligence have been used in the field of education and that of finance to make remarkable predictions. AI is used to develop better systems for the management of loan repayment using ML algorithms. The purpose of this research paper is to use machine learning algorithms to analyze the education loan repayment rate of the borrower that can be lucrative to both the borrower and the lender.

2 Literature Review

Student education loan debt holds a large proportion of debts among the young borrowers in America and is expanding rapidly over time. Researchers have been conducting detailed researches for years on how one's debts are interrelated to the overall finance and how much impact it has on the borrower's payments and general lifestyle.

In research [3], the authors work on the objective that whether granting a loan to a person is safe for the banks or not. This is determined by mining the big data of previous records of people to whom the loan was sanctioned. It tries to reduce the risk factor involved with approving loans and concludes that applicants with a poor credit history are less probable to get a loan because they might not be able to pay it back and also applicants with high income asking for a low amount of money are more likely to get their loans sanctioned as they are more likely to pay them back.

In the research [4], the authors analyze whether an applicant is a defaulter or not by applying data mining on the bank's data using R language. The random forest classification model is used to test and train the data. The analysis results in easy identification of required information for successful loan prediction and reduction of the number of bad loan problems.

In [5], the authors applied analysis on seven major elements considered while sanctioning the loan by the banks. It concludes that most numbers of loan applicants prefer short-term loans, and the loans applied were vastly for debt consolidation.

In [6], the author uses R language to find the prospect of default of a bank loan application to help banks prevent losses. The model uses the decision tree classification algorithm to predict the category of new borrowers.

Three different types of models are applied to the dataset in [7] to predict the loan approval of customers. It determines that the decision tree works the best on them and helps most accurately to classify the loan applicant.

In paper [14], the authors emphasize the use of a hybrid model for better accuracy as compared to traditional models. The author uses R, Python, and Waikato Environment for Knowledge Analysis (Weka) to introduce a new technique that can increase the progress of the banking sector by the use of classification models.

In research [15], the authors write about using an ensemble model for loan prediction analysis and how it performs as compared to stand-alone models. Different types of ensemble learning algorithms such as Bagging, Boosting, Stacking, and AdaBoost are described in the paper. Benefits of ensemble models such as better forecasting, a more constant model, better results, and reduced errors are also discussed.

In paper [16], the authors propose a classification task for differentiating between defaulters and non-defaulters to conduct credit risk management. They have used tools like Waikato Environment for Knowledge Analysis (Weka) and Konstanz Information Miner (KNIME) for examining the performance of the classifiers. Machine learning algorithms are applied on two different credit scoring datasets, and the results show that random forest performs the best while naïve Bayes exhibits the least accuracy.

In [17], the author discusses how feature selection and ensemble framework can be useful in getting more accurate results. Three credit scoring datasets are fit on five different models, and the output is obtained based on the weighted voting approach. As a result, the ensemble model outperformed the traditional feature selection techniques.

In research [18], the importance of data mining to extract vital information about customer behavior and loans from a huge amount of accumulated dataset is discussed. A classification model based on data mining is built. Three algorithms are used to build the model, and it is tested on the Weka application. The results indicate that J48 works the best with the highest accuracy and low mean absolute error followed by BayesNet and naïve Bayes algorithm.

In [19], a framework for risk evaluation based on the k-means clustering technique is discussed. Relevant attributes are selected using information gain theory. Three attributes are selected for primary risk and six for secondary risk. Results show that based on risk percentage, the customers are classified into low, medium, and high groups and that this model predicts higher accuracy and takes less time.

Through this review, it can be inferred that loan repayment is an essential domain on which machine learning can be applied to produce better and more accurate results and the algorithms used have proven to work well on them.

3 Researched Methodology

Several steps are followed to conduct the analysis (see Fig. 1).

3.1 Data Collection

The dataset used is the college scorecard project [8]. It was designed to increase transparency and to provide the students and their families the facility to access data without any discrepancies. The federal government released the dataset in 2010. The

Fig. 1 Methodology flow

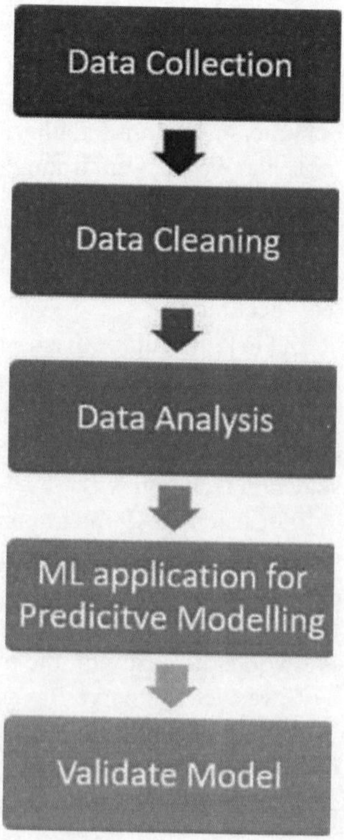

dataset contains information about thousands of colleges. Since it is a huge dataset and comprises a large number of features, only some useful features are selected that are renamed according to their properties.

3.2 *Data Cleaning*

Invalid and empty entries possess a big problem that hinders the correctness of the training model. It affects the accuracy of the model. The data is refined by analyzing the number of NULL values and is cleaned (See Fig. 2).

Since COST_P has the largest null values percentage (68.04%), it is dealt with first. The values in COST_P and COST_A contain a large number of null values as these features represent the average cost of attendance for program year institutions and academic year institutions, respectively. Thus, a common feature named 'TotalCost' is calculated to get the average cost of attendance irrespective of the institution type. Rows with a large number of null values are removed completely. The threshold is set to 30%, which means that any row containing more than 30% NULL values is removed (See Fig. 3).

There are two types of null values in the dataset that are present in the form of–'NULL' and 'PrivacySuppressed'. According to the data dictionary [8], the 'PrivacySuppressed' elements are the elements not revealed for privacy purposes and did not meet the reporting standards. These are either removed, replaced, or imputed to get a clean dataset free of null values. Features containing non-numeric values are also encoded. All duplicate institute entries are removed as well.

	Feature	Null Value Percentage
0	INST_NAME	0.000000
1	HIGH_DEG	0.000000
2	INST_TYPE	0.000000
3	PD_CS	11.313547
4	UGDS	11.240082
5	UGDS_NRA	11.240082
6	COSTA_A	49.588598
7	COSTA_P	68.042903
8	PCTFLOAN	11.636791
9	CDR3	13.620335
10	COMP_ORIG_YR4_RT	5.744931
11	GRAD_DEBT_MDN	4.231560
12	AGE_ENTRY	7.373845
13	MEDIAN_HH_INC	30.091096
14	MD_EARN_WNE_P8	19.144872
15	UGDS_WOMEN	11.240082
16	COMPL_RPY_3YR_RT	16.882163

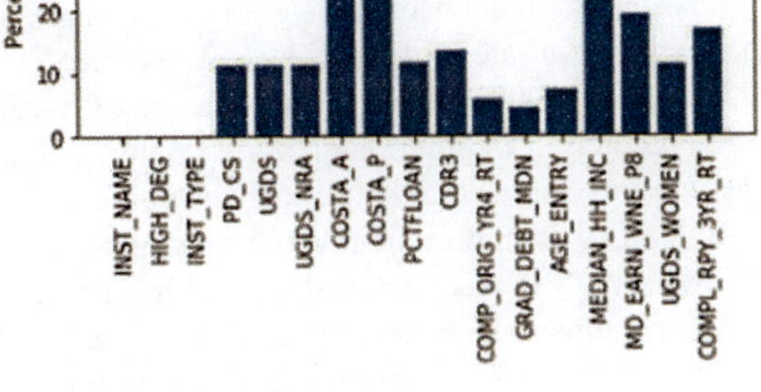

Fig. 2 Percentage of null values in each feature

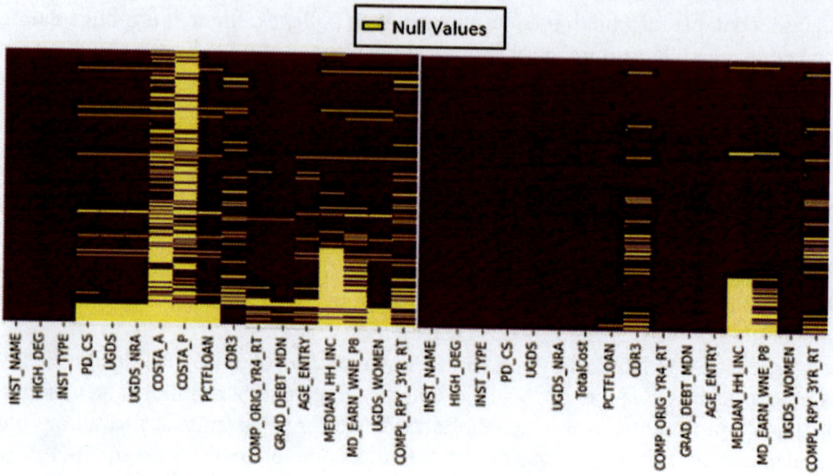

Fig. 3 Before and after cleaning of rows with excessive null values

3.3 *Data Analysis*

All the values in the dataset belong to different ranges. If some features are of a greater magnitude than others, it might have more influence on the predicted result than other equally important features. It may hamper the results and may fail in learning equally from all the features [9]. Therefore, all the values are standardized. Then, the dataset is analyzed to obtain the correlation between different features (See Fig. 4) and the correlation of each feature with the output feature (See Fig. 5).

As observed, the median earnings of students working and not enrolled eight years after entry (MD_EARN_WNE_P8) is the most closely related feature to the output feature, i.e., three-year repayment rate for completers (COMP_RPY_3YR_RT).

3.4 *Machine Learning Application for Predictive Modeling*

In this step, various machine learning models are built on the dataset to predict the loan repayment rate of the borrower. These are:

Multiple Regression. It is an extension of linear regression. In real-world problems, prediction of some value rarely depends on only one characteristic. Unlike linear regression where the response variable depends on a single feature, multiple regression takes into account various independent variables to make a prediction. It follows a many to one relationship. Multiple linear regression assumes that there is a linear relationship between the dependent and the independent variables and forms the baseline model to perform machine learning regression problems.

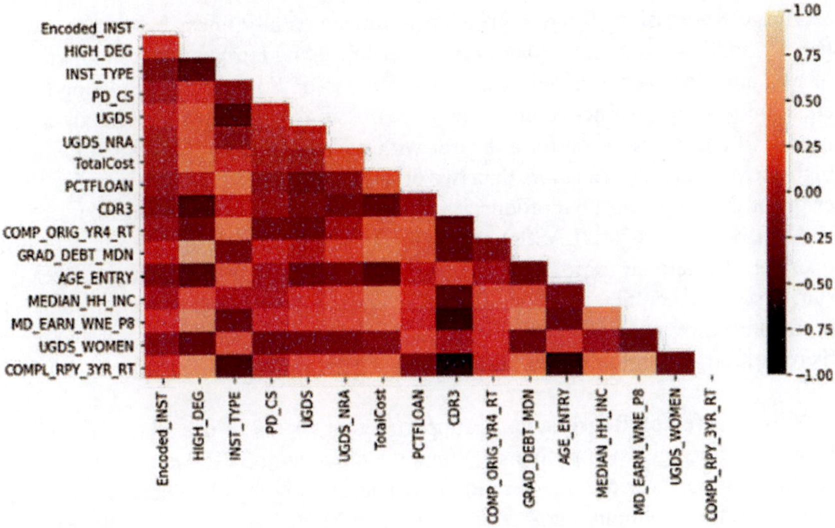

Fig. 4 Correlation among features

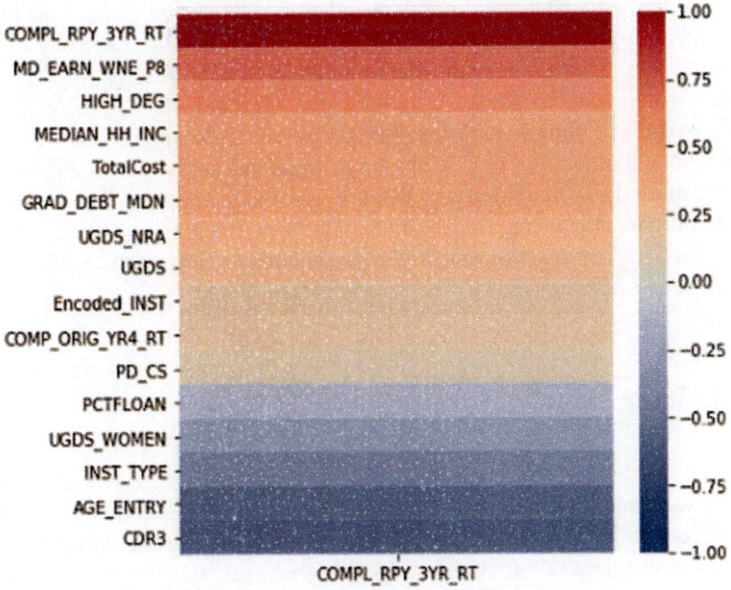

Fig. 5 Correlation of all the features with the output feature

Ridge Regression. Ridge regression is fundamentally linear regression with a penalty and is used to reduce variance in the dataset by introducing a small amount of bias and also used to avoid over fitting in the data. For answer estimation for an equation with no distinct solution, ridge regression is the most general algorithm used [10] and is also used for a dataset with multicollinearity, i.e., when there are high correlations between more than two predicted variables [11]. It might give some error for the test dataset but produces a generalized model for the whole dataset.

LASSO Regression. LASSO stands for least absolute shrinkage and selection operator. It is similar to ridge regression. Unlike ridge regression that penalizes the sum of squared coefficients, LASSO does so with the total of their absolute values. It is an alternative to the classical least square estimate used in linear regression when there is a large number of features and is used to avoid problems like over fitting [12].

Support Vector Regression. A support vector regressor uses the same rule as SVMs but for regression problems. When we move with SVR, we aim at basically considering points within the decision boundary. Our best fit line here is the hyper-plane with the maximum number of points [13]. This algorithm is complicated, but the idea always remains the same, i.e., to minimize the error.

Decision Tree Regression. A decision tree is a supervised machine learning model that is employed to predict the dependent variable based on decision rules of associated independent variables. It is a regression algorithm that has a tree structure and is used to predict target values in a nonlinear fashion. It employs a top-down approach without backtracking.

K-Nearest Neighbor Regression. KNN regressor works on the same rule as that of the KNN classifier, i.e., producing results based on k-nearest training points. It predicts a numerical value based on a distance formula. In a regression problem, KNN gives the output as the average of the values of k-nearest neighbors. The optimal value of k can be found by the elbow method where the model runs on the training set for different values of k, and the best-suited one is determined.

3.5 Validate Model

The data is divided into the ratio of 3:1 of training and testing data, respectively. The model is validated by fitting it on the training data, and the accuracy of the model is further determined by predicting the values of the test data and comparing them with the actual values. Then, R2 score is calculated using this test data that helps in determining how well the dataset fits the model.

4 Result and Discussion

Out of all the models, the support vector regressor shows the best results with the highest R2 score followed by the KNN regressor that performs the best with the value of $k = 5$ (See Fig. 6).

Multiple linear regression, ridge regression, and LASSO regression perform almost similarly on the dataset. Decision tree regression has the lowest R2 score. The R2 scores of all the models are specified in Table 1 with the bar graph representation (See Fig. 7).

.

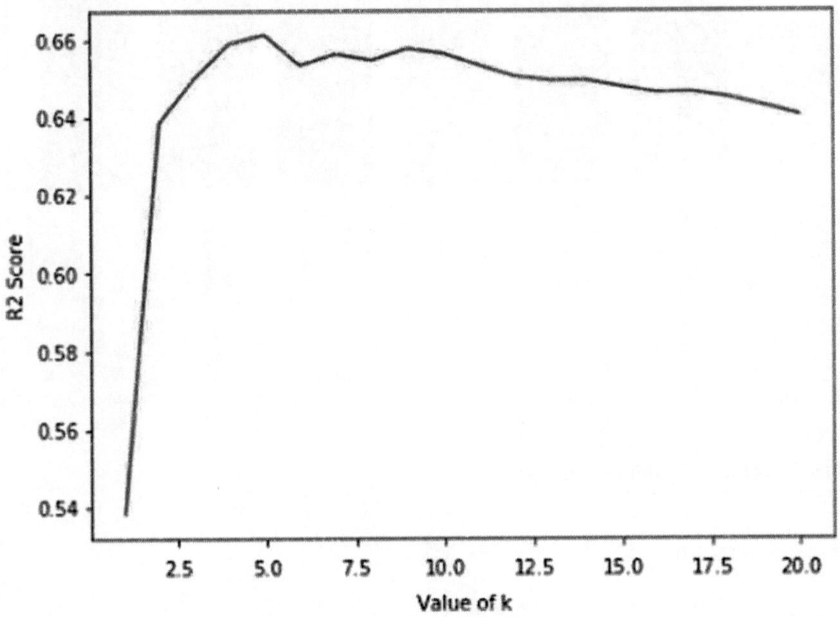

Fig. 6 R2 Score versus values of *k*

Table 1 Model type and their R2 scores

Model	R2 score (%)
Multiple linear regression	54.179
Ridge regression	54.121
LASSO regression	54.179
Support vector regression	69.225
Decision tree regression	47.829
KNN regression	66.116

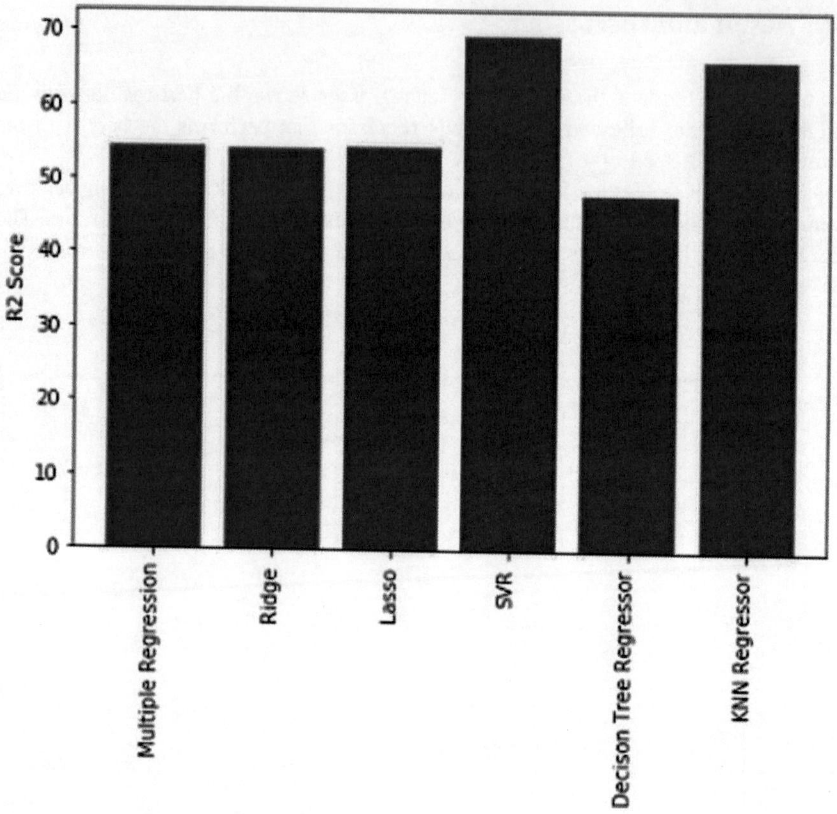

Fig. 7 R2 scores for different models

5 Conclusion

In this paper, six different models are applied to the dataset, and an analysis of the same is done. The dataset initially contained hundreds of features from which the most useful ones are selected to make an accurate prediction of the education loan repayment rate of students. It consisted of null values, incomplete information, redundant data that is cleaned, analyzed, and new features are derived from them as well. Data is divided into testing and training data, and the training set of this clean dataset is fitted on each model. The accuracy of each model is determined by calculating the R2 score. The results show that a support vector regressor works the best on the dataset.

References

1. U.S. Department of Education, Office of Federal Student Aid (2019) Federal student loan portfolio. https://studentaid.ed.gov/sa/about/data-center/student/portfolio. Last accessed Aug 2019
2. U.S. Department of Education, Office of Federal Student Aid (2019) Default rates. https://studentaid.ed.gov/sa/about/data-center/student/default. Last accessed Aug 2019
3. Supriya P, Pavani M, Saisushma N (2019) Loan prediction by using machine learning models. Int J Eng Tech 5(2):144–148
4. Rawate KR, Tijare PA (2017) Review on prediction system for bank loan credibility. Int J Adv Eng Res Dev 4(12):860–867.
5. Jency XF, Sumathi VP, Sri JS (2018) An exploratory data analysis for loan prediction based on nature of the clients. Int J Recent Technol Eng 7:176–179
6. Sudhamathy G (2016) Credit risk analysis and prediction modelling of bank loans using R. Int J Eng Technol 8:1954–1966
7. Kumar R, Jain V, Sharma PS (2019) Prediction of loan approval using machine learning. Int J Adv Sci Technol 28:455–460
8. U.S. Department of Education College Scorecard (2020) https://collegescorecard.ed.gov/assets/FullDataDocumentation.pdf. Last accessed Mar 2020
9. Scikit-learn Machine Learning in Python (2011) https://scikit-learn.org/stable/modules/preprocessing.html#standardization-or-mean-removal-and-variance-scaling, pp 2825–2830
10. Brilliant.org, Ridge Regression, https://brilliant.org/wiki/ridge-regression. Last accessed 14 Sept 2020
11. Mind Majix, Ridge Regression, https://mindmajix.com/ridge-regression. Last accessed 11 July 2019
12. Mind Majix, Lasso Regression, https://mindmajix.com/lasso-regression. Last accessed 25 Apr 2019
13. Analytics Vidhya, Support Vector Regression Tutorial for Machine Learning, https://www.analyticsvidhya.com/blog/2020/03/support-vector-regression-tutorial-for-machine-learning/. Last accessed 27 Mar 2020
14. Soni PM, Paul V (2019) Algorithm for the loan credibility prediction system. Int J Recent Technol Eng8(1S4):1080–1087
15. Goyal A, Kaur R (2016) A survey on ensemble model for loan prediction. Int J Eng Trends Appl 3(1):32–37
16. Torvekar N, Game PS (2019) Predictive analysis of credit score for credit card default. Int J Recent Technol Eng 7(5S2):283–286
17. Tripathi D, Edla DR, Kuppili V (2018) Credit scoring model based on weighted voting and cluster based feature selection. In: International conference on computational intelligence and data science 2018, Procedia Comput Sci 132:22–31, Elsevier, India
18. Hamid AJ, Ahmed TM (2016) Developing prediction model of loan risk in banks using data mining. Mach Learn Appl Int J 3:1–9
19. Kavitha K (2016) Clustering loan applicants based on risk percentage using K-means clustering techniques 6(2):162–166

Predicting the Result of English Premier League Matches

Ashutosh Ranjan, Vishesh Kumar, Devansh Malhotra, Rachna Jain, and Preeti Nagrath

Abstract Nowadays predictive models or prediction of results in any sports become popular in data mining community, and particularly, English Premier League (EPL) in football gains way much attention in the past few years. There are three main approaches to predict the results: statistical approaches, machine learning approaches, and the Bayesian approaches. In this paper, the approach used is machine learning and evaluating all features that influences the results and attempts to choose the most significant features that lead a football team to win, lose, or draw and even considering the top teams. This predictive model basically helps in betting areas and also for managers to have a knowledge how to set up their team by analyzing the results also companies like StatsBomb which use these kinds of tools for setting up scouting networks for searching of hidden gems throughout the world. These features help predict the best possible outcome of the EPL matches using these classifiers logistic regression, support vector machine, random forest, and XGBoost; the data used for prediction is selected from the Web site: datahub.io, and the model is based on the data of last ten seasons of EPL. K-fold cross-validation is used to describe the accuracy of the model.

Keywords Machine learning · English Premier League · Logistic regression · Support vector machine · Random forest · XGBoost

---

A. Ranjan (✉) · V. Kumar · D. Malhotra
Department of Electronics and Communication, Engineering, Bharati Vidyapeeth College of Engineering, New Delhi, India

R. Jain · P. Nagrath
Department of Computer Science and Engineering, Bharati Vidyapeeth College of Engineering, New Delhi, India
e-mail: rachna.jain@bharatividyapeeth.edu

P. Nagrath
e-mail: preeti.nagrath@bharatividyapeeth.edu

P. K. Singh et al. (eds.), *Proceedings of Second International Conference on Computing, Communications, and Cyber-Security*, Lecture Notes in Networks and Systems 203,
https://doi.org/10.1007/978-981-16-0733-2_30

1 Introduction

There are a number of predictive exemplars like academic performance prediction, stock market prediction, house price prediction, etc., and a number of anomalous approaches are made to extrapolate the result in order to get maximum possible accuracy. Similarly, the prediction of a winning team in any particular sport has become very popular in the last few decades and especially in a global sport like football among the football fans. The managers of football teams analyze the results which help them make and modify few strategies accordingly. And some organizations even have business related to prediction. But the major reason for popularity of sport prediction among the data scientists is the challenging complexity to predict the effective results or outcome as there can be plenty of factors based on which the result may depend. Some of them may be weather, skills, teamwork, strategies, and many more.

This paper proposes a model to predict the outcome of English Premier League (EPL) match-winning teams. The model is completely based on a machine learning approach. We have also used a concept of home and away that represents a team playing in the home ground and opponent ground, respectively, as this has been clearly visible that the percentage of a team winning a match escalates in their respective home grounds. The primary task for the model was to find a dataset having all the necessary feature like home goals, away goals, home corners, away corners, etc., that helped us go onto the next task, feature selection. For the feature selection process, we used the recursive feature elimination (RFE) classifier that helped us align all the feature in a rank-wise order with maximum to minimum influence on the model. Hence, we came out with final 13 features that had the maximum amount of influence on the prediction. Post the feature selection process, we used the predictive models like logistic regression, support vector machine, random forest, and XGBoost that gave us different accuracy in the prediction and XGBoost came out to give the maximum percentage of accuracy that is 58.73%. The accuracy of remaining classifiers: Logistic regression: 58.28%, support vector machine (SVM): 58.68%, and random forest: 57.84%. At the very end, to represent the result, we have used a confusion matrix that is basically a form of table matrix that gives us the correct and incorrect prediction.

Section 2 provides the information about some related researches in the field of predictive models using different approaches.

Section 3 explains about extraction of data, feature selection, and train dataset.

Section 4 discusses about the models used, results, and graph representation.

Section 5 concludes the research.

Last section has a list of references of research papers.

2 Related Work

A prediction model for predicting matches of football league in England using artificial intelligence and machine learning algorithms was built. They began by reducing three-class classification to the two-class classification whether a team will win or lose and for preliminary test, features like team which is home or away and the form are used, and later, they extended to three-class classification by implementing one-vs-all stochastic gradient descent algorithm on the same features and got the results same as SVM but with no overfitting in decision boundary. Using linear classifier, random forest, and SVM, they got error rates of 0.48, 0.50, and 0.50, respectively [1].

A prediction model is built to predict the football match results various machine learning algorithms were used, but performance of support vector machine (SVM) was not an appropriate technique for those sets of features as the artificial neural networks got the results by 85% on the same sets of features. Gaussian combination kernel type is used to generate 79 support vectors at 100,000 iterations. The prediction accuracy obtain was 53.3%. We slightly improved SVM prediction accuracy to 58.68% [2].

A prediction model to predict the possible outcome of National Basketball Association (NBA) league developed by testing several classification methods and selected naive Bayes method. Unlike other sports like football, basketball has only two-class classification win or lose, so it makes bit easier to predict outcomes. The prediction accuracy was highest in naive Bayes combined with normalization which was 67%. Model was tested on 778 games with 148 attributes [3].

A prediction model was created to predict the result of football league for 2014–16 season. Dataset of nine seasons (2005–2014) was used for training purpose and following two seasons, i.e., 2014–2016 as test data. They employed tenfold cross-validation for obtaining the optimal value of K, but the later obtained the best value of k which is 6 for all the models. Mean test accuracy score for different models was as follows NB = 0.51, linear SVM = 0.545, RBF SVM = 0.55, random forest = 0.57, and gradient boosting = 0.58 [4].

Acharya and Sinha [5] described a model to predict the students' performance using machine learning algorithms, and among all, they concluded that the decision tree (DT) was most convenient algorithm to generate the set. The training set contains 309 instances, whereas the testing set contains 104 instances and the efficiency obtained 79% on the training dataset whereas 66% on the testing dataset.

Khan et al. [6] described a model which predicts stock market using machine learning classifiers and social media, news. Both football results and stock market prediction are hard to predict as there are many external factors on which prediction depends. They perform prediction on data providing by social media and financial news and achieved accuracies, respectively, 80.53% and 75.16% by using deep learning and different classifier, but out of the classifiers, random forest classifier showed the highest accuracy of 83.22%.

A prediction model was built to predict football outcomes using Bayesian approach. There are three Bayesian algorithms, which are naive Bayes (NB), tree-augmented naive Bayes (TAN), and general Bayesian network (K2), applied on the data of three seasons of English Premier League (EPL) from 2015 to 2017. The data of 380 matches of each season is divided into ten equal set size, and each set is separated into two groups which is training set of 90% and testing set of 10%. TAN showed 90% which best results out of these three models, while NB and K2 algorithm showed 74.03% and 75.26%, respectively, in terms of average and seasonal accuracy for three seasons [7].

3 Methodology

3.1 Dataset

The dataset used in building this predictive model included results from 2009/10 to 2018/19 season.

This dataset is used for training of our model. Each team plays 38 matches throughout the season. Hence, for 10 seasons, total matches played were 3800 which accounts for the number of rows in our training dataset. Also, the number of columns present in the dataset was 78 with features like FTHG, FTAG, FTR, etc. We would look more into it in our 'feature selection' section.

For testing purpose, we have used the data of latest ongoing season that is 2019/20. Matches completed so far are 288.

3.2 Feature Selection

In football, home team has added advantage. The reason for this can be various, but one of the most important reasons considered is the crowd support. Being a home team, crowd turnout from that particular side will always going to be large. This in turn overwhelms the opponent side sometimes.

The domination of home team can be seen through following representation

This is the reason that our features are divided on the basis of home team and away team.

Also, from the following bar graph, we can observe how a team scores more at home rather than away from home.

Similarly, we can see how other features like goals against, crosses, etc., are affected if a team is playing at home or away.

Also, for feature selection, we made use of RFE classifier which ranks your input features lower the rank less its effect on model and hence that feature can be excluded (Table 1).

Table 1 Top 13 features used for training the model

Feature	Meaning
HAS	Home attacking strength
HDS	Home defensive strength
AAS	Away attacking strength
ADS	Away defensive strength
HST	Home shots on target
AST	Away shots on target
HC	Home corners
AC	Away corners
ARC	Away red cards
HRC	Home red cards
HYC	Home yellow cards
HF	Home fouls
AF	Away fouls

The domination of home team can be seen through following representation. We can clearly visualize that 46.3% of matches are won by those teams that are playing at their home. So, while predicting results, it is very helpful to know if the team that is playing is home team or away (Fig. 1).

This is the reason that our features are divided on the basis of home team and away team. Also, from the following bar graph, we can observe how a team scores more at home rather than away from home (Fig. 2).

Similarly, we can see how other features like goals against, crosses, etc., are affected if a team is playing at home or away.

Also, for feature selection, we made use of RFE classifier which ranks your input features lower the rank less its effect on model and hence that feature can be excluded.

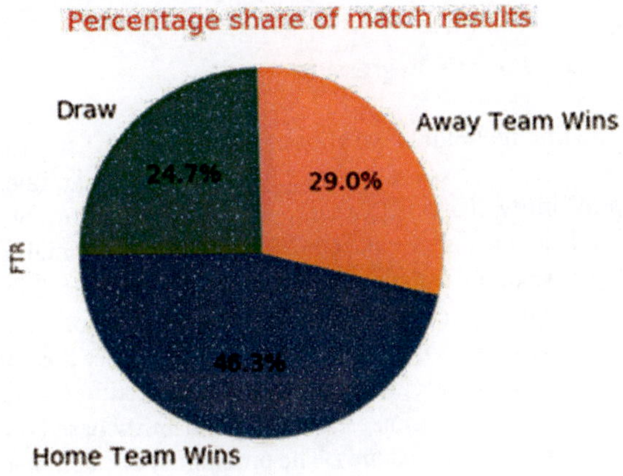

Fig. 1 Percentage share of match results found through dataset used (created using matplotlib)

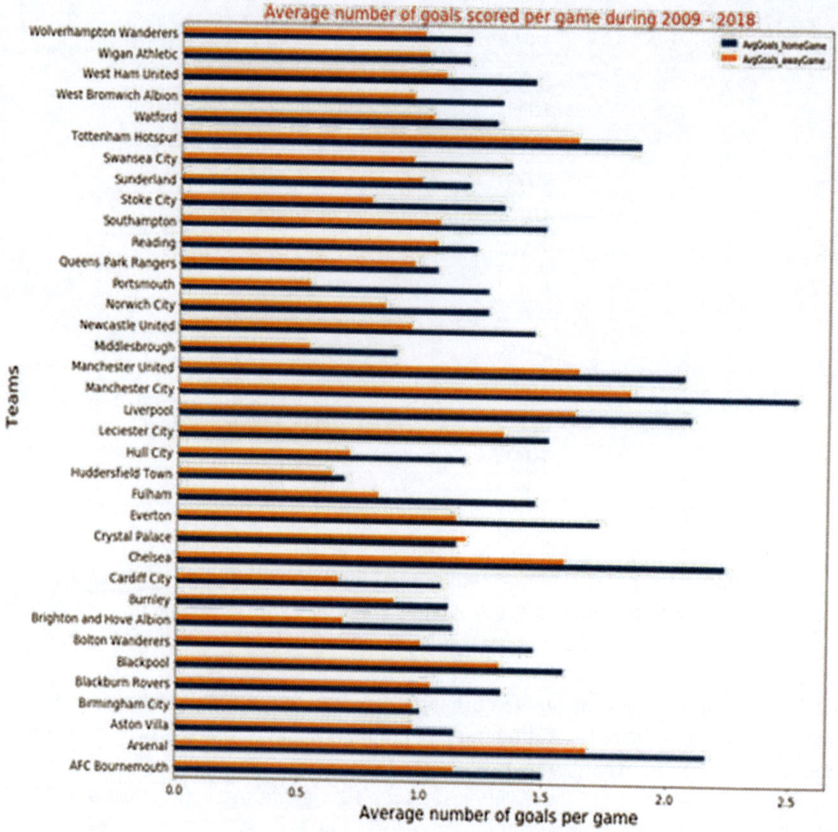

Fig. 2 Bar graph comparing home and away goals of various teams

3.3 Predictive Models

3.3.1 Logistic Regression

The first model that we chose to apply on our data was logistic regression. Logistic regression is a classification algorithm much like linear regression used to find the probability of a target variable where the target variable Y is a function of variable X. Logistic regression can be binomial and multinomial regression. In case of binomial regression, the output variable has two outcomes, 0 that represents failure or 1 that represents success. However, in case of multinomial regression, the target variable can have two or more than two outcomes as per the necessary. The problem at hand is a multinomial regression as we are predicting win (1), loss (−1), and draw (0).

Logistic regression estimates a probability based on the categorical output values and various input features. The probability is calculated via logistic function (Figs. 3 and 4).

Fig. 3 Sigmoid function

$$h_\theta(x) = \frac{1}{1 + e^{-\theta^T x}}$$

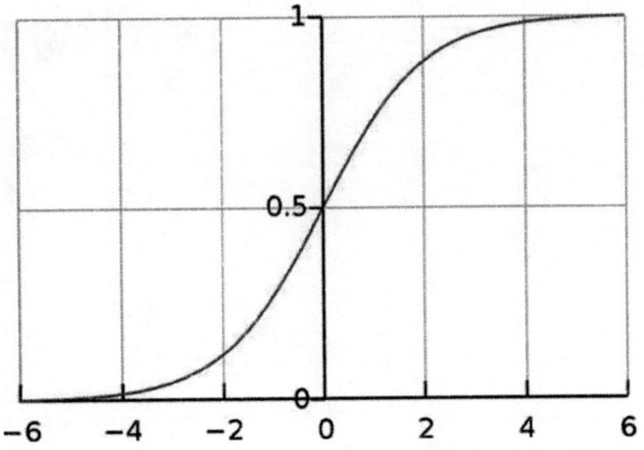

Fig. 4 Richards, Geoff, 'Standard logistic sigmoid function'

Fig. 5 Gradient descent equation

$$\theta_j := \theta_j - \alpha \frac{1}{m} \sum_{i=1}^{m} (h_\theta(x^{(i)}) - y^{(i)}) x_j^{(i)}$$

$$h_\theta(x) = \frac{1}{1 + e^{\theta^T x}}$$

Through, this sigmoid function probability is calculated for a weight to be classified. The cost function used in logistic regression is binary cross entropy.

Now, since we have to minimize the cost function, we use gradient descent, thus minimizing the error and providing us with the best fit line for our classification problem (Fig. 5).

$$\theta_j := \theta_j - \alpha \frac{1}{m} \sum_{i=1}^{m} (h_\theta(x^{(i)}) - y^{(i)}) x_j^{(i)}$$

3.3.2 Support Vector Machine

Next predictive model at hand is SVM. SVM is used for both types of problems be it classification or regression. However, it works better for classification problems. Here, each item is plotted on a *n*-dimensional space where *n* represent the combined number of features with each feature having some coordinate value (Fig. 6).

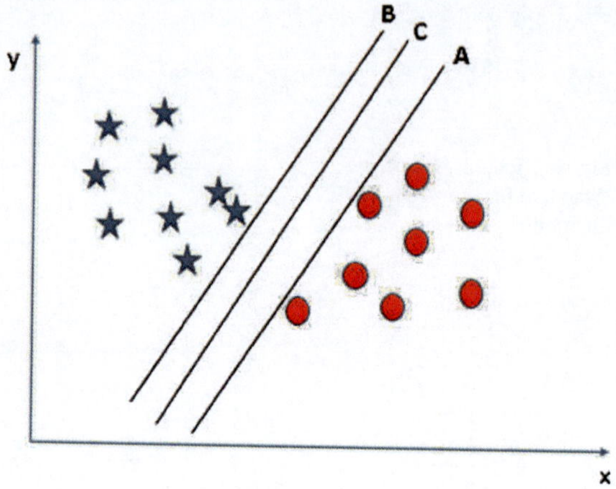

Fig. 6 Ray, Sunil, 'Identify the right hyperplane'

The main objective of support vector machine is to create a hyperplane that segregates the data points.

To understand better, we can have a perfect easy-to-understand example. We can see the above graph and find the best hyperplane that separates the two different classes from each other. It is simply visible that the hyperplane B separates the classes with highest efficiency which is what is actually done by a support vector machine. It searches for the best manner to classify a dataset for each feature. The hyperplane is selected, so the margin distance should be maximum because of which it can be more of a generalize model. These margins are drawn through the nearest data points, and these data points are known as ***Support Vectors***. Hyperplanes are basically decision boundary classifying the data points. Type of hyperplane depends upon the features; if there are two features, the hyperplane would be a line, if three it would become a 2D plane, and so on.

In SVM, margin would be maximized, so that it would work as our optimization function and will be known as hinge loss function.

$$c(x, y, f(x)) = \begin{cases} 0, & \text{if } y * f(x,) \geq 1 \\ 1 - y * f(x), & \text{else} \end{cases}$$

Parameter used

Kernel: It is used to decide the hyperplane that is used to separate data. We have used rbf kernel which is also the default parameter.

C = This parameter is used for regularization of data. For higher value of C, the separation of data is good, but if you have noise data present, then c should be low.

3.3.3 Random Forest

Random forest is a type of ensemble learning. It creates lots of individual decision trees on a training set, often on an order of tens or even thousands. The process of building decision trees is randomized. This is done by selecting data randomly, and also features are also randomly selected. Random forest algorithm can be very advantageous for our project because of its high efficiency and ability to work with large datasets efficiently. Adding to that, it is also robust to overfitting. Random Forest algorithm uses a tree-like-graph for the classification purposes (Fig. 7).

For creation of random forest, the no. of trees is decided. Each tree is built from different random sample of data. Hence, we will input the training data for last ten years along with the set of features shortlisted.

Parameters Used:

- n_estimators: Used to define the number of trees for the ensemble. The default value of this parameter is 100
- min_samples_leaf: Used to set the minimum value of samples to be called leaf nodes. It is very helpful in smoothing the model.

3.3.4 XGBoost

XGBoost is an implementation of gradient-boosted decision trees designed specifically for better speed and high performance. It is an algorithm that has recently been dominating applied machine learning because of being a multipurpose algorithm. It is also an ensemble learning. XGBoost works on boosting principle. It involves grouping classifiers and applying them on whole of the data sequentially [4].

It can be used for classification, regression, ranking, and predictions as well. Apart from that, XGBoost supports all kinds of languages like JAVA, C++, R, and Python which is why it stands apart.

Parameters used:

- learning_rate: XGBoost is likely to learn quickly and overfit, so to avoid this, learning_parameter can be used so as to slow down the learning rate.
- max_depth = Controls the max depth of the tree. Helps in avoiding overfitting as more the depth learning will be better.
- min_child_weight = Through this parameter, we can choose the sum of weights of observations for a child that is minimum. Higher values avoid overfitting as it prevents a highly specific learning relations for a particular sample.

4 Results and Graphs

For training purpose, we used dataset from season 2009/10 to 2018/19 and used different models on it out of which XGBoost gave highest accuracy. Two different

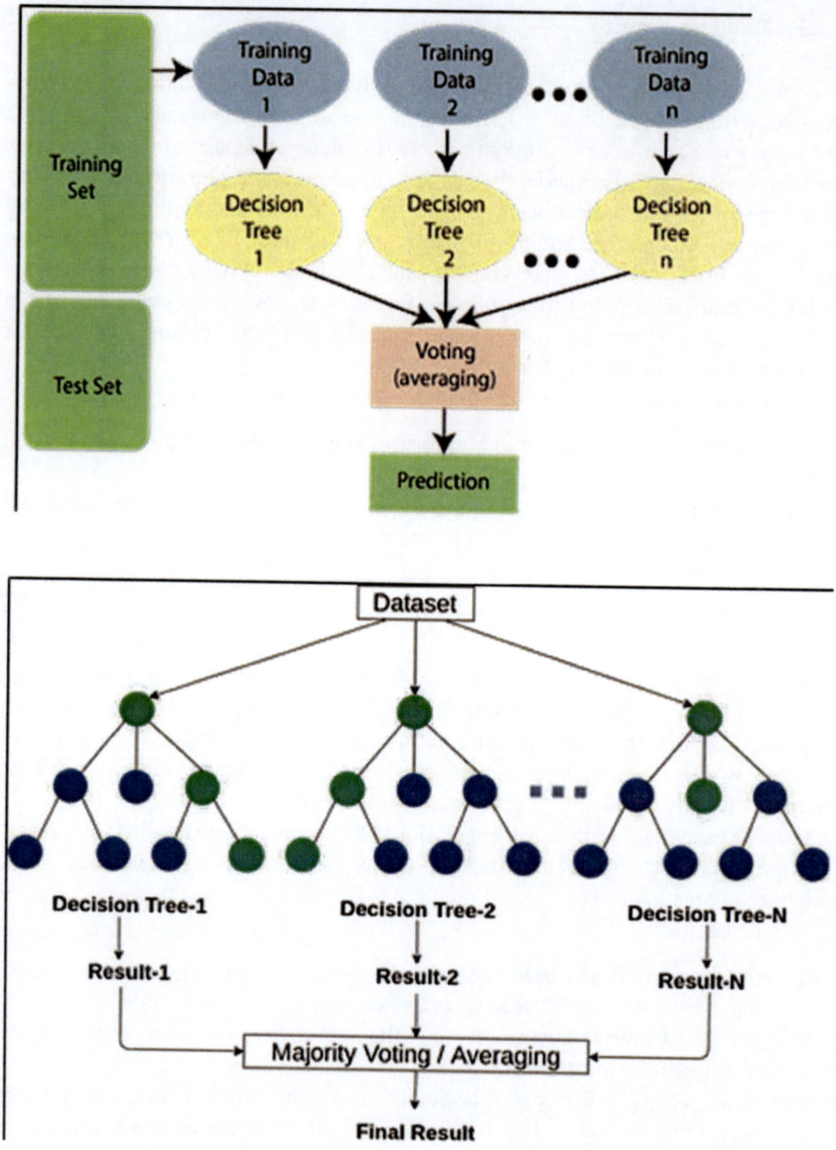

Fig. 7 Formation of random forest

training sets were used in the first dataset only those data were present whose values are known to us before match has started which will not change during matches. However, in second dataset, we had data like crosses and shots on target which can vary match to match (Fig. 8).

Now, a test dataset was taken which had 67 matches played in 2019/20 season on which XGBoost model was applied.

We got an accuracy of 66% which can be verified by the following confusion matrix (Fig. 9; Table 2).

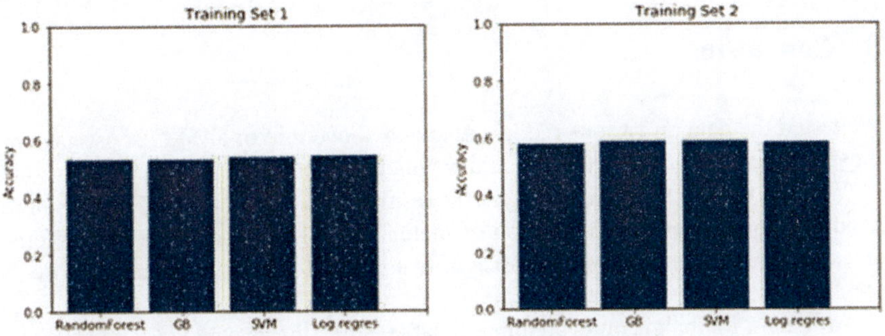

Fig. 8 Comparison of training dataset accuracy on different models

Fig. 9 Confusion matrix

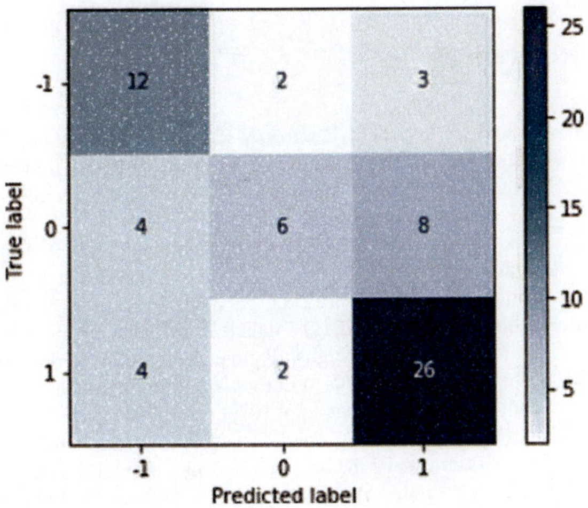

Table 2 Results of testing

	Precision	Recall	F1 score	Support
−1	0.60	0.71	0.65	17
0	0.60	0.33	0.43	18
1	0.70	0.81	0.75	32
Accuracy			0.66	67
Macro avg	0.63	0.62	0.61	67
Weighted avg	0.65	0.66	0.64	67

5 Conclusion

We tested the data on various models as stated above; out of all, XGBoost gave the most promising result with 66% accuracy which is good and better than results of [1, 4]. We used various features to analyze the result; some of features are known beforehand and some after the completion of match. We all know key role in predicting any sort of result lies in the importance of features and how major its effect is on the result.

Although, football is quite an unpredictable game; the accuracy can be increased further more. This can be done by finding out more features that majorly affect the outcome. Some of the features that can be included are injured players, few key players of that team, performance against a particular team, average league position, and many more. Also, deep learning algorithms can also be used to predict the results more accurately.

References

1. Ulmer B, Fernandez M, Peterson M (2013) Predicting soccer match results in the English Premier League. Doctoral dissertation, Ph. D. thesis, Doctoral dissertation, Ph. D. dissertation, Stanford
2. Igiri CP (2015) Support vector machine–based prediction system for a football match result. IOSR J Comput Eng (IOSR-JCE) 17(3):21–26
3. Miljković D, Gajić L, Kovačević A, Konjović Z (2010) The use of data mining for basketball matches outcomes prediction. In: IEEE 8th international symposium on intelligent systems and informatics. IEEE, pp 309–312
4. Baboota R, Kaur H (2019) Predictive analysis and modeling football results using machine learning approach for English Premier League. Int J Forecast 35(2):741–755
5. Acharya A, Sinha D (2014) Early prediction of student's performance using machine learning techniques. Int J Comput Appl 107(1)
6. Khan W, Ghazanfar MA, Azam MA et al (2020) Stock market prediction using machine learning classifiers and social media, news. J Ambient Intell Human Comput
7. Razali N, Mustapha A, Yatim FA, Ab Aziz R (2017) Predicting football matches results using Bayesian networks for English premier league (EPL). In: Iop conference series: materials science and engineering, vol 226, no 1, pp 012099. IOP Publishing

Comment Filtering Based Explainable Fake News Detection

Dilip Kumar Sharma and Sunidhi Sharma

Abstract Fake News Detection is one of the most currently researched areas over the globe; many methods have come to light using different features as their sources. Hence, there are also methods using existing comments on any news article which can be used to determine the credibility of the news article as fake or real. Here, we have introduced a hypothesis that uses a machine learning approach to check the credibility of comments before they can be analyzed for further fake news detection. So, we have used various text classification algorithms to check for our hypothesis that filtering comments since there is a high possibility that the comments used for any analysis can be useless and full of useless stuff. For example, the comments showing only the emotion of readers like 'Yesss' or 'Nooo!' and likewise or the comments built using only the curse words. Such comments would prove useless as a contributing factor for fake news detection and might also affect the results of fake news detection for any news article. These text classifiers are—Complement Naïve Bayes, Logistic Regression, Multinomial Naïve Bayes, and Support Vector Machine. Out of these, the best accuracy is provided by the MultinomialNB method of 75.7% and Decision Tree with 75.4% as opposed to the original algorithm with an accuracy of 73.3% using the same dataset. Since the MultinomialNB has provided the best improvement in all the metrics compared to the original method, and we are focusing our paper on this method. This hypothesis aims to classify comments as junky (useless) comments and utility (useful) comments. These utility comments will be further used for analysis to identify fake news. Also, since the size of comments per article may vary from a few tens to a few hundred or thousands, we have used the semi-supervised approach to classify the comments in junky or utility comments classes. We have also collected data from various sources and collaborated them to fetch ourselves from a usable dataset. It contains 415 records with contents or article data for each record, along with many comments for each record. Moreover, we have also classified those comments into junky and utility comment classes using the basic definition of spam filtering. This can be improvised for different uses using

D. K. Sharma · S. Sharma (✉)
Department of Computer Engineering and Applications, GLA University, Mathura, Uttar Pradesh, India
e-mail: sunidhi.sharma_mt18@gla.ac.in

P. K. Singh et al. (eds.), *Proceedings of Second International Conference on Computing, Communications, and Cyber-Security*, Lecture Notes in Networks and Systems 203,
https://doi.org/10.1007/978-981-16-0733-2_31

different criteria. Hence, eradicating the useless comments and only analyzing the useful comments for better identification of fake news is fake news detection.

Keywords Fake news detection · Comment analysis · Comment credibility check · Machine learning · MultinomialNB

1 Introduction

With the increasing developments in society, the risk of encountering fake news is also increasing. Considerably in current situations when there is a major 'Coronavirus Pandemic'. And even a simple hoax or myth or a fake news article may result in fear wave among public resulting in severe physical or mental harm. Another incident may be taken from a recent US President Donald Trump interview when he asked doctors to inject disinfectants to evade coronavirus [1]. Even though he later dismissed the statement as a part of a joke, it still claimed a significant portion of the doctor's efforts to calm the society. Hence, resulting in the need for a fake news detection system.

Fake News Detection being on a list of one of the immediately needed tools for this society has captured the much-needed focus of many researchers. Hence, many of them are coming up with various sorts of techniques for fake news detection. They can be coarsely identified in two categories, manual and automated. Manual fake news detection uses human resources and human time to identify the news as real or fake one by one. And automated fake news detection uses machine learning or such methods that do not need human interference, to identify the news as real or fake. For these automated fake news detection, the model is trained using a few or more of those news articles' features. Some use sources, and some use author, some use comments, some use words, etc. Hence, we are using our hypothesis over the comment feature as taken by the dEFEND system [2].

Here we have used the indirect human response to the articles to identify the news article as fake or real. The most prominent and detailed responses of readers can be considered as their comments to those articles. Hence, using those comments to identify fake news can be a liable option. Though there are many other ways to identify the fake news, there is still a long way to go for the automated methods to completely identify the fake news on all terms. There are various exceptional methods, but they all lack due to some or other conditions. For example—some methods identify only a specific type of fake news; there are many who cannot identify the hidden innuendos, there are many who have a specific requirement of data to check for validity of news, some even support a certain type of data be it text or image, etc. Apart from this, the organizations or companies using human resources to identify fake news also have a certain criterion to select the news that they analyze. Being selective, that is still not a very satisfying option. Hence, most existing solutions are dependent only on news articles and their features and not comprising the reader's point of view.

Hence, using the comments as identification of fake news can be one of the safest bets considering human psychology, where people do not hold back when correcting others. This also somewhere identifies the issue of temporal constraint. For example—if a news article is fake today but can be real tomorrow, recent comments can be used with a higher weightage to analyze the validity of news articles. Not to assume that those comments are completely true or written with genuine concerns, we have used supervised machine learning to identify genuine comments from fake. This will result in a refined set of comments that can be used to identify the news article as fake or real. Hence, using indirect human resources without any selective criterion for news articles.

The rest of this paper is as follows: Sect. 2 includes the related works; Sect. 3 includes the methodology of the proposed work. Section 5 refers to the results and analysis part. While Sect. 6 is the Conclusion and Future Work.

2 Related Works

Since, with the development of fake news detection systems, the notorious entities existing in the world are also developing in quantity as well as in quality, one should be prepared to identify all sorts of different fake news. The type of fake news one encounters in a day is also very difficult to determine. With the help of different basis, studies have identified many types of fake news. There is misinformation (created with no intent to deceive) and disinformation (created with an intention to deceive) [3]. There are also rumors, propaganda, clickbait, hoaxes, conspiracy theory, satire news, etc. [4]. Reference [5] has also differentiated the types of fake news based on velocity, veracity, and validity. There are many more factors to identify the types of fake news, but as there is no definite definition of fake news, its types can also be considered as indefinite.

As mentioned earlier, when there is manual fake news detection, many organizations or groups are involved in that. These organizations and groups that have taken the initiative can be categorized as either for-profit or non-profit. Most of those employ humans to check the claims manually. Some organizations target a particular topic/subject, while many target all subjects. But the biggest drawback is the requirement of huge manpower and time. Most of such sources are very selective about the claims that they check. So, not every claim/news article can be ensured of being checked in such cases. For example, sites like PolitiFact [6], Snopes [7], HoaxSlayer [8], etc. are a few of many such examples that have proven themselves to be efficient in manual fake news detection.

While automated fake news detection may refer to the detection techniques without much human interference, since there is lots of data to consider and process, automated approaches come as a refuge for fake news detection. Most of the algorithms revolve around artificial intelligence and machine learning. Yet some innovative researchers came up with out-of-the-box algorithms to identify fake news. Overall, these approaches include training a model on a given training dataset and then testing it for accuracy. This training is done considering a few of the articles' features. For text features, Natural Language Processing is used to fetch tokens for training purposes [9]. Some also use networking approaches to trace the source and forwarding patterns to identify whether the news is real or fake. Some use deep learning for fake news detection. In deep learning, researchers have deduced algorithms using Recurrent Neural Networks [2], Convolutional Neural Networks [10], Long Short-Term Memory, Event Adversarial Neural Networks [11], Geometric Deep Learning [12], etc. Reference [13] provides multi-modal checks for identification of fake news using the textual and visual data. They use the similarity between these features to identify the news as fake or true. Though it should be noticed that they are not using the video or network attributes. Reference [14], on the other hand, have used the textual data to extract features like news articles, creators, and subjects to use in the deep diffusion neural networks model. Here, they have not identified the temporal features or any other modal aspects of the news articles.

Reference [15] provides an approach that uses Bayesian interference to identify the fake news and jointly learns about users' flagging accuracy over time. Reference [16] taking on a different turn uses user profiles to identify the possibility of posting or identifying the fake or real news.

As a part of related work, one should also consider the comment verifications done using different methods. Those methods include [17] that consist of sarcasm and sentiments as different definitions and two artificial neural networks to identify the comments as fake or real. Another one includes [18], which generates the comments along with using the real posted comments to fix their model. [19] have paid focus on the toxicity of comments rather than the comments being real or fake. Though they have tried to provide a line between spammers, trolls, haters, and genuine users, their methods cannot be used due to their emphasis on toxicity rather than fakeness.

Hence, one can say, there are many methods of working fine as an individual entity. But it still proves as a matter of discussion that whether their constituent disadvantages can be catered to by using a compliment method sidewise to get better results. Our hypothesis is also somewhat working around a similar concept and tries to prove the better result theory.

3 Problem Statement

Working on Fake News Detection, our work incorporates the methodology of 'dEFEND: Explainable Fake News Detection' [2] and the functionality of the comment credibility check. The basic idea being the use of Recurrent Neural Network to identify the news article as fake (1) or real (0). In Layman's terms, the comments are identified to be related to news content. And hence, such explainable user comments can be used to deduce the credibility of news content. The working that goes inside this idea is that the article includes n statements, while each statement (s_i) is formed from m words (w_{ij}). This is taken care of by encoding words and then statements. On the other hand, the comments (c_i) are as well encoded in the forms of words, then statements. These encoded articles and comments are then used to formulate a co-attention matrix to find the article to be real or fake.

The original problem that may render the result of the article is the credibility of comments that go without any checks. Hence, making even the remotely close useless comment a participant in the deciding factor. Hence, there needs to be a check that can identify junky comments and eradicate them. At the same time, only use utility comments that contain some relevant information to contribute in fake news detection. For Example—For a news article, if there are many comments that can be considered vague or simply the statements used to indicate shock or surprise. Such comments may interfere in the comment statement co-attention matrix and therefore have a part in classification.

4 Proposed Solution

We intend to embed a semi-supervised machine learning classifier along with using deep learning model. To tackle the comment classification into junky (0) and utility (1), we use eight different text classifiers in machine learning and embed them into the system.

To classify the comments, we have identified the generic comments as junky. For example, comments like—"Right", "You are right", "LOL", "Oh my God", "Is it real?", etc. A simple hypothesis is that if a simple word used n number of times in various comments can play a game-changing role. Hence, to remove most of these comments, the accuracy of the system can be increased.

The comments (c_i) data item will be a consolidation of utility comments from both known comments (k_i) and unknown comments (u_i). Here, the known comments refer to those data items that have been labeled as junky or utility. While Unknown comments (u_i) refers to the data items that have not been labeled yet and need to be classified as junky or utility. In the end, comments used would be the utility comments from both known and unknown data items. Here, the known data item refers to the list of comments that have been classified and unknown as the list that has yet to be classified as junky or utility.

The system includes two levels of classification. That is, first being the classification of comments as utility or junky. Another being the classification of the news article as real or fake. Therefore, it would be safe to say that on both levels, the classification is between two classes with different labels on both levels. Also, on both levels, the data of consideration is different. Moreover, on the first level, we use semi-supervised machine learning, MultinomialNB (among many). While on the second level, we use deep learning RNN to counter the fake news detection problem (see Fig. 1).

In detail, as can be seen in Fig. 1, it depicts the working of the complete fake news detection system. Here, starting from the fetching of data from the dataset, we would be only using the two attributes, that is, contents and comments. The data fetched from the fields is preprocessed to remove stop words, lemmatize, tokenization, normalization, etc. These tokens or words can be used for further processing. Content data after preprocessing goes through the encoding or training phase where

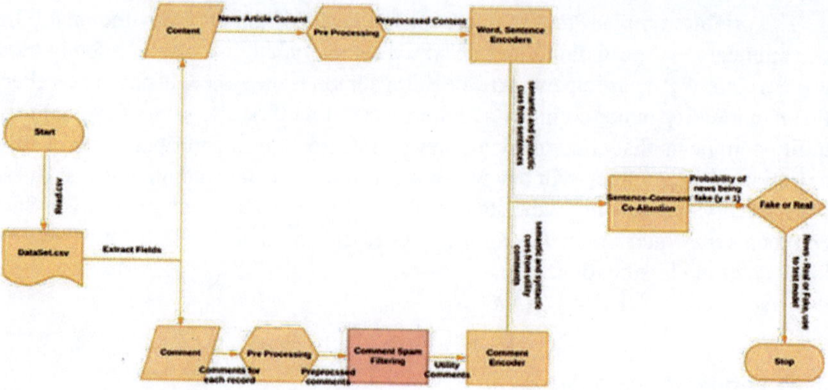

Fig. 1 Conceptual model of working method

the words are encoded into vectors, and on a coarser level, sentences are also are encoded that contain the encoded words. In other words, the sentences are used to form vectors of features, and so will be their constituent words. On the other hand, comments need to go through comment filtering before they can be encoded. Here, for each record, there are many comments. These comments go through classification themselves, and only the utility comments can be used further for encoding into a vector of features. Similar to sentences, comments are also encoded with the sole exception of the comments' constituent words not being encoded. Moving further, both the vectors of sentences and comments, a sentence-comment co-attention matrix is formed, which provides the probability of article being in the class of fake news.

4.1 Comment Filtering

Comment Filtering refers to differentiating comments on the basis of their usefulness. And using the useful or utility comments while removing the useless or junky comments. This is done using the semi-supervised Multinomial NB classifier. Though we are using many methods of Machine Learning that can be used for text classification, we are focusing on the Multinomial Naïve Bayes method for the sole reason of getting better results than the original method for all the performance metrics. This will require the dataset to be manipulated in such a way that there is an indication for all the comments (or at least most of them) to identify the class they fall, utility, or junky. Here, instead of adding another attribute to identify the comment class, we modify the comments with a token.

The comments for each article are double colon-separated (::). And for each comment, there is an indicator (0/1) that provide with the class of that comment, i.e., 0—junky comment, 1—Utility comment. Hence, the resulting comment would be of the format, $C = (0|1)*$.

Fig. 2 Comment partitions

We use two data items storing the comments. One (Known—*k*) with comments already identified in classes manually for training purposes. While second (Unknown—uk) with comments still to be identified in classes. Hence, there is a list of comments already classified into two classes. We used the Known comments (*k*) to train and fit the MultinonialNB classifier using supervised machine learning. Then, the predefined model was used to predict the Unknowns (uk) to identify the junky comments and utility comments.

The comment filtering starts with the preprocessed comments field. That data is separated using separator as '::' for each record to differentiate each comment from others, as shown in Fig. 2. Hereby, since there might be a huge number of comments, so there is a possibility that few or more comments are not classified as utility or junky. Hence, the known and unknown comments are separated into different data items. Once differentiated, the known comments will be used to train the model, while the same model will be used to predict the class of all the unknown comments. After the classification of all the comments in the unknown, we will separate the utility comments and junky comments in both known and unknown data items. These utility comments are used in further processing of encoding the comments. Moreover, it has been taken care that the comments of different records or articles are not mixed to ensure the explainability of the fake news detection method.

4.2 deFEND—Fake News Detection

The second step is the part of dEFEND fake news detection that can be identified to be using Deep Learning Recurrent Neural Networks. Here the working is divided into three parts:

First being the encoding of sentences and their constituting words using the Hierarchical Attention Neural Network to identify the semantic and syntactic features after the preprocessing of the data is done. Here, the sentences are encoded, and their constituting words are encoded as well.

The second step being comment encoding. Here, instead of using all the comments like was done in [2], we used our filtered comments such that only utility comments are used while junky comments or spam comments are removed. These filtered comments are encoded in a similar fashion as the sentences.

The third and final step is to create the sentence comment co-attention matrix, which will be used to identify the probability of a sentence or news article to be fake news ($y = 1$). As this is meant to be explainable fake news detection, more weight is to be given to those comments that represent any reasoning towards the sentences. Using the logic, a matrix is formed to map sentences to comments using the weights paramcters. This would provide the comment and sentence attention vectors as the weighted sum of sentence features and comment features.

This, after using these learned features and predicting, provides us with $y = [y_0, y_1]$, which is the predicted probability given y_0 represents the probability of real news while y_1 represents the probability of fake news.

5 Result Analysis

5.1 Dataset Used

Since we have modified an already existing model to improve the accuracy, we have compared the modified work with the original. Here, the modified work refers to the work, including the different methods. For each method, the work modified is different. Also, since the method's major requirement relies on the availability of datasets, it has been quite difficult to obtain a dataset that contains comments along with the news content. Although the only fields require for the method are the news article content and the comments for each content record. It has been assumed here that the minimum number of comments for each content section is 1. So, one needs to identify the comments as junky and utility as well. We obtained the dataset from [20, 21] and consolidated different tables to form one in the format we require for our work. In our work, the dataset we used has two major fields, namely contents and comments. For each content record, there are a different number of comments each.

5.2 Experimental Results

According to [20, 21], this dataset was fetched from PolitiFact containing 415 records. Hereby, we use 33% records for comment classification. We are using comments pertaining to 33% articles as test data, while training data contain the comments pertaining to rest 67% of articles. While we are dividing the records as 311 for training and 104 for testing in deep learning. That is, we are using 25% data as test data and 75% as train data. We have taken the random state as 42 to allow the repetitive production of the same data. Also, we are using 30 epochs for the same.

Table 1 Table to compare the original and modified methods

Method	Accuracy	Precision	Recall	F1_Score	AUC
dEFEND	0.733577	0.785187	0.839216	0.800759	0.689515
ComplementNB	0.753205	0.772599	0.899510	0.825289	0.688181
Logistic Regression	0.741667	0.754845	0.905392	0.820319	0.668900
MultinomialNB	0.757051	0.799562	0.856373	0.820263	0.712909
SVM	0.739103	0.774576	0.859804	0.810399	0.685458

Table 2 Table to compare the accuracy of methods used to modify the original work

Method	Accuracy (Machine Learning)
ComplementNB	0.9284356093344858
Logistic Regression	0.975695764909248
MultinomialNB	0.9596542783059637
Support Vector Machine	0.9755229040622299

After using the dataset for all the methods, we got the results as described below. All the data is provided below in Table 1. As shown in Table 1, the highest accuracy can be obtained by the Multinomial Naïve Bayes algorithm with 75.7% compared to the original dEFEND with accuracy 73.3%. On the other hand, all the metrics for MultinomialNB can be seen to have increased. The method ComplementNB also has an increased value of accuracy.

When it comes to machine learning algorithms for comment classification, Logistic Regression has the highest accuracy (97.57%) and Support Vector Machine not too far behind (97.5%), while MultinomialNB has medium accuracy (95.96%). This can be more identified with the help of Table 2.

In terms of precision, MultinomialNB has the highest precision and recall somewhat in the middle. The recall is highest for Logistic Regression. To elaborate on this point, we have used the graph to denote the accuracy comparison (Fig. 3).

The given above graph represents that the results for the modified methods provide with better accuracy, that is, a better chance to provide a result that is a true positive or a true negative. This also can be seen in Fig. 4 that the precision and recall have also changed a lot, especially the ComplementNB and Logistic Regression.

Here, as one can see, the precision has increased for MultinomialNB and not for any other methods, but the recall has increased for almost all the methods. Hence, one can say that fake news is detected by far more chance than before. So, we can also say that the resulting tilt towards terming any article that has even an ounce of possibility to be fake, the system depicts as fake which is a good option as it is needed for fake news to be identified as fake even on a miniscule portion of it being fake.

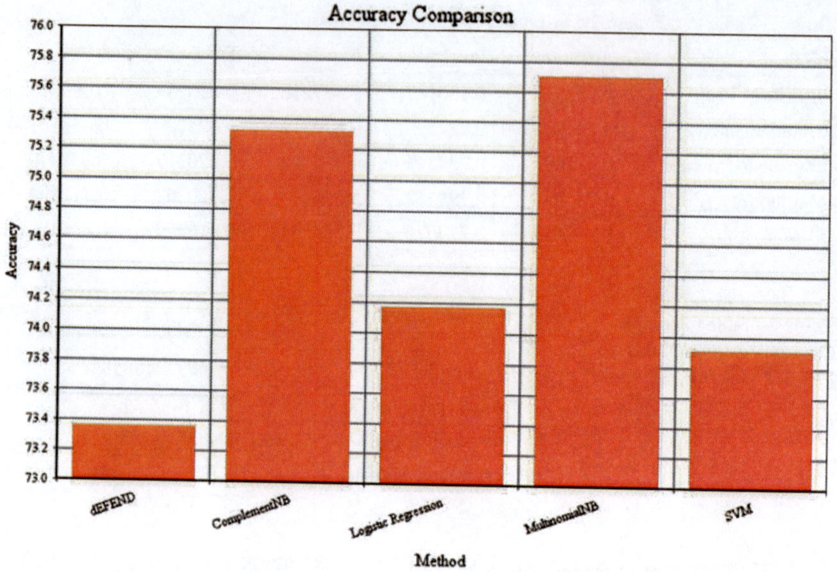

Fig. 3 Accuracy comparison for all the methods (including the original)

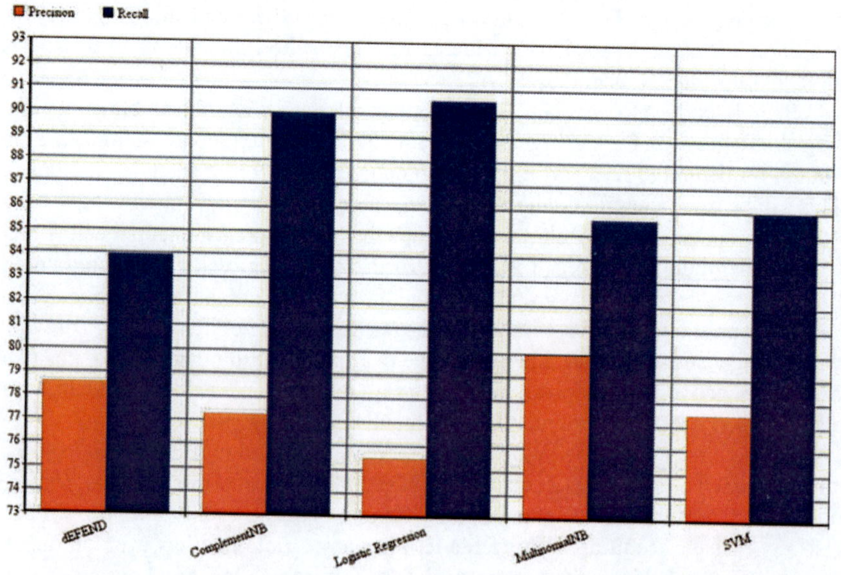

Fig. 4 Precision and recall comparison for all the methods (including the original)

6 Conclusion and Future Aspect

As is seen that with the mentioned modifications and including the comment filtering in dEFEND, the accuracy increases for all the given methods. Out of all the methods used to identify and filter useless comments, the MultinomialNB method has provided an increase in all the metrics considered and provided that the dataset available was minimal in size and hence limited the many possibilities of outcomes. Yet as the number of comments increases in news articles spreading like fire, these methods can be useful. Therefore, one can say that our hypothesis has been proven true, providing better accuracy for fake news detection by using comment filtering. Moreover, we have proven that using filtered comments for fake news detection is much better than using all the comments for similar fake news detection, even with the most common classifiers. This can be more refined by using different classifiers for different news threads or areas. Also, the models created for comment filtering can be different for various news records as compared to our generic model for all news records. Though the biggest challenge one may face is getting a hand on any dataset that includes a certainly good amount of comments along with the news articles. Apart from other such challenges of no early detection of fake news. This can only be achieved when once there are a considerable number of comments on any news article.

Since we ourselves graded the comments as junky or utility based on the basic concept of removing only the one or two-word comments that have minimal to none impact over information transfer, removing only the senseless comments, we could make a difference in the results. There can also be a scope of higher difference provided the criteria for classifying the comments can be more elaborate. One can also try other text classification methods to check for any difference in results or better results. The support of different languages in comments can also be a part of future work.

Acknowledgements This work was part of the study conducted for the project "Identification of Unreliability and Fakeness in Social Media Posts" funded by the Council of Science and Technology (CST), UP, India. We thank CST, UP, India, who provided us with the opportunity and support in this project.

References

1. https://www.bbc.com/news/world-us-canada-52407177. Accessed on 1 Aug 2020
2. Shu K, Cui L, Wang S, Lee D, Liu H (2019) dEFEND: explainable fake news detection. In: Proceedings of 25th ACM SIGKDD conference on knowledge discovery and data mining (KDD 2019). Anchorage, AK, USA. ACM, New York, NY, USA, 11 pages, August 4–8, 2019
3. Silva FCDD, Alves RVDC, Garcia ACB (2019), Proceedings of the 52nd Hawaii international conference on system sciences. https://doi.org/10.24251/HICSS.2019.332
4. Zannettou S, Sirivianos M, Blackburn J, Kourtellis N (2019) The web of false information: rumors, fake news, hoaxes, clickbait, and various other shenanigans

5. Zhang X, Ghorbani AA (2017) An overview of online fake news: characterization, detection, and discussion
6. https://www.politifact.com/. Accessed on 1 Aug 2020
7. https://www.snopes.com/. Accessed on 1 Aug 2020
8. https://hoaxy.iuni.iu.edu/. Accessed on 1 Aug 2020
9. Sharma S, Sharma DK (2019) Fake news detection: a long way to go. In: 2019 4th international conference on information systems and computer networks (ISCON). Mathura, India, pp 816–821. https://doi.org/10.1109/ISCON47742.2019.9036221
10. Oshikawa R, Qian J, Wang WY (2018) A survey on natural language processing for fake news detection. arXiv preprint arXiv:1811.00770
11. Wang Y, Ma F, Jin Z, Yuan Y, Xun G, Jha K, Su L, Gao J (2018) EANN: event adversarial neural networks for multi-modal fake news detection
12. Monti F, Frasca F, Eynard D, Mannion D, Bronstein MM (2019) Fake news detection on social media using geometric deep learning
13. Zhou X, Wu J, Zafarani R (2020) SAFE: similarity-aware multi-modal fake news detection. arXiv preprint arXiv:2003.04981
14. Zhang J, Dong B, Philip SY (2020) Fakedetector: effective fake news detection with deep diffusive neural network. In: 2020 IEEE 36th international conference on data engineering (ICDE). IEEE, pp 1826–1829
15. Tschiatschek S, Singla A, Gomez Rodriguez M, Merchant A, Krause A (2018) Fake news detection in social networks via crowd signals. In: Companion proceedings of the the web conference 2018, pp 517–524
16. Shu K, Wang S, Liu H (2018) Understanding user profiles on social media for fake news detection. In: 2018 IEEE conference on multimedia information processing and retrieval (MIPR). IEEE, pp 430–435
17. Gamova AA, Horoshiy AA, Ivanenko VG (2020) Detection of fake and provokative comments in social network using machine learning. In: 2020 IEEE conference of russian young researchers in electrical and electronic engineering (EIConRus). St. Petersburg and Moscow, Russia, pp 309–311. https://doi.org/10.1109/EIConRus49466.2020.9039423
18. Yanagi Y, Orihara R, Sei Y, Tahara Y, Ohsuga A (2020) Fake news detection with generated comments for news articles. In: 2020 IEEE 24th international conference on intelligent engineering systems (INES). Reykjavík, Iceland, pp 85–90. https://doi.org/10.1109/INES49302.2020.9147195
19. Risch J, Krestel R (2020) Toxic comment detection in online discussions. In: Deep learning-based approaches for sentiment analysis Springer, Singapore, pp 85–109
20. Cuilimeng/dEFEND-web. https://github.com/cuilimeng/dEFEND-web. Last accessed 25 May 2020
21. Cui L, Shu K, Wang S, Lee D, Liu H (2019) dEFEND: a system for explainable fake news detection. In: CIKM 2019—Proceedings of the 28th ACM international conference on information and knowledge management (international conference on information and knowledge management, proceedings). Association for Computing Machinery, pp 2961–2964. https://doi.org/10.1145/3357384.3357862

Comparative Analysis of Intelligent Solutions Searching Algorithms of Particle Swarm Optimization and Ant Colony Optimization for Artificial Neural Networks Target Dataset

Abraham Ayegba Alfa, Sanjay Misra, Adebayo Abayomi-Alli, Oluwasefunmi Arogundade, Oluranti Jonathan, and Ravin Ahuja

Abstract The optimizations approaches of ant colony optimization (ACO) and particle swarm optimization (PSO) were targeted at improving the outcomes of artificial neural networks for finding best solution in the space. Both ACO and PSO are derived from the artificial intelligence concept that imitate the natural behaviors of animals in finding best path to foodstuff location relative and back their nest. The artificial neural networks (ANNs) are reliant on estimated research scheme in which models are generated for unspecified function in order find suitable interrelationships in input and output datasets. These are not without challenges including large time of computation, expansive hidden layer size, and poor accuracy. This paper examines the effects of pretraining dataset with ACO and PSO prior training process of ANN in order to overcome the aforementioned problems of speed and accuracy through optimization of the local and global minima. The outcomes of the study revealed that the ACO outperformed PSO in conjunction with ANN in terms of RAE, MSE, RMSE, and MAPE utilized. The error rates of ANN pretrained with ACO and PSO

A. A. Alfa
Kogi State College of Education, Ankpa, Nigeria
e-mail: abraham.alfa@kscoeankpa.edu.ng

S. Misra (✉) · O. Jonathan
Covenant University, Otta, Nigeria
e-mail: sanjay.misra@covenantuniversity.edu.ng

O. Jonathan
e-mail: Jonathan.oluranti@covenantuniversity.edu.ng

A. Abayomi-Alli · O. Arogundade
Federal University of Abeokuta, PMB 2240 Abeokuta, Ogun State, Nigeria
e-mail: abayomiallia@funaab.edu.ng

O. Arogundade
e-mail: arogundadeot@funaab.edu.ng

R. Ahuja
Shri VikarmShilla University, Gurgaon, India

P. K. Singh et al. (eds.), *Proceedings of Second International Conference on Computing, Communications, and Cyber-Security*, Lecture Notes in Networks and Systems 203,
https://doi.org/10.1007/978-981-16-0733-2_32

distinctively are 62 and 73% accordingly. Benchmarking the results against the solution optimization studies, ACO and PSO algorithms are most preferred in finding the best solution or nearest-optimal in search spaces.

Keywords Artificial neural network · PSO · ACO · Solution search space · Best solution · Swarm intelligence · Optimization · Target dataset · Nearest-optima · Pretraining · Training dataset

1 Introduction

The quest to minimize time consumption and the cost implication of carrying out jobs based on guesses and to eventually deal with uncertainties of numerical and mathematical modeling gave rise to soft computing methods such as artificial neural network (ANN). ANNs have been deployed to diverse areas of human endeavors including engineering, nonlinear, and optimization problems. It is similar to the organic neurons of the human brain that carry signals or information in form of stimuli [1].

In practice, the training of ANNs makes use of dataset to enable the model experience or learn based on the instances obtainable for the purpose of generating forecasts [2]. There are serious weaknesses with ANN such as poor accuracy and large computational time due to extended training processes and local minima round trap [3].

Conventionally, ANN models are trained with well-defined database in order to produce outcomes for testing component of dataset. The point of training networks involves tasks learning and similarities discovery without afore computation. To strengthen the learning process of ANNs, there is need to minimize the error and improve the outcomes of predictive tasks. According to [4], several optimizations have developed to attain better performance of ANNs including PSO, ACO, genetic algorithm, etc.

This paper introduces the concept of pretraining dataset with optimization algorithms of ACO and PSO prior actual training of the ANNs. Both ACO and PSO simulate the natural behavior of moving toward foodstuff in the simplest route from the nest and back [4]. The goal is to generate actions on the basis of large datasets heuristics in arriving at the finest solution. This paper investigates the performance of ANNs pretrained with PSO and ACO searching algorithms for optimal solution in search space.

The remaining components of this paper are sectioned as follows: Sect. 2 is the concept of optimization. Section 3 is the methodology. Section 4 is the discussion of results. Section 5 is the conclusion.

2 The Concept of Optimization

The progression of resolving complex tasks by means of customary approaches has drastically decreased due many changes in the use of metaheuristic-based optimization algorithms. The main idea behind the advancements of metaheuristic algorithms is combination of several techniques in such that the new algorithm imbibes certain characteristics of the original algorithms. ACO and PSO algorithms have widespread usages because of their ability to solve variety of tasks [5]. During optimization, its processes are unbroken and comprehensive adding to daily tasks of individuals [6].

In its truest form, optimization is undertaken to select the finest alternative or alternative within a given preferences set. In many cases, optimization tasks are applicable to human areas of endeavor such as manufacturing, stock prediction, pattern recognition, agriculture, and engineering projects [7, 8]. Optimization is a quantitative tool in network decision making especially in one or more objectives optimization for some kinds of prearranged scenarios. However, optimization algorithms evolved by means of nature-driven evaluation that rely population about metaheuristics as general-purpose algorithms because of applications area cut across pool of issues [9, 10].

2.1 Artificial Neural Networks

A dataset for training the ANN enables its model to learn or acquire the features and relationships within data instances for the purpose of arriving at forecasts [3]. The outcomes of forecasts are usually optimized with diverse approaches such as fuzzy inference systems (FIS), PSO, genetic algorithm, ACO, and others. The main target of most optimizations is to minimize the time, errors, and costs of carrying out searches for solutions and deal with nonlinear or uncertain parameters about numerical and mathematical modeling problems, soft computing methods, and variety of engineering tasks [2].

There are diverse statistical approaches for moderating parameters such as weights, bias, and hidden neurons structures in ANN-based models. And, the combination of optimization processes to minimize errors in generally [3, 11]. However, ANNs have the prospects of being capable of carrying out near-accurate and reliable forecasts over massive dataset with divergent spread of attributes [12].

This paper adopted the key components of ANN-based models proposed by [13] including.

The Input: The input comprises binary attributes representations in the datasets. The hidden layer: This is the most complex structure of the ANNs, because it holds functions (differentiable, monotonic, and continuous) that connect weighted neuron of input to produce eventual solution. The output: It makes generalized information gathered from learning operations to produce appropriate outcomes within allowable boundary [13].

2.2 Particle Swarm Optimization

Researchers have continued to model the behavior of swarm for particles in terms of position and velocity in multidimensional space. The particles randomly move across the hyperspace and retain the finest position found in that process [14, 15]. The main operation of PSO starts by random swarm initialization in the search space for different iterations in a consecutive manner. Consequent upon this, it can be applied or deployed with ease to solve different optimization functional problems including nonlinear representations. By mathematical representation, particles initialization can be obtained with Eq. 1.

$$Y_c = y_{mi} + \text{rand}(y_{ma} - y_{mi}) \tag{1}$$

where y_{mi} and y_{ma} represent the minimum and maximum values in boundary of the solution search space, and Y_c is the different positions, c. Customarily, in every iterations, all particles search for the local best (LB) and later the global best (GB) in the quest to attain the solutions with optimal values. For that reason, every particle makes effort to maintain previous success and keep pace with the finest agent. The LB and GB are constantly updated whenever there is a difference between minimum the particles fitness values against present LB and GB values whose velocity and position can be determined using Eqs. 2 and 3.

$$s_{c+1} = \partial s_c + b_1 p_1(LB - Y_c) + b_2 p_2(GB - Y_c) \tag{2}$$

$$y_{c+1} = y_c + s_c + 1 \tag{3}$$

where p_1 and p_2 are values of random functions [0, 1], c_1 is the cognitive knowledge term, c_2 is the social knowledge term, and ∂ is the balancing function between local and global searches efforts. This paper utilizes the PSO solution searching algorithm to pretrain dataset prior to actual training of ANNs in order to direct search and minimize trials/errors.

2.3 Ant Colony Optimization

Ant colony optimization (ACO) makes use of heuristic means to easily determine or track the best solutions by searching the space with agents [16]. The ACO imitates the computational intelligence of swarm of several animals (or ants) in the seeking behavior for food which share similar characteristics with PSO [1]. The goal of optimization algorithms is to minimize the large dimension dataset within its subsets which is considered as a necessary phase of analytical prediction and accuracy.

Data is made up of several distinct features in terms of types and size known as structured, semi-structured, or unstructured form such as text, audio, and video [17]. According to a study in [5], the ant routing algorithm assigns and seeks the shortest (or top solution) routing in which a thorough search is carried out by the ant to select the shortest (or top solution) route. Often, this leads to defined function known as improved (or maximum) searching procedure.

2.4 Related Works

A hybrid scheme of PSO and ACO was introduced for optimal design procedure of truss structures. The PSO performs the complete space for design (optimize globally), while ACO determines a local search after the finest solution (optimize locally) was chosen with PSO [12]. The target was to construct high performance, simple and reliable forecasting scheme with prompt updates on parameters of fuzzy set. ACO was used to resolve problem of vehicle routing by [18].

In this case, fresh routing paths were generated by stepping up the concentration of pheromone or ant weight stratagem that serves as mutation agent for resolving the challenge of vehicle routing. Again, a tuning procedure on the rules lists with multitechnical factors was introduced into neurofuzzy system rule base by [19]. Again, the issues of landslide susceptibility mapping (LSM) forecast were resolved by optimizing ANN with PSO. The approach generated optimal network parameters and weights for the ANN that showed better reliability and accuracy for LSM [3].

PSO was chosen to generate optimal settings of the FLCs such as the membership functions to attain the finest localization reliability for indoor light-emitting diode positioning system in smart homes [20]. Similarly, the use of soft computing paradigm for eddy current braking system relied on PSO and FLC to improve effectiveness. The entire procedure employed PSO and FLC to optimize the design of control system in brakes in terms of peak overshoot and time. The PSO offered minimal peak overshoot for eddy current braking system than FLC [21].

Hybrid soft computing of PSO-ANN approach was used to forecast level of damage of conventional rubble mound breakwater of tandem breakwater. The role of PSO was to optimize bias and weights in the ANN through varied hidden neurons for improved accuracy and reliability [2]. A new approach for power electronics lifetime optimization and a faster speed response in a brushless DC motor drive was proposed by [22]. It made use of a fuzzy-PSO controller (i.e., FLC enhanced with PSO) whose outcomes were superior in terms of speed and lifetime efficiency of power electronics [23]. One study that considered pretraining of dataset to streamline the searching target of neural networks using ACO was proposed by [24].

3 Methodology

In this paper, the target attributes for outcomes in the dataset are determined before actual construction of the proposed neural networks. This approach makes use of PSO search capability to generate target solution or nearest-optima to optimize the trial, cost, and time for the ANN in generating forecast for Nigeria Stock Prices movement. The entire process of obtaining target solutions for the neural network optimization using PSO algorithm is illustrated in Fig. 1.

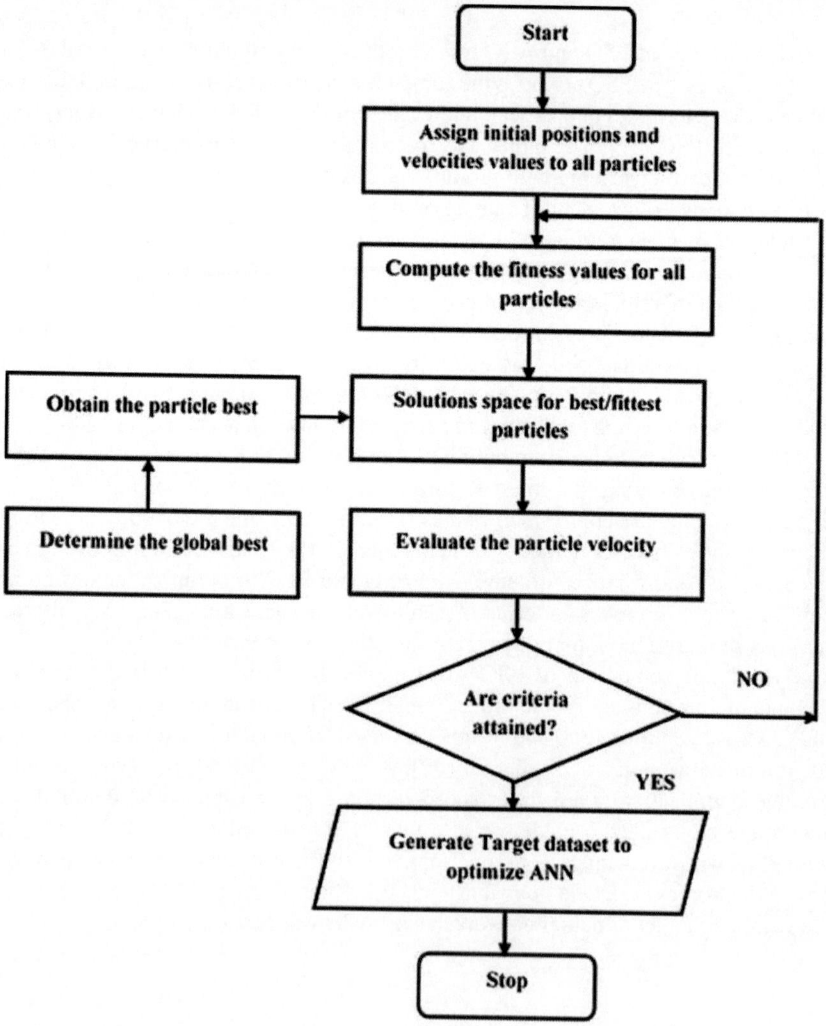

Fig. 1 ANN's target dataset optimization procedure of PSO

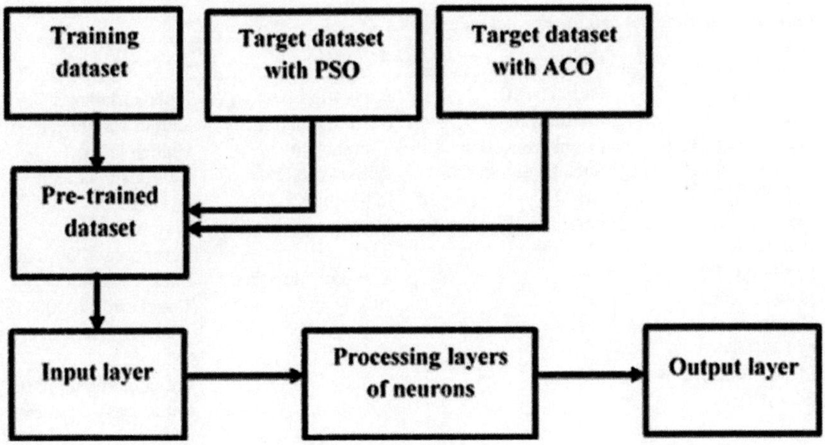

Fig. 2 Structure of feedforward ANN optimization

In Fig. 2, the entire PSO pretraining optimization involves allocation of initial positions and velocities to all participating particles, computation of fitness values of particles, solution space for best/fittest particles (the particles' best is obtained, and global best is determined), particle velocity is ascertained, and optimized target dataset is generated for ANN modeling.

The architecture of the proposed ANN based on PSO optimization for the target dataset is composed of the input, process, and outputs using appropriate parameter values. Similarly, the ACO optimization of target dataset for the ANN adopted that procedure present in [19]. The ANN makes use of the target generated from the solution space to generate reliable results autonomously. The original dataset utilized for this study is the UniLever Plc standard share prices for a period of 12 months which serves as inputs for the PSO algorithm and optimized feedforward ANN model as illustrated in Fig. 2.

From Fig. 2, the description of the proposed model is provided as follows:

Input Layer. This receives the *Pre-trained dataset* from the ACO and PSO procedure which optimize *Training dataset*. Training dataset serves as the input containing best solution sets for the ANN model.

Target dataset with ACO and target dataset with PSO. These provide the solution examples optimized by the procedures of ACO and PSO solution searching algorithms.

Processing layer of neurons. This block in ANN performs the various computations of seeking the best solutions using the knowledge of patterns drawn from *training dataset* and target datasets of ACO and PSO in order to improve the solution processing time and accuracy.

Output layers. This block shows the results of the best solutions generated by the complex computations of the processing layers of neurons, which is better, faster,

Table 1 Experimental parameters and their values

System	PSO	ACO [19]	ANN
Hardware HDD: RAM: 3.0 GHz CPU Speed: 2.0 GHz Software OS: Windows 8 Application: MATLAB R2013a Data collection Historical share prices of UniLever Plc–12/08/2014 to 06/25/2015	Particles: 10 Inertia weight: 0.78 Cognitive weight: 1.38 Social weight: 1.58 Arg: 1 Iteration: 1000	Population size: 80 Section function: Tournament Elitism preserved individual: 5 Mutation probability: 0.001 Size reduction factor: 100	Input variable: 3 Output variable: 1 Hidden layer: 1 Neurons in hidden layer: 50 Training: Levenberg–Marquardt Backpropagation Target error: 0.00001 Transfer function: Sigmoid Initial momentum: 0.1 Initial learning rate: 0.001

and accuracy with previous pretraining optimizations of ACO and PSO solution searching algorithms.

3.1 *Settings for the Experimentation*

The minimal values and parameters for the PSO, ACO, and ANN are presented in Table 1.

4 Discussion of Results

In this paper, the neural networks were distinctive trained with PSO and ACO algorithms before deploying each to search solutions in vector space. To ascertain the effectiveness of the solution searching algorithms, the outcomes of neural networks after utilizing trained data for diverse optimizations are presented in Table 2.

Table 2 Comparisons of the ACO and PSO solutions searching outcomes

Algorithms	ACO [24]	PSO
RAE	0.01	0.02
MSE	2.35	12.20
RMSE	1.53	3.49
MAPE (%)	0.06	0.13
Std. error of the estimate	0.62	0.73

In Table 2, the ACO and PSO algorithms performed differently during solutions searching operations. The ACO algorithm largely outperformed the PSO algorithm especially for MSE (2.35/12.20) and RMSE (1.53/3.49), respectively. However, the disparity in performance between ACO and PSO is relatively smaller for RAE (0.01/0.02), MAPE (0.06%/0.13%), and standard error of the estimate (0.62/0.73) accordingly. The reason for the improved performance of ACO algorithm against PSO algorithm is the capability of the former to attain nearest-optimal or finest solution in the space vector.

More so, the natural behavior offered by ACO makes it faster to direct search toward the finest solution with smallest errors as against PSO. Therefore, optimization of neural networks can be carried out preferably with ACO for better accuracy and minimal errors. The graphical illustration of the performances of the ACO and PSO is presented in Fig. 3.

In Fig. 3, the performance of the proposed ACO search algorithm outperformed contemporary searching algorithm of PSO in terms of the evaluation parameters MSE and RMSE. However, there are negligible differences between the two searching algorithms for evaluation parameters such as RAE, MAPE, and standard error of the estimate. The solution searching outcomes of this paper are compared to the soft computing approaches: heuristic procedure in [24], and fuzzy logic-based heuristic in [19] as shown in Table 3.

From Table 3, the metaheuristic solution searching procedure for ACO [24] was better than the PSO algorithm proposed in this paper by 62–73%. Similarly, the PSO search algorithm was better than neurofuzzy technique in [19] using performance metrics such as RMSE and MAPE. Whenever accuracy of searching pattern is to be considered, the ACO alone is most preferable to finding best solutions inside the local and global spaces.

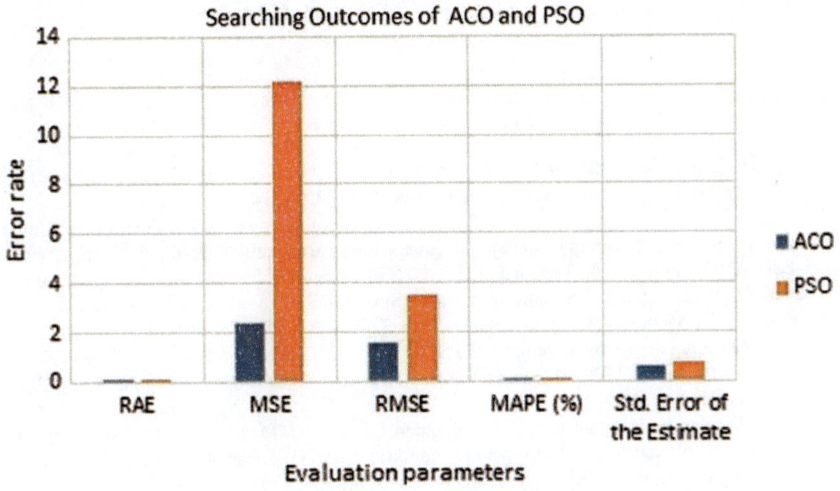

Fig. 3 Performance of ACO and PSO searching algorithms compared

Table 3 Benchmarking of searching algorithms

Metric	This paper	Alfa et al. [24]	Neurofuzzy system [19]
RMSE	3.49	1.534	30.995
MAPE (%)	0.13	0.06	0.35

5 Conclusion

This paper found that, the purpose of optimization is to determine the fittest solution within a wider set of solutions or search space, which is purely a heuristic approach. PSO is a key solution searching algorithm imitating the features of natural processes such as finding the best path to a solution. The swarm particles conduct search by means of their distinct experience as well as the location of the optimist particle in the solution space. Conversely, the ACO makes use of natural process of moving within shortest distance of the nest to the source of food.

These can be applied to the complex combinatorial problems in which agents in population generate diverse solutions using experiences from the past processes. In both cases, the entire searching activity is controlled by heuristic information probabilistically. Of all two optimization processes covered in this paper, ACO is the best solution searching algorithm against PSO by 62–73%. Comparatively, the performances of PSO and ACO (proposed models) and neurofuzzy system (benchmark work) are 9.69 and 4.26–86.05% for RMSE and 24.07 and 11.11–64.81% for MAPE, respectively.

It follows that the accuracy of ANN can best be improved upon through pretraining procedure especially with ACO. In future works, there is need to minimize the error rates of ANN with other kinds of optimization algorithms such as random forest and dragonfly.

References

1. Tian H, Shu J, Han L (2018) The effect of ICA and PSO on ANN results in approximating elasticity modulus of rock material. Eng Comput 35(1):305–314
2. Kuntoji G, Rao S, Nava E, Reddy B (2019) Prediction of damage level of inner conventional rubble mound breakwater of tandem breakwater using swarm intelligence-based neural network (PSO-ANN) approach. In: Bansal J et al (eds) Soft computing for problem solving, advances in intelligent systems and computing, vol 817. Springer, Singapore, pp 441–453
3. Moayedi H, Mehrabi M, Mosallanezhad M, Safuan A (2019) Modification of landslide susceptibility mapping using optimized PSO-ANN technique. Eng Comput 35(3):967–984
4. Pal A, Chakraborty D (2014) Prediction of stock exchange share price using ANN and PSO. Int J Eng Sci 80(1):62–70
5. Xu C, Gordan B, Koopialipoor M, Armaghani DJ, Tahir MM, Zhang X (2019) Improving performance of retaining walls in dynamic conditions developing an optimized ANN based on ant colony optimization technique. IEEE Access 7:94692–94700

6. Okewu E, Misra S (2016) Applying metaheuristic algorithm to the admission problem as a combinatorial optimization problem. In: Mizera-Pietraszko J et al (eds) Advances in digital technologies, ICADIWT 2016. IOS Press, Amsterdam, pp 53–64
7. Crawford B, Soto R, Johnson F, Misra S, Paredes F (2014) The use of metaheuristics to software project scheduling problem. In: Murgante B et al (eds) Computational science and its applications—ICCSA 2014, LNCS, vol 8583. Springer, Cham, pp 215–226
8. Crawford B, Soto R, Peña C, Riquelme-Leiva M, Torres-Rojas C, Misra S, Paredes F et al (2015) A comparison of three recent nature-inspired metaheuristics for the set covering problem. In: Gervasi O et al (eds) Computational science and its application—ICCSA 2015, LNCS, vol 9158. Springer, Cham, pp 431–443
9. Soto R, Crawford B, Galleguillos C, Misra S, Olguín E (2014) Solving Sudokus via Metaheuristics and AC3. In: 2014 IEEE 6th international conference on adaptive science and technology. IEEE, Otta, Nigeria, pp 1–3
10. Crawford B, Soto R, Johnson F, Vargas M, Misra S, Paredes F (2015) A scheduling problem for software project solved with ABC metaheuristic. In: Gervasi O et al (eds) Computational science and its applications—ICCSA 2015, LNCS, vol 9158. Springer, Cham, pp 628–639
11. Chakraborty R (2010) Fundamentals of neural networks: soft computing course lecture notes. Computer Science Department, Indian Institute of Technology, Madras, India
12. Gholizadeh S, Fattahi F (2012) Serial integration of particle swarm and ant colony algorithms for structural optimization. Asian J Civ Eng (Build Hous) 13(1):127–146
13. Su Y (2005) An investigation of continuous learning in incomplete environments. PhD dissertation, University of Nottingham, UK
14. Hamdi H, Regaya Ben C, Zaafouri A (2018) Real-time study of a photovoltaic system with boost converter using the PSO-RBF neural network algorithms in a MyRio controller. Sol Energ 183:1–16
15. Mohd Aras MS, Abdullah SS, Jaafar HI, Yusof AA, Mohd Tumari MZ, Yan HG (2019) Optimization of single input fuzzy logic controller using PSO for unmanned underwater vehicle. In: Md Zain Z et al (eds) Proceedings of the 10th national technical seminar on underwater system technology 2018, LNEE, vol 538. Springer, Singapore, pp 15–26
16. Eltamaly AM, Farh HMH (2019) Dynamic global maximum power point tracking of the PV systems under variant partial shading using hybrid GWO-FLC. Sol Energ 177:306–316
17. Manoj RJ, Preveena A, Vijayakumar K (2018) An ACO-ANN based feature selection algorithm for big data. Cluster Comput 22(2):395–3960
18. Hlaing SZS, Khine MA (2011) An ant colony optimization algorithm for solving traveling salesman problem. In: International conference on information communication and management, vol 16, pp 54–59
19. Rajab S, Sharma V (2017) An interpretable neuro-fuzzy approach to stock price forecasting. Soft Comput 23(3):921–936
20. Pau G, Collotta M, Maniscalco V, Choo KR (2019) A fuzzy-PSO system for indoor localization based on visible light communications. Soft Comput 23(14):5547–5557
21. Singh AK, Nasiruddin I, Sharma AK (2019) Implicit control of eddy current braking system using fuzzy logic controller (FLC) and particle swarm optimisation (PSO). J Discr Math Sci Crypt 22(2):253–275
22. López MG, Ponce P, Soriano LA, Molina A, José J, Rivas R (2019) A novel fuzzy-PSO controller for increasing the lifetime in power electronics stage for brushless DC drives. IEEE Access 7:47841–47855
23. Nguyen D (2019) Designing PSO-based PI-type fuzzy logic controllers: a typical application to load-frequency control strategy of an interconnected hydropower system. In: Proceedings of the 2019 3rd international conference on automation, control and robots. ACM, New York, USA, pp 61–66
24. Alfa AA, Misra S, Ahmed KB, Arogundade O, Ahuja R (2020) Metaheuristic-based intelligent solutions searching algorithms of ant colony optimization and backpropagation in neural networks. In: Singh P et al (eds) Proceedings of 1st international conference on computing, communications, and cyber-security—IC4S 2019, LNNS, vol 121. Springer, Singapore, pp 95–106

25. Bin AY, Zhong-Zhen Y, Baozhen Y (2009) An improved ant colony optimization for vehicle routing problem. Eur J Oper Res 196:171–176

An Online Planning Agent to Optimize the Policy of Resources Management

Aditya Shrivastava, Aksha Thakkar, and Vipul Chudasama

Abstract Reinforcement learning-based systems have received a lot of attention in various domains in recent years. In such domains, an autonomous agent learns from environment to provide a solution. Resource scheduling is considered as research challenge where such autonomous agent optimizes the solutions. This work is presented as an investigation on the effectiveness of various algorithms which drives actions associated with autonomous agent. We give a detailed contention between three differing algorithms—Q-learning, Dyna-Q, and deep-Q-network, given the task of effectively allocating the resources in an online basis. Among the mentioned algorithms, the Q-learning and deep-Q-network, which are model free algorithms, have remained in wide use for planning. However, this paper focuses on highlighting the effectiveness of lesser-known model-based algorithm, Dyna-Q. The experiment results show agent-based policy derived by the Dyna-Q algorithm which provides optimized resource scheduling for the current environment.

Keywords Dyna-Q framework · Resource allocation · Reinforcement learning · Model-based learning

1 Introduction

Consummate management of resources has been imperative for operating high performance computing tasks. In cloud computing applications, users make requests to the service providers to utilize multiple resources. Subsequently, systematic resource

A. Shrivastava (✉) · A. Thakkar · V. Chudasama
Department of Computer Science and Engineering, Institute of Technology, Nirma University, Ahmedabad, India
e-mail: 17bit014@nirmauni.ac.in

A. Thakkar
e-mail: 17bit003@nirmauni.ac.in

V. Chudasama
e-mail: vipul.chudasama@nirmauni.ac.in

P. K. Singh et al. (eds.), *Proceedings of Second International Conference on Computing, Communications, and Cyber-Security*, Lecture Notes in Networks and Systems 203,
https://doi.org/10.1007/978-981-16-0733-2_33

management has to be done to make efficient use of those resources. Scheduling tasks in cloud computing using traditional methods which are in use today have been meticulously tested and implemented [7, 9]. Resource management has also been essential to reduce energy consumption in data centers [8] and predicting relay selection [10]. However, the policies in use at present are generally manually defined keeping in mind the nature and working of any particular environment. Automating these decisions to allocate resources without any human governance for various applications is a difficult task.

Reinforcement learning is turning out be the most promising and optimistic field of research in the recent years. It pans out a very resolute and distinctive approach for making decisions in complex environments which seem implausible by other machine learning paradigms. An agent receives a reward on the basis of what actions it takes and learns what decisions to make by gaining experience from the defined environment. Reinforcement learning has been used for many applications like a network traffic signal control system in a simulated environment [1]. Latest breakthroughs in game-playing agents in AlphaGo and Atari games have substantiated that deep reinforcement learning is capable of solving complex decision-making problems in complicated environments [14, 16]. Reinforcement learning has proved to outperform conventional approaches in resource management in computer clusters [13] and in cooling datacenters [4]. Other applications include automating robotics [12], Web system auto-configuration [2], and so on. These applications involve obscure and continuously changing environments which need adaptive and automated solutions.

Dyna-Q and deep-Q-network are the extension of Q-learning method in reinforcement learning. Dyna-Q is a simple yet vigorous algorithm which compounds Q-learning and Q-planning. Here, *planning* refers to the generation of simulated experiences other than real experience which is used to improve the efficiency of our model. This way of improving value functions and policies is referred to as *indirect reinforcement learning*. We can see some reasons that this method could enhance the existing approach:

1. In environments where we have to improve the policy using insubstantial experience, Dyna-Q helps to attain results with better efficiency.
2. Dyna-Q speeds up the learning procedure and takes less number of iterations to achieve maximum amount of reward.
3. There are other variants of Dyna-Q algorithm like $Dyna - Q^+$ which has an added exploration bonus, and it is found to perform better.

One of the basic examples illustrating the working of Dyna-Q is Dyna-Maze [17]. Dyna-Q has shown good results in applications like dialog policy learning [15] and path planning for robots [19]. On the other hand, deep-Q-network (DQN), a deep reinforcement learning approach uses a neural network instead of the Q-table to approximate the Q-value functions. DQN is used to scale the Q-learning approach to complex environments, i.e., having large number of states and actions. DQN has become common in reinforcement learning applications as it is used in most of the models for game playing agents. These algorithms can help to get better results for the resource management problem as they provide better decision-making techniques.

In the subsequent sections, we discern the architecture of the above-mentioned techniques, their application in a simple exemplary model and results. Section 2 shows the related work in this area which was used as an inspiration for this paper. In Sect. 3, we understand more about these algorithms. In Sect. 4, we delineate the methodology of our experiments. In Sect. 5, we test them on sample data and observe results in Sect. 6, and finally, we ponder upon the results and conclude our hypotheses.

2 Related Work

Many industries and cloud computing applications require allocation of certain jobs without compromising the resource limitation. Reinforcement learning was first used for such a problem in a space shuttle payload processing problem (SSPP) for NASA to minimize the total duration of the final schedule [20]. The temporal difference methods used in this paper were observed to perform better than the standard heuristic approaches. This paper was the first to explore the potential of reinforcement learning algorithms to solve scheduling task. Our paper is particularly inspired from Hongzi Mao's paper [13] on resource management with deep reinforcement learning. Noteworthy results were observed as a reinforcement learning model using deep neural networks performed better than the standard approaches like shortest job first and Tetris. The reinforcement learning model used in this paper was based on Monte Carlo methods. Findings by an IBM research team also provide support on feasibility of such methods [18]. This paper demonstrated some positive results of a new hybrid reinforcement learning method which combines strength of both reinforcement learning and model-based policies for resource valuation estimates. Also insights from other papers having similar outlook to this problem [3, 11] and papers on cluster scheduling [6] and task scheduling [5] provided motivation for this paper.

3 Reinforcement Learning

Here, we first explain the general setting of a reinforcement learning problem and how it evaluates the problem in a basic Q-learning setting. Then, we shine some light on how indirect reinforcement learning methods like Dyna-Q will augment the direct reinforcement learning methods. Lastly, we will see how DQN replaces the Q-table in a Q-learning problem.

3.1 Problem Setting

In a standard reinforcement learning problem, there are two main elements, an active decision-making agent and the environment. The agent interacts with the environment and tries to achieve a goal with the maximum reward possible. Figure 1 shows a labeled diagram of a general reinforcement learning problem setting. Consider a sequence of discrete time steps $t = 0, 1, 2, 3, \ldots$. At every time step t, the agent arrives at a *state* in the environment, $s_t \in S$ and based on the policy selects an action, $a_t \in A(S_t)$ from the environment. Here, S is the set of possible states, and $A(S_t)$ is the set of actions available in state s_t. After completing the *action*, the agent receives a reward $r_{t+1} \in R$ where R is the set of rewards, and then, it moves on to the state s_{t+1}. The agents maps states to the probabilities of selecting the actions. This is known as the policy of the environment, and it is generally denoted by π_t, where $\pi_t(a|s)$ is the probability of selecting action a if the state is s. Our ultimate goal is to choose the states which lead to obtain the maximum rewards. However, for continuing tasks generally, we discount the rewards to get the cumulative discounted return. This is represented as $\sum_{k=0}^{\infty} \gamma^k r_{t+k+1}$, where $0 \leq \gamma \leq 1$. Here, γ is the discount rate. Hence, our goal would be to maximize the expected cumulative discounted reward [17].

Q-learning is a model free of policy reinforcement learning algorithm. Consequently, it does not have a transition probability function associated with the Markov decision process, and it does not have a policy function. Q-learning performs actions to learn policies that maximize the total reward. The optimal value function or the Q-function is approximated using experiences gained from the environment. So for a finite Markov decision process, Q-Learning can find the optimal action-selection policy. A Q-table is maintained to store the Q-values obtained after every episode. This table would help to get the best action a_t to perform in a state s_t using the Q-values. Let us see the update rule for $Q(s_t, a_t)$, i.e., the Q-value of action a_t at state s_t to approximate the Q-function:

$$Q(s_t, a_t) \leftarrow Q(s_t, a_t) + \alpha[r_{t+1} + \gamma \max_a Q(s_{t+1}, a) - Q(s_t, a_t)]$$

Here, α is the learning rate which basically decides how much importance should be given to new value as compared to the old value and γ being the discounting rate to discount future rewards. These updates are iterated many times to get optimal Q-values. Also for taking action, we have two options: to explore the environment

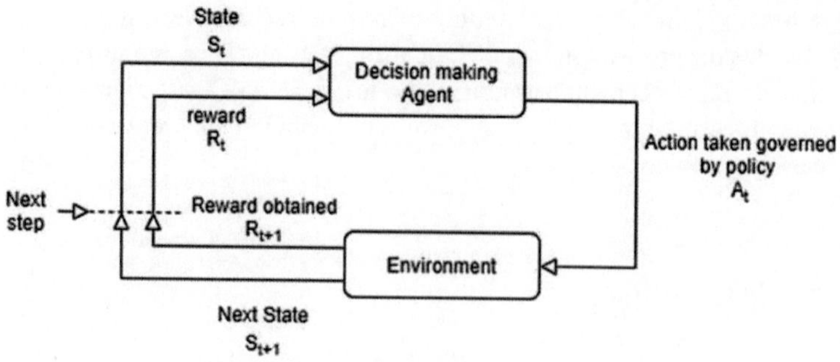

Fig. 1 General reinforcement learning problem setting

or to exploit by selecting the action with max value of those actions. This trade-off is decided by ϵ, where $0 \leq \epsilon \leq 1$. Once the agent gains enough experience, the algorithm converges if the hyperparameters are selected carefully. Hence, Q-learning handles the transitions and rewards which are stochastic.

3.2 *Dyna-Q*

Algorithm 1 Pseudocode of Dyna-Q algorithm

Initialize $Q(s, a)$ and model $M(s, a)$ for all the states $s \epsilon S$ and actions $a \epsilon A$ arbitrarily
Here S is the set of states, A is the set of actions at a particular state, Q is the set of Q-value pairs and M is the set of values of the model
while Q is not converged(for a certain number of episodes) **do**
  #Direct Reinforcement Learning starts
  $s \leftarrow$ current(non terminal) state
  $a \leftarrow \epsilon$-greedy(S, Q)
  Get reward $r \leftarrow R(s, a)$ and new state $s' \epsilon S$ by executing action a
  $Q(s, a) \leftarrow Q(s, a) + \alpha[r + \gamma \max_a Q(s', a) - Q(s, a)]$
  #Direct Reinforcement Learning ends
  #Model Learning starts
  $M(s, a) \leftarrow (s', r)$
  #Model Learning ends
  #Indirect Reinforcement Learning starts
  for i in range $1, \ldots, n$ **do**
    $s \leftarrow$ random previously observed state
    $a \leftarrow$ random action previously taken in s
    $(s', r) \leftarrow M(s, a)$
    $Q(s, a) \leftarrow Q(s, a) + \alpha[r + \gamma \max_a Q(s', a) - Q(s, a)]$
  #Indirect Reinforcement Learning ends
return Q

The Dyna-Q method is a combination of learning through real experiences and using a world model to generate hypothetical experience, i.e., planning. This combination of direct and indirect reinforcement learning along with model learning helps us to get better results. Figure 2 demonstrates the working of a generalized Dyna-Q architecture.

Along with direct updates using the update rule specified in Eq. 1, the experience also helps to build a model of the environment. Assuming a deterministic environment along with the Q-values $Q(s, a)$, we also maintain a model of the environment $M(s, a)$. This model helps to generate the simulated experiences, and search control selects the initial states and actions for the model to start with. The simulated experiences perform n number of planning steps where the Q-values are updated iteratively using the model. On performing this procedure on a number of episodes, Dyna-Q converges faster than the basic Q-learning problem. While this algorithm converges

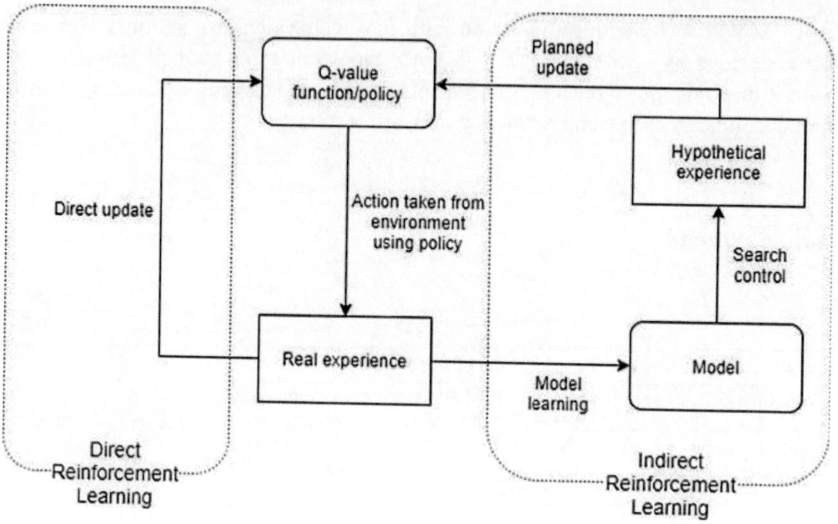

Fig. 2 General reinforcement learning problem setting

faster, it should be taken care of that number of epochs are limited as this algorithm takes up a lot of computing power. The algorithm shown below would make this clear.

3.3 Deep Q-Networks

Q-learning is a powerful algorithm but what happens when the environment has a large number of states and actions? The Q-values stored in the Q-table will take a lot of memory space, and the algorithm will take longer to run too. To solve this problem, we use neural networks replacing the Q-table to approximate the Q-function. The selection of actions is decided by the Q-network. The loss function for the neural network is defined as the mean squared error between the predicted and the target Q-values. The equation below shows the cost function used in deep-Q-networks. $\text{Cost} = [Q(s, a; \theta) - (r + \gamma \max_a Q(s', a; \theta))]^2$

Here, θ denotes the weights of the neural network. Also, $Q(s, a; \theta)$ represents the predicted Q-value, and $Q(s', a; \theta)$ represents the target Q-value. Using this equation, we train the network to learn Q-values. As the target is unstable and variable in the reinforcement learning problem generally, a separate network with similar architecture is used which estimates the target. This helps to stabilize the training. This is how the Q-learning problem is scaled using deep-Q-networks for large-scale applications.

4 Problem Description

In this section, we present the formulation of the problem of job scheduling as a reinforcement learning based on the algorithms delineated above. We show how to represent each these algorithms in a way suitable for job scheduling. Finally, we provide our approach and solution adhering to the apparatus prescribed above.

4.1 Proposed Model

Initially, we account a cluster that can contain jobs demanding at most d different resource types, for instance, CPU, memory, etc. These jobs arrive and make a request for the resources in a sequential manner at different timestamps. The job scheduler chooses from a pool of k jobs. Here, initially, the pool shall be empty, and scheduler shall start to accommodate the jobs in the pool as they arrive until a total of k jobs are present in the pool. The rest of the jobs are queued according to their timestamps as they arrive. Once any particular job is completed, the scheduler scans and chooses one or more of the waiting jobs for scheduling depending upon the availability in the pool after each timestep. We assume here that each job is aware of the resources that it needs to utilize. Here, the resource indication is that each job j is given by $r^j = (r_1^j, r_2^j, r_3^j, \ldots, r_d^j)$. For the convenience and evaluation purposes, note that scheduler allocates the job in a non-preemptive way, the duration of the job represented by T_j is known beforehand, and we treat cluster as single collection of resources instead of discrete resources fragmented across the system. While these aspects form an important portion when considering the practical scenario, we allow to derogate them to investigate and capture insights about the essential elements and provide a non-trivial setting. This can help capture the essential elements of multi-resource scheduling and provide a non-trivial setting to study the effectiveness of RL methods in this domain.

4.2 Scenario

4.2.1 State-Space

First we begin by providing the description for the state that our scheduler shall encounter. Note that here, the RL agent shall be the job scheduler which carries out the selection-action of most seemingly optimal job and allocates the resources to it. Thus, in all, the state-space from the job scheduler perspective may include the following:

1. Current jobs with allocated resources.
2. The jobs in the pool waiting to be scheduled.

Here, the jobs pool shall be initially represented by a vector comprised of job vectors each of length d—depicting values of each resources. Ideally, the vector containing vectors (no. of resources) can have indefinitely large number of vectors. However, as mentioned earlier, it can become impractical for the agent to choose from such large number of resources. Therefore, the agent shall maintain only the first k jobs. These k jobs shall either be chosen from directly in case of the algorithms Q-learning and Dyna-Q and shall be given as an input to the neural network in the case of deep-Q-Network.

4.2.2 Action-Space

After every time step, the agent is given the task of choosing the best job from the input space of k jobs. Thus, the full action space set of an agent shall be characterized by $\{\phi, 1, 2, \ldots, k\}$. Here, any element, for instance 3, means "schedule the 3rd job from the pool." and that ϕ signifies void selection of jobs meaning no job shall be selected. The scheduler examines the state after every time-step and gets to choose from the input space. Once any job from the input space is selected, the scheduler stays frozen until the time one of the scheduled process gets finished.

4.2.3 Rewards

We craft the rewards that can efficiently guide the agent to learn the optimal solution. In our experiments, the rewards for allocating a particular job would be the weighted sum of the values of the stipulated resources. (Please refer 4.2.4 for more on how to calculate the weighted sum). Thus, for every state, the agent shall be given the task to maximize the reward for which the optimal solution would be to select the resources that represent the highest priority in terms of highest rewards or weighted sum. Please note that the agent's job is not to maximize the immediate reward but the cumulative reward from the start till the time that the jobs are allocated and the agent does not receive any reward for intermediate decisions during a time-step or for any void action chosen.

4.2.4 Algorithms

It shall be impractical to give the vector representing the resources for a specific job as an input to one specific input neuron. Therefore, to counter this problem, we allow for a weighted sum of the resources to be computed. This means that we shall separately maintain a weight vector represented by,

$$W = \{w_1, w_2, w_3, \ldots, w_d\} \tag{1}$$

In this way, every job shall obtain its weighted sum value that shall account for the final value of the form that can be presented in the action space of the agent. We describe the complete process as below:

$$y = W^{\top} X \tag{2}$$

$$y = \begin{bmatrix} w_1 \\ w_2 \\ w_3 \\ \vdots \\ w_d \end{bmatrix} \begin{bmatrix} r_1^1 & r_2^1 & r_3^1 & \dots & r_d^1 \\ r_1^2 & r_2^2 & r_3^2 & \dots & r_d^2 \\ r_1^3 & r_2^3 & r_3^3 & \dots & r_d^3 \\ \vdots & \vdots & \vdots & \vdots & \vdots \\ r_1^k & r_2^k & r_3^k & \dots & r_d^k \end{bmatrix} \tag{3}$$

$$y = \begin{bmatrix} w_1 \cdot r_1^1 & w_2 \cdot r_2^1 & w_3 \cdot r_3^1 & \dots & w_d \cdot r_d^1 \\ w_1 \cdot r_1^2 & w_2 \cdot r_2^2 & w_3 \cdot r_3^2 & \dots & w_d \cdot r_d^2 \\ w_1 \cdot r_1^3 & w_2 \cdot r_2^3 & w_3 \cdot r_3^3 & \dots & w_d \cdot r_d^3 \\ \vdots & \vdots & \vdots & \vdots & \vdots \\ w_1 \cdot r_1^k & w_2 \cdot r_2^k & w_3 \cdot r_3^k & \dots & w_d \cdot r_d^k \end{bmatrix} \tag{4}$$

$$y = \begin{bmatrix} j^1 \\ j^2 \\ j^3 \\ \vdots \\ j^k \end{bmatrix} \tag{5}$$

Here, Eq. 2 describes the operation for eliciting weighted sum for all of the respective jobs. W here is a d dimension vector, and therefore, its transpose—$W^{\top}$ is of dimension $1 \times d$ and is multiplied with the jobs vector—X of dimension $d \times k$. Here, d represents the types of resources, and k represents the number of jobs. And finally, we get a vector of length k representing the effective value of each of the jobs. And this is the entity that shall be given as an action space for the user.

This preprocessed vector shall be an action space of three agents adopting three different respective algorithms whose contention we explore here—Q-learning, Dyna-Q, and deep-Q-network.

5 Experiments and Discussion

We simulate the environment discussed in 4 and train our RL agents using the three algorithms described in 3 namely—Q-learning, Dyna-Q, and deep-Q-network. Among the three methods discussed, our primary advocacy lies with the recently proposed Dyna-Q method. The other two methods namely Q-learning and deep-Q-network are the well established and experimented with methods, and therefore, we choose them for validation of the proposed Dyna-Q method. We run the experiments for 30,000 training steps, and in this section, we reflect on the behavior of our agents over the course of its learning to maximize its reward. It can be observed from Fig. 3a that the agent learns gradually when trained using Q-learning procedure. This suggests that conventional model free-based learning demands significant amount of time to train well. It is only after 2500 epochs that the agent's learning really starts to bump up. The Q-learning agent is seen to have consistently exploring and being affected by these exploring decisions. The Q-learning agent while exploring is faced by lesser rewards multiple times from 7500 to 12500 steps. However, it not until 20000 steps that the Q-learning agent counters the highest reward and stays consistent since then. Thus, for its application in job scheduling, the Q-learning can induce skepticism because of its inability to learn the optimal solution quickly.

Next, we explored the Dyna-Q procedure under identical circumstances. The Dyna-Q method learns to maximize its rewards in its early phase itself. Here, the agent takes an action for a few steps, and while waiting for the reward and shifting to the next phase, it simulates the identical environment to perform virtual actions and meanwhile also learns from it. This enables the agent, as observed from the figures to learn to make the optimal choices approximately ten times faster than its predecessor—Q-learning method. For instance, the optimal reward of 30 units is reached by a Q-learning agent at 25000th training step. However, for the case of Dyna-Q, it reaches this mark at just around 3000 steps. Such high yielding capacity of the Dyna-Q algorithm can be attributed to its ability to simulate a virtual environment in parallel. Employing this algorithm instead of Q-learning algorithm can help in marginally overcoming the problem of latency.

Lastly, we employed deep-Q-network for the allocation of resources. In this case, the policy of the algorithm was represented as neural network. The neural network shall take the input of the collection of resources as described in 4.2.4. And the neural network shall output the probability distribution over the action space, and then, the action with the highest probability shall be chosen and performed. And as shown in Fig. 3c, the deep-Q-network algorithm is the fastest one to learn. However, during the later stages, the algorithm's performance degrades. This means that the agent trained with deep-Q-network faces difficulty when it encounters unprecedented states. Fig. 3d shows its cost to be consistently reducing, and simultaneously, the agent loses its optimal performance (Fig. 3c). This behavior may introduce skepticism for employing the algorithms in real systems.

Thus, in general sense, considering the overall performance of agents with three different algorithms, we suggest that Dyna-Q is the most robust and reliable algorithm for the resource allocation technique. This technique can help in marginally overcoming the problem of both latency in reaching optimality that is faced by mainstream Q-learning algorithm and low efficacy displayed deep-Q-network when faced with high variance. Also the some works have known to derogate deep-Q-network to applied to real systems. This is because the computational complexity that it

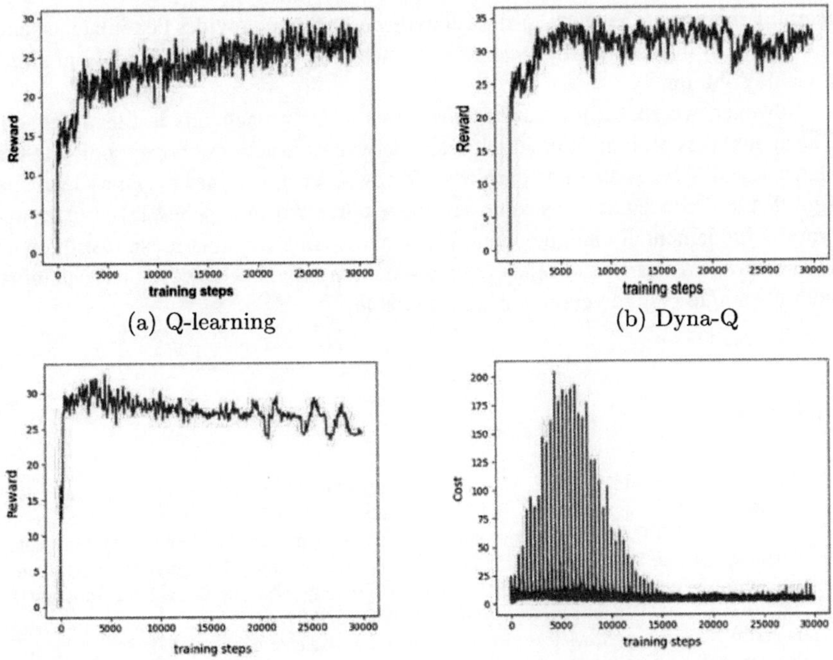

(a) Q-learning (b) Dyna-Q

(c) Deep-Q-Network (d) Deep-Q-Network cost

Fig. 3 Figure shows the contention between the performances of the three approaches considered. The vertical axis represents reward, and horizontal axis represents training steps. Since the third architecture opted for contention is deep-Q-network which consists of a neural network, we also include the graph showing improvement in learning cost of deep-Q-agent

involves. Thus, considering various perspectives, we suggest that Dyna-Q seems the most optimal algorithm for scheduling resources in cloud on an online basis. Also, this algorithm can be boosted even further, simulating the environment for virtual learning in an online fashion.

6 Conclusion

The need of managing the resources efficiently has been increasing with the new advancements in cloud computing and many other fields. However, making it autonomous has remained a challenge. In this work, we attempt to shed light on this challenging issue. The results achieved above provide conclusive evidence on how a reinforcement learning approach called Dyna-Q can prove to be a rather plausible alternative for resource management. This paper conducts experiments in a resource management problem setting where we introduce and compare the reinforcement learning methods. The Dyna-Q algorithm discussed here provides better latency and less difficulty in achieving optimal performance in a practical environment and offers a viable solution.

However, we spot a few further directions for improvements in this approach. The agent takes prolonged time in learning appropriate actions corresponding to the states. Dyna-Q faces the same problem. However, when it comes to Dyna-Q, we see significantly reduced times as compared to the other two approaches. Dyna-Q framework is fundamentally unique about its scalability in a way that it can distribute its learned progress across multiple systems and environments. However, its scalability within a single system yet remains questionable.

References

1. Arel I, Liu C, Urbanik T, Kohls AG (2010) Reinforcement learning-based multi-agent system for network traffic signal control. IET Intell Transp Syst 4(2):128–135
2. Bu X, Rao J, Xu CZ (2009) A reinforcement learning approach to online web systems auto-configuration. In: Proceedings of the 2009 29th IEEE international conference on distributed computing systems, ICDCS '09. IEEE Computer Society, USA, pp 2–11. https://doi.org/10.1109/ICDCS.2009.76, https://doi.org/10.1109/ICDCS.2009.76
3. Dutreilh X, Kirgizov S, Melekhova O, Malenfant J, Rivierre N, Truck I (2011) Using reinforcement learning for autonomic resource allocation in clouds: towards a fully automated workflow
4. Evans R, Gao J, Deepmind ai reduces google data centre cooling bill by 40%. https://deepmind.com/blog/article/deepmind-ai-reduces-google-data-centre-cooling-bill-40
5. Gawali MB, Shinde SK (2018) Task scheduling and resource allocation in cloud computing using a heuristic approach. J Cloud Comput **7**(1). https://doi.org/10.1186/s13677-018-0105-8, https://doi.org/10.1186/s13677-018-0105-8
6. Grandl R, Ananthanarayanan G, Kandula S, Rao S, Akella A (2014) Multi-resource packing for cluster schedulers. SIGCOMM Comput Commun Rev 44(4):455–466. https://doi.org/10.1145/2740070.2626334
7. Hameed K, Ali A, Jabbar M, Junaid M, Haider A, Naqvi M (2016) Resource management in operating system—a survey of scheduling algorithms
8. Heller B, Seetharaman S, Mahadevan P, Yiakoumis Y, Sharma P, Banerjee S, McKeown N (2010) Elastictree: Saving energy in data center networks, pp 249–264
9. Huang YF, Chao BW (2001) A priority-based resource allocation strategy in distributed computing networks. J Syst Softw 58(3):221–233. https://doi.org/10.1016/S0164-1212(01)00040-1, http://www.sciencedirect.com/science/article/pii/S0164121201000401
10. Jiang J, Das R, Ananthanarayanan G, Chou PA, Padmanabhan V, Sekar V, Dominique E, Goliszewski M, Kukoleca D, Vafin R et al (2016) Via: Improving internet telephony call quality using predictive relay selection. In: Proceedings of the 2016 ACM SIGCOMM conference, pp 286–299
11. Karthiban K, Raj JS (2020) An efficient green computing fair resource allocation in cloud computing using modified deep reinforcement learning algorithm
12. Kober J, Bagnell JA, Peters J (2012) Reinforcement learning in robotics: a survey. Int J Robot Res 32:1238–1274
13. Mao H, Alizadeh M, Menache I, Kandula S (2016) Resource management with deep reinforcement learning. In: Proceedings of the 15th ACM workshop on hot topics in networks, HotNets '16, Association for Computing Machinery, New York, NY, USA, pp 50–56. https://doi.org/10.1145/3005745.3005750, https://doi.org/10.1145/3005745.3005750

14. Mnih V, Kavukcuoglu K, Silver D, Graves A, Antonoglou I, Wierstra D, Riedmiller M (2013) Playing atari with deep reinforcement learning. arXiv preprint arXiv:1312.5602
15. Peng B, Li X, Gao J, Liu J, Wong KF (2018) Deep Dyna-Q: integrating planning for task-completion dialogue policy learning. In: Proceedings of the 56th annual meeting of the association for computational linguistics (volume 1: Long Papers). Association for Computational Linguistics, Melbourne, Australia, pp 2182–2192. https://doi.org/10.18653/v1/P18-1203, https://www.aclweb.org/anthology/P18-1203
16. Silver D, Huang A, Maddison CJ, Guez A, Sifre L, van den Driessche G, Schrittwieser J, Antonoglou I, Panneershelvam V, Lanctot M, Dieleman S, Grewe D, Nham J, Kalchbrenner N, Sutskever I, Lillicrap T, Leach M, Kavukcuoglu K, Graepel T, Hassabis D (2016) Mastering the game of Go with deep neural networks and tree search. Nature 529(7587):484–489. https://doi.org/10.1038/nature16961
17. Sutton RS, Barto AG (2018) Reinforcement learning: an introduction, 2nd edn. The MIT Press. http://incompleteideas.net/book/the-book-2nd.html
18. Tesauro G, Jong NK, Das R, Bennani MN (2006) A hybrid reinforcement learning approach to autonomic resource allocation. In: 2006 IEEE international conference on autonomic computing, pp 65–73
19. Viet H, An S, Chung T (2011) Extended dyna-q algorithm for path planning of mobile robots. J Meas Sci Instrum 2(3):283–287
20. Zhang W, Dietterich TG (1995) A reinforcement learning approach to job-shop scheduling. In: IJCAI, vol 95. Citeseer, pp 1114–1120

CNN-Based Approach to Control Computer Applications by Differently Abled Peoples Using Hand Gesture

Hitesh Kumar Sharma, Prashant Ahlawat, Manoj Kumar Sharma, Md Ezaz Ahmed, J. C. Patni, and Sahil Taneja

Abstract In today's world, the computer and digital technology have revolutionized every aspect of human life. From connecting with each other in a split second to sharing ideas sitting miles apart, the Internet and digital technology have changed the way humans used to look at things. But the advancement of technology has not been revolutionary for every part of human community, especially differently abled people. The learning curve of this new technology has been designed in such a fashion that it is near impossible for differently abled people to learn it, use it and interact with it. There is always a void between the differently abled and the modern digital technology. It has been always difficult for them not to only use the modern technology and derive its benefits, but also to interact with other people and share their ideas and thoughts with them.

Keywords Convolutional neural network (CNN) · Differently abled · Machine learning · HCI · Deep learning

1 Introduction

To enable differently abled users not only interact with the modern computers but also explain their thoughts with other people, we have aimed to build a whole new interaction and feedback complete system for differently abled people to help them not only interact with the system in their sign language but also simultaneously receive a feedback, which also helps them to interact with other people. We have

H. K. Sharma (✉) · J. C. Patni · S. Taneja
Department of Cybernetics, School of Computer Science, University of Petroleum and Energy Studies, EnergyAcres, Bidholi, Dehradun 248007, Uttarakhand, India

P. Ahlawat · M. K. Sharma
Manipal University, Jaipur, India

M. E. Ahmed
CS Department, Saudi Electronic University, Al Madina, Kingdom of Saudi Arabia
e-mail: m.ezaz@seu.edu.sa

P. K. Singh et al. (eds.), *Proceedings of Second International Conference on Computing, Communications, and Cyber-Security*, Lecture Notes in Networks and Systems 203,
https://doi.org/10.1007/978-981-16-0733-2_34

aimed to incorporate the sign language into our system, to detect what the person is aiming by recognizing his/her specific gestures and produce a meaningful response out of it. We have also aimed to incorporate voice inputs for the differently abled people for specific apps in the system like video players, which will detect their voice and append specific commands to it. It not only makes it easier to interact with the system apps but also reduces the hassle. Many other techniques has evolved in past for converting hand gesture actions to text.

1.1 Convolutional Neural Network (CNN)

Convolutional neural network (CNN)-based deep learning models are commonly used for image classification. CNN model is different from ANN models in no. of depth layers or hidden layers in neural network [1]. As we go deep in layer, model learns more detailed features for better classification. Since hand gesture image dataset is more diverse in nature due to its capturing complexity factors like light intensity, background, shape, size of hands, etc. Classify this dataset need more nos. of filters and hidden layers for identify each gesture without failure and we will be able to produce accurate results. Deep learning-based convolutional neural network (CNN) uses the following layers for training and testing of given image data (Fig. 1).

- Multiple sequential convolutional layer with filters
- Multiple polling layers
- Fully connected layer
- Output layer with softmax activation function.

In neural networks, convolutional neural network (Convents or CNNs) is one of the main categories to do image recognition and image classifications. Objects

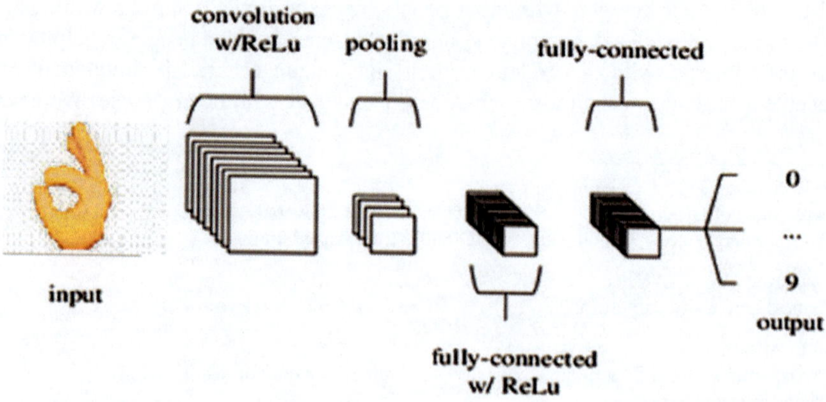

Fig. 1 Hidden layers in deep learning-based CNN model

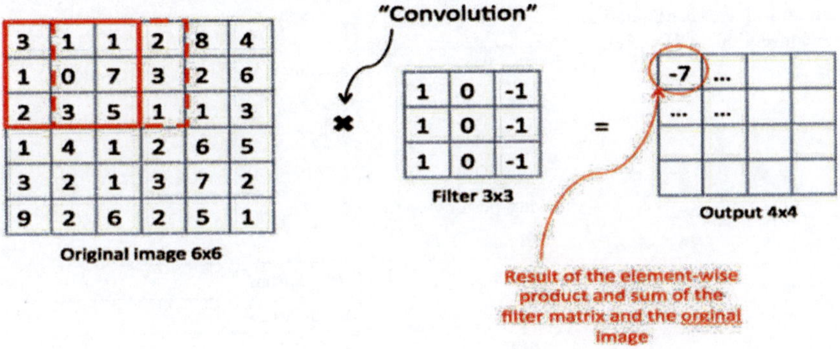

Fig. 2 Mathematical operators used in CNN model

detection, recognition faces, etc., are some of the areas where CNNs are widely used. CNN takes an image as input and classifies it under a certain category. This image is seen as an array of pixels and it depends on the image resolution (Fig. 2).

Based on the image resolution, it will see $h * w * d$ (h = height, w = width, d = dimension), e.g., an image of $6 \times 6 \times 3$ array of the matrix of RGB (3 refers to RGB values) and an image of $4 \times 4 \times 1$ array of a matrix of a grayscale image. This uses mathematical operation for matrix multiplication. Filter matrix is applied over image matrix to find features in this layer [2–4]. The brief description is given below.

- An image matrix (volume) of dimension $(h \times w \times d)$
- A filter $(f_h \times f_w \times d)$
- Outputs a volume dimension $(h - f_h + 1) \times (\mathrm{w} - f_w + 1) \times 1$ (Fig. 2).

They are having limited functionalities and the hardware components are required more for these past implementation. Gloves and sensors are additional components used to implement these complex and expensive techniques. Other limitation for previous methods was the restriction of background. To avoid noise from captured images is to use a specific background, the execution platform was limited to GPU and affording GPU is not possible for all users. As we have already mentioned that differently abled people are lacking in using new technologies.

There should be an efficient and a much easier system build for differently abled people to enable them and make them use and derive benefits of modern digital technology and computers and present their ideas to the rest of the world. Layard configuration of a conventional CCN is given in Table 1. The conventional CNN design contains 02 Conv2D layers, two max-pooling layers, and two dense layers. Three dropout layers also added in between Cov2D and max-pooling layers.

Table 1 Layer configuration of conventional CNN [Ref. 1]

Layers	Layer shape details
Conv2D	32 filters, 5 × 5, ReLU activation
Max-pooling	2 × 2
Dropout	20%
Conv2D	32 filters, 3 × 3, ReLU activation
Max-pooling	2 × 2
Dropout	20%
Flatten	800 neurons
Dense	128 neurons
Dropout	20%
Dense	64 neurons
Output	Softmax 6 classes

2 Objective

The main objective of this research work is to build a system which can increase the human–computer interaction (HCI) for the people who are differently abled and by which they can convey their message/thoughts to whom so ever they want to. The core objective of this work is to implement a lower complex system with very few restrictions. It should work on limiting computing powered resource like CPU. The solution is implemented in OpenCV. The image of hand gesture is captured and passes to the system for controlling application asked by differently abled peoples. Capture the image of the sign as input and convert it into hue, lightness, saturation (HLS) color space [5, 6].

The converted image is then recognized by the model which will be trained with such sign images and will do the assigned action. The recognized sign is then converted into audio format, and the action done can be acknowledged to the user in the form of audio, which enhance the user interaction with the machine and provides a completely new interface for the differently abled people to use and derive benefits of the modern technology and computers.

3 Hand Gesture Recognition Dataset Specifications

For this work, we have used the hand gesture recognition database given by T. Mantecón, C.R. del Blanco, F. Jaureguizar, N. García in their paper [2]. This dataset is available on Kaggle. It consists 20,000 images with different hand gestures of diverse peoples. 10 hand gestures of 10 different peoples are captured in this dataset. The different peoples are considered as subject in this dataset. So there are 5 Male subjects and 5 Female subjects taken for capturing their different hand gestures. The images were captured using the leap motion hand tracking device (Table 2).

Table 2 Hand gesture image dataset specification (LeapGestRecog) [2]

Total no. of images	20,000
Total peoples involved	10 (05 Female and 05 Male)
Hand gesture types recorded	10 (Table 3)
No. of classes	10
Image dimension	640 * 240
No. of sampled images	2010
Images used for training	1407
Images used for testing	603

From a large dataset of 20,000 image, we have sampled only ~10% images (2010) for training our CNN model. We have sampled 10% images due to limited computational resources. Out of 10 different people's samples, we have picked only one people sample. The sample dataset contains all types (10 types) of hand gestures of a single person. Each type contains ~200 images with different angles.

In Table 3, the different hand gestured is mapped with a label number. Ten different hand gestures are captured and labeled from 01 to 10.

The proposed model is trained to classify a given images into a class ranging from 01 to 10 as per the details given in the following table.

As we have mentioned above that only 10% images or 2010 images are sampled to train our model due to resource limitation. Six randomly selected images from sampled images are shown in Fig. 3. These images show the six different hand gestures of a single person.

Table 3 Classification used for every hand gesture

Hand gesture	Label used
Thumb down	00
Palm (horizontal)	01
L	02
Fist (horizontal)	03
Fist (vertical)	04
Thumbs up	05
Index	06
Ok	07
Palm (vertical)	08
C	09

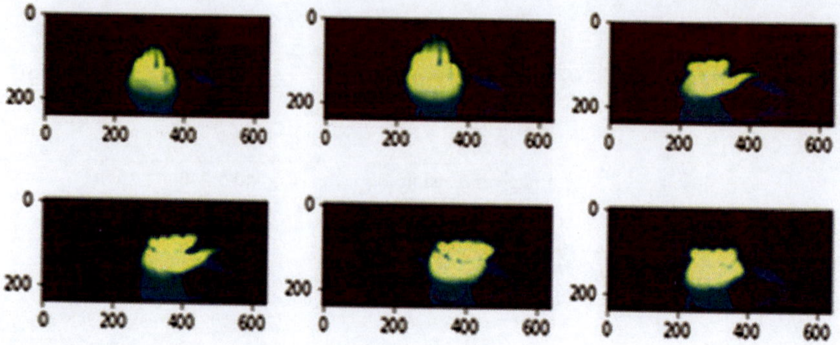

Fig. 3 Randomly selected images from training sample

4 Methodology

To prepare a binary mask for the work, we will use hand without gloves. The segmentation of hand mask will be done using skin color. OpenCV is used for segmentation [7]. We will convert BGR color space to HLS color space. First of all, we will create a binary mask of the hand. The actual color encoding is done by hue channel. We have to identify the hue value for hand skin. After that the adjustment for saturation and lightness range will be done. In our work, we have figured out hue range of 0–30° and a saturation range 5–60% using a simple color picker. Images will be filtered with a hue range.

From 0 to 15 as we have to divide it by 2. Blurring and smoothing operation will be applied on hand image to remove or minimize the noise. We will close with the following hand mask. Given in the following figure, our trained model is based on CNN model. In neural networks, convolutional neural network (Convents or CNNs) is one of the main categories to do image recognition and image [8, 9] classifications. Objects detection, recognition faces, etc., are some of the areas where CNNs are widely used. CNN takes an image as input and classifies it under a certain category. This image is seen as an array of pixels and it depends on the image resolution

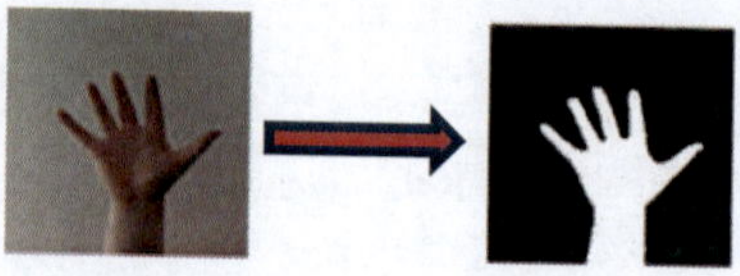

Based on the image resolution, it will see $h \times w \times d$ (h = height, w = width, d = dimension), e.g., an image of $6 \times 6 \times 3$ array of the matrix of RGB (3 refers to RGB values) and an image of $4 \times 4 \times 1$ array of a matrix of a grayscale image.

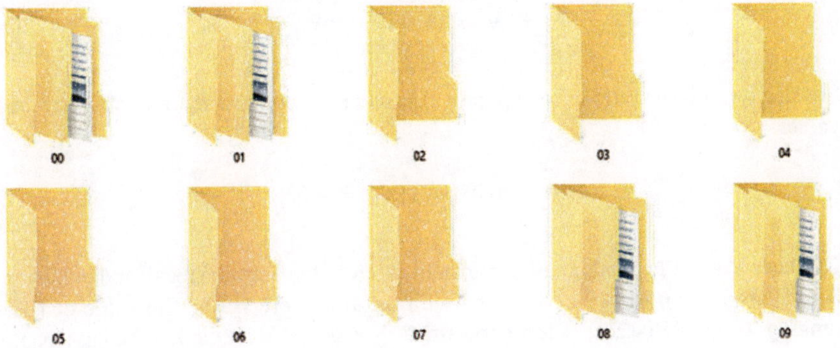

Fig. 4 Sampled image dataset of 10 types of hand gesture [refer Table 3 for label]

5 Implementation of Model

The proposed CNN model is implemented in Python using TensorFlow Keras library with OpenCV supporting library [10, 11]. 1407 images of 10 hand gestures were to the model during training phase. Binary mask for every gesture was created using HSL format [12].

Some of the masks are as follows:

Training the model using Keras library was used to implement CNN model for training to detect gestures.

02 Convolutional layers and 05 epoch were used for better accuracy. A final.h5 file was used in the main code. We trained the system for around 10 gestures each having a dataset of around 200 images (as shown in Fig. 4).

6 Experimental Results of Proposed CNN

In this section, the result analysis of implemented model is presented. To evaluate the accuracy and efficiency of model, several parameters need to be evaluated. These parameters are defined in Eqs. 1, 2, 3 and 4.

PPV: It is a fraction of positively correct predicted value to total positively predicted values.

$$PPV = TP/(TP + FP) \tag{1}$$

Sensitivity: It is the ratio of total positive correct predictions and no. of positive class values.

$$Sensitivity = TP/(TP + FN) \tag{2}$$

F1-score: As per the mathematical formula, it is relation between recall and precision. And F1-score is positively correlated with recall and precision. The value of F1 is ranging [0, 1]. If F1-score is more toward 1, than model is predicting the best results of classification.

$$F1 = 2 * \left[(PPV * Sensitivity)/(PPV + Sensitivity)\right] \tag{3}$$

Accuracy: It shows the percentage of perfect predication of classification.

$$Accuracy = 100 * [(TP + TN)/(TP + TN + FP + FN)] \tag{4}$$

where TP = true positive, TN = true negative, FP = false positive and FN = false negative. TP, TN, FP and FN are decided based on the following matrix (Table 4).

	Actual values		
Predicted values		Positive (1)	Negative (0)
	Positive (1)	TP	TN
	Negative (0)	FN	FP

CNN model training parameters are given in Table 5. The epochs called and the prediction performance parameters are given in the following table. The training time results are accurate for a good classification model.

To test the prediction output, we have given 12 images as test input to the new model and matched the output with labeled data. The results are shown in Fig. 5. The output generated by trained model is plotted using matplotlib library of Python.

As we can see that the title given to each image is 93% matched with the label assigned for each input image. The prediction score for all 12 test images is 93%.

The results shown in Fig. 6 are proving the accurate results for hand gesture classification by the proposed CNN model. The accuracy is almost 100%.

Table 4 Confusion matrix for proposed CNN model

	Predicted thumb down	Predicted palm (H)	Predicted L	Predicted fist (H)	Predicted fist (V)	Predicted thumbs up	Predicted index	Predicted OK	Predicted palm (V)	Predicted C
Actual thumb down	72	0	0	0	0	0	0	0	0	0
Actual palm (H)	0	53	0	0	0	0	0	0	0	0
Actual L	0	0	53	0	0	0	0	0	0	0
Actual fist (H)	0	0	0	48	0	0	0	0	0	0
Actual fist (V)	0	0	0	0	54	0	0	0	0	0
Actual thumbs up	0	0	0	0	0	58	0	0	0	0
Actual index	0	0	0	0	0	0	65	0	0	0
Actual OK	0	0	0	0	0	0	0	73	0	0
Actual palm (V)	0	0	0	0	0	0	0	0	69	0
Actual C	0	0	0	0	0	0	0	0	0	58

Table 5 Model training parameters

Training parameters	Value
Epochs	05
PPV	0.96
Sensitivity	0.96
F1-score	0.96
Accuracy (%)	99.73

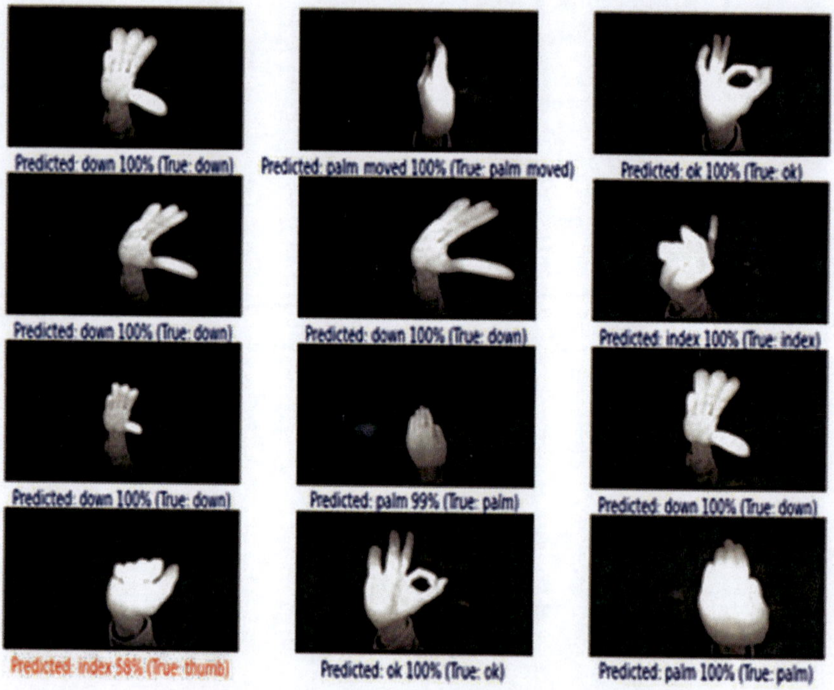

Fig. 5 Hand gesture prediction using the proposed CNN model

7 Application of Proposed CNN Model for Differently Abled Peoples

As we have seen in result section, our model is predicting accurate gesture given to the model as input. Almost 100% correct classification is performed by this model. Due to its highly accurate result, this model can be used for detecting hand gestures of differently abled peoples, who cannot see or speak and map these gestures with some pre-defined computer application and its operations. Based on the gesture provided by a differently abled, the system will detect it and run the application on his/her

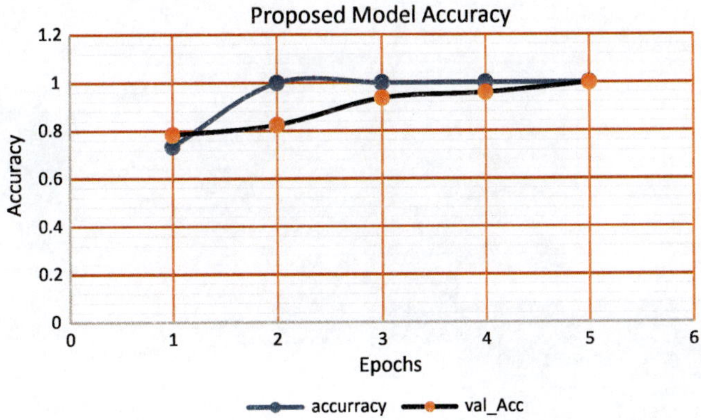

Fig. 6 Accuracy of the proposed CNN on train data and validation data

computer mapped with that gesture. It will help them to use the computing machine in same way as a normal person do.

Detecting the gesture from a real-time video, CV2 was used to capture the gesture real time. It is a Python-based library for image processing. As shown in Fig. 7, a real-time video is capture and a region of interest (ROI) is framed to capture the hand gesture. Capturing the hand gesture image and passed to implemented CNN model for classifying gesture image.

After detecting the gesture, a message related to the gesture is announced using pyttsx library (text to speech conversion) to make it more differently abled-friendly. Extending these capabilities to an application—VLC Media Player. Using these

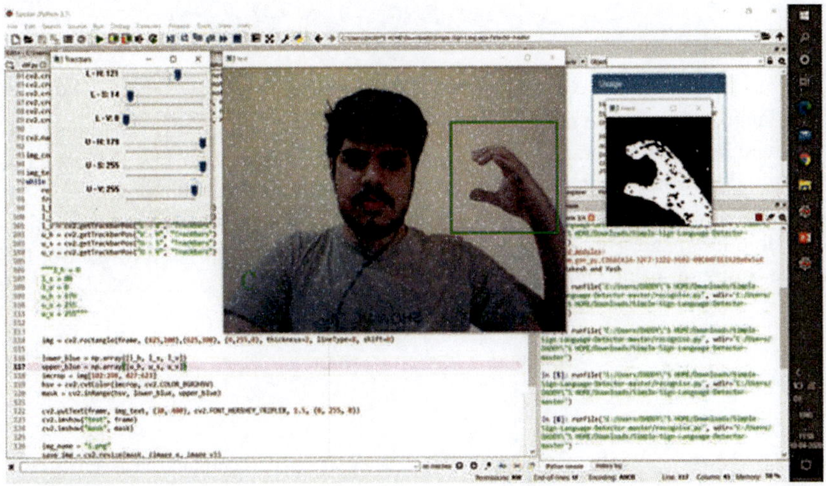

Fig. 7 Capturing and classifying hand gesture from a real-time video

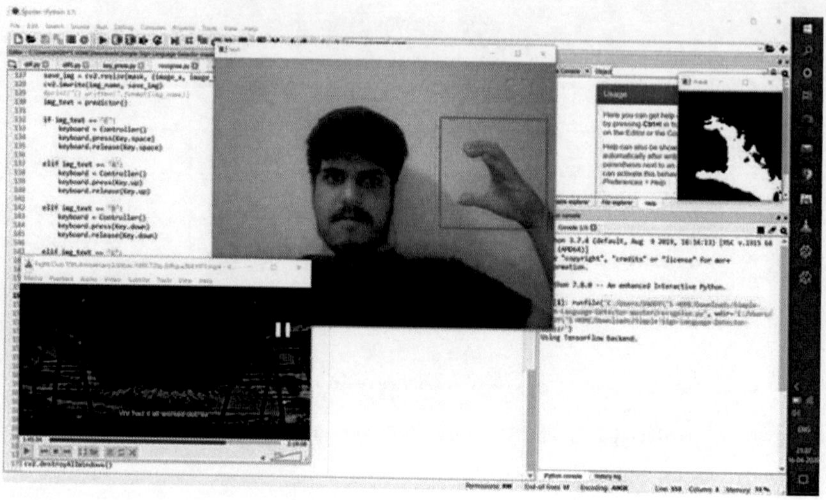

Fig. 8 Running a computer application (VLC player) based on gesture detected

hand gestures, the user can control this application and perform some of the basic controls like—announcing the gesture, making it friendly for them. After detecting the gesture, a message related to it is announced using pyttsx library (text to speech conversion) to make it more friendly (Fig. 8).

Based on different gesture provide by them the application features like play, pause, volume up, volume down, forward and backward can be controlled by differently abled peoples.

8 Conclusion

Sign language being the only communication means for differently abled community hampers their interaction with the computer. This system has the potential of minimizing this communication barrier by working as a mediator between the computer and the person and converting their sign language into a computer understandable language. By achieving higher accuracy, this can be used to control the whole computer using sign language.

References

1. Alani AA et al (2018) Hand gesture recognition using an adapted convolutional neural network with data augmentation. In: 24th IEEE international conference on information management
2. Mantecón T, del Blanco CR, Jaureguizar F, García N (2016) Hand gesture recognition using infrared imagery provided by leap motion controller. In: Int. Conf. on Advanced Concepts for

Intelligent Vision Systems, ACIVS 2016, Lecce, Italy, 24–27 October 2016, pp 47–57. https://doi.org/https://doi.org/10.1007/978-3-319-48680-2_5
3. Patni JC, Sharma HK (2019) Air quality prediction using artificial neural networks. In: ICACTM 2019, June 2019
4. Kumar Sharma H, Kshitiz K, Shailendra (2018) NLP and machine learning techniques for detecting insulting comments on social networking platforms. In: ICACCE 2018
5. Pardeshi V, Sagar S, Murmurwar S, Hage P (2017) Health monitoring systems using IoT and Raspberry Pi—a review. In: 2017 international conference on innovative mechanisms for industry applications (ICIMIA), Bangalore, pp 134–137
6. Navya K, Murthy MBR Dr (2013) A Zigbee based patient health monitoring system. Int J Eng Res Appl 3(5):483–486
7. Mathan Kumar K, Venkatesan RS (2014) A design approach to smart health monitoring using Android mobile devices. In: IEEE international conference on advanced communication control and computing technologies (ICACCCT), pp 1740–1744
8. Mukhopadhyay SC (2015) Wearable sensors for human activity monitoring: a review. IEEE Sens J 15(3):1321–1330
9. Djuknic GM, Richton RE (2001) Geolocation and assisted GPS. Computer 34(2):123–125
10. Sanders G, Thorens L, Reisky M, Rulik O, Deylitz S (2003) GPRS networks. WileyHoboken, NJ
11. Sharma HK, Shastri A, Biswas R (2013) A framework for automated database tuning using dynamic SGA parameters and basic operating system utilities. Database Syst J
12. Sharma HK, Shastri A, Biswas R (2015) Auto-selection and management of dynamic SGA parameters in RDBMS. Database Syst J

A Frequency-Based Approach to Extract Aspect for Aspect-Based Sentiment Analysis

Rahul Pradhan and Dilip Kumar Sharma

Abstract Data is king nowadays, and users worldwide express their views on different platforms to aggregate this data and analyze it. Sentiment analysis becomes a major tool for analysts. Sentiment analysis can be done on different levels. This will be discussing a more granular level of sentiment analysis using aspect-based sentiment analysis, which aims to predict the sentiment polarity of text for a specific target. The majority of work done in this field focuses on the extraction of aspect or feature and then finding their sentiments polarity and aggregating them to find the whole text's final polarity. Aspect extraction is the key to this process; our work will be focusing on aspect extraction. In this paper, we will address the issue of aspect extraction and then propose our approach to deal with it and show how it is better than these existing approaches.

Keywords Sentiment analysis · Deep learning · Natural language processing · Aspect-based sentiment analysis · Feature selection · Feature hierarchy

1 Introduction

Sentiment analysis is a text mining technique that dealt with textual content such as reviews, comments, posts on social media platforms such as Facebook and Twitter. Sentiment analysis assigns each piece of text (review, comment, or post) a sentiment based on textual properties; these sentiments are positive, negative, or neutral [1].

Figure 1 Shows the typical process that is followed by many sentiment analysis techniques mainly based on supervised learning algorithms [2–4]. Positive sentiments here mean that the user had said something good about the product, person, event, or place, while negative means the user said something in the text that is not

R. Pradhan (✉) · D. K. Sharma
GLA University, Mathura UP281406, India
e-mail: rahul.pradhan@gla.ac.in

D. K. Sharma
e-mail: dilip.sharma@gla.ac.in

P. K. Singh et al. (eds.), *Proceedings of Second International Conference on Computing, Communications, and Cyber-Security*, Lecture Notes in Networks and Systems 203,
https://doi.org/10.1007/978-981-16-0733-2_35

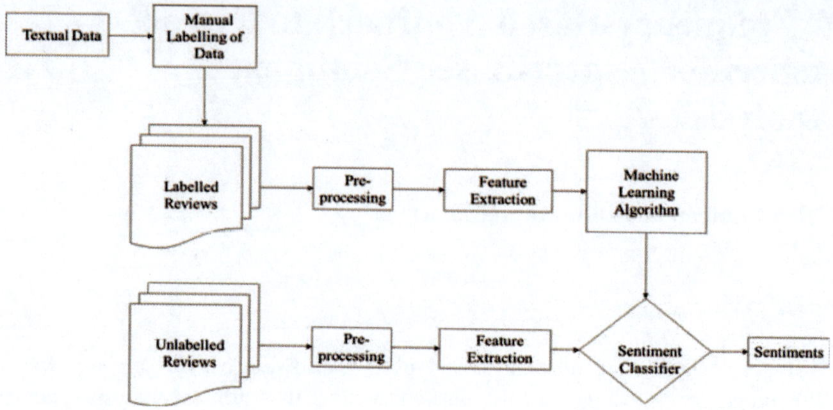

Fig. 1 Sentiment classification process

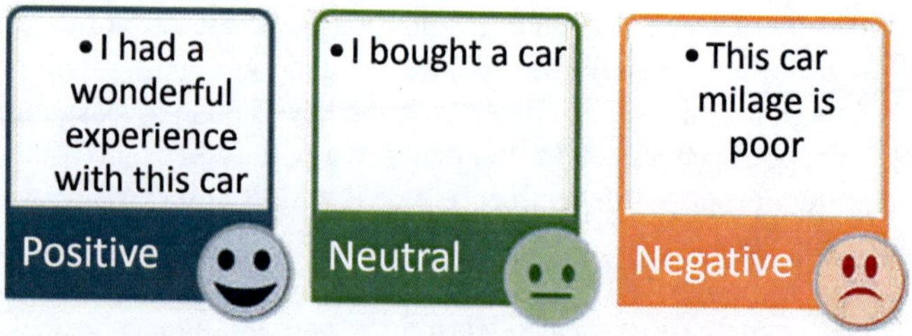

Fig. 2 Sentiments based on the opinion expressed by the user in text

in good light for that text's subject [5]. A third category says neutral when the user had not said anything good or bad in text. Figure 2 shows the three sentiment classes with opinion text[6, 7].

Sentiment analysis makes life easier for marketing executives. It will automate the task of analyzing comments, reviews, posts, which were earlier done by these people manually, and they wasted endless hours in this. Such analysis plays a significant role in making and monitoring brand image by gaining insights from user comments and reviews about products. [8, 9]. Many types of research had been done constantly and efforts are made to set standards that can be used to evaluate the reputations of brands such as in [10], they had shown various techniques that helps in evaluating reputations and what are the advantages of each of these approaches.

Deep learning approaches were explored by various researchers [11, 12] in the past, and these approaches have accuracy. However, they are mostly domain-dependent, while syntactic approaches are based on rules that identify the proper grammatical dependencies, as [13] used space and time to detect opinion sentiments. Simultaneously, [14] considers emoticons and other unusual nonverbal features of tweets to identify the polarity.

1.1 Sentiment Analysis

After introducing sentiment analysis in this section, the discussion will focus on the techniques used in detecting sentiment. Sentiment classification on granularity is done on three levels, which are 1. document level, 2. sentence level, and 3. aspect level [15]. A graphical representation of this taxonomy of sentiment analysis can be seen in Fig. 3.

In the first level, the focus will be on the whole document. It is considered to be expressing an opinion about a single topic, but since the document is long enough, this possibility that it is about a single topic is quite low, which is usually the case [5, 16]. Therefore, many researchers or analysts keen to use the second level, which identifies the text's polarity on the sentence level [17]. That is, they consider each sentence as a single text and assuming it will talk about a single topic, which usually the case as one will often express things about one thing in a sentence [18, 19]. Exceptions are always there many at times it is observed that the same sentence is talking about two or more subjects, topics and they are usually compared, and both have different sentiments, for example, "I love Nissan Micra, but I will never travel from Suzuki Alto." The result shows a general sentiment polarity based on the document or sentence keywords in both of the above levels [20, 21].

In the post or review, given by the user on either social media or an e-commerce Web site, their opinion text length can be of few words to a paragraph. This means for certain product users are very much interested in posting their reviews, so they write in a detailed manner. The detailed review usually discusses the product but also talks about various features of that product. Some features might be having positive sentiments; some might have negative sentiments.

This multifeature review was analyzed better by aspect-based sentiment analysis. Aspect-based sentiment analysis is a textual analysis technique that has two necessary steps: first, to identify the aspect that requires breaking down the text into different aspects, then assigning each aspect with a positive, negative, or neutral sentiment.

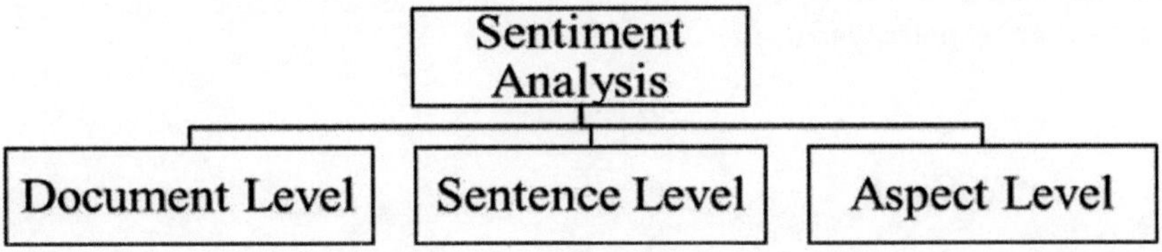

Fig. 3 Sentiment classification based on the granularity

1.2 Aspect-Based Sentiment Analysis

Sentiment analysis (SA) is a proven tool that is usually be employed by analysts while processing large databases of reviews, comments, and posts. SA gives the polarity of text that is positive or negative [22].

But for reviews that had mixed feelings can be dealt with aspect-based sentiment analysis (ABSA), in this, the assignment of polarity to each aspect as they are mentioned in the text under consideration. This technique helps in providing a complete picture of the customer's opinion.

For ABSA, first thing is to gather data and then analyzed it using machine learning algorithms. In aspect-based sentiment analysis, the main objective will be targeting a pair of aspects and text that describe its sentiment. Here rather having SA of the whole entity the prime focus will be on various aspects of the entity's entity or attributes. This helps customers or business people to understand and use information about sentiments more efficiently. This way, identification of what features or aspects need redesigning or upgradation or so on can be done.

Aspect detection is the field itself. There are various approaches available such as frequency-based methods, supervised, unsupervised, hybrid machine learning methods, and syntax-based methods. As its name suggested, in the frequency-based method, the approach is to try and identify the keywords or more prominent words in search of aspect in the hope that many reviewers must use these words in their review. Aspect can be a single word or something like a compound noun, bigrams or other methods need to be employed to identify them, such as "screen size," "noise canceling," "door lock," and many more. This method is useful and proved too powerful enough to capture majorly all aspects, but still, some detected nouns commonly used come out to be not an aspect for some product. Some aspects that are not commonly observed or discuss by many reviewers will go undetected in this system, and this is the major flaw in this method, but still, in comparison to others, it is fast and other trade-off is relatively low.

Figure 4 illustrates the difference between the two approaches. In this figure, one can see that ABSA analyzed the text to the granularity of detecting each feature and then assigning polarity to it.

2 Related Work

Our work is inspired by one of the problems discussed by Giachanou and Crestani [2]. Giachanou et al. [2] had discussed the process of opinion mining. They have clearly defined the aspect and its importance while analyzing the sentiment of the given problem. They discuss the problem of aspect-based Twitter sentiment analysis and the problem of aspect extraction and detection in short texts.

These approaches for aspect extraction are divided into four categories:

- Rule-based approaches

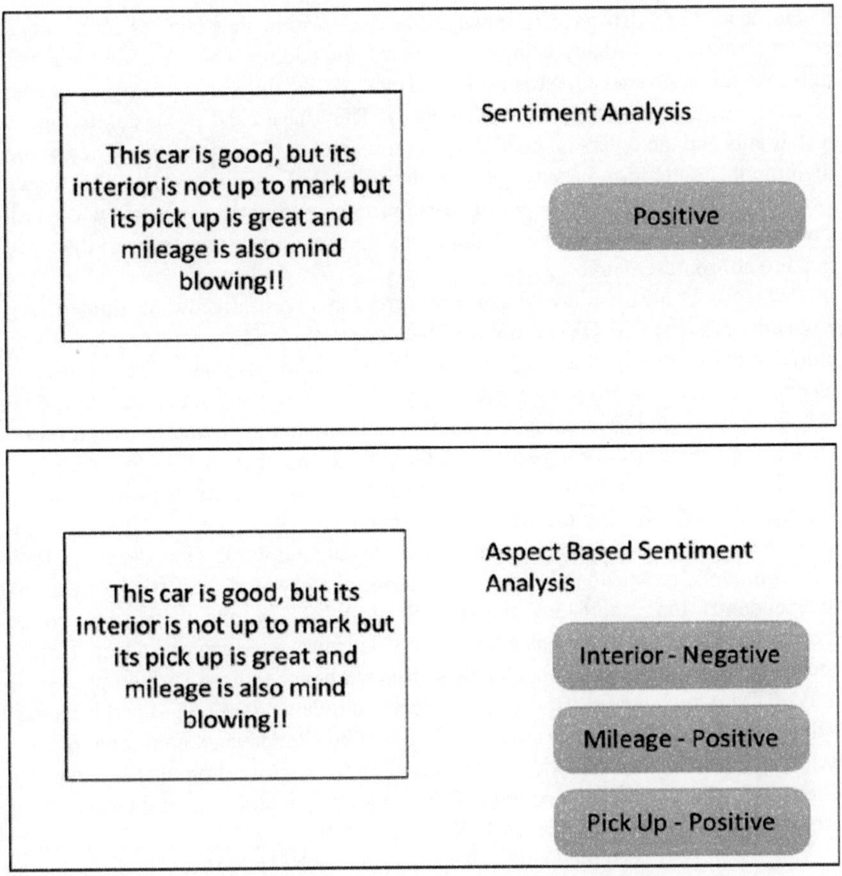

Fig. 4 Comparison between the granularity of analysis in both SA and ABSA

- Supervised approaches
- Unsupervised approaches
- Deep learning-based approaches

Liu et al. [37] presented the rule based approach but that have been crafted by then using hand pick rules and tune them so there will not much scope of errors or loop holes. This approach is faster, unsupervised and domain independent. Dragoni et al. [23] have discussed the approach under unsupervised machine learning. They have to provide a domain-independent approach. Most of these approaches are regularly mostly domain-dependent as features might be common between the domains, but their meaning and weight differ. They created a pairwise solution in which they try to detect pairs of aspects and their words of sentiments. This approach is good in the sense they use part of speech tagging and visualization to see relations.

He et al. [24] also gave an unsupervised algorithm that focuses more on the co-occurrence of words by using neural word embedding. They divided their work into two subtasks, first to extract all the aspect terms and second to create clusters of these terms so that they form a category. They have used two datasets, one on restaurants and the others on beer, both contain a total of 4400 reviews. They named their model as attention-based aspect extraction (ABAE). This ABAE creates a set of word embedding that uses the word co-occurrence to find the representative words. They use the attention mechanism to reduce the weight or importance of the words that are not representative.

Rezaeinia et al. [36] have discuss the problems with already pre-trained word embeddings such as GLOVE and Word2Vec, they improved the performance of word embedding by 2% using the combination of lexicons. Vargas et al. [25] used attention mechanisms for extracting the aspects they tried to find words that gain attention as per the domain and then using the category attribution approach to assign one of the desired categories to each sentence. They used similarity metrics to obtain these assignments and often have to average the score to assign a category to more than one sentence or assigning more than one category to a sentence.

Tran et al. [26] used deep learning approaches in their work. They discussed three LSTM models, independently long short-term memory (IndyLSTM), bidirectional independently long short-term memory (Bi-IndyLSTM), and their proposed Bi-IndyLSTM-CRF model for aspect extraction. This was the supervised deep learning approach. Their proposed model had first uses word embedding. They employed Bi-IndyLSTM, which was in itself contains forward independent LSTM and backward independent LSTM, and then finally, they used the CRF layer. Conditional random field (CRF) is a kind of statistical models used for predicting structures and patterns. They used a dataset [27] that contains 7686 reviews on laptops and restaurants. They used TensorFlow and NVIDIA Tesla K80 GPU [28].

Yang et al. [29] presented a deep learning approach. They focus on the attention score of aspect so that just using an average score for the attention. They proposed a weighted scheme for it. For this, they propose a co-attention mechanism that focuses on keywords to produce more effective context representation between both target and context. They use LSTM for this task. They also produce a weighted location function to capture the position effect on the co-attention mechanism.

Poria et al. [30] also work on a deep learning approach for aspect extraction. They used a seven-layer deep convolutional neural network to classify each term in review as aspect or non-aspect. They combine some linguistic patterns with CNN and produce an ensemble approach for aspect extraction. Since these sets of linguistic patterns are rule-based, this approach is a mixture of deep learning and rule-based approaches.

Deep2s [31] proposed an approach which is the mix of both rule-based approach and deep learning. They use a deep semantic representation of reviews with syntactic patterns. They exploit syntactic patterns to identifying the dependencies among terms in review, and for capturing the semantics, they use abstract meaning representation.

There are more hybrid approaches, such as [24], which mix supervised and unsupervised approaches. Therefore, it is more suitably comes under semi-supervised.

They took a seed word for each category and then extract aspect terms and cluster them in aspect categories. The proposed seeded aspect and sentiment model (SAS). SAS models both aspect and as well as the corresponding sentiment as well. They further improve SAS by adding maximum entropy to it so that they can differentiate between words that are aspect and words that are representing sentiments clearly identified using the POS tagging.

Textual knowledge is used to get a better aspect extraction. These approaches, as used by Khan and Jeong [32], integrates sentiment, and structure knowledge. This amalgamation of two produces better results when there is limited data for training.

3 Proposed Methodology

In this section, we will be discussing our proposed approach. The first thing we will be using reviews, and we need to preprocess them. In preprocessing, we remove all the hyperlinks, emoji, and extra spaces.

After preprocessing, we need to find the aspect. For aspect detection, we used the frequency-based method. We analyzed all the reviews and calculated the frequency in two patterns. Firstly, we calculate using the unigram approach, and then we use the bigram approach to identify aspects.

Figure 5 Shows the framework to calculate the frequency of each unigram present in reviews. We had used a similar framework to calculate the frequency of bigrams

Let us consider sentence $s = \{w_1, w_2, \cdots w_n\}$ where w_i is the$i$th word in sentence s. We take a pair of bigram w_i, w_{i+1} and add them to our frequency table if they

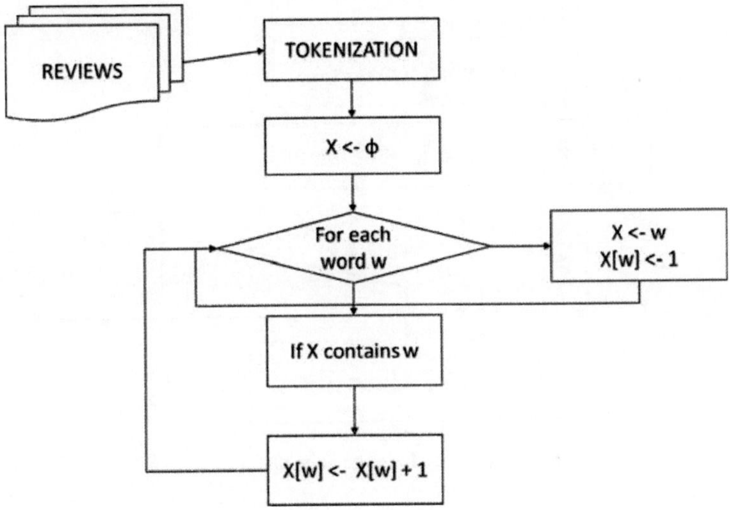

Fig. 5 Approach for calculating the frequency of unigrams

do not exist already and set their frequency f to 1. We continue his process for each bigram. If they already exist in the frequency table, we add 1 to their current frequency value f.

This approach helps to detect f of each bigram, which are compound nouns such as "Computer Processor." However, we could not capture the bigrams aspect, such as "Processor of this computer" for this approach, we need to consider each sentence and calculate bigrams within that sentence. However, they could not be placed adjacent to each other. For this, we consider each sentence s, and then we calculate the frequency of each pair of the word occur in a sentence. This way, we can capture all bigram even they are not occurring together. We have to remove stopwords and perform stemming before calculating the frequency to reduce the dictionary size. Figure 6 demonstrates the approach explained using a block diagram.

Stemming can be employed using Porter stemming. We had also performed lemmatization to find the correct root words as usually stemming over stem or under stem as data collection had a lot of Internet words so, this handles better by lemmatization. We had performed case folding as it will help us to calculate the correct frequency. We had employed the better machine learning-based case folding approach available with one of the Python package.

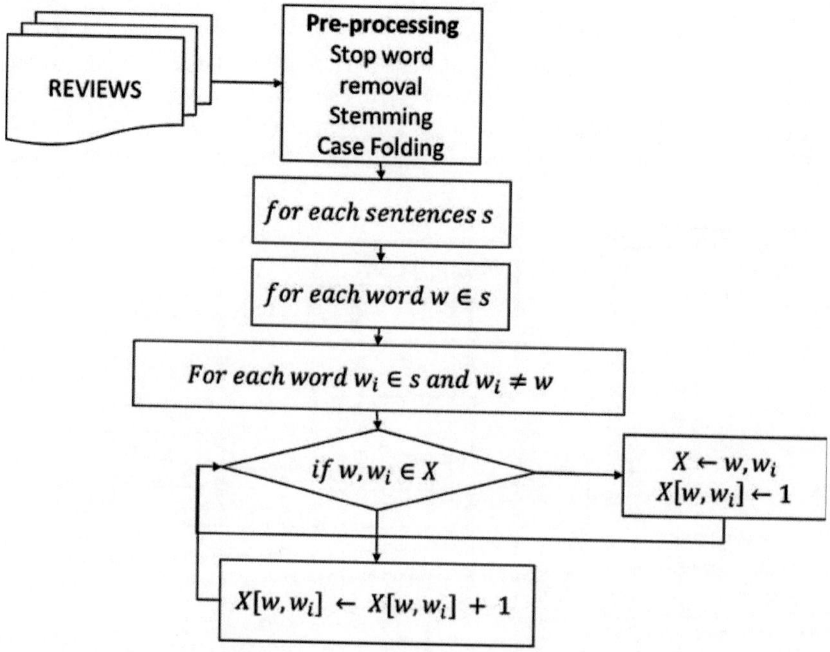

Fig. 6 Approach for calculating frequency for word pairs that not adjacent to each other

Table 1 Detail of dataset

Product name	Number of aspects
Digital camera 1	237
Digital camera 2	174
Cell phone	302
Mp3 player	674
DVD player	296

4 Experimental Setup

This section discusses the experimental setup we used to implement and simulate our approach. We had used the dataset to validate our results. This dataset is standard, and we also scrape few product pages to extract review from Amazon.com, but the annotation of aspect was not there and this makes things difficult, we annotate them manually, but we did not include that in our study.

4.1 Data Collection

Customer reviews are the source and inspiration behind the field of opinion mining and sentiment analysis. Various researchers [33, 34] had focused their research on customer reviews. We had used the customer review collections from Minqing Hu [35]. This collection contains a review of five electronics products in which two are cameras, one DVD player, one mp3 player, and one cell phone. They have collected their review from Amazon and CNET.

They have chosen these Web sites as they have large numbers of reviews available on their portal, and each review had a title and text in which the user or customer writes his or her review. Each product they select, they only store the first 100 reviews on it.

They had manually read all the reviews and tag each aspect on which the user or customer expresses their opinion. They also had a tagged aspect with sentiment polarity, but they did not include any aspect that they think manually. They are neutral. They do not tag any neutral sentences. Table 1 shows detail about the dataset.

5 Result Analysis

Most of the researchers use precision, recall, and F1-score to evaluate their approaches. Therefore, we had also employed these metrics to evaluate our approach, as well.

Table 2 Precision, recall, and F1-score of the proposed approach

	Precision	Recall	F1-score
Digital camera 1	0.63	0.78	0.69
Digital camera 2	0.58	0.82	0.68
Cell phone	0.61	0.81	0.67
Mp3 player	0.56	0.76	0.63
DVD player	0.57	0.79	0.65

The evaluation metrics are defined as follows:

$$\text{Precision} = \frac{\text{TP}}{\text{TP} + \text{FP}} \tag{1}$$

$$\text{Recall} = \frac{\text{TP}}{\text{TP} + \text{FN}} \tag{2}$$

$$\text{F1-Score} = \frac{2 \times \text{Precision} \times \text{Recall}}{\text{Precision} + \text{Recall}} \tag{3}$$

The results are summarized in the following Table 2.

From Table 2, we can see we had quality results on dataset FBS [35] precision and recall shows our approach select a good subset of features. These results are encouraging for a rule-based approach.

Figure 7 shows how the trend of precision, recall, and F1 measure of our approach across different products

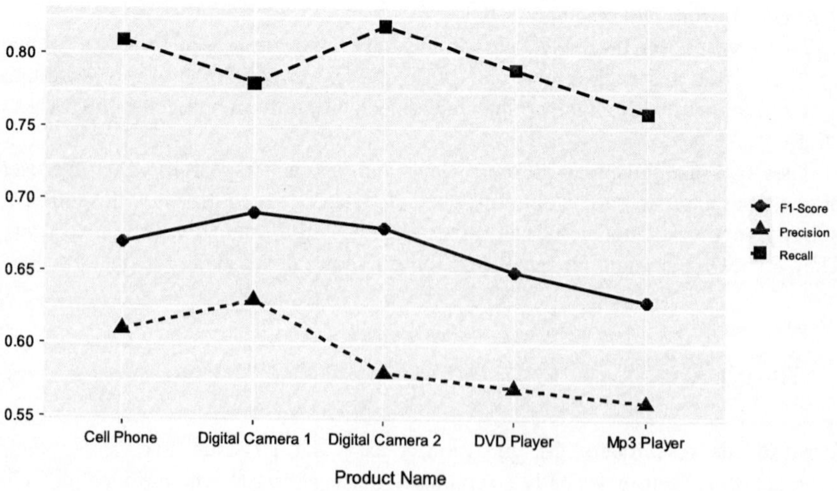

Fig. 7 Precision, recall, and F1-score plot of our proposed approach

This way, we can observe for product digital camera one our precision is highest. At the same time, for recall, Mp3 player has the lowest as people use aspect keywords as sound, loud as verb or adjective but useless nouns in reviews.

6 Conclusion

In this paper, we proposed a set of rules as a technique for aspect extraction. This work is the first step toward aspect-based sentiment analysis. Our experimental results are very encouraging, and the approach proposed effectively deals with the issue in aspect extraction. Aspect extraction is a very relevant problem in the current scenario as nowadays. Opinion mining is used in all fields from Peron relationship agencies, politicians, and companies to monitor product feedback. Aspect-based SA will be beneficial if the product developer wants to upgrade the product or want an idea about user needs and the gap in the market which they can tap.

In our future work, we will refine and tune our rule-based approach to get better results. As a rule-based approach cannot be improved further after an extended, we need to keep adding more rules or new sets in our lexicon to increase or capture more cases, but there is a limit to it. What is more, when we add a new rule, there are chances that it might capture this new case, but afterward, we find it starts conflicting with some other case. Therefore, it is better to tune them as much as we can.

References

1. Al-Smadi M, Talafha B, Al-Ayyoub M, Jararweh Y (2019) Using long short-term memory deep neural networks for aspect-based sentiment analysis of arabic reviews. 10:2163-2175
2. Giachanou A, Crestani F (2016) Like it or not: a survey of twitter sentiment analysis methods. 49:1-41
3. Xu H, Liu B, Shu L, Yu PS (2019) Bert post-training for review reading comprehension and aspect-based sentiment analysis
4. Do HH, Prasad P, Maag A, Alsadoon A (2019) Deep learning for aspect-based sentiment analysis: a comparative review. Expert Syst Appl 118:272–299
5. Vinodhini G, Chandrasekaran R (2012) Sentiment analysis and opinion mining: a survey. 2:282-292
6. Gupta C, Jain A, Joshi N (2019) A novel approach to feature hierarchy in aspect based sentiment analysis using OWA operator. 661–667
7. Pascual F (2019) A comprehensive guide to aspect-based sentiment analysis. MonkeyLearn. https://monkeylearn.com/blog/aspect-based-sentiment-analysis/ (2019). Accessed August 2020
8. Sun C, Huang L, Qiu X (2019) Utilizing BERT for aspect-based sentiment analysis via constructing auxiliary sentence
9. Peng H, Xu L, Bing L, Huang F, Lu W, Si L (2020) Knowing what, how and why: a near complete solution for aspect-based sentiment analysis. 8600–8607
10. Chiranjeevi P, Santosh DT, Vishnuvardhan B (2019) Survey on sentiment analysis methods for reputation evaluation. In: Cognitive informatics and soft computing, Springer, pp 53–66

11. Agarwal Y, Katarya R, Sharma DK (2019) Deep learning for opinion mining: a systematic survey. 782–788 (2019)
12. Rathi S, Shekhar S, Sharma DK (2016) Opinion mining classification based on extension of opinion mining phrases. 717–724
13. Samuel A, Sharma DK (2017) A spatial, temporal and sentiment based framework for indexing and clustering in twitter blogosphere. 32:3619–3632
14. Samuel A, Sharma DK (2018) A novel framework for sentiment and emoticon-based clustering and indexing of tweets. 17:1850013
15. Chauhan GS, Meena YK (2020) DomSent: domain-specific aspect term extraction in aspect-based sentiment analysis. In: Smart systems and IoT: innovations in computing, Springer, pp 103–109
16. Tsytsarau M, Palpanas T (2012) Survey on mining subjective data on the web. 24:478-514
17. Schouten K, Frasincar F (2015) Survey on aspect-level sentiment analysis. IEEE Trans Knowl Data Eng 28:813–830
18. Liu B (2012) Sentiment analysis and opinion mining. 5:1-167
19. Pang B, Lee L (2008) Opinion mining and sentiment analysis foundations and trends in information retrieval. 2
20. Tang H, Tan S, Cheng X (2009) A survey on sentiment detection of reviews. Expert Syst Appl 36:10760–10773
21. Zhao W, Guan Z, Chen L, He X, Cai D, Wang B, Wang Q (2017) Weakly-supervised deep embedding for product review sentiment analysis. IEEE Trans Knowl Data Eng 30:185–197
22. Ruder S, Ghaffari P, Breslin JG (2016) Insight-1 at semeval-2016 task 5: deep learning for multilingual aspect-based sentiment analysis
23. Dragoni M, Federici M, Rexha A (2019) An unsupervised aspect extraction strategy for monitoring real-time reviews stream. 56:1103-1118
24. Mukherjee A, Liu B (2012) Aspect extraction through semi-supervised modeling. 339–348
25. Suarez Vargas D, Pessutto LR, Pereira Moreira V (2020) Simple unsupervised similarity-based aspect extraction. . arXiv: 2008.10820
26. Tran TU, Hoang HT, Huynh HX (2020) Bidirectional independently long short-term memory and conditional random field integrated model for aspect extraction in sentiment analysis. In: Frontiers in intelligent computing: theory and applications, Springer, pp 131–140
27. SEMEVAL: Semeval task 4. https://alt.qcri.org/semeval2014/task4/. Accessed August 2020
28. NVDIA: Nvdia tesla. https://www.nvidia.com/en-us/data-center/tesla-k80/ (2020). Accessed July 2020
29. Yang C, Zhang H, Jiang B, Li K (2019) Aspect-based sentiment analysis with alternating coattention networks. 56:463-478
30. Poria S, Cambria E, Gelbukh A (2016) Aspect extraction for opinion mining with a deep convolutional neural network. Knowledge-Based Syst. 108:42–49
31. Li X, Wang B, Li L, Gao Z, Liu Q, Xu H, Fang L (2020) Deep2s: Improving aspect extraction in opinion mining with deep semantic representation. 8:104026–104038
32. Wu S, Xu Y, Wu F, Yuan Z, Huang Y, Li X (2019) Aspect-based sentiment analysis via fusing multiple sources of textual knowledge. Knowledge-Based Syst. 183:104868
33. Khan J, Jeong BS (2016) Summarizing customer review based on product feature and opinion. 1:158-165
34. Loh HT, Sun J, Wang J, Lu WF (2009) Opinion extraction from customer reviews. 48999:753-758
35. Hu M, Liu B (2004) Mining and summarizing customer reviews. 168–177
36. Rezaeinia SM, Rahmani R, Ghodsi A, Veisi H (2019) Sentiment analysis based on improved pre-trained word embeddings. Expert Syst Appl 117:139–147
37. Liu Q, Gao Z, Liu B, Zhang Y (2015) Automated rule selection for aspect extraction in opinion mining

Sentiment Analysis Techniques on Food Reviews Using Machine Learning

Shilpa Gite, Abhishek Udanshiv, Rajas Date, Kartik Jaisinghani, Abhishek Singh, and Prafful Chetwani

Abstract Review or opinion is a text which expresses the user's thought and response to the product or service he/she has availed or purchased. Processing this input and getting to know whether these sentiments are positive, negative, or neutral is called sentiment analysis. The reviews are then used by data analysts to perform evaluations about the product/service. We classify these sentiments in three principal types positive, negative, and neutral. Amazon Food reviews dataset would be used to train the classifier. Zomato, being the most popular food delivery site and restaurant aggregator provides information, menus, and user reviews of restaurants. Python language is used to conduct research on a carefully chosen data and apply a classification algorithm on it.

Keywords Sentiment analysis · Classification techniques · Amazon fine food reviews · Literature review · Future scope

S. Gite (✉) · A. Udanshiv · R. Date · K. Jaisinghani · A. Singh · P. Chetwani
Symbiosis Institute of Technology, Pune, Symbiosis International (Deemed University), Pune, Maharashtra, India
e-mail: shilpa.gite@sitpune.edu.in

A. Udanshiv
e-mail: abhishek.udanshiv@sitpune.edu.in

R. Date
e-mail: rajas.date@sitpune.edu.in

K. Jaisinghani
e-mail: kartik.jaisinghani@sitpune.edu.in

A. Singh
e-mail: abhishek.a.singh@sitpune.edu.in

P. Chetwani
e-mail: chetwani.prafful@sitpune.edu.in

P. K. Singh et al. (eds.), *Proceedings of Second International Conference on Computing, Communications, and Cyber-Security*, Lecture Notes in Networks and Systems 203,
https://doi.org/10.1007/978-981-16-0733-2_36

1 Introduction

1.1 Sentiment Analysis

Over the past few years, an interesting and popular research area emerging lately is sentiment analysis. The opinion or reviews which are held by any number of individuals are reviewed and analyzed using sentiment analysis. These reviews can be related to any event, brand, person, service, product, or current trending affairs. Earlier, magazines, newspapers, and other sources were used by the people to express their views, but not all of the public opinion was covered in these sources. However, with the advancement in technology, the people have commenced to express their feelings on different social networking and microblogging sites. Food bloggers and foodies have begun to review restaurants and dishes. Different aspects like the ambiance, hygiene, etc., are also a part of their review process which gives consumer an idea of how the food and service is offered by the restaurant. Consumer's review plays an important role in food industry as the restaurants are continuously trying to improve their standards.

1.2 Background

Sentiment analysis is a fresh and innovative field of research which is aimed at studying the sentiments of the people from the text they type in and classify them accordingly. Sentiment analysis analyzes people's sentiments, opinions, attitudes, evaluations, appraisals, and emotions toward services, restaurants, products, individuals, organizations, issues, topics events, and their attributes.

The text in the sentiment can be classified into the following:

- the polarity of the sentiment conveyed.
- Agreement or disagreement with respect to a topic (e.g., debates);
- News either good or bad;
- Pros and cons of a product.

To conduct sentiment analysis, various classifiers and methodologies can be used to seek more accuracy and results.

2 Literature Survey

Liu has defined [1] an opinion "is simply a positive or negative sentiment, view, attitude, emotion, or appraisal about an entity or an aspect of the entity" from an opinion holder at a specific time. The entity is different in each of the case. The entity could be an event, organization, a debate topic, a news which consists of aspects that convey both components and traits of the entity.

Alessia D'Andrea et al. [2] defined sentiment analysis as a "new field born in Natural Language Processing (NLP) aimed at detecting subjectivity in text or extracting and classifying opinions and sentiments. Sentiment analysis studies people's opinions, attitudes, evaluations, appraisals, and emotions towards services, products, individuals, organizations, issues, topics events and their attributes".

Machine learning algorithm was found to have the abilities of adaptation and creation of trained models for particular purposes and backgrounds. Machine learning-based algorithm was suitable for an application-specific need whereas lexicon-based analysis had a limitation on the number of lexicons and the assignation of a permanent sentiment.

Among the machine learning approaches, the most used were (i) Bayesian networks, (ii) Naive Bayes classification, (iii) maximum entropy, (iv) neural networks, and (v) support vector machine.

Tools for sentiment analysis were highlighted in the paper which helped to recognize and analyze the sentiment. Tools such as emoticons, which symbolized the face expression such as happy or sad emoticon, helped recognize the sentiment very easily although there have been many emojis which describe a different sentiment such as frustration. Tools like happiness index marked the norms for English words and score whereas positive and negative affect schedule (PANAS) define an eleven-sentiment psychometric scale. Tools like SentiStrength, Senti WordNet, and SenticNet were based on the lexical analysis. The paper emphasized the application areas such as business, politics, public actions, and finance where these tools and approaches could be used. The paper stressed the problem of irony and sarcasm in the text which made the task more difficult, problematic and less accurate.

The study presented by Tan et al. [3] is based on the Amazon Products and their ratings/reviews given by the customers. Traditional algorithms have been used here along with RNN. By applying all these classifiers, there was a better understanding of these algorithms with respect to the product aspect. Two types of feature techniques were applied: The first type was an old-style method. A dictionary, based on the everyday words and index of each word, was constructed. The limit for the word dictionary was set to be 6 occurrences and ended up gathering 4223 words from the entire dataset; then the transformation of each review into a vector, where each value represented how many times the word shows up. Another type of feature used was the 50-d glove2 dictionary which was pertained on Wikipedia. For this part, the advantage of the meanings of each and every word was taken. Each review was represented by the mean vector of 50-d glove method. Because of the way of characterization of each review, the features got undermined and the distance between distinctive reviews was not that precise.

For these two types of features, all the algorithms, i.e., Naive Bayes, support vector machine (SVM), K-nearest neighbor (KNN), long short-term memory (LSTM) were tested. From the outcomes, it was seen that the correctness on the test set is the best when LSTM on the first type of feature was used. One of the main reasons the precision was not high enough is because of the data imbalance. Resampling and different weighting techniques that were feedbacks of the audience were tried but that did not give any positive outcome. The lack of data points was considered as a drawback here.

In this paper by Wu and Ji [4], the author states the importance of sentiment analysis in natural language processing. Sentiment analysis is the process of assigning score to words which makes it simpler to analyze large datasets. Various methods are available for doing sentiment analysis on large datasets, some of the mainly used methods are bag of words and n-grams. With recent development in NLP, deep learning can also be used to solve sentiment analysis problems.

Before constructing a deep learning model, the authors have found the features of the given dataset. The authors use Twitter for obtaining a sentiment label for each word in the review. It was noticed that although many of the positive reviews have more positive words than negative words, it is astonishing to find that even around half of the negative reviews have more positive words than negative words.

In this paper, RNN method is used by the author to implement sentiment analysis on the Amazon Food Review dataset. Constructing a binary tree is very important in order to feed the RNN model. The author uses an RNN model with a one hidden layer recursive neural network. However, multiple sentences are to be to RNN. Therefore, the author proposes the model recursive neural network for multiple sentences (RNNMS). The baseline is a Naive Bayes classifier and the average of all word vectors of a review is used as the feature vector for a review.

The author concludes that RNNMC performs better than the baseline. Even when trees are labeled insufficiently, RNNMC still outperforms all other metrics we have than the baseline Naive Bayes classifier using averaged word vectors as input features, which means that understanding phrase-level structure helps sentiment analysis tasks.

Sasikala et al. [5] have proposed a method that classifies the reviews on a scale of 1 to 5 centered on the sentiments in the works. The words had been used to determine the rating of the reviews. Predictive techniques like Naïve Bayes were to be used to test the data. The values below score 3 are predicated as negative and the ones above 3 are predicated as positive. After performing stemming and pruning of the data, further algorithms were applied to predict the results. They proposed an approach for mining the food reviews based on score combined with existing text analyzing packages. The proposed system with score ratings had produced decent results. The limitation of this technique was that it worked better just for open opinions like rating or scores.

McAuley et al. [6] have proposed a latent factor recommendation system that proposed a qualitative analysis of the users who gave a review. In this system, the model assigned experience level (i.e., eui = E) to the expert and beginners were assigned the smallest level. (i.e., eui = 1). Here, it was found that novices and intermediate users have lower prediction accuracy. Here they argued that users do not become professionals via a smooth progression, but rather they evolve through several distinct stages. In experience progression, the study on how users progress through experience levels as a function of time was conducted. Observations on how much time a user spends at an experience level with "longest" meant covering the longest period and the greatest number of reviews were made. Also, a difference was established based on time spent on each level. It was found that despite several reviews that were required to progress through a level remained the same, a user who eventually become an expert spent more time on a particular level. Next, user retention was studied, and it was found that user who left the community had lower experience level compared to those who stayed. Also, it was found that the user who leaves the community was because of the inability to adapt to linguistic norms. Next acquired taste as a qualitative measure was studied, product bias terms between the most expert (level 5) and the least expert (level 1) for each time I was calculated, the positive di value indicated that a product is desired by experts over beginners. The research was successful in not only discovering acquired taste but also when the users acquired them.

Blair-Goldensohn et al. [7] have used a hybrid approach where they have combined a dynamic aspect extractor, where characteristics are determined from the review using text only, and a static extractor, where characteristics are predefined and extraction classifiers trained on a set of labeled data. The model which was used is also a hybrid model that uses lexicon-based and machine learning approaches. Static extractors leverage the fact that restaurants and hotels constitute a mass of online searches for local reviews. Thus, by constructing dedicated extractors for these domains, they improved the total accuracy of the structure.

3 Proposed System Architecture

Figure 1 presents the detailed system architecture for food sentiment analysis.

3.1 Data Extraction

The data source was collected from the reviews posted on Zomato. The Zomato API will be used to extract the reviews. A set of filtering parameters is set during the extraction of reviews from Zomato such that they can match any specific criteria. API is used to keep the query running after it has been generated. The output of this query will be all the relevant Zomato source data. In every record that is generated through a review, information such as id, text, username, and so on can be extracted. If any user makes his location public, the data relevant to the location from where a review is posted is generated in the form of latitude and longitude from the Zomato API. However, due to security and privacy reasons, users have stopped sharing their location. The location is used as a filtering parameter in the principle query later on Zomato. Thus, depending upon the settled set of locations, the reviews are extracted.

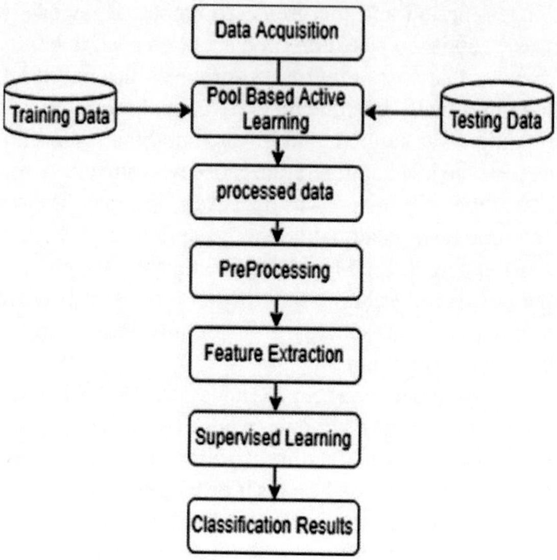

Fig. 1 Workflow diagram of the process

3.2 Data Preprocessing

There is a certain amount of irrelevant data available within the dataset that is extracted from Amazon Fine Food Reviews. Any kind of random characters or useless information needs to be filtered out from the reviews. The natural language processing tool is applied for filtering out this useless data.

Within the general natural language research, it is not useful to include certain advanced grammar available in the English language. The information analysts consider the main word relations as important even though linguistics defines a few other word relations within a sentence, due to which 50 dependencies have been defined in NLP [8–10]. The most used dependencies among these 50 are nsubj, amod, and dobj. The reviews that contain meaningful information are recognized using these relations [11]. The relations among nouns and adjectives or verbs are discovered using nsubj relation within any noun sentence. Irrespective of complementing a noun in a sentence or not, this is considered to be of high importance.

3.3 Data Analysis

Creation of Dataset and Data cleaning

The given dataset which is Amazon Fine Food Reviews is used to extract the data. Positive and negative classes are created for classifying the reviews. Redundant reviews are removed to get better results. Data points are created.

Preprocessing of Reviews

The extraction of keywords becomes difficult due to the presence of slangs and incorrect spellings in reviews. Thus, a preprocessing step is performed for filtering out the slang words and misspellings before extracting the features. Any slang words present in the reviews are replaced with their relevant meanings using the slang word dictionary. The slang word dictionary is created using the domain information.

Our dataset requires some preprocessing before we conduct further analysis and making the prediction model.

Now in the preprocessing phase, we perform the following activities on the dataset in the order below:-

1. Start by removing the HTML tags from the reviews.
2. We removed all the punctuation or only remove those punctuation which do not denote any sentiment like coma, period, hashtag, etc.
3. Check if the word is constructed using English letters and is not alphanumeric
4. The word length should be greater than 2. (there is no two-letter adjective).
5. Convert all words from uppercase to lowercase.
6. Remove stop words.

Creation of Feature Vector

The features are extracted from reviews in the next step. In the initial step, specific features like hashtags and emoticons are extracted. Based on the polarity of emotions, they depict, particular weights are assigned to the emoticons based on their polarity. The positive emoticons are assigned with weight "1" and negative emoticons with "−1" weight. A hashtag can be positive and negative. Within the vector of features, they are included as individual features.

Featurization Techniques

We have to encode the data before injecting it into the classifier. The assignment of textual data to real valued vectors is called as feature extraction. The following are some feature extractors:

1. Bag of Words (BoW) [12]—BOW is an algorithm that sums how many times a word appears in a document. The text in this model is represented as a bag of words ignoring the word order and grammar. Only the multiplicity of a word is counted. Multiplicity of a word is counted.
2. N-Grams & Bi-Grams [13]: N-gram is a sequence of N-words. It is fetched from the existing text. It is often used for predicting the next item in such sequence. The item here could be a word, letter, and syllable. Unigram/1-g is a unique set of words present in a sentence. Bi-Gram is a grouping of two words and so on.
3. TF-IDF [14]: This technique enumerates the word in the document, and a weight is allocated to each word which indicates the significance of that word. The idea behind TF-IDF is that the words that occur more often in one document and less regularly in other documents should be given more importance as they are more valuable for classification.
4. Word2Vec [15]: Word2vec takes a huge amount of input from corpus of text. The input is then placed in vector space. Each word has its own vector. Words

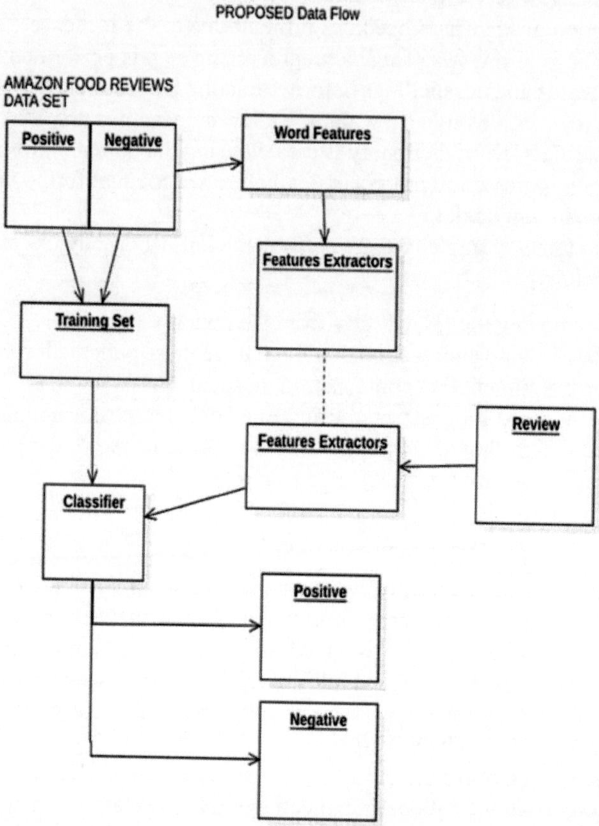

Fig. 2 Proposed system component and architecture diagram

are positioned in space in such a way that related words are close to each other. Words which share common meanings/context lie close to each other in space. (Fig. 2)

3.4 *Classification Techniques*

A text classification issue needed to be resolved is sentiment analysis. The machine learning approach and lexicon-based approach are the two broader categorizations of these classification approaches.

4 Machine Learning Approaches

The text is classified by the machine learning-based approach using classification techniques [15, 16]. The two broader categorizations of these machine learning techniques are:

(i) Unsupervised learning: There is no category involved and the targets are not provided by them at all. Thus, clustering is considered to be an important factor here.

(ii) Supervised learning: A labeled dataset is used to develop this method. When the classification approach is to be designed, the labels are provided to the model. For getting significant outputs when going through the decision making these labeled datasets are trained. The determination and extraction of particular sets of features such as the sentiments can be detected in the success of both of these learning techniques.

1. ***Naive Bayes Classifier*:-** A considerable number of features are utilized in feature vector through the Naïve Bayes classifier. Since these features are independent equally, analyzing them exclusively is important.
   A feature vector denoted by "*X*" is included here which is defined by $X = \{\times 1, \times 2, \ldots . \times m\}$. The class label is represented by *yj*. The classification of different types of independent features such as positive and negative keywords, emoticons and emotional keywords is done efficiently using Naïve Bayes. The relationships among features are not considered in the Naïve Bayes classifier. Thus, the relationships which exist among emotional keyword, negation words and speech tag are not utilized in it.
2. ***Support Vector Machine Classifier (SVM)*:-** Huge margin is used for classification through the SVM classifier. A hyperplane is used to differentiate the reviews. A discriminative function is utilized by SVM as:
   The feature vector is denoted in the above equation by "*X*", weights vector by "*w*" and the bias vector by "*b*". The nonlinear mapping which transforms information space to high-dimensional feature space is denoted by φ (). On the training set, "*w*" and "*b*" are recognized automatically. A linear kernel is applied for classification in this approach.
3. ***Decision Tree***
   Decision tree is a nonparametric approach in supervised machine learning. A decision tree is basically a flowchart like tree structure. The internal node present in this tree represents a test on a feature. Each leaf node represents a decision taken after computing all features. And the path present between them simply shows the classification rules. The time complexity of decision trees is a directly related to the number of files and number of features in the given data. The decision trees can handle great dimensional files with good accuracy.

4. ***K-nearest neighbors (KNN)***

K-nearest neighbors method assumes that all similar things are in close proximity, that is all the data points are actually near to each other. KNN shows the idea of similarity in terms of distance between the two points.

The two sets of data are used within machine learning approaches. The first set is used for training the classifier and the second set is used for testing. The training dataset is collected to initiate machine learning. The training data is used in the next step for training a classifier. Selecting the feature is an imperative decision to be made after the selection of a supervised classification approach. The representation of documents can be known through this. During sentiment classification, the most commonly used features included opinion word, speech information, negations, and term presence with a degree measure of frequency. When having an initial set of labeled opinions seems unrealistic for training the classifier, techniques such as semi-supervised and unsupervised are designed.

The sentiment dictionary which consists of opinion words is used by a lexicon-based approach. The polarity is determined by matching these words with the rest of the data. For understanding how much positive, negative, and objective the words comprised in the dictionary are, the sentiment scores are assigned to all opinion words. The sentiment lexicon which is an accumulation of known and precompiled sentiment phrases, idioms, and terms is used as a base for the lexicon-based approaches. For different traditional genres of communication, this approach is developed.

This approach has two subclassifications:

(i) Dictionary-based:- The terms which are collected normally and then annotated manually are utilized for this approach. The synonyms and antonyms of a particular word within the dictionary are searched for growing this set. WordNet is an example of one such dictionary using which a thesaurus called SentiWordNet is developed. The domain and context-based orientations cannot be managed by this method which is its major drawback [17].

(ii) Corpus-Based:- The dictionaries related to a particular domain are provided by the corpus-based approach. A set of seed opinion terms which grow from the search of relevant words using statistical or semantic techniques is created by these dictionaries [18].

4.1 Dataset Description

Amazon Food Reviews dataset comprises of reviews of fine foods from Amazon. Reviews include various users and products from various categories of Amazon. The dataset contains more than 500 K reviews with number of upvotes and total votes to those comments. This dataset has been obtained from Kaggle.

The data includes reviews in the span of 10 years with 568,454 number of reviews from 256,059 users for 74,258 products related to food industry with 260 users with more than 50 reviews. It has the following columns for each review.

1. Id:
2. Row Id:
3. Productid: (Unique identifier for the product)
4. Userid: (Unique identifier for the user)
5. Profile Name: (Profile name of the user)
6. HelpfulnessNumerator: (Users who find the review helpful/Relatable)
7. HelpfulnessDenominator: (Users who find the review not helpful/Not Relatable)
8. Score: (Rating between 1 and 5)
9. Time (Timestamp for the review)
10. Summary: (Brief summary of the review)
11. Text: (Text of the review)

5 Results and Discussion

We first define performance metrics to understand how a particular method performs in terms of accuracy and precision.

5.1 AUC-ROC Curve

ROC stands for receiver operating characteristic. It tells us good the model can distinguish between two things, for example, if a consumer likes the product or not. TPR stands for true positive rate and false positive rate (FPR). AUC stands for area under the curve. When the two curves in AUC are distinct and do not overlap it indicates that the model is performing well. And if the curves are overlapping, it means that the model is underperforming and its predictions are very random in nature. An excellent model will have an AUC near to 1 which indicates it has a good measure of separability and vice versa.

5.2 Confusion Matrix

Confusion matrix is a performance measurement graph; it is used when the output is two or more classes. It is presented in a tabular form with four different combinations of predicted and actual values namely true positive, false positive, true negative, and false negative. (Figs. 3, 4, 5, 6, 7, 8, 9, 10); (Tables 1, 2, 3)

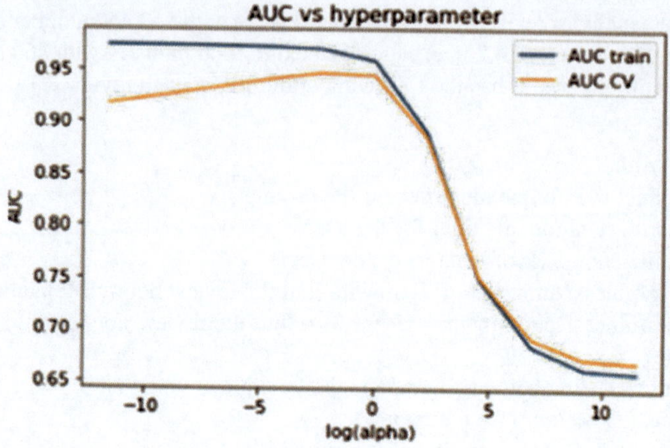

Fig. 3 AUC versus Hyperparameter for Naïve Bayes (SET 2: TF-IDF + FEATURE ENGINEERING)

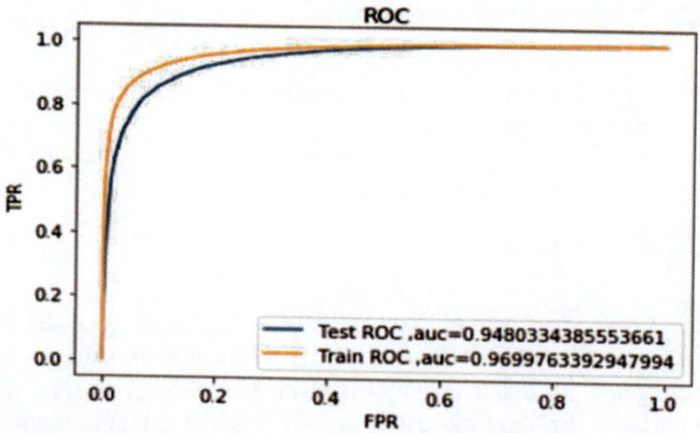

Fig. 4 TPR versus FPR curve: ROC curve for Naïve Bayes (SET 2: TF-IDF + FEATURE ENGINEERING)

6 Limitations

Majority of reviews provided by the user depict direct statements of the user, but sometimes the reviews do not contain any direct adjectives, it is then the process of analyzing sentiments from these reviews gets very difficult, almost impossible. sarcasm, euphemism, irony, negation, and exaggeration are some ways used by user to express their sentiment toward a particular product, service, or service provider [19]. It is very difficult to analyze these kinds of reviews as they rarely convey the same meaning as what the syllables contribute to, limiting the sentiment analysis

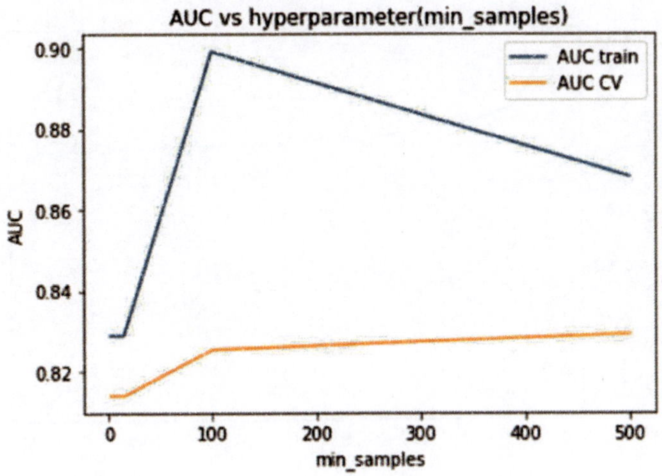

Fig. 5 AUC versus Hyperparameter curve (Min samples) for DECISION TREE (SET 3 AVG WORD2VEC)

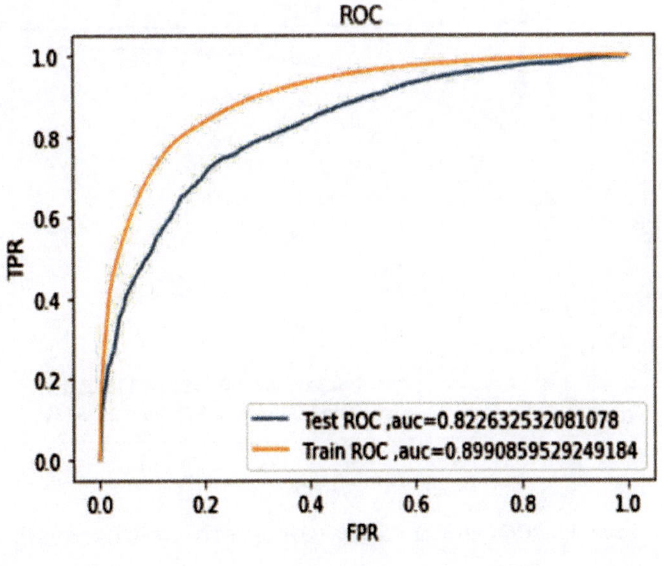

Fig. 6 TPR versus FPR, ROC Curve for DECISION TREE (Alpha = 1: For SET 3 AVG WORD2VEC)

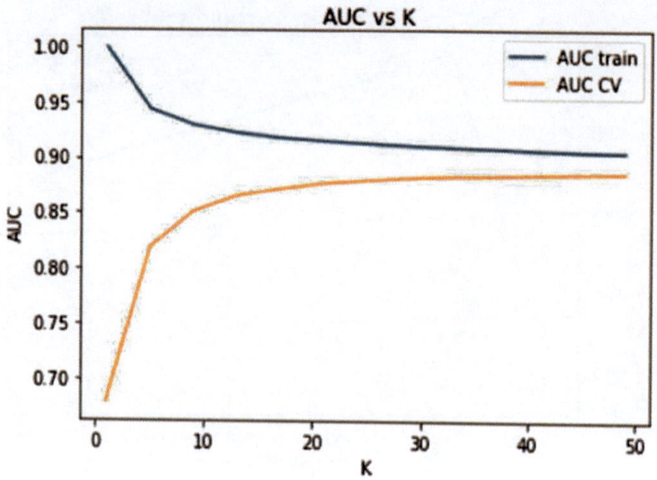

Fig. 7 AUC versus K curve for KNN (SET 3 AVG WORD2VEC)

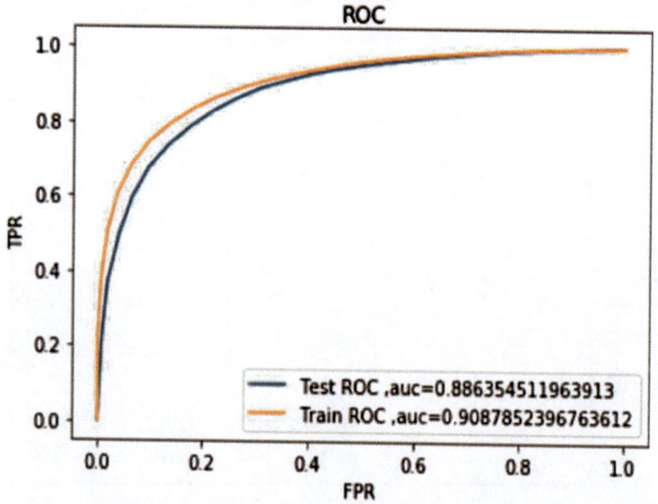

Fig. 8 TPR versus FPR ROC Curve for KNN ($k = 37$:For SET 3 AVG WORD2VEC)

algorithm. When there is an aggregate rating of 3 is observed it is very difficult to classify whether the depicted sentiment is positive or negative. It is one of the greatest limitations of sentiment analysis. User's idea of negative might differ from our idea of negative which is another big challenge in sentiment analysis. As ideas differ person to person, a wrong conclusion can supply a wrong perception toward the better good. Sometimes the users are biased toward the product, service, or the service provider. These users provide biased reviews for the same. Although sentiments

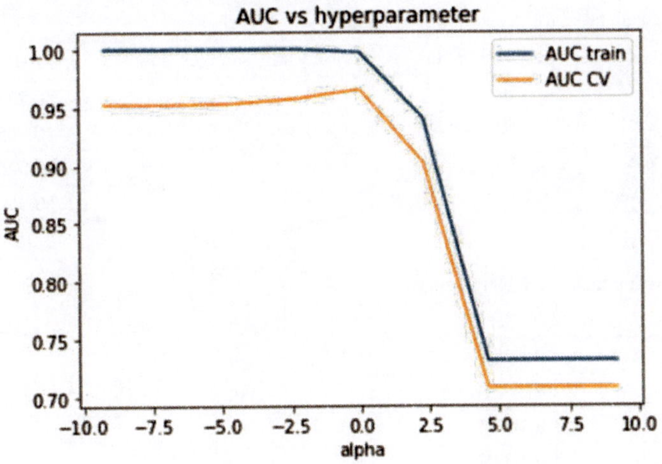

Fig. 9 AUC versus Hyperparameter curve for SVM (SET 1 TF-IDF)

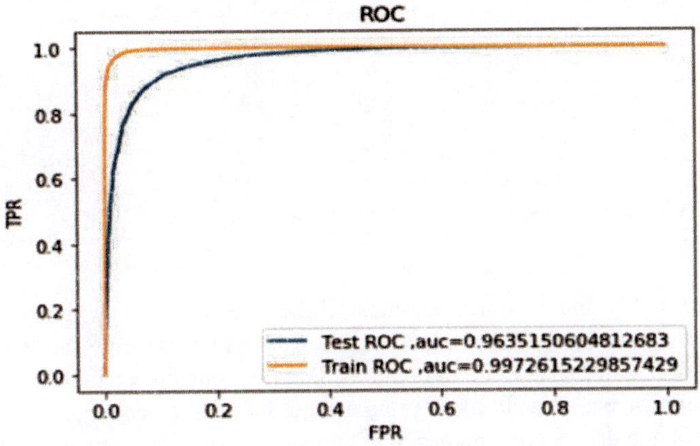

Fig. 10 TPR versus FPR, ROC Curve for SVM (SET 1 BOW)

Table 1 Result table for Naïve Bayes method

Vectorizer	Feature engineering	Hyperparameter	AUC
BOW	NO	1	0.91
TFIDF	NO	0.1	0.931
BOW	YES	1	0.939
TFIDF	YES	0.1	0.948

Table 2 Result table for decision tree method

Vectorizer	Best depth	Best min_samples_split	AUC
BOW	50	500	0.8218
TFIDF	50	500	0.8055
AVG W2Vec	10	500	0.8266
TDIDF W2Vec	10	500	0.7565

Table 3 Result table for KNN method

Vectorizer	Model		Hyperparameter	AUC
BOW	BRUTE		29	0.829
TDIDF	BRUTE		49	0.871
AVG W2V	BRUTE		37	0.886
TDIDF W2V	BRUTE		49	0.821
BOW	K D TREE		21	0.724
TFIDF	K D TREE		25	0.6
AVG W2V	K D TREE		45	0.775
TFIDF W2V	K D TREE		57	0.742
BOW	Linear	YES	1	0.931
TFIDF	Linear	YES	1	0.963
AVG W2V	Linear	YES	0.001	0.925
TFIDF W2V	Linear	YES	0.01	0.876

can be thoroughly analyzed through most of these reviews, they are biased so they alter the process output. There is no way we can analyze the true sentiments from these reviews, limiting the process output. Sentiment analysis needs to be evolved in qualitative aspect as well, users often are not habitual to new tastes and dishes, some dishes and drinks are a matter of experience progression which can be known as acquired taste. These reviews might obstruct the decision-making process.

7 Conclusion

In this research, we have proposed a supervised learning model to differentiate a massive amount of food review dataset which was unlabeled. We applied various machine learning algorithms on the dataset, we have been successful in determining that SVM outperformed Naïve Bayes, decision tree, and KNN. The featurization techniques also played a major role in improving a classifier's performance. When the classifier reaches its threshold, featurization and feature engineering techniques played a minor but really significant role in improving the capacity of the classifier.

The classifier's showed us very descriptive results through AUC-ROC and confusion matrixes. It was observed in almost every classifier that the TF-IDF SETS performed well with respect to other techniques. SVM along with TF-IDF and featured engineering has been the best approach so far in this project with a whopping AUC value of 96.3%. One major task in using sentiment classification approaches and tools for posts, blogs, and reviews in social media is to overcome the anomaly that really represents particular problem since it is not easy to make use of coreference information. Ordinarily, the analyzed posts may comprise of irony and sarcasm, which are particularly hard to detect. We have used a general yet wide approach on achieving our goal of sentiment analysis on food reviews. However, we need a more diverse approach to achieve a best model which eliminates all the problems and limitations. Sentiment analysis has a progressive solution and it needs to be continuously improvised as the analyzing a sentiment from a text requires a very intelligent system. Also, the overall sentiment of the text has to be studied in order to get to a conclusion. Poor quality of products and services results in reduced customer satisfaction. On the other hand, under the conditions of taut competition, there are no barriers for the consumer to change the supplier of goods and services. All these factors may result in loss of clients and a decrease of a company's competence indexes. Therefore, high-quality standards must be provided by effective executive decisions and based on opinion mining as a feedback. So, a progression of approaches and tools is essential to beat this constraint.

8 Future Scope

The era of getting meaningful insights has arrived as a result of massive increase in Web blogging. It is time for organizations to get statistical and business performance overview from the user's opinion. It is one of the finest ways to get an understanding on how the business performs. A need for hybrid combination of methods is needed to improve the performance of the model. A systemwise enough to recognize the acquired taste of the particular individual would be a step forward in this analysis approach as some reviews by people often ignore the acquired taste as a dependency. There are many dependencies when it comes to sentiment analysis on food and restaurants which cannot be compared to sentiment analysis of product reviews. There is a lot of scope in analyzing the video and images on the Web. These days people are expressing their thoughts and emotions through videos and vines on various social media platforms such as Facebook, Instagram, and YouTube along with text review. Sentiment analysis will have to pace up with this change. Tools which are helping companies to change strategies based on Facebook and Twitter will also have to accommodate the number of likes and retweets that the thought is generating on the social media.

References

1. Liu B, Zhang L (2012) A survey of opinion mining and sentiment analysis. In: Mining text data. Springer, Boston, MA, pp 415–463
2. Author, Alessia D, Ferri F, Grifoni P, Guzzo T (2015) Approaches, tools and applications for sentiment analysis implementation. Int J Comput Appl 125(3)
3. Tan, W., Wang, X., & Xu, X. Sentiment Analysis for Amazon Reviews.
4. Wu J, Ji T (2016) Deep learning for amazon food review sentiment analysis
5. Sasikala P, Sheela LMI (2018) Sentiment analysis of online food reviews using customer ratings. Int J Pure Appl Math 119(15):3509–3514
6. McAuley JJ, Leskovec J (2013) From amateurs to connoisseurs: modeling the evolution of user expertise through online reviews. In: Proceedings of the 22nd international conference on World Wide Web May, pp 897–908
7. Blair-Goldensohn S, Hannan K, McDonald R, Neylon T, Reis G, Reynar J (2008) Building a sentiment summarizer for local service reviews
8. Tyagi P, Chakraborty S, Tripathi RC, Choudhury T (2019) Literature review of sentiment analysis techniques for microblogging site. Available at SSRN 3403968
9. Sahayak V, Shete V, Pathan A (2015) Sentiment analysis on twitter data. Int J Innov Res Adv Eng (IJIRAE) 2(1):178–183
10. Shinde V, Pawar A, Ahirrao S, Phansalkar S (2019) Emotions identification by using unsupervised aspect category based sentiment classification
11. Gaind B, Syal V, Padgalwar S (2019) Emotion detection and analysis on social media. arXiv preprint arXiv:1901.08458
12. Pang B, Lee L, Vaithyanathan S (2002) Thumbs up?: sentiment classification using machine learning techniques. In: Proceedings of the ACL-02 conference on Empirical methods in natural language processing-vol 10. Association for Computational Linguistics, pp 79–86
13. Medhat W, Hassan A, Korashy H (2014) Sentiment analysis algorithms and applications: a survey. Ain Shams Eng J 5(4):1093–1113
14. Raghavendra TS, Mohan KG (2019) Web mining and minimization framework design on sentimental analysis for social tweets using machine learning. Proc Comput Sci 152:230–235
15. Cataldi M, Ballatore A, Tiddi I, Aufaure MA (2013) Good location, terrible food: detecting feature sentiment in user-generated reviews. Social Netw Anal Mining 3(4):1149–1163
16. Surjandari I, Naffisah MS, Prawiradinata MI (2015) Text mining of twitter data for public sentiment analysis of staple foods price changes. J Indus Intell Info Eng Technol Publ 3(3)
17. Tripathy A, Agrawal A, Rath SK (2016) Classification of sentiment reviews using n-gram machine learning approach. Expert Syst Appl 57:117–126
18. Gaspar R, Pedro C, Panagiotopoulos P, Seibt B (2016) Beyond positive or negative: qualitative sentiment analysis of social media reactions to unexpected stressful events. Comput Hum Behav 56:179–191
19. Liu B (2012) Sentiment analysis and opinion mining. Synthesis Lectures on Human Lang Technol 5(1):1-167

Parts of Speech (POS) Tagging for Dogri Language

Shivangi Dutta and Bhavna Arora

Abstract Parts of speech tagging is an important activity of natural language processing, information extraction, language translation, speech synthesis, question understanding and many more. Parts of speech tagging is basically the problem of assigning the parts of speech tags to the words in the text. As per English grammar, there are many parts of speech tags such as noun, adjective, pronoun, verb, adverb, preposition, conjunction and interjection. The methods of assigning tags to words are categorized into rule-based, stochastic and hybrid approach. In this paper, a rule-based parts of speech tagger for Dogri (regional language of Jammu) language along with the algorithm and the modular structure of the system is presented. The proposed system is evaluated over a number of corpus with six different parts of speech tags for Dogri, and hence, the evaluation is done on five datasets of Dogri corpus, and the corresponding results are also demonstrated in the paper.

Keywords Parts of speech tagging · Rule-based approach · Dogri language

1 Introduction

Natural language processing (NLP) is a part of computer science and artificial intelligence which deals with human languages. There is various application area of natural language processing, i.e., sentimental analysis, chatbot, speech recognition, machine translation, spell checking, keyword search, information extraction, advertisement matching, intellect disambiguation, information recovery, information handling, information analysis and interrogating, machine interpretation and many more. Parts of speech (POS) tagging is the important method of NLP in which labels are allotted to words. POS tagging can be done using various techniques which are broadly classified as supervised technique and unsupervised techniques which are further sub-divided into various categories [1]. There are various steps involved

S. Dutta (✉) · B. Arora
Central University of Jammu, Jammu and Kashmir, India

P. K. Singh et al. (eds.), *Proceedings of Second International Conference on Computing, Communications, and Cyber-Security*, Lecture Notes in Networks and Systems 203,
https://doi.org/10.1007/978-981-16-0733-2_37

in natural language processing such as tokenization, stemming, lemmatization, POS tagging, name entity recognition and chunking.

Tokenization is the process of breaking strings into tokens which in turn are small structures or units that can be used. Stemming, usually, refers to normalizing the words into their base form or root form. Stemming algorithm works by cutting off the end or the beginning of the word taking into account a list of common prefixes and suffixes that can be found in the inflected word. This indiscriminate cutting can be successful in some cases but not always. Lemmatization on the other hand takes into consideration the morphological analysis of the word. To do so, it is necessary to have a detailed dictionary which the algorithm can look through to link the form back to its original word or the root word which is also known as lemma. Basically, lemmatization groups together different inflected form of the word called lemma. It is somehow similar to stemming as it maps several words into one common root word. But the major difference between stemming and lemmatization is that the output of the lemmatization is the proper word.

POS tags: The grammatical type of the word is referred to as POS tags. It specifies how the word purposes in sense in addition to grammar within the sentence. The word can have more than one part of speech depending on the context in which it is used, this is called ambiguity of words. For example, "Google" something on the Internet. Google here is used as verb although it is a proper noun. There are basically eight parts of speech tags, namely noun, pronoun, verb, adverb, adjective, preposition, conjunction and interjection [2].

POS tagging: It is the initial step or a primal task in processing a natural language. POS tagging can be done by consideration of various linguistic rules, random pattern or by the combination of both [3]. Dogri language like any other Indian language being morphologically rich language having less linguistically distinctive patterns, so the development of POS tagger for Dogri language is a difficult task.

Problem: Ambiguous and unknown words tagging is the major problem in any POS tagging system. The main problem when dealing with Dogri language is that words may have different meaning in different contexts. In this scenario, we focus on the word rather than the context. Understanding these words is an easy task for humans but for a machine, it is very tedious job. Another problem is ambiguous words. Many words can have more than one tags. In this case, we focus on the context rather than the word [4].

In the initial sections, Dogri parts of speech tagging system is discussed along with the algorithm to tag the words as well as the basic modular structure of the tagger is also described. The latter section describes the tagging scheme, screenshots of the system along with the evaluated results of the system.

2 Literature Survey

Kumawat and Jain [4] have done the comparison on various datasets using different POS tagging techniques. Trigram and HMM models are applied on the Hindi text corpus of tourism (3000 sentences), health (1000 sentence) and general (1000 sentences) domain and results are generated. Average accuracy of trigram is 92.98%. For HMM model, the average accuracy is 95.45%.

Modi and Nain [5] proposed a parts of speech tagging system for Hindi language using the rule-based approach. The proposed system works in three steps. In the first step, the corpus is matched with the already existed trained data, and if match is found, tags are assigned to them. On untagged words second step is applied, i.e., regular expressions are searched in this step. Finally, in the third step, various lexical rules which are based on assumptions are applied, and the words are tagged. The system gives the precision of 91.84%.

Modi et al. [6] presented a combinational approach of tagging the words. Probabilistic approach was applied for tagging the words which were already known, and rule-based approach (based on regular expression) was applied for tagging the unknown words to acquire the average precision of 95.08%.

Antony and Soman [7] explored the numerous innovations in POS taggers and tags for various Indian languages. Different approaches for different languages over the period of time had been discussed by various researchers, and the corresponding results were shown. Study on various POS tagging systems and different techniques has been discussed by different authors. Study on various POS taggers for various Indian languages like Hindi, Bengali, Punjabi, Tamil, Telugu, Malayalam and Kannada has been done, and the results have been shown in the paper.

3 System Description

The designed system, proposed in the paper, uses a rule-based approach and six tags. The proposed system tags noun (N), adjective (ADJ), verb (V), helping verbs (HV), stop words (SW) and a special category called other tags (OT). Various Dogri rules are hence identified and implemented to tag the mentioned parts of speech [8, 9].

3.1 Algorithm [10]

Step 1: Input the Dogri text.

Step 2: Tokenize the input Dogri text.

Step 3: Normalize the text by separating the punctuation marks and symbols from the text.

Step 4: Recognize various Dogri POS tagging rules.

Step 5: Assign the various POS categories to the tokenized Dogri text.

Step 6: Check for the ambiguity of the words.

Step 7: Display the tagged data as the desired result.

3.2 Basic Module Structure

See Fig. 1.

3.2.1 Module Description: The Section Explains the Modification of Basic Module Structure for Parts of Speech Tagging for Hindi Corpus by Mishra and Mishra [11] into Parts of Speech Tagging for Dogri Language, as Both the Languages are Developed from the Devanagri Script

Input Dogri File:

This module accepts and reads Dogri (Unicode) corpus by browsing any text file from the drives with path if user browse the file.

Preprocessing:

Preprocessing Dogri text includes number of phases which could be used according to the model applied.

Word Tokenizer:

This tokenization is the process to separate word/tokens from input data text. The separation of input text into tokens is significant for POS tagging. This tokenization task is done by searching spaces between the words. The words separated from sentences are treated as single token so, we can deal with each word separately.

Stemmer:

Stemming is defined as the process of converting the words similar in morphology to their respective root words by removing the ending or suffixes from the words.

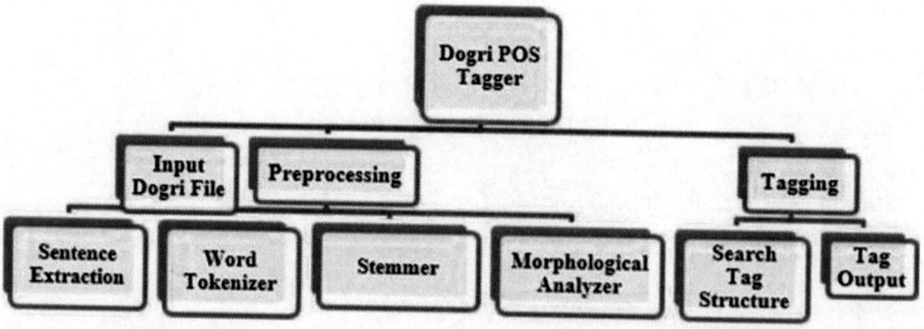

Fig. 1 Basic module structure of POS tagging. The motivation behind developing the model is taken from the referred paper

Like any other language, Dogri language is rich in morphology. A set of rules are made considering the language to accomplish the process of stemming. The resulting words after performing the process of stemming are not necessarily a root word. To overcome this problem, a database of the root words is needed to be maintained to provide the relevant output [12–14].

Morphological analyzer:

Morphological analysis aims to diagnose the inner structure of the word. The words after stemming are examined to check whether they are inflected or not. If stem word is modulated, then the root word is shaped by addition of replacement characters with stem word. A morphological analyzer is expected to yield root words for a given input document. There is need to project some standard rules called inflection rule which will permit the system to process the stem of words and discover the actual root word [15].

Tagging:

Tagging is the process of assigning the appropriate parts of speech to the words present in the corpus with the use of certain tagging techniques as discussed earlier. The module tags the Dogri words in the text with related tags.

Search Tag Structure:

The rule-based POS tagging technique has been used in the study. The module identifies the tag pattern as per the Dogri rules. In the designed system, in preprocessing phase, only the sentence extraction and word tokenization have been applied.

Tag Output:

The resulting tagged Dogri text is displayed as the output of the system.

4 Result

In this section, analysis of the proposed work is described along with its working and the screenshots of the working model. The performance parameters are evaluated, and graphical results are shown in this section. The tool used to perform tagging of Dogri text is Python which has the inbuilt library for natural language processing known as natural language toolkit.

4.1 Tagging Scheme

The proposed system tags noun (N), adjective (ADJ), verb (V), helping verbs (HV), stop words (SW) and a special category called other tags (OT). In natural language processing, non-useful information is called stop words. Stop words are considered irrelevant for search reasons as they often appear in the language. These phrases consume time and space, and they are often programmed to ignore. Natural language toolkit (NLTK) in Python has a list of stop words stored in 16 different languages. But there is no such pre-constructed list in Dogri language. So, in the study, stop words have also been tagged as SW. We can remove them later by storing the list of words that we consider as stop words. The designed system tags words other than noun, adjective, verb, helping verb and stop words as other tags. These are basically the words which fall in the categories other than the above mentioned. Other tags include the words which fall into the categories of conjunction, preposition, interjection and adverb.

4.2 Result and Demonstration of Work Performed

The overall result of Dogri part of speech tagger is described below.

Input Text

The first step of the model is to insert the Dogri text file. The input data is the Dogri (Devanagari script) text collected from social media. The model has been designed for demonstration. This module reads Dogri (Unicode) corpus by browsing any .text file from the drives. The output of the module is shown in Figs. 2, 3 and 4.

On clicking the "Show Dataset" button, the selected file will be displayed. The screenshot of the displayed input file is shown in Fig. 5.

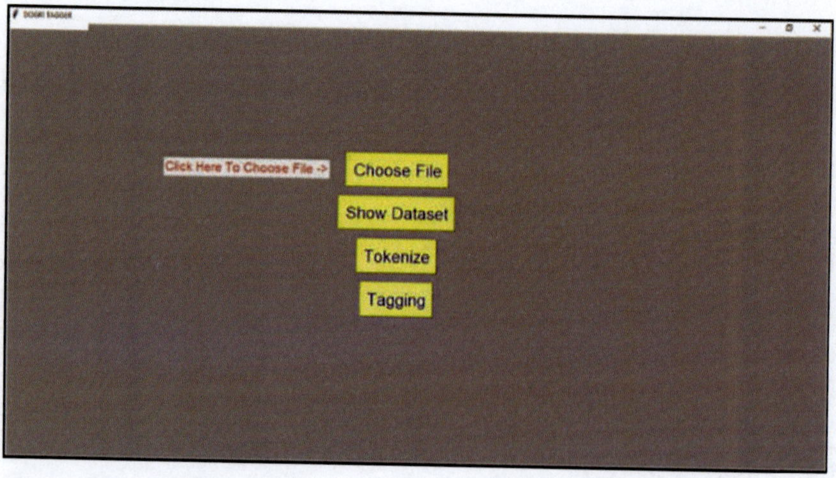

Fig. 2 Screenshot of main window

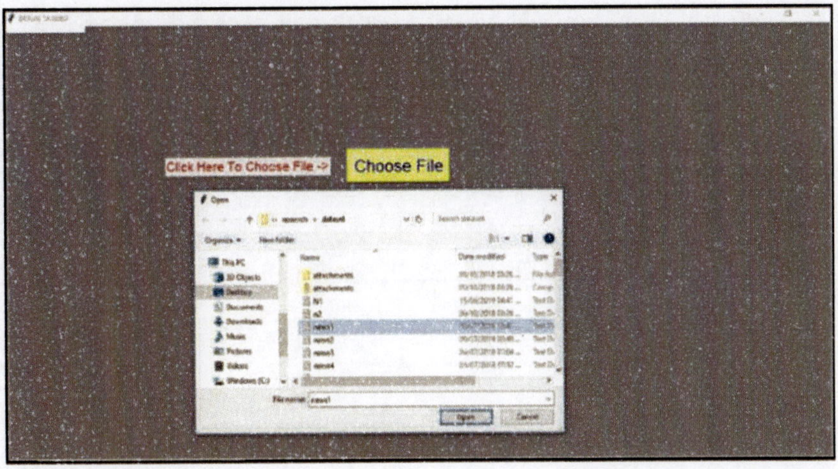

Fig. 3 Screenshot of inserting file

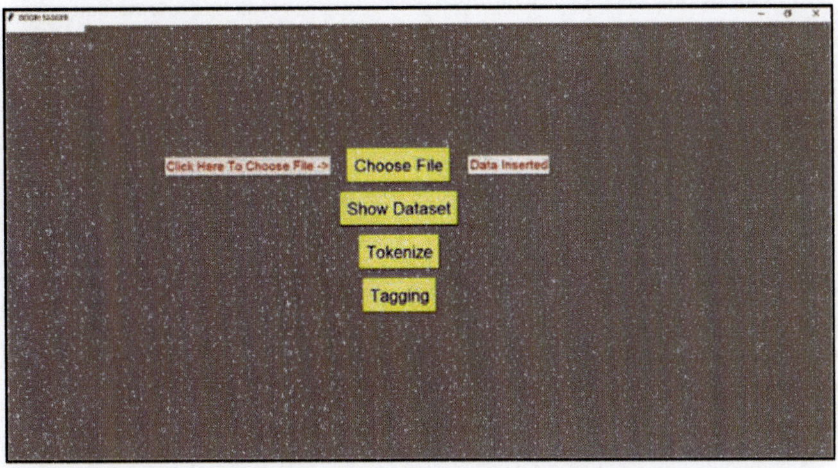

Fig. 4 Screenshot of inserted file

हिमाचल प्रदेश च घटदे जमीनै हेठ पानी दे स्तर दा कारण अन्नेबाह शैहरीकरण-दीपक गिरकर- पिछले दिनें हिमाचल प्रदेश दी प्रसिद्ध सैलसफा आह्ली थाहर शिमला च पानी गित्तै हाहाकार मचे दा हा। सरकार ने पिछले दिनें गित्तै स्कूल बंद करीदित्ते हे। प्रदेश च स्थितियां काबू थमां बाहर होने आह्ली बक्खी बधा करदियां न। जंगलें दी अग्ग ते जंगलें दी अन्नेबाह् कटाई कन्नै बी भू-जल स्तर तेजी कन्नै डिग्गदा जाकरदा ऐ। प्रदेश च पिछले किश सालें च पानी दे कुदरती स्रोतें दा बड्डे पैमाने उप्पर दोहन कीतागेआ ऐ। किश भ्रश्ट राजनीतिज्ञें, नौकरशाहें ते भू-माफिया दी संडगंड दे कारण हिमाचल प्रदेश दा चेहरा लगातार बिगड़दा जाकरदा ऐ। भू-माफिया ने कब्जे करियै बहुमंजला इमारतें दा निर्माण करियै हिमाचल दे जादातर शै गी कंक्रीट दे जंगलें च तबदील करी दित्ता ऐ जिसलै के हिमाचल प्रदेश मूंकप संवेदी खेतर ऐ। सिड़क चौड़ीकरण दा मलबा नदिये च सुट्टेआ जाकरदा ऐ। थ्होड़ी हारी बी बरखा होंदी ऐ तां हाड़े च अड़की दा एह्। मतलब सिड़कें उप्पर आई जंदा ऐ जिस कन्नै आओजाई च बिघन पेई जंदा ऐ। गड्डियें दी बदधी तदाद दे कारण बी हिमाचल प्रदेश दे शैहें च केईं बारी सिड़कें उप्पर जाम लगदे न। प्रदेश च शिमला, मनाली, धर्मशाला, डलहौजी ते कसौली प्रमुख अंतर्राश्ट्रीय सैलसफा केंद्र न।

Fig. 5 Input Dogri text file

```
['[', "'\\ufeffहिमाचल", 'प्रदेश', 'च', 'घटदे', 'जमीने', 'हेठ', 'पानी', 'दे', 'स्तर', 'दा', 'कारण', 'अन्नेबाह', 'शैहरीकरण-दीपक', 'गिरकर-',
'पिछले', 'दिनें', 'हिमाचल', 'प्रदेश', 'दी', 'प्रसिद्ध', 'सैलसफा', 'आह्ली', 'थाहर', 'शिमला', 'च', 'पानी', 'गित्तै', 'हाहाकार', 'मचे', 'दा', 'हा',
'""', '.', "'सरकार", 'ने', 'पिछले', 'दिनें', 'गित्तै', 'स्कूल', 'बंद', 'करीदित्ते', 'हे', '""', '.', "'प्रदेश", 'च', 'स्थितियां', 'काबू', 'थमां', 'बाहर', 'होने',
'आह्ली', 'बक्खी', 'बधा', 'करदियां', 'न', '""', '.', "'जंगलें", 'दी', 'अग्ग', 'ते', 'जंगलें', 'दी', 'अन्नेबाह्', 'कटाई', 'कन्नै', 'बी', 'भू-जल', 'स्तर',
'तेजी', 'कन्नै', 'डिग्गदा', 'जाकरदा', 'ऐ', '""', '.', "'प्रदेश", 'च', 'पिछले', 'किश', 'सालें', 'च', 'पानी', 'दे', 'कुदरती', 'स्रोतें', 'दा', 'बड्डे',
'पैमाने', 'उप्पर', 'दोहन', 'कीतागेआ', 'ऐ', '""', '.', "'किश", 'भ्रश्ट', 'राजनीतिज्ञें', '.', 'नौकरशाहें', 'ते', 'भू-माफिया', 'दी', 'संडगंड', 'दे',
'कारण', 'हिमाचल', 'प्रदेश', 'दा', 'चेहरा', 'लगातार', 'बिगड़दा', 'जाकरदा', 'ऐ', '""', '.', "'भू-माफिया", 'ने', 'कब्जे', 'करियै', 'बहुमंजला',
'इमारतें', 'दा', 'निर्माण', 'करियै', 'हिमाचल', 'दे', 'जादातर', 'शै', 'गी', 'कंक्रीट', 'दे', 'जंगलें', 'च', 'तबदील', 'करी', 'दित्ता', 'ऐ', 'जिसलै',
'के', 'हिमाचल', 'प्रदेश', 'मूंकप', 'संवेदी', 'खेतर', 'ऐ', '""', '.', "'सिड़क", 'चौड़ीकरण', 'दा', 'मलबा', 'नदिये', 'च', 'सुट्टेआ', 'जाकरदा', 'ऐ',
'""', '.', "'थ्होड़ी", 'हारी', 'बी', 'बरखा', 'होंदी', 'ऐ', 'तां', 'हाड़े', 'च', 'अड़की', 'दा', 'एह्', '""', '.', "'मतलब", 'सिड़कें', 'उप्पर', 'आई',
'जंदा', 'ऐ', 'जिस', 'कन्नै', 'आओजाई', 'च', 'बिघन', 'पेई', 'जंदा', 'ऐ', '""', '.', "'गड्डियें", 'दी', 'बदधी', 'तदाद', 'दे', 'कारण', 'बी', 'हिमाचल',
'प्रदेश', 'दे', 'शैहें', 'च', 'केई', 'बारी', 'सिड़कें', 'उप्पर', 'जाम', 'लगदे', 'न', '""', '.', "'प्रदेश", 'च', 'शिमला', '.', 'मनाली', '.', 'धर्मशाला', '.',
```

Fig. 6 Tokenized Dogri text

Preprocessing

After the dataset is inputted in the system, the next step is to perform the preprocessing. This step has number of phases depending upon the type of approach used to tag the text. The tokenization phase of preprocessing is used in the proposed algorithm. In this step, each word of the Dogri text needs to be separated so that the proposed system could tag each word in the file. The process which is used to separate the words from file is called tokenization. In this phase, each word of the Dogri text is tokenized, and the results are shown in Fig. 6.

Tagging

After each Dogri word is tokenized, the tagging of the words in the text is done. So, in this step, tagging has been done using various rules [9] described below:

- There is a high probability that noun follows an adjective.
- There is a probability that a postposition follows a noun.
- There is a probability that end word of each sentence is a stop word.
- There is a high probability of verb following a noun.
- There is a probability that the word preceding 'दा','दी', 'न', 'ने', 'दे'is a verb.
- In Dogri language, the main verb is followed by an auxiliary verb.

The module tags each Dogri word in the sentence with their related tags like noun (N), adjective (ADJ), verb (V), helping verb (HV), stop word (SW) and other tags (OT) for rest of the word categories. The screenshot of the tagged text is shown in Fig. 7.

List of Untagged Punctuation and symbols

Like any other language, Dogri language consists a number of punctuations and symbols. They have not been tagged as they are not considered as any part of speech. The untagged symbols have been removed from the final result. Table 1 shows the list of symbols that have not been tagged and are hence removed from the dataset. The same can be seen in the screenshot below (Figs. 8 and 9).

["\\ufeffहिमाचल(N)", 'प्रदेश(N)', 'च(OT)', 'घटदे(OT)', 'जमीनै(N)', 'हेठ(OT)', 'पानी(N)', 'दे(OT)', 'स्तर(N)', 'दा(OT)', 'कारण(N)',
'अन्नेबाह(ADJ)', 'शैहरीकरण-दीपक(N)', 'गिरकर-(N)', 'पिछले(ADJ)', 'दिनें(N)', 'हिमाचल(N)', 'प्रदेश(N)', 'दी(OT)', 'प्रसिद्ध(ADJ)',
'सैलसफा(N)', 'आह्री(SW)', 'थाहर(N)', 'शिमला(N)', 'च(OT)', 'पानी(N)', 'गित्ते(SW)', 'हाहाकार(v)', 'मचे(V)', 'दा(HV)', 'हा(SW)',
"'सरकार(N)", 'ने(OT)', 'पिछले(ADJ)', 'दिनें(N)', 'गित्तै(SW)', 'स्कूल(N)', 'बंद(V)', 'करीदित्ते(HV)', 'हे(SW)', "'प्रदेश(N)", 'च(OT)',
'स्थितियां(N)', 'काबू(N)', 'थमां(OT)', 'बाहर(N)', 'होने(v)', 'आह्री(SW)', 'बक्खी(OT)', 'बधा(V)', 'करदियां(HV)', 'न(SW)', "'जंगलें(N)",
'दी(OT)', 'अग्ग(N)', 'ते(OT)', 'जंगलें(N)', 'दी(OT)', 'अन्नेबाह(ADJ)', 'कटाई(N)', 'कन्नै(OT)', 'बी(SW)', 'भू-जल(N)', 'स्तर(N)',
'तेजी(ADJ)', 'कन्नै(N)', 'डिग्गदा(V)', 'जाकरदा(HV)', 'ऐ(SW)', "'प्रदेश(N)", 'च(OT)', 'पिछले(ADJ)', 'किश(N)', 'सालें(N)', 'च(OT)',
'पानी(N)', 'दे(OT)', 'कुदरती(ADJ)', 'सोतें(N)', 'दा(OT)', 'बड्डे(ADJ)', 'पैमाने(N)', 'उप्पर(OT)', 'दोहन(V)', 'कीतागेआ(HV)', 'ऐ(SW)',
"'किश(N)", 'भ्रश्ट(ADJ)', 'राजनीतिज्ञें(N)', 'नौकरशाहें(N)', 'ते(OT)', 'भू-माफिया(N)', 'दी(OT)', 'संडगंड(v)', 'दे(OT)', 'कारण(N)',
'हिमाचल(N)', 'प्रदेश(N)', 'दा(OT)', 'चेहरा(N)', 'लगातार(OT)', 'बिगड़दा(V)', 'जाकरदा(HV)', 'ऐ(SW)', "'भू-माफिया(N)", 'ने(OT)',
'कब्जे(N)', 'करियै(v)', 'बहुमंजला(ADJ)', 'इमारतें(N)', 'दा(OT)', 'निर्माण(N)', 'करियै(v)', 'हिमाचल(N)', 'दे(OT)', 'जादातर(N)',
'शै(N)', 'गी(OT)', 'कंक्रीट(N)', 'दे(OT)', 'जंगलें(N)', 'च(OT)', 'तबदील(N)', 'करी(v)', 'दित्ता(SW)', 'ऐ(SW)', 'जिसलै(OT)', 'के(SW)',
'हिमाचल(N)', 'प्रदेश(N)', 'मूंकप(N)', 'संवेदी(ADJ)', 'खेतर(N)', 'ऐ(SW)', "'सिड़क(N)", 'चौड़ीकरण(v)', 'दा(OT)', 'मलबा(N)',
'नदियें(N)', 'च(OT)', 'सुट्टेआ(V)', 'जाकरदा(HV)', 'ऐ(SW)', "'प्होड़ी(N)", 'हारी(SW)', 'बी(SW)', 'बरखा(N)', 'होंदी(SW)', 'ऐ(SW)',
'तां(OT)', 'हाड़े(N)', 'च(OT)', 'अड़की(V)', 'दा(HV)', 'एह(SW)', "'मतलब(N)", 'सिड़कें(N)', 'उप्पर(OT)', 'आई(v)', 'जंदा(SW)',
'ऐ(SW)', 'जिस(OT)', 'कन्नै(OT)', 'आओजाई(N)', 'च(OT)', 'बिघन(N)', 'पेई(V)', 'जंदा(HV)', 'ऐ(SW)', "'गड्डियें(N)", 'दी(OT)',
'बदधी(v)', 'तदाद(N)', 'दे(OT)', 'कारण(N)', 'बी(SW)', 'हिमाचल(N)', 'प्रदेश(N)', 'दे(OT)', 'शैहें(N)', 'च(OT)', 'केईं(OT)', 'बारी(OT)',

Fig. 7 Tagged Dogri text

Table 1 List of removed symbols

S. No.	Symbol	S. No.	Symbol	S. No.	Symbol
1	!	11	\	21	^
2	(	12	,	22	&
3	)	13	<	23	*
4	[	14	>	24	_
5	]	15	/	25	~
6	{	16	?	26	"
7	}	17	@		
8	;	18	#		
9	:	19	$		
10	'	20	%		

'उप्पर', 'आई', 'जंदा', 'ऐ', 'जिस', 'कन्नै', 'आओजाई', 'च', 'बिघन', 'पेई', 'जंदा', 'ऐ', '"', ':', "'गड्डियें", 'दी', 'बदधी', 'तदाद', 'दे', 'कारण',
'बी', 'हिमाचल', 'प्रदेश', 'दे', 'शैहें', 'च', 'केईं', 'बारी', 'सिड़कें', 'उप्पर', 'जाम', 'लगदे', 'न', '"', ':', "'प्रदेश", 'च', 'शिमला', ':', 'मनाली', ':',
'धर्मशाला', ':', 'डलहौजी', 'ते', 'कसौली', 'प्रमुख', 'अंतर्राश्ट्रीय', 'सैलसफा', 'केंद्र', 'न', '"', ':', "'इने", 'शैहें', 'च', 'पानी', 'दी', 'समस्या'
'दिनों', 'दिन', 'होर', 'बदधी', 'जा', 'करदी', 'ऐ', '"', ':', "'लगातार", 'डिग्गदा', 'भू-जल', 'स्तर', 'प्रदेश', 'दी', 'मुक्ख', 'समस्या', 'बनीगेआ',
'ऐ', '"', ':', "'रिआसता", 'च', '9524', 'कुदरती', 'जल', 'सप्लाई', 'योजनाएं', 'बिच्चा', 'पिछले', 'किश', 'साले', 'च', '1022', 'जल', 'सप्लाई',
'योजनां', 'सुक्की', 'चुकी', 'दियां', 'न', '"', ':', "'कसौली", 'च', 'अवैध', 'निर्माण', 'ते', 'अवैध', 'निर्माण', 'टाहने', 'गित्ते', 'गेई', 'दी',
'सरकारी', 'अधिकारी', 'दी', 'मौत', 'दे', 'मामले', 'च', 'सुप्रीम', 'कोर्ट', 'ने', 'संज्ञान', 'लेता', 'हा', 'ब', 'उसदे', 'बाद', 'बी', 'प्रदेश', 'च',
'अवैध',

Presence of Punctuation

Fig. 8 Tokenized sample data

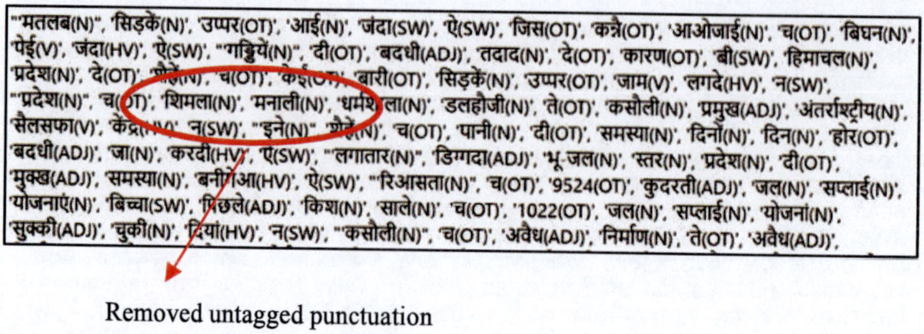

Fig. 9 Sample tagged data

4.3 *Performance Analysis Perimeter*

The designed system is evaluated on five datasets having 882, 158, 441, 162, 226 words, respectively, collected from social media, and the measure of evaluation in percentage is shown in Fig. 10. These are considered as the standard performance indicator of the system. Like any other language, Dogri also has lots of ambiguous words. Ambiguous words are the words which give different meaning in different context.

$$\text{Precision}(P) = \frac{\text{Total no. of correct word tagged by POS after applying rules}}{\text{Total no of words}}$$

$$\text{Recall}(R) = \frac{\text{No. of correct tag by POS}}{\text{Total no of correct word tag by POS after applying rules}}$$

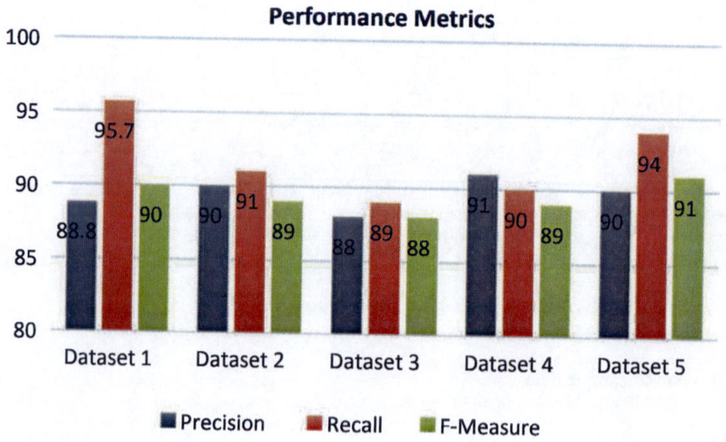

Fig. 10 Evaluated result

$$F\text{-Measure} = \frac{2PR}{P + R}$$

5 Conclusion and Future Scope

Parts of speech tagger designates the words in the sentence fragments to suitable tags. In the paper, rule-based technique to assign tags to words is used. Algorithm to assign tags is developed, and the basic modular structure of Dogri parts of speech tagger is also discussed in the paper. Five Dogri datasets have been evaluated, and the corresponding results of dataset 1 are 88.8% of precision, 95.7% of recall and 90% of F-measure; of dataset 2 are 90% of precision, 91% of recall and 89% of F-measure; of dataset 3 are 88% of precision, 89% of recall and 88% of F-measure; of dataset 4 are 91% of precision, 90% of recall and 89% of F-measure and of dataset 5 are 90% of precision, 94% of recall and 91% of precision, which are also shown in the paper with the screenshots. In the future, more tags can be added to the desired system as well as various rules can be recognized to remove ambiguity of words.

References

1. Bhatta S, Parmara K, Patelb M (2015) Sanskrit tag-sets and part-of-speech tagging methods—a survey. Int J Innov Emerg Res Eng (IJIERE) 2
2. Kumar S (2018) Developing POS tagset for Dogri. Lang India 18(1). www.languageinindia.com
3. Singh J, Joshi N, Mathur I (2013) Development of Marathi part of speech tagger using statistical approach. In: Proceedings of the 2013 international conference on advances in computing, communications and informatics, ICACCI 2013, 2013, pp 1554–1559
4. Kumawat D, Jain V (2015) POS tagging approaches: a comparison. Int J Comput Appl 118(6):32–38
5. Modi D, Nain N (2016) Part-of-speech tagging of Hindi corpus using rule-based method. Springer, India
6. Modi D, Nain N, Nehra M (2018) Part-of-speech tagging for Hindi corpus in poor resource scenario. J Multimed Inf Syst 5(3):147–154
7. Antony PJ, Soman KP (2011) Parts of speech tagging for Indian languages: a literature survey. Int J Comput Appl (0975-8887) 34(8)
8. Garg N, Goyal V, Preet S (2012) Rule based Hindi part of speech tagger. COLING (Demos) 2:163–174
9. Dutta S, Arora B (2019) Preprocessing for parts of speech (POS) tagging in Dogri language. Int J Innov Technol Expl Eng (IJITEE) 8(8S3):114–120
10. Bagul P, Mishra A, Mahajan P, Kulkarni M, Dhopavkar G (2014) Rule based POS tagger for Marathi text. Int J Comput Sci Inf Technol 5(2):1322–1326
11. Mishra N, Mishra A (2011) Part of speech tagging for Hindi corpus. In: International conference on communication systems and network technologies, pp 554–558
12. Dubey P (2018) The Hindi to Dogri machine translation system: grammatical perspective. Int J Inf Technol

13. Gupta V (2014) Hindi rule based stemmer for nouns. Int. J. Adv Res Comput Softw Eng 4(1):62–65
14. Pande BP, Tamta P, Dhami HS (2014) A Devanagari script based stemmer. Int J Comput Linguist Res 5(4):119–130
15. Pimpalshende A, Mahajan AR (2018) Extraction of root words using morphological analyzer for Hindi text. Int J Soft Comput 13(5):134–138

Deep Learning-Based 2D and 3D Human Pose Estimation: A Survey

Pooja Parekh and Atul Patel

Abstract In the real world, estimation of human pose has gained considerable consideration owed to its diverse application. Here, 2D pose estimation has remarkable research and achieves targeted output however challenges still remain in 3D pose estimation. As deep learning can improve the presentation of human pose estimation, it also brings very closest result. A literature review of deep learning methods for human pose estimation presented and analyzes the methodology used by this paper. It also includes real-world video with crowded scene pose estimation with latest research information. With a methodology-based taxonomy, we sum up and discuss recent works. It also addresses and compares the datasets used in this function. Thus, this survey makes interpretable each phase in the approximation pipeline and help to reader with easy comprehensive information. Future work and Challenges are detected.

Keywords Human pose estimation · Computer vision · Convolution neural network · 2D to 3D human pose

1 Introduction

Human pose estimation is defined because the problem of localization of human joints (also referred as key points—elbows, wrists, etc.) in images or videos. Along with the technological evolutions, new applications are constantly emerging because the great interest of various domains. Human pose estimation is not only an important com-putter vision problem but also plays a vital role in the following in a number of real-world applications, such as action recognition [1], gaming [2], augmented reality [3], pedestrian detection [4], automated lip reading [5], sports scenes [6],

P. Parekh (✉) · A. Patel
Charotar University of Science and Technology, Changa, Gujarat 388421, India
e-mail: poojaparekh.mca@charusat.ac.in

A. Patel
e-mail: atulpatel.mca@charusat.ac.in

P. K. Singh et al. (eds.), *Proceedings of Second International Conference on Computing, Communications, and Cyber-Security*, Lecture Notes in Networks and Systems 203,
https://doi.org/10.1007/978-981-16-0733-2_38

medical imaging [7], digital entertainment [8], human–computer interaction [9], video surveillance [10], face recognition [11], and gesture recognition [12]. There are various devices also built to capture motion and gesture capturing. This problem is solved by some researcher by dividing this problem in the subtasks such as estimating single-person pose, estimating multiple person pose, and finding pose of human in crowded places. To solve that subtask, two of the strategies are used in the previous researches are bottom-up and top-down. In top-down approaches, firstly it identifies and focuses single person using bounding box as object desector. After that estimate the pose of the single person, whereas in bottom-up approach detect different free semantic entities and group into single-person pose.

2 Methodology and Taxonomy

The above taxonomy described the overall models, approaches, and methods build in previous researches, which shows that the human pose estimation as whole problem divided into 2D and 3D pose estimation. The 2D pose estimation has two approaches single-person pose estimation and multiperson pose estimation with the different methods [13], such as direct regression and heat map-based approaches. Main goal of single-person approach is to locate the human joint location which is direct or indirectly given in image or video using direct regression method later than heat map approach direct predict the key points existence in that position. About the goal of multiperson pose estimation to locate individual human key points which are uncertain by top-down and bottom-up approaches. Bottom-up approach finds all of the key point first and then groups them to assign human joints key points while top–bottom approach has exactly reverse process it detects first human and then predicts single-person key points. 3D pose estimation [14] finds two approaches: generative and discriminative. Generative model is also known as top-down or model-based approach. This approach processing for restore pose comprises two separates parts, the modeling and the estimation. A probability function in the first stage is built for all dimensions of the question considered such as the descriptor of the image, the structure of human body, camera model, and also the constraints being added at a time. The second estimation part identifies the most hidden pose from the image observation and probability function. Another generative approach category found in the literature [14] is partial which is also known as bottom-up which follows a different path through representation of human skeleton as body part collection with the constraints. Pictorial structure model (PSM) is the most conspicuous example of part-based model. This model mainly used in 2D pose estimation [15, 16, 17]. This PSM model represents human body as series of deformable pieces arranged in configure. It is powerful model of the body that results in effectively inferring the respective part. And second approach is discriminative model which is also referred to as model-free. It is again classified into learning-based and example-based model [14]. In learning-based model learn a mapping function from image observation to pose space, which must be well-generalized from the test set for a new image [18].

A set of instances, in example-based model descriptors with their corresponding pose are stored and the final pose is estimated by the candidates being interpolated obtained from search for similarity [14].

3 Deep Learning-Based Methods Previously Used for Pose Estimation

The first top-down, deep learning method using FLIC dataset was proposed by Those and Szeged [19]; they show DNN-based regression and cascade of pose regressors which is formulated as DNN-based regression problem toward body joints. It achieves high precision pose estimation in holistic manner. But because of fixed input size of 220*220, the network has limited capacity to seem thoroughly. Second limitation during this model is low accuracy in high precision region. They used FLIC dataset which accommodates 4000 training and 1000 test images obtained from popular Hollywood movies and LSP dataset and its extension carries with it 11,000 training and 1000 test images. As evaluation metrics they used percentage of correct parts (PCP). And learning rate is 0.0005. This method is fast and clear, trained in an end-to-end fashion and without much modification, works with the 3D pose detection but it is difficult for mapping and for multiperson this method cannot be applied. So for the answer of this problem heat map method is employed by Jonathan Thompson [20] from New York University. For better accuracy, they used large and wider dataset MPII + 20 K images from YouTube into mulita regress or heat map regress or by increases pooling layers. He [20] proposes a ConvNet architecture which predicts human joint location in RGB images. This architecture implements a sliding window detector that overlaps context to come up with an output of a rough heat map. In general, a sliding window could be a box of rectangle with fixed size. The image slides through the box (Fig. 1).

This figure shows the complete model architecture using FLIC and MPII dataset. They [19] used the quality PCK on FLIC dataset and PCKH for MPII dataset to live the model performance. As compared to first approach it is easy to visualize

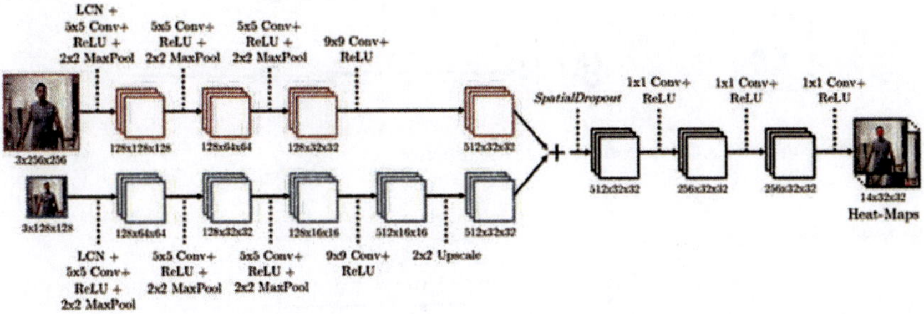

Fig. 1 Torch 7 full model architecture using FLIC and MPII dataset

and apply at complicated cases but it cannot be applied for 3D pose detection and it consumes high memory to produce high-resolution heat map (Tables 1 and 2).

In continuing the next framework [21], the hierarchical feature extractor expressive power is extended to incorporate input and output both spaces. It is a self-corrective model that uses feedback on the prediction of error and which does not predict output in one go. So it is called as iterative error feedback (IEF) that follows up-down feedback from author. Joao Carrera's IEF method demonstrates good novelty and performs very well (Fig. 2 and Table 3).

The Stacked Hourglass [22] is completely unique which repeats bottom-up and top-down processing with in-stating supervising by successive processes of pooling and unsampling. The architecture of the hourglass is intended to gather information of all scales. The network outputs prediction is pixelwise. The network setup features a layer of convolution and max-pooling layer which are used for processing apps.

Table 1 Results on MPII dataset

IEF	Head	Shoulder	Elbow	Wrist	Hip	Knee	Ankle	Body	Body
	96.1	91.9	83.9	77.8	80.9	72.3	64.8	84.5	82.0

Table 2 Results on FLIC dataset

IEF	Head	Shoulder	Elbow	Wrist
	92.6	75.8	57.1	60.4

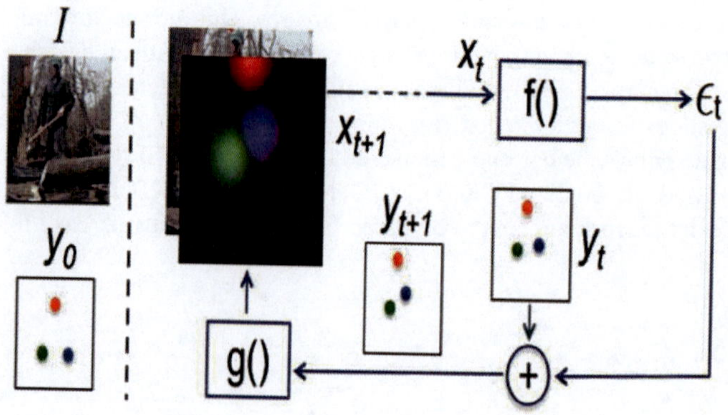

Fig. 2 IEF implements iterative error feedback function

Table 3 Results of IEF method

IEF	Head	Shoulder	Elbow	Wrist	Hip	Knee	Ankle	Body	Body
	95.5	91.6	81.5	72.4	82.7	73.1	66.9	81.9	81.3

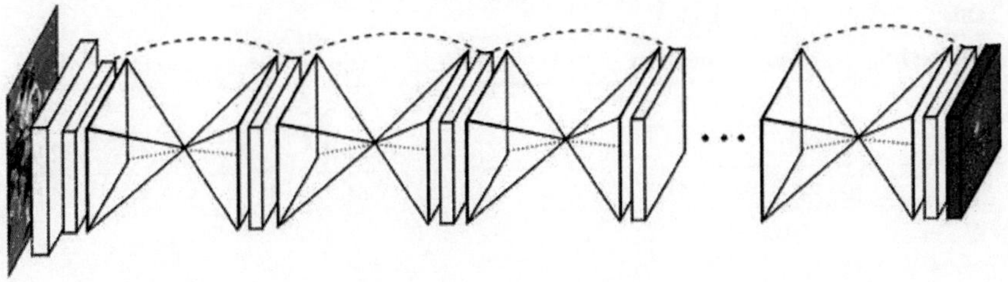

Fig. 3 Multiple stack hour glasses bottom-up top-down inferences

Table 4 Results of stacked hourglass method

Stack hourglass	Head	Shoulder	Elbow	Wrist	Hip	Knee	Ankle	Body
	98.2	96.3	91.2	87.1	90.1	87.4	83.6	90.0

The network productions heat maps predict specific joints occurring at each pixel level. This network was tested on FLIC and MPII and achieves 2% more accuracy in all told joints and 4 to 5% accuracy on ankle and knees. On every scale, the hourglass captures data. In this way, global and local data are fully captured and used by the network to recognize the predictions (Fig. 3 and Table 4).

Convolutional architectures [3] of **Convolution pose machine** displayed a sequential architecture composed of convolution networks which is capable automatically learning a spatial model for pose by communication increasingly between stages. This does not include the use of any inferences on graphical model. This architecture learns image features and image spatial models for structured prediction task. This architecture faces problem with multiple people in a single end-to-end architecture. But for single-person pose estimation with LSP and MPII dataset they got best results. Leeds Sports Pose dataset consists of 11000 images for training and 1000 for testing and used evaluation matric is Percentage Correct Key (PCK) shown in this paper [19]. And the PCKh-0.5 score from the model [19] achieves 87.95 and 78.28% PCKh-0.5 on the ankle implemented with Cafe. This method gives SOTA performance for single-person pose prediction on LSP, FLIC, and MPII dataset. The other method of deep learning is deep cut [23] which used for detecting poses in multiperson images. The model works by detecting the number of individuals in an image and then predicts the joint location of each image. The author [19] adapts Fast-Track for the task in order to get strong part detectors. They alter it in two ways: area size and identification of proposal production. This model is successful in predicting different parts of the body as shown below (Table 5).

As it could be suboptimal to use proposal for body parts identification, the authors use a full convolutionary VGG with a 32-px stage and reduce the stage to 8-px [19]. Then scale the image data to a standing height of 340 px and therefore the best values are obtained. They first try the softmax for the loss function, which brings out the probability of assorted body parts. At the moment they used sigmoid

Table 5 Results of CPM Method

CPMon LSP	Head	Shoulder	Elbow	Wrist	Hip	Knee	Ankle	Auc
	97.2	91.2	83.8	79.3	91.3	87.0	83.1	63.6
	96.1	91.9	83.9	77.8	82.6	75.5	68.4	82.4
	99.3	81.3	79.5	88.5	84.7	86	68.4	86.5

activation mechanism on the neurons within the output and also the lack of cross-entropy. Ultimately they determine that the function of sigmoid activation obtains better results than the function of softmax failure. The model is tested and focused on Leeds Sports Poses and LSP extended in described [24].

Simple baselines for human pose estimation and tracking this paper written by china authors from Microsoft research Asia and also the university of Electronics science and Technology of china. The approach utilizes in this network introduce a few deconvolutionary layers within ResNet architecture in the last convolution point. They pose calculation solution of this paper is contracted on deconvolutionary layers are placed on a ResNet. On a COCO test-dev break, the model reaches a 73.3 mAP and 57.8 in multiple item tracking accuracy (MOTA). The structure makes the generation of heat map from low resolution and in-depth image in no time. Three deconvolutionary layers are employed by choice, with batch normalization and ReLU activation (Fig. 4)

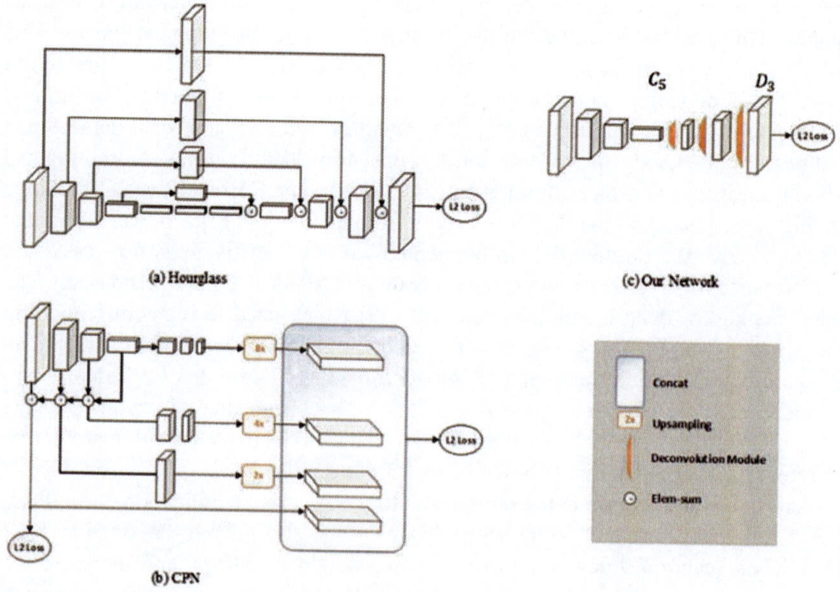

Fig. 4 Proposed flow of the tracking framework

Table 6 Results of **SPEE + STN** method

SPEE + STN	Head	Shoulder	Elbow	Wrist	Hip	Knee	Ankle	Fbody
	92.1	90.5	84.0	76.4	80.3	79.9	72.4	82.1

And results are ResNet-50 with input size 256*192 AP is 70.0 and ResNet-50 with input size 384*288 AP is 72.2.

Regional multiperson pose estimation (RMPE) [25] Proposed by shanghai Jiao Tong University, China and Tencent Youtu. The framework contains three components: symmetric spatial transformer network (SSTN), non-maximum suppression parametric pose (NMS), and pose guided proposals generator (PGPG). This system achieves a 76.6 mAP dataset with the MPII.

In this framework, the human detector bounding boxes are fed into the module "Symmetric STN + SPPE." Then, the pose proposals are automatically generated. To get the approximate human poses, these poses are modified by parametric pose NMS. "Parallel SPPE" is introduced at training to ignore the local least, which significantly outperforms the state-of-the-art multiperson methods [26]. Estimation of human posture throughout terms of accuracy and effectiveness (Table 6).

OpenPose [27] is a real-time, open source, multiperson pose estimation in 2D framework that includes body, bottom, and facemask key points. An approach to identifying 2D human poses in image and video in real time uses a nonparametric representations known as PAF which stands for part affinity field [27]. Some authors are from IEEE. CNN accepts input and predicts confidence map for component association identification of body parts and PAFs by this approach. It also provides the dataset for the foot with 15000 human foot examples. Iteratively the network construction predicts affinity fields encoding map of point-to-point association and confidence detection. OpenPose [27] is attached with related software and API so it is capable of fetching image from various sources such as camera feed, web cam, video, or images. This also supports different architecture like Linux, Mac OS and embedded system and hardware like CUDAGPU, Open CL GPU, and CPU. It also works for crowded scene as real world human pose estimation for this scenario one of the authors [27] form the KIT Fraunhofer Institute of Optonics and Karlsruhe Institute of technology KIT proposes methods for estimating human from crowd pose estimation. In such heavily populated environments, the difficulties of estimating person include individual next to each other, shared occlusion, and limited visibility. There are three blocks in OpenPose: body and bottom identification, arm identification, and body identification. The results shown on human multiperson dataset of the MPII, The COCO key points challenge dataset and the proposed foot dataset of the paper. The results are shown below (Table 7).

This paper suggests methods for estimating human crowds pose estimates. In such densely populated areas, the difficulties of estimating presents include very close person to each other, mutual obstructions, and limited perceptibility. As the ResNet50 network used single pose detector to optimize pose estimation for congested images. It also follows top-down approach that confines each person and then performs

Table 7 Results of OpenPose method

OpenPose	Head	Shoulder	Elbow	Wrist	Hip	Knee	Ankle	mAP
	92.1	91.3	82.3	72.6	70.9	66.8	79.0	79.0

an approximation of a single individual pose for each prediction. Occlusion-Net separations in the previous layers after two moved convolutions so that a combined illustration can be learnt. After one altered convolution, the Occlusion-Net cross-branch splits up. The detection networks for occlusion output two sets of heat maps per pose: one visible key heat map. One heat map for the visible key points, and the other for occluded key points. The results are shown below (Table 8).

From INRIA-CentraleSupelec and Facebook AI Research [28] which aims at mapping all human RGB image pixels to the human body's 3D surface. The DensePose-COCO dataset is also introduced. This is a dataset of 50,000 COCO images that have been manually annotated with image-to-surface correspondence. The authors [28] using this dataset they use to train CNN-based system. Dense correspondence is providing in the presence of variations in background, occlusions, and scale. Correspondence is basically a representation of how the pixels in another image correspond to the images in one image. As an input, a single RGB image is taken in this model to establish a correspondence between surface points and pixel of the image. Combining with the Mask-RCNN method has built up the methodology in this model. The model operates on a GTX 1080 GPU for a 240/320 image at 20–26 frames per second and on a 240/320 image at 4–5 frames per second. To create the DensePose-RCNN system, the authors dense regression framework paired with Mask-RCNN architecture. To assign and coordinates prediction, the model uses a fully convolutionary network dedicated to generate a classification and a regression head.

Those used the same architecture in the Mask-RCNN key point branch. It consists of an 8 alternating stack of 33 completely convolutionary and 512 channel ReLU layers. The authors are conducting tests on a 1.5k image test set containing 23000 humans and a 48000 human training set. The figure below is a comparison of their performance against other methods (Fig. 5)

The paper's writers [29] presents a box-free bottom-up method for estimating pose and segmenting instances for multiperson pictures. This means the authors identify parts of the body first and then organize those parts into individual instances. The average precision of COCO test-dev key point of 0.665 achieved by this approach using single-scale inferences 0.687 using multiscale inferences. The proposed model in this paper is a fully convolutionary box-free. The proposed model in this paper is a fully convolutionary box-free method that first predicts all of the main points in

Table 8 Results of crowd pose method

Crowd pose	AP	AP Easy	AP Med	AP Hard
	65.5	75.2	66.6	72.6

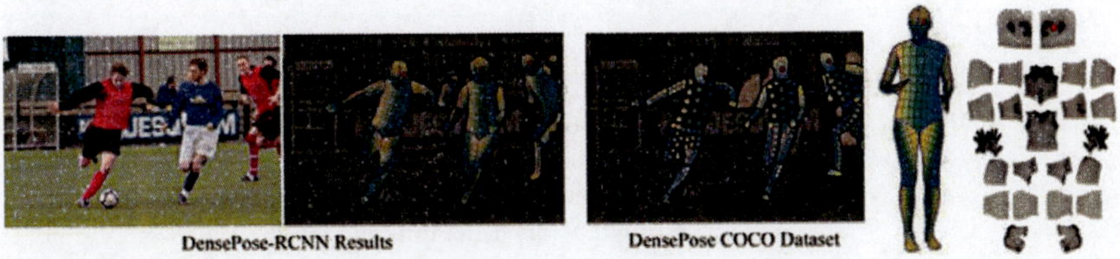

Fig. 5 DensePose method results on RCNN and COCO dataset

Table 9 Results of ResNet framework

ResNet101 (Single scale)	AP	AP 50	AP ?	AP M	AP L	ResNet101	AR	AR 50	AR ?	AR M	AR L
	0.655	0.871	0.714	0.613	0.715	(Single scale)	0.701	0.897	0.757	0.650	0.771
ResNet152 (Single scale)	0.66	0.880	0.720	0.624	0.723	ResNet152 (Single scale)	0.710	0.903	0.766	0.661	0.777
ResNet101 (Multiscale)	0.678	0.886	0.744	0.630	0.748	ResNet101 (Multiscale)	0.745	0.922	0.804	0.686	0.825
ResNet101 (Multiscale)	0.687	0.890	0.754	0.641	0.755	ResNet101 (Multiscale	0.754	0.927	0.812	0.697	0.830

an image for each person. The model is conditioned on the COCO key point data collection. During the key point detection stage, the model identifies visible key point of an individual in the picture. The Person Lab system [29] was accessed for the standard COCO key point's task and COCO instance segmentation. The findings are summarized below (Tables 9 and 10).

4 Dataset Used in 2D Pose Estimation and 3D Pose Estimation

Year	Dataset name	Description	Capacity
2008	Buffy [30]	This dataset contains data from TV show where line segment is used to indicate the position, size, and orientation of body parts	472 edges training 276 edges testing
2010	LSP [31]	This dataset contains only for specific sports category where images are scaled	1000 edges training 1000 edges testing

(continued)

(continued)

Year	Dataset name	Description	Capacity
2011	Shelf [32]	This dataset is more focused on camera orientation and annotated the body joints of four actors using different camera. It achieved 3D ground truth derivation though triangular using three cameras	1 video, 4 subjects, and multiple people
2012	MPII Cooking Activities [33]	A subset focusing on detecting fine-grained activities	44 no. of videos, 12 subjects, and 65 actions
2013	FLIC [34]	FLIC stands for frame labeled in cinema as name insists and it consists of the data from movies of Hollywood	3987 image frames training 1016 image frames testing
2013	KTH Multiview Football II [35]	This dataset is comprised of photographs of professional footballers after an Allsvenskan League Match. It consists of two parts: one with ground truth pose and the other with 2D and 3D ground truth pose	4 Videos 2 Characters and 4 actions
2014	Parse [36]	This is the smallest dataset with different annotation such as facial expression, gaze direction, and gender	10 image frame training 205 image frame testing
2014	MPII Human Pose [37]	This MPII dataset is set benchmark for evaluate of articulated human pose estimation. Data are systematically collected using taxonomy of everyday human activities	410 activities 2.5 ~ 10^4 images
2014	Poses in the wild [38]	In this dataset, data are from three Hollywood movies and generated 30 videos sequences	900 frames and 30 video sequences
2014	MSCOCO [39]	This dataset includes data directly form Internet which contains diverse activities	115 ~ 10^3images training 5 ~ 10^3images validation

(continued)

(continued)

Year	Dataset name	Description	Capacity
2014	Human3.6 M [40]	This is intended to train realistic human sensing system and evaluate various pose in typical human activities such as taking photos, greetings, and eating. It is mainly deployed for real complex environments using correct 3D geometry with synchronized Image and motion capture and depth data	1376 videos 11 Subjects 15 actions, 3.6 × 106 poses
2015	MARCOnI [41]	Multiview, indoor and outdoor, cameras varying in number and types and 3D condition pose estimation task	12 videos
2015	PosePrior Pose-Conditioned Joint Angle Limits- [42]	Prior based on Pose-Conditioned Joint Angle Limits for 3D pose estimation task	N/A
2016	CMU MoCAP [43]	The Mocap lab in Wean basement contains 12 Vicon Infrared MX-40 camera, each capable of capturing 120 Hz with 4-megapixel resolution image. The cameras are places	2605 videos 109 subjects 23 actions
2017	AI	The largest dataset among the all which include direct data from the images. Nowadays Facebook used this dataset	210 ~ 10^3 images training 30 ~ 10^3 images validation 60 ~ 10^3 images testing
2017	Challenger	This dataset gets the videos from MPII Human Pose dataset. It points on three aspects: (1) single-person pose estimation (2) multiperson pose estimation in videos (3) multiperson articulated tracking	514 videos, including 66, 374 frames, 300 videos training 50 videos validation 208 videos testing

(continued)

(continued)

Year	Dataset name	Description	Capacity
2017	[44]	The images for the datasets originate from the Leeds Sports Pose dataset and its extended version, as well as the single person tagged people from the MPII Human Pose dataset. Because we publish several types of annotations for the same images, a clear nomenclature is important: we name the datasets with the prefix "UP" (for Unite the **People**, optionally with an "i" for initial, i.e., not including the Fashion Pose dataset)	UPi-s1h 26,294 images 44.3 GB UP-3D, -P14h 8515 images 46 GB UP-S31 8515 images 1.8 GB UP-P14 8128 images 3.5 GB UP-P91 8128 images 3.5 GB
2015	Pose track [45]	Prior based on Pose-Conditioned Joint Angle Limits for 3D pose estimation task	N/A
2016	Unite the people dataset [46]	The Mocap lab in Wean basement contains 12 Vicon Infrared MX-40 Camera, Each capable of capturing 120 Hz with 4–megapixel resolution image. The cameras are places	2605 videos 109 subjects 23 actions
2017	PosePrior -[42]	The largest dataset among the all which include direct data from the images. Nowadays Facebook used this dataset	210 ~ 10^3 images training 30 ~ 10^3 images validation 60 ~ 10^3 images testing
2017	CMU MoCAP [43]	The images for the datasets originate from the Leeds Sports Pose dataset and its extended version, as well as the single person tagged people from the MPII Human Pose dataset. Because we publish several types of annotations for the same images, a clear nomenclature is important: we name the datasets with the prefix "UP" (for **U**nite the **People**, optionally with an "i" for initial, i.e., not including the Fashion Pose dataset)	UPi-s1h 26,294 images 44.3 GB UP-3D, -P14h 8515 images 46 GB UP-S31 8515 images 1.8 GB UP-P14 8128 images

(continued)

(continued)

Year	Dataset name	Description	Capacity
2017	AI	Dense human pose estimation aims at mapping all human pixels of an RGB image to the 3D surface of the human body. We propose Dense Pose-RCNN, a variant of Mask-RCNN, to densely regress part-specific UV coordinates within every human region at multiple frames per second	50 K COCO images Totaling > 5 million corresponding
2018	Challenger	It found on YouTube covering a broad range of activities and people, e.g., dancing, stand-up comedy, how-to, sports, disk jockeys, performing arts, and dancing sign language signers. Each video has been manually annotated with 2D locations of the upper body joints. BBC Pose recorded from BBC with an overlaid sign language interpreter	50 YouTube videos One hundred frames 20 BBC pose each 0.5–1.5 h in length
2019	[44]	Joint Track Auto (JTA) is a huge dataset for pedestrian pose estimation and tracking in urban scenarios created by exploiting the highly photorealistic video game Grand Theft Auto V developed by Rockstar North	~ 500 K frames ~ 10 M body poses 3D annotation, occlusion annotation

5 Conclusion

In this survey, we presented a comprehensive assessment method for human pose estimation based on deep learning with some features and limitations. Also, the current method for estimating human pose significantly increased, and they are continuously improved for better application in the complex environments. However, some areas need to be addressed such as occlusion and self-occlusion is one of the pose challenges are still remain to solve for estimating human poses. Human Prior are still under improvement before they can produce a satisfactory performance to the full body parts and the discussed dataset is larger in size and contains unbalanced dataset

Table 10 Comparative review of all these above methodology

Sr. no.	Methodology	Characteristics	Datasets
1	Deepose	is a DNN-based regression on the joints for single-person pose estimation	MPII, LSP, and FLIC datasets
2	IEF	It is top-down approach for 2D human pose	MPII and LSP
3	Stacked Hourglass	It follows both bottom-up and top-down	FLIC and MPII Human Pose
4	OpenPose	It is bottom-up approach used for multiperson pose estimation for 2D	MPII (multiperson) dataset
5	Deep cut	It is bottom-up approach used for multiperson pose estimation	Leeds Sports Poses (LSP), LSP extended (LSPET), and MPII Human Pose
6	RMPE(Alphapose)	It is top-down approach and it used three networks to detect pose estimation for mutiperson pose estimation	MPII (multiperson) dataset
7	Convolution pose machine	It is for 2D human pose estimation	MPII, LSP, and FLIC datasets
8	DensePose	It is from INRIA-CentraleSupelec and Facebook AI Research for 3D	DensePose-COCO dataset

and pose distribution is also unbalanced thus this three challenges need to be explore in future work.

References

1. Kim H, Lee S, Lee D, Choi S, Ju J, Myung H (2015) Real-time human pose estimation and gesture recognition from depth images using superpixels and SVM classifier. Sensors 15(6):12410–12427
2. Belghit H, Bellarbi A, Zenati N, Otmane S (2018) Vision-based pose estimation for augmented reality: a comparison study. arXiv preprint arXiv:1806.09316
3. Wei SE, Ramakrishna V, Kanade T, Sheikh Y (2016) Convolutional pose machines. In: Proceedings of the IEEE conference on computer vision and pattern recognition. pp 4724–4732
4. Aguilar WG, Luna MA, Moya JF, Abad V, Parra H, Ruiz H (2017) Pedestrian detection for UAVs using cascade classifiers with meanshift. In: 2017 IEEE 11th international conference on semantic computing (ICSC) January, IEEE, pp 509–514
5. Bowden R, Cox S, Harvey R, Lan Y, Ong EJ, Owen G, Theobald BJ (2013) Recent developments in automated lip-reading. In: Optics and photonics for counterterrorism, crime fighting and defence IX; and optical materials and biomaterials in security and defence systems technology X October, vol 8901. International Society for Optics and Photonics, pp 89010J

6. Thomas GA (2006) Real-time camera pose estimation for augmenting sports scenes
7. Doignon C, Nageotte F, Maurin B, Krupa A (2008) Pose estimation and feature tracking for robot assisted surgery with medical imaging. In: Unifying perspectives in computational and robot vision. Springer, Boston, MA, pp 79–101
8. Oh K (1997) U.S. Patent No. 5,616,078. Washington, DC, U.S. Patent and Trademark Office
9. Okada R, Stenger B (2008) A single camera motion capture system for human-computer interaction. IEICE Trans Inf Syst 91(7):1855–1862
10. Wang Y, Liu Y, Tao L, Xu G (2006) Real-time multi-view face detection and pose estimation in video stream. In: 18th International conference on pattern recognition (ICPR'06) August. vol 4. IEEE, pp 354–357
11. Zhang X, Gao Y (2009) Face recognition across pose: a review. Pattern Recogn 42(11):2876–2896
12. Noroozi F, Kaminska D, Corneanu C, Sapinski T, Escalera S, Anbarjafari G (2018) Survey on emotional body gesture recognition. IEEE Trans Affec Comput
13. Dang Q, Yin J, Wang B, Zheng W (2019) Deep learning based 2d human pose estimation: a survey. Tsinghua Sci. Technol. 24(6):663–676. Chicago
14. Sarafianos N, Boteanu B, Ionescu B, Kakadiaris IA (2016) 3d human pose estimation: a review of the literature and analysis of covariates. Comput Vis Image Underst 152:1–20
15. Felzenszwalb P, Huttenlocher D (2005) Pictorial structures for object recognition. Int J Comput Vision 61(1):55–79
16. Eichner M, Marin-Jimenez M, Zisserman A, Ferrari V (2012) 2D articulated human pose estimation and retrieval in (almost) unconstrained still images. Int J Comput Vision 99(2):190–214
17. Pishchulin L, Andriluka M, Gehler P, Schiele B (2013) Poselet conditioned pictorial structures. In: Proceedings IEEE conference on computer vision and pattern recognition, Portland, Oregon, pp 588–595
18. Huang J-B, Yang M-H (2009) Estimating human pose from occluded images. In: Proceedings 9th Asian conference on computer vision. vol Part I, Springer, Xian, China, pp 48–60
19. Toshev A, Szegedy C (2014) Deeppose: Human pose estimation via deep neural networks. In: Proceedings of the IEEE conference on computer vision and pattern recognition. pp 1653–1660
20. Tompson J, Goroshin R, Jain A, LeCun Y, Bregler C (2015) Efficient object localization using convolutional networks. In: Proceedings of the IEEE conference on computer vision and pattern recognition. pp 648–656
21. Carreira J, Agrawal P, Fragkiadaki K, Malik J (2016) Human pose estimation with iterative error feedback. In: Proceedings of the IEEE conference on computer vision and pattern recognition. pp 4733–4742
22. Newell A, Yang K, Deng J (2016) Stacked hourglass networks for human pose estimation. In European conference on computer vision October, Springer, Cham, pp 483–499
23. Pishchulin L, Insafutdinov E, Tang S, Andres B, Andriluka M, Gehler PV, Schiele B (2016) Deepcut: joint subset partition and labeling for multi person pose estimation. In: Proceedings of the IEEE conference on computer vision and pattern recognition. pp 4929–4937
24. Mwiti D (2019) A 2019 guide to human pose estimation. Heartbeat 5 August 2019. [Online] Available https://heartbeat.fritz.ai/a-2019-guide-to-human-pose-estimation-c10b79b64b73
25. Fang HS, Xie S, Tai YW, Lu C (2017) Rmpe: regional multi-person pose estimation. In: Proceedings of the IEEE international conference on computer vision. pp 2334–2343
26. Fabbri M, Lanzi F, Calderara S, Palazzi A, Vezzani R, Cucchiara R (2018) Learning to detect and track visible and occluded body joints in a virtual world. In: Proceedings of the European conference on computer vision (ECCV). pp 430–446
27. Cao Z, Simon T, Wei SE, Sheikh Y (2017) Realtime multi-person 2d pose estimation using part affinity fields. In: Proceedings of the IEEE conference on computer vision and pattern recognition. pp 7291–7299
28. Alp Güler R, Neverova N, Kokkinos I (2018) Densepose: dense human pose estimation in the wild. In: Proceedings of the IEEE conference on computer vision and pattern recognition. pp 7297-7306

29. Papandreou G, Zhu T, Chen LC, Gidaris S, Tompson J, Murphy K (2018) Personlab: person pose estimation and instance segmentation with a bottom-up, part-based, geometric embedding model. In: Proceedings of the European conference on computer vision (ECCV), pp 269–286
30. Ferrari V, Marin-Jimenez M, Zisserman A (2008) Progressive search space reduction for human pose estimation. In: Vision C, Recognition P (eds) Anchorage. USA, AK, pp 1–8
31. Johnson S, Everingham M (2010) Clustered pose and nonlinear appearance models for human pose estimation. In: Proceedings of the British machine vision conference, Aberystwyth, UK
32. Berclaz J, Fleuret F, Tretken E, Fua P (2011) Multiple object tracking using k-shortest paths optimization. IEEE Trans Pattern Anal Mach Intell 1806–1819
33. Rohrbach M, Amin S, Andriluka M, Schiele B (2012) A database for fine grained activity detection of cooking activities. In: Proceedings IEEE conference on computer vision and pattern recognition, providence. Rhode Island, pp 1194–1201
34. Sapp B, Taskar B (2013) Modec: Multimodal decomposable models for human pose estimation. In: Computer vision and pattern recognition (CVPR), Portland, OR, USA, pp 3674–3681
35. Kazemi V, Burenius M, Azizpour H, Sullivan J (2013) Multiview body part recognition with random forests. In: Proceedings 24th British machine vision conference, Bristol, United Kingdom
36. Antol S, Zitnick CL, Parikh D (2014) Zero-shot learning via visual abstraction. In: European conference on computer vision. Zurich, Switzerland, pp 401–416
37. Andriluka M, Pishchulin L, Gehler P, Schiele B (2014) 2d human pose estimation: new benchmark and state of the art analysis. In: Proceedings of the IEEE conference on computer vision and pattern recognition. Columbus, WI, USA, pp 3686–3693
38. Cherian A, Mairal J, Alahari K, Schmid C (2014) Mixing body-part sequences for human pose estimation. In: Proceedings of the IEEE conference on computer vision and pattern recognition. Columbus, WI, USA, pp 2353–2360
39. Lin T-Y, Maire M, Belongie S, Hays J, Perona P, Ramanan D, Dollar P, Zitnick CL (2014) Microsoft coco: common objects in context. In: European conference on computer vision, Zurich, Switzerland, pp 740–755
40. Ionescu C, Papava D, Olaru V, Sminchisescu C (2014) Human3.6m: large scale datasets and predictive methods for 3D human sensing in natural environments. IEEE Trans Pattern Anal Mach Intell 36(7):1325–1339
41. Elhayek A, de Aguiar E, Jain A, Tompson J, Pishchulin L, Andriluka M, Bregler C, Schiele B, Theobalt C (2015) Efficient ConvNet-based marker-less motion capture in general scenes with a low number of cameras. In: Proceedings IEEE conference on computer vision and pattern recognition. Boston, MA, pp 3810–3818
42. Akhter I, Black M (2015) Pose-conditioned joint angle limits for 3D human pose reconstruction. In: Proceedings IEEE conference on computer vision and pattern recognition, Boston, Massachusetts. pp 1446–1455
43. Rogez G, Schmid C (2016) Mocap-guided data augmentation for 3d pose estimation in the wild. In: Advances in neural information processing systems, pp 3108–3116
44. Wu J, Zheng H, Zhao B, Li Y, Yan B, Liang R, Wang W, Zhou S, Lin G, Fu Y et al. (2017) Ai challenger: a large-scale dataset for going deeper in image understanding. arXiv preprint arXiv:1711.06475
45. Andriluka M, Iqbal U, Milan A, Insafutdinov E, Pishchulin L, Gall J, Schiele B (2017) Posetrack: a benchmark for human pose estimation and tracking. arXiv preprint arXiv:1710.10000
46. Lassner C, Romero J, Kiefel M, Bogo F, Black MJ, Gehler PV (2017) Unite the people: closing the loop between 3d and 2d human representations. In: Proceedings of the IEEE conference on computer vision and pattern recognition, pp 6050–6059

Deepfake: An Overview

Anupama Chadha, Vaibhav Kumar, Sonu Kashyap, and Mayank Gupta

Abstract Recent advancements in digital technologies have significantly increased the quality and capability to produce realistic images and videos using highly advanced computer graphics and AI algorithms due to which it becomes difficult to distinguish between the real media and fake media. These computer-generated images or videos have useful applications in real life; however, these can also lead to various threats related to privacy and security. Deepfake is one of the ways which can lead to these threats. Term "Deepfake" is combination of two terms "deep learning" and "fake." Using deepfake, anyone can replace or mask someone else face on another person's face in an image or a video. Not only this, deepfake can change the original voice and facial expressions also in an image or a video. Nowadays, deepfake uses techniques like deep learning and AI to replace the original face, voice, or expressions. It is very hard for a human to detect that the content has been manipulated by deepfake techniques. This paper has attempted to introduce this concept of deepfake and has also discussed different types of deepfakes. The paper has also discussed methods to create and detect deepfake. The motivation behind this paper is to make the society aware about the deepfake tricks along with the treats offered by it.

Keywords Deepfake · Types of deepfake · Deepfake creation · Deepfake detection

1 Introduction

In the last decade, social media has become so popular that almost every single person is on at least on one or more than one social media platform. As a result, more and more people are uploading images and videos on social media. In a survey, it was found that around 2 billion photos and videos are uploaded on various social media. This tremendous increase in the digital images have led in the development of many images altering techniques and the development of software like Adobe Illustrator,

A. Chadha (✉) · V. Kumar · S. Kashyap · M. Gupta
MRIIRS, Faridabad, Haryana, India
e-mail: anupma.fca@mriu.edu.in

P. K. Singh et al. (eds.), *Proceedings of Second International Conference on Computing, Communications, and Cyber-Security*, Lecture Notes in Networks and Systems 203,
https://doi.org/10.1007/978-981-16-0733-2_39

Adobe Photoshop, etc. As this software is developed, some people have started altering real images to spread fake information. There is a field called digital image forensics research which works effortlessly in detection of these altered images, so that the circulation of the fake content can be reduced. Various attempts have been made to detect a fake image [1, 2], and most of them are done either by analyzing inconsistencies relative to that of a normal camera pipeline or by analyzing image noise [3]. Image noise is considered as a good indicator for detecting the slicing of one image over the other image.

The danger of the fake news being spread is increasing as more than 100 million hours of video content is watched over the Internet daily. The methods have developed and are successful in detecting the altered images, but detecting an altered video is still a challenge.

1.1 Deepfake

Deepfake is the combination of two terms, i.e., "deep learning" and "fake." Using deepfake, anyone can replace or mask someone else face on another person's face in an image or a video. Deepfake uses neural networks that analyze large datasets to learn how to mimic a person's facial expressions, voice, and inflections. A video of two different people is input to the deepfake algorithm, so that it can train on it to swap faces. In simple words, deepfake uses deep learning and AI to study a person's face movement and replaces the face of the person in the video by this face in the video using image mapping. As a result, a new fake video is produced with a replaced face. The first public deepfake incident was recorded in 2017 when on Reddit a user posted a fake video of a celebrity in a sexual act. It is very difficult to detect a deepfake video just by looking at it. This is because people use real footage to make deepfake videos.

2 Types of Deepfakes

2.1 Photo Deepfake

Face and Body Swapping

In this, the changes are done to the face and body by replacing or blending the body and the face with someone else's face or body. The result is a completely different person in the original image. Example of this approach can be seen in many applications using the aging filers. This can be useful for the customers to virtually try clothes, cosmetics, or hairstyles.

2.2 *Audio Deepfake*

Voice Swapping

In this, the voice of the person in the original audio is replaced by the voice of another person by imitating someone else's voice [4]. The approach was used by a fraudster who used AI to mimic the voice of the CEO and tricked the manager into transferring $243,000. It can be useful for adding voices for audiobooks. AI can mimic different voice of male and female in different accents to give listeners a good feel of the book.

Text-to-Speak

In this, the written text is translated into audio by the AI. Text can be translated into the audio of different voices and in different accents [5]. A real-life example is: Someone made some controversial recording in the voice of Jordan B. Peterson, a famous professor. This method can be used for correcting misspoken words in a script of a film without making a new recording.

Video Deepfake

Face-Swapping

In this, the face of the person in the original video is swapped with the face of another person. This is explained in the deepfake creation section where we have discussed how the face of the actress Amy Adams was swapped with the face of the actor Nicolas Cage. This can be used in film production where the face of the stunt man can be replaced with the face of the lead actor.

Face-morphing

In this, the face of one person changes into the face of another person through a seamless transaction. This can be used in video games. The player can upload his picture and the character's face in the game morphs into the face of the player, giving the player feeling of the environment of the game and significantly improving the overall gameplay experience.

2.3 *Audio and Video Deepfake*

Lip-Syncing

In this, not only the face is replaced but also the mouth of the person moves according to the creator's mouth, and the word spoken by the creator is translated in the voice of the victim. The readers can check this in action in "You Won't Believe What Obama Says in This Video!" which was created by Jordan Peele. This can be used in the instructions and ads videos that are needed to be translated into different languages using the same voice and original video (Table 1).

Table 1 Types of deepfake and their applications

Types	Method and description	Example	Application
Photograph Deepfake	**Face and Body Swapping** Changing the face and body of the person by blending	Mobile application using the aging filters	Customers can virtually try dress, cosmetics before buying them
Audio Deepfake	**Voice Swapping** The voice is replaced by someone else voice	It was used to trick the manager by mimicking CEO's voice and transferring $243,000	Mimicking good speakers for audio book
	Text-to-Speak Written text converts into audio	Someone made controversial recording in the voice of Jordan B. Peterson	Can be used in film industry to correct misspoken words
Video Deepfake	**Face-Swapping** Original face is swapped with another face	In the movie "fast and furious" Paul Walker's face was swapped with his brother's face	Swapping actor's face with stuntman, so that the actors can be safe
	Face-morphing Changing the face by morphing through seamless translation	Saturday Night Live star Bill Hader morphs in and out of actor Arnold Schwarzenegger on the talk show	In video games, players can give their face to their avatar
Audio and video deepfake	**Lip-Syncing** Mouth moves according to face and perfect audio is also mimicked	"You Won't Believe What Obama Say In This Video" is a perfect example	Advertisements and instruction videos can be converted to different languages without re-shooting

3 Deepfake Creation

Deepfake videos are so flawless that they can fool anyone. Various tools and applications are used to develop these deepfake videos. These applications mostly use deep learning techniques for developing these videos. The first deepfake video was created using the FakeApp which was developed by Reddit user. To understand it more clearly, let us take an example of this still image from the movie "Man of Steel" where the actress Amy Adams's face is replaced with another actor Nicolas Cage as shown in Fig. 1.

Fig. 1 shows the original image from the movie Man of Steel with actress Amy Adams's face on the left, and on the right is the frame of the deepfake replacing the face with Nicolas Cage. This example is showing how a female face is replaced with a male face. This is how it was done:

1. The region of the image showing Amy Adams face was extracted from the original video.

Fig. 1 Frame from the deepfake clip of "Man of Steel" movie

2. This extracted image is used as an input for the deep learning, a technique of AI that is used to automatically generate a matching image, Nicolas Cage.
3. The generated image is now swapped with the original face inside the original video and hence creating a deepfake video.

4 Deepfake Detection

The deepfake contents are rapidly increasing which is a great threat to privacy, social security, and integrity of the Internet. To prevent these deepfake contents, the detection is a vital containment tool. Over the years, different detection methods and approaches have proposed to tackle deepfakes. Earlier methods and approaches were based on detecting deepfakes by recognizing patterns. Recent methods to detect deepfakes are based on deep learning and artificial intelligence to examine images and videos to find the inconsistencies between the real content and deepfake content.

4.1 Temporal Sequential Analysis

Video manipulation is implemented on a frame by frame basis to make it hard to detect the alteration made on the video. David Guera and Edward Delp [6] described that a deepfake video contains discrepancies between the frames. To detect the temporal inconsistencies between the frames on the deepfake video, they proposed a method using a convolutional long short-term memory (LSTM) structure for processing frame sequences. Long short-term memory (LSTM) networks and convolutional neural networks (CNN) are two major parts of convolutional LSTM. The LSTM

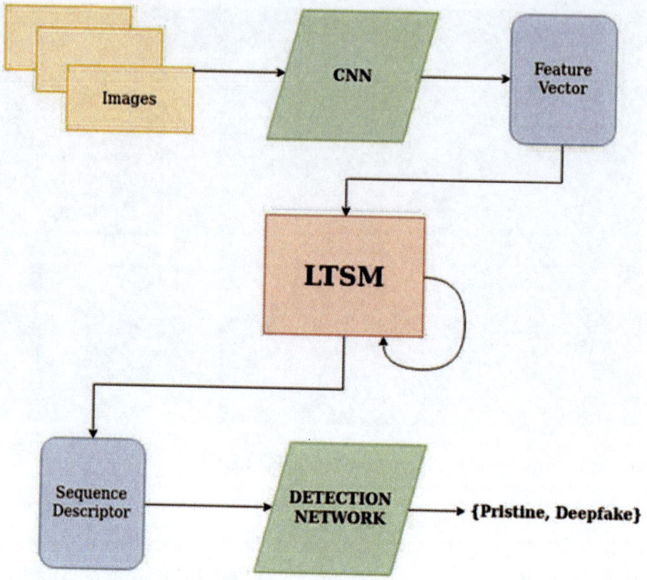

Fig. 2 Deepfake detection using CNN and LTSM

networks are special types of neural network architectures that are capable to analyze longer sequences of data. The LSTM networks can also be considered as improved versions of recurrent neural networks (RNN). In this method, they used convolution neural networks to extract a set of features for every frame and pass them into LSTM networks to produce a temporal sequence description. Then, they use a SoftMax layer to calculate the possibility of the frames being a deepfake as shown in Fig. 2. This method has been experimented with a dataset of 600 videos including 300 deepfake videos from different video hosting Web sites and 300 videos from the HOHA dataset [7]. The results obtained from experiments show promising performance with an accuracy greater than 97%.

4.2 *Eye Blinking Method Based on Physiological Signals*

Another method to detect deepfakes is an eye blinking method that is based on a physiological signal [8] that describes that a frequent blinking of a person's eyes does not well represent in deepfake videos. In other words, the eyes blinking rate of a person in synthesized fake videos are lower than the original videos. To differentiate a deepfake or untampered video, Li and Ming-Ching at [8] first convert a video into frames with the face region to extract eye areas.

The extracted frames are then further prepared for sequence processing. Then, these new eye area sequences are cropped and passed into long-term recurrent convolutional networks (LRCN) which predict the possibility of eyes either open or close. This method has been experimented with a dataset of 49 interview and presentation videos from the Web generated by deepfake algorithms. The obtained result from the experiments using this method shows promising performance.

4.3 Capsule Network

Hinton et al. in 2011 [9] first introduced capsule networks as CNNs have restricted applicability to "Inverse Graphics." Though in the beginning, they encountered a similar problem which was also faced by CNNs, i.e., restricted performance of the system and low accuracy in algorithms.

In 2017–18, further improvements were suggested. Dynamic routing algorithm [10] and its variance and maximum routing algorithm [11] were incorporated to conquer these problems. Due to these algorithms, capsule networks were able to produce better results.

In another field, Iesmantas and Alzbutas [12] used a capsule network to detect breast cancer by using binary classification [12]. Yang et al. used capsule network in the field of text domain [13]. Nguyen et al. [14] were experts in using capsule networks for digital media forensics. These researchers used capsule networks in multiple fields due to which they were motivated to maintain the development of "Capsule-Forensics" to detect fake images or videos that were generated by the computer.

Capsule-forensics overview

Capsule-forensics method is shown in Fig. 3. The first step depends on the input, if the input is video, then the system separates the frames. So, if the task is to identify the computer-generated frames, each frame is divided into small blocks. However, if the task is to identify the fake face or faces, the system crops the facial area using a face detection algorithm. In general, bigger the input results become more precise. Generally used image sizes are 100 × 100, 128 × 128, 256 × 256, and 299 × 299 [14–18] as these are big enough to provide data to detect fake content.

Then, separated frame is passed through the VGG-19 network [19] as it is trained on ImageNet Large-Scale Visual Recognition Challenge (ILSVRC) database [20] before it passes the capsule network.

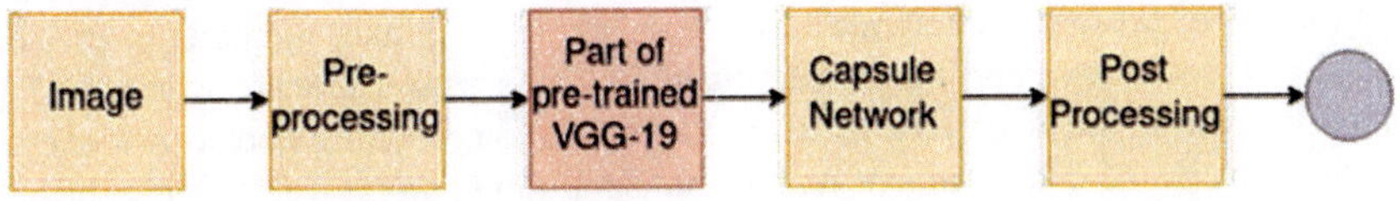

Fig. 3 Capsule-forensics pipeline

In the last step in which the task is to identify the computer-generated images, the score is obtained. However, if the input is video, then the score of all frames becomes average as it becomes the final output.

5 Various Other Deepfake Creation and Detection Methods Proposed

5.1 Digital Media Forensics

Digital media forensics use various kind of analysis and approaches to develop technologies and devices to detect deepfake content. The CNN based [21] is a counter-forensic method used for detecting modified images to ensure the authenticity of a camera image. A physiologically based detection method [22] is a digital forensic technique that finds the difference between a face generated by a computer and an actual human face. The low-level manipulations such as duplicated frames [23] in the video are detected through video-based digital forensics.

5.2 Face-Based Video Manipulation Methods

Face-based manipulation in videos is a most common method in creation of deepfake videos. Face2Face [24] is a novel approach proposed by J. Thies1 which can transfer real-time facial expression. The proposed approach can change face movements in video streams. An alternative approach to Face2Face has also been proposed by H. Averbuch-Elor [25]. Face aging with conditional generative adversarial networks (GAN) [26] proposed by G. Antipov is a method that can change the original person's age in an image by altering facial attributes. Similar methods that can change facial attributes such as skin color also have been proposed [27].

6 Conclusion

The significant improvement in the quality and capability of deep generative networks makes the deepfake content more realistic and flawless. In this paper, an effort has been made to present a comprehensive overview of deepfake, methods to create and detect deepfakes, and various applications of deepfake. Deepfake is used in many interesting ways in various fields like education, creating social awareness or in the world of entertainment and at the same time can be used to breach the security and privacy of people. The mentioned deepfake detection methods have been experimented with datasets of multiple manipulated and original content given

Table 2 Deepfake detection methods

Methods	Techniques used	Applied on	Dataset
Temporal sequential analysis [6]	LSTM and CNN	Videos	A dataset of 600 videos including 300 deepfake videos and 300 videos from HOHA dataset [7]
Eye blinking method based on physiological signals [8]	LRCN	Videos	A dataset of 49 interview and presentation videos from the Web generated by deepfake algorithms
Capsule-forensics [14]	Capsule networks	Videos/images	ImageNet Large-Scale Visual Recognition Challenge [20]

in Table 2, and result obtained from these methods is promising. These promising results have motivated authors for their future work. The future work in this field will be to generate efficient deepfake detection methods which will help to provide more secure social life. Hence, we can conclude that deepfake has its own set of pros and cons and is totally up to us how and where to use it.

References

1. Farid H (2009) A survey of image forgery detection. IEEE Signal Process Mag 26(2):25–26
2. Redi JA, Taktak W, Dugelay J-L (2011) Digital image forensics: a booklet for beginners. Multimedia Tools Appl 51(1):133–162
3. Julliand T, Nozick V, Talbot H (2015) Image noise and digital image forensics. In: 14th International workshop on digital-forensics and watermarking (IWDW 2015). vol 9569. Tokyo, Japan, October 2015 pp 3–17
4. Suwajanakorn S, Seitz SM, Kemelmacher-Shlizerman I (2017) Synthesizing Obama: learning lip sync from audio. ACM Trans Graph 36(4):Article 95
5. Kietzmann J, Lee LW, McCarthy IP, Kietzmann TC (2017) Deepfake: trick or treat. Elsevier 63(2):1–12
6. Guera D, Delp EJ, "Deepfake video detection using recurrent neural networks", 15th IEEE International Conference on Advanced Video and Signal Based Surveillance (AVSS) ,1–6, 2018.
7. Laptev I, Marszalek M, Schmid C, Rozenfeld B (2008) Learning realistic human actions from movies. In: Proceedings of the IEEE conference on computer vision and pattern recognition, 1–8, June
8. Li Y, Chang MC, Lyu S (2018) Exposing AI created fake videos by detecting eye blinking. In: IEEE International workshop on information forensics and security (WIFS), pp 1–7
9. Hinton GE, Krizhevsky A, Wang SD (2011) Transforming auto-encoders. In: International conference on artificial neural networks (ICANN). Springer, pp 44–51
10. Sabour S, Frosst N, Hinton GE (2017) Dynamic routing between capsules. In: Proceedings of conference on neural information processing systems (NIPS), pp 1–11
11. Hinton GE, Sabour S, Frosst N (2018) Matrix capsules with EM routing. In: International conference on learning representations workshop (ICLRW). pp 1–15
12. Iesmantas T, Alzbutas R (2018) Convolutional capsule network for classification of breast cancer histology images. In: International conference image analysis and recognition (ICIAR). Springer, pp 853–860

13. Yang M, Zhao W, Ye J, Lei Z, Zhao Z, Zhang S (2018) Investigating capsule networks with dynamic routing for text classification. In: Conference on empirical methods in natural language processing (EMNLP), pp 3110–3119
14. Nguyen HH, Yamagishi J, Echizen I (2019) Capsule forensics: using capsule networks to detect forged images and videos. In: International conference on acoustics, speech and signal processing (ICASSP), IEEE, pp 2307–2311
15. Rahmouni N, Nozick V, Yamagishi J, Echizen I (2017) Distinguishing computer graphics from natural images using convolution neural networks. In: International workshop on information forensics and security (WIFS), IEEE, pp 1–11
16. Nguyen HH, Tieu N-DT, Nguyen-Son H-Q, Nozick V, Yamagishi J, Echizen I (2018) Modular convolutional neural network for discriminating between computer-generated images and photographic images. In: International conference on availability, reliability and security (ARES). ACM, pp 1–10
17. R¨ossler A, Cozzolino D, Verdoliva L, Riess C, Thies J, Nießner M, (2018) Face forensics: a large-scale video dataset for forgery detection in human faces. pp 1–21. arXiv preprint arXiv: 1803.09179
18. R¨ossler A, Cozzolino D, Verdoliva L, Riess C, Thies J, Nießner M (2019) Face forensics++: learning to detect manipulated facial images. In: International conference on computer vision (ICCV). IEEE, pp 1–11
19. Simonyan K, Zisserman A (2015) Very deep convolutional networks for large-scale image recognition. In: International conference on learning representations (ICLR), pp 1–14
20. Russakovsky O, Deng J, Su H, Krause J, Satheesh S, Ma S, Huang Z, Karpathy A, Khosla A, Bernstein M, Berg AC, Fei-Fei L (2015) ImageNet large scale visual recognition challenge. Int J Comput Vision 115:211–252
21. Guera D, Wang Y, Bondi L, Bestagini P, Tubaro S, Delp EJ (2017) A counter-forensic method for CNN-based camera model identification. In: Proceedings of the IEEE conference on computer vision and pattern recognition workshops, pp 1840–1847
22. Conotter V, Bodnari E, Boato G, Farid H (2014) Physiologically-based detection of computer generated faces in video. In: Proceedings of the IEEE international conference on image processing. pp 248–252
23. Wang W, Farid H (2007) Exposing digital forgeries in interlaced and deinterlaced video. IEEE Trans Inf Forensics Secur 2(3):1–15
24. Thies J, Zollhofer M, Stamminger M, Theobalt C, Nießner M (2016) Face2Face: real-time face capture and reenactment of rgb videos. In: Proceedings of the IEEE conference on computer vision and pattern recognition. pp 2387– 2395
25. Averbuch-Elor H, Cohen-Or D, Kopf J, Cohen MF (2017) Bringing portraits to life. ACM Trans Graph 36(6):196:1–196:13
26. Antipov G, Baccouche M, Dugelay J-L (2017) Face aging with conditional generative adversarial networks. arXiv:1702.01983. 1–5 (2017)
27. Lu Y, Tai Y-W, Tang C-K (2017) Conditional cyclegan for attribute guided face image generation. arXiv:1705.09966 1–16 (2017)

Instinctive and Effective Authorization for Internet of Things

Nidhi Sinha, Meenatchi Sundaram, and Abhijit Sinha

Abstract Internet of Things (IoT) is currently deployed across applications, most of them connected to the Internet or at least connected to a gateway (superior processing capabilities) which is in turn connected to the Internet. The wireless sensor networks (WSNs) refer to a group of spatially dispersed and dedicated sensors for monitoring or recording data and collecting the same in a centralized location. Much research has been done to address the problem of security arising due to concern of authentication, avoidance of DOS attacks, identity hijacking, spoofing, etc. Some even went in depth to address issues related to authentication in a heterogeneous environment, i.e., solves authentication among devices of different make and model deployed in different networks and still trying to connect, addressing multiple authentication or certification (chain of) authorities. However, much less of research has focused on trying to address the true identity of the device. This paper proposes a scheme in post-authentication to explore and validate the identity of the device and later take a decision that needs to be done as necessary for the dynamic authorization phase. Here, we propose the post-authentication using dynamic authorization—Nonce (one-time credential) for a device which is not associated nor owned by system to perform limited use privilege operation on sensitive resource.

Keywords Authentication · M2M · Security · IoT · Identity · Authorization · Nonce

N. Sinha (✉) · M. Sundaram · A. Sinha
Garden City University, Bangalore 560049, India

M. Sundaram
e-mail: meenatchi.s@gardencity.university

P. K. Singh et al. (eds.), *Proceedings of Second International Conference on Computing, Communications, and Cyber-Security*, Lecture Notes in Networks and Systems 203,
https://doi.org/10.1007/978-981-16-0733-2_40

1 Introduction

Internet of Things loosely based on sensors or edge nodes which apart from performing the primary role of monitoring or collecting information from the environment also can connect with a local aggregator or directly to the central information accumulator (server) passing the data [26]. The ability to connect and pass information or the need to discover new nodes and connect has raised concerns at multiple levels of security. The security begins with a question on how to authenticate the device for machine-to-machine communication [11, 16], secure data in transit, secure data at rest, spoofing, etc., and all of these need to be done within the limited scope of a constrained device which is constrained by processing speed, memory, and power management—especially in case it is battery powered or low powered [10, 12]. Much of it is addressed and still pursued as a research topic for further optimization like secure communication and certificate-based trust as described [1, 14, 15]. WSN architecture supports both centralized and distributed IoT management [2, 5, 17]. In centralized managed nodes, the primary authentication challenge is handled mostly by certificate-based trust [1] as the certs installed in the device and managed by a single entity, and hence, once the authenticated identity of the device is established and understood well and more, trust [20] is exerted. Being decentralized nature of an ever-growing ecosystem, it is a necessity to be able to authenticate and have a secure connection for fairly unknown devices (build and make), and different nodes are managed by multiple servers including hosting their identity (different chain of PKI certificates). Much of the work to authenticate is covered in terms of authentication based on the trust model [3, 4]. Key developments are addressed in case if a node can authenticate a node in the same network or case nodes authenticating each other belonging to a heterogeneous network [5, 6, 13].

Despite being authenticated through the existing mechanism of authentication, the challenge posed in the above scenario is twofold, i.e.,

1. Establish the identity of the device which is not physically static (extended authentication)
2. Once identified as a vehicle; authorize it to perform an operation; for example—allow it to pass through the gate; where the authorization can be dynamic: as simple as it can be for one time only.

In paper [7], the proposed idea was a classical take on authentication and authorization; where authorization was handed over to two subcomponents named as policy decision point (PDP) and capability manager (CapM) to handle the authorization initially. Another paper recently published in 2019 [8] shares concerns regarding resource constrained RS components in active constrained environment ACE working on principle of authentication and authorization and actors as clients, resource servers, and authorization host communicating over CoAP. It focuses on bootstrapping of credentials and management of access using token life duration [23]. One paper published (2013) [9] talked about Capability based Access Contro (CAC) for future of IoT devices secured interaction.

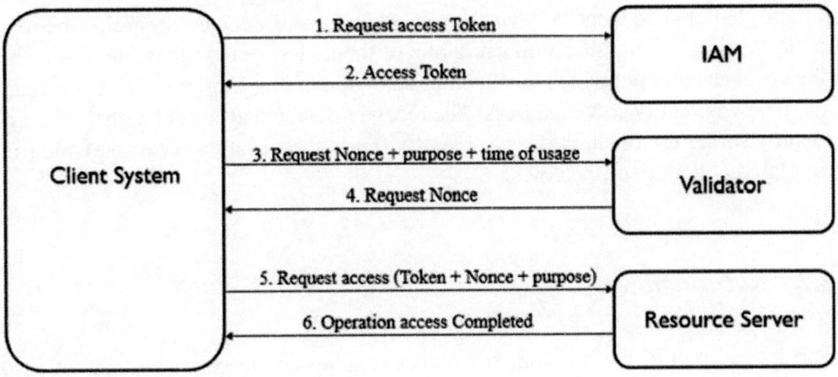

Fig. 1 Basic flow of request and permission

In above papers referenced, there is a scope of improvisation on security aspect by bringing more fine grain granular authorization and adding dynamicity of access which can be auto-granted for one time or limited use decoupled from access token. The proposal mentioned here focuses more on backend servers and communication to grant access to perform sensitive privileged operations. This paper attempts to address on how to provide authentication and authorization [19, 20] for devices which are directly not governed by access granter and relies on identity established by another OEM and is connected only for specific transaction. The device which needs permission is neither owned nor associated with the grant provider but only limited to specific and possibly one-time interaction.

Below figure highlights basic flow of request between client system which has at least an IoT device and client's identity authentication [9] and authorization server establishing identity of its device with system on right that is supposed to grant access to client system to perform operation (Fig. 1).

2 Problem Statement

2.1 Identity for Internet of Things (IoT) Framework

While all problems related to authentication focuses on establishing the authenticity of the device, little is done to establish the true identity of the device. However, the statement is not true for homogenous devices or devices using communication protocol like Universal Plug and Play (UPnP) which has a governing body to define different kinds of devices to maintain identity, versions and quickly understand the capability that each device offers. These are pre-coded in invoking/consuming devices to perform a full-fledged operation on the target device.

The challenge to identify devices goes beyond this. Imagine a scenario where the device initiates communication using one of the understood protocol, and the other device takes the input which must comprise of something similar to a CA certificate and its own identity server address. The receiver device reaches out to the initiator's identity server establishing its own identity (build trust using a certificate) and tries to fetch the identity of the device.

2.2 Authorization

The second part of the problem is whether to allow to perform a certain action or not. Once the identity of the device is established, the part to allow to perform an operation is yet to be addressed certainly. The challenge is posed in the following ways:

1. Devices authenticated; however, the action to be allowed has to be pre-configured as policies based on which decision needs to be taken.
2. Multiple devices are packaged together with the varied identity and authorization access.

Both the above points can be described hereby:

For point 1, the device's identity, generic name, etc., need to be pre-configured, and decision on access needs to be taken based on policies defined or through manual intervention at the need of time. The concern with this approach is multi-fold:

There is an ever-evolving list of devices sharing the same generic identifier like printer, camera, etc., that if granted access, all similar devices sharing the same generic name will have the same access level.

Improvising on above with second level classification like using brand or make or using again some generic identifier can narrow down the access for all devices but still open for many due to conflicting or unintentional impersonating identity based on which policies are defined and inadvertently grant access yet to many.

Further tightening of authorization policies based on absolute identities like a combination of unique device id, brand, make, model, etc., the authorization becomes robust addressing the concern shared in the above two points. However, this comes at a cost of maintenance on the below accounts:

1. The specific policy for each device that is comparable only to user-based access control (UBAC) will outgrow the capacity of any traditional identity and access management solution within a short period stressing on hardware and database, requiring redesign.
2. The management aspect which needs to be done only by humans becomes impossible, as every day numerous such devices need to be identified and added as per policy needed.
3. This in turn adds to a separate security risk. The policies added if not maintained properly as auto-deleted or invalidation, being managed by humans adds to the

complication that this error is bound to happen at a certain point leading to leaked policy and the device gaining access for an indefinite time.

3 Proposed Solution

Summarizing the problem statement as below:

1. Establishing identity for cross-manufactured devices effectively.
2. Assert the authority level of a device with a high level of granularity.
3. Have a maintainable list which can be made effective for only a short period.

3.1 Solution Overview

For the above-mentioned points that need to be addressed, we propose the below architecture. The components of the architecture are Nonce (one-time token), pre-configured IAM servers, and custom policy that can be managed.

Figure 2 highlights the use case as well as the major components involved as example. The use case is all about a vehicle presumably IoT enabled which needs to pass through the entrance of another system (physical). However, to identify and authorize the device, presumably both the IAM need to be configured and exchange the information beforehand the identity of the device. This will be verified once the device is challenged, and it submits its identity.

Figure 3 above highlights the components involved in transaction. The above depiction of solution tries to address problem where an IoT device client is associated with an organization and can be directly authenticated by them; however, the same needs access to perform operation on a different system, where the challenge becomes twofold viz., to establish identity and to allow perform operation for a limited usage. High-level overview of steps involved is

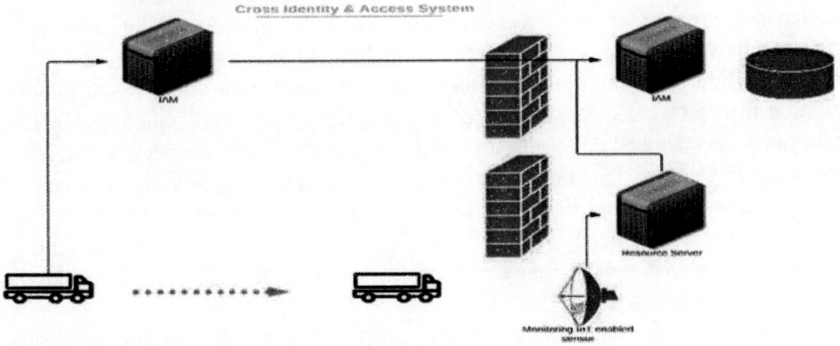

Fig. 2 High-level representational diagram [18]

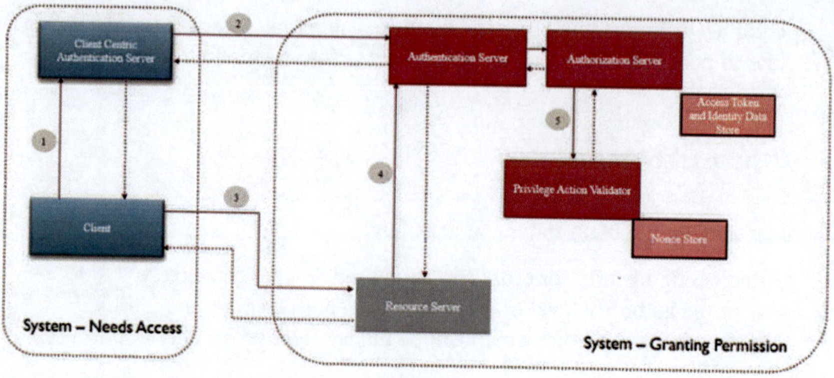

Fig. 3 Component diagram involved in transaction

1. Client requests access to its IAM server for operation.
2. Client IAM server establishes identity of device with target organization using pre-established trust between them.
3. At time of execution, client requests access to perform operation on resource server.
4. Resource server requests validation of access token from authentication and authorization server.
5. In case of sensitive operation as defined in authorization server, privilege action validator is requested to perform validation and grant final access.

The exact sequence of operation is detailed next.

3.2 Description

As shown in Fig. 4, which is all about pre-configuration or setup part which involves the system, we will henceforth call it as consumer system, and which needs its IoT device to be granted access for a certain action, we will henceforth call it as Assessor system. Hereby, both the consumer system and the Assessor system are assumed to be pre-configured to the point that they know and trust each other, much like the relationship that we see between supplier and factory. Point to be noted that is both the systems need not be pre-configured and can be dynamically connected as needed. However, this can be done only if the Assessor system enables inbound requests for general purpose from any IP (much like for public use). However, in the latter case, the consumer system needs to prove its identity via CA signed certificates and or in combination with the pre-registration process which eventually becomes like the earlier case (pre-configured).

The consumer system initiates a request to the Assessor by passing the device unique identity which needs to be authorized later along with details like purpose

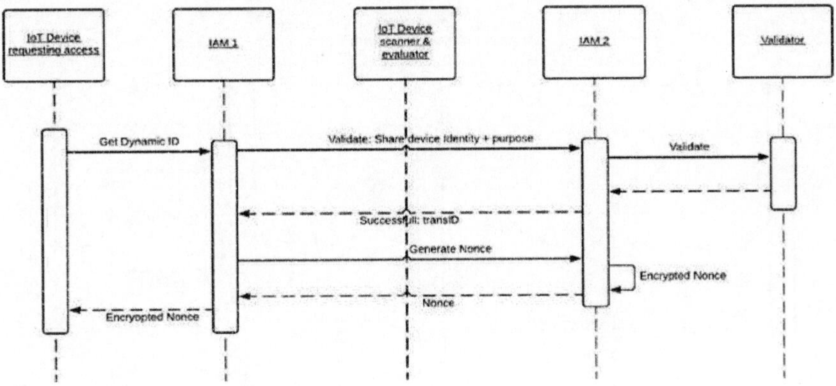

Fig. 4 (Setup) Pre-communication for AUTH

and possibly validity of dynamic identity needed. The Assessor system generates a dynamic identity by validating it. The validation process is where the policies and trust may be involved. For the pre-configured system in a controlled environment where consumer systems is on boarded in a tightly governed process, the validation can be automated easily. However, for public or auto/self—on boarding Assessor systems where the on boarding process is loosely governed, the validation may involve humans to understand and approve the generation of the token. This can still be automated and driven by policies, but it entirely depends on the nature of permission, the impact of a wrong decision, and the criticality of the mission.

Once validation is successful, a transaction ID is generated. With transaction ID—associated with the consumer identity, a unique short-lived Nonce is to be generated and stored with the consumer system. This is necessary as if Nonce needs to be refreshed, the transaction ID can be used to achieve it without the need for any intervention. The nonce in question is encrypted which can be done with a symmetric key, as nobody else other than the generator must be or intended to decrypt and use it. The encryption solves two purposes:

1. If Nonce is based on a random generator, it cannot be guessed.
2. The decryption of Nonce itself solves half of the identity issue.

Below are the properties of Nonce to be implemented/ensured:

1. The Nonce is supposed to be used only for one-time use with defined time-to-live (TTL).
2. Any refresh of Nonce must delete any of the existing ones and then create a new one.
3. During refresh, the validity or TTL can be increased, but none of the attributes like device identity or purpose can be modified.
4. If a nonce is not present in the database as it might have expired or used, it cannot be refreshed, and in that case, transaction ID itself needs to be invalidated to avoid any accidental reuse or leak.

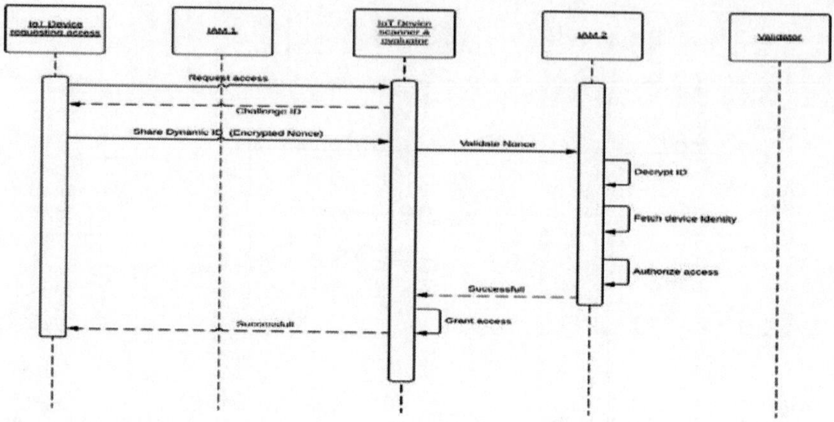

Fig. 5 Authentication and authorization process (in-action)

5. Once nonce is used, i.e., submitted for validation, irrespective of operation successful or not, both Nonce and transaction ID must be deleted from the system and invalidated to avoid further use.

Figure 5 details more about the usage post-preparation which was described above as represented in Fig. 1. What follows here is a simple and fast identity system. When the IoT enabled device needs access to the Assessor system, it simply passes the encrypted Nonce (token) upon challenge along with its identifier. Additionally, it may be enforced by the Assessor system to pass the digital signature to further ensure the static identity which is established as a device that is approved or part of the configured consumer system.

Once the encrypted Nonce is shared, the assessor system performs briefly below steps:

1. Decrypt the token. This is done to ensure that it was provided by the same system and has not tampered. This step is though optional, and it is used to provide a way, so that Nonce itself has not tampered. This step is normally used to avoid storing Nonce in the database for faster validation. However, in this context, it is advised to store as it is supposed to be limited for one-time use only that too within time validity as approved earlier during generation.
2. The token for which the record is fetched and cross-checked with the identity of the device to ensure cross usage due to accidental or intentional pass is avoided as passed in the request.
3. The most important usage is authorization which is done in the same step. This is done to ensure what was approved earlier is exactly to be allowed to be done in need. This is validated against the purpose recorded against the token in the database compared to the purpose of action being asked for by the consumer system IoT device.

If all above is successfully validated, the device is allowed to perform the operation. Point to be noted that is this can be further enhanced by giving it another dynamic token which can be used by the device to perform a series of operations much like a session token. However, for that, additional authentication is not needed, but just authorization for each step is what needs to cross-checked or verified.

4 Implementation and Evaluation

In this section, we walk over different perspective points and analyze how the proposal faired against the goals intends to solve. This section highlights two main points.

4.1 Implementation

To implement our solution, we had three setup configured. One is a Raspberry Pi 4 model b having 4 GB of RAM and 64 bit quad core processor running at 1.4 GHz. The device hosted Ubuntu operating system where application to seek permission was written in Python. The server side is simulated using two Java Spring Boot-based applications running in Docker container wherein one container acting as client-side authentication server to which the device is permanently associated with and second a single Docker container hosting server based on our proposed solution to which the client device is not associated and still needs to grant permission to device. The proposed solution was written in Java, and DB reference used was PostgreSQL. The base machine on which both Docker containers were running was having Intel i5 10th generation i5-10210U 4 core processor with base processing speed as 1.60 GHz and 16 GB of DDR4 RAM of speed 2666 MHz setup on Intel motherboard hosting Ubuntu 20.04. Client reference Docker instance was configured with 1 core and 1 GB of RAM. The Docker container hosting our solution-based authentication and authorization server Docker container was provided two cores and 2 GB of RAM sufficient enough to carry out operation and still roughly leaving more than approx. One GB of RAM free at any given time. Both Docker containers had base image of CentOS.

4.2 Analysis

Security Analysis. We performed threat risk modeling and analysis based on OWASP security [22, 25] top threats to system based on which evaluation was performed.

1. For a DOS or DDOS attack where the system can be brought down just by bombarding legitimate or illegitimate requests, it will only bring the system

down but not compromise on data or access of operation. The proposal in this paper assumes to use the best of security approaches for system implementation like using TLS v1.2 or better and not SSL for securing communication over network or have gateway and smart firewall [21, 24] configured to detect such attacks causing spike and bar the IP source once learnt about the attack without even letting the request reach authentication or authorization server to reduce drain.

2. Another aspect of session riding wherein the same request is mimicked—cloning the data over the network using man in the middle attack and replayed which is well handled by the system as the Nonce only used once and discarded can never be reused and access to such operation even using the exact same payload that will always result in forbidden operation response.
3. Critical to this is one very edge case scenario where the original request even before reaching the server and assuming the token and Nonce being leaked due to source system being compromised can allow the attacker targeting our system with hosted proposed solution to perform operation the way it wants. However, in this case as well, if the purpose in operation or identity while performing operation or if the time mismatch happens (example: the attacker wants to perform the operation too early or too late then the designated operation time) will result in denial as the data passed along with token and nonce must match with the data stored with our target system and found in case of such operation raises suspicion if the purpose of usage changed or being done by some other entity camouflaged as originator hence unauthorized.

Based on the discussion above, we see that the proposed system has better handling of security and still allow the legitimate requester to carry out operation. This solution will also deny operation to legitimate requester if done in wrongful means like at some other time instance or for some other purpose or even deny using it repeatedly.

Communication Overhead. As we see over multiple figures and explanation in main solution section, the core communication between IoT device and its server for securing access token or using the access token or Nonce requesting the resource server for performing operation, there is very marginal additional overhead respective to communication bytes (as it is assumed to use existing communication protocol like CoAP or HTTPS) or number of communication to perform as it still boils down to just two request to perform as any existing standard for authentication and authorization. The communication bytes which get added are due to addition of purpose while requesting access to perform operation which should match with the one that was used to generate nonce. However, due to additional verification introduced at server side, it adds latency which is needed to decrypt and lookup in database. Since it is done at server side, it can scale where we can have more control to ensure minimal latency SLA. The scale can be done horizontally to address faster turnaround and achieve higher throughput while handling very high amount of traffic. In our setup, we found the latency on an average of ~115 ms. In comparison, the overall authentication request latency was ~100 ms, without having Nonce validation, implying an increase of just ~15 ms.

5 Conclusion

This proposal intends to solve the problem of authorization and authentication both in an effective manner without compromising on speed while at the same time improves effective yet simple authorization mechanism. The solution proposed solves an issue of pre-configured or known devices, as this can be ever-expanding as much agreed by involved parties and does not need a central governing agency. Below are the pros of this proposal that we would like to cover:

5.1 Pros of This Proposal

1. Below are the benefits that this system provides vis—a—vis other solutions
2. Solves the authentication and authorization problem all in one solution.
3. The speed of processing is not compromised as multiple hops are not needed. Given some more processing capability, this solution can be further enhanced to IoT device, where the single IoT device can itself act as a complete Assessor system pre-configured to allow for a Nonce, and validation of Nonce through backend support system can be avoided.
4. This solution plugs the gap of intended usage and actual usage difference, where the purpose is validated before use.
5. Since Nonce is used, which is short-lived, a cynical issue like session riding or reuse is avoided.
6. In case of Nonce, itself seems to be compromised by the consumer system or lost or needs to be refreshed to extend the usage that can be done, and hence, additional approval is avoided.
7. The solution can be extended as mentioned earlier to have dynamic registration of the consumer system with the Assessor system for certain cases to make the process automated.
8. The validation process needed for Nonce generation can also be automated depending upon the criticality of access needed or as allowed by the process.
9. The devices and access requests need not be managed by the central governing body, rather it encourages the dynamic addition of devices as deemed fit by involved parties.

5.2 Future Work

This system though shortens the gap of machine-to-machine (M2M) authorization along with authentication; however, this is not yet complete and needs further enhancement to overcome certain aspects that it does not address good enough. Below are the points that need further work.

It assumes problems to be solved for larger entities like multiple companies that are involved and need to exchange. However, it does not solve the issue of a single random IoT entity that needs certain action to be done for on Assessor system. Though the solution can be extended to solve that, however, as of today, it is not scoped in solution nor verified to be fool proof enough.

The solution assumes that both the IoT devices communicate in common language or protocol. It does not address what if both of them use non-interoperable communication language.

Though the system designed was supposed to have minimal backend intervention for faster authentication and authorization process, however, some setup and pre-configuration using backend are enforced. If the IoT device of Assessor is pre-known and has less traffic (fewer data to store), the token itself can be stored in access IoT device with the right validation logic to avoid the round trip. However, this solution currently is not completely backend free and despite comparatively lighter, it needs work to make it truly lightweight. Furthermore, suppose the technologies like machine learning and artificial intelligence are integrated in the future. In that case, it will bring more maturity in the above-proposed scenario at the time of authorization (to make them SMART).

References

1. Cooper D, Santesson S, Farrell S, Boeyen S, Housley R, Polk W (2008) Internet X.509 public key infrastructure certificate and certificate revocation list (CRL) profile RFC 5280
2. Gubbi J, Buyya R, Marusic S, Palaniswami M (2013) Internet of things (IoT): a vision architectural elements and future directions. Futur Gener Comput Syst 29(7):1645–1660
3. Pacheco J, Hariri S (2016) IoT Security framework for smart cyber infrastructures. In: IEEE 1st International workshops on foundations and applications of self* systems (FAS*W). pp 242–247
4. Olesia V, Leonid Kupershtein, Olga Shulyatitska, Viktor Malyushytskyy, The authentication method in wireless sensor networks based on trust model. In: IEEE first Ukraine conference on electrical and computer engineering (UKRCON), pp 993–997
5. Porambage P, Schmitt C, Gurtov A, Gerdes S (2014) PAuthKey: A pervasive authentication protocol and key establishment scheme for wireless sensor networks in distributed IoT applications. Int J Distrib Sens Netw (357430)
6. Kim H (2017) Securing the internet of things via locally centralized, globally distributed authentication and authorization. In: EECS Department, University of California, Berkeley, Technical Report No. UCB/EECS-2017–139
7. Hernández-Ramos J, Pawlowski M, Jara AJ, Skarmeta A, Ladid L (2015) Toward a lightweight authentication and authorization framework for smart objects. IEEE J Sel Areas Commun 33:690–702
8. Echeverría S, Lewis GA, Klinedinst D, Seitz L (2019) Authentication and authorization for IoT devices in disadvantaged environments. In: IEEE 5th World forum on internet of things (WF-IoT), Limerick, Ireland, pp 368–373
9. Mahalle PN, Anggorojati B, Prasad NR, Prasad R (2013) Identity authentication and capability-based access control (IACAC) for the internet of things. J Cyber Secur Mobil 1(4):309–348
10. IEEE (2011) 802.15.4–2011 IEEE Standard for local and metropolitan area networks—Part 15.4: low-rate wireless personal area networks (LR-WPANs), pp 1–314

11. ZigBee Specification Version 1.0, ZigBee Alliance, https://www.zigbee.org/home.aspx(2008)
12. Kushalnagar N, Montenegro G, Schumacher C (2007) IPv6 over low-power wireless personal area networks (6LoWPANs): overview assumptions problem statement and Go. RFC 4919
13. Shin S, Shon T, Yeh H, Kim K (2014) An effective authentication mechanism for ubiquitous collaboration in heterogeneous computing environment. Peer-To-Peer Netw Appl 7(4):612–619
14. Liu Y, Li J, Guizani M (2012) PKC based broadcast authentication using signature amortization for WSNs. IEEE Trans Wireless Commun 11(6):2106–2115
15. Rescorla E, Modadugu N (2006) Datagram transport layer security. In: IETF RFC 4347
16. Lu R, Li X, Liang X, Shen X, Lin X (2011) GRS: the green, reliability, and security of emerging machine to machine communications. IEEE Commun Mag 49(4):28–35
17. Li CT, Hwang MS, Chu YP (2009) An efficient sensor-to sensor authenticated path-key establishment scheme for secure communications in wireless sensor networks. Int J Innov Comput Info Control 5(8):2107–2124
18. Icon of truck used in figure 2 made by Freepik from www.flaticon.com.
19. Trnka M, Cerny T, Stickney N (2018) Survey of authentication and authorization for the internet of things. Hindawi Secur Commun Netw **2018**(ID 4351603):1–17
20. Kim H, Lee EA (2017) Authentication and authorization for the internet of things. IT Professional 19(5):27–33
21. Moosavi SR, Gia TN, Rahmani AM, Nigussie E, Virtanen S, Isoaho J, Tenhunen J (2015) SEA: a secure and efficient authentication and authorization architecture for iot-based healthcare using smart gateways procedia computer science. pp 452–459
22. Humayed A, Lin J, Li F, Luo B (2017) Cyber-physical systems security—a survey. IEEE Internet Things J 4(6):1802–1831
23. Lee S-H, Huang K-W, Yang C-S (2017) TBAS: token-based authorization service architecture in internet of things scenarios. Int J Distrib Sens Netw 13
24. Tanwar S, Tyagi S, Kumar N (2019) Multimedia big data computing for IoT applications: concepts, paradigms and solutions. In: Intelligent systems reference library, Springer Nature Singapore Pte Ltd., Singapore, pp 1–425
25. Singh PK, Pawłowski W, Tanwar S, Kumar N, Rodrigues JJ, Obaidat MS (Eds) In: Proceedings of first international conference on computing, communications, and cyber-security (IC4S 2019). vol 121. Springer
26. Singh PK, Kar AK, Singh Y, Kolekar MH, Tanwar S (Eds) In: Proceedings of ICRIC 2019: recent innovations in computing. vol 597. Springer

Methodical Analysis and Prediction of COVID-19 Cases of China and SAARC Countries

Sarika Agarwal and Himani Bansal

Abstract COVID-19 pandemic has become a major challenge for all the countries of the world. No medicine has been developed till now to cure it. Coronavirus (COVID-19) is the family of viruses that causes illness and has symptoms like the common cold, influenza, and severe acute respiratory syndrome (SARS) that spread via breathing droplets. Proper analysis and prediction of the COVID-19 patients and its increasing rate of spread will help the government and people to mitigate its effect. This gives a reason to analyze, compare, and predict the cases in India, China, and SAARC countries to make early decision for taking preventive measures to combat its effects in a timely manner. In this paper, we have analyzed COVID-19 cases from January 21, 2020 to June 25, 2020 and have predicted the cases of COVID-19 for the period of next two weeks using multiple linear regression and polynomial regression models of machine learning.

Keywords COVID-19 · Linear regression · Polynomial regression · Coronavirus · Prediction · Machine learning

1 Introduction

Corona means crown. Coronavirus has a crown-like structure which is known to be initiated from the animal and transmitted to a human. This virus is new to the human immune system to fight. This virus can stick to almost any substance and is one-nine hundredth of a width of a hair in size. As of today, coronavirus disease (COVID-19) has spread in almost all the countries. More than 215 countries and 5,607,791 people are affected by coronavirus disease as on May 26, 2020 [1].

S. Agarwal (✉) · H. Bansal
Department of CSE/IT, Jaypee Institute of Information Technology, Noida, India
e-mail: sarikagarwal.it@gmail.com

H. Bansal
e-mail: singal.himani@gmail.com

P. K. Singh et al. (eds.), *Proceedings of Second International Conference on Computing, Communications, and Cyber-Security*, Lecture Notes in Networks and Systems 203,
https://doi.org/10.1007/978-981-16-0733-2_41

Pneumonia without any cause was first identified in Wuhan, China, and the same was reported to the World Health Organization (WHO) office in China on December 31, 2019. This disease has spread to every province of mainland China and other countries also. In January 2020, the UK and Russia found two corona positive cases. In January, only one case was found in Sweden and Spain. Canada reported 4 cases. After analyzing data about symptoms of patients, it was declared as a health emergency. Then, WHO named this disease as coronavirus disease (COVID-19). COVID-19 is an abbreviated term, where CO denotes Corona, VI denotes virus, D denotes disease, and 19 refer to the year it was discovered. The symptoms of COVID-19 disease are the same as normal viral diseases like fever, tiredness, loss of smell, loss of taste, and dryness. Less commonly seen symptoms are diarrhea, conjunctivitis, a rash on skin, shortness of breath, and loss of speech. The symptoms differ in age group. It can cause more severe health issues to those whose age is above 60 [2]. If the patients are already suffering from diabetes, lung disease, and heart disease then it may lead to death. The transmission rate of this virus is indicated by the reproductive number (R_0—R-nought). R_0 tells how many people are infected from one person with that disease. WHO estimated R_0 for COVID cases between 1.4 and 2.5 on January 23, 2020 [3]. It means on an average, one person transmits COVID to 1–3 people. The total number of confirmed cases in January was 8096 worldwide. The confirmed infected cases increased exponentially up to 7,102,957 and deaths were 406,343 on June 8, 2020. Total persons recovered were 3,466,581. Figure 1 shows the exponential growth of world's confirmed, death, and recovered COVID cases data from January 22, 2020 to June 20, 2020. Recovery cases are almost 40 lakhs out of 80 lakhs cases on June 20, 2020 which shows that recovery rate is more than 50% throughout the world.

The person who gets recovered from corona develops antibodies in their body that fights against the coronavirus. Antibodies may help people to move to work without being scared and can protect that person from the re-infection of COVID-19. But no evidence has been found till now. Scientists are not sure as how long these antibodies will protect against coronavirus. Immunity against corona disease is

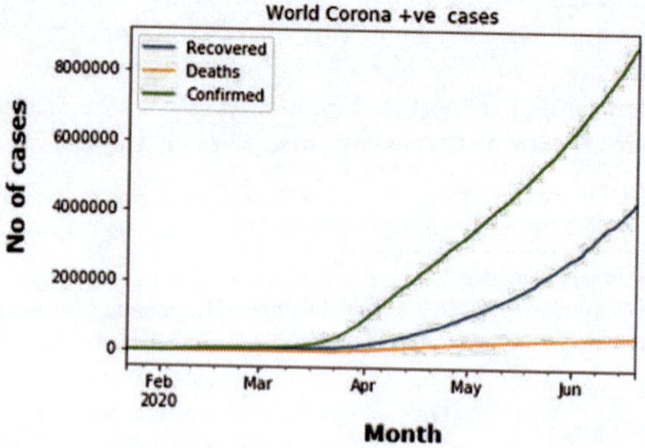

Fig. 1 World data statistics on COVID-19 from January 22, 2020 to June 20, 2020

developed in one or two weeks. The body starts fighting against viral without delay which is very natural. Three cells, namely macrophages, neutrophils, and dendritic, help to diminish the progress of the virus and prevent it from causing symptoms [4]. Analysis and prediction of such disease are necessary to combat its effect in a timely manner.

This paper is divided into six sections. Section 2 gives the motivation behind taking up this study and our contribution in this study. In Sect. 3, comparative analytical study of COVID-19 in China, India, and other SAARC countries is done. Section 4 has results of methodical prediction of COVID confirmed, recovered and death cases. Section 5 details the comparative study and Sect. 6 marks the conclusion of the paper.

2 Motivation and Contribution

Novel coronavirus was first found in China and spread in almost all the countries. A number of COVID-19 cases are increasing day by day and we have limited sources which has become a problem for the government to provide the medical aid to all the infected persons. Early prediction of COVID-19 cases might be helpful for making necessary arrangements. This generated the need to analyze, compare, and predict the coronavirus cases in India, China, and SAARC countries using the multiple linear regression and polynomial regression models of machine learning.

With this aim, the authors have used a dataset from Kaggle [5] that contains date-wise count of confirmed, death, and recovered cases of COVID-19 from different countries. The dataset contains datewise number of cases found, country of origin of the patient, exact count of confirmed, death, and recovered cases from January 22, 2020 to June 20, 2020. With this dataset, we got hold on the current situation of COVID-19 in the world and did comparative study of cases in China and SAARC countries. We then predicted COVID-19 cases in India, China, and SAARC countries from June 26, 2020 to July 10, 2020. South Asian Association for Regional Cooperation (SAARC) has Afghanistan, Bangladesh, Bhutan, India, Maldives, Nepal, Pakistan, and Sri Lanka as its member countries.

3 Comparison of COVID-19 Among India, China, and Other SAARC Countries

China was the first country to report the case of COVID-19 from Wuhan seafood wholesale market. It was reported that COVID-19 was transmitted from animals [6]. Though it was transmitted from animals, it was concluded that it can spread from human to human. The first COVID case in China was reported on November 20, 2019. The Chinese government did not support the doctors. They even criticized who were willing to intimate others the seriousness of the new SARS Virus in the city of Wuhan. After one month, in December 2019, 60 confirmed cases of SARS disease were found in Wuhan. The cases kept multiplying. On December 30, 2019, Health Commission in Wuhan asked local hospitals in Wuhan to report all the information about cases of pneumonia of unclear cause in the past week [7]. Dr. Li Wenliang also warned fellow doctors about the seriousness of the disease and advise them to wear protective clothing to avoid infection. Figure 2 shows the confirmed, recovered, and death cases of COVID-19 in China (till June 20, 2020).

We can easily predict from the graph that China has controlled the COVID-19 cases. Initially, the COVID-19 cases were rising simultaneously, the recovery rate was also increasing. By February 2020, China was able to control the increasing rate of COVID-19 and reduced the new cases by more than 90% [8]. Since, China did not stop the people to enter or move from the country; on January 13, 2020, the first case was confirmed outside China in Thailand. India also reported its first case on January 30, 2020. The patients had the travel history of China [9]. By June 11, 2020, India had total 286,579 confirmed cases. Out of them, 141,029 patients recovered (including 1 migration) while 8102 died. India currently has the largest number of confirmed cases in Asia. Figure 3 shows confirmed, death, and recovered cases in India till June 20, 2020. The graph shows that initially transmission rate was slow, but from the middle of April 2020, it is multiplying very fast.

All the SAARC countries (Afghanistan, Nepal, Bhutan, Sri Lanka, Bangladesh, India, Maldives, and Pakistan) have COVID-19 patients. Figure 4 shows India has the largest number of confirmed, recovered, and death cases in all the SAARC countries followed by Pakistan and Bangladesh. India has the highest population among all the SAARC countries, that may be one of the reasons for the fast transmission of COVID-19. Recovery cases are also fast in India followed by Pakistan and Bangladesh.

The death rate is 2.5% which means that patients are recovering from the disease. The effect of COVID-19 is mild in SAARC countries as compared with China. Figure 5 represents share of active, recovered, and dead patients in percentage across SAARC countries.

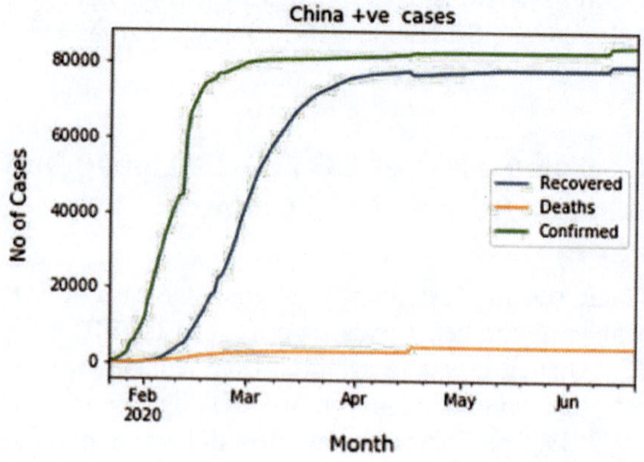

Fig. 2 Confirmed, recovered, and death cases of COVID-19 in China (till June 20, 2020)

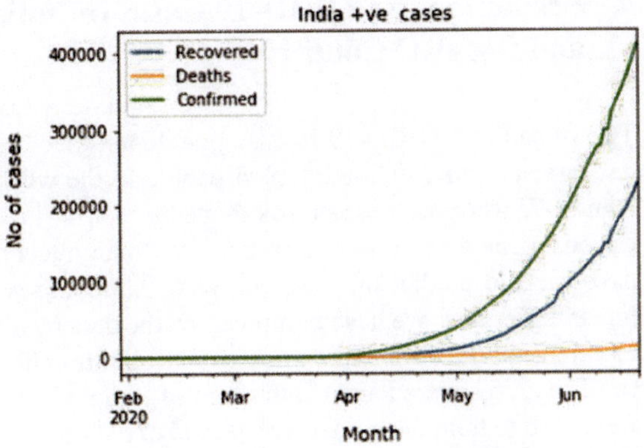

Fig. 3 Confirmed, recovered, and death cases of COVID-19 in India (till June 20, 2020)

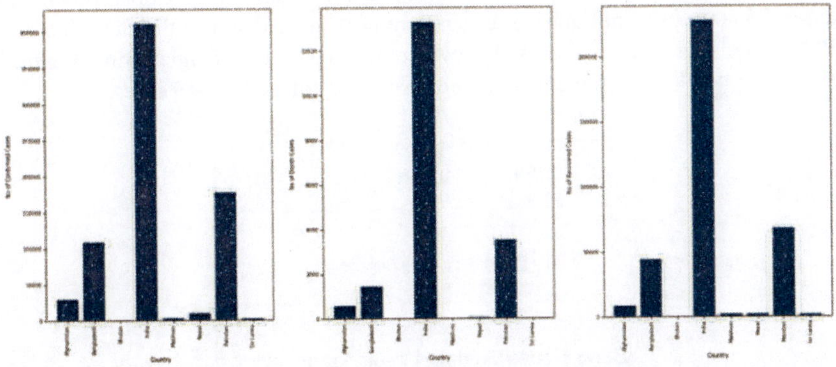

Fig. 4 COVID-19 confirmed, recovered, and death cases of SAARC countries till June 20, 2020

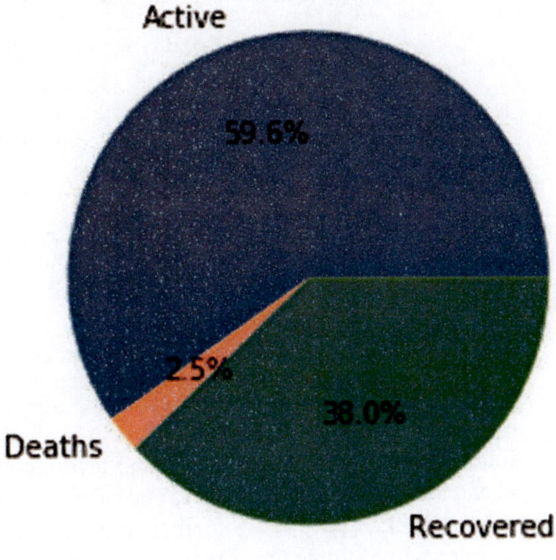

Fig. 5 Share of active, recovered, and dead patients in SAARC countries

4 Prediction of COVID-19 Cases in India, China and SAARC Countries

The spread of COVID-19 is unstoppable and has been declared as a pandemic. It has infected more than 8,007,804 people in the world by June 15, 2020 and more than 50% were recovered throughout the world. Prediction of confirmed cases of Corona virus disease were also done by many techniques like ARIMA [15]. We have tried to predict the cases in SAARC countries till July 20 through multiple linear regression. We have preprocessed the data by dividing the date column in day, month, year and eliminating the extra information like gender and symptoms. Date, confirmed, recovered and death cases of India, China, and other SAARC countries were used to train the model. The period of data considered is from January 22, 2020 to June 12, 2020. We have focused only on SAARC countries and China COVID-19 cases. We have used multiple linear regression to train and predict the COVID cases. Polynomial regression is also used to overcome the problem of underfitting. Figure 5 shows the flow of our implementation of prediction of COVID-19 cases.

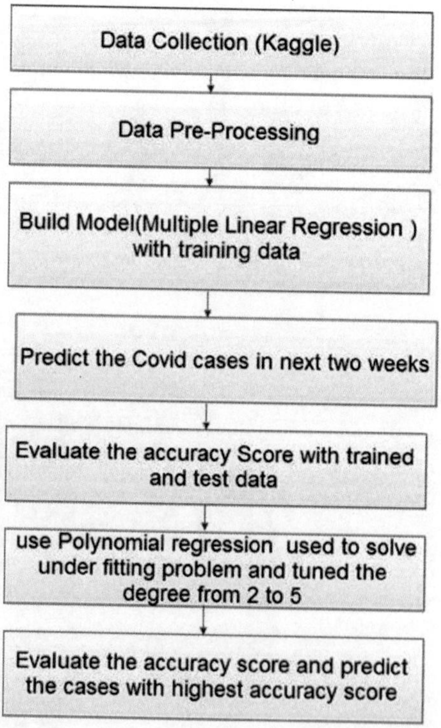

Multiple linear regression is a machine learning algorithm used to predict the output and take more than one feature. The calculation for multiple linear regression is shown in Eq. 1. β_0, β_1, β_n are coefficient, x_1, x_2, x_n are independent variable and $\in$ is an intercept. Multiple linear regression is used to find the relationship between the dependent and independent variables.

$$Y_i = \beta_0 + \beta_1 x_1 + \beta_2 x_2 + \ldots + \beta_n x_n + \in \tag{1}$$

Polynomial regression is also a kind of regression that is used to overcome the problem of underfitting. Underfitting is a situation when our model does not learn enough from training data resulting in unreliable prediction. To overcome underfitting, we increased the features of the model. The model in Eq. 1 is transformed into Eq. 2.

$$Y_i = \beta_0 + \beta_1 x_1 + \beta_2 x_2^2 + \beta_3 x_3^3 + \ldots + \beta_n x_n^n \tag{2}$$

Polynomial regression with degree 1 is same as linear regression. To increase the features of our model, we have to tune the degree parameter to the extent that gives maximum accuracy and minimum variance between the training and testing model. Selecting a degree is a challenging task [10]. If the degree of the polynomial is less, it will lead to the problem of underfitting, i.e., not be able to fit the model properly. If the value of the degree of the polynomial is greater than actual, it will lead to the problem of overfitting. Overfitting problem can be solved through regularization.

We have used multiple linear regression for predicting the confirmed, recovered, and death cases of COVID-19 in SAARC countries and China. The accuracy score for the trained model came to be 0.11623268358995764 and the accuracy score of the test model came as 0.0938602714826949. This shows that accuracy is not as desired. Since the model trained by us was showing the underfitting problem, polynomial regression was also used. Table 1 gives the predicted datewise (daily) confirmed cases of SAARC countries and China from June 26, 2020 to July 10, 2020 obtained from polynomial regression at degree 2.

Polynomial regression of degree 2, 3, and 4 was applied for the prediction made. Table 2 shows the training and testing accuracy score at different degrees of polynomial regression.

As degree is increasing, the variance is also increasing between training and testing models. By analyzing Table 2, polynomial regression at degree 2 and 3 is good to predict the confirmed cases.

5 Results

The COVID-19 virus originated from China and spread almost all over the world. The first COVID case in SAARC countries was on January 23, 2020, in Nepal. After that, it was reported across all the SAARC countries.

Figure 6 shows exponential growth of COVID cases in all the SAARC countries till June 20, 2020. All the countries are trying hard to control COVID. Preventive measures like lockdown, social distancing, and regular hand wash have been taken. In spite of all these, our prediction shows that there will be an exponential growth

Table 1 Two weeks prediction of COVID-19 patients in SAARC countries and China

Date	Afghanistan	Nepal	Bhutan	Srilanka	Bangladesh	India	Maldives	Pakistan	China
26/6/20	17,516	20,009	22,501	24,994	27,486	29,978	32,471	34,963	3336
27/6/20	17,770	20,263	22,755	25,247	27,740	30,232	32,725	35,217	3346
28/6/20	18,024	20,517	23,009	25,501	27,994	30,486	32,979	35,471	3356
29/6/20	18,278	20,770	23,263	25,755	28,248	30,740	33,233	35,725	3365
30/6/20	18,532	21,024	23,517	26,009	28,502	30,994	33,486	35,979	3375
1/7/20	18,303	20,796	23,288	25,780	28,273	30,765	33,258	35,750	3472
2/7/20	18,557	21,049	23,542	26,034	28,527	31,019	33,512	36,004	3482
3/7/20	18,811	21,303	23,796	26,288	28,781	31,273	33,765	36,258	3492
4/7/20	19,065	21,557	24,050	26,542	29,035	31,527	34,019	36,512	3502
5/7/20	19,319	21,811	24,304	26,796	29,288	31,781	34,273	36,766	3512
6/7/20	19,573	22,065	24,557	27,050	29,542	32,035	34,527	37,020	3521
7/7/20	19,826	22,319	24,811	27,304	29,796	32,289	34,781	37,273	3531
8/7/20	20,080	22,573	25,065	27,558	30,050	32,542	35,035	37,527	3541
9/7/20	20,334	22,827	25,319	27,812	30,304	32,796	35,289	37,781	3551
10/7/20	20,588	23,081	25,573	28,065	30,558	33,050	35,543	38,035	3560

Table 2 Training and testing accuracy score at different degree of polynomial regression

Degree	Testing accuracy sore	Training accuracy score
2	0.12756589341174684	0.16024107460679723
3	0.17943516161710302	0.22900462331233495
4	0.21600801519253746	0.3014880334951651
5	0.32105688124553533	0.40943900115182696

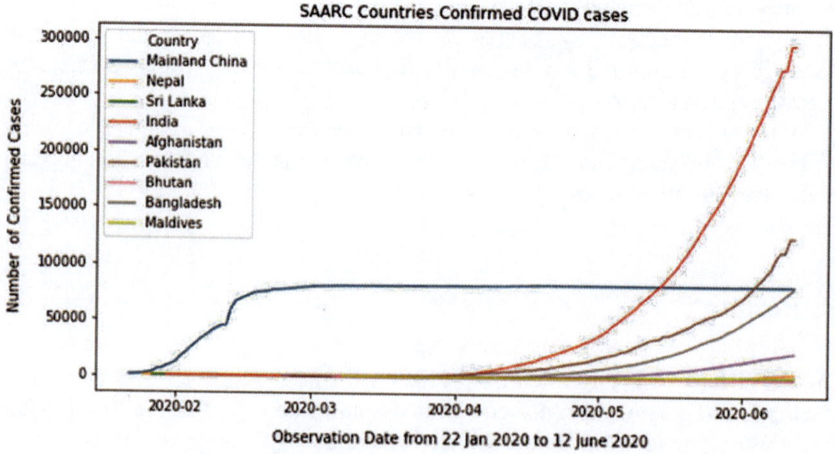

Fig. 6 Confirmed COVID cases in SAARC countries and China

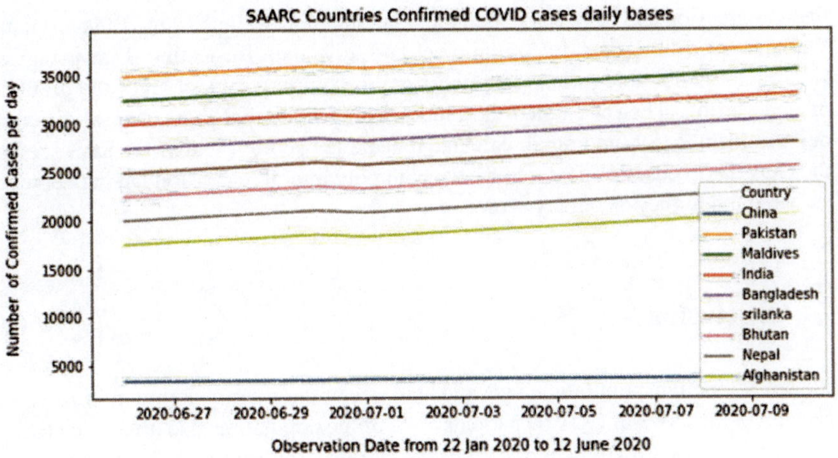

Fig. 7 Daily confirmed COVID cases prediction from June 26, 2020 to July 10, 2020

from June 26, 2020 to July 10, 2020. China have controlled the cases, but still cases will be found as per our prediction.

There is an exponential growth of COVID cases in India. Though Nepal was the first country that confirmed the COVID cases among all SAARC countries, Nepal COVID cases are not increasing exponentially. After India, Pakistan has shown exponential growth in COVID cases. The third country is Bangladesh where the COVID cases are increasing. The growth of COVID-19 cases is also controlled in China. Figure 7 shows the prediction of confirmed cases from June 26, 2020 to July 10, 2020 in SAARC countries and China.

6 Comparison with Other Schemes

Many models of machine learning have been used by the authors in literature to predict the cases of COVID-19. Some of the models used are traditional regression, multiple linear regression, polynomial regression and long short-term memory (LSTM). Jha et al. [11] predicted 7003 deceased cases by September 1, 2020 in Texas using Bayesian model. The author argues that prior distribution is set by the COVID experts and can be useful for small datasets. Tobias et al. [12] predicted confirmed cases of COVID-19 in Italy and Spain under lockdown using quasi-Poisson regression model. The quasi-Poisson model is a linear function of the mean. Gu et al. [13] applied cubic regression equations, which used the number of days as the input variable to predict the conformed COVID-19 cases in China and world. Pavlyshenko [14] used logistic curve to model COVID-19 spread. Different authors used different models but the efficiency of the model is based on the accuracy score of the training and testing data. No doubt error rate is also to be considered. We have used multiple linear regression and have calculated the accuracy score of both trained and test data. We have used 80% data for training set and 20% data for testing. Linear regression model is not able to learn enough from the training set and gives the problem of underfitting. Then we tuned the model with polynomial regression at different degrees from 2 to 5 and again calculated accuracy score. Finally, we have predict the COVID-19 cases of next two weeks with polynomial regression having degree 3 showing the highest accuracy score.

7 Conclusion

We did not include population, age, and land size of the country. Population size may also affect the cases of COVID patients. Age of the population also affects the cases, as if the patients are younger, the recovery rate may increase and vice versa. In our prediction, we realized that the cases in all SAARC countries will increase exponentially. More precautions must be taken. No doubt mental stress is also increasing among the people. COVID-19 has badly affected business, employment, and people who are living below the poverty line. The government is also doing well in providing medical assistance to COVID patients. It is time to ramp up the preventive measures and the precautions to be taken.

References

1. World Health Organization. Rolling updates on coronavirus disease (COVID-19). Accessed from https://www.who.int/emergencies/diseases/novel-coronavirus-2019/events-as-they-happen
2. How does COVID-19 affect different age groups. Accessed from https://www.nwhn.org/how-does-covid-19-affect-different-age-groups/
3. Worldometer. Accessed from https://www.worldometers.info/coronavirus/
4. World Health Organization. "Immunity Passports" in the context of COVID-19. Accessed from https://www.who.int/news-room/commentaries/detail/immunity-passports-in-the-context-of-covid-19
5. Kaggle. Accessed from https://www.kaggle.com/search?q=covid+19+dataset
6. Cascella M, Rajnik M, Cuomo A, Dulebohn SC, Di Napoli R (2020) Features, evaluation and treatment coronavirus (COVID-19). In: Statpearls [internet]. StatPearls Publishing. The Lancet
7. BBC News. Coronavirus: what did China do about early outbreak? Accessed from https://www.bbc.com/news/world-52573137
8. Remuzzi A, Remuzzi G (2020) COVID-19 and Italy: what next? The Lancet
9. Wikipedia. COVID-19 pandemic in India. Accessed from https://en.wikipedia.org/wiki/COVID-19_pandemic_in_India
10. Pandey G, Chaudhary P, Gupta R, Pal S (2020) SEIR and Regression Model based COVID-19 outbreak predictions in India. arXiv preprint arXiv:2004.00958
11. Jha PK, Cao L, Oden JT (2020) Bayesian-based predictions of COVID-19 evolution in Texas using multispecies mixture-theoretic continuum models. Comput Mech 1–14
12. Tobias A (2020) Evaluation of the lockdowns for the SARS CoV 2 epidemic in Italy and Spain after one month follow up. Sci Total Environ 725:

13. Gu C, Zhu J, Sun Y, Zhou K, Gu J (2020) The inflection point about COVID-19 may have passed. Sci Bull 5:98
14. Pavlyshenko BM (2020) Regression approach for modeling COVID-19 spread and its impact on stock market
15. Chakraborty T, Ghosh I (2020) Real-time forecasts and risk assessment of novel coronavirus (COVID-19) cases: a data-driven analysis. Chaos Solitons Fractals 109850

Coronary Artery Disease Prediction Techniques: A Survey

Aashna Joshi and Maitrik Shah

Abstract Machine learning has become a salient part of our life nowadays. It has a significant effect on the medical decision support system also. In the healthcare domain, it is beneficial to predict the disease and perform analysis to derive useful patterns from the electronic health records to reduce the toll. The primary cause of death worldwide is coronary artery disease (CAD), also known as atherosclerosis. It occurs when any of the arteries get blocked, resulting in weak or no blood flow to parts of a heart, leading to a heart attack. The prediction of CAD at an early stage is possible with the help of machine learning techniques like support vector machine, artificial neural network, k-nearest neighbors, decision trees, logistic regression, fuzzy rule-based methods, and many more. This paper gives insights into the research done on the prediction of this disease. We reviewed in-depth knowledge of the disease, various diagnostic techniques, and available datasets. Finally, we discussed and concluded how the machine learning technique creates an impact on predicting this disease.

Keywords Coronary artery disease · Machine learning techniques · Prediction

1 Introduction

Coronary heart disease continues to be a leading cause of morbidity and mortality among the different age groups [1]. Sometimes cholesterol, calcium, and other fatty materials present in the blood deposits into the wall of arteries make that artery narrowed, and gradually it gets blocked and resists blood flow reaching to other parts of a heart. This plaque can rupture at any time, which can lead to a heart attack. This is how coronary artery disease is occurred. According to a WHO report of 2016, 31% of total death is because of this disease, which is approximately 17.9 million people. By 2020 it would be more than 20 million people and will increase in the future [2] as demonstrated in Fig. 1. This depicts the importance of predicting CAD on time. Other previously done reviews concentrate on one kind of data only. This

A. Joshi (✉) · M. Shah
L.D. College of Engineering, Ahmedabad, India

P. K. Singh et al. (eds.), *Proceedings of Second International Conference on Computing, Communications, and Cyber-Security*, Lecture Notes in Networks and Systems 203,
https://doi.org/10.1007/978-981-16-0733-2_42

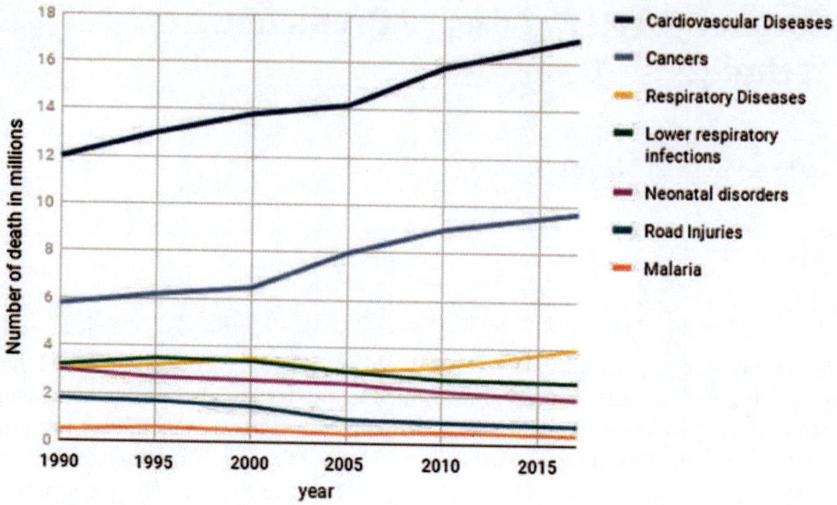

Fig. 1 Coronary artery disease statistics [33]

survey paper takes all the possible ways for prediction in consideration and provides an adequate comparative analysis on all of these.

The rest of the paper is organized as follows: Sect. 2 represents the prediction techniques of CAD, which can be categorized into two parts, manual and automatic. Section 3 describes the different state-of-the-art techniques used in prediction based on different types of data and compared the previous state of the art and dataset available. Section 4 provides the discussion of the literature survey done. Finally, we concluded with open research issues and challenges.

2 Prediction of Coronary Artery Disease

CAD can be predicted in two ways: manual prediction and automatic prediction. A cardiac specialist makes manual prediction through a specific procedure, and to predict it automatically, we have intelligent machine learning algorithms.

2.1 Manual Prediction

A conventional procedure for prediction is shown in Fig. 2. It starts with a symptomatic patient of cardiovascular disease where the patient gets diagnosis by a doctor. Based on the symptoms of patients, the doctor will record an electrocardiogram (ECG) as it is an easily accessible and widely used diagnostic tool to capture the

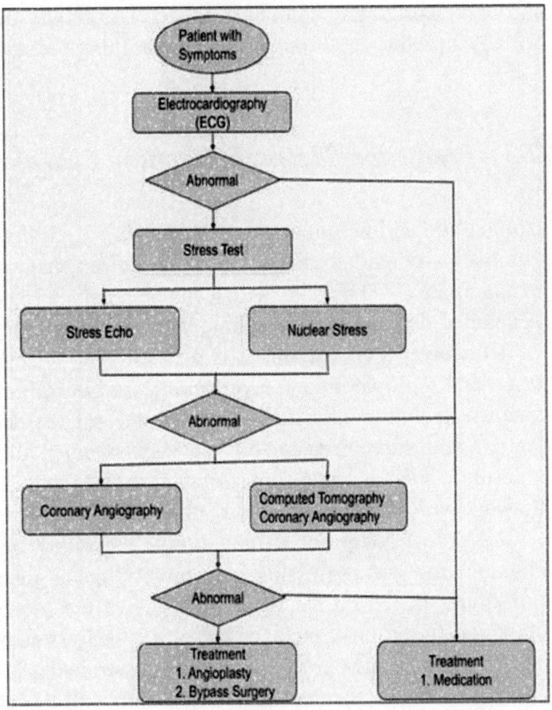

Fig. 2 Manual prediction for CAD

heart's abnormal activity [3, 4]. If it is reasonable, then medication will be suggested, and if it is not, we need to go to the further stage. In the next step, the doctor will perform a stress test to see his heart functioning healthily or not. This test generally can be of two types: stress echo and nuclear stress. In these tests, the patient's heart functioning is observed before and after some stress generating activities like the treadmill test. If it is normal, then medication will be suggested; otherwise, the patient is suggested to go for imaging procedure which is angiography. Most preferred imaging techniques are coronary angiography (invasive) and computed coronary tomography angiography (noninvasive). It takes photographs of the heart to find the exact location and size of the stenosis in the arteries. If unusual plaque growth is found in any of the arteries, then treatment like angioplasty and bypass surgery can be done of the patient; otherwise, medication will be suggested to reduce the plaque build-up and open up the artery.

Manual prediction is costly, time-consuming and also requires expert opinion to predict the disease. According to a report of National Rural Health Mission, approx 8% of primary health centers in rural India did not have a doctor, and as of March 2017, 61% of them were functioning with just one doctor. Moreover, prediction by doctors can be error-prone too. A local survey by Indira Gandhi Medical college says that 171 patients out of 335 patients, i.e., 48.9%, were diagnosed normal after performing angiography [5]. This shows how a huge number of people are going through angiography without the need. The above study and statistics also motivate for CAD prediction through machine learning techniques.

2.2 *Prediction Through Machine Learning Techniques*

Computational intelligent machine learning algorithms are the way to predict CAD automatically. With the help of machine learning techniques, a machine can be trained to anticipate CAD by data, which can be resolved issues of manual prediction [6–8]. A detailed description of machine learning algorithms is given below:

Multilayer Perceptron. It is a multilayer structure of the input layer, output layer, and multiple hidden layers comprised of artificial neurons. It learns from the data itself; that is why it is more suitable for the decision support system. It can learn in both supervised and unsupervised manner. To predict the CAD, any learning algorithm, such as backpropagation, is used to determine the weights of the neural network which defines the interconnection between neurons [9].

***K* Nearest Neighbor**. It is a simple algorithm that is supervised and used for classification and regression problems. Objects are classified through the voting procedure. Based on the majority votes of the k-nearest neighbors, the object is classified based on the class type and k-nearest samples in the space. Similarity among data points can be defined by metric like Minkowski in a data space. It calculates the distance between data points and the target variable by any distance measuring formula such as Euclidean distance [10].

Support Vector Machine. It is a supervised classifier but can also be used in regression analysis. It classifies objects by separating data points using hyperplanes by separating data points of the first class from second class [10]. It is done by nonlinear mapping of data points by support vectors and margins. If the CAD dataset is having linearly separable data, then two parallel hyperplanes can be used to classify the sample; otherwise, nonlinear classifiers are used by the kernel to transform attributes into bigger dimensional space to make them linear. It uses both linear and nonlinear kernel functions for prediction.

Decision Trees. Decision trees are easier to understand and implement. They fall under supervised learning criteria. It can be used to predict conditional probabilities and derive decision rules based on association rules with output variable. Frequently used algorithms to create a decision tree are a CART, ID3, C4.5, J48, and CHAID [11].

Naive Bayes. It is one of the well-known probabilistic data mining algorithms. It uses maximum likelihood method to estimate parameters and Bayes' probability theorem to classify objects [11]. For example, CAD can be predicted by symptoms of a patient such as a chest pain, cholesterol levels, blood pressure, stress level, smoking, and family history.

Logistic Regression. It is used to perform predictive analysis by finding relationships between the dependent variable, such as a symptom and one or more independent variables such as a person having the disease or not [12]. It estimates the parameters of binary logistic model. It follows the Bernoulli distribution and classifies objects into two groups.

Fuzzy Logic. It is an approach that follows fuzzy degrees of truth rather than crisp values just true or false (0 or 1). It generally has three stages: fuzzification, fuzzy inference system, and defuzzification. First input variables are fuzzified, and the membership function is calculated and then assigned based on the degree of the variables. In such kind of rule-based system, a disease can be predicted by formulating rules [13]. For example, if the age of a patient is greater than 65, and he has angina, then he is at high risk of having CAD.

Ensemble Methods. It is a technique of combining multiple single classifiers to improve the accuracy of a classifier. Better performance than individual classifiers is achieved by combining a weak learner with keen learners. It can be done by various methods such as bagging, boosting, stacking, and majority voting. CAD can be classified more accurately by using ensemble learning techniques [14].

3 State of the Art

For the automatic prediction of coronary artery disease, machine learning algorithms are the best option as they provide computational intelligence to the machine for predicting the disease accurately by itself. Machine learning techniques like feature selection and modeling can improve the performance to a great extent in CAD prediction [15, 16]. Apart from this, recently fog computing [1], cloud computing, artificial intelligence, and Internet of things play an vital role in healthcare 4.0 [17–21]. This section provides information about related work as per the type of the data for the prediction: text (numeric or categorical), signals, and images [22].

Prediction through signals can be made by electrocardiogram (ECG), phonocardiogram (PCG), and photoplethysmogram (PPG). Usually preferred imaging techniques are invasive coronary angiography (CA), intravascular ultrasound (IVUS) image, and noninvasive computed tomography coronary angiography (CTCA). Text data comprises of symptoms, demographics, and biomarkers. Various machine learning classification techniques are explained in the above section to process these types of data. A literature survey, according to data types, is shown in Fig. 3. A detailed discussion of papers is given below.

3.1 Signals

In a signal-based approach, a signal is processed first and then classified into normal and abnormal by some frequently used parameters in diagnosis such as ST elevation, ST depression, Q wave, and R wave progression. ECG signals are identified as CAD ECG signals by the convolutional neural network using long short-term memory (LSTM) network in [3]. A survey is done on ECG signals in [23], which contains

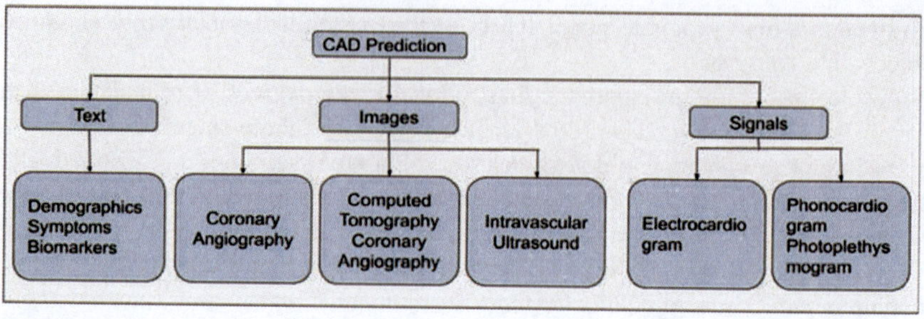

Fig. 3 Taxonomy

Table 1 Summary of the papers based on signal data

Source	Year	Dataset	Proposed approach	Accuracy
[34]	2019	Fortis Hospital in Kolkata (Private)	Rule-based + signal-based classification	–
[24]	2017	In-house experimental dataset	SVM	87%
[23]	2017	Multiple	Survey of various techniques	According to approaches
[35]	2018	From Fortis hospital in Kolkata (Private)	Metadata-based rule engine + SVM	–
[3]	2018	PhysioNet database (Public)	CNN with LSTM network	95.76%

an explanation of signal processing, available datasets, and information about how to identify different features of the signal. Some researchers have applied the fusion approach using both the signals PPG and PCG for the prediction [24]. Prediction in a multistage approach is also performed by few researchers, i.e., in the first stage, CAD probability is calculated, and if there is a greater chance of having CAD, then in next stage signals are processed to achieve surety of prediction. The summary of the papers based on signal data is shown in Table 1.

3.2 *Images*

Research on image processing is nowadays very booming because of its adaptive and accurate nature. Tianming Du, Xuqing Liu, Honggang Zhang, and Bo Xu have tried to find stenosis in arteries by image processing using CALD-Net on angiograms [25]. The authors proposed a new end-to-end framework for detecting lesions. Research

Table 2 Summary of the papers based on image data

Source	Year	Dataset	Proposed approach	Accuracy (%)
[28]	2018	Their institutional ethical review board (Private)	Recurrent convolution network	80
[25]	2018	–	Deep learning with CALD-Net	88
[27]	2018	dataset B from MIC-CAI challenge 2011 (Private)	Convolution neural network with ResNet101	99
[36]	2018	database of EVINCI study (Public)	ANN, SVM, and RF	85
[29]	2018	EVINCI study (Public)	SVM, ANN, RF, and J48	–
[26]	2018	–	Image processing (CNN)	83.08
[37]	2019	Database of EVINCI study (Public)	Logistic regression	83
[38]	2019	Institutional ethical review board (Private)	SVM	80

of vessel segmentation is also done by multiple CNNs in angiographic images [26], in which they have used two-channel CNN for coarse and one-channel CNN for fine segmentation of vessels. An effort is also made to detect calcification in coronary arteries on IVUS images by using ResNet101 [27]. Recent advancements in imaging techniques have introduced a new noninvasive method CTCA, it is a noninvasive imaging technique, but the patient is exposed to radiation that can be harmful. Research on CTCA images is also conducted by the convolutional neural network, logistic regression, and SVM to determine stenosis. By using recurrent convolutional neural network, automatic plaque and stenosis characterization is performed [28]. A comparative study with different classifiers such as J48, SVM, ANN, and RF on these images is also performed in [29, 30]. The summary of these papers is given in Table 2.

3.3 Text

Research has also been done on various textual datasets by using multiple classifiers and compared the result. A combined belief rule-based system is created to predict the severity of the disease [31]. ANN and KNN are applied on different datasets and compared with each other [32]. Some researchers have applied SVM, NB, and KNN on the Cleveland dataset and got good accuracy [10]. Fuzzy knowledge is also applied in the medical diagnostic system with an inference engine and used to extract rules from the result obtained [13]. Decision trees also give good accuracy in prediction of CAD [5, 11]. Many researchers [2] observe multiple classifiers as a single classifier

Table 3 Summary of the papers based on textual data

Source	Year	Dataset	Proposed approach	Accuracy
[9]	2018	5 hospitals of Guwahati (Private)	Recurrent neural network	81%
[10]	2018	Cleveland (Public)	SVM, Näive Bayes, and KNN	84%
[13]	2017	Combined	Fuzzy logic	–
[5]	2018	Department of Cardiology, Indira Gandhi Medical College (Private)	C4.5, NB tree, MLP	97.6%
[11]	2018	ERIC laboratory (Private)	J48, RF, and NB	100% (RF)
[12]	2019	Heart disease dataset (Public)	DT, LR, SVM, MLP, and NB	86% (RF without PCA)
[14]	2019	Cleveland (Public)	Ensemble technique	85%
[2]	2018	Z-Alizadeh Sani, South African, Heart disease dataset from UCI (Public)	NB, RF, KNN, MLP, and SVM and combined bagging, ensemble classifier	81.84% (Cleveland), 87.12% (Z-Alizadehsani)
[15]	2020	Z-Alizadeh Sani (Public)	Particle swarm optimization-based extreme learning machine	97.60%
[16]	2020	Extended Z-Alizadeh Sani (Public)	Decision trees, Naive Bayes, and deep learning	99% (NB, Deep learning)

and as a combined model through ensemble techniques. These textual data related papers are summarized in Table 3.

4 Discussion

For image-based studies, most of the researchers have used their own datasets, but EVINCI study has provided their supplementary data at their journal (EHJ). In 252 patients out of 697 patients, image data of myocardial perfusion scintigraphy (MPS), CTCA, and invasive coronary angiography (ICA) are given. For signal-based studies publicly available ECG dataset is available at PhysioNet database. Publicly available textual datasets: Cleveland dataset is from heart disease UCI machine learning repository. It has 303 records and 14 attributes. Statlog dataset is also from the UCI machine learning repository and has 303 records and the same 14 attributes as Cleveland. South African dataset is from the BigML dataset repository. It has 462 records and 10 attributes. These datasets have attributes of demographics, symptoms, and biomarkers of patients, but the Z-Alizadeh Sani dataset has a new kind of data, ECG

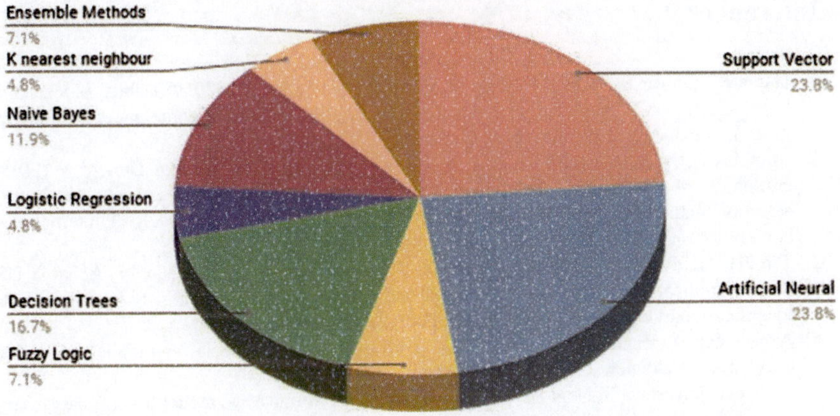

Fig. 4 Most commonly used machine learning techniques to predict coronary artery disease

features. It is also from the UCI machine learning repository having 303 records and 54 attributes. This dataset has been extended by adding three new attributes based on stenosis found in the main three kinds of arteries: the left main coronary artery, left circumflex artery, and the right coronary artery.

Though the imaging technique's prediction is very accurate, it requires an invasive procedure or involves radiation, which is harmful, time-consuming, and costly. Moreover, there is a need for expertise for the right prediction from the angiogram and requires a large number of resources like time, expensive tools, techniques, and laboratory setup [5]. That is why noninvasive prediction techniques from patient's demographics and symptoms related information are necessary to be sure before going to the final stage of imaging technique. The frequency of machine learning techniques used in the literature survey is shown in Fig. 4.

5 Conclusion

Coronary artery disease is the most crucial matter in the healthcare domain. According to a WHO report, it takes almost 30% of total death globally by leading any other cause and increasing year by year gradually. That is why early and accurate prediction of it is essential to save a valuable life. This paper provides in-depth knowledge of the disease, risk factors, and possible ways to predict it in a manual and machine learning approach. It also provides a detailed examination of different machine learning prediction techniques of CAD according to different data types available such as text, images, and signals. This survey provides all the information to create the noble medical system for accurate diagnosis owing to which a proper treatment could be delivered to patients to reduce the death toll.

References

1. Gupta R, Tanwar S, Tyagi S, Kumar N, Obaidat MS, Sadoun B (2019) Habits: blockchain-based telesurgery framework for healthcare 4.0. In: 2019 international conference on computer, information and telecommunication systems (CITS). IEEE, pp 1–5
2. Kolukisa B, Hacilar H, Goy G, Kus M, Bakir-Gungor B, Aral A, Cagri Gungor V (2018) Evaluation of classification algorithms, linear discriminant analysis and a new hybrid feature selection methodology for the diagnosis of coronary artery disease. In: 2018 IEEE international conference on big data (Big Data). IEEE, pp 2232–2238
3. Tan JH, Hagiwara Y, Pang W, Lim I, Oh SL, Adam M, Tan RS, Chen M, Acharya UR (2018) Application of stacked convolutional and long short-term memory network for accurate identification of cad ecg signals. Comput Biol Med 94:19–26
4. Ashishdeep A, Bhatia J, Varma K (2016) Software process models for mobile application development: a review. Comput Sci Electron J 7(1):150–153
5. Verma L, Srivastava S, Negi PC (2018) An intelligent noninvasive model for coronary artery disease detection. Complex Intell Syst 4(1):11–18
6. Jaykrushna A, Patel P, Trivedi H, Bhatia J (2019) Linear regression assisted prediction based load balancer for cloud computing. In: 2018 IEEE Punecon. IEEE, pp 1–3
7. Chauhan K, Jani S, Thakkar D, Dave R, Bhatia J, Tanwar S, Obaidat MS (2020) Automated machine learning: the new wave of machine learning. In: 2020 2nd international conference on innovative mechanisms for industry applications (ICIMIA). IEEE, pp 205–212
8. Vachhani H, Obiadat MS, Thakkar A, Shah V, Sojitra R, Bhatia J, Tanwar S (2019) Machine learning based stock market analysis: a short survey. In: International conference on innovative data communication technologies and application. Springer, Cham, pp 12–26
9. Talukdar J, Dewangan BK (2018) Analysis of cardiovascular diseases using artificial neural network. In: 2018 fifth international conference on parallel, distributed and grid computing (PDGC). IEEE, pp 132–137
10. Nassif AB, Mahdi O, Nasir Q, Talib MA, Azzeh M (2018) Machine learning classifications of coronary artery disease. In: 2018 international joint symposium on artificial intelligence and natural language processing (iSAI-NLP). IEEE, pp 1–6
11. Dhar S, Roy K, Dey T, Datta P, Biswas A (2018) A hybrid machine learning approach for prediction of heart diseases. In: 2018 4th international conference on computing communication and automation (ICCCA). IEEE, pp 1–6
12. Wu CM, Badshah M, Bhagwat V (2019) Heart disease prediction using data mining techniques. In: Proceedings of the 2019 2nd international conference on data science and information technology, pp 7–11
13. Ibrahim N, Mahadi LF, Mahmud F (2017) Initial study to evaluate fuzzy logic on diagnosis of generic atherosclerosis. In: 2017 international conference on vision, image and signal processing (ICVISP). IEEE, pp 123–129
14. Latha CBC, Jeeva SC (2019) Improving the accuracy of prediction of heart disease risk based on ensemble classification techniques. Inform Med Unlocked 16:100203
15. Shahid AH, Singh MP, Roy B, Aadarsh A (2020) Coronary artery disease diagnosis using feature selection based hybrid extreme learning machine. In: 2020 3rd international conference on information and computer technologies (ICICT). IEEE, pp 341–346
16. Ghasemi F, Neysiani BS, Nematbakhsh N (2020) Feature selection in pre-diagnosis heart coronary artery disease detection: a heuristic approach for feature selection based on information gain ratio and Gini index. In: 2020 6th international conference on web research (ICWR). IEEE, pp 27–32
17. Bhatia J, Patel T, Trivedi H, Majmudar V. HTV dynamic load balancing algorithm for virtual machine instances in cloud. In: 2012 international symposium on cloud and services computing (ISCOS). IEEE, pp 15–20
18. Kumari A, Tanwar S, Tyagi S, Kumar N (2018) Fog computing for healthcare 4.0 environment: opportunities and challenges. Comput Electr Eng 72:1–13

19. Shah NB, Shah ND, Bhatia J, Trivedi H (2019) Profiling-based effective resource utilization in cloud environment using divide and conquer method. In: Information and communication technology for competitive strategies. Springer, Singapore, pp 495–508
20. Bhatia J, Kumhar M (2015) Perspective study on load balancing paradigms in cloud computing. IJCSC 6(1):112–120
21. Singh PK, Kar AK, Singh Y, Kolekar MH, Tanwar S (2019) Proceedings of ICRIC 2019: recent innovations in computing, vol 597. Springer Nature
22. Chauhan K, Patel H, Dave R, Bhatia J, Kumhar M (2020) Advances in single image super-resolution: a deep learning perspective. In: Proceedings of first international conference on computing, communications, and cyber-security (IC4S 2019). Springer, Singapore, pp 443–455
23. Revathi J, Anitha J (2017) A survey on analysis of ST-segment to diagnose coronary artery disease. In: 2017 international conference on signal processing and communication (ICSPC). IEEE, pp 211–216
24. Choudhury AD, Banerjee R, Pal A, Mandana KM (2017) A fusion approach for non-invasive detection of coronary artery disease. In: Proceedings of the 11th EAI international conference on pervasive computing technologies for healthcare, pp 217–220
25. Du T, Liu X, Zhang H, Xu B (2018) Real-time lesion detection of cardiac coronary artery using deep neural networks. In: 2018 international conference on network infrastructure and digital content (IC-NIDC). IEEE, pp 150–154
26. Yang S, Yang J, Wang Y, Yang Q, Ai D, Wang Y (2018) Automatic coronary artery segmentation in X-ray angiograms by multiple convolutional neural networks. In: Proceedings of the 3rd international conference on multimedia and image processing, pp 31–35
27. Sofian H, Ming JTC, Mohamad S, Noor NM (2018) Calcification detection using deep structured learning in intravascular ultrasound image for coronary artery disease. In: 2018 2nd international conference on biosignal analysis, processing and systems (ICBAPS). IEEE, pp 47–52
28. Zreik M, Van Hamersvelt RW, Wolterink JM, Leiner T, Viergever MA, Išgum I (2018) A recurrent cnn for automatic detection and classification of coronary artery plaque and stenosis in coronary ct angiography. IEEE Trans Med Imaging 38(7):1588–1598
29. Kigka VI, Georga EI, Sakellarios AI, Tachos NS, Andrikos I, Tsompou P, Rocchiccioli S, Pelosi G, Parodi O, Michalis LK et al (2018) A machine learning approach for the prediction of the progression of cardiovascular disease based on clinical and non-invasive imaging data. In: 2018 40th annual international conference of the ieee engineering in medicine and biology society (EMBC). IEEE, pp 6108–6111
30. Vora J, Tanwar S, Tyagi S, Kumar N, Rodrigues JJPC (2017) FAAL: Fog computing-based patient monitoring system for ambient assisted living. In: 2017 IEEE 19th international conference on e-health networking, applications and services (Healthcom). IEEE, pp 1–6
31. Ahmed F, Chakma RJ, Hossain S, Sarma D et al (2020) A combined belief rule based expert system to predict coronary artery disease. In: 2020 international conference on inventive computation technologies (ICICT). IEEE, pp 252–257
32. Terrada O, Cherradi B, Raihani A, Bouattane O (2020) Atherosclerosis disease prediction using supervised machine learning techniques. In 2020 1st international conference on innovative research in applied science, engineering and technology (IRASET). IEEE, pp 1–5
33. Ritchie H (2018) Causes of death. Our world in data. https://ourworldindata.org/causes-of-death
34. Banerjee R, Ghose A, Sinha A, Pal A, Mandana KM (2019) A multi-modal approach for non-invasive detection of coronary artery disease. In: Adjunct proceedings of the 2019 ACM international joint conference on pervasive and ubiquitous computing and proceedings of the 2019 ACM international symposium on wearable computers, pp 543–550
35. Banerjee R, Bhattacharya S, Bandyopadhyay S, Pal A, Mandana KM (2018) Non-invasive detection of coronary artery disease based on clinical information and cardiovascular signals: a two-stage classification approach. In: 2018 IEEE 31st international symposium on computer-based medical systems (CBMS). IEEE, pp 205–210

36. Sakellarios A, Siogkas P, Georga E, Tachos N, Kigka V, Tsompou P, Andrikos I, Karanasiou GS, Rocchiccioli S, Correia J et al (2018) A clinical decision support platform for the risk stratification, diagnosis, and prediction of coronary artery disease evolution. In: 2018 40th annual international conference of the IEEE engineering in medicine and biology society (EMBC). IEEE, pp 4556–4559
37. Sakellarios AI, Tsompou P, Siogkas P, Kigka V, Andrikos I, Tachos N, Georga E, Kyriakidis S, Rocchiccioli S, Pelosi G et al (2019) Predictive models of coronary artery disease based on computational modeling: the smartool system. In: 2019 41st annual international conference of the IEEE engineering in medicine and biology society (EMBC). IEEE, pp 7002–7005
38. Zreik M, van Hamersvelt RW, Khalili N, Wolterink JM, Voskuil M, Viergever MA, Leiner T, Išgum I (2019) Deep learning analysis of coronary arteries in cardiac CT angiography for detection of patients requiring invasive coronary angiography. IEEE Trans Med Imaging 39(5):1545–1557

IoT Supported Healthcare (Or: Computer Aided Healthcare)

COVID-19 a "BIG RESET"—Role of GHRM in Achieving Organisational Sustainability in Context to Asian Market

Meenu Chaudhary and Loveleen Gaur

Abstract As per a report by UNCTAD, with outbreak of COVID-19 worldwide, human activities have stopped witnessing the revival of nature; CO_2 emission, global air traffic reduced by 25%, 60%, respectively. The effects on environment like improved air quality index, less resource consumption are short-lived as these are likely to rise to previous levels once economic activities pick up after crisis. Henceforth, it has become need of an hour to tackle the environmental concerns at organisational level by aligning environmental management, HRM and technology. Considering Asian economic advancement and environment adversity, it is imperative to explore GHRM in Asian context to move a step forward attaining organisational sustainability. Our findings, based on multimethod for collection of qualitative and quantitative data and analysis via NVIVO and IBM SPSS 23, respectively, confirm the degree of implementation of GHRM practices and also found the relationship between GHRM and organisational sustainability. Data was collected using survey questionnaire and interview from 107 h professionals of various sectors. However, more such studies are important for developing countries to address environmental concerns.

Keywords GHRM · GHRM and environmental management · GHRM and organisational sustainability · GHRM and sustainable development

1 Introduction

As per survey by KPMG, around 72 and 52% companies of N100 and G250, respectively, do not acknowledge climate change as financial risk in their annual reports [1]. With global interest being diverted to environmentalism, from specific treaties to cope up with climate change, for example Kyoto 1997 and Copenhagen 2009 [2],

M. Chaudhary (✉) · L. Gaur
Amity International Business School, Amity University, Sector 125, Noida, India

L. Gaur
e-mail: lgaur@amity.edu

P. K. Singh et al. (eds.), *Proceedings of Second International Conference on Computing, Communications, and Cyber-Security*, Lecture Notes in Networks and Systems 203,
https://doi.org/10.1007/978-981-16-0733-2_43

or from pollution consequential from high-profile industrial accidents, policies by government to slow-down the effect and reverse the exploitation of natural resources and its impact [3], expectations have risen from organisations to take accountability of environment management to (EM) globally [4]. The objectives of this paper are to find gaps by exploring the status of GHRM practices/activities in Asian-based organisations, addressing the gaps and finding correlation between GHRM practices and sustainability of organisation.

Need for awareness to value natural capital is not a new thing; however, it has become a common practice in corporate [5]. It has now become more important seeing the recent scenario of COVID-19 when human activities have stalled and reviving of nature has become quite evident. Though carbon emission has decreased due to pandemic, concern for climate change remains intact considering the previous record; after financial crisis 2008, drop was a minor fluctuation in long trend (Temple 2020) [6].

Besides, COVID-19 has led to global economic catastrophe. China's export fall by 17% in early 2020 and world trade is expected to fall between 13–32% (WTO 2020) [7]. As per Global Survey on Sustainability, less than half of the respondents worldwide know about UN Sustainable Development Goals (SDGs) [8]. Environmental performance has become the emerging issue for global business leaders, and thus, EM is of utmost importance for business world leaders [9, 10]. Economists say, due to increasing industrialisation, foreign investment and entrepreneurial nature, India needs to prioritise the environment management to promote the Indian economy. As stated by World Bank, cost of environmental degradation amounts to $80 bn a year and child deaths to 23% can be attributed to environmental aspects [11]. Hence, to avoid exploitation of natural resources leading to natural calamities and accidents at workplace, organisations must take responsibility of environment management. Researchers have suggested that environmental disruptions can be due to human activities. Hence, in order to reduce these environmental disruptions, human interference needs to be supervised [12, 13]. Studies show that only few organisations focus on internal factors such as participation and impact of human behaviour in protection of environment [14].

Concern for environment degradation has been witnessed from last two decades. From various options to reduce the burden, technology is considered as the most feasible one, but it seems to be unclear if technology can be alone sufficient. Usage of technology in achieving sustainability like use of pollution control devices, reforming existing technologies, using benign materials. However, shifting to complex technologies poses difficult task for policymakers as it not only involves change in technology but also some fundamental changes in organisation. Presently, with reference to management, becoming green is new norm [15], the literature on green marketing, green management, green accounting and green human resource management is increasing [16]. However, green human resource management (GHRM) is still in its primary level with most work in this area is theoretical [17]. GHRM, a subdivision of green management (GM), emphasizes on the role of human behaviour in EM [18] and sustainability [19]. GHRM is described as set of policies, system and practices that change its employees' attitude towards green which benefits the individuals, society, natural environment and business [20]. Under the umbrella of sustainability HRM, researchers are determining the relationship between certain set of HRM activities and environmental sustainability. Environmental management is as a side-shoot of a wider accounting schema, identified as "triple bottom line", that assimilates three aspects, i.e., social, environmental and financial facets [21]. Sustainability metrics with respect to technology involves three basics of sustainability: economic, environmental and social concerns, corresponding the triple bottom line (1997) [22]. Companies are now realising that responsibility of corporates towards sustainability is not just tool for building a brand but an important feature for business growth [14] by including greener options like telecommuting, flexible work schedules.

Though GHRM has been highlighted by researchers in promoting greener organisations many times, but it is still a less researched area [12, 23]. Also, the existing literature mostly provide insights of Western framework [16]. However, keeping in consideration the Asian economic progress and environmental issues, it has become paramount to study GHRM in Asian context. Though GHRM is a prerequisite, but it is still under-researched area in Asia.

The flow of the study is as follows. Section 1 includes the introduction followed by Sect. 2 that presents the literature review. Section 3 outlines the methodology, and Sect. 4 presents the findings and analysis.

2 Theoretical Background

Due to the rapid growth of green environment management, the role of "greening" concept is imperative in improvising the organisations' environmental performance [24]. A green workplace is socially responsible and environment sensitive [15], virtual workplace and green building being its characteristics. Google has positioned itself in the top position for not only following environmental practices but also in promulgation their environmental records [25]. However, human factor is the key factor for success of organisations' environmental activities. Green HRM focused on entire HRM practices like recruitment and selection (R&S) of personnel, training and development, compensation, performance management, employment engagement and organisational existence [26].

Green recruitment discusses the process of hiring people with behaviour, acquaintance and skill set of EMS in organisation. HRM practices may have additive properties when practices A and B have self-determining and nonoverlapping effects on organizational results (denoted by the formula $2 + 2 = 4$), substitution properties when practices A and B lead to similar results and addition of one practice has null effect on anticipated outcome (denoted by the formula; $2 + 2 = 3$), or interactive effects when their collaboration leads to outcome significantly greater than the total of distinct and autonomous outcomes (denoted by the formula $2 + 2 = 5$) [27]. On basis of this, there may be some influence of business's green reputation on green recruitment in numerous forms. Author supports the implication of green image on recruitment but no evidence of effect of green activities' information on website [28].

Creation and sustenance of an environment extensive firm also necessitates the hiring of employees who are keen to involve with EM goals. Some major multinationals are adopting GHRM in form of "employer banding" to improvise the attraction of environmentally conscious candidates [29]. Author found the positive influence of brand image on environmental commitment. Also, candidates also emphasise on implementation of environment-friendly plans [30].

Competitive advantage can be achieved via green training. More time and resource investment is necessary of content development of such trainings by focusing on gaps to fill and opportunities for application [31]. Author established positive relation between environmental training and environmental management maturity specifying that ET is not important of EM only but also for evolution with organisation [32]. Study identifies the positive correlation between green training and green supply chain management in procurement and purchase and dealing with customers [33].

The perceived value of green performance appraisal increases when individual behaviour is measured and hence compliance increases. Though performance management system (PMS) evaluates the green behaviours, green compensation system (CMS) confirms the integration of results of performance evaluation with rewards and benefits [34]. Performance management is an important human resource practice for encouraging environmental behaviour and sustainability, thus promoting green performance management (GPMS) [35]. Besides the barriers, adoption of green practices can help in many ways: better performance, improvement in organisational culture [16], cost reduction, effective utilisation and improving company goodwill.

Reward policies primarily focus on attracting, motivating and retaining employees for development of new abilities, knowledge and actions that pave way to achieve organisational goals [36]. Green compensation and reward system mean alignment of green behaviours with organisational green activities implemented in the organisation. Rewards and bonuses must be provided to employees for their efforts in creation of environment conducive culture [37]. Employees must be given rewards for their efforts in promotion of green behaviours in their lifestyle, workplace and reduction of carbon footprints [38].

Employees shall comply to a green behavioural motto if they are engaged with organisation for accomplishing the firm's resolution of implementation green policies [39]. Study revealed high positive correlation between employee involvement and green performance. Organisation can achieve green goals through employee participation [40]. Employees' green participation increases green business profit [41].

The advancement and diffusion of new and additional technology, particularly environment-friendly technology, is regarded as one of the ways to solve environmental and development problems. Philosophers of technology recommend that a better understanding of association between technology and society "is crucial for building a better world" [42]. As per The Special Report on Emission Scenarios (IPCC 2000, technology was of similar importance for future GHG emissions as population and economic growth combined. Leapfrogging is an idea that developing countries can bypass the dirty stages of industrial growth by moving to clean and modern technologies, using resources that generate less pollution [43]. Technology sustainability can be defined as a proposed evaluation of socio-technical factors that let technology develop, implement and maintain properly, considering the needs of all stakeholders, attract long-term users and create positive consequences corresponding to the purpose of the technology and initial intents of its developers (financial, social, etc) [44].

As per reviewed literature, adoption of GHRM can enhance the green image, help in employee branding, but inclination of employees towards goal of sustainable environment can be done by including skill development and green training, green-based performance management and reward management.

3 Methodology

To fulfil the purpose of the study, researcher has used multimethod as data is collected and analysed using two methods. In the first section of the study, archival method was used which includes study of past documents and texts to supplement research strategies [45]. This method gave understandings of GHRM from the existing literature. The study also includes categorising and classifying the relationship between EM and HRM practices. The second section of the research includes qualitative and quantitative techniques. The questionnaire was created with Google form and was forwarded to HR professional who have required knowledge for this research; henceforth, we used purposive sampling. Out of 180 distributed questionnaires, responses were received from 121 h professionals; however, only 107 questionnaires were fit for analysis due to the incomplete responses by other respondents. The flow of the study is depicted in Fig. 1.

Respondents were requested to rate the green human resource activities on five-point Likert scale (ranging from 1 = strongly disagree to 5 = strongly agree). The quantitative responses were analysed using SPSS, and qualitative responses were analysed using NVIVO to derive the outcome of study.

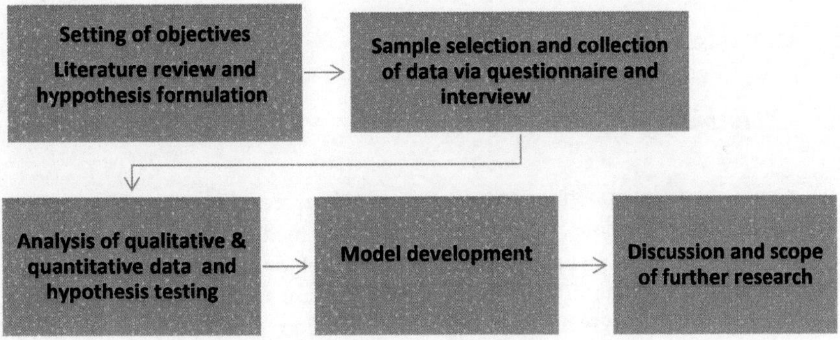

Fig. 1 Flow chart of research methodology used in this paper

Based on the literature review, list of green human resource practices was identified.

In order to study GHRM, variables were defined for green recruitment and selection, green training and development, performance management, compensation management and employee engagement as defined in Table 1. With a purpose to analyse the consequence of the above-listed HR practices on organisational sustainability, a study was conducted among human resource professionals of various sectors. As per the nature of study, purposive sampling was used. It is associated to the group of non-probability sampling techniques, in which sample population is selected on the parameters of their acquaintance and expertise in research topic. With reference to this paper, HR professionals were selected on convenience basis who have right information on this topic. The interview questions included were: Does your company assess green behaviours, is there any provision to keep track of paper used in recruitment cycle? Do you use job portal for hiring? Do you have budget to provide training on environmental issues? Can you provide the details of budget? Is there any dedicated green team? Does your company canteen have characteristics of green cafeteria? Who handles the green issues and what is the position in hierarchical structure? Which HRM initiatives have been taken to address environmental concerns? What impact did it had? Which environmental practices are important as per you? Which ISO certification company owns? Do you think GHRM is need of an hour? The distinctive features of the sampled population are mentioned in Table 2.

Out of 107 respondents, counts of male and females were 58 and 49, respectively, among which respondents with experience of less than 5 years, between 5 < 10, 10 < 15, 20 or more were 10, 39, 30, 19, and 9, respectively.

On the basis of the literature review and purpose of the study, the following hypotheses are tested in this research:

HP1: Environmental practices under green human resource management are well implemented in organisation.

HP2: There is significant correlation between green human resource practices and organisational sustainability.

4 Analysis

4.1 Reliability Analysis

Five-point Likert scale ranging from 1—strongly disagree to 5—strongly agree was used to evaluate the responses. Cronbach's alpha was included in evaluation process to determine the internal consistency of items in a study to measure its reliability [48]. The results of reliability test for each construct are summarised below.

Cronbach alpha for green recruitment and selection, green training and development, green performance management system, green compensation management and green employee engagement are 0.891, 0.843, 0.862, 0.826, 0.875, respectively.

Table 1 List of GHRM practices used in research [46, 47]

Construct	Label	Activities as measurement items
Recruitment and selection (GRS)	RS1	Informing job applicants, the commitment towards environment of organisation
	RS2	Display of environmental standards in job advertisement
	RS3	Verifying of candidates' environmental knowledge and skill set in recruitment
	RS4	Prioritising the candidates with skills and hands-on in ecological projects
	RS5	During orientation process, new employees are introduced with environmental norms
Training and development (GTD)	TD1	Identification and investigation of employees' need for environmental training
	TD2	Delivery of environmental training to employees and managers for development of ecological skills and knowledge
	TD3	Continuous improvement in environmental practices in organisation through training
	TD4	Creation of integrated training module for emotional participation of workforce in EM
Performance management system (GPMS)	PMS1	Usage of green performance objectives in performance appraisal and management
	PMS2	Establishment of green targets, objectives and accountability for managers and employees
	PMS3	Conduct of environmental assessments
	PMS4	Consequences attached to non-attainment or non-compliance with environment management goals
	PMS5	Continuous feedback to employees with regard to their performance in attainment of environmental goals
Compensation management (GCM)	CM1	Reward system for staff on innovative environmental performance
	CM2	Excellence award to personnel on environmental related performance and achievement

(continued)

Table 1 (continued)

Construct	Label	Activities as measurement items
	CM3	Incentives for encouraging an environmentally friendly practice
	CM4	Provision for loans or tax incentives (Usage of fuel-efficient cars, electric vehicles, bicycle loans)
Employee engagement (GEE)	EE1	Well-defined green mission to supervise the employees' conduct in EM
	EE2	Integration of learning climate with green behaviour for awareness among employees
	EE3	Availability proper communications channels (formal and informal) for spreading green culture at workplace
	EE4	Employee involvement in quality improvisation and sessions for solving problems related to green matters
	EE5	Participation of employees in EM, e.g. bulletins, recognition for low usage of carbon, green teams, feedback and suggestions from employees

Table 2 Specifications of population included in study using SPSS

		Work experience (in years)					Total
		1 < 5	5 < 10	10 < 15	15 < 20	20 or more	
Gender	Male	6	22	12	12	6	58
	Female	4	17	18	7	3	49
Total		10	39	30	19	9	107

The ideal scale measurement for reliability should be above 0.7. Since the readings of each construct is above 0.7, questionnaire is considered reliable. With the objective to identify the implementation of GHRM activities in the organisation, respondents evaluated the below mentioned items on the basis of degree of acceptance and implementation in their respective organisations. Summary of all surveyed activities is mentioned in Table 3.

The aggregate and mean value were calculated for variables of each green HR practices to find out the status of GHRM in surveyed organisations. The mean value ranges between 3 and 2.46. The mean value for each variable is approximately 3 which signifies that each green HR practice has moderate implementation. The topmost two score was found for green training and development (GTD) and green recruitment and selection (GRS). The lowermost grade of execution was found for green performance management system (GPMS) with lowest mean score for item

Table 3 Summary of reliability test of each construct

Construct	N of items	Cronbach's alpha
GRS	5	0.891
GTD	4	0.843
GPMS	5	0.862
GCM	4	0.826
GEE	5	0.875

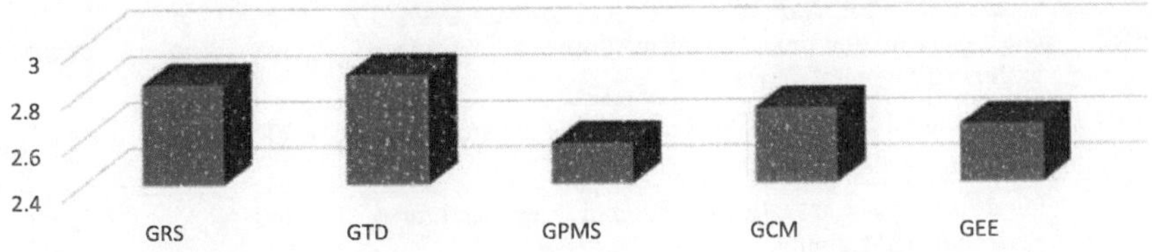

Fig. 2 Graphical representation of degree of implementation of GHRM practices

PMS4, consequences attached to non-attainment or non-compliance with environment management goals. The analysis of the above data shows that TD4 is the most widely practiced activity implying that organisations are trying to involve employees using connect to achieve the environmental goals. However, PMS4 is the least rated item signifying that environmental objectives do not have much importance with respect to performance management system of organisations (Fig. 2).

On the basis of mean score corresponding to extent of implementation of green HR practices, green training and development has topmost rank while green performance management system has lowest rank.

Mean, standard deviation and inter-correlation of variables used for each GHRM practice have been summarised using SPSS in the below mentioned Table 4.

Coefficient correlation ranges from −1 (perfect negative) to +1 (perfect positive correlation). Closer the coefficient values near −1 and +1, higher the correlation; closer to zero means weaker the correlation between variables. As we can see, inter-correlation values are near greater than and near to +1 signifying significant positive correlation.

Hypothesis 1 predicts that environmental practices under green human resource management are well implemented in organisation. The mean value of GHRM is 2.732, which specifies the moderate level of implementation of GHRM practices in surveyed organisations Since the mean value is less than 3 confirming the insignificant implementation of GHRM activities. Henceforth, Hypothesis 1 is not supported. Inter-correlation values for all constructs are stated in Table 5. It is quite evident from the results that all GHRM practices are correlated. Also, GHRM as a whole and its five GHRM practices have significant correlation with organisational sustainability. Therefore, Hypothesis 2 is supported.

Research further investigates the reasons for moderate implementation of GHRM practices in organisations. Respondents were asked to rank barriers (rank 1 for most

Table 4 Evaluation of implementation of green HRM activities

Label	Activities as measurement items	Aggregate value	Mean
TD4	Creation of integrated training module for emotional participation of workforce in EM	321	3.00
TD2	Delivery of environmental training to employees and managers for development of ecological skills and knowledge	319	2.98
RS2	Display of environmental standards in job advertisement	317	2.96
CM2	Excellence award to personnel on environmental related performance and achievement	317	2.96
RS5	During orientation process, new employees are introduced with environmental norms	308	2.88
RS1	Informing job applicants, the commitment towards environment of organisation	304	2.84
TD1	Identification and investigation of employees' need for environmental training	300	2.80
RS4	Prioritising the candidates with skills and hands-on in ecological projects	297	2.78
EE3	Availability proper communications channels (formal and informal) for spreading green culture at workplace	297	2.78
PMS5	Continuous feedback to employees with regard to their performance in attainment of environmental goals	295	2.76
EE1	Well-defined green mission to supervise the employees' conduct in EM	293	2.74
RS3	Verifying of candidates' environmental knowledge and skillset in recruitment	291	2.72
TD3	Continuous improvement in environmental practices in organisation through training	291	2.72
CM1	Reward system for staff on innovative environmental performance	289	2.70
CM4	Provision for loans or tax incentives (Usage of fuel-efficient cars, electric vehicles, bicycle loans)	285	2.66
EE2	Integration of learning climate with green behaviour for awareness among employees	280	2.62
PMS1	Usage of green performance objectives in performance appraisal and management	278	2.60
EE5	Participation of employees in EM, e.g., bulletins, recognition for low usage of carbon, green teams, feedback and suggestions from employees	278	2.60
PMS2	Establishment of green targets, objectives and accountability for managers and employees	276	2.58
CM3	Incentives for encouraging an environmentally friendly practice	276	2.58
EE4	Employee involvement in quality improvisation and sessions for solving problems related to green matters	270	2.52
PMS3	Conduct of environmental assessments	265	2.48

(continued)

Table 4 (continued)

Label	Activities as measurement items	Aggregate value	Mean
PMS4	Consequences attached to non-attainment or non-compliance with environment management goals	263	2.46

significant and rank 8 for most insignificant barrier) to evaluate the barriers that are affecting the acceptance and execution of GHRM at workplace. The result derived from SPSS is shown in Table 6.

As seen, lack of green culture has the lowest mean 2.12 which means that it is the main reason for moderate implementation of GHRM in sampled organisations. However, staff resistance is the most insignificant reason and less difficult to deal with mean value of 6.82. Mostly, every company consider social media sites like LinkedIn, Facebook, Twitter, blogs for attracting talent which may help employer in screening of candidates through his online activities regarding perspective towards environment and also save time and resources. Around sixty per cent organisations keep track of papers used during recruitment and selection. Survey revealed that only few assess green behaviours or personality of candidates during interview. However, it is found that it is very rare to keep track of employees' green behaviour and is not included in performance appraisal cycle. Besides training on environmental issues, some organisations are involved in few practices like internal meetings, campaigns, awareness programmes on environmental issues, NGOs and few others. Some in-house practices such as save water, close tap after use, switch off lights by the last person', posters are also followed. Around half of the organisations has features of green cafeteria like displaying the amount of food wasted the previous day near buffet/counter area so that people consciously serve themselves, less usage of plastic utensils.

Green teams work for aligning various departments to achieve green goal; however, very few organisations have conceptualised green teams. Few reported that it is managed by administration department with no separate team. Corporate Social Responsibility is often managed by HR team or is closely associated with HR department. Mostly, all organisations are involved in CSR practices such as finding different modes to reduce CO_2 emission, increasing green cover by planting trees, awareness programmes on environmental issues, association with NGOs for rural development, education, abolition of child labour.

Data was collected for ISO 9001 and 14001 certification as ISO 14001 includes Environmental Management System. The major difference between ISO 9001 and 14001 is that former is focused on quality management system and latter is focused on EMS. ISO 14001 is considered as GHRM practice as few functions of HRM such as performance management system, training and development, competency mapping are related with it. Very few organisations were ISO 14001 certified and others were ISO 9001 certified. In Asia, differences are found in subregions due to political stability of particular governments, financial resources, focus on training and development, abilities of individual countries to execute.

Table 5 Result of surveyed variables

S. No.	Variables	Mean	SD	1	2	3	4	5	6	7
1	GHRM	2.732	0.789	(0.968)						
2	GRS	2.836	0.865	0.842**	(0.917)					
3	GTD	2.875	0.838	0.921**	0.752**	(0.858)				
4	GPMS	2.576	0.912	0.913**	0.721**	0.788**	(0.921)			
5	GCM	2.725	0.812	0.926**	0.674**	0.785**	0.874**	(0.872)		
6	GEE	2.652	0.851	0.908**	0.689**	0.818**	0.798**	0.836**	(0.883)	
7	OS	3.131	0.682	0.479**	0.412**	0.462**	0.376**	0.453**	0.467**	(0.833)

Note **Correlations are significant at the 0.01 level (two-tailed)

Table 6 Ranking data of barriers in implementation of GHRM

Variables	N	Sum	Mean	SD
Lack of green culture	107	227	2.12	1.586
Lack of understanding green policy	107	310	2.90	1.581
Lack comprehensive plan to implement GHRM	107	357	3.34	2.056
Lack of organisational support	107	492	4.60	2.185
Complexity in implementing green technology	107	530	4.96	1.124
Lack of technical support	107	582	5.44	2.251
Implementation expense	107	623	5.82	1.758
Staff resistance	107	730	6.82	1.119

Willingness and involvement of employees is required for achieving GHRM goals [49]. Employee engagement can be developed by taking feedbacks, suggestions and implementing them. Respondents report various activities like reducing carbon footprints, plantation of trees, saving electricity and paper, waste management undertaken by organisations.

5 Alignment of Green HRM and Organisational Sustainability

Sustainability defined as the development that serves the present without staking the needs of future generation. New dimensions like culture, ethics, aesthetics, compassion, mutual help must be included which were neglected in previous approaches [50]. Sustainability poses both challenge and opportunity for a developing country like India where GHRM is still at nascent stage. "Green Business" is a mounting issue with regard to sustainability of economic advancement [51]. Organisational sustainability is a long-term business process that "the principle for business function to enhance societal, environmental and economic arrangements" [52].

Ongoing research has confirmed that implementation of green HRM practices has direct or indirect effect on environmental performance (EPF). To achieve organisational sustainability goal, it is imperative for green business and green HRM to work in synchronisation along with technology sustainability. Technically advanced unmanned aerial vehicles accompanied with connectivity can reduce human interaction like monitoring social distancing, announcement purposes, examining symptoms and sanitisation of infected areas [53] (Fig. 3).

Integrated acceptance and sustainability assessment model (IASAM) suggests integration of acceptance evaluation with socio-technical factors for framing multi-level framework for technology assessment. As per this model, sustainability of technology that leads to organisational sustainability depends of four factors: proper management of all assets, quality of technology (quality of

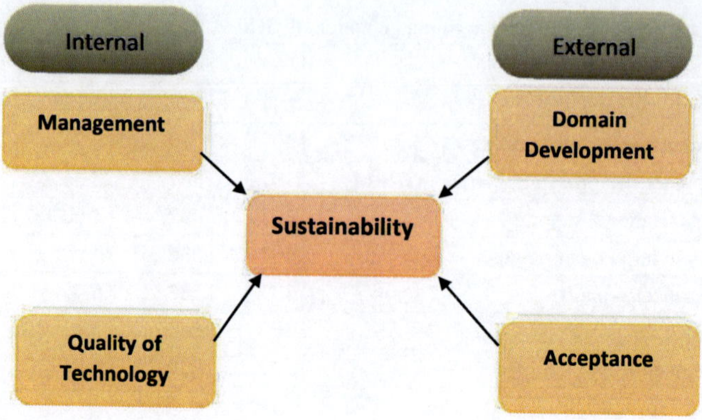

Fig. 3 IASAM model

product/production/supporting devices), impact of domain and societal development and acceptance and economic framework. IASAM model evaluates the existing technology as well as technology in development phase so that it can be used for sustainable purpose (Fig. 4).

A well-defined HR framework can drive human resource to introduce policies and practices. Encouraged managers and employees' involvement eases implementation of green HRM, and they may also generate ideas to go green promoting the EPF goals.

6 Discussion

When every organisation is struggling to make topmost position, it is very important make shift from strategic motives to sustainability. To translate strategy into operational practice, it is imperative to align environment and human resource practices. Organisations are still lacking the understanding and execution of GHRM and relevance of green behaviour. GHRM is still an underdeveloped area with less empirical studies been done. The study reveals that GHRM has moderate degree of implementation in surveyed organisations due to various barriers affecting the same with lack of green culture, insufficient understanding of green policy and lack of comprehensive plan to implement GHRM being the most common ones. Data analysis reflects that organisations have certain GHRM activities but these are not formally structured or diligently followed. Though they try to create awareness among employees through internal meetings, awareness campaigns, but none of them was able to reveal the environmental management budget. Qualitative research revealed that organisations do not have knowledge base of GHRM concept and its application, lack of approach in execution of HRM linked environmental practices, insufficient developed measurement and reporting process in context to GHRM. It can be further said

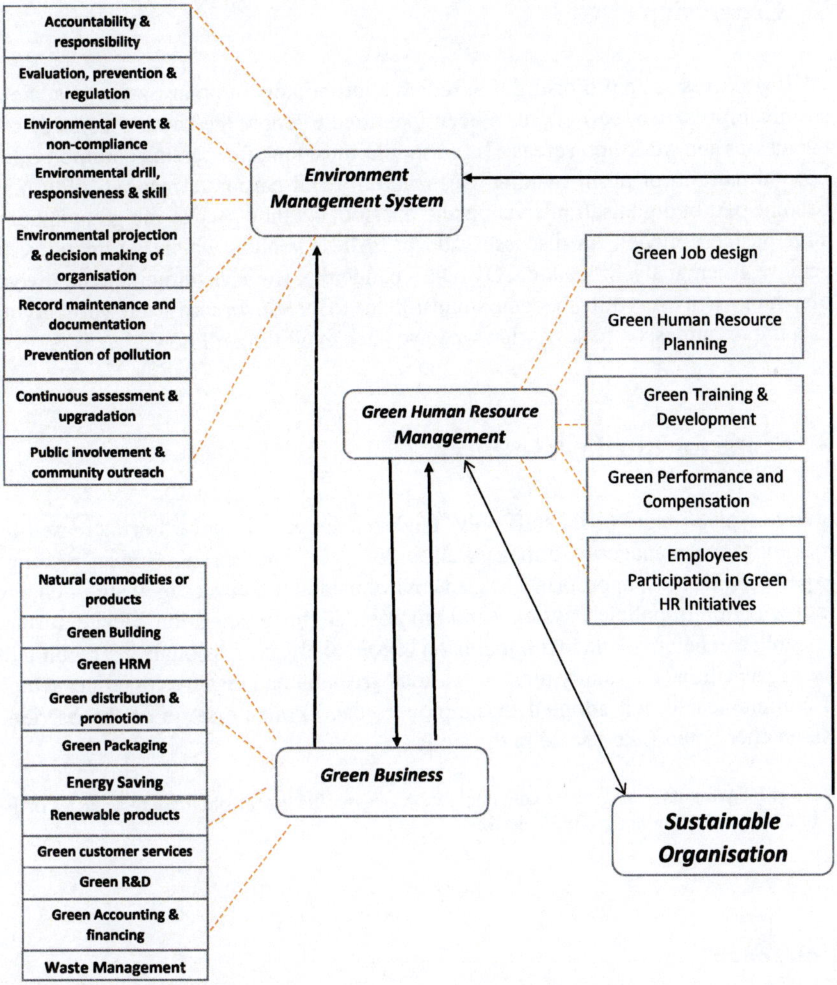

Fig. 4 Organisation sustainability model through integration of EMS and green human resource HRM. *Source* Created by Author

that some organisations make arrangements not to promote green behaviours, but to confine environmental deterioration because of production. However, companies can take initiatives to synchronise green initiatives with sustainability by identifying the present scenario of company and launches green plan (short-term initiatives), tracking performance and building environmental culture (middle-term initiative), incorporating environmental rational into business strategies (long-term initiative).

7 Conclusion

GHRM proves to be promising if executed appropriately in organisation to achieve sustainability. Also, barriers have been identified to know the current status of our businesses and work-upon areas. Organisations must look for creating common value beyond the aim of profit making. Sustainability and corporate responsibility may become part of organisation's vision and mission, but they need to integrate in operative business model. As discussed above, EMS is venture which is imperative to achieve green goals. Since the COVID-19 pandemic, we have witnessed the revival of nature when everything became standstill due to lockdown across the world giving us a important lesson to save what we have been exploiting till now.

8 Scope for Further Research

Study involved managerial staff only; however, scope can be further increased by including non-managerial staff as well with larger sample size. Human resource experts can be most appropriate to lead environmental practice as they are in-charge for regulation of policies, systems and processes in the organisation. Hence, further research can be broadened with inclusion of role of HR professionals in greening of the organisation. This study presents various green human resource practices which if communicated and adopted by employees can eliminate some of the barriers. Researcher opted Likert scale in the study.

Acknowledgements This research did not receive any specific grant from funding agencies in the public, commercial or not-for-profit sectors.

References

1. José Luis Blasco AK (2017) The road ahead: the KPMG survey of corporate responsibility reporting 2017
2. Victor DG (2001) The collapse of the kyoto protocol and the struggle to slow global warming. Princeton University Press, Princeton, NJ
3. Shrivastava P, Berger S (2010) Sustainability principles: a review and directions. Organ Manag J 7:246–261. https://doi.org/10.1057/omj.2010.35
4. Rondinelli DA, Berry MA (2000) Environmental citizenship in multinational corporations: social responsibility and sustainable development. Eur Manag J 18(1):70–84
5. Bocken NM, Short SW, Rana P, Evans S (2014) A literature and practice review to develop sustainable business model archetypes. J Clean Prod 65:42–56. https://doi.org/10.1016/j.jclepro.2013.11.039
6. Temple J (2020) Why the coronavirus outbreak is terrible news for climate change. MIT Technol Rev. https://www.technologyreview.com/2020/03/09/905415/coronavirus-emissions-climate-change/. Last accessed 27 April 2020

7. World Trade Organization (WTO) (2020) Trade set to plunge as COVID-19 pandemic upends global economy. Press Release April 8. https://www.wto.org/english/news_e/pres20_e/pr855_e.htm
8. Federal Ministry for Environment NC (2020) Global survey on sustainability and SDGs: awareness, priorities, need for action
9. BCG and MIT (2009) The business of sustainability: imperatives, advantages and actions. In: BCG Report, September. BCG and MIT, Boston, MA
10. McKinsey (2013) The business of sustainability. In: McKinsey Report, Summer. J Organ Behav Mental Behav
11. Mallet V (2013) Environmental damage costs India $80 bn a year. Financial Times, 17. Available at www.ft.com/content/0a89f3a8-eeca-11e2-98dd-00144feabdc0
12. Ones DS, Dilchert S (2012) Environmental sustainability at work: a call to action. Ind Organ Psychol 5(4):444–466
13. Oskamp S (2000) A sustainable future for humanity? How can psychology help? Am Psychol 55(5):496–509
14. Margaretha M, Saragih S (2013) Developing new corporate culture through green human resource practice. In: International conference on business, economics, and accounting, pp 1–10
15. Sathyapriya J, Kanimozhi R, Adhilakshmi V (2013) Green HRM—delivering high performance HR systems. Int J Mark Human Resour Manag 4(2):19–25
16. Renwick DWS, Redman T, Maguire S (2013) Green human resource management: a review and research Agenda. Int J Manag Rev 15(1):1–14. https://doi.org/10.1111/j.1468-2370.2011.00328
17. Jabbour CC (2013) Environmental training in organisations: from a literature review to a framework for future research. Resour Conserv Recycl 74:144–155
18. Jackson SE, Ones DS, Dilchert S (2012) Managing human resources for environmental sustainability, vol 32. Wiley
19. O'Donohue W, Torugsa N (2016) The moderating effect of 'green' HRM on the association between proactive environmental management and financial performance in small firms. Int J Human Resour Manag 27(2):239–261
20. Opatha HHP, Arulrajah AA (2014) Green human resource management: simplified general reflections. Int Bus Res 7(8):101
21. Elkington J (2006) Governance for sustainability. Corporate Governance Int Rev 14(6):522–529
22. Elkington J (1997) Cannibals with forks: the triple bottom line of 21st century business. Capstone, Oxford, p 402
23. Rimanoczy I, Pearson T (2010) Role of HR in the new world of sustainability. Ind Commer Train 42(1):11–17
24. Ambec S, Lanoie P (2008) Does it pay to be green? A systematic overview. Acad Manag Persp 45–62
25. Kaur H (2013) Today's success mantra-going green at functional areas of HRM. Int J Manag Bus Stud 3(1):96–99
26. Matlay H, Khandekar A, Sharma A (2005) Organizational learning in Indian organizations: a strategic HRM perspective. J Small Bus Enterprise Dev
27. Chadwick C (2010) Theoretic insights on the nature of performance synergies in human resource systems: toward greater precision. Human Resour Manag Rev 20(2):85–101. https://doi.org/10.1016/j.hrmr.2009.06.001
28. Guerci M, Longoni A, Luzzini D (2015) Translating stakeholder pressures into environmental performance—the mediating role of green HRM practices. Int J Human Resour Manag 27(2):262–289. https://doi.org/10.1080/09585192.2015.1065431
29. Ehnert I (2009) Sustainable human resource management. Springer, London
30. Grolleau G, Mzoughi N, Pekovic S (2012) Green not (only) for profit: an empirical examination of the effect of environmental-related standards on employees' recruitment. Resour Energ Econ 34(1):74–92

31. Touboulic A, Walker H (2015) Theories in sustainable supply chain management: a structured literature review. Int J Phys Distrib Logistics Manag 45(1/2):16–42. https://doi.org/10.1108/IJPDLM-05-2013-0106
32. Jabbour CJC (2015) Environmental training and environmental management maturity of Brazilian companies with ISO14001: empirical evidence. J Clean Prod 96:331–338
33. Teixeira AA, Jabbour CJC, de Sousa Jabbour ABL, Latan H, de Oliveira JHC (2016) Green training and green supply chain management: evidence from Brazilian firms. J Clean Prod 116:170–176. https://doi.org/10.1016/j.jclepro.2015.12.061
34. Mishra Pavitra (2017) Green human resource management: a framework for sustainable organizational development in an emerging economy. Int J Organ Anal 25(5):762–788. https://doi.org/10.1108/IJOA-11-2016-1079
35. Gholami H, Rezaei G, Saman MZM, Sharif S, Zakuan N (2016) State-of-the-art green HRM system: sustainability in the sports center in Malaysia using a multi-methods approach and opportunities for future research. J Clean Prod 124:142–163
36. Teixeira AD, Jabbour CJC, Jabbour SLB (2012) Relationship between green management and environmental training in companies located in Brazil: a theoretical framework and case studies. Int J Prod Econ 139(2):1–12
37. Liebowitz J (2010) The role of HR in achieving a sustainability culture. J Sustain Dev 3(4):50–57
38. Pillai R, Sivathanu B (2014) Green human resource management. Zenith Int J Multi Res 4(1):72–82
39. Robertson JL, Barling J (2013) Greening organizations through leaders' influence on employees' pro-environ-mental behaviors. J Organ Behav 34(2):176–194
40. Zutshi A, Sohal AS (2004) Adoption and maintenance of environmental management systems: critical success factors. Manag Environ Qual Int J 15(4):399–419
41. Benn S, Teo ST, Martin A (2015) Employee participation and engagement in working for the environment. Personnel Rev
42. Johnson D, Wetmore J (eds) (2009) Technology and society. Building our sociotechnical future. MIT Press, Cambridge, MA
43. Perkins R (2003) Environmental leapfrogging in developing countries: a critical assessment and reconstruction. Nat Resour Forum 27:177–188
44. Aizstrauta D, Ginters E (2013) Introducing integrated acceptance and sustainability assessment of technologies: a model based on system dynamics simulation. In: Springer LNBIP 145 series modeling and simulation in engineering, economics and management, Spain. Springer, Berlin Heidelberg, pp 23–30
45. John Mohr MV (2002) Archival research methods
46. Tang G, Chen Y, Jiang Y, Paillé P, Jia J (2017) Green human resource management practices: scale development and validity. Asia Pacific J Human Resour 56(1):31–55. https://doi.org/10.1111/1744-7941.12147
47. Chaudhary R (2019) Green human resource management in Indian automobile industry. J Global Responsib. https://doi.org/10.1108/jgr-12-2018-0084
48. Santos JRA (1999) Cronbach's alpha: a tool for assessing the reliability of scales. J Extension 37(2):1–5
49. Aragon-Correa JA, Martin-Tapia I, Hurtado-Torres NE (2013) Proactive environmental strategies and employee inclusion: the positive effects on information sharing and promoting collaboration and the influence of uncertainty. Organ Environ 40:1–23
50. Leal Filho W, Raath S, Lazzarini B, Vargas VR, De Souza L, Anholon R, Orlovic VL (2018) The role of transformation in learning and education for sustainability. J Clean Prod 199:286-295. https://doi.org/10.1016/j.jclepro.2018.07.017
51. Das SC, Raj KS (2016) Green HRM and organizational sustainability: an empirical review. Kegees J Soc Sci 8:227–236
52. CIPD (2012) A collection of thought pieces—responsible and sustainable business: HR leading the way. CIPD, London, UK

53. Gupta R, Kumari A, Tanwar S, Kumar N (2020) Blockchain-envisioned softwarized multi-swarming UAVs to tackle COVID-I9 situations. IEEE Netw

Anemia Multi-label Classification Based on Problem Transformation Methods

Bhavinkumar A. Patel and Ajay Parikh

Abstract The CBC report is considered to be the most important report for assessing the overall health of the human body which is an important test for the diagnosis of diseases such as anemia, cancer, infections, vitamin, and mineral deficiencies. Anemia is a common health problem among people worldwide. Anemia is not a disease but a sign of a serious illness. Therefore, it can be prevented from serious diseases diagnosed at an earlier stage. The pattern of CBC parameters is found to be very complex. Therefore, a multi-label classification of the types of IDA, Vitamin B12, aplastic, and sickle cell anemia has been made by the machine learning problem transformation method which gives an idea of the type of anemia that is likely to occur due to the abnormal condition of CBC parameters. The use of the model can help predict anemia at an earlier stage. And serious diseases can be avoided. Ultimately, people can be saved from financial costs, kept mentally healthy, and concentration and regularity in teaching children can also be improved. This multi-label classification inspires the application of methods to classify these anemia types. Binary relevance, classifier chains, and label power set methods are used with different base classifications, and the results are analyzed. The results obtained by the SVM model based on the classifier chains method of multi-label classification have proved to be superior to other methods.

Keywords Anemia · Multi-label classification · Machine learning

B. A. Patel (✉) · A. Parikh
Department of Computer Scince, Gujarat Vidyapith, Ahmedabad, Gujarat, India
e-mail: bhavin.patel@gujaratvidyapith.org

A. Parikh
e-mail: ajay@gujaratvidyapith.org

P. K. Singh et al. (eds.), *Proceedings of Second International Conference on Computing, Communications, and Cyber-Security*, Lecture Notes in Networks and Systems 203,
https://doi.org/10.1007/978-981-16-0733-2_44

1 Introduction

CBC plays a vital role in a thorough examination of the health of the body. It is useful for the diagnosis of various diseases such as anemia, infections, cancer, vitamin, and mineral deficiency. Anemia is a common problem in people all over the world which is likely to reason critical illness in the body. Therefore, it is very important to make a diagnosis at an early stage which can cause worldwide epidemics and mortality [1]. Anemia and its four types are classified based on CBC. It shows the type of anemia caused by the abnormal state of the CBC parameters. Many variations and complexities in the pattern of CBC parameters have been observed. Modern approaches to health care the prognosis of anemia is based on the basic blood test parameters of the clinical laboratory, and thus, it is very important to prevent anemia at an early stage which can improve their level of health and education to prevent the risk of serious diseases.

A proposed model has been predicted IDA, Vitamin B12, aplastic, and sickle cell using a multi-label classification problem transformation method by machine learning.

1.1 Types of Anemia

Anemia is a low level of hemoglobin (HB) and red blood cells (RBCs) in the human body. Here, Fig. 1 shows the types of anemia.

1.1.1 Iron Deficiency

Iron deficiency anemia which is caused due to a low amount of iron in the body is called iron deficiency anemia. Iron is the main and essential component of hemoglobin for its proper function. The main cause of low iron levels in the body is long-term blood loss. These basic parameters of the CBC report are used to predict iron deficiency anemia (HB-Low, PCV-Low, MCV-Microcytic, and MCHC-Low).

Case-1

IDA when anemia is present: Clinical parameters of iron deficiency anemia when anemia is present have low HB and RBC levels.

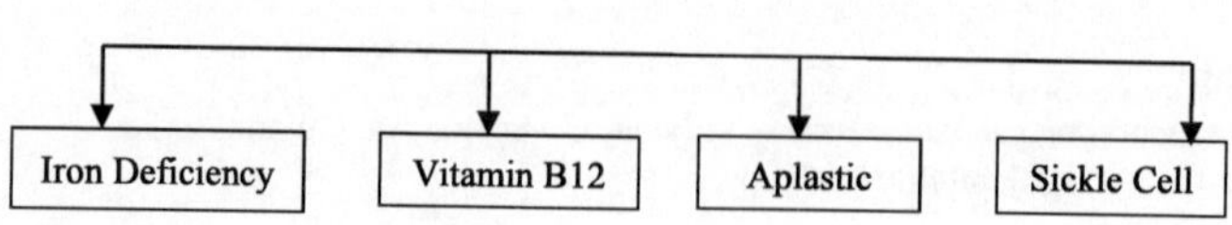

Fig. 1 Types of anemia

Case-2

IDA when anemia is not present: Iron deficiency anemia when anemia is not present the condition of clinical parameters is observed only in case of low hemoglobin level. It can be called latent iron deficiency anemia which is found in stage 3.

In stage 3, anemia (decrease in hemoglobin level) is present but the appearance of red blood cells remains normal.

1.1.2 Vitamin B12

Vitamin B12 deficiency occurs when the body has insufficient levels of Vitamin B12. This is an important vitamin for the construction of RBC and the hygienic functioning of the powerful system.

Vitamin B12 has low levels of both HB and RBC parameters, in which the condition of the clinical parameters of anemia is observed. The parameters of CBC (HB-Low, PCV-Low, MCV-Macrocytic, and MCHC-High) are also taken into account for its prediction. When there is no anemia but there is Vitamin B12, only HB is found to below.

1.1.3 Aplastic

Aplastic anemia is an autoimmune disease that fails to produce a sufficient number of blood cells in the body. It has the lowest proportion of major CBC parameters like RBC, WBC, and platelet. The parameters of CBC (HB-Low, RBC-Low, WBC-Low, PCV-Low, MCV-Macrocytic, MCH-High, and Platelet-Low) are also taken into account for its prediction [2].

1.1.4 Sickle Cell

A sickle cell is usually a faction of a blood distortion, and it is inherited. It indicates the result of the abnormality of protein in the conduction of oxygen and hemoglobin found in red blood cells. Sickle cell anemia is predicted based on the parameters of the CBC (HB-Low, RBC-Low, WBC-High, PCV-Low, MCV-Macrocytic, MCH-High, and Platelet-High) [3].

The effect of abnormal conditions of different parameters of CBC on the types of anemia is shown in Table 1.

Table 1 Differences between the laboratory parameters of the types of anemia

CBC parameter	IDA	Vitamin B12	Aplastic	Sickle cell
HB	↓	↓	↓	↓
RBC	↓	↓	↓	↓
WBC	Ordinary	Ordinary	↓	↑
PCV	↓	↓	↓	↓
MCV	↓ Microcytic	↑ Macrocytic	↑ Macrocytic	↑ Macrocytic
MCH	Ordinary	Ordinary	↑	↑
Platelet	Ordinary	Ordinary	↓	↑
MCHC	↓	↑	Ordinary	Ordinary

2 Materials and Methods

2.1 Dataset

For anemia multi-label classification, the CBC report of the newly admitted students of Gujarat Vidyapith for higher education from the last five years, i.e., 2014 to 2018 has been used as a dataset, in which a total of 2190 samples have been taken. In addition, CBC samples of people aged 21 to 32 years have been taken from a total of five clinical pathology laboratories, two from Ahmedabad district, and three from Gandhinagar district. It includes a total of 40,938 sample training datasets and 768 sample test datasets. There are many disparities due to different testing equipment. Thus, not all parameters of the CBC report can be said to be important for the prognosis of anemia. That can be said based on our previous research paper [1, 4]. For that only 16 main parameters have been taken.

Multi-label classification of anemia dataset using a total of 16 parameters as shown in Table 2.

2.2 Multi-label Classification

Multi-label classification are closely related to modern real-world applications, including disease diagnosis and genre classification.

Table 3 shows three out of four positive labels so this is called a multi-label classification.

Multi-label classification is divided into three methods, namely problem transformation, algorithm adaptation, and ensemble approach. Problem transformation methods convert a multi-label dataset into one or more single-label datasets so that the classification problem is solved by a single-label classification. To handle direct multi-label data, an algorithm adaptation method is used.

Table 2 Multi-label anemia classification dataset attributes

S. No.	Attribute	Description	S. no	Attribute	Description
1.	Age	Age Group 21–32	9.	RDW	Normal <11 >15 Abnormal ↓<11 ↑>15
2.	Sex	Male Female	10.	WBC	Normal <3400 >9600 Abnormal ↓<3400 ↑>9600
3.	HB	Male <13.5 Female < 12	11.	Platelet	Male < 13,500 Female < 157,000
4.	RBC	Male < 4.3 Female < 3.5	12.	Anemia_Status (Target) Lable-1	Male HB < 13.5, RBC <4.3 Female HB < 12, RBC < 3.5
5.	PCV (HCT)	Male < 41 Female < 36	13.	IDA (Target) Lable-2	Low HB, RBC, PCV, MCHC MCV = Microcytic
6.	MCV	Microcytic <80 Normocytic <80 >100 Macrocytic >100	14.	Vitamin B12 (Target) Lable-3	Low HB, PCV, MCV = Macrocytic MCHC = High
7.	MCH	Normal <25.4 >34.6 Abnormal ↓ <25.4 ↑ >34.6	15.	Aplastic (Target) Lable-4	Low HB, RBC, WBC, PCV, Platelet MCV = Macrocytic MCH = High
8.	MCHC	Normal <31 >36 Abnormal ↓<31 ↑>36	16.	Sickle cell (Target) Lable-5	Low HB, RBC, PCV High WBC, MCH, Platelet MCV = Macrocytic

Table 3 Multi-label classification

IDA	Vitamin B12	Aplastic	Sickle cell
Yes	Yes	No	Yes

Multi-label classification is evaluated using the problem transformation method to solve the problem.

2.2.1 Problem Transformation

The main purpose of the approach is to convert the actual multi-label into a set of single-label classification. Its function is not directly based on the classification method used so it is independent. In this research, group-related algorithms are used, with several problem transformation methods.

Table 4 Multi-label classification labels

X	Y1 (Anemic)	Y2 (IDA)	Y3 (VitB12)	Y4 (APA)	Y5 (sickle cell)
X1	1	1	0	0	0
X2	1	0	1	0	1
X3	1	0	1	1	0
X4	0	1	0	0	0
X5	0	0	1	0	1

X	Y1
X1	1
X2	1
X3	1
X4	0
X5	0

X	Y2
X1	1
X2	0
X3	0
X4	1
X5	0

X	Y3
X1	0
X2	1
X3	1
X4	0
X5	1

X	Y4
X1	0
X2	0
X3	1
X4	0
X5	0

X	Y5
X1	0
X2	1
X3	0
X4	0
X5	1

Fig. 2 Binary relevance of each label classification

In this research, three methods related to the problem transformation category have been used. Binary relevance, which converts multi-labels to 5 different single label. Classifier chains that are called binary relevance improvements. It chains each label one by one using the previous target variable as the input space. Label power set, which converts a multi-label into a single-label.

where X is the input variable, and Y is the target variable.

In Table 4, X1, X2, X3, …, Xn are the independent variables, and Y1 (Anemic), Y2 (IDA), Y3 (VitB12), Y4 (APA), Y5 (SickleCell) are the target variables [5], where Y1 is anemic or non-anemic, Y2 is an iron deficiency, Y3 is a Vitamin B12 deficiency, Y4 is aplastic, and Y5 is a sickle cell in which 0 indicates negative and 1 indicates positive.

Binary Relevance

Binary relevance is one of the most used problem transformation methods. BR treats each label's prediction as a free binary classification function. This is a simple technique that basically treats each label as a separate classification problem.

The binary relevance problem divides into five different single-label classification problems as shown in Fig. 2.

Experimental results have been evaluated using different evaluation measures, with example-based measures (hamming loss, accuracy), ranking-based measures (average accuracy, coverage, and one error), and label-based measures (macro-F1 and micro-F1) are detected.

Table 5 A Binary relevance model evaluation comparison

Classifier	SVM	Logistic R.	J48	Naïve Bayes	Decision table	Random forest
Accuracy ↑	1	0.988	0.982	0.815	0.786	0.658
Hamming loss ↓	0	0.003	0.004	0.046	0.052	0.093
Micro-F1 ↑	1	0.991	0.988	0.87	0.856	0.767
Macro-F1 ↑	0.421	0.421	0.421	0.405	0.413	0.413
One error ↓	0.579	0.579	0.579	0.581	0.579	0.579
AvgPrecision ↑	1	1	1	0.999	1	1

B Label wise accuracy comparison

Label	SVM	Logistic R.	J48	Naïve Bayes	Decision table	Random forest
Anemia status	1	0.999	1	0.951	1	0.927
IDA	1	0.995	0.982	0.909	0.982	0.807
Vitamin B12	1	0.993	1	0.943	0.999	0.863
APA	1	0.999	1	0.974	0.845	0.977
Sickle cell	1	1	1	0.993	0.915	0.958

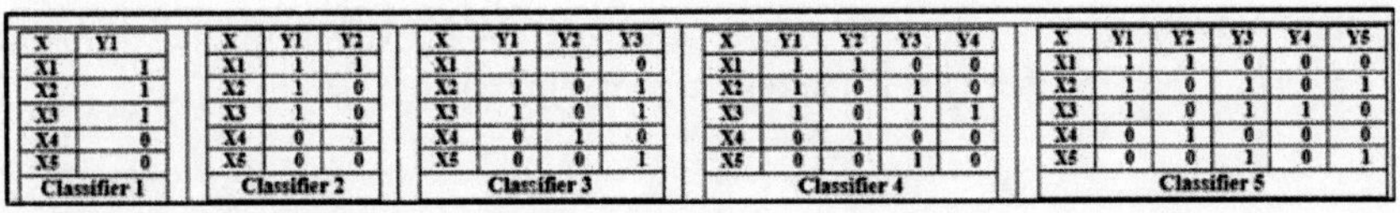

Fig. 3 Multi-label chain classifier each labels chain

Table 5A compares the multi-label classification evaluation model based on its different criteria using different classifiers with the help of the binary relevance (BR) method.

Table 5B shows an example base evaluation of five different labels using the binary relevance method of problem transformation of multi-label classification.

Classifier Chains

The classifier chains method based on the BR method eliminates the shortcomings of the BR and performs a high forecast, retaining the significant advantage of the BR, a very meaningful short-term complication [6]. CC provides a normal problem transformation method that has previously achieved BR efficiency and competes with higher accuracy of more computational complex methods [7].

Here, X is taken as the input and Y as the label (Fig. 3) [5].

Both this method and the binary relevance method are similar; the only difference is that they form a chain to maintain the relationship of the label.

Table compares the multi-label classification evaluation model based on its different criteria using different classifiers using the classifier chains (CC) method.

Table 6 A Chain classifier model evaluation comparison

Classifier	SVM	Logistic R.	Decision table	Random forest	J48	Naïve Bayes
Accuracy ↑	1	0.997	0.995	0.992	0.982	0.941
Hamming loss ↓	0	0.001	0.001	0.002	0.004	0.014
Micro-F1 ↑	1	0.997	0.997	0.995	0.988	0.958
Macro-F1 ↑	0.421	0.421	0.421	0.421	0.421	0.414
One error ↓	0.579	0.579	0.579	0.579	0.579	0.583
AvgPrecision ↑	1	1	1	1	1	0.997

B Label wise accuracy comparison

Label	SVM	Logistic R.	Decision table	Random forest	J48	Naïve Bayes
Anemia status	1	1	1	1	1	0.984
IDA	1	0.999	0.995	0.992	0.982	0.965
Vitamin B12	1	0.999	1	1	1	0.990
APA	1	1	1	1	1	0.993
Sickle cell	1	0.999	1	1	1	1

Table 6B shows an evaluation of accuracy based on an example of five different labels using the classifier chains method of problem transformation.

Label Power Set

The label power set method in the problem transformation approach creates a new class for each connection of labels and then uses the multiclass classification approach to solve the problem. Its disadvantage is that this approach leads to a fatal increase in the number of classes; as a result, many generated classes have very low labeled patterns that lead to overexploitation. In BR, this defect is eliminated by the label power set, also known as label cardinality [7].

Moreover, by using all the unique existing subsets (specific subsets of labels) of multi-labels in training patterns using class target values.

Here, X is taken as the input and Y as the label.

Figure 4 shows the same labels in X1 and X4, and the same set of labels in X3 and X6.

The label power set problem as shown in Table 7 has given each possible label combination a unique class.

In Table 8B, the outcome acquires by each label based on the classification method are classified on the basis of six classification algorithms. The outcome is presented on the basis of various evaluation measures. The optimal outcome achieved by the SVM classifier in each step are strong which is statistically significant.

In the research work, the hamming loss and accuracy of the example-based steps as well as in label-based measures, average precision, and one error are used to appraise the operation of the machine learning classifier.

X	Y1(Anemic)	Y2(IDA)	Y3(VitB12)	Y4(APA)	Y5(SickleCell)
X1	0	1	1	0	1
X2	1	0	0	0	0
X3	0	1	0	0	1
X4	0	1	1	0	1
X5	0	0	1	0	1
X6	0	1	0	0	1
X7	1	1	0	1	0

Fig. 4 Label power set classification

Table 7 Label power set each label classification

X	X1	X2	X3	X4	X5	X6	X7
Y1	1	2	3	1	4	3	5

Table 8 A Label power set model evaluation comparison

Classifier	SVM	Logistic R.	Random forest	J48	Naïve Bayes	Decision table
Accuracy ↑	1	0.997	0.992	0.982	0.979	0.583
Hamming loss ↓	0	0.002	0.002	0.004	0.004	0.092
Micro-F1 ↑	1	0.995	0.995	0.988	0.986	0.748
Macro-F1 ↑	0.421	0.421	0.421	0.421	0.42	0.381
One error ↓	0.579	0.579	0.579	0.579	0.579	0.599
AvgPrecision ↑	1	1	1	1	1	0.989

B Label wise accuracy comparison

Label	SVM	Logistic R.	Random forest	J48	Naïve Bayes	Decision table
Anemia_Status	1	0.999	1	1	0.999	0.952
IDA	1	1	0.992	0.982	0.986	0.965
Vitamin B12	1	0.996	1	1	0.996	0.621
APA	1	1	1	1	0.997	1
Sickle cell	1	0.997	1	1	1	1

3 Result and Discussion

The dataset contains 40,938 samples in the training set and 768 samples in the test set where five class labels are taken in each component. Different types of multi-label classifications have been performed on anemia datasets using the Meka-Release-1.9.3-SNAPSHOT tool. Methods of problem transformation are used to build a model and evaluate their performance using measures such as average precision, one error, accuracy, hamming loss.

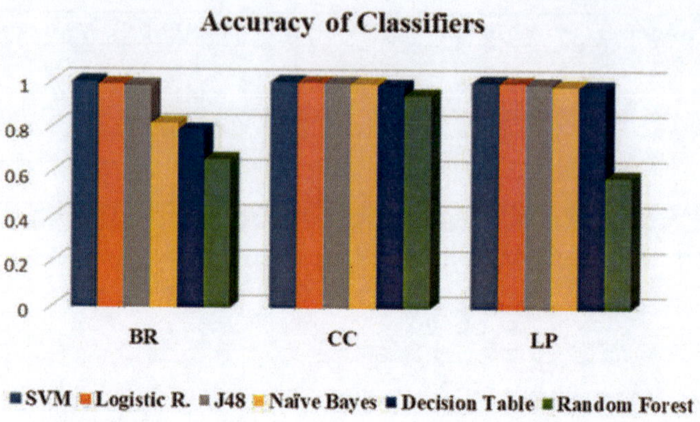

Fig. 5 Accuracy of classifiers difference

The operation of multi-label classifiers with various base classifiers, including SVM, J48, random forest, logistic regression, decision tree, and naive Bayes, is present in the previous table (Fig. 5).

Classifier chains with SVM classifier depending on the method, the accuracy of the machine learning classifier is more than any other classifier [7].

The best classification indicates that the value of the hamming loss is less than or equal to zero. From Fig. 6, it can be seen that the SVM is showing the good performance of the CC-based model with base classifier in which the hamming loss value is found below.

Like the hamming loss metric, the value of one error must below for an efficient classifier. As shown in Fig. 7, better performance of both BR and CC-based classification is observed.

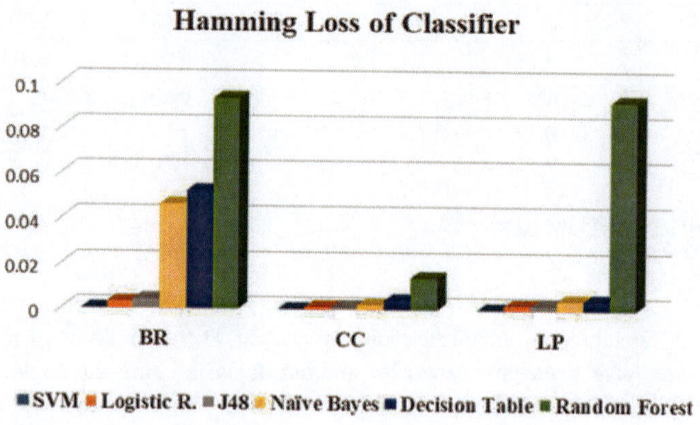

Fig. 6 Hamming loss classifiers difference

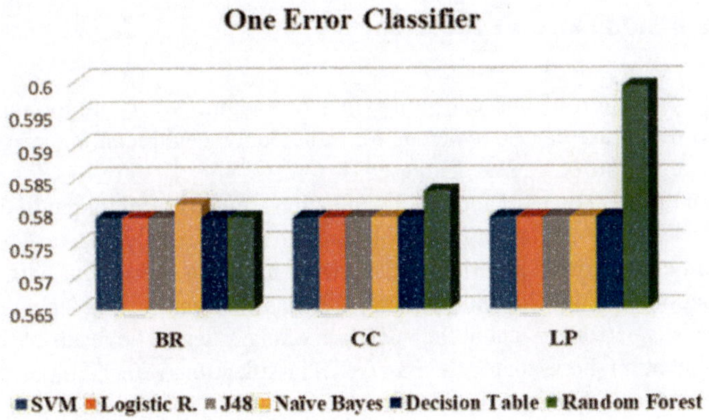

Fig. 7 Difference between a one error of classifiers

The accuracy and other multi-label classification parameters found in the model evaluation indicate that B.R and CC-based SVM classifiers perform best. Therefore, multi-label classification can work very well in classification functions of anemia type in the CBC test (Fig. 8).

The results suggest that the chain classifier method using SVM-based classification provides a better classification of the type of anemia than other methods. An analogical analysis of multi-label classifiers based on various operational steps is presented in the diagrams [7].

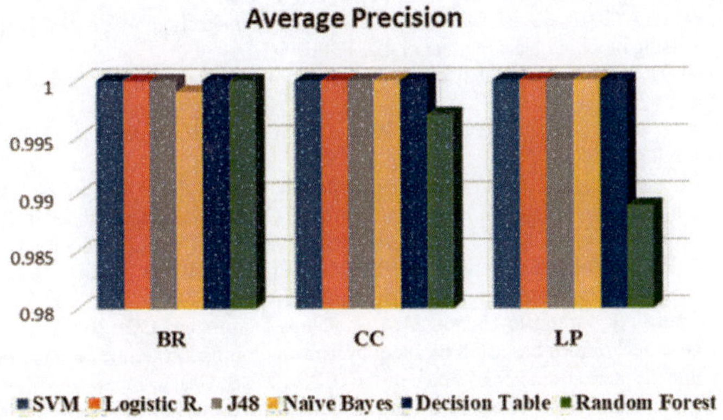

Fig. 8 Difference between an average precision of classifiers

4 Conclusions and Future Work

Thus, the conclusion of that early diagnosis of anemia is very important through which severe disease can be avoided. An Individual's health can be corrected and the drop out proportion can be decreased.

The four different types of anemia classification are also useful for predicting many serious diseases. IDA, Vitamin B12, Aplastic, and sickle cell anemia have been classified using classification algorithms. This work has been carried out with problem transformation methods like binary relevance, classifier chains, and label power set. The analysis is performed using the operation of classifiers. The result of the multi-label classification shows that SVM-based CC classifications offer significantly better performance than other classifications.

As research opportunities and future work, the task of classifying different types of anemia could be expanded using other machine learning algorithms to improve performance. Chronic anemia, hemolytic anemia, thalassemia as well as other serious diseases and infections can be detected by other future parameters of CBC.

References

1. Patel B, Parikh A (2020) Impact Analysis of the complete blood count parameter using Naive Bayes. In: IEEE 5th international conference on inventive computation technologies (ICICT 2020)
2. Wikipedia. https://en.wikipedia.org/wiki/Aplastic_anemia
3. Wikipedia. https://en.wikipedia.org/wiki/Sickle_cell_disease
4. Patel B, Parikh A (2018) Search for an essential parameter and technique for effective prediction of disease using machine learning technique. JETIR 5(10)
5. Jain S (2017) Solving multi-label classification problems (Case studies included). Analyticsvidhya
6. Jesse R, Bernhard P, Holmes G, Frank E (2011) Classifier chains for multi-label classification. Mach Learn 85:333–359
7. Pushpaa M, Karpagavalli S (2017) Multi-label classification: problem transformation methods in Tamil phoneme classification. In: 7th International Conference on Advances in Computing and Communications, ICACC-2017, 22–24 August 2017. Elsevier, Science Direct
8. Read J (2013) Multi-label Classification. Universidad Carlos III de Madrid. Department of Signal Theory and Communications Madrid, Spain (2013)
9. Giraldo-Forero AF, Jaramillo-Garzon JA, Castellanos-Domınguez CG (2013) A comparison of multi-label techniques based on problem transformation for protein functional prediction. In: 35th annual international conference of the IEEE EMBS, Osaka, Japan
10. Santos AM, Canuto AMP, Feitosa NA (2011) A comparative analysis of classification methods to multi-label tasks in different application domains. IJCISIM 3:218–227. ISSN 2150-7988
11. Giraldo-Forero AF, Jaramillo-Garzon JA, Castellanos-Domınguez CG (2015) Evaluation of example-based measures for multi-label classification performance. In: Ortuno F, Rojas I (eds) IWBBIO 2015, Part I, LNCS 9043. Springer International Publishing Switzerland, pp 557–564
12. Raed A, Thabtah F, Al-Radaideh Q (2015) A multi-label classification approach based on correlations among labels. Int J Adv Comput Sci Appl (IJACSA) 6(2)
13. Li L, Liu H, Ma Z, Mo Y, Duan Z, Zhou J, Zhao J (2014) Multi-label feature selection via information gain. In: Luo X, Yu JX, Li Z (eds) ADMA 2014, LNAI 8933. Springer International Publishing Switzerland, pp 345–355

14. Agarwal AM (2014) Diagnostic approach to Anemia. University of Utah School of Medicine, Department of Pathology, ARUP Laboratories
15. Christine AH (2019) Anemia assessment. Springer Nature, Switzerland AG
16. Grigorios T, Ioannis K, Ioannis V (2010) Mining multi-label data. In: Maimon O, Rokach L (eds) Data mining and knowledge discovery handbook, 2nd ed. Springer Science Business Media, LLC
17. Grigorios T, Ioannis K (2007) Multi-label classification: an overview. IGI Publishing
18. de Carvalho A, Freitas A (2009) A tutorial on multi-label classification techniques. In: Foundations of computational intelligence, vol 5

Automated Disease Detection and Classification of Plants Using Image Processing Approaches: A Review

Shashi and Jaspreet Singh

Abstract For preventing damages in agriculture field plant monitoring is necessary. Plant monitoring or plant disease detection at primary stage can enhance the productivity of crops in terms of quality and quantity both. Field monitoring can be possible in many ways like farmers can take the help of experts or they can use the pesticides also for removing unwanted plants and diseases from plant but experts presence is not possible at every place second problem about pesticides in how much quantity farmers should use it. So these traditional approaches not suitable as much or they takes lots of time. For effective growth of yield and for increment the farmers benefit there is need of automated plant disease detection. Automated plant disease detection can be possible by many techniques such as by Image processing, computer vision, machine learning and through neural network, etc. In this paper, we discussed the Image Processing technique with its approaches such as Image Acquisition, Preprocessing, Segmentation, Feature Extraction and Classification. This paper shows the potential of plant disease detection system that offers the favorable opportunity in agriculture field. The review presents in this paper showing the detailed discussion about existing studies with their strengths and limitation and also giving the information about uncovered research issues on which future scope be there.

Keywords Image processing · Segmentation · Classification · KNN · ANN · SVM

1 Introduction

Plant disease is a big challenge for agriculture field; this affects the crop yield as well as crop production. Plant disease always vary from one season to another depends on environmental condition. Farmers grow variety of crops according to their field capability and available sources but there are several types of disease attacks that

Shashi (✉) · J. Singh
G D Goenka University, Gurugram, India

J. Singh
e-mail: jaspreet.singh@gdgu.org

P. K. Singh et al. (eds.), *Proceedings of Second International Conference on Computing, Communications, and Cyber-Security*, Lecture Notes in Networks and Systems 203,
https://doi.org/10.1007/978-981-16-0733-2_45

leads to the destruction of plants. These types of diseases can reduce the productivity of plants. Plant infection mainly caused by bacteria, fungi, viruses, or other types of agents.

Some common plant and fruit diseases are:

- Rust—found on the surface of lower leaves in form of reddish-orange spore.
- Leaf Spot—These leaves contain dark water spots, sometimes with yellow color.
- Leaf Curl—Effected leaves shape destroys in this.
- Gray Mold—It looks like gray soft, mushy spots on leaves, stems, and flowers.
- Apple Scab—It's scabby spots appears on fruit and leaves.
- Canker—Disease mostly appear in Cherries, Strawberries, and plums.
- Black Knot—Mainly relates to fruit and leaves attack mainly plum, cherry, apricot trees.
- Blossom End Rot—Tomato and pepper mostly affected by this disease appears a large spot on the bottom end.
- Brown Rot—a disease mainly affecting almond, apricot, cherries, plum etc.
- Cedar Apple Rust—Disease affected the upper surface of the leaves with yellow, pinhead-sized spots.
- Club Root—Disease affects the cabbage family with clubbed roots.
- Downy Mildew—This disease contains yellow to white patches on the upper surfaces of older leaves.
- Fire Blight—It is a type of bacterial disease that infects plant leaves.
- Early Blight—Lower surface of leaves contain concentric rings of brown spots.
- Many more.

Nowadays various ways are present to remove these plant disease like by manually, through mechanical cultivation or with the help of pesticides [1]. Removing the affected plants by manually is very time-consuming, costly and laborious process second method Usage of pesticides but over limit of pesticides can damage the crop. In such situation, automatic detection of disease is easier as well as cheaper task.

Disease detection is an automatic form by just looking at the indications on leaves makes it easier and cost-effective. This provides support for computer vision to give image-based automated supervision and process direction. Whereas if automatic disease detection is used then it will give more accurate results, within a less time and less efforts [2]. The presence of automated system for detection and diagnosis of plants disease, it may offer a useful assistance to the agronomist through optical observation of leaves of infected plants [3].

The aim of this paper is to summarize the different components that are used in a typical smart agriculture field starting from image acquisition to classification and then finally building the models. The main scope of this existing work is to perform a deep survey by integrating the approaches of image processing technique.

1.1 Image Processing Approaches

Above figure represents the generalized framework for disease detection and classification. The basic disease detection system includes the following phases:

1.1.1 Image Acquisition

The first task is to capture or collect the images through various types of digital cameras, digital scanner, mobile phones, etc. we can also collect image from web. Acquisition of image not a simple task there are many challenges to deal with this process environmental condition, stability in image, etc. Present dataset available on web contains both healthy and affected plants images. Image is represented in the pixels [4]. Now a day's many datasets are available like Quantitative-plant.org, kaggle.com, data.mendley.com, openweather.org, etc. from where we can directly collect thousands of images.

1.1.2 Preprocessing

In general plants, image experiences from low resolution, pixel disturbance, and external things which appear on images are essential to eliminate before applying any automated technique or algorithm. The per-processing step helps to improve the image visual quality through contrast enhancement and noise elimination. The pre-processing also contains some techniques such as resizing, image clipping, etc. to make it suitable for the subsequent processing [5].

1.1.3 Image Segmentation

After removing the undesirable factors from image next step is to segmentation. In segmentation process image gets breakdown into small regions on the basis of same pixel setting and attributes values. Aim of image segmentation is to understand the overall image description such as whether the image is a real thing image, a scene or moving scene [6]. Segmentation algorithms are selected on the basis of intensity values or gray level properties which able to detect similar or discontinue pixels. This process can be done with the help of many techniques like Thresholding-means segmentation, gray scale, histogram, RGB images conversion into HIV model, etc. Another aim of segmentation is to recognize image object accurately so that the related information of similar pixel are labeled together [7].

1.1.4 Feature Extraction

After segmentation of image there is a need to diagnose the specific region of image. Feature extraction helps to detect the specific region of image. In this features of image can be extracted on the basis of texture, color, shapes, size, etc. these features are collected information about any image for further processing. Method carried out for plant defected region after that features of image like shape, color, boundary, and texture are extracted for the disease spots to identify healthy or diseased plants. In [8] the proposed work related to cotton plants, there are three main parts of the cotton leaf spot, cotton leaf color segmentation and edge-based detection segmentation, analysis, and disease classification.

1.1.5 Pattern Matching and Classification

Pattern matching is a process in which sequence of patterns is collected and includes the matching of same pattern of tokens with the help of regular expression or regular matrices.

Pattern is a class that contains a set of patterns sharing common attributes. The collective index features of any image present in database by the help of algorithm pattern recognition and matching take place. After matching patterns the next method is classification.

Classification is a method in which using features and learned model assign a pattern to a category. Classification can be two types supervised classification or unsupervised classification [9].

1. Supervised classification contains three steps—Finding training area, generate file, classify.
2. Unsupervised Classification contains two steps—Creating clusters, Assign Classes.

Classification method contains number of techniques for classifying the images like Naïve Bayes classifiers, K-nearest neighbors algorithms, support vector machine (SVM), Decision tree, KNN, ANN, etc. [10].

2 Literature Review

M. K. Singh et al. proposed image enhancement using increase the contrast of images, RGB images to gray images with the help of histogram equalization and cumulative distribution. Color co-occurrence method for feature extraction simultaneously global color histogram (GCH) and local binary pattern (LBP) are used. Proposed work is very helpful for finding different types of leaf disease.

Fachao Qin et al. represented leaf segmentation using simple linear iterative clustering (SLIC) algorithm with good performance in superpixel generation for optical images but SLIC degrade its result in case of noisy images but this problem overcome with polarimetric synthetic aperture radar (PolSAR), AirSAP and ESAR L-band. Superpixel method provided great computational efficiency with fast running time [11].

Proposed work related to grape disease detection and leaf disease detection, work used unsupervised segmentation process based on K-mean clustering, $l * a * b$ color space model, iterative color clustering was conducted using Squared Euclidian distance, etc. [12, 13].

SR Dubey et al. focused on defect segmentation, feature extraction, and classification. Existing paper worked on an improved sum and difference histogram (ISADH) and support vector machine for the fruit disease detection [14]. Peng Guo et al. presented an automatic method for cucumber disease detection. In this study they used color moment and texture feature with good segmentation accuracy [15].

Gizelle K. Vianna et al. proposed work is related to the detection of tomato crop disease using Artificial neural network (ANN) classifier. ANN proved helpful for finding the damage surface of tomato efficiently [16].

In paper [17] authors introduced techniques in which after image acquisition, by creating color values, color transformation structure are converted into space values. K-means method used for image segmentation and GLCM calculations used for feature extraction, the existing study described about the leaf disease detection. This approach enhanced the productivity of crops and provided better algorithm accuracy.

In paper [18] authors described the work on fruit disease detection most of times classifies the apple disease. Existing work focused on apple scab, black knot and blotch disease mainly. Image segmentation with the help of K-means method after that recognition and classification took place with support vector machine to identify infected part.

In paper [19] authors described techniques for pomegranate fruit disease detection, common diseases relate to pomegranate are: Alterneria, Bacterial Blight and anthracnose. Image Preprocessing involved image resizing, filtering and morphological operations. Color-based segmentation with the help of RGB to HSV, YCbCr, $l * a * b$ etc. for best performance. Feature extraction includes color, morphology for obtaining boundary of images.

Bhumika S. Prajapati et al. proposed work aim to identify cotton leaf disease. Images captured for this work through digital cameras. In Pre-processing step they used background removal technique using RGB to HSV, Otsu threshold method used in image segmentation provided better results as compare to color removal technique. On the basis of color, shape, and texture features of images were extracted. Existing work used Support Vector machine for classification result good accuracy and reliability [20].

Sowmya GM et al. proposed a system for diagnosis the leaf disease using Matlab. This work was based on image processing approaches for pomegranate leaf disease detection. Segmentation techniques like thresholding-mans clustering and region-growing used here, feature extraction is done through GLCM technique. This work proved user friendly, accurate, fast, efficient, and effective for disease detection. This work was very costly and needs continuous monitoring of experts [21].

Sudhir Rao Rupanagudi et al. [22] described a methodology to perform tomato grading and with high level of accuracy and low cost. Different type of tomato ripening was considered using color and texture of tomato. Tomato maturity estimator used $L * a * b$ color space and color grading based on machine vision. Presented work is capable for tomato grading not for tomato disease detection.

S. N. Ghaiwat et al. presented survey on plant leaf disease detection and classification. In this work different classification method was discussed like K-nearest classifier, artificial neural network, support vector machine, generic algorithm, probabilistic neural network, fuzzy logic, etc. with their advantages and disadvantages over time, accuracy and other factors [23].

Gui et al. [24] proposed a method for soybean leaf disease detection based on salient regions. In this work low-level feature of luminance and color, combined with multi-scale analysis and K-means algorithm was used.

Mohammad et al. [25] described work on cotton plant disease identification. In this work, they discriminate healthy and diseased cotton crops using SVM classifier. They used K-means clustering for plant leaves. They discovered two main diseases which affect the healthy leaf, these are: Leaf spot and Leaf minor. The study described the difference between diseased and non-diseased leaf on the basis of color, structure, and size.

3 Research GAP/Future Directions

- In segmentation techniques we need to extract some features manually.
- The selection of appropriate features and their extraction is very time-consuming and laborious process.
- Existing Classifiers are not able to recognize each and every disease.
- There is a need for recognize more kind of diseases through classifier.
- To overcome above problems, we need to develop the new model that would automatically classify and detect the diseases.

4 Discussion

In our entire work by the help of exiting study tried to cover the different techniques with their performances. The main focus of this survey is to discover the ideas behind different techniques how the healthy and unhealthy images can be distinguished with how much accuracy and reliability. This paper is constructed in different sections. Figure 1 shows the Image Processing Approaches with detail Summary about each and every methodology covered in its subsections. Each phase perform a different task for disease detection and identification. In Fig. 2 there are some diseases pictures are given. Table 1 shows the comparisons of different techniques of segmentation and classification by different authors as some popular techniques described with

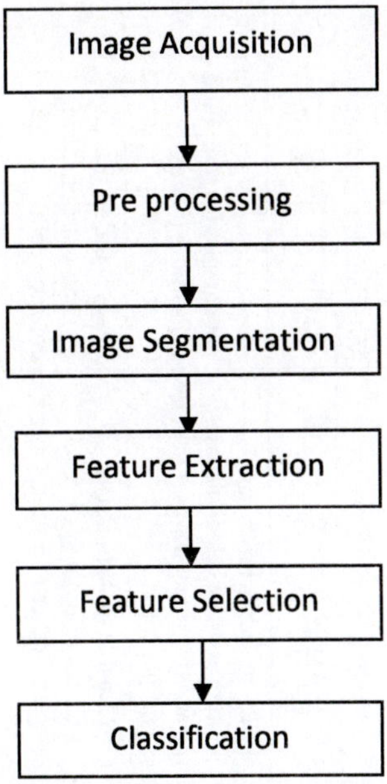

Fig. 1 Image processing approaches

Fig. 2 Some common plant diseases with their name

Table 1 Analyses of various techniques and algorithm of image processing

Authors N Year	Title	Techniques	Advantages	Disadvantages
Gilbert Gutabaga Hungilo et al. [26]	Image Processing Techniques for Detecting and Classification of Plant Disease: A Review	Convolution Neural Network (CNN) Approach	Convolution method is very Effective to diagnose disease of plant	This approach needs extra time as lot of trail and errors occur in its procedure
Guiling Sun et al. [27]	Plant diseases recognition based on image processing technology	Threshold Segmentation Methods-Otsu method, Iterative method, Image segmentation	Appropriate Threshold automatically, more scientific reliable and efficient	As the disease situation becomes gradually complex the no. of errors gradually increased
Khirade et al. [28]	Plant Disease detection using image processing	K-means method, ANN, SVM for image classification	Better accuracy power of SVM and ANN	Work needs more time and complex procedure
Mamta Mittal et al. [29]	An Efficient Edge Detection Approach To Provide Better Edge Connectivity For Image Analysis	Edge enhancement/segmentation, Thresholding as detection system	Proposed methods can successfully detect disease and this method was less noise proportion	The proposed algorithm and not capable for blur, time consuming process
Gurleen Kaur Sandhu et al. [30]	Plant disease Detection Techniques: A Review	K-Nearest neighbor (KNN), support vector machine (SVM) artificial neural network (ANN) etc	Capable to distinguish diseased and non-diseased plant leaves and fruit	There is the lot that can still be done in this field for enhancement

(continued)

Table 1 (continued)

Authors N Year	Title	Techniques	Advantages	Disadvantages
Pantazi et al. [31]	Automated leaf Disease detection in different crop species through image feature analysis and one class classifier	Support vector strategy, One class support vector machine (OCSVMs)	This Method was capable for detecting various health conditions including healthy, downy, mildew, powdery mildew and black spot easily	Proposed algorithm is not suitable for all environment conditions
Jagadeesh D. Pujari et al. [32]	Image processing waste detection of fungal diseases in plants	K-means clustering, gray level co-occurrence matrix, nearest neighbor etc	For Different classes of agriculture crops: Fruit crops, vegetable crops, curial crops and commercial crops by fungal disease easily diagnose	Presented work is complex and challenging in terms of outdoor conditions variability and general symptoms
Mohammad Naved Qureshi et al. [33]	An improved method for image segmentation using K-means clustering with neutrosophic logic	Based on K-means clustering using neutrosophic logic	Proposed method described better results on real as well as synthesis images	This work is very time consuming and complex process
Vijai Singh et al. [34]	Detection of plant leaf diseases using image segmentation and soft computing techniques	K-means clustering for image segmentation and ANN, Bayes classifier	Proposed work gives optimum result with very less computational efforts	Recognition rate in classification process is not so good. It can be improved

advantages and disadvantages. On the basis of Table 1 in Sect. 3 we discussed some research gap/future direction. In future, we can work on these points and overcome existing drawbacks. In Sect. 5 conclusion part and future aspect are there.

5 Conclusion

The present surveys about image processing techniques for finding the different types of plants and their related diseases. Some popular diseases are mentioned in the introduction part which is related to plant leaves and fruits. In this survey, we study about every step of image processing. There are many challenges occur during image acquisition and pre-processing. Most used Techniques in existing work are: K-Mean method, Thresholding, Segmentation, Otsu method, Support Vector Machine, Artificial neural network etc are described their role and efficiency level. All these techniques proved helpful to analyze distinction between healthy or infected plants. Literature Review part showing the work of different authors with their different techniques and application areas. The Review gives us ideas about research gap and future scope also.

References

1. Gutte VS, Gitte MA (2016) A survey on recognition of plant disease with help of algorithm. Int J Eng Sci 7100
2. Butale NM, Kodavade DV (2019) Detection of plant leaf diseases using image processing and soft-computing techniques
3. Mohanty SP, Hughes DP, Salathé M (2016) Using deep learning for image-based plant disease detection. Frontiers Plant Sci 7:1419
4. Mohammad MB, Srujana RN, Jyothi AJN, Sundari PBT (2016) Int J Appl Sci Eng Manag 5(02):84–88
5. Perez-Sanz F, Navarro PJ, Egea-Cortines M (2017) Plant phenomics: an overview of image acquisition technologies and image data analysis algorithms. Giga Sci 6(11):gix092
6. Unay D, Gosselin B, Kleynen O, Leemans V, Destain MF, Debeir O (2011) Automatic grading of Bi-colored apples by multispectral machine vision. Comput Electron Agric 75(1):204–212
7. Pujari JD, Yakkundimath R, Byadgi AS (2013) Classification of fungal disease symptoms affected on cereals using color texture features. Int J Sig Process Image Process Pattern Recogn 6(6):321–330
8. Dubey SR, Jalal AS (2012) Detection and classification of apple fruit diseases using complete local binary patterns. In: 2012 third international conference on computer and communication technology. IEEE, pp 346–351
9. Dubey SR, Jalal AS (2016) Apple disease classification using color, texture and shape features from images. Sig Image Video Process 10(5):819–826
10. Revathi P, Hemalatha M (2012) Advance computing enrichment evaluation of cotton leaf spot disease detection using Image Edge detection. In: 2012 third international conference on computing, communication and networking technologies (ICCCNT'12). IEEE, pp 1–5
11. Qin F, Guo J, Lang F (2014) Superpixel segmentation for polarimetric SAR imagery using local iterative clustering. IEEE Geosci Rem Sens Lett 12(1):13–17

12. Revathy R, Chennakesavan SA (2015) Threshold based approach for disease spot detection on plant leaf. Trans Eng Sci 3(5):72–75
13. Li G, Ma Z, Huang C, Chi Y, Wang H (2010) Segmentation of color images of grape diseases using K_means clustering algorithm. Trans Chin Soc Agric Eng 26(1):32–37
14. Dubey SR, Jalal AS (2014) Fruit disease recognition using improved sum and difference histogram from images. Int J Appl Pattern Recogn 1(2):199–220
15. Guo P, Liu T, Li N (2014) Design of automatic recognition of cucumber disease image. Inf Technol J 13(13):2129–2136
16. Vianna GK, Oliveira GS, Cunha GV (2017) A neuro-automata decision support system for the control of late blight in tomato crops. World Acad Sci Eng Technol Int J Comput Electr Autom Control Inform Eng 11(4):455–462
17. Mainkar PM, Ghorpade S, Adawadkar M (2015) Plant leaf disease detection and classification using image processing techniques. Int J Innov Emerg Res Eng 2(4):139–144
18. Varughese S, Shinde N, Yadav S, Sisodia J (2016) Learning-based fruit disease detection using image processing. Int J Innov Emerg Res Eng 3(2)
19. Khot ST, Supriya P, Gitanjali M, Vidya L (2016) Pomegranate disease detection using image processing techniques. Pune Int J Adv Res Electr Electron Instrum Eng
20. Prajapati BS, Dabhi VK, Prajapati HB (2016) A survey on detection and classification of cotton leaf diseases. In: 2016 international conference on electrical, electronics, and optimization techniques (ICEEOT). IEEE, pp 2499–2506
21. Deshpande T, Sengupta S, Raghuvanshi KS (2014) Grading and identification of disease in pomegranate leaf and fruit. Int J Comput Sci Inform Technol 5(3):4638–4645
22. Sowmya GM, Chandan V, Kini S (2017) Disease detection in pomegranate leaf using image processing technique. Int J Sci Eng Technol Res (IJSETR) 6(3):396–400
23. Rupanagudi SR, Ranjani BS, Nagaraj P, Bhat VG (2014) A cost effective tomato maturity grading system using image processing for farmers. In: 2014 international conference on contemporary computing and informatics (IC3I). IEEE, pp 7–12
24. Ghaiwat SN, Arora P (2014) Detection and classification of plant leaf diseases using image processing techniques: a review. Int J Recent Adv Eng Technol 2(3):1–7
25. Gui J, Hao L, Zhang Q, Bao X (2015) A new method for soybean leaf disease detection based on modified salient regions. Int J Multimedia Ubiquit Eng 10(6):45–52
26. Malathi M, Aruli K, Nizar SM, Selvaraj AS (2015) A survey on plant leaf disease detection using image processing techniques. Int Res J Eng Technol (IRJET) 2(09)
27. Hungilo GG, Emmanuel G, Emanuel AW (2019) Image processing techniques for detecting and classification of plant disease: a review. In: Proceedings of the 2019 international conference on intelligent medicine and image processing, pp 48–52
28. Sun G, Jia X, Geng T (2018) Plant diseases recognition based on image processing technology. J Electr Comput Eng
29. Khirade SD, Patil AB (2015) Plant disease detection using image processing. In: 2015 international conference on computing communication control and automation, pp 768–771
30. Mittal M, Verma A, Kaur I, Kaur B, Sharma M, Goyal LM, Roy S, Kim TH (2019) An efficient edge detection approach to provide better edge connectivity for image analysis. IEEE Access 7:33240–33255
31. Sandhu GK, Kaur R (2019) Plant disease detection techniques: a review. In: 2019 international conference on automation, computational and technology management (ICACTM), pp 34–38
32. Pantazi XE, Moshou D, Tamouridou AA (2019) Automated leaf disease detection in different crop species through image features analysis and one class classifiers. Comput Electron Agric 156:96–104
33. Pujari JD, Yakkundimath R, Byadgi AS (2015) Image processing based detection of fungal diseases in plants. Procedia Comput Sci 46:1802–1808
34. Qureshi MN, Ahamad MV (2018) An improved method for image segmentation using K-means clustering with neutrosophic logic. Procedia Comput Sci 132:534–540

Heart Disease Prediction Using Machine Learning

Jaydutt Patel, Azhar Ali Khaked, Jitali Patel, and Jigna Patel

Abstract Early Prediction of Chronic Heart Disease (CHD) makes use of the data collected on people over the span of 10 years. The data contains various information corresponding over 4000 individuals, attributes such as sex, age, diabetes, total cholesterol along with 11 more are used to train various machine learning models to predict if a certain individual has a high likelihood of suffering from CHD in the next ten years. This investigation will be used by health care institutions and societies to predict CHD beforehand and suggest the necessary preventive precautions required. In the article, the authors evaluate machine learning techniques: Support Vector Machines (SVM), Decision Tree, Artificial Neural Network (ANN), Naive Bayes on the Framingham data set by using parameters such as accuracy, recall, precision, and F-score for comparison. The authors use these classification approaches to determine which approach would work the best and under what scenarios based on experiment study.

Keywords Heart disease · Machine learning · SVM · ANN · Decision tree · Gaussian · Normalization

1 Introduction

Early prediction of Diseases provides the power to help prevent a disease from happening or to enable health insurance organizations to only cover those diseases which are most likely to be faced by their customer. An early disease prediction model would help evaluate many new patterns and trends which may lead to a lifestyle disorder or disease. Early disease prediction models function very similarly to the models used by banking companies, which are used to determine an individual's credit score based on their financial records.

Early prediction can be performed using either Binary classification, which would evaluate the likelihood of the disease happening to an individual or not and return a

J. Patel · A. A. Khaked · J. Patel (✉) · J. Patel
Institute of Technology Nirma University, Ahmedabad, India
e-mail: jitali.patel@nirmauni.ac.in

P. K. Singh et al. (eds.), *Proceedings of Second International Conference on Computing, Communications, and Cyber-Security*, Lecture Notes in Networks and Systems 203,
https://doi.org/10.1007/978-981-16-0733-2_46

yes or no answer. A binary classification usually provides a probability of a certain event of happening on the basis of the data given to it, the event in our case being the development of an illness. Binary classification generally falls under the area of supervised learning. Binary classification, as the name suggests, is a special type of classification problem where there are only 2 possible classes. A data set in binary classification is either tagged as 0 or 1, representing the two possible classes. Such classification problems fall under the domain of supervised learning, which requires a significant amount of domain knowledge and a hefty data set. The more refined and large the data set provided to the model, the better would be the results.

2 Related Work

Purushottam et al. [1] used a data mining approach for heart disease prediction. This study used a ten-fold cross-validation. The accuracy achieved was 86.3% in the testing and 87.3% in training. The database used in this study is the UCI database.

Gavhane et al. [2] surveyed multiple machine learning methods being used along with their accuracies. The survey thin out the methods with best result to be Recursive Neural Network having 92% accuracy and Decision Tree with 91% accuracy.

Jabbar et al. [3] explored the possibility of using hidden naive bayes classifiers to detect CHD. getting results close to 100% hidden Naive Bayes classifier relieved itself from the assumption of feature independence in Naive Bayes classification.

Noura et al. [4] used an artificial neural network for heart disease diagnosis. Feed-forward back propagation neural networks are used in this study as the classification algorithm. The model has 13 neurons in the input, 20 neurons in the hidden layer, and one neuron in the output layer. UCI heart disease dataset is separated in the test, and the target is used in this study. The accuracy achieved is 88% in this study.

Tomov et al. [5] used Deep Neural Network (DNN). This study has proposed a DNN called Heart Evaluation for Algorithmic Risk-reduction and Optimization five (HEARO-5). HEARO-5 has regularization and also deals with missing data and/or outlier data. K-means cross-validation and Metthews Correlation Coefficient (MCC) function is used in evaluating the model. The HEARO-5 has shown 99% accuracy and 0.98 MCC in the UCI dataset.

Dutta et al. [6] used a neural network with a convolution layer to classify class-imbalanced clinical data. Dataset used is the National Health and Nutritional Examination Survey (NHANES). To achieve better accuracy in class imbalanced data, this study has used a two-step approach. First, they used the Least Absolute Shrinkage and Selection Operator (LASSO) based feature weight assessment and then majority-voting based identification of essential features. Even though there is a 35:1 ration in the dataset, this study has shown 77% accuracy in the positive cases and 81.8% accuracy in the negative cases that is 85.7% of the total dataset.

Jahromi et al. [7] explain the use of a Gaussian naive Bayes classifier on features having fewer dependencies on each other. It uses principle component analysis to improve the conditional independence of features.

Najeeb Abbas Al-Sammarraie et al. describe the working of a neural network and back-propagation in a classification problem. A study on how the number of neurons in an artificial neural network affects the learning process helped them describe a relation between the number of neurons in the hidden layer versus the number of iterations needed to learn the classification.

Pouriyeh et al. [8] explore a number of methods, concluding that SVM with boosting outperforms all the other methods such as K-Nearest Neighbor (KNN), Multilayer Perceptron (MLP), Decision Tree.

Shimpi et al. [9] tested out Support Vector Machine, Random Forest, Logistic Regression, and KNN on the UCI Machine Learning Repository dataset of cardiac arrhythmia, which uses ECG signals to detect possibilities of cardiac arrhythmia. The paper offered a detailed comparative study between the four suggested machine learning approaches and their accuracies.

Shaikhina et al. [10] trained a classifier using artificial neural networks to determine the compressive strength of the osteoarthritic trabecular bone from its structure and biological properties. They achieved this on a particularly small dataset and explained how by using transfer learning techniques, they managed to train their model by fine-tuning a model previously trained for determining the compressive strength of concrete.

Khourdifi et al. [11] used the Fast Correlation-Based Feature Selection (FCBF) method to remove redundant features; this will improve the accuracy of the classification algorithm. Classification is done using different machine learning algorithms like KNN, SVM, Naive Bayes, Random Forest, and MLP, Artificial Neural Network optimized by Particle Swarm Optimization (PSO) combined with Ant Colony Optimization (ACO). Maximum accuracy of 99.65% was achieved using the combination of ACO and PSO with FCBF. Similarly in Bashir, Saba, et al. [12] used Naive Bayes, Support vector machines and Random Forest methods to extract the most relevant features from the data sets.

Saqlain et al. [13] used AFIC data. First, data mining techniques were used to extract the useful data, and after that, machine learning multiclass classification techniques were used for the prediction. Accuracy and the area under the curve (AUC) of this model is 86.7% and 92.4%.

Ambekar et al. [14] signified the importance of data sanitization. Data collected from surveys and health organisations is not always complete or organised. Various data cleansing and techniques were used and verified over k-nearest neighbour and Naive Bayes classification methods.

Repaka et al. [15] used the Naive Bayes approach for the classification. They divided data into 80% training data and 20% testing data. In this work, 89.77% accuracy was achieved.

Seema et al. [16] used two datasets from UCI, one for heart disease and second for diabetes. Different classification techniques were used like Naive Bayes, SVM, Decision Tree, and ANN. SVM gives the highest accuracy of 95.556% in the heart disease dataset, and Naive Bayes gives the highest accuracy of 73.588% in the diabetes dataset.

3 Proposed Approach

In this article, we shall be using various classification techniques on a single data set which has been formed on the Framingham Heart Study conducted by the National Heart, Lung, and Blood Institute of the United States. The procedure shall follow these steps (i) Load the data set, (ii) fill in the 'na' values, (iii) balance out the two classes, (iv) feature selection, (v) normalizing the data, (vi) training the model (v) Evaluating the model.

3.1 *Load the Data*

We upload the framingham data set; this data set is classified into two classes, which can either be a one or zero. One represents the individuals who developed a CHD over a period of ten years, while zero represents the individual who did not develop any CHD.

The data set consists of various features for every individual, which are (i) Sex (ii) Age (iii) Education (iv) Current smoker, (v) Cigarettes per day (vi) On blood pressure medication (vii) Prevalent stroke (viii) Prevalent hypertension (ix) Diabetes (x) Total cholesterol (xi) Systolic blood pressure (xii) Diastolic blood pressure (xiii) Body mass index (xiv) Heartbeat rate (xv) Glucose in blood.

3.2 *Fill 'na' Values*

There are a few instances in the data set which have a few feature values marked as 'na', which implies that that particular value was not extracted during the framingham study. To train the model, it's necessary to have every feature to contain an acceptable value; therefore, the 'na' is replaced with the average of the values in that particular feature.

3.3 *Balance Two Classes*

The framingham data set has an imbalanced data, which means that there are more negative instances (instances belonging to class 0) and fewer positive instances (instances belonging to class 1). To balance this, we make copies of the positive instances and shuffle the data set to achieve a data set with both negative and positive instances, having a number similar to each other. This is necessary; otherwise, the model iterates more over the negative instances, and overtime gains a bias towards the negative instances; if the data set is balanced, the model iterates over the positive and negative instances an equal number of times, leading to better accuracy.

3.4 Feature Selection

Framingham data set has a few features which don't hold much relevance to the prediction of a CHD development in 10 years. Such features can be dropped to improve the accuracy of the model.

3.5 Normalizing the Data

Various features in the data set have a different range of magnitude. Features such as heartbeats per minute range from 50 to 120, while values such as age are in the range 20–80. On the other hand, we have many binary values which are either in the form on 0 and 1. If the model were to be trained without normalization, then the values with higher magnitude would overwhelm the binary values and values with low magnitude during training, causing the values of low magnitude insignificant. We, therefore, normalize all these values by setting them between the range of 0–1.

3.6 Training the Model

There are four models that have been trained for the prediction of CHD for comparative study in this article. The methods used for classification are (i) Artificial Neural Network (ANN), (ii) Support Vector Machine (SVM), (iii) Decision Tree, (iv) Gaussian Naive Bayes. Each of these methods has its own pros and cons, which would be studied using the framingham data set.

Artificial neural network The ANN model used for our classification is a serial neural network with each neuron using a rectified linear unit (ReLU). The network has a total of sex ReLU layers with dimensions 1024, 512, 124, 64, 32, 16, respectively, while the output layer has a layer of 2 softmax neurons, which give the probability of the two possible classes for binary classification.

Support Vector Machine SVM classifiers form a separating hyperplane to separate the two classes, which allows the classification. SVM is best suited for binary classification as a hyperplane can only separate two sets of data, which works perfectly in this scenario.

Decision Tree A decision tree is a machine learning tool that evaluates the various features of a data set and creates a tree representing the flow of decisions to be made on a top-down basis. The tree can then take an instance and evaluate the features using the conditions mentioned along the tree.

Gaussian Naive Bayes Gaussian Naive Bayes classifier is an extension of the Naive Bayes classifier. Naive Bayes classifiers are simple and use Bayes's probability theorem for predicting the probability of an instance belonging to a certain class. Naive Bayes is called so as it is overly simplified because of the assumption that each feature is independent of each other, which is not true in real-life situations. By using Gaussian distribution, we can implement Naive Bayes classification with real-life data such as in our case where there may be a relation between features.

3.7 Evaluating the Model

The models are all evaluated using the k-folds cross-validation method. The models have been validated over four-folds, and for each validation method, (i) Precision (ii) Recall (iii) F1 score and (iv) Accuracy have been calculated. These values will be used to compare the various models to obtain the most efficient model.

4 Results

The Framingham dataset being used in this investigative study has a total of 4238 instances, of which 644 instances are positive, which means 644 instances did develop a CHD over ten years, and 3594 instances are labeled negative. Since there are more negative values than positive, the data set is augmented to have 3864 positive values and 3594. This is achieved by making six copies of the positive instances.

Another result derived is the covariance of all the features to their class, as shown in Fig. 1. To evaluate this result, we use the original dataset and compare the column representing the feature to the column representing the class by using the formula:

$$\mathrm{cov}_{x,y} = \frac{\sum_{i=1}^{N} (x_i - \bar{x}) + (y_i - \bar{y})}{N} \tag{1}$$

Covariance helps in providing an idea of how a feature affects the classification. If the covariance is positive, it means that the relationship is directly proportional, whereas a negative value signifies an inversely proportional relation.

Figure 1 shows the effect of each attribute on the result which is nothing but the covariance of the attribute. Higher positive value shows that this attribute plays a major role in determining the output. Hence attributes like male, age, prevalentHP are more important than the other attributes.

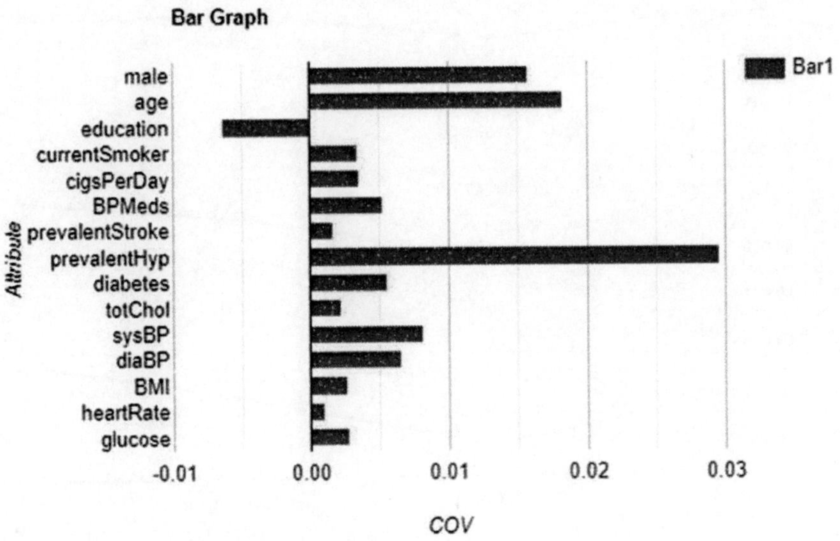

Fig. 1 Co-variance of different attributes of data

4.1 Results and Evaluation

The performance of the models is evaluated on the evaluation scores mentioned above. The final dataset size is of 7458 instances, a confusion matrix is formed of the results, and from them, the precision, recall, f1 score, and accuracy are calculated as depicted in Table 1. Figure 2 shows the comparison between all the models trained using the dataset In this general comparison, there is no k-folds cross-validation, and the validation set is picked out from the dataset in the ratio of 2:8, which means 20% of the instances in the dataset are used for validation.

Table 1 Values of different parameters for different algorithms

Parameter	ANN	SVM	Decision tree	Gaussian
Accuracy	88.89	67.42	92.59	56.57
Precision	83.38	67.76	86.7	81.39
Recall	99.2	77.94	100	33.26
F1-Score	90.6	72.06	94.08	47.22

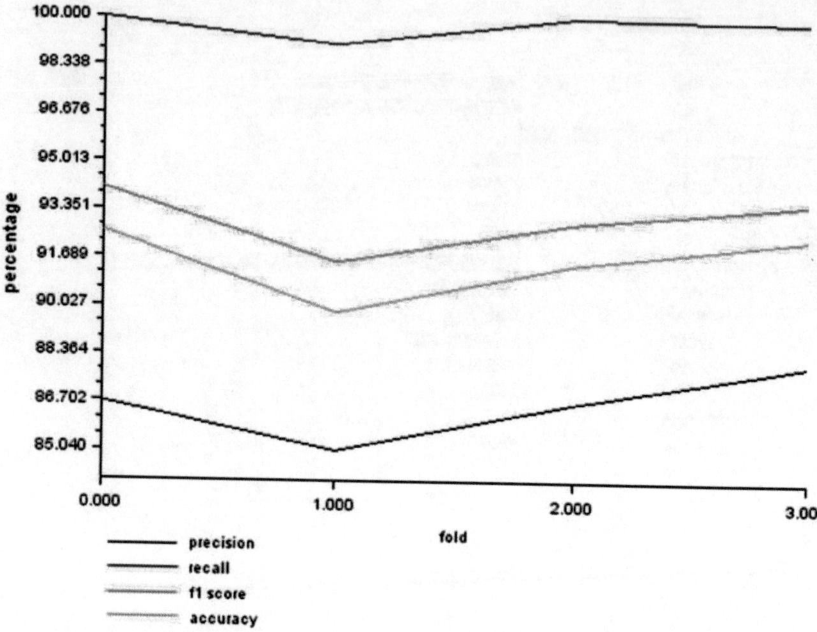

Fig. 2 Comparison of the result of the decision-tree algorithm

4.2 Decision Tree

As depicted in Fig. 2, the decision tree has the highest performance compared to the other methods in the parameters.

Figure 2 shows the change in the value with change in no. of folds applied for Decision tree algorithm. Recall value is highest and precision value is least. F1 score and accuracy have nearly similar values (Table 2).

Table 2 Results of Decision Tree

Fold	Precision	Recall	F1-Score	Accuracy
No Fold	86.7	100	94.08	92.59
2	85.04	99.06	91.51	89.78
3	86.68	100	92.86	91.44
4	88.03	99.84	93.56	92.35

4.3 SVM

SVM (Support vector machine) has 67.53% accuracy. Like the Gaussian technique in the current model having different accuracy measures that do not increase with the increase in the number of folds. It remains the same (Table 3).

Figure 3 shows the change in the value with change in no. of folds applied for SVM algorithm. Recall value is highest, but unlike decision tree where precision value was least, here accuracy value is least. Accuracy and precision value are close to each other.

Table 3 Results of SVM

Fold	Precision	Recall	F1-Score	Accuracy
No Fold	67.76	77.94	72.06	67.42
2	68.65	76.55	72.38	67.5
3	68.57	77.17	72.61	67.62
4	68.66	76.81	72.5	67.58

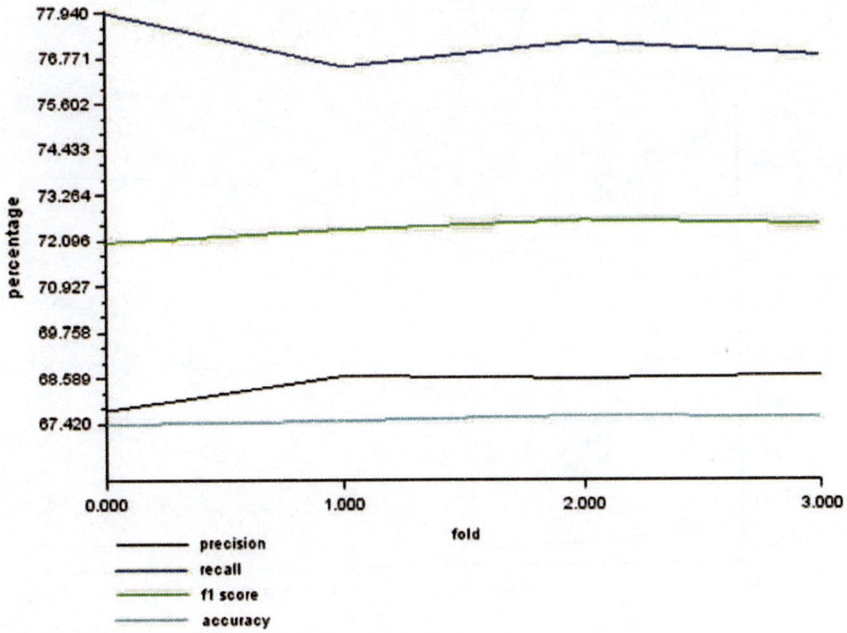

Fig. 3 Comparison of the result of the SVM algorithm

4.4 ANN

While training, the ANN number of epochs used was 100, and learning rates were 0.001, and shuffle was also enabled. The accuracy achieved after the training was around 88%, which is higher than the SVM or Gaussian techniques. As the number of epochs is increased, the accuracy also increases. It is also evident from the graph that when using the k-fold to test the accuracy as the number of fold increases all the measures of the accuracy also increases. Because when the number of fold k-1 is higher, and we have more data to train our model; hence accuracy increases (Table 4).

Figure 4 shows the change in the value with change in no. of folds applied for ANN algorithm. Recall value is highest and precision value is least which is similar to decision tree algorithm graph.

Table 4 Results of ANN

Fold	Precision	Recall	F1-Score	Accuracy
No Fold	83.38	99.2	90.6	88.89
2	82.54	93.89	87.84	85.53
3	85.75	96.85	90.95	89.28
4	88.11	97.38	92.51	91.22

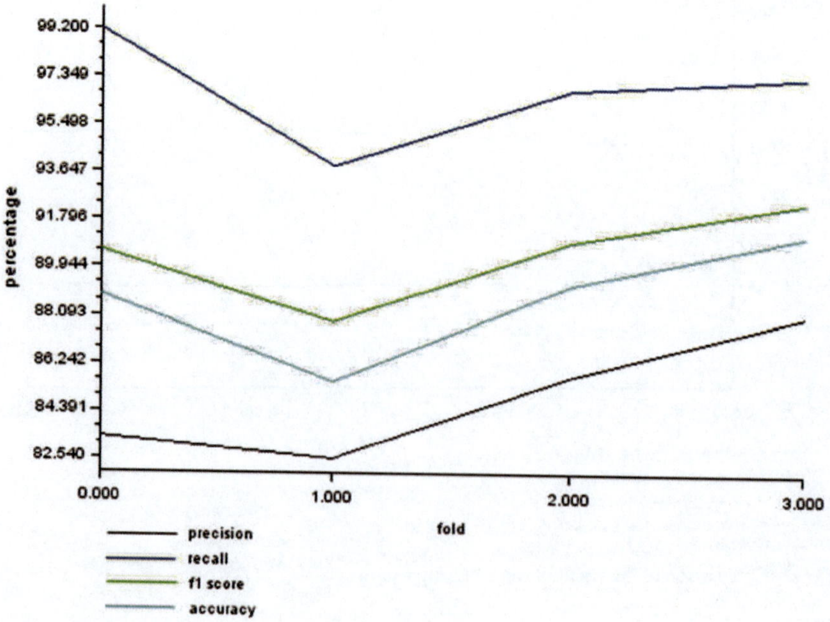

Fig. 4 Comparison of the result of the ANN algorithm

Table 5 Results from gaussian algorithm

Fold	Precision	Recall	F1-Score	Accuracy
No Fold	81.39	33.26	47.22	56.57
2	77.35	31.25	44.51	56.66
3	77.64	31.28	45.12	57.02
4	77.34	31.5	44.75	56.78

4.5 *Gaussian*

Gaussian techniques have the least accuracy of all. Even after normalization and data, augmentation accuracy is still at 56.75\%, which is the least of all techniques. While using k-fold to test the accuracy of the technique in other techniques with the number of fold accuracy increases but in the Gaussian techniques, it remains the same regardless of the number of folds as it can be seen from the graph (Table 5).

Figure 5 shows the change in the value with change in no. of folds applied for Gaussian algorithm. Unlike any other algorithm where the recall value was the highest here precision value is highest and recall value is least.

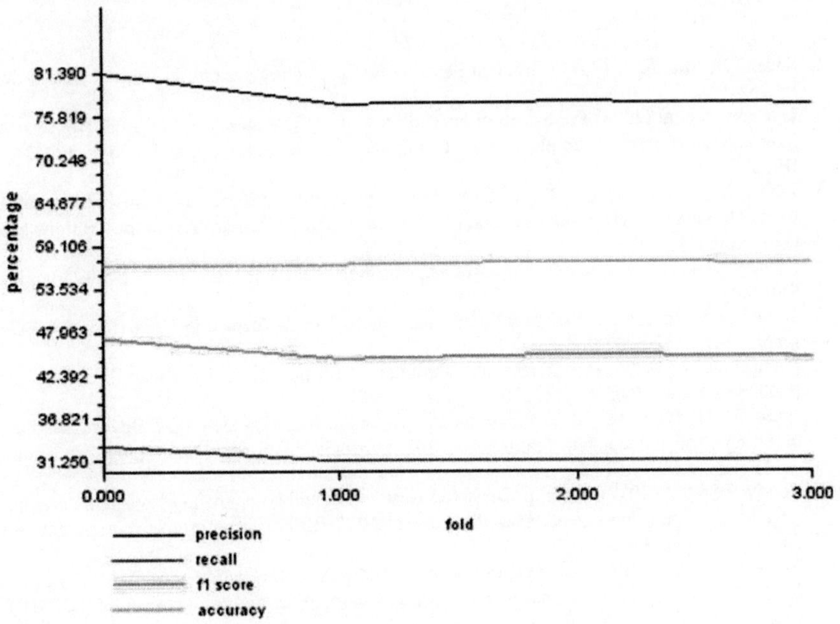

Fig. 5 Comparison of the result of the Gaussian algorithm

5 Conclusion

Based on the accuracy achieved after training the model and testing it on different k-folds, we come to a conclusion signifying the fact that methods such as ANN and Decision Tree are more suitable for binary classifications having a data set with parameters varying between categories and numeric values. Although SVM happens to be a well-known method for binary classification, it only seems to perform well when it is provided exclusively with numeric parameters. However, in the framingham dataset, categorical parameters do exist, such as prevalent stroke, prevalent hypertension, current smoker, and sex. Unlike SVM, Decision Tree happened to handle such categorical parameters a lot more efficiently. ANN also performed flexibly, providing high accuracy. The poorest performance was noted when using the Gaussian approach. This can primarily be hinged onto the fact that the Gaussian method considers all the parameters to be independent of each other. In contrast, all the parameters of the framingham dataset are highly interrelated. This could be the primary reason as to why the method performed so poorly with an accuracy approaching 50%.

References

1. Saxena K, Sharma R (2016) Efficient heart disease prediction system. Procedia Comput Sci 85:962–969
2. Gavhane A et al (2018) Prediction of heart disease using machine learning. In: 2018 Second international conference on electronics, communication and aerospace technology (ICECA). IEEE
3. Jabbar MA, Samreen S (2016) Heart disease prediction system based on hidden naïve bayes classifier. In: 2016 International conference on circuits, controls, communications and computing (I4C). IEEE
4. Ajam N (2015) Heart diseases diagnoses using artificial neural network. IISTE Netw Complex Syst 5(4)
5. Tomov N-S, Tomov S (2018) On deep neural networks for detecting heart disease. arXiv preprint arXiv:1808.07168
6. Dutta A et al (2020) An efficient convolutional neural network for coronary heart disease prediction. Expert Syst Appl (2020):113408
7. Jahromi AH, Taheri M (2017) A non-parametric mixture of Gaussian naive Bayes classifiers based on local independent features. In: 2017 Artificial intelligence and signal processing conference (AISP). IEEE
8. Pouriyeh S et al (2017) A comprehensive investigation and comparison of machine learning techniques in the domain of heart disease. In: 2017 IEEE symposium on computers and communications (ISCC). IEEE
9. Prajwal S et al (2017) A machine learning approach for the classification of cardiac arrhythmia. In: 2017 International conference on computing methodologies and communication (ICCMC). IEEE
10. Shaikhina T, Khovanova NA (2017) Handling limited datasets with neural networks in medical applications: A small-data approach. Artif Intell Med 75:51–63
11. Khourdifi Y, Bahaj M (2019) Heart disease prediction and classification using machine learning algorithms optimized by particle swarm optimization and ant colony optimization. Int J Intell Eng Syst 12(1):242–252

12. Bashir S et al (2019) Improving heart disease prediction using feature selection approaches. In: 2019 16th International bhurban conference on applied sciences and technology (IBCAST). IEEE
13. Saqlain M et al (2016) Identification of heart failure by using unstructured data of cardiac patients. In: 2016 45th International conference on parallel processing workshops (ICPPW). IEEE
14. Ambekar S, Phalnikar R (2018) Disease risk prediction by using convolutional neural network. In: 2018 Fourth international conference on computing communication control and automation (ICCUBEA). IEEE
15. Repaka AN, Ravikanti SD, Franklin RG (2019) Design And Implementing Heart Disease Prediction Using Naives Bayesian. In: 2019 3rd International conference on trends in electronics and informatics (ICOEI). IEEE
16. Deepika K, Seema S (2016) Predictive analytics to prevent and control chronic diseases. In: 2016 2nd International conference on applied and theoretical computing and communication technology (iCATccT). IEEE

Future of Augmented Reality in Healthcare Department

Gouri Jha, Lavanya shm. Sharma, and Shailja Gupta

Abstract In today's world, augmented reality (AR) is a highly challenging and immerging technology which presents some additional information to the existing real world. This is done by using special glasses like Google glasses or with help of advanced devices. This technology is an advance version of virtual reality. As, in VR, we have to work in a completely virtual environment, but in this technology, we do not have to work in virtual world, but being in the real world, we are getting some additional information. This paper provides a brief description about the architecture of AR, possible solutions provided by several researchers, and academicians, their challenging issues and real-time application in medical or emergency department.

Keywords Augmented reality · Virtual reality · Medical field · HOLO- BLSD

1 Introduction

Augmented reality (AR) which is a self-defined by its name as augmented means adding something more to anything and reality means real world. So, adding something to the existing real world defines augmented reality. It is an upgrade version of virtual reality (VR). As, in VR, we are completely in a blind condition due to VR glasses, AR glasses are like normal glasses; it just add some additional information to the real world. In medical world, new technologies are introducing day by day [1–3]. These technologies are resulting in ease of medical operations and education. With the help of these technologies, it becomes very easy to learn and perform complicated tasks. Among these emerging technologies, there is a technology known as

G. Jha (✉) · L. Sharma
Amity Insititute of Information Technology, Amity University, Noida, Uttar Pradesh, India

S. Gupta
Department of Computer Science and Technology, Manav Rachna University, Faridabad, India

P. K. Singh et al. (eds.), *Proceedings of Second International Conference on Computing, Communications, and Cyber-Security*, Lecture Notes in Networks and Systems 203,
https://doi.org/10.1007/978-981-16-0733-2_47

Fig. 1 Applications of augmented reality [1]

augmented reality (AR) [4–6]. This technology not just contributed in the development of medical but also other fields too. Not just in the medical field but also in other fields like architecture, gaming military, etc., these technologies are contributing lot as shown in Fig. 1.

It just changed both the perspective of learning as well as applying it in real world. Augmented reality is an emerging technology, and its contribution in the medical field results in life-saving operations. Operations like spinal surgery and sinus are a bit complex before augmented reality. Augmented reality had done two significant tasks; the first one is "reduction of stress" and the other one is "reduction of time limit." There are some challenges with this technology because it is a newly emerging one [7, 8]. People are unfamiliar with the correct use of AR, and as a result, many life-taking incidents had occurred. One of the examples of this type of case is Pokémon Go game [9, 10]. This game had resulted in many serious accidents which had risked some lives.

This paper is categorized into seven sections. Section 1 deals with introductory part of AR in medical department, whereas in Sect. 2, related work is discussed. Section 3 deals with architecture of AR. In the next Sect. 4, real-time applications of AR in medical department are discussed. In Sect. 5, challenges of AR are discussed. In Sect. 6, some more fields other than medical are discussed where AR is used. In the next Sect. 7, future scope of AR in medical department is discussed. In the last, conclusion of the work is discussed.

2 Literature Review

This section deals with the work done by various researchers in the domain of augumented reality for health care. Rodriguez and Huang [2] discussed about the usefulness of AR/GIS and also the way through which student can do independent study of AR/GIS. Deshmukh et al. [11] discussed a 3D manual system for advances and effective learning using AR. Çolak and Yünlü [3] discussed the use of augmented reality and virtual reality in engineering education. Sharma and Garg [12] discussed about enabling opportunities and challenges of e-health system applications. Bottino et al. [13] discussed a self-directed life support system using AR; using this, humans can get life-saving education without any expert. Wasenmuller et al. [14] discussed an application for discrepancy check for industrial purpose. Mizell [15] discussed a HUD set used for manufacturing many products like aircraft manufacturing, form board diagram and other masking devices. Azuma et al. [16] discussed a survey potential of AR applications in medical, designing and repair of complex equipment. Dünser et al. [4] discussed a survey report based on the analysis of published paper since 1993–2007. Atkuri et al. [17] discussed technique, scope and status of AR in medicine. Sharma and Lohan [5] discussed about visual surveillance systems and conventional methods in different suspectable environments.

3 Architecture

The AR glasses or the AR devices have four basic requirements. As like in the AR glasses, these components are as follows:

- *The Display*: This AR display is also known as combiner. As it combines the eye glasses with the digital LED or OLED display, the computer-generated images can be sent to the eyes [18]. So, when we wear AR glasses, we are analyzing two things; the first is the real world and the other one is the computer-generated images.
- *The Camera:* The second component is camera. This camera is placed in your AR glasses, and if you are using the AR app, then your phone camera is used. This camera is used to capture the images real-world images as your eyes cannot capture it. [10, 19].
- *The Registration:* The registration consists of some icons, and these icons are not visible to the user. These icons help the glasses to place a virtual object in the real world. That is why you can see a car placed in your garage or a sofa pops up in your room when you are using home décor app. These icons use various things like the corners of wall and length of wall for guidance and placing of virtual object [8, 20].
- *The Computer Vision:* This is the stage where the magic of AR takes place. Here, both the images which are taken by the camera and the registration are combined and placed on the display which are sent to our eyes [21] (Fig. 2).

4 Applications of Augmented Reality

There are many surgeries, checkups and medical subjects existing where augmented reality is used [23]. As augmented reality is used for creating medical applications for surgeries and for study purpose too, here are some fields of medicine where AR is used:

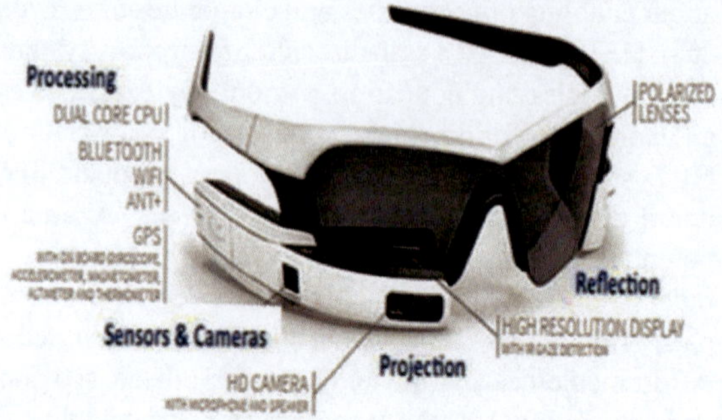

Fig. 2 AR glass composition [22]

1. *Spinal Surgeries:* The old medical system is a bit complex for performing spinal surgeries. As spinal is the back bone for our body and its very sensitive, the doctors have worked with extra alertness. The old system for spinal surgery requires high cost CT scan, and it is very difficult to determine the anatomy of spinal. With this scan and difficulty in anatomy, there is one more big problem that is doctor' s have to see the screen many times while performing the surgery. So, all these problems are solved with help of augmented reality in the following way [24]:

   - *AR Helped in cost effectiveness:* As it requires a lot of money for the CT scan of spine, now with help of Microsoft HoloLens and AR technology which cost just $3000, it is basically one-time investment (Fig. 3).
   - *AR helped in building the anatomical structure of spine:* Just by using the AR, there is an inaccuracy in depth perception, so the combination of AR and VR was introduced [18]. Now, by this combination, doctors can get the distance between instruments and organs in a single window.
   - *Reduction of stress and helped in projection of spine without any physical cut on body:* As in old spinal surgeries, doctor' s have to watch the operation screen, and the patient at the same time which creates much stress and complications. But now, with the help of Microsoft HoloLens, the doctors

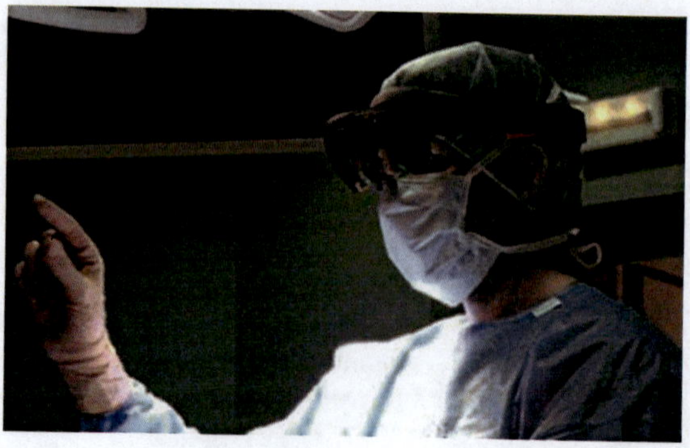

Fig. 3 Doctor using AR glasses [17]

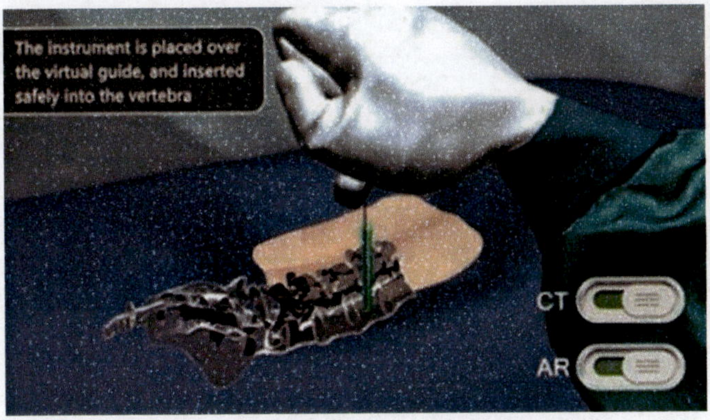

Fig. 4 Inserting screw in spinal using AR glasses [5]

can watch the Spin CT through the skin and can perform the drill and pre-plan to place the screw before surgery [25, 26] (Fig. 4).

2. *AR in Bone Tumor Resection:* The physicians of South Korea have developed an AR system that work as navigational device, and it focuses on the navigational imaging of bone tumor [27]. As in old techniques, CT scans and X-rays are used to identify the tumor area which is not so much accurate. But, with this system, the accuracy is increased. In this study, physicians had used Microsoft Surface Pro 3 for the tracking as well as for workstation too. The camera is used to track the distance between the target and the instruments [26, 22]. The image data which was received from the camera was filled in the software, and as a result, the system provides an image with the virtual bar which contain five sections. In this section, the blue colored area shows the normal bone reason, the green colored area shows the safety margins and the red color area indicates the bone tumor (Fig. 5).
3. *AR in sinus surgery:* Sinus surgeries are traditionally done through endoscopy, but in this surgery, there is a risk of damaging the other body tissues and nerves. So, an AR-based system was introduced which produces computer-generated images with real-time surgery [24]. This system shows the distance between the sinus and instrument in mm. This system also prevents physical damages of untargeted organs like optic nerves and arteries. It provides an alert sound if the instrument is very close to the target or damaging an untargeted organ (Fig. 6).
4. *Augmented reality in virtualizing 3D Radiology images:* Augmented realty is used for making the 3D radiology images. These images create a 3D model of the organs and present all the different part of that organ with different colors. With the help of this 3D model, doctors can see the conditioning of the organ more precisely. The main advantage of this 3D model is that now, doctors can see the infected area in layering means with the help of AR we can see the different layer of any body part [18, 30] (Fig. 7).

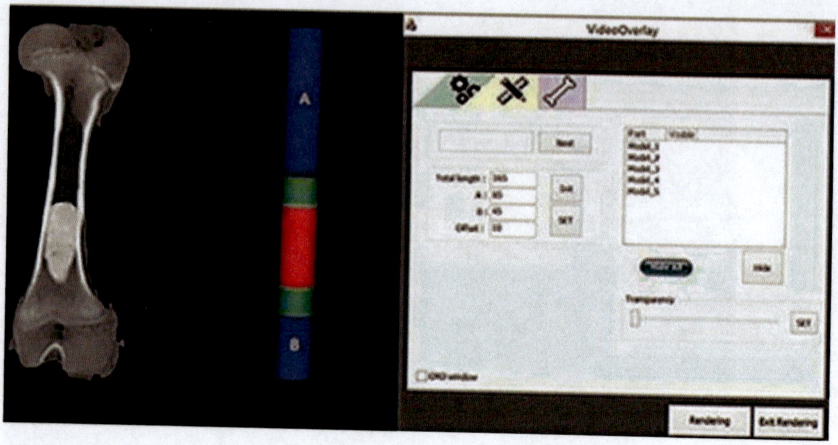

Fig. 5 Software showing the bone analysis [28]

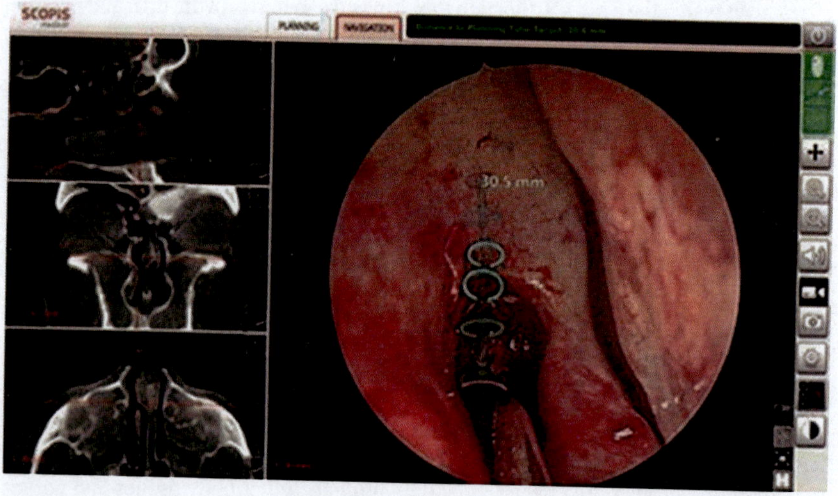

Fig. 6 Live screening of sinus operation using AR [29]

Fig. 7 View of 3D radiology image of liver using AR glasses [31]

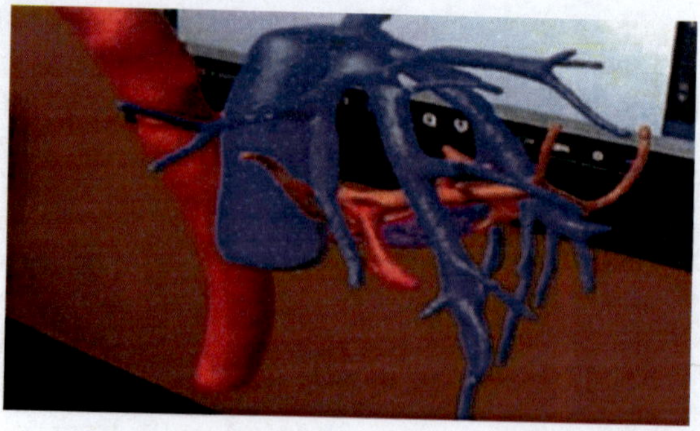

This image shows the 3D radiology image of liver. In this, the purple area indicates the tumor and the red area shows the arteries. So, like this, we can create 3D imaging of any body part and watch it layer by layer.

5. *Augmented Reality in Veins and allergy detection:* Augmented reality helps to detect the veins through the skin [18, 32]. Now, the nurses can see the veins of a baby or adult through the skin with help of AR glasses. This will save time and can help to reduce the number of mistakes (Fig. 8).

Not only this but also AR helps to detect weather a person has some kind of allergy or not before an operation. This can be done by watching the relevant data on the AR screen of the glasses.

6. *Use of Augmented Reality in Medical Education:* Medical students are getting a lot of help by using augmented reality. As of now, students can see the detailed design of any body part like 3D model of radiology imaging [20] (Fig. 9).

Students can learn without the help of an expert. Students can tackle with the future technical difficulty if they had already gone through whiles their practice. It helps in just-in time and just-in place learning.

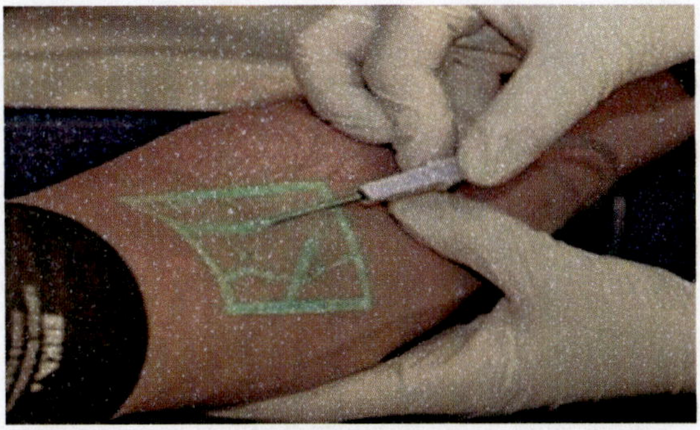

Fig. 8 Watching veins of hand using AR glasses [21]

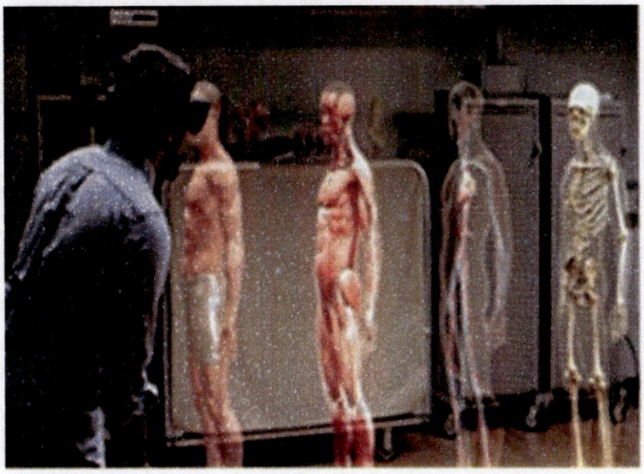

Fig. 9 Human body structure using AR glasses [33]

5 Challenges in Augmented Reality

This technology is an advance version of virtual reality. Some of the major challenging issues of AR are listed below:

- *Lack of Knowledge:* As augmented reality is an emerging technology, there are very less people who know the correct use of it. So, it is the biggest challenge while using it [7, 34].
- *Security Issue:* Augmented reality has a breach of security issue because while using augmented reality, the user is in two worlds: one is real and other is virtual. So, sometimes, they forget one world and start working in virtual one. This led to serious accidents. The biggest example of this breech is Pokémon Go game. We all know that while using this game, many humans had lost their life [10].
- *Not Well-Developed Technology:* Augmented reality is a new concept, and there are very less developers and tools available for its development [7, 10, 35].
- *Lack of Business Models:* Augmented reality had taken a kick start, but many investors still do not want to invest a huge amount of money because there is not a lack of proper business model as we saw in Pokémon Go that people gone crazy because of the brand Pokémon, but after some time, the hype goes down and no business strategy helps it to rise again [7, 35].

6 Some More Fields of Augmented Reality

There are several fields in which augmented reality is playing and could play a vital role. These are listed below:

1. *Architecture field:* Architecture field is using AR technology in a very interactive way [8, 36]. Now, engineers can just design the architecture of any building in paper, and with the help of that design, AR can show how the house will look after completion of construction.
2. *Military:* AR has become an important part in modern military training [20, 22, 30]. It is helping militaries in many ways, and one of them is training the

new bees with the help of augmented reality war field. In this field, trainees have special glasses and special equipments; now, they can kill the virtually created terrorists and can make their performance better. The second is AR sand table; this table helps soldiers to understand the battle field more easily and efficiently. The explainer can add more graphical things like airplanes and ships movements while explaining a war situation.

3. *Gaming Industry:* Gaming industry is one of the main industries who is using AR as well VR at a very high level [10, 25]. Now, we can play AR games on our mobiles without any AR glasses. In AR games, you can create your own track in your own room or at any place where you want, and not only this, you can also have a race on this track. The battle games are the category in which we

can play more interesting AR games. In this, we can make teams and can play multiplayer games in any area of real world. You can detect which had killed or how much power is still remaining of that player [25].

7 Future Scope of AR in Medical Field

The scope of augmented reality in medical department is growing day by day [37, 38]. As it is the second most highly money invested field of AR. There are many medical operations where radiology is used and AR has become a very efficient technology for the radiological data. So, for making and analyzing the anatomy of different body parts, AR can be used [39].

- AR is also contributing in educational field. So, AR technology can be used for learning complex medical operations with or without experts or we can say that it can result in an independent study for the medical students [19, 40].
- AR can also assist the patients for having their medicines and how and when to take it. Such as if a patient wants to inject an injection for pain relief and no doctor is available their, so he can get assisted by an AR application for proper injecting location and technique [27].
- AR can also be used for scanning and displaying a detailed 3D structure of a tumor. As all the details and the infected area will be known to the doctors, then they can handle the operation very easily [4]. AR is used for detecting the blood veins and differentiating the blood vessels and soft tissue [8, 40]. AR can be used for detecting the infection of blood with the help of old relevant data.
- AR in cosmetics or skin operations. AR can be used for cosmetic operations; for example, we can create a virtual 3D image of the patient with the help of AR and AR application should show the after image that how the patient looks after getting surgery (Fig. 10).

Fig. 10 Augmented reality in cosmetic surgery [41]

8 Conclusion and Future Work

Augmented reality is a very effective technology in medical as well as in other sectors. This technology has a very huge potential, and it can do more innovative things as it had done in the past. This technology had contributed in medical world which has resulted in life-saving processes. For those operations which take a lot of time and had lot of complex task, AR had reduced its complexness and time limit. This technology has some breeches, but with time, those will be filled. Moreover, AR is a new and emerging technology which will help a lot in our future, and we should encourage its development.

References

1. Applications of augmented reality available at: https://vstream.ie/wp-content/uploads/2018/10/Screen-Shot-2018-10-17-at-12.30.02.png
2. Rodriguez J, Huang CY (2017) An emerging study in augmented reality & geographical information system. Int J Comput Theory Eng 9(6)
3. Çolak O, Yünlü L (2018) A review on augmented reality and virtual reality in engineering education. J Ducational ND Instr Stud World 8
4. Andreas Dünser, Raphaël Grasset, Mark Billinghurst (2008) A survey of evaluation techniques used in augmented reality studies. Human Interface Technology Laboratory, New Zealand, Sept 2008
5. Sharma L, Lohan N, Yadav DK (2017) A study of challenging issues on video surveillance system for object detection, J Basic Appl Eng Res, vol. 4, Issue 4, pp 313–318
6. Future of AR: http://www.tekshapers.com/blog/FutureofAugmented-Reality
7. AR challenges: https://theappsolutions.com/blog/development/augmented-reality-challenges/
8. Jha G, Singh P, Lavanya Sharma, (2019) Recent advancements of augmented reality in real time applications. Int J Recent Technol Eng 8(2S7):538–542
9. Challenges: https://www.chrp-india.com/blog/biggestchallenges-facingin-augmented-reality/
10. Sharma L, Lohan N (2019) Internet of things with object detection. In: Handbook of research on big data and the IoT, IGI Global, pp 89–100, Mar 2019. ISBN: 9781522574323. https://doi.org/10.4018/978-1-5225-7432-3.ch006
11. Deshmukh SS et al (2018) 3D Object tracking and manipulation in augmented reality. Int Res J Eng Technol 05(01)
12. Sharma L, Garg PK, Smart E-healthcare with Internet of things: current trends challenges, solutions and technologies. In: From visual surveillance to Internet of Things, Taylor & Francis, CRC Press, vol 1, p 215
13. Bottino A et al (2018) Holo-BLSD: an augmented reality self-directed Learning and evaluation system for effective basic Life support defibrillation training. IMSH 2018, At Los Angeles
14. Wasenm¨uller O, Meyer M, Stricker D (2016) Augmented reality 3D discrepancy check in industrial applications. In: IEEE international symposium on mixed and augmented reality (ISMAR), At Mexico
15. Caudell TP, Mizell DW (1992) Augmented reality: an application of heads-up display technology to manual manufacturing processes. IEEE Kauai, HI, USA, USA
16. Ronald A et al (2001) Recent advances in augmented reality. IEEE Comput Graph Appl
17. Atkuri P et al (2018) Augmented reality in medicine: technique, scope and status. Int J Sci Res 7(2)
18. Sharma L, Garg P (2020) From visual surveillance to internet of things. Chapman and Hall/CRC, New York. https://doi.org/10.1201/9780429297922

19. Makkar S, Sharma L (2019) A face detection using support vector machine: challenging issues, recent trend, solutions and proposed framework. In: Third international conference on advances in computing and data sciences (ICACDS 2019, Springer), Inderprastha Engineering College, Ghaziabad, 12–13 Apr 2019
20. Sharma L, Lohan N (2019) Performance analysis of moving object detection using BGS techniques, Int J Spatio-Temporal Data Sci, Inderscience, vol 1(1), pp 22–53
21. Ar in military: https://www.techradar.com/news/deathbecomesar-how-the-military-is-usinga ugmented-reality
22. Shubhankar S, Mohit, Sharma L, Use of motion capture in 3D animation: motion capture systems, challenges, and recent trends. In: 1st IEEE international conference on machine learning, big data, cloud and parallel computing (Com-IT-Con), India, pp 309–313, 14th–16th Feb
23. Sharma L (2020) Human detection and tracking using background subtraction in visual surveillance, Towards smart world: homes to cities using internet of things, Taylor & Francis, CRC Press, pp 317–329
24. Liu Y, Hou R (2010) About the sensing layer in internet of things. Comput Study 5:55
25. AR scope: https://www.ncbi.nlm.nih.gov/pmc/articles/PMC4948614/
26. Anand A, Jha V, Sharma L (2019) An improved local binary patterns histograms techniques for face recognition for real time application. Int J Recent Technol Eng 8(2S7):524–529
27. Sharma L, Garg PK, Future of internet of things. From visual surveillance to internet of things, Taylor & Francis, CRC Press, vol 1, pp 245
28. AR in Information: https://en.wikipedia.org/wiki/Augmented_reality
29. AR in Medical field: https://www.archersoft.com/en/blog/howaugmented-reality-usedme dicine
30. AR Future in medical: https://bmcmededuc.biomedcentral.com/articles/, https://doi.org/10.1186/s12909-018-1244-9
31. AR in information: https://www.autodesk.com/redshift/whatisaugmented-reality/
32. Sharma L, Yadav DK (2016) Histogram based adaptive learning rate for background modelling and moving object detection in video surveillance. Int J Telemedicine Clin Practices Inderscience. ISSN: 2052-8442. https://doi.org/10.1504/ijtmcp.2017.082107
33. Body Anatomy: http://scarfedigitalsandbox.teach.educ.ubc.ca/anatomy-4d-using-ar-to-teach-humanbodystructure/
34. Future work: https://www.ijarse.com/images/fullpdf/1508652201_3109_IJARSE.pdf
35. Sharma L, Introduction: from visual surveillance to internet of things. From visual surveillance to internet of things, Taylor & Francis, CRC Press, vol 1, pp 14
36. AR Ftuture: https://haptic.al/vr-and-ar-inmedicaleducation-cd1c90cc3de3
37. Sharma L, Garg PK, Block based adaptive learning rate for moving person detection in video surveillance. From visual surveillance to internet of things, Taylor & Francis, CRC Press, vol 1, pp 201
38. Sharma L (2020) Towards smart world: homes to cities using internet of things. Taylor & Francis, CRC Press. (ISSN: 9780429297922)
39. Sharma L, Garg PK, Agarwal N, A foresight on e-healthcare trailblazers. From visual surveillance to internet of things, Taylor & Francis, CRC Press, vol 1, pp 235
40. Sharma L, Garg PK, IoT and its applications. From visual surveillance to internet of things, Taylor & Francis, CRC Press, vol 1, pp 29
41. AR glass: http://www.arreverie.com/blogs/wpcontent/uploads/2017/11/ar-devices-arreverie-768x271.jpg
42. Kumar A, Jha G, Sharma L (2019) Challenges, potential & future of IOT integrated with block chain. Int J Recent Technol Eng 8(2S7):530–536
43. Sharma L, Singh A, Yadav DK (2016) Fisher's linear discriminant ratio based threshold for moving human detection in thermal video. Infrared physics and technology, Elsevier
44. Sharma L , Sengupta S, Kumar B (2021) An improved technique for enhancement of satellite image. J Phys: Conf Ser 1714:012051

45. Singh S, Sharma L, Kumar B (2021) A machine learning based predictive model for coronavirus pandemic scenario, J Phys: Conf Ser 1714 012023
46. Sharma L (2020) The rise of internet of things and smart cities, Towards smart world: homes to cities using internet of things, Taylor & Francis, CRC Press, pp 1–19
47. Sharma L (2020) The future of smart cities. Towards smart world: homes to cities using internet of things, Taylor & Francis, CRC Press, pp 1–19

E-health in Internet of Things (IoT) in Real-Time Scenario

Gourav Jha, Lavanya Sharma, and Shailja Gupta

Abstract Internet of things (IoT) has drawn much attention in recent years, and it has a very significant role in the IT industry. The IoT helps to modernize the healthcare system with promising technologies and economic prospects. This paper presents working of IoT in e-health, IoT-based technologies, and communication range in the e-health sector. Furthermore, this paper also analyzed the IoT security and privacy in e-health, security risk which is involved in the e-health and the real-time application which is used in health care. The IoT application is the utmost necessary part of daily life, and it is used in almost every sector of human and industry activity. Communication standards of IoT will be provided along with some real-time applications.

Keywords Internet of things · Health care · E-health · Real-time tracking location · Virtual health · MIoT

1 Introduction

Augmented The internet of things (IoT) ensures to changes our livelihood to make them simpler, efficient and most importantly smart. This paper aims to provide information on e-health IoT universe in a different perspective and also reflects the importance of this kind of technology in the medical sector. IoT technologies are rapidly increased in terms of device installation, this trend shows that IoT became the most important part in the e-health sector, and also, we can rely on that with no hesitation [1–6].

IoT comes in as the interfacing stage for all the particular elements associated with a common social insurance framework. Besides it supplies the unobtrusive and installed intensity of processing which separates the information from nature and

G. Jha · L. Sharma
Amity Insititute of Information Technology, Amity University, Noida, Uttar Pradesh, India

S. Gupta (✉)
Department of Computer Science and Technology, Manav Rachna University, Faridabad, India

P. K. Singh et al. (eds.), *Proceedings of Second International Conference on Computing, Communications, and Cyber-Security*, Lecture Notes in Networks and Systems 203,
https://doi.org/10.1007/978-981-16-0733-2_48

trades it commonly for a universal data framework, making it continuous for an omnipresent smart framework. As we know that the world technology is growing much rapidly and intelligently in every sector like industries and business, the main reason of growing technology is because of revolution of the Internet of things as we know that is IoT. The global market of IoT in e-health alone is 26.34 billion dollars in 2016 and expected to reach [7–13].

The global market of IoT may reach 148.76 billion dollars approximately in the year 2025. The healthcare IoT also called the Internet of medical and things (IOMT) provides many facilities like smart medical care, improved public health care and saves thousands of lives across the world, and nowadays, health care in IoT is playing a very vital role in everyone's life. Internet of things also includes many components of a healthcare system such as data collection which works as the collection of real-time data comprises IoT-driven sensors to gather ongoing checking information from smart sensors, the another component is big data which really became an important component of IoT health care in recent times, and any smart devices that are connected with users or patient and generate related data about the personal health and send back to the cloud will be a part of the IoT (Fig. 1).

The last component is data analysis, the mechanism of the data analysis articulation by machine learning, to acquire relevant data from past stored result data machine learning is a key component in the health care.

All components have their own role and process in health care and also played a very important role in e-health care [14–17]. There are also many fundamental areas where the e-health is used such as health records data, diagnosis, post-medical phase, and monitoring of data.

This paper is categorized into seven chapters. Section 1 deals with introductory part of e-health in IoT, whereas Sect. 2 is related work. Section 3 deals with IoT leading a smart medical revolution. In the next Sects. 4 and 5, communication standards of e-health and challenging issues are discussed. In Sect. 6, communication

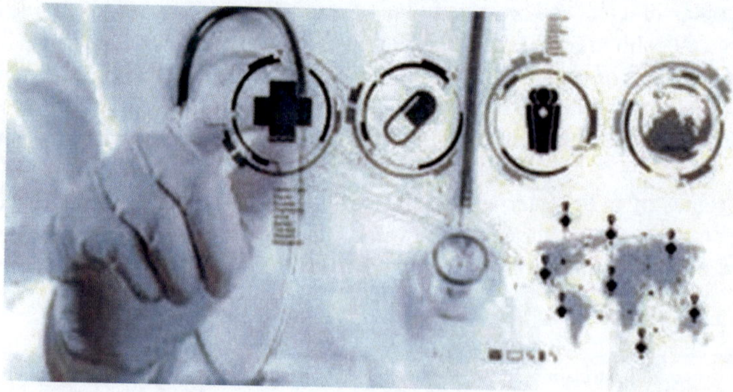

Fig. 1 E-health in IoT [2]

standards of IoT are provided. In next Sect. 7, real-time applications are discussed. In the last, conclusion of the work is discussed.

2 Literature Review

This section deals with the work done by various researchers in this specific domain to date. Scarpato et al. [1] discussed about the cloud environments that help to share and collect data directly from the devices with the using of IoT, and it is also providing a vast amount of data or input to be stored and mechanism of the data analytics process. Because of healthcare application, there is an increased in life expectancy of patients. This paper also showed the privacy and security during the sharing of real-time data. Xang et al. [18] discussed that, in recent years, IoT technology draws much attention because of its promising alleviation of the strain on health care, and thus, this paper presented overall suitability for a sensor-based IoT health care system. Challenges that healthcare IoT faces include security and privacy.

Sharma and Lohan [19] discussed about object detection in IoT field and in surveillance system alongside with methods, applicational areas, challenges, and advantages of IoT technology in surveillance system. Devendan et al. [14] discussed that IoT is a connecting network interfacing thing which has naming, detecting, and processing capacity. IoT helps to alter the data from the physical world to the digital world. The IoT has a variety of application domains, and IoT in health care is gone for engaging individuals to live more beneficial life by using smart gadgets. This paper presents couple of uses of IoT in rural health care and approaches to improve essential health needs of the creating countries and innovation utilized in IoT. Sharma et.al [15] discussed about e-health system alongside with challenges in implementing IoT in medical department. Sebestyen et al. [20] discussed and break down the effect of 'IoT—Internet of Things' on the structure of new e-health arrangements. The authors expected to illustrate that the communication models, conventions, and advances advanced under the IoT idea have an incredible potential in the usage of Internet-based healthcare frameworks.

Azzawi et.al. discussed (IoT) ascends as an incredible area where capable gadgets and sensors can interface and trade data over the Internet. The significance of IoT gadgets and information can be basic, so security limitations are required to safeguard the IoT information from intruders; verification is one of fundamental and imperative intends to affirm information protection and security.

3 IoT Leading a Smart Medical Revolution

Here are the means by which IoT is realizing an intelligent medicinal revolution, changing the e-health industry incredibly. Some of the important are listed below.

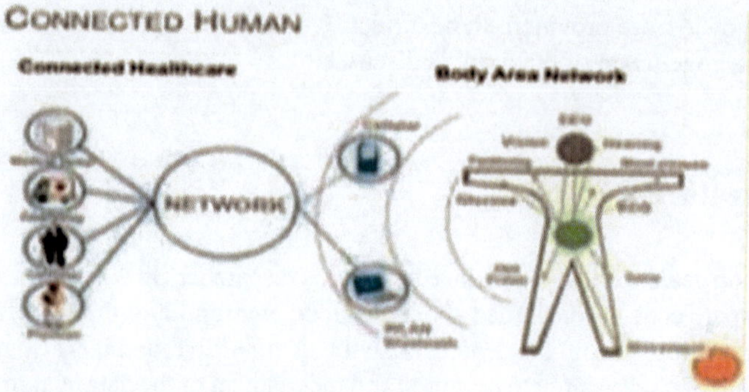

Fig. 2 IoT for E-health system project concept [24]

- *Telehealth*: The another name of telehealth is remote patient monitoring system. This technology is used in a such a way that the health care is provided to the patient who are in a distance. The telehealth is beneficial in many ways such as it increases the access of patients, improves the real-time data analysis, and also, it is cost effective. The working of remote patient monitoring system: The patient is connected to the monitoring system, and vitals are recorded through probes. The tablet sends life and recorded data to the cloud using WI-FI [21–23].
- *Real-Time Tracking Location (RTLS)*: The real-time location tracking uses the technology of the location sensor which is attached or connected to the device and people. Nowadays, health care is likely to be 52%, and the organization uses the real-time location services in their daily lives to improve their healthcare result and status within the patient. RTLS also comprehends ordinary difficulties by following important resources, staff, and patients on virtual maps, helping emergency clinic representatives and directors save time (Fig. 2).

- *E-health Application*: The introduction of e-health application creates many possibilities, and because of this application, the health care is very easily accessible to everyone [17]. Some of the healthcare groups developed mobile application to start round the clock message passing with the patients and users and also increase the healthcare awareness among them (Fig. 3).

- Virtual health assistant—Virtual health assistant helps the health care system in many ways, most of the hospitals installed chat-based and virtual-based assistant in their organization which helps their medical staff or administrator to enhance the result and produce a better outcome, and it also increases the patient experience [26, 27] (Fig. 4).

Fig. 3 Healthcare app development [25]

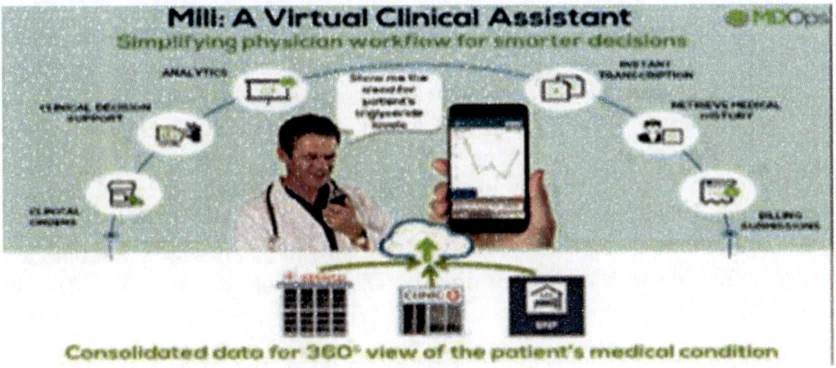

Fig. 4 Virtual health assistant

4 Communication Standards of E-health in IoT

In the health care in IoT, the communication area is related to it and is divided into two categories. The first is short-range communication, and the second is long-range communication. In advancing of e-health in IoT, we used the wireless body area network (WBAN) to communicate between the two devices that are connected to the Internet of things. Firstly, we have to know what is WBAN, so basically, it is a wireless network of wearable computing device that connected to a cloud network and also in or around the human body to exchange of data and also serve the variety of application. Both communication standards are equally necessary in e-health in IoT [28].

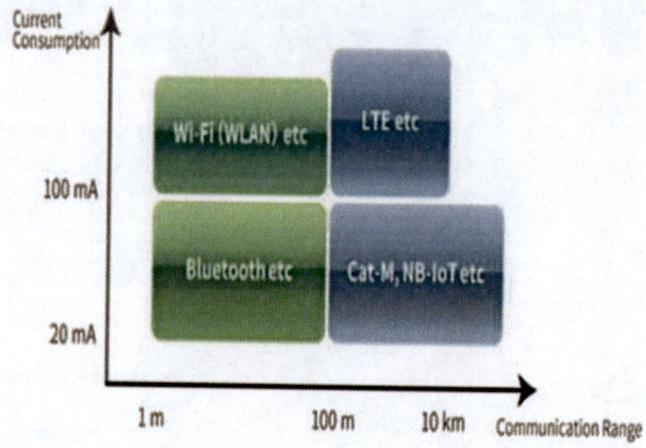

Fig. 5 Range and consumption scale of SRM [32]

1. *Short-Range Communication*: The short-range communication application of network technologies is growing very rapidly in recent years, short-range communication also helps to expand network easily, because in this type of communication, we do not need predefined and wired infrastructure or framework. In the short-range communication, new sensor and input devices can easily be configured and easily joined or connected [29–31].
2. *The Bluetooth low energy*: It is also known as Bluetooth 4.0, it is introduced in 2010, and with the introduction of the Bluetooth 4.0, there is a rapid increase in wearable devices mostly in the fitness sector, i.e., flue band, Fitbit, flex, and shine. Star topology technique is used by BLE which results often in the best for the healthcare application and devices and that is why sensor does not communicate with each other directly. The range of BLE is 150–1570 m in widely open area, and high data rate is 1 MBPS (Fig. 5).

3. *Long-Range Communication*: The reason of the higher success of e-health in IoT is because of the introduction of the long-range communication standards, and it is also very suitable for the IoT applications. LPWA is a subset of the long-range communication definitive [30]. In general, the LRC has the long reach of communication as compared to the traditional communication such as the Bluetooth or WI-FI. Long-range communication has another advantage and is that it supports the 3G network that they are intended to help short blasts of information rarely [26, 32, 33]. This is appropriate for a vast number of social insurance applications. The most famous or important standard of long-range communication is SigFox. The description of SigFox is given below:

For, e.g., SigFox—SigFox provides just a limited number of functionality, but it is expanded very vast as compared to any other communication standards. The SigFox base framework is very similar to the cellular—antenna mount on tower, and SigFox uses the star topology which is suitable for healthcare devices.

5 IoT E-health Challenging Issues

This is a rapid growth of IoT in e-health, and there are many challenges that rise which must be addressed carefully before it reaches out of our zone. Some of the important e-health open challenges are listed below:

- *Cross Domain*: In IoT e-health, many fields intersect with each other at some point, and some of the fields are bio-engineering, embed system, network design, and data analysis. Therefore, to maintain that level of framework structure and also the design verification requires a very vast amount of knowledge in that field.
- *Heterogeneous*: IoT measures the cyber and physical world, and because of that, it includes the hardware and software component. Thus, it is important to give detailed or proper consideration to interfacing and compatibility of such an all-encompassing framework. Of associated gadgets, arrange segments, calculation frameworks that must deal with information volume, assortment, speed, and accuracy [34].
- *Data Management*: In health care, IoT tackles the data management challenges in recent times because of the increasing number of patients which are connected to the IoT e-health. The data is taken from the human body via wearable devices or sensor from which human is connected so that device sends the real-time data, but human body constantly changes and so the data of an individual, and thus, it becomes very difficult to maintain a very large amount of ongoing flux data.
- *Scalability*: In order to design a human services Internet of things on a very scaled-down, every client should have direct access to healthcare administrations from convenient gadgets, for example, cell phones. These administrations will require them very own sensors particularly for information gathering, alongside secure focal servers for dealing with client requests [28, 35].
- *Human-factor engineering and interfaces*: The interface between front-end advancements, for example, sensors, PCs, tablets, and other cell phones gives one of the quickest difficulties for IoT e-health improvement. End clients will be required to self-train so as to utilize the gadgets effectively. Also, a large number of the gadgets will be conveyed in remote areas; old populaces specifically will be probably the most remarkable IoT clients, featuring a reasonable requirement for e-health frameworks that can be conveyed basically and self-sufficiently (Fig. 6).

- Another big challenge is device update needs to be managed effectively as well security patches to firmware and software will have a number of challenges. Also, over the air updates may not be possible with all kind of IoT devices. Many a times, the device owners may also not show much interest in applying an update to the system.
- The communication channel through which data is going from cloud devices or vice versa also needs to be secure and also uses the transport encryption to adopt standards like TLS.

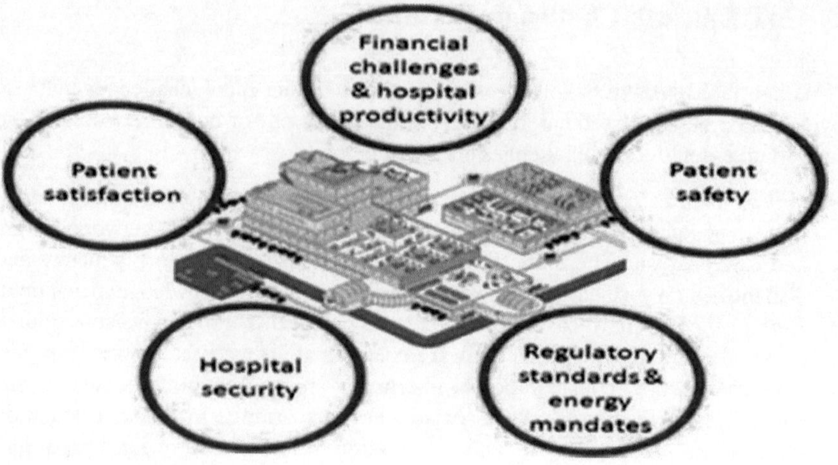

Fig. 6 Challenges in e-health sector [3]

Security and the Privacy Issue on the Medical Internet of Things (MIoT)

MIoT is subdivided into three layers which are the perception layer, network layer, and the another is application layer, and all these layers perform their specific task, such as to collect healthcare real-time data from different types of devices is a work of perception layer, the network is subdivided into wired and wireless system and middleware which mechanize and send the data or input which is taken by perception layer, it also ensures the privacy and security, and the third layer is application layer that unified the healthcare information resources to provide personalized medical services. Following are the security and the privacy issues on the medical Internet of things:

(1) *Data Usability*: Data usability is to guarantee that information or information frameworks can be utilized by approved clients. Enormous information brings incredible advantages as well as vital difficulties, for example, wrong information and nonstandard information.
(2) *Data Integrity*: Data integrity, or 'data quality,' alludes to the way toward keeping up the precision, reliability, and consistency of data over its whole 'life-cycle.' Applied to health, this can incorporate (yet is not restricted to) keeping up the exactness of patient's personal health, details, summary, and many more that are very credential to the patients.
(3) *Data Auditing:* Review of medical data is a successful way to screen the utilization of assets and a typical measure for finding and following unusual occasions. Also, cloud service organizations more often play entrusted roles, which require sensible auditing techniques [27].
(4) *Patient information privacy*: The data of the patient's or users can be divided into two categories, one is normal record, and the another is sensitive records which include patient's mental status, infectious diseases, drug addiction,

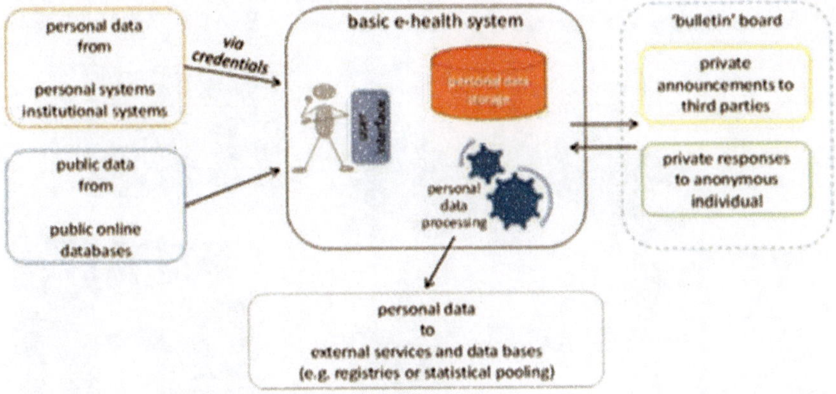

Fig. 7 Basic privacy and security of e-health system [36]

genetic information, and personal identification data. So, we have to make sure that any unauthorized users do not get these types of information or neither the data can be modified (Fig. 7).

6 Real-Time Applications / System Components of E-health Sensor

In healthcare system, sensory devices play a very important role, and because of that devices, we can easily detect or know any problem within the patients and user, these devices are shown the real-time data of the patients which helps to easily overcome the problem [37,38]. Some real-time application/sensory devices are as follows

- *ECG electrodes*—An ECG cathode is a gadget connected to the skin on specific pieces of a patient's body —by and large the arms, legs, and chest—during an electrocardiogram methodology. It recognizes electrical driving forces delivered each time the heart pulsates. The number and situation of anodes on the body can change, yet the capacity continues as before [37]. The power that an anode recognizes is transmitted by means of this wire to a machine, which makes an interpretation of the power into wavy lines recorded on a bit of paper. The ECG records, in an extraordinary detail, are utilized to analyze a wide scope of heart conditions.
- Blood pressure (BP) sensor—It is a smart device that helps to measure the pressure of the blood in arteries which pumped in our body through heart when our heart thumps, it contracts and pushes blood, and due to this, a pressure arises on the arteries. BP is recorded as two numbers, one is systolic pressure (sp), and another is diastolic pressure (dp). Moreover, a regular circulatory strain sensor can store 80 estimations data with time and date [28].

Fig. 8 Sensor devices in health care [19]

- Pulse oximetry sensors—Heartbeat oximetry measures the level of oxygen in the blood. Like heartbeat, blood oxygen level is positively not a basic sign, however that as it might, fills in as a pointer of respiratory limit and can help in diagnosis of particular conditions, for instance, hypoxia (low oxygen including the body's tissues) [35]. All things considered, and beat oximetry is an essential augmentation to a general medicinal service observing system. By getting PPG signals, beat oximetry measures the blood oxygen [38] (Fig. 8).

7 Conclusion and Future Work

In this paper, we analyzed that the emergency department part is embracing the IoT very rapidly. This exponential development of the wearable gadgets, capable advancements, and cloud-based information logical techniques are giving new era of healthcare system frameworks. Despite, the extensive number of applications of IoT in medicine numerous issues is as yet open, and they need ingenious answers to be solved. This paper presents advances in IoT-based future healthcare or medical services for different use case scenarios, explores the ongoing and developing communication technologies and standards in IoT. The security issues are also involved in the e-health sector, and it is very crucial to take some important measure before any big adversity. This paper also discussed some real-time application/sensor devices which are used in the e-health sector, and due to this, there is a big improvement in healthcare sector.

References

1. Scarpato N et al (2017) E-health-IoT universe: a review. Int. J Adv Sci Eng Inform Technol 7(6)
2. Barnagh P et al (2012) Semantics for the internet of things: early progress and back to the future. Centre for Communication Systems Research, University of Surrey, Guildford, UK
3. Sharma L, Lohan N (2019) Performance analysis of moving object detection using BGS techniques. Int J Spatio-Temporal Data Sci
4. E-Health in IOT. Available at: https://encrypted-tbn0.gstatic.com/images?q=tbn:ANd9GcSQjBqW3Yd5CnK35LtivM6gnVcLESTRoZwWwm xwIOpSp4lXvigKzA
5. IoT for E-health system project concept. https://www.iotforall.com/how-iot-enables-tomorrows-telemedicine/
6. IoT for E-health system project concept. https://dzone.com/articles/iot-find-and-track-the-next-generation-of-healthcare
7. https://study.com/academy/lesson/what-are-ehealth-mhealth-applications-definition-uses.html
8. Virtual health assistant. Available at: https://www.medstartr.com/uploads/images/projects/100144/solutionimg_59e2488306e82.png
9. https://searchhealthit.techtarget.com/news/2240022287/Bluetooth-standard-could-advance-use-of-wireless-medical-devices
10. IoT devices. Available at: https://engineering.eckovation.com/sigfox-vs-lora-one-prefer-iot-device/
11. Impact of IoT. Available at: https://blog.eai.eu/possible-impact-of-iot-in-health-care-in-developing-countries-cannot-be-underestimated/
12. Anand A, Jha V, Sharma L (2019) An improved local binary patterns histograms techniques for face recognition for real time application. Int J Recent Technol Eng 8(2S7):524–529
13. Jha G, Singh P, Sharma L (2019) Recent Advancements of augmented reality in real time applications. Int J Recent Technol Eng 8(2S7):538–542
14. Devendran T, Agnes Archana DA, Suseela S (2018) Challenges and issues of healthcare in internet of things (IOT). Int J Latest Trends Eng Technol
15. Sharma L, Garg PK (2019) Smart E-healthcare with internet of things: current trends challenges, solutions and technologies. In: From visual surveillance to internet of things, vol 1. Taylor & Francis, CRC Press, Boca Raton, p 215
16. Chacko N, Hyanjneh T (2017) Security and privacy issues with IoT in healthcare, vol 4, issue 14. Ordham Center for Cybersecurity, Fordham University, New York, NY, USA
17. Sensor devices in healthcare. Available at: https://www.cooking-hacks.com/blood-pressure-sensor-mysignals-ehealth-medical
18. Baker S et al (2017) Internet of things for smart healthcare: technologies, challenges, and opportunities. James Cook University, Townsville
19. Sharma L, Lohan N (2019) Performance analysis of moving object detection using BGS techniques in visual surveillance. Int J Spatiotemporal Data Sci Inderscience 1:22–53
20. Sebestyen G et al (2014) eHealth solutions in the context of internet of things. In: IEEE international conference on automation, quality and testing, robotics. Abdullah M et al (2016) A review on internet of things (IoT) in healthcare, vol 11. University Kembangan Malaysia, pp 10216–10221
21. Security issues of IoT. Available at: https://techcrunch.com/2016/03/01/iot-security-needs-scalable-solutions/

22. Liu Y, Hou R (2010) About the sensing layer in internet of things. Computer Study 5:55
23. Sharma L, Yadav DK (2016) Histogram based adaptive learning rate for background modelling and moving object detection in video surveillance. Int J Telemedicine Clinical Practices. ISSN: 2052–8442. https://doi.org/10.1504/IJTMCP.2017.082107
24. IoT for E-health system project concept. Available at: https://image.slidesharecdn.com/iot-150630175330-lva1-app6891/95/iot-for-ehealth-system-project-concept-1-638.jpg?cb=1435687353
25. Virtual Health. Available at: https://www.pharmacytimes.com/publications/issue/2015/may2015/next-it-virtual-health-assistant-engaging-patients-and-improving-their-outcomes
26. Kumar A, Jha G, Sharma L (2019) Challenges, potential & future of IOT integrated with block chain. Int J Recent Technol Eng 8(2S7):530–536
27. Sharma L (2020) Introduction: from visual surveillance to internet of things. In: From visual surveillance to internet of things, vol 1. Taylor & Francis, CRC Press, Boca Raton, p 14
28. Sharma L, Garg PK (2020) Future of internet of things. From visual surveillance to internet of things, vol 1. Taylor & Francis, CRC Press, Boca Raton, p 245
29. Range and consumption scale of SRM. Available at: https://dl.cdn-anritsu.com/images/tm/technologies/iot/tec
30. Sharma L, Garg P (eds) (2020) From visual surveillance to internet of things. Chapman and Hall/CRC, New York. https://doi.org/10.1201/9780429297922
31. Sharma L, Lohan N (2019) Internet of things with object detection. In: Handbook of research on big data and the IoT. IGI Global, pp 89–100. ISBN: 9781522574323. https://doi.org/10.4018/978-1-5225-7432-3.ch006
32. Challenging issues of IoT. Available at: https://www.softwebsolutions.com/resources/challenges-in-healthcare-analytics.html
33. Shubham S, Shubhankar V, Sharma L (2019) Use of motion capture in 3D animation: motion capture systems, challenges, and recent trends. In 1st IEEE international conference on machine learning, big data, cloud and parallel computing (Com-IT-Con), India, , 14–16 Feb 2019, pp 309–313
34. Sharma L, Garg PK (2020) Block based adaptive learning rate for moving person detection in video surveillance. In: From visual surveillance to internet of things, vol 1. Taylor & Francis, CRC Press, Boca Raton, p 201
35. Sharma L, Garg PK (2020) IoT and its applications. In: From visual surveillance to internet of things, vol 1. Taylor & Francis, CRC Press, Boca Raton, p 29
36. Sharma L, Garg PK, Agarwal N (2020) A foresight on e-healthcare Trailblazers. In: From visual surveillance to internet of things, vol 1. Taylor & Francis, CRC Press, Boca Raton, p 235
37. Makkar S, Sharma L (2019) A face detection using support vector machine: challenging issues, recent trend, solutions and proposed framework. In: Third international conference on advances in computing and data sciences (ICACDS 2019), Inderprastha Engineering College, Ghaziabad, 12–13 April 2019. Springer, Berlin
38. Sharma L, Singh A, Yadav DK (2016) Fisher's linear discriminant ratio based threshold for moving human detection in thermal video. Infrared Phys Technol

Diagnosis of Heart Disease Using Internet of Things and Machine Learning Algorithms

Amit Kishor and Wilson Jeberson

Abstract In the current scenario of the digital world, the healthcare industry generates a huge amount of patient data. Manual handling of these produced data becomes very difficult for doctors. The Internet of things (IoT) is very effectively handling the produced data. The IoT captured huge amounts of data, and with the machine learning algorithms, it can detect the disease and diagnose the disease. The work aims to apply various machine learning methods to the produced data. A machine learning framework has proposed for early prediction of heart disease in conjunction with IoT. The developed model is evaluated with k-nearest neighbor (K-NN), decision trees (DTs), random forest (RF), multilayer perceptron (MLP), Naïve Bayes (NB), and linear-support vector machine (L-SVM). The model achieved the diagnostic accuracy for 82.4%, 81.3%, 92.3%, 88.2%, 89.6%, and 82.4% for K-NN, DT, RF, MLP, NB, and L-SVM, respectively. As per the experimental results, random forest has the highest prognosis rate of 92.3%.

Keywords Machine learning · Internet of Things · Cloud computing · Health care

1 Introduction

In the current scenarios of human lives, the lifestyles of humans are changing day by day. The changes occurred in their lives due to eating habits, social changes, and technological changes,. In the busy schedule of lifestyles, the human age becomes shorter than the prior one. The human being starts suffering from different diseases due to careless behavior and lifestyle. Heart disease is one of the major diseases found in today's life. As per the report of World Health Organization (2020), in the year 2016, 17.9 million people died due to cardiovascular disease, and this cardiovascular disease is the main reason for death with 31% of overall death. Out of which, 85% of deaths (around 15.2 million) are due to heart attack and stroke.

A. Kishor (✉) · W. Jeberson
Department of Computer Science and Information Technology, Sam Higginbottom University of Agriculture, Technology and Sciences, Allahabad, U.P., India

P. K. Singh et al. (eds.), *Proceedings of Second International Conference on Computing, Communications, and Cyber-Security*, Lecture Notes in Networks and Systems 203,
https://doi.org/10.1007/978-981-16-0733-2_49

Advancement in the era of the medical field is helping to predict the risk of heart disease patients who get affected in the future. It is strongly recommended that all people, including those who feel healthy, see a cardiologist twice a year to see if there is any evidence that it may cause heart disease [1]. Variety of test related to heart issues can be conducted by well-equipped health organizations and that will provide valuable information that help clinician to diagnose the patient's level of risk of heart disease and provide their views [2].

Now, technology changes day by day, and in this section, we present a significant impact of technological changes in the healthcare field. Machine learning overview is provided here with their benefits and how it will be implemented in health care. In the year of late 1970, the provision of health care is started, and it is considered as Healthcare 1.0. In this era of health care, there is very less technological development, and resources were available. Due to the lack of coordination with digital systems, the medical practitioner used paper-based prescriptions. The period of years from 1991 to 2005 was considered as Healthcare 2.0 in the healthcare industry. In this era, the combination of health with information comes into this system. The coordination with the digital system of doctors with patients starts in this duration, and doctors start to use cloud servers to store their data. From the years 2006 to 2015, this period is considered as Healthcare 3.0. The electronic healthcare records were introduced. The sharing of data made patient's and doctor's online communication easy in Healthcare 3.0. From 2015 to the current day, we have experienced Healthcare 4.0. Healthcare 4.0 is full of technological experiences such as cloud, fog computing, artificial intelligence, the internet of things (IoT), and machine learning. The adoption of technological advancements in healthcare industries is the key challenge in our society for both patients and doctors.

Machine learning entered health care with different features, and it helps both patients and doctors as well in the prognosis of the disease. Machine learning uses digital medical data for processing. In the past decades, researchers explored various machine learning approaches in health care. These approaches are used to segmentation and classification of data. The IoT devices are used to send health data to the cloud; after that, the large volume of patient's data is processed by the cloud and provides the ability to analyze big data. The collection of patient's data is processed through different body sensors. Then, the collected data is stored on cloud servers via IoT devices. The patient data is processed here with efficiency. Remotely, doctors can take action on data after processing, and they can also store their response on the cloud. The fog computing can be used for fast processing and response. Often, the prediction is based on the understanding and experience of the doctors, which can sometimes be wrong and lead to undesirable consequences. Therefore, an automated medical information analysis model is needed to be developed. Due to the remote physical situation, it is sometimes not possible to get to the medical services on-time, and therefore, it is life-threatening. The main motivation for using the IoT-based healthcare system is to use the latest wireless technology in health care to achieve real-time care and early diagnosis for life-threatening diseases. The developed system will work on collected information and will take decisions accordingly. The developed model can help diagnose the disorder, often with fewer medical tests and before the patient develops serious symptoms. To meet this need, we used health information systems in hospitals. To diagnose the disorder, with a large amount of collected data and historical data of the disease, a machine learning approach based on an intelligent disease prediction model has been developed. Hence, machine learning techniques are used to efficiently process a high volume of patient data and extract inalienable, previously hidden, and widely useful information for the e-healthcare system. The main contributions of the work are as follows:

1. Develop a machine learning-based model for the prognosis of heart disease.
2. Measuring the performance of the model with different machine learning techniques.

The remaining research work is outlined in different sections as follows. Section 2 describes the previous work done. Section 3 presents the materials and methods used. The proposed diagnostic model is presented in Sect. 4. Section 5 represents the significance of research work. Result and discussion are presented in Sect. 6. Section 7 concludes the presented work.

2 Related Work

Hameed et al. [3] represented a cloud-based e-healthcare system where all patient data is recorded in a single database. The proposed structure depends on a service-oriented architecture (SOA), and the system offers improvement in cost management, the maintenance of the patient profile, and the selection of the right specialist.Verma and others [4] proposed a k-means algorithm with particle swarm to determine the risk factors using the treatment for coronary heart disease (CAD). They extracting the data by implementing a variety of learning algorithms, including multilayer perceptron, multi-nomial logistic regression, fuzzy rule algorithms, and C4.5. They use data set provided by a Medical College in India, that contain 26 and 335 features and instances, respectively. According to the test results, the maximum accuracy of the MLR is 88.4%. Forkan et al. [5] proposed a ViSiBiD predictive model for daily monitoring and prevention of diseases, which is based on the analysis of the vital symptom of the patient. To study the cloud platform, machine learning methods were used, as well as some implementations of map reduction. They used 4893 publicly available patient records and found that six biosignals were different from normal and distinct characteristics for observation of data events. The collection of data events is performed at intervals of 1–2 h. According to the test results, in comparison with the other one, the random forest shows the maximum accuracy of 95.85%. Osman and Aljahdali [6] introduced a method with improved feature extraction, and they used k-means and SVM methods as classification techniques for detecting diabetics. From experimental results, they show that the developed technique has a better classification rate when compared to other available methods. Zhang et al. [7] developed a new cancer prediction method based on statistical learning theory. The used method is suitable for binary classification and multi-class problems. The support vector machine (SVM) method creates large hyperplanes in multidimensional space. They used to maximize the difference between the data points. The hyperplane is created with support vectors. The SVM offers the best accuracy but takes longer computational time. Devi and Shyla [8] applied data mining techniques for the early prediction of diabetes disease. 768 instances have been used to measure accuracy. The used data set is collected from PIMA India. Their study proves that as compared to other techniques, J48 classifier has superior accuracy. Hsu et al. [9] proposed a health-based model for assessing breast cancer risk. Sampling and dimensionality reduction were used to preprocess the test data in the work. Hence, the prediction of risk is done based on used different classifiers. The developed model was cost-sensitive. Benjamin et al. [10] analyze the factors that become a reason for heart disease. The authors considered smoking, high cholesterol, a symptom of diabetics, and high pressure of blood as the prime factors for heart disease. The authors also discussed the heart disease statics for stroke and other heart issues. Rizvi [11] predicted heart disease based on machine learning and deep learning. They used some machine learning data mining methods and deep learning methods to predict heart disease. The authors have surveyed heart disease work. After the analysis, the authors said that neural networks have better efficiency rather than other methods. Uyar and Ilhan [12] presented a method to diagnose heart disease. The diagnosis is based upon the genetic algorithm and neural networks. They used a total of 297 samples of patient data, and it is divided into 85 and 15% for training and testing work. The model has given 97.78% accuracy. Doupe et al. [13] highlighted the use of different machine learning approaches to the healthcare sector. They presented that decision tree, deep learning, and ensemble methods for identification subpopulation experiences, identification of nonlinear patterns, and improvement in the use of different machine learning approaches in combination. Nilashi et al. [14] proposed a machine learning-based model for heart disease prediction. The model used fuzzy SVM. They highlighted the improvement in the accuracy of the data classification and diagnostic time of disease. Mohan et al. [15] proposed hybrid machine learning techniques for cardiovascular disease. They combined different machine learning techniques and also combined the features. The model achieved an accuracy of 88.7%.

3 Materials and Methods

The used methods are as k-nearest neighbor (K-NN), decision trees (DTs), random forest (RF), multilayer perceptron (MLP), Naïve Bayes (NB), and linear-support vector machine (L-SVM). From UCI, the patient data for heart disease is collected as shown in Table 1, and also, description of data with their attributes is given in Table 1. The data set is downloaded from https://archive.ics.uci.edu/ml/datasets/Heart+Disease. We use MATLAB tools for simulation.

Table 1 Data set for heart disease

S. No.	Name of attribute	Description
1	Age	Age
2	Sex	Gender
3	CP	Chest pain
4	Trestbps	Blood pressure (resting)
5	Chol	Cholesterol
6	FBS	Blood sugar (fasting)
7	Restecg	ECG (resting)
8	Thalach	Heart rate (maximum)
9	Exang	Heart rate (maximum)
10	Oldpeak	ST depression induced by exercise
11	Slope	The slope of peak exercise segment
12	CA	Major vessels number
13	Thal	Heart rate (resting)
14	NUM	Heart disease status

4 Proposed System Model

The Internet of Things connects everything to the Internet. The IoT has been ported all over the Internet, and the main role of IoT is to provide new management system that offers new opportunities to increase the activities of production, agriculture, finance, and health care. The Internet comprises an important role in developing IoT-based healthcare monitoring system. The maintenance of records for any doctor is very difficult because it contains enormous data. However, physicians must use this historical data to predict a patient's health status. Various machine learning technologies have been used in medical application field over the decades. Although machine learning methods have problems with tuning parameters, after effectively tuning these different parameters, it can improve the prediction efficiency and the performance of current machine learning methods. There is a long history of studying information, finding hidden connections, and predicting future patterns. This is sometimes referred to as "database disclosure research." We are planning to develop an Internet of Things (IoT)-based remote online health monitoring framework supported by cloud computing. These current situations are expected to take advantage of machine learning in the cloud while maintaining a health services database, with the specific end goal of continuously monitoring patient health and overseeing remediation for each client in the infrastructure. We proposed a framework for a machine learning model that could be applied to cloud data and tested its performance. The required facilities for patients that are available in remote locations or at some distances can make easily available on their homes with some sensors. The third type of client includes patients who visit the clinic and laboratory, including all modern facilities, but do not have medical doctors but have all the necessary medical equipment and a support group to act as intermediaries. The patient's health information is deployed on a cloud server from where doctors can access the information and respond accordingly. All medical information collected is first sent to mobile devices via a network of sensors. Mobile networks act as the operator of the IoT and are used to send patient health information to the cloud. So far in this research implementation has been carried out including the application of leading different machine learning methods (such as K-Nearest Neighbor (K-NN) method, Decision Trees (DT), Random Forest (RF), Multilayer Perceptron (MLP), Naïve Bayes (NB), and Linear-Support Vector Machine (LSVM) method) in a data set of diseases namely heart disease. Random forests are unmatched in precision, like other learning algorithms that are controlled by classification and work efficiently with large databases. The random forest classifier generates a series of decision trees from a randomly selected subset of the training set. Then, it collects the results from all different decision trees that are combined to determine the final test object. In this research work, the extraction of information from databases random forest classifier is used. The block diagram of research work is presented in Fig. 1. In the proposed model, the methodology is divided into six stages.

The first stage is preprocessing of data, and the patient data is collected through sensors, and the sensor input data is preprocessed. In preprocessing, missing data is removed. The second stage is feature selection of data, and the heart disease attributes like chest pain, cholesterol, and blood sugar, etc., are selected for further processing. The third stage is the splitting of data. In this stage, the complete data of the patient is split into the ratio of 80 and 20% for training and testing purposes. In stage four, the six machine learning algorithm is applied to train the model. The data is trained according to each used machine learning techniques. In stage five, the testing of the model is done. The model is tested for each used machine learning techniques. And the last stage is prediction through the model; finally, according to the training of the data set, the model is tested, and it predicts the output as per the area under curve method as positive and negative. Positive indicates the presence of heart disease and negative indicates the non-presence of heart disease in the supplied input.

5 Significance of the Proposed Model

Six different data mining algorithms are used named as K-NN, DT, RF, MLP, NB, and L-SVM for the prediction heart disease. The effectiveness of these methods is compared with the goal of providing a better quality of service as a health goal.

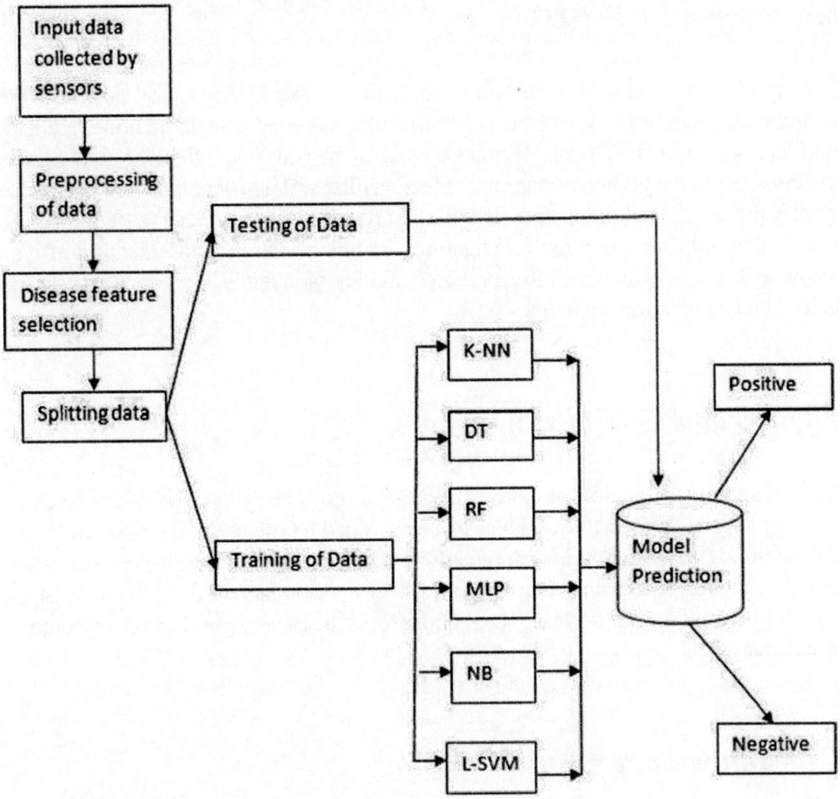

Fig. 1 Block diagram of the proposed system model

5.1 K-Nearest Neighbor (K-NN)

The K-NN algorithm is a supervised and non-parametric learning approach to machine learning. The K-NN method is based on the data points of the nearest neighbor, finds unrecognized data points, and arranges the data points according to a voting system. Supervised learning indicates that data is labeled, and the algorithm learns from input data to predict the output. It even works great even if the training data is huge. Implementation of the K-NN method is easy but requires a lot of memory because of its alertness to noise and slow testing.

5.2 *Decision Tree (DT)*

The decision tree method is based on a tree shape diagram. This method calculates the conditional probability of various attribute nodes, where the attribute nodes are leaf, non-leaf, and branch. The upper node is used as the root node, the leaf node acts as the class label, the branch nodes are the test results, and non-leaf hubs are utilized to signify the test. The domain knowledge is not required for the decision trees method. In this method, interpretation and handling of numerical and classification of data are easy. The performance of the method completely depends upon the data sets and is limited to one attribute outpu

5.3 *Random Forest (RF)*

Leo Breiman is the person who developed the random forest method of the machine learning approach. Overfitting issues can be avoided by the use of the random forest. This method is used for various applications. The applications involve classification, predictions, and selections. The random forest method has various characteristics. This can be used for multi-class. The random forest has a very good quality prediction of accuracy.

5.4 *Multilayer Perceptron (MLP)*

Multilayer perceptron (MLP) is an advanced feeding technology for artificial neural networks. MLP consists of a minimum of three (input, hidden, and output) layers nodes. MLP is based on a supervised learning approach. In this method, the neural network layer considers the error in the output and tries to restore the output to the hidden layer to revise the internal weights. The modification in the network can be also done in the time training time. All nodes on this network are sigmoid. In comparison with other methods, the MLP method has a high degree of predictive accuracy.

5.5 *Naïve Bayes (NB)*

Naïve Bayes classification is coming under the category of the supervised machine learning algorithm; this method is a kind of probabilistic nature, and it is based on Bayes theorem and having strong (naïve) independence assumptions. Naïve Bayes is an accurate, effective, and high accuracy even if the data set is large. The impact of a particular feature in a class is not dependent on the other feature in the class. Naïve

Bayes used a small set of training data set to train the model and predict the disease. Computational cost is low for Naïve Bayes, and it works for multi-class prediction problems.

5.6 *Linear-Support Vector Machine (L-SVM)*

L-SVM is a supervised learning method and based on the statistical learning theory (S-LT). This method prefers to solve binary classification problems, and it is also used for multi-class relevant vector problems. The L-SVM method predicts heart disease with high-dimensional space by creating a large hyperplane. The hyperplane classifies the heart data into two parts (existence and non-existence) of the disease. L-SVM works on the concept of maximization of the hyperplane where the margins between two classes become high. SVM offers the best precision but requires more computation time. Electronic health records and healthcare at home, etc., are different applications of IoT in the healthcare field. The use of IoT in health care reduces the cost of the prognosis in healthcare applications and improves the quality of the application.

6 Result and Discussion

In this section, the results of the developed prognosis model using six machine learning methods are presented, and the data set is divided into training and testing purpose. Coronary heart disease contains an aggregate of 303 samples and 14 attributes. The important features of data set for heart dieases are selected through feature selections. We use MATLAB for the access to performance analysis of used methods. The heart disease prognosis of the accuracy of comparison is shown in Table 2. Table 3 is used to show the area under the curve (AUC) of used machine learning methods.

Table 2 Comparison of accuracy with various machine learning methods

Machine learning methods	Accuracy (in %)
K-nearest neighbor	75.73
Decision tress	72.45
Random forest	75.73
Multilayer perceptron	67.54
Naïve Bayes	76.26
Linear-support vector machine	77.37

Table 3 AUC for various machine learning methods

Machine learning methods	AUC (in %)
K-nearest neighbor	82.4
Decision tress	81.3
Random forest	92.3
Multilayer perceptron	88.2
Naïve Bayes	89.6
Linear-support vector machine	82.4

The developed prognosis model achieves the highest accuracy of 77.37% with L-SVM in comparison to other selected machine learning methods, whereas the K-NN, DT, RF, MLP, and NB achieved 75.73, 72.45, 75.73, 67.54, and 76.26% and also achieved prognosis rate of 92.3% with random forest machine learning method whereas the K-NN, DT, MLP, NB, and L-SVM achieved 82.4, 81.3, 88.2, 89.6, and 82.4%. The graphical representation of the results is shown in Figs. 2 and 3.

The developed model has the accuracy 92.3%, and it has been compared with some exiting models that are selected the on heart disease parameters such as sugar, chest pain, etc., of the patients). Vembandasamy et al. [16] presented the 86.4% prediction accuracy of heart disease, Purushottam et al. [17] demonstrated prediction model with 86.3% accuracy, and Mohan et al. [15] presented the model with 88.7% accuracy for cardiovascular disease.

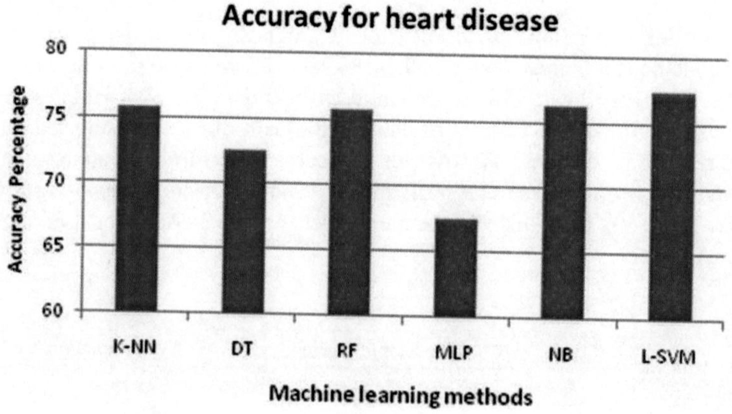

Fig. 2 Heart disease accuracy

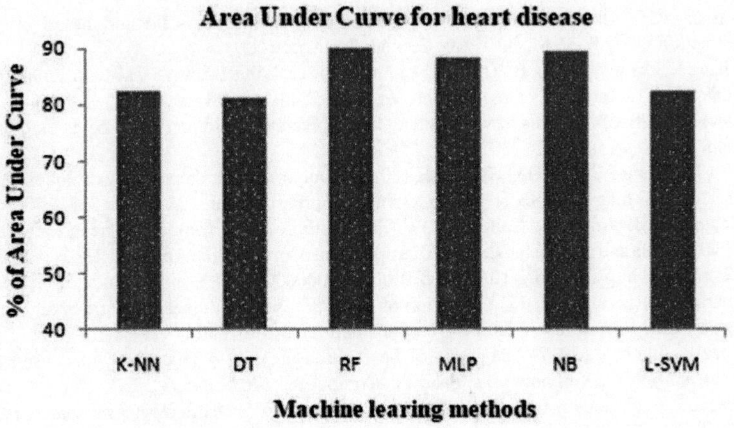

Fig. 3 AUC for machine learning

7 Conclusion

In this research, a framework is developed for health care based on IoT with machine learning. In recent times, heart disease is the main cause of death from the disease. The developed system detects and prognoses heart disease. It makes interaction easier for both patients and doctors. Six machine learning methods (k-nearest neighbor, decision trees, random forest, multilayer perceptron, Naïve Bayes, and linear-support vector machine) are applied to the developed model with heart disease data set. The maximum prognosis rate and maximum accuracy of 92.3 and 77.37% are achieved for heart disease with random forest and L-SVM machine learning methods. In the future, the proposed work can be extended to detect and diagnose other life-threatening diseases. This works can also be lengthened to different applications such as forensics observations and generation of weather report.

References

1. WHO—World Health Organization (2020). https://www.who.int/health-topics/cardiovascular-diseases/#tab=tab_1
2. AHA: American Heart Association website (2017). https://www.heart.org
3. Hameed RT, Mohamad OA, Hamid OT, Tapus N (2015) Design of e-healthcare management system based on cloud and service oriented architecture. In: 2015 E-health and bioengineering conference (EHB), pp 1–4. IEEE
4. Verma L, Srivastava S, Negi PC (2016) A hybrid data mining model to predict coronary artery disease cases using non-invasive clinical data. J Med Syst 40(7):178
5. Forkan ARM, Khalil I, Atiquzzaman M (2017) ViSiBiD: a learning model for early discovery and real-time prediction of severe clinical events using vital signs as big data. Comput Netw 113:244–257

6. Osman AH, Aljahdali HM (2017) Diabetes disease diagnosis method based on feature extraction using K-SVM. Int J Adv Comput Sci Appl 8(1)
7. Zhang L, Zhou W, Wang B, Zhang Z, Li F (2018) Applying 1-norm SVM with squared loss to gene selection for cancer classification. Appl Intell 48(7):1878–1890
8. Devi MR (2016) Analysis of various data mining techniques to predict diabetes mellitus. Int J Appl Eng Res 11(1):727–730
9. Hsu JL, Hung PC, Lin HY, Hsieh CH (2015) Applying under-sampling techniques and cost-sensitive learning methods on risk assessment of breast cancer. J Med Syst 39(4):40
10. Benjamin EJ, Virani SS, Callaway CW, Chamberlain AM, Chang AR, Cheng S et al (2018) Heart disease and stroke statistics—2018 update: a report from the American Heart Association. Circulation. https://doi.org/10.1161/CIR.0000000000000558
11. Sharma H, Rizvi MA (2017) Prediction of heart disease using machine learning algorithms: a survey. Int J Recent Innov Trends Comput Commun 5(8):99–104
12. Uyar K, İlhan A (2017) Diagnosis of heart disease using genetic algorithm based trained recurrent fuzzy neural networks. Procedia Comput Sci 120:588–593
13. Doupe P, Faghmous J, Basu S (2019) Machine learning for health services researchers. Value Health 22(7):808–815
14. Nilashi M, Ahmadi H, Manaf AA et al (2020) Coronary heart disease diagnosis through self-organizing map and fuzzy support vector machine with incremental updates. Int J Fuzzy Syst 22:1376–1388. https://doi.org/10.1007/s40815-020-00828-7
15. Mohan S, Thirumalai C, Srivastava G (2019) Effective heart disease prediction using hybrid machine learning techniques. IEEE Access 7:81542–81554
16. Vembandasamy K, Sasipriya R, Deepa E (2015) Heart diseases detection using Naive Bayes algorithm. Int J Innov Sci Eng Technol 2(9):441–444
17. Saxena K, Sharma R (2016) Efficient heart disease prediction system. Procedia Comput Sci 85:962–969

Diabetes Prediction Using Machine Learning

Harsh Jigneshkumar Patel, Parita Oza, and Smita Agrawal

Abstract Diabetes is a chronic disease which is characterized by the rise of sugar level in blood. There are many complications of the disease when it remains undetected and untreated. This disease, most of the time, gets identified by various symptoms. With the adverse effects of this disease on the patient's entire life, it is crucial to take the necessary actions to mitigate the results. Hence, this disease needs to be identified as soon as possible. The growth of machine learning technology helps to identify such problems. The motive behind this research paper is to build a machine learning model that can identify the probability of a person testing positive for diabetes based on the features. Thus, various machine learning algorithms are used to make a comparative study through which best ML technique has been identified. Model uses random forest, SVM, logistic classification, naive Bayes, KNN and decision tree which are implemented on Pima Indians Diabetes Dataset. Evaluation is done on three different accuracy measures which are accuracy, precision and recall. Along with classification algorithms, the use of gradient boosting and bootstrapping has been used for the improvisation of the results of the evaluation metrics and the classification process. Bootstrapping used avoids overfitting of the classification algorithm used, whereas the boosting uses the weak learners to learn from their errors by learning from the stronger ones. The results of the evaluation metrics compared showed a huge improvement in the entire process of the prediction system based on important features.

Keywords Diabetes · Logistic classification · SVM · Naïve Bayes · KNN · Random forest · Decision tree · PCA · Boosting · Bootstrapping

H. J. Patel (✉) · P. Oza · S. Agrawal
CE Department, Institute of Technology, Nirma University, Ahmedabad, India
e-mail: 17bit028@nirmauni.ac.in

P. Oza
e-mail: parita.prajapati@nirmauni.ac.in

S. Agrawal
e-mail: smita.agrawal@nirmauni.ac.in

P. K. Singh et al. (eds.), *Proceedings of Second International Conference on Computing, Communications, and Cyber-Security*, Lecture Notes in Networks and Systems 203,
https://doi.org/10.1007/978-981-16-0733-2_50

1 Introduction

Diabetes is a disease which causes high blood sugar. Hormone insulin is used to take sugar from blood into your cells which is to be used for energy. When suffering from diabetes, the body is unable to produce enough insulin. This disease comes with symptoms such as frequent urination, increased hunger, tiredness, weight loss and increased thirst.

The two types of diabetes are type 1 and type 2. In type 1 diabetes, immune system attacks and destroys cells of pancreas, where insulin is made. The cause of this type of diabetes is still unknown. This type is rare and most commonly found in children. Type 2 diabetes is found to be most common amongst the patients suffering from diabetes. In this type, pancreas produces some insulin but that amount is not sufficient for the body needs. This type of diabetes can be treatable but can also cause major health complications. There is also one another type of diabetes known as gestational diabetes. This type of diabetes only triggers by pregnancy.

People, nowadays, living in a stressful environment, surrounded by pollution and have an unhealthy lifestyle, the risk of having such diseases increases. Obesity, lack of exercise, junk food, etc., all such are now becoming serious problems amongst people. As the risk increases, the methods to prevent such diseases also increased.

Modern technologies are coming into the medical sector with great speed. Technologies which can detect problems and can cure them as soon as possible are needed. Another thing which is needed is to predict the likelihood of having such diseases beforehand, which can help someone to take care of himself and makes necessary changes into his/her lifestyle to reduce the risk of getting such diseases. Machine learning, here, plays a huge role in prediction of some diseases with the help of existing data. This field in computer science provides system the ability to automatically learn and improve from upcoming experiences without being explicitly programed to do so. Here, this technology is used to predict the happening of the diabetes based on the training of the model on the dataset and then gives the outcome for different set of data. Many researchers are conducting experiments on predicting the likelihood of having diabetes in future with the help of various classification algorithms like SVM [1], random forest [2] and naive Bayes [3] as they prove that machine learning algorithms works better in diagnostic many diseases. This technology gains its popularity because of its capability of handling large amount of data. Now, we will go through work of various researchers who had experimented with various algorithms for medical purpose.

Bootstrapping is a resampling method used by independently sampling with replacement from an existing sample data with the size N.

2 Literature Survey

This objective of this section is to review the existing literature works, provides the details of their model and compares the results.

Sisodia et al. [4] proposed a model which can predict the probability of diabetes with the most accurate results. Three machine learning algorithms were used in this model, namely decision tree, SVM and naive Bayes. Experiment is performed on the Pima Indians Diabetes Dataset. Algorithms are compared on the various measures like accuracy, precision, F-measure and recall. Accuracy of naive Bayes was highest with 76.30%. Therefore, naïve Bayes outperformed both SVM and decision tree here.

Indoria et al. [5] discussed the survey of various research papers regarding the work on diabetes prediction and the model they used. Berina et al. [6] gave the classification model used for prediction of diabetes and other cardiovascular diseases using the algorithms—artificial neural networks (ANNs) and Bayesian networks (BNs). Dinu et el. [7] again performed the work on diabetes prediction but using only two algorithms, which are decision tree J48 and naïve Bayes. J48 gives the accuracy of 76.95% by use of cross-validation. Naïve Bayes gives the accuracy of 79.56% using percentage split. Zahed Soltani et al in his research work focused on the artificial neural network approach to diagnose diabetes type 2 disease. They make use of Pima Indians Diabetes Dataset, and their model uses PNN which is implemented in MATLAB. The training and testing accuracy of this model is 89.56% and 81.49%, respectively.

Mir et al. [8] discussed the classification model which uses Waikato Environment for Knowledge Analysis (WEKA) tool for processing different algorithms on the dataset of Pima Indians. Algorithms which are proposed in this model are: support vector machine, naïve Bayes, random forest and simple CART algorithm. These four classifiers are compared on the basis of their training and testing time and their accuracies. The accuracy of SVM, in this model, is highest with 79.13% followed by naïve Bayes (77%) and then random forest and simple CART with accuracy of 76.5%.

Yuvaraj et al. [9] researched for the diabetes prediction by the following methodology. Using Hadoop clustering with R is for the analysis and prediction of the data with the help of machine learning algorithms. The components of Hadoop used are MapReduce and Hadoop distributed file systems. The machine learning techniques used for the classification process are decision tree, naïve Bayes and random forest. The R packages used for the modelling are rhdfs, rhbase, plyrmr, rmr2 and ravro. Feature selection methods such as information gain are be used for the creation of feature vectors which is based on the entropy of the data points in the dataset.

Aishwarya et al. [10] worked on a diabetes prediction model using the following methodology. The different algorithms used for the comparison are K-nearest neighbour, decision tree, naïve Bayes, support vector machine, logistic regression and random forest. Pre-processing of the dataset is done in order to reduce the errors of the missing values and duplicate data in the dataset. Data is processed through various algorithms with the corresponding features and values as input with parameters if needed. Comparison of the evaluation metrics of each algorithm is done to check the best prediction algorithm amongst the others.

Swapna et al. [11] worked on the diabetes prediction modelling in the following way. The extraction of the features takes place using the CNN and LSTM neural networks, and the support vector machine is used for the classification of the data according to the features. A particular CNN 5-LSTM network is used which is an example of ensemble learning. The data input points are fed into the five-layer CNN model which converts them into a feature map which is then fed into the LSTM network model with the dropout regularization function. The kernel used by SVM for classification is the RBF kernel.

Mahabub et al. [12] worked on a diabetes prediction system which is based on the machine learning algorithms. The prediction modelling is done with the help of 11 classification algorithms which are naïve Bayes, KNN, SVM, random forest, artificial neural network, logistic regression, gradient boosting, AdaBoosting, boosting tree, SVC and multilayer perceptron. The entire work is performed on the WEKA software. The pre-processing of the data takes place which includes normalizing and indexing the data points according to the missteps observed in the dataset. It also includes finding out missing data and deleting unnecessary columns.

Hyperparameter tuning is then done to reduce the cost and loss of the computation process. The cross-validation set is converted into tenfold dataset. The classification results are then compared on the dataset to find the best accuracy result. This is followed by hard and soft voting classification. Soft voting depends also on the anticipated probability of the classifier.

Kumari et al. [13] developed a model which only uses SVM as a classifier. Here, to evaluate the robustness of this model, a tenfold cross-validation was performed in training dataset. The dataset here used is of Pima Indians. The training dataset is first divided into 10 equal-sized subsets. Each subset was tested on a model trained with remaining nine subsets. This cross-validation was repeated 10 times. Then the performance of this model was assessed. This model gives the accuracy of 78% (Table 1).

The related work gives an overview of the various machine learning tools and algorithms being used for building of diabetes prediction systems. The algorithm with the highest accuracy is the K-nearest neighbours algorithm. The algorithms used for the prediction modelling system have shown approximately same results but the KNN algorithm used has shown the highest result for the PIMA dataset. The other observation is related to the disadvantage that some of the machine learning models do not work efficiently on large datasets as compared to the KNN algorithm. At the same time, the problem of overfitting and data inconsistency is not dealt with specifically which shows a reduction in the results. With the analysis of the research results from the table, it helped in deciding which machine learning model to select for the initial process of the building of the prediction model.

Table 1 Study of prediction of diabetes

[4]	Compare different diabetes prediction algorithms	K-Nearest neighbours, decision tree, Naïve Bayes, SVM, logistic regression	Machine learning algorithms based	Logistic regression gives the direction of association as well	Logistic regression works well only on linearly separable data
[5]	Improvisation of diabetes prediction algorithms	CNN 5-LSTM, Support vector machine	Deep learning methodology	Feature extraction is done by both CNN and LSTM	Does not work well on large datasets
[6]	Diabetes prediction using Hadoop clusters and ML algorithms	Hadoop, MapReduce, decision tree, Naïve Bayes and random forest	Hadoop cluster based along with machine learning	Works on large datasets and compatible with cloud	Diversified method brings cost of computation
[7]	Diabetes prediction using optimal features	Decision tree, SVM and Naïve Bayes	Machine learning	Decision tree and random forest has higher accuracy	Random forest works like a black box with no control over working
[14]	Improvisation of diabetes prediction algorithms	KNN, Naïve Bayes, SVM, ANN, Random forest, Gradient boosting, AdaBoosting, SVC	Machine learning	Ensemble learning increases accuracy with use of a greater number of features	Computation is expensive due to use of multiple algorithms
[8]	Diabetes prediction improvement using ML algorithms	Gradient boosting, Naïve Bayes and Logistic regression	Machine learning	Improves the accuracy with less cost of computation	Does not work well on large datasets
[15]	Prognostication of likelihood of diabetes	Decision tree, SVM, Naïve Bayes	Machine learning	Naïve Bayes out[performs decision tree and SVM	The overall accuracy is low in comparison to deep learning model
[9]	Prediction of diabetes and cardiovascular diseases	ANN, decision tree, Bayes networks, PNN, Cross-validation Naïve Bayes	Machine learning	PNN outperforms all other methodologies	Does not work well on large datasets

(continued)

Table 1 (continued)

[10]	Processing of different prediction algorithms	WEKA software, SVM, Naïve Bayes, random forest and cart	WEKA and machine learning	SVM works the best of all	Does not include all the features efficiently
[11]	SVM diabetes prediction for	Support vector machine	Machine learning	Works well on the tenfold cross-validation dataset	This does not work well on large datasets or large number of features
[12]	Decision tree for diabetes prediction	J48 decision tree using WEKA	Machine learning and WEKA software	Prediction accuracy is high using tenfold dataset	Is restricted to one environment for the implementation
[2]	Naïve Bayes for diabetes prediction	Naïve Bayes, gradient boosting	Machine learning	The accuracy increased steeply	Does not work well with large datasets
[13]	SVM classifier used for diabetes prediction	Support vector machine	Machine learning	The training time decreased	The accuracy is very low compared to KNNs
[1]	J48 decision tree with WEKA	Decision tree algorithm, WEKA	Machine learning and WEKA software	The number of evaluation metrics and methodologies have increased	The algorithm does not show higher accuracy compared to the rest of them
[16]	Compare different diabetes prediction algorithms	K-Nearest neighbours, decision tree, Naïve Bayes, SVM, logistic regression	Machine learning algorithms based	Logistic regression gives the direction of association as well	Logistic regression works well only on linearly separable data

3 Methodology

Procedure of our model is explained below in the form of a flowchart. The flow of the entire work is depicted which is used for building of the model.

3.1 Dataset Used

We evaluated this model on dataset, namely Pima Indians Diabetes Dataset. This dataset comprises of details of 768 female patients. There are 8 sets of numeric attributes on which 9th attribute is targeted. 9th column is a binary attribute, where

0 means patient is not diabetic and 1 means patient is diabetic [17]. The description of the dataset is given in the following Table 2.

We have chosen 260 samples for training data and 508 samples for the test data for the machine learning model as the training accuracy reaches the peak comparatively faster as compared to the test dataset [1, 6, 7, 10, 18]. The true positives here represent the number of negative cases predicted negative by the algorithm, false positives were representing the positive diabetes when the prediction is negative, true negative is showing the number of cases with diabetes predicted as positive and the false negative to be the number of cases with the absence of diabetes predicted to be present. We have restricted the data to pregnant women over 20 years of age [16–20] (Fig. 1).

Table 2 List of attributes of the dataset used

S. No.	Attributes used	Attribute type	Attribute description
1	Pregnant	Numeric	No. of times pregnant
2	Plasma	Numeric	Plasma glucose concentration
3	Pressure	Numeric	Blood pressure (mm Hg)
4	Skin	Numeric	Triceps skin fold thickness
5	Insulin	Numeric	2-Hour serum insulin
6	Mass	Numeric	Body mass index
7	Pedi	Numeric	Diabetes pedigree function
8	Age	Numeric	Age of patient in years
9	Class	Binary	No for tested negative and Yes for tested positive

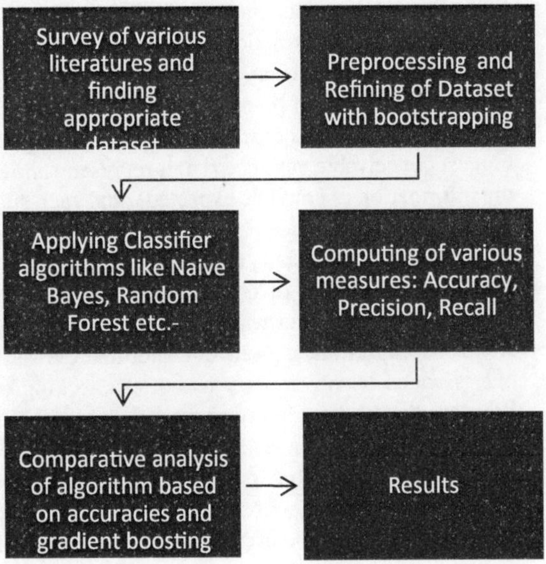

Fig. 1 Flow diagram showing the process for construction of the model

3.2 Algorithm Used

- **Random Forest**: Random forest model is an example of ensemble learning algorithm. Ensemble learning algorithms are those which consolidate multiple machine learning algorithms of similar or distinctive kind for characterizing objects. Random forest classifier makes a lot of decision trees from an arbitrarily chosen subset of training set. It is at that point totals the votes from various choice trees to choose the last class of the test object [6, 10].
- **Logistic Classifier**: Logistic classifier is a 'statistical learning' system arranged in 'supervised' machine learning (ML) strategies devoted to 'classification' jobs. It has increased a huge demand for most of the recent two decades particularly in money-related division because of its conspicuous capacity of identifying defaulters.
- **Decision Tree Classifier**: Decision tree algorithm falls under the class of supervised learning. The decision tree utilizes the tree portrayal to take care of the issue in which each leaf node relates to a class name and qualities correspond to the interior node of the tree. A decision tree is where every node speaks about a component (quality), each connection (branch) speaks to a decision made (rule) and each leaf represents to a result (categorical or continuous value) [6, 7, 9, 15, 18].
- **Naïve Bayes**: The naive Bayes classifier is a machine learning algorithm which uses Bayes hypothesis as its core idea of computation. This algorithm also works on the independence of the amongst the properties of the information nodes. Famous employments of these classifiers incorporate spam channels, text investigation and medical analysis. They are broadly utilized for AI since they are easy to execute. It is mainly useful when the input dimensionalities are high [3, 21].
- **PCA**: Principal component analysis (PCA) is a statistical procedure that employs an orthogonal transformation to change over a lot of observations of perhaps correlated factors (entities every one of which takes on different numerical values) into a lot of estimations of linearly uncorrelated factors called principal components. This transformation is characterized so that the principal component part has the biggest conceivable variance (i.e. it represents, however, much of the variance in the information as could be expected), and each succeeding segment thus has the most elevated difference conceivable under the constraint that it is orthogonal to the previous components [12, 14, 22].
- **Boosting:** Two types of gradient boosting are used for the diabetes prediction using machine learning which are AdaBoost and gradient boosting algorithms. AdaBoost algorithm is used to customize the sample distribution. The weaker one is added to the stronger one after checking of the performances [23, 24]. Whereas the gradient boosting works such as the weaker learner is trained with the errors of the stronger learner. The entire process is based on the gradient boosting escalating process. The results of the paper also state that the gradient boosting used with the KNN algorithm shows the greatest results of 92.0384%. XGBoost is the method of gradient boosting used in this paper. The tuning of the parameters is comparatively easy with the XGBoost methodology.

- The bootstrap sampling method is implemented in the following sequence of steps:
  - A sample of size n is selected from the population with replacement with size *n*, replacing the sample *P* times with each resampled sample known as the bootstrap sample.
  - Estimate the parameters for each bootstrap sample with total *P* number of estimates being calculated with further construction of the sample distribution with the following *P* estimates for statistical inference such as estimation of the standard error of the statistic for the particular parameters followed by obtaining of confidence interval for the particular parameter [17, 25, 26].

  (a) Standard error of the mean of the sample:

$$\sigma_{\bar{x}} = \sigma/\sqrt{n}$$

Here, σ stands for the standard deviation of the entire population, whereas n is the sample size of the sample being selected and replaced for *P* times during the process [3, 21]. Most of the times, the standard deviation of the population is unknown and not accurate so the only substitute for the same is to calculate the estimated standard error. This is calculated using the sample standard deviation *S* [19, 20] (Fig. 2).

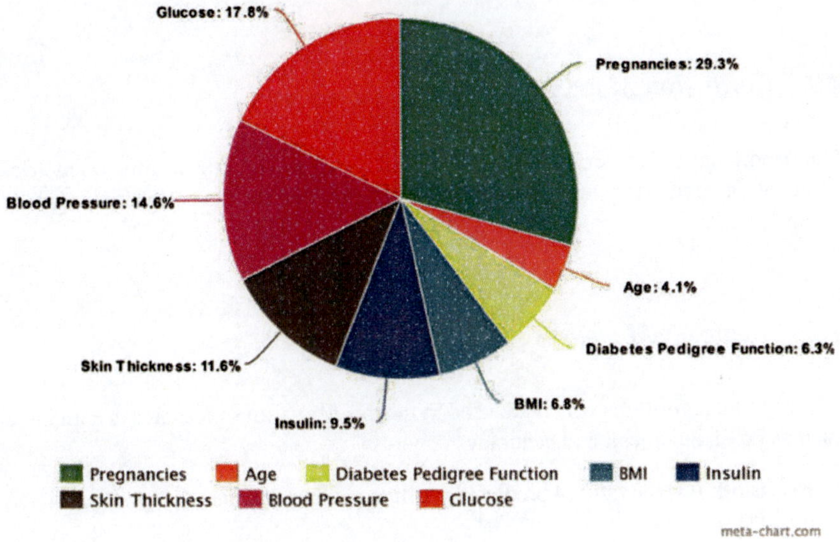

Fig. 2 Chart showing variance of attributes using PCA

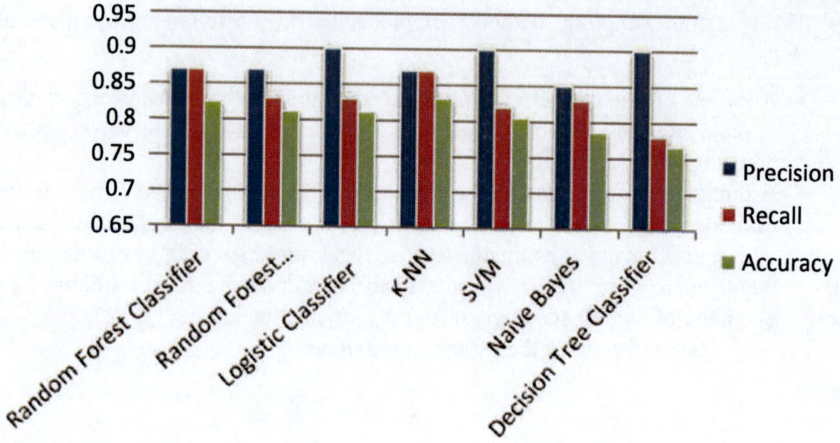

Fig. 3 Structure of confusion matrix

4 Experimental Analysis

This section describes the experiments which are done in order to obtain results after the training of dataset using all the algorithm described above.

4.1 Confusion Matrix

Confusion matrix is an evaluation metric used for calculating the evaluation metrics of the algorithms which are used [17, 19]. The structure of the confusion matrix is given (Fig. 3).

4.2 Accuracy Measures

Using all the algorithms which are used in this model various measures is computed such as precision, recall and accuracy.

- **Precision**: It is the ratio of predicted positive instances to the total of all predicted positive instances.

$$\text{Precision} = \frac{\text{True Positive}}{\text{True Positive} + \text{False Positive}}$$

- **Recall**: It is the ratio of number of predicted positive instances to the actual total number of positive instances.

$$\text{Recall} = \frac{\text{True Positive}}{\text{True Positive} + \text{False Negative}}$$

- **Accuracy**: It is the ratio of sum of true predicted instances which are true positive and true negative, to the total number of instances.

$$\text{Accuracy} = \frac{\text{True Positive} + \text{False Negative}}{\text{Total Number of Instances}}$$

The following table represents the accuracy measure value of all the classification algorithms which are used in this model.

5 Result Analysis

This research is dedicated to make a comparative study on different machine learning algorithms on Pima Indians Diabetes Dataset. The model applies several machine learning algorithms, and analysis has been made by tuning different parameters and hyperparameters. PCA is used to finding out the most correlated attributes which affects the classification. With percentage split of 80:20, we have found that the accuracy of algorithm KNN is highest with 83.11% followed by random forest (82.46%), logistic classifier (81.16%), SVM (80.51%), naïve Bayes (78.57%) and decision tree (76.62%). This outcome might help in enhancing accuracy measures for different models and to provide a simple and early identification of diabetes which would help in reducing future complications (Fig. 4).

Fig. 4 Bar graph showing the accuracy measures of different algorithms used in the model

	Class 1 Predicted	Class 2 Predicted
Class 1 Actual	True Positive	False Negative
Class 2 Actual	False Positive	True Negative

6 Results and Conclusion

Artificial intelligence plays an important role not only in the field of computer science but also in various other fields like health care, agriculture and entertainment. With the blessing of artificial intelligence, we are able to identify and solve the problems faced in different fields. One such problem is taken out from the field of healthcare. Diabetes is one of the major diseases which many people are diagnosed with. Hence, early identification of this disease is a necessity. This identification has been possible with different algorithms which machine learning has to offer.

After applying various algorithms, we found that the KNN algorithm was able to outperform the other machine learning algorithms because of the hyperparameter tuning performed on number of neighbours to be considered, the algorithm is used to find the nearest neighbour and the weight function is used in the algorithm. After the application of gradient boosting along with KNN algorithm, the prediction accuracy is depicted a steep increase to 92.0834%. The major reason for the increase in the accuracy of the prediction algorithm was the use of gradient boosting and bootstrapping. The future work in the following research could deal with finer level of feature extraction from the dataset along with scaling the visualization and analysis of the dataset on a greater level so it becomes more precise regarding which features are really crucial to the prediction and which are the ones which actually create noise and deteriorate the results. Dealing with the missing data could be improved by using different algorithms in the form of ensemble learning as well as transfer learning to increase the efficiency of the entire model.

References

1. Sivansena R, Dhivya KDR (2017) A review on diabetes mellitus diagnosing using classification on pima Indian diabetes dataset. Int J Adv Res Comput Sci Manag Stud 5(1). ISSN: 2321-7782
2. Birjais R, Mourya AK, Chauhan R, Kaur H (2019) Prediction and diagnosis of future diabetes risk: a machine learning approach. SN Appl Sci 1(9):1–8
3. Saru S, Subashree S (2019) Analysis and prediction of diabetes using machine learning. Int J Emerg Technol Innovative Eng (54). ISSN: 2394-6598
4. Sisodia D, Sisodia DS (2018) Prediction of diabetes using classification algorithms. In: International conference on computational intelligence and data science (ICCIDS), Procedia Comput Sci 132(2018):1578–1585
5. Indoria P (2018) A survey: detection and prediction of diabetes using machine learning techniques. Int J Eng Res Technol (IJERT) 7(03). ISSN: 2278-0181
6. Alić B, Gurbeta L, Badnjević A (2017) Machine learning techniques for classification of diabetes and cardiovascular diseases. In: 2017 6th mediterranean conference on embedded computing (MECO) (pp 1–4). IEEE
7. Dinu AJ, Ganesan R, Joseph F, Balaji V (2017) A study on deep machine learning algorithms for diagnosing of diseases. Int J Appl Eng Res 12(17):6338–6346. ISSN 0973-4562
8. Mir A, Dhage S (2018) Diabetes disease prediction using machine learning on big data of healthcare. In: International conference on computing communications control and automation (ICCUBEA)

9. Yuvaraj N, SriPreethaa KR (2019) Diabetes prediction in healthcare systems using machine learning algorithms on Hadoop cluster. Cluster Comput 22(1):1–9
10. Jakka A, Vakula Rani J (2019) Performance evaluation of machine learning models for diabetes prediction. Int J Innov Technol Exloring Eng 8(11):1976–1980
11. Swapna G, Vinayakumar R, Soman KP (2018) Diabetes detection using deep learning algorithms. ICT Express 4(4):243–246
12. Mahabub A (2019) A robust voting approach for diabetes prediction using traditional machine learning techniques. SN Appl Sci 1(12):1–12
13. Kumari A, Chitra R(2013) Classification of diabetes disease using support vector machine. Int J Eng Res Appl (IJERA) 3(2). ISSN: 2248-9622
14. Soltani Z, Jafarian A (2016) A new artificial neural networks approach for diagnosing diabetes disease type II. Int J Adv Comput Sci Appl 7(6)
15. Sneha N, Gangil T (2019) Analysis of diabetes mellitus for early prediction using optimal features selection. J Big Data 6(1):1–19
16. Islam MA, Jahan N (2017) Prediction of onset diabetes using machine learning techniques. Int J Comput Appl 180(5). ISSN: 00975-8887
17. Iyer A, Jeyalatha S, Sumbaly R (2015) Diagnosis of diabetes using classification mining techniques. IJDKP 5(1):01–014
18. Ellis AC (2014) Increased fracture risk in patients with type-2: an overview of the underlying mechanisms and the usefulness of imaging modalities and fracture risk assessment tools. Maturitas 79(3)
19. Newsfeed on World Health Organization on Diabetes: https://www.who.int/news-room/fact-sheets/detail/diabetes
20. Tanwar S, Bhatia Q, Patel P, Kumari A, Singh PK, Hong WC (2020) Machine learning adoption in blockchain-based smart applications: the challenges, and a way forward. IEEE Access, 8:474–488
21. Qiu A (2019) An improved prediction method for diabetes based on a feature based least angle regression algorithm. In: ICMLSC 2019: proceedings of the 3rd international conference on machine learning and soft computing, Jan 2019
22. Zia UA, Khan N (2017) Predicting diabetes in machine datasets using machine learning techniques. Int J Sci Eng Res 8(5). ISSN: 2229-5518
23. Kadhm M (2018) An accurate diabetes prediction system based on K-means clustering and proposed classification work. Int J Appl Eng Res 13
24. Predicting diabetes disease using mixed data and supervised machine learning algorithms. In: SCA '19: Proceedings of the 4th international conference on smart city applications, Oct 2019
25. Zafar F (2019) Predictive analysis in healthcare for diabetes prediction. In: ICBET' 19: proceedings of the 2019 9th international conference on biomedical engineering and technology, Mar 2019
26. Mir A (2018) Diabetes disease prediction using machine learning on big data of healthcare. In: IEEE, 2018 fourth international conference on computing communication control and automation (ICCUBEA)

Myocardial Infarction Detection Using Deep Learning and Ensemble Technique from ECG Signals

Hari Mohan Rai, Kalyan Chatterjee, Alok Dubey, and Praween Srivastava

Abstract Automatic and accurate prognosis of myocardial infarction (MI) from electrocardiogram (ECG) signals is a very challenging task for the diagnosis and treatment of heart diseases. Hence, we have proposed a hybrid convolutional neural network—long short-term memory network (CNN-LSTM) deep learning model for accurate and automatic prediction of myocardial infarction using ECG dataset. The total 14552 ECG beats from "PTB diagnostic database" are employed for validation of the model performance. The ECG beat time interval and its gradient value ID are directly considered as the feature and given as the input to the proposed model. The used data is unbalanced class data, hence synthetic minority oversampling technique (SMOTE) & Tomek link data sampling techniques are used for balancing the data classes. The model performance was verified using six types of evaluation metrics and compared the result with state-of-the-art method. The experimentation was performed using CNN and CNN + LSTM model on both imbalance and balance data sample, and the highest accuracy achieved is 99.8% using ensemble technique on balanced dataset.

Keywords CNN · Deep learning · ECG · LSTM · MI · SMOTE

1 Introduction

Myocardial infarction (MI) is the most common and fatal cardiovascular disease (CVD), which impedes blood flow to the heart muscle due to partial or complete blockage of the coronary arteries. The oxygenated blood to the cardiac muscle is supplied by coronary arteries, if there is any obstruction into it, the heart muscle segment may die due to lack of blood flow into it [1]. The damage or death of cardiac muscle tissue causes the change in the normal cardiac conduction system, resulting

H. M. Rai (✉) · A. Dubey · P. Srivastava
Department of ECE, Krishna Engineering College, Ghaziabad, India

K. Chatterjee
Department of Electrical Engineering, Indian Institute of Technology (ISM), Dhanbad, India

P. K. Singh et al. (eds.), *Proceedings of Second International Conference on Computing, Communications, and Cyber-Security*, Lecture Notes in Networks and Systems 203,
https://doi.org/10.1007/978-981-16-0733-2_51

the life-threatening arrhythmias which may leads to sudden cardiac arrest. There are several symptoms may be seen in case of MI such as chest pain, breathing problems, and unconsciousness but many individual do not experience any symptoms, that why it is also called "silent heart attack". According to one estimate, it has been found that about 22–65% of MIs do not show any symptoms, which means that they are silent. Therefore, patients do not get time to prepare themselves, which makes the disease more dangerous and fatal, the mortality rate is very high as a result, and the mortality rate of MI is very high [2].

Hence, the early detection of the MI is very much important to provide timely treatment and reduce the mortality rate. The electrocardiogram (ECG) signal analysis is the most appropriate technique for detecting the MIs at early stage. But the manual and incorrect detection of cardia abnormality may lead to loss of life of the patients suffering from MIs. The main objective of our work is to assist the medical practitioner by automatic, accurate and quick detection of MI from ECG signals.

The use of machine learning for computational computing has been enormously increased in every field of computation encluding medical and health care [3–5]. From past few years, the deep learning models have exponentially rised especially in the field of medical and health care. Many researchers, academicians, and scientists have proposed different techniques for the detection of cardiac arrhythmias such as MI. For the automatic classification of arrhythmias, some researchers have utilized neural network (NN) [6], support vector machine (SVM) [7], decision tree [8], radial basis function (RBF) [9], K-nearest neighbors (KNN) [10], and hybrid classifiers [11] for detecting the arrhythmias into different classes. Reasat et al. [12] developed an automated method for the detection of inferior MI using inception frame-based shallow CNN and attained 84.54% of average classification accuracy. Liu et al. [13] presented a myocardial infarction (MI) detection algorithm using convolutional neural network (on) on multi-lead ECG signals. Their ML-CNN model uses sub 2D convolution which can use complete character of total leads, and it also uses 1D filter to generate local optimal features. The algorithm was evaluated on PTB diagnostic ECG dataset, and it achieved 95.40% of sensitivity, 97.37% of specificity and 96% of accuracy. Kayikcioglu et al. [14] proposed a technique to categorize the ST segments using distribution of time frequency-dependent features for the detection of myocardial infarction (MI) from multi-lead ECG waveforms. The weighted KNN algorithm provided best performance with average sensitivity of 95.72%, accuracy of 94.23% and specificity of 98.18% using Choi–Williams frequency distribution (CWFD) features. Liu et al. [15] presented a new hybrid RNN (LSTM)-based method for the prediction of MI from 12 lead ECG and obtained overall 93.08% accuracy.

Despite these excessive efforts, still there are few scope of improvements based on the review carried out on the MI detection from ECG signals. Data imbalance issue is the most common in medical signal and image which has not been addressed by the researchers. To resolve the issue of imbalance dataset, we have come up with the solution by applying SMOTE and tomek link jointly, along with hybrid CNN-LSTM model. Also, we have detected the MI from PTB database with the accuracy of 99.8%.

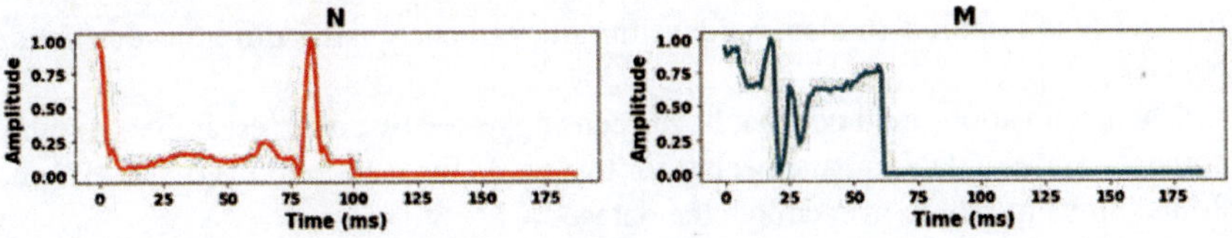

Fig. 1 Visual representation of normal (left) and MI beats (right) from PTBDB

The structure of this paper is arranged as follows: Sect. 2 briefs about the materials and methods employed in this work. Section 3 explains about the proposed methodology, and Sect. 4 describes about the result and discussion. The final Sect. 5 provides concluding remarks and future work of this paper.

2 Materials and Methods

2.1 ECG Dataset

The PTBDB (PTB diagnostic database) comprises 549, 12 lead ECG recording from 290 individuals publically available at PhysioNet Web site to download [16]. Especially, this database contains 148 individuals diagnosed with MI in 368 recording and 52 individuals with healthy control in 80 recordings, apart from this other records are diagnosed with seven different types of arrhythmias [16]. For this work, we used ECG lead-II signals of only two classes, myocardial infarction (MI) and healthy ECG beats (N). The total number of ECG beats utilized for MI detection is 14,552 including 4046 healthy beats (N) and 10,506 MI beats, in other words, MI beats are 72%, and normal betas are ony 28% of the total utilized beats. ECG signals are segmented and down sampled of size 187, and also zero padded to make all the beats of same size, also all the signals are filtered and preprocessed, and this dataset is available publically at kaggle respiratory [17, 18]. The visualization of normal and myocardial infarction (MI) beats sample is shown in Fig. 1.

2.2 Data Balancing

If the difference between the two classes is large in terms of the number of samples, then the data is said to be imbalanced. In case of data imbalance, one class (majority) has more dominance or influence over the other classes [19].

Hence, the imbalance datasets mostly impacted the classifier performance because classifiers are mainly designed for balanced class problems. The another issue is that this imbalance data may not be visualized in the overall accuracy of the classification for that it is required to also validate the performance using different evaluation metrics.

There are various methods that have been suggested by many researchers to overcome the issue of data imbalance, but for this work, we have employed SMOTE and Tomek links jointly to oversample the dataset.

2.3 SMOTE and Tomek Links Sampling

The method applied to oversample the minority class data (MI) is synthetic minority oversampling technique (SMOTE) which creates the synthetic samples close to the minority class instead of simply generating the multiple copies. The SMOTE generates the minority data samples based on the K-nearest neighbors algorithm, it selects nearest neighbor from the minority data samples, and then based on linear interpolation new samples are generated [20].

Tomek links are the technique of under sampling which reduces the number of instances from majority classes based on the data samples which belongs to the borderline. This process keeps on repeating and creates the gap between the borderline of majority and minority classes by deleting the majority samples [21].

2.4 Convolutional Neural Network (CNN) Model

CNN was initially designed for the pattern recognition tasks in the field of computer vision for edge detection, segmentation and object detection task, but because of its versatile applicability, it is being used in almost every application. The commonly used layers of CNN are convolution (Conv) layer, ReLU layer, pooling layer, and batch normalization. The convolution layer is the first layer of the CNN which does the convolution operation, and it is supported by the kernel size, filter number, and padding. ReLU does the nonlinear operation which stands for "rectified linear unit", and it is a type of activation function which introduces nonlinearity into the network. Mainly, two types of pooling operation are available: max pooling and average pooling; it reduces the number of parameters in the training, and it avoids the overfitting problem and increases the efficiency of the network. The batch normalization process normalizes the every layer's input by applying the variance and mean of the present batch, during training process.

2.5 Long Short-Term Memory Network (LSTM)

LSTM network is the widely and most preferred DL model used for sequential and time series data, providing a solution to short-term memory (STM) as to why it is named "long- 'short-term memory'". The basic concepts of LSTM operation are in gates and cell states, which are forget gate, input gate, cell state, output gate, and hidden state [22]. In forget gate (f_t) system, the present input (x_t) is concatenated with the previous hidden state (h_{t-1}) and passes through the sigmoid activation function to produce the output between 0 and 1. Forget gate is mathematically expressed as follows:

$$f_t = \sigma\left(w_f\left[h_{t-1}, x_t\right] + b_f\right) \tag{1}$$

Input gate (i_t) is mainly used for updating the memory or cell state, and in this gate also sigmoid function is used between concatenated value of current input (x_t) and previous hidden state (h_{t-1}) and expressed by the equation.

$$i_t = \sigma\left(w_i\left[h_{t-1}, x_t\right] + b_i\right) \tag{2}$$

Cell state (C_t) is also known as the memory state, and the operation in this state takes place in three step. First step, the previous cell state value (C_{t-1}) gets multiplied (point wise) with the forget gate (f_t) value (from Eq. 1), and in the second step candidate (C_t) vector is point wise multiplied with the input gate (i_t) vector (from Eq. 2). In the final step, both of these values get added together to create the memory cell or current state value (C_t) to be fed to the next stage, which is given by the equation:

$$C_t = f_t * C_{t-1} + i_t * \widehat{C}_t, \text{Where, } \widehat{C}_t = \tan h\left(w_c\left[h_{t-1}, x_t\right] + b_c\right) \tag{3}$$

Output gate (O_t) is also calculated in a similar way as input and forget gate but with different weight and bias value, here also the sigmoid activation function is used for providing the output between 0 and 1. The output gate operation is mathematically given by

$$O_t = \sigma\left(w_o\left[h_{t-1}, x_t\right] + b_o\right) \tag{4}$$

Hidden state (h_t) and cell state memory is only passed to the next LSTM stage, and all the gates are used for regulating and computing the value of these states. It is the point wise vector multiplication of output gate value (from Eq. 4) and cell state value (from Eq. 3) after applying tanh activation, and it is presented in Eq. 5. The overall visualization of signal stage LSTM network, including all gates and states, is shown in Fig. 2.

$$h_t = O_t * \tan hC_t \tag{5}$$

Fig. 2 Basic structure of single-stage LSTM network

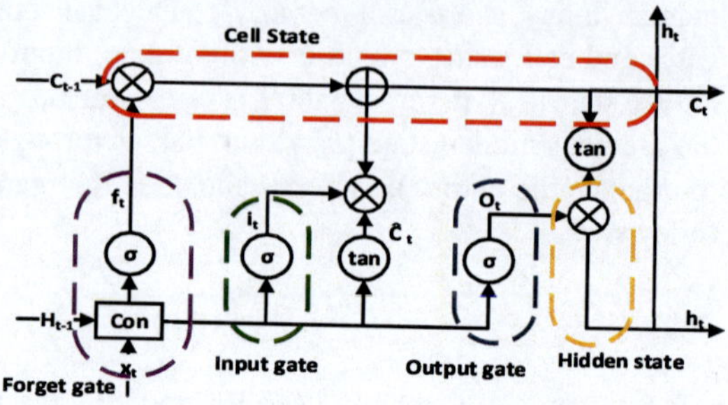

3 Proposed Methodology

The methodology employed for the predication of myocardial infarction (MI) from ECG signals, using CNN, CNN-LSTM and ensemble technique, is presented in this section. The proposed methodology includes two deep learning models with different architecture: One is CNN model and another is CNN-LSTM model along with ensemble technique in the final stage.

In this work, the automatic MI detection and also oversampling methods have been employed in the ECG signals. The ECG signals are first preprocessed by filtering and segmenting it, and then the time interval and gradient of these time series data were calculated. In the next step, the preprocessed imbalance data is directly trained on the training dataset using CNN model and also CNN-LSTM model. In the final step, the imbalanced dataset is balanced using SMOTE-Tomek link method and then trained using the CNN and CNN-LSTM model. The predicted result from both the model is ensemble by averaging them, and final model performance is obtained. Instead of data augmentation technique, data oversampling using SMOTE-Tomek link is applied for generating more number of sample in minority instances. Since the ECG signal is time series and nonlinear data, we have preferred LSTM in combination with CNN among many deep learning models.

3.1 Data Splitting (Distribution)

The data distribution also plays a very important role in the prediction task using classifiers. In this work, total 14,552 ECG beats from PTB database consisting MI beats 10,506 and normal beats 4056 are utilized for MI detection. The ratio of distribution is 80:20 for train: test dataset, respectively, i.e., 11,641 beats for training and 2911 beats for testing the model. Further, the train dataset has been divided into two parts, training and validation in the ratio of 80:20, respectively. Finally, 9312 ECG beats are used for learning the proposed models, and 2329 beats are used for validating the training performance.

3.2 Proposed CNN Model

The 21 layer CNN deep learning model has been proposed to accurately and automatically detect the MI from ECG signal, and the structure of the proposed CNN model layer wise is illustrated in Fig. 3. The input is the 1D ECG signal of sample size (87×1) and reshapes it to the ($187 \times 2 \times 1$) to make suitable for 2D convolutional (Con2D) layer. The complete structure is designed into four parts (stage), where first two stages contain the convolutional layer followed by batch normalization, ReLu non-linearization and max pooling with dropout (0.5) layer. Throughout the model, stride is constant ($s = 1$) and filter size of Con2D layer ($f = 3, 1$) in each segment, except first, ($f = 3, 2$) is used. Second stage with 128 filter size is repeated for extracting the deep features, and third and fourth stages do not contain max pool layer to maintain the feature map and do not allow it to shrink further. Con2D layer varies in the filter number from first to fourth stage as 256, 128, 64, and 64, respectively.

3.3 Proposed CNN-LSTM Model

The 19 layer hybrid CNN-LSTM model is designed for detection of MI from ECG signals and demonstrated the structure of the model, layer by layer in Fig. 3. The complete design is built-up into four segments, first two segments are exactly same as CNN model, but the third and fourth segments are modified to optimize the better result. In this model, also the filter size ($f = 3, 1$) was kept constant for con2D layer except first and fourth layer, where we have used ($f = 3, 2$) and ($f = 6,1$) to match the feature size with the forward layers. The stride and padding throughout the model are one ($s = 1$) and valid, respectively. The segments third and fourth are entirely different from the CNN model, and it is indicated by the red box, similar to CNN model, and this segment does not contain max pooling layer for avoiding further reduction of feature size. The last fourth segment consists of LSTM layer trailed by reshape which is combined with the CNN module. First to third segments represent the CNN model, and fourth segment is added extra for embedding the LSTM layer into it and completed with FC + softmax layer similar to CNN model.

3.4 Performance and Parameter Evaluation

There are several methods to evaluate the performance of the classifier, but we have used confusion matrix and computed various evaluation metrics to verify the model classification result. Six types of performance evaluation metrics, recall (Re%), specificity (Sp%), precision (Pr%), accuracy (Acc), and F1-score (F1%), and classification error rate (CER%) is used for validation of the classifier performance.

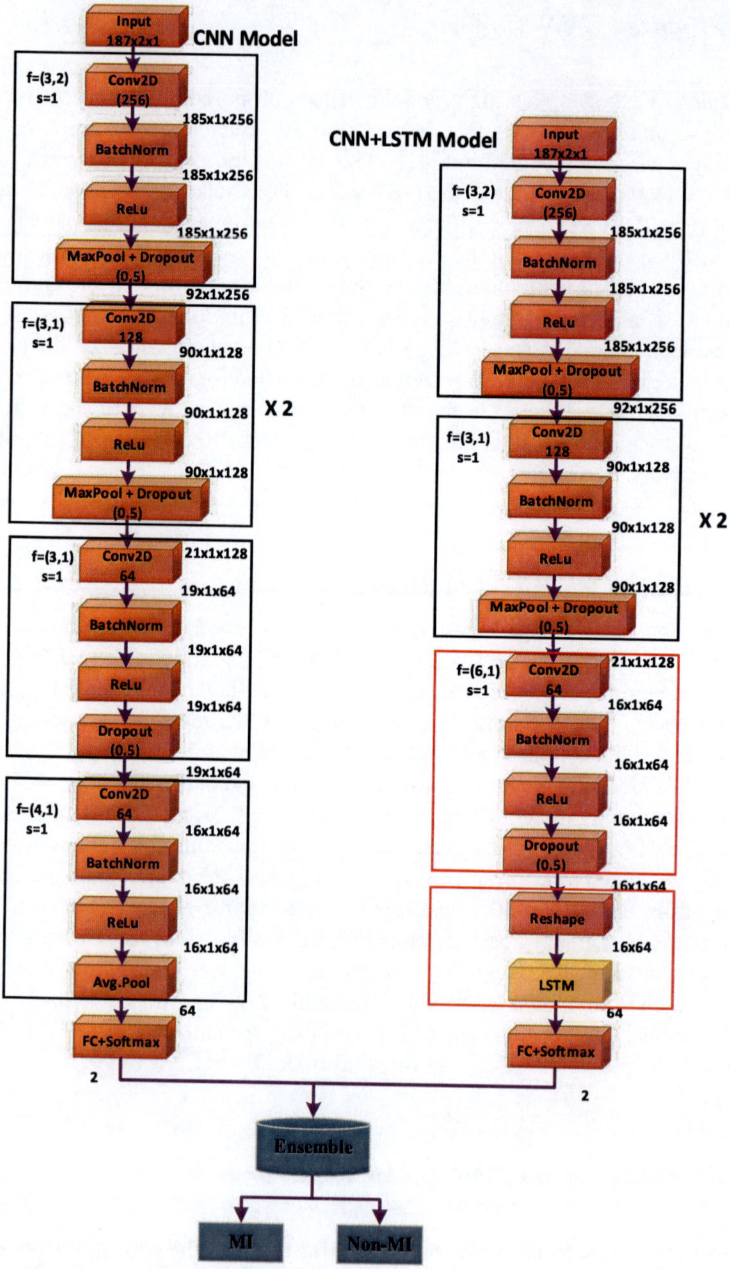

Fig. 3 Block diagram of the proposed methodology

$$\text{Precision(Pr \%)} = \frac{\text{TP}}{\text{TP} + \text{FP}} * 100, \text{Recall(Re\%)} = \frac{\text{TP}}{\text{TP} + \text{FN}} * 100$$

$$\text{Specificity(Sp\%)} = \frac{\text{TN}}{\text{TN} + \text{FP}} * 100, F1 - \text{Score}(F1\%) = \frac{2 * \text{Pr} * \text{Rec}}{\text{Pr} + \text{Rec}} * 100$$

$$\text{Error Rate(ER\%)} = \frac{\text{FP} + \text{FN}}{\text{Total beats}} * 100, \text{Accuracy(Acc\%)} = \frac{\text{TP} + \text{TN}}{\text{Total beats}} * 100$$

$$\text{TP} = \text{True Positive}, \text{FP} = \text{False Positive},$$
$$\text{TN} = \text{True Negative}, \text{FN} = \text{False Negative}$$

4 Result and Discussion

The experiment was performed on total 14,552 ECG beats from PTB database consisting MI beats 10,506 and normal beats 4056. The Python environment has been used for validating the model with TensorFlow and keras library on Kaggle respiratory with GPU support. Laptops with hardware configuration, 8 GB RAM, 1 TB HDD, i5 core Pentium processor, and NVIDIA graphics card have been used for the experimentation. The hyperparameters used during the training process were as, Adam optimizer, batch size 128, momentum 0.9, learning rate 1e-3, epoch 100, and 1e-7 decay. The prediction was made on all three datasets training, validation and test, where training and validation data were already used for learning but the test data was unexposed to the training process.

The first experimentation was performed on imbalance dataset using CNN and CNN + LSTM models. The detection was executed on test dataset, and the prediction results from both the models were ensemble to attain the final result. The accuracy and loss history of training and validation on the original dataset using CNN model are visualized in Fig. 4, and the training history using proposed CNN + LSTM model is shown in Fig. 5. It was observed from both the figures that the model learned well using both the proposed model, but there is slight improvement noticed on the CNN + LSTM model training history.

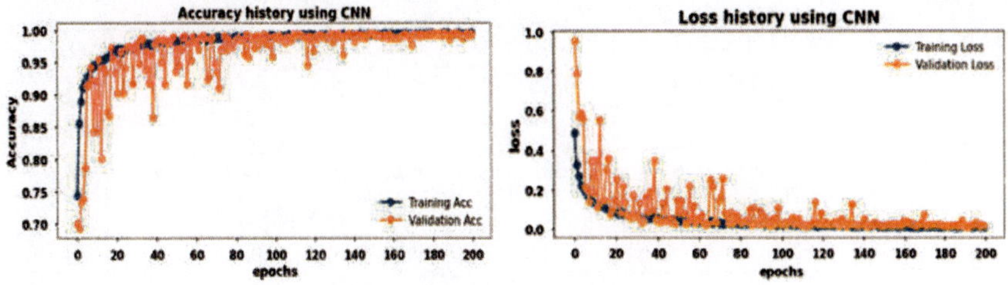

Fig. 4 Accuracy and loss history using CNN model during training

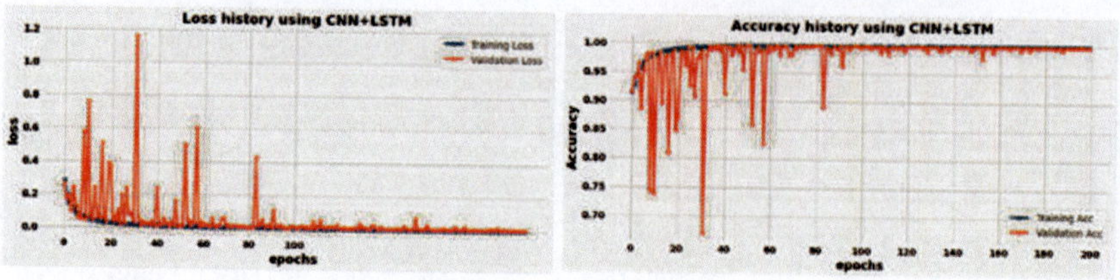

Fig. 5 Accuracy and loss history using CNN + LSTM model during training

The training was performed on the original dataset (imbalance), and after training the model was validated on test dataset (2911 beats). From the prediction result, it was found that the ensemble results from both the model provide better result as compared to CNN and CNN + LSTM model. The overall accuracy achieved using CNN, CNN + LSTM, and ensemble model is 99.3%, 99.5%, and 99.6% respectively. The performance evaluation metrics of each models are presented in Table 1.

The second experiment was performed on the balanced training dataset by SMOTE + Tomek link resampling method. The total 11,641 ECG beats from both the classes were oversampled to 16,798 beats where 8399 beats are from each class. Hence, after balancing the datasets, it was trained using both the models and the prediction was made on the reserved test dataset only. The training performance using CNN and CNN + LSTM model on the balanced dataset is depicted in Figs. 6 and 7, respectively.

The prediction result in Table 2 demonstrates that the class (Cl) accuracy has improved as well as overall accuracy is also improved compared to the imbalanced dataset. The overall accuracy obtained using CNN, CNN + LSTM, and ensemble is 99.5%, 99.7%, and 99.8%, respectively, on balanced dataset. Hence by balancing the dataset using SMOTE +Tomek link sampling methods, not only the overall

Table 1 Evaluation metrics of the proposed model on imbalance test dataset

Predicted ECG beats						% Pr	% e	% Sp	% F1	% ER	% Acc
Model	Cl	TP	FN	FP	TN						
CNN	N	793	16	4	2098	99.5	99.2	98	98.8	0.7	99.3
	M	2098	4	6	793	99.2	99.5	99.8	99.5		
Average						99.4	99.4	98.9	99.1		
CNN + LSTM	N	98	11	3	2099	99.6	99.5	98.6	99.1	0.5	99.5
	M	2099	3	11	798	99.5	99.6	99.9	99.7		
Average						99.6	99.6	99.2	99.4		
Ensemble	N	801	8	4	2098	99.5	99.6	99	99.3	**0.4**	**99.6**
	M	2098	4	8	801	99.6	99.5	99.8	99.7		
Average						**99.6**	**99.6**	**99.4**	**99.5**		

Bold value indicates the proposed method result which is better as compared to others two

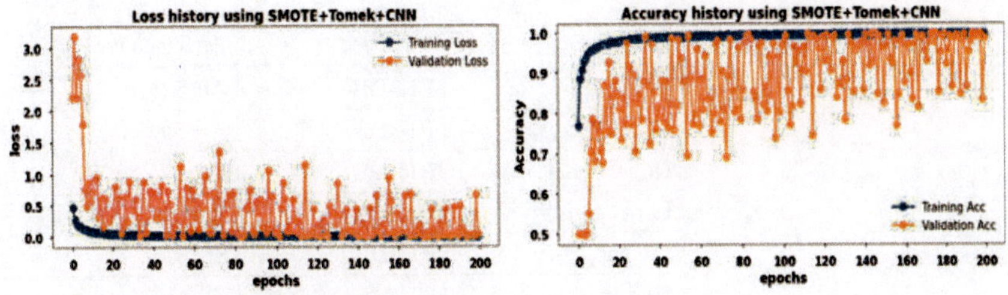

Fig. 6 Training history using CNN model on SMOTE + Tomek balanced data

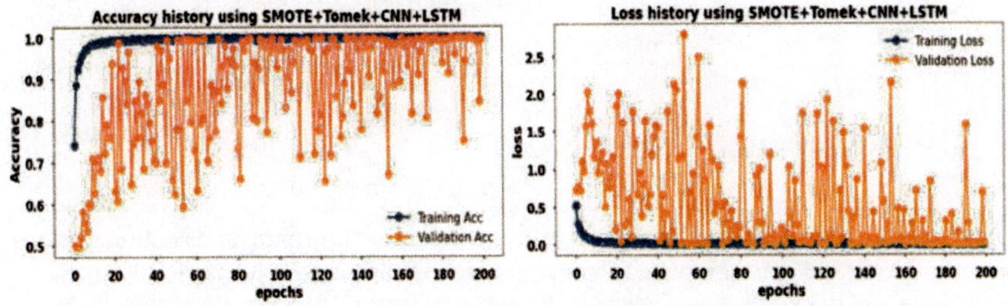

Fig. 7 Training history using CNN model on SMOTE + Tomek balanced data

Table 2 Evaluation metrics of the proposed model on balanced test dataset

Predicted ECG beats						% Pr	% Re	% Sp	% F1	% ER	% Acc
Model	Cl	TP	FN	FP	TN						
CNN	N	800	9	5	2097	99.4	99.6	98.9	99.1	0.5	99.5
	M	2097	5	9	800	99.6	99.4	99.8	99.7		
Average						99.5	99.5	99.3	99.4		
CNN + LSTM	N	803	6	4	2098	99.5	99.7	99.3	99.4	0.3	99.7
	M	2098	4	6	803	99.7	99.5	99.8	99.8		
Average						99.6	99.6	99.5	99.6		
Ensemble	N	806	3	2	2100	2100	99.8	99.9	99.6	**0.2**	**99.8**
	M	2100	2	3	806	806	99.9	99.8	99.9		
Average						**99.8**	**99.8**	**99.8**	**99.8**		

Bold value indicates the proposed method result which is better as compared to others two

accuracy increases but each individual class prediction values also increases. The obtained result was also compared with state-of-the-art techniques to validate the proposed model performance, as presented in Table 3.

Table 3 Comparison with state-of-the-art techniques

Literature	Database	Performance (%)
Kora [23]	PTBDB	Acc: 96.7
Liu et al. [13]	PTBDB	Acc: 96.00
Dohare et al. [24]	PTBDB	Acc: 96.66
Liu et al. [25]	PTBDB	Acc: 94.82
Lui and Chow [2]	PTBDB	Sp: 97.7
Sharma et al. [26]	PTBDB	Acc: 94.4
Savostin et al. [10]	PTBDB	97.3%
Kayikcioglu et al. [14]	EDB, MITDB, LSTDB	Acc: 94.23
Proposed	PTBDB	**Acc = 99.8**

5 Conclusion and Future Scope

The hybrid CNN-LSTM-based deep learning model for automatic and accurate detection of myocardial infarction (MI) was proposed in this paper. The 14,552 ECG beats were used for the verification of model performance from PTB diagnostic dataset. The two models are proposed for the detection of MI and normal beats from ECG signals using two proposed models, CNN and CNN + LSTM. The data resampling method SMOTE + Tomek link is used for balancing the data classes. The CNN-LSTM model performance was tested on 2,911 ECG beats along with training and validation dataset and obtained 99.8% overall accuracy which proves the model supremacy compared to all state-of-the-art techniques.

In the future scope of this work and the generation adverse network (GAN) for data enrichment, the large dataset will be used with the fastest intensive learning model to reduce the computation time.

References

1. Acharya UR, Fujita H, Oh SL, Hagiwara Y, Tan JH, Adam M (2017) Application of deep convolutional neural network for automated detection of myocardial infarction using ECG signals. Inf Sci (Ny) 415–416:190–198. https://doi.org/10.1016/j.ins.2017.06.027
2. Lui HW, Chow KL (2018) Multiclass classification of myocardial infarction with convolutional and recurrent neural networks for portable ECG devices. Inform Med Unlocked 13:26–33. https://doi.org/10.1016/j.imu.2018.08.002
3. Singh PK, Kar AK, Singh Y, Kolekar MH, Tanwar S (2019) Recent innovations in computing. Springer Nature, Switzerland AG
4. Singh PK, Panigrahi BK, Suryadevara NK, Sharma SK, SAK (eds) (2020) Proceedings of ICETIT 2019. Springer International Publishing, Cham. https://doi.org/10.1007/978-3-030-30577-2

5. Singh PK, Pawłowski W, Tanwar S, Kumar N, Rodrigues JJPC, Obaidat, MS (eds) (2020) Proceedings of first international conference on computing, communications, and cybersecurity (IC4S 2019). Springer Singapore, Singapore. https://doi.org/10.1007/978-981-15-3369-3
6. Banerjee S, Mitra M (2013) ECG beat classification based on discrete wavelet transformation and nearest neighbour classifier. J Med Eng Technol 37:264–272. https://doi.org/10.3109/03091902.2013.794251
7. Khalaf AF, Owis MI, Yassine IA (2015) A novel technique for cardiac arrhythmia classification using spectral correlation and support vector machines. Expert Syst Appl 42:8361–8368. https://doi.org/10.1016/j.eswa.2015.06.046
8. Alarsan FI, Younes M (2019) Analysis and classification of heart diseases using heartbeat features and machine learning algorithms. J Big Data 6:1–15. https://doi.org/10.1186/s40537-019-0244-x
9. Singh R, Mehta R, Rajpal N (2018) Efficient wavelet families for ECG classification using neural classifiers. In: Procedia computer science, pp 11–21. Elsevier B.V. https://doi.org/10.1016/j.procs.2018.05.054
10. Savostin AA, Ritter DV, Savostina GV (2019) Using the K-Nearest neighbors algorithm for automated detection of myocardial infarction by electrocardiogram data entries. Pattern Recognit Image Anal 29:730–737. https://doi.org/10.1134/S1054661819040151
11. Hernandez-Matamoros A, Fujita H, Escamilla-Hernandez E, Perez-Meana H, Nakano-Miyatake M (2020) Recognition of ECG signals using wavelet based on atomic functions. Biocybern Biomed Eng 40:803–814. https://doi.org/10.1016/j.bbe.2020.02.007
12. Reasat T, Shahnaz C (2018) Detection of inferior myocardial infarction using shallow convolutional neural networks. In: 5th IEEE Region 10 humanitarian technology conference 2017, R10-HTC, pp 718–721. Institute of Electrical and Electronics Engineers Inc. https://doi.org/10.1109/R10-HTC.2017.8289058
13. Liu W, Zhang M, Zhang Y, Liao Y, Huang Q, Chang S, Wang H, He J (2018) Real-time multilead convolutional neural network for myocardial infarction detection. IEEE J Biomed Heal Inform 22:1434–1444. https://doi.org/10.1109/JBHI.2017.2771768
14. Kayikcioglu İ, Akdeniz F, Köse C, Kayikcioglu T (2020) Time-frequency approach to ECG classification of myocardial infarction. Comput Electr Eng 84. https://doi.org/10.1016/j.compeleceng.2020.106621
15. Liu W, Wang F, Huang Q, Chang S, Wang H, He J (2020) MFB-CBRNN: a hybrid network for MI DETECTION using 12-Lead ECGs. IEEE J Biomed Heal Inform 24:503–514. https://doi.org/10.1109/JBHI.2019.2910082
16. Bousseljot R, Kreiseler D, Schnabel AN (1995) The PTB diagnostic ECG database. Biomed Tech 40, 317. https://doi.org/10.13026/C28C71
17. Fazeli S (2020) ECG heartbeat categorization dataset. https://www.kaggle.com/shayanfazeli/heartbeat. Last accessed 2020/08/21
18. Kachuee M, Fazeli S, Sarrafzadeh M (2018) ECG heartbeat classification: a deep transferable representation. In: Proceedings—2018 IEEE international conference on healthcare informatics, ICHI 2018. pp. 443–444 (2018). https://doi.org/10.1109/ICHI.2018.00092
19. Somasundaram A, Reddy US (2016) Data imbalance: effects and solutions for classification of large and highly imbalanced data. In: Proceeding 1st international conference on research in engineering, computers and technology (ICRECT 2016), pp 28–34
20. Chawla NV, Bowyer KW, Hall LO, Kegelmeyer WP (2002) SMOTE: synthetic minority oversampling technique
21. Younes Charfaoui: resampling to properly handle imbalanced datasets in machine learning. https://heartbeat.fritz.ai/resampling-to-properly-handle-imbalanced-datasets-in-machine-learning-64d82c16ceaa. Last accessed 2020/08/28
22. Phi M (2020) Illustrated guide to LSTM' s and GRU' s : a step by step explanation. https://towardsdatascience.com/illustrated-guide-to-lstms-and-gru-s-a-step-by-step-explanation-44e9eb85bf21. Last accessed 2020/08/23

23. Kora P (2017) ECG based myocardial Infarction detection using Hybrid Firefly algorithm. Comput Methods Programs Biomed 152:141–148. https://doi.org/10.1016/j.cmpb.2017.09.015
24. Dohare AK, Kumar V, Kumar R (2018) Detection of myocardial infarction in 12 lead ECG using support vector machine. Appl Soft Comput J 64:138–147. https://doi.org/10.1016/j.asoc.2017.12.001
25. Liu W, Huang Q, Chang S, Wang H, He J (2018) Multiple-feature-branch convolutional neural network for myocardial infarction diagnosis using electrocardiogram. Biomed Signal Process Control 45:22–32. https://doi.org/10.1016/j.bspc.2018.05.013
26. Sharma M, Tan RS, Acharya UR (2018) A novel automated diagnostic system for classification of myocardial infarction ECG signals using an optimal biorthogonal filter bank. Comput Biol Med 102:341–356. https://doi.org/10.1016/j.compbiomed.2018.07.005

BL_SMOTE Ensemble Method for Prediction of Thyroid Disease on Imbalanced Classification Problem

Rajshree Srivastava and Pardeep Kumar

Abstract The imbalanced classification problem is one of the most challenging problems in various domains such as in machine learning and data mining. In this state of an imbalanced dataset, each class associated with a given dataset is distributed unevenly. This case arises when the positive class is smaller than the negative class. To overcome this problem, oversampling and undersampling techniques are used. Undersampling leads to the problem of information loss. In this paper, borderline_synthetic minority oversampling technique (BL_SMOTE) ensemble method is used for the prediction of thyroid disease to solve imbalanced classification problems using the oversampling technique. For the ensemble, we have used decision tree and random forest classifier. The proposed method for detection of the thyroid has achieved 98.88% accuracy, 99.12% specificity, 98.93% F-measure, and 98.66% sensitivity on thyroid UCI repository dataset. The proposed method is competitive to the other methods proposed in the literature for prediction of thyroid disease on an imbalanced classification problem.

Keywords Thyroid · Imbalanced dataset · Borderline_SMOTE · Decision tree · Ensemble · Random forest · Classification · SMOTE first section

1 Introduction

The thyroid is a small butterfly-shaped gland located at the base of the neck. Hormones perform a major role in controlling the flow of blood to maintain metabolism in the human body. There are three types of hormones that are produced by the thyroid gland to maintain bloodstream for the regulation of metabolism, namely T4 (thyroxin), thyroid-stimulating hormone (TSH), and tri-iodothyronine

R. Srivastava (✉)
Department of Computer Science and Engineering, JUIT, Himachal Pradesh, Waknaghat, Solan, India

P. Kumar
JUIT, Himachal Pradesh, Waknaghat, Solan, India

P. K. Singh et al. (eds.), *Proceedings of Second International Conference on Computing, Communications, and Cyber-Security*, Lecture Notes in Networks and Systems 203,
https://doi.org/10.1007/978-981-16-0733-2_52

(T3). An increase in the production of hormones leads to hyperthyroidism, whereas a decrease in the production of hormones leads to hypothyroidism. Goiter is also one of the thyroid disorders that enlarge the thyroid gland. There are many symptoms of thyroid disease like obesity, weakness, hair fall, muscle ache, fatigue, etc., in the early stage. Therefore, its diagnosis is important for better survival of life. There is degradation in the performance of machine learning (ML) and data mining due to classification imbalance problems [1]. Some of the reasons includes insufficient data, i.e., less number of positive classes than negative classes, distribution of class, the cost of error is not even [2], etc. Credit card fraud detection [3], detection of an oil spill in satellite images [4], intrusion detection [5], medical diagnosis [6], etc., are some of the real-life example. To overcome the problem of classification imbalance, undersampling and oversampling techniques are mostly used. The undersampling technique works by randomly eliminating the majority class, leading to form an imbalance ratio of the dataset. One of the major disadvantages of this technique is that there are chances of loss of useful data. To address this problem, the oversampling technique is used. It works by generating the same replica of an existing data point of the minority class. Some of its techniques are synthetic minority oversampling technique (SMOTE) proposed by Chawla et al. [7]. This works by creating new minority data along the line of joining at each minority point and one of its neighbors. The second technique is the BL_SMOTE technique proposed by Han et al. [8]. This technique mostly focuses on strengthening the data points nearby decision boundary and oversampling its borderline points, whereas adaptive synthetic (A.DASÝN) technique works by integrating the concept of adaptively synthesizing the new data points, depending on the ratio of majority class samples in k-neighborhood of the point [9]. The ensemble method is defined as a method that raises the overall performance of a single classifier. It is frequently used to solve the classification imbalance problem [10]. The decision tree works by dividing the instances and features. It helps us in finding estimated outcomes and in taking decisions for future outcome especially in medical diagnosis, forecasting revenue, sale, etc. There are many machine learning tools available for decision tree analysis that can be used for the work. One of the advantages of the decision tree is that it does not require normalization of data and scaling of data. Random forest is also known as ensemble random forest. It is said as trees of trees in which many trees support to decide on the problem statement. It provides a good distinction of all attributes of the medical data, etc. Larger the tree, the more accurate will be the result. It also solves the problem of over fitting and does automatic feature selection. Due to the above advantages, we have ensemble the two along with the BL_SMOTE technique and found a better prediction result.

This paper is summarized as follows, Section I focuses on introduction, Section II covers with related work, Section III explains dataset description, Section IV proposed work, Section V covers experiment and result analysis, and Section VI conclusion followed by future work.

2 Related Work

We have studied some of the methods proposed to handle the imbalanced classification problem. Tarek et al. [11] achieved 95.3% accuracy with the Type-2 fuzzy logic system (FLS) approach. The proposed approach helps to handle the uncertainty of data and imprecision. Kotsiantis et al. [12] proposed cascade generalization reweighting (CWGM) approach and achieved 95% accuracy. Fereshteh et al. [13] present the multiclass support vector machine techniques (MCSVM) for detection and classification of hypothyroid problems. The result shows that one-against-all-support vector machines (OAASVM) approach performs better than one-against-one-support vector machines (OAOSVM) approach with 95.30% accuracy. Astha et al. [14] proposed the SMOTE and clustering undersampling technique (SCUT) technique and achieved 98.2% accuracy. This approach firstly oversamples the minority class examples by generating synthetic examples and then clustering analysis is done to undersample the majority class. Pan et al. [15] proposed an ensemble method based on random forest and principal component analysis (PCA) to reduce the feature dimension. The method achieved 96.16% accuracy. Shreela et al. [16] achieved 95.36% accuracy using ranker search algorithm + Naïve Bayes approach. In this approach, the ranker search algorithm solves the problem of redundancy and noisy data.

Mustafa et al. [17] proposed farthest distance-based synthetic minority oversampling technique (FD-SMOTE) technique, and it achieved 84.129% with SVM and Naïve Bayes classifier; FD-SMOTE with multiperceptron (MP) gives 82.72%; FD-SMOTE with N neighbor gives 77.12%; FD-SMOTE with bagging gives 84.11%; FD-SMOTE with random forest (RF) gives 83.21%. The work shows that FD_Smote and PCA significantly reduce the dimensionality and also balances the minority class. Syed et al. [18] proposed an ensemble gain ratio feature selection (EGFS) model to solve the classification imbalance problem and achieved 96.45% with EGFS model + KNN (K-nearest neighbor) model; EGFS model + LR (linear regression) 94.94% and EFGS model + NB (Naïve Bayes) 93.33%. The model ensembles random forest and gain ratio algorithm to find the most efficient features and then aligned it with LR, Naïve Bayes, and KNN. They have achieved the good result by using four features in their work. Rekha et al. [19] proposed distance-based bootstrap sampling model using bagging method on imbalanced dataset and achieved 98.19% accuracy. In this work, the authors show that the use of a distance-based approach with ensemble bagging provides a efficient result to address the problem of an imbalanced classification problem. There is an improvement in the performance of classification accuracy from their work.

3 Dataset Description

3.1 Dataset Description

The thyroid disease dataset is taken from university of California, Irvine (UCI) repository dataset [20]. The original dataset is having 2800 records and 30 features including the class. After preprocessing, i.e., after removal of missing data, we have used 1947 records and 29 features. Table 1 shows dataset. Out of 29 features, six features are continuous namely age, TSH, TT4, T3 T4U, FTI, and rest are categorical. Class is divided into two groups sick/negative class.

4 Proposed Work

To overcome the problem of imbalanced dataset, oversampling technique is used in the proposed model. The proposed method works in three phases, preprocessing phase, ensemble method phase, and prediction phase. In the first phase, preprocessing of the data is done by cleaning of data, removal of missing data and noisy data. This step is considered important for each dataset as it affects the result. After preprocessing step, we have balanced the dataset by using BL_SMOTE oversampling technique. Then, in the second phase, we have ensemble decision tree and random forest classifier to improve the prediction result. There are two types of methods that

Table 1 Thyroid disease dataset [20]

Features	
Age	TSH
TT4	T3
T4U	FTI
Pregnant	Thyroid surgery
I131 treatment	Query hyperthyroid
Query hypothyroid	Tumor
Goitre	Lithium
Hypo pituitary	Psych
TSH measured	Sex
T3 measured	Query on thyroxine
TT4 measured	On thyroxine
T4U measured	On antithyroid medication
FTI measured	Sick
TBG	Referral source
Class	

are used for combining models namely majority voting and average method. In an average method, it takes the average prediction of each class, whereas in majority voting final prediction is done based on maximum votes of each classifier. In this work, majority voting is used. Figure 1 shows the proposed method. Figure 2 shows process of building of balanced ensemble classifier. The proposed algorithm works by applying the BL_Smote technique on the imbalanced dataset, after that model 1, i.e., decision tree classifier and model 2 as random forest classifier are ensemble together in model based on voting method. At last prediction of thyroid disease is done based on performance metrics, i.e., accuracy, sensitivity, F-measure, and specificity.

Algorithm for BL_SMOTE ensemble method.

Input:

Let D: Dataset; Db: Balanced dataset; X: Training set; Y: Testing set; DT: Decision tree; RF: Random forest; Model_1: DecisionTreeClassifier; Model_2: RandomForestClassifier.

Output

P: Prediction of model

Begin

Load D

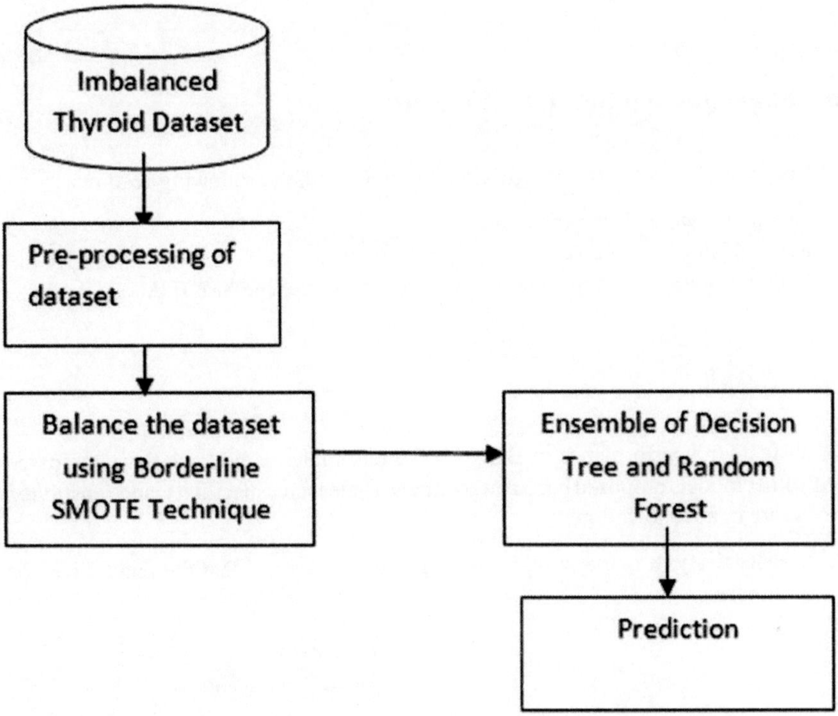

Fig. 1 Proposed BL_SMOTE ensemble method

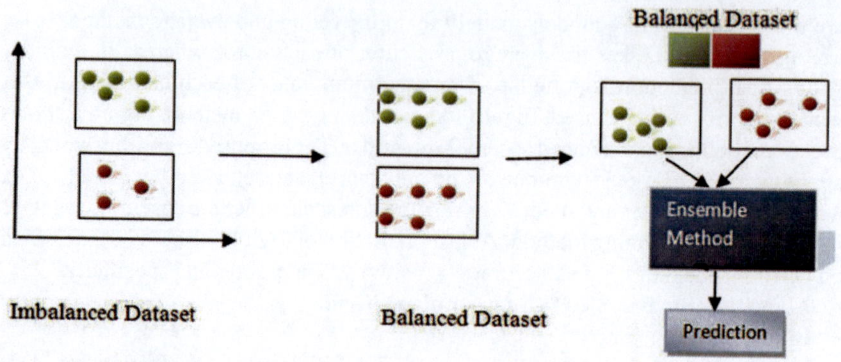

Fig. 2 Building of a balanced ensemble classifier

D ← Borderline SMOTE
Db ← model1
Db ← model2
Model ← (model1, model2, voting)
Prediction P
End

5 Experimental Setup and Result

The code was executed on Google Co laboratory with the following features:

- Runtime: Python 3, i5 processor
- RAM: 12.6 GB available
- Python libraries: Numpy, pandas, sckit-learn, Borderline SMOTE.

5.1 *Result*

To evaluate the performance of BL_SMOTE ensemble method, we have compared with other models proposed based on accuracy, F-measure, specificity, and sensitivity. These are defined as follows:

(a) **Sensitivity**: It is the probability of positive test given that the patient has the disease.

$$\text{Sensitivity} = \frac{\text{True positive}}{\text{True positive} + \text{False negative}} \tag{1}$$

(b) **Specificity**: It is probability of negative test given that the patient is well.

Table 2 Result of balanced dataset by applying BL_SMOTE technique

Dataset	Number of classes	Total number of records	Sick class	Negative class
Actual dataset	2	1947	158	1789
After BL_SMOTE technique	2	3578	1789	1789

$$\text{Specificity} = \frac{\text{True negative}}{\text{True negative} + \text{False negative}} \quad (2)$$

(c) **Accuracy**: It is probability of the results that are classified correct.

$$\text{Accuracy} = \frac{\text{True positive} + \text{True negative}}{\text{True positive} + \text{False positive} + \text{True negative} + \text{False negative}}$$

(d) F-measure: It is defined as test accuracy of the model.

$$F\text{ - measure} = 2 * \frac{\text{Precision} * \text{Recall}}{\text{Precision} * \text{Recall}} \quad (4)$$

From the analysis, it is found that the BL_SMOTE ensemble method performs better than other models. From Table 2 and Fig. 3, it can be observed that the actual

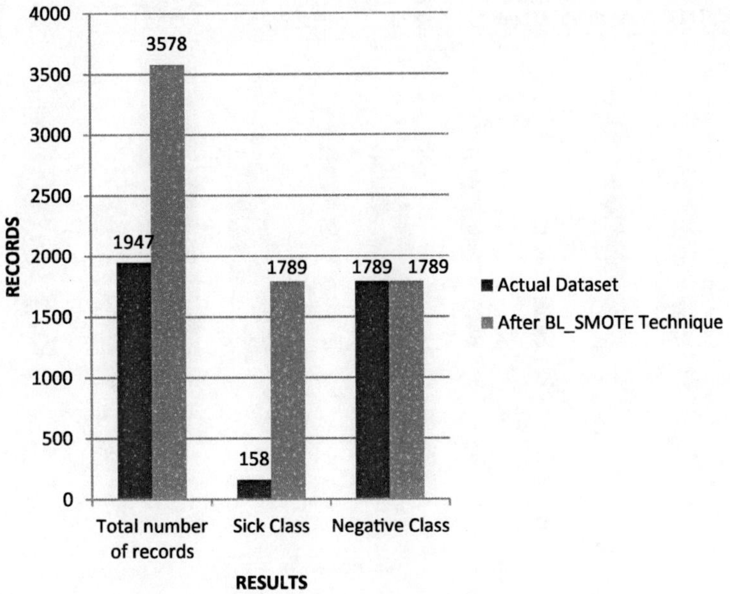

Fig. 3 Result of balanced dataset by applying BL_SMOTE technique

dataset is having 1947 total no. of records, 158 sick classes, and 1789 negative class. After balancing the dataset with the BL_SMOTE technique, we have 3578 total no. of records, 1789 sick classes, and 1789 negative classes. Table 3, Fig. 4, and Fig. 5 show the comparative analysis of the models and it is found that the BL_SMOTE ensemble method performs better than the rest of the methods in terms of specificity and accuracy. In terms of sensitivity, it is better than EGFS model + KNN model, rankers search + Naïve Bayes model and Type 2 FLS model. In comparison with F-measure, the proposed method had achieved the best result from EGFS model + KNN, distance-based bootstrap + bagging method, rankers search + Naïve Bayes model, and SCUT technique. Table 4 shows the confusion matrix. Confusion matrix is important for evaluation of model, it consists of true positive

Table 3 Comparative analysis of the models

Model	Accuracy	Sensitivity	Specificity	*F*-measure
EGFS model + KNN	96.45	96.50	–	96.50
Distance-based bootsrap + bagging	98.19	–	–	92.81
FD_SMOTE	84.12	–	–	–
Ensemble method	96.16	–	–	–
Ranker search + Naïve Bayes	95.36	95.4	–	94.6
SCUT technique	98.2	–	–	97.4
Type-2 FLS approach	95.3	94.5	–	–
OAASVM	95.30	–	–	–
BL_SMOTE ensemble method	**98.88**	**98.66**	**99.12**	**98.93**

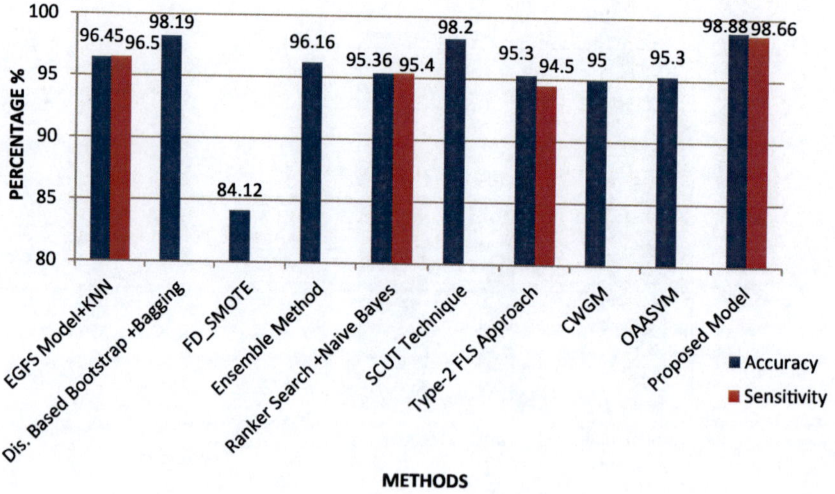

Fig. 4 Comparative analysis of the result based on accuracy and sensitivity

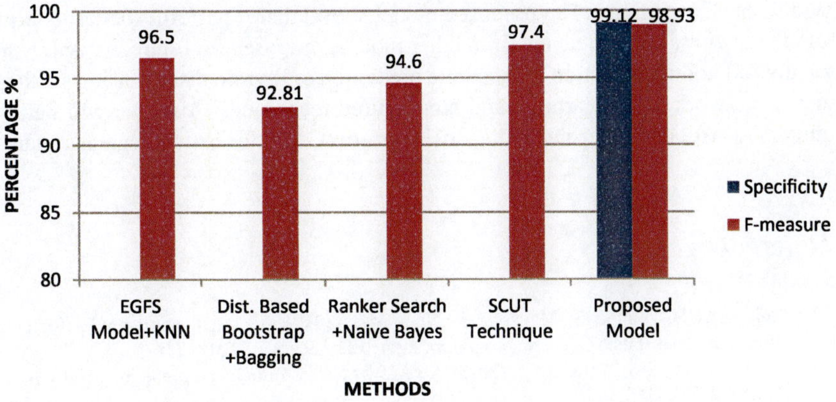

Fig. 5 Comparative analysis of result based on specificity and *F*-measure

Table 4 Confusion matrix

Predicted Values	True Values: Positive	True Values: Negative
Negative	True Positive = 371	False Negative = 4
Positive	False Positive = 4	True Negative = 337

(TP), false negative (FN), true negative (TN), and false positive (FP). From the comparative result analysis, it is found that the BL_SMOTE ensemble method has achieved a better result in terms of accuracy 98.88%, specificity 99.12%, F-measure 98.93%, and sensitivity 98.66%.

6 Conclusion and Future Work

Undersampling and oversampling techniques are the two methods that are used to resolve the problem of imbalanced classification of a dataset. This paper proposes BL_SMOTE ensemble method which solves the imbalance classification problem of thyroid disease detection. The performance of this model is evaluated on four parameters and achieved 98.88% accuracy, 98.93% *F*-measure, 99.12% specificity, and 98.66% sensitivity. The proposed model shows that it performs better in terms of effectiveness and efficiency from other models proposed. In future, we will work on thyroid nodule detection for accurate prediction of thyroid disease using artificial neural network, deep learning, and bio-inspired techniques. Also, we will build a more powerful computer-aided diagnosis system (CAD) for better diagnosis results.

References

1. Tahir MAUH, Asghar S, Manzoor A, Noor MA (2019) A classification model for class imbalance dataset using genetic programming. IEEE Access 7:71013–71037
2. Abd Elrahman SM, Abraham A (2013) A review of class imbalance problem. J Netw Innov Comput 1:332–340
3. Awoyemi JO, Adetunmbi AO, Oluwadare SA (2017) Credit card fraud detection using machine learning techniques: a comparative analysis. In: 2017 international conference on computing networking and informatics (ICCNI), pp 1–9. IEEE (2017)
4. He H, Ma Y (2013) Imbalanced learning: foundations, algorithms, and applications. Wiley
5. He H, Garcia EA (2009) Learning from imbalanced data. IEEE Trans Knowl Data Eng 21:1263–1284
6. Li DC, Liu CW, Hu SC (2010) A learning method for the class imbalance problem with medical data sets. Comput Biol Med 40:509–518
7. Chawla N, Bowyer K, Hall L, Kegelmeyer W (2002) SMOTE: synthetic minority oversampling technique. J Artif Intell Res 16:321–357
8. Han H, Wang W, Mao B (2005) Borderline-SMOTE: a new oversampling method in imbalanced data sets learning. In: International conference on intelligent computing, pp 878–887
9. He H, Bai Y, Garcia EA, Li S (2008) ADASYN: adaptive synthetic sampling approach for imbalanced learning. In: Proceedings of the international joint conference on neural networks, pp 1322–1328
10. Lin W, Wu Z, Lin L, Wen A, Li J (2017) An ensemble random forest algorithm for insurance big data analysis. IEEE Access 5:16568–16575
11. Helmy T, Rasheed Z, Al-Mulhem M (2011) Adaptive fuzzy logic-based framework for handling imprecision and uncertainty in classification of bioinformatics datasets. Int J Comput Methods 8(3):513–534
12. Kotsiantis SB (2011) Cascade generalization with reweighting data for handling imbalanced problems. Comput J 54:1547–1559
13. Chamasemani FF, Singh YP (2011) Multi-class support vector machine (SVM) classifiers—an application in hypothyroid detection and classification. In: 6th international conference of bio-inspired computer theory and application, pp 351–356
14. Agrawal A, Viktor HL, Paquet E (2015) SCUT: multi-class imbalanced data classification using SMOTE and cluster-based undersampling. In: 7th international joint conference on knowledge discovery, knowledge engineering and knowledge management (IC3K), vol 1, pp 226–234
15. Pan Q, Zhang Y, Zuo M, Xiang L, Chen D (2016) Improved ensemble classification method of thyroid disease based on random forest. In: 8th international conference on information technology in medicine and education (ITME), pp 567–571
16. Dash S, Das MN, Mishra BK (2016) Implementation of an optimized classification model for prediction of hypothyroid disease risks. In: International conference on invention computer technology (ICICT), vol 2, pp 4–7
17. Mustafa N, Memon RA, Li JP, Omer MZ (2017) A classification model for imbalanced medical data based on PCA and farther distance based synthetic minority oversampling technique. Int J Adv Comput Sci Appl 8:61–67

18. Pasha SJ, Mohamed ES (2020) Ensemble gain ratio feature selection (EGFS) model with machine learning and data mining algorithms for disease risk prediction. In: International conference on inventive computation technologies (ICICT), pp 590–596. IEEE
19. Rekha G, Reddy VK, Tyagi AK, Nair MM (2020) Distance-based Bootstrap sampling in bagging for imbalanced data-set. In: International conference on emerging trends in information technology and engineering (IC-ETITE), pp 1–6. IEEE
20. https://archive.ics.uci.edu/ml/datasets/Thyroid+Disease

Computer-Aided-Diagnosis System for Symptom Detection of Breast and Cervical Cancer

Piyushi Jain, Drashti Patel, Jai Prakash Verma, and Sudeep Tanwar

Abstract As witnessed,' Cancer metastasis is the leading cause of death worldwide, lots of efforts done for understanding the pathology of cancer for prognosis and diagnosis. Cancer treatment at an early stage can increase the chances of survival of the sufferer considerably. This research aims to contribute to the detection of breast and cervical cancer in the early stages. A comparative study of classification techniques includes Support Vector Classifier (SVC), Multi-layer Perceptron (MLP), and Random Forest (RF) has done to identify the best model for cancer prediction. These models are differentiated for different symptoms collected from electronic health care data. A correlation matrix with a heat map is used for symptoms/feature selection from the results of biopsy examinations applied on a dataset collected from the UCI repository. The predictive model based on Random forest techniques achieves the highest testing accuracy 98.83% in the case of cervical cancer and 96.50% for breast cancer with the selected symptoms/feature.

Keywords Classification · Random forest · Support vector classifier · Multi-layer perceptron · Cancer research

---

P. Jain · D. Patel · J. P. Verma (✉) · S. Tanwar (✉)
Institute of Technology, Nirma University, Ahmedabad, Gujarat 382381, India
e-mail: jaiprakash.verma@nirmauni.ac.in

S. Tanwar
e-mail: sudeep.tanwar@nirmauni.ac.in

P. Jain
e-mail: 17bit081@nirmauni.ac.in

D. Patel
e-mail: 17bce075@nirmauni.ac.in

P. K. Singh et al. (eds.), *Proceedings of Second International Conference on Computing, Communications, and Cyber-Security*, Lecture Notes in Networks and Systems 203,
https://doi.org/10.1007/978-981-16-0733-2_53

1 Introduction

The healthcare industry manages a large amount of data every day from operational information systems and clinical, such as electronic health records. Data analytics includes techniques, methodologies, practices, technology, and application for a large amount of data. It is useful for better understand and make a timely decision for the domain applications. The use of information technology and medical data to build a medical support system can use to detect cancer early. It increases the treatment chances and decreases the death rate of a cancer patient. Medical image processing is the best method for cancer detection. There are many different types of medical imaging modalities available like magnetic resonance imaging (MRI), digital mammogram (DM), ultrasound (US), microscopic image, and infrared thermography (IRT) [1, 2]. These modalities produce an image. Using these images, we can detect breast cancer and decrease mortality rates by 30–70%. Using some a feature classification and extraction formulated model as computer-aided detection can be helpful for physicians and experts in detecting cancer [3, 4].

Cancer is the rampant growth of abnormal cells [5]. Instead of apoptosis, old cells start growing maniacally and hence forming a cluster or tumor in that part of the body. Machine learning is prevalent in the field of healthcare today. Reliable models developed can give higher accuracy by learning from past datasets and improving efficiency after each prediction. Various symptoms or factors can be identified to classify each healthcare issue uniquely. For instance, Human papillomavirus, smoking, oral contraceptives, and multiple pregnancies are some of the major risk factors for cervical cancer [6]. Similarly, alcohol, DES exposure, estrogen exposure, family history, and breast density are some common causes of breast cancer. Methods like Cancer screening, imaging tests, and Fine-needle aspiration biopsy (FNAB) are used to detect cancer. FNAB is a diagnostic technique in which a thin hollow needle is allowed to pass through the tissue for the sampling of cells. This sample after being stained is examined under a microscope for diagnosis and a compilation of such results is used for breast cancer diagnosis in this project. Classification of cancer is done based on the tissue of origin [7]. Most of the time cancer is detected too late that the treatment becomes challenging and hence less likely to succeed so there is a need to detect it in the early stages. The accuracy of the detection is significant for the treatment of the sufferer.

Due to the advancement in computer technology, our lives are becoming easier [8]. Technology nowadays has a solution to every problem which existed and many more to come [9]. It can be used to predict cancer. Other than the cancer prediction ML can be used to determine relevant features that help to detect cancer or can be used to determine any risk related to cancer [10].

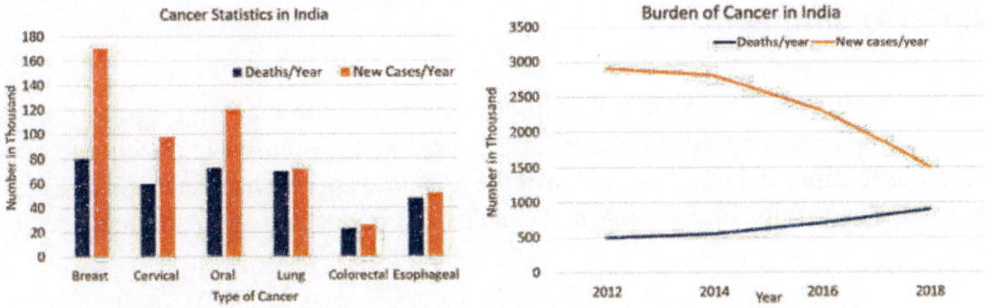

Fig. 1 Cancer survivor in India **a** Comparison of different cancer based on deaths and new cases. **b** Count of new cases and death due to cancer in India

1.1 Motivation

India ranks third in cancer cases after China and the USA. Every year more than one million new cases are registered, shown in Fig. 1b. More than half a million deaths are caused due to ignorance and patients being deprived of early diagnosis. For every two women newly diagnosed with breast cancer, one dies of it. And in every eight minutes, one woman died of cervical cancer in India [11]. Even after observing a great decline in the number of new cases of cancer per year, the number of deaths goes on increasing each year as illustrated in Fig. 1b. A reason behind this could be the negligence of early diagnosis and determination of the disease [12].

1.2 Contribution

In this paper, breast cancer and cervical cancer are predicted using three machine learning techniques SVC, MLP, and RF. Using a correlation matrix and a heatmap graph representation, two different sets of symptoms are created and used for prediction. The accuracy achieved is as high as 96.50% with Wisconsin dataset for 32 features computed from a digitized image of a fine needle aspirate (FNA) of a breast mass and an accuracy of 98.83% with 36 attributes which include demographic information, habits, and significant medical records of patients of cervical cancer. Following are the contributions of this paper:

- To explore different features of Breast cancer and Cervical cancer and their significance in causing the disease.
- To provide a comparative study of the results of a predictive model based on SVC, RF, and MLP for two different models with different features set.

1.3 Organisation

The remaining paper is organized as follows. Section 2 summarizes the related work in the field of cancer prediction. Section 3 provides the proposed approach. Methodology, execution, and implementation of the proposed approach are included in Sects. 4 and 5. The results, discussions, and conclusions in Sects. 6 and 7, respectively.

2 Related Work

In this subsection, we will try to summarise most of the research works that have been done on the diagnosis of various cancers including breast and cervical cancer (refer Tables 1 and 2). Machine learning models were proved useful in identifying it in early stages at least 3 months in advance [13, 14]. The work is done by Tseng et al. [14] showed a random forest model to predict breast cancer metastasis at least 3 months in advance. The prediction was made based on multiple parameters and attributes which include: Demographic data (age of the diagnosed), Tumor information (TNM stage), Pathology data, and Laboratory data, i.e., serum biomarkers.

Singh et al. [13] determined relevant biomarkers to predict Breast Cancer in early stages using different feature selection methods like filter method and wrapper method and tested the same features on different models like kernel-based support vector machine, Naïve Bayesian, Linear discriminant, Quadratic discriminant, Logistic regression, K-nearest neighbors and Random forest. He proposed that features like age, glucose, insulin, HOMA, and resisting have the potential to be used as reliable biomarkers for the detection of breast cancer. For determining

Table 1 Abbreviations table

S. No.	Abbreviations	Full form	S. No.	Abbreviations	Full form
1	DT	Decision tree	2	IB	Instance-based
3	NN	Nearest neighbour	4	SVM	Support vector machine
5	BN	Bayesian network	6	ANN	Artificial neural networks
7	QD	Quadratic discriminant	8	GRU-SVM	Gated recurrent unit-support vector machine
9	RF	Random forest	10	KNN	K-nearest neighbor
11	LR	Logistic regression	12	MLP	Multi layer perceptron
13	LASSO	Least absolute shrinkage and selection operator	14	LD	Linear discriminant

Table 2 Comparative Analysis of work in the area of Cancer Research

Approach	Year	Objective	Pros	Cons	DDT	DIB	NN	SVM	DBN	ANN	QD	GRU-SVM	RF	DKNN	LR	MLP	LASSO	LD
Data Min-ing [9]	2009	To provide a broad review of different techniques of classification	Relation b/w DT, BN, and RC was found	No clear picture of which method is best	✓	✓	✗	✗	✓	✗	✗	✗	✗	✓	✗	✗	✗	✗
Data Min-ing [5]	2013	To build a predictive model for breast cancer patients who were followed up for 2 years	SVM model with high accuracy (0.957) was developed	Important attributes like S-phase fraction and DNA index were not included	✓	✗	✗	✓	✗	✓	✗	✗	✗	✗	✗	✗	✗	✗
Machine learning [8]	2019	To check the risk associated with the cancer patients of PICC-related Vein Thrombosis due to chemotherapy	Drinking and malnutrition is strongly related to thrombosis	Dataset was not distributed and was limited to a smaller number	✗	✗	✗	✗	✗	✗	✗	✗	✓	✗	✗	✗	✓	✗
Machine learning [7]	2019	To predict the breast cancer using 3 machine learning algorithms	SVM achieved the highest accuracy (92.7%)	More research is required to increase the accuracy of the predictive models	✗	✗	✗	✓	✗	✗	✗	✗	✗	✓	✓	✗	✗	✗

(continued)

Table 2 (continued)

Approach	Year	Objective	Pros	Cons	DDT	DIB	NN	SVM	DBN	ANN	QD	GRU-SVM	RF	DKNN	LR	MLP	LASSO	LD
Machine learning [2]	2018	Comparison of different machine learning algorithms for prediction of breast cancer	MLP achieved the highest accuracy (99.0384%)	K-fold cross-validation should be used to substantiate the results	✗	✗	✓	✓	✗	✗	✗	✓	✗	✗	✓	✓	✗	✗
Machine learning [17]	2019	To build a prediction model using serum biomarkers and other 7 clinical features	Cancer metastasis was predicted 3 months in Advance	The accuracy obtained was not as expected	✗	✗	✗	✓	✓	✗	✗	✗	✓	✗	✓	✗	✗	✗
Machine learning [12]	2019	To determine relevant biomarkers for the prediction of breast cancer	Age, glucose, insulin, HOMA, resistin have the potential to be relevant biomarkers for breast cancer	The dataset was limited so more research is needed to verify it with a greater dataset	✗	✗	✗	✗	✓	✗	✓	✗	✓	✓	✓	✗	✗	✓

the cancer type (benign or malignant) many machine learning models are used like gated recurrent unit-Support Vector Machine (GRU-SVM), Linear Regression, Multilayer Perceptron (MLP), Nearest Neighbour (NN) search, SoftMax Regression, and Support Vector Machine (SVM) as proposed by Agarap et al. [15].

The study in the cancer research using machine learning has mainly focused on comparing different models for the prediction and finding the best one out of it [15–18]. Houby et al. [18] provided a comprehensive review and comparison of different classification techniques including Decision tree induction, Bayesian networks, K-Nearest Neighbor classifiers, and Instance-based learning. The study concluded that whenever the results of the Bayesian network are accurate, the Decision tree's results are not and vice versa. On the other hand, decision tree and rule classifiers show similar results. Ahmad et al. [16] compared data mining techniques: Decision tree, Support vector machine, and Artificial neural network. 24 variables were used for recurrence modeling of breast cancer and it was found that SVM gave the best results on the test set with the least error rate and highest accuracy.

Liu et al. [17] provided the risk associated with the cancer patients of PICC-related Vein Thrombosis due to chemotherapy and supportive care therapy using machine learning models like Random Forest and LASSO. He also proposed that drinking and malnutrition are strongly associated with PICC-related thrombosis. Kourou et al. [19] compared three different supervised machine learning algorithms Logistic Regression, Nearest neighbor and support vector machine for prediction of Breast Cancer. 32 features were used and out of the three Support Vector Machine gave the highest accuracy.

Agarap et al. [15] compared different machine learning models for the diagnosis of Breast Cancer. He used models like gated recurrent unit-Support Vector Machine (GRU-SVM), Linear Regression, Multilayer Perceptron (MLP), Nearest Neighbour (NN) search, SoftMax Regression, and Support Vector Machine (SVM) and it gave satisfactory results by classifying them as a benign tumor or malignant tumor. MLP gave the highest accuracy. Other models also gave satisfactory results.

3 Proposed Work

Cancer prediction with correct accuracy is crucial for its treatment and survival of the patient. Many applications that already exist and are being used by people worldwide. These apps have features like adding medications, prescriptions, interacting with the community with stories, track of treatment records, and others. Research has shown that the most frequently faced problems by cancer patients include: anxiety for their future, economic burden-financial difficulty, unhappiness, or depression. Hence now we know that treatment of cancer not only requires proper medications but also relief from psychological stress. There is no app existing that could help the patient with both services mentioned above.

The alert will be raised with a mobile application for patients suffering from cancer. This application will help them by keeping a record of their medications and prescriptions and giving them remainders accordingly. The main feature that this application is the common platform it will provide to the patients seeking job offers and companies that are willing to provide the same. The app will provide them advertisements and links for the same.

4 Methodology

Data Preprocessing is one of the most important steps of data mining where we prepare the data before mining. Data in the real world is inconsistent, incomplete, and is likely to contain error, hence there is a need to modify it before processing and applying any of the predictive models.

4.1 Handling Null Values or Missing Values

We need to handle null values as they have no significance and can not be processed with other datatypes. We handled null values by removing some of the rows and by adding mean values to the null values columns in some cases.

4.2 Feature Selection

It is a method of selecting a subset from the relevant set of features for use in model construction. Feature selection includes removing rows of missing values, selecting which influence the result and have a greater possibility of deviating and giving wrong output if ignored. We used the Correlation Matrix with Heatmap. Heatmap helps us to correlate the data and shows how much it is related to each other. Here for breast cancer data, we used this technique from which we chose the top 10 features for model 2. While model 1 kept as it is with all the features.

4.3 Prediction

Data split into two parts: Test set and Training set. The training set was used for training the model and the test set is used to test the trained model. Stratified splitting is done to ensure that each split is similar for some features. We have used 3 of the supervised machine learning algorithms and compared them. Random forest, Support Vector Classifier, and Multi-Layer Perceptron.

4.4 Performance Measurement

The performance was measured using the confusion matrix and calculating and comparing the precision, accuracy, recall, f1-score, roc auc score.

Precision (p) is defined as tp/(tp + fp). It is a ratio of total patients having cancer and predicted correctly to the patients the model predicted that they have cancer and they have cancer or they have cancer and the model predicted it wrong i.e. they don't have cancer. The recall (r) is defined as tp/(tp + fn). It is a ratio of total patients having cancer and predicted correctly to the patients the model predicted that they have cancer and they have cancer or they don't have cancer and the model predicted it wrong i.e. they have cancer.

F1 score is defined as 2 * (p * r)/(p + r). It shows the balance between precision and recall. We also have accuracy for that but if we have large no of patients who do not have cancer than the value will differ and it may look better.

Algorithm1: Prediction

Input: Dataset D, n independent variable xi=x1,x2,.., xn
Output: dependent output variable y
1. Obtain training set and test set using stratified split
a. X train=x1, x2,.., xn where xi!=y
b. y train=y
Similarly $X_testandy_test$
2. Apply a predictive model
3. Calculate precision := tp/(tp+fp)
4. Calculate recall := tp/(tp+fn)
5. Use another model and repeat step 3 and 4.

---

5 Execution and Implementation

We implemented these models on the cervical cancer dataset having a total of 36 features and 858 records. First, all the data were preprocessed as it contained null values or missing values. These values were simply made 0. Data was split into two parts training set and testing set. The splitting was done using stratified splitting and with ratio 8:2. Model 1 was designed using all the 36 features while model 2 was designed using only the top 31 features. As per Table 3, These top 31 features were selected using the correlation graph showing how each feature is related to the output. There was a minor difference in the accuracy of both the models so we can say that most of the features even if not taken does not deviate from the prediction (Table 4).

For breast cancer, we had a total of 31 features and 569 records. It contained no null values so we implemented the model with all features. Heatmap correlation was used to select the features out of 31. From this, we selected 10 features and the

Table 3 Hyperparameters

Algorithm	Hyperparameter	
Random Forest	criterion = ‘gini’, n *estimators* = 10, *random state* = 42	
MLP Classifier	activation = ‘relu’, alpha = 0.0001, beta_1 = 0.9, beta_2 = 0.999, hidden_layer_sizes = (5, 10, 5), learning_rate = ‘constant’, learning_rate_init = 0.001, solver = ‘adam’, validation_fraction = 0.1	0.999, *hiddenlayer sizes* =
SVC	kernel = ‘rbf’, C = 5, gamma = 10, degree = 3	

models were created as shown in Table 5. When only 10 features were taken the accuracy increased by a small percentage.

6 Result and Discussion

Our analysis for cervical cancer shows that accuracy for Random Forest, Support vector classifier, and MLP classifier is, respectively, 98.83%, 97.09%, and 95.38%. The results were best when all the 36 features used in the model-1.

Our analysis for breast cancer shows that accuracy for Random forest, support vector classifier, and MLP respectively 95.10%, 93.20%, and 95.80% when all the features took (refer Table 6 Overall the models with all features gave better results) (Fig. 3).

A precision-recall curve drawn taking recall on x-axis and precision on the y-axis. Higher the area under the curve, the higher the recall and precision show better the model. If the curve show straight from (0, 1) to (1, 1) and to straight down, then it is said to be a perfect plot. Results of model-1: uses all 36 features. The discussions are showing in Fig. 2.

Cervical Cancer, Model-1, RF micro-average precision-recall calculated is 0.986, shows that out of 1000 results predicted 986 results were true and remaining others we wrongly predicted as true. The results of RF are the best among the three. 2. Roc curve is drawn as 1-specificity on the x-axis and sensitivity on the y-axis. Roc curve specifies how good the model is to separate each class, here it is whether there is cancer or there is no cancer. For the perfect model, this curve goes from bottom left to top left and top right. Here we can measure the performance as if the curve is closer to the perfect curve the model is having good separability.

In model-2: where we used the top 31 features, RF performed the with ROC curve value equal to 0.79702. Usually, the AUC of other classifiers should be greater than that of random as it is expected for them to perform better. In this model, the AUC for the random forest is 0.89 which indicates that random forest performed better than the other two. In other words, there were more negative values predicted in positive class and positive values predicted in negative class in SVC and MLP.

Table 4 List of features used in different model for breast cancer

Feature	Model1D	Model2D	Feature	Model1D	Model-2	Feature	Model1D	Model-2	Feature	Model1D	Model2D
Radius mean	✓	✓	Texture mean	✓	✗	Fractal dimension worst	✓	✗	Concave points worst	✓	✓
Perimeter mean	✓	✓	Area mean	✓	✓	Symmetry worst	✓	✗	Compactness worst	✓	✗
Smoothness mean	✓	✗	Compactness mean	✓	✓	Concavity worst	✓	✗	Area worst	✓	✓
Concavity mean	✓	✓	Concave points mean	✓	✓	Smoothness worst	✓	✗	Texture worst	✓	✗
Symmetry mean	✓	✗	Fractal dimension mean	✓	✗	Perimeter worst	✓	✓	Radius worst	✓	✓
Radius se	✓	✗	Texture se	✓	✗	Concave points se	✓	✗	Symmetry se	✓	✗
Texture se	✓	✗	Perimeter se	✓	✗	Fractal dimension se	✓	✗	Concavity se	✓	✗
Area se	✓	✗	Smoothness se	✓	✗	Compactness se	✓	✗			

Table 5 List of features used in different model for cervical cancer

Feature	Model1D	Model2D	Feature	Model1D	Model2D	Feature	Model1D	Model2D	Feature	Model1D	Model2D
Age	✓	✓	Number of sexual partners	✓	✓	Biopsy	✓	✓	STDs: HIV	✓	✓
First sexual intercourse	✓	✓	Num of pregnancies	✓		Schiller	✓	✓	STDs: Hepatitis B	✓	✓
Smokes	✓	✗	Smokes (years)	✓	✗	Hinselmann	✓	✓	STDs: molluscum contagiosum	✓	✓
Smokes (packs/year)	✓	✗	Hormonal contraceptives	✓	✗	Dx	✓	✓	STDs: AIDS	✓	✓
Hormonal Contraceptives (years)	✓	✗	IUD	✓	✓	Dx:CIN	✓	✓	STDs: Number of diagnosis	✓	✓
IUD (years)	✓	✓	STDs	✓	✓	Citology	✓	✓	STDs: genital herpes	✓	✓
STDs (number)	✓	✓	STDs: condylomatosis	✓	✓	STDs: Time first diagnosis	✓	✓	STDs: cervical condylomatosis	✓	✓
STDs: vaginal condylomatosis	✓	✓	STDs: vulvo-perineal condylomatosis	✓	✓	Dx:HPV	✓	✓	STDs: HPV		
STDs: syphilis	✓	✓	STDs: pelvic inflammatory disease	✓	✓	Dx:Cancer	✓	✓			

Table 6 Performance measurement—results

Model	Cancer	Approach	Recall	Precision	F1 score	ROC score	Accuracy (%)
Model1	Cervix	Random Forest	0.8	0.8	0.8	0.89	98.83
Model1	Cervix	SVC (kernel = 'rbf')	0.2	1.0	0.33	0.6	97.09
Model1	Cervix	MLP Classifier	0.6	0.5	0.54	0.79	95.38
Model2	Cervix	Random Forest	0.8	0.75	0.66	0.79	98.25
Model2	Cervix	SVC (kernel = 'rbf')	0.2	1.0	0.33	0.6	97.09
Model2	Cervix	MLP Classifier	0.8	0.8	0.8	0.89	96.56
Model1	Breast	Random Forest	0.86	1.0	0.92	0.93	95.10
Model1	Breast	SVC (kernel = 'rbf')	0.81	1.0	0.89	0.90	93.20
Model1	Breast	MLP Classifier	0.92	0.96	0.94	0.95	95.80
Model2	Breast	Random Forest	0.90	1.0	0.95	0.95	96.50
Model2	Breast	SVC (kernel = 'rbf')	0.96	0.92	0.94	0.95	95.80
Model2	Breast	MLP Classifier	0.90	0.96	0.93	0.94	95.10

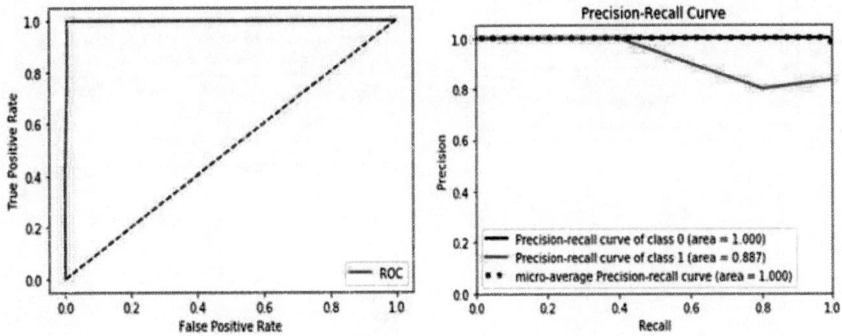

Fig. 2 Cervical cancer Random Forest—model-1

In Breast Cancer, Fig. 4 shows the analysis graph for one of the algorithms which outperformed the best. SVC performed the best in both the models. From Fig. 4, the precision-recall curve area calculated is 0.997. When we created the model with ten features, we see that SVC here also performed better. In SVC, the precision-recall curve area calculated is 0.994 4. For a model with all the features, the results show in Fig. 4.

From Fig. 4, the area under the ROC curve value is 0.9920 for SVC. From Fig. 4, the area under the ROC curve value for SVC model 2 is 0.9907. So we can say that the more probability is for SVC.

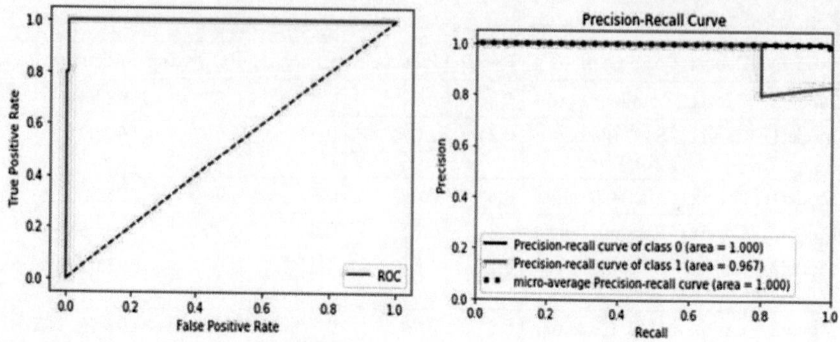

Fig. 3 Cervical cancer Random Forest—model-2

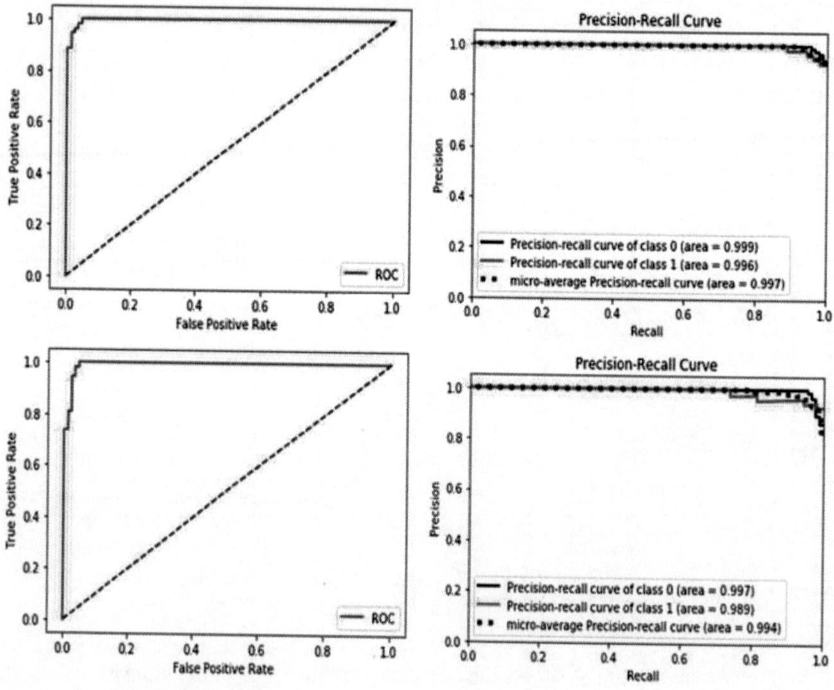

Fig. 4 Breast cancer SVC model-1

7 Conclusion

The study of cancer has become an extensive concern due to the growing number of patients and deaths in a developing country like India. This study aimed at the prediction of cancer in the early stages using different machine learning algorithms. The machine learning algorithms after feature selection and data splitting were implemented and the results were examined. Three machine learning algorithms were used SVC, RF, MLP out of which for cervical cancer Random forest gives the highest accuracy and recall whereas Support vector classifier gives higher precision and MLP gives the least accuracy among all the three. For breast cancer when all the features were taken, MLP gives the highest accuracy and recall while Random forest and SVC give the higher precision and SVC gives the least accuracy among all the three. When only 10 features were taken random forest gives the highest accuracy and precision while MLP gives the least accuracy among all the three. For cervical cancer, the dataset was limited and had fewer patients having cancer so the result may vary for greater datasets. Using a larger dataset with better quality, accuracy and recall can be improved for future work. As stated earlier in the paper more research and development are required to overcome financial stress on cancer patients.

References

1. Mostert B, Sleijfer S, Foekens JA, Gratama JW (2009) Circulating tumor cells (ctcs): detection methods and their clinical relevance in breast cancer. Cancer Treatment Rev 35(5):463. https://doi.org/10.1016/j.ctrv.2009.03.004. https://www.sciencedirect.com/science/article/pii/S0305737209000395
2. Tunali I, Gray JE, Qi J, Abdalah M, Jeong DK, Guvenis A, Gillies RJ, Schabath MB (2019) Novel clinical and radiomic predictors of rapid disease progression phenotypes among lung cancer patients treated with immunotherapy: an early report. Lung Cancer 129:75–79
3. Tian Y, Shi Y, Chen X, Chen W (2011) Auc maximizing support vector machines with feature selection. Proc Comput Sci 4:1691–1698 (Proceedings of the international conference on computational science, ICCS 2011). https://doi.org/10.1016/j.procs.2011.04.183. https://www.sciencedirect.com/science/article/pii/S1877050911002419
4. Yassin NI, Omran S, Houby EME, Allam H (2018) Machine learning techniques for breast cancer computer aided diagnosis using different image modalities: a systematic review. Comput Methods Programs Biomed 156:25–45
5. Davis CP (2019) Cancer facts. https://www.medicinenet.com/cancer/article.htm
6. Forsyth AW, Barzilay R, Hughes KS, Lui D, Lorenz KA, Enzinger A, Tulsky JA, Lindvall C (2018) Machine learning methods to extract documentation of breast cancer symptoms from electronic health records. J Pain Symptom Manage 55(6):1492–1499
7. Tsai CJ, Riaz N, Gomez SL (2019) Big data in cancer research: real-world resources for precision oncology to improve cancer care delivery. Seminars Radiation Oncol 29(4):306–310 (Big Data in Radiation Oncology)
8. Singh PK, Kar AK, Singh Y, Kolekar MH, Tanwar S (2019) Proceedings of ICRIC2019: recent innovations in computing, vol 597. Springer
9. Singh PK, Panigrahi BK, Suryadevara NK, Sharma SK, Singh AP (2019) Proceedings of ICETIT 2019: emerging trends in information technology, vol 605. Springer
10. Brownlee J (2019) Supervised and unsupervised machine learning algorithms. https://machinelearningmastery.com/supervised-and-unsupervised-machinelearning-algorithms/
11. Annual report 2018 (2018), https://nicpr.icmr.org.in/
12. Parry N, Comment LA (2019) What world cancer day told India about health. https://www.healthissuesindia.com/2019/02/08/what-world-cancer-daytold-india-about-health/
13. Singh B (2019) Determining relevant biomarkers for prediction of breast cancer using anthropometric and clinical features: a comparative investigation in machine learning paradigm. Biocybern Biomed Eng 39:393–409. https://doi.org/10.1016/j.bbe.2019.03.001

14. Tseng YJ, Huang CE, Wen CN, Lai PY, Wu MH, Sun YC, Wang HY, Lu JJ (2019) Predicting breast cancer metastasis by using serum biomarkers and clinicopathological data with machine learning technologies. Int J Med Inform 128:79–86
15. Agarap AF (2017) On breast cancer detection: an application of machine learning algorithms on the wisconsin diagnostic dataset. CoRR abs/1711.07831. https://arxiv.org/abs/1711.07831
16. Eshlaghy A (2013) Using three machine learning techniques for predicting breastcancer recurrence
17. Liu S, Zhang F, Xie L, Wang Y, Xiang Q, Yue Z, Feng Y, Yang Y, Li J, Luo L, Yu C (2019) Machine learning approaches for risk assessment of peripherally inserted central catheter-related vein thrombosis in hospitalized patients with cancer. Int J Med Inform 129:175–183
18. Houby E (2018) A survey on applying machine learning techniques for management of diseases. J Appl Biomed 16(3):165–174
19. Kourou K, Exarchos TP, Exarchos KP, Karamouzis MV, Fotiadis DI (2015) Machine learning applications in cancer prognosis and prediction. Comput Struct Biotechnol J 13:8–17

Blockchain Adoption for Trusted Medical Records in Healthcare 4.0 Applications: A Survey

Umesh Bodkhe, Sudeep Tanwar, Pronaya Bhattacharya, and Ashwin Verma

Abstract Healthcare 4.0 allows monitoring of electronic health record (EHR) at distributed locations, through wireless infrastructures like Bluetooth, ZigBee, near-field communication (NFC), and GPRS. Thus, the private EHR data can be tampered by malicious entities that affect updates through different stakeholders like patients, doctors, laboratory technicians, and insurance agencies. Hence, there must be a notion of trust among aforementioned stakeholders. Moreover, the accessed volume of data is humongous; thus, to ensure security and trust, blockchain (BC)-based solutions can handle timestamped volumetric data as chronological ledger. Motivated from the same, the paper presents a systematic survey of BC applications in Healthcare 4.0 ecosystems. The contribution of the paper is to conduct a systematic survey of BC adoption in Healthcare 4.0. The survey identifies tools and technologies to support BC-based healthcare applications and addresses open challenges for future research of integrating BC to secure EHR in Healthcare 4.0 ecosystem.

Keywords Healthcare 4.0 · BC · Decentralized EHR

U. Bodkhe (✉) · S. Tanwar (✉) · P. Bhattacharya · A. Verma
Department of Computer Science and Engineering, Institute of Technology, Nirma University, Ahmadabad, Gujarat, India
e-mail: umesh.bodkhe@nirmauni.ac.in

S. Tanwar
e-mail: sudeep.tanwar@nirmauni.ac.in

P. Bhattacharya
e-mail: pronoya.bhattacharya@nirmauni.ac.in

A. Verma
e-mail: ashwin.verma@nirmauni.ac.in

P. K. Singh et al. (eds.), *Proceedings of Second International Conference on Computing, Communications, and Cyber-Security*, Lecture Notes in Networks and Systems 203,
https://doi.org/10.1007/978-981-16-0733-2_54

1 Introduction

Recently, the focus has shifted towards efficient storage of personal health information (PHI) and clinical trials from manual to electronic forms. Patient EHR data are normally collected through sources like medical sensors, laboratory records, health prescriptions, and insurance companies. The collected data are mainly heterogeneous; thus, to maintain EHR, industry ecosystem shifted gradually from Healthcare 1.0 to Healthcare 4.0. Records were manual and stored in files in Healthcare 1.0. Thus, any intruder could access those manual records and breach privacy and confidentiality of patients [1]. Healthcare 2.0 reduced manual intervention, and EHR is stored in electronic forms at centralized servers. The servers were vulnerable to network attacks to gain access to patient critical and sensitive information. Also, the servers were single point of contact; hence, proper load balancing was required. Healthcare 3.0 saw transitions towards decentralizing EHR by mobile apps for efficient management and retrieval of EHR. Hence, cost was reduced but the apps lacked intelligence for personalized care. Healthcare 4.0 introduced decentralized intelligence to support real-time monitoring and decision analytics. However, Healthcare 4.0 suffers from limitations of trust among medical stakeholders, record fragmentation, heterogeneous locations, complex recommender models to enhance personalization, and logistics handling.

Thus, to handle aforementioned issues in Healthcare 4.0, a decentralized trust is required between industry stakeholders. Hence, BC in Healthcare 4.0 is a seeming revolution that can provide anonymous trust as an agreed set of truth among all participating stakeholders [3, 4]. A BC forms a chronological and distributed ledger that addresses the information sharing. At the international front, Onik et al. [2] proposed the projected growth of patients worldwide by 2020 to reach 1.6 million. Figure 1a depicts the details of technological revolution in healthcare industry. Figure 1b presents usage of BC in Healthcare 4.0 [5].

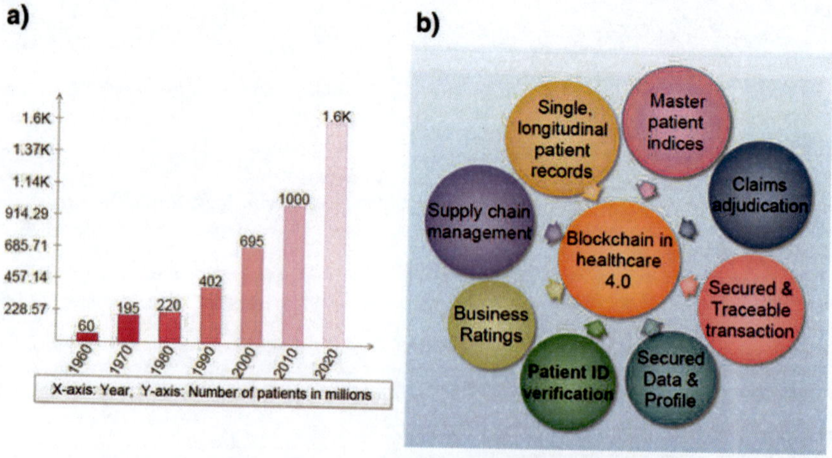

Fig. 1 **a** Projected worldwide growth of patients by 2020 [2] and **b** pros of BC in Healthcare 4.0

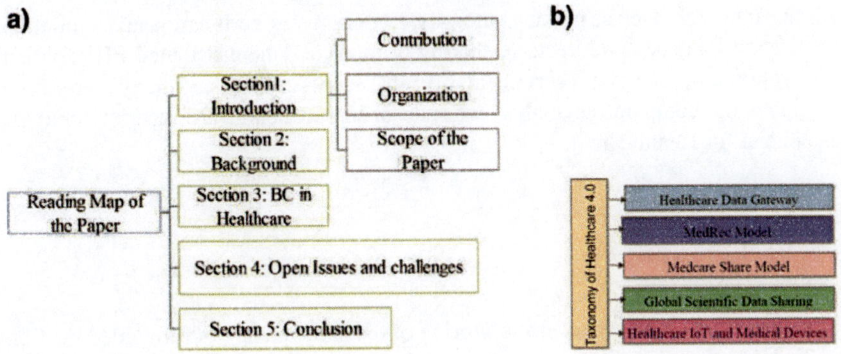

Fig. 2 **a** Flow graph of the paper and **b** solution taxonomy of Healthcare 4.0 ecosystem

1.1 Contribution

1. The role of security in managing EHR in the healthcare environment is presented in detail.
2. Comparative analysis of several existing approaches to secure Healthcare 4.0 concerning different parameters is discussed.
3. Finally, open issues, research opportunities, and directions for future research in Healthcare 4.0 are discussed.

1.2 Organization

Figure 2a shows the flow graph of the entire paper. The manuscript is organized as follows as Background knowledge and importance of smart healthcare is presented in Sect. 2.

In Sect. 3, we outlined the solution taxonomy of BC in healthcare application. Section 4 presents research challenges and open issues in smart health care. Finally, Sect. 5 discusses the concluding remarks of the proposed survey paper.

1.3 Scope of the Paper

Till date, many surveys have been conducted that explored various security aspects of EHR, PHR, MHR, and other healthcare-related data. However, as per our surveyed literature, many of these surveys primarily focused on secured and reliable approaches, techniques, and mechanisms. Most of the researchers have considered some parameters such as privacy, integrity, EHR security, verification and validation, and network latency architecture, sharing of medical data, distributed EHR, patient encryption key, privacy-preserving algorithms, access control, and many more. Table 1 depicts the comparative analysis of state-of-the-art healthcare security standards beneficial for Healthcare 4.0.

2 Background

Health diagnosis and proposer monitoring of the patient using the unguided medium is the prime functionality of Healthcare 4.0. Relevant heath-related data of different patients are collected from various IoT sensors [25, 26]. Heterogeneous devices generated humongous amount of data which require in-depth analysis and real-time monitoring. The patient's data are very confidential which can be susceptible by the Internet attacks [27, 28]. Moreover, we need to secure the same for maintaining the privacy of every patient. The analysed data are shared among medical stakeholders such as hospitals, surgeons, physician doctors, patients, and medicals. Hence, secured and reliable communication among the various healthcare stakeholders for various primary decisions is the necessity in the Healthcare 4.0 [29]. The primary decisions include planning of new services in the hospital, doctor recommendation, symptom analysis of various health-related issues, and overall system improvement.

3 BC in Health care

At present, centralized healthcare client–server-based platforms are used to store many healthcare databases. In such platforms, all the permissions and decisions are carried out by single party/administrator. Also, authentication of all the stakeholders in the centralized healthcare system is done by single entity. Single point of failure, central server crash, security, and privacy are some of the common issues in centralized healthcare systems. To counter the challenges of these conventional centralized systems, we can use BC-based solutions.

A BC is an immutable linked list of blocks where data are stored and shared in a tamper-resistant, distributed, secured, and transparent way [30, 31]. If one of the blocks is modified in the chain, it breaks cryptographic links which disrupt the whole BC. Hence, it provides security and maintains transaction record in verifiable manner. It eliminates requirement of trusted parties, and even, the untrusted individual/device can interact in a secured manner. Process integrity, traceability, disintermediation security, automation, immutability, trust, costs, traceability, and faster processing are the advantages of BC. There have been several evolutions of BC. For example, the first generation of BC started with the bitcoin network in 2009 (also known as BC 1.0). In this generation, the key concept was about payment and cryptocurrency. In BC 2.0, smart contract (i.e., Ethereum) and the hyperledger frameworks were

Table 1 Secured Healthcare 4.0 state-of-the-art BC-based techniques

Authors	Years	Objective	1	2	3	4	5	6	7	8	Pros	Cons
Shae et al. [30]	2017	To design BC-based precision medicine decentralized platform	✔	✔	✔	✔	✔	✔	✔	✘	Flexibility in data sharing, transparent, and scalable	Mortality rate is very high
Zhang et al. [42]	2017	To propose attack-free BC-based social healthcare network	✔	✔	✔	✔	✔	✔	✔	✔	Faster settlement	Required more computational power
Xia et al. [40]	2017	To propose fully trusted healthcare data sharing model using BC	✔	✔	✔	✔	✔	✔	✔	✔	Tamper-proof transactions	Lack of scalability and data interoperability
Ri fi et al. [28]	2017	To explore use of BC technology in an eHealth data access	✔	✔	✔	✔	✔	✔	✘	✘	Secure data sharing	Data exchange in humongous quantity, lack of scalability and interoperability
Magyar et al. [23]	2017	To tackle with privacy issues in HER by the integration of BC with HER database	✘	✔	✔	✔	✔	✔	✘	✘	Trust, security, speed, disintegration	Interchangeability and data integrity
AlHadhrami et al. [1]	2017	To understand the feasibility of BC in health care	✔	✔	✔	✔	✔	✔	✘	✘	Easily organized data and consent management	Sybil attacks
Jiang et al. [16]	2018	To propose health data exchange, BC-based system	✔	✔	✘	✔	✘	✔	✔	✔	High-level privacy	Minimum throughput of the system
Theodouli et al. [38]	2018	To design a system for smooth healthcare data sharing	✘	✔	✔	✘	✘	✔	✔	✘	Automation and accountable workflow	Pseudo-anonymity, single point of failure

(continued)

Table 1 (continued)

Authors	Years	Objective	1	2	3	4	5	6	7	8	Pros	Cons
Li et al. [21]	2018	To understand the importance of healthcare-related data and its preventive mechanism	✔	✘	✔	✔	✔	✔	✘	✔	Secured cryptographic solutions	Damage issue during data storage
Fan et al. [13]	2018	To propose efficient and tamper-resistant health data sharing with BC network	✘	✔	✔	✘	✔	✔	✔	✔	Secured maintenance of EMR systems, privacy preservation	Huge computation power
Griggs et al. [14]	2018	To propose BC-based EHR system and smart contracts for monitoring the patients residing at remote places	✘	✔	✔	✔	✔	✔	✔	✘	Real-time patient monitoring	Response time is minimum
Sun et al. [32]	2018	To perform in-depth survey to understand novelty of attribute-based signature for health care using BC	✘	✔	✔	✔	✔	✔	✔	✔	Anonymity and nonrepudiation	Less storage capacity
Nikoloudakis et al. [25]	2019	To provides seamless assessment system for all existing and newly introduced network entities	✔	✘	✔	✔	✔	✘	✔	✔	Testing was done on a large number of nodes	Time duration to do assessment was not focused
Saia et al. [29]	2019	To formalize communication between entities and tracker through BC	✔	✔	✔	✔	✔	✘	✘	✔	Internet of Entities a new paradigm for secure and trusted communication	Enabling all devices currently to shift to IoE would be a challenge

(continued)

Table 1 (continued)

Authors	Years	Objective	1	2	3	4	5	6	7	8	Pros	Cons
Hathaliya et al. [15]	2019	a bio-metric-based authentication algorithm with effective communication and computational cost	✔	✔	✔	✔	✔	✔	✘	✔	Comparison with 3 other algorithm on 15 + attacks	Crypt-analysis not done to prove strength of proposed algorithm
McGhin et al. [24]	2019	Exploring the usefulness of BC as per usability in different domains	✔	✔	✘	✔	✔	✔	✔	✔	Unexplored things many survey papers explained	Real-world evaluation not done
Aloqaily et al. [2]	2019	To design trust-based cluster of vehicles to prevent IDS for communicating vehicles in smart city	✔	✘	✘	✔	✔	✔	✔	✔	Good experimental results with different attributes focused	No testing done for real-world scenario
Khezr et al. [19]	2019	To study BC deployment in health care	✔	✔	✔	✘	✔	✔	✔	✔	A very good comprehensive work	Detailing not done for some domains
Otoum et al. [27]	2019	Evaluation of IDS for WSN-based critical monitoring infrastructure using ML/DL	✔	✔	✔	✘	✔	✔	✔	✔	Implemented multiple algorithms and shown comparative results	Very less results
Bhattacharya et al. [4]	2019	BC integration with deep learning to strengthen HER security	✔	✔	✔	✘	✔	✔	✔	✘	Novel deep learning-based recommender algorithm for patient illness	Experimental evaluation needs to be validated in real-world scenarios
Bodkhe et al.	2020	Decentralized trust building using BC in Healthcare 4.0 applications: A comprehensive survey	✔	✔	✔	✔	✔	✔	✔	✔	A comprehensive survey	–

1—Architecture, 2—Integrity of data, 3—Data exchange, 4—Access control mechanism, 5—Distributed HER, 6—Encryption key for the patient, 7—Simulation tool, 8—Algorithm/pseudo-code introduced [30, 32]. It was only in BC 3.0 that the utility of BC in a healthcare EHR by mobile apps. BC 4.0 is the latest age of BC innovation. It is used in business-usable condition for making and running application in Healthcare 4.0. In summary, aforementioned risks associated with the centralized control system can be eliminated using decentralized systems, such as BC-based systems. BC stores the data and builds the structural data storage which makes the network more tamper-proof.

The security standards are differentiated using various parameters. It includes cost reduction, access control mechanism, security, privacy, and many more. Exchange or sharing of EHR, PHR, or MHR is the necessity in the healthcare sector for dictating the disease symptoms, possible appropriate treatments, medicines, and for various healthcare-related research works. The existing patient's data access control methods are not adequate from the security perspective. In some situations, doctors are not having adequate/enough patients database due to lack of patient's history for the proper treatment. Due to improper maintenance of the patient's data/record, sometime doctors do not receive a proper patient's history even in a medical emergency. These challenges can be resolved by secured maintenance of EHR, PHR, and MHR by using BC.

Figure 3 discusses an idea about the BC-based healthcare ecosystem. Figure 2b presents a solution taxonomy of BC in healthcare applications. The authors of [33] developed BC-based healthcare data gateway (HDG) platform from where the secured sharing of EHR by the doctors and patients can be achieved. The proposed platform has data usage, storage, and data management layer. Data usage consists of users who directly used the data such as pharma companies, doctors, pharmacist, laboratory technicians, and government. Data confidentiality and integrity is maintained by reliable and trust-based storage layer.

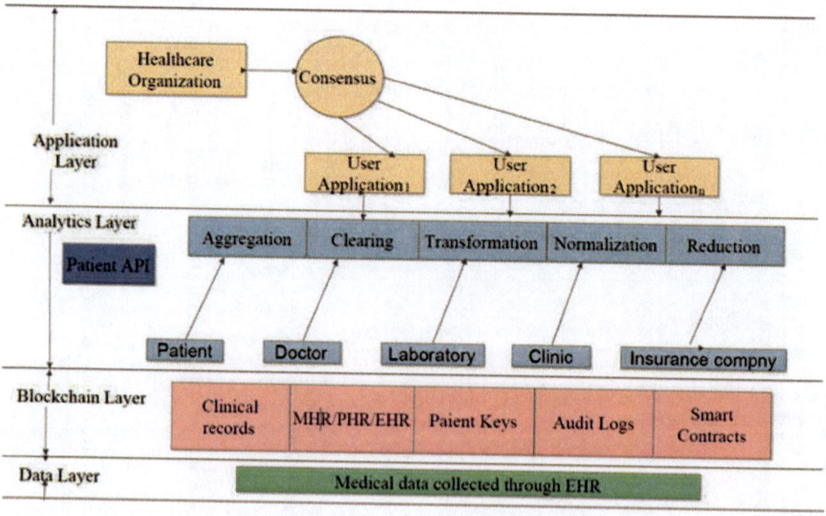

Fig. 3 BC-based healthcare ecosystem

Management of metadata, indexing mechanism, and schema is properly maintained by the data management layer. The authors in [34] developed a BC-based MedRec model to access permission for medical data using BC. The proposed model securely shares the medical data through BC. The authors of [6] designed BC-based precision medicine decentralized platform for clinical trial. It examines the data and maintains the integrity of data and identity privacy.

The authors of [7] proposed attack-free BC-based social healthcare network by using pervasive social network (PSN) using different protocols. Healthcare database is shared among PSN nodes using the extended version of the protocol. The proposed model has a limitation with respect to transport performance. The authors of [35] focused on a BC-based healthcare system, and they also discussed a framework for managing and sharing electronic medical records, especially for the cancer patients. In this way, it will reduce an overall time of sharing EMR and the cost of the system. Xia et al. [8] proposed and developed a system called MedShare which is totally BC based which is used to handle the large amount of medical information on big data. The author focused on data privacy issue, but few issues such as data interoperability, key management, and scalability were not discussed. Rifi et al. [9] focused on the problem such as interoperability and scalability and focused on the advantage of using BC-based technology for sharing medical records to achieve better performance. Liang et al. [36] developed a system, which solves the problem of identity management and privacy issue in sharing health information. Jiang et al. [12] developed a blockchain-based model for exchanging healthcare information, which somehow resolved the above-mentioned issues.

In this technological revolution, BC has been used extensively in research towards the medical field to store, share, and maintain health records securely. Theodouli et al. [13] focused on the use of this data for further research and innovation in the medical healthcare system as well as industry. By exploring the needs of the medical healthcare system, the authors of [13] presented an architectural design of a system which ensures permission and management of healthcare data with the help of BC. The three-layer system consists of platform, middleware, and BC network layer, where the first two layer is purely cloud based, and the combination of these three layers enhances the integrity and security of medical health records. It is also used to verify the exchange and interoperability of the data in the system which also gave the workflow automation, auditing, accountability, and data integrity as the additional benefit. In [19], the nodes of infrastructure which are connected wireless-based are used to communicate, and this communication without any measure is insecure. The author here proposed a BC-based approach to secure this communication and also setup a trust among them. This could also prevent communication with any unidentified entity and compromise the confidentiality. The authors in [37] emphasize on edge computing in the same scenario that reduces latency compared to cloud-based infrastructure. They proposed BC-based approach for securing this edge computing on data produced by mobile nodes, i.e. mobile edge computing. The framework has good amount of security and trust as it is based on BC, and proof-of-work was also given for the same.

The authors in [22, 24] focused on finding intrusion detection in the wireless-based systems like smart cities. They observed this approach could be possible approach for attack at hospitals were data collected of patients are shared through wireless setup in the premise between laboratory, doctor, reception and patient present in premise. The authors in [21] have surveyed on BC use in Healthcare 4.0 beyond just authentication. The focus was specifically to find use of BC in fraud detection, medical incentives, scalability, and standardization process of BC in health care. The paper also focused on its application use like GEM network, OmniPHR, MedRec, and PSN. The authors of [3, 5, 38, 39] have explored various dimensions of usability of BC and provided its validations. They have majorly focus on financial transactions happening and securing them with BC, while [3] provided it in tourism along with hospitality. They have also derived the equations of mining incentives and its distribution to miners who contribute in maintaining BC infrastructure by providing computational power. Table 2 provides a detailed overview of various approaches to integrate BC in Healthcare 4.0 applications using different parameters.

4 Open Issues and Challenges

As discussed in the above sections, BC can leverage decentralized trust among healthcare stakeholders and add chronology to health records. This adds accountability for all users in the Healthcare 4.0 ecosystem. Thus, healthcare sector has employed potential use cases of analysing the benefits of BC-leveraged health systems, but still the large-scale deployments and mainstream usage are far from reality. The challenges of adoption of BC in Healthcare 4.0 are due to lack of technical professionals, legal issues of cryptocurrency adoptions and lack of global conversion standards, trained personals for business promotions, and overall trust of healthcare stakeholders over complete adoption [42]. Due to this, despite numerous advantages, BC-envisioned healthcare systems are not deployed at practical fronts. Table 3 provides a detailed overview of different research challenges and possible countermeasures by BC technology in Healthcare 4.0.

5 Conclusion and Future Scope

Healthcare 4.0 is rapidly evolving towards deployment of structured decentralized, automated and secure solutions for storage of patient health records. In a similar direction, trust and immutability of stored EHR is a prime concern. Integration of BC in Healthcare 4.0 can leverage efficient use cases through chronology and trust in stored transactions. It also provides automation of payments via smart contracts without third-party intermediaries. The survey highlighted the possible key features and functionalities of BC-based EHR storage and presented the challenges in the main-stream adoption of BC in healthcare ecosystems. The survey also provided

Table 2 Approaches to integrate BC in Healthcare 4.0 applications

Authors	Years		Objective	1	2	3	4	5	6	7	8	Simulator/dataset/methodology
[3]	2017		To access permission for medical data using BC	✘	✔	✔	✔	✔	✔	✔	✘	Ethereum
[12, 28, 30]	2017		To propose BC-based framework for the clinical trials and precision medicine	✔	✔	✔	✔	✔	✔	✔	✘	Neo
[1, 22, 23]	2017		Data sharing BC-based mobile health usage system/framework	✔	✔	✔	✔	✔	✔	✘	✘	EthereumM
[38]	2018		To design tamper-proof and attack-resistant healthcare data sharing system	✔	✘	✔	✔	✔	✔	✘	✘	Lisk
[16, 21]	2018		To propose BC-based framework for health care	✔	✘	✔	✔	✔	✔	✔	✘	Ark
[18]	2018		To discuss security aspects of BC-based heterogeneous HER in the cloud	✔	✔	✔	✔	✔	✔	✘	✔	Eos
[14, 39]	2018		To design smart contracts to secure remote patient monitoring	✘	✔	✔	✔	✔	✔	✔	✘	Ethereum
[32]	2018		To understand the importance of attribute-based signature and BC for healthcare	✘	✔	✔	✔	✔	✔	✔	✔	Lisk
[15, 25, 29]	2019		Ethereum	✘	✔	✔	✔	✔	✔	✔	✘	Stratis
[24]	2019		Exploring the usefulness of BC as per usability in different domains	✘	✔	✘	✔	✔	✔	✘	✔	Wainchain

(continued)

Table 2 (continued)

Authors	Years		Objective	1	2	3	4	5	6	7	8	Simulator/dataset/methodology
[2]	2019		To design trust-based cluster of vehicles to prevent IDS for communicating vehicles in smart city	✔	✔	✔	✔	✔	✘	✔	✔	Waves
[19]	2019		To study BC deployment in health care	✔	✔	✔	✘	✔	✔	✔	✔	QTUM
[4]	2019		BC integration with deep learning to strengthen HER security	✔	✔	✘	✔	✔	✔	✔	✔	CordaDApp

1—Architecture, 2—Integrity of data, 3—Data exchange, 4—Access control mechanism, 5—Distributed HER, 6—Encryption key for the patient, 7—Simulation tool, 8—Algorithm/pseudo-code

Table 3 Open challenges in Healthcare 4.0

Parameters	Challenges	Implications	BC-leveraged solutions
Master patient indices	Data redundancy and complexity in HER schema, resulting in inconsistent normalization	Inconsistency in healthcare records, different EHR schema for every field	Efficient hashed data structure with multi-keys linked to single patient identifier that results in efficient search and access mechanisms
EHR management	Access to EHR records to untrusted third-party stakeholders	Security and privacy concerns as malicious intruder may tamper HER records	Valid health blocks added post-authorization of EHR by patient, and access is provided to only authorized stakeholders via proper credentials
Data integrity	Complex interrelationships among stored data that results in less responsive applications	Difficulty in creation of efficient trained healthcare analytics and responsive queries	Chronological entries, simplification of timestamped ledgers, fetch, and query of EHR records are persistent
Clinical trials	Lack of global standards in EHR records and solutions are proprietary	Unstructured and different formats with interconversion issues that to the overall conversion complexity	Data transaction through structured formats like JavaScript object notation (JSON) that allows lightweight exchange and is compatible with BC
Drug traceability	Drug counterfeiting	Low customer satisfaction	Chronological, immutable and timestamped ledger that forms auditability in added transactions
Data enrichment	Accurate and understandable formats of stored EHR	High latency in processing of EHR data	Single patient identity for storage of patient EHR in BC, with multiple associations of healthcare stakeholders mapped to the same identifier
Security HER access	Lack of cryptographic primitives	Challenges of data privacy	Efficient public and certificateless cryptographical primitives can be embedded with the chain structure

useful insights to the readers about the importance of BC as a facilitator to ensure trust and validation in health care. The survey intends to serve as a guideline to medical stakeholders, industry personals, and researchers in similar domain. As part of future scope, deeper insight of efficient consensus mechanisms and efficient EHR storage patterns needs to be explored for challenges for uniformity and efficient access.

References

1. Bodkhe U, Tanwar S, Parekh K, Khanpara P, Tyagi S, Kumar N, Alazab M (2020) Blockchain for industry 4.0: a comprehensive review. IEEE Access 8:79764–87980. https://doi.org/10.1109/ACCESS.2020.2988579
2. Onik MMH, Aich S, Yang J, Kim CS, Kim HC () Blockchain in healthcare: challenges and solutions. In: Big data analytics for intelligent healthcare management. Elsevier, pp 197–226
3. Bodkhe U, Bhattacharya P, Tanwar S, Tyagi S, Kumar N, Obaidat MS (2019) Blohost: blockchain enabled smart tourism and hospitality management. In: 2019 International conference on computer, information and telecommunication systems (CITS), pp 1–5
4. Bodkhe U, Tanwar S (2020) A taxonomy of secure data dissemination techniques for iot environment. IET Softw 1–12 (2020). https://doi.org/10.1002/SEN-2020-0006
5. Bhattacharya P, Tanwar S, Bodke U, Tyagi S, Kumar N (2019) Bindaas: blockchain-based deep-learning as-a-service in healthcare 4.0 applications. IEEE Trans Netw Sci Eng 1–1. https://doi.org/10.1109/TNSE.2019.2961932
6. Shae Z, Tsai JJP (2017) On the design of a blockchain platform for clinical trial and precision medicine. In: 2017 IEEE 37th international conference on distributed computing systems (ICDCS), pp 1972–1980
7. Zhang J, Xue N, Huang X (2016) A secure system for pervasive social network-based healthcare. IEEE Access 4:9239–9250
8. Xia Q, Sifah E, Omono Asamoah K, Gao J, Du X, Guizani M (2017) Medshare: trust-less medical data sharing among cloud service providers via blockchain. IEEE Access 1–1
9. Rifi N, Rachkidi E, Agoulmine N, Taher NC (2017) Towards using blockchain technology for ehealth data access management. In: 2017 fourth international conference on advances in biomedical engineering (ICABME), pp 1–4
10. Magyar G (2017) Blockchain: Solving the privacy and research availability tradeoff for EHR data: a new disruptive technology in health data management. In: 2017 IEEE 30th Neumann Colloquium (NC), pp 000135–000140
11. Alhadhrami Z, Alghfeli S, Alghfeli M, Abedlla JA, Shuaib K (2017) Introducing blockchains for healthcare. In: 2017 international conference on electrical and computing technologies and applications (ICECTA), pp 1–4 (2017)
12. Jiang S, Cao J, Wu H, Yang Y, Ma M, He J (2018) Blochie: a blockchain-based platform for healthcare information exchange. In: 2018 IEEE international conference on smart computing (SMARTCOMP), Taormina, Italy, pp 49–56. https://doi.org/10.1109/SMARTCOMP.2018.00073
13. Theodouli A, Arakliotis S, Moschou K, Votis K, Tzovaras D (2018) On the design of a blockchain-based system to facilitate healthcare data sharing, pp 1374–1379
14. Li H, Zhu L, Shen M, Gao F, Tao X, Liu S (2018) Blockchain-based data preservation system for medical data. J Med Syst 42:1–13
15. Fan K, Wang S, Ren Y, Li H, Yang Y (2018) Medblock: efficient and secure medical data sharing via blockchain. J Med Syst 42
16. Griggs K, Ossipova O, Kohlios C, Baccarini A, Howson E, Hayajneh T (2018) Healthcare blockchain system using smart contracts for secure automated remote patient monitoring. J Med Syst 42

17. Sun Y, Zhang R, Wang X, Gao K, Liu L (2018) A decentralizing attribute-based signature for healthcare blockchain. pp 1–9
18. Nikoloudakis Y, Pallis E, Mastorakis G, Mavromoustakis CX, Skianis C, Markakis EK (2019) Vulnerability assessment as a service for fog-centric ICT ecosystems: a healthcare use case. Peer-To-Peer Netw Appl 12(5):1216–1224
19. Singh PK, Panigrahi BK, Suryadevara NK, Sharma SK, Singh AK (eds) Proceedings of ICETIT 2019, emerging trends in information technology, lecture notes in electrical engineering (LNEE), Springer Nature Switzerland AG. https://doi.org/10.1007/978-3-030-30577-2
20. Hathaliya JJ, Tanwar S, Tyagi S, Kumar N (2019) Securing electronics healthcare records in healthcare 4.0: a biometric-based approach. Comput Electr Eng 76, 398–410
21. McGhin T, Choo KKR, Liu CZ, He D (2019) Blockchain in healthcare applications: research challenges and opportunities. J Netw Comput Appl 135:62–75
22. Aloqaily M, Otoum S, Ridhawi IA, Jararweh Y (2019) An intrusion detection system for connected vehicles in smart cities. Ad Hoc Netw 90:101842
23. Khezr S, Moniruzzaman M, Yassine A, Benlamri R (2019) Blockchain technology in healthcare: a comprehensive review and directions for future research. Appl Sci 9(9)
24. Otoum S, Kantarci B, Mouftah HT (2019) On the feasibility of deep learning in sensor network intrusion detection. IEEE Netw Lett 1(2):68–71
25. Ladha A, Bhattacharya P, Chaubey N, Bodkhe U (2020) Iigpts: Iot-based framework for intelligent green public transportation system. In: Singh PK, Paw lowski W, Kumar N, Tanwar S, Rodrigues JJPC, Obaidat MS (eds) Proceedings of first international conference on computing, communications, and cybersecurity (IC4S 2019), vol 121. Springer International Publishing, Berlin, pp 183–195
26. Tanwar S, Agarwal B, Goyal L, Mittal M (2019) Energy conservation for IoT devices: concepts, paradigms and solutions. Springer, Berlin
27. Bodkhe U, Tanwar S, Shah P, Chaklasiya J, Vora M (2020) Markov model for password attack prevention. In: Singh PK, Paw lowski W, Kumar N, Tanwar S, Rodrigues JJPC, Obaidat MS (eds) Proceedings of first international conference on computing, communications, and cyber-security (IC4S 2019), vol 121. Springer International Publishing, pp 831–843
28. Singh PK, Kar AK, Singh Y, Kolekar MH, Tanwar S (2019) Recent innovations in computing, vol 597. Springer Nature, Switzerland AG
29. Tanwar S (2020) Fog computing for Healthcare 4.0 environments. Springer, Berlin
30. Bodkhe U, Tanwar S, Bhattacharya P, Kumar N (2020) Blockchain for precision irrigation: opportunities and challenges. Trans Emerg Telecommun Technol 1–35. https://doi.org/10.1002/ett.4059
31. Bodkhe U, Tanwar S (2020) Secure data dissemination techniques for iot applications: Research challenges and opportunities. Softw Pract Experience 1–23. https://doi.org/10.1002/spe.2811
32. Tanwar S, Singh P, Kar A, Singh Y, Kolekar M (2020) Proceedings of ICRIC 2019-recent innovations in computing. Springer, Berlin
33. Yue X, Wang H, Jin D, Li M, Jiang W (2016) Healthcare data gateways: found healthcare intelligence on blockchain with novel privacy risk control. J Med Syst 40:218
34. Azaria A, Ekblaw A, Vieira T, Lippman A (2016) Medrec: using blockchain for medical data access and permission management. In: 2016 2nd international conference on open and big data (OBD), pp 25–30
35. Dubovitskaya, A, Xu Z, Ryu S, Schumacher M, Wang F (2017) How blockchain could empower eHealth: an application for radiation oncology, pp 3–6
36. Liang X, Zhao J, Shetty S, Liu J, Li D (2017) Integrating blockchain for data sharing and collaboration in mobile healthcare applications. In: 2017 IEEE 28th annual international symposium on personal, indoor, and mobile radio communications (PIMRC), pp 1–5
37. Bhattacharya P, Tanwar S, Shah R, Ladha A (2020) Mobile edge computing-enabled blockchain framework-survey. In: Singh PK, Kar AK, Singh Y, Kolekar MH, Tanwar S (eds) Proceedings of ICRIC 2019. Springer International Publishing, Cham, pp 797–809
38. Kabra A, Bhattacharya P, Tanwar S, Tyagi S (2020) Mudrachain: blockchain-based framework for automated cheque clearance in financial institutions. Future Gener Comput Syst 102:574–587. ISSN 0167-739X. https://doi.org/10.1016/j.future.2019.08.035

39. Srivastava A, Bhattacharya P, Singh A, Mathur A, Prakash O, Pradhan R (2018) A distributed credit transfer educational framework based on blockchain. In: 2018 second international conference on advances in computing, control and communication technology (IAC3T), Allahabad, India, pp 54–59. IEEE
40. Kaur H, Alam A, Jameel R, Mourya A, Chang V (2018) A proposed solution and future direction for blockchain-based heterogeneous medicare data in cloud environment. J Med Syst 42:156
41. Uddin MA, Stranieri A, Gondal I, Balasubramanian V (2018) Continuous patient monitoring with a patient centric agent: a block architecture. IEEE Access 6:32700–32726
42. Tanwar S, Tyagi S, Kumar N (2020) Security and privacy of electronics healthcare records. IET, UK

The Amalgamation of Blockchain and IoT: A Survey

Jignasha Dalal

Abstract Blockchain, the form of Distributed Ledger Technology, is getting momentum. Data in the blockchain is in the form of transactions. Blockchains are transparent, immutable and auditable distributed ledgers with peer to peer connection, cryptography and consensus algorithm. All the parties involved have the same copy of data (transparency), data cannot be modified or deleted (immutability) and a full history of transactions is available (auditable). Based on history, future transactions are validated by all the parties (consensus). Internet of Things means connecting lightweight devices to share information among themselves and to enable some functionalities based on the shared information. This exchange of information among multiple devices must be secure as it contains sensitive and safety-critical data. Because of the scale and distributed nature of IoT, security and privacy are major concerns. Traditional security solutions are not suitable for IoT devices as they have limited resources in terms of memory and processing power. Blockchain can bridge the gap. The information exchange among these devices can be stored on the blockchain to increase transparency among the devices. This paper provides a review of literature in the area of IoT and blockchain.

Keywords IoT · Blockchain · Ethereum · Security · IOTA · Tangle

1 Introduction

A blockchain is a distributed, peer to peer, immutable, append-only ledger where all the transactions are visible to all the parties. Transactions can be in the form of currency, land records, patient records, Personal Identifiable Information (PII) and so on. The transactions are grouped into blocks and blocks are cryptographically linked. These chains of blocks grow continuously as the number of transactions grows. The Blockchain is stored on all the nodes and all nodes required to validate

J. Dalal (✉)
K. J. Somaiya Institute of Engineering and Information Technology, Mumbai University, Mumbai, India
e-mail: jignasha@somaiya.edu

P. K. Singh et al. (eds.), *Proceedings of Second International Conference on Computing, Communications, and Cyber-Security*, Lecture Notes in Networks and Systems 203,
https://doi.org/10.1007/978-981-16-0733-2_55

Table 1 Distributed database versus distributed ledger

Distributed database	Distributed ledger
Data can be deleted or modified by authorised parties	Data cannot be deleted, modified by any of the parties involved
Data can be added anywhere by the authorised party	Data can be appended only after verification by all the parties
Client–server architecture	Peer to peer architecture

the transactions so that they can agree on a single global state. This is done through a Consensus algorithm. Blockchains also provide a way to store logic. This is done through Smart Contracts. Smart contracts are the executable program codes which are triggered by some events. They are similar to traditional contracts in the real world. There are three kinds of blockchains: (i) Public or Permissionless, where anyone can join the blockchain, (ii) Consortium or permissioned, where only authorised parties can join the blockchain and (iii) Private with centralized authority. The difference between Distributed ledger and Distributed database is shown in Table 1.

IoT is the seamless interconnection of heterogeneous devices to provide services in the sectors of social media, businesses, intelligent transports and smart cities. These connected devices generate a huge amount of data traffic. The sensitivity of the data depends on the application where IoT is used. The security of this data while in storage and in the transmission is very important. Blockchain provides immutability, transparency, auditability and decentralization. The security of the shared data among the devices can be significantly improved using Blockchain. Ferrag et al. [1] provided a detailed overview of Blockchain usage in various application areas of IoT. They have also shown when to use Blockchain model for IoT applications. When there is a peer to peer communication among the devices, synchronization is needed among the devices and no centralized server, then Blockchain technology can be incorporated with an IoT application. They also classified the IoT applications in different domains like the Internet of Vehicles, access management, Internet of cloud, SDN, Edge Computing, Distributed P2P applications, Data Storage, etc. and provided the overview of existing applications with Blockchain in these domains. They analysed vulnerabilities in Blockchain and how those vulnerabilities are applicable in Blockchain-IoT applications. They provided a detailed analysis of security and privacy for Blockchain-based IoT applications.

In the existing IoT applications, most devices are connected through the central server. The central server becomes a single point of failure. Blockchain can bring decentralization and also can help in transferring the value of massive IoT data. Yu et al. [2] suggested a 3-layer distributed network architecture with PBFT-DPOC consensus algorithm. The Network architecture consists of DAPP layer, Blockchain layer and the intelligent device layer. To reduce the pressure on devices with limited resources they have used DPOC (Delegated Proof of Contribution) consensus algorithm. This algorithm shows better throughput than consensus algorithm used in

Bitcoin and Ethereum. Lo et al. [3] did a systematic literature review for incorporating Blockchain in IoT applications. They identified some challenges in the IoT domain. The challenges are:

- Lack of standards
- Limited bandwidth and computation capability
- The integrity of states of devices
- Interoperability among devices.

They analyzed different solutions addressing the above challenges and found some interesting insights:

- Most of the solutions were tested in the test net.
- During the testing, very few devices were connected.
- Some tests have used Ethereum test nets which are using Proof of Authority consensus algorithm.
- Public Blockchain like Ethereum is not suitable for IoT as latency is very high, throughput is less and also transaction fees are involved.
- Permissioned Blockchains with PBFT are more suitable for IoT.
- The performance should be analysed for the whole process starting from transaction submission to transaction confirmation.
- IOTA means "small". It is specifically designed for incorporating Blockchain concept in IoT applications. But it suffers from 33% attack. Bitcoin and Ethereum suffer from 51% attack.

Casino et al. [4] presented an in-depth analysis of applications using Blockchain technology. The IoT is a domain where the use of Blockchain can bring a revolution. For Blockchain to be used with IoT applications to make it more transparent and secure, Blockchains need to be scalable, privacy-preserving and must have low latency. Various consensus algorithms are developed to increase scalability. For IoT applications, it is also needed to reduce resource consumption. Kumar and Jain [5] have proposed a PoG (Proof of Game) consensus algorithm. Multi-round and Multi-bit challenges are more suitable for the devices with limited resources to confirm the Block in stipulated time.

The contributions of this paper are as follows:

1. This paper reviews different IoT domains where Blockchain can be incorporated. It also discusses the challenges associated with the implementation of Blockchain in these domains.
2. The paper also reviews the IOTA (means small), a DAG-based, fee less Blockchain architecture and how it is suitable for incorporating with IoT devices.

The paper is organized as follows: Sect. 2 reviews the research work in different IoT domains with Blockchain. The domains are Access Control, Blockchain, IoT and Machine learning, Data storage and sharing, Health, Supply Chain and VANETs. Section 3 provides Analysis and Summary of the literature review and Sect. 4 concludes the paper.

2 IoT Domains

2.1 Access Control

In IoT, devices need to share their resources with other IoT devices. They need to have the local policy which defines who can access their resources. But the devices havc limited storage to define these access management policies. If the devices are static, they can be managed with a centralized server where access policies are stored and defined. In the case of dynamic IoT scenario, where IoT devices are mobile and joining and leaving any time, access management cannot be done with a centralized server. Novo [6] has suggested an access management system with Blockchain. In his system, IoT devices do not write any information on the blockchain, so high latency problem with Blockchain is not present here. Instead, IoT devices read access control information from the Blockchain. Read operation from Blockchain system is inexpensive as it does not require consensus. The set of IoT devices need to register with a manager who is responsible for interacting with Blockchain. The manager can be a single point of failure. The Ethereum Blockchain with a single smart contract is used for defining access control policies. The agent node is the node on the blockchain. A service provider can act as an agent node and the owner of IoT devices can act as a manager node. Then the system has been compared with existing centralized IoT access management systems. The proposed system provides good scalability.

IoT devices are needed to share data with external parties. These external parties must have proper authorisation to access the data. Traditional access control schemes do not work in constrained IoT environment. Attribute-based access control is more appropriate for IoT [7–10]. Fully decentralization systems are not possible, even with Blockchain. Permissioned Blockchain requires all the participants to be authenticated before joining the network. In the dynamic IoT environment, when IoT devices are joining and leaving the network any time, they need to get authenticated each time they connect to the network. The authentication of the participant is done by some central authority.

2.2 Blockchain, IoT and Machine Learning

Combination of Blockchain, IoT and Machine learning can be a great solution to Industrial IoT. Liu et al. [11] suggested the use of Deep Reinforcement Learning (DRL) and private Ethereum blockchain for Industrial IoT data storage and sharing. The smart portable mobile terminals (MTs) include smartphones, UAVs with sensors, cameras, gyroscope and GPS. The system is suggested for Industrial IoT environment like a manufacturing plant. The deep learning technology is used for efficient collection of data and Blockchain is used for securely storing and sharing the data. The MTs move around the plant and collect data and they submit this data to the

Blockchain node in the form of transactions. To prevent the malicious behaviour of MTs, a Certificate Authority (CA) is used to verify the authenticity of MTs and data submitted by them. The system is tested against malicious behaviour of MTs, Eclipse attack and majority attacks. The system provides better security than a traditional centralized database.

The system is not fully decentralized as it is using the CA. the private Ethereum network is set up, but the roles of a network user are not specified. DRL is a combination of traditional reinforcement learning and deep learning. Smart cities are using IoT in the areas of transportation, energy distribution, security, manufacturing and agriculture. These IoT systems generate a huge amount of data and processing this huge amount of data efficiently and fruitfully is a big challenge. Machine learning can help with the classification of data. One of the classification methods is SVM (Support Vector Machine). This classifier needs training data for the classification. As the size of the training data grows, the accuracy of classification increases. It is difficult to get such a huge training data from one entity. The solution is to combine the data from multiple entities. The entities are reluctant to share data because of privacy (in health care), ownership and trust. Shen et al. [12] suggested SVM based training on Blockchain-based encrypted data. IoT data providers encrypt the data collected from the IoT devices and store it on the Blockchain. The analyst process and analyse encrypted data. The Paillier cryptographic system is used for this purpose. It is an additive homomorphic encryption technique which works on encrypted data. The security and efficiency of the system were evaluated using different experiments. The system can be extended for other classification models.

The combination of IoT, Blockchain and ML makes it possible to analyse and perform an audit of the IoT data.

2.3 Data Storage and Sharing

Zhou et al. [13] suggested a Blockchain-based threshold IoT system Beekeeper. The system consists of servers, devices and a leader. Devices send encrypted data to the server,the server will perform homomorphic computations on encrypted data and sends back the results. The leader will have the decryption key and will be able to decrypt the result. All the communication among the participants is done through Blockchain. The record nodes manage the Blockchain. The beekeeper was tested on Ethereum Blockchain. In this system, IoT devices need to perform cryptographic computations and it is not feasible. So, Edge Computing/fog computing is introduced to reduce the computational load on IoT devices [14–18]. Hyperledger Fabric and Ethereum were used as Blockchain platform. Edge devices are placed in the same network as IoT devices. They help IoT devices to perform cryptographic computations.

Wang et al. [16] proposed a hierarchical Blockchain architecture called CHAIN SPLITTER. The old blocks are stored on the cloud and only a few recent blocks are stored on the IoT devices. A centralized database is used in the local IoT network to store all the data in the local network.

Data storage and sharing of IoT data on Blockchain can provide transparency, immutability and auditability. It also provides decentralization, so there is no single point of failure. Full Decentralization is not possible, somewhere you will have some kind of central authority to verify the authenticity of the devices and validity of the data.

IoT devices generate a huge amount of data at very high speed. Cloud computing is the solution to process this large amount of data as IoT devices have limited storage and computational abilities. Transmitting such large data to centralized cloud servers at very high speed through the Internet is expensive and not secure. Sharma et al. [19] proposed a software-defined fog node based distributed cloud architecture to solve the cost and security issue with centralized cloud-based architecture. The fog node is a collection of SDN controllers with Blockchain. This architecture ensures high security, real-time deliveries, High scalability and low latency. The performance of the system is better than the traditional centralized cloud architecture.

Pan et al. [20] suggested an architecture EDGECHAIN which combines IoT, edge computing and Blockchain. A permissioned Blockchain is used for resource allocation to IoT devices from the Edge computing resources. All the IoT devices transactions and activities are stored on the Blockchain, which will help to analyse and audit the behaviour of IoT devices. Edge chain is placed between centralized cloud services and IoT devices. It does resource management for IoT devices as they are resource-constrained. Edge chain processes the massive data generated from IoT devices and only processed output is sent to the cloud servers.

Peña and Fernández [21] suggested an IoT architecture SAT-IOT consisting of different entities to manage data flow from IoT devices. In the case of dynamic IoT environment, the topology management entity will manage the topology of the IoT devices. The IoT visualization entity is incorporated to get the visual of the IoT system. The concept "Edge-Cloud Computing Location transparency" lets computation nodes, in an IoT network topology change dynamically (without administrator intervention) to fulfil the efficiency criteria defined for the IoT system. The "IoT Computing Topology Management" concept integrates the hybrid networks (cloud, edge, devices and their wireless or wired links) as part of the IoT Platforms. This gives an IoT system global view, from the hardware and communication infrastructures to the software deployed on them. The Embedded IoT Visualization System concept offers a mechanism to check the deployment of the new IoT system in the platform.

Guo et al. [22] have suggested a Blockchain-based Edge computing system to improve the efficiency of authentication in the IoT system. An optimized PBFT algorithm is designed and used in the Blockchain. A distributed authentication system based on name resolution strategy and Elliptic curve cryptography is used. To reduce delay a caching mechanism is introduced. The 3-layer architecture consists of the Physical layer, Blockchain edge layer and Blockchain node layer. In optimized PBFT algorithm, there is one speaker and other peers are congressmen. Speaker runs the consensus process for other peers. Each round a speaker is selected by the peers.

Speaker sends pre-prepare messages to the congressmen and if congressmen agree, they send prepare messages to the speaker. If the speaker receives prepare messages from $2f + 1$ peer, where $f = \mid (N - 1) /3\mid$, the speaker sends commit messages to all the peers. When it receives the response from the $f + 1$ peer, the consensus is achieved. Caching strategy minimizes the download latency.

Lei et al. [23] have proposed GROUP CHAIN—a 2 level chain to enhance the performance of Blockchain in IoT architecture. There is a group chain and vice chain. The group chain contains the leaders who can generate vice blocks without PoW. Miners compete for membership in the leader group using PoW. Once the leaders are selected, they can generate any number of vice blocks without PoW, which is added to the Vice chain after getting a signature from all the leaders. This enhances the Bitcoin mining mechanism. The leader is required to deposit some amount, if it behaves honestly during an inspection period, the deposit will be refunded. This will prevent the leader from signing invalid transactions and creating a denial-of-service attack by generating vice blocks continuously.

IoT devices are resource-constrained. They generate massive data which need to be processed and analysed. The processing and transmission of data need an infrastructure along with IoT devices. This infrastructure can be a centralized cloud computing architecture, where the data from IoT devices are stored and processed. But transmitting data to the centralized data servers through the internet is expensive and makes data vulnerable. The data generated by IoT devices contain sensitive information most of the time. So, the concept of fog computing/edge computing is being used where the processing and computation devices are at the edge of the IoT devices' network. These edge devices form an edge network and they facilitate IoT devices with authentication, access control, information sharing, and passing the processed information or computation to cloud servers. It is possible to use Blockchain in edge computing layer and even in cloud computing layer to make the processing, computing and sharing of the data more transparently and securely. Most of the research has used permissioned Blockchain. This combination of edge computing and blockchain with IoT can help in developing many real-life applications for smart cities and industries.

2.4 Health

Kumar et al. [24, 25] proposed a smart health care system based on Ethereum Blockchain. The authors have given a comprehensive review of existing healthcare systems with blockchain. There are separate Blockchains for Doctor, supplier, patient, hospital, staff and insurance. These Blockchains are integrated into one smart healthcare blockchain. It is a full-fledged and comprehensive healthcare system. Kumar et al. [24, 25]] have proposed three PoG based consensus algorithms for wearable kidney devices. The authors also reviewed the existing consensus algorithms and their limitations for using them in implementation of IoT based application with Blockchain.

Xu et al. [26] have suggested HEALTHCHAIN, a Blockchain-based healthcare data privacy-preserving scheme. In the scheme, the patient medical data is encrypted. Patients have the right to revoke or grant access to their medical data. It is not advisable to store the full patient's health data on the Blockchain. So, the patients' health data is stored on the Interplanetary File System (IPFS) in encrypted form. It does not have a central server. The file is stored on different peers in parts. A unique hash string is associated with each file. This hash is stored on the Blockchain. There are two chains involved, Userchain and Doc chain. Userchain is a public Blockchain and Docchain is consortium Blockchain. Encryption keys and encrypted data are separated to achieve flexibility in key management.

Miners in the Blockchain need continuous power for mining. Mobile IoT devices request microgrids to supply power to them. Miners also can request nearby MECs to compute the hash for the mining. These MECS need the power to compute the hash on behalf of miners. Microgrids can provide efficient energy allocation. Li et al. [14, 15] proposed a microgrid based energy supply system for powering IoT mobile devices for mining computation. The microgrids provide real-time scheduling and decision making based on the energy consumption of miner. The energy allocation is formed as a Stackelberg game to optimize the profit for microgrids and cost-effectiveness for miners.

2.5 Supply Chain

There are multiple participants involved in supply chain management. Blockchain can bring transparency and create trust among these participants. Tsang et al. [27] have proposed a Blockchain-based IoT system with fuzzy logic to ensure the Traceability and quality of food by assessing the shelf life of perishable food. The system combined cloud technology and blockchain technology and fuzzy logic. They are using the concept of Blockchain vaporization to increase the efficiency of the system. For a particular batch of food, batch id, container id and IoT id are stored on the Blockchain. One the batch of food reaches the endpoint or deal is completed, this data will be removed and stored on the cloud for future reference. The data from the blockchain is used as input to the fuzzy evaluation technique for quality assessment.

2.6 VANETS (Vehicular Adhoc Networks)

The modern transportation system is intelligent as it incorporates IoT devices with internet connectivity. 5G technology will reduce latency and increase throughput. SDN (Software Defined Network) simplifies the management of IoT devices in the vehicle and also of VANETS. Security and privacy are very important for VANETS. Incorporating Blockchain in VANET will help to achieve the security and privacy in VANET. Xie et al. [28] have proposed a 5G-SDN enabled Blockchain base system for

VANETS. All vehicles and base stations are involved in maintaining the Blockchain. Each vehicle is assigned an ID. The vehicle is required to collect the videos and images of road conditions and broadcast it to other vehicles and base stations. This data will help the SDN controller to monitor the position of the vehicle and also helps in traffic management. SDN controller is the centralized system which is responsible for all kinds of policies. The message sharing among the vehicles is maintained on the Blockchain to detect malicious nodes. It is not possible to avoid malicious nodes. To detect them is the only solution. In case of an accident, the transactions on the Blockchain can be checked and originated vehicle ID can be identified. Vehicle Id does not reveal any information about the vehicle owner. The mapping between Vehicle ID and Vehicle Number is stored in the DMV database.

Zhang et al. [29] have proposed architecture of VANET using blockchain and Mobile Edge Computing (MEC). It has three layers-Service layer, Perception layer and Edge computing layer. The blockchain is used in Perception layer and Service layer to ensure the security of data transmission and security of data respectively. All vehicles will run wallets and Perception layer will perform the tasks of blockchain. All vehicles will be able to communicate with the blockchain in Perception layer.

3 Analysis and Summary

The paper has provided a detailed literature review of recent research in IoT and Blockchain. The Summary of IoT domains with Blockchain Technology is shown in Table 2. The major challenges in incorporating IoT systems in daily life and industry are:

(i) Massive data generation by IoT.
(ii) Storage and security of the data.
(iii) Configuration and management of IoT data.
(iv) Detection of malicious Behaviour.
(v) Defining and applying Access control policies for inter-device communication and external request.

Incorporation of Blockchain with IoT has great advantages. It can provide decentralization, improve the performance and increase the security of IoT system. Massive data storage problem can be solved by using cloud storage, but centralized cloud storage is inefficient in terms of transmission, processing of data and real-time delivery of data. It is also a single point of failure. Edge computing along with cloud storage is a better solution. Blockchain can be incorporated in the Edge computing layer to record the data exchange between cloud and IoT devices as well as the sharing of the data with other devices or external entity.

It is seen from Table 2 that a High throughput and low latency Blockchain platform is needed to make the real implementation of IoT with Blockchain. All proposed systems are developed on Ethereum and Hyperledger Fabric. Hyperledger Fabric

Table 2 IoT Domains and Proposed Systems using Blockchain

IoT domain	Requirements	Proposed systems
Access control	Low read latency	ACL is stored on the blockchain, the device only needs to read ACL
Blockchain, IoT and machine learning	Data storage, high throughput, low write latency	Blockchains are not designed to store data. Use ethereum and hyperledger fabric
Data storage and sharing	Low read latency, high throughput	Edge devices are used to reduce computational load on IoT devices
Health	Data storage, high throughput, Low write latency, scalability	PoG based consensus algorithms are used. Multibit and Multi round challenges are more suitable for IoT devices. IPFS is used for storing private data
Supply chain	Low read and write latency, high throughput, Scalability	Hyperledger fabric and ethereum are used
VANETS	Low read and write latency, High throughput, scalability, data storage	Involves real-time transactions

uses PBFT (Practical Byzantine Fault Tolerance) consensus algorithm, which is not scalable. For VANETs, a high throughput and low latency blockchain is must as they involve real time transactions and high latency transactions may have serious consequences. For small scale IoT system, Hyperledger Fabric is suitable. But for industry level IoT system, a scalable Blockchain platform is needed.

Ethereum is a public Blockchain and transaction fees are involved [9, 10, 30]. Transactions are grouped into the block and blocks are added by the miners. Confirmation of transaction takes at least six-block times and it consumes a lot of energy as Proof of Work algorithm is used. These problems can be solved by using DAG-based blockchain architecture. IOTA is DAG-based Blockchain. Transactions are arranged as the vertices of DAG and each new transaction needs to approve two old transactions [31]. It is a feeless architecture. The latency is very low and throughput is high compared to Ethereum. It is a public Blockchain and Highly scalable.

4 Conclusion

Combining IoT and Blockchain is rewarding in terms of transparency, privacy and security. But real-time implementation has many challenges. Most of the proposals, models and architectures suggested are Proof of Concept. They are not tested in the real environment. Majority of implementations are based on Ethereum and Hyperledger fabric. Hyperledger Fabric is suitable for small scale IoT system. To increase the efficiency and performance of the IoT system new Blockchain architectures need to be examined. DAG-based architecture IOTA can be an alternative to Ethereum for IoT implementation.

References

1. Ferrag MA, Derdour M, Mukherjee M, Derhab A, Maglaras L, Janicke H (2019) Blockchain technologies for the internet of things: research issues and challenges. IEEE Internet Things J 6(2):2188–2204
2. Yu S, Lv K, Shao Z, Guo Y, Zou J, Zhang B (2018) A high performance blockchain platform for intelligent devices. In 2018 1st IEEE international conference on hot information-centric networking (HotICN), pp 260–261. IEEE
3. Lo SK, Liu Y, Chia SY, Xu X, Lu Q, Zhu L, Ning H (2019) Analysis of blockchain solutions for IoT: a systematic literature review. IEEE Access 7:58822–58835
4. Casino F, Dasaklis TK, Patsakis C (2019) A systematic literature review of blockchain-based applications: current status, classification and open issues. Telematics Inform 36:55–81
5. Kumar A, Jain S (2019) Proof of Game (PoG): a game theory based consensus model. International conference on sustainable communication networks and application. Springer, Cham, pp 755–764
6. Novo O (2019) Scalable access management in IoT using blockchain: a performance evaluation. IEEE Internet of Things J 6(3):4694–4701
7. Ding S, Cao J, Li C, Fan K, Li H (2019) A novel attribute-based access control scheme using blockchain for IoT. IEEE Access 7:38431–38441
8. Islam MA, Madria S (2019) A permissioned blockchain based access control system for IOT. In: 2019 IEEE international conference on blockchain (blockchain). IEEE, pp 469–476
9. Liu H, Han D, Li D (2020) Fabric-IOT: a blockchain-based access control system in IoT. IEEE Access 8:18207–18218
10. Liu Y, Hei Y, Xu T, Liu J (2020) An evaluation of uncle block mechanism effect on ethereum selfish and stubborn mining combined with an eclipse attack. IEEE Access 8:17489–17499
11. Liu CH, Lin Q, Wen S (2019) Blockchain-enabled data collection and sharing for industrial IoT with deep reinforcement learning. IEEE Trans Ind Inform 15(6):3516–3526
12. Shen M, Tang X, Zhu L, Du X, Guizani M (2019) Privacy-preserving support vector machine training over blockchain-based encrypted IoT data in smart cities. IEEE Internet of Things J 6(5):7702–7712
13. Zhou L, Wang L, Sun Y, Lv P (2018) Beekeeper: a blockchain-based iot system with secure storage and homomorphic computation. IEEE Access 6:43472–43488
14. Li R, Song T, Mei B, Li H, Cheng X, Sun L (2019a) Blockchain for large-scale internet of things data storage and protection. IEEE Trans Serv Comput 12(5):762–771
15. Li J, Zhou Z, Wu J, Li J, Mumtaz S, Lin X, Gacanin H, Alotaibi S (2019) Decentralized on-demand energy supply for blockchain in internet of things: a microgrids approach. IEEE Trans Comput Soc Syst 6(6):1395–1406
16. Wang G, Shi Z, Nixon M, Han S (2019) Chainsplitter: towards blockchain-based industrial iot architecture for supporting hierarchical storage. In: 2019 IEEE international conference on blockchain (blockchain). IEEE, pp 166–175
17. Truong HTT, Almeida M, Karame G, Soriente C (2019) Towards secure and decentralized sharing of IoT data. In 2019 IEEE international conference on blockchain (blockchain). IEEE, pp 176–183

18. Bajoudah S, Dong C, Missier P (2019) Toward a decentralized, trust-less marketplace for brokered IoT data trading using blockchain. In: 2019 IEEE international conference on blockchain (blockchain). IEEE, pp 339–346
19. Sharma PK, Chen MY, Park JH (2017) A software defined fog node based distributed blockchain cloud architecture for IoT. IEEE Access 6:115–124
20. Pan J, Wang J, Hester A, AlQerm I, Liu Y, Zhao Y (2019) EdgeChain: an edge-iot framework and prototype based on blockchain and smart contracts. IEEE Internet of Things J 6(3):4719–4732
21. Peña MAL, Fernández IM (2019) SAT-IoT: an architectural model for a high-performance fog/edge/cloud IoT platform. In: 2019 IEEE 5th World Forum on Internet of Things (WF-IoT). IEEE, pp 633–638
22. Guo S, Hu X, Guo S, Qiu X, Qi F (2019) Blockchain meets edge computing: a distributed and trusted authentication system. IEEE Trans Ind Inf
23. Lei K, Du M, Huang J, Jin T (2020) Groupchain: towards a scalable public blockchain in Fog computing of IoT services computing. IEEE Trans Serv Comput 13(2):252–262
24. Kumar A, Krishnamurthi R, Nayyar A, Sharma K, Grover V, Hossain E (2020a) A novel smart healthcare design, simulation, and implementation using Healthcare 4.0 processes. IEEE Access 8:118433–118471
25. Kumar A, Kumar Sharma D, Nayyar A, Singh S, Yoon B (2020b) Lightweight Proof of Game (LPoG): A Proof of Work (PoW)'s extended lightweight consensus algorithm for wearable kidneys. Sensors 20(10):2868
26. Xu J, Xue K, Li S, Tian H, Hong J, Hong P, Yu N (2019) Healthchain: a blockchain-based privacy preserving scheme for large-scale health data. IEEE Internet of Things J 6(5):8770–8781
27. Tsang YP, Choy KL, Wu CH, Ho GTS, Lam HY (2019) Blockchain-driven IoT for food traceability with an integrated consensus mechanism. IEEE Access 7:129000–129017
28. Xie L, Ding Y, Yang H, Wang X (2019) Blockchain-based secure and trustworthy Internet of Things in SDN-enabled 5G-VANETs. IEEE Access 7:56656–56666
29. Zhang X, Li R, Cui B (2018) A security architecture of VANET based on blockchain and mobile edge computing. In: 2018 1st IEEE international conference on Hot Information-Centric Networking (HotICN). IEEE, pp 258–259
30. Wood G (2014). Ethereum: a secure decentralised generalised transaction ledger. Ethereum Project Yellow Paper 151(2014):1–32
31. Gal A (2018) The tangle: an illustrated introduction. Retrieved from https://blog.iota.org/the-tangle-an-illustrated-introduction-4d5eae6fe8d4

C2B-SCHMS: Cloud Computing and Bots Security for COVID-19 Data and Healthcare Management Systems

Vivek Kumar Prasad, Sudeep Tanwar, and Madhuri Bhavsar

Abstract Technologies play an essential role in mitigating the physical human need and replacing this with robots (bots). Hence reducing social involvement results in a reduction in COVID-19 patients. This proofs to be safe for the human generation and humanity too. The technological aspects of cloud computing resource management and bots can be used for the management and security of the patient's data and incorporating intelligent decision support in case of the massive reporting of the patients. As the healthcare sector continues to offer life-critical services while working to improve treatment and patient care with new technologies, criminals and cyber threat actors look to exploit the vulnerabilities that are coupled with this expertise. Healthcare organizations collect and store vast amounts of personal information, making them a primary target for cyber-criminals. In this paper, we will explore and discover the security implications and privacy issues of these health care technologies related to the management of patient's data. We also describe various security breaches in medical data and used a framework called as C2B-SCHMS which usages machine learning-based isolation graph for handling anomaly.

Keywords Cloud computing · Bots · Healthcare systems · Security · Decision support system · Machine leaning · Isolation graph

V. K. Prasad (✉) · S. Tanwar (✉) · M. Bhavsar
Computer Science Department, Institute of Technology, Nirma University, Ahmedabad, India
e-mail: vivek.prasad@nirmauni.ac.in

S. Tanwar
e-mail: sudeep.tanwar@nirmauni.ac.in

M. Bhavsar
e-mail: madhuri.bhavsar@nirmauni.ac.in

P. K. Singh et al. (eds.), *Proceedings of Second International Conference on Computing, Communications, and Cyber-Security*, Lecture Notes in Networks and Systems 203,
https://doi.org/10.1007/978-981-16-0733-2_56

1 Introduction

The COVID-19 pandemic's impact is felt across industries, indicating the emergence of clear and irreversible changes to exist. Despite full support from communities and governments; private and public healthcare systems fight an uphill battle against a largely unknown enemy, while also coping with a lack of adequate operator/staff and resources. Hence this paper deals with managing and securing the data for the healthcare systems during the periods of COVID19 pandemic [1]. Globally, it is expected that the healthcare cloud computing market will grow. Cloud offers tremendous potential to transform the healthcare landscape by reinterpreting systems, processes, and computer intelligence, enabling innovation, the flow of communication from momentous data sets, reducing costs, and accumulative efficiency and customization while adhering to rigorous expectations of security, adherence, and privacy [2]. Digitization in the healthcare sector and patient data is undergoing many problems such as data breaches of patients, phishing hacks and many more. These issues scale from malware that compromises the integrity of systems and privacy of patients to distributed denial of service (DDoS) attacks that disrupt facilities' that provide services to a patient. This valuable data can be used for identity theft, says Peter Carlisle, head of EMEA at cloud and data security company Thales eSecurity [3]. Healthcare sectors have been propagated from paper-based record systems to Electronic Medical Record (EMR) and Electronic Health Record (EHR) systems to improve patient care quality.

Body Sensor Network [4] is a booming technology that is being discovered. Both industries and academic organizations are emerging sensor based integrated systems for remote patient monitoring. For example, recently, Intel's Integrated Digital Hospital (IDH) has the intention at the enhancement of health care worldwide by connecting people, processes and technologies together in one platform. The IDH system comprises all mobile point-of-care (MPOC) and other information technologies to integrate patient and administrative data into a comprehensive, digital view of a patient's health.

Such innovations will also have many opportunities for the delivery of health care, there are a range of protection and privacy concerns that need to be addressed to uphold and preserve the standards of medical ethics, values and societal expectations. These implications include access rights to data from patients, how, whether and when data is stored, data transmission protection through wireless media.

Figure 1 shows how the COVID-19 cases in India are arising [5] and hence it reflects that technological ideas are a must required things that need to be incorporated for the management and security of the healthcare systems. Motivated by these aspects, we proposed a framework for the management of healthcare services using cloud computing and bots security services for reducing the impact of COVID-19 pandemic and is named as C2B-SCHMS (Cloud Computing and Bots Security for COVID-19 Data and Healthcare Management Systems).RPM (Remote Patient Monitoring), also known as home care tele health, allows daily monitoring devices

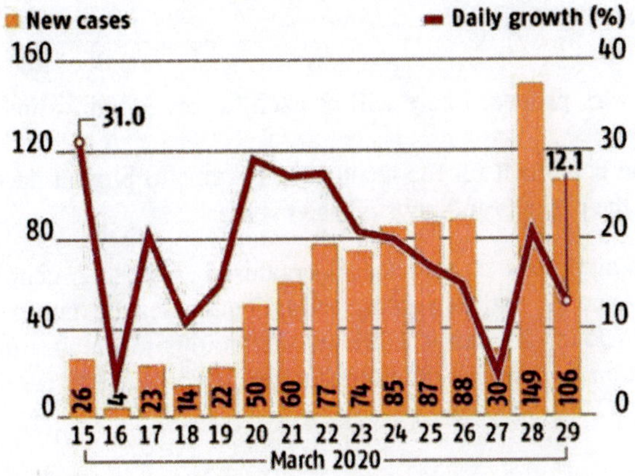

Fig. 1 COVID-19 cases on the rise in INDIA in the month of March 2020. *Source* Ministry of health and family welfare, business standard calculations

such as Diabetes (glucose level), blood pressure for the patient having cardiac problems. This patient data is sent and stored in a relational database so the doctors can view the data as a specific instance or as a trend. In case of heterogeneous methodology the sensor's data (medical sensor devices) can be merged into the RPM. Both the concepts such as remote patient monitoring systems and heterogeneous methodologies can be mapped to the bot systems to make things easier. There are endless instances in which a personal digital assistant or bot may support physicians, nurses, patients or their families. Better organization of patient routes, handling medicine, assisting in emergency situations or in first aid, providing a solution to basic medical problems: all these are potential scenarios for bots to step in and relieve the stress on medical professionals [6]. The same has been depicted in the Sect. 5 with the help of the proposed framework.

1.1 Motivation

- A framework has been proposed for managing the COVID-19 patient's data where cloud computing and bots services are used.
- The security features have been discussed while maintaining healthcare information.
- The proposed scheme monitors and identifies any anomaly in the dataset of the CC using bots.

1.2 Contribution

- The framework proposed here will be used for the identification of the security breaches.
- This will be beneficial for the healthcare systems to provide faster and secured services to the patients and physicians via bots.

The remaining portion of the paper is organised as; Sect. 2 mentions the various attack scenarios on the patient's data. Section 3 indicates the multiple data breaches present in the Cloud. Section 4 discusses the various techniques used against the attackers. Section 5 mentions the proposed framework and its results followed by the SWOT of CC and conclusion.

2 Various Attacks on Patient's Data

With the everyday unstable risk terrain and introduction of new emerging threats and vulnerabilities, security violations are normal to grow in the coming years. Table 1 shows various attacks and its descriptions [7]. Patient data is stored in data centers with volatile levels of security. Several hospitals across the country have reportedly been compromised with ransomware by the use of the obsolete JBoss server program. In these cases, the attacker uploaded malware without any interaction from

Table 1 Possible attacks on the healthcare data

Types of attack	Description
XML signature attacks/flooding	Sending many requests to a victim machine
Flooding attacks by worms, viruses/malicious programs by hackers	The malicious script is written by the hacker and this script will be spread all over the network
Denial of service/distributed Denial of service	The attacker tries to make network resources unavailable to legitimate users
Data breaches	Unauthorized users have accessed data
Phishing attacks	Often used to steal sensitive information via an illegal website
Spoofing attacks	The person or program masquerades as another by falsifying data to gain illegitimate access
IP/port scanning attack	An attacker has the goal of finding the active ports and exploits a known vulnerability
Man-in-the-middle attack	Attack where the attacker secretly sniffs and possibly modify the communication between two parties
Eavesdropping attack	Someone tries to steal information that electronic devices transmit over a network

the victim to the out-of date server, which in turn will start infecting the hospital's devices through familiar workstations used by everyday staff. Presbyterian Hospital in California was among the affected hospitals, and this reflected in a situation like delayed patient care, the hospital eventually charged $17,000 to get back access to the data and their network. Actors used an open-source method, JexBoss to scan the Internet or compromised JBoss servers, and infected networks, irrespective of the size in which they operate. Another example, this was the case in 2014 associated with Boston Children's Hospital.

The expansion of polymorphic attack vectors is tormenting healthcare at higher rates than other industries and organization. For example, FortiGuard Labs stated that in the year 2017 healthcare precept an average of almost 32,000 intrusion attacks per day as compared to over 14,300 attacks per organization in other industries.

A phishing attack on the Aultman Health Foundation based in Ohio has potentially infringed data from 42,600 patients of its Ault Works division of occupational medicine, hospital, and 25 physician practices for more than a month [8]. After the attack was discovered on March 28, officials led an analysis that found hackers gained access to multiple email accounts in mid-February, which continued until late March. Although the email accounts were not on the machines that store EHR data, some patient data was included in the breached emails.

Baltimore-based Life Bridge Health and Life Bridge Potomac Practitioners have been targeted by a ransomware attack that has potentially exposed some 500,000 patients' personal data for more than a year, officials said. They also said they discovered the breach on March 18, with a malware attack on their server hosting the EHR of Life Bridge Potomac Specialist and the patient registration and billing systems [9] for Life Bridge Safety. However, on September 27, 2016, the subsequent investigation reported that the hackers first obtained access to the EHR and servers. And the data compromised included demographic details, birth dates, medical records, details about hospitals and care, insurance data including Social Security numbers for individual patients. Chatbots are of exceptional help—especially in industries such as healthcare safety. This area includes dealing with sensitive personal health knowledge. However, there are a range of security issues concerning personal health records and Health Insurance Portability and Accountability Act (HIPAA) laws resulting from electronic data transfer and chatbot use. What happens if a patient is sending chatbot messages and providing very personal medical history? Such security aspects must be handled.

3 Data Breaches in Cloud

Cloud computing data protection has become a big problem as the data is located at various locations. The primary issue of user interest in cloud automation is maintaining data privacy. The protection of data security thus becomes a primary aim for the potential application of cloud computing technology in healthcare institutions. According to Gartner, security violations of 95 per cent will be seen in the cloud by 2020.

There are many good examples of data breaches. An antivirus organization known as Bitdefender in year 2015 had security attack due to which its customers' credentials such as username and password were stolen [10]. This attack happened in its public cloud which was deployed on AWS cloud. After the attack, hacker demanded $15,000 to the company. TalkTalk is the British telecommunication provider who found many incidents of a data breach in the years 2014 and 2015, respectively. It involved the theft of individual information of its 4million customers. Later on, these data were used to do scam calls to extract the sensitive and important information of customers.

The cloud-based file locker called Intralinks became a victim of a data breach with the Google AdWords. The vulnerability came into the picture, when users share links to share files, then the link was copied to the search box in the browser. This allows links are shared among various users in the network [11]. The chatbot is based on an AI-powered platform full of background knowledge of the healthcare. Patients describe their symptoms, and a chatbot then makes diagnoses achievable. The app is efficient because a video connection with a doctor is optional. But what if the background knowledge is a false one (on the attacked data)/not the actual data. Thus the bot must be aware of the security breaches at each level of the cloud computing; as the from here the main analysis happens.

4 Techniques Used Against Attacks

To protect health care organizations from these cyber threats, healthcare organizations have taken several procedures to tackle. But not all technique doesn't apply to every healthcare institute. Security requirements and implementations differ based on how the organization has set up its technical resources, and what is considered to be critical to the business and patient caution. Some of the solutions are described below, which are currently used by various health sectors to overcome the attacks.

Some organizations have adopted different techniques based on their requirements and infrastructure of the hospital and its facilities [12].

A. Automate Software Updates and Software Patching [13]: Many healthcare organizations have adopted decentralized software patching and updates. Use of automation to improve the speed and accuracy of the software updates process to eliminate vulnerabilities that intruder can exploit the system.
B. Implemented Access Control [14]: Data should be protected with a role-based access control system so that employees can only access the system and perform their job tasks accordingly. Many organizations have also implemented technologies for tracking and analyzing data access to help unusual patterns and its identification.
C. Implement Data Loss Prevent Mechanism DLP [15] solution is one of the effective ways to prevent data breaches by ensuring that sensitive information isn't lost, or accessed by unauthorized entities. Data Loss Prevent software

helps you control end-to-end activities, filter data on networks, and monitor data in the cloud to protect data in use.

D. Use of Data Encryption Techniques [16] Data Encryption makes the information unreadable in the network, thus making it difficult for hackers/intruders to access the data even if any mobile device is stolen or tampered. Healthcare institutes have started implemented encryption for both data at rest (which is stored in the database) as well as data in motion (which is sent to the end-user) to ensure data is not leaked in any situation.

E. The ability to detect bot traffic is, however, more complicated than it ever was. Bot developers are continually finding new ways to circumvent the functions of standard security solutions for bot detection. And as they are now beginning to make extensive use of artificial intelligence and deep learning, effective bot detection without truly specialized know-how—and without artificial intelligence/deep learning is waste.

In this research paper we are focusing on the concept of automatically detecting bad bots by making use of intelligent AI and deep learning based bots to identify and handle any attacks on the cloud computing based healthcare system. Its architecture and procedure is described in Sect. 5.

5 Proposed Framework

Figure 2 indicates about the proposed architecture where the bot are used to manage the information between the patients and the hospital's stakeholders for the smooth management of the services. New chat bot/bots buddies who aim to make patients' lives better are in demand now a days. Bots, computer programs or advanced analytic that conduct conversations using auditory or textual methods are becoming increasingly common and popular. Bots are an Intelligent application-powered, text-based or voice-based frameworks have expanded and take their place in healthcare, too. The Medical Futurist claims that they can relieve the pressure on physicians and help patients learn to care properly about their well being. Numerous activities can be done using bots in the CC based medical systems as remote patients care, providing information for the patients on search, providing virtualized essential services, ability to scale (through bot's intelligence) when patients increases, offers better storage, this also shares data with the hospital managers/stakeholders and data scientists [17]. Here we are focusing on the security aspects of the intelligent bots using the concepts of the Long short term memory and isolation forest for the anomaly detection. It is obvious that the bots will be making the decisions based on the "dataset available for the analysis purpose" in the cloud computing environment. Hence if the genuine/original dataset are available, the intelligent bots can pass the correct information to the patients, doctors, and other management persons of the hospitals without the involvement of the human need. Hence the dataset.

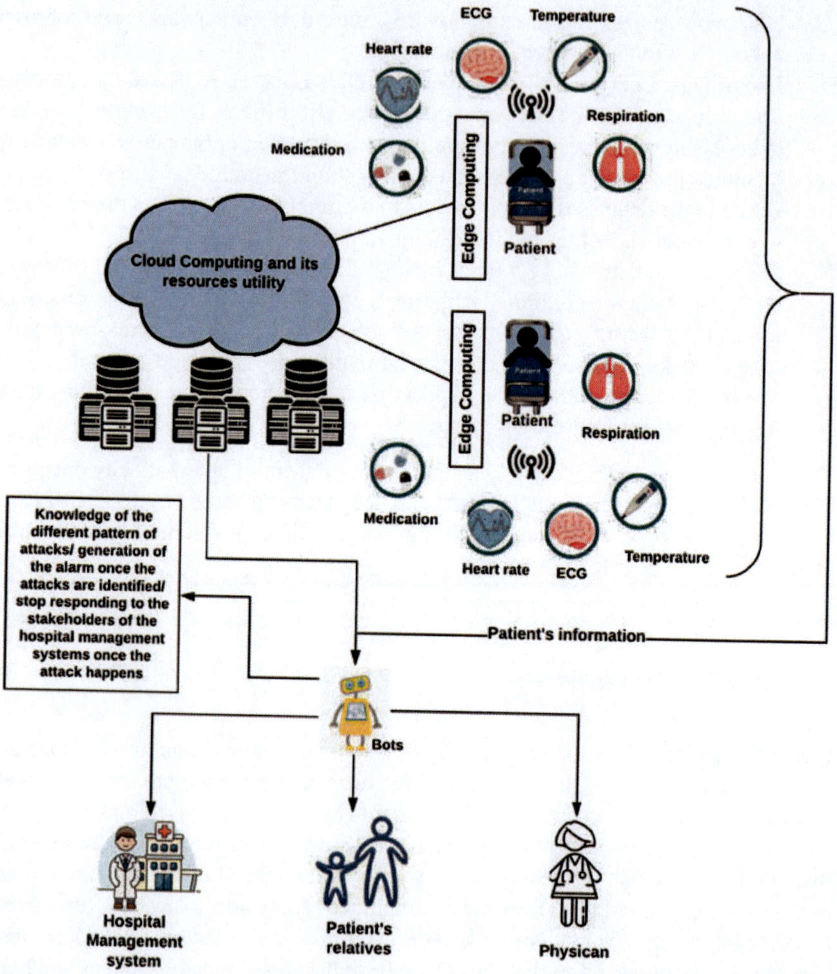

Fig. 2 Security and Cloud Computing services from where the bots are making the decisions play an important role [18]. But when the dataset itself is anomaly-based. Then the resultant information passing to the patients/doctors and others will be of no use; instead, this will create havoc in the system and improper management of the computing resources [19]. Hence the bots have to identify the anomalies along with concerning their own jobs. Therefore in this research paper, we are focusing on the CPU parameter for anomaly detection using deep learning and isolation forest concepts. The dataset contains COVID-19 X-ray image consisting of 750 VMs (Virtual Machines), and the parameter that we are considering here is the CPU utilization. Figure 3 shows the reading of the dataset of the CPU without the anomaly/no attack has happened and the data is clean. Figure 4 is a variant of the Fig. 3 and shows the reading from 2000 to 3500 timestamp

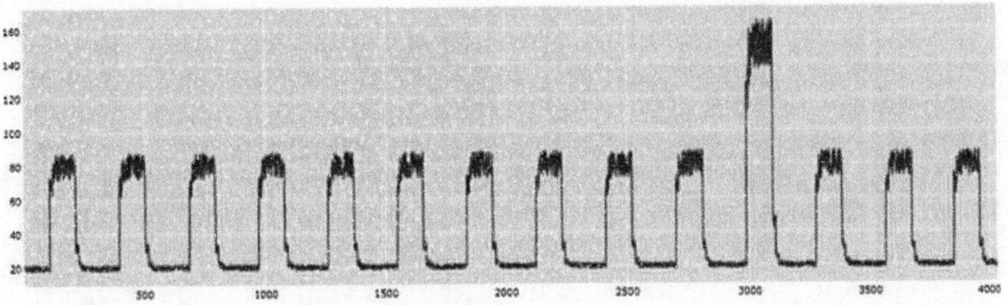

Fig. 3 CPU utilization, In graph x-axis: timestamp (ms) and in Y-axis: CPU utilization in MHz

Figure 5 show the implementation of the machine learning based isolation graph, that detects the anomaly which identified in between the timestamp 3000–3100 ms and in the range of the CPU utility from 100 to 165 MHz.

Hence if the anomaly is identified, then the bots will take appropriate action to mitigate from the anomaly/glitch. If there is no anomaly, then bots can pass the information safely to the end-users. The normal observation value is calculated as 0.96 and the outlier accuracy was calculated as 0.91. The scores values can be classified as; if the values are close to 1, then we can conclude that this is an anomaly and if the score are smaller than 0.5 then its a normal observation. If the scores are close to 0.5 then the entire sample doesn't seems to have clearly identified the distinct anomalies. In our experimental setup our values were near to 1 and much lesser than 0.5. These results and analysis can be helpful for the bots to recognize whether the data that is given to the patients/doctors and others is a valid one. [Note: Here, we are assuming that the CC resource manager has already made the workload patterns and if the model's mismatches with the present one, then this will be identified by the machine learning-based isolation graphs and thus intimated to the bots. Bots in turn will stop communicating the messages with the stakeholders of the healthcare system till the anomaly is prevented].

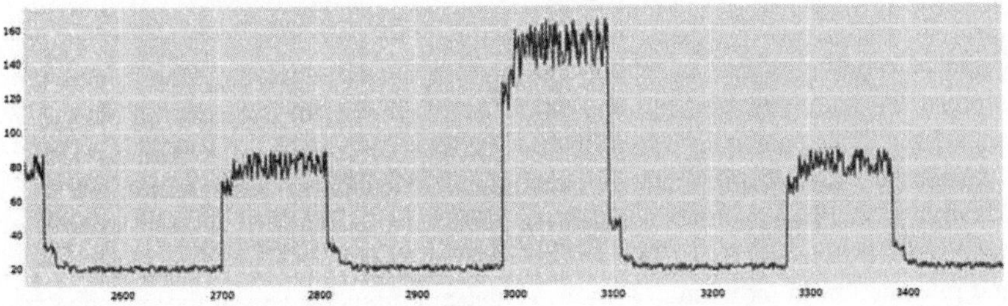

Fig. 4 CPU utilization, in graph x-axis: timestamp (ms) and in Y-axis: CPU utilization in MHz. Reading shown from timestamp 2500–3500 ms

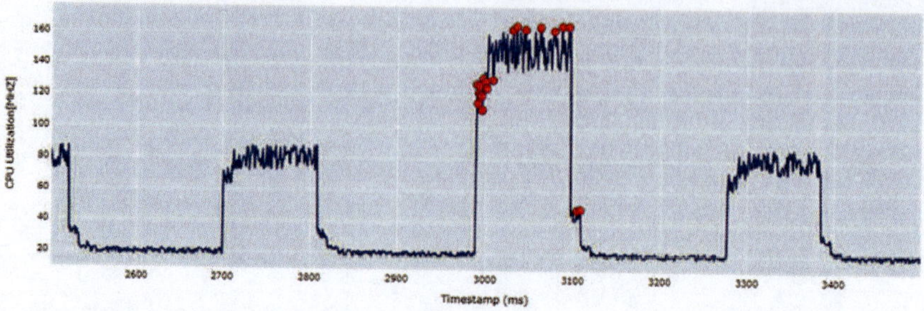

Fig. 5 Readings and highlighting of the anomaly points i.e. from the timestamp 3000–3100 ms and the utilisation of the CPU is 100–165 MHz

Table 2 SWOT analysis

Strength	*Weakness*
Energy efficient, flexible, easy to customize, cost efficacy	Privacy laws, blending with local software, QoS
Opportunity	*Threat*
Elastic to new requirements Adoption of new technology	Loss of data Lack of standards Data privacy

6 SWOT Analysis for Implementing Cloud in Healthcare System

SWOT helps us to find out the efficiency of deployment of evolving technology in the Healthcare domain (Table 2).

7 Conclusion and Future Work

Cloud computing plays an important roles in the healthcare systems, and bots can provide details of the appointment to health care professionals and help them update medical records into the Cloud. The service combines integrated medical intelligence with natural language capabilities, extensibility tools, and compliance constructs, enabling healthcare organizations such as providers, payers, pharmaceuticals, HMOs, telehealth to provide access to trusted and relevant healthcare services and information to individuals. Anomaly detection mechanism in the bots can add advantages over the medical healthcare systems. Here in this research paper, the framework called as C2B-SCHMS has been used and works with the methods of Machine learning based isolation graph to detect any anomaly in the dataset. This will make cloud computing trustable for medical services.

References

1. Keesara S, Jonas A, Schulman K (2020) Covid-19 and health care's digital revolution. N Engl J Med 382(23):e82
2. https://cio.economictimes.indiatimes.com/news/cloud-computing/7-ways-howcloud-computing-is-shaping-healthcare-in-2019/68614849
3. https://www.forbes.com/sites/kateoflahertyuk/2018/10/05/why-cyber-criminalsare-attacking-healthcare-and-how-to-stop-them/66e97de57f69.
4. Lin K, Li Y, Sun J, Zhou D, Zhang Q (2020) Multi-sensor fusion for bodysensor network in medical human–robot interaction scenario. Inf Fusion 57:15–26
5. Ministry of health and family welfare, business standard calculations. https://www.business-standard.com/article/current-affairs/statsguru-covid-19-cases-on-the-rise-in-india-maharashtra-worst-affected-120033000061.html
6. Bhuyan SS, Kabir UY, Escareno JM, Ector K, Palakodeti S, Wyant D, Kumar S, Levy M, Kedia S, Dasgupta D, Dobalian A (2020) Transforming healthcare cybersecurity from reactive to proactive: current status and future recommendations. J Med Syst 44:1–9
7. Qi J, Yang P, Min G, Amft O, Dong F, Xu L (2017) Advanced internet of things for personalised healthcare systems: a survey. Pervasive Mob Comput 41:132–149
8. https://www.healthcareitnews.com/news/phishing-hack-ohio-provider-breaches-data-42000-patients-month
9. https://www.healthcareitnews.com/news/lifebridge-health-reveals-breach-compromised-health-data-500000-pa
10. https://businessinsights.bitdefender.com/spate-of-ransomware-attacks-on-healthcare-providers-raises-serious-health-concerns
11. Prasad VK, Bhavsar M (2017, August) Efficient resource monitoring and prediction techniques in an IaaS level of cloud computing: survey. In: International conference on future internet technologies and trends. Springer, Cham, pp 47–55
12. Gupta R, Tanwar S, Tyagi S, Kumar N (2020) Machine learning models forsecure data analytics: a taxonomy and threat model. Comput Commun 153:406–440
13. Mell P, Bergeron T, Henning D (2005) Creating a patch and vulnerability management program. NIST Spec Publ 800:40
14. Prasad VK, Bhavsar MD, Tanwar S (2019) Influence of montoring: Fog and edge computing. Scalable Comput Pract Experience 20(2):365–376
15. Shah K, Prasad V (2017) Security for healthcare data on cloud. Int J Comput Sci Eng (IJCSE) 9(5)
16. Chauhan K, Prasad V (2015) Distributed denial of service (ddos) attack techniques and prevention on cloud environment. Int J Innov Adv Comput Sci 4:6
17. Prasad VK, Bhavsar MD (2020) Monitoring IaaS Cloud for Healthcare systems: healthcare information management and cloud resources utilization. Int J E-Health Med Commun (IJEHMC) 11(3):54–70
18. Alesanco A, Sancho J, Gilaberte Y, Abarca E, García J (2017). Bots in messaging platforms, a new paradigm in healthcare delivery: application to custom prescription in dermatology. In: EMBEC NBC 2017. Springer, Singapore, pp 185–188
19. Prasad VK, Mehta H, Gajre P, Sutaria V, Bhavsar M (2017, August) Capacity planning through monitoring of context aware tasks at IaaS level of Cloud computing. In: International conference on future internet technologies and trends. Springer, Cham, pp 66–74

FemtoCloud for Securing Smart Homes—An Edge Computing Solution for Internet of Thing Applications

Abhinav Rawat, Avani Jindal, Akshat Singhal, and Abhirup Khanna

Abstract Increasing capabilities of mobile devices have manifested great potential for the feasibility of edge computing to aid IoT applications toward building more efficient and smart systems of the future world. This paper is intended to propose a multiparameter and secured FemtoCloud solution for securing smart homes from coercions. The architecture focuses on securing mainly smart buildings and smart residential locations. On identification of any threat, the data and suggestions are provided to the user's mobile device (User Node) via Femtocell. Our system is bifurcated into two brief machine learning models which are trained and deployed at both the levels, i.e., mobile and cloud. Edge devices perform the rudimentary computations at the edge level itself to increase the response time factor. On identification of a serious threat, Femtocell is triggered to send the data of mobile device to the cloud for performing comprehensive processing to estimate the intensity of the threats. We use modified open-source machine learning models to detect and determine the situations which can be of potential threat. The objective is to leverage mobile devices and the cloud to analyze the most appropriate solution. Finally, we aim to provide challenges and future opportunities to present a wide scope of research in the same sphere.

Keywords FemtoCloud · Mobile computing · Edge computing · Deep learning

1 Introduction

Smart devices [1] and data produced by them have grown exponentially over the past few years. The capable computing platforms and processors to process data at the edge have become the need of the hour. Cloud computing is the key to process computations, but latency has always been a concern in the field where performance in terms of time is a decisive factor.

A. Rawat (✉) · A. Jindal · A. Singhal · A. Khanna
University of Petroleum and Energy Studies (UPES),, Dehradun, Uttarakhand 248007, India

P. K. Singh et al. (eds.), *Proceedings of Second International Conference on Computing, Communications, and Cyber-Security*, Lecture Notes in Networks and Systems 203,
https://doi.org/10.1007/978-981-16-0733-2_57

Internet of Thing Applications [2] requires high processing at the edge level to accomplish the requirements and goals of the system. Smart homes, smart cities [3], smart hospitals, smart grids, etc. are the potential use cases. However, all IoT applications require cloud to compute and offload the data. Cloud itself is self-sufficient to meet all the requirements and derive the most efficient solution but constraints like response time, bandwidth cost minimization and overall cost minimization hinders the smart solutions [4]. Since all IoT applications are heavily dependent on high-end computations for performance, it heavily relies on its own servers or cloud.

Gradually, the concept of edge computing comes to the rescue which has already shown the potential to address the problem of response time factor along with the bandwidth cost minimization [5] and also the data privacy. FemtoCloud has the potential to integrate edge computing with IoT applications to leverage the edge/mobile devices and orchestrate cloud services along with minimizing the computation load. Femto-cells access points [6] (FAPs) are equipped with multiple mobile connectivity potentials. The aim is not only to minimize the latency but also to avoid the remote cloud costs while improving the quality of services (QoS). Femto-cells [7] is a smart wireless solution for connecting mobile devices over the network. It is built to improve wireless reception inside a small location. Mobile devices work closely with the Femtocells to dynamically offload or migrate workloads to the cloud in case of high processing needs. FAPs uses a Secure Gateway (SeGW) for migrating data from Femtocell to the cloud which is collectively called as home node base station gateway (HNB-GW). HNB-GW's aim to connect all the HNBs, i.e., Femtocells in the vicinity or which are interconnected over a network. Femtocell acts as the HNB station which operates in a licensed spectrum connected to Internet Service Provider (ISP) via wired/wireless broadband. The components for the proposed architecture are:

1. Home node base station (HNB)
2. Internet connectivity
3. Mobile devices (user node)
4. Surveillance cameras (SC).

We term this model as Femtocell-based Home Monitoring System (FCHMS). The goal of FCHMS is to perform rudimentary computations and migrate data to cloud for comprehensive computations if and when required.

The key contributions of our paper are:

1. A FemtoCloud-based architecture for monitoring and securing smart homes using FAPs and mobile cloud technology.
2. Mobile computations determine the need for migrating data to the cloud for comprehensive processing. This increases the response time, cost efficiency and decreases the bandwidth cost minimization. Offloading also reduces the power consumption and overutilization of mobile devices.
3. Machine learning models at the two levels increase the trustworthiness measurement, correctness of decisions, and decreases the workload of the mobile devices.
4. Enabling cloud increases the consistency and availability of the system.

2 Traditional Cloud Versus FemtoCloud

2.1 *Benefits and Pitfalls of Current Cloud Computing Model*

The conventional cloud model is centralized and works remotely from a data center. Cloud offerings like (1) Pay-as-You-Use; (2) 99.999999% guaranteed uptime (AWS Datacenters); (3) Elasticity between servers, storages, etc. (4) Orchestration, (5) Running Costs, (6) Security [8], (7) Automation, (8) Insight has made the cloud a popular and favorable IT solution, eventually changing the way we used to process and work on the data. Current Cloud Computing Model [9] has been demonstrated to be a massive achievement for the existing Cyberspace scenario. However, a lot of challenges are faced by IoT applications [1] due to centralized cloud mode, we have explained a few of them as follows:

1. Bulk and Velocity of Data Storage and Migration: IoT tends to produce a lot of significant data. Due to the presence of fewer numbers of large-sized remote data centers, only wealthy firms can invest in it. Further, IoT devices carry and generate tremendous amounts of data which eventually needs to be stored over the cloud. However, due to fewer data centers, it becomes very difficult to transfer data at a higher velocity which in turn increases complexities.
2. Latency owing to the remoteness between edge IoT devices and data centers: The existing remote cloud infrastructure lacks a high transfer rate of data. Keeping in mind about the near future when data accumulation will be a great concern, it is very much imaginable that high latency will be a big challenge for quite a number of edge devices that involve end-to-end encryption.
3. Monopoly versus nascent IoT competition: The most major concern is that the existing cloud infrastructure is very extortionate to construct and is only affordable to those giant organizations that tend to define proprietary protocols.

2.2 *FemtoCloud Computing Model*

Mobile device usage continues to increase at a faster pace as they have evolved to be powerful and extensive computational tools. Most of the mobile devices in a crowded area or work-specific areas are either idle or underutilized whose computational capabilities can be used for offloading the computational processes. This will help in improving the user experience and decreasing the latency of the system. FemtoCloud computing is an amalgamation of Femtocell network grids and cloud computing.

In the FemtoCloud model, the Femtocells which are small low power cellular stations with computational and storage power are deployed to implement the cloud. In FemtoCloud, cluster formation is used for cooperating Femtocells together. VMs in FemtoCloud can be used by users to offload their computations. The computations are distributed on the resources available in a particular cluster that are nearest to the server. The information processed is sent back to the server.

1. It helps with faster data processing at the edge as they are readily available at the nearest.
2. With the reduction in time taken for data travel, latency gets reduced.
3. Most of mobile devices have under-utilized resources which will be put to use
4. Mobile devices are becoming smart and powerful with computational capabilities which will help in deploying edge networks and reducing the costs.
5. Since mobile devices are available almost everywhere, they will help in deploying a better and more distributed edge network.

2.3 *Comparative Study*

After the comprehensive study of both the technologies, we have tried our best to formulate the below-mentioned table for a brief understanding about the differences between the various parameters which tends to prove to use the FemtoCloud Computing Model for the proposed architecture (Table 1).

3 Literature Review

We have made our related work focused primarily on the same technologies and different kinds of solutions for different use cases. We have also made our best

Table. 1 Brief evaluation between traditional cloud computing and FemtoCloud computing

Parameters	Traditional Cloud computing [10]	FemtoCloud computing
Availability	Less number of large sized data centers	Highly available
Latency	High due to remote data centers	Low due to proximity of users
Proximity of services	Usually remote and distant from users	At edge, local to the user
Performance	Capable of performing complex operations	Capable of basic computations
Cost optimization	Highly efficient	More efficient than Cloud computing
Bandwidth consumption	High due to transmission of data to data centers	Low due to edge computing
Security	Data attacks during long transmissions	Low due to edge processing
Scalability/flexibility	Scalable at data centers	Scalable both at edge and Datacenter
Applications	Most of the Cloud native applications	Applications on IoT, smart homes, smart solutions

efforts to explain various technologies that together constitute our FCHMS proposed architecture. Computer applications and networks [11] are one of the most important factors for efficient working of our model. With the high technological shift taking place, we have a great Internet-enabled IoT system with billions of devices and mobile devices that are now capable of performing huge computations. The data is being created at a very faster pace than ever before and hence needs to be processed at the same time. There are numerous technologies available, but all lack some standardization which can lead to missing some information. The authors have tried to explain a new edge computing-based cloud system for IoT applications. The advancement of the mobile and Internet of Things (IoT) [12] technologies over the last years has created high latency and low bandwidth efficiency a big challenge. The traditional cloud is centralized but with the support of edge computing can help in reducing both of the challenges. In addition to this, FemtoCloud technology (mobile edge computing) is of great use to reduce the dependency on centralized cloud servers.

The future IoT is being rapidly evolved around the Internet for the fact that billions of devices are being added to the system and the data generated needs to be processed at a very high speed [1]. With traditional clouds, low latency is expected, servers are centralized and future IoT needs processing to be very fast for the matter of fact that automated systems cannot be delayed with instructions that are being processed from the data collected by the near edge IoT devices. To overcome this problem, we can use the edge technology to process the information in nearby servers to not only reduce the latency but also to reduce the bandwidth consumption as data has a lesser path to travel. Such prospect is empowered by a progression of developing advancements, including system work virtualization and programming characterized organizing. The authors did a comprehensive work in the field of computing capabilities of mobile devices in terms of processing power and memory are increasing at a very fast rate. But it has been significantly observed that most of these resources are underutilized. A collection of collocated devices can be formed which can be used as the edge cloud service. For this, the FemtoCloud [13] architecture was proposed to be a mobile cloud from the mobile devices cluster which will be dynamic, self-configuring and multi-device. This architecture will have coordinated cloud service formed out of multiple mobile devices. The aim of the FemtoCloud is to improve the Quality of Experience (QoE) so as to avoid remote cloud costs. In order to achieve the most of Femto Access Points (FAPs), their computation capacity should be exploited at maximum. Computational efficiency is limited in mobile devices due to limited communication and computational capabilities [5]. This can be improved with a collaboration of traditional cloud and edge computing technologies, where tasks will be processed partially at both the ends. First, we will compute the latency of all the devices in the edge-cloud system. Then we will divide the problem into two subproblems. First one is to create a communication resource allocation with the edge devices, and second is to computational resource allocation among the edge and cloud. With this, we find lesser latency and high computation as processes are distributed all over the network.

One of the FemtoCloud cloud solutions is Precision Agriculture. PA [14] is growing at a faster rate as IoT and cloud computing offers great connectivity. Although the system is not useful in all regions for the obvious environmental reasons but with some provided platforms it offers great efficiency. The system is based on a multi-tier platform: (i) physical layer to interact with the devices in touch with crop and collect the information, (ii) the edge network to process all the information provided, (iii) the cloud network to store past records and perform the required predictions based on past. Network virtualization can be used to increase the deploying flexibility. Edge computing has great potential in reducing the challenges of traditional cloud and meanwhile ensuring the data security and privacy [2]. There are several use-cases available where cloud offloading and collaborative edge have been very beneficial to the system. With this not only that the traditional cloud is still supported but also other distant networks are connected together which increases the data sharing and collaboration because of data closeness. Also, there are many opportunities in the field that have a lot to extract.

4 Proposed Architecture

4.1 Conceptual Architecture

We have termed our proposed architecture as FCHMS model. It is proposed to evaluate initial threats and offload data [15] to the cloud from the mobile device via Femtocell. Femtocell is the home node base station (HNB) which is responsible for connecting mobile devices in a specific geographic location. FCHMS is an abbreviation for Femtocell-based Cloud Solution for effective home monitoring system. Femtocell (HNB) itself is connected to the Internet via Internet Service Provider (ISP) through a broadband connection. Since Femtocell is a smart wireless solution for connecting mobile devices; it is effective for offloading tasks from mobile devices to the cloud.

Uu interface [6, 7] is used to connect HNB with the mobile devices; Iuh interfaces are used to connect HNB with the HNB-GW while being connected through a Secured Internet Gateway (Se-IGW) for secured transfer of data. Iups/Iucs interface, i.e., Iu-circuit Switched/ Iu-packet Switched interface [16], is used to connect HNB-GW with the Core Network (Fig. 1).

The above-mentioned figure depicts the proposed methodology of this paper. The user node performs the rudimentary operations based on the data sent by the surveillance cameras (SC) and notifies the registered Femto Cell in case of threat detection. In such scenario, Femtocell forwards the raw and processed data to the cloud for comprehensive operations via the secured channels and Se-IGW. The operations undergone in cloud are followed by the updating of the database for maintaining consistency and also inform the emergency departments as and when instructed by

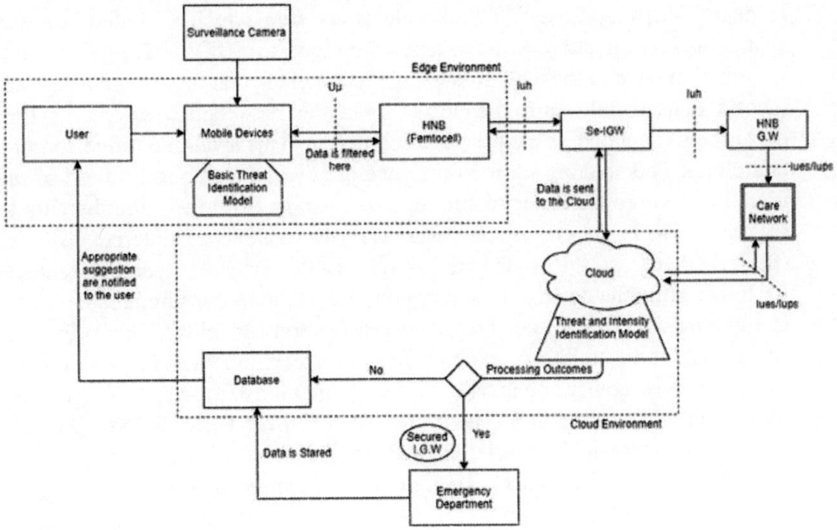

Fig. 1 Conceptual architecture

the user node. However, the sole functionality of the architecture lies on the hands of the users.

Functionality of Architecture: The functionality of the architecture is bifurcated into the following five phases:

1. Image captured by the cameras and transfer to mobile devices (MS): Image is considered as the data which is to be transferred.
2. Basic Threat Identification Model for Mobile Device: Using image classification techniques, we can classify an image as either threatening or safe. There are many models hosted by TensorFlow Lite for image classification which can be made use of for edge computing devices. For our application, we will be using the Efficient Net-Lite model because of its high accuracy in contrast with the other advanced models, and it also minimizes latency. Increasing the lite version means increasing the accuracy with a slight increase in latency. For our application, we will use the Lite-2 model. It has a size of approximately 6 MB, accuracy 77% and CPU latency of 12 ms.
3. Transfer of Data from MS to the HNB: After the above-mentioned model has derived the results, the raw data, results of the model and the unique Id (IP Address or IMEI Number) of the mobile device are sent to the HNB station.
4. Data Transmission from HNB to the Cloud: HNB stations are equipped with storage capacities which maintain a database of the parameters of data which has been sent to the cloud. For instance, if the initial threat model identifies a potential risk, the raw data on which the model was used along with the results and the information of the mobile device is sent to the cloud via HNB with a record of metadata stored in the database of the HNB station. This metadata

helps to re-identify the user, i.e., mobile device from which the initial data was sent. This identification, in turn, helps the cloud to notify and send the vital information back to the user. It is visible to conclude that multi-parameters are identified, used and sent to the cloud for detailed processing. However, not all data is being sent to the cloud and is discarded at all the levels, i.e., edge, mobile, and cloud. This is done so as to improve the overall cost optimization of the model and to reduce the load and also the storage of transferring the data to the cloud. The basic security parameters consist of a Secured Internet Gateway (Se-IGW) and also the HNB gateway (HNB-GW) which is specifically made for transferring the data from Femtocell to the cloud in a secured manner.

5. Comprehensive Threat and Intensity Identification Model for Cloud: For this component of the architecture, object detection techniques can be used to identify suspicious activity or behavior. There is a plethora of models by TensorFlow for custom object detection. Therefore, we will use a Faster RCNN (Region-based Convolutional Neural Network) Resnet50 model which has already been trained on the COCO dataset. This model has a speed of 106 ms and a mAP (mean Average Precision) of 32. If that dataset is found to be insufficient, we can use transfer learning and train it again for our custom dataset. This is done to acknowledge all the possible risks and to make the model more comprehensive and reliable.

Security aspects of the Architecture: Security of the FCHMS model heavily depends upon the two types, internal and external. Internal security covers aspects like the security of data inside the cloud and also the secured data transmission. External security supervises the factors which can compromise the user's data and the associated processing of it. Internal security [6, 11] may/may not be completely secured, but the existing technologies of Secured IGW and HNB GW are sufficient to manage and deal with the secured data transmission. User data is transmitted to the Femtocell it is registered under. Femtocell is an independent node which is connected through an Internet gateway. This channel is secured for the transmission of data to maintain data integrity and confidentiality. As predicted by Gartner, 95% security breaches in the cloud are expected due to the customer's fault.

External security is heavily dependent on the client-side validation. Multiple steps are taken to maintain the client-side security. However, there exists no fault-tolerant technique which works independently of clients' participation. Participation, as well as vigilance, are equally essential attributes which a client must follow and practice to ensure client-side security breaches.

The FCHMS model considers the factor of external security breaches, and below mentioned steps are the way to achieve it.

1. Secured Internet Gateway ensures secured transmission of data over the Internet, keeping it inaccessible from the reach of unauthorized access.
2. On every new data being sent to the cloud, a replica of the database is saved in another region of the cloud.

3. User must use an ID and password which is provided to him on successful registration of his mobile device under a Femtocell.
4. MFA or Multiple Factor Authentication, phone printing is used to send the OTP or One-Time-Password to the user. Phone printing is an effective MFA method to bypass the knowledge-based authentications and also the diversion of SMS to an unauthorized individual.
5. Apart from the above-mentioned steps, users will also be notified to update their passwords once every 5 times they access the data stored in the cloud.

Deep Learning Models: For the FCHMS model, we use two deep learning models at each level, respectively, i.e., edge (Image Classification Model) and cloud (Object Detection Model). The first one is a classification model on the edge which classifies an image as either a potential threat or non-threat. The second one, or the object detection model, will be deployed directly on the cloud, and it performs the processing on the data which is sent by the mobile device via Femtocell only on identification of the potential threat.

The Image Classification Model serves the purpose of classifying the images as of potential threat or non-threat. For instance, a human or unidentified moving object is found in the image, it will be considered as a potential threat and is sent forward for the comprehensive step of processing which is the object detection model deployed on the cloud. For this, associated Femtocell is triggered and the data along with the processing is offloaded to the cloud. For another instance, the image will be classified as a non-threat if some domestic animal or an identified non-moving object is detected by the edge deep learning model. The models are pre-trained and thus, the identified objects can include a ball or a vehicle passing by. Since the non-threatening elements won't have any detrimental effects, data associated with them will not be taken into further consideration and thus discarded.

The Object Detection Model is invoked when the edge-level classification model detects the image to be of potential threat. It works in 2 phases. First, the data is again considered for a re-evaluation of potential threat. On the successful determination of humans (which are a potential threat), the second phase is initiated. Only when an image is classified and determined as a potential threat, the model will further process to determine the intensity of the threat. As discussed before, the images with threat have the presence of either a human or an unidentified moving object. If the unidentified object is a human, further processing will be performed or else will be termed as non-threat and the associated data will be discarded. In the case of the image detecting the presence of a human, the level of threat will be computed. For initial processing, we propose a three-level threat intensity determination approach. The three levels are understated as following:

1. Level 1: This level is considerably safe. Threat level 1 signifies the presence of a human with no identification of any threat. For instance, a human or nothing suspicious like a delivery box or a letter, threat level 1 is triggered and is notified to the user. Here, the user is the mobile device from where the raw data was offloaded to the cloud for detailed processing.

2. Level 2: This level is triggered when a moderate threat is determined. For instance, the presence of a group of people or a human with objectionable objects such as a stick or baton. A group of humans might refer to a gang or mob which can be a potential threat. In such a scenario, all the devices registered under the associated Femtocell from where the user's mobile device offloaded the data to the cloud will be notified.
3. Level 3: This is the highest level of threat. On identification of a human with a suspicious object like a knife or a gun or anything which is fatal or lethal, threat level 3 will be triggered. In any such scenario, not only all the registered devices will be notified but a piece of a processed image along with a prompt suggesting informing the nearest police station will be sent to the user's mobile device.

Since this model plays a very important yet significant role for the FCHMS, high accuracy is also a challenging factor. We not only propose to use detailed datasets like Granada [17] but also to prepare our personalized and modified datasets.

Comparison of various models considered for our FCHMS model: The edge level model is a classification model which classifies an image as either a potential threat or non-threat. For this, we use the [18], Efficient Net-Lite model with version 2. It is a pre-trained model provided by TensorFlow. This is deployed and auto-operated on the edge device. This model is felicitous as it fulfills all the thresholds of the factors which determine the efficiency of the FCHMS Model. The following table demonstrates models from different series and versions which are available for image classification [19] (Table 2).

The values of latency and accuracy specified are of the quantized version (INT8) of the model using a CPU. In the above models, it is observable and evident that high accuracy comes at a cost of high latency. Especially in the case of the Inception model, the accuracy achieved is 79.5% with a large latency of 268 ms which cannot fulfill our determining factor of response time optimization. The Mobile net model has a very low latency of 3.1 ms but is accompanied by an extremely poor accuracy of 55.9%, which again makes this model unfit for the FCHMS. Therefore, we propose to use the EfficientNet-Lite2 model which has an optimal balance between latency of 12 ms with a considerable accuracy of 77%.

The second model serves the functionality for object detection. It is used to validate the results of that model and also provide additional insights pertaining to the level of threat. The model used for serving the above-mentioned purpose is the

Table 2 Comparison of various models available for image classification

Model name	Latency (ms)	Accuracy (%)		Model size (MB)
EfficientNet-Lite2	**12**	**77**		**~6**
EfficientNet-Lite3	18	79		~10
Mobilnet_V1_0.75_128_quant	3.1	55.9		2.6
Inception_V4_quant	268	79.5		41

Table 3 A comparison of various models available for object detection model

Model name	Response time	COCO mAP
faster_rccn_resnet50_coco	**89 ms**	**30**
faster_rccn_nas	1833 ms	43
ssd_mobilenet_v1_0.75_depth_quantized_coco	29 ms	16
faster_rcnn_inception_resnet_v2_atrous_coco	620 ms	37

faster_rccn_resnet50 model which is pre-trained on the COCO dataset provided by TensorFlow. The COCO dataset covers a lot of objects in its 80 classes and therefore, is best suited for the FCHMS. This model is further modified by adding more classes as required. The approach of transfer Learning and further implemented to train the model. The other models which were taken into consideration are mentioned as follows. While keeping our determining factors in account, we propose the following comparative study of all the models [20].

The below-mentioned table depicts the comparison of the processing speed and mAP of a few object detection models provided by TensorFlow. The proposed model for the purpose of Object Detection has a response time of 106 ms and 32 mAP (mean Average Precision) which is a measure of performance. The faster_rccn_nas model has a great performance mAP of 43 but also a high response time of 1833 ms which is not efficient (Table 3).

On the other side, ssd_mobilenet_v1_0.75_depth_quantized_coco model has a very low response time of 29 ms, but it does not have a good enough mAP. The faster_rcnn_inception_resnet_v2_atrous_coco model has a considerable mAP and a response time lesser than faster_rccn_nas. But, a latency of 620 ms might not be an optimal solution for our FCHMS. The motive of minimizing delay in the computations would fail reasonably. Undoubtedly, we tend to consider the faster_rccn_resnet50_coco model.

4.2 Physical Architecture

The physical architecture comprises all the physical aspects of the FCHMS model. These components work independently yet simultaneously in harmony to perform all the functionalities of the system efficiently. The most prominent and the most important is the Femtocell which forms the backbone of the system. However, all other components work under the authority of the Femtocell. The Internet and network play the most vital role for the safe and secure transmission of the data. Femtocell uses the HNB gateway for connecting the mobile devices to the cloud. HNB gateway is a dedicated technology specifically built for connectivity of the Femtocells.

The main functionality of the components is to capture the data and perform the necessary transmission of data to the predefined destinations via Secured Internet Gateway. The major components are defined as follows:

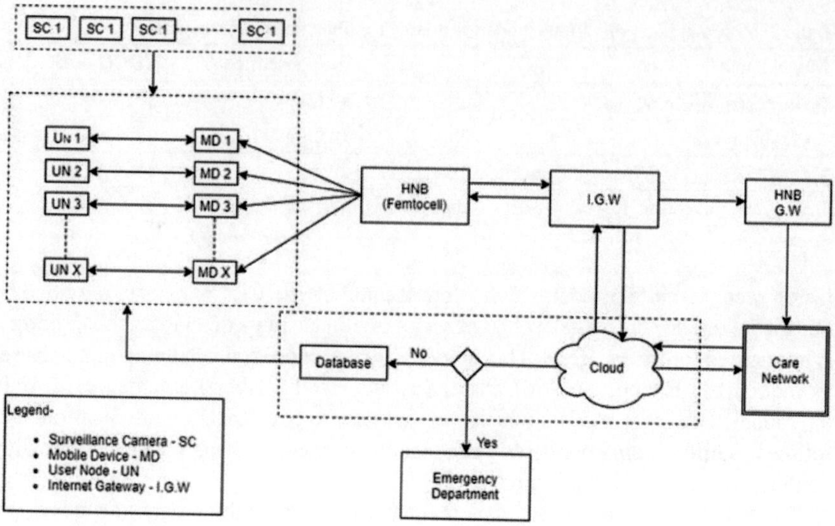

Fig. 2 Physical architecture

1. Home node base station (HNB), i.e., Femtocell-Femtocells are basically wireless devices which are cloud-enabled and are responsible for connecting a number of mobile devices to the cloud. A basic Femtocell can register around 4–5 devices while an enterprise Femtocell can register around 16–18 devices in it.
2. Internet connectivity: This is the basic necessity of the whole architecture and responsible for maintaining the interconnectivity among the devices, Femtocells and the cloud.
3. Mobile devices (user node): This node establishes a direct connection from the surveillance cameras and is the first store where the raw data arrives.
4. Surveillance cameras (SC): These are the edge devices which are placed in a vicinity with the objective to capture the data and send it to the user node, i.e., the mobile device.

Figure 2 is the pictorial illustration which explains the connectivity of all the components of the architecture and also how the data is defined to flow from each component.

4.3 *Case Study*

The Video Surveillance Management Service incurs huge storage and network costs. With the advancement of technology, people want streaming the CCTV footage and getting the notifications of suspicious activities on the tap without any delay. Traditionally with Internet-connected CCTV systems, live streaming came as an option but the constraints put a setback on the technology. Eventually with the advancement of cloud computing, the cloud technology most of the constraints got resolved and were also very cost effective.

The cloud-based video surveillance is named as Video Surveillance as a Service (VSaaS) [21] which gives many advantages over the traditional systems by reducing the IT costs, less infrastructure investments, low power costs, reliability, etc. With most of the things being managed at the cloud end, managing cameras becomes flexible. But these solutions are provided by the companies using their cloud services having common servers to provide the notifications for the suspicious activities which causes delay as servers are always busy processing all the streams coming from the cameras.

In our model, for this latency, FemtoCloud will be used to process the threat level at the edge network so that there is less delay in identifying the threat and notifying the security and the owner, making the architecture more reliable and available. Our aim is to shift the operationalities as close to the edge devices as possible. With FemtoCloud, it becomes easier for the users to control the functionality of the system related to them, giving them the opportunity to master their security measures.

5 Challenges and Perspectives

Based on various parameters, there exist numerous challenges which can hinder the efficiency of the FCHMS model. We have tried to cover various aspects in terms of hardware limitations [22, 23], software limitations and security support of IoT devices with cloud.

1. Hardware Limitations: Many factors determine the feasibility and cost optimization of implementing the security mechanisms. Bandwidth, latency and response time factor determine the efficiency of the hardware. Not only the device's specifications but the climatic conditions also affect the efficiency, for instance, radio signal propagation.
2. Software Limitations: We propose the optimal deep learning models for minimizing the probability of errors. However, there exist certain loopholes in the existing solution such as hidden objects which are non-detectable by simple surveillance cameras. Deployment on cloud ensures 99.9999999% of model availability (AWS Cloud). But the potential of edge devices is still neglected in terms of availability and reliability.
3. Security Support of Cloud: With the best security practices and approaches, cloud is still one of the secured platforms for accessing the IT services over the Internet. With the trends and facts, it is predicted that major security breaches tend to happen due to lack of attention from the client's side. Major data integrity loss also tends to happen from the client's device rather than from the cloud.

6 Conclusion

In this paper, we anticipated a multiparameter and secured FemtoCloud solution aimed at securing smart homes from coercions. The architecture focuses on the data captured by the edge devices, namely surveillance cameras to capture data and send it to the nearest mobile device in vicinity to identify the potential threats in a preserved parameter of any geographical location (A Smart Residential or a Smart Building Solution). On identification of any threat, the data and suggestions are provided to the user node or the user's mobile device via Femtocell. Technically, Femtocell is a master node station which guides, commands, connects all mobile devices (nodes) with the cloud, providing good coverage in a limited geographical location. Edge devices work on capturing data and sending data to mobiles in the vicinity, performing basic computations at the edge level itself. On identification of a serious threat, Femtocell is triggered to send the data of mobile (user node) to the cloud for performing comprehensive processing to estimate the intensity of the threats. Our system is bifurcated into two brief machine learning models which are trained and deployed at both the levels, i.e., mobile and cloud. Rudimentary computations are performed by a pre-trained Efficient Net-Lite Model for determining the possible threat. We use a modified open-source Faster RCNN Resnet50 COCO-trained comprehensive model to detect and determine the objects which can be a potential threat.

References

1. Pan J, McElhannon J (2017) Future edge cloud and edge computing for internet of things applications. IEEE Internet Things J 5(1):439–449
2. Shi W, Cao J, Zhang Q, Li Y, Xu L (2016) Edge computing: vision and challenges. IEEE Internet Things J 3(5):637–646
3. Khanna A, Goyal R, Verma M, Joshi D (2018, February) Intelligent traffic management system for smart cities. In: International conference on futuristic trends in network and communication technologies. Springer, Singapore, pp 152–164
4. Singh PK, Panigrahi BK, Suryadevara NK, Sharma SK, Singh AK (eds) Proceedings of ICETIT 2019, Emerging trends in information technology. Lecture notes in electrical engineering (LNEE), Springer Nature Switzerland AG. https://doi.org/10.1007/978-3-030-30577-2
5. Ren J, Yu G, He Y, Li GY (2019) Collaborative cloud and edge computing for latency minimization. IEEE Trans Veh Technol 68(5):5031–5044
6. De D, Mukherjee A (2015) Femto-cloud based secure and economic distributed diagnosis and home health care system. J Med Imag Health Inform 5(3):435–447
7. De D (2016) Mobile cloud computing: architectures, algorithms and applications. CRC Press
8. Ullah F, Naeem H, Jabbar S, Khalid S, Latif MA, Al-Turjman F, Mostarda L (2019) Cyber security threats detection in internet of things using deep learning approach. IEEE Access 7:124379–124389
9. Singh PK, Kar AK, Singh Y, Kolekar MH, Tanwar S (2019) Recent innovations in computing. Springer Nature, Switzerland AG, vol 597. ISBN 978-3-030-29406-9
10. Tomar R, Khanna A, Bansal A, Fore V (2018) An architectural view towards autonomic cloud computing. In: Data engineering and intelligent computing. Springer, Singapore, pp 573–582
11. Subashini S, Kavitha V (2011) A survey on security issues in service delivery models of cloud computing. J Netw Comput Appl 34(1):1–11

12. Li H, Ota K, Dong M (2018) Learning IoT in edge: deep learning for the Internet of Things with edge computing. IEEE Netw 32(1):96–101
13. Habak K, Ammar M, Harras KA, Zegura E (2015, June) Femto clouds: leveraging mobile devices to provide cloud service at the edge. In: 2015 IEEE 8th international conference on cloud computing. IEEE, pp 9–16
14. Zamora-Izquierdo MA, Santa J, Martínez JA, Martínez V, Skarmeta AF (2019) Smart farming IoT platform based on edge and cloud computing. Biosys Eng 177:4–17
15. Khanna A, Kero A, Kumar D (2016, October) Mobile cloud computing architecture for computation offloading. In: 2016 2nd international conference on next generation computing technologies (NGCT). IEEE, pp 639–643
16. Deb P, Mukherjee A, De D (2018) A study of densification management using energy efficient femto-cloud based 5G mobile network. Wirel Personal Commun 101(4):2173–2191
17. Lim J, Al Jobayer MI, Baskaran VM, Lim JM, Wong K, See J (2019, November) Gun detection in surveillance videos using deep neural networks. In: 2019 Asia-Pacific signal and information processing association annual summit and conference (APSIPA ASC). IEEE, pp 1998–2002
18. Higher accuracy on vision models with Efficient Net-Lite. (n.d.). TensorFlow Blog. Retrieved June 27, 2020, from https://blog.tensorflow.org/2020/03/higher-accuracy-on-vision-models-with-efficientnet-lite.html
19. Hosted models | TensorFlow Lite (2020). TensorFlow. https://www.tensorflow.org/lite/guide/hosted_models
20. T. (n.d.-b). TensorFlow/models. GitHub. https://github.com/tensorflow/models/blob/master/research/object_detection/g3doc/detection_model_zoo.md
21. VSaaS: 11 Reasons Cloud Video Surveillance is Moving to the Cloud. (2020, June 9). Eagle eye networks. https://www.een.com/vsaas-video-surveillance-moving-to-cloud/
22. Batalla JM, Vasilakos A, Gajewski M (2017) Secure smart homes: opportunities and challenges. ACM Computing Surveys (CSUR) 50(5):1–32
23. Ferrer AJ, Marquès JM, Jorba J (2019) Towards the decentralized cloud: survey on approaches and challenges for mobile, ad hoc, and edge computing. ACM Comput Surv (CSUR) 51(6):1–36
24. Balasubramanian V, Zaman F, Aloqaily M, Alrabaee S, Gorlatova M, Reisslein M (2019, April) Reinforcing the edge: autonomous energy management for mobile device clouds. In IEEE INFOCOM 2019-IEEE conference on computer communications workshops (INFOCOM WKSHPS), pp 44–49
25. Vijayalakshmi V, Vimal S (2019) A new edge computing based cloud system for IoT applications. Int J Recent Technol Eng (IJRTE) 8(2). ISSN: 2277–3878
26. Domb M (2019) Smart home systems based on internet of things. In Internet of Things (IoT) for automated and smart applications. IntechOpen
27. Gedawy H, Harras KA, Habak K, Hamdi M (2020) FemtoClouds beyond the edge: the overlooked data centers. IEEE Internet Things Mag 3(1):44–49

Security and Privacy Issues

Global Intrusion Detection Environments and Platform for Anomaly-Based Intrusion Detection Systems

Jyoti Snehi, Abhinav Bhandari, Manish Snehi, Urvashi Tandon, and Vidhu Baggan

Abstract The defense is the critical element of the computer system, and the most challenging issues are detecting the intrusion attacks. The IDS is the most critical cyber-security factor which can detect intrusion before, during, and after an attack. This paper provides an overall IDS benchmarking which quantifies different IDS properties, types of anomaly-based IDS that are deployed in different environments or platforms, and comparison among them based on methods used, their details, and advantages of each method. We have analyzed the different IDS techniques based on anomaly and various issues associated with anomaly-based IDSs. We addressed global environments for intrusion detection and framework for behavioral or anomaly-based intrusion detection systems and discussed the challenges facing anomaly-based IDSs. After reviewing the various anomaly-based IDS techniques, we have analyzed that successful detection rates could not be achieved by a single technique. To lower the false prediction rate and decreased the complexity of the process, an efficient automated hybrid technique is suggested for achieving accurate detection rates to enhance anomaly detection.

Keywords Intrusion detection system (IDS) · Anomaly-based IDS · Behavior-based IDS

J. Snehi (✉) · V. Baggan
Institute of Engineering and Technology, Chitkara University, Chitkara University, Punjab, India
e-mail: Jyoti.snehi@chitkara.edu.in

A. Bhandari
Department of Computer Science and Engineering, Panjabi University, Patiala, India
e-mail: bhandarinitj@gmail.com

M. Snehi
Engineering Services, Infosys Limited, Chandigarh, India
e-mail: snehi.manish@outlook.com

U. Tandon
Chitkara Business School, Chitkara University, Punjab, India
e-mail: urvashi.tandon@chitkara.edu.in

P. K. Singh et al. (eds.), *Proceedings of Second International Conference on Computing, Communications, and Cyber-Security*, Lecture Notes in Networks and Systems 203,
https://doi.org/10.1007/978-981-16-0733-2_58

1 Introduction

Cybercriminals across the globe are taking advantage of vulnerabilities and stealing information in unsecured networks. Different defensive strategies are used to defend the front access points of linked networks from internal and external attacks. While the detection of internal attacks by anti-threat applications alone is ineffective, firewalls and anti-malware software alone is not enough to protect a whole network from any attack. Intrusion detection systems (IDSs) are hardware and software systems that automatically detect and respond to attacks on computer systems and ensure integrity, availability, and confidentiality. An IDS is a framework for the security management of individual computer systems or computer networks. IDS collects traffic data from a computer or network system and are known as audit data. This audit data is analyzed to detect any system security policy violations, and a security break is concluded if any security breach is identified. This security breach is possible from two ends, one being inside the system/network known as misuse or outside the system/network. Network traffic data is used as the audit trail for intrusion detection. NIDS offers real-time detection via the hardware sensors that are usually placed at different ends in the network or along with the installed software on the network-connected computer systems. NIDS has network-level intrusion detection capabilities and is usually the standalone hardware tool. Intrusions can be classified as:

1. Single Intruder Single Terminal: It is an Intrusion situation in which a single intruder at one terminal attached directly or remotely begins an attack.
2. Single Attacker Multiple Terminal: It is an intrusion situation in which an intruder uses multiple windows with each window targeting a different target device and establishing multiple connections to the same target, to control a specific computer session.
3. Multiple Intruders Multiple Terminal: It is an intrusion case where multiple intruders simultaneously engage in one or more intrusions, one or more target machines, and intruders try to spread suspicious activity through several simultaneous sessions.
4. An intrusion detection system: It is a type of computer network protection system and a central component for the safety of production systems and protection tools designed to automatically alert administrators when something is trying to compromise the information system by malicious activities [1].

IDS supports by monitoring changes in network behavior, inspecting system operation, distinguishing between normal and abnormal activities with a restriction that often gives false alarms, takes time, and is not 100% safe from attacks.[1] A global IDS is a network consisting of several individual IDSs (IIDSs) consisting of a connection layer, a layer of Individual IDS, and a layer of detection schemes addressing all possible attacks [2, 3]. Intrusion detection attacks can be Denial-of-Service (DOS) Attacks, i.e., flooding and manipulation of defects. Eavesdropping attacks, spoofing attacks, intrusion attacks, or root attack users (U2R), logon misuse, and application-level attacks. IDS features include operating the machine without human interference,

Table 1 Types of attacks

S. No.	Categories of attacks	Types of attacks
1	Attacks associated with unauthorized access	Cracking password spoofing
		Trojan horses
		Network packet listening
		Unauthorized access to resources
		Misuse of authorized privileges
		Scanning ports and services
2	Misuse through Unauthorized access	Altering of information or deletion identity imitation
		Unauthorized creation of false data
		Illegal configuration changes
3	Denial of service	Flooding
		Exploiting the weakness of the system

fault tolerance, self-recovery during device crashes, and resilience to subversion, efficiency, observation of deviations from normal behavior, scalability, adaptability, power, and convenience for detection and reusability of attacks. The attacks can be divided into various categories according to their source [4]. Some of the attacks are (Table 1).

This paper provides a compiled and consolidated list of various anomaly-based methods used for the intrusion detection system along with evaluation metrics in Sect. 2. The study of methods of IDS is presented in Sect. 3 and models and methods for anomaly-based intrusion detection in Sect. 4. The last section consists of issues related to anomaly-based IDSs and finally, the work is concluded.

2 Evaluation Metrics

There are various metrics used for IDS benchmarking which quantifies different IDS properties. The evaluation metrics aim to ensure a fair assessment of the efficiency of the established framework. One may distinguish the metric groups as:

1. Performance-related metrics: Metrics quantifying non-functional properties of test-based IDSs, such as efficiency, overhead rate, and resource usage, etc., are considered under performance-related metrics.
2. Security-related metrics: Metrics that calculate IDS properties related to security problems, such as coverage of attacks, the accuracy of attack detection, etc. are covered under safety-related metrics.

There are a variety of assessment criteria available that can research and analyze the results of different classifiers. Table 2 shows the confusion matrix representation for IDS. The confusion matrix is often used for representing the effects of a

Table 2 Confusion matrix

		Predicted (P)	
		Normal	Anomalous
Actual (A)	Normal	True negative (1)	False positive (0)
	Anomalous	False negative (0)	True positive (1)

process of classification. The matrix of uncertainties serves as the basis for different metric measurements. The uncertainty matrix indicates the exact solution to a classification problem. A confusion matrix as shown in the table provides explanations of a classifier's actual and expected classifications made and includes details of the classifications that a classifier made and predicted [5].

True Negatives: Number of instances accurately determined to be non-attacks represented by TN.

False Negatives: Number of wrongly predicted instances determined to be as non-attacks represented by FN.

False Positives: Number of cases identified erroneously determined to be as attacks represented by FP.

True Positives: Number of instances corrected represented as TP [7].

The term false positive refers to security systems that mistakenly view legitimate requests as spam or breaches of security. Essentially, the IDS can sense something that shouldn't be. Instead, the IDSs are vulnerable to false negatives when the device fails to detect a request it should have. These are troublesome IDS-related problems. Nonetheless, vendors spend a lot of time working on them, and thus, IDS is not assumed to detect a large percentage of false positives or negatives. Table 2 shows various types of alarms based on a confusion matrix. Elements are developed from the IDS classification model.

Table 3 explains the usefulness of certain elements under various situations. The uncertainty matrix comprises four components, i.e., true positive, true negative, false positive, and false negative [6].

The evaluation metrics focus on the following major attributes of performance:

- Accuracy of results
- Performance of the developed system
- Fault tolerance
- Completeness of results

Table 3 Alarms based on the confusion matrix

Alarm	Description
True negative	No alarm raised when no attack happened
False negative	A scenario of actual attack when no alarm is raised
True positive	Correctly alarms an attack
False positive	An alarm is generated when no attack has happened

- Timeliness.

A strong false-positive number raises the false alarm number. Consequently, the NIDS will drop the false warning, which contains a valid packet. As a result, the IDS may drop a huge amount of normal packet and the network performance may be significantly impacted. The popular IDS metrics include:

(1) Accuracy (A):

$$A = \frac{\mathrm{TP} + \mathrm{TN}}{\mathrm{TP} + \mathrm{TN} + \mathrm{FP} + \mathrm{FN}} \tag{1}$$

(2) False alarm rate: Defined as protective, the speed at which usual connections occur. It is represented as FAR.

$$\mathrm{FAR} = \frac{\mathrm{FP}}{\mathrm{FP} + \mathrm{TN}} \tag{2}$$

(3) Reminder (R): Reminder is the proportion of properly predicted attack cases to the actual size of the attack class.

$$R = \frac{\mathrm{TP}}{\mathrm{TP} + \mathrm{FN}} \tag{3}$$

(4) False-Negative Rate: False-negative rate estimation using the FNR is done by calculating

$$\mathrm{FNR} = 1 - R = \frac{\mathrm{FN}}{\mathrm{TP} + \mathrm{FN}} \tag{4}$$

(5) Precision (P): Precision is the proportion of attack cases calculated to be correct following attack class size:

$$P = \frac{\mathrm{TP}}{\mathrm{TP} + \mathrm{FP}} \tag{5}$$

(6) Specificity (S): Specificity is the percentage of True Negative points in which negative elements.

$$S = \frac{\mathrm{TN}}{\mathrm{FP} + \mathrm{TN}} \tag{6}$$

(7) False-Positive Rate (FPR): This is correlated with the wrongly predicted positives.

$$\mathrm{FPR} = \frac{\mathrm{FP}}{\mathrm{FP} + \mathrm{TN}} \tag{7}$$

(8) F-score: It tests the balance between accuracy (A) and recall (R), and it is a measure of the test's accuracy.

$$F\text{-Score} = \frac{2 * P * R}{P + R} \tag{8}$$

In Eqs. 1–8, accuracy is the ratio of the total number of correct predictions the size of the set of individual data. False alarm rate defines protective, the speed at which usual connections occur. A reminder is the proportion of properly predicted attack cases to the actual size of the attack class, false-negative rate estimation using the FNR. Precision is the proportion of attack cases calculated to be correct following attack class size. Specificity is the percentage of true-negative points in which negative elements are calculated by the equation. False-positive rate is correlated with the wrongly predicted positives. F-score tests the balance between accuracy and recall, and it is a measure of the test's accuracy [6, 7].

3 Study of Methods of IDS

IDS tracks gather and analyze the network traffic and user actions in the way suspicious actions are detected. IDS can submit early alarm regarding the exposure risks resulting from any attack. It is to warn system administrators to perform accurate response measurements, thereby reducing the potential for significant system damage. Two architectures may be used to create global intrusion detection schemes. A centralized system where the central manager ensures the correlation layer work by installing specifically dedicated hosts and communicating with IIDS and performing correlation tasks based on the various alarm reports created by IIDs. As part of individual IIDs, a hierarchical architecture is introduced by contact between each IID to perform correlation tasks and to report warnings to the administrator if an intrusion is detected.

An IDS consists of several components such as a sensor that generates security events, a monitoring screen, warnings, and sensor controls, and central engine recording events that the sensor has entered into a database and updates the security events received. IDS groups can be divided into several classes to define and detect attacks, depending on the methods used to detect attacks and based on input data collected from various devices, such as audit reports, user or device activity, system requests, the application process, and network traffic [8].

(A). **IDS Methods based on Deployment**:

(1) HIDS: A host-based intrusion detection system is deployed to detect suspicious activities. They collect the information from a given host and analyze it. HIDS analyzes the information when any system or application activity changes; it instantly notifies the network manager that the computer is at risk. HIDS is primarily used to protect the integrity of systems [7].

(2) NIDS: A network-based intrusion detection system monitor network traffic and analyze traffic to identify potential intrusions such as DoS attacks, port scanning, and so on. NIDS collects the network packets and points out how they react to known attack signatures or compare the user's actual behavior to known attacks in real time [9].

(A). **IDS Methods based on Detection**

(a) Knowledge/Signature/ Misuse detection technique:
Signature-based recognition and identification are obtained by comparing the data found with the database of signatures. A signature may have a set of rules or patterns about known attacks beforehand. The first part gathers and analyzes the intrusion detection packets for the network. The second component declines the packets immediately and checks whether or not it corresponds to the block table rules, if packets with no autonomous element manager, autonomous coordinator, and correspondence to those rules are forwarded to the alert clustering module which generates alert for suspicious packets. The third component blocks suspicious packets and sends alerts to other IDSs. The last component acquires alerts and makes the decisions about packets. By implementing the above IDS, we can protect the system from a single point of fault attack.

(b) Behavior/Anomaly/Heuristic Detection Techniques: Detection technique based on anomaly or behavior measures the actual user operation against the user logs previously loaded. It causes many false alarms because of irregular network and device behavior. It also needs large datasets that can be identified at different rates to train the system for user profiles which are common Unknown attacks within an anomaly detection network. Large-data intrusions at different cloud levels are hard to monitor [7].

4 Methods for Anomaly/Behavior-Based Intrusion Detection

IDS plays a significant role in protecting the networks from malicious activity. Over the past few years, data mining methods have become increasingly important in tackling security issues within networks. Intrusion detection systems are designed to recognize low and high false alarm intrusion detection rates. Detection based on abnormalities concentrates on recognizing the behavior of the network. The behavior of the network is consistent with the predefined behavior; otherwise, it will be recognized or cause anomalies to be observed in the event. The agreed behavior in the network is planned or studied by the network administrator's directives [10]. The crucial step in assessing network conduct is the IDS engine's ability to break at all speeds through various protocols. The engine must be capable of interpreting the protocols and understanding its meaning. Since this protocol review is computationally expensive, the benefit it offers is that the set of rules results in less false positive alarms. Anomaly-based IDS tracks device behavior and detects the attack. Based on rules, events are classified as either natural or anomalous or heuristic rather than signatures. Anomalous means behavior which is unusual or which deviates from natural. A heuristic is nothing but a tool for practical thinking, exploration, problem-solving, etc. The system, therefore, works quickly toward seeking an acceptable solution using these methods and also identifies incorrect computer activities [11].

The anomaly detection system comprises three components, such as the semi-supervised, monitored, and controlled discovery of anomalies. The supervised method of anomaly detection aims at developing the model and constructs the anomalous records and usual records separately [12]. The semi-supervised methods for detecting anomalies are aimed at building the model, based on normal data. It is an exception because the documents do not fit the pattern. The semi-supervised methods can now construct the model according to abnormal data, but all the anomalies can be very difficult to classify. Semi-supervised detection, for this reason, includes a labeled database, which often reflects the high false-positive rate [13]. This problem is solved using the unsupervised anomaly detection method, and several new phenomena are found. This method did not involve a named dataset and was implemented without a change in every program [34].

(A) **Pros related to anomaly-based IDS**

- Database upgrades are not required to identify the latest attacks.
- Upon installation of a program, some maintenance time is required.
- It tracks parallel network operations and builds network operating profiles.
- Use a broader framework to recognize risks most efficiently.
- New attacks with no signature can be identified.

(B) Cons related to anomaly-based IDS

- Irregular conduct of network in regular traffic does not cause a warning to be sent to an administrator.
- False positives in the anomaly-based setup can get even more [9].
- The key limitation of detecting phenomena is the identification of a rule set [10].

IDS-based anomaly detection consists primarily of three methods, i.e., machine learning, arithmetic, and knowledge-based detection. The mathematical models suggest Internet-traffic activities. The anomaly detection mechanism employs statistical methods to traffic two datasets on the network. The current network profiles are contained in the first time-based study, and the second collection of data is the statistic profile previously eligible.[14] The network activities were performed, the current profile was determined and the anomaly score was calculated by comparing two activities. Usually, the score represents the degree of an anomaly for a given procedure. The knowledge-based approach is extensively utilized in IDS. A set of rules is used to check the input data label, and two stages for a set of rules are used. The input training data defines the various classes and network behaviors and extracts event classification from regular activities [15]

The following are numerous methods used for the identification of anomaly-based intrusion as compiled in Table 4.

5 Major Issues in Anomaly-Based IDS

An IDS focusing on network anomalies should lower the false alarm rate. Even then, the false warning cannot be reduced. The various challenges for the intrusion detection system are:

1. The creation of a general methodology or parameter set that can be used to test the framework for intrusion detection is a challenging problem.
2. Current intrusion detection systems aren't easy to track. Potential attack situations are difficult to anticipate, and it is difficult to reproduce known attacks. A lack of a standard audit trail format impedes testing and comparison of the efficacy of current programs with specific scenarios of attack.
3. Another challenge when new patterns are found in a NIDS is to upgrade the database without any output losses and the issue to tackle is the removal of pre-processing data technical problems during preparation and also during the delivery phase.
4. The main challenge is to establish an effective method of selecting the attributes for each attack type, and it is much harder to pick the best classifier from a collection of unrelated and unbiased classifiers to create an efficient anomaly-detection ensemble approach.
5. Compared to signature-based detection, anomaly-based detection is intended to recognize any activity that deviates from the normal pattern/profile of usage and is considered an intrusion.
6. While detection of anomalies can detect unknown attacks, it can generate large amounts of false alarms.
7. Weaknesses of current intrusion detection systems are the total cost of introducing a system for detecting the intrusion is high.
8. The definition of the rules of the system of experts and the choice of the underlying statistical metrics must be made by someone who is not only a protection expert but who is also familiar with the language of design of the rules of the system.

Detection systems for intrusion that are designed for one environment are difficult to use in other environments where similar policies and issues can occur. That's because most of the software needs to be unique to the monitoring context, and standards need to be set. That software is designed ad-hoc and tailored in a manner that suits its purpose. It is difficult to reuse and retarget when the device is constructed in such a structured way as to be unreliable or with restricted strength [43].

Table 4 Methods used for identification of anomaly-based intrusion

S. No.	Type of methods	Details	Advantages
1	Artificial Intelligence-based Methods [16–21]	It is a subgroup of machine learning that resembles biological neural networks' learning patterns	1. Helpful in tackling rule-based IDS deficiencies 2. Detects Known behavioral attacks and anomalies 3. Versatile, and easy to adapt
2	Fuzzy logic-based [21, 22]	It involves the fuzzification of input variables based on variables like spoofing, fraud, repudiation, disclosure of information, denial of service, privilege advancement, rule determination, and aggregation Outputs rule	1. This helps smooth out the sudden separation between normal behavior and abnormal activity 2. The resolution of complex problems causes high warnings when anomalous network activity has been detected
3	Fusion-based Combination Methods [23, 24]	It requires the fusion of multiple classifiers using a separate representation of Patterns apps	1. Ensemble methods perform well when many classifiers are combined. 2. More effective, as multiple base-level classifiers harness the diversity of predictions 3. Improve trade-off between a false alarm and identification rates for attacks
4	Machine learning techniques based on IDS [25, 26]	It focuses on pattern detection and building dataset-based intrusion detection framework	1. Improved accuracy 2. Fewer requirements for human knowledge
5	Genetic Algorithm-based IDSs [21, 27–32]	Computational models and search algorithms whose functioning is based on the principles of natural selection and a programming technique, which mimics biological evolution as a problem-solving approach	1. Provide better and faster classification 2. Takes less time for training 3. Possess capabilities for parallel processing 4. High attack detection rate and low false-positive rate
6	Ensemble-based methods and systems [13, 23]	Multiple individual classifiers are combined to get an overall classifier that can outperform each	1. It weighs the individual opinions combined to reach a final decision

(continued)

Table 4 (continued)

S. No.	Type of methods	Details	Advantages
7	Distributed and Collaborative IDS [33–35]	It allows information collected from various sources to detect network device attacks such as doorknob attacks and DDoS attacks	1. Scalability and greater versatility 2. Detects High-Speed Network DOS attacks 3. Reduce costs for computation 4. Its simpler and speedier to track, evaluate
8	Agent-based Intrusion detection [36]	They use the concept of automatic computing to track network traffic and system operations using autonomous sensors to identify suspected accidents	1. With less human involvement the system has the properties of self-management 2. During runtime, the agents are reconfigurable without the need to restart them 3. To achieve the necessary characteristics a layered management model is used
9	Statistical methods based on IDS [13, 23]	This involves gathering and evaluating data records in object sets, and developing a standard statistical model of user behavior. Statistical IDS utilizes the Univariate, Multi-Variate, and Time Series model	1. It does not warrant advanced awareness of the goal of the system's normal activities 2. It provides correct warning of malicious activities or generates an alarm 3. It analyzes traffic using the sudden transition hypothesis
10	Cloud-based intrusion detection [8, 37, 38]	It consists of a single controller used to handle IDS instances allocated to each Cloud user and provider	1. A large volume of data is managed by a multi-threaded method with a single node. 2. Low CPU, memory usage, and loss of packets to increase the overall cloud IDS performance. 3. It is capable of running simultaneous data analysis processing.

(continued)

Table 4 (continued)

S. No.	Type of methods	Details	Advantages
11	Virtual Machine Monitor-based IDS [39]	It encapsulates the monitoring system within a virtual machine which is controlled outside of the intruders' control and the detection and response mechanism can be implemented as host system processes	1. It allows processes to be analyzed separately, anomalous activities observed, and intrusions from compromised processes impeded. 2. Disruptions to legitimate procedures of visitors are minimized. 3. The virtual machine does not need to be suspended for confirmation of intrusion.
12	IDS for Grid and Cloud Computing IDS [11]	It has an evaluation framework that is intended to cover threats that can't be handled by network and host-based systems. GCCIDS incorporates an application of information and behavior to identify unique intrusions	1. Increasing threat reporting helps to detect newly identified threats by the expert system
13	Bayesian Classifier-based IDS [40, 41]	The Bayesian IDS is built from a Bayesian naïve classifier that works by realizing that features are likely to occur in attacks and regular TCP traffic	1. Improve R2L attack accuracy with Bayesian methods
14	Wireless-based IDS [42]	A wireless IDS is capable of recognizing ad-hoc networks, a can configuration that would theoretically enable hackers to manipulate a wireless system	1. It provides honesty, confidentiality, and accessibility 2. It deals with wide coverage and unrestricted access suggesting transparency against attacks 3. Wireless networks are flexible and independent of the framework of the structure

(continued)

6 Conclusion and Future Work

IDS are the most significant cyber-security factors able to detect intrusion before or after an attack. It plays an important role as a computer and network protection framework. An ID monitors the data on a device and analyzes it to detect any threat.

Table 4 (continued)

S. No.	Type of methods	Details	Advantages
15	Hypervisor-based IDS	Hypervisors are similar to VMMs and are a software layer that interposes Isolation, Inspection, and Interposition between the operating system and the underlying hardware based on three VM capabilities	1. It manages and implements different security measures for every VM 2. Availability of knowledge 3. This eliminates the requirement that software is installed in a virtualized cloud environment on the host computer or virtual machine 4. A hypervisor performs control and supervisory functions on the operating system

Intrusion detection has progressed dramatically over time, notably in the last few years, thanks to the latest technologies. This paper provided an overall overview and comparison between the types of anomaly-based IDS deployed in various environments or platforms. It illustrated the features, benefits, and disadvantages of every form. We have introduced the classification of IDS categories based on some parameters including a computer, network, host, virtual machine, and input data. The information presented provides an important starting point for discussion of IDS research and development based on anomalies. This study addresses numerous limitations and anomaly-related IDS problems, such as high false alarm rates, difficult to implement in large databases, heavy network traffic, time complexity in the training and testing phase, etc. After reviewing the various anomaly-based IDS techniques, we concluded that successful detection rates could not be achieved by a single technique. An efficient automated hybrid technique is suggested for achieving accurate detection rates to enhance anomaly detection. This also helps lower the false prediction rate and decreases the complexity of the process.

References

1. Saxena AK, Sinha S, Shukla P (2017) General study of the intrusion detection system and survey of agent-based intrusion detection system. In: Proceeding—IEEE international conference on computing, communication and automation, ICCCA 2017. 2017-Janua, 417–421 (2017). https://doi.org/10.1109/CCAA.2017.8229866
2. Labiod H, Boudaoud K, Labetoulle J (2000) Towards a new approach for intrusion detection with intelligent agents. Netw Inf Syst J 2:701–739
3. Agarwal N, Hussain SZ (2018) A closer look at intrusion detection system for web applications. Secur Commun Netw. https://doi.org/10.1155/2018/9601357
4. Bhandari A, Sangal AL, Kumar K (2014) Characterizing flash events and DDoS attacks—an empirical investigation. https://doi.org/10.1002/sec

5. Azwar H, Murtaz M, Siddique M, Rehman S (2019) Intrusion detection in secure network for cybersecurity systems using machine learning and data mining. In: 2018 IEEE 5th international conference on engineering technologies and applied sciences, ICETAS 2018, pp 1–9. https://doi.org/10.1109/ICETAS.2018.8629197
6. Xin Y, Kong L, Liu Z, Chen Y, Li Y, Zhu H, Gao M, Hou H, Wang C (2018) Machine learning and deep learning methods for cybersecurity. IEEE Access 6:35365–35381. https://doi.org/10.1109/ACCESS.2018.2836950
7. Singhal A (2007) Intrusion detection systems. Adv Inf Secur 31:43–57. https://doi.org/10.4018/978-1-59904-168-1.ch007
8. Yassin W, Udzir NI, Muda Z, Abdullah A, Abdullah MT (2012) A Cloud-based intrusion detection service framework. In: Proceedings 2012 international conference on cyber security, cyber warfare and digital forensic, CyberSec 2012, pp 213–218. https://doi.org/10.1109/CyberSec.2012.6246098
9. Verma J, Bhandari A, Singh G (2020) Review of existing data sets for network intrusion detection system. Adv Math: Sci J 9(6):3849–3854. https://doi.org/10.37418/amsj.9.6.64
10. Satam P (2017) Anomaly based Wi-Fi intrusion detection system. In: Proceedings—2017 IEEE 2nd international workshops on foundations and applications of self* systems, FAS*W 2017, pp 377–378. https://doi.org/10.1109/FAS-W.2017.180
11. Vieira K, Schulter A, Westphall C, Westphall CM (2010) Intrusion detection for grid and cloud computing. IT Prof 12:38–43. https://doi.org/10.1109/MITP.2009.89
12. Arshad J, Azad MA, Amad R, Salah K, Alazab M, Iqbal R (2020) A review of performance, energy and privacy of intrusion detection systems for IoT. Electronics (Switzerland). 9:1–24. https://doi.org/10.3390/electronics9040629
13. Khraisat A, Gondal I, Vamplew P, Kamruzzaman J (2019) Survey of intrusion detection systems: techniques, datasets and challenges. Cybersecurity 2. https://doi.org/10.1186/s42400-019-0038-7
14. Anton SDD, Sinha S, Dieter Schotten H (2019) Anomaly-based intrusion detection in industrial data with SVM and random forests. In: 27th international conference on software, telecommunications and computer networks, SoftCOM 2019. https://doi.org/10.23919/SOFTCOM.2019.8903672
15. Nascimento G, Correia M (2011) Anomaly-based intrusion detection in software as a service. In: Proceedings of the international conference on dependable systems and networks, pp 19–24. https://doi.org/10.1109/DSNW.2011.5958858
16. Alrajeh NA, Lloret J (2013) Intrusion detection systems based on artificial intelligence techniques in wireless sensor networks. Int J Distrib Sens Netw. https://doi.org/10.1155/2013/351047
17. Ali A, Hu Y, Hsieh CG, Khan M (2017) A comparative study on machine learning algorithms for network defense. 68:1–19. https://doi.org/10.25778/PEXS-2309
18. Kanimozhi V, Jacob TP (2019) artificial intelligence based network intrusion detection with hyper-parameter optimization tuning on the realistic cyber dataset CSE-CIC-IDS2018 using Cloud computing. ICT Express 5:211–214. https://doi.org/10.1016/j.icte.2019.03.003
19. Yin C, Zhu Y, Fei J, He X (2017) A deep learning approach for intrusion detection using recurrent neural networks. IEEE Access 5:21954–21961. https://doi.org/10.1109/ACCESS.2017.2762418
20. Kumar G, Kumar K, Sachdeva M (2010) The use of artificial intelligence based techniques for intrusion detection: a review. Artif Intell Rev 34:369–387. https://doi.org/10.1007/s10462-010-9179-5
21. Napanda K, Shah H, Kurup L (2015) Artificial intelligence techniques for network intrusion detection. Int J Eng Res V4:357–361. https://doi.org/10.17577/ijertv4is110283
22. Elhag S, Fernández A, Altalhi A, Alshomrani S, Herrera F (2019) A multi-objective evolutionary fuzzy system to obtain a broad and accurate set of solutions in intrusion detection systems. Soft Comput 23:1321–1336. https://doi.org/10.1007/s00500-017-2856-4
23. Bhuyan MH, Bhattacharyya DK, Kalita JK (2014) Network anomaly detection: methods, systems and tools. IEEE Commun Surv Tutorials 16:303–336. https://doi.org/10.1109/SURV.2013.052213.00046

24. Giacinto G, Roli F, Didaci L (2003) Fusion of multiple classifiers for intrusion detection in computer networks. 24:1795–1803. https://doi.org/10.1016/S0167-8655(03)00004-7
25. Amrita, Kant, S.: Machine learning and feature selection approach for anomaly based intrusion detection: A systematic novice approach. International Journal of Innovative Technology and Exploring Engineering. 8, 434–443 (2019).
26. Sedjelmaci H, Senouci SM, Ansari N (2017) Intrusion detection and ejection framework against lethal attacks in UAV-aided networks: a bayesian game-theoretic methodology. IEEE Trans Intell Transp Syst 18:1143–1153. https://doi.org/10.1109/TITS.2016.2600370
27. Anand Sukumar JV, Pranav I, Neetish MM, Narayanan J (2018) Network intrusion detection using improved genetic k-means algorithm. In: 2018 international conference on advances in computing, communications and informatics, ICACCI 2018, pp 2441–2446. https://doi.org/10.1109/ICACCI.2018.8554710
28. Goyal A (1999) GA-NIDS: a genetic algorithm based network intrusion detection system. Electr Eng 2–5
29. Srinivasa KG (2012) Application of genetic algorithms for detecting anomaly in network intrusion detection systems. Lecture notes of the Institute for Computer Sciences, Social-Informatics and Telecommunications Engineering, LNICST, vol 84, pp 582–591. https://doi.org/10.1007/978-3-642-27299-8_61
30. Majeed PG, Kumar S (2014) Genetic algorithms in intrusion detection systems: a survey. Int J Innov Appl Stud 5:2028–9324
31. Rai N (2014) Genetic algorithm based intrusion detection system. Int J Comput Sci Inf Technol (IJCSIT) 5:4952–4957
32. Uppalaiah B, Anand K, Narsimha B, Swaraj S, Bharat T (2012) Genetic algorithm approach to intrusion detection system. Engineering. 3(2):156–160. ISSN 0976-8491
33. Othman SM, Alsohybe NT, Ba-alwi FM, Zahary AT (2018) Survey on intrusion detection system types. Int J Cyber-Sec Digit Foren 7:444–462
34. Aggarwal P, Kumar S (2015) Analysis of KDD Dataset attributes—class wise for intrusion detection. Procedia Procedia Comput Sci 57:842–851. https://doi.org/10.1016/j.procs.2015.07.490
35. Snehi, M.: Security management in SDN using Fog computing: a survey. In: Strategies for e-Service, e-Governance, and Cyber Security. CRC Press (2020).
36. Kene SG, Theng DP (2015) A review on intrusion detection techniques for cloud computing and security challenges. In: 2nd international conference on electronics and communication systems, ICECS 2015, pp 227–232. https://doi.org/10.1109/ECS.2015.7124898
37. Shelke MPK, Sontakke MS, Gawande AD (2012) Intrusion detection system for cloud computing. Int J Sci Technol Res 1:67–71
38. Mehta A, Panda SN (2016) Comparative analysis of cloud simulators and authentication techniques in Cloud computing. J Today's Ideas Tomorrow's Technol 4:181–191. https://doi.org/10.15415/jotitt.2016.42010
39. Laureano M, Maziero C, Jamhour E (2004) Intrusion detection in virtual machine environments. In: Conference proceedings of the EUROMICRO, vol 30, pp 520–525. https://doi.org/10.1109/eurmic.2004.1333416
40. Altwaijry H, Algarny S (2012) Bayesian based intrusion detection system. J King Saud Univ Comput Inf Sci 24:1–6. https://doi.org/10.1016/j.jksuci.2011.10.001
41. Mukherjee S, Sharma N (2012) Intrusion Detection using naive bayes classifier with feature reduction. Procedia Technol 4:119–128. https://doi.org/10.1016/j.protcy.2012.05.017
42. Peng K, Leung VCM, Zheng L, Wang S, Huang C, Lin T (2018) Intrusion detection system based on decision tree over big data in Fog environment. Wirel Commun Mobile Comput. https://doi.org/10.1155/2018/4680867.
43. Lundin E, Jonsson E (2000) Anomaly-based intrusion detection: privacy concerns and other problems. Comput Netw 34:623–640. https://doi.org/10.1016/S1389-1286(00)00134-1

Image Steganography Using Bit Differencing Technique

Mudasir Rashid and Bhavna Arora

Abstract The swift progression of evidence correspondence in contemporary time demands protected communication of data. Steganography obscures personal material in numerous record organisations, for example, photograph, content, sound, and audio–visual, impalpability, payload and vigour are the key complications on the way to steganography. This proposed work would give a nascent procedure that could be used for storing imperative data inside some cover metaphors by means of most significant bits (MSB) of the cover metaphors that have been presented. Firstly, the bits are numbered in increasing order, and bit number 6 is concerned over storing the confidential data which is based on the difference of bits 6 and 7. The result is generated on the basis of the difference whether the resultant bit is same as that of secret bit or not. If the resultant bit is distinct as compared to confidential pixel, the bit number 6 is modified and updated in the original bit. The outcome communicates the suggested model that must give good percentile of improvements in signal-to-noise ratio. Here, picture is fragmented into red, green and blue components where red components used for signalling on the way to store information in green or blue components of the portrait. The proposed method built on most significant bits (MSB) is used for safeguarding the organisation from unauthorised admittance, and the interlopers would not be intelligent to admittance the intimate data.

Keywords Bit differencing · Steganography · Image hiding

1 Introduction

Steganography is a substantial zone of investigation in image processing together with numerous outcomes. Steganographic techniques are the ways of using original pictures for storing essential and important information though there is not any change in the original image's appearance. Steganography and cryptography are the procedures for making certain about the cataloguing and anonymous information.

M. Rashid (✉) · B. Arora
Central University of Jammu, Jammu, Jammu and Kashmir 181143, India

P. K. Singh et al. (eds.), *Proceedings of Second International Conference on Computing, Communications, and Cyber-Security*, Lecture Notes in Networks and Systems 203,
https://doi.org/10.1007/978-981-16-0733-2_59

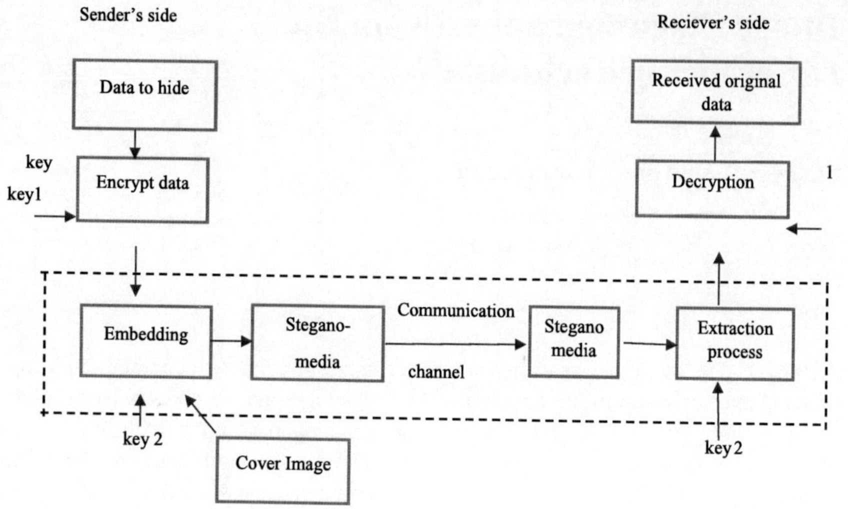

Fig. 1 Block diagram of image steganography

In cryptography, anonymous content is reformed into figure content, even though in steganography, the anonymous content endures as formerly nevertheless it is set in one more association of information. Currently, within the view of absurd correspondence outlines, defending the anonymous information from the interlopers is a problematic task. The anonymous data that is to be taken from the developers is a tough task. Steganography obscures existence of the information and safeguard anonymous information from an unapproved get to. The information hiding structure encompasses the following subcomponents: the important message to be stored, an original image in which to store the data and the resulted hidden image or text [1]. Mystery data to be encoded is known as plain content. Original document can be content, picture, sound or video in which information is hidden. Steganographic records the yield of the steganographic framework that contains the concealed data. The three central merits of a steganographic structures are as follows: (1) safekeeping, (2) payload capacity and (3) strength (Fig. 1).

1.1 Image Steganography Techniques

Keeping overview of the scrutiny of steganography implemented procedures, all tools are panelled into two groupings:

- Spatial domain-based steganography
- Transform domain-based steganography.

- ***Spatial Domain-Based Steganography***

Spatial steganography fundamentally integrates least significant bit (LSB), and this form of addition is a representative, straight forward mode to pact with implanting information in a concealment portrait. The least significant bit (the eighth bit) of a rare or the total of the bytes inside a portrait is altered to a bit of the secret memorandum.

Pixel: (10101111 11101001 10101000)
(10100111 01011000 11101001)
(11011000 10000111 01011001)

Secret message: 01000001
Result: (10101110 11101001 10101000)
(10100110 01011000 11101000)
(11011000 10000111 01011001).

- ***Transform Domain-Based Steganography:***

Principally, there are frequent categories of intensity level changes that occur to transfer a portrait to a re-appearance range, some of which are discrete cosine transform, KL transform and wavelet transform. Transform domain policies have an ideal location over LSB structures which is nice, and they obscure information in regions of the portrait that are a lesser amount offered to pressure, trimming and picture preparing.

The Discrete Cosine Transform (DCT)

These measured transforms translate the pixels so as to stretch the effect of "scattering" the portion of the pixel over representation [3]. The DCT fluctuates a symbol from a portrait illustration into a replication illustration, by gathering the pixels into 8×8-pixel squares and altering the pixel barricades into 64 DCT. DCTs are operated in steganography as—images are fragmented into 8×8 squares of pixels. Working from left to right and top to bottom, the DCTs are useful to each square. Each square is jam-packed through quantization table to scale the DCT coefficients, and message is implanted in DCT coefficients.

The Discrete Wavelet Transform (DWT)

Wavelets transformation (WT) varies over spatial space information to the re-appearance range information. Wavelets are useful in the portrait steganographic model on the estates that the wavelet alteration purely parts the high-relapse and low-relapse statistics on a pixel by pixel premise. This structure injects important files or information in the control and point areas of the portrait. When all is said and done, the human eye is more and more quick-tempered to disorder in the horizontal quarters of a portrait.

2 Literature Survey

In the paper [4], summary of image steganography, its applications and procedures are presented. Most of the strong and weak points related to important image files with diverse procedures of storing messages are given in which one method is deficient in payload dimensions, whereas other is deficient in robustness but together upsurges the possibility recognition.

The authors in [5] provide new and efficient ways of steganography using least significant bit (LSB) of image pixels. This paper [5] also provides the detection capability for the number of bits being used for embedding prior the user can make difference between hidden image and cover image.

The paper [6] presents the overview investigation to discuss the various effective parameters of the given techniques. The effectiveness of the parameters is assessed using mean square error (MSE) and peak signal-to-noise ratio (PSNR), processing time and security. The results disclosed that discrete cosine transform is superlative technique.

The paper [7] gives some important standards and guidelines computed using the overview investigation and provides up-to-the-minute appraisal and scrutiny of the methods used by steganography. The conclusion of the paper provides acclamation for the object-oriented technique.

This paper [8] presents the procedures used for discrete cosine transform (DCT)-based steganography and discrete wavelet transform (DWT), and the authors have provided various practices used for hiding information or some undisclosed files in the image file formats. The evaluation of results has been performed to know which procedure is good for image hiding.

The authors in [9] have presented technique for hiding essential information using most significant bits (MSB) of image pixels. The difference between bit number 5 and bit number 6 is calculated and if the outcome is unlike from that of secret data bit. Then, value of bit number 5 is altered. The consequences generated from the above investigation reveal that the projected method advances signal-to-noise ratio.

The paper [10] assesses several methods and techniques used in stegano scrutiny, notions and techniques used in spatial representation. Altogether the probable imminent exploration drifts related to steganography safekeeping and substantiation abilities are summarised.

The authors in [11] recommend novel technique of image coding by loading data in the carefully chosen pixel and on the succeeding value of the designated pixel. Selected pixel is used to accumulate the primary bit of the data, and pixel+1 value is used to store another bit of the data. The pixel+1 variable is generated by applying mathematical function on the 7th bit of image data.

This paper [2] analyses the existing image steganographic techniques and its types. Moreover, contribution of various modalities of these techniques is concerned. General procedure, necessities, different characteristics, different categories and their routine estimations regarding to image steganography are overviewed.

The authors in [12] overview all the steganographic approaches (current steganographic methods, such as F5, Outguess, Steghide, JPhide and Jsteg) used for several JPEG images, and their suitable arrangement has been done. Furthermore, the cataloguing of wholly varieties of steganographic images has been performed by the proposed algorithm to detect 109 features and trans SVM.

Coloured steganography created on DCT and a universal adaptive region (GAR) is presented in paper [13]. In this method, the same area under unique image coefficients and mystery data is realised. The main advantages of using this method increased payload capacity among most steganographic techniques which are gained.

Histogram technique is used for data hiding in the paper [14]. In this paper, pixels are mainly characterised based on image characteristics. From the smooth region, a smaller number of pixels are selected, whereas huge fidelity requires large number of pixels.

The authors in paper [15] have proposed called as speeded up robust features (SURF) used for identifying the important areas in the cover image. Steganography process is completed by modulating the secret data by using wavelet coefficients.

A genetic method is for detecting the highest quality vicinity from the original image [16]. Then mystery bit is embedded randomly using on LSB replacement.

In the paper [17], the reduction in the comparison in between original data and steganographic data is done by using pixel modification algorithm, whereas LSB method is used for embedding private file.

The new and amended method has been proposed in the paper [18] which is used for addressing various problems related to data hiding which were not addressed yet like step of security, key size and payload capacity.

The authors in the paper [19] have proposed a novel algorithm known as Blowfish encryption algorithm which is used for improving proficiency and safekeeping. The undisclosed data is encoded using blowfish algorithm formerly implanting the message.

A novel technique has been proposed in the paper [20] which is resulted from the grouping of steganography and cryptography. Cipher scheme is used for coding the data, and the resultant data is implanted inside image via LSB method.

In the paper [21], the author has proposed a new method by grouping steganography and cryptography. In this technique, Jamal encryption algorithm has been used as cryptographic algorithm, and LSB technique with 128 bit sego-key has been used as steganographic algorithm.

A new technique for hiding the confidential data has been proposed in the paper [22] which addresses the pixel values of every image that has been used in the experiment.

3 Problem Statement

In the new era of communication and expertise, perception of steganography is a very decisive and vital idea to be understood by the innovators and programmers. To explore the security and substantiation concerns in the communication, first concept known as LSB-based hiding was introduced due to its easiness of usage. Though the idea has several numbers of drawbacks regarding to the security, this concept is applicable all over the world to get the evidence conveyed from one user to other. To upsurge the efficiency and security of the system, most significant bit (MSB)-based image steganography procedure has been proposed which uses 16-bit image file to store and hide the significant data. By the applications and calculations of the proposed system, the hidden data could be shifted from one user to other user lacking the intervention of any other unauthorised third party. In the prosed system, the surreptitious data would be warehoused in the calculated MSB of the cover file.

Demerits of LSB-based image steganography

1. The information being transferred using this concept could be easily attacked by the attackers. The attackers could get the LSBs of the original file to generate image.
2. This concept was used since other efficient techniques were not discovered yet, and it is very less efficient as compared to other techniques.
3. In case of LSB-based image hiding, if several numbers of LSBs are used for storing the important information, the quality may get decreased so it is not preferably used.

Merits of MSB-based image steganography

1. The MSB-based image hiding is preferred as compared to LSB-based image hiding because the interlopers cannot decrypt the image or file simply by generating MSB of the picture.
2. This encryption technique is more difficult to use in image hiding because it was made to remove the complexity of the LSB-based image hiding so the interlopers could not attack this type of encoding.
3. This technique removes the drawback of the LSB-based image hiding because more than one MSBs can be used to store the data files without degrading the original image's quality.

Based on the existing solutions, we highlight the advantages and disadvantages of classification methods in Table 1 and various image steganography techniques in Table 2. These solutions make balance between different attributes like security, payload capacity, data capacity, etc. So, depending on the situations, the best feasible solution must be chosen.

4 Proposed Work

In this system, a brand-new approach is provided wherein most significant bits (MSB) of the given file are calculated on the way to conceal mystery pixels of the file. The confidential message is saved with inside the MSB of the image. The concept is used to increase the efficiency and security of the data communication.

Table 1 Comparison of various steganographic classification methods [2]

References	Classification technique	Method explanation	Advantages	Disadvantages
Rabie and Kame [13]	Region-based transform domain DCT	Coloured steganography is built on DCT and universal adaptive region (GAR)	Sophisticated payload dimensions Adequate noiselessness	Deprived security Reduced amount of corroboration for pictures being used
Cheng et al. [14]	Region grounded spatial domain histogram alteration	Histogram move technique is used for data hiding. The estimate model used is based on image inpainting	As the reference points on the receiving end are correctly identified, so accuracy is guaranteed	Lower payload capacity Geometric and compression attacks are less concerned
Hamid et al. [15]	Region-based transform domain DWT	Detecting the most important regions in the cover image, the method called as speeded up robust features (SURF) is used	Even if the image is modified by the attacks, the SURF can be used to identify the points	The total keeping data capacity is not as per requirement
Shah and Bichkar [16]	Spatial domain LSB genetic algorithm	A genetic model for detecting the highest quality vicinity from the original picture and then private pixel embedded normally on LSB replacement	Higher state of being imperceptible A number of cryptographic techniques are used to increase security	High complexity
Bandyopadhyay et al. [17]	A spatial domain LSB GA	The reduction in the difference between cover image and steganographic image is done by using pixel modification algorithm, whereas LSB method is used for embedding secret data	Higher state of being imperceptible More robust against histogram and noise attacks	It is least concerned about the geometric modifications

Table 2 Review of various image steganography techniques [23]

	Technique	Advantages	Disadvantages
Spatial domain	Least significant bit (LSB) Pixel value detector (PVD) Deviation extension	Simpler to use for encoding and decoding of data Good payload capacity long encoding capability Efficient inserting capability Efficient visual quality compared to past methodologies	Minimum safety during geometric, compression and statistical targeting Missing of safety equipment and techniques Minimum safety during geometric, compression and statistical targeting Need large location data for extracting secret data Poor control of capacity
Transform domain	Discrete fourier transform (DFT) Discrete cosine transform (DCT)	Normal transform domain accustomed in file hiding model Better visual quality than DFT	Minimum inserting capability Minimum visual quality Missing of safety Less inserting capability Missing of safety techniques Minimum robustness against attack

4.1 Encryption Algorithm for Image Steganography

Step 1: Interpret the quilt picture and textual content that is to be kept hidden with inside the cowl picture.
Step 2: Transform the textual content to computer understandable form.
Step 3: Return MSB of every byte of quilt picture.
Step 4: Restore MSB of cowl picture by every bit of private text one after another.
Step 5: Print steganographic picture.
Step 6: Display the steganographic image.

4.2 Decryption Algorithm for Image Steganography

Step 1: Interpret the steganographic file.
Step 2: Return most significant bit (MSB) of each and every bit of steganographic image.
Step 3: Get pixels and transform each and every 8 bits into characters.
Step 4: Display the original cover image (Tables 3 and 4).

Table 3 Efficiency analysis of the proposed system (256 * 256) 2 KB pictures [11]

Picture name	Textual width (KB)	PSNR	MSE
Lena	2	57.4020	0.1894
Baboon	2	57.4176	0.1756
Home	2	57.4785	0.1874
Girl	2	57.3572	0.1924
Clock	2	57.3333	0.1898
Cameraman	2	57.4505	0.1874

Table 4 Efficiency analysis of the proposed system using (256 * 256) 4 KB images [11]

Image name	Message size	PSNR	MSE
Lena	Four kilobytes	52.4146	0.3755
Baboon	Four kilobytes	52.3867	0.3653
Home	Four kilobytes	52.4108	0.3875
Girl	Four kilobytes	52.3231	0.3976
Clock	Four kilobytes	52.3892	0.3876
Cameraman	Four kilobytes	52.3855	0.3755

5 Experimental Results and Analysis

The performance is reviewed on the premise of two arguments [24], that is, peak noise-to-signal ratio (PSNR) and mean square error (MSE). Returned results display the maximum efficiency of the proposed system.

$$\text{Mean Square Error} = \frac{1}{R * C}\sum_{i=1}^{R}\sum_{j=1}^{c}\left(x_{ij} - \chi'_{ij}\right)$$

whereas R and C portray the properties of the picture matrix, x_{ij} portrays the quilt picture, and x'_{ij} portrays the steganographic pictures.

$$\text{PSNR} = \log_{10} + \sum_{n=1}^{\infty}\left[\frac{\mathrm{i}^2}{\text{MSE}}\right](dB)$$

whereas i portrays the highest possible value of the pixel in a quilt picture. PSNR is calculated in decibel (Table 5).

Table 5 Evaluation of the proposed system with distinctive methods on PSNR by hiding 8 KB of pixels in picture of resolution (256 ∗ 256)

Picture name	Traditional LSB system [25]	SCC system [25]	PIT system [25]	FMM system [25]	CST system [25]	Proposed system [25]
Lena	40.51	41.64	41.35	42.57	53.94	50.45
Baboon	65.87	58.88	56.90	55.66	59.89	50.70
House	53.40	49.77	49.10	65.44	53.20	48.40
Couple	51.39	45.88	47.54	48.39	53.88	47.43
Trees	67.27	50.77	47.63	48.12	39.46	45.30
Moon	65.10	52.32	44.40	55.78	49.50	47.55

6 Conclusion

In this work, we have studied fundamentals of steganography, several classification approaches and dissimilar steganographic procedures. In addition, this effort presents a steganographic policy using grey colour image as a concealment medium. MSBs are used to obscure surreptitious message inside cover image to upsurge the safekeeping of the procedure as an alternative of entrenching message inaugural after maximum leftward crook or lowest right, and dominant port is carefully chosen for implanting, providing more robustness to the technique.

References

1. Khan Z, Shah M, Naeem M, Mahmood T, Khan S, Amin N, Shahzad D (2016) Threshold based steganography: a novel technique for improved payload and SNR. Inte Arab J Inf Technol 13(4):380–386
2. Kadhim IJ, Premaratne P, Vial PJ, Halloran B (2019) Comprehensive survey of image steganography: techniques, evaluations, and trends in future research. Neurocomputing 335:299–326
3. Mehta AM, Lanzisera S, Pister KSJ (2016) Steganography 802.15.4 wireless communication
4. Morkel T, Eloff JHP, Oliver MS (2015) An over view of image steganography. Information and Computer Security Architecture (CSA) Research Group
5. Chandramoul R, Memon N (2014) Analysis of LSB based image steganography techniques. 0-7803-6725- 1/0 1/$10.00 IEEE
6. Kaur G, Kochhar A (2012) A steganography implementation based on LSB & DCT. Int J Sci Emerg Technol Latest Trends 4(1):35–41
7. Cheddad A, Condell J, Curran K, Mc Kevitt P (2010) Digital image steganography: survey and analysis of current methods. Sig Process 90:727–752
8. Kaur N, Bansal A (2014) A review on digital image steganography. (JCS T) Int J Comput Sci Inf Technol 5(6):8135–8137
9. Islam AU, Khalid F, Shah M, Khan Z, Mahmood T, Khan A, Al U, Naeem M (2016) An improved image steganography technique based on MSB using bit differencing. 978-1-5090-2000-3/16/$31.00 ©2016 IEEE

10. Li B, He J, Huang J, Shi YQ (2017) A survey on image steganography and steganalysis. J Inf Hiding Multimedia Sig Process 2(2)
11. Joshi K, Gill S, Yadav R A new method of image steganography using 7th bit of a pixel as indicator by introducing the successive temporary pixel in the gray scale image. Hindaw J Comput Netw Commun 2018(9475142):10. https://doi.org/10.1155/2018/9475142
12. Pan X, Yan BT, Niu K (2010) Multiclass detect of current steganographic methods for JPEG format based re-stegnography. 978-1-4244-5848-6/10/$26.00 ©2010 IEEE
13. Rabie T, Kamel I (2017) High-capacity steganography: a global-adaptive-region discrete cosine transform approach. Multimedia Tools Appl 76:6473–6493
14. Cheng PH, Chang KC, Liu CL (2017) A reversible data hiding scheme for VQ indices using histogram shifting of prediction errors. Multimedia Tools Appl 76:6031–6050 https://doi.org/10.1007/s11042-015-3142-z
15. Hamid N, Yahya A, Ahmad RB, Al-Qershi O (2012) Characteristic region based image steganography using speeded-up robust features technique. In: International conference on future communication networks
16. Shah PD, Bichkar RS (2018) A secure spatial domain image steganography using genetic algorithm and linear congruential generator. In: Dash SS et al (eds) International conference on intelligent computing and applications, advances in intelligent systems and computing, vol 632. https://doi.org/10.1007/978-981-10-5520-1_12
17. D Bandyopadhyay, Dasgupta K, Mandal JK, Dutta P, Ojha VK, Snášel V (2014) A framework of secured and bio-inspired image steganography using chaotic encryption with genetic algorithm optimization (CEGAO). In: Krömer P et al (eds) Proceedings of the fifth international conference on innovations in bio-inspired computing and applications IBICA. Advances in intelligent systems and computing, vol 303. https://doi.org/10.1007/978-3-319-08156-4_27
18. Rahna E, Govindan VK, "A novel technique for secure, lossless steganography with unlimited payload and without exchange of stegoimage. Int J Adv Eng Technol. ISSN: 22311963
19. Barhoom TS, Mousa SMA (2015) A steganography LSB technique for hiding image within image using blowfish encryption algorithm. Int J Res Eng Sci (IJRES) 3(3):61–66. ISSN (Online): 2320-9364, ISSN (Print): 2320-9356. www.ijres.org
20. Laskar SA, Hemachandran K (2012) High capacity data hiding using LSB steganography and encryption. Int J Database Manag Syst (IJDMS) 4(6)
21. Al-Qwider WH, Salameh JNB (2017) Novel technique for securing data communication systems by using cryptography and steganography. Jordanian J Comput Inf Technol (JJCIT) 3(2)
22. Khan Z, Shah M, Naeem M, Mahmood T, Khan SNA, Amin NU, Shahzad D Threshold-based steganography: a novel technique for improved payload and SNR. In: IAJIT first online publication
23. Gayathri C, Kalpana V (2013) Study on image steganography techniques. Int J Eng Technol (IJET) 5(2). ISSN: 0975-4024
24. Joshi K, Gill S, Yadav R (2018) A new method of image steganography using 7th bit of a pixel as indicator by introducing the successive temporary pixel in the gray scale image. Hindaw J Comput Netw Commun 2018(9475142):10. https://doi.org/10.1155/2018/9475142
25. Khan M, Muhammad S, Irfan M, Seungmin R, Sung BW (2015) A novel magic LSB substitution method (M LSBSM) using multilevel encryption and achromatic component of an image. Springer, Berlin, Germany

Detection and Prevention of DoS and DDoS in IoT

Meetu Sharma and Bhavna Arora

Abstract Internet of Things (IoT) is a network of interconnected devices embedded with software, sensors and essential electronics that allow us to gather and exchange data between them. Through IoT, it is difficult to guarantee the privacy and protection of the users due to various artifacts linked to the Internet. Denial of Service (DoS) and Distribution Denial of Service (DDoS) are among the main security issues in IoT. DoS is a type of attack where attackers try to prevent access by legitimate users to the service. A DDoS is where multiple systems target a single, DoS attack system. This occurs when several systems overload a target system's bandwidth or resources, normally at one or more servers. This is because of resource-constrained IoT network characteristics that have become a big victim. The early detection of DoS and DDoS attacks will prevent the resource-constrained devices from becoming a target and early death. This paper focuses on vulnerabilities in IoT such as Distributed Denial of Services (DDoS). Many privacy-conserving mechanisms have been discovered (such as automatic solution learning, and DDoS warning mechanisms). And, related work is under way. The goal of this paper is to present the detection and prevention of DDoS in IoT and privacy issues faced by the IoT environment and current mechanisms for its security.

Keywords Denial of Service (DoS) · DDoS · Security · Internet of Things (IoT) · Constrained

1 Introduction

IoT is an advanced analytical and automation system that takes advantage of processing, cloud computing, collaboration and machine intelligence technology to create a complete product or service framework. These devices require greater transparency control and effectiveness when added to any industrial environment. The Internet of Things has been introduced in recent decades as an groundbreaking

M. Sharma (✉) · B. Arora
Department of Computer Science and Information Technology, Central University of Jammu, Jammu, Jammu and Kashmir 181143, India

P. K. Singh et al. (eds.), *Proceedings of Second International Conference on Computing, Communications, and Cyber-Security*, Lecture Notes in Networks and Systems 203,
https://doi.org/10.1007/978-981-16-0733-2_60

technology which has a significant effect on human life. The Internet of Things is about combining the real and digital world into one ecosystem. It has, however, been a common concern that such revolutionary ideas will cause safety problems. To eliminate the possible risk of people revealing their private information, users need to grasp the concept of multiple attack tactics to eavesdrop the details of the person, with the DoS attack being considered one of the most common methods of attack. There Is also a expensive range of requirements for IoT devices. Monitoring is one of the clearest benefits of IoT. Through this, the exact quantity of equipment, water delivery and use, intelligent energy storage and protection delivery conveniently obtained gives an benefit in understanding items in advance IoT System Architecture. Denial of Service (Dos) is a digital assault that tries to make a computer or associate asset unaccessible to its expected customers by momentarily or unconclusively disrupting Internet-related host administrations. Refusal of administration is typically promoted by overwhelming computer or asset-focused people with needless demands attempting to overburden structures and preventing a few or any specific requirements from being fulfilled. For Distributed Denial of Service DDoS is short. DDoS is a kind of DoS assault in which different negotiated frameworks, which are regularly contaminated with a Trojan, are used to focus on a solitary framework that causes a Denial of Service (DoS) assault. Casualties of a DDoS assault consist of both the end based on the frame and all structures malignantly used and limited by the programmer in the attack conveyed [1]. In a DDoS assault, the approaching traffic flooding the unfortunate casualty starts from a wide range of sources—possibly many at least thousands. This viably makes it difficult to stop the assault just by obstructing a solitary IP address; additionally, it is hard to recognize authentic client traffic from assault traffic when spread across such a significant number of purposes of cause. To maintain a strategic distance from the potential hazard we use Denial of administration (DoS) a sort of assault where aggressors endeavor to keep real clients from getting to the administration. In DoS assault, the aggressor generally sends extreme messages asking the system or server would not have the option to discover the arrival locations of the assailants when sending the validation endorsement, making the server hold up before shutting the association, the aggressor sending more confirmation messages with invalid bring addresses back. Henceforth, the procedure of confirmation and serve holds up will start again helping the system or server occupied. There are various classes of DoS assault happening at sensors and mist hubs of IoT engineering. At mist hubs, there are six regular classes of dos assault that exist at mist layer of IoT design are:

Smurf flooding of ICMP reverberation answer.
Neptune flooding of synchronizing on port(s).
The dark mentioning of URL having numerous backslashes from a webserver.
Tear causing framework reboot or crash utilizing misfragmented UDP bundle.
Case pinging with deformed bundles causing reboot or crash.
Land-sending UDP parcel having a similar source and goal address to a remote host.

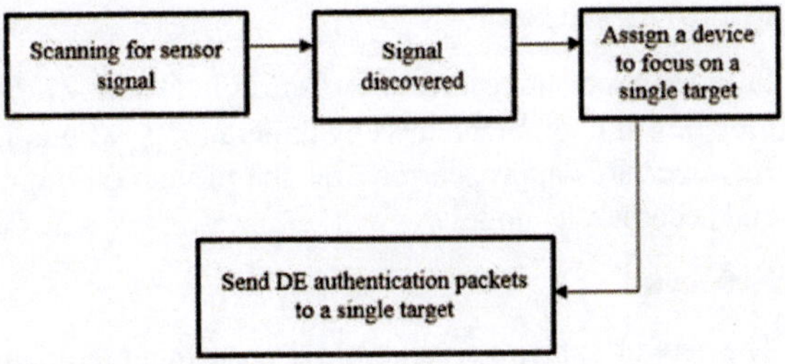

Fig. 1 Process of denial of service

Fatigue assault. A Denial of Service (DoS) assault is not the same as a DDoS assault. Dos assault ordinarily utilizes one PC and one web association with a flood a focused on framework asset.
DDOS utilizes numerous PCs and web associations with flood the objective asset (Fig. 1).

In the DDoS attack, the victim's incoming traffic flood originates from possibly hundreds of thousands or more from several separate outlets. This essentially renders it hard to avoid the assault by merely blocking a specific IP address; however, when scattered over too many points of origin, it becomes very difficult to differentiate valid user traffic from attack traffic.

1.1 Security Concerns in IoT

Internet of Things is a platform of real-world devices that communicate in real time. There are various threats involves in IoT security is shown (Fig. 2).

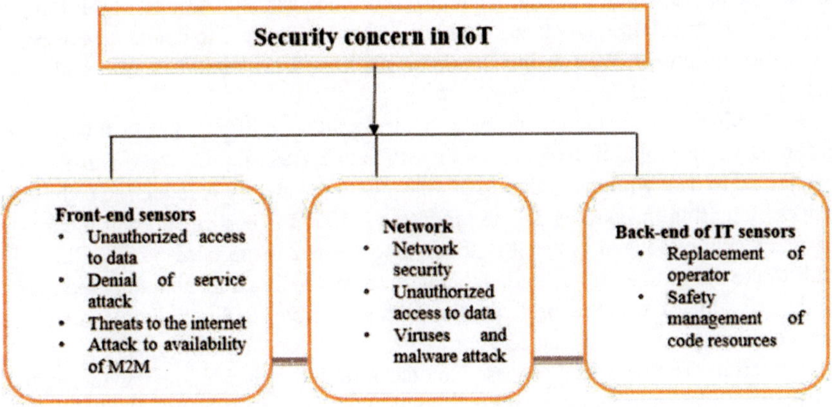

Fig. 2 Security concern in IoT

Front-end sensors

Front-end sensors receive sensors to collect data. The data is then transmitted through modules or computers, thereby undertaking various sensor networking services [1]. However, this approach requires the protection with business installing computers and access to its nodes.

Network

The network plays a crucial role in ensuring interconnection as well as the efficiency of IoT operation. If a large number of machines send data causing congestion in the network, there are a significant range of IoT nodes and classes that may cause service attacks to be declined.

Back-End sensors

These sensors have high-security, middleware and gathering specifications, analyzing sensor data in actual time to improve business understanding. At any time, IoT security has numerous extreme standards of confidentiality, security, safety, data integrity, data confidentiality, and availability.

1.2 Detection of DDoS at Application Layer

There are various faces involved in preventing DDoS attacks in which four phases involved as follows were explained in previous studies.

Prevention. The protection process focuses on shielding a network from attack by installing appropriate security equipment in various locations. In fact, mitigation also preserves server capital and guarantees that the actual client is able to access online services. DoS attacks sent by robotic tools allow multiple programs to approach those Web pages without any human interference. Probable protection of this kind of attack by software design is to grant only authentic consumer to connect web server tools and equipment. Web design should be successful, which the attacker could not delay.

Reduction. The reduction process is also said to mitigation, and this step is enforced when violation happens, with sufficient protection. Countermeasures are performed to deal with the violation or slow down the attack. A reduction technique works by halting the assault. DDoS reduction creation is best regarded if the traffic of attack acknowledged as usual is small, also known as the false-positive limit. In addition to the mitigation technique that is supposed to block an unauthorized traffic source IP address which causes an attack, this method would explicitly guarantee real consumer access to a web service.

Detection. The detection process includes a running machine review to find bad traffic which leads to DDoS attacks. Detection requires a innovatory technique for detecting broad illegal traffic on GET requests opposed to web server. The bulk of the detection strategies was used to form DDoS identification such as matching trends, clustering, predictive analysis, examining variations, correlations, and similarity. Detection usually development utilizes data background as the primary source to train the data to generate a threshold that will be applied to a parameter using a particular procedure for counting the GET request obtained. The wrong-positive rate.

Monitoring. The monitoring process required, the use of devices, such as network monitoring software, obtains the requisite information about a host or network. Monitoring is carried out in real time, as it is necessary to detect DDoS attacks. If the intruder began a DDoS assault using a botnet installed at various locations around the globe, tracking method becomes complicated. According to, dynamic monitoring is required to shape defenses for attacks. This chart provides a schematic view of the protection's life cycle.

1.3 Limitations and Challenges Faced by DDoS Attack

Associating an identity to a single person is a hazard because this can result in profiling and monitoring. Therefore, disallowing these activities in IoT and taking some preventive steps is one of the biggest challenges. Localization and monitoring attempt to establish and monitor the location of the person through space and time. The major challenge is developing protocols that inhibit IoT interactions such activity. In e-commerce applications, profiling information relating to a specific person to infer preferences through correlation with other profiles and data is very common. The major challenge is to align business interests in profiling and data collection with the privacy requirements of users. Many difficulties of maintaining privacy in IoT include distributing data safely via a shared channel without shielding the general network users, avoiding unwanted collection of information regarding the nature and characteristics of personal items.

2 Recent Detection Methods for DDoS in IoT

Hasan et al. [2] presented a paper which uses machine learning approaches to predict an attack and anomaly in IoT sensors. The algorithms for machine learning were used which are logistic regression (LR), support vector machine (SVM), logistic regression (LR), and artificial neural network (ANN). The measurement criterion used in the performance comparison is precision, precision, f1, and field under the characteristic curve controlled by the receiver. The program obtained test accuracy of 99.4% for the decision tree, random forest, and ANN. While the [3] accuracy of these techniques is the same, another metric shows that random forest performs comparatively better.

Bakhtiar et al. [3] introduced an IDS with a lightweight algorithm for DoS detection. J48 learning machine algorithm has been checked as reliable for use in restricted applications, so in this study, we have fitted the middleware with a lightweight J48-based IDS to solve the DoS threat. The test results said 75% of network packets could be identified by the IDS. Kajwadkar et al. [4] introduced a novel method, early prevention and detection algorithm for DoS and DDoS attacks. The algorithm was designed to fit in with the limited setting. In addition, the proposed algorithm can be equated with more research, and deep analysis can be carried out.

de Lima Filho et al. [5] has introduced the intelligent identification program (smart detection system) algorithm, the web solution for detecting DDoS attacks. He employed various machine learning methods. He observed Denial of Service attacks utilizing specific algorithms. The software used the random forest tree algorithm to classify network traffic based on flow protocol samples taken directly from network computers. Several experiments were conducted to calibrate the unit and to calculate its performance. Evidence suggested the approach proposed is possible and shows better performance compared with some recent and applicable approaches to literature. The proposed system was tested on three intrusion detection systems. While the system has achieved significant results within its reach, some improvements are needed, such as improved hit rates between attack classes and an automated parameter adjustment mechanism to maximize the rate of attack detection.

Daud et al. [6] concluded that the purpose of this paper has been accomplished with success. The results of this experiment show he is vulnerable to DoS attack by the IoT sensor node. Therefore, taking other security measures to counter the below susceptibility, for example, the deployment of numerous intrusion detection systems to identify DoS attack trends and signatures, and the clustering of sensor nodes to maximize the lifetime of the network, ensuring the efficiency of the IoT sensor node. To increase the lifespan of the IoT network for more study and creation on the use of clustering techniques for the IoT sensor node.

Santosh Kumar et al. [7] presented a paper on the identification of Dos attacks. He suggested a topology management method (TMM) in his paper for the recovery of the attack. The proposed TMM used the network to recover and compared it to current approaches. Further analyzes are to identify stealth denial of service attacks. Additionally, it is possible to evaluate several other fields of the IEEE 802.11 protocol frame and extract new features to reduce the time taken to detect the attack and increase analytical performance.

Cui Y et al. [8] demonstrated that the online hacking and attacking on Dos attack evaluated in IoT devices posed a threat to net health. The Internet of Things (IoT) will cause serious losses of property as an important component of the information era once it is targeted. This experiment aims to use three devices to replicate the Denial of Service (DOS) assault concept. Kali Linux initiated the attack in a variety of different ways. In his paper, he described the experiment's modified variables and showed how they can affect the effect.

Guleria A et al. [9] presented a paper for the given idea of also getting their influence for community holding webpage guests concerning certain DDOS attacks. Here, a DDOS assault was investigated to investigate the transmission of believing group holding website guests would specifically capture such general party movement. Only this paper needs to claim flood hit. This paper also mentioned a strategy with reminiscent anomalies clinched alongside visitors to the group's website, fundamentally based on an unrestrained α-strong model.

Ladislav Huraj et al. [10] suggested these IoT devices and their rapid development of the Internet would cause many security issues. This article presents the effectiveness of selected real-world IoT devices in demonstrating the UDP-based distributed reflective DoS attack, as a particular form of DDoS attack led to the layer of transport. The experiments display this type of attack on four heterogeneous IoT platforms representatives: IP camera, Raspberry Pi single-board machine, network printer, and smart lights. The experiment findings indicate the ability to be used in the DRDoS (Distribution Reflection Denial of Service) attack on all investigated IoT devices.

3 Comparison of Techniques of Different Researchers for Dos and DDos Based on Parameters

See Table 1.

4 Detection and Prevention Methods for DDoS in IoT

4.1 Detection Techniques

Detection of a DDoS attack is performed in network context by different strategies to prevent the serious injury. DDoS detection techniques for attacks have a workflow which tends to diagnose the impact of DDoS assaults [1] (Fig. 3).

Honey net cloud is a diverse group of various sub-networks. It includes honeypots. Honeypots are usually traffic controlled and alert HTTP, FTP, and UDP protocols. Toggle Bridge sends question that comes after passing dynamic supply board. The IP address is given to it, and it varies at regular time intervals for each and every honeypot and board. Stop fingerprinting methods. By utilizing this strategy, the intruder thereby is confused. What is it? Any request from a suspect node shall enter honeynet, dynamic framework of provisioning defines the sum of malicious order comes in request analysis is as loading is opposed to preset load threshold honeypot [2].

Fog computing-based security system (FOCUS)

From the recent improvement on fog computing, FOCUS has been developed and is a protection framework focused on fog computing. Fog computing is close the computers and end-users dependent on IoT. FOCUS offers a security mechanism on two occasions. A VPN is implemented at first level to secure the contact channel, and then, a challenge answer authentication mechanism is used to identify the unlawful traffic from DDoS attack [3]. FOCUS is a great strategy because it has less reaction

Table 1 Comparison study of DoS and DDoS in IoT

No.	Title	Parameter	DDOS-level identification	Evaluation method	Data set	Matrix performance
1	HADEC: live DDoS detection system, based on Hadoop [11]	Timestamp, source network, IP address, packet protocol, and packet header	DDoS high rate: TCP SYN, http post, UDP, and ICMP	Experiment	Dataset experiments	Measure utilization, processor and memory
2	D-FACE: a collaborative anomaly-based approach to early identification of DDoS threats and flash events [12]	Time window size, packet header, and generalized parameter	High intensity and low rate assault and a bunch of memories	Experiment	MTT Lincoln, CAIDA and FIFA	Precise classification, false-positive rate, *F*-measurement and precision
3	Defending HTTP web servers against attacking DDoS by detection of duration dependent attack flow [13]	Threshold whitelist and blacklist	High rate DDOS attack	Simulation (OPNET experiment)	Experiment dataset	Detection speed
4	Real-time prevention of DDoS threats using FPGA [14]	Origin IPs, variable indexing of origin IPs, and volume of packets	High rate HTTP DDoS	Experiments	CAIDA, TUIDS, and DARPA	Level of identification, accuracy, false positive and false negatives
5	FHSD: an advanced tool for spoof detection of network DDoS attacks [15]	Hop count, source MAC address, OS passive fingerprinting	High rate HTTP DDoS	Experiments	DARPA LLDOS inside 1.0 and experiment dataset	Detection rate, accuracy, and false positive
6	Detection of cloud based HTTP DDoS attacks using matrix covariance approach [16]	List of covariances and the TCP channel header	High rate HTTP DDoS	Simulation (MATLAB)	KDD cup 99 and experiment dataset	Level of identification, accuracy, false positive and false negatives

Fig. 3 Detection and prevention techniques for DDoS in IoT

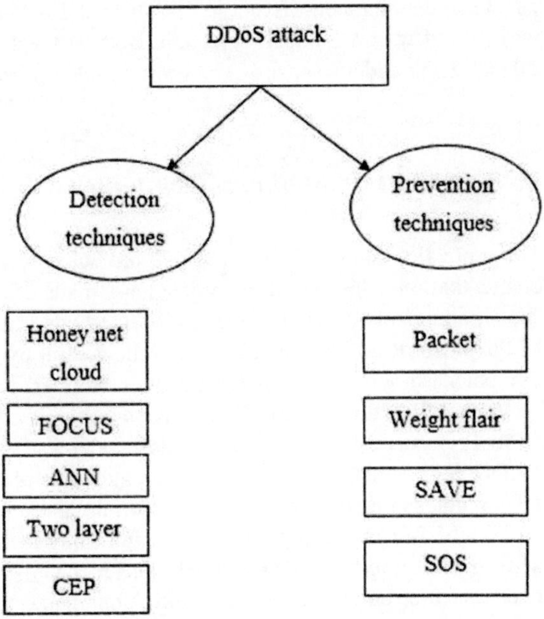

time and less use of bandwidth. However, it includes precise description of network traffic from traffic analysis device.

ANN IDS-based artificial neural network is used to evaluate the risks IoT faces. To capture and interpret information from several IoT devices and identify a DDoS attack inside the IoT network, it is implemented as an offline framework for detecting some kind of intrusion. They suggested an intrusion prevention method focused on a neural network to identify DDoS assaults. The identification or reconnaissance method was focused on classifying normal traffic patterns and malignant patterns. For this ANN model, the presentation demonstrated more than 99% precision. It effectively identifies DDoS attacks with greater precision for unauthorized IoT network traffic. It also increases network reliability but is not particularly successful in real-time response.

Two-Layer approach There are two major forms of DDoS attacks, high rate traffic that triggers massive traffic spikes and low-rate traffic attacks that are more equal to regular real traffic attacks. Detecting them all at the same time is difficult because this technique uses two-layer approach to identify all threats. There are three levels to complete. At first point, the device named detection with average filters (DAF) is passed through to filter high-intensity DDoS attack metrics. The remaining metrics are transferred by (DDFT), which is identification of low-rate DDoS attacks with differential Fourier transformation. It detects both low-rate and high-rate DDoS attacks. However, it is hard to distinguish when low-rate and high rate are similar. The CEP architecture consists of three primary layers: event generator, event processor and action system. The incident detector scans and tracks the network traffic as soon as an accident happens. Event generator contains two modules: (i) packet analyzer and (ii) software for attack detection. Both of these modules evaluate the form of DDoS attack and also examine incoming packet properties.

4.2 Prevention Methods/Techniques

For protection against DDoS attacks, protective measures are often ideal. It is a shame that once the assault is initiated and made effective it will seriously damage the computer of the victim. Prevention methods often aim to handle the majority of threat traffic and hence aid to avoid the assault by DDoS. In this manner, victim computer is not impacted by assault and continues its normal operations.

Packet filtering Every counteractive measure is equivalent to a patch in any situation. Preventive approaches aim to address defense vulnerabilities that are regulated by DDoS assaults. Packet filtering strategy is one of the strategies for avoidance of DDoS attacks that decrease harmful incoming packets.

Weight-fair throttling: Weight-fair throttling mechanism prevents a web server at upstream router from DDoS attack. This mechanism is weight-fair since the leaky bucket at the router controls the traffic anticipated for the server. On the basis of connection count, congestion control algorithm regulates the bucket count of network traffic capacity sent for the traffic server. In this mechanism, even if some of the routers are compromised, then system can still be in working condition. The routers are disabled, but the device will still run.

SAVE: In this process, the source position sends messages regularly to all locations with valid IP addresses. This approach helps routers to easily identify specific paths and IP address ranges as well. Router already knows the intended ranges of IP addresses, routers take valid addresses from routing tables, and then routers block the packets with one based on that knowledge. Routers block address packets that are not within predefined IP address set. The paradigm being introduced is constructive as it avoids packets with null addresses. It filters inappropriately presented packets properly, but legitimate packets may also be lost during the transient time because it is not efficient against intelligent IP spoofing.

5 Conclusion

This paper provides a study of recent methods of identification in the application layer detecting and preventing DoS and DDoS attacks. Research related to the DDoS attack has gained significant interest, particularly those occurring at the application layer. DDoS attack identification is very difficult because the traffic occurs because of other system types which can be influenced by botnet, such as IoT devices and the presence of DDoS as utilities may be considerably complex in detecting such an attack. The latest methods used to detect an assault on DDoS in IoT have developed various strategies for detection purposes. The DoS and DDoS attacks can lead to jamming of networks and can disrupt any network environment. Early detection and prevention of these attacks can lead to a better network service environment. As future work, the proposed methods can be compared with more works and its deep analysis can be done.

References

1. Singh K, Singh P, Kumar K (2017) Application layer HTTP-GET flood DDoS attacks: research landscape and challenges. Comput Secur 65:344–372
2. Hasan M, Islam M, Zarif II, Hashem MMA (2019) Internet of things attack and anomaly detection in IoT sensors in IoT sites using machine learning approaches. Internet Things 7:
3. Bakhtiar FA, Pramukantoro ES, Nihri H (2019) A lightweight IDS based on J48 algorithm for detecting DoS attacks on IoT middleware. In: 2019 IEEE 1st global conference life sciences technology, pp 41–42
4. Kajwadkar S (2018) A novel algorithm for DoS and DDoS attack detection in internet of things. In: 2018 Conference on information communication and technology, pp 1–4
5. de Lima Filho FS, Silveira FAF, de Medeiros Brito Junior A, Vargas-Solar G, Silveira LF (2019) Smart detection: an online approach for DoS/DDoS attack detection using machine learning. Secur Commun Networks 2019:1–15
6. Daud M, Rasiah R, George M, Asirvatham D, Rahman AFA, Halim AA (2018) Denial of service: (DoS) impact on sensors. In: 2018 4th International conference on information management ICIM, pp 270–274
7. Santhosh Kumar S (2017) An anomaly behavior-based detection and prevention of DoS attack in IoT environment. In: 2017 Ninth international conference advanced computing, pp 287–292
8. Guo K, Wang D, Zhi H, Lu Y, Jiao Z (2020) A trusted resource-based routing algorithm with entropy estimation in integrated space-terrestrial network. IEEE Access 8:122456–122468
9. Irum A, Khan MA, Noor A, Shabir B (2020) DDoS detection and prevention in internet of things
10. Alqahtani H, Sarker IH, Kalim A, Hossain SM, Ikhlaq S, Hossain S (2020) Cyber intrusion detection using machine
11. Cui Y, Liu Q, Zheng K, Huang X (2018) Evaluation of several denial of service attack methods for IoT system. In: 2018 9th International conference on information technology in medicine and education, pp 794–798
12. Guleria A, Kalra E, Gupta K (2019) Detection and prevention of DoS attacks on network systems. In: 2019 International conference on machine learning, big data, cloud parallel computing, pp 544–548
13. Huraj L (2018) IoT measuring of UDP-based distributed reflective DoS attack. In: 2018 IEEE 16th international symposium on intelligent systems and informatics, pp 209–214
14. Ali U (2018) Open access HADEC : hadoop-based live DDoS detection framework
15. Behal S, Kumar K, Sachdeva M (2018) D-FACE: an anomaly-based distributed approach for early detection of DDoS attacks and flash events. J Netw Comput Appl 111:49–63
16. Fox MR (1981) Cover letter PC-24(4)

Approach for Ensuring Fragmentation and Integrity of Data in SEDuLOUS

Anand Prakash Singh and Arjun Choudhary

Abstract With the rapid adoption of cloud services, more and more data are being uploaded on the cloud platform. These data are under threat from various threat actors constantly researching to steal, corrupt, or get control over the data. The threat actors are not only limited to malicious attackers, but these also include curious service providers, social activists, business entities, and nations. They pose a serious risk to cloud services. There are various approaches to protect data at various levels. Division and replication is one such approach where data are divided into chunks and spread over the cloud to reduce the risk of data leakage and simultaneously increase the accessibility. Under division and replication approach, SEDuLOUS provides a heuristic algorithm for data placement in a distributed cloud environment. In this research work, we have provided a comprehensive analysis on cloud data storage services and associated security issues, analysis of SEDuLOUS algorithm, and methodology to improve the SEDuLOUS by specifying minimum fragments to ensure fragmentation of all files and hashing of each chunk to identify the compromised storage nodes.

Keywords SaaS · Data security · Data leakage · Data loss · Fragmentation and replication · SEDuLOUS

1 Introduction

Cloud computing is a sophisticated on-demand computing service that can be rapidly provisioned from anywhere on the network using any platform with minimal service provider interaction [1]. It has so evolved and become the necessity of our daily life

---

A. P. Singh (✉) · A. Choudhary
Sardar Patel University of Police Security and Criminal Justice, Jodhpur, India
e-mail: spu18cs01@policeuniversity.ac.in

A. Choudhary
e-mail: a.choudhary@policeuniversity.ac.in

P. K. Singh et al. (eds.), *Proceedings of Second International Conference on Computing, Communications, and Cyber-Security*, Lecture Notes in Networks and Systems 203,
https://doi.org/10.1007/978-981-16-0733-2_61

that it has been termed as the fifth utility after water, gas, electricity, and communication [2]. Cloud computing has evolved from many different technologies such as virtualization, autonomic computing, grid computing, and many other technologies [3]. It offers various benefits to the user such as reduced cost, scalability, on-demand service, and anywhere access. One of the major advantages of cloud computing is that it helps users in focusing on their core business rather than on managing the information technology infrastructure. To harness the advantages offered by cloud computing, more and more data owners are migrating their data on the cloud platform [4]. Since the cloud is generally accessed through the public network, cloud service provider is not fully trusted by the user, due to various security concerns. These security concerns need to be addressed to provide a better ecosystem for accessing the cloud services and assimilating with current information technology infrastructure. The remainder of this paper is organized as follows. Section 2 describes the cloud storage services and benefits, Sect. 3 provides a brief overview of cloud security issues, Sect. 4 covers related work including strength and weaknesses of SEDuLOUS, Sect. 5 gives suggested improvements to SEDuLOUS, Sect. 6 explains the experimental setup, Sect. 7 contains results and analysis, and at the end, Sect. 8 provides conclusion and future scope of the paper.

2 Cloud Storage

Cloud storage is a storage service on Internet offered by the cloud service provider and known as "Storage as a Service" (SaaS) [5] in cloud computing domain, for example, Google Drive, Dropbox, Microsoft's OneDrive formerly known as SkyDrive, Amazon's S3 (simple storage service), and elastic file system (EFS), etc. The cloud storage service providers have virtually unlimited storage capacity and rent in as a pay-as-you go model. The user does not need to buy or manage storage infrastructure while enjoying scalability and durability with anywhere, anytime access for their use. The cloud storage can be accessed through traditional storage protocols or via an application programming interface (API) providing compatibility and ease. Many service providers also offer complementary services designed to help user in collection, management, security, legal compliance, and analysis of data at massive scale. There are three types of cloud data storage namely block storage, file storage, and object storage.

2.1 *Block Storage*

In block storage, data are broken or organized into fixed size chunks called blocks. Each block is assigned a unique identifier and handed over to the storage system for storage. This makes the job of storage system easier as some of the blocks can be placed on one node and some blocks can be stored on another node. This makes its performance better in terms of latency. However, as no higher-level metadata like data format, access/creation date, type or ownership, etc., are stored with the data, it requires these aspects to be taken care by application or data bases. This is similar to direct-attached storage (DAS) or storage area network (SAN). The major offerings for this type of cloud storage solutions are Amazon's EBS (elastic block storage), Google persistent disks, Microsoft Azure Premium Storage, Rackspace Cloud Block Storage, etc.

2.2 *File Storage*

It provides simplest and primitive way of storage. Almost, all operating systems support file storage. It is based on filing system for paper-based documents. The files are given names, tagged with metadata and then organized in folders having directories and sub-directories. It has a hierarchical system which is best suited for large content repositories, media stores, operating system files, or user home/office directories. The major cloud offering for this type of storage is Google Drive, Dropbox, Microsoft's SkyDrive, and Amazon EFS.

2.3 *Object Storage*

It is also known as object-based storage. The data are arranged in the discrete units called objects and stored in flat memory. Each object has data itself, allotted a unique identifier and metadata. Object-based storage provides access through API calls or simple http/https protocols. There is no need to mount the disk. Object-based storage provides capability to easily add or remove metadata to the data. It is best suited for large amount of indexed data. The major offerings for object-based cloud storage solutions are Amazon's S3, Rackspace Cloud Files, Azure BLOB, Google cloud storage, etc.

3 Cloud Data Security Issues

Cloud computing is an amalgamation of various technologies such as virtualization, utility computing, service-oriented architecture, networking, and World Wide Web, etc. There are multiple threats and security issues at each abstraction level. Various researchers have studied and presented their studies in different ways. Even the business entities like Gartner [6] in their published report have specified top seven threats, whereas the Cloud Security Alliance (CSA) in its report named "The Treacherous twelve" pointed out 12 security concerns [7]. Varghese and Buyya in

[8] found privacy and denial of service as major concern. A summary of the major issues related to cloud data security is presented below.

3.1 Data Leakage

Data leakage or data breach is one of the top security concerns. It happens when the data get exposed to the wrong entity while it is being transferred, stored, or audited. It may happen due to careless handling of data, human error, accidental issues like hardware failure, application vulnerabilities, or a deliberate attempt of an attacker. Architecture, the services of application are advertised or used using application program interfaces (APIs). These APIs, if not properly secured, pose serious threats which can be utilized by the attacker to compromise the security of data [9, 10]. The damage due to data leakage depends on the value of data which varies from entity to entity like financial, health, and personal information that are valuable to cyber criminals, whereas business or trade secrets are valuable for business adversaries.

3.2 Data Loss

It occurs when data become permanently inaccessible for the user. This happens due to various reasons such as malicious attack, accidental or deliberate deletion by service provider, hardware failures, physical catastrophe like fire or earthquake, or loss of encryption keys, [11]. To avoid data loss, proper backup mechanism should be followed.

3.3 Lack of Control

Once the data are transferred to the third-party infrastructure, the data owner loses their control over the data [4]. If something happens to the cloud service providers infrastructure, like malware attack, power outage, or legal restriction, etc., that affects directly the data, the data owner has no option other than to rely on the service provider for the corrective and preventive measures.

3.4 Data Privacy

Handling the data privacy is a complex issue due to various administrative, technical, legal, and regulatory reasons [12]. Sometimes, the data owner outsources the encryption and key management infrastructure to the cloud service provider. Even if the data are properly encrypted, the various other information such as access pattern and data sizes reveals various information about the data. Due to legal or regulatory requirements, the service provider needs to share the user data with law enforcement or regulatory authorities [13, 14].

3.5 Data Scavenging

In most of the implementation, while deleting the data, only links from the registry/record are removed, and it is possible to recover the data using various disk or file forensic tools. The faulty disks discarded by the service provider may be used by the malicious entities to recover the data [15]. Also, it becomes relatively easier for the attacker to brute force the encryption schemes used by the data owner. To avoid this issue, it is required to properly destroy the faulty media, or data are completely wiped out by rewriting the used data blocks.

3.6 Shared Technology Vulnerability

The cloud infrastructure is shared between multiple users also called tenant. The attackers take advantage of the same to launch various attacks discussed as under:

Cross VM Attack. The side channel attacks launched using one VM on adjacent VM residing on same physical machine are called cross VM attack. The side channel attack [16] refers to mechanism that is used to gain the sensitive information for abusing the encryption framework. These attacks include cache-based side channel attack, timing attack, power-monitoring attack, electro-magnetic attack, acoustic and eavesdropping cryptanalysis, and differential fault analysis.

VM Migration. For energy efficiency in data centers [17–19] or servicing of the VMs, the VMs are migrated from one physical machine to another. These VMs contain user data in attached virtual storage. Using the data forensic tools, it is possible to retrieve the other user's data in unused blocks [20].

VM Roll Back attack. The attacker can utilize the roll back feature of virtual machine to brute force even if the guest OS had restriction on number of failed attempts [21].

VM Escape Attack. In this attack, the hypervisor vulnerability is exploited to break the isolation layer. This allows the successful attacker to gain control of host machine and all other virtual machines [9, 22].

VM Jumping Attack. If the attacker is not able to attack the target VM, host, or hypervisor directly, then it attacks another VM and use it as platform for further attack [9, 22].

4 Related Work

Data security is very challenging and important aspect in outsourcing the data over cloud. There are various approaches to deal with the security issues. Data fragmentation and replication has been widely discussed approach to achieve data security. Medina et al. [23] conducted a survey of articles focused on fragmentation and replication between year 2010 and 2018. They found 83 paper published during the period covering problems such as fragmentation, replication, application to cloud, ease of implementation, performance, completeness, and cost model. Based on survey on metaheuristic approaches proposed between year 2006–2019, Mansouri et al. [24] concluded that there is no metaheuristic replication algorithm fulfilling all the required factors such as response time, and resource utilization, latency. Further, they concluded that security is one of the main factors neglected by most of replication algorithm.

In paper, Santos et al. [25] presented performance comparison of fragmentation vs encryption to secure data in cloud environment. The results shown that better performance is achieved in fragmentation, whereas better security is achieved using encryption. They concluded that fragmentation is plausible approach in the environment such as big data, where privacy and performance are the major concern.

In paper, Ali et al. [26] suggested DROPS methodology which used fragmentation of files along with strategic distribution across the available nodes ensuring that no nodes have multiple fragments to minimize the risk of significant data leakage. The paper also suggested controlled replication so that only desired number of copies are replicated to ensure availability. In paper, Pandithurai et al. [27] proposed number theory research unit (NTRU) encryption algorithm to further secure the fragmented chunks with slight performance lag. Khatod et al. [28] presented a hybrid approach named as Enigma where the fragments are made using Jigsaw puzzle strategy and encryption was applied on the selective chunks only to club the benefit of replication and encryption.

In paper, Hudic et al. [29] applied fragmentation on data base, wherein the various tables in normalized data base were considered as fragments for applying fragmentation. They further classified tables as various confidentiality levels to apply encryption on the selective tables to ensure better confidentiality.

In paper, Kang et al. [30] presented a heuristic approach for data placement in distributed cloud environment named security-awarE data placement mechanism for cLOUd storage systems (SEDuLOUS). SEDuLOUS provides heuristic method for security-aware data placement to achieve high performance while satisfying security requirement. Here, the security requirements define that the storage node being selected to store a chunk of file must be minimum how much hops away from the other storage nodes having chunk of data from the same file. Here, we provide brief methodology, and readers are encouraged to read paper [30] for details. In SEDuLOUS, user submits the file to controlling node. The network graph of storage nodes is converted to meet the security requirement. Thereafter, the T-coloring algorithm is applied to select set of candidate nodes. Then, set of candidate storage nodes are evaluated based on heuristic approach to identify the set of storage nodes having minimum storage/access time.

5 Improvement to SEDuLOUS.

We found that in SEDuLOUS [30], the chunk size (size of fragments on any node) is decided based on maximum chunk size defined by administrator to be stored on any node. Decreasing the maximum chunk size for the system results in requirement of more fragments of files, and hence, chances of rejecting the large size files in the system due to security constraints increase. Whereas, increasing the maximum chunk size for the system results in more files being stored in a single node. Due to these limitations, the SEDuLOUS fits well if the variation in the size of files is less. However, if the user has to store files with large variation in file sizes, then SEDuLOUS fails, for the files having size smaller or close to the maximum chunk size defined for the system, i.e., the complete file will be stored on the single node defeating the objective of the algorithm. We propose to correct this issue by defining minimum number of chunks for a file in the system.

In addition to the above, SEDuLOUS advocated that the encryption will add additional processing overhead and hence proposed storing data in plain on storage nodes. This provides an opportunity to the attacker to modify the data on the compromised nodes. The same may not be noticed by the user. We propose to correct this issue by hashing each chunk on storing and using the hash for comparison with the hash of data chunk on retrieval to identify possible modification/manipulation of data chunk and compromised storage node.

6 Experimental Environment

6.1 Algorithm Implementation

We implemented the algorithm and conducted the experiment in simulated environment as shown in Fig. 1 using work station with 8 core Intel Core i5 -8250U CPU @ 1.60 GHz CPU and 16 GB memory. The files submitted to access node with required minimum number of fragments (default 3 Nos fragments) are divided into chunks as per constraints (minimum number of fragments and max chunk size defined for system). Then, as per security constraints, the candidate storage nodes are selected using SEDuLOUS or RSN. Thereafter, the chunks of file are stored in selected storage nodes. The code was implemented using LAMP stack. For simulation of storage cloud network, we used the NKN-topolgy as given in article [31]. All

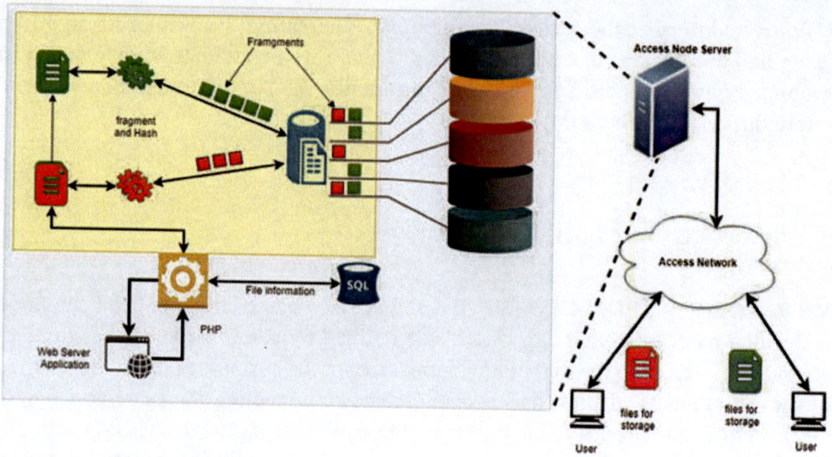

Fig. 1 Simulation environment

the links were given BW of 10 Gbps, and Delhi node was selected as access node. We presumed that there is no delay in data read and write operations for all the storage nodes.

6.2 Data Sets

To perform the measurements for our problem statement, it was required to collect benchmark data (files) representing different data structures such as text, pdf, html, xls, biblography, paper, news, jpg, and png. Accordingly, we googled out and found compression benchmark data sets [32] named Calgary and Canterbury compression corpora, enwik8 and Silesia compression corpus. The total benchmark data sets used by us contained 49 files amounting to 330 MB.

7 Results and Analysis

7.1 Restriction on Minimum Number of Chunks

We implemented the dynamic fragmentation in our application by providing an option to the user to select minimum number or fragments. We provided value of minimum fragments as 3 and defined the maximum chunk size of the system as 1 MB. This resulted in minimum 3 fragmentation of files having size lesser than 1 MB. Hence, we could achieve fragmentation for the files sizes smaller than the maximum chunk size defined for the system.

7.2 Use of Hashing to Avoid Unnoticed Modification of Data

To maintain the integrity of stored data chunks, we performed hashing of data chunks, and the hashes were stored in database at access nodes. We observed that this resulted into the identification of nodes where data were modified either due to errors or due to compromise of the node.

7.3 Performance of SEDuLOUS Over NK Topology

The plot of file size vs time taken to store date over NKN topology using SEDuLOUS and random selection node methods is given in Graph 2. The uploaded file sizes in

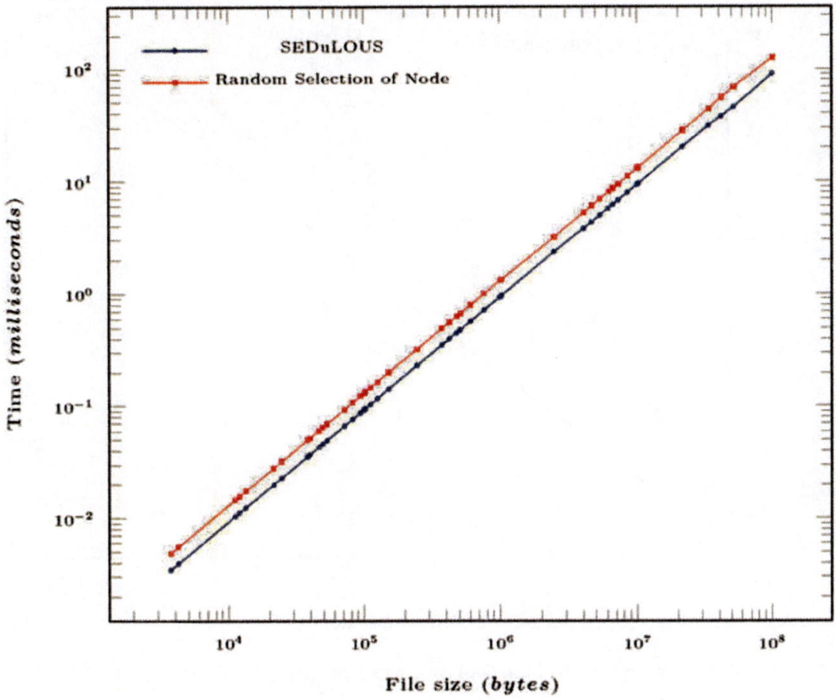

Graph 2 File size versus storage time for SEDuLOUS and random selection of node

logarithmic scale are represented on x-axis, and time taken in milliseconds for storage using SEDuLOUS (in blue) and RSN (in red) is represented in logarithmic scale on y-axis. It may be noticed that the performance of SEDuLOUS over NKN is 28% better than random selection of nodes. The authors of SEDuLOUS found SEDuLOUS 20 and 19% better on random network and Internet2 topology, respectively. Hence, we may conclude that the performance of SEDuLOUS is also dependent on the network topology of the storage network.

7.4 Performance Overhead Due to Hashing

The plot of file size vs time for hashing during storage, hashing during retrieval, and SEDuLOUS is shown in Graph 3. The uploaded file sizes in logarithmic scale are represented on x-axis, and time taken in milliseconds for computation of hash during storage, i.e., upload (in red), computation of hash during retrieval, i.e., download (in orange) and for storage using SEDuLOUS (in blue) is represented in logarithmic

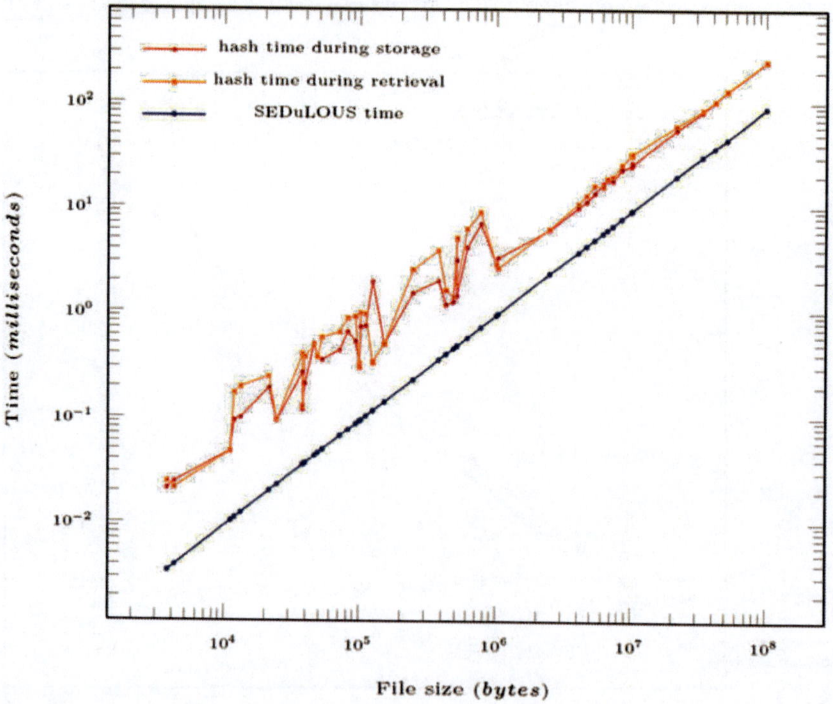

Graph 3 File size versus time for hash (storage), hash (retrieval), and SEDuLOUS

scale on y-axis. It may be noticed that the hashing time is approximately 5 times higher than the storage time using SEDuLOUS. However, the ratio may further improve if we take into account the time taken by storage hardware or reduced the available BW (currently 10 Gbps).

8 Conclusion

In our work, we proposed defining minimum or dynamic number of chunks as an improvement of SEDuLOUS, which provides adequate security to the data sizes lesser than or comparable to the maximum chunk size defined for the system. This made SEDuLOUS usable for user data/files having large variation in size.

The hashing of data chunks was proposed to identify unnoticed modification of data on compromised storage nodes. However, hashing increases the storage or retrieval time by fivefold due to hashing of each chunk based on selected link BW and topology. This may further reduce if we consider various network equipment latency such as router, firewall, and switches. In the information security world, the CIA triad is a basic principle. The`three tenets depend on each other, i.e., if we increase the level of one, the other two are adversely affected. Here also, hashing surely increases integrity but on the cost of availability, i.e., longer storage/access time.

During the performance study, we found that the performance of SEDuLOUS was 28% better on the NKN topology network as compared to random selection of nodes.

This work may be extended to study the performance of SEDuLOUS on inclusion of other parameters such as storage device performance, and network equipment performance. It may also be extended to study the performance over data center networks such as DCell, FAT tier, and Three tier. The concept of security constrains in SEDuLOUS may also be extended to replication strategy proposed in the DROPS methodology.

References

1. Mell P, Grance T (2011) The NIST definition of cloud computing. Special Publication, National Institute of Standard and Technology, pp 145–800
2. Buyya R, Yeo CS, Venugopal S, Broberg J, Brandic I (2009) Cloud computing and emerging IT platforms: vision, hype, and reality for delivering computing as the 5th utility. Futur Gener Comput Syst 25(6):599–616
3. Alani MM (2016) Elements of cloud computing security. In: Computer science. DOI https://doi.org/10.1007/978-3-319-41411-9_1. Springer, Berlin
4. Henze M, Matzutt R, Hiller J, Mühmer E, Ziegeldorf JH, Giet JVD, Wehrle K (2017) Practical data compliance for cloud storage. In: 2017 IEEE international conference on cloud engineering (IC2E), pp 252–258
5. Buyya R, Broberg J, Goscinski AM (2010) Cloud computing: principles and paradigms, vol 87. Wiley, London

6. STAMFORD Gartner Identifies the Top Seven Security and Risk Management Trends for 2019 (2019). https://www.gartner.com/en/newsroom/press-releases/2019-03-05-gartner-identifies-the-top-seven-security-and-risk-ma. Last Accessed 22 July 2020
7. T. T. W. Group (2016) The treacherous 12: cloud computing top threats in 2016. Cloud Security Alliance. Last Accessed 20 June 2020
8. Varghese B, Buyya R (2018) Next generation cloud computing: new trends and research directions. Futur Gener Comput Syst 79:849–861
9. Ouffoué G, Ortiz AM, Cavalli AR, Mallouli W, Domingo-Ferrer J, Sánchez D, Zaidi F (2016) Intrusion detection and attack tolerance for cloud environments: the clarus approach. In: 2016 IEEE 36th international conference on distributed computing systems workshops (ICDCSW), pp 61–66
10. Torkura KA, Sukmana MIH, Meinig M, Kayem AVDM, Cheng F, Graupner H, Meinel C (2018) Securing cloud storage brokerage systems through threat models. In: 2018 IEEE 32nd international conference on advanced information networking and applications (AINA), pp 759–768
11. Kajal N, Ikram N (2015) Security threats in cloud computing. In: International conference on computing, communication and automation. IEEE, pp 691–694
12. Tari Z, Yi X, Premarathne US, Bertok P, Khalil I (2015) Security and privacy in cloud computing: vision, trends, and challenges. IEEE Cloud Comput 2(2):30–38
13. Choo KKR, Sarre R (2015) Balancing privacy with legitimate surveillance and lawful data access. IEEE Cloud Comput 2(4):8–13
14. Quick D, Choo KKR (2016) Big forensic data reduction: digital forensic images and electronic evidence. Clust Comput 19(2):723–740
15. Khan N, Al-Yasiri A (2016) Identifying cloud security threats to strengthen cloud computing adoption framework. Procedia Comput Sci 94:485–490
16. Anwar S, Inayat Z, Zolkipli MF, Zain JM, Gani A, Anuar NB, Khan MK, Chang V (2017) Cross-VM cache-based side channel attacks and proposed prevention mechanisms: a survey. J Netw Comput Appl 93:259–279
17. Patel D, Gupta RK, Pateriya R (2019) Energy-aware prediction-based load balancing approach with VM migration for the cloud environment. In: Data, engineering and applications. Springer, pp 59–74
18. Xiao X, Zheng W, Xia Y, Sun X, Peng Q, Guo Y (2019) A work load aware VM consolidation method based on coalitional game for energy saving in cloud. IEEE Access 7:80421–80430
19. Xiao X, Xia Y, Zeng F, Zheng W, Sun X, Peng Q, Guo Y, Luo X (2019) A novel coalitional game-theoretic approach for energy aware dynamic VM consolidation in heterogeneous cloud datacenters. In: International conference on web services. Springer, pp 95–109
20. Jordon M (2012) Cleaning up dirty disks in the cloud. Netw Secur 2012(10):12–15
21. Xia Y, Liu Y, Chen H, Zang B (2012) Defending against VM rollback attack. In: IEEE/IFIP international conference on dependable systems and networks workshops (DSN 2012). IEEE, pp 1–5
22. Singh A, Chatterjee K (2017) Cloud security issues and challenges: a survey. J Netw Comput Appl 79:88–115
23. Castro-Medina F, Rodríguez-Mazahua L, Abud-Figueroa MA, Romero-Torres C, Reyes-Hernández LÁ, Alor-Hernández G (2019) Application of data fragmentation and replication methods in the cloud: a review. In 2019 International conference on electronics, communications and computers (CONIELECOMP). IEEE, pp 47–54
24. Mansouri N, Javidi MM (2020) A review of data replication based on meta-heuristics approach in cloud computing and data grid. In: Soft computing, pp 1–28
25. Santos N, Lentini S, Grosso E, Ghita B, Masala G (2019) Performance analysis of data fragmentation techniques on a cloud server. Int J Grid Util Comput 10(4):392–401
26. Ali M, Bilal K, Khan SU, Veeravalli B, Li K, Zomaya AY (2018) Drops: division and replication of data in cloud for optimal performance and security. IEEE Trans Cloud Comput 6(2):303–315
27. Pandithurai O, Shenbagalakshmi R, Sindujha AU (2019) A noval approach of drops with NTRU in cloud. In: 2019 5th international conference on science technology engineering and mathematics (ICONSTEM), vol 1, pp 261–265

28. Khatod V, Ingale S, Gund K, Gorde S, Joshi R, Khengare R (2020) Enigma: a hybrid approach to file security in cloud. In: Proceedings of ICETIT 2019. Springer, Cham, pp 1005–1015
29. Hudic A, Islam S, Kieseberg P, Rennert S, Weippl ER (2013) Data confidentiality using fragmentation in cloud computing. Int J Pervasive Comput Commun
30. Kang S, Veeravalli B, Aung KMM (2016) A security-aware data placement mechanism for big data cloud storage systems. In: 2016 IEEE 2nd international conference on big data security on cloud (BigDataSecurity), IEEE international conference on high performance and smart computing (HPSC), and IEEE international conference on intelligent data and security (IDS), pp 327–332
31. Raghavan S (2014) E-science infrastructure: national knowledge network (NKN) initiative. CSI Trans ICT 2(3):207–215
32. Peazip. Compression benchmark. https://www.peazip.org/peazip-compression-benchmark.html. Last Accessed 22 July 2020

Efficient Classification of True Positive and False Positive XSS and CSRF Vulnerabilities Reported by the Testing Tool

Monika Shah and Himani Lad

Abstract Security testing is essential for website and web applications in current days. It is easy for an attacker to invade the security and do malicious activities through web applications if they are not properly protected against known attacks. General practice in web application before release is using testing tools to recognize the possible set of vulnerabilities, which can be true-positive or false-positive. Then the developer team will be asked to revise code to protect against true-positive vulnerabilities. For that, the testing team needs to classify each reported vulnerability into true-positive and false-positive individually, which is very time-consuming. This article suggests innovation in this practice to reduce the time of recognizing true-positive vulnerabilities. It presents a novel approach to classify reported multiple vulnerabilities of different attacks using a single script and in single go. These attacks should share common triggering events or testing process. Cross-Site Scripting (XSS) and Cross-Site Request Forgery (CSRF) attacks are chosen to illustrate the approach.

Keywords Security testing · False-positive · XSS · CSRF · Website security · Performance · Testing script

1 Introduction

Web Applications are becoming very essential mode for business, shopping, communications, booking or hiring resources, bill payments, online banking, to name a few to support online service requests. Ideally, Web applications should be designed to make users securely share confidential data for various types of services. Unfortunately, statistics say that web applications are the most common mode for attackers to steal users' confidential information. Attackers scan the web applications and

M. Shah (✉) · H. Lad
Nirma University, Ahmedabad, Gujarat, India
e-mail: monika.shah@nirmauni.ac.in

H. Lad
e-mail: 17mcei06@nirmauni.ac.in

P. K. Singh et al. (eds.), *Proceedings of Second International Conference on Computing, Communications, and Cyber-Security*, Lecture Notes in Networks and Systems 203,
https://doi.org/10.1007/978-981-16-0733-2_62

find out vulnerabilities in it to access confidential data or do malicious activities. The attacker performs many attacks like SQL Injection, Cross-Site Scripting, Cross-Site Request Forgery, Identity theft, broken authentication, sensitive data exposure, security misconfiguration, to name a few. OWASP (Open Web Application Security Project) is an organization working on the improvement of the security of software. Among all the security attacks, Cross-Site Scripting (XSS) and Cross-Site Request Forgery (CSRF) are in the list of the top Attacks declared by OWASP. OWASP has also provided the techniques for identifying possible vulnerabilities of these attacks. Looking toward the necessity to secure software against unauthenticated access, researchers and developers have put their hard efforts to provide security testing tools to report vulnerabilities present in the web application.

Some of the security tools are signature-based and they detect only those attacks which match the signatures stored in it. So, if the time to time new signatures of attacks is not updated, the presence of such attacks could not be identified. The major issue is there does not exist any security tool in our best knowledge that generates zero False positive alarms. As a result, the testing team needs to perform manual testing of each reported vulnerabilities to eliminate false-positive alarms. This helps to improve the quality of software and gain the trust of end-users but at the cost of delayed software release. There are some efforts to automate the analysis of each reported vulnerability and classify it into false-positive or true-positive. But, analysis of each reported vulnerability individually makes the process much slower for high false-positive vulnerabilities rate. Henceforth, this paper proposes a novel approach to reduce time to identify false-positive alarms. The main concept adopted here to identify a set of attacks category having a similar flow of triggering attack, identify common vulnerable locations reported for these attacks, and generate a joint script for all these attacks. To illustrate our approach, popular top attacks Cross-Site Scripting (XSS) and Cross-Site Request Forgery (CSRF) have been chosen in this work.

This paper is organized into five sections. In this survey, Sect. 2 illustrates the traditional approach used for security testing in web applications, Sect. 4 talks about topmost security attacks, Sect. 3 talks about the tools used to test these attacks, Sect. 6 proposes an approach to reduce is some scripts to be tested to classify true positive and false-positive vulnerability alarm generated for XSS and CSRF by testing tools, and the last section is the conclusion of our experiment.

2 Traditional Approach for Security Testing

In traditional security testing of a web application, testing software scan the whole application to list the possible vulnerabilities and prepare scripts for attacking these vulnerabilities, get them executed, predict vulnerabilities, and report alerts for them. Ailure to detect the existence of vulnerabilities is known as false-negative reporting. Ideally, no vulnerabilities to be left out open. In other words, all expect zero negative results. In this regard, the web application needs to be tested using multiple security tools designed to detect different types of vulnerabilities effectively. Vulnerability reports generated by all security tools are integrated into a single list.

Anomaly detection and Zero-Day Attack detection approach used by security tools makes the probability of false-positive vulnerability report by the security tool. If false-positive vulnerabilities are also forwarded to the developer, a large amount of time would be wasted to find alternate protection solutions for properly designed code. As a result, the application release would be delayed unnecessarily. To overcome this issue, the testing team needs to spend more man-hours in the verification of each reported vulnerabilities. This process includes injecting an attacking script for each reported vulnerability, analyze the GET/POST URL and response. If the footprint of this script is not observed, then that vulnerability is marked as false-positive and eliminated from the list. Figure 1 presents this traditional method of classifying into true-positive and false-positive. This process bothers much more when the false-positive to the true-positive ratio is high. Many of the testing software tools tried to follow the OWASP standards. Some of the tools are licensed, and some of them are open source tools. One of the most used tools is OWASP ZAP. Mburano and Si [1] presents statistics of different scanners according to the OWASP benchmark in Table 1, which shows the footprint of false-positive for XSS and SQL injection attacks report. Security testing using OWASP ZAP was also performed on six different versions and builds of a web application. The comparative statistics of true-positive and false-positive vulnerabilities for XSS and CSRF reported by the OWASP ZAP testing tool are presented in Table 2. Both statistics show a significant amount of false-positive vulnerability reports for various attacks like XSS, CSRF, SQL injection, and to name few. Such a high rate of false-positive vulnerability justify efforts done by the testing team to eliminate false-positive vulnerabilities and reduce the overhead of developers. But, it demands expert testing engineers. On the other side, cross-checking each reported vulnerabilities to classify it into true-positive and false-positive is a big overhead especially when the true-positive to false-positive ratio is high.

3 Security Testing Tools

Security testing tools list out the possible vulnerable areas from the application. This section compares the popular licensed tools and open-source tools for security testing.

3.1 WebInspect Tool

WebInspect is the most used tool among all the licensed tools because it covers maximum attacks declared by OWASP [2]. It asks a separate license for each application. It is a signature-based tool, which uses predefined attack signatures [3].

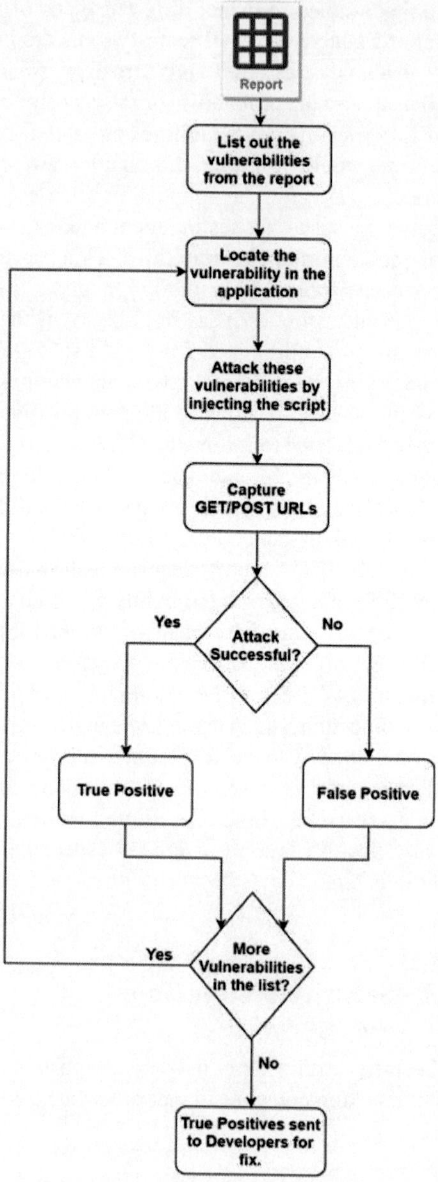

Fig. 1 Traditional method of classifying true positive and false-positive

WebInspect is an automated vulnerability scanning tool that is used to scan a web application. WebInspect has an external viewpoint on web applications, so at first, it will spider the application to find all the accessible pages and forms. After that, it will launch all the suitable attacks on all the possible access points and will determine all the vulnerable points in the application. Based on this scanning, it will generate the

Table 1 Empirical statistics for different scanners according to OWASP benchmark

	Command injection		LDAP injection		SQL injection		Cross site scripting		Path traversal	
	ZAP	Arachni	ZAP	Arachni	ZAP	Arachni	ZAP	Arachni	ZAP	Arachni
TP	41	39	8	20	158	136	158	136	0	0
FN	85	87	19	19	114	136	114	136	133	133
TN	125	125	32	32	224	227	224	227	135	135
FP	0	0	0	0	8	5	8	5	0	0
TPR%	32.54	30.96	29.63	74.07	58.09	20	58.09	20	0	0
FPR%	0	0	0	0	3.45	2.16	3.45	2.16	0	0
Score%	33	31	30	74	55	48	76	64	0	0

Table 2 Comparative statistics for true positive and false-positive detected in scanning of different versions of a web application

	Total vulnerabilities	XSS vulnerabilities			CSRF vulnerabilities		
		Total	TP	FP	Total	TP	FP
Report-1	326	54	42	12	43	33	10
Report-2	315	62	44	18	38	26	12
Report-3	443	53	36	17	36	30	6
Report-4	536	66	49	17	34	25	9
Report-5	268	44	36	8	37	22	15
Report-6	593	72	59	13	56	40	16

report of all the vulnerabilities of the application. The result has high true-positive and high false negatives because it can't detect newly discovered attacks.

3.2 *OWASP ZAP Tool*

OWASP introduced a tool Zed Attack Proxy (ZAP) to detect the top 10 types of web application attacks including XSS and CSRF. OWASP ZAP is highly popular among open-source security testing software. It is developed and maintained by Security Expert Engineers of OWASP through community lead open source projects. These engineers keep updating the tool from time to time according to the discovery of new attack signatures. So, this tool can detect a high number of true-positive as compared to other tools [4]. Unfortunately, like other security testing tools, ZAP also generates false-positive reports.

Vega et al. [5] have evaluated testing tools and report deficiencies in identifying vulnerabilities of web applications. It also shows that OWASP ZAP has more success with less testing time in compare to WebInspect. OWASP ZAP is found more popular due to its attractive features like open source, free, regular update of attack patterns identified by the OWASP community. Therefore, the authors have chosen here to perform security testing of web applications using ZAP.

4 Cross-Site Scripting and Cross-Site Request Forgery Attacks

OWASP has declared the topmost and dangerous attacks as mentioned in the previous section. Two of these attacks are XSS and CSRF. Both of these attacks are very dangerous to any web application as they directly or indirectly attack the server, cookies, and sessions of the authenticated user.

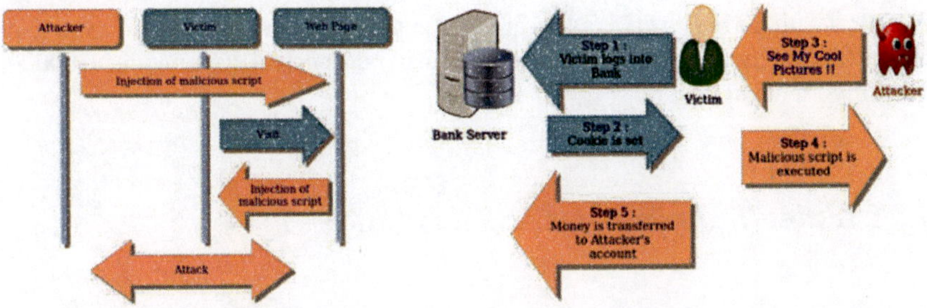

(a) Cross-Site scripting(XSS) (b) Cross-Site Request Forgery(CSRF)

Fig. 2 Flow of XSS and CSRF attacks

4.1 Cross-Site Scripting (XSS)

Cross-Site Scripting is almost like client-side attacks because this attack uses all client-side languages, most often Javascript and HTML [6]. The attacker injects malicious script in the application input to steal users' confidential information [6]. On the basis of the type of attack, the malicious script is either reflected at the victim's browser or stored at the server. An attacker saves the malicious script on the victim's browser and whenever the victim tries to load that particular web page or link, it gets executed and performs malicious activities [7]. An XSS stored attack can be harmful to all users [8, 9]. This attack can be succeeded because of poor user input validation, where the attacker injects malicious script into the code of the web server and the user will not be able to judge whether it is malicious code or not [10, 11]. Figure 2a displays the flow of a Cross-Site Scripting attack.

4.2 Cross-Site Request Forgery (CSRF)

In the Cross-Site Request Forgery attack, the attacker forces an authenticated user to send a forged HTTP request without his/her knowledge. Here, the attacker takes privileges of all the authorities that a victim has and uses the victim's authenticated session to do malicious activities [12, 13].

CSRF usually targets the functionalities of state changes on the server like changing passwords, online ordering etc. [14]. Figure 2b illustrates a flow of Cross-Site Request Forgery attack. For an example, if any user logged in and authenticated to a banking website, and the attacker manages to send a link or a pop-up to the user which can be also 1*1 pixel image [12]. If the user clicks on that link, the malicious code behind that link will get executed, and it might use the session which is authenticated by the bank website to send money to the attacker. Suppose XYZ is transferring money to ABC. Then the URL on XYZ should be displayed like https://bank.com/transfer.do?acct=XY Z&amount=100,000 But, the attacker will replace this URL with the below URL when clicked on the forged link or image. Example of image link: ha href = "https://bank.com/transfer.do?acct=ABC&amount=100000i View my Pictures! h/ai.

4.3 Testing Techniques for XSS and CSRF

Testing the vulnerability of web application against XSS attack Preventive measures for cross-site scripting attack suggest developers to design the code such that it escapes untrustable data from HTML context includes body, javascript, attribute, CSS, URL, etc. where data can be placed [15]. The most popular way to test the Cross-Site Scripting attack is'Black Box Testing'. In traditional manual black-box testing for XSS attack, all possible vulnerable parts of the web applications are identified, and attacks are performed by injecting script at these points. If these attacks are successful, then white box testing will be done to review code and protect the vulnerability by appropriate code patch. In 'white box' testing, the source code is analyzed manually or using code analytis tools [16]. On otherside, aliero et al. discuss issues with white box testing to detect vulnerabilities of web applications.

Automated scanner injects XSS attack script into http body or request parameters, and check http response for reflection of script in http body or header parameter [17], if reflected, it reports vulnerability. Gowda et al. [17] mentions that if 'contenttype' is other than "application/javascript", then it is false-positive.

Testing the vulnerability of web application against CSRF attack To test this attack, Black box testing and grey box testing are two techniques that are used in common. In traditional manual black-box testing of CSRF, the tester will intentionally create one Html page that will have the URL which needs to be tested. Then the tester observes POST URLs on click to the link. If the same session is used to access that URL as well, then this is a vulnerability [14].

In Gray Box Testing, the tester will check if the session management is dependent on the client-side values i.e. all the information available to the browser, then the application is vulnerable to access client-side attributes like cookies and HTTP authentications (an application-level authentication). So, it is really easy for an attacker to steal the session from the user [14].

Automated scanner search CSRF token in Http request/response to find CSRF vulnerability. Each of these reported CSRF vulnerabilities will be verified manually for false-positive alarm identification. The reported vulnerability is false-positive if it is not changing state. If Http-request is for state change, analyze 'content-type' [17].

5 Related Work

False-Positive Rate (FPR) is one of the important criteria for the evaluation of Security testing tools. It is observed that false-positive count may increase while scanning large web application [18]. Mohammadi et al. [19] propose a grammar-based XSS attack tests generation technique to detect cross-site vulnerabilities caused by incorrect encoder usage. It is aimed to reduce the false-positive rate. Similar efforts have been seen at Webguardia [20, 21] to reduce false positive vulnerabilities of top attacks like XSS. But, their empirical results conclude that it is technically infeasible to completely avoid generation of false-positives and false-negatives. With advancement, many security tools like ZAP facilitate several features to reduce or handle false-positive vulnerability alarms. It includes features [22] like (i) Tagging confidence level of reported falsepositive alert as false-positive while using ZAP manually, (ii) configuration of threshold level off, low or high, (iii) configuration of strength low or high alarm (iv) Apply Alert filters to specify criteria to have false-positive confidence level, (v) ignore alerts on specific URLs, to name a few. This demands for a skilled person with sufficient knowledge and experience to configure these settings. There are still issues with such security tool configuration: (i) These different type of configuration may not be generalized for each web applications, (ii) Configuration like High threshold, low strength may miss some real vulnerabilities (false negative) (iii) Configuration like Low Threshold or High Strength may increase the number of false-positives. As a result, the testing team of a web application needs to classify false-positive and true-positive alerts manually.

Some intelligence efforts [17, 18] are observed to make this classification automated. Gowda et al. [17] have compared accuracy of various machine learning approach aiming classify given vulnerability into true-positive and false-positive, where decision tree is found more accurate for CSRF and Random forest is found more accurate for XSS. NETSPARKER [18] reports confirmed true-positive by automating the process of injecting script and verifying response. It generates a probable vulnerability list, but each of these probable vulnerability need to be verified individually. To conclude, most of existing efforts are either focusing on automating procedure of classifying each reported vulnerability into true-positive and false-positive. Verifying each reported vulnerability individually is time-consuming or train tool to reduce false-positive. This paper proposes a novel approach, where multiple vulnerabilities will be tested using single script and in single go.

6 Proposed Approach

This paper proposes a novel approach where a set of attacks will be identified, which has high similarity in process of classifying it into false-positive and truepositive. Then, a joint single script will be designed to test all those attacks considering differences among them. For example, XSS and CSRF both share some common properties

Table 3 Similarity between XSS and CSRF

	Similarity in XSS and CSRF
Client/server side script	Client side script
Application domain	Website, web application
Requirement to trigger an attack	Need some form of user activity
Way of trigger an attack	Trick victims to execute some script

as described in Table 3. From Fig. 3, it can be concluded that the flow of XSS and CSRF is much similar. A similar way of triggering attacks is the motivation for this approach. In both the attacks, the attacker tricks victim to click on the image of a link or popup which has malicious code in the background. So, both attacks could be tested using the same technique'on click event'. In XSS, an attacker steals confidential information from the victim like cookies, session ids, etc. In a CSRF attack, the attacker gets control of the victim's authenticated session-id and does malicious activity using it. The article [23] also describes that some XSS attacks can be prevented through use of CSRF tokens. Looking toward much overlapping between XSS and CSRF, this paper proposes to design a joint script for classifying reported XSS and CSRF vulnerability into true-positive and false-positive for a scenario.

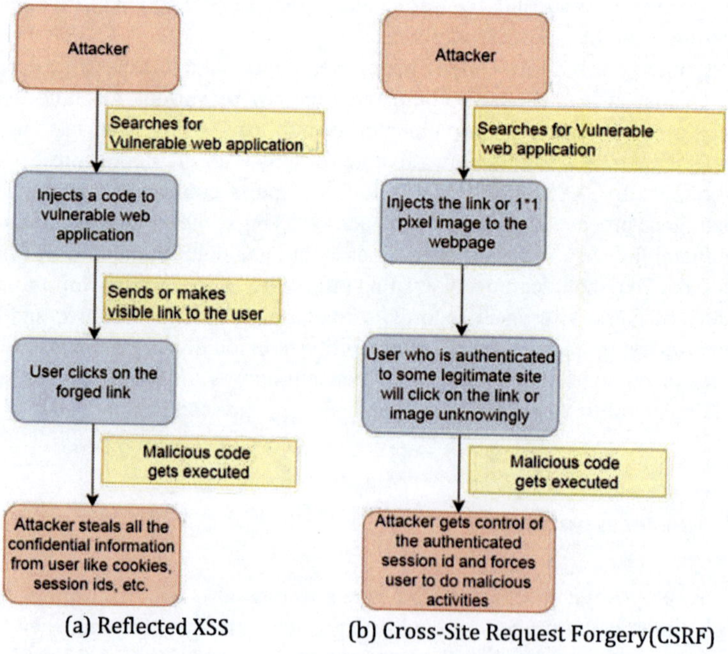

Fig. 3 Similar flow of reflected XSS and CSRF in traditional approach

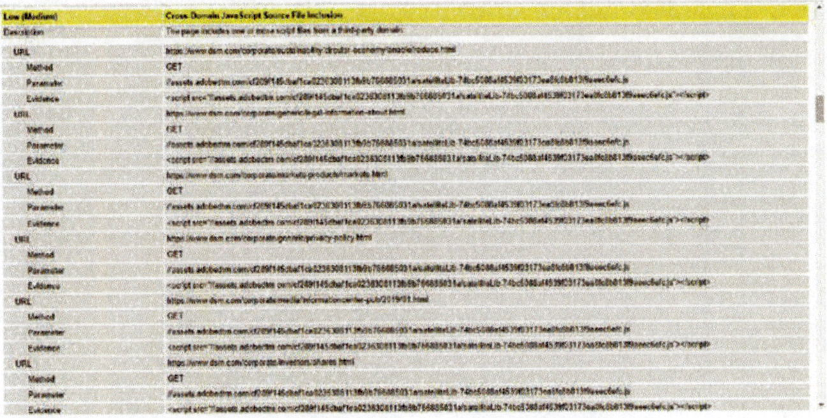

Fig. 4 Vulnerability report generated by ZAP

Here, the process starts when the report from the testing tool is ready. This report contains information include attack type, the Risk Level of the attack, location of the attack, method(GET/POST) used to perform the attack, and the parameters as shown in Fig. 4. To demonstrate the proposed approach.

XSS and CSRF attacks are extracted from the report. This report will be sorted as per the location attribute. All reported vulnerability with the same location would be tested using a single script and injected the script. At the same time, any network packet scanning tool scans all the GET and POST URLs for the web browser. The URLs scanned by this tool and impact keywords are compared with the parameters of the attack and if they match, the vulnerability is classified as a True positive of that attack type. This process is summarized in Fig. 5. This approach also shows the parallel process to verify the existence of XSS as well as CSRF to improve the performance of the process further. In this example, three vulnerability of types (stored XSS, reflected XSS, and CSRF) are combined for the vulnerability in the testing report share a common location. In brief, instead of design, execution, and analysis of three different vulnerabilities from the report, only one script would make the process near to 3 times more efficient. But, the tool may not report all these three types of vulnerability for all locations. For example, the sixth build in Table 1 shows 72 vulnerability of XSS (stored + reflected), which is not double of CSRF vulnerability. In the worst case, if all vulnerabilities (72 + 56 in the sixth build) do not share different locations, then 128 scripts need to be designed, executed and analyzed separately. But, looking toward the similar nature of triggering XSS and CSRF, this probability is much less. Considering the best case, where only 56 scripts need to be designed and tested. This approach will help to reduce the classification of reported vulnerability by half.

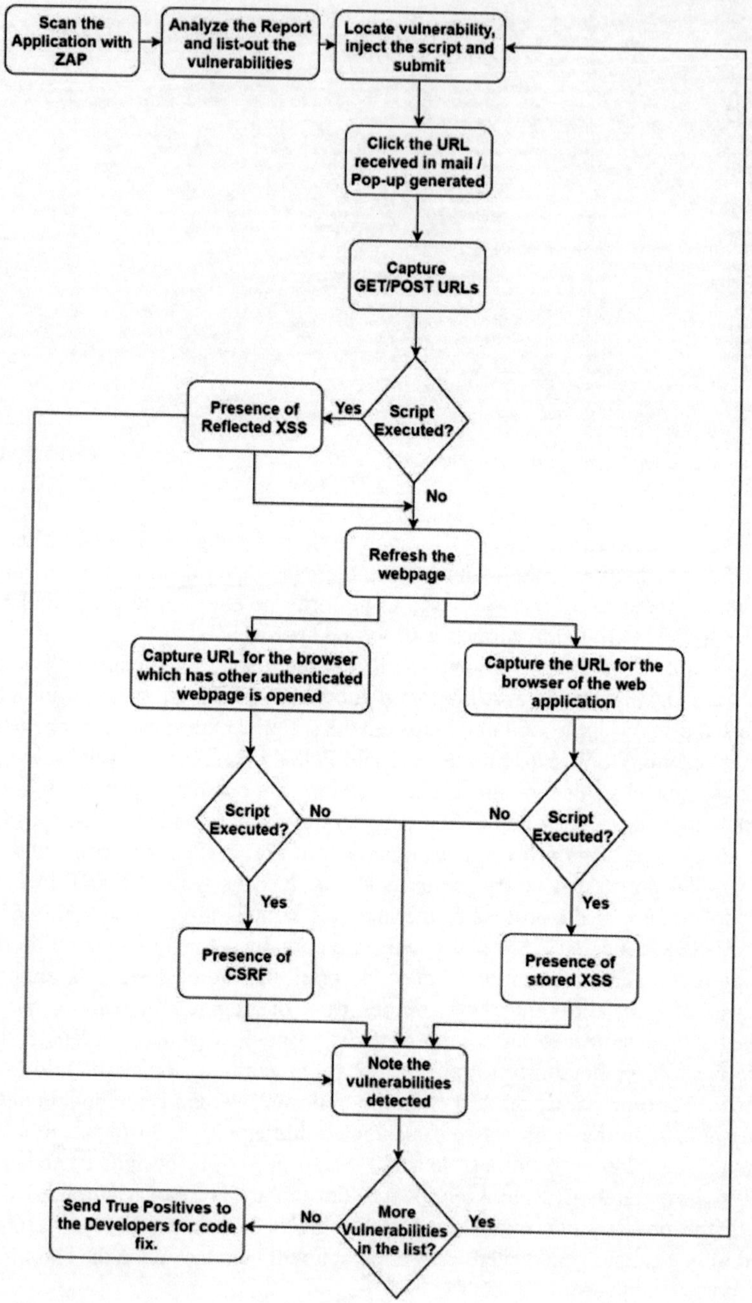

Fig. 5 Proposed approach

7 Conclusion

Statistics and previous analysis show that there is no such testing software that assures zero false-positive vulnerability, and the rate of false-positive vulnerability report is also remarkable. Unfortunately, due to the different contexts of web applications, existing tools are still not able to stop reporting zero false-positive vulnerabilities. This causes immense investment in manual verification of each reported vulnerability and delays the software revision process. The approach presented here shows a novel approach to identify different types of vulnerabilities sharing common ways of triggering an attack or process of security testing, generate a joint script for them to classify reported vulnerabilities in a single go. It can reduce overall testing time. This paper has illustrated an approach to design a joint script to classify XSS and CSRF vulnerabilities into true-positive and false-positive identification. This shows us the direction to identify other attacks also with similar flow and generate joint scripts for them. This work throw lights on some more research scopes like finding flow similarity among reported possible vulnerabilities using computing intelligence, automatic generation of the joint script for attacks with similar flow and its reported vulnerabilities, and most important is automatic filtering out false-positive vulnerabilities.

References

1. Mburano B, Si W (2018) Evaluation of web vulnerability scanners based on owaspbenchmark. In: 2018 26th international conference on systems engineering (ICSEng). IEEE, pp 1–6
2. Mohammed R (2016) Assessment of web scanner tools. Int J Comput Appl 133(5):0975–8887
3. Auronen L (2002) Tool-based approach to assessing web application security. HelsinkiUniversity Technol 11:12–13
4. Sagar D, Kukreja S, Brahma J, Tyagi S, Jain P (2018) Studying open source vulnerability scanners for vulnerabilities in web applications. IIOAB J 9(2):43–49
5. Muñoz FR, Armas Vega EA, Villalba LJG (2018) Analyzing the traffic of penetration testing tools with an ids. J Supercomput 74(12):6454–6469. https://doi.org/https://doi.org/10.1007/s11227-016-1920-7
6. XSSTypesPrevention (2019) Cross site scripting (XSS) attack tutorial with examples, types and prevention. https://www.softwaretestinghelp.com/cross-site-scriptingxss-attack-test/. Accessed 2020–02–20
7. Kaur D, Kaur P (2017) Cross-site scripting attack and their prevention during development. Int J Eng Dev Res 5(3)
8. XSSOWASP: Cross-Site Scripting (XSS) OWASP. https://www.owasp.org/index.php/Cross-site_Scripting_(XSS)/. Accessed 20 Feb 2020
9. Mahmoud SK, Alfonse M, Roushdy MI, Salem AM (2017) A comparative analysis of cross site scripting (XSS) detecting and defensive techniques. In: 2017 8th international conference on intelligent computing and information systems (ICI-CIS), pp 36–42. https://doi.org/10.1109/INTELCIS.2017.8260024
10. Kaur G (2014) Study of cross-site scripting attacks and their countermeasures. Int J Comput Appl Technol Res 3(10):604–609
11. XSSTesting: Testing for Cross site scripting. https://www.owasp.org/index.php/Testing_for_Cross_site_scripting/. Accessed 20 Feb 2020

12. CSRF: Cross-Site Request Forgery(CSRF). https://www.tutorialspoint.com/securitytesting/cross_site_request_forgery.htm/ (2020). Accessed 20 Feb 2020
13. Gupta J, Gola S (2016) Server side protection against cross site request forgery usingcsrf gateway. J Inform Tech Softw Eng 6(182):2
14. CSRFOWASP: Cross-Site Request Forgery(CSRF) OWASP. https://www.owasp.org/index.php/Cross-Site_Request_Forgery_(CSRF)/. Accessed 20 Feb 2020
15. XSS: Testing Cross-Site Scripting. https://www.tutorialspoint.com/security_testing/testing_cross_site_scripting.htm/. Accessed 20 Feb 2020
16. Fonseca J, Vieira M, Madeira H (2007) Testing and comparing web vulnerabilityscanning tools for SQL injection and XSS attacks. In: 13th Pacific Rim international symposium on dependable computing (PRDC 2007), pp 365–372. https://doi.org/10.1109/PRDC.2007.55
17. Gowda S, Prajapati D, Singh R, Gadre SS (2018) False positive analysis of softwarevulnerabilities using machine learning. In: 2018 IEEE international conference on cloud computing in emerging markets (CCEM). IEEE, pp 3–6
18. NETSPARKER: How Netsparker ensures false positives free web vulnerability scans. https://www.netsparker.com/blog/web-security/false-positives-thedirty-secret-of-the-web-security-scanning-industry/. Accessed 20 Feb 2020
19. Mohammadi M, Chu B, Lipford HR (2017) Detecting cross-site scripting vulnerabilities through automated unit testing. In: 2017 IEEE international conference on software quality, reliability and security (QRS). IEEE, pp 364–373
20. Srinivas (2017) Reduce false positives in application security testing. https://www.castsoftware.com/blog/reduce-false-positives-in-applicationsecurity-testing. Accessed 20 Feb 2020
21. Vithanage NM, Jeyamohan N (2016) Webguardia-an integrated penetration testingsystem to detect web application vulnmserabilities. In: 2016 international conference on wireless communications, signal processing and networking (WiSPNET). IEEE, pp 221–227
22. ZAPFAQ: How do I handle a false positive. https://www.zaproxy.org/faq/how-doi-handle-a-false-positive/. Accessed 20 Feb 2020
23. CSRFVsXSS: Cross-Site Request Forgery (CSRF) vs Cross-site Scripting, howpublished= "https://portswigger.net/web-security/csrf/xss-vs-csrf", note = . Accessed 20 Feb 2020
24. Aliero MS, Ghani I, Qureshi KN, Rohani MF (2020) An algorithm for detectingsql injection vulnerability using black-box testing. J Ambient Intell Humaniz Comput 11(1):249–266

A Survey on Hardware Trojan Detection: Alternatives to Destructive Reverse Engineering

Archit Saini, Gahan Kundra, and Shruti Kalra

Abstract System security has always been associated with the software being used on it. The hardware has been by default considered trusted. This root of trust for hardware has been violated after the emergence of Hardware Trojan (HT) attacks. Such attacks can be used by an adversary to leak important or secret information or to conduct a system failure. This paper gives a broad overview on different techniques that can be used to detect HT inside a circuit in place of destructive reverse engineering. Further a detailed literature survey has been presented which gives an overview of the efficiency of the detection techniques used in the literature.

1 Introduction

A Hardware Trojan attack is a malicious modification done during IC fabrication in any foundry, with intentions of affecting the behavior of an IC during field operation, such as creating undesired functionality or for creating backdoors for information leakage. The goal for HT insertion in a circuitry should be such that is not detected during conventional post manufacturing testing and has a rare trigger with a malicious payload [1–5]. Hardware Trojans are one of the biggest threats to have emerged in this decade for integrated circuitry, and economic trends have played a vital role to magnify the problem where manufacturing processes have started to increasingly rely on untrusted fabrication facilities [4, 6–8]. This makes Hardware Trojan detection significantly important. In view of the perspective of an attacker, the purpose of HT attacks can be wide ranging. These attacks can be planned to bring down the reputation of a company so as to achieve a competitive benefit in the business place [9–11]. Hardware Trojans can also be inserted in electronics equipment used in some national missions and caused major harm, by leaking confidential information through the backdoors created using these Trojans [12]. Many military mishappenings have also been linked to the existence of a hardware modification in the electronics used. These

A. Saini · G. Kundra · S. Kalra (✉)
Jaypee Institute of Information Technology, Noida, India
e-mail: shruti.kalra@jiit.ac.in

P. K. Singh et al. (eds.), *Proceedings of Second International Conference on Computing, Communications, and Cyber-Security*, Lecture Notes in Networks and Systems 203,
https://doi.org/10.1007/978-981-16-0733-2_63

Hardware Trojans can achieve more than one can think of, from leaking secret information, or achieving access to security systems, to even causing system failure, and thus trust validation of system on chip design becomes extremely important. These Hardware Trojans can be extremely difficult to detect as an intelligent attacker can plan these attacks by inserting just a few transistors in a multimillion transistor SoC design. Another way could be to change the doping concentration of the transistors, thus decreasing the reliability of the IC [13–15].

Any of these modifications can be detected in pre-silicon or post-silicon testing. Although in pre-silicon verification process, a golden integrated circuit (IC) is required to compare the functionality of ICs, thus validating the functionality of the IC, but the golden IC is not always available. Moreover, this does not validate that an IC is free from Trojan, as Trojan insertion might not affect the functionality of the IC (Trojan attack for leaking information), or the rare conditions at which the trojan is triggered to cause functional errors are not activated in the pre-silicon verification. Post-silicon testing methods include de-packaging of the IC and reverse engineering, but both of these methods are neither cost effective or time efficient [16–19]. Thus, Hardware Trojan imposes a major threat to security of a SoC design, and reliable detection methods become extremely crucial.

Any Hardware Trojan attack has two components: (1) Payload (2) Trigger. Payload is defined as the effect of a Hardware Trojan on a SoC design when it is activated. It defines the purpose of a Trojan, whether it will be affecting the functionality of a design, leaking private information, causing system failure, reducing the life of an IC, increasing path delay or power dissipation, or any other malicious intent. Trigger is defined as the rare set of conditions and/or sequence of events at the nodes which activate a Trojan. Only after a Trojan has been triggered, Trojan payload is experienced. The Hardware Trojan detection approach can be divided into two categories: destructive approach and non-destructive approach which are further divided into techniques such as logic testing, side-channel analysis and hardware IP trust verification, as shown in Fig. 1.

Following contributions are made in this paper:

1. An overview of HT highlighting its importance in the IC industry has been presented.

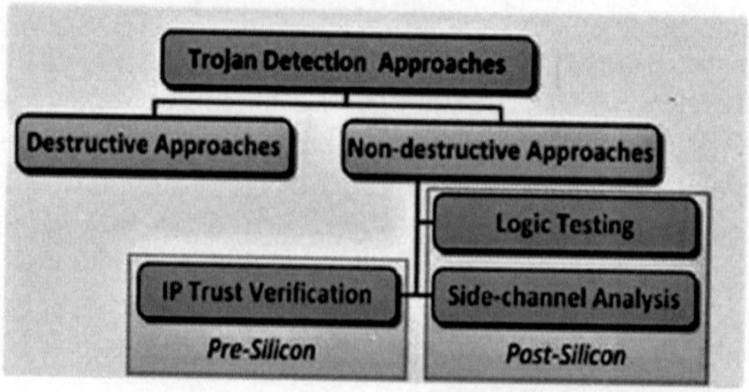

Fig. 1 Overview of different detection techniques [1]

2. A survey of different HT detection techniques available in the literature that can be utilized for detecting HT in a circuit has been analyzed.
3. The efficiency of various detection techniques presented above has been compared on parameters such as performance, implement ability, time need, success chance, infrastructure need and coverage scope.
4. The techniques are grouped on the basis of test modality for detection of HT.

This paper is organized as follows. Section 2 describes the destructive reverse engineering technique for Hardware Trojan detection. Section 3 highlights non-destructive approaches for Hardware Trojan detection which includes techniques such as logic testing, side-channel analysis and IP trust verification. Section 4 presents the comparison of different HT detection techniques on parameters such as performance, implement ability, time need, success chance, infrastructure need and coverage scope. Section 5 is the conclusion which is followed by Table 2 in which the literature survey of 25 papers is described.

2 Destructive Reverse Engineering

The destructive reverse engineering techniques are not considered efficient and feasible. After the fabrication and packaging of an IC are done, there is little to no visibility of the components inside it. Destructive reverse engineering can be used to de-package the IC and to get microscopic view of each layer to rebuild the design for trust validation of the final product. The destructive techniques [40, 41] use a sample of the manufactured ICs which are subject to de-metallization using chemical mechanical polishing (CMP) followed by scanning electron microscope (SEM) image reconstruction and analysis [15]. For an IC of appropriate complexity, it would take several weeks or months to do this, but this technique guarantees of 100% assurance that any unwanted modification in the IC will be caught. The IC used in this process cannot guarantee trust for all ICs, and there will only be assurance of a single IC. There is a possibility that the attacker has only infected few samples in an entire group of ICs. Therefore, the destructive techniques are not seen as a viable method to detect Hardware Trojan.

3 Non-Destructive Approaches

- Logic Testing—To detect the Hardware Trojan, it is important that the trigger condition is satisfied and also to observe the effect of such event on the output node [18]. One of the most popular logic testing techniques is multiple excitation of rare occurrence (MERO) [19]. The aim of this technique is to obtain a set of compact test patterns and maximizing Trojan detection coverage. It is done by detecting low probability conditions at the internal nodes and then by obtaining a set of vectors that should trigger each low probability nodes to their rare logic values multiple times (e.g., M times, M is a parameter defined by the user). The increased toggling rate of nodes improves the chances of activating a Trojan in contrast to random patterns. Logic testing is used on small-sized circuits, because if a number of input nodes are large, then larger will be the possible combinations of the trigger nodes.

- Side-Channel Analysis—Power consumed by a circuit depends upon the type and number of circuit elements, and the path delay relates to the number and type of logic levels or to the capacitive load in a path. So, side-channel analysis is evaluating the effect of a Trojan on a side-channel parameter, where static and transient current profiles are some of these parameters. It is used in post-silicon testing to overcome the limitations of logic testing approaches [42–44]. Any malicious inclusion to a circuit would lead to a change in the power consumed by a circuit, due to the addition/alteration of the current circuit elements. Presence of a Hardware Trojan is bound to have a delay impact due to the additional logic level, or increased capacitive load. Moreover, every Trojan will constantly monitor its activation condition or trigger which will impact the power consumption of the circuit [16]. All these inferences show that the impact of an unknown Trojan can be evaluated in a circuit by observing several physical parameters between a golden circuit and a Trojan infected one. This side-channel approach is directly dependent on two key features: (1) Signal-to-Noise Ratio (SNR)—Noise can be present due to process and environmental variations, and it can mask the impact of a Trojan on a physical parameter. Several statistical approaches have been used to isolate the Trojan from the noise. (2) Trojan to circuit Ratio (TCR)—In the absence of noise, TCR becomes the most significant factor, as in a multimillion gate SoC TCR is very small and effect of the Trojan on the physical parameters is under the negligible range for any circuit leading to failure in Trojan detection. Side-channel analysis can be done using one or more of the following parameters:
  - Static Current—Static CMOS gates are subject to leakage current in idle mode, so any change in the current is drawn from the power supply, even when no switching is happening helps in Trojan detection [43]. This is highly dependent on the TCR.
  - Transient Current—Main aim is to notice any switching activity [42]. Natural parameter variations in both chip to chip (inter-die) and within chip (intra-die) are crucial to take into account for this mechanism to work. Item Path Delay—It can be minimal for small Trojan circuit, and important vector selection sensitizing paths affected by Trojans is very vital. This technique is capable enough of detecting a single extra gate in a circuit [45, 12].
  - Multiple parameters to isolate the effect of noise, multiple parameters like transient current and Fmax can be used. Moreover, test vectors can be taken as such to improve the coverage spectrum of Trojan detection by manifolds. Following are the limitations for side channel analysis:
  - It requires a golden model or golden chip.

- Golden models are absolutely vital part of side-channel analysis as simulation-based analysis may have potential inaccuracies in the simulation model.
- Golden model can be obtained by two methods: (1) destructive reverse engineering and (2) exhaustive testing to verify trustworthiness. And both of these methods are largely cost and time withdrawing, making them highly infeasible.

- IP Trust Verification—A large numbers of IP cores used in todays SoC design are from different IP vendors, with different vendor giving different degree of reliability. Hence, trust on these third parties IP is crucial to ensure dependable SoC. Trojans can be easily inserted into different IPs by a dishonest designer or an untrusted CAD tool in a design house. The trust verification of the IPs can be done through directed test and verification. But the lack of golden design makes these direct tests infeasible. Another way to ensure trust is based on proof-carrying code method. A set of security-related properties is formed, and the designer crafts the proof of these properties. If there is any undesired modification in the IP, it is likely to violate the proofs. The IP user carries out the validation of security-related properties to make sure that no hardware description language (HDL) code was modified. This technique cannot ensure complete trust because the IP vendor who crafted the proof of these properties can be a dishonest person. But still it offers a line of defense against malicious alteration.

4 Comparison of Different Hardware Trojan Detection Techniques

In Table 1, different Hardware Trojan detection techniques are distinguished to understand the efficiency of these methods. Factors such as performance, coverage scope, implement ability, infrastructure need, time need and success chance are considered. Table 2 explains briefly about the recent work done on Hardware Trojan detection which uses the non-destructive detection techniques mentioned in this paper.

Table 1 Comparison between different Hardware Trojan detection techniques [5]

	Performance	Implementability	Time need	Success chance	Infrastructure need	Coverage scope
Reverse engineering	Low	Low	Very long	High	High	Low
Logic testing	Middle	High	Middle	Low	Middle	High
Side-channel analysis	High	Middle	Middle	Middle	Middle	High

Table 2 Summary of recent work on Trojan detection

S. No.	Paper	Detection method	Trojan model	Golden model	Test modality	Experiment trojan coverage %
1	[6]	Logic testing	Unknown trojan payload	C432	Logic implications	84% accuracy
2	[11]	Side-channel analysis	1-gate	Simulation	Static power, delay	98–100% accuracy
3	[20]	Side-channel analysis	1, 3, 5 added 2-input gates	Simulation	Static power	80–90% accuracy
4	[7]	Side-channel analysis	Combinational, comparator, Mux	Synthesized SEA IP core	Path delay	Increased delay detection is difficult Decreased delay detection is 100%
5	[9]	Side-channel analysis	Combinatorial and a sequential	Spartan 3AN FPGA	Path delay	100% accuracy
6	[10]	Side-channel analysis	Invertor chain	Simulation	Path delay	Does not guarantee trojan detection, can be used with other techniques
7	[12]	Side-channel analysis	2,4-bit comparator	Simulation	Path delay	100% accuracy
8	[8]	Side-channel analysis	8-bit comparator, combination trojan	C7552	Transient current	100% accuracy
9	[21]	Side-channel analysis	2, 3, 4-input gates	Simulation	Transient power	90–100% accuracy
10	[22]	Side-channel analysis	Comparator	Simulation	Transient power	90–100% accuracy
11	[4]	Side-channel analysis	AES-T100	AES-128 bit crypto core	Transient power	100% accuracy
12	[16]	Side-channel analysis	Counter: 16-bit Comparators: 3, 8-bit	Invasive characterization	Transient power	100% accuracy

(continued)

Table 2 (continued)

S. No.	Paper	Detection method	Trojan model	Golden model	Test modality	Experiment trojan coverage %
13	[23]	Side-channel analysis	Counter: 1, 3, 7, 9-bit Comparators: 3, 5, 20 bit	Simulation	Transient power	Easily detect trojans as small as 0.1% the circuit area
14	[24]	Side-channel analysis	2-input gates activated by trojans	Simulation	Functional	100% accuracy
15	[25, 26]	Side-channel analysis	Varying flip-flop numbers	Simulation	Transient power, pattern generation	80–90% accuracy
16	[27]	Side-channel analysis	2-input NAND gate, combination of an AND, XOR gate	ISCAS benchmark circuits C17 and S27	Leakage power	Very accurate
17	[4]	Side-channel analysis	Trojan model given in [28, 29] is inserted in AES-128 bit crypto core	AES-128 bit crypto core	Static and dynamic power	100% accuracy
18	[30]	16-ROs used (ring oscillator based)	64-bit LFSR is used as HT	128-bit LFSR generates test patterns for AES encryption blocks	Power supply noise	Very low trojan gate switching frequency may lead to an unsuccessful detection, as the trojan induced PSN decreased as the trojan gate switching frequency decreases
19	[31]	TeSR, a temporal self-referencing-based side-channel analysis	Small, rarely activated sequential trojans	Simulation model, no golden model required	Temporal variations in transient current signature of sequential hardware trojans [32, 25]	100% accuracy

(continued)

Table 2 (continued)

S. No.	Paper	Detection method	Trojan model	Golden model	Test modality	Experiment trojan coverage %
20	[33]	Scanning electron microscopy [SEM] and multiple image alignment	Combinational HT within a 40 K gates IC	Golden design IC required	Fully automated image processing	100% accuracy
21	[34]	Side-channel analysis for symmetric paths (covering entire IC using symmetric paths)	Non-functional trojan of two gates	Detection metric (DM) of a suspect IC compared with the detection threshold (DT), no golden IC required. (ISCAS-85 c17 benchmark circuit), HSPICE simulation	Symmetric path delays	Detection rate of 100% is achievable with maximum of 8% intra-die and 10% inter-die variation
22	[35]	Logic testing	Failure-type hardware trojans	Timing logic hardware chips and combinational logic hardware chips	State transition diagram	100% ideally, actual success rate of 92.8% and false retrieval rate of 7.14%
23	[36]	Side-channel analysis or even direct triggering	Tri-state buffer addition	s386, s5378, s9234a, s13207a and s38417 ISCAS89 benchmark circuits	Increasing transition probability of nets in ICs, the transition probabilities are increased to their maximum possible values by inserting tri-sate buffer [37]	Benchmark circuits have a number of nets with low transition probabilities even after adding tri-state buffer, and these nets should be checked as they are most vulnerable to being accessed by an adversary
24	[38]	Ring oscillator network (RON)-based hardware trojan detection method [30] with effective data analysis algorithm for trojan detection	Sequential—four 64-bit LFSRs, four 64-bit shift registers; combinational—ten 32-bit ripple carry adders, ten 32-bit ripple carry adders	Simulation	Dynamic power and static power	Static power analysis alongside the dynamic allows for even better trojan detection efficiency

(continued)

Table 2 (continued)

S. No.	Paper	Detection method	Trojan model	Golden model	Test modality	Experiment trojan coverage %
25	[39]	Power supply transient signal analysis with a statistical model	One gate trojans to nine gate trojans	ISCAS 85 benchmark circuit	Power supply transient currents	Trojan detection for trojans as few as of four gates, trojan activation is not required

5 Conclusion

Ever since the emergence of hardware-related Trojan attacks, different people came up with different Hardware Trojan detection techniques. In this paper, important methods are discussed that can be used in place of destructive reverse engineering which is a highly inefficient method. There is further discussion about the advantages and disadvantages of each method and then finally compared all the methods with the help of various parameters to get a better understanding of them. Thus, following points can be concluded:

- Logic testing and side-channel analysis are the major subclasses of Hardware Trojan detection mechanisms.
- Logic testing can be used for Trojans with unknown payloads. C432 golden model is tested for Trojan with unknown payload. Logic implications allow us to detect Trojans. This gives an accuracy of Trojan detection of 84% for this case.
- Dynamic power, static power, transient current, static current, path delay all come under side-channel analysis and can play a crucial role in Trojan Detection.
- Static power is sufficient for Trojan detection with an accuracy of more than 80–90% for gated Trojan models. Both static and dynamic power are considered to achieve an accuracy of almost 100% for AES 128 crypto core and sequential—four 64-bit LFSRs, four 64-bit shift registers, Combinational—ten 32-bit ripple carry adders and ten 32-bit ripple carry adders.
- Path delay detection criteria can be used for combinational and sequential Trojans like invertor chain, 2,4-bit comparators, MUX with a detection accuracy of more than 80%.
- Transient power analysis gives a detection accuracy of more than 90% for 2, 3, 4-input gates, Comparator, AES-T100, Counter: 16-bit Comparators: 3, 8-bit.
- s386, s5378, s9234a, s13207a and s38417 ISCAS89 benchmark circuits with Trojans can be handled by increasing transition probability of nets in ICs, and the transition probabilities are increased to their maximum possible values by inserting tri-sate buffer.
- For a non-functional trojan of two gates, detection metric (DM) of a suspect IC is compared with the detection threshold (DT). In this case, no golden IC is required (ISCAS-85 c17 benchmark circuit), which is handled using HSPICE simulation.
- Combinational HT within a 40 K gates IC can be dealt using scanning electron microscopy [SEM] and multiple image alignment techniques. This is a fully automated image processing method.

References

1. Bhunia S, Hsiao MS, Banga M, Narasimhan S (2014) Hardware trojan attacks: threat analysis and countermeasures. Proc IEEE 102(8):1229–1247

2. Ngo XT, Hoang VP, Le Duc H (2018, November) Hardware trojan threat and its countermeasures. In: 5th NAFOSTED conference on information and computer science (NICS). IEEE, pp 35–40
3. Wang SJ, Wei JY, Huang SH, Li KSM (2016, December) Test generation for combinational hardware trojans. In: 2016 IEEE Asian hardware-oriented security and trust (AsianHOST). IEEE, pp 1–6
4. Shende R, Ambawade DD (2016, July) A side channel based power analysis technique for hardware trojan detection using statistical learning approach. In: 2016 Thirteenth international conference on wireless and optical communications networks (WOCN). IEEE, pp 1–4
5. Sharifi E, Mohammadiasl K, Havasi M, Yazdani A (2015) Performance analysis of hardware trojan detection methods. Int J Open Inf Technol 3(5)
6. Cornell N, Nepal K (2017, August) Combinational hardware trojan detection using logic implications. In: 2017 IEEE 60th international midwest symposium on circuits and systems (MWSCAS). IEEE, pp 571–574
7. Kumar P, Srinivasan R (2014, March) Detection of hardware trojan in SEA using path delay. In: 2014 IEEE students' conference on electrical, electronics and computer science. IEEE, pp 1–6
8. Hou B, He C, Wang L, En Y, Xie S (2014, August) Hardware trojan detection via current measurement: a method immune to process variation effects. In: 2014 10th International conference on reliability, maintainability and safety (ICRMS). IEEE, pp 1039–1042
9. Exurville I, Zussa L, Rigaud JB, Robisson B (2015, May) Resilient hardware trojans detection based on path delay measurements. In: 2015 IEEE international symposium on hardware oriented security and trust (HOST). IEEE, pp 151–156
10. Li J, Lach J (2008, June) At-speed delay characterization for IC authentication and trojan horse detection. In: 2008 IEEE international workshop on hardware-oriented security and trust. IEEE, pp 8–14
11. Potkonjak M, Nahapetian A, Nelson M, Massey T (2009, July) Hardware trojan horse detection using gate-level characterization. In: 2009 46th ACM/IEEE design automation conference. IEEE, pp 688–693
12. Jin Y, Makris Y (2008, June) Hardware trojan detection using path delay fingerprint. In: 2008 IEEE international workshop on hardware-oriented security and trust. IEEE, pp 51–57
13. Adee S (2008) The hunt for the kill switch. IEEE Spectr 45(5):34–39
14. Kumagai J (2000) Chip detectives [reverse engineering]. IEEE Spectr 37(11):43–48
15. Collins DR (2008) Trust in integrated circuits. Defense advanced research projects agency Arlington VA microsystems technology office
16. Agrawal D, Baktir S, Karakoyunlu D, Rohatgi P, Sunar B (2007, May) Trojan detection using IC fingerprinting. In: 2007 IEEE symposium on security and privacy (SP'07). IEEE, pp 296–310
17. Tehranipoor M, Koushanfar F (1900) A survey of hardware trojan taxonomy and detection. IEEE Ann Hist Comput 01:1
18. Chakraborty RS, Narasimhan S, Bhunia S (2009, November) Hardware trojan: threats and emerging solutions. In: 2009 IEEE international high level design validation and test workshop. IEEE, pp 166–171
19. Chakraborty RS, Wolff F, Paul S, Papachristou C, Bhunia S (2009, September) MERO: a statistical approach for hardware trojan detection. In: International workshop on cryptographic hardware and embedded systems. Springer, Berlin, pp 396–410
20. Alkabani Y, Koushanfar F (2009, November) Consistency-based characterization for IC trojan detection. In: Proceedings of the 2009 international conference on computer-aided design, pp 123–127
21. Banga M, Hsiao MS (2009, July) VITAMIN: voltage inversion technique to ascertain malicious insertions in ICs. In: 2009 IEEE international workshop on hardware-oriented security and trust. IEEE, pp 104–107
22. Rad RM, Wang X, Tehranipoor M, Plusquellic J (2008, November) Power supply signal calibration techniques for improving detection resolution to hardware trojans. In: 2008 IEEE/ACM international conference on computer-aided design. IEEE, pp 632–639

23. Wang X, Salmani H, Tehranipoor M, Plusquellic J (2008, October) Hardware trojan detection and isolation using current integration and localized current analysis. In: 2008 IEEE international symposium on defect and fault tolerance of VLSI systems. IEEE, pp 87–95
24. Wolff F, Papachristou C, Bhunia S, Chakraborty RS (2008, March) Towards trojan-free trusted ICs: problem analysis and detection scheme. In: 2008 Design, automation and test in Europe. IEEE, pp 1362–1365
25. Banga M, Hsiao MS (2008, June) A region based approach for the identification of hardware trojans. In: 2008 IEEE international workshop on hardware-oriented security and trust. IEEE, pp 40–47
26. Banga M, Hsiao MS (2009, January) A novel sustained vector technique for the detection of hardware trojans. In: 2009 22nd International conference on VLSI design. IEEE, pp 327–332
27. Karunakaran DK, Mohankumar N (2014, July) Malicious combinational hardware trojan detection by gate level characterization in 90 nm technology. In: Fifth international conference on computing, communications and networking technologies (ICCCNT). IEEE, pp 1–7
28. https://www.trust-hub.org
29. Lin L, Kasper M, Gneysu T, Paar C, Burleson W (2009, September) Trojan side-channels: lightweight hardware trojans through side-channel engineering. In: International workshop on cryptographic hardware and embedded systems. Springer, Berlin, pp 382–395
30. Qin Y, Xia T (2017, October) Sensitivity analysis of ring oscillator based hardware trojan detection. In: 2017 IEEE 17th international conference on communication technology (ICCT). IEEE, pp 1979–1983
31. Narasimhan S, Wang X, Du D, Chakraborty RS, Bhunia S (2011, June) TeSR: a robust temporal self-referencing approach for hardware trojan detection. In: 2011 IEEE international symposium on hardware-oriented security and trust. IEEE, pp 71–74
32. Lin L, Burleson W, Paar C (2009, November) MOLES: malicious off-chip leakage enabled by side-channels. In: 2009 IEEE/ACM international conference on computer-aided design-digest of technical papers. IEEE, pp 117–122
33. Courbon F, Loubet-Moundi P, Fournier JJ, Tria A (2015, August) SEMBA: a SEM based acquisition technique for fast invasive hardware trojan detection. In: 2015 European conference on circuit theory and design (ECCTD). IEEE, pp 1–4
34. Vaikuntapu R, Bhargava L, Sahula V (2016, May) Golden IC free methodology for hardware trojan detection using symmetric path delays. In: 2016 20th International symposium on VLSI design and test (VDAT). IEEE, pp 1–2
35. Xia D, Zhu YF (2012, November) A research on detection algorithm of failure-type hardware trojan. In: 2012 Fourth international conference on multimedia information networking and security. IEEE, pp 918–921
36. Dhar T, Roy SK, Giri C (2019, January) Hardware trojan detection by stimulating transitions in rare nets. In: 2019 32nd international conference on VLSI design and 2019 18th international conference on embedded systems (VLSID). IEEE, pp 537–538
37. Salmani H, Tehranipoor M, Plusquellic J (2011) A novel technique for improving hardware trojan detection and reducing trojan activation time. IEEE Trans Very Large Scale Integr (VLSI) Syst 20(1):112–125
38. Qu K, Wu L, Zhang X (2015, December) A novel detection algorithm for ring oscillator network based hardware Trojan detection with tactful FPGA implementation. In: 2015 11th International conference on computational intelligence and security (CIS). IEEE, pp 299–302
39. Rad R, Plusquellic J, Tehranipoor M (2008, June) Sensitivity analysis to hard-ware trojans using power supply transient signals. In: 2008 IEEE international workshop on hardware-oriented security and trust. IEEE, pp 3–7
40. Chipworks Inc. Semiconductor manufacturing reverse engineering of semiconductor components, parts and process. [Online]. Available: http://www.chipworks.com
41. Kash JA, Tsang JC, Knebel DR (2002) US Patent No 6,496,022. Washington, DC, US Patent and Trademark Office
42. Narasimhan S, Du D, Chakraborty RS, Paul S, Wolff FG, Papachristou CA, … Bhunia S (2012) Hardware trojan detection by multiple-parameter side-channel analysis. IEEE Trans Comput 62(11):2183–2195

43. Aarestad J, Acharyya D, Rad R, Plusquellic J (2010) Detecting trojans through leakage current analysis using multiple supply pad I_{DDQ}s. IEEE Trans Inf forensics and Secur 5(4):893–904
44. Du D, Narasimhan S, Chakraborty RS, Bhunia S (2010, August) Self-referencing: a scalable side-channel approach for hardware trojan detection. In: International workshop on cryptographic hardware and embedded systems. Springer, Berlin, pp 173–187
45. Rai D, Lach J (2009, July) Performance of delay-based trojan detection techniques under parameter variations. In: 2009 IEEE international workshop on hardware-oriented security and trust. IEEE, pp 58–65

Comparative Study of Various Intrusion Detection Techniques for Android Malwares

Leesha Aneja and Jaspreet Singh

Abstract The spread of digital crimes have increased with the expansion in the use of smartphones. Especially, the major security threats have been seen in the case of android devices as android is the most famous working framework among smart phones. As these gadgets store confidential data of clients like private information, monetary data, thus malwares are being produced for stealing data. The reason behind why android OS is progressively prone toward malware assaults is that it does not put restrictions on its clients to download from unreliable sites. For understanding the risks to the Android clients' data, it is relevant to comprehend the distinction in the conduct of genuine and pernicious applications and study mobile malware detection. There are various methodologies for these Intrusions' identification, for example, static investigation, dynamic investigation and hybrid investigation which have been covered in this paper along with their functionalities. The benefits and constraints of each classification of android malware detection systems are also discussed. Therefore, this paper fundamentally focuses on the comparative study of these techniques.

Keywords Malwares · Intrusion detection · Static analysis · Dynamic analysis · Hybrid analysis · Comparative study

1 Introduction

There are lots of smart phones available in the market using different sorts of operating systems (OS). Android is one of the intensely used OS for mobile phones and is prominent in the market. It is built on modified version of the Linux kernel and holds a tremendous market share. Its open-source nature, performance and good user interface have made users to select cell phones that utilize android. With the

L. Aneja (✉) · J. Singh
GD Goenka University, Gurugram, India

J. Singh
e-mail: jaspreet.singh@gdgu.org

P. K. Singh et al. (eds.), *Proceedings of Second International Conference on Computing, Communications, and Cyber-Security*, Lecture Notes in Networks and Systems 203,
https://doi.org/10.1007/978-981-16-0733-2_64

expansion of smartphones, there is a dynamic growth in the production of malwares. An expanding number of softwares is being utilized for harming users by stealing users' private information like bank credentials, location, camera, etc., thus leading to major cyber crimes [1].

There are various third-party application stores from where the users can download applications for android. When users download from some unreliable sites, it becomes easier for malware authors to repackage android applications with pernicious code to enter into their systems and fetch their data. Malware contamination are giving billions of budgetary effect around the world. So, the identification and examination of android malware is becoming a significant area for research.

This paper discusses three different forms of malware intrusion detection system (MIDS). These three forms of MIDS include static, dynamic, and hybrid. Static examination is managed without executing the application, the application is dissembled, and manifest file is analyzed for dangerous permissions or signatures. It is quicker than dynamic examination yet experiences issues like code obfuscation. Dynamic investigation analyzes malicious behavior of application during runtime. It expects applications to be run and observed in a virtualized environment. It cannot cover pernicious code that is not executed during runtime. In the hybrid examination, features acquired from the static and dynamic investigation are utilized to distinguish vindictive conduct. This paper focuses on some analytical techniques and tools utilized for these forms.

Malwares have become a constant issue in computer security, and a couple of countermeasures ought to be taken to beat this issue. To address this, we have examined how to identify the pernicious application and how to analyze them. As declared above, for analyzing malware, we have examined the diverse sort of methods. Our study overviewed current researches with the expectation to have a better comprehension in building up a powerful answer to safeguard cell phones from malware. Thus, this study is based on exploring current research for improving malware recognition by surveying and classifying the current mobile malware identification strategies.

1.1 Contributions of the Paper

The valuable contributions of this work are mentioned below:

(a) This study reviewed different researches on android malware recognition with the expectation to have a better comprehension in building up a viable answer to safeguard cell phones from malware.
(b) Benefits and constraints of each categorization of MIDS have been explored and comparative analysis of these different intrusion detection techniques; i.e., static, dynamic, and hybrid have been undertaken.

The organization of this paper is as follows. Section 2 discusses the different approaches to malware detection in android. Section 3 provides comparative analysis. Section 4 concludes this paper and provides a way forward.

2 Different Approaches to Android Malware Detection

Various frameworks for perceiving malware have been proposed. In any case, mobile malware assessment still needs improvement. Though there are different characteristics of the available strategies for recognition of malware, but they have their shortcomings as well. Mobile phones and desktops, the two devices, have same working in terms of the data, applications, etc. Thus, similar security system can be applied to both of the devices. However, one needs to consider that phones have a limited space, memory, CPU rather than PCs. In this manner, applying any mechanism to cell phones needs to consider these limitations. In spite of the accomplishment of deploying IDS in the desktop environment, adjusting IDS in mobile environment requires a well plan design so as to conquer the constraint of the cell phones. Shabtai et al. [2] recorded few threats in Android like maliciously using the permissions granted to an installed application, draining resources, and exposing private content.

Shaerpour et al. have looked into a few patterns of android malware recognition and highlighted the following difficulties in android malware detection: issue of false negatives and false positive in recognizing android malware, concern of gathering enormous measure of information in powerful investigation which exhausts the limited storage resources of mobile devices. To defeat all the difficulties and distinguish the android malware detection inside and out, it is required to have a progressively precise and comprehensive review on existing android malware identification.

A lot of research has proposed techniques for investigating and identifying android malware. Current standard android malware discovery techniques fall into following classes: static investigation, dynamic examination, and hybrid. As discussed, the static investigation is done without executing the application and it decompiles android application source codes so as to make sense of malicious codes.

The dynamic examination gathers android application runtime information to see if the application executes with pernicious conduct. Static examination is not considered to be precise and productive as it is incapable of detecting intrusions during runtime of an application. It cannot uncover every malignant element to accomplish extensive detection. In reality, the dynamic examination can recognize update attacks. But, this sort of strategies frequently consumes resources. Hybrid analysis is another procedure which is a blend of both static and dynamic examinations.

2.1 Static Approach

In the static investigation, the examination of the apps is done and the permissions are extricated just by analyzing the code. Thus, it leads to lesser utilization of resources. This examination cannot identify runtime mistakes, logical irregularities, and conceivable security infringement, which is considered as a significant drawback of static investigation. Permission and API calls are generally used features.

Some current static examination techniques centers around stratifying and detecting various kinds of malicious softwares. RiskRanker [3] check applications' investigate features like authorizations and API calls to recognize malware. Some works included extracting API calls from binary files along with API frequencies as features in static examination. There have been different other features which are utilized for static investigation, for example, size of file [4]. Also, different features can be separated from APK document including Manifest.xml, classes.dex file and can be utilized for investigation. Such static technique was introduced by Sayfullina [5]. Shen et al. proposed a topology diagram dependent on android parts so as to identify malware and oppose against basic muddling utilized by programmers [6]. Features under networking such as TCP/UDP ports, destination IP, and HTTP request can also be taken into consideration. The static examination is a fast and reasonable methodology. Two primary methodologies under static examination include:

2.1.1 Signature-Based Analysis

Here, the application signatures are compared with a database of known malwares. So, it is unsuccessful in detecting unknown malwares because of the limited database of malware signatures as new malwares are being produced every day. The same needs to be updated in the database.

2.1.2 Permission Based Analysis

Here, the code is analyzed without running the application. It performs reverse engineering and analyzes permissions requested by apps in AndroidManifest.xml file. Mainly, it opts manifest file into consideration which is the downside of this approach.

Static analysis has certain drawbacks as the attackers use some anti-forensic techniques by which analyst is unable to find the suspicious code. Here are some of the techniques utilized by the attackers or malware authors:

(a) *Self-define communication Protocol*: Some malware is executing remotely that implies they connect with the code and execute remotely on explicit occasion. They take private data from the gadget and send to some server.
(b) *String Encapsulation*: The string is significant for data perspective. Numerous malware designers utilize solid encryption to evade the plaintext identification. Algorithms like DES or AES are applied for strong encryption with the goal that it becomes difficult for an examiner to decode the cypher text.
(c) *Code Obfuscation*: Malware authors use obfuscation code in which all the classes, bundles, strategies are renamed to single letter set in Java, the language in which android is written. So, it is exceptionally difficult to recognize various parts of the code and even it hard to comprehend the functions of the code.
(d) *Environment check*: Some kind of malevolent code is run in a particular type of environment or some particular sort of android cell phone. The code is

verified before execution. So, if a malware runs on a different device, it will stop executing; thus, an expert will not be able to comprehend the conduct of the malware.

2.2 Dynamic Approach

Dynamic analysis, or behavioral analysis, checks for API calls, system calls, IP address, network traffic, and so forth, while executing the application at runtime. It is fundamentally the testing and assessment of an application by executing it continuously on emulator. The goal is to discover errors in a program, as opposed to analyzing the code offline. The resource utilization in this examination method is more when contrasted with static investigation. This method becomes significant when static examination cannot fulfill the need because of the previously mentioned disadvantages.

Here are some notable strategies which can be discussed. TaintDroid gives system-wide dynamic taint tracking for android [7]. It features any malevolent information that begin from delicate sources, for example, location, microphone, camera, and so forth. This examination strategy is utilized to observe and obtain sensitive information before the transmission to the network interface of a cell phone; hence, this technique keeps away from data leakages. It centers exclusively on dataflow. Thus, it is unable to recognize such network intrusions. DroidTrace was proposed by Zheng et al. [8], a ptrace-based powerful investigation framework to observe nominated system calls of a target process that are executing dynamic payloads.

Zhai et al. proposed a recognition structure by exploring and analyzing features from API calls, system calls and native code execution [9]. In [6], the researchers have used authorizations and control flow graphs alongside support vector machines (SVMs) to separate malware from genuine applications. Mimran et al. [2] examined a conduct-based anomaly detection framework for identifying deviations in applications conduct by recognizing malware with self-upgrade capabilities.

Androdialysis [10] investigates the intents of every application as parameters for the classification. In their work, Schmidt et al. [11] proposed intrusion discovery framework that tracked the system activities through process list of open files, network traffic, etc. A. malware detection framework was discussed using KNN, SVM, J48, random forest, etc., [12]. Huang et al. likewise utilized AI method for order of android application and have a most extreme precision of 81% with J48 [13]. Malik et al. [14] investigated the conduct through framework call trace of 345 noxious applications utilizing AI algorithms.

The dynamic analysis approach is further classified into the following.

2.2.1 Anomaly-Based Detection

Any pattern or observation that does not fall in the expected region that is considered for normal behavior is termed as an anomaly or an outlier. Anomalies when go unnoticed in an android device can leak private information to the malware author. This method utilizes AI algorithms to recognize the vindictive conduct of the applications. To prepare a model for a new malware, features from the current malware are utilized. To identify the pernicious application, applications are introduced on the clients' device. The tools which utilizes this identification strategy give good results and consequently requires a great deal of resources. AI-based algorithms are being referred as promising approaches for identifying malware.

2.2.2 Taint Analysis Detection

This methodology utilizes a logical procedure called "dynamic taint analysis." TaintDroid gives system-wide data stream tracking for android. It can all the while track various wellsprings of delicate information, for example, camera, GPS, receiver, and so forth. Thus, it can distinguish the information discharge in various applications which are downloaded from different sources other than Google play. TaintDroid can label delicate data or information naturally, as long as the labeled data leaves the framework through any channel; it is recorded by TaintDroid.

Nowadays, there are lots of framework available for analysis of the malware. A hybrid analysis framework has been discussed in the upcoming section which includes both static and dynamic analysis.

2.3 *Hybrid Approach*

There are certain weaknesses of the above two approaches. For instance, static identification invests a great deal of time in identifying malevolent applications. But, dynamic recognition typically consumes many computing resources. Because of the above reasons, we have to talk about a solution to deal with android malware which is hybrid analysis. It is strategy that can combine run-time information removed from dynamic investigation along with code from static examination to recognize noxious conduct of the applications. This technique includes integrating static permissions received while breaking down the code of the application with dynamic permissions or data extricated while the application is executed.

Below are the examples of tools utilizing hybrid analysis approach:

(a) *Mobile Sandbox*: In this, the manifest.xml document is parsed for recognizing the suspicious code. In powerful investigation, they utilize emulator for executing the malicious application and behavior of application is observed. This sandboxed environment is utilized to isolate running projects and provide security to the host. To understand the conduct of apps, system traffic and local calls are checked.

(b) *Andrubis*: In Andrubis system, for dynamic examination, test results of static investigation are utilized. In static examination, this system is concentrated on an android manifest.xml record and byte code. The data that originates through static investigation is utilized for dynamic examination and produces viable outcomes. In this powerful examination, they do following investigation to be specific: taint tracing, method tracing, system level analysis, etc.

3 Comparative Analysis

Malware identification methods need to be improvized as the malware applications are expanding with the growth of cell phone users. Malware detection that utilizes static detection techniques face the problem of code obfuscation, and also, this technique cannot detect update attack as it does not analyze the code at run time. But, dynamic analysis solves this problem as it involves analyzing the behavior of application in real time. But, this method often consumes good amount of operating resources. So, these methods are consolidated to beat the disadvantages in distinguishing malevolent applications. From the examination, it is presumed that hybrid investigation, which is a blend of both static and dynamic investigation, is more productive approach and gives better outcomes when contrasted with static and dynamic examination independently. The hybrid examination is gaining popularity on the grounds because it produces precise outcomes in detecting malware application. Different tools used by these techniques have already been discussed in Sect. 2. The following Table 1 exhibits the comparison of these techniques on the basis of the parameters like time, input, resource utilization, etc.

4 Conclusion and Future Scope

Our paper presents a systematic literature survey of the MIDS approaches in android. Approaches like static, dynamic, and hybrid have been presented. The advantages and disadvantages of each method have been discussed with its functionalities. Our work compared and analyzed the malware detection approaches according to various essential factors such as time undertaken, code obfuscation, resource utilization, inputs utilized, etc. Malware discovery methods that use static examination can discover definitely known malware with high precision and proficiency. In any case, they are ineffective in discovering interruptions occurring at runtime. However, the dynamic

Table 1 Comparative interpretation of malware detection techniques

Factors	Static analysis	Dynamic analysis	Hybrid analysis
Time taken	Less	More	More
Input	Manifest file, APK file, etc	System calls, Runtime API data	Data obtained from both static and dynamic analysis
Code obfuscation	Yes	No	No
Resource utilization	Less	More	More
Effectiveness and accuracy	Less as compared to dynamic	Better than static analysis	Better than static and dynamic analysis
Target code execution	Not possible	Possible	Possible
Advantages	Low cost and requires less time for analysis	Provides deep analysis and higher detection rate with unknown malware detection	Extracts features of static and dynamic analysis both, providing more accurate results
Limitations	Limited signature database and can detect only known malware types	More time and power consumption	High cost

investigation can recognize update attacks. Although, dynamic examination techniques frequently consume huge operating resources, these strategies are consolidated to defeat the disadvantages in perceiving malevolent applications. From the examination, it is contemplated that hybrid assessment, which is a blend of both static and dynamic assessment, is progressively powerful and gives increasingly precise outcomes when contrasted with static and dynamic investigation separately. We have also studied researches, where dynamic analysis and static analysis have been used in training and detection phases subsequently.

Future work can include use of anomaly detection in different phases of static, dynamic, and hybrid analysis to examine their performance. More in-depth features and parameters can be explored to have a more comprehensive comparative analysis.

References

1. Singh PK, Kar AK, Singh Y, Kolekar MH, Tanwar S. Recent innovations in computing, vol 597. Springer, Switzerland. ISBN: 978-3-030-29406-9
2. Shabtai A, Tenenboim-chekina L, Mimran D, Rokach L, Shapira B, Elo-vici Y (2014) Mobile malware detection through analysis of deviations in application network behavior. In: Digital investigation. Elsevier
3. Grace M, Zhou Y, Zhang Q, Zou S, Jiang X (2012) RiskRanker: scalable and accurate zero-day android malware detection. In: Proceedings of the 10th international conference on mobile systems, applications, and services (MobiSys '12), ACM, pp 281–294

4. Kolbitsch C, Comparetti PM, Kruegel C, Kirda E, Zhou XY, Wang X (2009) Effective and efficient malware detection at the end host. In: USENIX security symposium, pp 351–366
5. Sayfullina L, Eirola E, Komashinsky D, Palumbo P, Miche Y, Lendasse A, Karhunen J (2015) Efficient detection of zero-day Android malware using normalized bernoulli naive bayes. In: 2015 IEEE Trustcom/BigDataSE/ISPA, pp 198–205
6. Shen T, Zhongyang Y, Xin Z, Mao B, Huang H (2014) Detect android malware variants using component based topology graph. In: 2014 IEEE 13th international conference on trust, security and privacy in computing and communications, pp 406–413
7. Enck W, Gilbert P, Chun BG (2008) TaintDroid: an information-flow tracking system for real-time privacy monitoring on smartphones. In: 9th USENIX symposium on operating systems design and implementation, pp 393–407
8. Zheng M, Sun M, Lui JC (2014) DroidTrace: a ptrace based android dynamic analysis system with forward execution capability. In: Wireless communications and mobile computing conference (IWCMC), pp 128–133
9. Li J, Zhai L, Zhang X, Quan D (2014) Research of android malware detection based on network traffic monitoring. In: Industrial electronics and applications (ICIEA), pp 1739–1744
10. Feizollah A, Anuar NB, Salleh R, Suarez-Tangil G, Furnell S (2017) Androdialysis: analysis of android intent effectiveness in malware detection. Comput Secur 65:121–134. https://www.sciencedirect.com/science/article/pii/S016740481630160
11. Schmidt AD, Schmidt HG, Clausen J, Yuksel KA, Kiraz O, Camtepe A, Albayrak S (2008) Enhancing security of Linux-based android devices. In: Proceedings of 15th international Linux Kongress, pp 1–16
12. Aneja L, Babbar S (2019) Malware detection in android devices using system calls under dynamic analysis. IJIET 13(3)
13. Aneja L, Babbar S (2017) Research trends in malware detection on android devices. Springer
14. Malik S, Khatter K (2016) System call analysis of android malware families. Indian J Sci Technol (IJST) 9(21)

Error Detection Using Syntactic Analysis for Air Traffic Speech

Narayanan Srinivasan and S. R. Balasundaram

Abstract Air traffic controllers and pilots communicate primarily through voice/speech to perform their day-to-day operations. Automatic speech recognition when applied in this domain can reduce the workload of both controller and pilot. The speech processing has several challenges in this domain like poor quality of radio channel, faster speaking rate and very strict vocabulary with infinite accent combinations. As errors have impact over the speech recognition process, the focus is on error detection in air traffic controllers speech when using automatic speech recognition. In this line, various error detection techniques available in normal English speech recognition are compared, and specific techniques which can help in air traffic controllers speech domain are discussed. Also, this paper emphasizes on syntactic analysis as a major component in the post-processing. Syntactic analysis along with phonetic string distance analysis helped to obtain close to 10% overall improvement in word error rate and 10–15% improvement in concept recognition rate for the experiments conducted over air traffic speech data considered for discussions.

Keywords Air traffic control (ATC) speech · Automatic speech recognition · Error analysis · Error rate · Phonetic matching · Error correction · Syntactic analysis

1 Introduction

Air traffic controller (ATC) and pilots use voice communication to a larger extend in their day-to-day operations and for safety of the flight. Despite the advancement in the technologies, voice control over radio is the primary means of giving instructions to aircraft within an airport's vicinity. Most of the operational, safety instructions pertaining to takeoff, taxi, landing and approach are given through voice commands. Claudiu [1] mentions that 80% of all pilot radio messages contain at least 1 error and 23% of flight level incursions and 40% runway incursions are due to communication errors. Deploying a speech recognition engine (ASR) in this domain has

N. Srinivasan (✉) · S. R. Balasundaram
National Institute of Technology, Tamil Nadu, Tiruchirappalli 620015, India
e-mail: nbasandroid@gmail.com

P. K. Singh et al. (eds.), *Proceedings of Second International Conference on Computing, Communications, and Cyber-Security*, Lecture Notes in Networks and Systems 203,
https://doi.org/10.1007/978-981-16-0733-2_65

been attempted by many, but in a constrained environment. No global deployment of ASR systems for ATC speech has been done so far. Even though ASR systems are matured and provide higher accuracy for general English applications (like Google Voice, SIRI, Alexa), due to the complexity and challenges of this domain, the usage of ASR systems is very limited. One of the basic requirements in this domain is to detect errors in the speech decoded text and apply appropriate error correction techniques which retain the concept, intent and semantics of each command. This way of error detection, error correction and concept extraction would help in bringing improvement in the overall speech recognition accuracy.

1.1 Automatic Air Traffic Speech Recognition

Van Nhan Nguyen [2] talks about evolution of various modules in ASR like feature extraction, language model, lexical model,and acoustic models. Techniques like hidden Markov model (HMM), state vector machines (SVM) and artificial neural networks (ANN) are already used by several authors to implement these models. Most of the work for ATC ASR started off with adapting ATC domain utterances into an off the shelf speech engine. CMU Sphinx was adapted with ATC corpus data from ATCOSIM corpus. A single speaker corpus was chosen to understand the errors in a detailed way. All the new modules recommended in this work are added to the post-processing step of the speech processing architecture which is described in Fig. 2. Future work is planned to perform multi-speaker analysis covering accent issues.

2 Error Analysis in Air Traffic Speech Recognition

A typical instruction from a controller or pilot has the following syntax Table 1.

Few examples of instructions are

"<swissair nine three five two> <climb> <flight level three five zero> <set> <course to gotil>"
"<hapag lloyd six five three> <climb> <flight level two nine zero> <set> <course trasadingen>"
"<sobelair two five five seven> <maintain> <heading> <contact> <rhein one three two decimal four>"
"<alitalia four zero one> <climb> <flight level two nine zero>"

Instructions can be a single instruction or a combination of instructions. The instruction may appear simple as it has high syntax control with small vocabulary.

Table 1 ATC instruction format

<CALL SIGN>	<ACTION COMMAND>	<DETAILS OF ACTION COMMAND>

Some of the challenges in converting the above speech instructions into text by a speech engine are varied accent, faster speech rate, mixing of non-standard vocabulary, named entity recognition (NER) like call sign or way points, ambiguity in saying numbers as both altitude, speed and heading are numbers and using fillers, sounds while speaking.

3 Literature Survey

A basic application of speech is a decoded speech display application. More advanced applications which rely heavily on accurate speech decoding are call sign detection, concept extraction, read back/hear back verification, ambiguity detection in instruction, command prediction and airport/controller/pilot efficiency improvements. Joakim [3] talks about various recognition errors in the ASR decoded ATC speech that are given below.

1. A correct utterance decoded and identified as correct utterance
2. A wrong utterance decoded and identified as wrong utterance
3. A correct utterance decoded and not identified as wrong utterance
4. A substituted utterance decoded and identified as substituted utterance
5. An inserted utterance decoded and identified as inserted utterance.

Hering [4] analyses the ATC utterances to understand the errors that come out of speech decoding. Claudiu et al. [1] mention about the need for reducing pilot/ATC communication errors by applying ASR in controller console and then use link systems to transmit to aircraft. Their argument is that, as the ASR is deployed in controller console, it is less subject to noise and other radio disturbances which can occur in cockpit. Chen et al. [5] discuss in detail on the need of strong acoustic and language model for eliminating errors in read back transcriptions. The authors also emphasize on concept accuracy than word level accuracy for higher applications. Recognition rates and word error rate of aircraft identifications (ACID) alias call sign were discussed. The authors conclude that call sign identification is a critical step in recognition process. They performed a top-level classification of controller and pilot instruction and then subcategories like CLIMB, TAKEOFF, HOLD SHORT, LINE UP AND WAIT instruction types. For the corpus and experiments done, they quoted 16% WER for controller instructions with adapted language model (LM), and 15% WER for controller instructions with trained model were achieved. Very high WER rates were observed by them for pilot spoken instructions. Based on the error analysis, there were issues in recognition of aircraft identification or call sign because of word errors. 79% concept recognition has been achieved for aircraft identification or call sign from the controller spoken instructions. Use of more discriminant aircraft identification or call sign detection techniques to achieve higher identification rate was suggested. This is our motivation to focus on determining more errors and possible corrections for call sign. We performed syntactic analysis to split the call sign from the complete utterance and then applied various similarity algorithms to identify more call signs correctly.

To understand the intricacies, we experimented and captured results in three categories

1. Analysis with similarity algorithms using full utterance
2. Analysis with similarity algorithms using syntactic split utterances
3. Word by word error categorization and analysis.

4 Overview of the Architecture

A typical speech processing system has the architecture as given in Fig. 1 with the following components.

4.1 Proposed Architecture

Most of the ASR in ATC architectures follow the pattern described in Fig. 1 ignoring the need for syntactic analysis. We propose the following architecture which includes additional steps in post-processing steps to support efficient error detection. The architecture can be depicted as below (Fig. 2).

In this architecture, syntactic analysis is defined as part of post-processing step. In this step, the speech output for ATC syntax is analysed, and then, it is used to determine errors in the output. Applying the error detection techniques at the pre-processing stage would affect the portability of the system to different airport contexts or countries. Hence, post-processing techniques are focussed in this work. Analysing

Fig. 1 Speech processing reference architecture

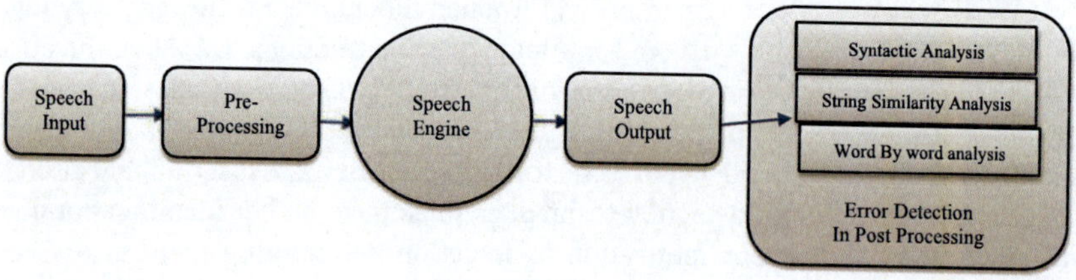

Fig. 2 Proposed architecture

the errors with syntactic analysis results in more errors to be determined and thus reduces errors in decoded output. The step wise algorithm is given below.

4.2 Methodology

In the proposed architecture, syntax analysis of the command utterances at various levels in the context of error detection is driven through the following algorithms.

i. SyntANL—applied to either full or parts of communication and the results populated
ii. Error detection—parsing the utterances and applying sequence of SyntANL calls
iii. HS,JS,ES,FS,LS are Hamming, Jaccard, EdiTex, Fuzzy, Levenshtein similarity algorithms, respectively.

Algorithm ErrorDetection (ST)

Input : ST - communication string
Output : Comparison of different algorithms with Error Detection %

```
Begin
        CS ← Separate(ST,1) /*Separate callsign from ST
        RCS ←Separate(ST,2) /*Separate rest of call sign from ST
        Call SyntANL(CS, HS)
        Call SyntANL(CS,JS)
        Call SyntANL(CS,ES)
        Call SyntANL(CS,FS)
        Call SyntANL(CS,LS)
        Call SyntANL(RCS, HS)
        Call SyntANL(RCS, JS)
        Call SyntANL(RCS, ES)
        Call SyntANL(RCS, FS)
        Call SyntANL(RCS, LS)
End.
```

5 Speech Corpora and Experiment Setup

We used ATCOSIM air traffic control [6] as our speech corpus. This corpus was recorded as ATC real-time simulation and covers the ATC operator speech. Total of ten hours is available as part of this corpus, and this is available free of charge. We considered seven sessions of German native male speaker for the ATC sector Solingen. Carnegie Mellon University (CMU) Sphinx speech engine which is available for public usage was used for the experiments. Default acoustic model from

CMU Sphinx and adapted ATC text were used for creating the language model and dictionary.

6 Error Detection Approaches

The baseline model was established using adapted text. Word error rate (WER) for the baseline models is given in Table 2.

As expected, without any improvisation, the baseline WERs were high. Experiments were conducted using syntactic algorithms, string similarity algorithms and phonetic matching algorithms. Further, word by word error analysis was also done and compared with baseline metrics.

6.1 Error Detection Using String Similarity Algorithms—Full Utterances

Table 3 provides the list of sample utterance that is used for decoding.

Table 2 Baseline metrics with adapted model

Method	WER (%)
Full utterance	27
Call sign only	29
Rest of call sign	17

Table 3 Full text utterances of ATC communication

AERO LLOYD FIVE NINE ZERO CLIMB FLIGHT LEVEL THREE ONE ZERO
AIR MALTA ZERO ZERO FOUR DESCEND TO FLIGHT LEVEL THREE HUNDRED RATE ONE THOUSAND FIVE HUNDRED OR MORE
ALITALIA FOUR EIGHT FIVE RHEIN RADAR IDENTIFIED
BRITANNIA TRIPLE TWO ALFA RHEIN RADAR IDENTIFIED
EUROPA THREE SIX THREE RHEIN IDENTIFIED PROCEED DIRECT FRANKFURT
FOXTROT SIERRA INDIA CLIMB FLIGHT LEVEL TWO SEVEN ZERO
HAPAG LLOYD THREE FIVE FIVE ONE RHEIN RADAR IDENTIFIED
LUFTHANSA FOUR FOUR ONE SIX CLIMB FLIGHT LEVEL THREE HUNDRED SET COURSE TO TRASADINGEN
NETHERLANDS AIR FORCE THREE FIVE CONTACT RHEIN ONE THREE TWO DECIMAL FOUR

The string-matching algorithms also have a similarity measure which provides us better metric to determine errors in the utterances. To compare the performance of the different similarity algorithms, the values were plotted in a histogram chart. For comparing the concept recognition, we considered only the top results from histogram. The results are populated in Table 4,

Hamming similarity was able to reduce the WER% by 9% when compared with the baseline. This is very close to Chen et al. [5] Adapted WER of 16%. If we consider top five values from the histogram, WER would go down further. Also, we observed that concept recognition rate was higher when compared to the results observed by Chen et al. [5].

6.2 *Error Detection Using Syntactic Analysis—Call Sign and Rest of Call Sign*

In this section, a new approach syntactic analysis combined with string distance analysis has been proposed to improve the error detection rate in ATC speech processing. The approach used the syntactic analysis of the output utterances and then applied the string similarity analysis. ATC utterances referred in Table 5 typically adhere to the following syntax.

<call sign><action word><details on action>

In syntactic error analysis, we split the instruction into two parts as below

[<call sign>] – [<action word><details on action>]

The syntax followed is ICAO standard and used by ATC personnel and pilots. After segregating the utterances, the instructions are given in Table 5.

Syntactic Analysis—call sign only

String similarity algorithms were also used, and the results were plotted in histogram to understand these algorithms. Two experiments were conducted. One with call sign only and other with rest of call sign. Table 6 gives the histogram chart of the various similarity algorithms experimented.

Syntactic Analysis—rest of call sign only

The second experiment of using string similarity with rest of call sign is given in Table 7.

Out of these analyses, FuzzyWizzy algorithm was able to match more utterances than any other algorithm. Hamming similarity has a very less WER than other algorithms. Comparison (best observed with our experiments) with the baseline metrics on the three approaches discussed so far is given in Table 8.

Table 4 Error detection performance of similarity algorithms

Similarity algorithm	Histogram	Max utterance count	WER (%)	Command concept recognition (%)
Hamming	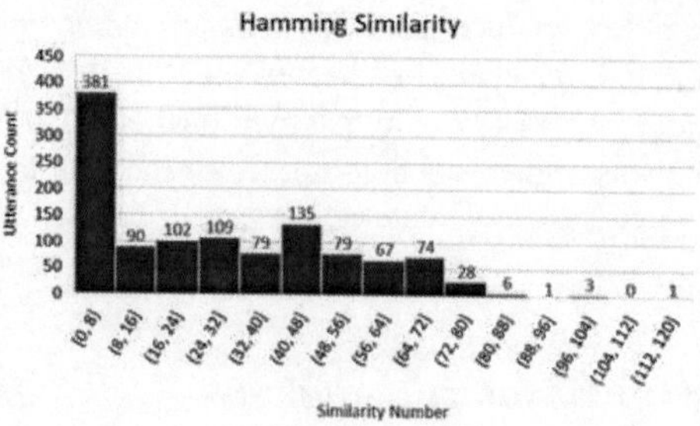	381	16	84
Jaccard	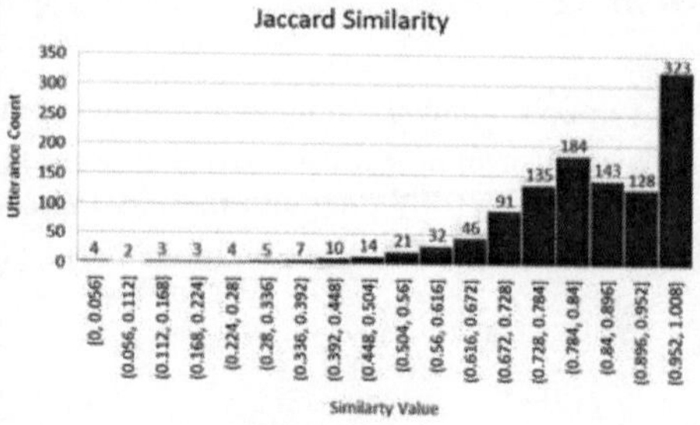	325	37	92
EdiTex	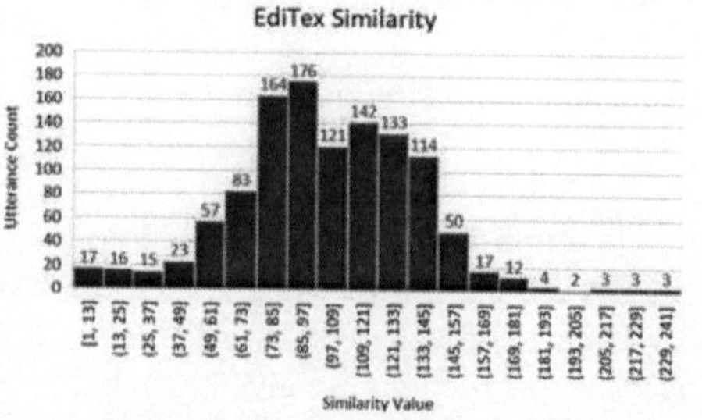	176	33	88
Fuzzy Wizzy	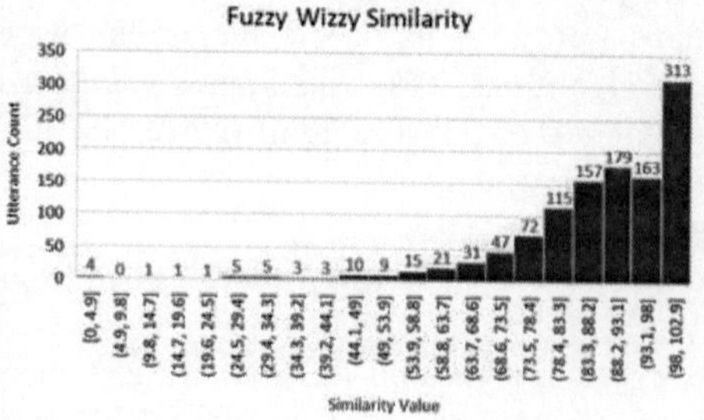	313	37	89

(continued)

Table 4 (continued)

Similarity algorithm	Histogram	Max utterance count	WER (%)	Command concept recognition (%)
Levenshtein	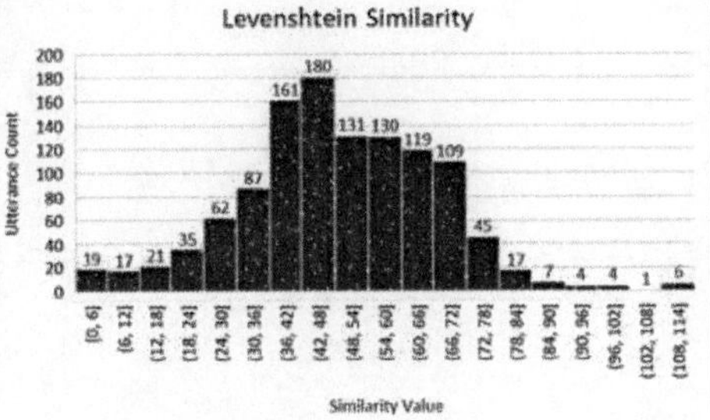	180	33	90

Table 5 Utterances with syntax split

Call sign	Instruction and additional info
AERO LLOYD FIVE NINE ZERO	CLIMB FLIGHT LEVEL THREE ONE ZERO
AIR MALTA ZERO ZERO FOUR	DESCEND TO FLIGHT LEVEL THREE HUNDRED RATE ONE THOUSAND FIVE HUNDRED OR MORE
ALITALIA FOUR EIGHT FIVE	RHEIN RADAR IDENTIFIED
BRITANNIA TRIPLE TWO ALFA	RHEIN RADAR IDENTIFIED
EUROPA THREE SIX THREE	RHEIN IDENTIFIED PROCEED DIRECT FRANKFURT
FOXTROT SIERRA INDIA	CLIMB FLIGHT LEVEL TWO SEVEN ZERO
HAPAG LLOYD THREE FIVE FIVE ONE	RHEIN RADAR IDENTIFIED
LUFTHANSA FOUR FOUR ONE SIX	CLIMB FLIGHT LEVEL THREE HUNDRED SET COURSE TO TRASADINGEN
NETHERLANDS AIR FORCE THREE FIVE	CONTACT RHEIN ONE THREE TWO DECIMAL FOUR

6.3 *Error Detection Using Word by Word Error Analysis*

ATC has a very strong vocabulary, and the usage of out of domain words in ATC is limited. There are numerous named entities which are specific only to this domain and not applicable in general English language. It is understood that categorization of words that belong to aviation specific or general English also helps. The following analysis was done using word by word

Table 6 Error detection using syntactic analysis—call sign only

Similarity algorithm (call sign)	Histogram	Max utterance count	WER (%)	Command concept recognition (%)
Hamming	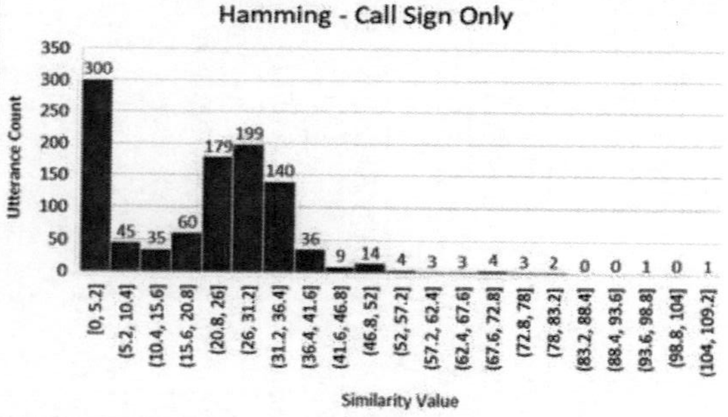	300	17	85
Jaccard	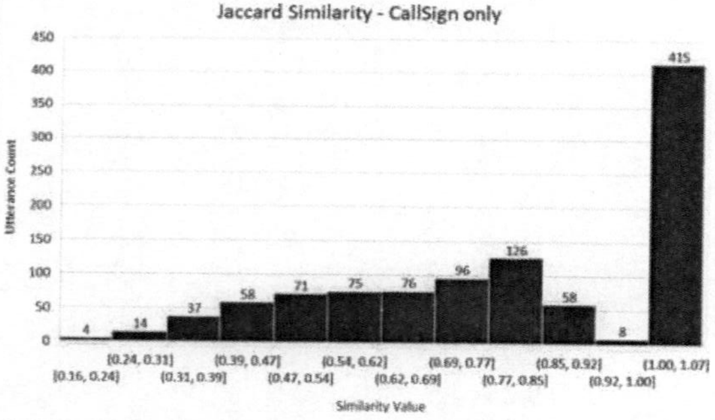	415	48	92

(continued)

Table 6 (continued)

Similarity algorithm (call sign)	Histogram	Max utterance count	WER (%)	Command concept recognition (%)
EdiTex	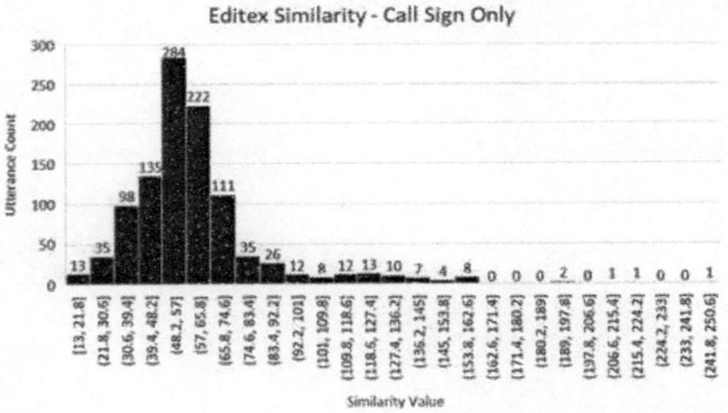	284	35	94
FuzzyWizzy	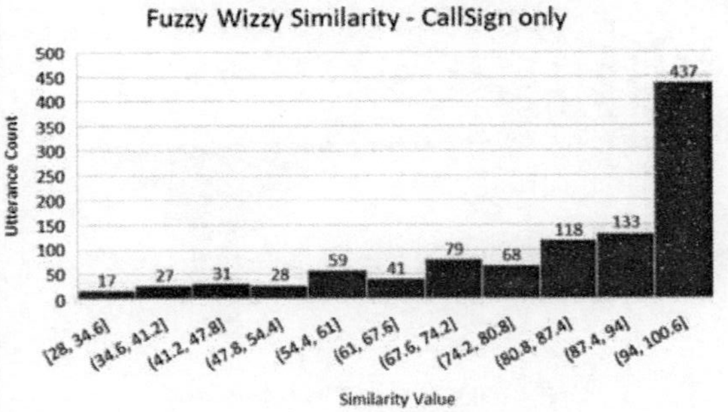	437	48	87

(continued)

Table 6 (continued)

Similarity algorithm (call sign)	Histogram	Max utterance count	WER (%)	Command concept recognition (%)
Levenshtein	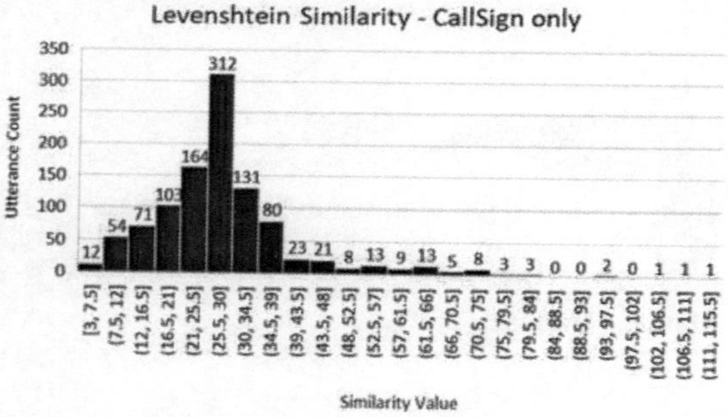	312	35	89

Table 7 Error detection using syntactic analysis—rest of call sign only

Similarity algorithm (Call Sign)	Histogram	Max utterance count	WER (%)	Command concept recognition (%)
Hamming	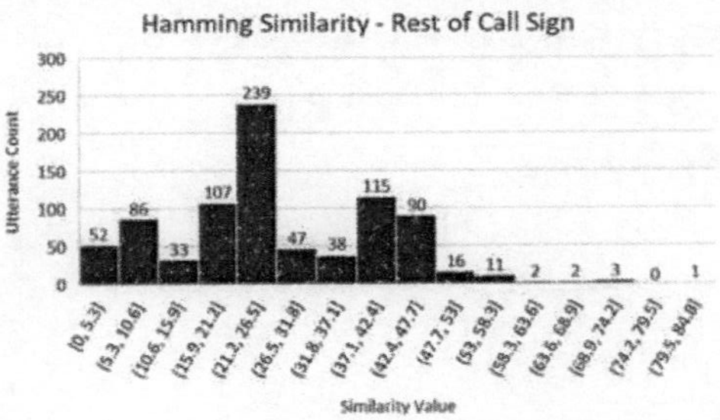	239	22	79
Jaccard	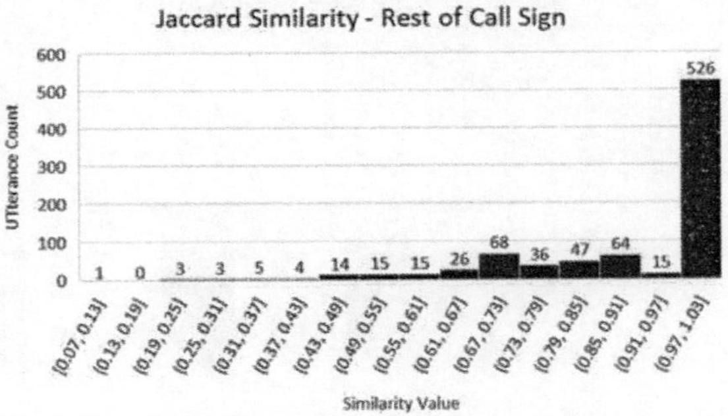	526	49	87

(continued)

Table 7 (continued)

Similarity algorithm (Call Sign)	Histogram	Max utterance count	WER (%)	Command concept recognition (%)
EdiTex	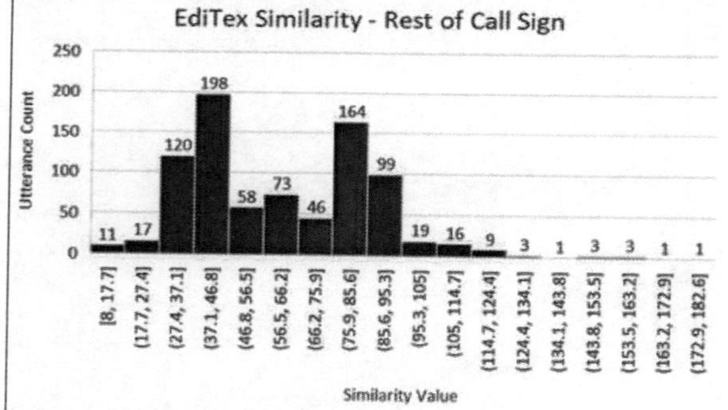	198	22	79
FuzzyWizzy	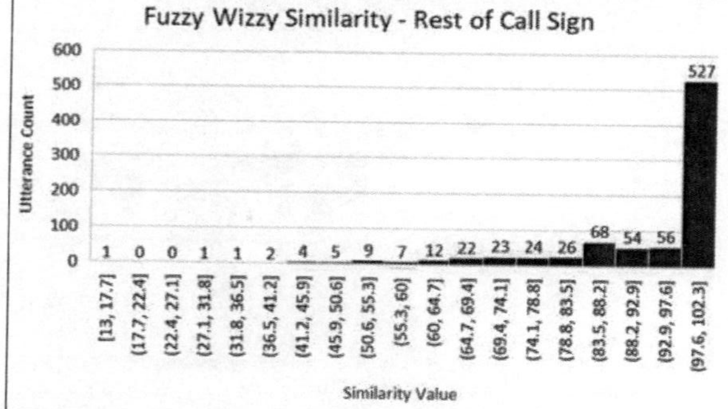	527	49	87

(continued)

Table 7 (continued)

Similarity algorithm (Call Sign)	Histogram	Max utterance count	WER (%)	Command concept recognition (%)
Levenshtein	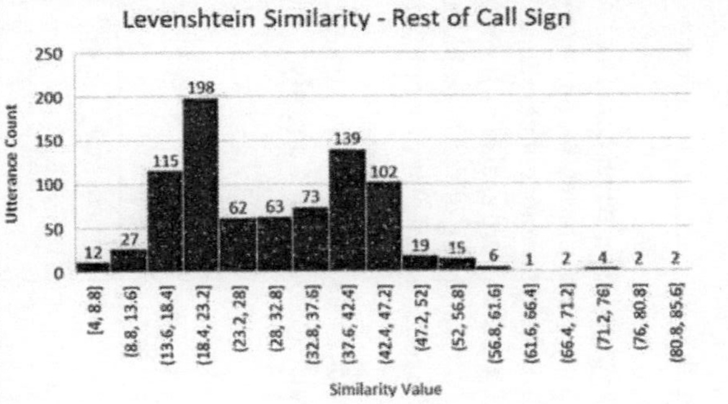	198	22	79

Table 8 Syntactic analysis comparison with baseline metrics

Approaches	Full utterance WER (%)	Call sign only WER (%)	Rest of call sign only WER (%)
Baseline (adapted LM)	27	29	17
Similarity algorithm	16	17	22
Concept command recognition	92	94	87

1. Number of words which were decoded correctly and decoded incorrectly.
2. Number of occurrences of words in the entire set of utterances.
3. Number of times a word was decoded correctly and decoded incorrectly.
4. Categorization of each word as "English" and "Domain Specific" to understand the contributing factor.

Table 9 Provides the error detection % with respect to one or more words in each utterance.

It is easily noticeable that error detection % is lesser when compared with other techniques discussed earlier in this work. The value is based on one or more word errors in the utterances, and hence, it may cover larger number of utterances and yield a lesser error detection %. On the other hand, word by word analysis can lead different ways to address errors in a much precise way by applying different techniques based on the category of words. Table 10 gives the individual count of words which were correctly decoded and incorrectly decoded.

Total number of occurrences of each word was counted and divided into two categories namely "Correctly Decoded" and "Not Decoded Correctly". The top ten results are given in Table 11

Table 9 Error detection % with word by word analysis

Error analysis word by word comparison	
Number of utterances with 1 or more wrong words	753
Number of utterances with all correct words	402
Error detection %	65%

Table 10 Words correctly and incorrectly decoded

Word error analysis	Count
Number of words wrongly recognized	165
Number of words correctly recognized	55
Total number of words	220

Table 11 Words "found" and "not found" count

Words not found count		Word found count	
Word	Not found count	Word	Found count
SAME	251	ONE	854
ME	121	THREE	751
PROCEED	97	TWO	728
MORNING	83	FOUR	541
CAN	61	SEVEN	541
AGAIN	58	RHEIN	537
DINKELSBUHL	57	FIVE	442
SAY	56	SIX	403
AFTER	46	IDENTIFIED	389
IT	43	DECIMAL	315

Table 12 Words "found" compard with "not found" numbers

Number of same words found > not found	87
Number of same words found < not found	78

Table 13 Domain words versus English words

Number of domain words in row 1 of Table 12 (out of 87)	61
Number of general English words in row 1 of Table 12	26

The following Table 12 is a summary of ratio of words found to not found.

From the above table, 87 words can be easily mapped to the correct entries of the same occurrence of the word. This significantly improves the concept recognition rate. Finally, the categorization of words based on domain or English was done to understand if there is any scope to improve concept recognition further (Table 13).

Domain words are a major contribution to the word errors, and hence, any focused algorithm for domain using string distance/similarity could help to detect and correct the errors in a better way. Moreover, word by word error is much easier to implement in post-processing than implementing costly string-matching algorithms. We observed that syntactic analysis along with word by word error analysis would help in making efficient error detection and thus error correction.

7 Conclusion

We were able to observe, experiment and illustrate that in-depth analysis of error in ATC ASR could help in making the overall speech engine accuracy better. Performing higher-level analysis helps to improve higher-level modules involved in ATC ASR, but in-depth analysis helps in making the rest of the higher-level modules perform better. Understanding the pattern of ATC ASR errors for a specific airport or a context could be considered as future work and help in attaining better read back error detection and correction or reducing concept error ratio. This would help adaptation of ATC adapted speech engines rather than training the engines from scratch with the challenges in getting the corpus for this domain. Phonetic comparison yields better result when combined with syntactic analysis and word by word analysis. Further work is planned to use machine learning methods to detect and correct errors in ATC domain.

References

1. Geacăr CM (2010) Reducing pilot/atc communication errors Using voice recognition. In: 27th international congress of the aeronautical sciences, 2010
2. Van Nguyen N, Holone H (2015) Possibilities, challenges and the state of the art of automatic speech recognition in air traffic control. World Acad Sci Eng Technol Int J Comput Electr Automation Control Inf Eng 9(8)
3. Karlsson J (1990) The integration of automatic speech recognition into the air traffic control system, MIT
4. Hering H (2001) Technical analysis of ATC controller to pilot voice communication with regard to automatic speech recognition systems. Eurocontrol Experimental Centre
5. Chen S, Kopald HD, Chong R, Wei Y, Levonian Z (2017) Read back error detection using automatic speech recognition. 25th USA/Europe air traffic management research and development seminar (ATM2017)
6. ATCOSIM Corpus Document. https://www.eurocontrol.int/eec/gallery/content/public/document/eec/conference/paper/2008/005_ATCOSIM_corpus.pdf. Last Accessed by 10 Aug 2019

Road Segmentation from Satellite Images Using Custom DNN

Harshal Trivedi, Dhrumil Sheth, Ritu Barot, and Rainam Shah

Abstract Recently, with the enhancement in the field of remote sensing and computation techniques, road detection from satellite images is getting possible. In these days, precise extraction of the lane from satellite images has become one of the major important fields of research in both remote sensing and transportation. The road network performs an imperative role in the traffic system, urban planning, route planning, and self-driving. In this paper, technique for road segmentation from the satellite images has been introduced. In the proposed method, a custom deep neural network (DNN) has been used for the detection of the road from satellite images. We have used a simple and custom neural network which is computationally faster and as accurate as a traditional deep neural network like Inception, YOLO, and ResNet-50 for road detection in the satellite images. In the initial stage, images are preprocessed with the help of OpenCV and morphology. We have annotated each pixel value as 0 for non-lane pixels and 1 for lane pixels. With this annotated data, we have trained our custom DNN model. The road region is denoted by white pixels, and black pixel denotes a non-road region. In the final result, the noise removal technique is used to remove distorted white pixels to improve the accuracy further.

Keywords Road extraction · Satellite images · OpenCV · Deep neural network

H. Trivedi · D. Sheth
Softvan Pvt. Ltd, Ahmedabad, India
e-mail: harshal@softvan.in

D. Sheth
e-mail: dhrumil@softvan.in

R. Barot
LD Engineering College, Ahmedabad, India

R. Shah (✉)
Gandhinagar Institute of Technology, Ahmedabad, India

P. K. Singh et al. (eds.), *Proceedings of Second International Conference on Computing, Communications, and Cyber-Security*, Lecture Notes in Networks and Systems 203,
https://doi.org/10.1007/978-981-16-0733-2_66

1 Introduction

There is such a huge number of astonishing ways machine learning and AI are utilized behind the scenes to affect our everyday existences. The term "artificial intelligence" is frequently used to portray machines that copy "intellectual" works that humans relate with the human personality, for example, "learning" and "critical thinking". It has become an urgent piece of day-by-day human lives today, and it aids pretty much every situation—regardless of whether you understand it or not. AI has the favorable circumstances over the normal intelligence as it is increasingly lasting, steady, and affordable, has the simplicity of duplication and scattering, can be recorded, and can perform certain undertakings a lot quicker and superior to the human. AI has been of tremendous use in different areas like information retrieval, computational linguistics, and language translation as stated by *Pannu Avnit* [1]. Road segmentation is significant for self-driving and pedestrian identification. Recuperating the 3D structure of street scenes gives significant logical data to improve their comprehension. Research has been undertaken for road segmentation using deep learning, which is an application of artificial intelligence, for proper training of data and accurate results.

Computer vision (CV) helps in mechanizing tasks that the human visual framework can do. Applications of CV range from undertakings, for example, modern machine vision frameworks to an investigation into AI and robots that can grasp their general surroundings. For acquiring road segmentation with higher accuracy, another important prospect is to gain elevated level interpretation from images or videos which can be provided by CV. Satellite images are one of the most dominant and significant devices. Satellites pictures give a decent portrayal of what is going on at each point on earth. There are a variety of ways in which satellite images can be utilized for research in different fields. There is an abundance of error-free data obtained from satellite images that if processed and analyzed properly can provide required results for road segmentation. Our method uses the satellite images and divides the image between in the lane and the non-lane part, which can help for the town planning. Also, this segmentation can help in order to uplift the rural areas.

2 Related Work

The need for road segmentation from remote sensing images has been increasing since the past few years. Many attempts have been made in order to get accurate results of road segmentation. Jiang et al. [2] have proposed an optimized convolutional neural network algorithm to segregate road from remote sensing images and utilized wavelet packet algorithms in order to increment the precision of division. Wei Xia et al. [3] used deep convolutional neural network as a tool for abstract feature extraction from a huge amount of data on which semi-supervised labeling technique is applied, and precise road boundaries through post-processing are accomplished. Henry et al. [4] present the substantial implementation of tree fully convolutional neural networks, namely FCN-8s, deep residual U-Net, and DeepLabv3+ . These methods resolve the major issue of distinguishing slender objects from the spotted environment and identifying numerous road patterns regardless of notable visual differences. A fully convolutional neural network is one of the most successful methods to apply pixel-wise segmentation. Identically sized prediction map can be produced by the method proposed by Long et al. [5]. Semantic segmentation of the satellite SAR images can be achieved with the help of the FCNN. Buildings, land use, bodies of water, and other natural areas were classified by applying off-the-shelf pre-trained FCNNs on SAR images. This method provided by Yao et al. [6] satisfactorily results for land use and natural classes but not for buildings. Ashwani Kumar et al. [7] focus on the object detection strategy to recognize objects on any gadget running the model and in any environment. In the given method, convolutional neural systems are utilized to build up a model which is made of multiple layers to group the given items into any of the characterized classes. Multi-scale feature maps have been used for object recognition. To accelerate the computational performance, single shot multi-box detector calculation has been utilized. Ashwani Kumar et al. [8] proposed an object identification strategy for blind people in real time to detect objects on all device with the help of this model. Convolutional neural networks have been used with single shot multi-box detector algorithm. The single shot multi-box detector algorithm uses standard VOC and COCO datasets. This model incorporates the sound gadget, which will be useful for the blind people.

According to Sandeep Reddy et al. [9], the proposed method for automatic road extraction from satellite images utilizing adaptive global thresholding and morphological operations has been proposed. Adaptive global thresholding is applied on satellite images to convert images into grayscale image. By applying histogram analysis on that image, road region is segmented. Morphological operation is implemented to expel the non-road region from the satellite image.

Yadav et al. [10] proposed a technique for road detection and extraction of satellite images. In this approach, the input image is preprocessed. The threshold of the preprocessed image is calculated by using Otsu's method. With the help of this threshold, a binary image is extracted from the gray image. Road network is extracted from the image by morphological operators. Wang, Weixing, et al. [11] observed distinct road features and road models and then after the road detection methods have been labeled into different categories such as knowledge-based methods, classification-based methods, dynamic programming, active contour model, and mathematical morphology. Merits and demerits of the above-listed methods and research accomplishments were highlighted.

Xu, Yongyang, et al. [12] suggested that the segmentation model was created based on densely connected convolutional networks (DenseNet). The developed model was trained with the help of the dataset of the preprocessed remote sensing images. The designed deep neural network GL-Dense-U-Net model was used for the road segmentation. All the preprocessed samples were given as the input, and the output of the trained model was classified in two categories: road and non-road.

Mnih, Hinton [13] have explained an approach for road extraction. In the proposed method, system is trained to identify road from expert-labeled data. Expert-labeled data and high-resolution aerial images are easily available; therefore, this method is well suited. Deepthy Mary Alex [14] has used low-resolution satellite images in this research paper to reduce the cost. These images are enhanced using discrete wavelet transform (DWT) and high-frequency sub-bands. The road is extracted from enhanced images with the help of level set and mean shift method. Alexander Buslaev et al. [15] proposed a technique for road segmentation based on a fully convolutional neural network of the U-net family. This architecture contains ResNet-34 pre-trained on ImageNet and decoder implemented from vanilla U-Net. This system incorporates restrained memory that facilitates the use of just one GTX 1080 or 1080TI video cards to perform whole training and makes extraction fast.

Xiaolong Liu et al. [16] suggested a method for road identification using the RPP model for monocular vision based on the amalgamation of fully convolutional network, residual learning, and pyramid pooling. In this method, the 112-layer RPP model has been used. Vladimir Khrayashchdev et al. [17] proposed an approach for recognition of geo objects by convolution neural network using satellite images from DSTL, Landsat-8, and PlanetScope databases. Three moderations of convolutional neural network were implemented. Yao Wei et al. [18] explained method road network extraction using CNN with multiple starting point tracers. This method has two stages. The first stage includes initial pixel-wise semantic segmentation, and in the second stage, road network maps are constructed using road centerline tracing and post refinements.

Rui Fan et al. [19] proposed a method for lane segmentation using ResNet. In this paper, incongruity of road is displayed as a linear model in the v-disparity map. In further procedure, proposed architecture is used to decrease the repetitive information. After that disparity maps and images are merged to create 3D images to train neural network. Jinsoo Kim et al. [20] suggested method for real-time object detection with the help of YOLO by merging information from RGB camera and LIDAR. Zhu Sun et al. [21] explained method for traffic congestion detector, which contains two modules: attention proposal module and deeply supervised Inception network. A very deep structure based on the Inception network was used to detect traffic congestion. As compared to previously used approaches YOLO, ResNet, and Inception, our model deep neural network provides same accuracy but it is a lightweight model and works faster. Syed Aley Fatima et al. [22] provided and have given a short writing review of object identification in remote sensing. This paper gives a short rundown of various object recognitions in remote sensing images and also examines about their merits and demerits. The principle focal point of this survey paper is on the satellite image and likewise examined the issues and advancement of current scenario. Bhardwaj et al. [23] have discussed various techniques used for image enhancement. Image enhancement techniques mentioned in this paper are contrast stretching and image sharpening, nonlinear image enhancement technique, genetic algorithm, generalized fuzzy enhancement, wavelet transform technique, multi-scale and single scale retinex improvement technique, etc.

3 Proposed Method

3.1 Task Description

In our method, images are preprocessed using OpenCV, and after that morphology is implemented to the image. In the RGB image, black and white pixels are manually assigned where white pixel represents "lane". The RGB image pixels which contain lane are stored in CSV as 1, and non-lane pixels are stored as 0. As per the above given configuration, training model was prepared with the assist of a deep neural network (DNN). We have used aerial imagery for roof segmentation (AIRS) dataset. Post-processing method, median blur algorithm, is implemented to the output image to eliminate noise and increment efficiency.

3.2 Proposed Flow

3.2.1 Training Phase

Input image
We have used aerial imagery for roof segmentation (AIRS) dataset having image size 10,000 * 10,000. It is a public dataset that targets at benchmarking the algorithms of roof segmentation from very-high-resolution aerial imagery, and hence, we used this dataset (Qi Chen 2018) [24]. The main features of AIRS are that it covers 457 km^2 orthorectified aerial images, and dataset contains 16 GB of training images and 1.6 GB of test images (Fig. 1).

Preprocessing
Crop the image into a size of 500 $\times$ 500, to get a better visualization of the lane and the non-lane pixels of the satellite image. For preprocessing, we used morphological image processing. Binary images may have various defects like noise and texture. Morphological image processing seeks after the objectives of casting off those deformities by representing the form and structure of the image. This method is an assortment of nonlinear features identified with the form or morphology of features in the image. Morphological operations depend on the relative ordering of pixel values, not on their numerical values. Morphological procedures check an image with a structuring detail. Some operations check whether the element fits within the neighborhood, whereas others check whether it hits the neighborhood (Fig. 2).

Perform annotation We manually annotated black and white pixels where white pixels denoted lane, whereas black pixels denoted non-lane areas. We created comma separated value (CSV) file which includes lane and non-lane pixels of RGB image. We have annotated the images with "lane" as white pixels and other parts of the image as black pixels. The RGB image pixels which contain lane are stored in CSV as 1, and non-lane pixels are stored as 0. Then we store the data for training (Fig. 3).

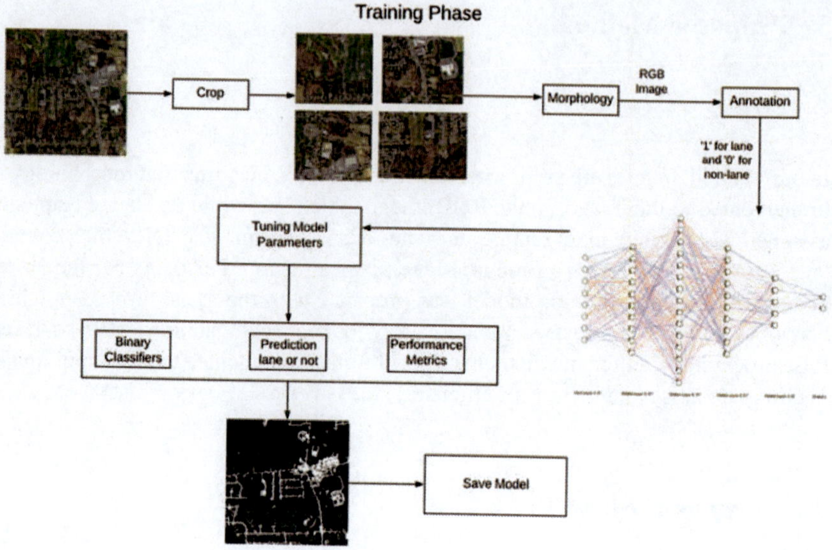

Fig. 1 Training phase of the proposed method

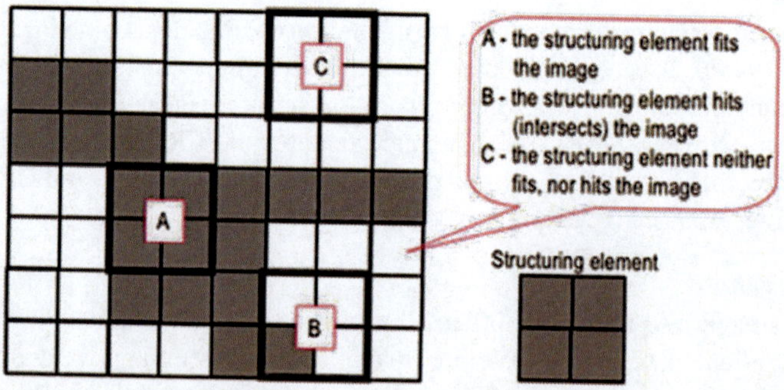

Fig. 2 Probing of an image with the structuring

Custom DNN

Train using our custom DNN model. A DNN is a group of neurons organized in a sequence of a couple of layers, wherein neurons receive as input the neuron activations from the preceding layer and carry out a simple computation. The neurons of the network jointly implement a complex nonlinear mapping from the input to the output. This mapping is learned from the data by adapting the weights of each neuron by applying a method named error back propagation.

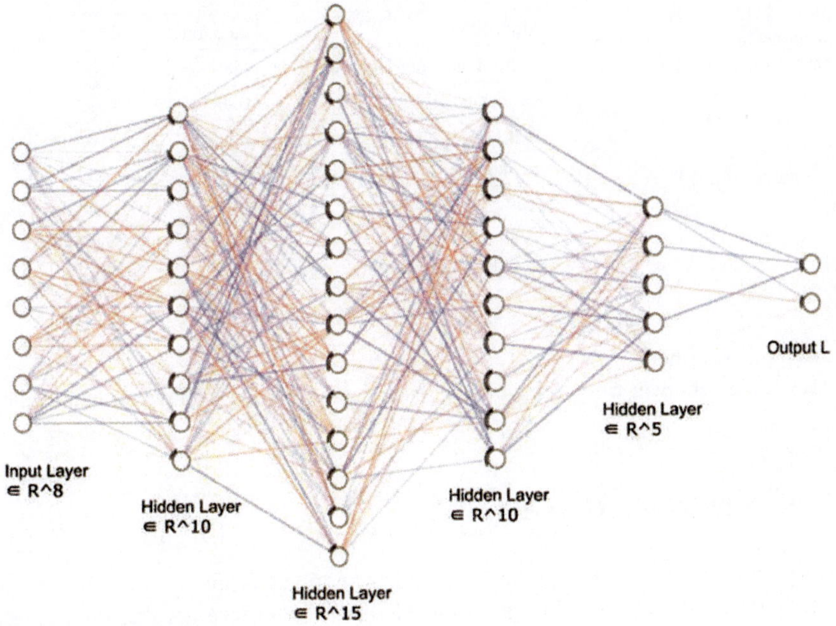

Fig. 3 Deep neural network

3.2.2 Evaluation Phase

Input image
We have each RGB image of size 10,000 × 10,000 from the AIRS dataset. We have used this image for training.

Preprocessing
We cropped the image into the size of 500 × 500, to get a better visualization of the lane and the non-lane pixels of the satellite image. Then we loaded the model for prediction.

Evaluation
Predict each pixel for 0 and 1 and create an image with the non-lane pixel as black pixels and lane predicted pixels as white pixels.

Post-processing For removing the noise, in post-processing, we have used the median filter algorithm. Median filter is valuable in diminishing irregular noise, particularly when the noise amplitude probability thickness has enormous tails and occasional patterns. Using this algorithm, we replace every gray pixels with the value of median of the neighboring pixels instead of using average (Lizhe Tan 2019) [25]. After noise removal using median filter algorithm, the accuracy of the lane segmentation is increased by approximately 5%.

Table 1 Comparison of accuracy and time complexity among different algorithms

Algorithms	Accuracy	Time (ms)
Custom deep neural network	0.86	700
ResNet-50	0.93	1500
YOLO	0.88	1100
Inception	0.91	1600
SSD	0.88	1000
SVC	0.78	600
Random forest	0.81	600

Final output image
After noise removal, we will have the output image which will show predicted lane.

4 Computational Complexity

We performed the road segmentation using various algorithms. From the above results, the ResNet, YOLO, Inception, and SSD have better accuracy than our custom DNN but time complexity is high. SVC and random forest are faster but less accurate. To keep the perfect balance between accuracy and time complexity, custom DNN is the best option to choose from. In our method, we have tuned hyperparameters into a traditional DNN architecture for faster and more accurate result. Our architecture is a four-layer custom DNN architecture having 120, 150, 120, and 70 nodes in the first, second, third, and fourth layer, respectively. Additionally, we have also implemented preprocessing method like morphological image processing and post-processing method, i.e., median filter algorithm for better result (Table 1).

5 Experimental Setup

We have split the dataset with 80% train images, 10% validation images, and 10% test images. On NVIDIA 1080ti GTX, we have trained for 50 epochs with each image of 10,000 × 10,000. We have cropped those images into 500 × 500. We have used a total of 95 test images and 900 train images, and after cropping we have obtained 1900 test images and 18,000 train images (Table 2).

Table 2 Custom DNN nodes

Hidden layer	Nodes
First	120
Second	150
Third	120
Fourth	70

6 Result Analysis

The results of the proposed method of road identification and extraction are shown in the below figures. The proposed technique has been implemented to satellite images. Figure 4 indicates the original colored image, it is given as input, and Fig. 7 shows the output image where white pixel denotes lane and black pixel denotes non-road region.

Accuracy: After performing post-processing, accuracy is ~86%.

Accuracy Formula:

$$\text{Precision} = \frac{\text{TP}}{\text{TP} + \text{FP}}$$

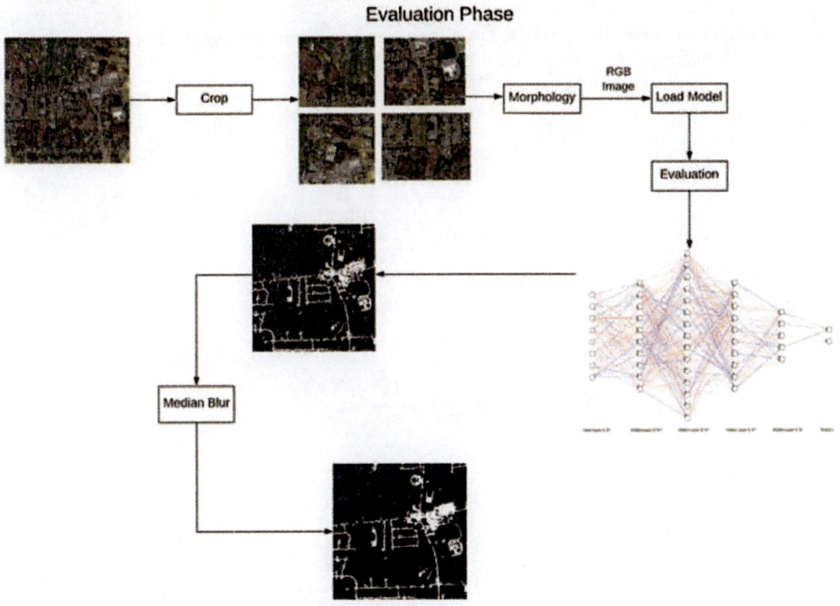

Fig. 4 Evaluation phase of the proposed method

Fig. 5 Sample input image which is used for training and evaluation purposes. This image is further trained by our custom DNN architecture

$$\text{Recall} = \frac{\text{TP}}{\text{TP} + \text{FN}}$$

$$\text{Precision} = \frac{\text{TP}}{\text{TP} + \text{FP}}$$

$$\text{F}_1 = \frac{2 * \text{precision} * \text{recall}}{\text{precision} + \text{recall}}$$

$$\text{Accuracy} = \frac{\text{TP} + \text{TN}}{\text{TP} + \text{FN} + \text{TN} + \text{FP}}$$

TP: True Positive
TN: True Negative
FP: False Positive
FN: False Negative

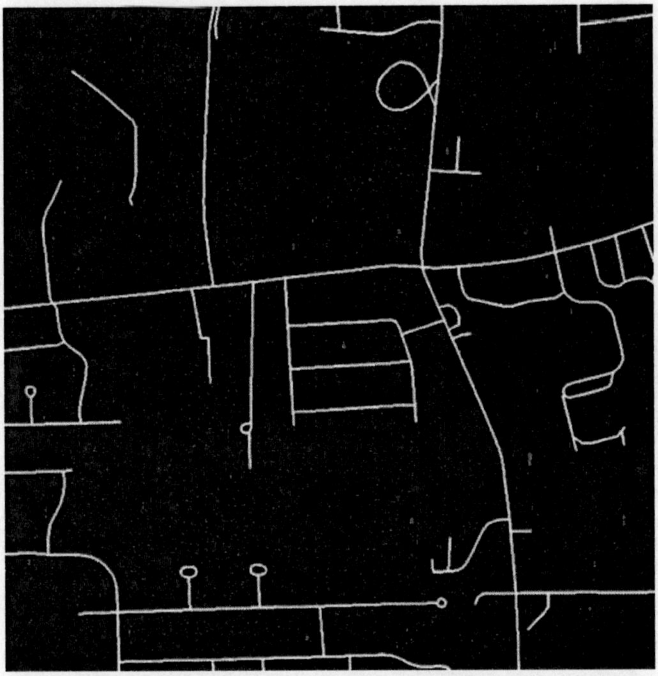

Fig. 6 Annotated sample image. During the training after applying morphology, this image was annotated by marking 1 for lane pixels and 0 for non-lane pixels. This image will be compared with the trained model for measuring the accuracy

Training accuracy: 91%
Testing accuracy: 86%

Test:
Recall—0.85
Precision—0.78
F_1—0.81.

We have calculated accuracy by summation of all the corrected classified pixel labels divided by the total number of pixels in the testing dataset. For calculating our model accuracy in road detection from the satellite image, this is the simplest and robust way. From the labeled lane and non-lane pixels, we have calculated how many of the test pixels were correctly classified by our custom DNN model. Other than accuracy, we have also calculated precision, recall, and F1-score (Figs. 5, 6 and 8).

Some other samples of road segmentation

See Figs. 9, 10, 11 and 12.

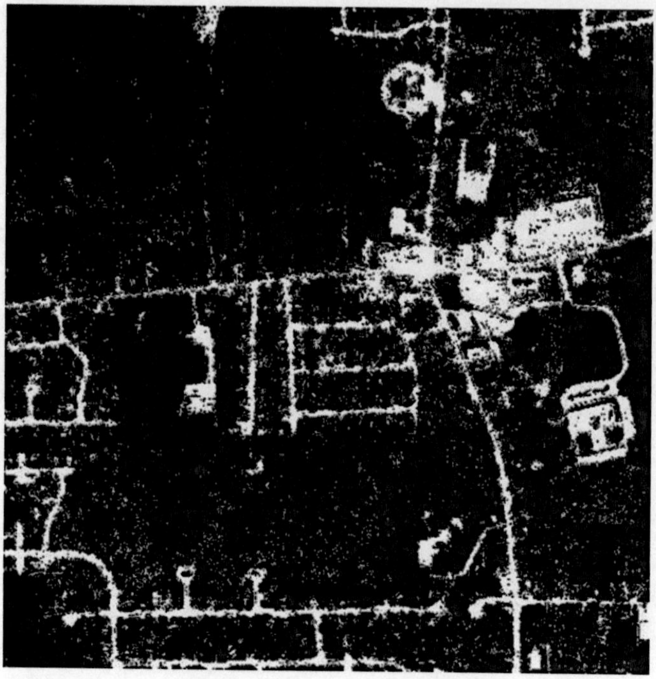

Fig. 7 Obtained after the model is trained, and this is the output image that classifies the lane from an input image

7 Conclusion and Future Work

We have presented the implementation and results of lane detection from the satellite images. We have implemented various algorithms, and our custom DNN shows the best balance between time complexity and accuracy. The first step is to perform preprocessing on the image. Then image pixels are stored in a CSV file and trained with DNN. Post-processing is performed using median filter algorithm after training to increase accuracy and to remove noise. We have achieved 86% accuracy with our custom DNN. We will try different algorithms for better accuracy while maintaining time complexity.

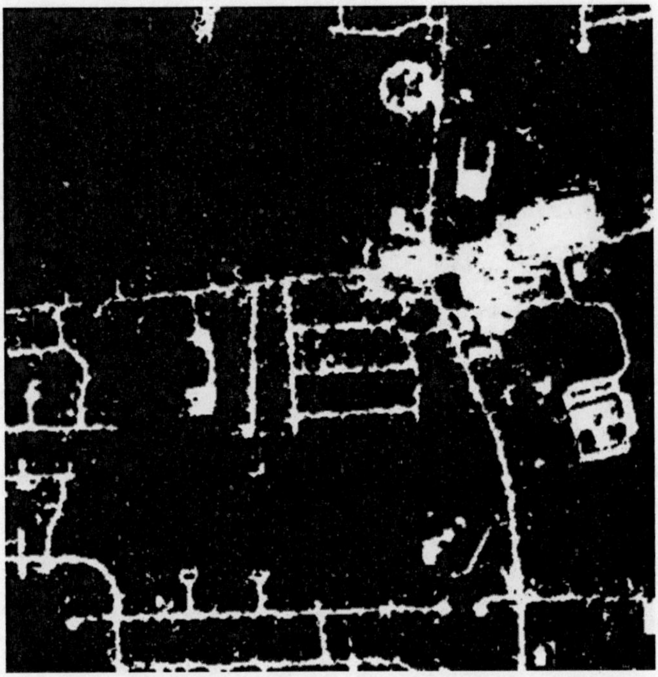

Fig. 8 Obtained after applying median blur to remove false positive pixels (noise) to improve accuracy

Fig. 9 Input image 1

Fig. 10 Output image 1

Fig. 11 Input image 2

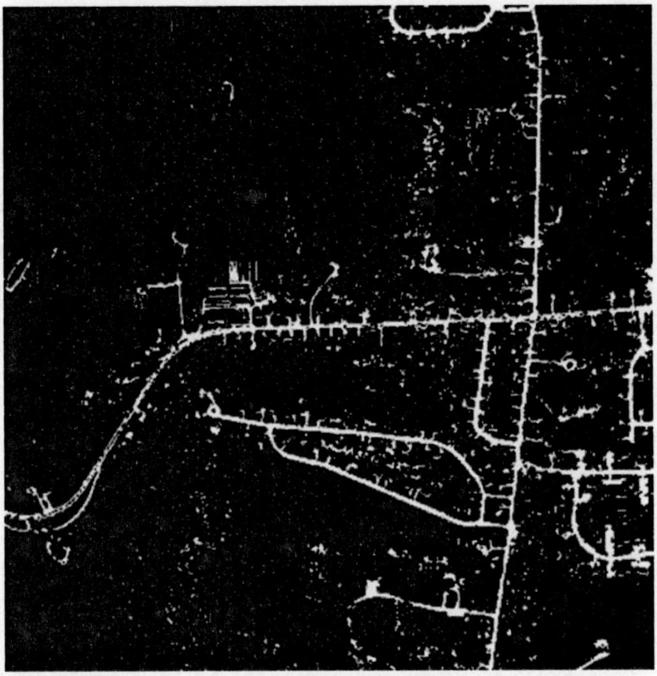

Fig. 12 Output image 2

References

1. Pannu A (2015) Artificial intelligence and its application in different areas. Artif Intell 4(10):79–84
2. Jiang Y (2019) Research on road extraction of remote sensing image based on convolutional neural network. EURASIP J Image Video Process 2019(1):31
3. Xia W et al (2018) Road extraction from high resolution image with deep convolution network—a case study of GF-2 image. Multi Dig Publishing Inst Proc 2(7)
4. Henry C, Azimi SM, Merkle N (2018) Road segmentation in SAR satellite images with deep fully convolutional neural networks. IEEE Geosci Rem Sens Lett 15(12):1867–1871
5. Long J, Shelhamer E, Darrell T (2015) Fully convolutional networks for semantic segmentation. In: Proceedings of the IEEE conference on computer vision and pattern recognition
6. Yao W, Marmanis D, Datcu M (2017) Semantic segmentation using deep neural networks for SAR and optical image pairs. 1–4
7. Kumar A, Srivastava S (2020) Object detection system based on convolution neural networks using single shot multi-box detector. Procedia Comput Sci 171:2610–2617
8. Kumar A, Reddy SSSS, Kulkarni V (2019) An object detection technique for blind people in real-time using deep neural network. In: 2019 fifth international conference on image information processing (ICIIP). IEEE
9. Sandeep Reddy D, Padmaja M (2016) Extraction of roads from high resolution satellite images by means of adaptive global thresholding and morphological operations. Int J Sci Eng Res 7(10)

10. Yadav P, Agrawal S (2018) Road network identification and extraction in satellite imagery using Otsu's method and connected component analysis. Int Arch Photogrammetry Rem Sens Spat Inform Sci
11. Wang W et al (2016) A review of road extraction from remote sensing images. J Traffic Transp Eng (English edition) 3(3):271–282
12. Xu Y et al (2018) Road extraction from high-resolution remote sensing imagery using deep learning. Rem Sens 10(9):1461
13. Mnih V, Hinton GE (2010) Learning to detect roads in high-resolution aerial images. In: Daniilidis K, Maragos P, Paragios N (eds) Computer vision ECCV 2010. Number 6316 in lecture notes in computer science. Springer, Berlin/Heidelberg, Germany, pp 210–223
14. Alex DM, Bindu KR, Reemamol PK (2013) Resolution enhancement and roadextraction for urban and sub-urban management. Int J Sci Eng Res 4(8):1640. ISSN 2229-5518
15. Buslaev A et al (2018) Fully convolutional network for automatic road extraction from satellite imagery. In: CVPR Workshops
16. Liu X, Deng Z (2018) Segmentation of drivable road using deep fully convolutional residual network with pyramid pooling. Cogn Comput Springer 10(2):272–281
17. Khryashchev V, Ivanovsky L, Pavlov V, Ostrovskaya A, Rubtsov A (2018) Comparison of different convolutional neural network architectures for satellite image segmentation. In: 2018 23rd conference of open innovations association (FRUCT), 13 November 2018. IEEE, pp 172–179
18. Wei Y, Zhang K, Ji S (2019) Road network extraction from satellite images using CNN based segmentation and tracing. In: IGARSS 2019-2019 IEEE international geoscience and remote sensing symposium. 28 July 2019. IEEE, pp 3923-3926
19. Fan R et al (2019) PT-ResNet: perspective transformation-based residual network for semantic road image segmentation. arXiv preprint arXiv:1910.13055
20. Kim J, Kim J, Cho J (2019) An advanced object classification strategy using YOLO through camera and LiDAR sensor fusion. In: 2019 13th international conference on signal processing and communication systems (ICSPCS). IEEE
21. Sun Z et al (2020) Exploiting deeply supervised inception networks for automatically detecting traffic congestion on freeway in china using ultra-low frame rate videos. IEEE Access 8:21226-21235
22. Fatima SA et al (2020) Object recognition and detection in remote sensing images: a comparative study. In: 2020 international conference on artificial intelligence and signal processing (AISP). IEEE
23. Bhardwaj N, Kaur G, Singh PK (2018) A systematic review on image enhancement techniques. In: Sensors and Image Processing. Springer, Singapore, pp 227-235
24. Chen Q et al (2018) Aerial imagery for roof segmentation: a large-scale dataset towards automatic mapping of buildings. arXiv preprint arXiv:1807.09532
25. Tan L, Jiang J (2018) Digital signal processing: fundamentals and applications. Academic Press

Performance Analysis of SoC and Hardware Design Flow in Medical Image Processing Using Xilinx Zed Board FPGA

Neol Solanki, Chintan Patel, Neel Tailor, and Nadimkhan Pathan

Abstract The requirement of the real-time implementation of the image processing algorithm compels FPGA adoption due to parallelism, reconfigurability, and pipelining architecture. This paper presents the coherent design of advanced edge detection algorithms using two design methodologies with FPGA: SoC design flow and hardware flow for the medical image processing purposes. Vivado HLS and Vivado IP integrator implement the SoC design flow, while system generator realizes the hardware flow. We have implemented Canny–Deriche edge detection and Laplacian of Gaussian (LoG) edge detection and practiced several brain tumor images with distinct mathematical parameters such as threshold and standard deviation. Thus, this paper aims to examine two edge detection algorithms in terms of noise reduction, edge response characteristics, and edge localization, and two design methodologies in three parameters: power consumption, resource utilization, and timing constraints. We present a real-time image processing method utilizing a pipeline structure that emphasizes medical image enhancements using this system on the Zed board SoC FPGA platform.

Keywords Medical image processing · SoC design flow · Hardware design flow · Marr–Hildreth filer · Canny–Deriche filter · xfOpenCV · Zynq-7000 all programmable · Zed board

1 Introduction

The management of digital images has become a premise of great importance in various sectors, such as medicinal, defenses, technological applications, and numerous other divisions. There are several instances where image processing aids

N. Solanki (✉) · C. Patel · N. Tailor · N. Pathan
Birla Vishvakarma Mahavidyalaya, Vallabh Vidyanagar 388120, Gujarat, India
e-mail: neolsolanki@gmail.com

C. Patel
e-mail: cspatel@bvmengineering.ac.in

P. K. Singh et al. (eds.), *Proceedings of Second International Conference on Computing, Communications, and Cyber-Security*, Lecture Notes in Networks and Systems 203,
https://doi.org/10.1007/978-981-16-0733-2_67

in examining, inferring, and arriving in conclusions [1–4]. Image processing's prime focus is to increase the image quality for human elucidation or the machines' recognition independently [2–4]. We have first preprocessed images to subdue noise and blur, i.e., filtering. After that, we recognized the image structure, extract meaningful data from the images, and obtain the image for better visualization.

The course can be challenging for image enhancement techniques to biomedical images [5–7]. The primary focus of the enhancement technique is to proffer a desirable visual aspect of medical images. Due to the high frame rate and image resolution in real time, biomedical image processing is an arduous job. Hardware implementation is always preferred compared to software implementation because it allows a fast throughput rate and higher performance [1, 2, 4]. Hardware implementation also minimizes latency and delivers the smooth operation of verifications and debugging. It also provides parallelism that decreases processing time.

There is a dearth of literature regarding the usage of FPGA for medical image processing. Intensely few researchers have focused on developing the image processing algorithms with hardware flow. Iuliana Chiuchisan suggested the real-time configurable system for medical image processing using FPGA [5]. They had implemented the low-level image processing algorithms such as contrast filter, brightness filter, inverting filter, and pseudo-filter on Virtex-6 FPGA ML605. Their conclusion proved that the design using FPGA provides versatile, modular, and scalable, reconfigurable architecture. V. Kasik and Z. Chvostkova demonstrated the acceleration of the image processing techniques using the FPGA platform Spartan 3AN [6]. Their work showed that the FPGA could meet the requirements of HDTV (1920 $\times$ 1080) with a frame rate of 60 Hz simultaneously consuming limited resources. They concluded that using FPGA can decrease the need for large video memory, high parallelism, and fast combinatorial graphics. Vladimir Kasik, Martin Cerny, Marek Penhaker, Václav Snášel, Vilem Novak, and Radka Pustkova suggested the advanced CT and MR image processing with FPGA [7]. They accomplished counting the ratio of intracranial fluid in the skull using the real parallel hardware of the FPGA using the Virtex-4 platform. They processed the image with 150 MHz, taking the only fraction of milliseconds to complete for 800 $\times$ 600 resolution images.

Today's lots of FPGA have the hardcore processor embedded with them. Designing with the programmable processor and configurable hardware is acknowledged as system on chip (SoC) design [8]. The SoC design flow reduces the design complexity and design time contrasted to the conventional hardware flow. However, SoC design flow necessitates the knowledge of industry standard protocol such as AXI for transferring information between processor core and hardware part. The Xilinx Zynq-7000 all programmable SoC devices come with both programmable ARM core (PS-processing system) and configurable hardware (PL-programmable logic) section. These two parts, PS and PL, can be used autonomously in Zynq-7000 devices, empowering engineers to practice both SoC design flow and hardware flow [8].

This paper aims to analyze the performance of SoC design flow and hardware design flow in medical image processing by examining the distinct phases of resource consumption, timing constraints, and power requirements (Table 1) on the Zed board.

Table 1 Analyses of the design flow

Algorithm	Hardware design flow			SoC design flow		
	Max delay(ns)	Min delay (ns)	Power (w)	Max delay (ns)	Min delay (ns)	Power (w)
Marr–Hildreth	4.011	0.333	0.226	6.349	0.027	1.719
Canny–Deriche	3.254	0.478	0.240	6.598	0.034	1.74

Table 2 Analysis of edge detection with the previous work [13]

Algorithm	Robert (threshold = 25)	Prewitt (threshold = 25)	Marr–Hildreth	Canny–Deriche
PSNR (dB)	8.6199	8.1932	10.6728	12.0389

We have also analyzed the edge detection algorithms in noise content (Table 2), edge response characteristics, and accuracy.

2 Tools and Design Flow

2.1 Vivado HLS

In these days, broadcast, clinical, and military apply very involved algorithms than formal time. The industry mainstay tool for algorithm developments is C, C++, and SystemC. Previous to the notion of high-level synthesis, the adaptation from high-level language (C, C++) to hardware description language (Verilog, VHDL) was vulnerable to error and consumed a lot of development time [9]. Vivado high-level synthesis (HLS) automates this course of action for designers. Furthermore, Vivado HLS provides diverse interface protocols (AXI4 Lite, AXI4 Stream, and FIFO) and a directive-based compiler that produces the greatest feasible quality of results (QoR) [9, 10]. For verification of the C/C++ module, it presents C simulation and automatic VHDL/Verilog co-simulation for synthesized hardware [10]. With the entrance of the xfOpenCV library, the Vivado HLS has simplified the image processing algorithms development for FPGA and SoC devices [11] (Fig. 1).

2.2 Vivado IP Integrator

Vivado IP integrator provides a higher-level GUI environment and also tcl command-based environment for the SoC and hardware design. With the Vivado IP integrator,

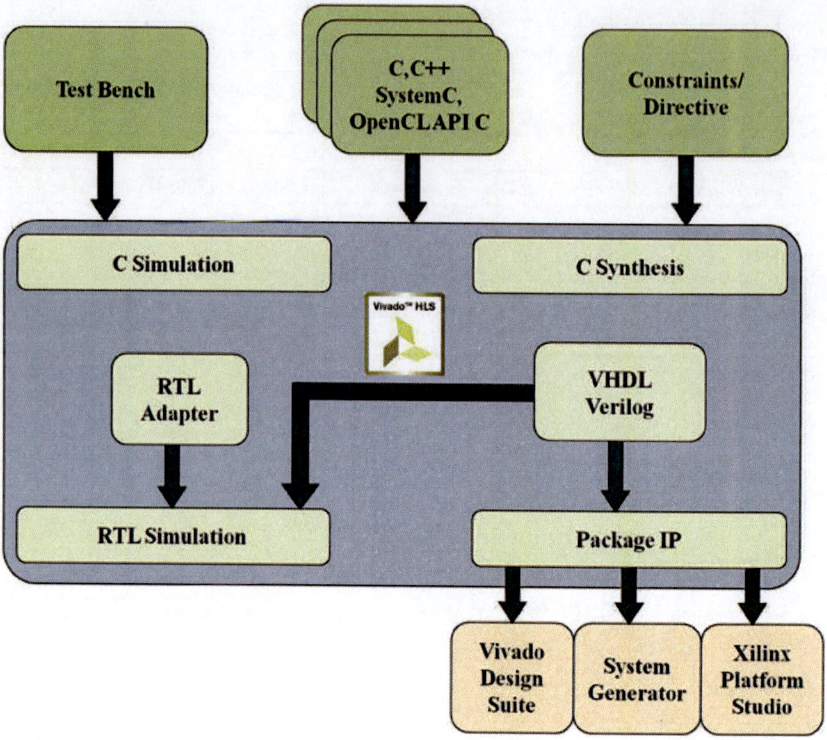

Fig. 1 Vivado HLS design flow

designers operate the design at the interface level rather than the signal level. This interface abstraction level increases design productivity and decreases design time. The Vivado atmosphere also supports commonly used interface protocol other than AXI [9].

2.3 System Generator

Xilinx system generator permits the same background for ISE Design Suite and MATLAB/Simulink. It delivers a coherent way of creating conundrum designs—the benefits of using XSG are because it gives flexibility, reusability, and does not expect profound knowledge [9]. XSG also offers a library of Simulink blocks, DSP functions, and memories. XSG generates VHDL/Verilog code using a code generator from the created design. It also performs VHDL/Verilog synthesis, mapping hardware, and floor plan [9, 12]. Furthermore, it also generates a user constraint file, test vectors, and test bench file. The hardware implementation uses XSG, known as Xilinx blocks, work with Boolean values with a fixed point. On the other side,

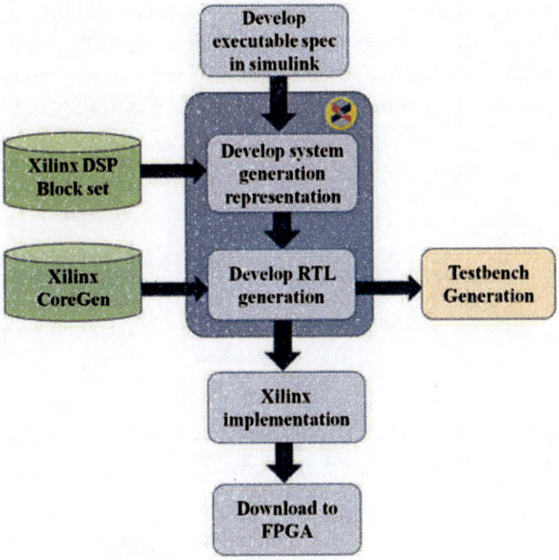

Fig. 2 System generator design flow

the Simulink block runs on a continuous and floating format. For converting the Simulink block data format to the Xilinx block compatible format, gateway blocks are necessary [12] (Fig. 2).

3 Advanced Edge Detection Methods

The initial edge detection techniques such as Robert, Prewitt, and Sobel comprise the convolution process with two masks and then calculating the magnitude response. These methods act on the notion of gradient and determining the maximum variation in the pixel neighborhood. On the other hand, these techniques do not reflect on noise substance and edge characteristics [13, 14]. In this paper, we have implemented the two advanced edge detection techniques, Marr–Hildreth and Canny–Deriche, on the Zynq-7000 all programmable SoC device: Zed Board.

3.1 Marr–Hildreth Edge Detection

Marr and Hildreth are one of the pioneer engineers for doing into the in-depth analysis of edge detection algorithms. They came up with two crucial arguments concerning the edges. First, intensity difference depends on the image scale ensuing in the usage of dissimilar sizes of masks for convolution. Second, an abrupt intensity value change produces zero-crossings in the second derivative of the image. These arguments implicate that the edge detection operator should have these two qualities: First, it should be proficient in computing discrete differential of the first and second order of each pixel. Second, it should be able to be "tuned" to any desired scale [14]. The most promising operator pleasing above all constraints is $\nabla^2 G$, where ∇^2 is the Laplacian operator, and G is the two-dimensional Gaussian function,

$$G(x, y) = e^{\frac{-(x^2+y^2)}{2\sigma^2}} \quad (1)$$

Using some essential calculus, we can prove that

$$\nabla^2 G(x, y) = \left[\frac{x^2 + y^2 - 2\sigma^2}{\sigma^4}\right] e^{\frac{-(x^2+y^2)}{2\sigma^2}} \quad (2)$$

Equation 2 is called the Laplacian of Gaussian (LOG). Figures 3 and 4 show the three-dimensional plot of—$-\nabla^2 G(x, y)$ and its cross section for $\sigma = 4$. Due to this

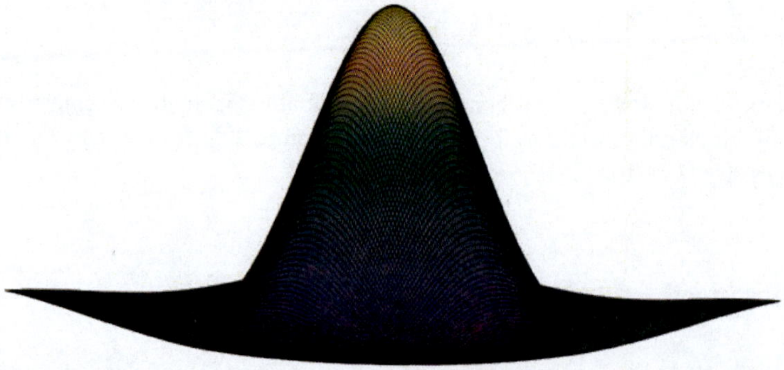

Fig. 3 LOG 3D plot

Fig. 4 LOG cross section

shape, it is also famous as a Mexican hat operator. The cross section shows the circle with the equation $x^2 + y^2 = 2\sigma^2$.

Sampling Eq. 2 generates the LOG mask of the desired dimensions. Still, it is necessary to scale coefficient after sampling operation such that the coefficient's addition becomes zero (essential condition for edge detection). The more efficient methodology is to sample Eq. 1 and generate the $n \times n$ Gaussian filter and convolve with the input image and another time performing convolution with the Laplacian mask. The Laplacian mask used is shown in Fig. 5. After the Laplacian operator, we perform zero-crossing detection to find out edges. 99.7% of the volume in the Gaussian filter lies under the range of ±3σ above the mean. Thus, as a rule, the size of the Gaussian filter should be such that n is the smallest odd integer greater than $6\sqrt{2}\sigma$ or 8.485, considering the mean to be zero [14, 15]. Figure 6 exemplifies these conditions for $w = 2\sqrt{2}\sigma$. In our work, we have practiced a Gaussian filter with $\sigma = 2$. Figure 7 depicts the mask after scaling the floating point to integers (Figs. 8 and 9).

Fig. 5 Laplacian kernel

+1	+1	+1
+1	−8	+1
+1	+1	+1

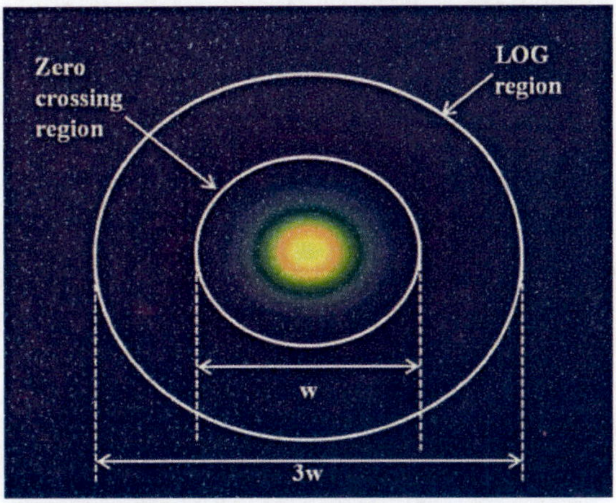

Fig. 6 LOG filter size condition

0	0	+1	+1	+2	+2	+2	+1	+1	0	0
0	+1	+2	+3	+5	+5	+5	+3	+2	+1	0
+1	+2	+4	+8	+11	+13	+11	+8	+4	+2	+1
+1	+3	+8	+15	+21	+24	+21	+15	+8	+3	+1
+2	+5	+11	+21	+31	+35	+31	+21	+11	+5	+2
+2	+5	+13	+24	+35	+40	+35	+24	+13	+5	+2
+2	+5	+11	+21	+31	+35	+31	+21	+11	+5	+2
+1	+3	+8	+15	+21	+24	+21	+15	+8	+3	+1
+1	+2	+4	+8	+11	+13	+11	+8	+4	+2	+1
0	+1	+2	+3	+5	+5	+5	+3	+2	+1	0
0	0	+1	+1	+2	+2	+2	+1	+1	0	0

Fig. 7 Gaussian filter

For obtaining the zero-crossings in the LOG image, we have processed the 3 × 3 region. The zero-crossing is recognized at the center pixel when at least one of its neighborhoods alters their signs, and their absolute difference surpasses the threshold [14, 15] (Figs. 10, 11, 12, 13, 14 and 15).

3.2 Canny–Deriche Edge Detection

The output of Canny–Deriche algorithms is excellent compared to other edge detection algorithms due to the complexity entailed in the algorithm. Three crucial peculiarities of this algorithm are good detection, sound localization, and single edge responses [15, 16]. Canny recommended the use of the first derivative of the Gaussian given by,

$$g(x) = -\frac{x}{\sigma^2} e^{\frac{-x^2}{2\sigma^2}} \tag{3}$$

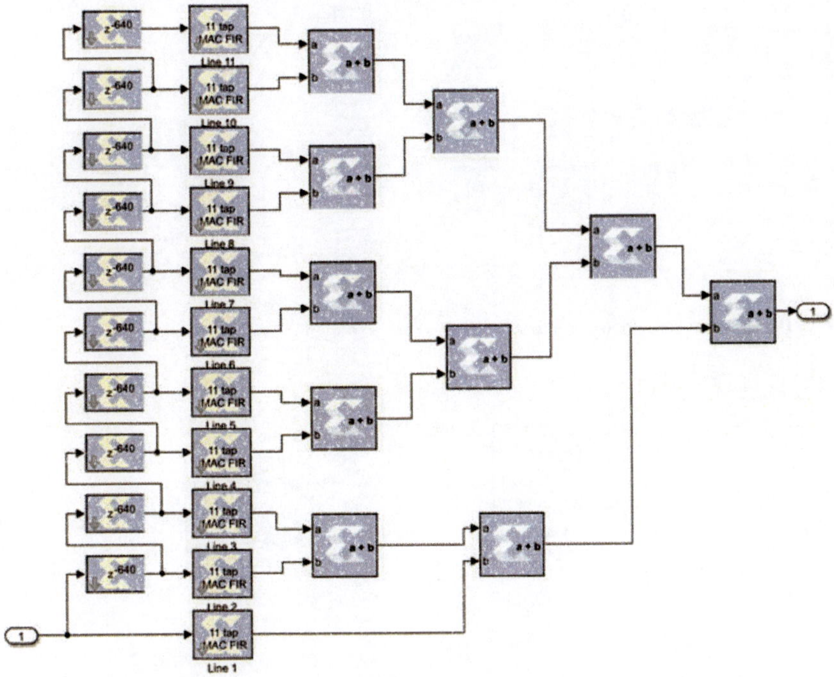

Fig. 8 Gaussian filter design

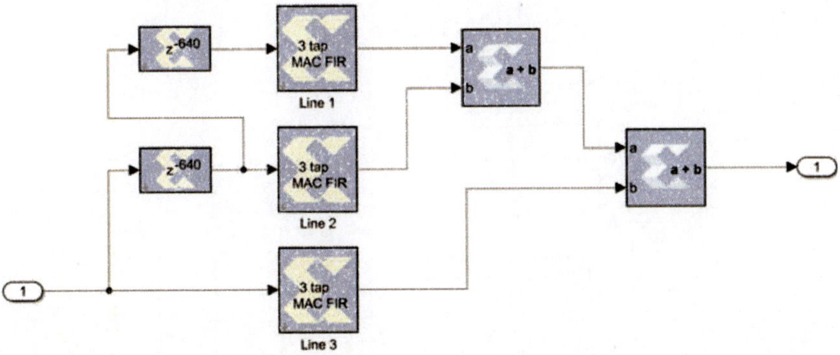

Fig. 9 Laplacian filter design

Nevertheless, later on, Deriche submitted a better optimal edge detector $f(x)$; it follows that $f(x)$ performs better than $g(x)$. It is mild and merely depends on the α. The value of α determines the SNR and localization of the edges [16]. The α performs the role of σ in the Canny–Deriche algorithm, but in an inverted way, i.e., $\alpha = 1/\sigma$. The maximum of $f(x)$ lies under the $x = 1/\alpha$. Increasing α will increase

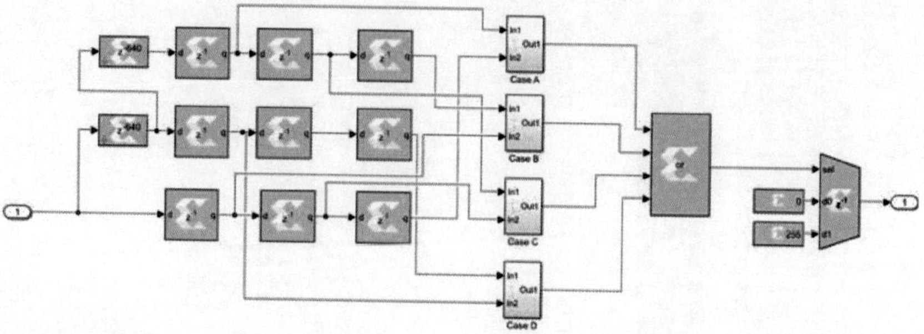

Fig. 10 Zero-crossing detection design

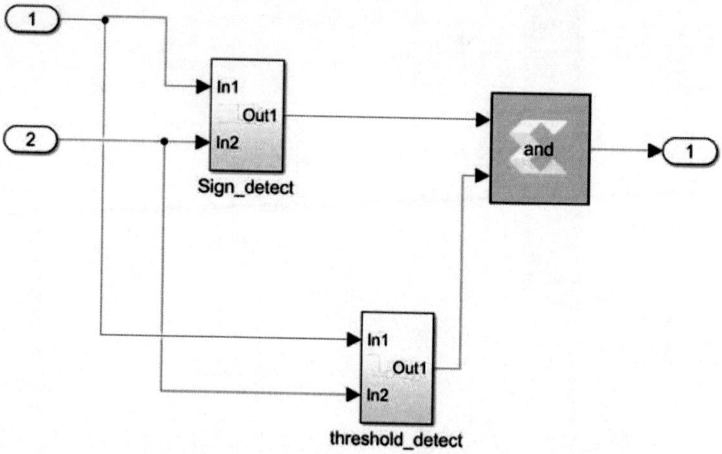

Fig. 11 Zero-crossing => case A

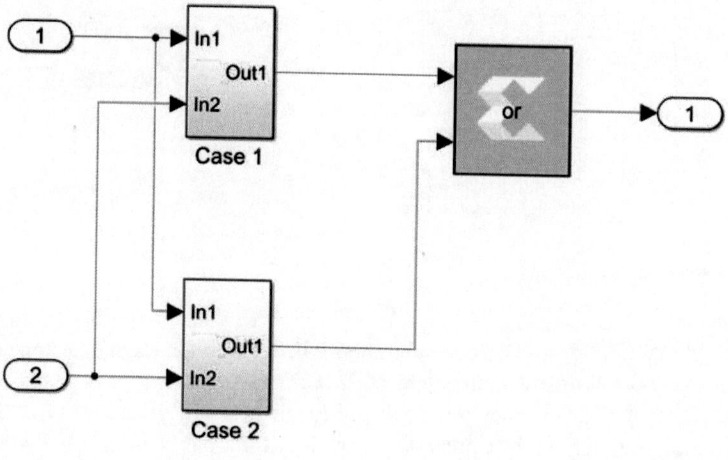

Fig. 12 Case A => sign-detect

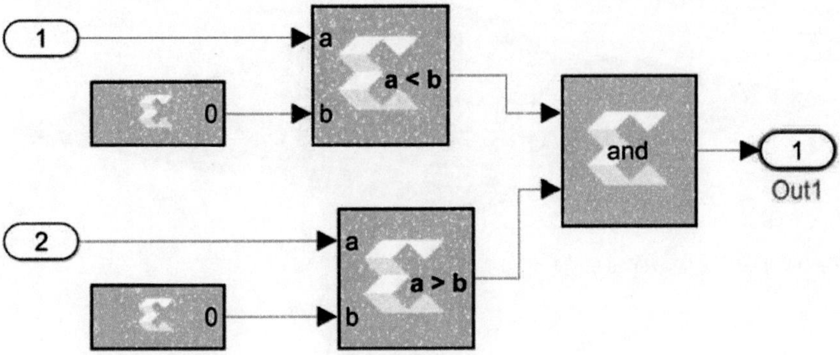

Fig. 13 Sign-detect => case 1

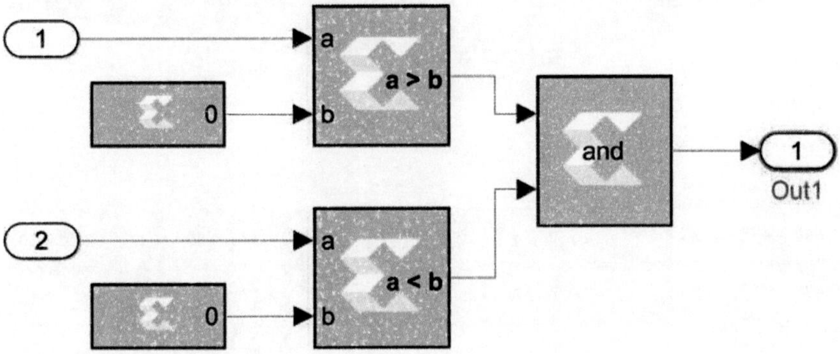

Fig. 14 Sign-detect => case 2

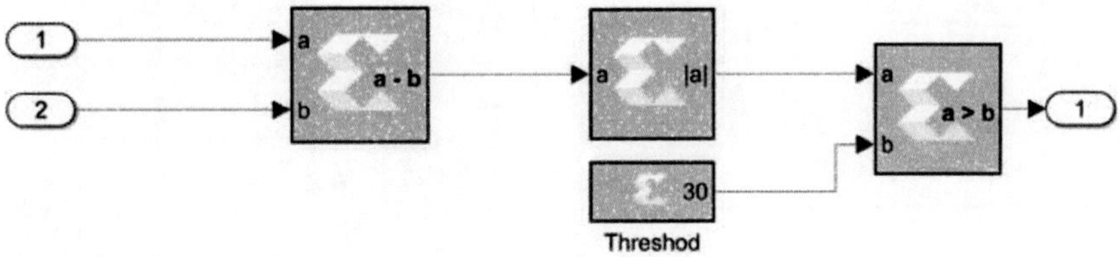

Fig. 15 Case A => threshold-detect

localization but also lowers SNR [16].

$$f(x) = -c \cdot x \cdot \mathrm{e}^{-\alpha|x|} \tag{4}$$

Equations 5 and 6 show a two-dimensional version of the filters. Figures 16 and 17 portray the three-dimensional plot for the horizontal and vertical filters, and Fig. 18

Fig. 16 Deriche horizontal 3D plot

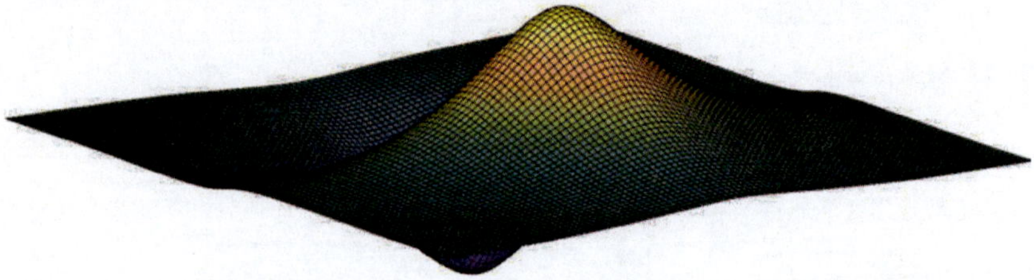

Fig. 17 Deriche vertical 3D plot

+10	+18	+25	0	−25	−18	−10
+20	+37	+50	0	−50	−37	−20
+37	+67	+91	0	−91	−67	−37
+50	+91	+124	0	−124	−91	−50
+37	+67	+91	0	−91	−67	−37
+20	+37	+50	0	−50	−37	−20
+10	+18	+25	0	−25	−18	−10

+10	+20	+37	+50	+37	+20	+10
+18	+37	+67	+91	+67	+37	+18
+25	+50	+91	+124	+91	+50	+25
0	0	0	0	0	0	0
−10	−20	−37	−50	−37	−20	−10
−18	−37	−67	−91	−67	−37	−18
−25	−50	−91	−124	−91	−50	−25

Fig. 18 Deriche horizontal (left) and vertical (right) filters

displays the masks used in this project with $\alpha = 1$.

$$X(x, y) = \frac{\left[-c \cdot x \cdot e^{-\alpha \cdot |x|}\right] \cdot \left[k \cdot (\alpha \cdot |y| + 1) \cdot e^{-\alpha |y|}\right]}{\alpha^2} \tag{5}$$

$$Y(x, y) = \frac{\left[-c \cdot y \cdot e^{-\alpha \cdot |y|}\right] \cdot \left[k \cdot (\alpha \cdot |x| + 1) \cdot e^{-\alpha |x|}\right]}{\alpha^2} \tag{6}$$

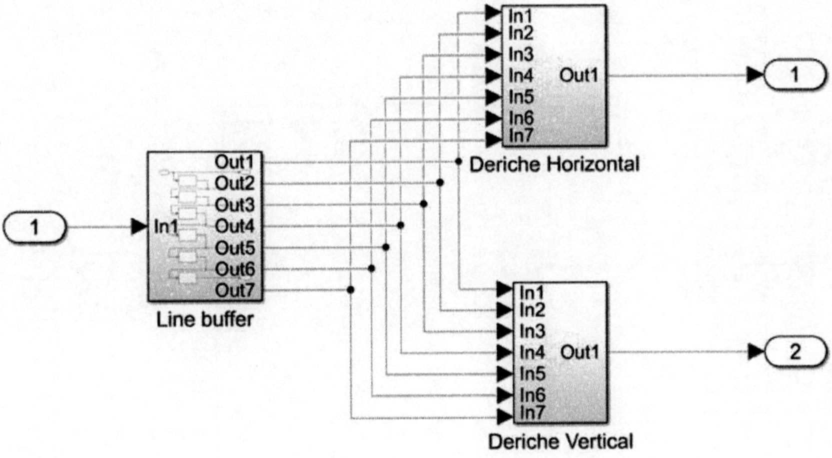

Fig. 19 Deriche filter

$$c = \frac{\left[1 - \mathrm{e}^{-\alpha}\right]^2}{\mathrm{e}^{-\alpha}} \tag{7}$$

$$k = \frac{\left[1 - \mathrm{e}^{-\alpha}\right]^2 \cdot \alpha^2}{1 + 2 \cdot \alpha \cdot \mathrm{e}^{-\alpha} - \mathrm{e}^{-2\alpha}} \tag{8}$$

Firstly, we convolved the input image with these two filters. After completing the convolution with the horizontal and vertical masks, we get two images, $r(x, y)$ and $s(x, y)$, respectively. $A(x, y)$ (Eq. 9) and $\theta(x, y)$ (Eq. 10) show the approximated calculation of gradient and phase, respectively (Figs. 19 and 20).

$$A(x, y) \approx |r(x, y)| + |s(x, y)| \tag{9}$$

$$\theta(x, y) = \tan^{-1} \frac{s(x, y)}{r(x, y)} \tag{10}$$

After calculating the gradient and phase response, we performed non-maxima suppression, followed by hysteresis. In non-maxima suppression, firstly, the phase is categorized as shown in Fig. 22. After that, we processed 3 × 3 regions and suppressed the pixel if its value is less than at least one of its neighborhoods in the direction dictated by the phase response of center pixel [14] (Fig. 21).

After doing the hysteresis with two threshold levels, we get three kinds of edges, sharp edges, weak edges, and zero edges (Figs. 23 and 24).

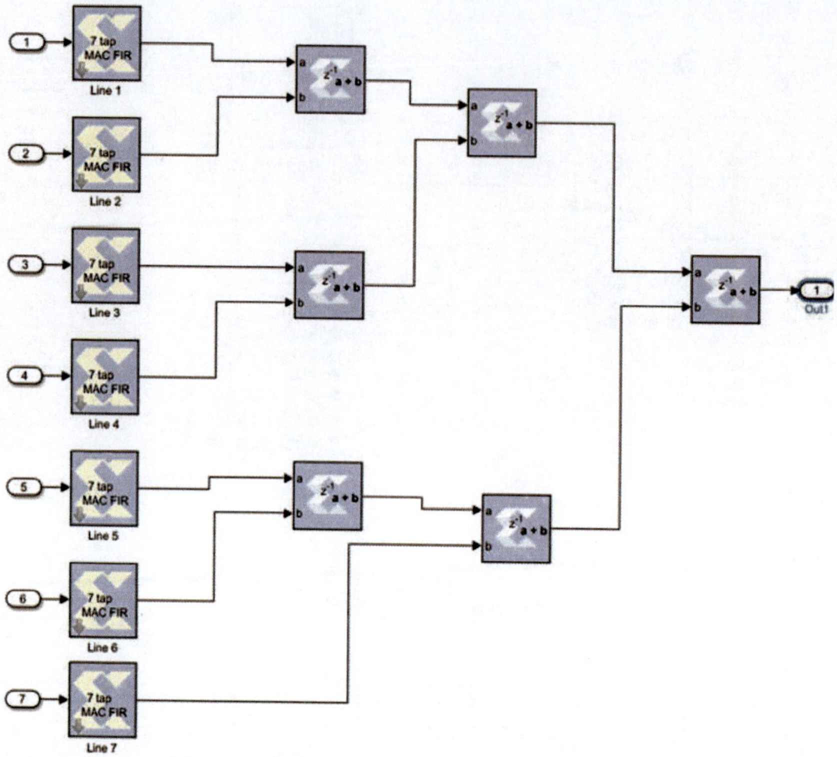

Fig. 20 Deriche internal design

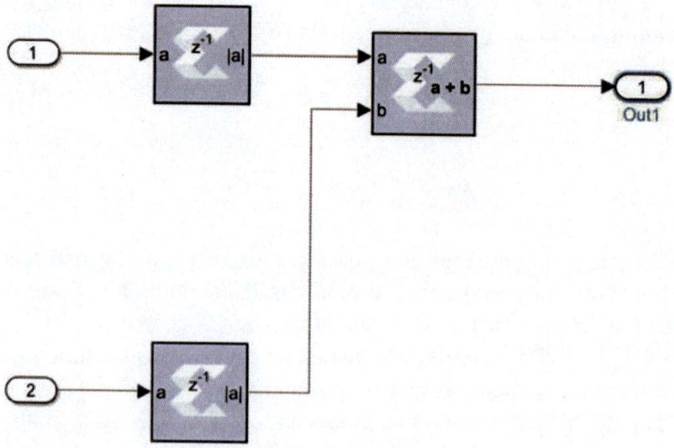

Fig. 21 Gradient calculation

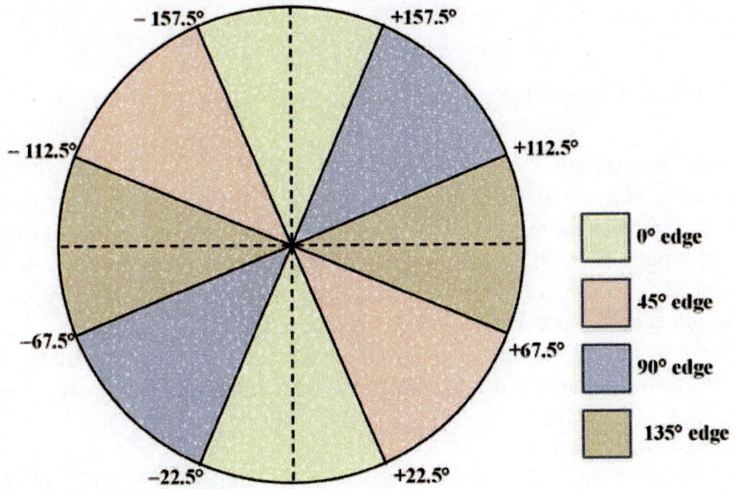

Fig. 22 Edge classification

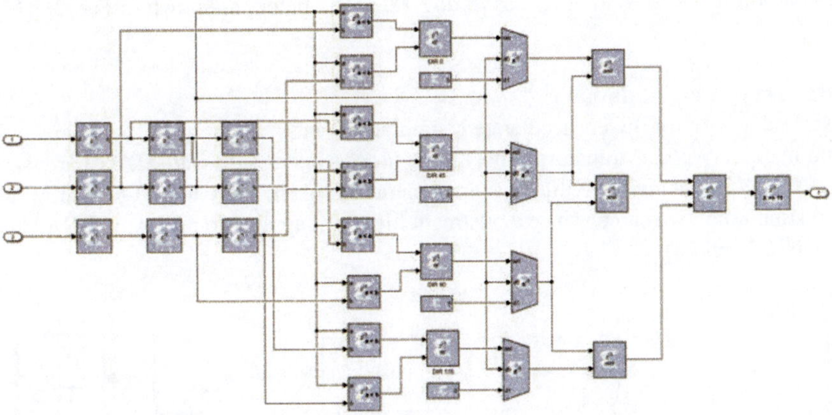

Fig. 23 Approximated Non-maxima filter

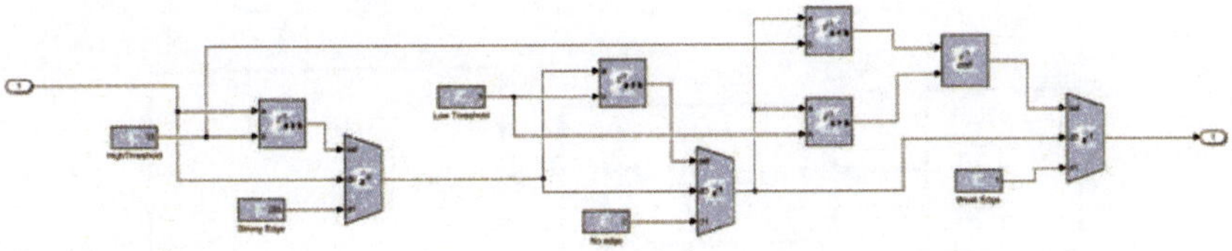

Fig. 24 Hysteresis

4 Implementation

4.1 System Generator

Image Preprocessing

Every image is a 2D array, and the range of an array is row × column. Image processing with hardware requires 1D image data. Image preprocessing block transforms an image from 2D to 1D. Figure 25 explains the configuration of image preprocessing. The resize block helps to convert image in desirable size, and this data, which is in 2D, is converted to 1D for processing [12, 17]. Frame conversions facilitate regenerating output signal to frame-based data, and after that frame, data is forwarded to the unbuffer block to convert into scalar output at a high sampling rate.

Image Post-processing

For transforming the scalar data into the floating (double) data type, we require image post-processing. We transformed those scalar-based data into frame-based data using a buffer block. Following that, we can convert those one-dimensional data to two-dimensional output by converting 1D to 2D block, as shown in Fig. 26 [12, 17].

Hardware Co-simulation

System generator gives accelerated simulation with the help of hardware co-simulation. We can transform any image processing design into JTAG hardware co-simulation using the Xilinx system generation token, and using JTAG hardware co-simulation, we can produce a bitstream file (.bit), applied to program FPGA [12] (Figs. 27 and 28).

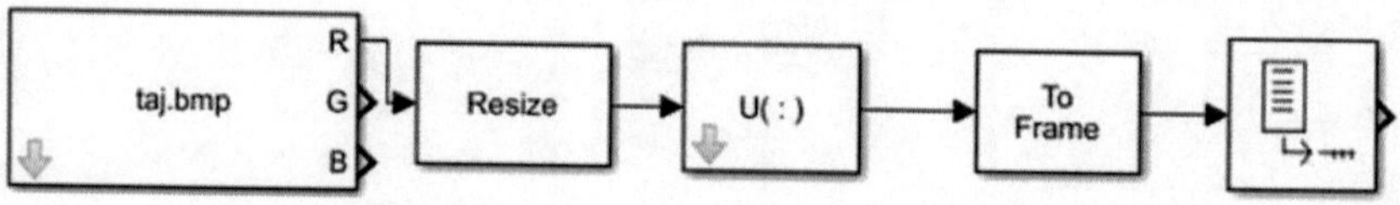

Fig. 25 Image preprocessing

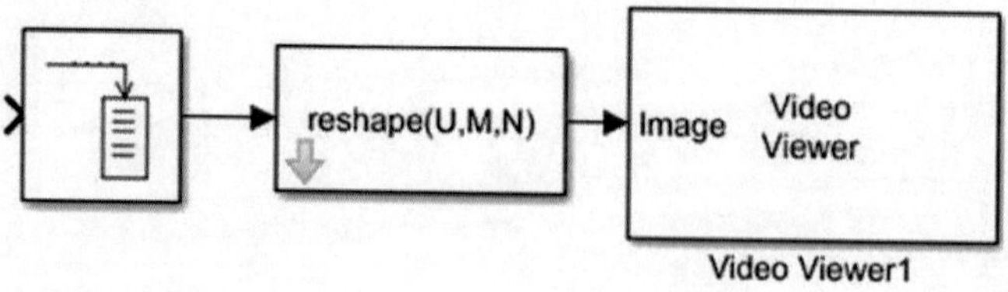

Fig. 26 Image post-processing

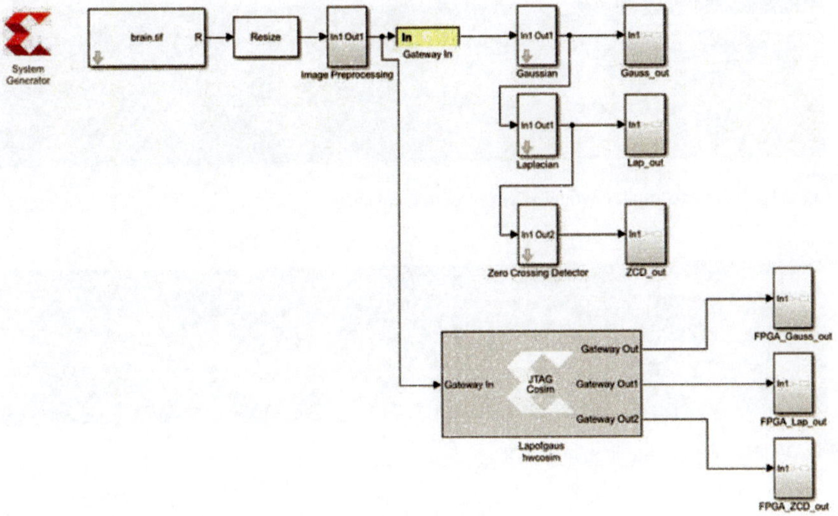

Fig. 27 Marr–Hildreth hardware co-simulation

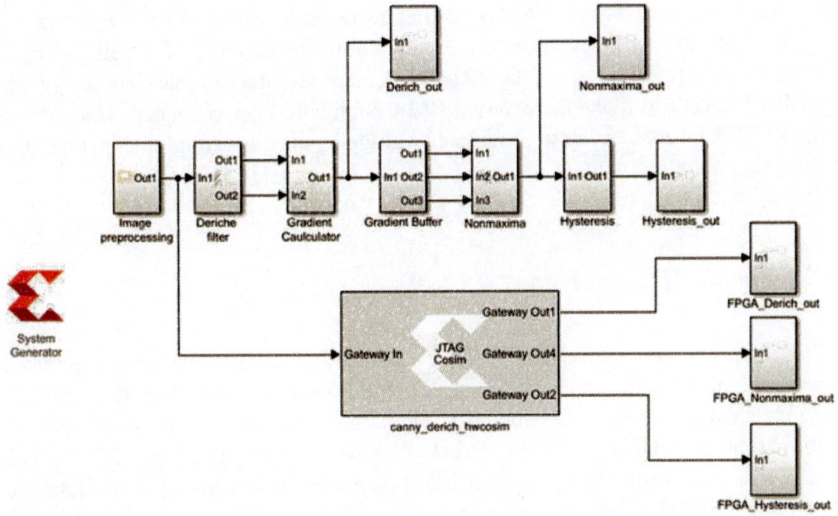

Fig. 28 Canny–Deriche hardware co-simulation

4.2 Vivado HLS

RTL Co-simulation

It is not prudent to review all filters' RTL simulation results in one paper. For better perception of the AXI protocol handshake in the module, this paper examines the

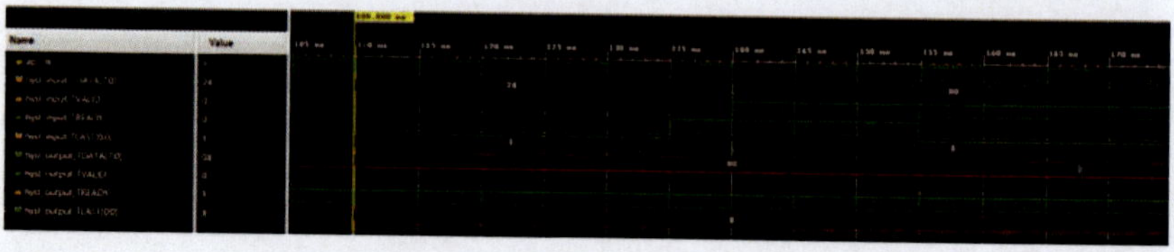

Fig. 29 Hysteresis input channel AXI protocol handshake

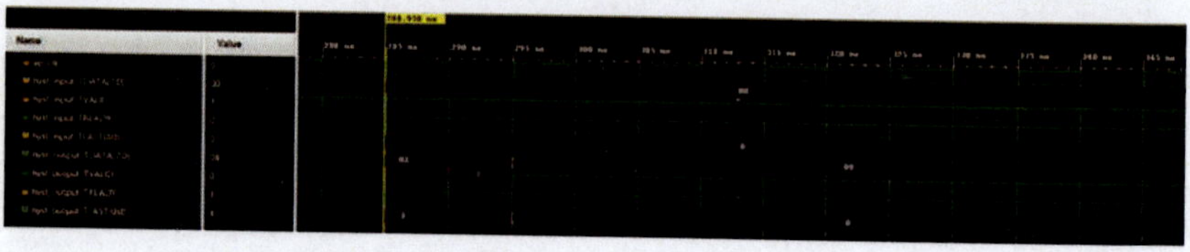

Fig. 30 Hysteresis output channel AXI protocol handshake

hysteresis filter. The AXI protocol contains multiple handshaking signals, but the primary signals are READY, VALID, and LAST [10, 18]. Both input and output channels have these interface signals. First, hysteresis asserts signals READY, and after some period, it asserts VALID and begins receiving the data from the external DMA engine. It continues accepting the data until the signal LAST gets asserted, as shown in Fig. 29. After IP starts operating, it asserts the output channel signals VALID, READY to make the external DMA engine to receive the data as shown in Fig. 30. DMA keeps accepting the data until the hysteresis output channel and the LAST signal are maintained.

4.3 *Vivado IP Integrator*

The Zynq processing system functions as a master and slave by the general-purpose master and high-performance slave port. First, the Zynq processing system acquires the pixels data from the DDR memory. Then, the Zynq system forwards the pixel information to the HLS generated IP via the AXI direct memory access (DMA). The AXI DMA transforms the memory-mapped data into the streaming of the data and vice versa [19]. Once the edge detection is complete, the AXI DMA receives the pixel streaming and maps it to the DDR memory address. The AXI broadcast transmits the same steam of data to the two separate IPs simultaneously [20]. AXI lite protocol helped us to share the threshold information to IPs. For measuring timing precisely, we employed AXI timer in 64-bit mode [21] (Figs. 31 and 32).

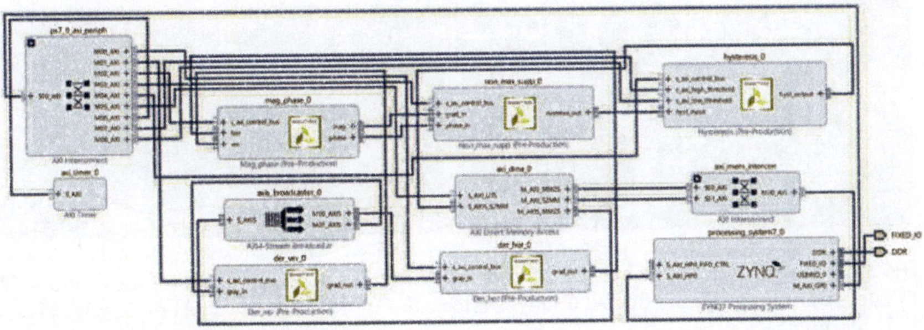

Fig. 31 Canny–Deriche filter design

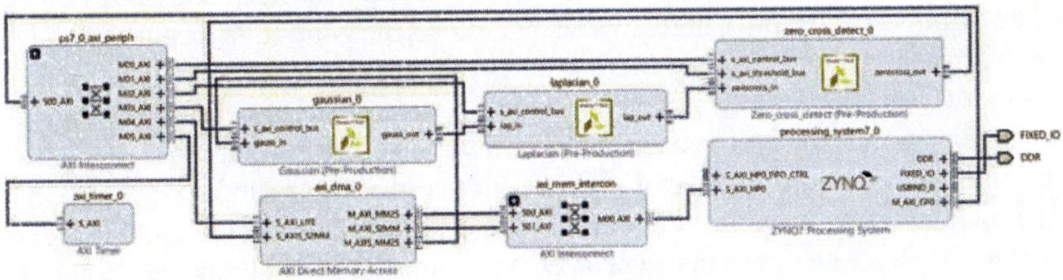

Fig. 32 Marr–Hildreth filter design

Fig. 33 640 × 640 brain tumor image (Image courtesy MedPix)

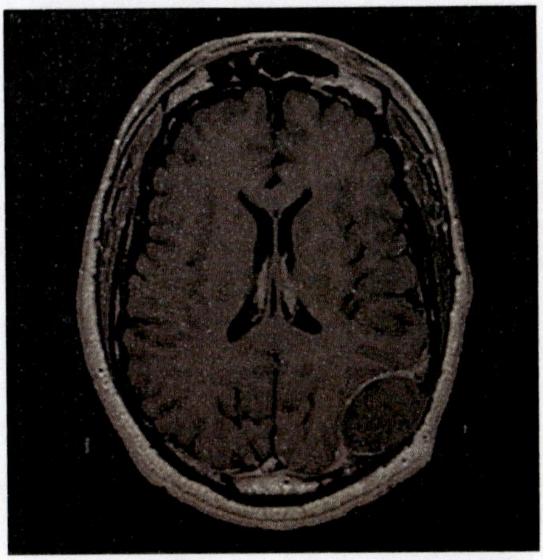

5 Results and Discussion

We have applied these both algorithms to the brain tumor image for comparative analysis. From the output images (Fig. 34), it is evident that Canny–Deriche delivers the better than the Marr–Hildreth algorithms. Due to the noise immunity of Deriche

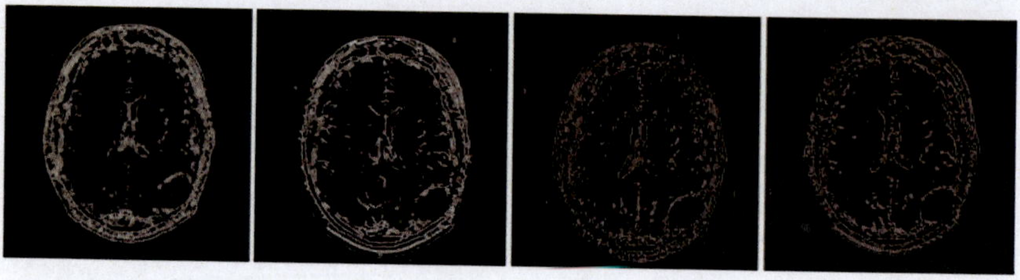

Fig. 34 **a** Marr–Hildreth, SoC flow, **b** Marr–Hildreth, hardware flow; with $\sigma = 2$, threshold = 15. **c** Canny–Deriche, hardware flow, **d** Canny–Deriche, SoC flow; with $\alpha = 1$ and low threshold = 1, high threshold = 5

filter, the results are visually more pleasant than the Gaussian smoothing in the Marr–Hildreth. Selecting a higher value of α allows the good edge localization while increasing the amount of σ in Marr–Hildreth reduces the edge localization due to reciprocal behavior of both parameters. Non-maxima filter assists in producing only a single edge response as evident from the thin edges in the Canny–Deriche filter in contradiction to the edges of the Marr–Hildreth that has multiple edge response. The optimal value of the threshold in both algorithms filters out the undesired edge content. We have analyzed the edge detection algorithm by applying peak signal-to-noise ratio (PSNR) (Table 2). A higher value of PSNR implicates the lower noise content in the image relative to the noiseless image constructed using MATLAB.

Figure 35 reflects the relative usage of resource consumption for all these algorithms. Vivado (Soc design flow) utilizes more resources compared to system generator (hardware flow) due to the auto-generation of FSM, BRAM, registers, and LUTs from the high-level code C/C++. Using the line buffer causes the image to shift down depending upon the kernel size, as indicated in the output of system generator (Fig. 34). However, having complete dominance over the behavior of the system with just a few lines of C/C++ code can resolve this issue in the SoC design flow. Conversely, building FSM using the hardware flow requires a more design time. For this reason, we were able to implement the non-maxima filter more accurately in SoC implementation compared to the hardware implementation in the same amount of time (Fig. 33).

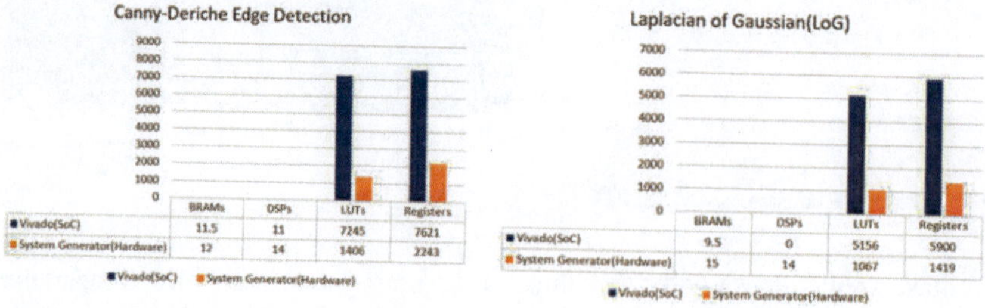

Fig. 35 Comparative analysis of resources

6 Conclusion

This paper studies two different design methodologies for two intricate edge detection methods: Marr–Hildreth and Canny–Deriche filter and analyzes in the course of resource consumption, power requirements, and timing constraints. Pipelining decreased the latency to the optimal value. We tested these systems on real-time images of brain tumors. The design time reduces, and accurate results are achievable in a shorter period without in-depth knowledge of lower-level abstraction details using SoC design flow with high-level synthesis. Although it reduces the design time, resource consumption may be higher than expected. On the other hand, hardware flow requires more design time and voluntary design of all algorithms and profound knowledge of system details at the lowest level. Apart from this, at the application level, these two algorithms are incredibly accurate edge detection compared to the Sobel, Robert, and Prewitt due to the complexity. Real-time video processing systems can use these IPs with other supplementary IPs such as AXI subset, AXI VDMA, and VGA core. Different image processing algorithms can use these designs at a hierarchical level, such as face recognition and object detection.

References

1. He X, Tang L (2019) FPGA-based high definition image processing system. In: Jia M, Guo Q, Meng W (eds) Wireless and satellite systems. WiSATS 2019. Lecture notes of the institute for computer sciences, social informatics and telecommunications engineering, vol 281. Springer, Cham. https://doi.org/10.1007/978-3-030-19156-6_20
2. van der Vlugt S, Alizadeh Ara H, de Jong R, Hendriks M, Marin RG, Geilen M, Goswami D (2019) Modeling and analysis of FPGA accelerators for real-time streaming video processing in the healthcare domain. J Sig Process Syst 91:75–91. https://doi.org/10.1007/s11265-018-1414-3
3. Chinchwadkar RM, Ingale VV, Gokhale A (2020) Hardware implementation of histogram-based algorithm for image enhancement. In: Iyer B, Rajurkar A, Gudivada V (eds) Applied computer vision and image processing. Advances in intelligent systems and computing, vol 1155. Springer, Singapore. https://doi.org/10.1007/978-981-15-4029-5_6
4. Sharif U, Mirzaei S (2018) High level synthesis implementation of object tracking algorithm on reconfigurable hardware. In: Voros N, Huebner M, Keramidas G, Goehringer D, Antonopoulos C, Diniz P (eds) Applied reconfigurable computing. architectures, tools, and applications. ARC 2018. Lecture notes in computer science, vol 10824. Springer, Cham. https://doi.org/10.1007/978-3-319-78890-6_48
5. Chiuchisan I (2013) A new FPGA-based real-time configurable system for medical image processing. In: 2013 E-health and bioengineering conference (EHB), Iasi, pp 1–4. https://doi.org/10.1109/EHB.2013.6707301
6. Kasik V, Chvostkova Z (2013) FPGA in technical resources of medical imaging. In: 2013 IEEE 11th international symposium on applied machine intelligence and informatics (SAMI), Herl'any, pp 193–196. https://doi.org/10.1109/SAMI.2013.6480973
7. Kasik V, Cerny M, Penhaker M, Snášel V, Novak V, Pustkova R (2012) Advanced CT and MR image processing with FPGA. In: Yin H, Costa JAF, Barreto G (eds) Intelligent data engineering and automated learning—IDEAL 2012. IDEAL 2012. Lecture Notes in computer science, vol 7435. Springer, Berlin, Heidelberg. https://doi.org/10.1007/978-3-642-32639-4_93

8. Crockett LH, Elliot RA, Enderwitz MA, Stewart RW (2014) The Zynq book: embedded processing with the ARM Cortex-A9 on the Xilinx Zynq-7000 all programmable SoC, 1st ed. Strathclyde Academic Media
9. Xilinx Inc Xilinx backgrounder, 9 reasons why the Vivado design suite accelerates design productivity
10. Xilinx Inc Vivado design suite user guide, high-level synthesis. UG902(v2018.3)
11. Xilinx Inc Xilinx OpenCV user guide. UG1233(v2019.1)
12. Xilinx Inc Vivado design suite user guide, model-based DSP using system generator. UG958(v2019.2)
13. Šušteršič T, Milovanović V, Ranković V, Filipović N, Peulić A (2020) Medical image processing using Xilinx system generator. In: Filipovic N (eds) Computational bioengineering and bioinformatics. ICCB 2019. Learning and analytics in intelligent systems, vol 11. Springer, Cham. https://doi.org/10.1007/978-3-030-43658-2_10
14. Gonzalez RC, Woods RE (2007) Digital image processing, 3rd ed. Prentice-Hall, pp 748–976
15. Huertas A, Medioni G (1986) Detection of intensity changes with subpixel accuracy using Laplacian-Gaussian masks. IEEE Trans Pattern Anal Mach Intell PAMI-8(5):651–664. https://doi.org/10.1109/tpami.1986.4767838
16. Deriche R (1987) Using Canny's criteria to derive a recursively implemented optimal edge detector. Int J Comput Vis 1:167–187. https://doi.org/10.1007/BF00123164
17. Mohapatra SK, Swain BR, Mahapatra SK (2015) Optimized approach of Sobel edge detection technique using Xilinx system generator. In: 2015 2nd international conference on electronics and communication systems (ICECS), Coimbatore, pp 338–341. https://doi.org/10.1109/ecs.2015.7124919
18. Gaikwad N, Patil VN (2018) Verification of AMBA AXI on-chip communication protocol. In: 2018 fourth international conference on computing communication control and automation (ICCUBEA), Pune, India, pp 1–5. https://doi.org/10.1109/iccubea.2018.8697587
19. Xilinx Inc AXI DMA v7.1-LogiCORE IP product guide, Vivado design suite (PG021)
20. Xilinx Inc AXI4-stream infrastructure IP suite v3.0 LogiCORE IP product guide, Vivado design suite (PG085)
21. Xilinx Inc AXI Timer v2.0-LogiCORE IP product guide, Vivado design suite (PG079)

SDN Firewall Using POX Controller for Wireless Networks

Sulbha Manoj Shinde and Girish Ashok Kulkarni

Abstract Recently, high-speed broadband services, social networking and cloud environments have increased Internet penetration significantly. Due to this, there is large amount of users data available in the Internet including personal data, enterprise data and financial data. This leads to serious threats from malicious users. Researchers have proposed various security threats for protecting this data from unknown threats. Most of the security solutions are employed on traditional networking techniques. They are very complex, and it is very difficult to manage them. Traditional networking techniques depend on manual configuration of devices. Each device may have different policy, hence leading to policy conflicts in managing the resources over network. Network security may be compromised in such case. Software-defined networking (SDSN) is a new paradigm addressing this issue. SDN provides various advantages including network wide visibility, centralized control, flexible network architecture and ease of management. The control plane (network controller) is separated from the data plane (forwarding devices). The controller is responsible for monitoring, management and controlling the behavior of the forwarding devices. OpenFlow protocol is used by the SDN controller. In this paper, we have proposed the SDN-based firewall. POX Python-based controller of SDN is used for control and management of the network. Using wireless network topology based on SDN controller, we have evaluated the performance of the network in terms of delay, TCP bandwidth and UDP jitter. Different wireless network topologies have been created for evaluating the performance.

Keywords Software-defined networking · Firewall · Mininet-WiFi · OpenFlow · POX controller

S. M. Shinde (✉) · G. A. Kulkarni
Department of Electronics & Telecommunication Engineering, HSM'S Shri Sant Gadge Baba College of Engineering & Technology, Bhusawal, India

P. K. Singh et al. (eds.), *Proceedings of Second International Conference on Computing, Communications, and Cyber-Security*, Lecture Notes in Networks and Systems 203,
https://doi.org/10.1007/978-981-16-0733-2_68

1 Introduction

Internet has become the integral part of human needs. It has reached almost every house in the world. It is observed that about 54% of worlds population are using Internet. As high-speed Internet can be easily accessed and different IoTs have been invented, there is huge number of IP address generated. The scope of IPv4 cannot accommodate this volume of IP address requests. Hence, organizations and institutions are preferring privately managed networks so as to make effective use of available addresses. Traditional networks are being used to implement these private networks for providing seamless connectivity to users and protecting users. Traditional networks are designed based on vendor-specific devices. This limits the addition and removal of specific device from the network because the network design team need to agree with vendor-specific commands to configure devices such as firewall, intrusion detection system (IDS) and IPSec [1] for implementation of security policies. Also annual configuration of devices may lead to configuration error and security threats.

Software-defined networking is a new network design paradigm, wherein the control logic of the network is separated from the data plane [2]. Controller is responsible for the control logic of the network plane. Controller makes the switching and routing devices as simple as the forwarding devices. Open standards for SDN have been developed by investments of famous organizations including Microsoft, Google, Yahoo, Facebook and Verizon [2]. Communication between the controller and data plane in SDN environment is enabled by OpenFlow protocol. OpenFlow protocol is used by the controller for passing switching, routing, load balancing or firewall policies onto data plane devices [3].

SDN architecture

The typical SDN architecture is shown in Fig. 1. SDN architecture is divided into three layers.

- Application layer
- Control layer
- Infrastructure layer.

The application layer deals with different network applications and functions used by the organization. Typical applications include intrusion detection systems, load balancing or firewalls. Firewall and load balance are implemented in traditional networks using specialized module. In SDN, the controller replaces the separate appliance which deals with management of the data plane.

Northbound and southbound APIs are used for communication between three layers of SDN architecture.

The controller resides in the control plane of the SDN. It is the brain of the SDN. The centralized controller is responsible for managing network policies and network flow.

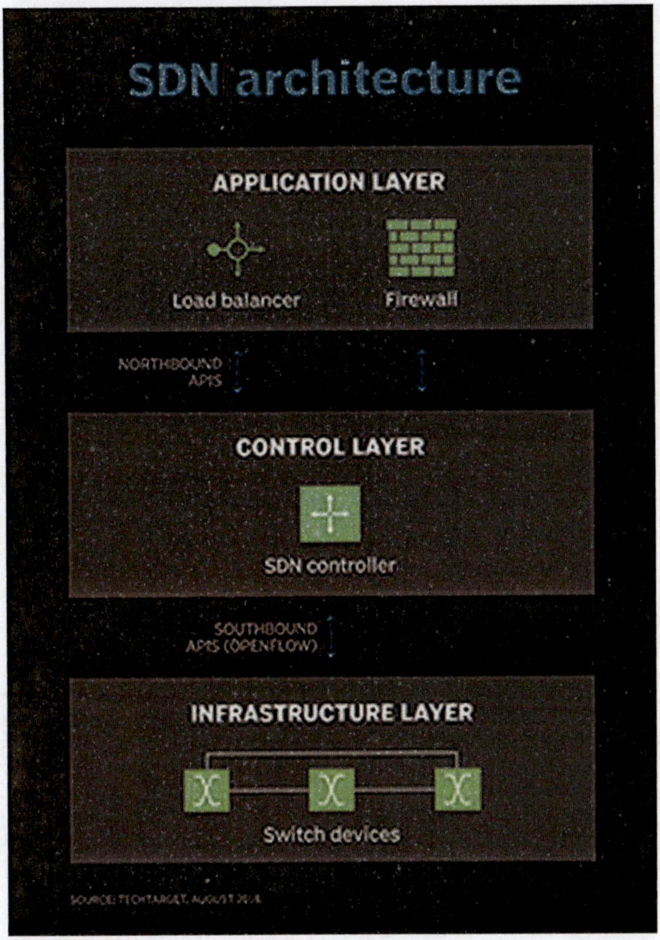

Fig. 1 SDN architecture

The infrastructure layer contains the physical switches. Northbound and southbound programming interfaces (APIs) are responsible for communication between these three layers. Communication between applications to the controller occurs through northbound interface. The southbound API is responsible for communication between controller and switches. An example of southbound protocol is OpenFlow protocol.

1.1 Motivation

Incoming and outgoing packet traffic from the network can be monitored and controlled by setting security rules in terms of firewall. A firewall can act as a barrier between an internal trusted network and an external untrusted network such as the Internet. In SDN as the centralized control enforces network wide security and prevents policy collision, it is preferable to implement firewall over the controller. This paper deals with implementation of POX-based firewall for providing network wide security. During this firewall implementation, some devices are secured from communication with other devices in the network. Rest of the paper is organized as below: Sect. 2 deals with related work done in the field of SDN security and firewall. Section 3 deals with the flow and methodology of the proposed firewall application. Section 4 deals with a basic model for firewall solution, followed by building blocks of the proposed solution. Section 5 explains performance analysis of firewall in terms of TCP and UDP delay, throughput and jitter. Conclusion and future work are covered in Sect. 6.

2 Related Work

Nowadays, traditional network systems are replaced by software-defined networking (SDN). The networks are more flexible with the use of network virtualization (NVF). It is possible to implement firewall in SDN. But it was not possible to implement stateful version of firewall in earlier version of OpenFlow protocol used in SDN. Authors in [4] have implemented stateful firewall in SDN switch. The Open vSwitch is used. The performance of SDN stateful firewall is also evaluated. SDN stateful firewall was able to work with small overhead increased in SDN switches.

Authors in [5] have focused on replacement of physical switches by the virtual networking domain. Also firewall and load balancing algorithm have been developed based on OpenFlow and SDN. The hardware firewall devices are replaced by simple firewall application. Floodlight controller of SDN is used as it displays the SDN network in GUI.

Authors in [6] have proposed a firewall application for SDN-based on MAC address. The average packet delay performance of both IP and MAC addressing is evaluated.

Authors in [7] have used POX, a Python-based controller. Layer 3-based firewall security is implemented using a full mesh topology with one controller and six switches with one host per switch. Learning switch coding of POX controller is modified using Mininet. Flow of the packets between different hosts is controlled based on the firewall rules written over learning switch.

Authors in [8] have studied UDP traffic for different bandwidths ranging from 1 Mbit/sec to 50 Mbits/sec in SDN environment designed with and without firewall topology, respectively. SDN-based firewall is installed on POX controller. Performance of SDN-based firewall is evaluated, and it is observed that percentage packet loss increases with increasing bandwidth.

Authors in [9] have introduced firewall for SDN for securing network attached devices and providing access control in SDN.

Authors in [10] have implemented some firewall functionalities on SDN by writing firewall applications. SDN POX controller is used here. The firewall is able to work on layer 2, layer 3 and layer 4. The firewall is filtering packets according to header, and they are matched against predefined policies. A packet is blocked if the matching is found, otherwise forwarded. POX controller of SDN is used over Mininet. Wireshark and Iperf have been used for analyzing the performance of firewall module.

Traditional networks have been replaced by SDN. Firewall is one of the essential components of SDN. But firewall links suffer from speed limitation as speed of firewall's link is slower than supported network interface. This results in packet loss or delay. Hence, for overcoming this issue, authors in [11] have implemented duplicate instances of firewall, as performance evaluation use of multiple controllers in networks showed improved performance.

Data-centric computing devices require to transfer files over long distances at high throughput. Stateful firewalls result in considerable packet loss. Hence, for dealing with this problem, authors in [12] have proposed a novel extension to the Science DMZ design, which uses an SDN-based firewall. NFShunt is introduced here which is a firewall based on Linux's Netfilter combined with OpenFlow switching. Thee bypass-switching policy is used by OpenFlow 1.0 controller. The performance is evaluated using TCP throughput and packet loss.

Authors in [13] have developed a firewall application for OpenFlow-based SDN. The application demonstrates that the firewall application can be built on software without dedicated hardware. Authors have built the firewall on POX controller and created SDN topology using VirtualBox and Mininet. In this study, authors cover the implementation detail of our firewall application, as well as the experimentation result [13].

Authors in [14] have described intrusion detection mechanism for OpenFlow-based SDN. Packet filtering firewall is developed over floodlight controller of SDN. Data passing through firewall is analyzed using association rules. Different patterns recorded can be used as motivation for development of an anomaly-based intrusion detection mechanism.

Firewall is responsible for monitoring incoming and outgoing traffic of the network and deciding whether to allow or block specific traffic based on security policy. It is very complicated for managing the firewall rules for large enterprise networks. Authors in [15] have developed a network application for SDN. The application aims at resolving different anomalies in the firewall rules set. Floodlight SDN OpenFlow controller is used for experimentation. The performance of network application is compared with the external floodlight firewall module.

3 Methodology

The firewall application implemented in this paper is responsible for filtering of packets according to their parsed headers. The packets are permitted or dropped based on the specified rules in the firewall. Hence, the firewall application needs to run with the learning switch module of POX controller. The learning switch module enforces the OpenFlow switch to act as a type of L2 learning switch. The two modules are required to run at the same time. But running the two modules at the same time may cause following two issues:

- Running of these two modules exactly at the same time may lead to OpenFlow error messages since both of them will be accessing the same buffer.
- Also policy conflict may occur as both modules might install different rules, flow entries to the OpenFlow switch.

To deal with these issues, there are two methods:

- Create one "Master" class which will include both firewall module and switch module. The "Master" class will be responsible for organizing the order of calling module's functions.
- In the another method both modules are separated. The switch module is to be modified so that it will listen to events fired by our firewall module in instead of listening to events triggered by OpenFlow protocol.

In our work, we have considered the second approach. L2 learning switch of POX controller is chosen. The learning switch is run along with the firewall application. The L2 learning switch listens to triggered events of firewall application. Thus, both module will work together in harmony.

Algorithm for the firewall application is depicted herewith.

Algorithm:

1. Start the POX controller with l2_learning switch
2. Start the firewall application.
3. Parse the firewall policies
4. Receive new event from switch.
5. Take our packet from the event and decode ethernet header.
6. Learn/update switch port associated with source MAC address/IP address.
7. If IPv4 packet then
    a. Check for ICMP messages and match the rule.
    b. Check if it is TCP traffic
        If TCP traffic then parse TCP packet and match the rules
    c. Check if it is UDP packet
        If it is UDP packet then parse UDP datagram and match the rules.

    Else
    check for ARP request and flood if it is ARP else drop.
8. If the rule is matched then drop or install flow entry for similar packets.
    Else
    Drop.

As depicted in the algorithm, the firewall application parses the firewall policies. The firewall application then listens to catch the new event packet from the switch. After receiving the packet, the packet is parsed by the firewall application, and Ethernet header is taken out. Table containing switch ports associated with MAC address/IP address is learned or updated. The firewall application then checks whether it is IP packet or ARP request. In case of "ping", the firewall application checks for ICMP, otherwise it checks for TCP segment or UDP datagram. According to the rules specified, the packet is blocked or allowed. For ARP request, it will store the flow table entry on switches. Due to this, packet processing will be speeded up for further packets, and bandwidth consumption of controller switch link is reduced.

4 Pox Firewall Experimental Setup

For carrying out this experiment, we have used Mininet-WiFi emulator and POX controller of SDN. POX is a Python-based open-source controller of SDN. The wireless topology as depicted in Fig. 2 was created using Mininet-WiFi emulator tool. Mininet-WiFi is a Python-based emulator able to create small network as well as large network. The code written on Mininet can be straightforward used in real network without any modification. Predefined topologies of Mininet include linear, single and tree. It is possible to create custom topologies using Mininet-WiFi. POC controller contains various SDN applications including hub application, L2 learning switch application and L3 learning application.

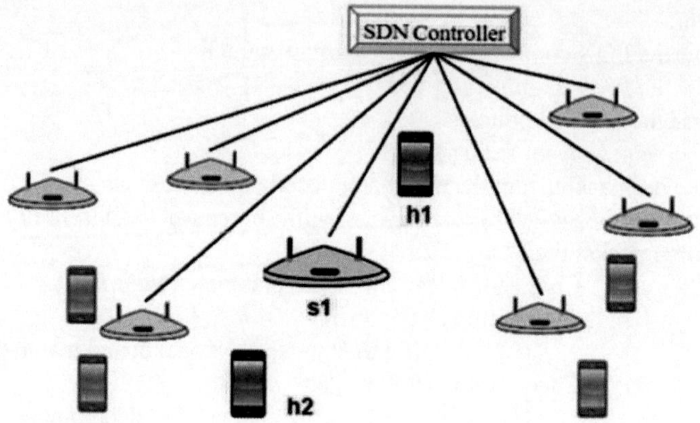

Fig. 2 Network topology

Table 1 depicts the MAC address and IP address for sample stations of the custom topology.

During this experiment, we have created two Python scripts. The first Python script is for creating custom topology, and the second script is firewall application. The custom topology created consists of one remote controller, two access points and four hosts. A number of hosts in the network is increased from 4, 6, 8, 12 and 20, and a number of access points in the network are 2 and 4, respectively. The performance of the network is evaluated in terms of bandwidth and jitter.

The firewall application is responsible for dropping the packets from predefined IP address of source and destination.

Command to run the custom topology is

sudo python sdn_handover.py

In another terminal, command to run firewall application along with L2 learning switch of POX controller is shown in Fig. 3.

The firewall application is able to parse three policies:

Table 1 Record of MAC and IP address for example topology

Net node	MAC address	IP address
Sta1	00:00:00: 00:00:02	10.0.0.2
Sta2	00:00:00: 00:00:03	10.0.0.3
Sta3	00:00:00: 00:00:04	10.0.0.4
Sta4	00:00:00: 00:00:05	10.0.0.5
Sta5	00:00:00: 00:00:06	10.0.0.6

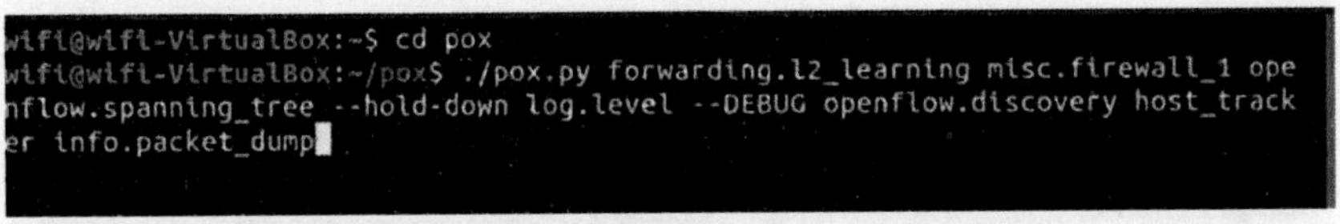

Fig. 3 Command to run firewall application along with POX controller

- MAC address: In this, the Ethernet header is cached for checking if it belongs to specific hosts, e.g., host 1 (00:00:00: 00:00:01), and it is to blocked in both directions.
- IP rule: In this application, layer 3 firewall is represented which is able to detect IP packet and checks if it matches the rule. It matches the rule, and then it is not allowed to forward. We have specified list blocking IP address of source and destination pair.
- TCP and UDP rule: Port security is performed for catching layer 4 security. Once the TCP segment or UDP datagram is cached, the rules are executed accordingly.

5 Experimental Evaluation

We have evaluated the performance of proposed firewall application using different wireless network scenarios including

- varying number of access points in networks, viz 2 and 4
- varying number of moving stations in networks from 4, 8, 12, 16 and 20.

We have compared the performance in terms of delay, TCP bandwidth and jitter. Also the comparison of delay, TCP bandwidth and jitter is done for running the same topology with and without firewall.

ICMP test

Figure 4 shows the testing of ping reachability after executing the firewall application. Figure 4 shows that all traffic from sta1 having IP address 10.0.0.2 to sta4 with IP address 10.0.0.5 is blocked as if it is considered a suspicious or malicious host. Thus, sta1 cannot ping sta4, and similarly sta4 cannot ping it back. Figure 4 displays the POX output of blocking the traffic coming from the source sta1 to destination sta4 (10.0.0.5).

Since there is no firewall rule installed between sta1 (10.0.0.2) and sta2 (10.0.0.3), both can communicate with each other using ping as shown in Fig. 5.

TCP test

TCP/UDP layer 4 rules are re-added, and TCP/UDP traffic is checked and matched with firewall policies for deciding whether to forward or drop the packet. We have evaluated the performance of the firewall application for TCP bandwidth. Iperf is

```
mininet-wifi> sta1 ping 10.0.0.5
'PING 10.0.0.5 (10.0.0.5) 56(84) bytes of data.
From 10.0.0.2 icmp_seq=1 Destination Host Unreachable
From 10.0.0.2 icmp_seq=2 Destination Host Unreachable
From 10.0.0.2 icmp_seq=3 Destination Host Unreachable
From 10.0.0.2 icmp_seq=4 Destination Host Unreachable
From 10.0.0.2 icmp_seq=5 Destination Host Unreachable
From 10.0.0.2 icmp_seq=6 Destination Host Unreachable
From 10.0.0.2 icmp_seq=7 Destination Host Unreachable
From 10.0.0.2 icmp_seq=8 Destination Host Unreachable
From 10.0.0.2 icmp_seq=9 Destination Host Unreachable
From 10.0.0.2 icmp_seq=10 Destination Host Unreachable
^C
--- 10.0.0.5 ping statistics ---
13 packets transmitted, 0 received, +10 errors, 100% packet loss, time 12277ms
pipe 4
```

Fig. 4 Ping unreachability for firewall installed rules

```
*** Unknown command: clear
mininet-wifi> sta1 ping 10.0.0.3
PING 10.0.0.3 (10.0.0.3) 56(84) bytes of data.
64 bytes from 10.0.0.3: icmp_seq=1 ttl=64 time=1.55 ms
64 bytes from 10.0.0.3: icmp_seq=2 ttl=64 time=1.15 ms
64 bytes from 10.0.0.3: icmp_seq=3 ttl=64 time=1.62 ms
64 bytes from 10.0.0.3: icmp_seq=4 ttl=64 time=1.47 ms
64 bytes from 10.0.0.3: icmp_seq=5 ttl=64 time=1.95 ms
64 bytes from 10.0.0.3: icmp_seq=6 ttl=64 time=1.54 ms
64 bytes from 10.0.0.3: icmp_seq=7 ttl=64 time=1.96 ms
64 bytes from 10.0.0.3: icmp_seq=8 ttl=64 time=1.53 ms
64 bytes from 10.0.0.3: icmp_seq=9 ttl=64 time=1.41 ms
64 bytes from 10.0.0.3: icmp_seq=10 ttl=64 time=1.47 ms
^C
--- 10.0.0.3 ping statistics ---
10 packets transmitted, 10 received, 0% packet loss, time 9012ms
```

Fig. 5 Output for hosts connectivity after running the firewall

used for establishing the connection between sta2 (10.0.0.3) and sta4 (10.0.0.5). As depicted in Fig. 6, Iperf server is initiated at sta1.

Similarly at sta4, we initiate the Iperf client to communicate with Iperf server at 10.0.0.3 (sta2).

As depicted in Fig. 7, sta4 client sends TCP packets to sta2 for 10 s.

As shown in Fig. 6, TCP bandwidth utilization at server is 9.75 Mbits/sec.

RTT Evaluation

Sta2 (10.0.0.3) is able to ping with sta4 (10.0.0.5) since there is no firewall rule installed. RTT evaluation based on layer 3 policy is installed, and sta2 can ping sta4t4. Figure 8 shows that sta2 successfully sent ICMP messages to sta4 and got responses. The round trip time required to send 10 ICMP messages was 9015 ms.

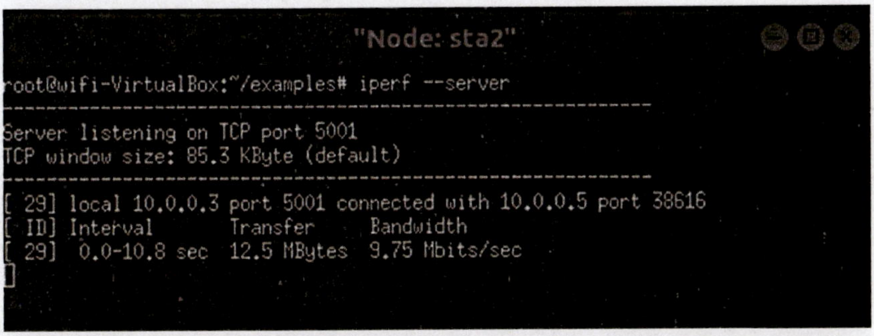

Fig. 6 TCP server initiated at sta2 using Iperf

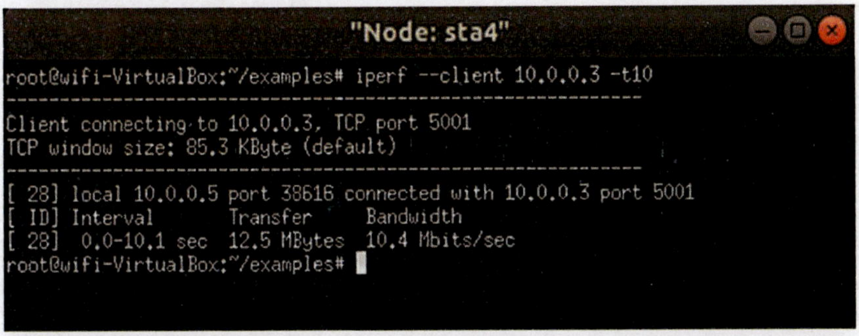

Fig. 7 TCP client sending packets to server

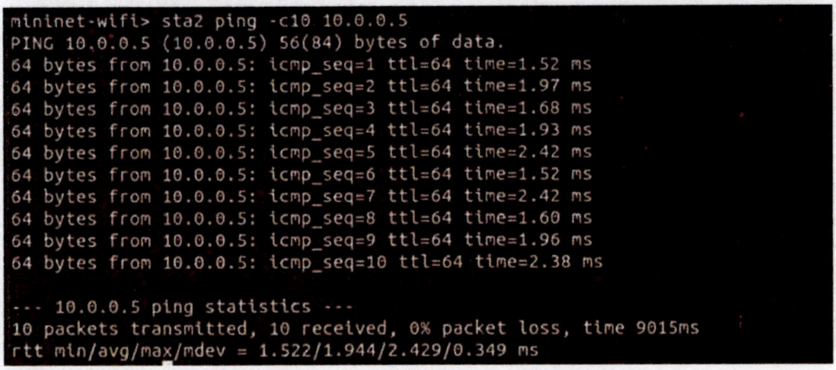

Fig. 8 ICMP traffic allowed from sta2 and sta4 with firewall running

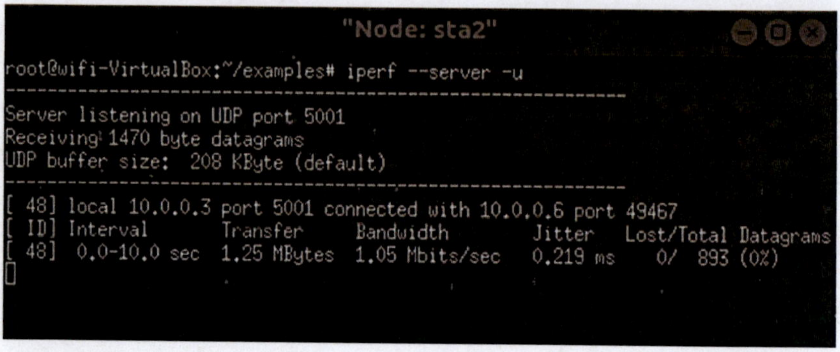

Fig. 9 UDP server

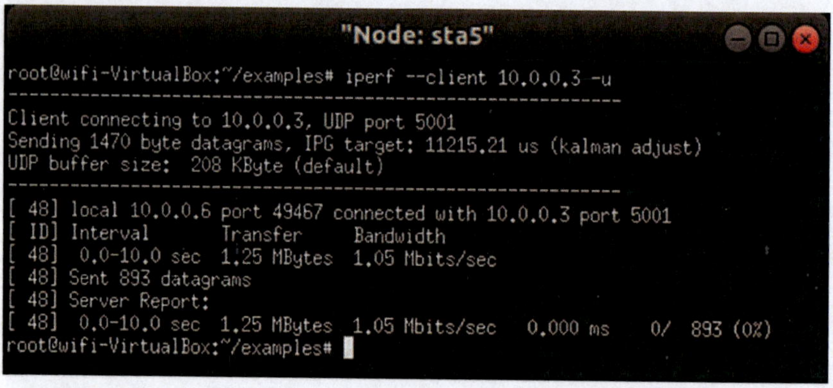

Fig. 10 UDP client

UDP Test

Iperf is used for establishing the connection between UDP server and client. As depicted in Fig. 9, UDP server is established at sta2 with IP address 10.0.0.3. The Iperf client sta5 sends UDP datagrams to server. Latency variation (jitter) is delay amount of time to transmit data from a source to a destination that varies over time.

Figure 9 shows measured bandwidth and jitter of UDP traffic with firewall installed on POX controller, whereas Fig. 10 shows UDP client sending the UDP datagrams.

6 Results and Discussions

The performance of the proposed firewall is evaluated in various scenarios.

6.1 Performance Evaluation with and Without Firewall Application

Table 2 shows the TCP bandwidth for different network topology under consideration. We have considered varying number of stations and access points.

As shown in Table 2, the average TCP bandwidth utilization for running the POX controller without firewall is more as compared with firewall application. The firewall application reduces the bandwidth utilization.

Table 3 depicts the performance in terms of Jitter.

Table 2 TCP bandwidth for varying network conditions

2AP and 2 R	Number of stations	4	8	12	16	20	Average
	With firewall	9.28	10.2	10.1	4.16	6.3	8.008
	Without firewall	9.24	10.4	9.34	10.1	6.47	9.11
2AP and 4 R	Number of stations	4	8	12	16	20	Average
	With firewall	9.33	10.1	10.3	10.2	5.02	8.99
	Without firewall	9.24	10.4	9.34	10.1	6.47	9.11
4AP and 2 R	Number of stations	4	8	12	16	20	Average
	With firewall	10	10.3	6.43	9.41	9.45	9.118
	Without firewall	9.79	9.91	8.95	10.4	10.4	9.89
4AP and 4 R	Number of stations	4	8	12	16	20	Average
	With firewall	10.2	10.4	9.9	9.52	10.3	10.064
	Without firewall	9.79	9.91	8.95	10.4	10.4	9.89

Table 3 Performance of the firewall in terms of jitter

2AP and 2 R	Number of stations	4	8	12	16	20	Average
	With firewall	0.245	0.178	0.169	0.257	0.236	0.217
	Without firewall	0.105	0.289	0.163	0.241	0.204	0.2004
2AP and 4 R	Number of stations	4	8	12	16	20	Average
	With firewall	0.199	0.279	0.223	0.337	0.216	0.2508
	Without firewall	0.105	0.289	0.163	0.241	0.204	0.2004
4AP and 2 R	Number of stations	4	8	12	16	20	Average
	With firewall	0.202	0.189	0.213	0.201	0.217	0.2044
	Without firewall	0.179	0.291	0.294	0.282	0.229	0.255
4AP and 4 R	Number of stations	4	8	12	16	20	Average
	With firewall	0.204	0.226	0.258	0.237	0.236	0.2322
	Without firewall	0.179	0.291	0.294	0.282	0.229	0.255

Table 4 Performance in terms of delay

Number of AP	2AP				
Number of stations	4	8	12	16	20
With firewall	9013	9012	9012	9013	9011
Without firewall	9014	9013	9014	9013	9012
	4AP				
Number of stations	4	8	12	16	20
With firewall	9012	9014	9017	9009	9012
Without firewall	9011	9012	9014	9017	9017

As shown in Table 3, the average jitter is increased for topology with two access points for POX controller with firewall. For four access points, the jitter is reduced for POX controller with firewall.

Table 4 shows the performance in terms of delay. As depicted in the table, the delay performance of the network does not affect significantly with the firewall application. Addition of firewall module to the POX controller does not affect the performance in terms of delay.

6.2 *Performance Evaluation with Varying Number of Access Points*

The performance is evaluated by increasing the number of access points in the network. As depicted in Fig. 11, TCP bandwidth is increased with four access point networks as compared with two access point networks.

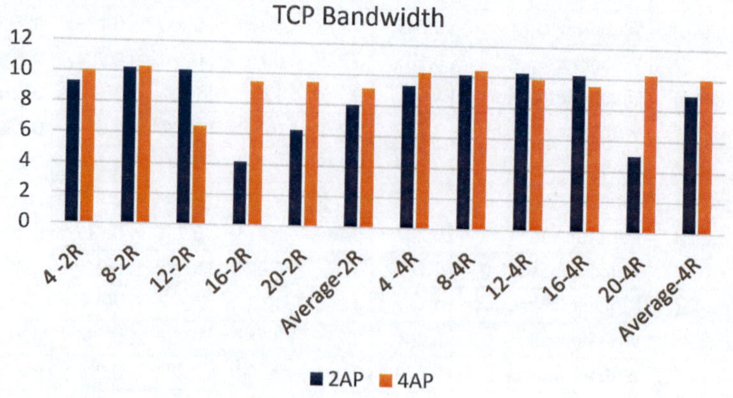

Fig. 11 TCP bandwidth for various APs

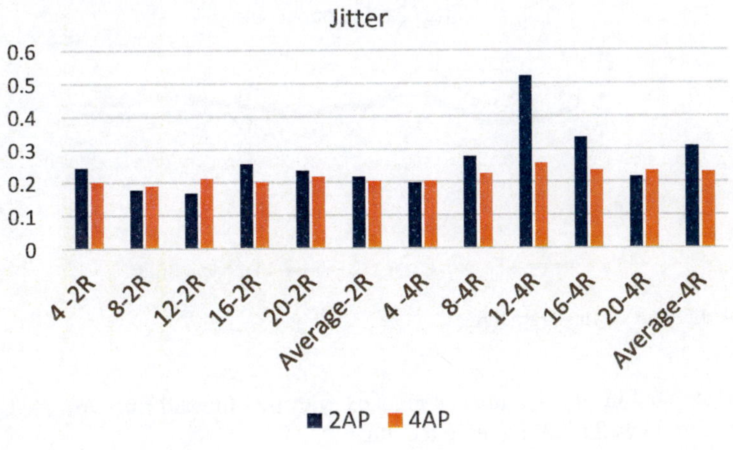

Fig. 12 Jitter for varying AP

Figure 12 depicts the jitter by varying AP in network.

As depicted in Fig. 12, the network with 4AP has reduced jitter as compared to network with 2AP.

6.3 *Performance Evaluation by Increasing the Number of Firewall Rules in the Network*

We have evaluated the performance of the network by increasing number of firewall rules.

Figure 13 depicts the comparison of TCP bandwidth for varying number of stations. TCP bandwidth is increased with four firewall rules installed as compared with two firewall rules installed.

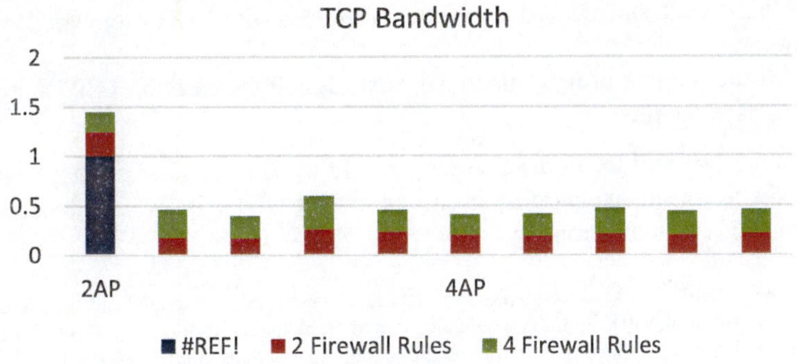

Fig. 13 TCP bandwidth for various firewall rules installed

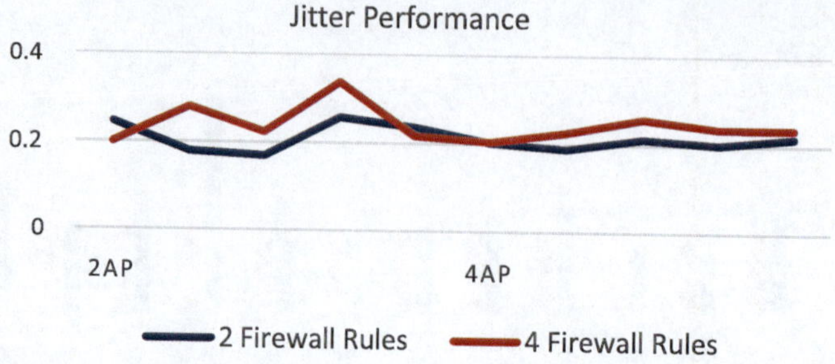

Fig. 14 Jitter for varying firewall rules

As depicted in Fig. 14, jitter is reduced with two firewall rules installed. With increase in firewall rules, increase the jitter.

7 Conclusion

One of the essential components of network security is firewall. It is responsible for monitoring and governing all network traffic. It identifies and blocks undesired traffic. In traditional networks, it was costly to deploy physical firewall. It was difficult to update the system with firewall rules installed. In this paper, we have designed SDN-based firewall for wireless networks using POX controller. We have evaluated the performance of the firewall application for ICMP, TCP and UDP messages. SDN firewall application is run along with L2 learning switch of SDN controller. We have evaluated the performance of the firewall application in terms of delay, TCP bandwidth and jitter by varying number of stations in the networks and for varying number of AP in wireless network.

- Overall TCP bandwidth is reduced with firewall application ads compared with simple POX controller without firewall application.
- By increasing the number of access points in the network can result in increased TCP bandwidth and reduced jitter. Jitter is reduced with 4AP as compared to 2AP networks.
- Increasing number of firewalls in the network will increase the TCP bandwidth and reduce the jitter.

In future, we will try to develop our firewall to be distributed so that it would reduce the overhead on controller in case of much traffic. Also we might add some functionalities including routing.

Acknowledgements Authors would like to welcome the reviewer's comments and suggestions to carry out this research work. Authors would also like to thank the authorities of SSGBT College of Engineering, Bhusawal, for providing the research laboratory to carry out this research.

References

1. Ahmad I, Namaly S, Ylianttilaz M, Gurtov A (2015) Security in software defined networks: a survey. IEEE Commun Surv Tutorials 17(4):2317–2346
2. Kreutz D, Ramos FMV, Verissimo P, Rothenberg CE, Azodolmolky S, Uhlig S (2014) Software-defined networking: a comprehensive survey. In: Proceedings of the IEEE, pp 14–76
3. Shieha A (2014) Application layer firewall using openflow. In: Interdisciplinary telecommunications graduate theses and dissertations, Paper 1, unpublished
4. Krongbaramee P, Somchit Y (2018) Implementation of SDN stateful firewall on data plane using open vSwitch. In: 2018 15th International joint conference on computer science and software engineering (JCSSE), Nakhonpathom, pp 1–5. https://doi.org/10.1109/jcsse.2018.8457354
5. Zope N, Pawar S, Saquib Z (2016) Firewall and load balancing as an application of SDN. In: 2016 Conference on advances in signal processing (CASP), Pune, pp 354–359. https://doi.org/10.1109/casp.2016.7746195
6. Rengaraju P, Kumar SS, Lung C (2017) Investigation of security and QoS on SDN firewall using MAC filtering. In: 2017 International conference on computer communication and informatics (ICCCI), Coimbatore, pp 1–5. https://doi.org/10.1109/iccci.2017.8117772
7. Kumar A, Srinath NK (2016) Implementing a firewall functionality for mesh networks using SDN controller. In: 2016 International conference on computation system and information technology for sustainable solutions (CSITSS), Bangalore, pp 168–173. https://doi.org/10.1109/csitss.2016.7779417
8. Monir MF, Akhter S (2019) Comparative analysis of UDP traffic with and without SDN-based firewall. In: 2019 International conference on robotics, electrical and signal processing techniques (ICREST), Dhaka, Bangladesh, pp 85–90. https://doi.org/10.1109/icrest.2019.8644395
9. Satasiya D, Raviya R, Kumar H (2016) Enhanced SDN security using firewall in a distributed scenario. In: 2016 International conference on advanced communication control and computing technologies (ICACCCT), Ramanathapuram, pp 588–592. https://doi.org/10.1109/icaccct.2016.7831708
10. Othman WM, Chen H, Al-Moalmi A, Hadi AN (2017) Implementation and performance analysis of SDN firewall on POX controller. In: 2017 IEEE 9th international conference on communication software and networks (ICCSN), Guangzhou, pp 1461–1466. https://doi.org/10.1109/iccsn.2017.8230351
11. Saad Waheed M, Al Mufarrej M, Sobhieh M, Al Barrak A, Baig A, Al Mazyad A (2017) Implementation of virtual firewall function in SDN (software defined networks). In: 2017 9th IEEE-GCC conference and exhibition (GCCCE), Manama, pp 1–9. https://doi.org/10.1109/ieeegcc.2017.8447955
12. Miteff S, Hazelhurst S (2015) NFShunt: a linux firewall with openflow-enabled hardware bypass. In: 2015 IEEE conference on network function virtualization and software defined network (NFV-SDN), San Francisco, CA, pp 100–106. https://doi.org/10.1109/nfv-sdn.2015.7387413
13. Suh M, Park SH, Lee B, Yang S (2014) Building firewall over the software-defined network controller. In: 16th International conference on advanced communication technology, Pyeongchang, pp 744–748. https://doi.org/10.1109/icact.2014.6779061
14. Sayeed MA, Sayeed MA, Saxena S (2015) Intrusion detection system based on software defined network firewall. In: 2015 1st International conference on next generation computing technologies (NGCT), Dehradun, pp 379–382. https://doi.org/10.1109/ngct.2015.7375145
15. Morzhov SV, Nikitinskiy MA (2018) Development and research of the PreFirewall network application for floodlight SDN controller. In: 2018 Moscow workshop on electronic and networking technologies (MWENT), Moscow, pp 1–4. https://doi.org/10.1109/mwent.2018.8337255

Latest Electrical and Electronics Trends

Output Load Capacitance Scaling-Based Energy-Efficient Design of ROM on 28 nm FPGA

Pankaj Singh, Bishwajeet Pandey, Neema Bhandari, Shilpi Bisht, and Neeraj Bisht

Abstract In this paper, we have proposed the design of an energy-efficient ROM by scaling down the output load capacitance. The ROM is implemented on a 28 nm-based CPG236 package and Artix-7 FPGA. To achieve the energy efficiency of ROM, we have reduced capacitance from 30 to 0 pF by reducing 15 pF capacitance at each step to exhibit the consequence of reduced capacitance on the total power consumption of ROM. Reduction of 35.88% in I/O power consumption, 26.35% in static power consumption, 35.44% in total power consumption, and 23.04% in the junction temperature of ROM is observed when capacitance is reduced from 30 to 15 pF. Subsequently reducing capacitance from 15 to 0 pF results in a reduction of 55.96% in I/O power, 17.89% in static power, 54.64% in total power consumption, and 29.76% in junction temperature. From experimental results, it has been observed that signal power and logic power are not varying with variation in capacitance. By scaling output load capacitance, not only energy efficiency is achieved, but also thermal management has also improved. Reduction in junction temperature results in low static power consumption and an increase in device reliability and lifespan. The design is implemented on Artix-7 FPGA using Vivado Design Suite and Verilog. Because of low power consumption, this design of ROM can be ingrained into a microprocessor setup or can be assembled as ROM ASIC.

Keywords Energy efficient · Output load capacitance · FPGA · ROM design · I/O power · Leakage power · Thermal management

P. Singh (✉) · B. Pandey · S. Bisht · N. Bisht
Birla Institute of Applied Sciences, Bhimtal, Uttarakhand, India
e-mail: pankaj@birlainstitute.co.in

N. Bhandari
G.B. Pant University of Agriculture and Technology, Pantnagar, Uttarakhand, India

P. K. Singh et al. (eds.), *Proceedings of Second International Conference on Computing, Communications, and Cyber-Security*, Lecture Notes in Networks and Systems 203,
https://doi.org/10.1007/978-981-16-0733-2_69

1 Introduction

Read-only memory (ROM) is an integrated chip which is preprogrammed with a specific set of instruction at the time of manufacturing. It is also known as firmware. It is a nonvolatile memory; because of this property, it retains data even when the power is switched off. ROM is used not only in computers but in many other electronic devices. ROM has become an integral part of many electronic devices. Data stored in a ROM can only be read but cannot be altered as shown in Fig. 1.

With the rapid use of ROM in electronic and battery-operated devices, its power consumption must be considered while designing. Energy efficiency has become an important aspect in the design of ROM. As a consequence, power consumption estimation before ROM fabrication has become necessary. As the submicron technologies have developed, challenges in designing portable applications, digital signal processors, and ASIC implementation with low power consumption have grown significantly [1]. Applications that rely on batteries for power sources require low power consumption for a long lifetime [2]. Therefore, it is essential to design ROM in an energy-efficient way which is used in such applications.

A capacitor can store electrical energy like a small rechargeable battery. It behaves like electric charge storage when voltage is applied across its plates, and it gives up the stored charge when required. The charge storing capability of the capacitor is measured in a unit called capacitance (Fig. 2).

By scaling output load capacitance, we can ensure low power consumption of ROM in applications. The uses of output load scaling are discussed in [4–7].

While designing energy-efficient ROM, all factors of power consumption must be considered. Cumulative of static power consumption and dynamic power consumption represents the total power consumption of any CMOS-based device. Static power consumption is due to the repercussion of leakage in the transistor. Static power

Fig. 1 Read-only memory

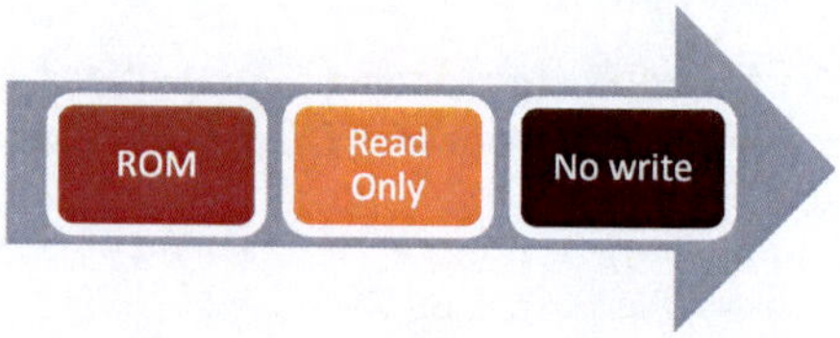

Fig. 2 Parasitic capacitor [3]

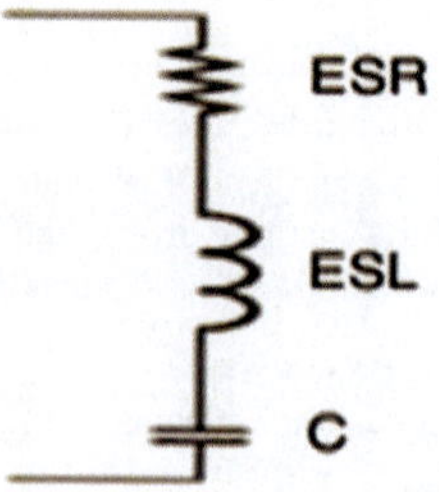

consumption is an important and non-avoidable aspect of total power consumption as it is the constant power that will be consumed by the device as long as it is powered on. Dynamic power consumption is comprised of signal power, logic power, and I/O power. Dynamic power consumption can be expressed using Eq. 1.

$$P = cv^2 f \tag{1}$$

In Eq. 1, *P* represents the dynamic power of ROM and its unit is watt; *C* represents output load capacitance of ROM and its unit is Farad; *V* represents the operating voltage of ROM and its unit is volts; and *F* represents operating frequency of ROM and its unit is HZ. By scaling output load capacitance, voltage, and frequency, we can reduce the power consumption. In this paper, we have considered output load capacitance for energy efficiency.

FPGA is a semiconductor device that has configurable logic blocks (CLB) connected via programmable interconnect and configurable I/O cell as shown in Fig. 3. CLB is made up of four components: lookup table (LUT), multiplexer, full adder, and D flip flop.

FPGA is designed to be reprogrammable using a hardware description language. VERILOG and VHDL are two popular choices for hardware description language. The fact that FPGA is reprogrammable distinguishes it from application-specific integrated chip (ASIC). ASIC is designed to do a specific task whereas FPGA can be reprogrammed to do any logic function. Due to the programmable nature of FPGAs, they can be used in many different applications such as defense and aerospace, automotive assistance, prototyping of ASIC, artificial intelligence, cloud computing, and telecommunication. FPGA has been in use for device prototyping for a long time

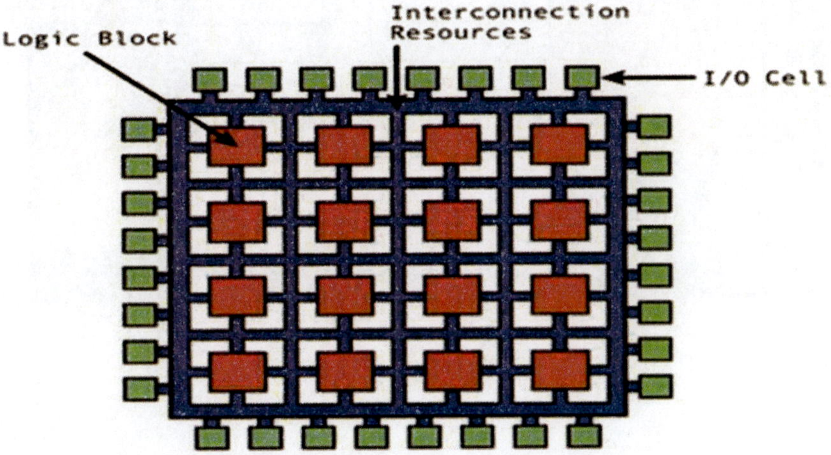

Fig. 3 Generic FPGA architecture [8]

because it can verify RTL replica of a design. A working prototype can be used for product demonstration and field testing.

To implement energy-efficient ROM, we have used 28 nm-based Artix-7 FPGA. Here, 28 nm represents the size of the transistor. We have chosen Artix-7 FPGA for the energy-efficient ROM implementation because it has the lowest power and cost over other 7 series FPGA. 7 series FPGA have the advantage of its 28 nm architecture which results in a substantially reduced static power consumption (65%), 30% reduced IO power consumption, low power High-K-Metal-Gate process, 25% reduced dynamic power consumption. All these considerations in return lower total power consumption by 50% over 40 nm FPGA. 7 series provide system design flexibility by catering the option of either reducing power budget to 50% or taking leverage of increased accessible performance and capacity at the preceding power budget as shown in Fig. 4. Artix-7 has 236 I/O pin, 106 IOB, 20,800 LUT, 41,600 flip flop, 50 Block RAM, 90 DSP which provide an optimum price to performance ratio [9].

The main aspiration of this paper is to propose a design of an energy-efficient ROM and analyze the power consumption at different output load capacitances. This paper is formed as follows: The previous related work is presented in Sect. 2, power analysis of ROM is described in Sect. 3, the results are represented and discussed in Sect. 4, and the Sect. 5 concludes the contributions of this paper.

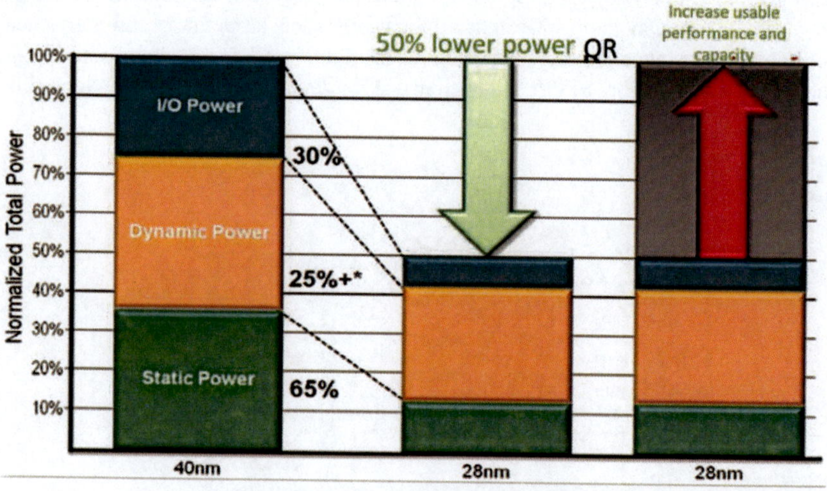

Fig. 4 Energy efficiency of 28 nm FPGA over 40 nm FPGA [10]

2 Related Work

In Reference [11], the author has designed a power-efficient ROM based on capacitance output load scaling for i7 processor operating at varying frequencies. In [12], the author designed low power consumption and high-performance ROM which is based on NOR type. The ROM design can be used in integrated embedded application or system on chip. In [13], the author proposed three methods for power-efficient ROM based on recycling and sharing of charge. The first method for power efficiency is based on the concept of recycling of charge for predecoder, the second method is based on the concept of recycling of charge for wordline decoder, and the third method is based on the concept of sharing of charge for bitline. In [14], the author proposed a ROM design on a custom layout with low power dissipation. The proposed ROM is based on CMOS technology, and the cell size for storage is optimized. In [15], the author designed a low power dissipation ROM based on the concept of charge sharing for capacitor. ROM reduces the power dissipation of bitlines. Capacitance scaling-based energy-efficient design of register is proposed and implemented on 28 nm Artix-7 FPGA [16]. A Unicode reader for the Punjabi language processing is designed on virtex-6 FPGA using capacitance scaling [17]. A similar work based on output load capacitance scaling is proposed for designing energy-efficient content addressable memory [18]. In [19], the author proposed a power-efficient ALU design for Wi-Fi ah channel. Power efficiency of ALU is achieved by varying capacitance and voltage. Low power dissipation FIR filter design for communication that is based on capacitance scaling is proposed in [20]. Low power dissipation comparator based on the concept of capacitance scaling on 28 nm FPGA is proposed in [21]. Low power consumption Unicode reader for the Malayalam language processing based on output load scaling is designed in [22]. The maximum capacitance of a polysilicon-gate MOSFET against oxide thickness is studied for different gate and substrate doping levels [23]. An energy-efficient image inverter is designed using capacitance and frequency scaling [24]. A biomedical wrist watch based on capacitance scaling energy efficient is proposed in [25].

3 Power Analysis of ROM

We have analyzed I/O power, static power, signal power, logic power, total power, and junction temperature for three different output load capacitances 0, 15, and 30 pF. The results of experimental work are demonstrated in Figs. 5, 6, and 7.[1]

There is a reduction of 23.04% in junction temperature, 35.88% in I/O power consumption, 26.35% in static power consumption, and 35.44% in total power consumption as a consequence of reducing output load capacitance from 30 to

[1]Fig. 5, 6, and 7 are the screenshots taken from Vivado design suite and a part of our original work.

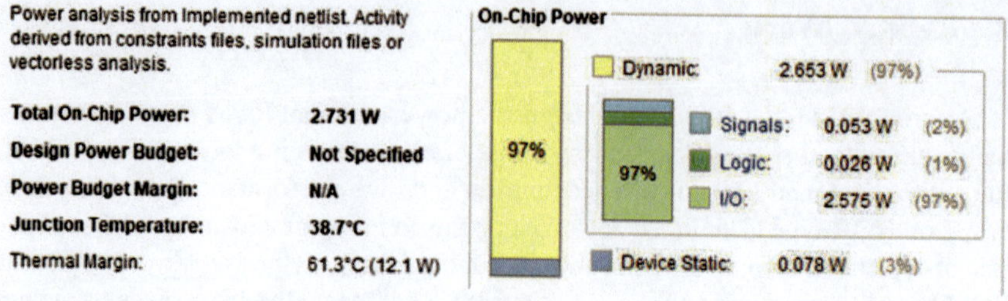

Fig. 5 Power analysis of ROM at 0 pF capacitance

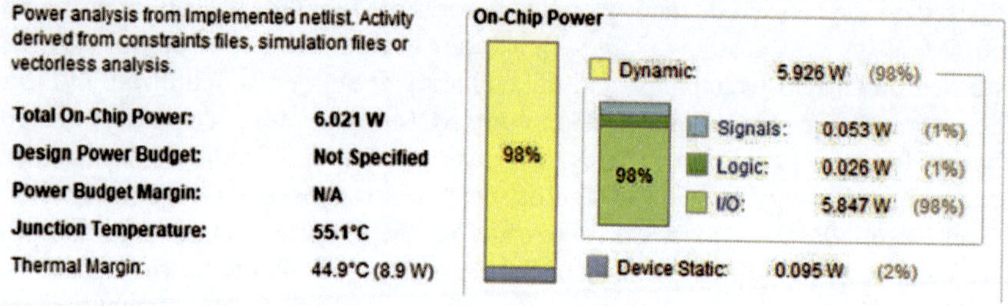

Fig. 6 Power analysis of ROM at 15 pF capacitance

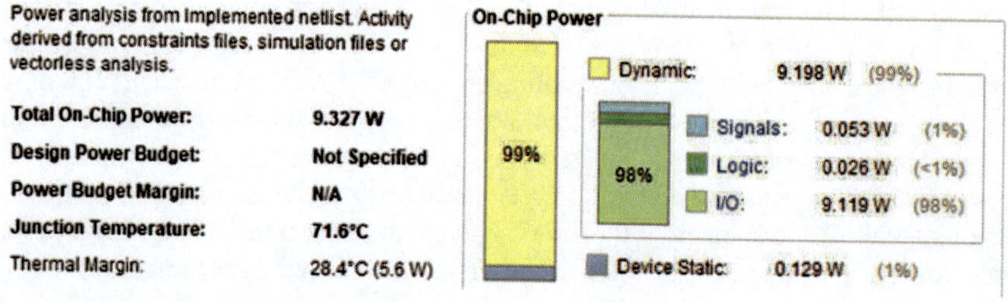

Fig. 7 Power analysis of ROM at 30 pF output load capacitance

15 pF. Similarly, there is a reduction of 29.76% in junction temperature, 55.96% in I/O power consumption, 17.89% in static power consumption, and 54.64% in total power consumption when output load capacitance is reduced from 15 to 0 pF as demonstrated in Table 1.

No variance in signal power and logic power is observed when output load capacitance is reduced. I/O power consumption, static power consumption, total power consumption, and junction temperature are corresponding to the variation in output load capacitance as illustrated in Figs. 8 and 9.

We have compared our results with the result of [26] which uses SSTL I/O standards for making energy-efficient ROM. Table 2 provides a comprehensive comparison between our model and the one used in [26]. Our ROM design is 28.43% more power-efficient.

Table 1 Power consumption with the different output load capacitances

Power/capacitance	30 pF	15 pF	0 pF
I/O power(W)	9.119	5.847	2.575
Signal power (W)	0.053	0.053	0.053
Logic power (W)	0.026	0.026	0.026
Static power (W)	0.129	0.095	0.078
Junction temperature (°C)	71.6	55.1	38.7
Total power (W)	9.327	6.021	2.731

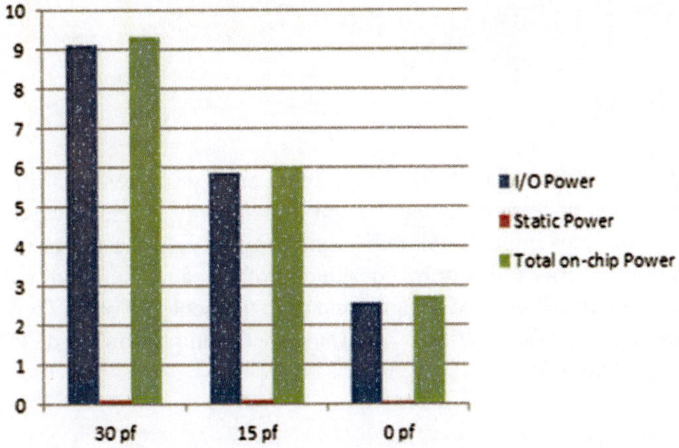

Fig. 8 Comparison of I/O power consumption, static power consumption, and total power consumption at the varying capacitance

Fig. 9 Reduction in junction temperature

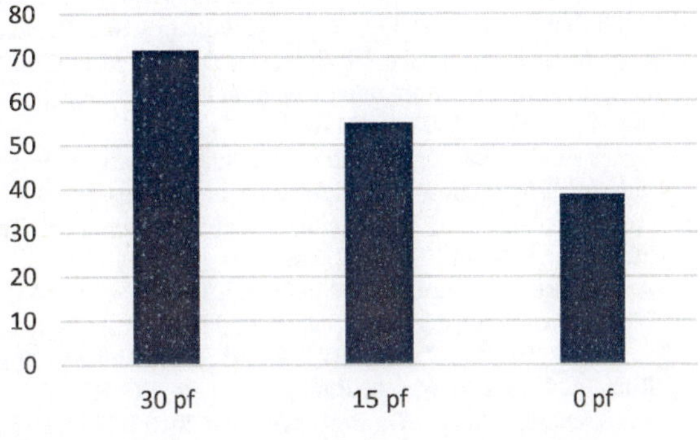

Table 2 Comparison of energy consumption at minimum operating parameters

	Minimum power consumption of the proposed ROM	Minimum power dissipation with SSTL2_II_DCI I/O standard
Parameter used	Output load capacitance (0 pF)	Output load capacitance (0 pF)
Power consumption (W)	2.731	3.816

4 Results and Discussion

By scaling output load capacitance, reduction in I/O power, static power, total on-chip power, and junction temperature is observed. However, variation in output load capacitance does not affect the signal power and logic power. When output load capacitance is reduced from 30 to 0 pF, there is a contraction of 71.76% in I/O power, 39.53% in static power, 70.71% in total power consumption, and 45.94% in junction temperature. Therefore, in the proposed ROM design, energy efficiency and thermal management have improved.

5 Conclusion

In this paper, we have proposed the design of an energy-efficient ROM by scaling output load capacitance. By making a ROM more energy efficient, we can ensure minimum power consumption, reliability, and longevity. In the future, we can try to make it more energy efficient by considering other parameters like voltage and ambient temperature. This ROM design could be implemented in other 7 series FPGA like Kinetix-7, Spartan-7, and Virtex-7, and energy consumption in these FPGA could be analyzed and compared.

References

1. Chandrakasan AP, Sheng S, Brodersen RW (1992) Low-power CMOS digital design. IEEE J Solid-State Circ 27(4):473–484. https://doi.org/10.1109/4.126534
2. Seok M, Hanson S, Seo JS, Sylvester D, Blaauw D (2008) Robust ultra-low voltage ROM design. In: IEEE custom integrated circuits conference. San Jose, CA, pp 423–426. https://doi.org/10.1109/cicc.2008.4672110
3. Series FPGAs packaging and pinout. https://www.xilinx.com/support/documentation/user_guides/ug475_7Series_Pkg_Pinout.pdf. Last Accessed 20 Aug 2020
4. Aoki T, Yokoi K (1997) Capacitance scaling system. IEEE Trans Instrum Meas 46(2):474–476. https://doi.org/10.1109/19.571889
5. Aguado-Ruiz J, Hernandez-Alvidrez J, Lopez-Martin AJ, Carvajal RG, Ramirez-Angulo J (2009) Programmable capacitance scaling scheme based on operational transconductance amplifiers. Electron Lett 45(3):159–161. https://doi.org/10.1049/el:20093596
6. Ruiz JA, Lopez-Martin A, Ramirez-Angulo J (2010) Three novel improved CMOS capacitance scaling schemes. In: Proceedings of 2010 IEEE international symposium on circuits and systems. Paris, pp 1304–1307. https://doi.org/10.1109/iscas.2010.5537257
7. Pandey B, Kumar T, Das T, Yadav R, Pandey OJ (2014) Capacitance scaling based energy efficient FIR filter for digital signal processing. In: International conference on reliability optimization and information technology (ICROIT). Faridabad, pp 448–451. https://doi.org/10.1109/icroit.2014.6798382
8. All about FPGA. https://www.eetimes.com/all-about-fpgas/. Last Accessed 25 Aug 2020
9. Series FPGAs configurable logic block user guide. https://www.xilinx.com/support/documentation/user_guides/ug474_7Series_CLB.pdf. Last Accessed 30 Aug 2020

10. series FPGA overview. http://www.eng.ucy.ac.cy/theocharides/Courses/ECE408/. Last Accessed 26 Aug 2020
11. Bansal M, Bansal N, Saini R, Kalra L, Mohan Singh P, Pandey B, Akbar Hussain DM (2014) FPGA based low power ROM design using capacitance scaling. Adv Mater Res 1082:471–474. https://doi.org/10.4028/www.scientific.net/ amr.1082.471
12. Chang CR, Wang JS, Yang CH (2001) Low-power and high-speed ROM modules for ASIC applications. IEEE J Solid-State Circ 36(10):1516–1523. https://doi.org/10.1109/4.953480
13. Yang BD, Kim LS (2003) A low-power ROM using charge recycling and charge sharing techniques. IEEE J Solid-State Circ 38(4):641–653. https://doi.org/10.1109/JSSC.2003.809516
14. Cui W, Wu S (2007) Design of small area and low power consumption mask ROM. In: IEEE international conference on integrated circuit design and technology. Austin, TX, pp 1–4. https://doi.org/10.1109/icicdt.2007.4299585
15. Yang BD, Kim LS (2006) A low-power ROM using single charge-sharing capacitor and hierarchical bit line. IEEE Trans Very Large Scale Integr (VLSI) Syst 14(4):313–322. https://doi.org/10.1109/tvlsi.2006.874303
16. Banshal SK, Pandey B, Brenda SJ (2014) Capacitance scaling aware power optimized register design and implementation on 28 nm FPGA. In: International conference on computer communication and informatics. Coimbatore, pp 1–4. https://doi.org/10.1109/iccci.2014.6921838
17. Kaur A, Singh G, Pandey B, Fazili F (2015) Capacitance scaling based Gurumukhi Unicode reader design for natural language processing. In: 2nd International conference on computing for sustainable global development (INDIACom). New Delhi, pp 1479–1483
18. Kaur T, Singh S, Pandey B (2015) Capacitance scaling based energy efficient internet of things (IoTs) enable CAM design on FPGA. Int J Eng Res Technol (IJERT) ICNTE. 3(01):1–4
19. Singh S, Agarwal M, Agrawal N, Kumar A, Pandey B (2015) Simulation and verification of voltage and capacitance scalable 32-bit Wi-Fi Ah channel enable ALU design on 40 nm FPGA. In: International conference on computational intelligence and communication networks (CICN). Jabalpur, pp 1363–1366. https://doi.org/10.1109/cicn.2015.264
20. Pandey B, Pandey N, Kaur A, Akbar husain DM, Das B, Tomar GS (2019) Scaling of output load in energy efficient FIR filter for green communication on ultra-scale FPGA. Wireless Pers Commun 106:1813–1826. https://doi.org/10.1007/s11277-018-5717-2
21. Saxena A, Gaidhani S, Pant A, Patel C (2016) Capacitance scaling based low power comparator design on 28 nm FPGA. Int J Comput Trends Technol (IJCTT) 42(2):72–76
22. Kaur A, Fazili F, Singh S, Sharma V, Singh A, Hashim Minver M (2015) Capacitance scaling based energy efficient and tera hertz design of malayalam unicode reader on FPGA. Int J u- and e- Serv Sci Technol 8(8):151–158. http://dx.doi.org/10.14257/ijunesst.2015.8.8.15
23. Versari R, Ricco B (1998) Scaling of maximum capacitance of MOSFET with ultra-thin oxide. Electron Lett 34(22):2175–2176. https://doi.org/10.1049/el:19981471
24. Das T, Pandey B, Rahman MA, Kumar T, Siddiquee T (2013) Capacitance and frequency scaling based energy efficient image inverter design on FPGA. In: International conference on communication and computer vision (ICCCV). Coimbatore, pp 1–5. https://doi.org/10.1109/icccv.2013.6906736
25. Madhok S, Verma G, Bhardwaj A, Verma H, Singhm I, Shekhar S (2015) Capacitance scaling with different IO standard based energy efficient bio-medical wrist watch design on 28 nm FGPA. Int J Bio-Sci Bio-Technol 7(4):145–158
26. Bansal M, Bansal N, Saini R, Pandey B, Kalra L, Akabar Hussain DM (2014) SSTL I/O standard based environment friendly energy efficient ROM design on FPGA. In: 3rd International symposium on environmental friendly energies and applications (EFEA). St. Ouen, pp 1–6. https://doi.org/10.1109/efea.2014.7059947

Image Correction and Identification of Ishihara Test Images for Color Blind Individual

Himani Bansal, Lalit Bhagat, Satyam Mittal, and Ayush Tiwari

Abstract Color blindness is when someone is unable to see color in a normal way or make out the differences in certain colors. The color-sensitive cells in our eyes react differently to various wavelengths of light giving our brain the information needed to create the 'perceived' color in our vision. There is no way to rectify the cells. In other words, color blind people cannot be treated. This paper uses image processing techniques to make Ishihara Test (test for color deficiency) images visible to partially and totally color blind people. Authors have targeted the following types of dichromatic color blindness—deuteranopia, protanopia, tritanopia and also monochromatic color blindness. All of this is achieved by applying various state-of-the-art image processing and mathematical operations in the image pixels to allow the user to 'perceive' the color as it was intended. To give a kind of proof of concept of the image processing for complete color blindness, a machine learning model has been implemented and trained on the widely known MNIST dataset.

Keywords Vision defects · Dichromacy · Convolutional neural networks · Image color analysis · Monochromacy

1 Introduction

Color blindness occurs to some extent in approximately 8% men and 0.5% women worldwide. People suffering from color blindness have to face so many challenges in their everyday tasks. Most commonly, color blindness occurs between red and green colors. It is often witnessed when someone sees the 'dot' on a packaging denoting whether the food is vegetarian or non-vegetarian. Another example is when someone cannot tell if the light which tells whether a washroom is occupied or not is red or green. The unfortunate part is that it is a genetic condition which occurs during the development of one or more kind of color-sensitive cells known as cones in the retina of our eye. To our knowledge, there is no medicine or cure for it. Modern

H. Bansal (✉) · L. Bhagat · S. Mittal · A. Tiwari
Department of CSE & IT, Jaypee Institute of Information Technology, Noida, India

P. K. Singh et al. (eds.), *Proceedings of Second International Conference on Computing, Communications, and Cyber-Security*, Lecture Notes in Networks and Systems 203,
https://doi.org/10.1007/978-981-16-0733-2_70

computers with their processing power have made it possible to implement various state-of-the-art image conversion techniques for the benefit of color blind people. The primary objective of this research is to demonstrate how daily life images can be color corrected. This paper makes use of the famous Ishihara Test (standard test for detecting color deficiency) images in order to perform the right color calibrations for three types of color blindness: deuteranopia, protanopia and tritanopia as well as for total color blindness, i.e., monochromacy. For deuteranopia, protanopia and tritanopia, a color conversion algorithm, according to the type of color blindness, is applied, and for monochromacy, a widely adopted image processing library called OpenCV [1] is used to perform various mathematical operations on the pixel values like contrast adjustment, median and Gaussian blurs, Skeletonization, dilation, etc. The information is directly extracted so that the person can read the Ishihara Test images. The processed image for monochromacy is tested on a computer vision, convolutional neural network model which has been trained on the MNIST dataset.

The work below is categorized in sections which are as follows: Sect. 2 presents the related work. Background is explained in Sect. 3. Section 4 presents the proposed methodology. Section 5 explains the experimental setup and analysis. Section 6 talks about the future scope and concludes the paper.

2 Related Work

Some research has been done on simulation of images for color blind people. The light that can discriminate in three-dimensional space with axis as signals of long-wavelength, middle-wavelength and the short-wavelength cones, i.e., LMS is presented in [2]. The algorithm of projecting wavelengths of one side of cone to different side of cone is given in [3]. This was done by transforming RGB space into LMS space of a LCD monitor and then mapping to the region which is visible to color blind person. Then, again, it is converted back to the modified RGB space. Later, in 2007, researchers in [4] came up with a method of recognition of traffic light through a camera using some algorithms for processing images. Researchers in [5] proposed a computer vision technique on a color blindness plate. Authors in [6] deal about how CNN can be used to read handwritten digits. References [7–9] presents some basic research for color blindness and color simulation.

3 Background

National Eye Institute categorizes color blindness in broadly three main categories which are listed under Sects. 3.1, 3.2 and 3.3. Section 3.4 talks about convolutional neural networks.

3.1 Red– Green Color Blindness

This is the most common category of color blindness which makes it difficult for a person to identify the differences between the shades of green and red color. It can further be classified into two types:

(a) **Deuteranopia/Deuteranomaly**—which makes green look similar to red. This is caused due to missing/malfunctioning *M*-cones (green). 'M' stands for medium wavelength light.
(b) **Protanopia/Protanomaly**—which makes it difficult to recognize red color. This is caused due to missing/malfunctioning *L*-cones (red). 'L' stands for long-wavelength light. A very common problem in this type is that purple seems to look more like blue and greens/yellows/oranges/reds/browns look more similar to each other.

3.2 Blue–Yellow Color Blindness

Tritanopia/tritanomaly falls under this category of color blindness. This is caused due to missing/malfunctioning *S*-cones (blue). 'S' stands for small wavelength light. It is quite rare and is generally caused due to aging of the eye or some medical condition. It could also just be a birth defect.

3.3 Complete Color Blindness

Monochromacy and achromatopsia consist of a range of conditions which range from partial to complete color blindness. Achromatopsia is also often linked with other eye conditions such as glare sensitivity, photophobia or light sensitivity. It is most prominent in low-light conditions. This is caused when either none of the cones are present in the retina or only one of the three is present.

3.4 Convolutional Neural Networks (CNN)

During the analysis of the image correcting algorithms, a pre-trained convolutional neural network is used. A typical convolutional network is a set of one or more layers of matrices containing layer weights. Each cell or point in the layer receives input from a small predefined area or window of the previous layer. These layers essentially form a feature map which stores abstract information about the input image. Adding multiple layers (as done in this paper) allows the network to learn multiple features of the input image. These layers are known as convolutional layers. Since when a feature is 'learned' by a layer, it becomes less relevant, we generally include another layer which does local subsampling as well as averaging of the aforementioned cells. This entire network is trained with the typical gradient descent and backpropagation algorithms.

4 Methodology

The algorithm presented in the research paper involves a RGB to LMS transformation. It begins by taking an image as an input and a matrix is created which consists of RGB values of every pixel in the input image. These values obtained are transformed into the three-dimensional LMS space. The LMS values of the input image are obtained by multiplying the RGB values of the matrix with a matrix of constant values using a chromatic adaptation matrix explained in [2]. The LMS values can be interpreted as those values which are experienced at the level of the retina. After this, a conversion is made which takes the missing cone and removes all the color information that was required by the missing cone, effectively removing the need for it. Hence, we obtain the changed LMS values L"M"S". At last, multiplication is performed on the obtained L"M"S" values with inverse matrix of those constants, and hence, the R"G"B" values are obtained which in turn represent values of that image which is recognized by the color blind person. These R"G"B" are perceived by a color blind person. The error matrix that is to be defined is essentially the loss that has occurred during the aforementioned conversion process of RBG to LMS which is calculated by finding the difference between the current R"G"B" image and the original RGB image. Now, for RGB to LMS transformation, we have used the algorithm given by the researchers in [2]. This linear transformation is done by multiplying the matrix suggested in [2]. Once data is transformed to LMS space, data related to one of the cone types, (selected according to the type of color blindness a person is suffering from) is deleted.

Figure 1 depicts the LMS color space. Let the letters—B, K, W, R, G, C, M, Y stand for blue, black, white, red, green, cyan, magenta and yellow, respectively. A normal person can see all colors. The Eq. (1) of KBWY plane is

$$\omega_1 L + \omega_2 M + \omega_3 S = 0 \tag{1}$$

After solving the equation, we get (2), (3), (4)

$$\omega_1 = M_W S_B - M_B S_W \tag{2}$$

$$\omega_2 = S_W L_B - S_B L_W \tag{3}$$

$$\omega_3 = L_W M_B - L_B M_W \tag{4}$$

Fig. 1 LMS color space

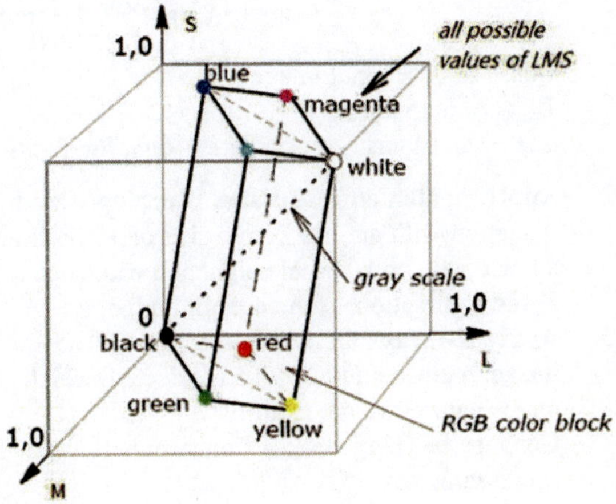

where ω_1, ω_2, ω_3 are coefficients for a scaler equation of a plane.

Using the above equation, the value of L_P and M_D are found for protanope (5) and deuteranope (6), respectively

$$L_P = \frac{-(\omega_2 M + \omega_3 S)}{\omega_1} \tag{5}$$

$$M_D = \frac{-(\omega_1 L + \omega_3 S)}{\omega_2} \tag{6}$$

Now, the calculation of error matrices is done which consist of values that were not visible to color blind person. The information is obtained by finding out the difference between R"G"B" values and RGB values, i.e., this part of image is lost. For example, if the M-cone (medium wavelength) is missing, i.e., deuteranopic, that person would not be able to correctly perceive the green part of the spectrum, making it hard for them to identify green sections from other sections. Due to this, the error picture that will be calculated would primarily consist of shades of green. The above transformation will then map this color information which was previously hard to perceive by the person to the more easily perceived blue side of the spectrum. On adding this newly found information to the original image, we would obtain the corrected image which the color blind person can easily see and be able to distinguish between colors of the image.

Now, to tackle monochromacy, a completely different set of algorithms is applied. Since the person is missing one or all the cones in the retina, simply mapping the color information to other parts of the spectrum is not possible. Thus, various techniques have been applied to make the original image clearer and recognizable. The steps are

1. **Increase the Contrast**—Using the weighted sum of arrays to increase the contrast of the image (also known as alpha blending), according to the Eq. (7):

$$I_{\text{output}} = \left(I_{\text{input}} \times \alpha + I_{\text{input}} \times \beta + \gamma\right) \tag{7}$$

where I_{input} = input image
I_{output} = output image
α, β = constants, γ = scalar constant for addition.

2. **Apply median and Gaussian Blurring**—Median blurring is applied for noise reduction with an odd kernel size of 13. Following this, Gaussian blurring is applied with both kernel height and width equal to 5. The standard deviation in X and Y directions is the default border.
3. **Apply K-means Color Clustering**—Identifies the top dominant colors of an image. A cluster number of 5 is selected (which gives the most consistent results on the dataset we are using).
4. **Convert to Gray scale**—Converting value of every pixel of the image to a single sample.
5. **Apply Thresholding**—To help in image segmentation. It fixes the problem where due to errors in clustering, the whites and blacks get inverted (numbers in black and background in white). It compares the percentage of white and blacks, and if white is more than black (which is undesirable), they both get inverted. An arbitrary range of 10–30% is selected as the optimal percentage of white.
6. **More Blurring and Thresholding**—To get clearer images by reducing a lot of noise and irregularities that may have made it into the image at this stage.
7. **Morphology Open, Close, Erosion**—Morphology open is just another name for erosion after dilation. It removes the irregularities from the borders of the foreground of an image. A kernel size of 6 × 6 is used for this operation. Morphology close removes small holes in the foreground which may have occurred due to some outliers or simply noise. Again, a kernel of 6 × 6 is used for this. Lastly, morphology erosion removes jagged pixels or edges on forefront object boundaries once again to output an even clearer image. Here, a kernel size of 7 × 7 is used.
8. **Skeletonizing**—It is a process for reducing foreground regions in a binary image.
9. **Dilation**—This operation essentially adds more pixels to the forefront object boundaries which cause it to 'grow' making it more readable by the machine learning model.

5 Experimental Results and Analysis

The algorithm was verified by two of the authors of this paper, who are color blind, and they were able to read the Ishihara Test images. Figure 2 presents the corrected images for the three types of dichromatic color blindness—deuteranopia, protanopia, tritanopia.

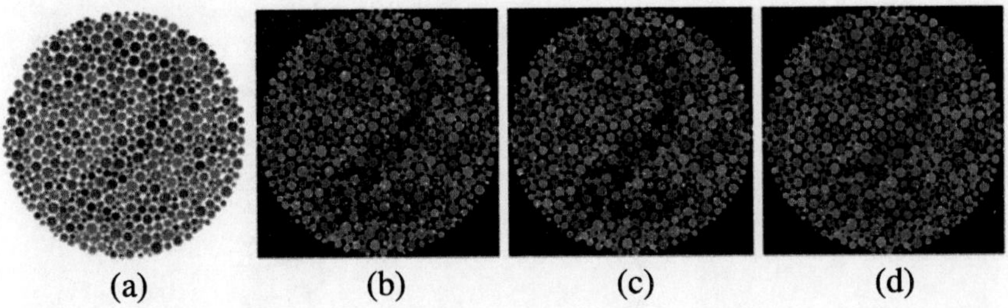

Fig. 2 **a** Input image, **b** corrected image for deuteranopia, **c** corrected image for protanopia, **d** corrected image for tritanopia

In order to demonstrate the readability of the above processed image after all the nine steps, as shown in Fig. 3, the output form around 50 Ishihara Test images has been fed into a pre-trained machine learning model trained on MNIST images. As it can be seen in Fig. 3j, that image processed for monochromatic color blindness is visible. We have also tested our processed images for monochromacy on a model trained with MNIST dataset. Since a modern MNIST model has an accuracy of around 98%, it provides a great way to test our corrected images as they are quite similar to the black and white images found in the MNIST dataset. We achieved an accuracy of 80%, which is the F1 score and weighted sum of diagonal elements in Fig. 4.

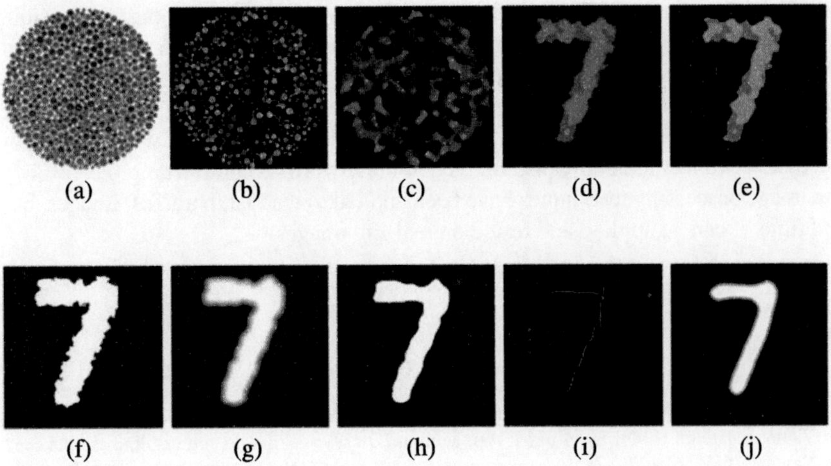

Fig. 3 **a** Input image, **b** increase the contrast, **c** apply median and Gaussian blurring, **d** apply K-means color clustering, **e** convert to gray scale, **f** apply thresholding, **g** more blurring and thresholding, **h** morphology open, close and erosion, **i** skeletonizing, **j** dilation

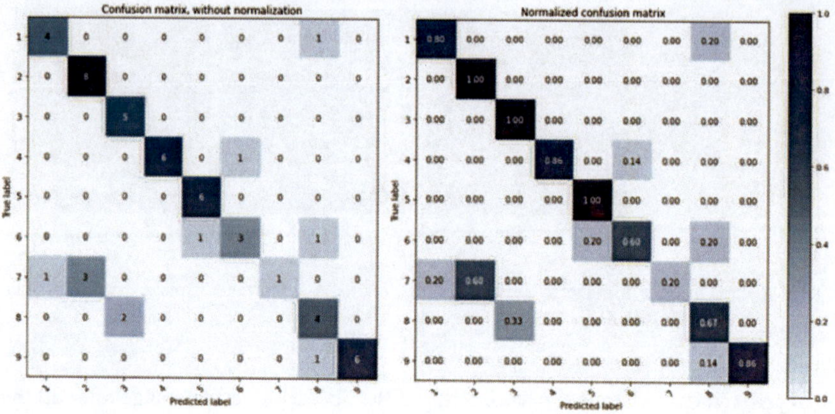

Fig. 4 Confusion matrix

6 Conclusion and Future Scope

The color palette of an input image has been adjusted, and visible color maps have been created for the three types of dichromatic color blindness—deuteranopia, protanopia, tritanopia. This is done by converting the color space into LMS and replacing the colors that are not visible to a color blind person by the color that is visible. For monochromatic color blindness, the information is extracted from Ishihara Test images using image processing techniques. All processed images are visible to color blind people. Also, the CNN model trained on MNIST data is able to read the Ishihara test images that had been processed for monochromatic color blindness using image processing techniques. Around 50 images for testing were used, and the model is able to classify the images with an accuracy of 80% which does not necessarily reflect the real world but is definitely worth accomplishing. As of now, the image processing techniques have been applied to read Ishihara Test images, but in future, it can be applied and tested on real-life images.

References

1. Bradski G (2000) The openCV library. Dr. Dobb's J Softw Tools
2. Viénot F, Brettel H, Ott L, M'barek AB, Mollon JD (1995) What do colour-blind people see? Nature 376(6536):127–128
3. Brettel H, Viénot F, Mollon JD (1997) Computerized simulation of color appearance for dichromats. JOSA A 14(10):2647–2655
4. Kim YK, Kim KW, Yang X (2007, August) Real time traffic light recognition system for color vision deficiencies. In: International conference on mechatronics and automation. IEEE, pp 76–81
5. Chauhan R, Ghanshala KK, Joshi RC (2018, December) Convolutional neural network (CNN) for image detection and recognition. In: First international conference on secure cyber computing

and communication (ICSCCC). IEEE, pp 278–282
6. Viénot F, Brettel H (2000, December) Color display for dichromats. In: Color imaging: device-independent color, color hardcopy, and graphic arts VI, vol 4300, pp 199–207. International society for optics and photonics
7. Capilla P, Dıez-Ajenjo MA, Luque MJ, Malo J (2004) Corresponding-pair procedure: a new approach to simulation of dichromatic color perception. JOSA A 21(2):176–186
8. Capilla P, Luque MJ, Diez-Ajenjo MA (2004) Looking for the dichromatic version of a colour vision model. J Opt A: Pure Appl Opt 6(9):906
9. Fidaner O, Ozguven N (2006) Analysis of color blindness. Stanford University, SCIEN Lab, 2006. Web. 4 June 2012. http://scien.stanford.edu/pages/labsite/2005/psych221/projects/05/ofidaner/colorblindness_project.htm

Gradient Feature-Based Classification of Patterned Images

Divya Srivastava, B. Rajitha, and Suneeta Agarwal

Abstract Image classification is the task of assigning a class to an image. It has a wide range of applications: image and video retrieval, object tracking, object recognition, Web content analysis, number plate recognition, OCR in banking systems, etc. Color, texture, gradient, shape, keypoint descriptors, etc. are the various features being used for the image classification. A patterned image is an image in which selected pattern is repeated, for example, horizontal stripes, vertical stripes, polka dots, geometric shapes, etc. Gradient feature plays a vital role in distinguishing the different patterns. Therefore, in the proposed approach, gradient features are used for the classification of patterned images like cloth patterns (vertical stripes, horizontal stripes, polka dots, etc.), English characters (capital and small alphabets) and numerals (0–9) and geometric shapes (square, triangle, etc.). The different patterns recognized in the present paper show the versatility of the approach. It can be applied to many of the real-time applications like number plate recognition, cloth pattern recognition and retrieval. The proposed approach achieves the accuracies of 95.4, 93.5, 91.4 and 92% on standard datasets describable texture dataset (vertical stripes, polka dots), EnglishImg dataset (small and capital English alphabets), numerals dataset (0–9) and geometric shapes (triangle, square) dataset, respectively.

Keywords Gradient · Image classification · Patterned images · Support vector machine

D. Srivastava (✉)
Assistant Professor, Bennett University, Greater Noida, UP, India

B. Rajitha
Assistant Professor, Motilal Nehru National Institute of Technology, Allahabad, Prayagraj, India
e-mail: rajitha@mnnit.ac.in

S. Agarwal
Professor, Motilal Nehru National Institute of Technology, Allahabad, Prayagraj, India
e-mail: suneeta@mnnit.ac.in

P. K. Singh et al. (eds.), *Proceedings of Second International Conference on Computing, Communications, and Cyber-Security*, Lecture Notes in Networks and Systems 203,
https://doi.org/10.1007/978-981-16-0733-2_71

1 Introduction

Image classification is the task of assigning a class to an image based on its features. Although, manually, classifying an image is an easy task as just by seeing the image one can classify it as rose or table or chair, etc., it becomes a challenging task when to be done automatically. A machine is required to be trained thoroughly to identify the relationship between the image and the class it belongs to. Selection of features, their representation, selection of suitable classifier and its training are important tasks for image classification.

Researchers have used several features, shape [11], texture [12, 14] and color [17] for image classification. Support vector machine (SVM), convolutional neural networks (CNN), K-nearest neural networks(KNN), etc. are the popular classifiers in the literature for image classification. Comparison of various classifiers is made in [1, 7] concluding that SVM gives better results for image classification as compared to other classifiers.

Texture features are preferred over color and shape for classification and retrieval of images. This is so because the color histogram does not give any information about the spatial layout of the image. Two completely different images may have the same color histogram [8]. Shape feature is generally used for the objects having a fixed shape like logos, trademarks and traffic signs. Therefore, due to the aforementioned limitations of color and shape features, researchers prefer texture features over color and shape. Haralick et al. [4] have used texture features for classification of satellite images and aerial photographs. Porebski et al. [10] classified the color-texture image set combining color-texture histograms. A pipelined approach using color and texture [13] has been used for classification and retrieval of color images. Classification and retrieval of stripped pattern clothes have been discussed in [15]. Pawening et al. [9] used the combination of GLCM and local binary pattern (LBP) for the classification of textile images. Susistra et al. [16] used wavelet sub-bands for the recognition of various patterns in cloths. Ishak et al. [5] have used Gabor wavelet and gradient feature for the classification of weed images. Therefore, it is observed that the selection of features for classification highly depends upon the types of images to be classified.

A patterned image is an image in which selected pattern is repeated, for example, horizontal stripes, vertical stripes, polka dots, geometric shapes, etc. Gradient feature plays a vital role in distinguishing the different patterns. Therefore, in the proposed approach, gradient features are used for the classification of patterned images like cloth patterns (vertical stripes, horizontal stripes, polka dots, etc.), English characters (capital and small alphabets) and numerals (0–9) and geometric shapes (square, triangle, etc.). The different patterns recognized in the present paper show the versatility of the approach. The proposed approach achieves the accuracies of 95.4, 93.5, 91.4 and 92% on standard datasets describable texture dataset (vertical stripes, polka dots), EnglishImg dataset (small and capital English alphabets), numerals dataset (0–9) and geometric shapes (triangle, square) dataset, respectively. Figure 1 shows the sample images from the datasets.

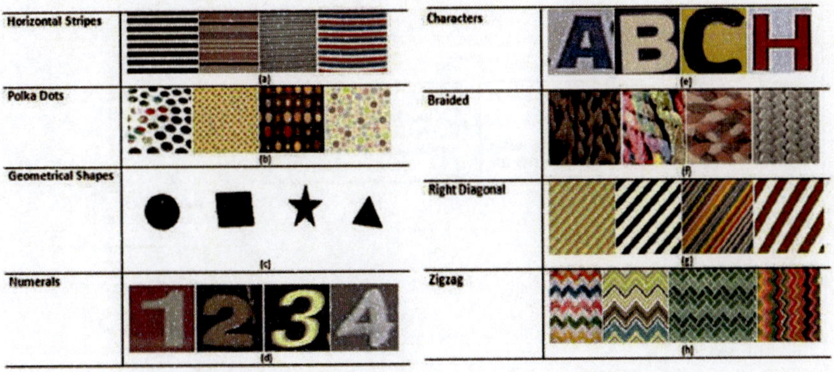

Fig. 1 Sample images from dataset. **a** horizontal stripes, **b** polka dots, **c** geometric shapes, **d** numerals, **e** characters, **f** braided, **g** right diagonal and **h** Zigzag

The rest of the paper is organized as: Sect. 2 discusses the proposed approach. Section 3 describes the datasets used for the experiments. Result analysis and effectiveness of the proposed approach are discussed in Sect. 4. Section 5 is the concluding section.

2 Proposed Approach

In the proposed approach, gradient feature is used to classify the patterned images as shown in flowchart in Fig. 2. Algorithm 1 train SVM is for the processing of various blocks. The proposed approach is discussed in the following subsections:

2.1 Image Preprocessing

An image of the dataset (having N images) is converted to grayscale (if not, already). It is then resized to 198 * 300. The preprocessed image (i) is then divided into m blocks (here, $m = 6600$) $B^i[j]$, of size 3×3, where $j = 1, 2, \ldots, 6600$. Thereafter, standard deviation ($Std(B^i[j])$) of gray values of each block is computed. Gradient features are extracted for all the blocks having $Std(B^i[j]) >$ threshold. A threshold of 10 has been set experimentally.

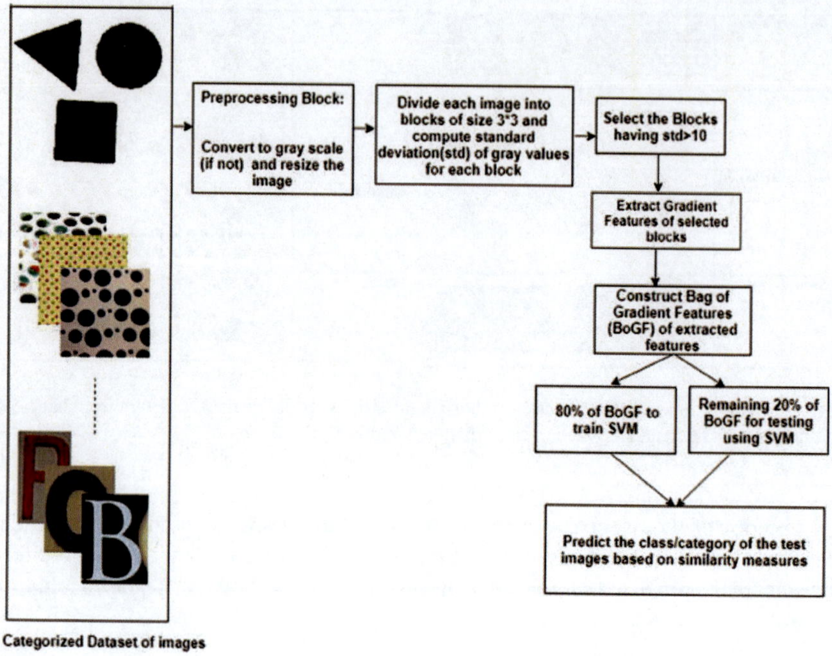

Fig. 2 Overview of proposed approach

2.2 *Gradient Feature Extraction*

For each block of *i*th image having $\text{Std}(B^i[j]) >$ threshold, eight gradient values G[1, 2, …, 8] for angles 0°, 45°, 90°, 135°, 180°, 225°, 270° and 315° are computed. These gradients are labeled as 1, 2, 3, …, 8, respectively. Highest value among these is taken as the main gradient (MG) of that block. Therefore, for each block, 9 values are computed (its eight gradient values followed by one main gradient). Thus, GFV^i (gradient feature vector of *i*th image) is of size $r \times 9$, where r is the number of blocks having the standard deviation greater than 10. GFV^i is represented as:

$$GFV^i \leftarrow [(G[1, 2, \ldots, 8], MG)_1; (G[1, 2, \ldots, 8], MG)_2; \ldots; (G[1, 2, \ldots, 8]MG)_r]$$

Bag of gradient features of *i*th image ($BoGF^i$) is constructed by splitting the GFV^i into eight vectors with respect to eight gradients values: $GFV^i{}_1, GFV^i{}_2, \ldots, GFV^i{}_8$. Let, $\mu^i{}_1, \mu^i{}_2, \ldots, \mu^i{}_8$ be the mean values of the respective vectors. Let the count of each vector be: $C^i{}_1, C^i{}_2, \ldots, C^i{}_8$. $BoGF^i$ is constructed by combining means and counts of ith image as shown below:

$$BoGF^i \leftarrow [(\mu^i{}_1, C^i{}_1), (\mu^i{}_2, C^i{}_2), (\mu^i{}_3, C^i{}_3), (\mu^i{}_4, C^i{}_4), (\mu^i{}_5, C^i{}_5), (\mu^i{}_6, C^i{}_6), (\mu^i{}_7, C^i{}_7), (\mu^i{}_8, C^i{}_8)]$$

The same procedure is repeated for all the images i = 1,2,3, …, N of the dataset to construct dataset feature vector (DFV) as:

$$\text{DFV} \leftarrow [\,\text{BoGF}^1, \text{BoGF}^2, \text{BoGF}^3, \ldots, \text{BoGF}^N]$$

SVM is trained and tested using 80% and 20% (selected randomly) of DFV, respectively.

3 Datasets for the Experiments

In the proposed approach, various patterns of cloths(horizontal stripes, vertical stripes, etc.), English alphabets (capital and small) and numerals (0–9) and geometric shapes (triangle, square, etc.) are taken from describable texture dataset (DTD) [2], EnglishImg [3], numerals [3] and geometric shapes [6], respectively. Some sample images of these are shown in Fig. 1. DTD dataset consists of ten categorized cloth patterns having 50 images in each. The English alphabet dataset [3] consists of 52 categorized data having 50 to 101 images of each category. Numbers dataset consists of ten categorized data having 50–101 images of each category. Geometrical_Shapes dataset consists of five different types of geometric shapes(circle, star, square, triangle, rectangle) with 101 different orientations of each. From each dataset, randomly 80 and 20% was taken to train and test SVM.

4 Result Analysis

Table 1 is the confusion matrix constructed for eight categories of cloth patterns. The diagonal elements show the number of images correctly classified in the respective category. For instance, in category 1 (braided), out of 11 test images, 10 were correctly classified to Class 1 (i.e., Braided), and 1 was miss classified to Class 3 (left diagonal). Table 2 shows the tabulation of the performance metrics: precision, recall, specificity, F-measure and class accuracy calculated using formulae 1, 2, 3, 4 and 5, respectively.

$$\text{Precision} = \frac{\text{Tp}}{(\text{Tp} + \text{Fp})} \tag{1}$$

$$\text{Recall/Sensitivity} = \frac{\text{Tp}}{(\text{Tp} + \text{Fn})} \tag{2}$$

$$\text{Specificity} = \frac{\text{Tn}}{(\text{Tn} + \text{Fp})} \tag{3}$$

Table 1 Confusion matrix for cloth patterns

Cloth Patterns	Braided	Horizontal Stripes	Left Diagonal	Paisley	Polka Dots	Right Diagonal	Vertical Stripes	Zigzag
Braided	10	0	1	0	0	0	0	0
Horizontal-Stripes	0	10	0	0	1	0	0	0
Left-Diagonal	1	0	23	0	0	2	0	0
Paisley	0	0	0	10	0	0	0	0
Polka Dots	0	0	0	0	9	0	0	0
Right-Diagonal	0	0	0	0	0	22	0	1
Vertical-Stripes	0	0	0	0	0	0	10	0
Zigzag	0	0	0	0	0	0	0	9

Table 2 Performance metrics for cloth patterns

Categories	Precision	Recall	Specificity	F-measure	Class accuracy (%)
Braided	0.58	1	0.12	0.73	90.9
Horizontal stripes	0.58	1	0.12	0.73	90.9
Left diagonal	0.79	0.95	0.25	0.86	92
Paisley	0.58	1	0	0.73	100
Polka dots	0.60	0.90	0	0.72	100
Right diagonal	0.78	0.91	0.14	0.84	95.7
Vertical stripes	0.58	1	0	0.73	100
Zigzag	0.56	0.90	0	0.69	100
Average	0.63	0.95	0.07	0.75	95.4

$$F - \text{measure} = \frac{2 * (\text{Precision} * \text{Recall})}{(\text{Precision} + \text{Recall})} \tag{4}$$

$$\text{ClassAccuracy} = \frac{\text{Tp}}{\text{Tp} + \text{Tn}} \tag{5}$$

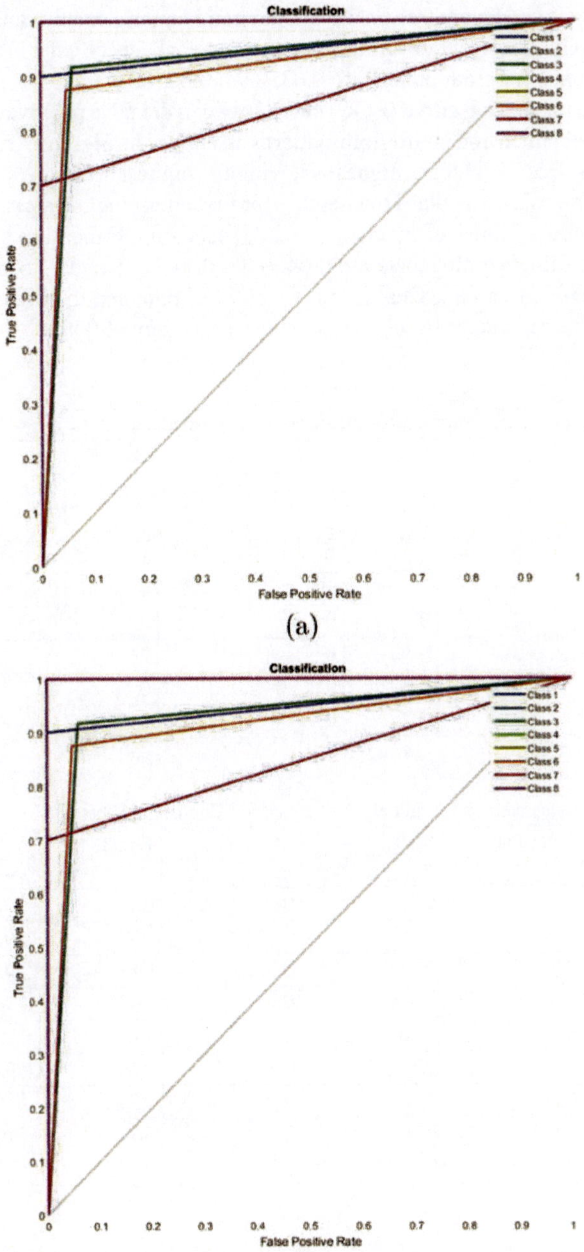

(a)

(b)

Fig. 3 **a** ROC plot for eight categories of clothes pattern: Class 1: braided, Class 2: horizontal stripes, Class 3: left diagonal, Class 4: paisley, Class 5: polka dots, Class 6: right diagonal, Class 7: vertical stripes and Class 8: zigzag, **b** ROC plot for first eight categories of English alphabets: Class 1: A, Class 2: B, Class 3: C, Class 4: D, Class 5: E, Class 6: F, Class 7: G and Class 8: H where Tp, Tn, Fp and Fn are true positive, true negative, false positive and false negative, respectively. The average values of precision, recall, specificity, F-measure and accuracy obtained for cloth patters are 0.63, 0.95, 0.07, 0.75 and 95.4, respectively.

The receiver operating curve (ROC) curve between the false positive rate and true positive rate obtained for eight cloth patterns is shown in Fig. 3a. Area under the curves classes 4, 5, 7 and 8 are highest showing the highest accuracy.

Table 3 is the confusion matrix generated for first eight English alphabets (capital A–H). The average values of precision, recall, specificity, F-measure and accuracy found for eight English alphabets are 0.64, 0.93, 0.11, 0.75 and 93.5, respectively, and are tabulated in Table 4. Characters ‘G’ and ‘H’ obtained the highest accuracy of 100%, and the character ‘A’ obtained the least accuracy of 90%.

Table 3 Confusion matrix for first eight categories of English alphabets

Alphabets	A	B	C	D	E	F	G	H
A	9	0	1	0	0	0	0	0
B	0	10	0	0	1	0	0	0
C	1	0	22	0	0	1	0	0
D	0	0	0	10	1	0	0	0
E	0	0	0	0	8	0	0	0
F	0	0	0	0	0	23	0	1
G	0	0	0	0	0	0	10	0
H	0	0	0	0	0	0	0	9

Table 4 Performance metrics for first eight categories of English alphabets

Categories	Precision	Recall	Specificity	F-measure	Class accuracy (%)
A	0.60	0.90	0.12	0.72	90
B	0.58	1	0.12	0.73	91.7
C	0.81	0.91	0.28	0.85	91.7
D	0.58	1	0.12	0.73	90.9
E	0.61	0.80	0	0.69	92
F	0.79	0.95	0.25	0.69	92
G	0.58	1	0	0.73	100
H	0.60	0.90	0	0.72	100
Average	0.64	0.93	0.11	0.75	93.5

Table 5 Comparison with other state-of-the-art approaches

	Pawening et al. (2015) [9]	Susithra et al. (2016) [16]	Proposed approach
Accuracy (%)	76.15	88.68	95.4

Table 5 shows the comparison of results of proposed approach with the state-of-the-art approaches [15, 16]. Pawening et al. [9] obtained the accuracy of 76.15% for textile images using the combination of GLCM and LBP, while Susithra et al. [16] obtained the accuracy of 88.68% for cloth patterns using wavelet subbands. The present approach obtained the accuracy of 95.4% for cloth patterns using gradient features. The proposed approach has also been tested on various other patterns like English alphabets and numerals and geometric shapes. For these patterns, accuracies obtained were 93.4%, 91.4% and 92%, respectively. Thus, the proposed approach outperforms the state-of-the-art approaches [9, 16].

5 Conclusion

In the proposed approach, gradient feature has been used to recognize the various patterned images: cloth patterns, English alphabets and numerals and geometric shapes having high directional properties. The proposed approach has a wide range of applications in image and video retrieval, object recognition and Web content analysis. It is useful for e-commerce sites for filtering the cloth patterns, recognizing number plates, recognizing optical character, etc. The average accuracies for cloth patterns, English alphabets and numerals and geometric shapes are 95.4, 93.5, 91.4 and 92, respectively. It is concluded that the proposed approach outperforms the existing state-of-the-art approaches by obtaining the accuracy of 95.4%.

References

1. Akata Z, Perronnin F, Harchaoui Z, Schmid C (2013) Good practice in large-scale learning for image classification. IEEE Trans Pattern Anal Mach Intell 36(3):507–520
2. Cimpoi M, Maji S, Kokkinos I, Mohamed S, Vedaldi A (2014) Describing textures in the wild. In: Proceedings of the IEEE conference on computer vision and pattern recognition, pp 3606–3613
3. De Campos TE, Babu BR, Varma M et al (2009) Character recognition in natural images. VISAPP (2) 7
4. Haralick RM, Shanmugam K, Dinstein IH (1973) Textural features for image classification. IEEE Trans Syst Man Cybern 2(6):610–621
5. Ishak AJ, Hussain A, Mustafa MM (2009) Weed image classification using gabor wavelet and gradient field distribution. Comput Electron Agric 66(1):53–61
6. Kaggle: Shapes Dataset (2020). https://www.kaggle.com/smeschke/four-shapes. Online; accessed 19 Sept 2020
7. Kamavisdar P, Saluja S, Agrawal S (2013) A survey on image classification approaches and techniques. Int J Adv Res Comput Commun Eng 2(1):1005–1009
8. Pass G, Zabih R (1999) Comparing images using joint histograms. Multimedia Syst 7(3):234–240
9. Pawening RE, Dijaya R, Brian T, Suciati N (2015) Classification of textile image using support vector machine with textural feature. In: 2015 International conference on information & communication technology and systems (ICTS). IEEE, pp 119–122

10. Porebski A, Vandenbroucke N, Macaire L, Hamad D (2014) A new benchmark image test suite for evaluating colour texture classification schemes. Multimedia Tools Appl 70(1):543–556
11. Ramesh B, Xiang C, Lee TH (2015) Shape classification using invariant features and contextual information in the bag-of-words model. Pattern Recogn 48(3):894–906
12. Singh S, Srivastava D, Agarwal S (2017) Glcm and its application in pattern recognition. In: 2017 5th International symposium on computational and business intelligence (ISCBI). IEEE, pp 20–25
13. Srivastava D, Goel S, Agarwal S (2017) Pipelined technique for image retrieval using texture and color. In: 2017 4th International conference on power, control & embedded systems (ICPCES). IEEE, pp 1–6
14. Srivastava D, Rajitha B, Agarwal S (2017) An efficient image classification using bag-of-words based on surf and texture features. In: 2017 14th IEEE India council international conference (INDICON). IEEE, p. 1–6
15. Srivastava D, Rajitha B, Agarwal S, Singh S (2018) Pattern-based image retrieval using glcm. In: Neural computing and applications, pp 1–14
16. Susithra K, Sujaritha M (2016) Clothing pattern recognition based on local and global features. Int J Sci Eng Res 7(3):106–110
17. Zou J, Li W, Chen C, Du Q (2016) Scene classification using local and global features with collaborative representation fusion. Inf Sci 348:209–226

Corrosion Estimation of Underwater Structures Employing Bag of Features (BoF)

Anant Sinha, Sachin Kumar, Pooja Khanna, and Pragya

Abstract According to World Corrosion Organization (WCO), the estimated annual cost of damage due to corrosion across the globe is approximately US$2.5 trillion which contributes 3–4% GDP of developed countries. Minimizing losses due to corrosion and to ensure longer life of structures is thus a major concern. Paper presents a technique employing bag of features (BoF) for underwater structural corrosion recognition. BoF methods are based on an unorganized grouping of image features and it is conceptually simpler than various other alternatives. The model is trained on three labeled datasets corroded, un-corroded, and damaged obtained from underwater structures along the Gomti River in Lucknow. Dataset of around 2000 images is used to train the model. Trained prototype BoF learning model is capable of efficiently classifying pure and corroded images and achieves an accuracy of 82.38% demonstrating the feasibility of this method. The technique proposed and its deployment on handheld and autonomous devices provide an efficient and intelligent method for underwater structure corrosion recognition.

Keywords Underwater structures · Corroded · BoF · SIFT · Otsu's method · MLSTSVM

A. Sinha (✉) · S. Kumar · P. Khanna
Amity University, Lucknow Campus, Lucknow, India

S. Kumar
e-mail: skumar3@lko.amity.edu

P. Khanna
e-mail: pkhanna@lko.amity.edu

Pragya
MVPG College, Lucknow University, Lucknow, India

P. K. Singh et al. (eds.), *Proceedings of Second International Conference on Computing, Communications, and Cyber-Security*, Lecture Notes in Networks and Systems 203,
https://doi.org/10.1007/978-981-16-0733-2_72

1 Introduction

Corrosion on the surface or underwater has been a potential area of investigation and research, and corrosion can be defined as the deterioration of structures due to diversified reasons such as chemical reactions, aging, and accidental damages. To ensure longer life of structures, it is necessary to conduct maintenance exercise regularly; however, regular maintenance is not an easy task, and it involves danger and risk factor when it comes to investigating areas which are under extreme conditions like underwater environment. Power plants such as thermal power plants or hydroelectric power plants involve underwater infrastructure, its smooth functioning, productivity, and safety depend only on regular examination; hence, underwater corrosion estimation of such part of the infrastructure which is submerged in water plays an important role. Major processes for underwater investigations are performed using non-destructive testing (NDT) methodologies such as the pulse-echo and ultrasonic angle beam method. Lately, military and civil industries have started using image processing technique for regular investigation of underwater structures along with devices which are operated remotely [1, 2].

Due to aging and flows in construction, the concrete structures submerged under the water are vulnerable to destruction [3], repair and restoration of such structures are extremely important, more specifically for the submerged substructures of flyover [4, 5]. Thus, a need for the development of an integrated system for monitoring and assessment of underwater structures is essential. Sensing and assessing in an underwater environment are applied in various applications such as applications related to oil exploration, global weather forecast, monitoring of water quality, studies related to aquatic life, defense operations, etc. [6, 7]. Underwater pipelines are the backbone of the offshore oil and gas industry, to ensure smooth transportation of material, and to avoid leakage due to corrosion it is necessary that periodic monitoring is done [8]. Among all significant incidents in offshore structures, oil, and gas pipelines, it is seen that corrosion is the major cause as depicted in Fig. 1 [9].

Estimated loss is of around $5000 billion USD to the global economy every year because of corrosion. In the report issued by National Association of Corrosion

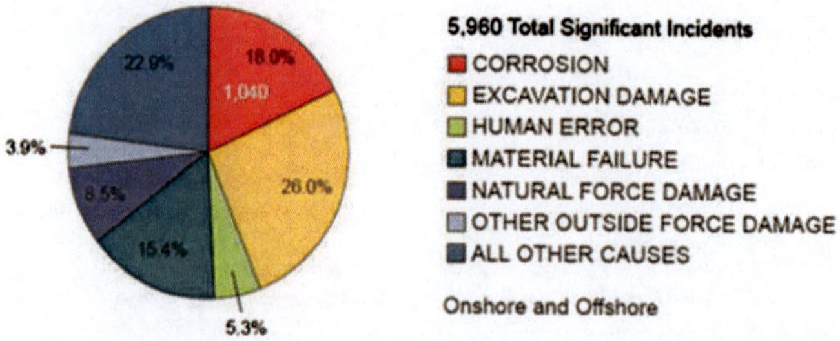

Fig. 1 Significant incidents in underwater environment [9]

Table 1 Estimated loss owing to corrosion [10]

Economic regions	Agriculture CoC (USD billion)	Industry CoC (USD billion)	Services CoC (USD billion)	Total CoC (USD billion)	Total GDP (USD billion)	CoC % GDP
USA	2.0	303.2	146.0	451.3	16,720	2.7
India	17.7	20.3	32.3	70.3	1670	4.2
European region	3.5	401	297	701.5	18,331	3.8
Arab World	13.3	34.2	92.6	140.1	2789	5.0
China	56.2	192.5	146.2	394.9	9330	4.2
Russia	5.4	37.2	41.9	84.5	2113	4.0
Japan	0.6	45.9	5.1	51.6	5002	1.0
Rest of the world	52.4	382.5	117.6	552.5	16,057	2.5
Global	152 7	1446.7	906.0	2505.4	74,314	3.4

Engineering, the cost of corrosion in different sectors of nine countries is estimated [10], and data is depicted in Table 1.

The paper is organized as follows; Sect. 2 presents the motivation for taking up the work, Sect. 3 provides the methodology adopted for identification and estimation of level of corrosion, Sect. 4 presents experimental setup utilized and processes employed, Sect. 5 explores about open issues related with the work, and finally paper concludes with optimum choices for design in Sect. 6.

2 Motivation

There are several processing steps involved while developing an image processing-based corrosion detection algorithm due to hostile underwater environmental conditions encountered during the image acquisition process. A raw or unprocessed image taken in an underwater environment is not of good quality and hence gives rise to serious problems during analysis. There are various factors responsible for the degradation of image quality in the underwater environment such as low contrast, poor visibility due the absence of natural light or non-uniformly distributed light, pepper noise, etc.; sometimes, the presence of dust particles in water also hamper the clarity of an image, actually underwater dust particle leads to backscattering of light [11, 12]. Furthermore, when the depth of water is increased, several colors are absorbed by the surrounding medium according to wavelengths. Most of the time, blue/green color is dominating in the underwater environment and the phenomenon is known as color cast as the depth increases the colors such as red and green disappears. Another major challenge regarding image processing is associated with water density in the sea which is approximately 800 times denser mediums than air. So, when light rays from the air enter the water, it is partially reflected and partly enters the water at the same time. Hence, for any visual testing (VT)-based corrosion inspection technique, there is a need for pre-processing. The number of research articles for various enhancement and de-blurring algorithms has been proposed till date. Yang Wang and Ching have proposed a convolutional neural network (CNN)-based framework called UIE-Net for underwater image enhancement. UIE-Net comprises of two sub-networks. CC-Net and HR-Net. CC-Net is utilized for the correction of color distortion of underwater images. HR-Net is for the light attenuation transmission map. It is utilized for the contrast enhancement of underwater images [13]. Tatsuya Baba and Keishu Nakamura have developed discrete cosine eigenbasis transformation (DCET)-based underwater image enhancement technique. Underwater images are achromatized by DCET, by this input images are transformed into images having discrete cosine eigenbasis (DCE) [14]. Though the number of algorithms for identification and estimation of corrosion are in place supported by diversified technological innovations, yet maintenance and repair is a costly issue. Vibration-based approaches have also been utilized and have become increasingly common for structural health monitoring scheme and they are generally, practiced broadly in bridges, railways, and tunnels. Marine structures are complex when compared to the listed ones; therefore, the problem is more challenging and it will be a new attempt and search front in structural health monitoring methods [15].

The limited resources for automated monitoring of underwater structure created a need to design integrated systems to estimate and synthesis corrosion/damage to structures. The main focus should be to enhance the safety and integrity of marine structures.

3 Methodology

The proposed prototype consists of HD self-stabilizing underwater camera with anti-noise feature to collect data with minimum noise, due to poor lighting conditions and visible suspended impurities noise is still inherent in sample images, the data is, therefore, will be subjected to cleaning process before feature extraction and bag of feature analysis is executed, and the information obtained is utilized to develop knowledge-based decision-making system for the parameters that are related to safety and integrity of the structural system. Figure 2 depicts the functional block diagram of the prototype proposed for underwater corrosion/damage estimation [16].

a. The first phase starts with data collection of the structure details under investigation, the data acquired comprises of:

- The material used for construction
- Age of the structure
- Location/depth of underwater
- Identification of the area to be investigated.

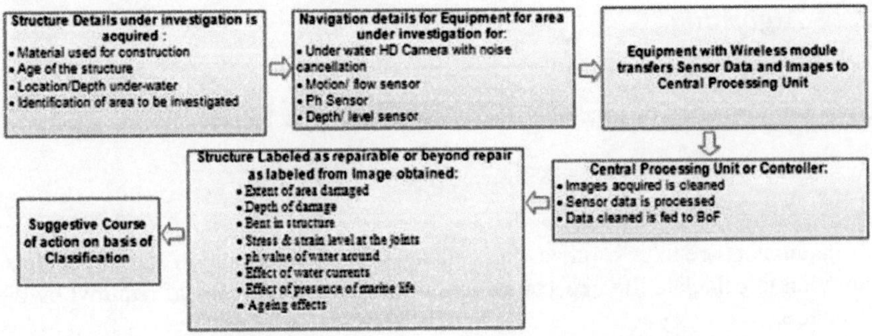

Fig. 2 Functional flow diagram for underwater corrosion/damage estimation

b. The second phase requires navigation details for equipment for the area under investigation for positioning equipment mounted with underwater HD camera equipped with noise cancelation feature, motion/flow sensor for stabilizing the equipment, level sensor to estimate the pressure.
c. The third phase deals with the wireless transfer of the data to the central unit.
d. The fourth phase deals with cleaning of image data, synthesis of image data, and information deduction from an image, simultaneously sensor data is mapped to estimate factors contributing to corrosion, and the process is achieved through a BoF classifier.
e. The fifth and final phase involves labeling the structure as repairable or beyond repair as deducted from the result obtained from BoF operation, and the following parameters are addressed.

- The extent of the area damaged
- Depth of damage
- Bent in structure
- Stress and strain level at the joints
- pH value of water
- Effect of water currents
- Effect of the presence of marine life
- Aging effects.

To develop knowledge-based decision-making system and to provide quick remedial solutions in case of emergency, the system will generate alert messages in case of a serious condition, and the following tasks would be addressed by the prototype.

Areas needing supervision should be highlighted.

- Identification of critical zones needing replacement
- The task of controlling the network of sensors and empirical evaluation of scaled images for the above parameters will be accomplished by the equipment though the prototype is still a handheld device the same would be extending to a self-propelled auto-bot in future course of work. The task of logic monitoring will be performed based on the predictive model build using BoF classifier.

3.1 Algorithm

Step 1. Information of structures under investigation is obtained and analyzed:

- Portion of structure submerged underwater
- Climatic conditions.

Parameters are to be estimated to evaluate natural deterioration and forced deterioration to calculate life expectancy and immediate reinforcement required by the structure.

Step 2. Location identification to set navigation course for finalizing:
Region or area of pillars/oil pipes/steel bases to be analyzed.

- Depth to dive
- Probable navigation
- Illumination details in the water
- Prevailing conditions current of flow
- Marine life, the density of marine plants in the region of investigation
- GPS-based navigation may be employed for remote structures /bridges/pipes/ship.

Step 3. Image acquisition and sensor data (pH value of water) of entire region of investigation with proper illumination and noise cancelation.

Step 4. Images observed are preprocessed for enhancement and restored through inverse filtering, i.e., estimation of degradation function.

Step 5. Once the image obtained attains appropriate quality; it is analyzed to evaluate the extent of deterioration the structure has undergone during the life elapsed.

Step 6. Parameters approximated depending on type of structure—and therefore type of corrosion/damage; to determine the extent of damage is wirelessly transferred to central unit for processing.

Step 7. At central unit for processing extent of damage based on the set of parameters that determine the health of the structure is worked out.

Step 8. Predefined mapping of acceptable and non-acceptable values of factors contributing to corrosion are mapped to classify structures/bridges/pipes/ship as—normal/damaged with the possibility of repair/damaged beyond possibility of repair—fitness is examined.

Step 9. Soft margin is used to accommodate the variability of parameters.

Step 10. Classification of damage/corrosion for deciding on the extent of damage/corrosion, labels the structure as repairable/not repairable.

Step 11. Depending on the label appropriate actions can be taken.

4 Experimental Setup and Process

The experiment was conducted at different bridge structures over Gomti River and adjoining areas of district Lucknow, supporting pillars submerged underwater were scanned, and test samples were collected for the labeling process. Figure 3 depicts one of the sample sites, the spread of locations were taken to accommodate structures of different age and maintenance record.

Experimental setup comprised of extended camera stick with a waterproof camera embedded with wireless module for test sample transfer to control unit, approximately 22 sites were located and per site, approximately 90 images were obtained for the labeling process. The total sample space had almost 2000 test samples.

The stick employed is sufficiently long to capture images of the corroded region of structures. With the help of this setup, target areas were identified, and the dataset was collected for analysis; however, more advanced self-automated ROV's can be employed for the purpose of data collection. ROV's are costly but capable of acquiring images more efficiently.

(a). Monkey Bridge, Lucknow **(b).** Daliganj Bridge

(c). Pillar Number 3 at Daliganj Bridge **(d).** Pillar Number 3 at Monkey Bridge

Fig. 3 Sample dataset from collection points

4.1 Hardware Components

a. **Waterproof camera**—The waterproof camera which is used for capturing underwater images is GoPro HDX-701. It is a 12-megapixel camera capable of capturing optimum quality underwater images with sufficient brightness and contrast support.
b. **Extending Stick**—Stick is capable of pairing with GoPro camera, it is equipped with a remote. With help of remote various tasks can be performed like, shots, zoom, flashlight, etc., the camera is mounted on the stick. The stick can be extended up to maximum length of 42 inch in this case and put inside the water near the corroded structure, and random images were clicked for region of interest finalization. Once finalized, a number of samples were clicked with varying lighting, visibility, and angle conditions.
c. **Wireless module**—Camera is equipped with wireless module which enables wireless transfer of data to control unit; however, owing to presence of impurities in water, marine life, and current, errors might appear during transfer. Under such circumstances, the concurrency of data can be verified with memory storage from camera.

4.2 Dataset and Pre-processing

The experimental setup was deployed at different bridges across the Gomti River in areas and adjoining to Lucknow, bridges like Monkey Bridge, Daliganj Bridge, New Pakka Pull, Old Pakka pull, and many others toward outskirts of city were examined, and the pillars connecting the bridge across the river were examined for cracks, damage, and corrosion. Regions of interest for analysis were identified for processing; Fig. 5 depicts the sample dataset location where the experiment was conducted. Dataset of about 2000 images was acquired, labeled, and segregated in three classes un-corroded, corroded, and damaged. The data obtained is often corrupted owing to the presence of poor and difficult visible conditions as it travels in the liquid medium; hence, images captured underwater are hazy and are of poor contrast further presence of other elements such as organic matter dissolved in medium or minute floating particles also degrade data samples. Range of visibility can be enhanced through artificial lighting, still, these sources will suffer from difficulties like scattering and absorption of light.

Figure 4(a) depicts acquired colored image from the corroded dataset, sample image acquired is first processed for noise removal and image degradation, this is achieved by the adaptive median filter and Wiener filter as depicted in Fig. 4b, after noise removal dataset is further processed to improve image contrast, Dualistic sub-image histogram equalization is employed for the process as depicted in Fig. 4c, Image is segmented to segregate the region of interest this is achieved by region growing depicted in Fig. 4d.

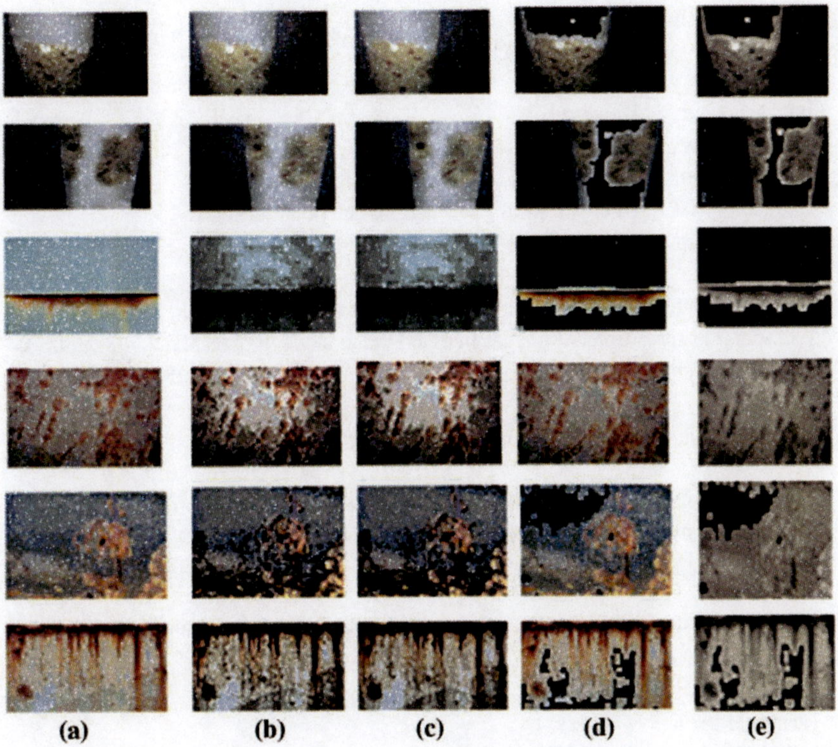

Fig. 4 Depicts sample images after pre-processing steps **a** data sample **b** noise cancelled image **c** contrast enhanced image **d** region of interest **e** grey scale image

Color images are often built of several stacked color channels, it is easier to deal with the single color channel as compared to multiple colored channels, and hence the inherent complexity of gray-level images is lower as compared to a colored image.

Algorithms employed for analysis were finalized, based on results that were optimal for human visualization. Adaptive thresholding is performed to have greyscale image on an image. Thresholding is achieved by applying Otsu's method, and this algorithm provides clustering-based image thresholding. This algorithm is based on selecting two classes or peaks. The Otsu technique extensively iterates and estimates the threshold with minimum intra-class variance, expressed as an algebraic weighted summation of two variances for the two defined classes.

4.3 Bag of Features

Bag of feature has been employed for corrosion classification of the structures; three datasets of un-corroded, corroded, and damaged labeled images are passed to the

Creating Bag-Of-Features.

----

* Image category 1: Corroded
* Image category 2: Damaged
* Image category 3: Un Corroded
* Selecting feature point locations using the Grid method.
* Extracting SURF features from the selected feature point locations.
** The GridStep is [8 8] and the BlockWidth is [32 64 96 128].

* Extracting features from 108 images in image set 1...done. Extracted 389268 features.
* Extracting features from 92 images in image set 2...done. Extracted 286768 features.
* Extracting features from 101 images in image set 3...done. Extracted 319928 features.

* Keeping 80 percent of the strongest features from each category.

* Balancing the number of features across all image categories to improve clustering.
** Image category 2 has the least number of strongest features: 229414.
** Using the strongest 229414 features from each of the other image categories.

Fig. 5 Feature extraction of corroded, damaged, and un-corroded dataset with BoF

Using K-Means clustering to create a 500 word visual vocabulary.
* Number of features : 688242
* Number of clusters (K) : 500

* Initializing cluster centers...100.00%.
* Clustering...completed 31/100 iterations (~4.05 seconds/iteration)...converged in 31 iterations.

* Finished creating Bag-Of-Features

Fig. 6 K-means clustering of corroded, damaged, and un-corroded dataset with BoF

training model to train the network. Once trained, the classifier will be capable of efficiently classifying pure and corroded images. Bag of features provides a compact description of images in the form of histograms of local descriptors. Bag of features algorithm primarily involves five steps.

Step 1. Extracting the data: The experimental setup described was employed at different underwater structure locations along Gomti River in Lucknow, to collect the image dataset. Dataset obtained was cleaned and labeled. Un-corroded, corroded, and damaged sets of data were employed for training the classifier.

Step 2. Feature identification and segregation: Objective is to segregate a group of images with the most significant information in the data. After identifying the critical features in each dataset, computation of vector takes place which will describe the

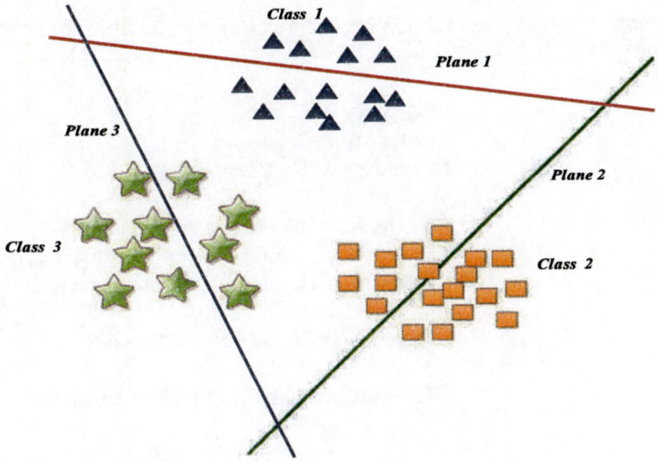

Fig. 7 Geometrical view of one-versus-all linear MLSTSVM classifier

features. The tasks of object recognition and image categorization are completed using SIFT descriptor. SIFT descriptor consists of feature extraction and detection both.

Step 3. Formulation of codebook and quantization: In encoding phase, the local descriptor is obtained in a new form by utilizing visual vocabulary.

Step 4. Classification phase: During categorization of image, the primary objective is to automatically annotate images with predefined groups. Image labels are predicted as soon as descriptors are extracted using a set of classifiers. To carryout classification, BoF has employed SVM classifier from machine learning toolbox. For the task, we have three categories of labeled dataset un-corroded, corroded, and damaged. The geometrical representation OVA MLSTSVM classifier is shown in Fig. 7.

Step 5. Performance of model: The algorithm trained the image category classifier for three categories—corroded, damaged, and un-corroded, following which encoding was done for each category, finished training the category classifier. Figure 8 depicts the confusion matrix and the accuracy achieved with BoF.

5 Open Issues

Underwater light attenuates in order of negative exponents leading to poor and difficult visible conditions as it travels in a liquid medium; hence, images captured underwater are hazy and are of poor contrast. This degradation puts a limit on the range of visibility to approximately about 20 m in clear water and 5 m or even less in

Fig. 8 Confusion matrix and accuracy achieved with BoF

```
Evaluating image category classifier for 3 categories.
-------------------------------------------------------

* Category 1: Corroded
* Category 2: Damaged
* Category 3: Un Corroded

* Evaluating 108 images from category 1...done.
* Evaluating 92 images from category 2...done.
* Evaluating 101 images from category 3...done.

* Finished evaluating all the test sets.

* The confusion matrix for this test set is:

                    PREDICTED
KNOWN        | Corroded  Damaged  Un Corroded
-----------------------------------------------------
Corroded     | 0.60      0.20     0.19
Damaged      | 0.02      0.90     0.08
Un Corroded  | 0.04      0.04     0.92

* Average Accuracy is 0.81.

ans =    0.8083
```

contaminated turbid water. Attenuation of light is due to phenomena called absorption and scattering, i.e., energy dissipation and change in the direction of light path, respectively. Absorption and scattering characteristics of light in liquid medium deteriorate the quality and performance of imaging techniques utilized for underwater applications.

6 Conclusion

This paper presents bag of features (BoF)-based methodology for the detection and estimation of underwater structural corrosion. The proposed model is capable of distinguishing between pure and corroded images and has achieved an accuracy of more than 83%. Three datasets of un-corroded, corroded, and damaged images are passed to the training model to train the network. Test samples were collected from different bridge structures over Gomti River and adjoining areas of district Lucknow. Approximately 22 sites were located and per site, approximately 90 images were obtained for the labeling process. The total sample space had almost 2000 test samples. Bag of feature algorithm primarily consists of five stages which extraction of feature, creation of codebook, features encoding, pooling features, learning and classification phase. The primary objective is identification of features through quantization of the local descriptors of images in the data according to visual vocabulary. A 500 words visual vocabulary is formulated through clustering of a large volume of local descriptors employing the K-means algorithm. Approximately 688,242 features are extracted, and 500 clusters are formed. Various non-destructive techniques have been used worldwide such as eddy current method, ultrasonic detection of corrosion, and infrared detection. The major drawback of these techniques is that they are expensive and require extreme expertise to perform operation and analysis on fetched data. Hence, the proposed model can be considered as a better alternative solution. The future course of work will focus at integrating the proposed model into an underwater remotely operated vehicle (ROV) for real-time monitoring of underwater structures.

References

1. C. I. Committee 546, Guide to Underwater Repair of Concrete, America Concrete Institute, Colo, USA, 2006
2. Lasa RP, Kessler R (1997) Practical application of cathodic protection systems for reinforcing steel substructures in marine environment. In: Proceedings of the international seminar on repair and rehabilitation of reinforced concrete structures, pp 16–31, Maracaibo, Venezuela, May 1997
3. Farid Uddin KM, Ohtsu M, Hossain KMA, Lachemi M (2007) Simulation of reinforcement–corrosion induced crack propagation in concrete by acoustic emission technique and boundary element method analysis. Canadian J Civ Eng 34:1197–1207
4. Sun X, Conglian Z (2011) Casing corrosion prediction based on grey support vector machine. In: 2011 International conference on electric information and control engineering, pp 4831–4834
5. Wu X, Ren J, Ye Y, Ren H (2016) Prediction of the laws of carbon steel erosion corrosion in sour water system based on decision tree and two kinds of artificial neural network model. In: 2016 Chinese control and decision conference (CCDC), pp 3872–3876
6. Bondada V, Pratihar DK, International conference on robotics and smart manufacturing, Elsevier
7. Naladala I, Raju A (2018) Corrosion damage identification and lifetime estimation of ship parts using image processing. In: International conference on advances in computing, communications and informatics (ICACCI)
8. Diaz JAI, Ligeralde Jr MI (2017) Rust detection using image processing via Matlab. In: TENCON 5z2017—2017 IEEE region 10 conference
9. Shi P, Fan X (2016) A detection and classification approach for underwater dam cracks. Sage
10. Bonnín-Pascual F, Rodriguez AO (2014) 'Corrosion detection for automated visual inspection. In: Developments in corrosion protection. InTech, Rijeka, Croatia (2014)
11. Pan B, Qian K, Xie H, Asundi A (2009) Two-dimensional digital image correlation for in-plane displacement and strain measurement: a review. Measure Sci Technol 20(6):27
12. Singh PK, Kar AK, Singh Y, Kolekar MH, Tanwar S (eds) (2019) Proceedings of ICRIC 2019: recent innovations in computing, vol 597. Springer Nature, Berlin
13. Wang Y, Chin (2017) A deep CNN method for underwater image enhancement. In: IEEE international conference on image processing, (ICIP)
14. Baba T, Nakamura K (2017) Image enhancement method for underwater images based on discrete cosine eigenbasis transformation. In: IEEE international conference on image processing, (ICIP)

15. Tanwar S (2020) Fog data analytics for IoT applications—next generation process model with state-of-the-art technologies, studies in big data. Springer International Publishing, Berlin, pp 1–550
16. Singh PK, Pawłowski W, Tanwar S, Kumar N, Rodrigues JJ, Obaidat MS (eds) (2020) Proceedings of first international conference on computing, communications, and cyber-security (IC4S 2019), vol 121. Springer, Berlin

On Performance Enhancement of Molecular Dynamics Simulation Using HPC Systems

Tejal Rathod, Monika Shah, Niraj Shah, Gaurang Raval, Madhuri Bhavsar, and Rajaraman Ganesh

Abstract The proposed work aims to enhance performance of molecular dynamics (MD) simulation code using various high-performance computing (HPC) approaches. The two-dimensional (2D) legacy code is parallelized using message-passing interface (MPI). Parallelization strategies when deployed with HPC platform, the performance and scalability improve with reduction in required computational time. Simulation experiments included two different numbers of atoms deployment keeping step size, time step, initial and boundary condition constant. Various profiling tools have been applied for identifying the hot spots that consume most of the execution time in the code. MD code is optimized employing following four approaches namely (1) force decomposition, (2) force decomposition with data organization, (3) intra- and inter-force decomposition with data organization and (4) intra- and inter-force decomposition with data organization and grid management. The output of these approaches is tested for the required accuracy by comparing its results with original standard MPI parallelized code. Simulation results for these approaches are found satisfactory from performance aspect. A comparative study is carried based on various performance metrics like execution time, speedup ratio and efficiency with multiprocessors. These approaches, when deployed on various platforms, are

T. Rathod · M. Shah · N. Shah · G. Raval (✉) · M. Bhavsar
Institute of Technology, Nirma University, Ahmedabad, India
e-mail: gaurang.raval@nirmauni.ac.in

T. Rathod
e-mail: tejal.rathod_srf@nirmauni.ac.in

M. Shah
e-mail: monika.shah@nirmauni.ac.in

N. Shah
e-mail: niraj.shah@nirmauni.ac.in

M. Bhavsar
e-mail: madhuri.bhavsar@nirmauni.ac.in

R. Ganesh
Institute of Plasma Research, Gandhinagar, India
e-mail: ganesh@ipr.res.in

P. K. Singh et al. (eds.), *Proceedings of Second International Conference on Computing, Communications, and Cyber-Security*, Lecture Notes in Networks and Systems 203,
https://doi.org/10.1007/978-981-16-0733-2_73

found better than standard MPI parallelized code except for the data organization approach. When the code is reformed implementing all approaches, the maximum speedup is found in the range of 2.5–4.5 times based on use of number of processors. Enhancement of code performance by saving computation time helps to solve the large-scale problems more efficiently.

Keywords Multipotential molecular dynamics simulation (MPMD) · High-performance computing (HPC) · Scalability

1 Introduction

Simulation is strategic compared to theoretical aspects, as it overcomes the hypothetical and experimental constraints. Particle simulation is widely used in many applications as there are various methods available for particle simulations. One of the methods is molecular dynamics (MD) method that solves the problems out of reach of classical models based on flow continuity of matter, momentum and energy. MD evolves a transient finite-sized molecular configuration [1]. There are limits on the typical length scales and time scales that can be investigated, and the consequences must be considered in analyzing the results along with validation of results.

1.1 Motivation

Simulation runs typically range from thousands to million time steps, corresponding to 2D or 3D regime. Some early results of such simulations have been followed by other works showing need for enlargement of simulation scale [2–4]. This is possible by parallelizing such computationally expensive code, so simulation time decreases or larger size problem can be solved. This parallelization is performed using algorithmic approaches on HPC platforms. This needs motivated to target the problem of reducing simulation time of molecular dynamics code.

1.2 Brief Literature Review

There are certain parallel MD algorithms that have been proposed in the past [5, 6]. Message-passing interface (MPI) plays an important role for parallelization. There has been a delusion that pure MPI can often outperform hybrid approach, but ample exceptions can be seen, and results may tend to vary with data size, input data, etc., even for a given code [7]. It is found that due to its molecular approach of MD simulations, it is an extremely time-consuming application and takes weeks or months to complete a single simulation [8]. The simulation time of such simulations can be reduced by (i) Load balancing: Dynamic load balancing is applied to the distribution of data and computation [9] which is time consuming, so finding an optimal partitioning to fully balance the load between all available distributed resources [10] will overcome the load balancing problem. (ii) Cache miss: When the data requested for processing by an element or application is not found in the cache memory, it causes execution delays by requiring the program or application to fetch the data from other cache levels or the main memory [11]. It can be resolved using block mechanism concept. (iii) Handle large-scale heterogeneous data and perform load balancing: MD simulation deals with large amount of data. It is difficult to give better performance for two phase simulation [5]. (iv) Data movement: Organization of data and dealing with large amount of data is a tedious task. MD simulation needs to consider data movement at each and every time step, so it affects communication time and increases communication overhead [2].

To overcome these issues, many researchers presented various techniques which are recited here. Liem et al. [12] proposed an efficient method for performing large-scale molecular dynamics simulation using the parallel link-cell algorithm on 2D space. There are some constraints like usage of network bandwidth, disk space, available memory and computation time, while dealing with force calculation between particles lies in a different processor. Putz et al. [13] introduced linked cell optimization techniques and used a domain decomposition approach which improves the performance by up to 45%. Number of force calculation reduced by adaptively restrained molecular dynamics (ARMD) [14]. It is used with LAMMPS which increases speedup more than 60%. Malakar et al. [15] designed a model for large-scale molecular dynamics simulation which reduces data movement time. They used optimal 3D decomposition on high-performance computers for code optimization and performance enhancement. After studying scalability challenges on large-scale molecular dynamics, they demonstrated the benefits of space-shared analysis [16] using LAMMPS code on supercomputers. Yang et al. [17] presented an end-to-end MD system on a single FPGA featuring online particle-pair generation; force evaluation on range-limited, long range and bonded interactions; motion update and particle data migration. They compare FPGA performance with GPU and achieve maximum throughput. Wang et al. [18] gave machine learning approach for data generated in running even a microsecond long MD simulation human comprehensible. They deal with interpretability and transferability challenge and also efficiently sample the underlying free energy surface and kinetics.

1.3 Objectives and Contributions of Present Work

From domain analysis, it is observed that MD simulation always deals with numerous calculations and increases communication overhead while calculating force between particles. It is also observed that due to load unbalancing problem and cache miss, execution time is increased, and scalability of code can be improved using multicore environments. In the present work, we used standard MPI-based parallelized code to run 2D MD simulations. The main objective of the present work is to verify efficiency of the 2D MD code and enhance its scalability on message-passing programming environments. We emphasize to minimize core saturation in MPI as execution time and core utilization can be improved. Hence, a couple of approaches have been implemented to solve the problems of slow computing speed and parallel scalability issues on HPC platform. Promising results are obtained in the present work.

1.4 Organization of Paper

Section 1 provides information on motivation, contributions, objectives and organization of the paper. Section 2 includes information about basics of molecular dynamics and various methods details that are used in present study. The Sect. 3 discusses about obtained results and its analysis, and Sect. 4 is conclusion of the present work. An acknowledgment and references details are included at the end of paper.

2 Algorithmic Approaches and Derived Parameters

This section describes necessary phases considered during simulation of molecular dynamics along with parameters, steps, computational resource used using various approaches to enhance the performance of the MD code.

2.1 Basics of Algorithm

In the present work, simulations are performed using already parallelized classical molecular dynamics (MD) code. The code was developed in structural programing and then parallelized using message-passing interface (MPI). The Leapfrog method is taken for simulation from numerous algorithms available [1]. Leapfrog is the simplest numerical scheme that is widely used for MD simulations. Key benefits are minimum storage requirements, and coordinates are accurate to third order in ΔT [1]. The algorithm uses intermolecular interactions defined by 12/6 Lennard–Jones pair potential. The computational plane is a classical 2D plane. The periodic boundary condition (PBC) is applied in all directions. The steps of MD simulations are (1) randomly initialize molecules in a computational domain and assign velocities to these molecules (2) determine forces using intermolecular potential data (3) solve Newton's equation of motion to determine acceleration and subsequently velocity and new position of molecule (4) repeat steps 2–4 for decided time period so that equilibrium is reached and (5) calculate necessary output. To decrease the computation time, cutoff radius concept is used so that limited interactions of molecules

Table 1 State fixed and input parameters

Input parameters	Value
N_X (number of atoms in X-direction)	91 and 271
N_Y (number of atoms in Y-direction)	91 and 271
N number of atoms ($N_x * N_y$)	8281 and 73,441
n (number density)	0.318
Γ	150
r_{Cut}	20
K	2
Time step (ΔT)	0.01
Step average	10
Step equilibrium	3000
Step limit	6000

are computed instead of all molecule computations. The cutoff radius in the present simulations is kept as 20. Simulations with the code are performed on Intel Parallel Studio environment with appropriate GSL library. Table 1 pertains to input parameters with values used in the code. Output parameters are kinetic energy, potential energy, total energy, velocity magnitude and pressure and code execution time.

2.2 Approaches to Improve Code Performance

In order to enhance the performance, computational time of the code can be reduced by applying various approaches. Sequential code of 2D MD which is developed using structural programming language is taken as a case study for applying the devised algorithms targeted for performance enhancement. Reverse engineering process analyzes the structure and design of code and provides information about computational time consumed by different parts of the code. The most time-consuming portion is known as hot spots. Thus, hot spot analysis helps to determine the health of the code and gives idea about the nature of all bottlenecks. Once hot spots are identified, computational capability of code is enhanced using different approaches that are discussed as follows. As per Newton's third law, force acting between two molecules is equal and opposite. Based on it, once force is calculated between molecules first and second, same output with opposite sign can be used while calculating force acting between second and first. Thus, it is possible to eliminate the redundant calculation that is nearly half of the total force calculation [19]. This approach is named as force decomposition in the present study. Algorithm steps to implement this approach in the code are as follows. (i) In the step (2) of MD simulation mentioned in Sect. 2.1, calculate the force between selected two molecules and store it. (ii) In the same step (2), the force acting between same molecules is required to be calculated once again and at that time force calculated in step (i) with opposite sign to be used. This approach is further improved by allocating intra-processor and inter-processor concept. The intra-processor concept is used for the calculation of the interaction between molecules resides in the same processor, while in case of inter-processor forces, interaction is calculated between molecules reside in the different processors [5]. The processor is allotted to the code in such a way that intra-processor and inter-processor force computation time reduces. The next approach reduces the catch miss by organizing data of molecules. Initially, molecules are arranged grid wise in proper sequence in a computational domain. However, after each time step, molecule moves, and its grid position may change. Due to this, as simulation proceeds, molecules in a particular grid are not in a proper sequence but randomly distributed. To carry out further simulation, certain data related to molecules of a grid is required, but due to random distribution of molecules, this leads to catch miss memory issue. If molecules data is arranged properly in an array, data can be fetched simultaneously, and it is possible to reduce cache miss time consumption. The standard code is parallelized horizontally, i.e., divides work amongst processes column wise. This limits utilization of number of processors. Similarly, limited processors can be used when the code is parallelized vertically, i.e., processors divide work row wise. If the code is modified in such a way that it uses combination of horizontal and vertical parallelization, i.e., dividing work combination of row and column wise, then computational work can be divided amongst many processors. This approach is termed in present work as grid approach and divides work heterogeneously in processors. Performance enhancement study with these approaches is measured and analyzed by execution time, data movements, memory usage and communication time without compromising accuracy of physical aspect of the code.

3 Results and Discussion

This section presents the results obtained with the approaches that are used to improve the performance of the parallelized code. Output of all approaches adopted in the present work has been compared with standard MPMD simulation approach output.

3.1 Determination of Hot Spots

Profiling is the process that provides information of CPU time consumed by different parts of the code. Using this information, time-consuming hot spots in the code are identified. Tools used for hot spot analysis are mentioned in Table 2.

Results of time consumed by various functions of the code have been collected from Gprof profiling tool as presented in Fig. 1. It has been identified that compute forces cells (blue color in Fig. 1) function acquires maximum time (71.31%). The libm_atan_19 (orange color in Fig. 1), inbuilt library function, also consumes substantial time, but it is not possible to reduce its computation time as it is an inbuilt

Table 2 Profiling tool features

Name of tool	Features
Gprof	Give collect timing information of functions [20]
Tuning and analysis utilities (TAU)	Aggregates the time spent in each routine-phase profiling [21]
Jumpshot	Graphical visualization tool use for post-mortem performance analysis [22]

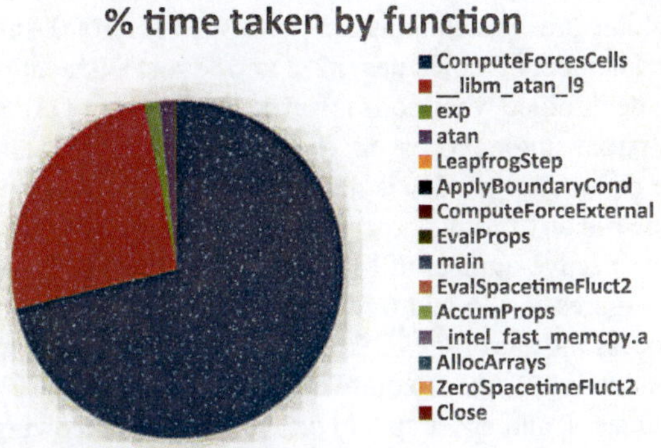

Fig. 1 Computational time comparison by various functions

function. Similarly, it is found using profiling tools such as TAU and Jumpshot that MPI_Bcast & MPI_Reduce function takes maximum time while communicating as shown in Fig. 2.

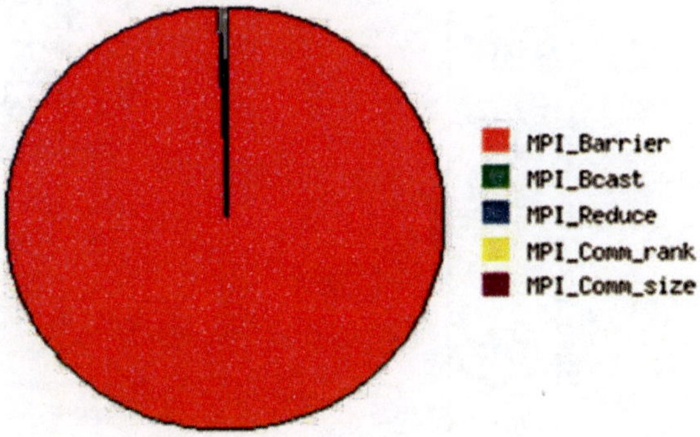

Fig. 2 Computational time used by various communication functions

3.2 MPI Results

To determine the performance improvement, results of standard MPI parallelized code (SMPC) are compared with different approaches discussed in Sect. 2.2. These approaches are force decomposition (FD), force decomposition with organization of data (FDOD), intra-grid and inter-grid force decomposition with organization of data (IIFDOD) and intra-grid and inter-grid force decomposition with organization of data and grid management (IIFDODGM). The workstation configuration used in the present work is 32 GB, Intel Xeon(R) CPU E5-1630 v4 @ 3.70 GHz × 8 architecture. Simulations are performed for 1000 time steps with two variations of total number of molecules (8281 and 73,441). Execution time, speedup and efficiency are determined when code is run on the number of processors, and it is compared with output of single processor. The execution time is total simulation time consumed by the code. The speedup is defined as ratio of execution time of serial code to parallel code run on different processors. Efficiency has been calculated by dividing speedup to respective number of MPI rank.

Figures 3, 4, 5 and 6 represent performance of a particular approach on various processors. It is observed from Figs. 3, 4, 5 and 6 that when the code related computation job is allocated to more computing elements (CPU processors), simulation time decreases, and speedup and efficiency increase owing to division of work between different processors. It can also be observed from these figures that computational time reduces when one by one various approaches of performance improvement are implemented in the code.

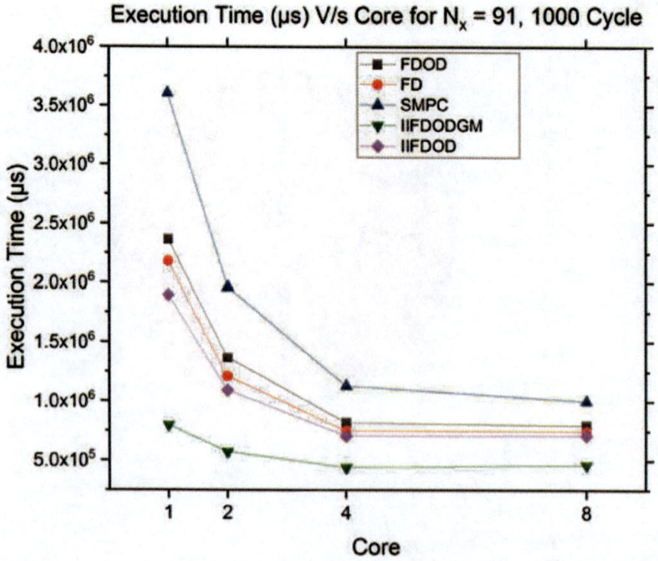

Fig. 3 Comparison of execution time for various approaches on different cores for 8281 molecules

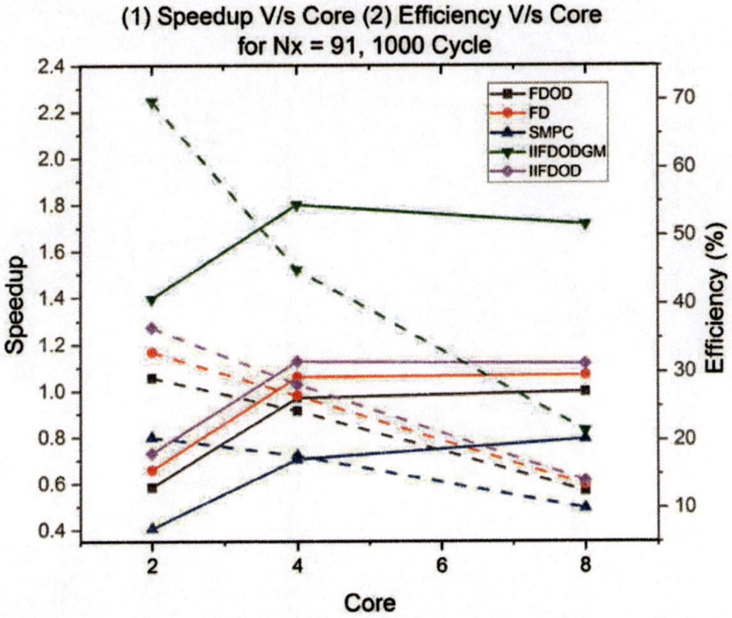

Fig. 4 Comparison of speedup and efficiency for various approaches on different cores for 8281 molecules

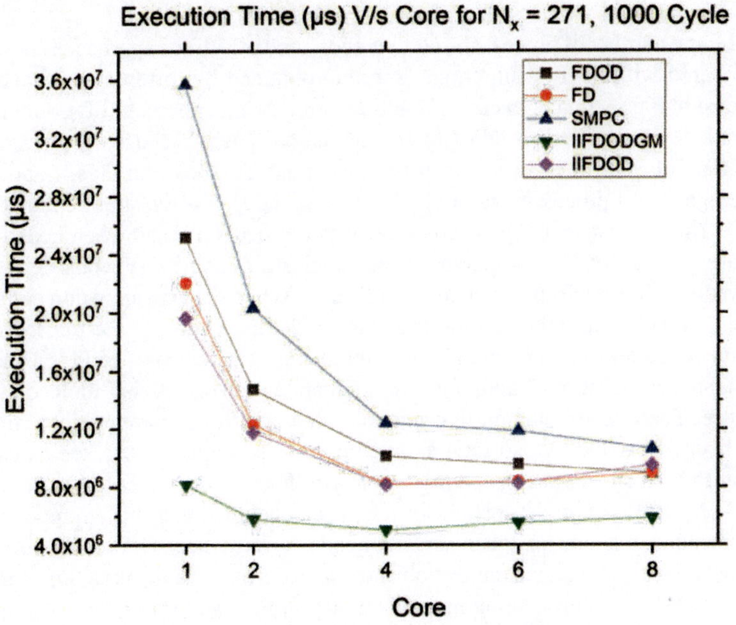

Fig. 5 Comparison of execution time for various approaches on different cores for 73,441 molecules

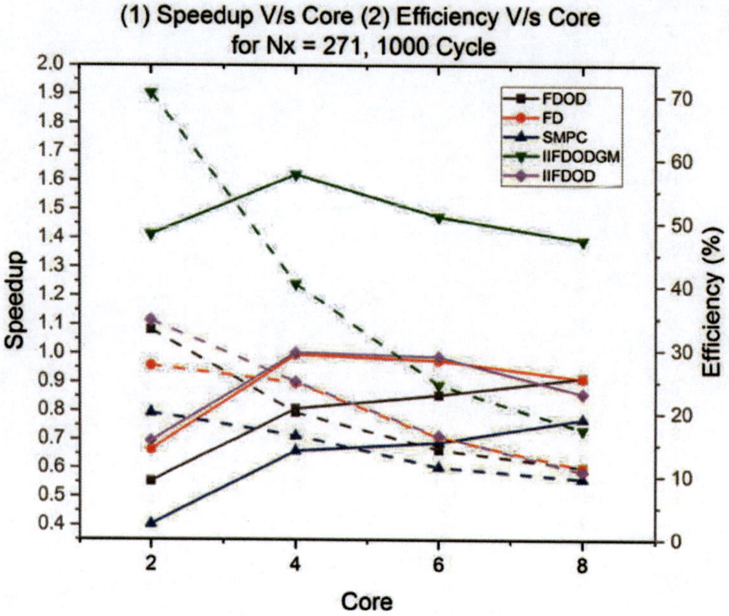

Fig. 6 Comparison of speedup and efficiency for various approaches on different cores for 73,441 molecules

For same number of processors, computation time of various approaches obtained is compared with SMPC output, and speedup obtained by different approaches is presented in Figs. 7 and 8. In case of force decomposition approach (FD), simulation time reduces due to reduction of force calculation. Also, based on profiling, it is found that this part of code is most time consuming. As shown in Figs. 7 and 8, on an average for all processors around 1.5 times, speedup is achieved as compared to SMPC. The highest speedup is achieved for two processors, and when code is run on more processors, due to increased communication latency, speedup decreases; however, still it is performing better than SMPC. When data organization approach (FDOD) is included, target is to improve the code in addition to the force decomposition adaptation, but computational time increases as per plot of Figs. 7 and 8. This happened because of addition of computations for organizing molecules in a sequence. There is a reduction in cache miss, but additional computational time is more than this reduction, so total simulation time is slightly increased as can be seen in Figs. 3 and 5, and speedup as well as efficiency is decreased as shown in Figs. 4 and 6. Based on these analysis, it can be concluded that this approach is not recommended to adopt for enhancement of code performance. The concept of intra-grid and inter-grid force decomposition and organization of data (IIFDOD) approach improves results with reduction in total simulation time and improves speedup and efficiency as compared to SMPC. This benefit is achieved due to proper allocation of processor for intra- and inter-grid force calculation. Referring Fig. 2, it may be noted

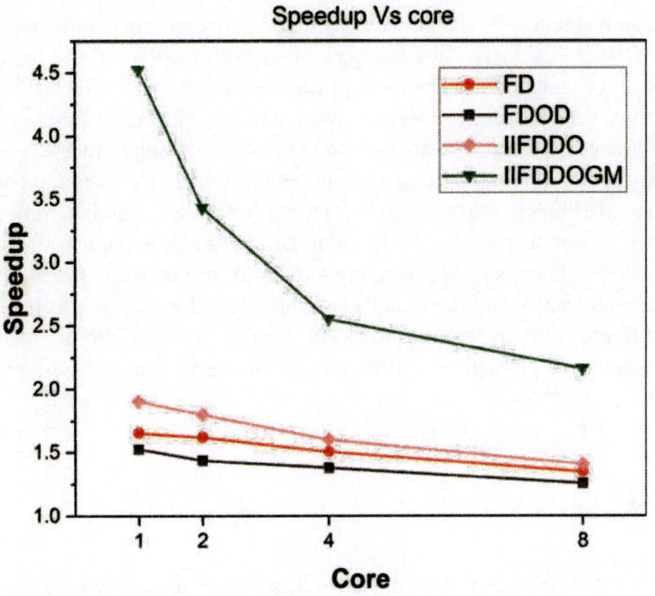

Fig. 7 Various approaches of speedup on different cores with respect to SMPC for molecules 8281

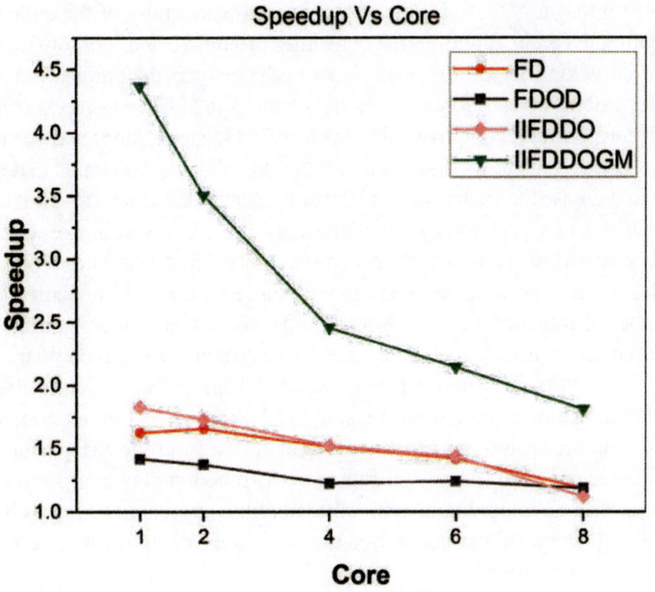

Fig. 8 Various approaches of speedup on different cores with respect to SMPC for molecules 73,441

that certain advantage of this approach is utilized in overcoming computational time loss of FDOD approach. The last approach of intra-grid and inter-grid force decomposition with organization of data and grid management (IIFDODGM) is applied, computational time reduces further, and substantial speedup and efficiency improvement are observed. The performance graph in Figs. 7 and 8 shows that least execution time is required compared to other approaches discussed. The speedup obtained as compared to SMPC is in the range of 2–4.5 times (refer Figs. 7 and 8). Similar results are obtained in Borstnik et al. [6] when simulations are carried out with 14,026 and 54,212 particles. They achieved speedup 4 and 4.6, respectively, for small number of molecules and large number of molecules; however, the code, platform, approach, etc., are different than the present work. In present study, as number of processor increases, there is a gradual reduction in speedup due to increased communication amongst the processors.

4 Conclusion

The present work describes performance enhancement of already MPI-based parallelized code on HPC platform employing four approaches. Parallelized implementation of the code improves performance of standard MPI code as well as all four approaches up to 4 times. Profiling method and hot spot study of the code are identified as compute forces function that consumes around 70% of computational time. This issue is tackled by different approaches. The force decomposition approach that reduces calculations of force between molecules to 50% and speedup obtained is 1.25–1.5 times. Further, computation time reduction strategy adopted by dealing with cache miss issues. Due to increase in computations, this method leads to decrease in speedup. To increase the scalability and performance efficiency, intra- and inter-force decomposition and a grid management approach are employed in the code. Results analysis suggests that intra- and inter-force decomposition results are found similar to force decomposition when code is run with more number of molecules; however, small number of molecules resulted in slight improvement in the performance. The grid management approach gives reasonably better performance as compared to other methods. The results are improved from 2 to 4.5 times. This is due to decrease in computation of force between molecules. In nutshell force decomposition, intra- and inter-force decomposition and grid management can improve the performance of already parallelized code. As part of future work, speedup may be improved further for the parallelized standard code with other possible approaches such as hybridization of MPI and OpenMP parallelization methods, optimizing the force computation domain, CUDA programming.

Acknowledgements We are grateful to the Institute of Plasma Research (IPR) for providing us the work environment and constant support. This work was fully funded through the Regular Research Project (RP) by the Board of Research in Nuclear Sciences (BRNS).

References

1. Rapaport DC (2004) The art of molecular dynamics simulation, 2nd edn. Bar-Ilan University, Israel
2. Chris M, Meyer R (2017) A hybrid algorithm for parallel molecular dynamics simulation computer physics communications. Comput Phys Commun 219:196–208
3. Brown WM, Wang P, Steven J, Arnold N, Tharrington AN (2011) Implementing molecular dynamics on hybrid high performance computers—short range forces. Comput Phys Commun 182(4):898–911
4. Anirban P, Abhishek A, Soumyendu R, Baidurya B (2014) Performance metrics in a hybrid MPI–OpenMP based molecular dynamics simulation with short-range interactions. J Parallel Distrib Comput 74(3):2203–2214
5. Shrinidhi H, Shrihari H, Raghu H, Yashonath S, Mohan TS (2012) Parallelizing molecular dynamics solutions for high performance. In: ATIP'12: proceedings of the ATIP/A*CRC workshop on accelerator technologies for high-performance computing: does Asia lead the way? Article No: 32, pp 1–4
6. Urban B, Benjamin T, Bernard R, Dusanka J (2012) Implementation of the force decomposition machine for molecular dynamics simulations. J Mol Graph Model 38:243–247
7. Loft R, Thomas S, Dennis J (2001) Terascale spectral element dynamical core for atmospheric general circulation models. In: SC'01: proceedings of the 2001 ACM/IEEE conference on Supercomputing. Denver, USA, pp 18
8. Wenqian D, Letian K, Keqin L, Ziyu H, Xiang-Hui X (2016) Implementing molecular dynamics simulation on sunway Taihu Light system. In: IEEE 18th international conference on high performance computing and communications. Sydney, Australia
9. Terry W, Reinhard V, Hanxleden J, Andrew M, Scott LR (1994) Parallelizing molecular dynamics using spatial decomposition. In: Proceedings of IEEE scalable high performance computing conference, vol 1. Knoxville, TN, USA, pp 95–102
10. Steffen H, Dirk P (2016) Towards understanding optimal load-balancing of heterogeneous short-range molecular dynamics. In: 2016 IEEE 23rd international conference on high performance computing workshops (HiPCW). Hyderabad, India. ISBN 978-1-5090-5774-0
11. Techopedia Homepage: https://www.techopedia.com/definition/6308/cache-miss. Last accessed 2020/02/24
12. Liem SY, Brown D, Clarke JHR (1991) Molecular dynamics simulations on distributed memory machines. Comput Phys Commun 67(2):261–267. ISSN 0010-4655. https://doi.org/10.1016/0010-4655(91)90021-C
13. Ptitz M, Kolb A (1998) Optimization techniques for parallel molecular dynamics using domain decomposition. Comput Phys Commun 113(2-3):145–167
14. Krishna S, Dmitriy M, Stephane R (2017) Parallel adaptively restrained molecular dynamics. In: International conference on high performance computing & simulation. Genova, Italy, pp 308–314
15. Preeti M, Venkatram V, Christopher K, Todd M, Michael P (2016) Optimal execution of co-analysis for large-scale molecular dynamics simulations. In: International conference for high performance computing, networking, storage and analysis Salt Lake City, UT
16. Preeti M, Christopher K, Todd M, Venkatram V, Michael P (2017) Scalable in situ analysis of molecular dynamics simulations. In: ISAV'17: proceedings of the in situ infrastructures on enabling extreme-scale analysis and visualization, pp 1–6
17. Chen Y, Tong G, Tianqi W, Rushi P, Qingqing X, Ahmed S, Chunshu W, Jiayi S, Charles L, Vipin S, Woody S, Martin H (2019) Fully integrated FPGA molecular dynamics simulations. In: SC'19: proceedings of the international conference for high performance computing, networking, storage and analysis, Article No: 67, pp 1–31
18. Yihang W, Joao M, Pratyush T (2020) Machine learning approaches for analyzing and enhancing molecular dynamics simulations. Curr Opin Struct Biol 61:139–145
19. Shu JW, Wang B, Chen M, Wang JZ (2003) Optimization techniques for parallel force-decomposition. Comput Phys Commun 154(2):121–130

20. Integrated performances monitoring homepage. https://ipm-hpc.sourceforge.net/. Last accessed 22 Feb 2019
21. Lawrence Livermore National Laboratory. Livermore computing center high performance computing homepage. https://hpc.llnl.gov/software/development-environment-software/tau-tuning-and-analysis-utilities. Last accessed 3 Apr 2019
22. Performance visualization for Parallel Programs. https://www.mcs.anl.gov/research/projects/perfvis/software/viewers/index.html. Last accessed data 10 Mar 2019

Author Index

A
Abayomi-Alli, Adebayo, 213, 227, 459
Abayomi, Funmilayo, 227
Agarwal, Rekha, 121
Agarwal, Sarika, 581
Agarwal, Suneeta, 1007
Agbaegbu, JohnBosco, 213
Agrawal, Smita, 703
Ahlawat, Prashant, 485
Ahmed, Md Ezaz, 485
Ahuja, Ravin, 213, 227, 459
Akinwale, Adio, 213
Alfa, Abraham Ayegba, 459
Alonge, Christianah, 227
Anand, Piyush, 409
Aneja, Leesha, 899
Arogundade, Oluwasefunmi, 213, 227, 459
Arora, Ashish, 183
Arora, Bhavna, 529, 833, 845
Asthana, Pallavi, 361

B
Baggan, Vidhu, 817
Balasundaram, S. R., 909
Balyan, Vipin, 141
Bansal, Anushree, 423
Bansal, Himani, 581, 997
Barot, Ritu, 927
Bhadana, Vartika, 241
Bhagat, Lalit, 997
Bhandari, Abhinav, 817
Bhandari, Neema, 987
Bhattacharya, Pronaya, 759
Bhatt, Bhupesh, 183
Bhat, Zahid A., 53
Bhavsar, Madhuri, 787, 1031
Bisht, Neeraj, 987
Bisht, Shilpi, 987
Bist, Divyanshu, 183
Bodkhe, Umesh, 759

C
Chadha, Anupama, 557
Chatterjee, Kalyan, 717
Chaudhary, Meenu, 607
Chauhan, Uttam, 23
Chetwani, Prafful, 511
Choudhary, Arjun, 857
Chudasama, Vipul, 471

D
Dalal, Jignasha, 775
Date, Rajas, 511
Dave, Rutvij, 339
Dubey, Alok, 717
Dutta, Shivangi, 529

G
Gaharwar, Gauravsingh, 163
Ganesh, Rajaraman, 1031
Ganotra, Sanjog, 81
Garg, Hitendra, 241, 265
Gaur, Loveleen, 607
Ghanchi, Asif, 3
Ghosh, Shilpa, 279
Gite, Shilpa, 153, 511

P. K. Singh et al. (eds.), *Proceedings of Second International Conference on Computing, Communications, and Cyber-Security*, Lecture Notes in Networks and Systems 203,
https://doi.org/10.1007/978-981-16-0733-2

Gupta, Lalit Mohan, 265
Gupta, Mayank, 557
Gupta, Shailja, 667, 679
Gupta, Sindhu Hak, 141

J
Jain, Piyushi, 743
Jain, Rachna, 183, 393, 409, 435
Jain, Rishit, 393
Jain, Satbir, 279
Jaisinghani, Kartik, 511
Jeberson, Wilson, 691
Jha, Gourav, 679
Jha, Gouri, 667
Jindal, Avani, 799
Jonathan, Oluranti, 459
Joshi, Aashna, 593

K
Kalra, Shruti, 885
Kashyap, Sonu, 557
Khaked, Azhar Ali, 653
Khanna, Abhirup, 799
Khanna, Pooja, 1017
Khatavkar, Hrituja, 153
Kishor, Amit, 691
Kotecha, Ketan, 3, 163, 339
Krishna, Achyut, 409
Kulkarni, Girish Ashok, 967
Kumar, Anil, 361
Kumar, Ashwani, 203
Kumar, Mohit, 203
Kumar, Pardeep, 731
Kumar, Sachin, 1017
Kumar, Vaibhav, 557
Kumar, Vishesh, 435
Kundra, Gahan, 885

L
Lad, Himani, 871
Limbachiya, Kuldeep, 3

M
Mahak, Mahapara, 293
Maheshwari, Priyam, 153
Malhotra, Devansh, 435
Malhotra, Parushi, 307
Malik, Praveen Kumar, 109
Malik, Shahid A., 53
Mishra, Sumita, 361
Misra, Sanjay, 213, 227, 459
Mistry, Mayur, 3, 163, 339
Mittal, Satyam, 997

N
Nagrath, Preeti, 39, 183, 393, 409, 435
Nayyar, Anand, 39
Nigam, Ritu, 279

O
Obhalia, Khusali, 97
Olaleye, Taiwo, 227
Oza, Parita, 377, 703

P
Panchal, Naitik, 23
Pandey, Bishwajeet, 987
Pandey, Neerav, 153
Pandya, Sharnil, 3, 163, 339
Parah, Shabir A., 53
Parekh, Pooja, 541
Parikh, Ajay, 627
Parikh, Pramit, 163, 339
Parmar, Rahil, 23
Patadiya, Vedant, 3
Patel, Ankit C., 323
Patel, Atul, 541
Patel, Bhavinkumar A., 627
Patel, Chintan, 945
Patel, Dhruval, 23
Patel, Drashti, 743
Patel, Harsh Jigneshkumar, 703
Patel, Jaydutt, 653
Patel, Jigna, 653
Patel, Jitali, 653
Patel, Samir, 377
Pathan, Nadimkhan, 945
Patni, J. C., 485
Pradhan, Rahul, 499
Pragya, 1017
Prasad, Vivek Kumar, 787
Pratap, Ajay, 203

Q
Qudus, Rauf, 213

R
Rai, Hari Mohan, 717
Rajitha, B., 1007

Ranjan, Ashutosh, 435
Rashid, Mudasir, 833
Rathod, Tejal, 1031
Raval, Gaurang, 1031
Rawat, Abhinav, 799
Rehman, Raqeebur, 53

S

Saini, Archit, 885
Sajad, Mir Saqlain, 253
Samad, Abdus, 265
Shah, Kashish, 163, 339
Shah, Khurshed A., 53
Shah, Maitrik, 593
Shah, Monika, 871, 1031
Shah, Niraj, 1031
Shah, Rainam, 927
Sharma, Deepak Kumar, 279
Sharma, Dilip Kumar, 447, 499
Sharma, Hitesh Kumar, 485
Sharma, Lavanya, 667, 679
Sharma, Manoj Kumar, 485
Sharma, Meetu, 845
Sharma, Paawan, 377
Sharma, Ritik, 393
Sharma, Sunidhi, 447
Shashi, 641
Sheikh, Javaid A., 53
Sheikh, Zakir Ahmad, 67
Sheth, Dhrumil, 927
Shinde, Sulbha Manoj, 967
Shivam, Anand, 3
Shrivastava, Aditya, 471
Siddiqui, Farheen, 253
Singh, Abhishek, 511
Singhal, Akshat, 799
Singh, Anand Prakash, 857
Singh, Jaspreet, 641, 899
Singh, Pankaj, 987
Singh, Pramod, 121
Singh, Shikha, 423
Singh, Yashwant, 67, 293, 307
Sinha, Abhijit, 567
Sinha, Anant, 1017
Sinha, Nidhi, 567
Snehi, Jyoti, 817
Snehi, Manish, 817
Solanki, Neol, 945
Srinivasan, Narayanan, 909
Srivastava, Divya, 1007
Srivastava, Praween, 717
Srivastava, Rajshree, 731
Srivastava, Shilpi, 153
Sundaram, Meenatchi, 567
Sur, Anirban, 3, 163, 339
Swadas, Prashant B., 97

T

Tailor, Neel, 945
Tandon, Urvashi, 817
Taneja, Sahil, 485
Taneja, Soham, 39
Tanwar, Sudeep, 361, 743, 759, 787
Thakkar, Aksha, 471
Thakur, Vrunda, 323
Tiwari, Ayush, 997
Tiwari, Bhawna, 141
Tiwari, Praveen, 109
Trivedi, Harshal, 927

U

Udanshiv, Abhishek, 511

V

Varshney, Yash, 409
Vegad, Mayur M., 97
Verma, Ashwin, 759
Verma, Jai Prakash, 743
Vividha, 39

Y

Yadav, Ankit, 361